KIRK-OTHMER

ENCYCLOPEDIA OF CHEMICAL TECHNOLOGY

Third Edition

VOLUME 24

Vitamins
to
Zone Refining

EDITORIAL BOARD

HERMAN F. MARK
Polytechnic Institute of New York

DONALD F. OTHMER
Polytechnic Institute of New York

CHARLES G. OVERBERGER
University of Michigan

GLENN T. SEABORG
University of California, Berkeley

EXECUTIVE EDITOR

MARTIN GRAYSON

ASSOCIATE EDITOR

DAVID ECKROTH

KIRK-OTHMER

ENCYCLOPEDIA OF CHEMICAL TECHNOLOGY

THIRD EDITION

VOLUME 24

VITAMINS
TO
ZONE REFINING

A WILEY-INTERSCIENCE PUBLICATION
John Wiley & Sons
NEW YORK • CHICHESTER • BRISBANE • TORONTO • SINGAPORE

Copyright © 1984 by John Wiley & Sons, Inc.

All rights reserved. Published simultaneously in Canada.

Reproduction or translation of any part of this work
beyond that permitted by Sections 107 or 108 of the
1976 United States Copyright Act without the permission
of the copyright owner is unlawful. Requests for
permission or further information should be addressed to
the Permissions Department, John Wiley & Sons, Inc.

Library of Congress Cataloging in Publication Data:

Main entry under title:
　Encyclopedia of chemical technology.

　　At head of title: Kirk-Othmer.
　　"A Wiley-Interscience publication."
　　Includes bibliographies.
　　　1. Chemistry, Technical—Dictionaries. I. Kirk, Raymond
Eller, 1890–1957. II. Othmer, Donald Frederick, 1904–
　　　III. Grayson, Martin. IV. Eckroth, David. V. Title:
Kirk-Othmer encyclopedia of chemical technology.

TP9.E685　　1978　　660′.03　　77-15820
ISBN 0-471-02077-X

Printed in the United States of America

CONTENTS

Vitamins, 1

Wastes, industrial, 227
Water, 257
Waterproofing and water/oil
    repellency, 442
Waxes, 466
Weighing and proportioning, 482
Welding, 502
Wheat and other cereal grains, 522
Wine, 549
Wood, 579
Wool, 612

Xanthates, 645
Xanthene dyes, 662
X-ray technology, 678
Xylenes and ethylbenzene, 709
Xylylene polymers, 744

Yeasts, 771

Zinc and zinc alloys, 807
Zinc compounds, 851
Zirconium and zirconium
    compounds, 863
Zone refining, 903

EDITORIAL STAFF FOR VOLUME 24

Executive Editor: **Martin Grayson**
Associate Editor: **David Eckroth**
Production Supervisor: **Michalina Bickford**
Editors: **Joyce Brown   Caroline L. Eastman   Carolyn Golojuch
            Anna Klingsberg   Mimi Wainwright**

CONTRIBUTORS TO VOLUME 24

M. A. Amerine, *University of California at Davis, Consultant, San Francisco, Wine Institute, Saint Helena, California,* Wine

A. J. Baker, *United States Department of Agriculture, Madison, Wisconsin,* Wood

Robert Bakish, *Bakish Materials Corporation, Englewood, New Jersey,* Water, supply and desalination

B. A. Bendtsen, *United States Department of Agriculture, Madison, Wisconsin,* Wood

Helmut Beschke, *Degussa Company, Wolfgang 6451, Federal Republic of Germany,* Vitamins, nicotinamide and nicotinic acid (B_3)

W. S. Boston, *CSIRO, Belmont, Victoria, Australia,* Wool

J. J. Brenden, *United States Department of Agriculture, Madison, Wisconsin,* Wood

Frank A. Brigano, *Olin Chemicals, New Haven, Connecticut,* Water, treatment of swimming pools, spas, and hot tubs

David L. Coffen, *Hoffmann-La Roche, Nutley, New Jersey,* Vitamins, pyridoxine (B_6)

W. E. Eslyn, *United States Department of Agriculture, Madison, Wisconsin,* Wood

Russell E. Farris, *Sandoz Colors & Chemicals, Martin, South Carolina,* Xanthene dyes

J. Philip Faust, *Olin Chemicals, New Haven, Connecticut,* Water, treatment of swimming pools, spas, and hot tubs

CONTRIBUTORS TO VOLUME 24

Heinz Friedrich, *Degussa Company, Wolfgang 6451, Federal Republic of Germany,* Vitamins, Nicotinamide and nicotinic acid (B$_3$)
Wilhelm Friedrich, *Universität Hamburg, Hamburg, Federal Republic of Germany,* Vitamins, vitamin B$_{12}$
Peter W. Fryth, *Hoffmann-La Roche, Inc., Nutley, New Jersey,* Vitamins, survey
Karl F. Graff, *The Ohio State University, Columbus, Ohio,* Welding
Guy H. Harris, *Consultant, Concord, California,* Xanthates
J. F. Harris, *United States Department of Agriculture, Madison, Wisconsin,* Wood
Mason Hayek, *E. I. du Pont de Nemours & Co., Inc., Wilmington, Delaware,* Waterproofing and water/oil repellency
David C. Herting, *Eastman Kodak Company, Rochester, New York,* Vitamins, vitamin E
Arnold L. Hirsch, *A. L. Laboratories, Inc., Chicago Heights, Illinois,* Vitamins, vitamin D
J. L. Howard, *United States Department of Agriculture, Madison, Wisconsin,* Wood
Otto Isler, *F. Hoffmann La Roche & Co., Basel, Switzerland,* Vitamins, vitamin A
Gerald M. Jaffe, *Hoffmann-La Roche, Inc., Nutley, New Jersey,* Vitamins, ascorbic acid
Ronald Jenkins, *Philips Electronic Instruments, Inc., Mahwah, New Jersey,* X-ray technology
Frank Kienzle, *F. Hoffmann La Roche & Co., Basel, Switzerland,* Vitamins, vitamin A
Axel Kleemann, *Degussa Company, Wolfgang 6451, Federal Republic of Germany,* Vitamins, nicotinamide and nicotinic acid (B$_3$)
Stuart M. Lee, *Ford Aerospace and Communications Corp., Palo Alto, California,* Xylylene polymers
C. Scott Letcher, *Petrolite Corporation, Tulsa, Oklahoma,* Waxes
Thomas B. Lloyd, *Gulf and Western National Resources Group, Bethlehem, Pennsylvania,* Zinc and zinc alloys; Zinc compounds
K. H. Mancy, *University of Michigan, Ann Arbor, Michigan,* Water, analysis
Christina Matter-Müller, *Swiss Federal Institute of Technology, Zurich, Switzerland,* Water, properties
Gerald Meyer, *U.S. Geological Survey, Reston, Virginia,* Water, sources and utilization
Olaf Mickelsen, *Consultant, Lula, Georgia,* Wheat and other cereal grains
R. B. Miller, *United States Department of Agriculture, Madison, Wisconsin,* Wood
Nelson L. Nemerow, *University of Miami, Coral Gables, Florida,* Wastes, industrial
Henry Nielsen, *Teledyne Wah Chang Albany, Albany, Oregon,* Zirconium and zirconium alloys
Ralph H. Nielsen, *Teledyne Wah Chang Albany, Albany, Oregon,* Zirconium and zirconium compounds
Heribert Offermanns, *Degussa Company, Wolfgang 6451 Federal Republic of Germany,* Vitamins, nicotinamide and nicotinic acid (B$_3$)
Yoshikazu Oka, *Takeda Chemical Industries, Ltd., Yodogawaku, Osaka, Japan,* Vitamins, thiamine (B$_1$)
R. C. Pettersen, *United States Department of Agriculture, Madison, Wisconsin,* Wood
James R. Pfafflin, *Consultant, Gillette, New Jersey,* Water, reuse; Water, sewage
Muthiah Ramanathan, *Roy F. Weston, Inc., West Chester, Pennsylvania,* Water, pollution
Derek L. Ransley, *Chevron Research Company, Richmond, California,* Xylenes and ethylbenzene
Gerald Reed, *Amber Laboratories, Inc., Milwaukee, Wisconsin,* Yeasts
J. W. Rowe, *United States Department of Agriculture, Madison, Wisconsin,* Wood

CONTRIBUTORS TO VOLUME 24

R. M. Rowell, *United States Department of Agriculture, Madison, Wisconsin,* Wood

James H. Schlewitz, *Teledyne Wah Chang Albany, Albany, Oregon,* Zirconium and zirconium alloys

Walter Showak, *The New Jersey Zinc Company, Palmerton, Pennsylvania,* Zinc and zinc alloys

W. T. Simpson, *United States Department of Agriculture, Madison, Wisconsin,* Wood

J. E. Singly, *Environmental Science & Engineering, Inc., Gainsville, Florida,* Water, municipal water treatment

R. B. Smart, *West Virginia University, Morgantown, West Virginia,* Water, analysis

Werner Stumm, *Swiss Federal Institute of Technology, Zurich, Switzerland,* Water, properties

† **S. Sussman,** *Olin Water Services, Stamford, Connecticut,* Water, industrial water treatment

Herbert Tanner, *Degussa Company, Wolfgang, 6451, Federal Republic of Germany,* Vitamins, nicotinamide and nicotinic acid (B_3)

Milan R. Uskoković, *Hoffmann-La Roche, Inc., Nutley, New Jersey,* Vitamins, biotin

Theodore H. Wegner, *United States Department of Agriculture, Madison, Wisconsin,* Wood

William W. Wells, *Michigan State University, East Lansing, Michigan,* Vitamins, inositol

William R. Wilcox, *Clarkson College of Technology, Potsdam, New York,* Zone refining

John A. Wojtowicz, *Olin Chemicals, New Haven, Connecticut,* Water, treatment of swimming pools, spas, and hot tubs

Fumio Yoneda, *Kumamoto University, Kumato, Japan,* Vitamins, riboflavin (B_2)

D. F. Zinkel, *United States Department of Agriculture, Madison, Wisconsin,* Wood

† Deceased.

NOTE ON CHEMICAL ABSTRACTS SERVICE REGISTRY NUMBERS AND NOMENCLATURE

Chemical Abstracts Service (CAS) Registry Numbers are unique numerical identifiers assigned to substances recorded in the CAS Registry System. They appear in brackets in the *Chemical Abstracts* (CA) substance and formula indexes following the names of compounds. A single compound may have many synonyms in the chemical literature. A simple compound like phenethylamine can be named β-phenylethylamine or, as in *Chemical Abstracts*, benzeneethanamine. The usefulness of the *Encyclopedia* depends on accessibility through the most common correct name of a substance. Because of this diversity in nomenclature careful attention has been given the problem in order to assist the reader as much as possible, especially in locating the systematic CA index name by means of the Registry Number. For this purpose, the reader may refer to the CAS Registry Handbook-Number Section which lists in numerical order the Registry Number with the *Chemical Abstracts* index name and the molecular formula; eg, **458-88-8,** Piperidine, 2-propyl-, (S)-, $C_8H_{17}N$; in the *Encyclopedia* this compound would be found under its common name, coniine [*458-88-8*]. The Registry Number is a valuable link for the reader in retrieving additional published information on substances and also as a point of access for such on-line data bases as Chemline, Medline, and Toxline.

In all cases, the CAS Registry Numbers have been given for title compounds in articles and for all compounds in the index. All specific substances indexed in *Chemical Abstracts* since 1965 are included in the CAS Registry System as are a large number of substances derived from a variety of reference works. The CAS Registry System identifies a substance on the basis of an unambiguous computer-language description of its molecular structure including stereochemical detail. The Registry Number is a machine-checkable number (like a Social Security number) assigned in sequential order to each substance as it enters the registry system. The value of the number lies in the fact that it is a concise and unique means of substance identification, which is independent of, and therefore bridges, many systems of chemical nomenclature. For polymers, one Registry Number is used for the entire family; eg, polyoxyethylene (20) sorbitan monolaurate has the same number as all of its polyoxyethylene homologues.

Registry numbers for each substance will be provided in the third edition cumulative index and appear as well in the annual indexes (eg, Alkaloids shows the Registry Number of all alkaloids (title compounds) in a table in the article as well, but the intermediates have their Registry Numbers shown only in the index). Articles such as Analytical methods, Batteries and electric cells, Chemurgy, Distillation, Economic evaluation, and Fluid mechanics have no Registry Numbers in the text.

Cross-references are inserted in the index for many common names and for some systematic names. Trademark names appear in the index. Names that are incorrect, misleading or ambiguous are avoided. Formulas are given very frequently in the text to help in identifying compounds. The spelling and form used, even for industrial names, follow American chemical usage, but not always the usage of *Chemical Abstracts* (eg, *coniine* is used instead of *(S)-2-propylpiperidine*, *aniline* instead of *benzenamine*, and *acrylic acid* instead of *2-propenoic acid*).

There are variations in representation of rings in different disciplines. The dye industry does not designate aromaticity or double bonds in rings. All double bonds and aromaticity are shown in the *Encyclopedia* as a matter of course. For example, tetralin has an aromatic ring and a saturated ring and its structure appears in the

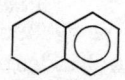

Encyclopedia with its common name, Registry Number enclosed in brackets, and parenthetical CA index name, ie, tetralin, [*119-64-2*] (1,2,3,4-tetrahydronaphthalene). With names and structural formulas, and especially with CAS Registry Numbers the aim is to help the reader have a concise means of substance identification.

CONVERSION FACTORS, ABBREVIATIONS, AND UNIT SYMBOLS

SI Units (Adopted 1960)

A new system of measurement, the International System of Units (abbreviated SI), is being implemented throughout the world. This system is a modernized version of the MKSA (meter, kilogram, second, ampere) system, and its details are published and controlled by an international treaty organization (The International Bureau of Weights and Measures) (1).

SI units are divided into three classes:

BASE UNITS

length	meter[†] (m)
mass[‡]	kilogram (kg)
time	second (s)
electric current	ampere (A)
thermodynamic temperature[§]	kelvin (K)
amount of substance	mole (mol)
luminous intensity	candela (cd)

SUPPLEMENTARY UNITS

plane angle	radian (rad)
solid angle	steradian (sr)

[†] The spellings "metre" and "litre" are preferred by ASTM; however "-er" are used in the Encyclopedia.
[‡] "Weight" is the commonly used term for "mass."
[§] Wide use is made of "Celsius temperature" (t) defined by

$$t = T - T_0$$

where T is the thermodynamic temperature, expressed in kelvins, and $T_0 = 273.15$ K by definition. A temperature interval may be expressed in degrees Celsius as well as in kelvins.

xiv FACTORS, ABBREVIATIONS, AND SYMBOLS

SUPPLEMENTARY UNITS

plane angle radian (rad)
solid angle steradian (sr)

DERIVED UNITS AND OTHER ACCEPTABLE UNITS

These units are formed by combining base units, supplementary units, and other derived units (2–4). Those derived units having special names and symbols are marked with an asterisk in the list below:

Quantity	Unit	Symbol	Acceptable equivalent
*absorbed dose	gray	Gy	J/kg
acceleration	meter per second squared	m/s²	
*activity (of ionizing radiation source)	becquerel	Bq	1/s
area	square kilometer	km²	
	square hectometer	hm²	ha (hectare)
	square meter	m²	
*capacitance	farad	F	C/V
concentration (of amount of substance)	mole per cubic meter	mol/m³	
*conductance	siemens	S	A/V
current density	ampere per square meter	A/m²	
density, mass density	kilogram per cubic meter	kg/m³	g/L; mg/cm³
dipole moment (quantity)	coulomb meter	C·m	
*electric charge, quantity of electricity	coulomb	C	A·s
electric charge density	coulomb per cubic meter	C/m³	
electric field strength	volt per meter	V/m	
electric flux density	coulomb per square meter	C/m²	
*electric potential, potential difference, electromotive force	volt	V	W/A
*electric resistance	ohm	Ω	V/A
*energy, work, quantity of heat	megajoule	MJ	
	kilojoule	kJ	
	joule	J	N·m
	electron volt†	eV†	
	kilowatt-hour†	kW·h†	
energy density	joule per cubic meter	J/m³	

† This non-SI unit is recognized by the CIPM as having to be retained because of practical importance or use in specialized fields (1).

Quantity	Unit	Symbol	Acceptable equivalent
*force	kilonewton	kN	
	newton	N	kg·m/s^2
*frequency	megahertz	MHz	
	hertz	Hz	1/s
heat capacity, entropy	joule per kelvin	J/K	
heat capacity (specific), specific entropy	joule per kilogram kelvin	J/(kg·K)	
heat transfer coefficient	watt per square meter kelvin	W/(m^2·K)	
*illuminance	lux	lx	lm/m^2
*inductance	henry	H	Wb/A
linear density	kilogram per meter	kg/m	
luminance	candela per square meter	cd/m^2	
*luminous flux	lumen	lm	cd·sr
magnetic field strength	ampere per meter	A/m	
*magnetic flux	weber	Wb	V·s
*magnetic flux density	tesla	T	Wb/m^2
molar energy	joule per mole	J/mol	
molar entropy, molar heat capacity	joule per mole kelvin	J/(mol·K)	
moment of force, torque	newton meter	N·m	
momentum	kilogram meter per second	kg·m/s	
permeability	henry per meter	H/m	
permittivity	farad per meter	F/m	
*power, heat flow rate, radiant flux	kilowatt	kW	
	watt	W	J/s
power density, heat flux density, irradiance	watt per square meter	W/m^2	
*pressure, stress	megapascal	MPa	
	kilopascal	kPa	
	pascal	Pa	N/m^2
sound level	decibel	dB	
specific energy	joule per kilogram	J/kg	
specific volume	cubic meter per kilogram	m^3/kg	
surface tension	newton per meter	N/m	
thermal conductivity	watt per meter kelvin	W/(m·K)	
velocity	meter per second	m/s	
	kilometer per hour	km/h	
viscosity, dynamic	pascal second	Pa·s	
	millipascal second	mPa·s	
viscosity, kinematic	square meter per second	m^2/s	
	square millimeter per second	mm^2/s	

Quantity	Unit	Symbol	Acceptable equivalent
volume	cubic meter	m^3	
	cubic decimeter	dm^3	L(liter) (5)
	cubic centimeter	cm^3	mL
wave number	1 per meter	m^{-1}	
	1 per centimeter	cm^{-1}	

In addition, there are 16 prefixes used to indicate order of magnitude, as follows:

Multiplication factor	Prefix	Symbol	Note
10^{18}	exa	E	
10^{15}	peta	P	
10^{12}	tera	T	
10^{9}	giga	G	
10^{6}	mega	M	
10^{3}	kilo	k	
10^{2}	hecto	h[a]	[a] Although hecto, deka, deci, and centi are SI prefixes, their use should be avoided except for SI unit-multiples for area and volume and nontechnical use of centimeter, as for body and clothing measurement.
10	deka	da[a]	
10^{-1}	deci	d[a]	
10^{-2}	centi	c[a]	
10^{-3}	milli	m	
10^{-6}	micro	μ	
10^{-9}	nano	n	
10^{-12}	pico	p	
10^{-15}	femto	f	
10^{-18}	atto	a	

For a complete description of SI and its use the reader is referred to ASTM E 380 (4) and the article Units and Conversion Factors which appears in Vol. 23.

A representative list of conversion factors from non-SI to SI units is presented herewith. Factors are given to four significant figures. Exact relationships are followed by a dagger. A more complete list is given in ASTM E 380-79(4) and ANSI Z210.1-1976 (6).

Conversion Factors to SI Units

To convert from	To	Multiply by
acre	square meter (m^2)	4.047×10^3
angstrom	meter (m)	1.0×10^{-10}†
are	square meter (m^2)	1.0×10^{2}†
astronomical unit	meter (m)	1.496×10^{11}
atmosphere	pascal (Pa)	1.013×10^5
bar	pascal (Pa)	1.0×10^{5}†
barn	square meter (m^2)	1.0×10^{-28}†
barrel (42 U.S. liquid gallons)	cubic meter (m^3)	0.1590

† Exact.

To convert from	To	Multiply by
Bohr magneton (μ_β)	J/T	9.274×10^{-24}
Btu (International Table)	joule (J)	1.055×10^3
Btu (mean)	joule (J)	1.056×10^3
Btu (thermochemical)	joule (J)	1.054×10^3
bushel	cubic meter (m^3)	3.524×10^{-2}
calorie (International Table)	joule (J)	4.187
calorie (mean)	joule (J)	4.190
calorie (thermochemical)	joule (J)	4.184†
centipoise	pascal second (Pa·s)	1.0×10^{-3}†
centistoke	square millimeter per second (mm^2/s)	1.0†
cfm (cubic foot per minute)	cubic meter per second (m^3/s)	4.72×10^{-4}
cubic inch	cubic meter (m^3)	1.639×10^{-5}
cubic foot	cubic meter (m^3)	2.832×10^{-2}
cubic yard	cubic meter (m^3)	0.7646
curie	becquerel (Bq)	3.70×10^{10}†
debye	coulomb·meter (C·m)	3.336×10^{-30}
degree (angle)	radian (rad)	1.745×10^{-2}
denier (international)	kilogram per meter (kg/m)	1.111×10^{-7}
	tex‡	0.1111
dram (apothecaries')	kilogram (kg)	3.888×10^{-3}
dram (avoirdupois)	kilogram (kg)	1.772×10^{-3}
dram (U.S. fluid)	cubic meter (m^3)	3.697×10^{-6}
dyne	newton (N)	1.0×10^{-5}†
dyne/cm	newton per meter (N/m)	1.0×10^{-3}†
electron volt	joule (J)	1.602×10^{-19}
erg	joule (J)	1.0×10^{-7}†
fathom	meter (m)	1.829
fluid ounce (U.S.)	cubic meter (m^3)	2.957×10^{-5}
foot	meter (m)	0.3048†
footcandle	lux (lx)	10.76
furlong	meter (m)	2.012×10^{-2}
gal	meter per second squared (m/s^2)	1.0×10^{-2}†
gallon (U.S. dry)	cubic meter (m^3)	4.405×10^{-3}
gallon (U.S. liquid)	cubic meter (m^3)	3.785×10^{-3}
gallon per minute (gpm)	cubic meter per second (m^3/s)	6.308×10^{-5}
	cubic meter per hour (m^3/h)	0.2271
gauss	tesla (T)	1.0×10^{-4}
gilbert	ampere (A)	0.7958
gill (U.S.)	cubic meter (m^3)	1.183×10^{-4}
grad	radian	1.571×10^{-2}
grain	kilogram (kg)	6.480×10^{-5}
gram force per denier	newton per tex (N/tex)	8.826×10^{-2}
hectare	square meter (m^2)	1.0×10^4†

† Exact.
‡ See footnote on p. xiv.

To convert from	To	Multiply by
horsepower (550 ft·lbf/s)	watt (W)	7.457×10^2
horsepower (boiler)	watt (W)	9.810×10^3
horsepower (electric)	watt (W)	$7.46 \times 10^{2\dagger}$
hundredweight (long)	kilogram (kg)	50.80
hundredweight (short)	kilogram (kg)	45.36
inch	meter (m)	$2.54 \times 10^{-2\dagger}$
inch of mercury (32°F)	pascal (Pa)	3.386×10^3
inch of water (39.2°F)	pascal (Pa)	2.491×10^2
kilogram-force	newton (N)	9.807
kilowatt hour	megajoule (MJ)	$3.6^\dagger$
kip	newton (N)	4.48×10^3
knot (international)	meter per second (m/s)	0.5144
lambert	candela per square meter (cd/m^2)	3.183×10^3
league (British nautical)	meter (m)	5.559×10^3
league (statute)	meter (m)	4.828×10^3
light year	meter (m)	9.461×10^{15}
liter (for fluids only)	cubic meter (m^3)	$1.0 \times 10^{-3\dagger}$
maxwell	weber (Wb)	$1.0 \times 10^{-8\dagger}$
micron	meter (m)	$1.0 \times 10^{-6\dagger}$
mil	meter (m)	$2.54 \times 10^{-5\dagger}$
mile (statute)	meter (m)	1.609×10^3
mile (U.S. nautical)	meter (m)	$1.852 \times 10^{3\dagger}$
mile per hour	meter per second (m/s)	0.4470
millibar	pascal (Pa)	1.0×10^2
millimeter of mercury (0°C)	pascal (Pa)	$1.333 \times 10^{2\dagger}$
minute (angular)	radian	2.909×10^{-4}
myriagram	kilogram (kg)	10
myriameter	kilometer (km)	10
oersted	ampere per meter (A/m)	79.58
ounce (avoirdupois)	kilogram (kg)	2.835×10^{-2}
ounce (troy)	kilogram (kg)	3.110×10^{-2}
ounce (U.S. fluid)	cubic meter (m^3)	2.957×10^{-5}
ounce-force	newton (N)	0.2780
peck (U.S.)	cubic meter (m^3)	8.810×10^{-3}
pennyweight	kilogram (kg)	1.555×10^{-3}
pint (U.S. dry)	cubic meter (m^3)	5.506×10^{-4}
pint (U.S. liquid)	cubic meter (m^3)	4.732×10^{-4}
poise (absolute viscosity)	pascal second (Pa·s)	$0.10^\dagger$
pound (avoirdupois)	kilogram (kg)	0.4536
pound (troy)	kilogram (kg)	0.3732
poundal	newton (N)	0.1383
pound-force	newton (N)	4.448
pound force per square inch (psi)	pascal (Pa)	6.895×10^3
quart (U.S. dry)	cubic meter (m^3)	1.101×10^{-3}
quart (U.S. liquid)	cubic meter (m^3)	9.464×10^{-4}
quintal	kilogram (kg)	$1.0 \times 10^{2\dagger}$

† Exact.

To convert from	To	Multiply by
rad	gray (Gy)	1.0×10^{-2}†
rod	meter (m)	5.029
roentgen	coulomb per kilogram (C/kg)	2.58×10^{-4}
second (angle)	radian (rad)	4.848×10^{-6}
section	square meter (m^2)	2.590×10^6
slug	kilogram (kg)	14.59
spherical candle power	lumen (lm)	12.57
square inch	square meter (m^2)	6.452×10^{-4}
square foot	square meter (m^2)	9.290×10^{-2}
square mile	square meter (m^2)	2.590×10^6
square yard	square meter (m^2)	0.8361
stere	cubic meter (m^3)	1.0†
stokes (kinematic viscosity)	square meter per second (m^2/s)	1.0×10^{-4}†
tex	kilogram per meter (kg/m)	1.0×10^{-6}†
ton (long, 2240 pounds)	kilogram (kg)	1.016×10^3
ton (metric)	kilogram (kg)	1.0×10^3†
ton (short, 2000 pounds)	kilogram (kg)	9.072×10^2
torr	pascal (Pa)	1.333×10^2
unit pole	weber (Wb)	1.257×10^{-7}
yard	meter (m)	0.9144†

Abbreviations and Unit Symbols

Following is a list of commonly used abbreviations and unit symbols appropriate for use in the *Encyclopedia*. In general they agree with those listed in *American National Standard Abbreviations for Use on Drawings and in Text (ANSI Y1.1)* (6) and *American National Standard Letter Symbols for Units in Science and Technology (ANSI Y10)* (6). Also included is a list of acronyms for a number of private and government organizations as well as common industrial solvents, polymers, and other chemicals.

Rules for Writing Unit Symbols (4):

1. Unit symbols should be printed in upright letters (roman) regardless of the type style used in the surrounding text.

2. Unit symbols are unaltered in the plural.

3. Unit symbols are not followed by a period except when used as the end of a sentence.

4. Letter unit symbols are generally written in lower-case (eg, cd for candela) unless the unit name has been derived from a proper name, in which case the first letter of the symbol is capitalized (W,Pa). Prefix and unit symbols retain their prescribed form regardless of the surrounding typography.

5. In the complete expression for a quantity, a space should be left between the numerical value and the unit symbol. For example, write 2.37 lm, *not* 2.37lm, and 35 mm, *not* 35mm. When the quantity is used in an adjectival sense, a hyphen is often used, for example, 35-mm film. *Exception:* No space is left between the numerical value and the symbols for degree, minute, and second of plane angle, and degree Celsius.

6. No space is used between the prefix and unit symbols (eg, kg).

7. Symbols, not abbreviations, should be used for units. For example, use "A," not "amp," for ampere.

8. When multiplying unit symbols, use a raised dot:

$$N \cdot m \text{ for newton meter}$$

In the case of W·h, the dot may be omitted, thus:

$$Wh$$

An exception to this practice is made for computer printouts, automatic typewriter work, etc, where the raised dot is not possible, and a dot on the line may be used.

9. When dividing unit symbols use one of the following forms:

$$m/s \; or \; m \cdot s^{-1} \; or \; \frac{m}{s}$$

In no case should more than one slash be used in the same expression unless parentheses are inserted to avoid ambiguity. For example, write:

$$J/(mol \cdot K) \; or \; J \cdot mol^{-1} \cdot K^{-1} \; or \; (J/mol)/K$$

but *not*

$$J/mol/K$$

10. Do not mix symbols and unit names in the same expression. Write:

$$\text{joules per kilogram} \; or \; J/kg \; or \; J \cdot kg^{-1}$$

but *not*

$$\text{joules/kilogram} \; nor \; \text{joules/kg} \; nor \; \text{joules} \cdot kg^{-1}$$

ABBREVIATIONS AND UNITS

A	ampere	AIP	American Institute of Physics
A	anion (eg, HA); mass number	AISI	American Iron and Steel Institute
a	atto (prefix for 10^{-18})	alc	alcohol(ic)
AATCC	American Association of Textile Chemists and Colorists	Alk	alkyl
ABS	acrylonitrile–butadiene–styrene	alk	alkaline (not alkali)
		amt	amount
abs	absolute	amu	atomic mass unit
ac	alternating current, *n*.	ANSI	American National Standards Institute
a-c	alternating current, *adj*.		
ac-	alicyclic	AO	atomic orbital
acac	acetylacetonate	AOAC	Association of Official Analytical Chemists
ACGIH	American Conference of Governmental Industrial Hygienists	AOCS	American Oil Chemists' Society
ACS	American Chemical Society	APHA	American Public Health Association
AGA	American Gas Association		
Ah	ampere hour	API	American Petroleum Institute
AIChE	American Institute of Chemical Engineers	aq	aqueous
AIME	American Institute of Mining, Metallurgical, and Petroleum Engineers	Ar	aryl
		ar-	aromatic
		as-	asymmetric(al)

ASHRAE	American Society of Heating, Refrigerating, and Air Conditioning Engineers	coml	commercial(ly)
		cp	chemically pure
		cph	close-packed hexagonal
ASM	American Society for Metals	CPSC	Consumer Product Safety Commission
ASME	American Society of Mechanical Engineers	cryst	crystalline
		cub	cubic
ASTM	American Society for Testing and Materials	D	Debye
		D-	denoting configurational relationship
at no.	atomic number		
at wt	atomic weight	d	differential operator
av(g)	average	d-	*dextro-*, dextrorotatory
AWS	American Welding Society	da	deka (prefix for 10^1)
b	bonding orbital	dB	decibel
bbl	barrel	dc	direct current, *n.*
bcc	body-centered cubic	d-c	direct current, *adj.*
BCT	body-centered tetragonal	dec	decompose
Bé	Baumé	detd	determined
BET	Brunauer-Emmett-Teller (adsorption equation)	detn	determination
		Di	didymium, a mixture of all lanthanons
bid	twice daily		
Boc	*t*-butyloxycarbonyl	dia	diameter
BOD	biochemical (biological) oxygen demand	dil	dilute
		DIN	Deutsche Industrie Normen
bp	boiling point	*dl*-; DL-	racemic
Bq	becquerel	DMA	dimethylacetamide
C	coulomb	DMF	dimethylformamide
°C	degree Celsius	DMG	dimethyl glyoxime
C-	denoting attachment to carbon	DMSO	dimethyl sulfoxide
		DOD	Department of Defense
c	centi (prefix for 10^{-2})	DOE	Department of Energy
c	critical	DOT	Department of Transportation
ca	circa (approximately)		
cd	candela; current density; circular dichroism	DP	degree of polymerization
		dp	dew point
CFR	Code of Federal Regulations	DPH	diamond pyramid hardness
		dstl(d)	distill(ed)
cgs	centimeter–gram–second	dta	differential thermal analysis
CI	Color Index		
cis-	isomer in which substituted groups are on same side of double bond between C atoms	(*E*)-	entgegen; opposed
		ε	dielectric constant (unitless number)
		e	electron
cl	carload	ECU	electrochemical unit
cm	centimeter	ed.	edited, edition, editor
cmil	circular mil	ED	effective dose
cmpd	compound	EDTA	ethylenediaminetetraacetic acid
CNS	central nervous system		
CoA	coenzyme A		
COD	chemical oxygen demand	emf	electromotive force

emu	electromagnetic unit	grd	ground
en	ethylene diamine	Gy	gray
eng	engineering	H	henry
EPA	Environmental Protection Agency	h	hour; hecto (prefix for 10^2)
		ha	hectare
epr	electron paramagnetic resonance	HB	Brinell hardness number
		Hb	hemoglobin
eq.	equation	hcp	hexagonal close-packed
esca	electron-spectroscopy for chemical analysis	hex	hexagonal
		HK	Knoop hardness number
esp	especially	hplc	high pressure liquid chromatography
esr	electron-spin resonance		
est(d)	estimate(d)	HRC	Rockwell hardness (C scale)
estn	estimation	HV	Vickers hardness number
esu	electrostatic unit	hyd	hydrated, hydrous
exp	experiment, experimental	hyg	hygroscopic
ext(d)	extract(ed)	Hz	hertz
F	farad (capacitance)	^{i}(eg, Pr^i)	iso (eg, isopropyl)
F	faraday (96,487 C)	i-	inactive (eg, i-methionine)
f	femto (prefix for 10^{-15})	IACS	International Annealed Copper Standard
FAO	Food and Agriculture Organization (United Nations)		
		ibp	initial boiling point
		IC	inhibitory concentration
fcc	face-centered cubic	ICC	Interstate Commerce Commission
FDA	Food and Drug Administration		
FEA	Federal Energy Administration	ICT	International Critical Table
		ID	inside diameter; infective dose
FHSA	Federal Hazardous Substances Act	ip	intraperitoneal
		IPS	iron pipe size
fob	free on board	IPTS	International Practical Temperature Scale (NBS)
fp	freezing point		
FPC	Federal Power Commission	ir	infrared
FRB	Federal Reserve Board	IRLG	Interagency Regulatory Liaison Group
frz	freezing		
G	giga (prefix for 10^9)	ISO	International Organization for Standardization
G	gravitational constant = 6.67×10^{11} N·m^2/kg^2		
		IU	International Unit
g	gram	IUPAC	International Union of Pure and Applied Chemistry
(g)	gas, only as in H$_2$O(g)		
g	gravitational acceleration	IV	iodine value
gc	gas chromatography	iv	intravenous
gem-	geminal	J	joule
glc	gas-liquid chromatography	K	kelvin
g-mol wt; gmw	gram-molecular weight	k	kilo (prefix for 10^3)
		kg	kilogram
GNP	gross national product	L	denoting configurational relationship
gpc	gel-permeation chromatography		
		L	liter (for fluids only)(5)
GRAS	Generally Recognized as Safe	l-	*levo*-, levorotatory

(l)	liquid, only as in NH$_3$(l)	ms	mass spectrum
LC$_{50}$	conc lethal to 50% of the animals tested	mxt	mixture
		μ	micro (prefix for 10^{-6})
LCAO	linear combination of atomic orbitals	N	newton (force)
		N	normal (concentration); neutron number
LCD	liquid crystal display		
lcl	less than carload lots	N-	denoting attachment to nitrogen
LD$_{50}$	dose lethal to 50% of the animals tested		
		n (as n_D^{20})	index of refraction (for 20°C and sodium light)
LED	light-emitting diode		
liq	liquid	n (as Bu^n), n-	normal (straight-chain structure)
lm	lumen		
ln	logarithm (natural)	n	neutron
LNG	liquefied natural gas	n	nano (prefix for 10^9)
log	logarithm (common)	na	not available
LPG	liquefied petroleum gas	NAS	National Academy of Sciences
ltl	less than truckload lots		
lx	lux	NASA	National Aeronautics and Space Administration
M	mega (prefix for 10^6); metal (as in MA)		
		nat	natural
M	molar; actual mass	NBS	National Bureau of Standards
$\overline{M}_w$	weight-average mol wt		
$\overline{M}_n$	number-average mol wt	neg	negative
m	meter; milli (prefix for 10^{-3})	NF	*National Formulary*
		NIH	National Institutes of Health
m	molal		
m-	meta	NIOSH	National Institute of Occupational Safety and Health
max	maximum		
MCA	Chemical Manufacturers' Association (was Manufacturing Chemists Association)		
		nmr	nuclear magnetic resonance
		NND	New and Nonofficial Drugs (AMA)
MEK	methyl ethyl ketone		
meq	milliequivalent	no.	number
mfd	manufactured	NOI- (BN)	not otherwise indexed (by name)
mfg	manufacturing		
mfr	manufacturer	NOS	not otherwise specified
MIBC	methyl isobutyl carbinol	nqr	nuclear quadruple resonance
MIBK	methyl isobutyl ketone	NRC	Nuclear Regulatory Commission; National Research Council
MIC	minimum inhibiting concentration		
		NRI	New Ring Index
min	minute; minimum	NSF	National Science Foundation
mL	milliliter	NTA	nitrilotriacetic acid
MLD	minimum lethal dose	NTP	normal temperature and pressure (25°C and 101.3 kPa or 1 atm)
MO	molecular orbital		
mo	month		
mol	mole	NTSB	National Transportation Safety Board
mol wt	molecular weight		
mp	melting point	O-	denoting attachment to oxygen
MR	molar refraction		

xxiv FACTORS, ABBREVIATIONS, AND SYMBOLS

o-	ortho	ref.	reference
OD	outside diameter	rf	radio frequency, n.
OPEC	Organization of Petroleum Exporting Countries	r-f	radio frequency, adj.
		rh	relative humidity
		RI	Ring Index
o-phen	o-phenanthridine	rms	root-mean square
OSHA	Occupational Safety and Health Administration	rpm	rotations per minute
		rps	revolutions per second
owf	on weight of fiber	RT	room temperature
Ω	ohm	^{s} (eg, Bu^s); sec-	secondary (eg, secondary butyl)
P	peta (prefix for 10^{15})		
p	pico (prefix for 10^{-12})		
p-	para	S	siemens
p	proton	(S)-	sinister (counterclockwise configuration)
p.	page		
Pa	pascal (pressure)	S-	denoting attachment to sulfur
pd	potential difference		
pH	negative logarithm of the effective hydrogen ion concentration	s-	symmetric(al)
		s	second
		(s)	solid, only as in $H_2O(s)$
phr	parts per hundred of resin (rubber)	SAE	Society of Automotive Engineers
p-i-n	positive-intrinsic-negative	SAN	styrene–acrylonitrile
pmr	proton magnetic resonance	sat(d)	saturate(d)
p-n	positive-negative	satn	saturation
po	per os (oral)	SBS	styrene–butadiene–styrene
POP	polyoxypropylene	sc	subcutaneous
pos	positive	SCF	self-consistent field; standard cubic feet
pp.	pages		
ppb	parts per billion (10^9)	Sch	Schultz number
ppm	parts per million (10^6)	SFs	Saybolt Furol seconds
ppmv	parts per million by volume	SI	Le Système International d'Unités (International System of Units)
ppmwt	parts per million by weight		
PPO	poly(phenyl oxide)		
ppt(d)	precipitate(d)	sl sol	slightly soluble
pptn	precipitation	sol	soluble
Pr (no.)	foreign prototype (number)	soln	solution
pt	point; part	soly	solubility
PVC	poly(vinyl chloride)	sp	specific; species
pwd	powder	sp gr	specific gravity
py	pyridine	sr	steradian
qv	quod vide (which see)	std	standard
R	univalent hydrocarbon radical	STP	standard temperature and pressure (0°C and 101.3 kPa)
(R)-	rectus (clockwise configuration)		
		sub	sublime(s)
r	precision of data	SUs	Saybolt Universal seconds
rad	radian; radius		
rds	rate determining step	syn	synthetic

^{t} (eg, But^t), t-, tert-	tertiary (eg, tertiary butyl)	Twad	Twaddell
		UL	Underwriters' Laboratory
		USDA	United States Department of Agriculture
T	tera (prefix for 10^{12}); tesla (magnetic flux density)	USP	*United States Pharmacopeia*
		uv	ultraviolet
t	metric ton (tonne); temperature	V	volt (emf)
		var	variable
TAPPI	Technical Association of the Pulp and Paper Industry	*vic*-	vicinal
		vol	volume (not volatile)
TCC	Tagliabue closed cup	vs	versus
tex	tex (linear density)	v sol	very soluble
T_g	glass-transition temperature	W	watt
tga	thermogravimetric analysis	Wb	weber
THF	tetrahydrofuran	Wh	watt hour
tlc	thin layer chromatography	WHO	World Health Organization (United Nations)
TLV	threshold limit value		
trans-	isomer in which substituted groups are on opposite sides of double bond between C atoms	wk	week
		yr	year
		(Z)-	zusammen; together; atomic number
TSCA	Toxic Substance Control Act		
TWA	time-weighted average		

Non-SI (Unacceptable and Obsolete) Units

		Use
Å	angstrom	nm
at	atmosphere, technical	Pa
atm	atmosphere, standard	Pa
b	barn	cm^2
bar†	bar	Pa
bbl	barrel	m^3
bhp	brake horsepower	W
Btu	British thermal unit	J
bu	bushel	m^3; L
cal	calorie	J
cfm	cubic foot per minute	m^3/s
Ci	curie	Bq
cSt	centistokes	mm^2/s
c/s	cycle per second	Hz
cu	cubic	exponential form
D	debye	C·m
den	denier	tex
dr	dram	kg
dyn	dyne	N
dyn/cm	dyne per centimeter	mN/m
erg	erg	J
eu	entropy unit	J/K
°F	degree Fahrenheit	°C; K
fc	footcandle	lx
fl	footlambert	lx
fl oz	fluid ounce	m^3; L
ft	foot	m

† Do not use bar (10^{5}Pa) or millibar (10^{2}Pa) because they are not SI units, and are accepted internationally only for a limited time in special fields because of existing usage.

xxvi   FACTORS, ABBREVIATIONS, AND SYMBOLS

Non-SI (Unacceptable and Obsolete) Units		*Use*
ft·lbf	foot pound-force	J
gf den	gram-force per denier	N/tex
G	gauss	T
Gal	gal	m/s^2
gal	gallon	m^3; L
Gb	gilbert	A
gpm	gallon per minute	(m^3/s); (m^3/h)
gr	grain	kg
hp	horsepower	W
ihp	indicated horsepower	W
in.	inch	m
in. Hg	inch of mercury	Pa
in. H$_2$O	inch of water	Pa
in.-lbf	inch pound-force	J
kcal	kilogram-calorie	J
kgf	kilogram-force	N
kilo	for kilogram	kg
L	lambert	lx
lb	pound	kg
lbf	pound-force	N
mho	mho	S
mi	mile	m
MM	million	M
mm Hg	millimeter of mercury	Pa
mμ	millimicron	nm
mph	miles per hour	km/h
μ	micron	μm
Oe	oersted	A/m
oz	ounce	kg
ozf	ounce-force	N
η	poise	Pa·s
P	poise	Pa·s
ph	phot	lx
psi	pounds-force per square inch	Pa
psia	pounds-force per square inch absolute	Pa
psig	pounds-force per square inch gauge	Pa
qt	quart	m^3; L
°R	degree Rankine	K
rd	rad	Gy
sb	stilb	lx
SCF	standard cubic foot	m^3
sq	square	exponential form
thm	therm	J
yd	yard	m

BIBLIOGRAPHY

1. The International Bureau of Weights and Measures, BIPM (Parc de Saint-Cloud, France) is described on page 22 of Ref. 4. This bureau operates under the exclusive supervision of the International Committee of Weights and Measures (CIPM).
2. *Metric Editorial Guide* (*ANMC-78-1*) 3rd ed., American National Metric Council, 5410 Grosvenor Lane, Bethesda, Md. 20814, 1981.
3. *SI Units and Recommendations for the Use of Their Multiples and of Certain Other Units* (*ISO 1000-1981*), American National Standards Institute, 1430 Broadway, New York, N. Y. 10018, 1981.
4. Based on *ASTM E 380-82* (*Standard for Metric Practice*), American Society for Testing and Materials, 1916 Race Street, Philadelphia, Pa. 19103, 1982.
5. *Fed. Regist.*, Dec. 10, 1976 (41 FR 36414).
6. For ANSI address, see Ref. 3.

R. P. LUKENS
American Society for Testing and Materials

V *continued*

VITAMINS

Survey, 1
Ascorbic acid, 8
Biotin, 41
Inositol, 50
Nicotinamide and nicotinic acid (B$_3$), 59
Pyridoxine (B$_6$), 94
Riboflavin (B$_2$), 108
Thiamine (B$_1$), 124
Vitamin A, 140
Vitamin B$_{12}$, 158
Vitamin D, 186
Vitamin E, 214

SURVEY

Vitamins do not constitute a homogeneous chemical group of compounds. Therefore, the term is a functional designation and can be defined in reference to the principal common function. Vitamins are essential chemicals needed in minute quantities to support metabolism; their activity is catalytic in nature and they must generally be provided by exogenous sources. They are supplied by natural foodstuffs, synthetic preparations, or in the form of provitamins, ie, vitamin precursors that are transformed in the organism.

The name vitamin was coined by Funk (1) in 1911 (as "vitamine"), who believed that the antiberiberi factor (thiamine) was an amine essential to the maintenance of life (*vita*). Although subsequent discoveries showed that not all vitamins are amines, the historical term has been retained; however, without the terminal e.

2 VITAMINS (SURVEY)

Solubilities

Vitamins are classified by solubility characteristics into the fat-soluble group and the water-soluble group containing the vitamins of the B series, including niacin, pantothenic acid, folic acid, biotin, and vitamin C. The fat-soluble vitamins are derivatives of partially cyclized isoprenoid polymers (A, E, K) or of sterol derivatives (D). Their activity appears to be a function of lipid solubility in various cell membranes affecting permeability or transport, or of their chemical groups as redox agents (A, E, K), coenzymes or enzyme activators (A, D, K), or enzyme inhibitors (E). The water-soluble vitamins are derivatives of carbohydrates (C) or pyridine, purines, and pyrimidines (B_1, B_2, B_6, niacin, folic acid), or complexes, eg, porphyrin–nucleotide (B_{12}), and amino acids–carboxylic acids (biotin, pantothenic acid). They function as enzyme activators and coenzymes (B_1, B_6, B_{12}, niacin, biotin, folic acid, pantothenic acid), as redox agents on enzyme reactions (B_2, B_{12}, niacin, folic acid, C), in nucleic acid synthesis (B_{12}, biotin, folic acid, C), and possibly as mitochondrial agents (B_2, niacin, C) (2).

In addition to their essential role in the physiologic processes of human and animal organisms, vitamins are used in the prophylaxis and treatment of various diseases including infectious and neoplastic conditions (3–5). Moreover, they are important industrial chemicals used for the fortification of foods, as stabilizing agents, and discoloration inhibitors because of their antioxidant properties (see also Food additives; Antioxidants and antiozonants).

The elucidation of the chemical nature of vitamins has led to the development of derivatives. The principal vitamins and structurally related compounds with similar activity called vitamers and their basic characteristics are given in Table 1.

So-Called Vitamins

The table does not include several substances that do not fit even the broad definition of the term vitamin and are not generally recognized as such, although they are called vitamins U, T, P, and L.

Vitamin U has been isolated from cabbage and other green vegetables and is probably L-methylioninemethylsulfonium chloride [1115-84-0]. It is said to have anti-ulcer activity and has been assessed in the treatment of peptic ulcer (6).

Vitamin T, so-called because of its original isolation from termites, has not been chemically defined. It is thought to be a complex of some vitamins and related substances and has been reported to have growth-promoting activity (7).

Vitamin P complex is the name given to a group of compounds known as bioflavonoids because of their flavonoid structure. They can be extracted from various fruits and plants and are claimed to exert a beneficial effect upon the integrity of the peripheral vasculature. Their possible pharmacologic utility is under investigation (8).

Vitamin L [118-92-3], o-aminobenzoic acid, so-called because it was claimed to be a lactation factor, contains as its active component (L_2), an adenine derivative. A description can be found in the Japanese patent literature (9).

Table 1. Vitamins, Vitamers, and Their Characteristics

Vitamin	CAS Registry No.	Other names	Vitamers Natural	CAS Registry No.	Synthetic	CAS Registry No.	Characteristics
A	[68-26-8]	vitamin A$_1$, retinol, axerophthol	α-carotene, β-carotene, γ-carotene, vitamin A esters	[7488-99-5] [7235-40-7] [472-93-5]	vitamin A acid, vitamin A esters	[302-79-4]	skin and mucosal integrity, maintenance of vision
C	[50-81-7]	ascorbic acid, antiscorbutic vitamin	dehydroascorbic acid	[490-83-5]	6-desoxyascorbic acid, isoascorbic acid	[528-81-4] [89-65-6]	antioxidant, supports capillary integrity, collagen formation
D	[520-91-2]	antirachitic vitamin	vitamin D$_2$ (ergocalciferol), vitamin D$_3$ (cholecalciferol)	[50-14-6] [67-47-0]	irradiated vitamin D$_4$	[511-28-4]	regulation of calcium and phosphorus metabolism, bone growth
E	[59-02-9]	α-tocopherol	β-tocopherol, γ-tocopherol, tocopherol esters	[148-03-8] [7616-22-0]	racemic α-tocopherol, tocopherol esters	[10191-41-0]	biological antioxidant, cell respiration
K	[84-80-0]	vitamin K$_1$, phylloquinone, coagulation vitamin	vitamin K$_2$	[11032-49-8]	menadione, menadiol	[58-27-5] [481-85-6]	blood clotting
B$_1$	[67-03-8][a]	thiamine, aneurin	thiamine pyrophosphate, cocarboxylase; thiamine orthophosphate	[582-37-6] [582-38-7]	thiamine disulfide, acetylated thiamine	[67-16-3] [1563-71-9][b]	carbohydrate metabolism coenzyme in pyruvate metabolism, antineuritic
B$_2$	[83-88-5]	riboflavin, vitamin G, lactoflavin, hepatoflavin, ovoflavin, verdoflavin	riboflavin mononucleotide, FMN; riboflavin dinucleotide, FAD	[146-17-8] [146-14-5]	7, 8, and/or 10-methyl or ethyl compounds		coenzyme in respiratory enzyme systems; cellular redox systems

Table 1 (continued)

Vitamin	CAS Registry No.	Other names	Vitamers Natural	Vitamers CAS Registry No.	Vitamers Synthetic	CAS Registry No.	Characteristics
B_6	[58-56-0][a]	pyridoxine, pyridoxol	pyridoxal, pyridoxamine, pyridoxal phosphate	[66-72-8] [524-36-7] [54-47-7]			coenzyme in some phases of amino acid metabolism
B_{12}	[68-19-9]	cobalamin, cyanocobalamin	hydroxycobalamin, aquocobalamin	[13422-51-0] [13422-52-1]	nitrocobalamin	[20623-13-6]	coenzyme in lipid, protein, nucleic acid synthesis antipernicious extrinsic factor
biotin	[58-85-5]	vitamin H, anti-egg-white-injury factor	desthiobiotin	[533-48-2]	biotinsulfoxide	[3376-83-8]	coenzyme for some carboxylations for pyruvic oxidase
folic acid	[59-30-3]	folacin, pteroylglutamic acid, antianemia factor	tetrahydrofolic acid, pteroyltriglutamic acid, citrovorum factor, leucovorin	[135-16-0] [89-38-3] [51057-63-7] [58-05-9]	pteroic acid, xanthopterin	[119-24-4] [119-44-8]	synthesis of nucleic acid, coenzyme in purine, pyrimidine metabolism, antianemia requirement
niacin	[59-67-6]	nicotinic acid, antipellagra factor, vitamin B_3[c]	niacinamide, NAD (CoI), NADP (CoII)	[98-92-0] [53-84-9] [53-59-8]	niacin esters, nikethamide, niacinamide	[59-26-7]	coenzyme in dehydrogenase and other systems, pellagra prevention, vasodilation
pantothenic acid	[79-83-4]	chick antidermatitis factor			pantothenic acid esters		component of coenzyme A in carbohydrate, acetyl transferase system

[a] For the HCl.
[b] For the N-acetyl.
[c] Vitamin B_3 has also been used to name pantothenic acid.

Physiologic Importance and Toxicity

Studies of the physiologic importance of the vitamins have demonstrated their participation in a large variety of enzymic processes, some of which are mentioned in Table 1. The essential role of vitamins in the animal organism is further proven by the deficiency symptoms caused by inadequate intake (10). Some vitamins also exhibit a certain species specificity. For example, humans, monkeys, and guinea pigs require vitamin C to be provided by exogenous sources, whereas dogs and rats are capable of synthesizing ascorbic acid.

The Food and Nutrition Board of the NAS–NRC periodically publishes a guide for the maintenance of good nutrition (see Table 2). The recommended allowances are based on studies dealing with vitamin deficiencies (hypovitaminosis), total lack of vitamins (avitaminosis), and toxic manifestations of vitamin excesses (hypervitaminosis). Some of the pathologic conditions resulting from deficiencies or lack are

Table 2. Recommended Daily Dietary Allowances[a]

Age, yr	Vitamin A, µg	Vitamin D, µg	Vitamin E, mg	Vitamin C, mg	Vitamin B$_1$, mg	Vitamin B$_2$, mg	Vitamin B$_6$, mg	Vitamin B$_{12}$, µg	Niacin, mg	Folacin, µg
0–½	420	10	3	35	0.3	0.4	0.3	0.5	6	30
½–1	400	10	4	35	0.5	0.6	0.6	1.5	8	45
1–3	400	10	5	45	0.7	0.8	0.9	2.0	9	100
4–6	500	10	6	45	0.9	1.0	1.3	2.5	11	200
7–10	700	10	7	45	1.2	1.4	1.6	3.0	16	300
males										
11–14	1000	10	8	50	1.4	1.6	1.8	3.0	18	400
15–18	1000	10	10	60	1.4	1.7	2.0	3.0	18	400
19–22	1000	7.5	10	60	1.5	1.7	2.2	3.0	19	400
23–50	1000	5	10	60	1.4	1.6	2.2	3.0	18	400
51–	1000	5	10	60	1.2	1.4	2.2	3.0	16	400
females										
11–14	800	10	8	50	1.1	1.3	1.8	3.0	15	400
15–18	800	10	8	60	1.1	1.3	2.0	3.0	14	400
19–22	800	7.5	8	60	1.1	1.3	2.0	3.0	14	400
23–50	800	5	8	60	1.0	1.2	2.0	3.0	13	400
51–	800	5	8	60	1.0	1.2	2.0	3.0	13	400

[a] Ref. 11.

Table 3. Toxicity of Vitamins

Vitamin	Species	LD$_{50}$, mg/kg	Route of administration
B$_1$	mouse	125	intravenous
	rat	250	intravenous
	rabbit	300	intravenous
	dog	350	intravenous
B$_2$	rat	560	intraperitoneal
folic acid	mouse	600	intravenous
	rat	500	intravenous
K	mouse	500	oral

mentioned in Table 1. Intake of vitamins A, B_2, C, and E is recommended for chemoprevention of cancer (12). Vitamins in excess of the body's ability to utilize them are generally excreted, but some untoward effects of hypervitaminosis have been shown to occur. However, these effects are primarily a function of the fat-soluble vitamins which are more readily stored in the organism (13). Thus, excess vitamin A has been the cause of certain skin lesions, irritability, and other symptoms (14), vitamin D has caused calcification of soft tissues (15), and increased clotting times have been associated with excess vitamin E in patients on anticoagulant therapy (16). As a class, vitamins are of a low order of toxicity (see Table 3).

Standards and chemical, biological, and animal assay methods have been developed. Fluorimetry and colorimetry are utilized as well as microbial assays. The biological activity of specific weights of pure vitamin preparations has become the standard unit of activity, called the International Unit (IU). However, weight units are now preferred and have been adopted by the USP (10).

Production

The commercial production of vitamins may be by chemical synthesis, isolation from natural sources, or microbial fermentation. The choice of process is usually determined by the availability of the starting materials and the economic feasibility of the process. Most commercial vitamins are produced by chemical synthesis. However, vitamins B_2 and B_{12} are also conveniently and economically produced by bacterial fermentation and the fat-soluble vitamins are also readily isolated from natural sources, eg, vitamin A from fish liver, vitamin E from soybean oil, vitamin K from fish meal, and vitamin D from liver oil or irradiated yeast (17).

Although the required vitamins are adequately available in the normal diet, uses such as in foodstuffs, medical applications, preservatives, and components of cosmetic formulations require the continually increasing production of vitamins, as shown in Table 4 (18–19).

Table 4. Production and Sales of Vitamins

Year	Production, metric tons	Sales, 10^6 $	Retail sales value, 10^6 $
1950	1,567	58.714	
1960	5,018	68.684	334
1965	7,392	65.366	271
1970	10,394	84.063	321
1975	13,586	123.551	452
1979	18,812	220.554	621
1980	19,313	239.923	916

BIBLIOGRAPHY

"Vitamins, Survey" in ECT 1st ed., Vol. 14, pp. 777–791, by H. R. Rosenberg and M. G. Weller, E. I. du Pont de Nemours & Co., Inc., "Vitamins, Survey" in ECT 2nd ed., Vol. 21, pp. 484–490, P. W. Fryth, Hoffmann-LaRoche, Inc.

1. C. Funk, *J. Physiol. (London)* **43,** 395 (1911).
2. R. J. Kutsky, *Handbook of Vitamins and Hormones*, 2nd ed., Van Nostrand Reinhold Co., New York, 1981.
3. W. Bollag and A. Matter, *Ann. N.Y. Acad. Sci.* **359,** 9 (1981).
4. R. E. Olson, *Circulation* **48,** 179 (1973).
5. S. S. Charleston and K. M. Clegg, *Lancet* **1,** 1402 (1972).
6. Z. Szabo and G. Vargha, *Arzneim. Forsch.* **10,** 23 (1960).
7. A. Koch and co-workers, *Naturwissenschaften* **38,** 339 (1951).
8. M. Foldi, *Angiologica* **9,** 133 (1972).
9. Jpn. Pat. 69 16,015 (July 16, 1969), T. Sato, R. Ishido, and A. Hosono.
10. *Vitamin Compendium*, F. Hoffmann-La Roche, Basel, Switzerland, 1976.
11. *Recommended Dietary Allowances*, 9th ed., National Research Council, Washington, D.C., 1980.
12. *Report on Diet, Nutrition, and Cancer*, National Research Council—National Academy of Sciences, Washington, D.C., 1982.
13. *Vitamin–Mineral Safety, Toxicity, and Misuse*, National Nutrition Consortium, Inc., The American Dietetic Association, Chicago, Ill., 1978, pp. 1–20.
14. D. R. Miller and K. C. Hayes in J. N. Hathcock, ed., *Nutritional Toxicology*, Vol. 1, Academic Press, New York, 1982, pp. 81–133.
15. R. Ziegler and co-workers, *Dtsch. Med. Wochenschr.* **100,** 415 (1975).
16. J. J. Corrigan, Jr., and F. I. Marcus, *J. Am. Med. Assoc.* **230,** 1300 (1974).
17. C. S. Sodano, *Vitamins Synthesis, Production and Use: Advances Since 1970*, Noyes Data Corp., Park Ridge, N.J., 1978.
18. *Statistical Abstract of the United States*, 102nd ed., U.S. Department of Commerce, Washington, D.C., 1981.
19. *Synthetic Organic Chemicals, U.S. Production and Sales, 1980*, U.S. Government Printing Office, Washington, D.C., 1981.

PETER W. FRYTH
Hoffmann-LaRoche, Inc.

VITAMINS (ASCORBIC ACID)

ASCORBIC ACID

Ascorbic acid [50-81-7] (L-ascorbic acid, L-*xylo*-ascorbic acid, L-*threo*-hex-2-enonic acid γ-lactone) (**1**) is the name recognized by the IUPAC–IUB Commission on Biochemical Nomenclature for vitamin C (1). The name implies the vitamin's antiscorbutic properties, namely, the prevention and treatment of scurvy. L-Ascorbic acid is widely distributed in plants and animals. The pure vitamin ($C_6H_8O_6$, mol wt 176.13) is a white crystalline substance derived from L-gulonic acid, a sugar acid, and

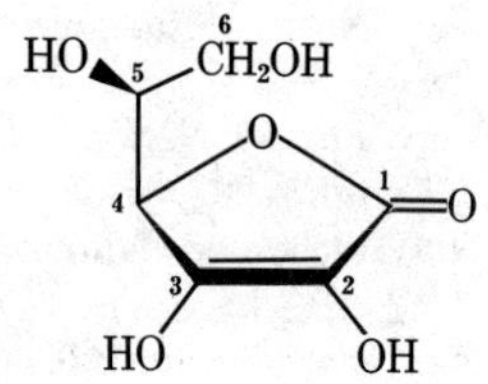

(**1**) L-ascorbic acid

synthesized both biologically and chemically from D-glucose. L-Ascorbic acid and its main derivatives are officially recognized by regulatory agencies and included in compendia such as the *United States Pharmacopeia* (USP), the *National Formulary* (NF), and *Food Chemicals Codex* (FCC). The most significant characteristic of L-ascorbic acid (**1**) is its oxidation to dehydro-L-ascorbic acid (L-*threo*-2,3-hexodiulosonic acid γ-lactone) (**2**). Both substances are active antiscorbutic agents; they act in concert as a redox system. The precise biochemical mechanisms of L-ascorbic acid's actions are somewhat obscure; nevertheless, this redox system appears to be the basis for the vitamin's primary physiological activities and technical applications. It is important in the food industry as a preservative in processed foods (see Food processing; Food additives). It also is used in other industries, eg, agriculture and photography (qv). The importance of vitamin C in nutrition and the maintenance of good health is well-documented. Over 22,000 references relating only to L-ascorbic acid have appeared since 1966–1967.

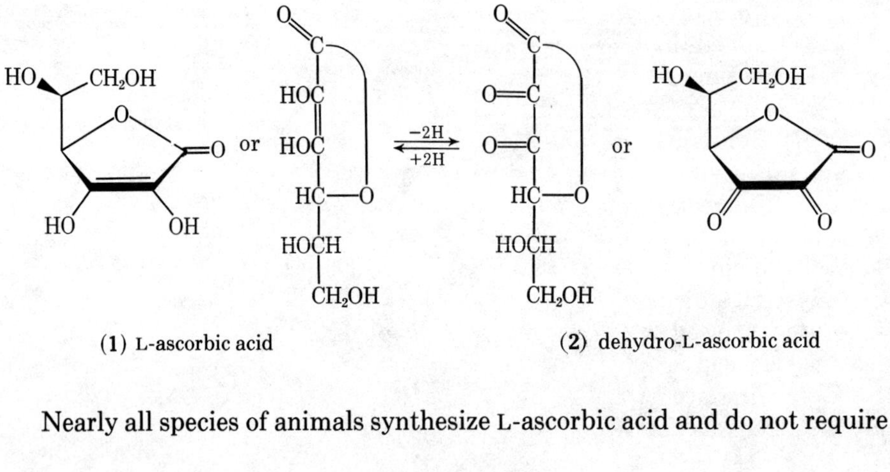

(**1**) L-ascorbic acid   (**2**) dehydro-L-ascorbic acid

Nearly all species of animals synthesize L-ascorbic acid and do not require it in their diets. Humans, other primates, the guinea pig, fruit-eating bats, some birds, including the red-vented bulbul, and related passeriformes and fish, eg, the coho salmon, rainbow trout, and carp, lack a liver enzyme, L-gulono-γ-lactone oxidase, and cannot synthesize the vitamin (2). When deprived of a dietary source for a sufficient length of time, they develop scurvy, the first deficiency disease to be recognized. Potentially fatal, it is characterized by anemia and alteration of protein metabolism; weakening of collagenous structures in bone, teeth, and connective tissues; swollen, bleeding gums with loss of teeth; fatigue and lethargy; rheumatic pains in the legs and degeneration of the muscles; skin lesions; and capillary weakness, massive hematomas in the thighs, and hemorrhages in many organs, including the eyes. Marginal deficiencies of the vitamin can occur under stressful situations from enhanced metabolism or poor utilization and can cause physiological changes that are not typical of scurvy but are detrimental to good health. Small quantities (10–60 mg/d) of L-ascorbic acid are sufficient to reverse the trend of both subclinical and clinical scurvy and alleviate their symptoms; however, larger quantities may be required for those whose marginal deficiencies are aggravated by environmental conditions, physiological stress, chronic disease, and other factors.

L-Ascorbic acid was isolated by 1921 and characterized and synthesized in 1933; however, its early history is associated with the etiology, treatment, and prevention of scurvy (3–5). Scurvy was observed in ancient times in Egypt, Greece, and Rome. It influenced the course of history, spontaneously ending many military campaigns and long ocean voyages by its outbreak resulting from inadequate amounts of vitamin C in rations. In the Middle Ages, scurvy was endemic in northern Europe when fresh fruits and vegetables were unavailable during winters. It was treated as a venereal disease with disastrous results.

The first clues to the treatment of scurvy occurred in 1535–1536 when Jacques Cartier, on advice from Newfoundland Indians, fed his crew an extract from spruce tree needles to cure an epidemic. In 1570, scurvy was prevented in sailors by the daily administration of two jiggers of lemon juice. Subsequently, it was proven in 1753–1757 by Dr. James Lind in his famous clinical experiment that the disease was associated with diet, was caused by lack of fresh vegetables and could be treated by feeding lemon juice. Evidence has shown that even with undefined scorbutic symptoms, vitamin C levels can be low, causing marked diminution in resistance to infections and slow healing of wounds.

The history of scurvy research leading to the discovery of vitamin C began in 1907, when it was observed that guinea pigs were as susceptible to scurvy as man, and the disease was experimentally produced in these animals (6). The findings led to the development of an assay for the biological determination of antiscorbutic activity of foodstuffs (7). Subsequent experimental diet studies were difficult because of the inactivation of the vitamin by air oxidation.

Between 1910 and 1921, the vitamin was isolated in almost pure form from lemons, and some of the physical and chemical properties were determined (8). It was discovered that vitamin C is easily destroyed by oxidation and best protected by reducing agents, and that 2,6-dichloroindophenol is reduced by solutions of the vitamin. Subsequent studies showed that the indophenol test measured only the vitamin and not dehydroascorbic acid which also has antiscorbutic activity. The actual isolation of vitamin C was first accomplished by Szent-Györgyi (9). Its relationship to the antiscorbutic factor was demonstrated in 1932 by Szent-Györgyi (9) and by King and

10   VITAMINS (ASCORBIC ACID)

Waugh (10). Subsequent isolations, structure determination, and synthesis are reported in refs. 10–11. The synthetic product was shown to have the same biological activity as the natural substance. It is reversibly oxidized in the body to dehydro-L-ascorbic acid (2) (L-*threo*-2,3-hexodiulosonic acid γ-lactone), a potent antiscorbutic agent with full vitamin activity.

Distribution

Ascorbic acid exists as a pool of ascorbate, which is distributed throughout the body with high concentrations in specific tissues (12). For example, the plasma in normal individuals contains 0.8–1.4 mg/100 mL, but the pituitary, adrenal, thymus, corpus luteum, and retina have concentrations several hundred times higher; brain, testicle, thyroid, small intestinal mucosa, lymph glands, lung, liver, spleen, white blood cells, pancreas, and salivary glands have concentrations 10–100 times that of plasma; and kidney, skeletal, smooth and cardiac muscle, and erythrocytes have ca 10 times that of plasma. Healthy male adults have an ascorbate body pool in excess of 1.5 g, which increases to 2.3–2.8 g with intakes of 200 mg/d; depletion of the pool to 600 mg initiates physiological changes and levels below 300 mg are associated with clinical signs of scurvy (13). The pharmacokinetic data indicate a normal ascorbate turnover of ca 60 mg/d; however, little is known about the factors controlling its rate, which is related to the total pool and is constant among individuals. It varies depending not only on daily intake but also on such external situations as smoking, stress, chronic diseases, diabetes, and other conditions (13–14). Smokers, for example, have enhanced metabolic turnovers of ca 100 mg/d and lower body pools, unless compensated for by increased daily intakes of ascorbic acid (15). Other factors, eg, impaired utilization, absorption, and storage, may account for lower body pools in the elderly, and daily intakes of 150 mg may be required to maintain a 1.0 mg/100 mL plasma level (16). In some animals, eg, the guinea pig, there appears to be evidence that ascorbic acid may exist in the liver in a bound form which may be protected against degradation by liver enzymes.

It is also present in vegetables and fruit, as illustrated in Table 1 (17–19). Plants rapidly synthesize L-ascorbic acid from carbohydrates; some of the acid is metabolized into carbohydrates and reconverted into L-ascorbic acid for storage in the metabolic pool. Its content in most plants rarely exceeds 100 mg % of fresh weight, except for a few species that accumulate it in their tissues. Fresh tea leaves, some berries, guava, and rose hips are accumulators of L-ascorbic acid and are rich sources. For practical purposes, raw citrus fruits are good daily sources of L-ascorbic acid, since appreciable amounts in other foods can be destroyed when cooked in the presence of air and when in contact with traces of copper.

Physical Properties

The physical properties of both synthetic L-ascorbic acid and natural vitamin C have been examined, and the data were comparable for both substances (20). Table 2 contains a summary of L-ascorbic acid's physical properties. The mass spectrum of L-ascorbic acid obtained by electron impact ionization shows fragmentations at m/e: 116, 85, 71, 70, and 61 (26). Properties relating to the structure of ascorbic acid were recently reviewed (27). Stabilization of the molecule by delocalization of the π electrons

Table 1. Content of L-Ascorbic Acid in Representative Foods

Food substances	L-Ascorbic acid, mg/100 g
Meat, fish, and milk	
meat, beef, pork, fish	≤2
liver, kidney	10–40
cow's milk	1–2
Vegetables	
asparagus	15–30
brussel sprouts, broccoli	90–150
cabbage	30–60
carrots	9
cauliflower	60–80
kale	120–180
leek	15–30
onion	10–30
peas, beans	10–30
parsley	170
peppers	125–200
potatoes	10–30
spinach	50–90
tomatoes	20–33
Fruit	
apples	10–30
bananas	10
grapefruit	40
guava	300
hawthorne berries	160–800
oranges, lemons	50
peaches	7–14
pineapples	17
rose hips	1000
strawberries	40–90

over the conjugated enediol system causes the H atom on O-3 to become highly acidic. The protolytic property of this H atom has been confirmed by x-ray and neutron diffraction studies (28) (see Table 3). The five-membered ring containing the enediol grouping is almost planar. The conformation of the side chain in the crystal is shown by structure (3).

(3) L-ascorbic acid

The structure of crystalline dehydro-L-ascorbic acid (2) was recently reported to be a dimer (4) (29). In water, it is hydrated to a hemiacetal monomer (5), which eventually undergoes ring opening to free the side chain as in L-ascorbic acid. Electrochemical studies indicate that dehydro-L-ascorbic acid (nonhydrated form) and L-ascorbic acid form a perfectly reversible couple (30).

Table 2. Physical Properties of L-Ascorbic Acid

Property	Characteristics	Ref.
appearance	white, odorless, crystalline solid with a sharp acidic taste	
formula; mol wt	$C_6H_8O_6$; 176.13	
crystalline form	monoclinic; usually plates, sometimes needles	
mp, °C	190–192 (dec)	21
density, g/cm^3	1.65	21
optical rotation	$[\alpha]^{25}$ + 20.5° to +21.5° (c = 1 in water)	21
	$[\alpha]^{23}$ + 48° (c = 1 in methanol)	
pH		21
5 mg/mL	3	
50 mg/mL	2	
pK_1	4.17	21
pK_2	11.57	21
redox potential	first stage: E_0^1 + 0.166 V (pH 4)	21
solubility, g/mL		21
water	0.33	
95 wt % ethanol	0.033	
absolute ethanol	0.02	
glycerol USP	0.01	
propylene glycol	0.05	
ether	insoluble	
chloroform	insoluble	
benzene	insoluble	
petroleum ether	insoluble	
oils	insoluble	
fats	insoluble	
fat solvents	insoluble	
Spectral properties		
uv	pH 2: E_{max} (1%, 1 cm) 695 at 245 nm (nondissociated form)	22
	pH 6.4: E_{max} (1%, 1 cm) 940 at 265 nm (monodissociated form)	
ir (KBr)	characteristic wavelengths, cm^{-1}	23
	3455, 3405, 3155 ν OH groups	
	2570 associated OH groups	
	1770, 1670 carbonyl lactone	
	1254 C—O—C lactone	
	1057 δ OH groups	
nmr[a]	^{1}H nmr (D$_2$O)	24
	δ 4.97 (d, 1 H, $J_{4,5}$ = 2 Hz, H-4), 4.10	
	(ddd, 1 H, $J_{5,6}$ = 5 and 7 Hz, $J_{4,5}$ = 2 Hz, H-5),	
	3.78 (m, 2 H, C-6)	
	^{13}C nmr (D$_2$O)	25
	δ 174.0 (C-1), 118.8 (C-2), 156.3 (C-3),	
	77.1 (d, $J_{C\text{-}4,\,H\text{-}4}$ = 158.8 Hz, C-4),	
	69.9 (d, $J_{C\text{-}5,\,H\text{-}5}$ = 145.0 Hz, C-5),	
	63.2 (t, $J_{C\text{-}6,\,H\text{-}6}$ = 145.0 Hz, C-6)	

[a] d = doublet; ddd = doublet of doublet of doublet; m = multiplet; and t = triplet.

(4) dehydro-L-ascorbic acid dimer

(5) dehydro-L-ascorbic acid hemiacetal

Table 3. X-Ray Studies of Crystalline L-Ascorbic Acid, Crystal Data[a]

Space group	P 21
a	1.7299 nm
b	0.6353 nm
c	0.6411 nm
β	102° 11'
V	0.68859 nm^3
Z	4
d_{calcd}	1.699 g/cm^3
μ (Cu K$_\alpha$)	13.9 cm^{-1}

[a] Ref. 28.

Chemical Properties

The reversible oxidation–reduction with dehydro-L-ascorbic acid is L-ascorbic acid's most important chemical property and the basis for known physiological activities and stabilities (31). Dehydro-L-ascorbic acid has been prepared by uv irradiation and by oxidation with halogens, ferric chloride, hydrogen peroxide, 2,6-dichlorophenol, indophenol, neutral potassium permanganate, selenium oxide, and many others. It has been reduced to L-ascorbic acid by hydrogen iodide, hydrogen sulfide, etc, without affecting the lactone ring. Solid L-ascorbic acid is stable when dry and gradually darkens on exposure to light.

The degradation of L-ascorbic acid in aqueous solution depends on many factors, eg, pH, temperature, presence or absence of oxygen, and presence or absence of metals. Comprehensive reviews of degradation reactions and mechanisms have been published (32–33). In aqueous solution, it is more sensitive to alkalies than to acids; it is most stable at pH 4–6. L-Ascorbic acid is sensitive to heat, and in the presence of oxygen and heat, it is oxidized at a rate proportional to the temperature rise. On standing, dehydro-L-ascorbic acid undergoes irreversible hydrolysis to 2,3-dioxo-L-gulonic acid (*threo*-2,3-hexodiulosonic acid) (**6**) and then oxidation to oxalic acid (**7**) and L-threonic acid ([R-(R^*,S^*)]-2,3,4-trihydroxybutanoic acid) (**8**) (see Fig. 1). In acidic solution, the degradation proceeds further, forming furfural, furfuryl alcohol, 2-furoic acid, 3-hydroxyfurfural, furoin, 2-methyl-3,8-dihydroxychroman, ethylglyoxal, etc. Most metals, especially copper, catalyze the oxidation of L-ascorbic acid. Oxalic acid exerts an inhibitory effect by chelating the copper in a nonionic form, which prevents the initial formation of a complex with L-ascorbic acid. Other stabilizers are metaphosphoric acids, amino acids, 8-hydroxyquinoline, glycols, etc. Another catalytic reaction, which accounts for loss of the vitamin, occurs with enzymes, eg, L-ascorbic acid oxidase, which is a copper–protein-containing enzyme. A synthetic enzyme prepared from calcium acetate, alkali phosphate, and a trace of copper ion behaves like the natural enzyme.

Biochemistry. Ascorbic acid's biochemistry and physiological functions have been extensively reviewed (17). It is an important reducing agent that functions primarily as an electron carrier. By giving up two electrons, it is converted to dehydro-L-ascorbic acid. Loss of one electron by interaction with oxygen, metals, or both, produces the reactive monodehydro-L-ascorbate free radical which is reduced to L-ascorbic acid by various enzymes in animals and plants (34). In some model systems, reactions occur because of its redox potential and not its coenzyme activity.

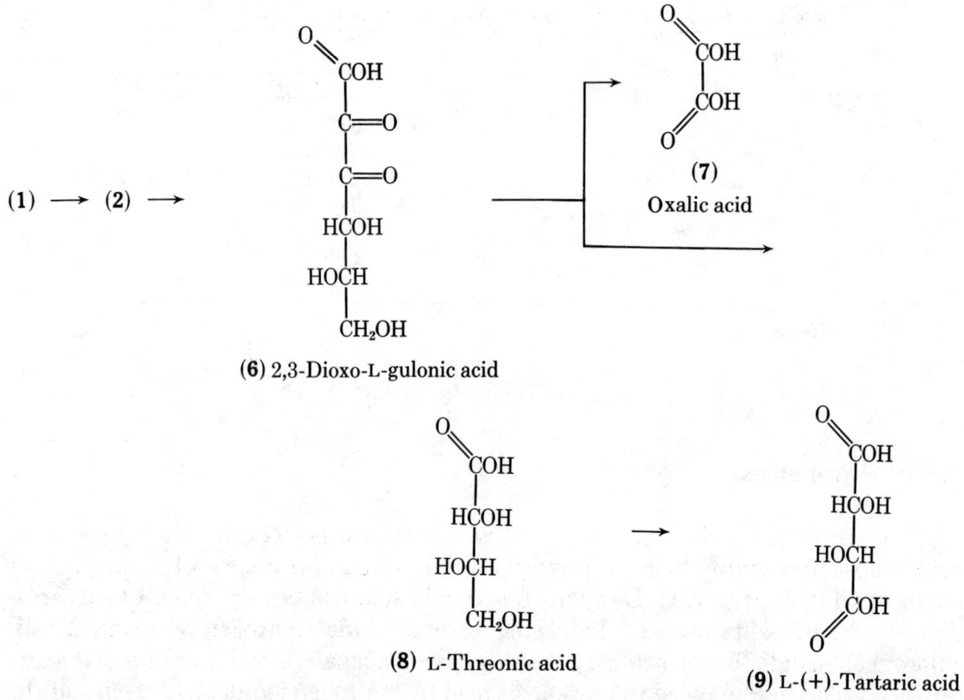

14 VITAMINS (ASCORBIC ACID)

(1) → (2) →

(6) 2,3-Dioxo-L-gulonic acid

(7) Oxalic acid

(8) L-Threonic acid

(9) L-(+)-Tartaric acid

Figure 1. Degradation of L-ascorbic acid.

Biosynthesis in Animals. Animals produce L-ascorbic acid by the glucuronic acid pathway in the livers of mammals and in the kidneys of other vertebrates, eg, amphibians and reptiles (35). The reactions involved in rats are illustrated in Figure 2;

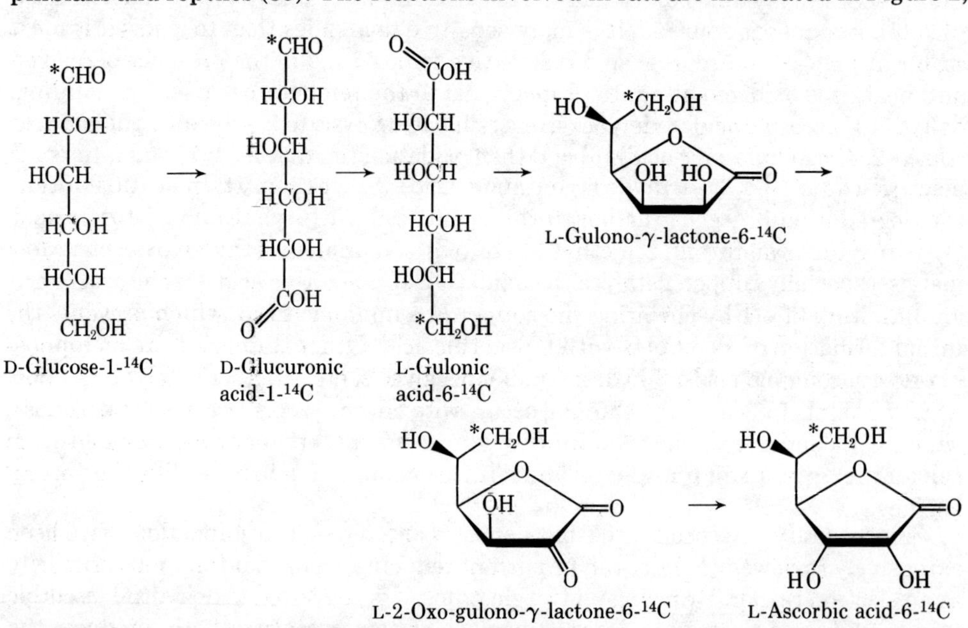

D-Glucose-1-^{14}C → D-Glucuronic acid-1-^{14}C → L-Gulonic acid-6-^{14}C → L-Gulono-γ-lactone-6-^{14}C → L-2-Oxo-gulono-γ-lactone-6-^{14}C → L-Ascorbic acid-6-^{14}C

Figure 2. Pathway for the biosynthesis of L-ascorbic acid in rats using C-1 labeled D-glucose; * indicates position of ^{14}C.

the incorporation of ^{14}C from D-glucose-1-^{14}C into urinary L-ascorbic acid in the C-6 position to L-ascorbic acid-6-^{14}C was measured. D-Glucose-6-^{14}C in similar experiments becomes L-ascorbic acid-1-^{14}C, and feeding either D-glucurono-γ-lactone and L-gulono-γ-lactone increases the amount of urinary L-ascorbic acid (36). In this pathway, the D-glucose chain remains intact, the C-1 and C-6 of D-glucose become the C-6 and C-1, respectively, of L-ascorbic acid, and the sequence of carbon-chain numbering is inverted. Humans, other primates, guinea pigs, fruit bats, some birds, and certain fish cannot synthesize L-ascorbic acid; thus, they must consume vitamin C from exogenous sources to survive (37). They lack L-gulono-γ-lactone oxidase, an essential oxidizing enzyme in liver, which converts L-gulono-γ-lactone into 2-oxo-L-gulono-γ-lactone, a tautomer of L-ascorbic acid, which presumably transforms spontaneously into the vitamin. The conversion of exogenous myo-inositol by carp into L-ascorbic acid may also proceed via D-glucuronic acid (38).

Glucuronic acid pathway. In animals, the glucuronic acid pathway (see Fig. 3) is an important route for glucose utilization, for the formation of glucuronides and mucopolysaccharides, and a principal detoxification process. Hexoses are transformed by a series of biochemical processes via D-glucuronic acid to L-ascorbic acid, L-xylulose, and D-glucaric acid. It was largely elucidated from studies of the biosynthesis of L-ascorbic acid. The pathway is important in the metabolism of sugars under normal and diseased conditions and in regulating physiological functions (39). Sugars entering the cycle at various positions are subject to different metabolic processes, including the requirement for insulin only between D-glucose and D-glucose-6-phosphate. The myo-inositol oxidation pathway has been implicated in the formation of free D-glucuronic acid and in the conversion of D-glucose into D-galacturonic acid in plants (40) (see Vitamins, inositol). However, there is no evidence linking the D-glucuronic acid biosynthesis in plants to L-ascorbic acid formation. L-Gulonic acid is the precursor of L-ascorbic acid and L-xylulose in animal tissue only.

The biosynthesis of L-ascorbic acid is inhibited by deficiencies of certain vitamins, eg, vitamin A, vitamin E, biotin, etc, but is stimulated by certain drugs, eg, barbiturates, chlorobutanol, aminopyrine, and antipyrine, and by carcinogens, eg, 3-methylcholanthrene and 3,4-benzpyrene (41). Aminopyrine and antipyrine also increase the excretion of L-xylulose in pentosuria patients who cannot synthesize 3-oxo-L-gulonic acid or reduce L-xylulose because of a genetic defect. The stimulation effects from exposure to xenobiotic compounds are attributed to induced biosynthesis of enzymes involved in the glucuronic acid pathway and are part of the body's drug detoxification process. It has been proposed that the excretion of D-glucaric acid can be used to diagnose both the exposure to body foreign substances and the drug metabolic capacity of the liver (42).

Biosynthesis in Plants. As in animals, L-ascorbic acid is also the product of hexose phosphate metabolism in plants, but its biosynthesis is more complicated. Studies with ripening strawberries and germinating cress seedlings suggest two mutually exclusive biosynthetic pathways for the conversion of D-glucose or D-galactose to L-ascorbic acid in plants (43). Both conserve the six-carbon chain and involve either C-1 or C-6 oxidation of the hexose. D-Glucose-1-^{14}C yields L-ascorbic acid with 70–80% of the activity in C-1 and 20–30% in C-6. Correspondingly, D-glucose-6-^{14}C gives ca 70% of the activity in C-6 of the L-ascorbic acid and 24% in the C-1 position. D-Galactose-1-^{14}C gives similar results. The main pathway is postulated as involving retention of configuration by successive oxidation at C-1, lactonization, oxidation at

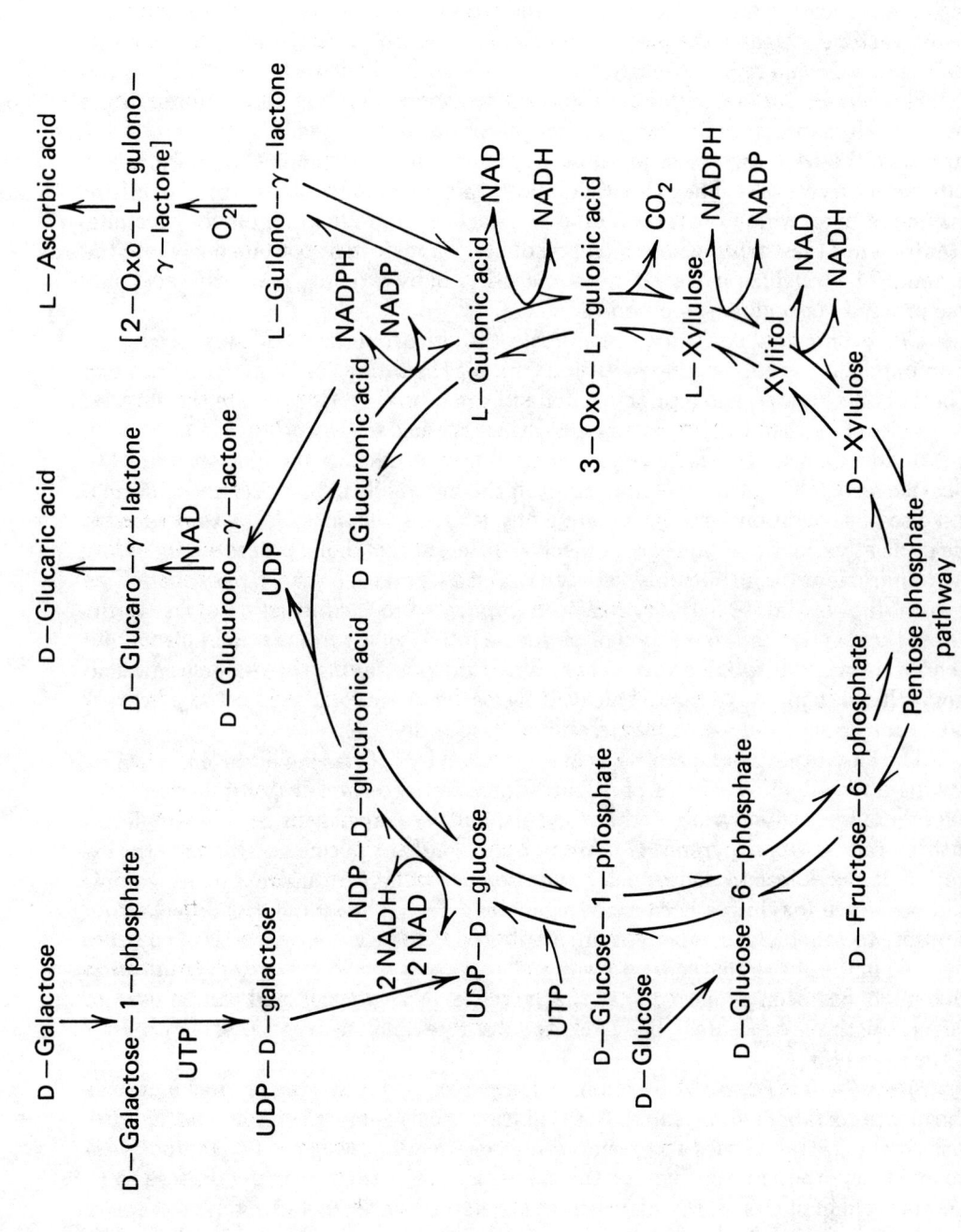

Figure 3. Glucuronic acid pathway. NAD = nicotinamide-adenine dinucleotide; NADH = reduced nicotinamide-adenine dinucleotide phosphate; NADPH = reduced nicotinamide-adenine dinucleotide phosphate; NDP = nucleoside diphosphate; UDP = uridine diphosphate; and UTP = uridine triphosphate.

C-2 or C-3, and epimerization at C-5 (see Fig. 4). Supporting evidence was obtained with D-glucose-1-^{14}C, -2-^{14}C, -6-^{14}C, and with D-glucose-6-^{3}H. The epimerization appears to involve hydrogen exchange between C-5 and the medium as D-glucose-5-^{3}H-6-^{14}C completely loses the ^{3}H label at C-5 (44). The epimerization mechanism is not yet fully known.

The other biosynthetic pathway from D-glucose and D-galactose may be postulated similarly to Figure 2 to account for configurational inversion (45). Both lower and higher plants form predominately L-ascorbic acid-6-^{14}C from D-glucose-1-^{14}C, D-galactose-1-^{14}C, D-glucurono-γ-lactone-1-^{14}C, and D-galacturonic acid-1-^{14}C and its methyl ester (43,46). The main precursor of L-ascorbic acid is L-galactono-γ-lactone, rather than L-gulono-γ-lactone which is less active (45–46), and is rationalized as being epimerized to L-galactono-γ-lactone prior to oxidation to L-ascorbic acid. This alternative pathway for the D-galactose series involves oxidation at C-6, reduction at C-1, and lactonization and oxidation at C-2 or C-3. But, in the case of the glucose series, epimerization at C-3 to L-galactono-γ-lactone must occur before the final oxidation (see Fig. 5). In some plants, eg, *Euglena gracilis*, the biosynthesis from D-glucose is further complicated; the intermediate D-glucuronic acid is also epimerized to D-galacturonic acid which is converted to L-ascorbic acid by a scheme similar to that in Figure 5 (47).

Metabolism. Studies with labeled L-ascorbic acid in rats and guinea pigs disclosed that the vitamin is oxidized to respiratory CO_2 (35). When given to humans in physiological doses of 50–100 mg/d, urinary oxalic acid is the principal metabolite; ca 30–50 mg/d form. But when given in large doses up to 10 g/d, the vitamin is excreted largely unmetabolized in urine and feces, and the urinary oxalate is increased by only 10–30 mg/d (15). The half-life of ascorbic acid is inversely related to the daily intake and is 13–40 d in humans and 3 d in guinea pigs, which correlates with the longer time for

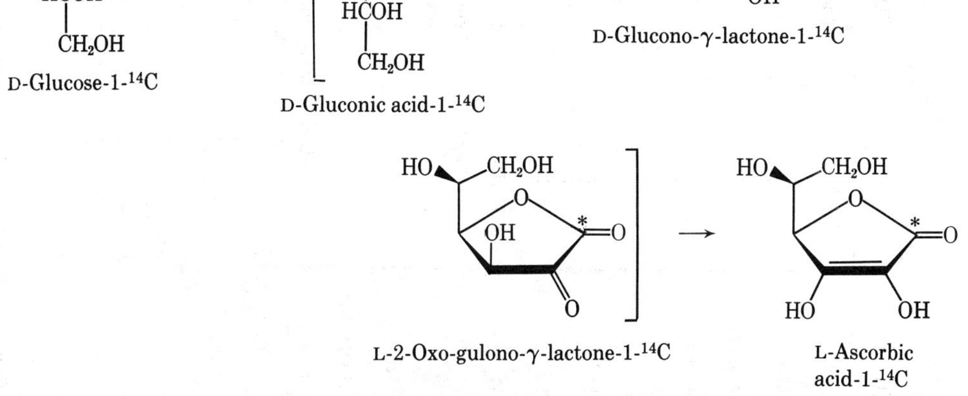

Figure 4. Suggested pathway for the biosynthesis of L-ascorbic acid (with retention of configuration) in higher plants based on D-glucose-1-^{14}C experiments; * indicates position of the label.

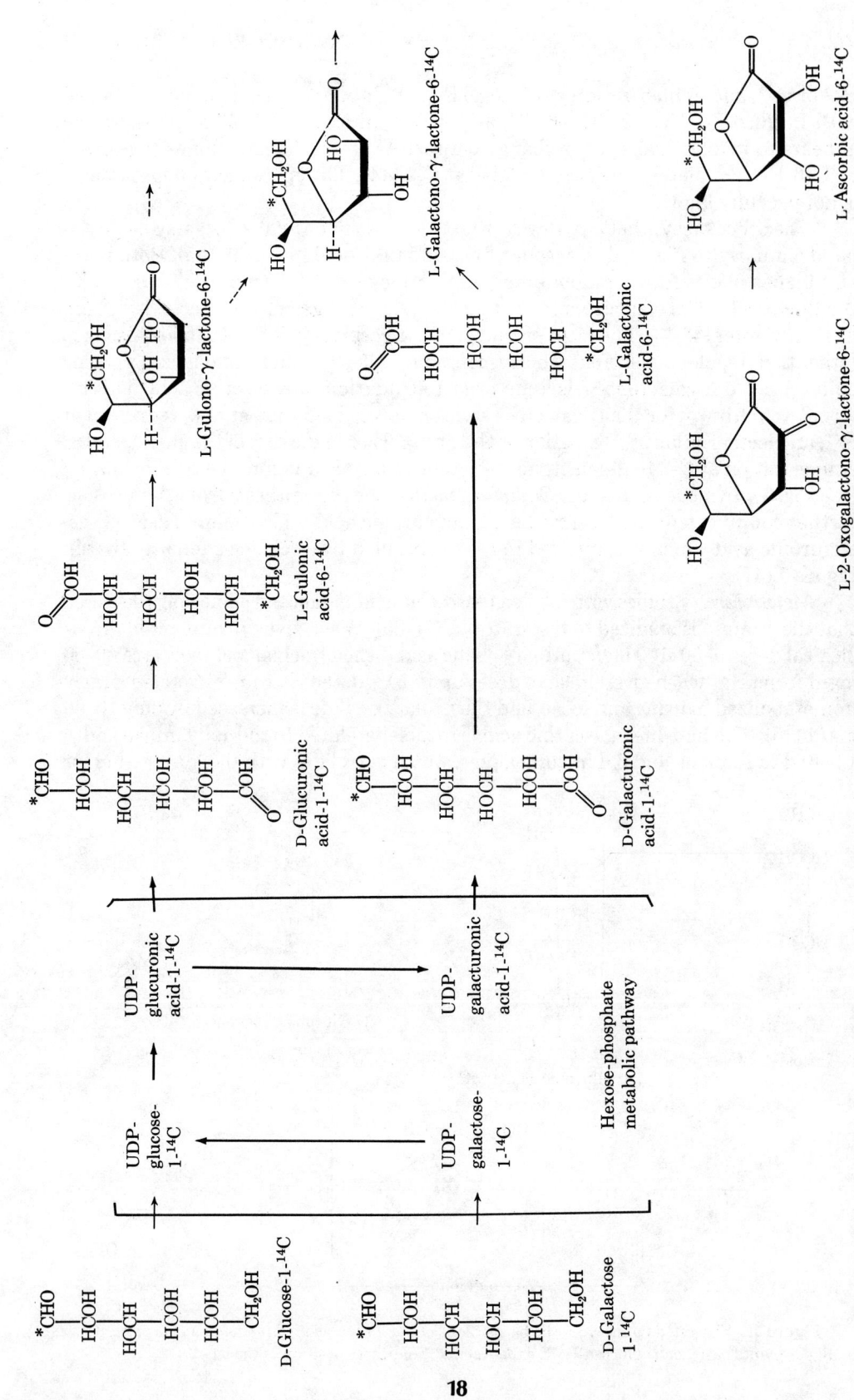

Figure 5. Proposed pathway for the biosynthesis of L-ascorbic acid (with inversion of configuration) in plants using C-1 labeled D-glucose or C-1 labeled D-galactose; * indicates position of the ^{14}C.

humans to develop scurvy, ie, 3 mo on a vitamin C-free diet compared to 3 wk for the guinea pig.

The steps in the metabolism of L-ascorbic acid are illustrated in Figure 6 (48). The formation of L-ascorbic acid 2-hydrogen sulfate (10) is a metabolic product in humans. The principal pathway involves loss of two successive electrons; the intermediate free radical immediately and irreversibly forms dehydro-L-ascorbic acid (2) which, as the monohydrated substance, forms a reversible electrochemical couple with L-ascorbic acid (30). Subsequent irreversible hydrolysis yields the biologically inactive 2,3-dioxo-L-gulonic acid (6) which undergoes degradation by alternative pathways: decarboxylation to carbon dioxide and five-carbon fragments as L-xylose, L-xylonic acid, and L-lyxonic acid and formation of oxalic acid and four-carbon fragments as L-threonic acid. Oxidation of L-ascorbic acid at the C-6 position to L-saccharoascorbic acid (L-*threo*-hex-2-enaric acid 1,4-lactone) (11) was reported to be an additional metabolic process in monkeys (49).

In plants, L-ascorbic acid metabolism parallels that observed in animals but is

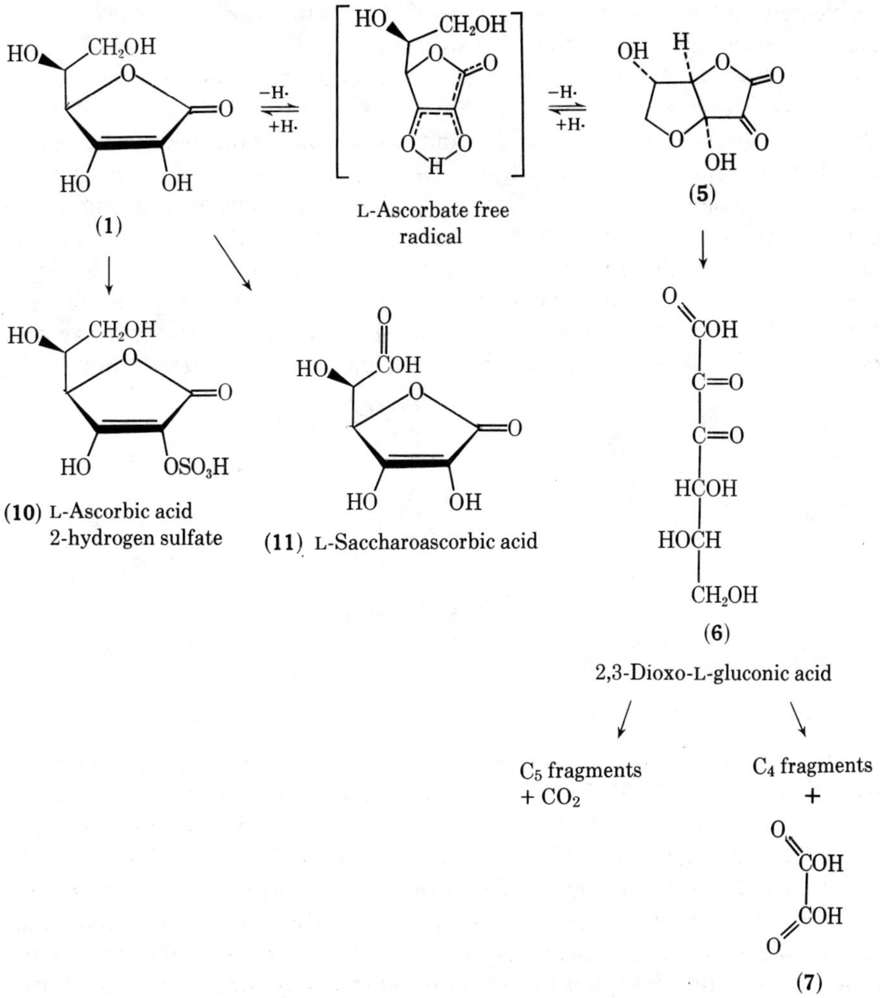

Figure 6. Metabolism of L-ascorbic acid.

further complicated by the formation of tartaric acid and two carbon fragments other than oxalic acid (43,50). Oxidative rupture of 2,3-dioxo-L-gulonic acid (6) occurs by two pathways, either between C-2 and C-3 or between C-4 and C-5. Tartrate-accumulating plants utilize both pathways to form L-(+)-tartaric acid (9); the former also yields oxalic acid (7), which can be metabolized to CO_2, and the latter pathway also yields a two-carbon fragment as glycolaldehyde, which is recycled by metabolism to hexoses. In some oxalate-accumulating plants, cleavage also occurs between C-2 and C-3, yielding oxalic acid (7) and L-threonic acid (8) instead of L-(+)-tartaric acid; six-carbon metabolites, eg, L-idonic acid, are also detected in plants and may be intermediates to tartaric acid (51).

Structure Determination and Synthesis

Chemical Constitution. The elucidation of the stereochemistry of L-ascorbic acid played a large role in its structure determination and in the development of syntheses and manufacturing processes. During the early studies, the configurational assignment of ascorbic acid to the L-series was confirmed by synthesis from L-xylose (13) (10–11,31,52–53). Subsequently, ascorbic acid was synthesized from D-glucose (12) because the chiral centers at C-2 and C-3 were in the correct configuration to become C-5 and C-4, respectively, of L-ascorbic acid (54).

The biosynthesis of L-ascorbic acid in animals and plants proceeds from D-glucose (12), D-galactose, and other carbohydrates. Its efficient, practical chemical synthesis starts with D-glucose (12), which is a readily available, inexpensive, pure enantiomer. These syntheses involve retention of the intact carbon chain and two alternative routes, whereby the C-1 of D-glucose becomes either the C-1 or the C-6 of L-ascorbic acid. Other carbohydrates, eg, L-gulose, L-idose, L-galactose, and L-talose, have the desired chirality at C-4 and C-5 to be considered potential starting materials for the synthesis of L-ascorbic acid, but their unavailability and high cost make them impractical.

(12) D-glucose   (13) L-xylose   (1) L-ascorbic acid

Many reviews include descriptions of structure determination studies and synthesis of L-ascorbic acid (17,31,53,55–56). The primary oxidation product, dehydro-L-ascorbic acid (2), becomes acidic as its lactone ring opens to form 2,3-dioxo-L-gulonic acid (6) with loss of vitamin activity (10,57). Further oxidation gives quantitative yields of oxalic acid (7) and L-threonic acid (8). The latter gives L-(+)-tartaric acid (9) on further oxidation (Fig. 1). X-ray studies with crystalline L-ascorbic acid reveal a cyclic, almost planar molecule with a perpendicular pseudoplane (58). Ultraviolet absorption studies performed in the mid-1930s led to the proposal that L-ascorbic acid was 3-oxo-L-sorbosone (14), which exists in different tautomeric forms (1), (15), and (16); several are represented. including L-ascorbic acid (1).

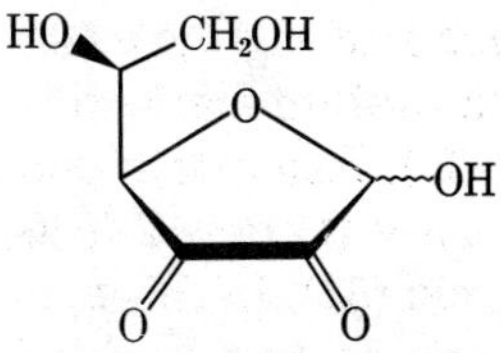

(14) 3-oxo-L-sorbosone 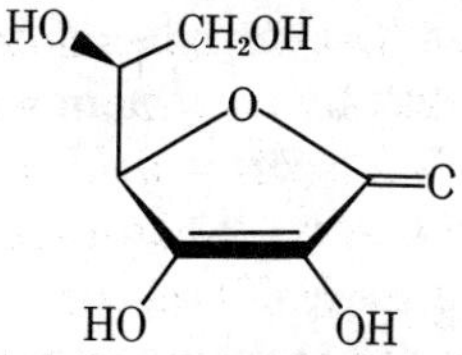   (1) 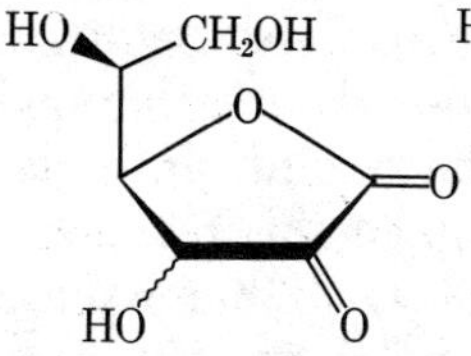   (15) 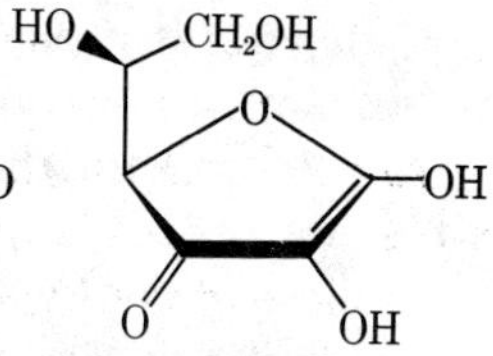   (16)

Syntheses. Prior to the completion of the structure elucidation studies, the first synthesis of L-ascorbic acid was reported in 1933 (11,52–53,57) (see Fig. 7). L-Xylose (13) was converted by way of its osazone (17) into L-xylosone (18), which reacted with hydrogen cyanide forming L-xylonitrile (19). Lactonization and enolization yielded L-ascorbic acid (1). Similar to the synthesis of L-ascorbic acid from L-xylose or L-xylosone, one of its enantiomers, D-araboascorbic acid (D-erythorbic acid, D-*erythro*-hex-2-enonic acid γ-lactone, D-isoascorbic acid) (22) was also prepared in 1933 by a similar series of reactions from D-arabinose (20) via D-arabinosone (21). The L-xylosone route to L-ascorbic acid was never commercialized, but it was valuable for structure determination and for synthesizing derivatives. It was lengthy, gave low yields (ca 40% from L-xylosone), and its key intermediate, L-xylosone, was too expensive and not readily available.

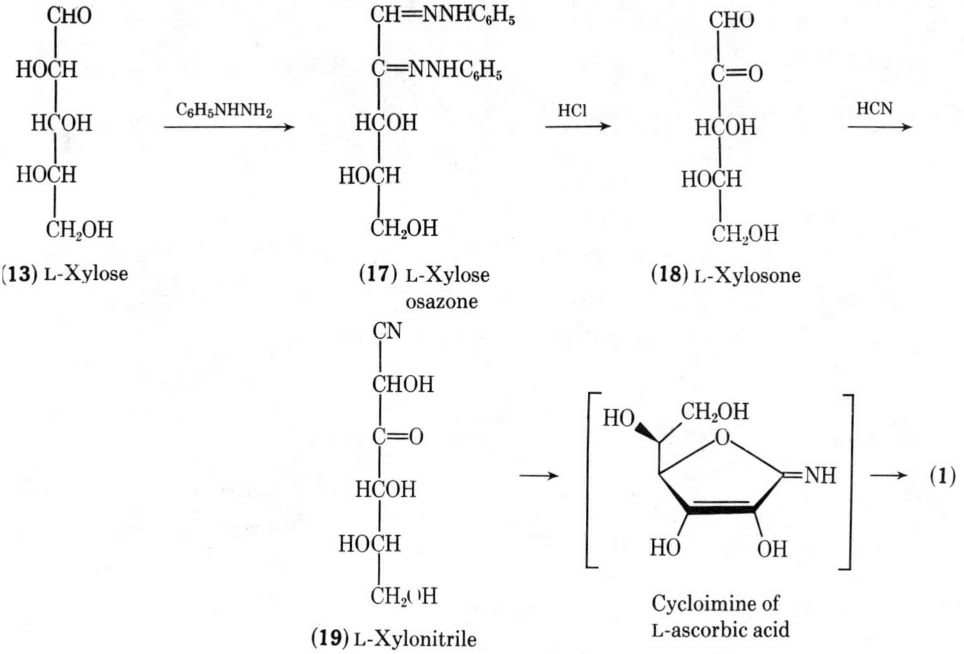

(13) L-Xylose → (17) L-Xylose osazone → (18) L-Xylosone → (19) L-Xylonitrile → [Cycloimine of L-ascorbic acid] → (1)

Figure 7. First reported synthesis of L-ascorbic acid.

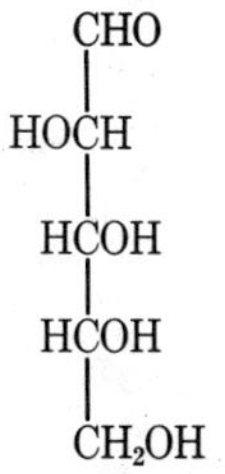

(20) D-arabinose 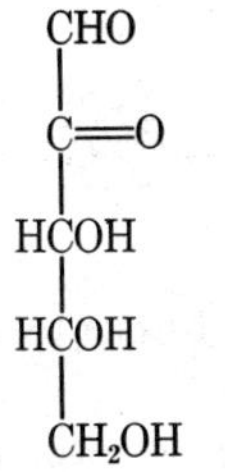   (21) D-arabinosone 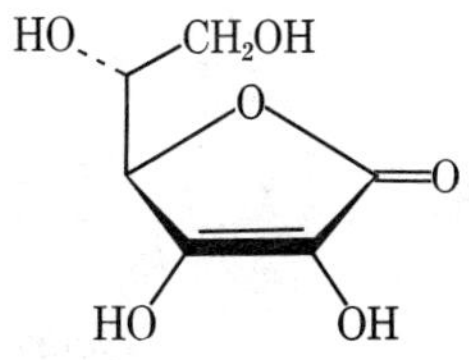   (22) D-erythorbic acid

22   VITAMINS (ASCORBIC ACID)

Reports that methyl-2-oxo-D-gluconate (D-*xylo*-2-hexulosonic acid methyl ester) is transformed into D-araboascorbic acid (**22**) by treatment with sodium methoxide were the basis for the second synthesis of L-ascorbic acid (59–60). Reichstein and Grüssner readily applied this reaction to methyl 2-oxo-L-gulonate (L-*xylo*-2-hexulosonic acid methyl ester) (**28**) for the synthesis of L-ascorbic acid (54). In this synthesis (see Fig. 8), D-glucose (**12**) was quantitatively hydrogenated to D-sorbitol (D-glucitol) (**23**), which was microbiologically oxidized to L-sorbose (**24**) with *Acetobacter xylinum* in ca 60% yield. L-Sorbose, on treatment with acetone and sulfuric acid, gave a mixture of 2,3-*O*-isopropylidene-α-L-sorbose and 2,3:4,6-bis-*O*-isopropylidene-α-L-sorbofuranose (**25**). The latter, after isolation, was oxidized with potassium permanganate in alkaline solution to 2,3:4,6-bis-*O*-isopropylidene-2-oxo-

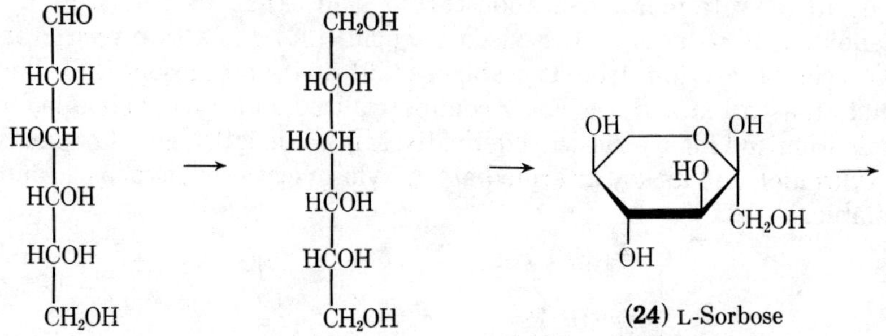

(**12**) D-Glucose   (**23**) D-Sorbitol   (**24**) L-Sorbose

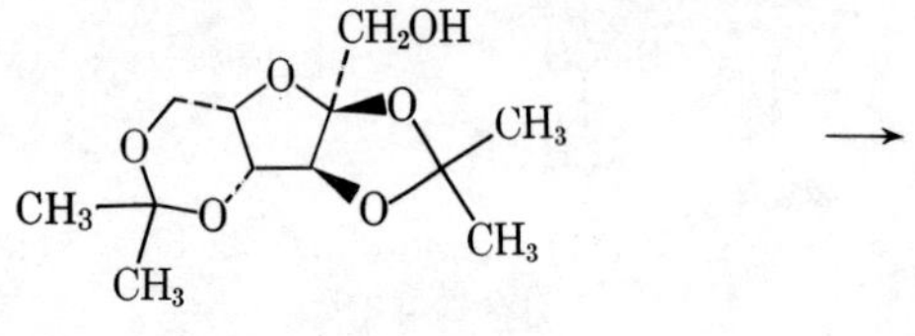

(**25**) 2,3:4,6-Bis-*O*-isopropylidene-α-L-sorbofuranose

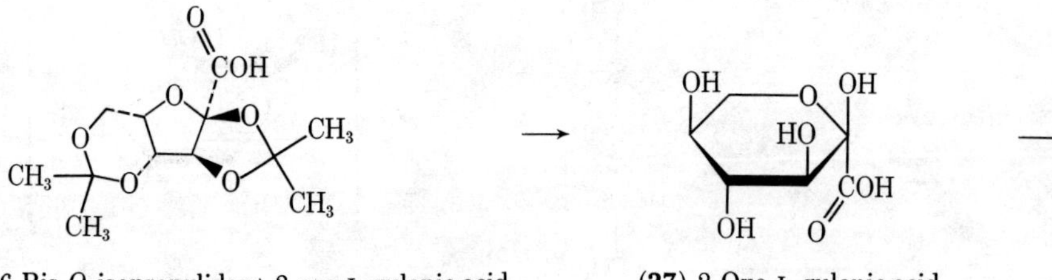

(**26**) 2,3:4,6-Bis-*O*-isopropylidene-2-oxo-L-gulonic acid   (**27**) 2-Oxo-L-gulonic acid

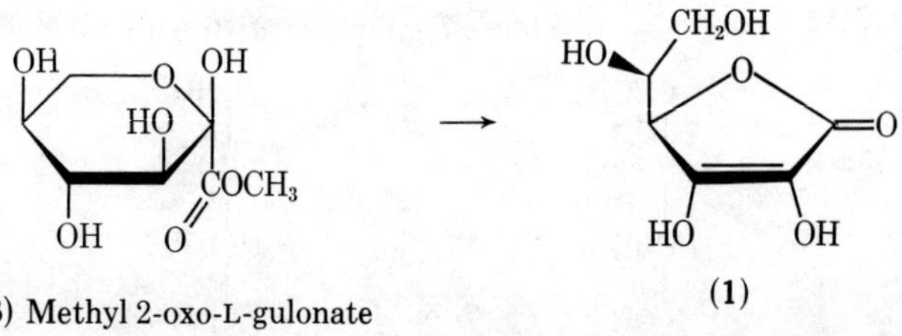

(**28**) Methyl 2-oxo-L-gulonate

(**1**) L-Ascorbic acid

Figure 8.   Second synthesis of L-ascorbic acid.

L-gulonic acid (α-L-*xylo*-2-hexulofuranosonic acid 2,3:4,6-bis-*O*-1-methylethylidene) (**26**), which was hydrolyzed in water to 2-oxo-L-gulonic acid (L-*xylo*-2-hexulosonic acid) (**27**). The latter was converted in 20% yield to L-ascorbic acid by heating in water. However, esterifying the acid first with methanol and hydrochloric acid gave the methyl ester (**28**), which on treatment with sodium methoxide and neutralization with hydrochloric acid, gave L-ascorbic acid in ca 70% yield.

The overall yield of L-ascorbic acid from D-glucose was ca 15–18% and from L-sorbose it was ca 30%. The method was efficient and became the basis for the industrial route to L-ascorbic acid. Its two main features are the use of low cost, readily available D-glucose as the starting compound and the use of the fully protected intermediate 2,3:4,6-bis-*O*-isopropylidene-α-L-sorbofuranose (**25**) for the second oxidation, which precludes overoxidation and other side reactions.

Various attempts to develop a superior, more economical L-ascorbic acid process have been reported since 1934. These approaches are summarized in ref. 55 and are outlined below.

Syntheses from D-glucose include (1) Inversion of the glucose chain (ie, C-1 of D-glucose becomes C-6 of L-ascorbic acid):

Reduction at C-1 followed by oxidation at C-5 and C-6
    Reichstein's second synthesis
    direct oxidation of D-sorbitol
    direct oxidation of L-sorbose
    oxidation of methyl α-L-sorbofuranoside

Oxidation at C-6 followed by reduction at C-1 and oxidation at C-5
    oxidation of D-glucurono-γ-lactone
    oxidation of L-gulonic acid, which is derived from D-glucuronic acid, and

(2) Noninversion of the glucose chain (ie, C-1 of glucose remains as C-1 of L-ascorbic acid):

Oxidation to 5-oxo-D-gluconic acid (D-xylo-5-hexulosonic acid) and reduction to L-idonic acid
    oxidation of L-idonic acid to 2-oxo-L-gulonic acid

Oxidation to 2,5-dioxo-D-gluconic acid (D-threo-2,5-hexodiulosonic acid) and reduction to 2-oxo-L-gulonic acid
    and direct fermentation to L-ascorbic acid.

Syntheses from other monosaccharides include synthesis from D-galactose, from L-xylose, and from L-threose.

Although no synthesis has reached the commercial importance of Reichstein's method, current developments in synthetic and biotechnological methods may provide the basis for a future advance; many workers view the direct synthesis of 2-oxo-L-gulonic acid from D-sorbitol, L-sorbose, L-idonic acid, or 2,5-oxo-D-gluconic acid as possibilities (55,61).

Manufacture

Although natural and synthetic vitamin C are chemically and biologically identical, in recent years, a limited amount of commercial isolation from vegetable sources, eg, rose hips, persimmon, citrus fruit, etc, has been carried out to meet the preference of some persons for vitamin C from natural sources. L-Ascorbic acid was the first vitamin to be produced in commercial quantities, and manufacture is based on Reich-

24 VITAMINS (ASCORBIC ACID)

stein's second synthesis (54). Many chemical and technical modifications have improved the efficiency of each step, enabling this multistep synthesis to remain the principal, most economical process (55). L-Ascorbic acid is produced in large, integrated, and automated facilities, involving both continuous and batch operations. The process steps are outlined in Figure 9. Procedures require ca 1.7 kg L-sorbose per kilogram of L-ascorbic acid with ca 66% overall yield (62).

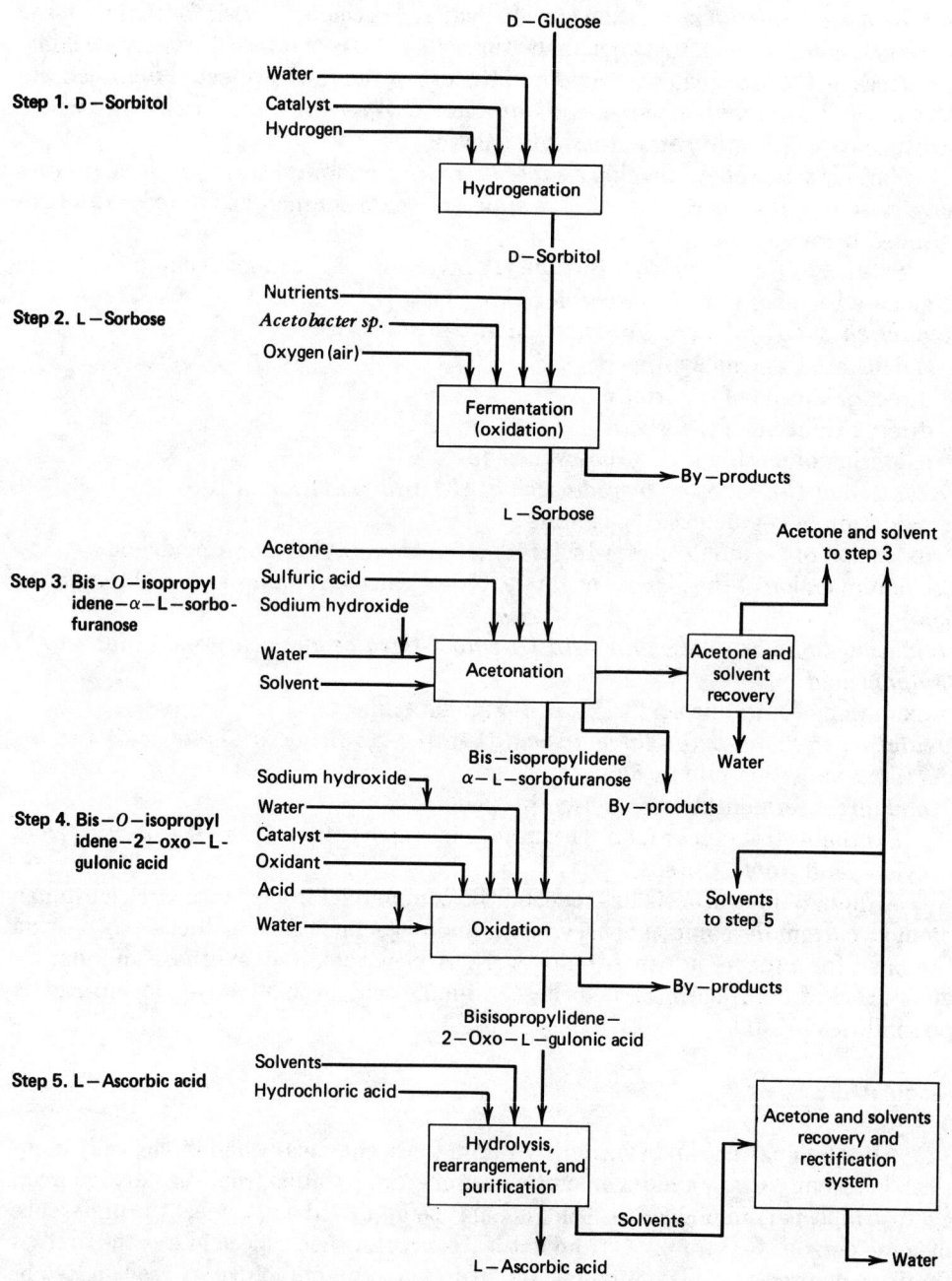

Figure 9. L-Ascorbic acid manufacture by the Reichstein and Grüssner synthesis (54).

The catalytic hydrogenation of D-glucose to D-sorbitol is carried out at elevated temperature and pressure with hydrogen in the presence of nickel catalysts in both batch and continuous operations with >97% yield (63–64). The cathodic reduction of D-glucose to L-sorbitol has been practiced (65). D-Mannitol is a by-product (66).

Sterile aqueous D-sorbitol solutions are fermented with *Acetobacter suboxydans* in the presence of large amounts of air to complete the microbiological oxidation. The L-sorbose is isolated by crystallization, filtration, and drying. Various methods for the fermentation of D-sorbitol have been reviewed (67). The microbiological oxidation of D-sorbitol to L-sorbose with *Acetobacter xylinum* was reported in 1904, and a procedure giving ca 75% yields was reported in 1933 (59,68–69). Other workers reported the application of different *Acetobacter*, *Bacterium*, and *Streptomyces* species for this reaction. *Acetobacter suboxydans* is the organism of choice as it gives L-sorbose in >90% yield (70). Large-scale fermentations can be carried out in either batch or continuous modes. In either case, sterility is important to prevent contamination with subsequent loss of product.

In the third step, L-sorbose reacts with acetone and excess sulfuric acid at low temperatures. The sorbose dissolves on conversion into the isopropylidene derivatives. The reaction mixture is worked up by dilution, neutralization, and extraction to separate the 2,3:4,6-bis-O-isopropylidene-α-L-sorbofuranose from the mono-O-isopropylidene-L-sorboses. Mother liquors are distilled to recover acetone and solvents, which are recycled. The original Reichstein and Grüssner process has been optimized by other workers to define practical reaction conditions giving ca 85% yields (62,71). The reaction of L-sorbose with acetone and sulfuric acid results in an equilibrium mixture of ca 80 wt % 2,3:4,6-bis-O-isopropylidene-α-L-sorbofuranose and ca 20% mono-O-isopropylidene-L-sorboses. The sulfuric acid acts as a catalyst and a dehydrating agent. Other acidic catalysts have been described, eg, zinc chloride–phosphoric acid, p-toluenesulfonic acid, copper(II) sulfate, and cationic exchange resins. Ferric chloride and perchloric acid are active catalysts that give >90% yields of bis-O-isopropylidene-L-sorbofuranose when azeotropic distillation is used for dehydration (72). Other methods involving acetone dimethylacetal, which eliminates the dependence on the removal of water to drive the reaction to completion, have been reported (73). The use of other ketonic and aldehydic protective agents for L-sorbose, eg, cyclohexanone and its derivatives, formaldehyde, benzaldehyde, etc, have also been described (54,74).

2,3:4,6-Bis-O-isopropylidene-α-L-sorbofuranose is oxidized at elevated temperatures in dilute sodium hydroxide in the presence of a catalyst: nickel chloride for sodium hypochlorite or electrolytic anodic oxidations or palladium–carbon for air oxidation. After completion of the reaction, the mixture is worked up by acidification to 2,3:4,6-bis-O-isopropylidene-2-oxo-L-gulonic acid, which is washed and dried. In the original Reichstein process and early modifications, potassium permanganate was the oxidant and almost 90% yields were obtained. Less expensive sodium hypochlorite with catalytic amounts of nickel chloride increased the yields to >90%; nickel peroxide may be the active oxidant (75). The oxidation with air was reported, and a practical process was developed with palladium–carbon or platinum–carbon as the catalyst (76–77). The practical electrolytic oxidation with nickel salts as the catalyst has also been described (78).

2,3:4,6-Bis-O-isopropylidene-2-oxo-L-gulonic acid is treated with ethanol and hydrochloric acid in the presence of inert solvents, eg, chlorinated solvents, hydrocarbons, ketones, etc. At the end of the reaction, crude L-ascorbic acid is filtered and purified by recrystallization from dilute ethanol. The pure product is isolated and dried. The mother liquors from the crude and recrystallization steps are treated in the usual manner to recover the L-ascorbic acid and solvents contained therein. The solvents and acetone liberated in the initial hydrolysis are recovered by distillation and fractionation for recycling. Many solvent systems have been reported for the acid-catalyzed conversion to L-ascorbic acid (55). Nonaqueous systems give >80% yields of high purity crudes (79). The L-ascorbic acid precipitates from the mixture as it forms, minimizing its decomposition, which can be significant during the reaction (80). The initial Reichstein and Grüssner process (54) for the conversion of 2,3:4,6-bis-O-isopropylidene-2-oxo-L-gulonic acid to L-ascorbic acid was base-catalyzed. Methyl 2-oxo-L-gulonate was first prepared, then converted to sodium L-ascorbate on treatment with sodium methoxide, and finally acidified. As with the acidic conversion to L-ascorbic acid, various solvents and bases have been reported (55).

Packaging. L-Ascorbic acid is screened or pulverized to a variety of particle sizes. It is usually packaged in 50-kg quantities in standard, polyethylene-lined containers, eg, fiber drums, corrugated boxes, etc. The recommended storage conditions are low humidity and temperatures of $\leq 23°C$.

Environmental Issues. The environmental concerns of an ascorbic acid manufacturing facility are typical of a chemical processing plant. Its operating design must be patterned to conform to environmental protection regulations. Measures must be taken to contain solvents and to keep emissions within official guidelines. Special condensers, continuous instrumental monitoring, and emergency containment and cleanup systems are required. Wastewater-treatment facilities may have to be provided to remove by-product organics and inorganics from effluent streams before disposal. The extent of these treatment facilities depends upon the location of the plant and the local tolerances. Usually, they are like secondary treatment facilities for organic removal; at some plant sites, additional treatment may be required to remove inorganic salts.

Economic Aspects

By the end of the 1970s, L-ascorbic acid accounted for 58% of the total domestic vitamin production and 37% of all vitamin imports (weight bases). U.S. production reached ca 10,900 metric tons per year, and world production was >30,000 t/yr (81). Table 4 contains a historical summary of production landmarks and resulting price changes. The declining L-ascorbic acid prices prior to 1974 reflect the successful development of Reichstein's second synthesis into a high-yield manufacturing process; implementation in well-engineered, efficient manufacturing plants; and economies of scale resulting from large facilities capable of producing >10,000 t/yr. The subsequent price increases are the result of worldwide inflation.

Price adjustments for inflation were made from the Producer Price Index. Based on 1966–1967 = 100 and the June 1982 index of 278.8, the $4.50 price in 1966 was equal to $12.55 in June 1982. The $12.00 selling price shown in Table 4 for 1982 further attests to the efficiency of the process and manufacturing operations. The rapid growth during the 1970s is attributed to L-ascorbic acid's growing popularity as a food supplement as a result of increasing awareness of its beneficial physiological effects, application of antioxidant properties in canned and frozen food preservation, and the expanding health-food industry.

Table 4. Historical Review of Vitamin C Production and Selling Prices

Year	Event	Selling price, $/kg
1933–1934	first productions from natural sources; announcements of Reichstein's second synthesis; production was ca 50 kg/yr	7000.00
1937	synthetic material produced commercially	127.00
1940		100.00
1945	annual U.S. production ca 17 t	25.00
1960	annual U.S. production ca 3000 t	6.00
1966	L-ascorbic acid demand increasing at ca 14%/yr; production ca 5000 t	4.50
1968–1973		3.25–4.25
1974	pressures from inflation and rapidly increasing raw material and energy costs gradually forced price increases	7.15–8.30
1979	U.S. production ca 10,000 t; world production ca 32,750 t	10.40
1982	future U.S. forecasts indicate a 6–8%/yr growth rate to ca 15,900 t/yr in 1984	12.00

Specifications

L-Ascorbic acid and derivatives are included in compendia such as USP, FCC (*Food Chemicals Codex*), etc. Specifications correspond to the physical properties listed in Table 2. The official assay is an iodimetric titration with standard iodine solution and starch as the indicator.

Analytical Methods

The first chemical method for assaying L-ascorbic acid was titration with 2,6-dichlorophenol–indophenol solution (82). This method and other chemical and physiochemical methods are based on the reducing character of L-ascorbic acid (83). Other oxidants, eg, iodine, potassium hexacyanoferrate(III), methylene blue, chloramine, and many others have been used for the assay, but they are not specific because other reducing substances in foods and natural products interfere (84). Photometric methods based on color formation with 2,4-dinitrophenylhydrazine, diazotized aromatic nitramines, etc are more specific. Polarographic methods have been widely used because of their accuracy and ease of operation. More recently, both gas–liquid (glc) and high performance liquid chromatographic (hplc) methods have been used for the assay of ascorbic acid. Sensitive hplc methods have been described for the assay of L-ascorbic acid in plants (85). Comprehensive reviews of all analytical methods, including the extraction of ascorbic acid from foods and biologic tissue, have been published (86).

Biochemical Functions

L-Ascorbic acid deficiencies involve metabolic effects that can be related to subclinical stages of classical scurvy. Abnormal effects are observed in collagen formation, fatty-acid metabolism, brain function, drug detoxication, infection, and fatigue. In plants, L-ascorbic acid is involved in cellular respiration, growth, and main-

VITAMINS (ASCORBIC ACID)

taining the carbon balance. Definitive hypotheses on its activities have not yet been formulated (43,50). It plays important roles in mixed-function oxidation reactions involving molecular oxygen, whereby only one of the two oxygen atoms is incorporated into the substance. L-Ascorbic acid appears to play a classical co-factorlike role as part of the active site of hydroxylating enzymes, whereas other functions with hydroxylases in collagen biosynthesis suggest a protective role. Its relation to transition-metal ions, its reducing properties, and its reaction with free-radical derivatives of oxygen may be central to its biological functions. Much still must be learned of its actions on a molecular level and the vitamin C requirements of man under various conditions. The nutritional aspects of L-ascorbic acid are reviewed in ref. 87.

Hydrogen and Electron Transport. L-Ascorbic acid's metabolism in plants via a two-step, biological, reversible oxidation reaction and an irreversible lactone-ring opening makes it an important agent for cellular respiration and plant growth. L-Ascorbic acid oxidase, a copper-containing enzyme present in plant tissue, effects the transfer of hydrogen from L-ascorbic acid to molecular oxygen. Glutathione (**29**) is part of the system reversing the L-ascorbic acid–dehydro-L-ascorbic acid interaction and may help maintain the plant's carbon balance (43). However, the factors that control L-ascorbic acid's turnover rate are not known.

(**29**) glutathione

Collagen Synthesis. The polypeptide collagens are the main component of skin and connective tissue, the organic substance of bones and teeth, and the ground substance between cells. Impaired syntheses of collagens, together with other factors, result in scurvy. The relationships of ascorbic acid to collagen formation and to overcoming high collagen turnover from tissues have been reviewed (48,88). The syntheses of collagens involve enzymatic hydroxylations of proline and of lysine; the former produces a stable extracellular matrix and the latter is needed for glycosylation and formation of cross-linkages in the fibers. The requirement for ascorbic acid is specific; it probably prevents inactivation of the hydroxylase enzymes by oxidation of the ferrous ions and thiol groups present. Some collagen synthesis occurs in the absence of L-ascorbic acid, but in such a case the fibrils are abnormal and fewer.

L-Ascorbic acid is also involved in the synthesis of fibroblasts and various microsomal and polysomal structures. It is also required for the synthesis of repair collagen but may not be needed for growth collagen. There is evidence that specific collagens are required for different tissues and organs and that other reducing agents can replace L-ascorbic acid in several of the hydroxylations and protective reactions against oxidation. In general, beneficial effects have been claimed for ascorbic acid in synthesizing the repair collagen, which in turn stimulates connective-tissue formation.

Carnitine Biosynthesis. Carnitine (γ-amino-β-hydroxybutyric acid, trimethyl betaine) (**30**) is a component of heart muscle, skeletal tissue, liver, and other tissues. It is involved in the transport of fatty acids into mitochondria, where they are oxidized to provide energy for the cell and animal (89). It is synthesized in animals from lysine and methionine by two hydroxylases, both containing ferrous iron and L-ascorbic acid. The substrate and co-factor requirements appear similar to those of collagen synthesis. Deficiencies of L-ascorbic acid could reduce the formation rate of carnitine, which can account for accumulation of triglycerides in blood, physical fatigue, lassitude, and other early scurvy symptoms.

(**30**) carnitine

L-Tyrosine Metabolism and Catecholamine Biosynthesis. L-Tyrosine metabolism and catecholamine biosynthesis occur largely in the brain, central nervous tissue, and endocrine system (48). L-Tyrosine metabolism involves two mixed-function oxidases which depend on the presence of L-ascorbic acid. In one, tyrosine is metabolized to homogentisic acid by *p*-hydroxyphenylpyruvate oxidase; ascorbate is involved in the oxidation and decarboxylation of the intermediate *p*-hydroxyphenylpyruvic acid (**31**). Without L-ascorbic acid, the intermediate substances *p*-hydroxyphenylpyruvic acid and *l*-*p*-hydroxyphenyllactic acid (**32**) are excreted as metabolites of the amino acid, such as in scorbutic patients and individuals with genetic defects in their tyrosine metabolism, eg, phenylketonuria (PKU) in the newborn.

(**31**)   (**32**)

L-Tyrosine is also metabolized to catecholamine, a neurotransmitter, by hydroxylation and decarboxylation, resulting in the formation of dopamine, which is then converted to noradrenaline, adrenaline, and adrenochrome (see Epinephrine and norepinephrine). L-Ascorbic acid is involved indirectly in the formation of dopamine and directly in the dopamine-β-hydroxylase reaction to noradrenaline. The ascorbate free radical (see Fig. 6) may be the primary product of the oxidations, as with other copper-containing enzymes such as ascorbate oxidase and peroxidases, but different from collagen hydroxylases. The catecholamine biosynthesis occurs in tissues, eg, adrenal, brain, etc, that have large pools of L-ascorbic acid. The L-ascorbic acid may enhance the biosynthesis by recycling 5,6,7,8-tetrahydrobiopterin required for the rate-limiting step of tyrosine hydroxylation (90). It may also protect the catecholamines by direct interaction and help remove adrenochrome, the toxic product of the oxidation which has been linked to mental disease (91). Various observations suggest complex interactions among catecholamines, their tissue receptors, and ascorbic acid. The precise role of ascorbic acid has not been completely elucidated (48).

Hydroxylation Reactions. Ascorbic acid has important biochemical functions with various hydroxylase enzymes in steroid, drug, and lipid metabolism (see Steroids). The involvement occurs during the initial stages of the interactions. Microsomal-cytochrome-P450-dependent hydroxylases catalyze the conversion of cholesterol to bile acids and the hydroxylation or demethylation of aromatic drugs and other xenobiotics, eg, carcinogens, pollutants, pesticides, etc, in the presence of L-ascorbic acid (48). Mitochondrial and microsomal steroid hydroxylases in the adrenal gland account for the formation of corticosteroids, which is impaired in scorbutic animals but reversed in the presence of L-ascorbic acid (92). In related important interactions, animals show diminished plasma glucocorticoid responses when exposed to stress situations (48).

Metal-Ion Metabolism. The reducing and chelating properties of L-ascorbic acid affect the absorption of metal ions from the diet and their mobilization and distribution throughout the body (see Chelating agents). Ascorbic acid tends to alleviate the effects of transition metals in the body. L-Ascorbic acid deficiency interferes with iron mobilization from the spleen but not from the liver; supplementary L-ascorbic acid increases its mobilization from body stores during the treatment of iron overload. Increased levels of iron in blood have been associated with increased turnover of L-ascorbic acid, decreased leucocyte ascorbate levels, and clinical signs of scurvy (93). This property of vitamin C is important in the treatment of anemias. Several recent comprehensive reviews on the interactions of ascorbic acid and metal ions have been published (94).

Endocrine and Allergic Responses. The effects of L-ascorbic acid on histamine metabolism related to scurvy and anaphylactic shock have been investigated and are varied (95) (see also Histamine and histamine antagonists). Increased histamine levels and excretion, as associated with allergies, are observed during the onset of scurvy when plasma ascorbate is low. Studies indicate that L-ascorbic acid prevents the accumulation of histamine and contributes to its degradation and elimination in the presence of copper ions (96). L-Ascorbic acid may also act by modulating prostaglandin syntheses which may mediate histamine sensitivity and cause relaxation by favoring the prostaglandin E series instead of the F series (97) (see Prostaglandins, Supplement Volume). It may also control histamine function by modulating cyclic nucleotide metabolism by inhibition of phosphodiesterases (98). Other interactions involving L-ascorbic acid in hypersensitivity reactions have been reviewed (99).

Interaction with Free Radicals. An important biological function of L-ascorbic acid is the destruction of free radicals derived from oxygen, eg, hydroxyl, singlet oxygen, and superoxide (100–102). These free-radical scavenging reactions are important in the eye and in the extracellular fluid of the lung as L-ascorbic acid protects against oxidizing agents, including ozone (103–104). The destructive oxidative effects, such as lipid peroxidation and cytotoxicity resulting from hydrogen peroxide generation, DNA damage, and mutagenesis, which have been attributed to vitamin C *in vitro*, are nullified by the removal of metals from the media and are held in check *in vivo* by the synergistic action of protective enzymes, eg, superoxide dismutase, glutathione peroxidase, and catalase and by the interaction of vitamin E in the lipid phase (102,105–106). The role of vitamin C in aging may be related to control of these interactions with the destruction of unwanted cell types (see Memory enhancing and antiaging drugs).

Immunological and Antibacterial Functions. Ascorbic acid is involved in the immunological and antibacterial functions of white blood cells by increasing their mobility, stimulating the energy-producing hexose monophosphate shunt within the cell, and protecting leucocyte membranes from oxidative damage (107). It also may play a role in increased interferon formation (108).

Uses

Vitamin C is an essential nutrient for health maintenance (109). The first uses of the vitamin were to prevent and treat scurvy. Subsequent uses developed from biochemical studies which delineated its extra-antiscorbutic activities related to maintenance of good general health. A daily intake of 60–100 mg of L-ascorbic acid is considered adequate, depending on body weight and rate of metabolism (see Table 5). Better methods and interpretations of results are needed to determine the optimum consumption of the vitamin for all people, especially those individuals who may have marginal deficiencies resulting from enhanced metabolic turnovers and lower body vitamin pools caused by exposure to environmental situations, stress, dietary inadequacies, and physiological and biochemical changes (90,110–113). Such people may require increased daily intakes of the vitamin (12–13,42,112). Alcoholics, smokers, the elderly, and surgical and cancer patients also fall into this category. Improved assessment methods can also furnish information on the mechanisms of the vitamin's actions in the body. Although the mechanisms are not completely defined, the vitamin does interact with essential nutrients and metabolites, thus accounting for its biochemical functions (114–115).

Most health uses of vitamin C depend upon administration of large amounts. Although its usefulness in these quantities is being debated, its safety and tolerance in humans at levels as high as 10 g/d has been demonstrated (116). At oral intakes up to 180 mg, 70% is absorbed, but at intakes of 1–12 g, the relative absorption decreases from ca 50% to ca 16% (117). The unabsorbed ascorbic acid is excreted *per se* in urine and feces and not as urinary oxalate (116). Reported adverse effects, eg, the metabolism of electrolytes and metals, interaction with vitamin B_{12} body stores, mutagenic activity,

Table 5. 1980 Recommended Vitamin C Daily Dietary Allowances[a]

Age group	Amount, mg
male	
adults, 19–51 yr and over	60
females	
adults, 19–51 yr and over	60
pregnancy, second half	80
lactation period	100
children	
<1 yr	35
1–3 yr	45
4–6 yr	45
7–10 yr	45
11–14 yr	50
15–18 yr	60

[a] Ref. 13.

urinary acidification, etc, were determined to be insignificant or nonexistent once definitive studies were conducted *in vivo* in the presence of protective enzymes (106).

Important uses of vitamin C relate to the prevention of megaloblastic anemia in formula-fed infants and other macrocytic anemias. L-Ascorbic acid increases the absorption of iron from the intestines and improves iron nutrition (93,118). The increased iron absorption was higher than expected with the ascorbic acid supplements tested; this factor proved important in raising the recommended daily allowance (RDA) for the vitamin from 45 mg to 60 mg (13).

Vitamin C requirements for wound repair, normal healing processes, and trauma are based upon its role in collagen synthesis and fiber cross-linking (119). Both processes are necessary to form new tissue of high tensile strength, yet the amounts required in surgical patients are still undefined (120). In recent studies, the ascorbic acid content of white blood cells was observed to be 42% lower on the third postoperative day (121). The results did not correlate with the extent of tissue trauma nor the volume of blood transferred, but could be related to leucocytosis and release of low ascorbate-containing white blood cells from the bone marrow. The results suggest a need for increased ascorbic acid during and after surgery. Other studies show that pressure sores, burns, and periodontal disease respond favorably to increased ascorbic acid intake (122–124).

Stimulation of the immune system by vitamin C accounts for its use in the prevention and treatment of infections (107). Mononuclear leucocytes are important to immunocompetence and normally contain the highest levels of ascorbate among the various cellular components of blood, despite variations in the plasma's ascorbate content (12). The mobility of white blood cells is also increased by the vitamin; their autoxidation is inhibited, serum levels of immunoglobulins are higher, and more antibodies can form (125). The extracellular ascorbic acid resulting from high daily supplements may also be important in the immune response by protecting leucocyte membranes from oxidative damage. The use of increased amounts of the vitamin with infections appears justified.

The use of ascorbic acid in ameliorating and speeding recovery from the common cold is based on immunostimulation and is recommended by Pauling (126). The vitamin's metabolism may be altered during the cold's early stages by losses of ascorbate from leucocytes, decreases in both plasma and urinary ascorbate, and increases in blood histamine levels (127–129). Reports from several studies of the effects of supplementation reveal only a small moderating effect on the severity and duration of colds (130). However, other studies claim positive results (131).

Vitamin C's possible role in cancer prevention, therapy, and management is being studied intensively. It has been related to immunostimulation, collagen encapsulation of tumors, deficiencies in tissue, activity as an antioxidant, and interaction with free radicals (132). In early studies, ascorbic acid protected mice against bladder malignancies from implanted 3-hydroxyanthranilic acid, a known carcinogen (133). It inhibited the growth of transplanted solid sarcoma-180 in mice and exhibited a synergistic effect with copper ions by inhibiting tumor growth (134–135). Nitrosamines and other *N*-nitroso compounds (which are known genotoxins and carcinogens) derived from foods, metabolic processes, smoking, and environmental sources have been prevented from forming by L-ascorbic acid; however, the vitamin has no effect on preformed nitrosamines or related substances (136). Large supplements of vitamin

C were reported to improve the quality and length of life of cancer patients and to cause remissions (137–139). Confirmation of these findings by others has not been completely successful, perhaps because of differences in experimental conditions, such as prior suppression of the immune system (140). Ascorbic acid may kill certain tumor cells, increase the effect of some tumor therapeutic agents, and stimulate the host's immune system against residual tumor cells (141). These actions may be related to enhanced hydrogen peroxide formation in tumor cells. In some tumors, it may increase the biological effectiveness of the red blood cells' normal oxygen dissociation of hemoglobin by increasing the formation of 2,3-diphosphoglycerate (142). However, intravenous ascorbic acid changes the electron spin resonance (esr) spectrum of erythrocytes so that the spectra resemble those from certain leukemias (143). The semidehydroascorbate radical has been detected, suggesting a diminished content of ascorbate oxidase or comparable enzymes followed by uncontrolled buildup of ascorbic acid in the cell and unregulated interactions with some blood constituents, eg, Cu^{2+} proteins, which produces the radical. *In vitro* experiments show that ascorbate oxidase reverses the effect, causing the leukemic erythrocytes to have normal esr spectra. A possible chemoprevention role has been suggested for ascorbic acid in blocking the transformation by carcinogens of normal cells into precancerous ones. This may be in addition to other associated roles, such as blocking both tumor cell initiation and formation of carcinogenic nitrosamines (144). Evidence that cancers may be influenced by dietary patterns has been reviewed and affirmed by a National Research Council committee, which recommended that dietary guidelines be periodically reviewed to provide new insights on the disease (145).

The role of vitamin C in lipid metabolism and its possible usefulness in cardiovascular disease were recently reviewed (146–147). Experimental, clinical and epidemiological evidence demonstrated L-ascorbic acid mediation in lowering the incidences of hypercholesterolemia and atherosclerosis. High levels of cholesterol in blood plasma are reduced by its metabolism to bile acids in the liver by mixed-function hydroxylases. Triglycerides are metabolized by lipolytic enzymes in the carnitine biosynthesis pathway to lower their level in serum. Damage to the normal structure and permeability of blood-vessel walls can be avoided by vitamin supplementation, which increases collagen and mucopolysaccharide syntheses within the walls. Reduced incidence and severity of thrombotic episodes, which are related to increased fibrinolytic activity, were reported. Indirect evidence that vitamin C deficiency may be a factor in arterial disease is indicated by an inverse relationship between L-ascorbic acid intake and human mortality rates (147–148). More recently, lower serum and leucocyte ascorbate levels have been reported in patients with coronary artery disease and cerebrovascular accidents (149). Although supplementary ascorbic acid appears beneficial, further studies are needed to substantiate its role in the etiology of cardiovascular disease.

Nearly all studies of alcoholism indicate widespread vitamin C deficiencies in alcoholics and lower levels of ascorbate in plasma and leucocytes (150). It has not been possible to conclude whether these observations result from dietary deficiency or from changes in the gastrointestinal absorption of the vitamin. Smokers also have decreased plasma ascorbate levels, which are shown in kinetic studies to be caused by increased quantitative metabolism and not by changes in absorption and excretion of ascorbate (15,151). Smoking has been considered as a type of stress and as a consumption of highly noxious substances, eg, nitrosamines, cyanides, etc, which affect the respiratory track. These substances may be prevented from forming or be detoxified by L-ascorbic acid (94,136,152). A proposal was made that the recommended daily allowances (RDAs) be broadened for further protection against all pollutants to which man is exposed (153). The RDA Commission officially recognizes acute emotional and environmental stresses as conditions for increased supplementation of L-ascorbic acid to maintain normal plasma levels (13). The effects of physical, physiological, emotional, and environmental stresses on lowering plasma ascorbate levels have been reviewed (154). Key biochemical functions involved are carnitine biosynthesis, catecholamine biosynthesis, and histamine degradation. Fatigue and lassitude are also symptomatic of vitamin C deficiency resulting from decreased carnitine biosynthesis.

Health-related uses of vitamin C originate with its involvement in many bodily biochemical functions. These are complicated by the vitamin's many interactions in specific enzymes and extracellular redox systems. It is thought that vitamin C helps in the prevention and treatment of chronic diseases other than scurvy that are aggravated by adverse external factors, eg, smoking, alcohol, pollution, drugs, and stress. However, some data have been interpreted differently, raising questions about vitamin C's actual roles in some health areas. Much work remains to resolve these questions. Fundamental knowledge of the vitamin's metabolism, rates of utilization, and the effects of its metabolites in various tissues and fluids will be helpful in assessing the reasons why low ascorbate levels occur and are associated with disease.

Industrial uses of L-ascorbic acid relate to its antioxidant properties. L-Ascorbic acid is widely accepted as a preservative and is GRAS for this application (155). It is also used as an antioxidant in the commercial preparation of beer (qv), fruit juices (qv), and canned and frozen vegetables and fruits; in the curing of meat; and in the flour industry to enhance baking qualities and appearance of bread. Its fatty acid esters, eg, L-ascorbyl palmitate, are used to protect fats and oils. It also prevents discoloration of the food during cooking and storage, and it has been demonstrated to aid in blocking the formation of N-nitroso compounds in bacon that is cured with nitrites. D-Erythorbic acid (22) (D-araboascorbic acid) is also used in the food industry as a preservative. L-Ascorbic acid is also used in agriculture as an abscission agent for fruit, in photography (qv) as a developing agent, and in metallurgy as a reducing agent.

Derivatives and Analogues

Various derivatives and analogues of ascorbic acid have been prepared in attempts to find substances with increased activity (see Table 6) (17,156–157). Only salts and C-6 substituted esters have full activity; they readily form ascorbic acid in the body

Table 6. Isomeric Ascorbic Acids

Substance	Structure	Mp, °C	$[\alpha]_D$ (H$_2$O)	Activity
L-ascorbic acid	(1)	192	+24°	1
D-xyloascorbic acid (D-ascorbic acid, D-*threo*-hex-2-enonic acid γ-lactone)	(33)	192	−23°	0
D-araboascorbic acid (erythorbic acid, D-*erythro*-hex-2-enonic acid γ-lactone)	(22)	174	−18.5°	1/20–1/40
L-araboascorbic acid (L-*erythro*-hex-2-enonic acid γ-lactone)	(34)	170	+17°	0

and are reversibly oxidized to dehydroascorbic acid. Several, eg, sodium L-ascorbate, calcium L-ascorbate, and L-ascorbyl palmitate, are commercially significant. However, ascorbic acid 2-hydrogen sulfate occurs in brine shrimp and is bioavailable to fish but not to guinea pigs, primates, or humans. L-Ascorbic acid sulfate has been described and its biological and pharmacological activities reviewed (158).

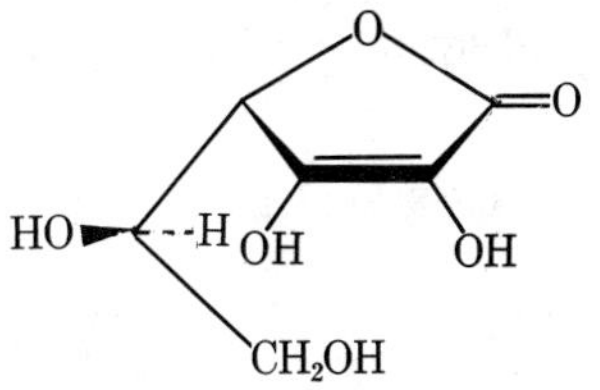

(33) 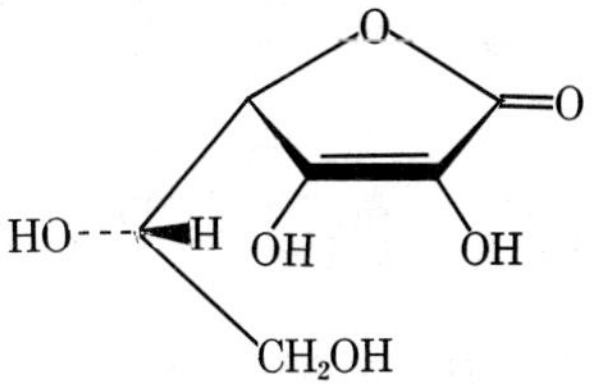   (34)

Derivatives, eg, 6-deoxy-L-ascorbic acid, 3-O-methyl ether of L-ascorbic acid, 2-amino-2-deoxy-L-ascorbic acid, etc, have been prepared, and their respective activities compared to L-ascorbic acid are 1/3, 1/25–1/50, and 0. The various heptono-L-ascorbic acids have also been prepared and exhibit 1/40 to 1/100 the activity of L-ascorbic acid. No analogue or derivative other than those convertible into L-ascorbic acid has activity comparable to or higher than L-ascorbic acid. Many are without antiscorbutic activity.

It appears that the highest vitamin C activity correlates with the D-configuration for the C-4 hydroxyl group, ie, lactone ring formation, at least a two-carbon substituent on the C-4 carbon, two enolic groups, L-configuration for the C-5 hydroxyl group, and a carbon chain no longer than six carbons. Some analogues may be inactive because of rapid elimination (35). D-Araboascorbic acid is inactive in single doses because of rapid elimination, but it is active in scorbutic guinea pigs when administered in frequent repeated small doses.

A bound form of ascorbic acid, ascorbinogen, occurs in cabbage, and it has been identified as a β-substituted indole derivative of L-ascorbic acid (21,157). Ascorbic acid polymer, which is a reaction product of polymerized ascorbic acid and polymerized formaldehyde and is known as P-2 Virucide, has been used with some success as an antiviral agent for bovine leukemia (159).

Methods for the preparation of L-ascorbic acids having isotopic C, H, and O in various positions have been described and reviewed (55,160–161).

BIBLIOGRAPHY

"Ascorbic Acid" in *ECT* 1st ed., Vol. 2, pp. 150–163, by Peter Regna, Chas. Pfizer & Co., Inc.; "Ascorbic Acid" in *ECT* 2nd ed., Vol. 2, pp. 747–762, by J. J. Burns, Burroughs Wellcome & Co. (U.S.A.), Inc.

1. *Biochim. Biophys. Acta* **107**, 4 (1965).
2. C. G. King, *World Rev. Nutr. Diet.* **18**, 47 (1973); J. E. Halver, R. R. Smith, B. M. Tolbert, and E. M. Baker, *Ann. N.Y. Acad. Sci.* **258**, 81 (1975).
3. R. E. Hodges and E. M. Baker in R. S. Goodhart and M. E. Shils, eds., *Modern Nutrition in Health and Disease*, 5th ed., Lea & Febiger, Philadelphia, Pa., 1973, p. 245; U. Wintermeyer, H. Lahann, and R. Vogel, *Vitamin C: Entdeckung, Identifizierung und Synthese—Heutige Bedeutung* in *Medizin und Lebensmitteltechnologie*, Deutscher Apotheker Verlag, Stuttgart, FRG, 1981.

4. I. Stone, *The Healing Factor—Vitamin C Against Disease*, Grosset & Dunlop, Inc., New York, 1972, p. 7.
5. I. M. Sharman in G. G. Birch and K. J. Parker, eds., *Vitamin C: Recent Aspects of Its Physiological and Technological Importance*, John Wiley & Sons, Inc., New York, 1974, p. 1.
6. A. Holst and T. Fröhlich, *J. Hyg.* **7**, 634 (1907).
7. H. Chick and E. M. Hume, *J. R. Army Med. Corps* **29**, 121 (1917); *J. Biol. Chem.* **39**, 203 (1919).
8. A. Harden and S. S. Zilva, *Biochem. J.* **12**, 259 (1918); S. S. Zilva and Welles, *Proc. R. Soc. London Ser. B* **90**, 505 (1917–1919); S. S. Zilva, *Biochem. J.* **21**, 689 (1927).
9. A. Szent-Györgyi, *Biochem. J.* **22**, 1387 (1928); J. L. Svirbely and A. Szent-Györgyi, *Biochem. J.* **26**, 865 (1932).
10. C. G. King and W. A. Waugh, *Science* **75**, 357 (1932).
11. J. L. Svirbely and A. Szent-Györgyi, *Nature* **129**, 576, 690 (1932). E. L. Hirst, R. W. Herbert, E. G. V. Percival, R. J. W. Reynolds, and F. Smith, *Chem. Ind. (London)*, 221 (1933); R. W. Herbert, E. L. Hirst, E. G. V. Percival, R. J. Reynolds, and F. Smith, *J. Chem. Soc.*, 1270 (1933); E. L. Hirst, E. G. V. Percival, and F. Smith, *Nature* **131**, 617 (1933); W. N. Haworth and E. L. Hirst, *J. Chem. Soc. Ind.* **52**, 645 (1933); T. Reichstein, A. Grüssner, and R. Oppenhauer, *Helv. Chim. Acta* **76**, 561 (1933).
12. W. E. Knox and M. N. D. Goswami, *Adv. Clin. Chem.* **1**, 121 (1961); R. M. Evans, L. Currie, and A. Campbell, *Br. J. Nutr.* **47**, 273 (1982).
13. *Recommended Dietary Allowances*, 9th ed., National Academy of Sciences, Washington, D.C., 1980, p. 72.
14. M. I. Irwin and B. K. Hutchins in M. I. Irwin, ed., *Nutritional Requirements of Man: A Conspectus of Research*, The Nutrition Foundation, New York, 1980, p. 247; T. K. Basu and C. J. Schorah, *Vitamin C in Health and Disease*, Avi Publishing Co., Westport, Conn., 1982, p. 61.
15. A. Kallner in J. N. Counsell and D. H. Hornig, eds., *Vitamin C (Ascorbic Acid)*, Applied Science Publishers, London, 1981, p. 63; D. H. Hornig and U. Moser in J. N. Counsell and D. H. Hornig, eds., *Vitamin C (Ascorbic Acid)*, Applied Science Publishers, London, 1981, p. 225.
16. P. J. Garry, J. S. Goodwin, M. C. Hunt, and B. A. Gilbert, *Am. J. Clin. Nutr.* **36**, 332 (1982).
17. W. H. Sebrell and R. S. Harris, eds., *The Vitamins, Chemistry, Physiology, Pathology, Methods*, 2nd ed., Vol. 1, Academic Press, Inc., New York, 1967, p. 305.
18. J. Marks, *A Guide to the Vitamins, Their Role in Health and Disease*, Medical and Technical Publishing Co., Lancaster, UK, 1975.
19. S. Nobile and J. M. Woodhill, *Vitamin C. The Mysterious Redox-System—A Trigger of Life*, MTP Press, Boston, Mass., 1981, p. 38.
20. J. P. Scannell, H. A. Ax, and W. Morris, *J. Agr. Food Chem.* **22**, 538 (1974).
21. R. J. Kutsky, *Handbook of Vitamins and Hormones*, Van Nostrand Reinhold Co., New York, 1973, p. 71; *The Merck Index*, 9th ed., Merck & Co., Inc., Rahway, N.J., 1976, p. 110.
22. J. S. Koevendel, *Nature* **180**, 434 (1957); Y. Ogata and Y. Kosugi, *Tetrahedron* **26**, 4711 (1970).
23. *Sadtler Standard Spectra*, Spectrum No. 5424, Sadtler Research Laboratory, Inc., Philadelphia, Pa., 1972.
24. *High Resolution NMR Spectra Catalog*, Vol. 2, Spectrum No. 464, Varian Assoc., Palo Alto, Calif., 1963.
25. V. F. Johnson and W. C. Jankowski, *Carbon-13 NMR Spectra*, Spectrum No. 171, Wiley-Interscience, New York, 1972; S. Berger, *Tetrahedron* **33**, 1587 (1977); T. Ogama, J. Uzawa, and M. Matsui, *Carbohydr. Res.* **59**, C32 (1977); T. Radford, J. G. Sweeny, and G. A. Jacobucci, *J. Org. Chem.* **44**, 658 (1979).
26. I. A. Al-Meshal and M. M. A. Hassan, *Anal. Profiles Drug Subst.* **11**, 45 (1982).
27. B. M. Tolbert, M. Downing, R. W. Carlson, M. K. Knight, and E. M. Baker, *Ann. N.Y. Acad. Sci.* **258**, 44 (1975); B. M. Tolbert, *Int. J. Vit. Nutr. Res. Suppl.* **19**, 127 (1979).
28. J. Hvoslef, *Acta Cryst.* **B24**, 23, 1431 (1968); **B25**, 2214 (1969).
29. W. Muller-Mulot, *Hoppe-Seyler's Z. Physiol. Chem.* **351**, 52,56 (1970); J. Hvoslef, *Acta Chem. Scand.* **24**, 2238 (1970); *Acta Crystallogr.* **B28**, 916 (1972); J. Hvoslef and B. Pedersen, *Acta Chem. Scand.* **B33**, 503 (1979); *Carbohydr. Res.* **92**, 9 (1981).
30. G. Dryhurst, K. M. Kadish, F. Scheller, and R. Renneberg, *Biological Electrochemistry*, Vol. 1, Academic Press, Inc., New York, 1982, p. 256.
31. W. N. Haworth, *Chem. Ind. (London)* **52**, 482 (1933); W. N. Haworth and E. L. Hirst, *Ergeb. Vitamin-u, Hormonforsch.* **2**, 160 (1939).
32. K. Mikova and J. Davidele, *Chem. Listy* **68**, 715 (1974).

33. F. E. Huelin, *Food Res.* **18**, 633 (1953); J. Campbell and W. G. Tubb, *Can. J. Res.* **28E**, 19 (1947).
34. W. H. Kalus and W. G. Filby, *Z. Naturforsch.* **36C**, 1088 (1981).
35. J. J. Burns in D. M. Greenberg, ed., *Metabolic Pathways*, Vol. 1, Academic Press, Inc., New York, 1968, p. 394.
36. F. A. Isherwood, Y. T. Chen, and L. W. Mapson, *Biochem. J.* **56**, 11 (1954); J. J. Burns and C. Evans, *J. Biol. Chem.* **23**, 897 (1956).
37. J. J. Burns, *Am. J. Med.* **26**, 740 (1959).
38. S. Ikeda and M. Sato, *Nippon Suisan Yukkaishi* **32**, 530 (1966).
39. G. J. Dutton, ed., *Glucuronic Acid—Free and Combined: Chemistry, Biochemistry, Pharmacology and Medicine*, Academic Press, Inc., New York, 1966; B. L. Horecker in G. Ritzel and G. Brubacher, eds., *Monosaccharides and Polyols in Nutrition, Therapy and Dietetics*, Hans Huber, Bern, Switz., 1967, p. 1.
40. O. Touster in E. H. Stotz and M. Florkin, eds., *Comprehensive Biochemistry*, Vol. 17, Elsevier, Inc., New York, 1969, p. 219; M. W. Loewus and F. A. Loewus, *Carbohydr. Res.* **82**, 333 (1980); *Ann. Rev. Plant Physiol.* **34**, 137 (1983); L. Sasaki and F. A. Loewus, *Plant Physiol.* **69**, 220 (1982).
41. A. H. Conney, G. A. Bray, C. E. Evans, and J. J. Burns, *Ann. N.Y. Acad. Sci.* **92**, 115 (1961).
42. W. R. F. Notten, *Alteration in the D-Glucuronic Acid Pathway and Drug Metabolism by Exogenous Compounds: D-Glucaric Acid Level as an Indication of Exposure to Xenobiotics*, thesis, Stichting Studentenpers, University of Nijmegen, Nijmegen, The Netherlands, 1975; W. R. F. Notten and P. T. Henderson, *Int. Arch. Occup. Environ. Health* **38**, 197, 209 (1977).
43. F. A. Loewus and M. M. Baig, *Methods Enzymol.* **18**, 22 (1970); F. A. Loewus, G. Wagner, and J. C. Yang, *Ann. N.Y. Acad. Sci.* **258**, 7 (1975); F. A. Loewus in J. Preiss, ed., *Biochemistry of Plants*, Vol. 3, Academic Press, Inc., New York, 1980, p. 77; C. H. Foyer and B. Halliwell, *Phytochemistry* **16**, 1347 (1977).
44. M. Grün, B. Rentstrøm, and F. A. Loewus, *Plant Physiol.* **70**, 1233 (1982); F. Celardin, F. J. Castillo, and H. Greppin, *J. Biochem. Biophys. Met.* **6**, 89 (1982).
45. F. A. Isherwood, Y. T. Chen, and L. W. Mapson, *Biochem. J.* **56**, 1 (1954).
46. M. M. Baig, S. Kelley, and F. A. Loewus, *Plant. Physiol.* **46**, 277 (1970); J. P. Helpser, L. Kagan, C. L. Hilby, T. M. Maynard, and F. A. Loewus, *Plant. Physiol.* **69**, 465 (1982).
47. S. Shigeoka, Y. Nakano, and S. Kitaoba, *J. Nutr. Sci. Vitaminol.* **25**, 299 (1979).
48. C. J. Bates in ref. 15, p. 1.
49. B. M. Tolbert, R. J. Hardrader, D. O. Johnsen, and B. A. Joyce, *Biochem. Biophys. Res. Comm.* **71**, 1004 (1976).
50. F. A. Loewus and J. P. F. J. Helsper, *Adv. Chem. Ser.* **200**, 249 (1982).
51. J. P. Helsper and F. A. Loewus, *Plant Physiol.* **69**, 1365 (1982); K. Saito and Z. Kosai, *Plant Cell Physiol.* **23**, 499 (1982).
52. T. Reichstein, A. Grüssner, and R. Oppenauer, *Nature* **132**, 280 (1933); *Helv. Chim. Acta* **16**, 1019 (1933); *Helv. Chim. Acta* **17**, 510 (1934); U.S. Pat. 2,056,126 (Sept. 29, 1936), T. Reichstein (to Hoffmann-LaRoche, Inc.).
53. R. G. Ault, D. K. Baird, H. C. Carrington, W. N. Haworth, R. W. Herbert, E. L. Hirst, E. G. V. Percival, F. Smith, and M. Stacey, *J. Chem. Soc.*, 1419 (1933); W. N. Haworth and E. L. Hirst, *Chem. J. Ind. (London)* **52**, 645 (1933); *Helv. Chim. Acta* **17**, 520 (1934).
54. T. Reichstein and A. Grüssner, *Helv. Chim. Acta* **17**, 311 (1934).
55. T. C. Crawford and S. A. Crawford, *Adv. Carbohydr. Chem. Biochem.* **37**, 79 (1980).
56. P. Karrer, K. Schopp, and F. Zehnder, *Helv. Chim. Acta* **16**, 1161 (1933).
57. R. W. Herbert and E. L. Hirst, *Nature* **130**, 205 (1932); E. L. Hirst, R. W. Herbert, E. G. V. Percival, R. J. W. Reynolds, and F. Smith, *Chem. Ind. (London)*, 221 (1933).
58. E. G. Cox, *Nature* **130**, 205 (1932); J. D. Bernal, *Nature* **129**, 721 (1932).
59. H. Ohle and R. Wolter, *Ber.* **63**, 843 (1930); H. Ohle, *Angew. Chem.* **46**, 399 (1933); H. Ohle, H. Erbach, and H. Carls, *Ber.* **67**, 324 (1934).
60. K. Maurer and B. Schiedt, *Ber.* **63**, 1054 (1933).
61. T. Sonoyama, H. Tani, K. Matsuda, B. Kageyama, M. Tanimoto, K. Kobayashi, S. Yagi, H. Kyotani, and K. Mitsushima, *Appl. Environ. Microbiol.* **43**, 1064 (1982).
62. I. T. Strukov and N. A. Kopylova, *Farmatsiya* **10**(3), 8 (1947); *Drug Cosmet. Ind.*, 40 (Aug. 1977).
63. W. Ipatiew, *Ber.* **45**, 3218 (1912).
64. L. W. Cover, R. Connor, and H. Adkins, *J. Am. Chem. Soc.* **54**, 1651 (1932).
65. A. T. Kuhn, *J. Appl. Electrochem.* **11**, 261 (1981).
66. W. E. Cake, *J. Am. Chem. Soc.* **44**, 859 (1922).

67. M. Kulhanek, *Adv. Appl. Microbiol.* **12,** 11 (1970).
68. G. Bertrand, *Ann. Chim. Phys.* **3,** 181 (1904).
69. H. H. Schulbach and J. Vorwerk, *Ber.* **66,** 1251 (1933).
70. P. A. Wells, J. J. Stubbs, L. B. Lockwood, and E. T. Roe, *Ind. Eng. Chem.* **29,** 1385 (1937); E. I. Fulmer and L. A. Underkofler, *Iowa State Coll. J. Sci.* **21,** 251 (1974).
71. R. G. Kristallinskaya, *Proc. Sci. Inst. Vit. Res. (USSR)* **3,** 78 (1941); Y. M. S. Slobodin, *J. Gen. Chem. (USSR)* **17,** 485 (1947).
72. U.S. Pat. 3,622,560 (Nov. 23, 1971), N. Halder, M. J. O'Leary, and N. C. Hindley; U.S. Pat. 3,598,804 (Aug. 10, 1971), N. Halder, N. C. Hindley, G. M. Jaffe, M. J. O'Leary, and P. Weinert; U.S. Pat. 3,607,862 (Sept. 21, 1971), G. M. Jaffe, W. Szkrybalo, and P. Weinert (all to Hoffmann-LaRoche, Inc.).
73. R. S. Glass, S. Kwoh, and E. P. Oliveto, *Carbohydr. Res.* **26,** 181 (1973).
74. U.S. Pat. 2,039,929 (May 5, 1936), T. Reichstein, U.S. Pat. 2,301,811 (Nov. 10, 1942), T. Reichstein (both to Hoffmann-LaRoche, Inc.); V. F. Kazimirova, *J. Gen. Chem. (USSR)* **24,** 1559 (1955).
75. J. W. Weijlard, *J. Am. Chem. Soc.* **67,** 1031 (1945); U.S. Pat. 2,367,251 (Jan. 16, 1945), J. Weijlard and J. B. Ziegler (to Merck and Co.); K. Nakagawa, R. Konaka, and T. Nakata, *J. Org. Chem.* **27,** 1597 (1962).
76. L. O. Shnaidman and I. N. Kushehinskaya, *Tr. Vses. Nauchno-Issled. Vitam. Inst.* **8,** 13 (1961).
77. U.S. Pat. 3,832,355 (Aug. 27, 1972), G. M. Jaffe and E. Pleven (to Hoffmann-LaRoche, Inc.).
78. U.S. Pat. 3,453,191 (July 1, 1969), G. J. Frohlich, A. J. Kratavil, and E. Zrike (to Hoffmann-LaRoche, Inc.); Eur. Pats. 40331 (Nov. 25, 1981) and 40709 (Dec. 2, 1981), R. Wittmann, W. Wintermeyer, and J. Butzka (to Merck GMBH); P. M. Robertson, P. Berg, H. Reimann, K. Schleich, and P. Seiler, *Chimia* **36,** 305 (1982).
79. F. Elger, *Festschr. Emil Barell*, 229 (1936); Brit. Pat. 428,815 (May 20, 1935); and Brit. Pat. 466,548 (May 31, 1937), T. Reichstein (to Hoffmann-LaRoche, Inc.).
80. P. P. Regna and B. P. Caldwell, *J. Am. Chem. Soc.* **66,** 246 (1944).
81. *Chem. Mark. Rep.*, 23 (Aug. 25, 1980); *Chem. Week*, 21 (Nov. 1, 1978).
82. J. Tillmans, *Z. Unters. Lebensmittel* **54,** 33 (1927).
83. R. Strohecker, *Wiss. Veroff. Dtsch. Ges. Ernahr.* **14,** 157 (1965).
84. J. R. Cooke in ref. 5, p. 31; T. Tono and S. Fujita, *Agric. Biol. Chem.* **45,** 2947 (1981).
85. M. Grün and F. A. Loewus, *Anal. Biochem.* **130** (1983).
86. J. R. Cooke and R. E. D. Moxon in ref. 15, p. 167; H. E. Sauberlich, M. D. Green, and S. T. Omaye, *Adv. Chem. Ser.* **200,** 199 (1982); R. C. Rose and D. L. Nahrwold, *Anal. Biochem.* **123,** 389 (1982).
87. M. Brin, *Chem. Technol.*, 428 (July 1982).
88. B. Peterkofsky and S. Udenfriend, *Proc. Nat. Acad. Sci.* **53,** 335 (1965); M. J. Barnes and E. Kodicek, *Vitam. Horm.* **30,** 1 (1972).
89. R. E. Hughes in ref. 15, p. 75.
90. K. F. Stone and B. J. Townsley, *Biochem. J.* **131,** 611 (1963).
91. S. Lewin, *Vitamin C. Its Molecular Biology and Medical Potential*, Academic Press, Inc., New York, 1976.
92. D. Liakakos, N. Doulas, D. Ikkos, C. Anoussakis, P. Vlachos, and G. Jouramani, *Clin. Chim. Acta* **65,** 251 (1975); V. A. Wilbur and B. L. Walker, *Nutr. Rep. Int.* **16,** 789 (1977).
93. S. R. Lynch and J. D. Cook, *Ann. N.Y. Acad. Sci.* **355,** 32 (1980); L. Hallberg, *Ann. Rev. Nutr.* **1,** 126 (1981); T. A. Morck and J. D. Cook, *Cereal World* **26,** 667 (1981).
94. O. A. Levander and L. Cheng, *Ann. N.Y. Acad. Sci.*, 355 (1980); E. J. Calabrese, *Nutrition and Environmental Health—The Influence of Nutritional Status on Pollutant Toxicity and Carcinogenicity*, Vol. 1, John Wiley & Sons, Inc., New York, 1980, p. 249.
95. W. Dawson and G. B. West, *Br. J. Pharmacol.* **24,** 725 (1965).
96. N. Subramanian, *Agents Actions* **8,** 484 (1978); C. A. B. Clemetson, *J. Nutr.* **110,** 662 (1980).
97. J. S. Douglas in A. Bouhuys, ed., *Living Cells in Disease*, Elsevier, Amsterdam, The Netherlands, 1976, p. 245; E. Pocwiardowska and L. Puglisi, *Acta Physiol. Pol.* **31,** 2 (1980).
98. D. Malamud and Y. Kroll, *Proc. Soc. Exp. Biol.* **164,** 534 (1980).
99. S. C. Sharma and C. W. M. Wilson, *Int. J. Vitam. Nutr. Res.* **50,** 163 (1980).
100. R. W. Fessenden and N. C. Verma, *Biophys. J.* **24,** 93 (1978).
101. R. S. Bodannes and P. C. Chan, *FEBS Lett.* **105,** 195 (1979).
102. B. E. Leibovitz and B. V. Siegel, *J. Gerontol.* **35,** 45 (1980).
103. S. D. Varma, S. Kumar, and R. Richards, *Proc. Natl. Acad. Sci. U.S.A.* **76,** 3504 (1979).

104. R. N. Matzen, *J. Appl. Physiol.* **11,** 105 (1957); C. C. Kratzing and R. J. Willis, *Chem. Biol. Interact.* **30,** 53 (1980).
105. H. F. Stich, L. Wei, and R. F. Whiting, *Cancer Res.* **39,** 4145 (1979); B. Peterkofsky and W. Prather, *J. Cell Physiol.* **90,** 61 (1976).
106. L. H. Chen and H. M. Chang, *Int. J. Vitam. Nutr. Res.* **49,** 87 (1979); B. E.. Leibovitz and B. V. Siegel, *J. Gerontol.* **35,** 45 (1980); E. P. Norkus, W. Kuenzig, and A. H. Conney, *Mutat. Res.* **117,** 183 (1983).
107. C. W. M. Wilson, *Ann. N.Y. Acad. Sci.* **258,** 355 (1975); R. Anderson in ref. 15, p. 249; W. R. Thomas and P. G. Holt, *Clin. Exp. Immunol.* **32,** 370 (1978).
108. W. F. Geber, S. Lefkowitz, and C. Hung, *Pharmacology* **13,** 228 (1975); B. V. Siegel and J. I. Norton, *Int. J. Nutr. Res. Suppl.* **16,** 245 (1977).
109. R. E. Hodges in R. B. Alfin-Slater and D. Kritchevsky, eds., *Human Nutrition*, Plenum Press, New York, 1980, p. 73.
110. R. E. Hughes, *Vitamin C: Some Current Problems*, British Nutrition Foundation, London, 1981.
111. M. Brin in ref. 15, p. 359; *Chemtech*, 428 (July 1982).
112. H. Kraut, *Wiss. Veroff. Dtsch. Ges. Ernahr.* **14,** 89 (1965).
113. M. Brin, *Adv. Chem. Ser.* **200,** 369 (1982).
114. S. Nobile and J. M. Woodhill in ref. 19, p. 66.
115. V. Siegel and B. Leibovitz, *Int. J. Vitam. Nutr. Res. Suppl.* **23,** 9 (1982).
116. W. R. Koerner and F. Weber, *Int. J. Vitam. Nutr. Res.* **42,** 528 (1972); L. A. Barness, *Ann. N.Y. Acad. Sci.* **258,** 523 (1975); A. Hanck, *Int. J. Vitam. Nutr. Res. Suppl.* **23,** 221 (1982); D. H. Hornig and U. Moser in ref. 15, p. 225.
117. W. Kuebler and J. Gehler, *Int. J. Vitam. Res.* **40,** 442 (1970).
118. L. Hallberg in ref. 15, p. 49.
119. W. V. Robertson, *Ann. N.Y. Acad. Sci.* **92,** 159 (1961); B. S. Gould, *Ann. N.Y. Acad. Sci.* **92,** 168 (1961); *Ann. N.Y. Acad. Sci.* **92** (1961); T. T. Irwin in ref. 15, p. 283; T. T. Irwin, *Int. J. Vitam. Nutr. Res. Suppl.* **23,** 277 (1982).
120. G. H. Bourne, *Proc. Nutr. Soc.* **4,** 204 (1946); N. Stone and A. Meister, *Nature* **194,** 555 (1962).
121. F. P. McGinn and J. C. Hamilton, *Br. J. Surg.* **63,** 505 (1976); T. T. Irwin, D. K. Chattopadhyay, and A. Smythe, *Surg. Gyn. Obst.* **147,** 49 (1978).
122. T. Hunter and K. T. Rajan, *Paraplegia* **8,** 211 (1971); T. V. Taylor, S. Rimmer, B. Day, J. Butcher, and W. Dymock, *Lancet* **2,** 544 (1974).
123. D. H. Klasson, *N.Y. State J. Med.* **51,** 2388 (1951); F. R. Klenner, *J. Appl. Nutr.* **23,** 61 (1971).
124. J. Carneiro and F. Fava de Moraes, *Arch. Oral Biol.* **10,** 833 (1965); A. Cowan, *Ir. J. Med. Sci.* **145,** 273 (1976).
125. W. Prinz, R. Bortz, B. Bregin, and M. Hersch, *Int. J. Nutr. Res.* **47,** 248 (1977); S. Banic, *Int. J. Vitam. Nutr. Res. Suppl.* **23,** 49 (1982).
126. L. Pauling, *Vitamin C and the Common Cold*, W. N. Freeman Co., San Francisco, Calif., 1970.
127. R. Hume and E. Weyers, *Scott. Med. J.* **18,** 3 (1973).
128. A. R. Schwartz, Y. Togo, R. B. Hornick, S. Tominaga, and R. A. Gleckman, *J. Infect. Dis.* **128,** 500 (1973); J. E. W. Davies, R. E. Hughes, E. Jones, S. E. Reed, J. W. Craig, and D. A. Tyrell, *Biochem. Med.* **21,** 78 (1979).
129. C. A. B. Clemetson, *J. Nutr.* **110,** 662 (1980).
130. T. W. Anderson, *Ann. N.Y. Acad. Sci.* **258,** 498 (1975); T. C. Chambers, *Am. J. Med.* **58,** 532 (1975); A. B. Carr, R. Einstein, L. Y. C. Lai, N. G. Martin, and G. A. Starmer, *Acta Genet. Med. Gemellol.* **30,** 249 (1981).
131. B. H. Sabiston and M. W. Radomski, *DCIEM Report No. 74/R/1012*, Defense Research Board (Def. Civ. Inst. Environ. Med., Downsview, Ont.), 1974, pp. 1–10; B. J. Luberoff, *Chemtech*, 76 (Feb. 1978).
132. E. Cameron, L. Pauling, and B. Leibovitz, *Cancer Res.* **39,** 667 (1979); J. W. T. Dickerson in ref. 15, p. 349; E. Cameron, *Int. J. Vitam. Nutr. Res. Suppl.* **23,** 115 (1982).
133. J. U. Schlegel, P. E. Pipkin, R. Mishimiera, and G. N. Schultz, *Trans. Am. Soc. Genito-Urinary Surg.* **61,** 85 (1969).
134. K. Yamafuji, Y. Nakamura, H. Omura, T. Sveda, and K. Gyotoku, *Z. Krebsforsch.* **76,** 1 (1971).
135. S. Bram, P. Froussard, M. Guichard, C. Jasmin, Y. Augery, F. Sinoussi-Barre, and W. Wray, *Nature* **248,** 629 (1980); R. L. Ikonopisov in W. H. McCarthy, ed., *Melanoma and Skin Cancer*, International Union Against Cancer, Sydney, Australia, 1972, p. 223.

136. J. J. Kamm, T. Dashman, A. H. Conney, and J. J. Burns, *Ann. N.Y. Acad. Sci.* **258,** 169 (1975); H. Newmark and W. J. Mergens in J. Solms and R. L. Hall, eds., *Criteria of Food Acceptances,* Foster Verlag AG, Switz., 1981, p. 379; S. S. Mirvish in M. S. Zedeck and M. Lipkin, eds., *Institution of Tumor Induction and Development,* Plenum Publishing Corporation, New York, 1981, p. 101; H. Okshiuma and H. Bartsch in ref. 15, p. 215; Y. K. Kim, J. S. Wishnok, and S. R. Tannenbaum, *Adv. Chem. Ser.* **200,** 571 (1982); D. Schmahl and G. Eisenbrand, *Int. J. Vitam. Nutr. Res. Suppl.* **23,** 91 (1982).
137. E. Cameron and L. Pauling, *Biol. Interact.* **9,** 273 (1974); **11,** 387 (1975); E. Cameron and L. Pauling, *Proc. Natl. Acad. Sci.* **75,** 4538 (1978).
138. A. Murta, F. Morishige, and H. Yamaguchi, *International Conference on Nutrition, Tokyo, 1981,* Medical News Group, London, 1982, p. 11.
139. A. Murta, F. Morishige, and H. Yamaguchi, *J. Vitam. Nutr. Res. Suppl.* **23,** 103 (1982).
140. E. T. Creegan, C. G. Moertel, J. R. O'Fallon, A. J. Schult, J. J. O'Connel, J. Rubin, and S. N. Frytak, *N. Engl. J. Med.* **301,** 686 (1979); E. T. Creegan and C. G. Moertel, *N. Eng. J. Med.* **301,** 1399 (1979).
141. K. N. Prasad, *Life Sci.* **27,** 275 (1980).
142. L. A. Wood and E. Beutler, *J. Hematol.* **25,** 611 (1973); l. G. Moore, G. J. Brewer, F. J. Oelshlegel, L. F. Brewer, and E. B. Schoomaker, *J. Pharm. Exp. Ther.* **203,** 722 (1977).
143. W. Lohmann, K. G. Bensch, E. Mueller, and S. O. Kang, *Z. Naturforsch.* **36C,** 1 (1981); W. Lohmann, *Z. Naturforsch.* **36C,** 804 (1981).
144. W. F. Benedict and P. A. Jones in M. S. Arnott, J. van Eys, and Y-M. Wang, eds., *Molecular Interrelations of Nutrition and Cancer,* Raven Press, N.Y., 1982, p. 351.
145. National Research Council, *Report of the Committee on Diet, Nutrition and Cancer,* National Academy Press, Washington, D.C., 1982.
146. S. D. Turley, C. E. West, and B. J. Horton, *Atherosclerosis* **24,** 1 (1976); E. Ginter, *World. Rev. Nutr. Diet.* **33,** 167 (1978); E. Ginter and P. Bobek in ref. 15, p. 299; T. K. Basu and C. J. Schorah, *Vitamin C in Health and Disease,* Avi Publishing Co., Westport, Conn., 1982, p. 95.
147. E. Ginter, *Ann. N.Y. Acad. Sci.* **258,** 410 (1975); *Adv. Lipid Res.* **16,** 167 (1978).
148. A. Hanck and H. Weiser in A. Hanck and G. Ritzel, eds., *Reevaluation of Vitamin C,* Verlag Hans Huber, Bern, Switz., 1977, p. 67.
149. J. Ramirez and N. C. Flowers, *Am. J. Clin. Nutr.* **33,** 2079 (1980); E. Cheraskin and W. M. Ringsdorf, Jr., *J. Electrocardiol.* **12,** 441 (1979); R. Hume, B. D. Vallance, and M. M. Muir, *J. Clin. Pathol.* **35,** 195 (1982).
150. P. J. Bonjour, *Int. J. Nutr. Res.* **49,** 434 (1979); A. Kallner in ref. 138, p. 15; M. Baines, *Int. J. Vitam. Nutr. Res. Suppl.* **23,** 287 (1982).
151. A. Kallner, D. Hartmann, and D. H. Hornig, *Am. J. Clin. Nutr.* **32,** 530 (1979).
152. H. Sprince, G. G. Smith, C. M. Parker, and D. A. Rinehimer, *Nutr. Rep. Int.* **25,** 463 (1982).
153. E. J. Calabrese, *Med. Hypothesis* **5,** 1273 (1979).
154. I. Stone in ref. 4, p. 152, 163, 178; M. Brin in ref. 15, p. 371; A. Kallner in ref. 138, p. 16.
155. N. Chamberlain in ref. 15, p. 87; M. D. Ranken in ref. 15, p. 105; J. D. Henshall in ref. 15, p. 123; H. Klaui in ref. 5, p. 16; G. G. Birch, B. M. Bionton, E. J. Rolfe, and J. D. Selman in ref. 5, p. 40; J. D. Henshall in ref. 5, p. 104; M. D. Ranken in ref. 5, p. 121; D. M. Gresswell in ref. 5, p. 136; B. N. Thewlis in ref. 5, p. 150.
156. F. Smith, *Adv. Carbohydr. Chem. Biochem.* **2,** 79 (1946).
157. J. Stanek, M. Cerny, J. Kocourek, and J. Pacak, *The Monosaccharides,* Academic Press, Inc., New York, 1963, p. 735.
158. G-Q Tao, *Sheng Lik'o Hseuh Chin Chan* **12,** 26 (1981).
159. L. MartinAponte, N. M. De Lopez, M. Szabuniewicz, and R. Rodriguez, *IRCS Med. Sci. Libr. Compend.* **8,** 522 (1978).
160. M. Williams and F. A. Loewus, *Carbohydr. Res.* **63,** 149 (1978).
161. S. L. von Schuching and A. F. Abt, *Methods Enzymol.* **18,** 1 (1970).

General Reference

P. A. Seib and B. M. Tolbert, eds., "Ascorbic Acid: Chemistry, Metabolism, and Uses," *Adv. Chem. Ser.* **200** (1982).

GERALD M. JAFFE
Hoffmann-La Roche, Inc.

BIOTIN

Biotin [58-85-5] (**1**) is one of the water-soluble B vitamins. In bound form, it is distributed widely as a cell constituent of animal and human tissues. The main sources of biotin are liver, kidney, pancreas, egg yolk, yeast, and milk. A high content of biotin in cows' milk occurs early in lactation. It is in different plant materials, especially in seeds, pollen, molasses, rice, mushrooms, fresh vegetables, and some fruits. Most fish contain biotin in small amounts. Biochemically, biotin functions as a cofactor for enzymes principal to carboxylation reactions. These reactions are involved in important biochemical processes, eg, gluconeogenesis and fatty-acid synthesis.

(**1**)
D-biotin

Isolation and Structure Determination

Biotin was isolated in 1936 from egg yolk, in 1941 from beef liver, and in 1942 from milk concentrates (1–3). The empirical formula for biotin, $C_{10}H_{16}N_2O_3S$, was established in 1941 and the full structure in 1942 (4–5). This structure was confirmed by the first total synthesis of biotin in the Merck Laboratories (6). The absolute configuration was established more than 20 years later by an x-ray crystallographic analysis (7).

Synthesis

In the first synthesis of racemic biotin in 1944, the starting material L-cysteine (**2**) was converted to the dicarboxylic ester (**3**) which, on Dieckmann condensation and decarboxylation, gave the thiophane ketone (**4**) (see Fig. 1). The activation of α-methylene protons by the oxo group on one side and the sulfide on the other led to the efficient aldol condensation with the aldehyde ester (**5**) to form (**6**). This reaction serves to attach the whole properly functionalized C-5 side chain in one step. The second nitrogen was then introduced to give the conjugated oxime (**7**). This substance, upon reduction with zinc, led to the enamide (**8**) with the double bond in the proper position so that catalytic hydrogenation would create two additional chiral centers with the desired all-cis configuration. Although this hydrogenation was less than stereospecific, it gave mainly the precursor (**9**), which was then easily transformed to biotin. Unfortunately, epimerization during the conversion of (**3**) to (**4**) led to optically inactive biotin.

From this early work, it became clear that any valuable synthesis of biotin requires an efficient and stereocontrolled formation of three chiral centers in the all-cis configuration. This was first accomplished in the Sternbach synthesis of optically active

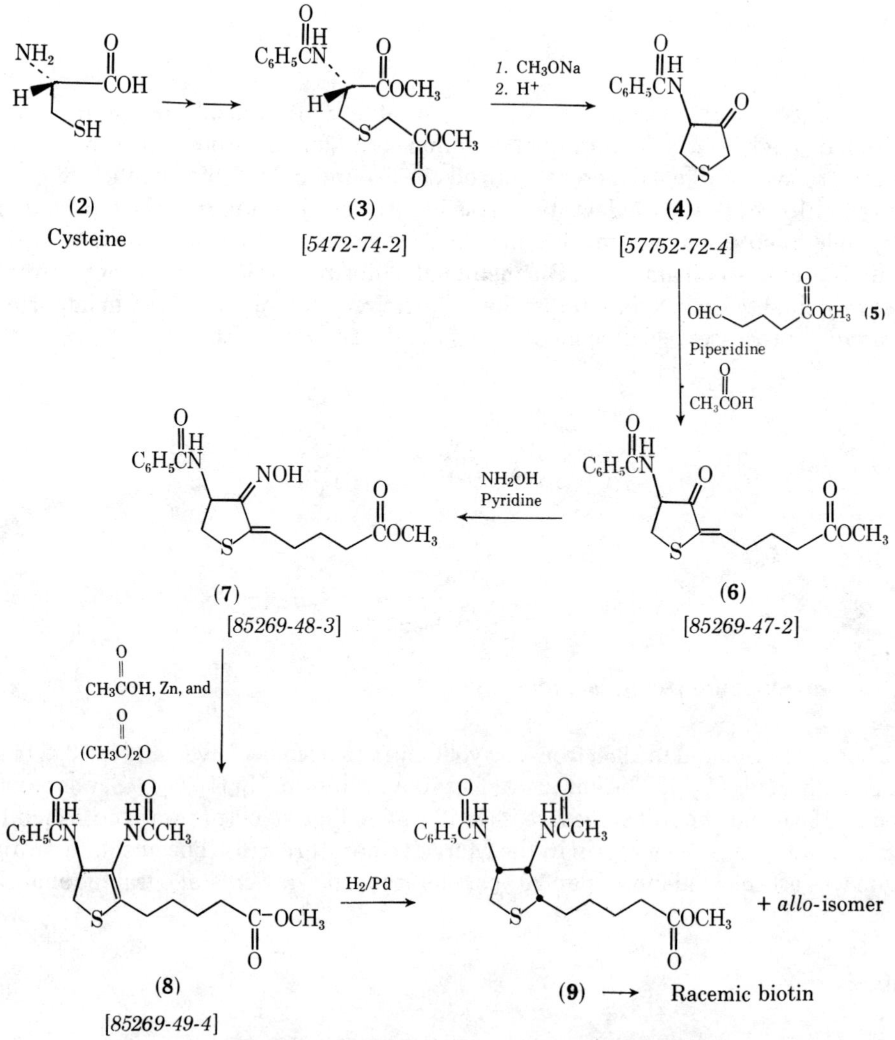

42   VITAMINS (BIOTIN)

Figure 1. Synthetic pathway for racemic biotin.

biotin, which later was known as the Hoffmann-La Roche industrial synthesis of biotin (8). The improved version of the original Sternbach synthesis is shown in Figure 2. Fumaric acid (10) serves as the starting material. The sequence of bromination, replacement of dibromide with benzyl-bromide, and ring closure with phosgene gives the imidazolone *cis*-dicarboxylic acid (11). The corresponding anhydride (12) is opened with cyclohexanol to the racemic monoester (13), which is resolved with (+)-ephedrine in high yield. The enantiomer of (14) is recycled back to the anhydride (12). Lithium borohydride reduces only the ester group of (14), thus producing the lactone (15) with the desired absolute configuration. Sulfur is then introduced by treatment with potassium thioacetate to give the thiolactone (16). The side chain is introduced in two phases. The first three carbons are attached by a Grignard reaction. Dehydration and hydrogenation over Raney nickel establishes the third chiral center stereospecifically as in (18). The last two carbons are then added by reaction of the cyclic sulfonium cation (19) with sodium dimethylmalonate. Hydrolysis of the ester groups of (20),

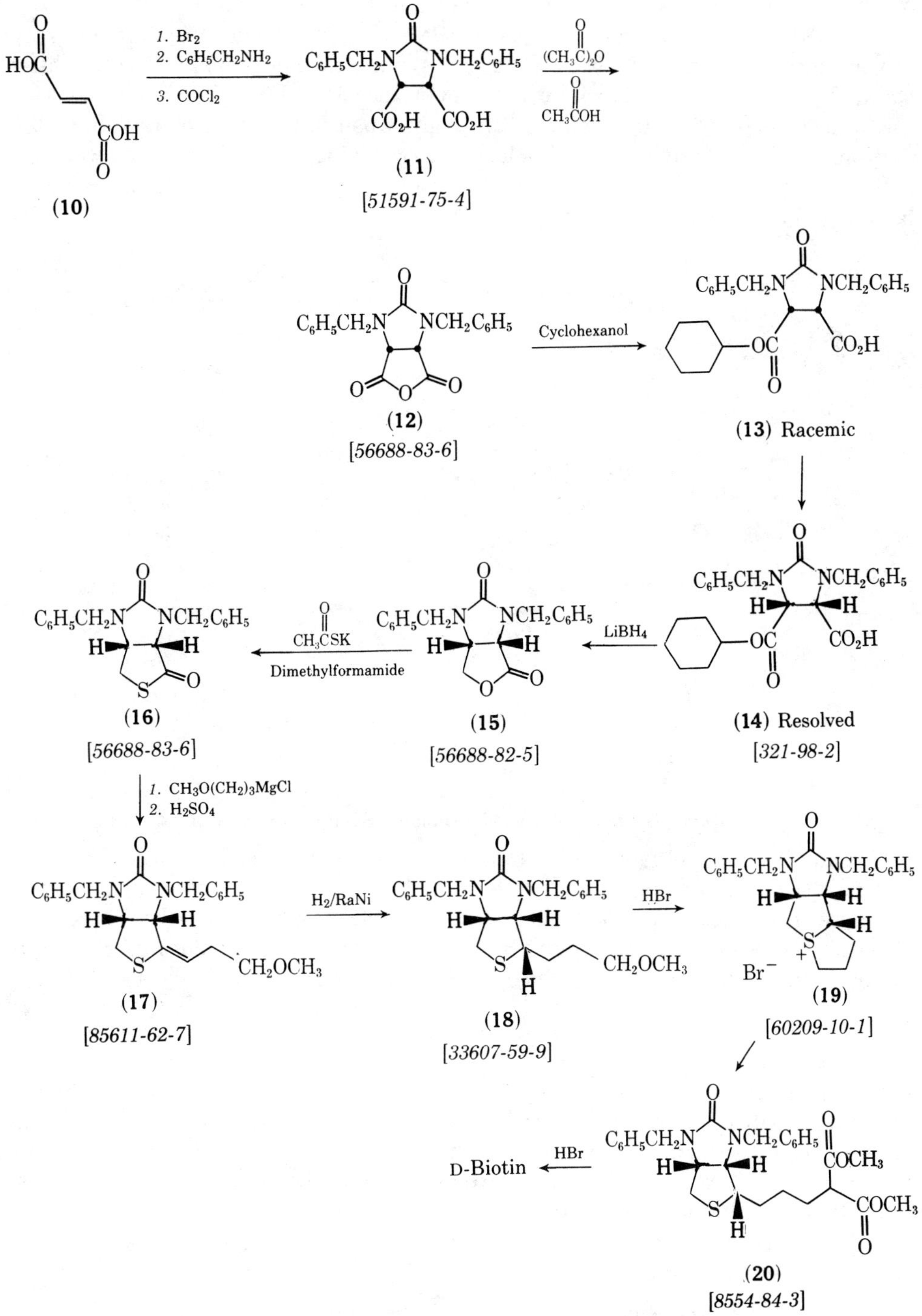

Figure 2. Synthetic pathway for D-biotin (Sternbach synthesis).

decarboxylation, and didebenzylation occur during heating with aqueous HBr to produce the optically pure biotin in a more than 25% overall yield.

In 1975, Sumitomo chemists replaced the optical resolution–reduction sequence of the Sternbach synthesis by an efficient asymmetric conversion of the prochiral cis-acid (11) to the optically active lactone (15) (see Fig. 3) (9). The acid reacts with the optically active dihydroxy amine (21) to give quantitatively the chiral imide (22). Sodium borohydride reduces stereoselectively the pro-R carbonyl group to give, after recrystallization, the optically pure hydroxy amide (23). Hydrolysis then yields the lactone (15).

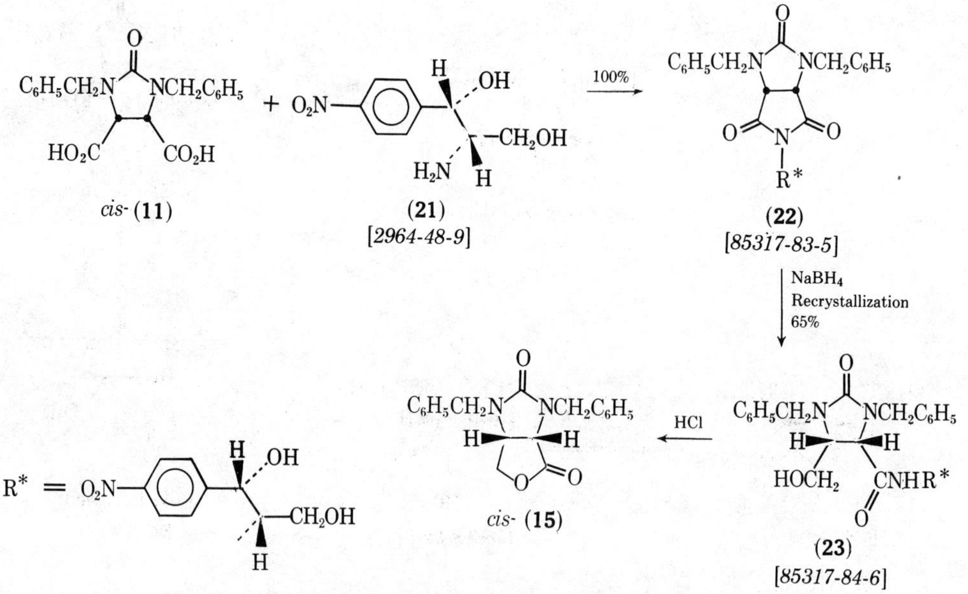

Figure 3. Synthetic pathway for D-biotin (Sumitomo synthesis).

A number of novel syntheses of biotin have been developed in the last ten years. One starts from cysteine, one from cystine, and three from carbohydrates; none require an optical-resolution step (10–12). The stereocontrolled formation of all chiral centers of biotin can be achieved in three syntheses by means of 1,3-dipolar nitrone-olefin cycloadditions and in two syntheses by [2 + 2] cycloaddition methods (11,13–14).

Biosynthesis

A number of fungi and bacteria synthesize biotin from pimelic acid by a metabolic pathway, whose last step is the conversion of dethiobiotin to biotin. This pathway has been thoroughly investigated. All the intermediates from pimelic acid to dethiobiotin are known and are formed by classical biochemical reactions (15). The only remaining problem in this biogenetic scheme is the elucidation of the mechanism for the transformation of dethiobiotin to biotin.

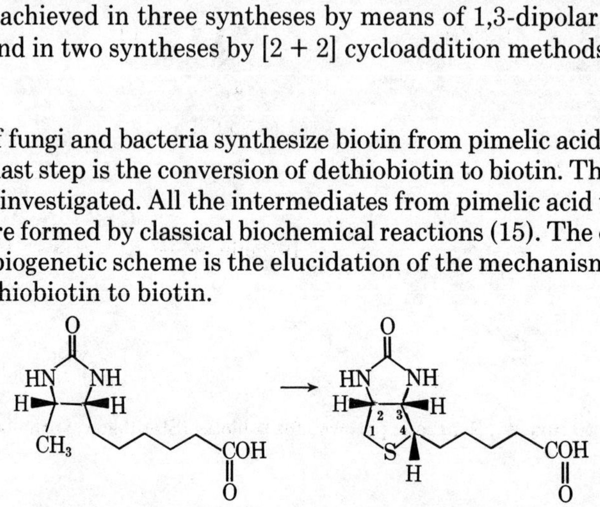

Evidence has been presented that the biosynthesis of biotin in *Aspergillus niger* and *E. coli* proceeds by the introduction of sulfur at C-1 and C-4 of dethiobiotin without apparent involvement of C-2 and C-3 (16–17). A more recent study clearly demonstrates that sulfur is introduced at C-4 of dethiobiotin with loss of the 4-pro S hydrogen atom (17). Since the configuration of biotin at C-4 is S, it follows that sulfur is introduced with retention of configuration at the C-4 prochiral center of dethiobiotin (18).

Enzymatic Reactions

Biotin plays its biochemical role as a prosthetic group of carboxylases, the enzymes that catalyze energy-dependent fixation of CO_2 to various biochemical substrates (19). The first biotin enzyme discovered was acetyl-CoA carboxylase (20). It effects the adenosine triphosphate dependent carboxylation of acetyl-CoA to malonyl-CoA.

$$\text{acetyl-CoA} + HCO_3^- + \text{adenosine triphosphate (ATP)} \xrightarrow{Mg^{2+}} \text{malonyl-CoA}$$

$$+ \text{ adenosine diphosphate (ADP)} + HPO_4^{2-}$$

This reaction is the incipient process in the pathway of fatty-acid biosynthesis. Fatty-acid synthetase is a cytosolic multienzyme system involved in the formation of palmitate from malonyl-CoA.

Propionyl-CoA carboxylase catalyzes the ATP-dependent carboxylation of propionyl-CoA to methylmalonyl-CoA.

$$\begin{array}{c} CH_3 \\ | \\ CH_2 \\ | \\ COSCoA \end{array} + ATP + HCO_3^- \xrightarrow{Mg^{2+}} \begin{array}{c} CH_3 \\ | \\ CHCO_2^- \\ | \\ COSCoA \end{array} + ADP + HPO_4^{2-}$$

Under biological conditions, propionate is produced by oxidation of odd-numbered fatty acids and by degradation of branched-chain amino acids. Methylmalonyl-CoA formation is the first step in metabolic conversion of propionate to succinyl-CoA and oxaloacetic acid; the latter is involved in the biosynthesis of glucose.

Pyruvate carboxylase catalyzes formation of oxaloacetate from pyruvate.

$$\begin{array}{c} CH_3 \\ | \\ C=O \\ | \\ CO_2H \end{array} + HCO_3^- + ATP \xrightarrow{Mg^{2+}} \begin{array}{c} CO_2H \\ | \\ CH_2 \\ | \\ C=O \\ | \\ CO_2H \end{array} + ADP + HPO_4^{2-}$$

This enzyme is present in human and animal liver, kidney, adipose tissue, intestinal mucosa, mammary glands, brain, adrenal glands, and other organs, and it is localized in the mitochondrial matrix. In the liver and kidney, it is an integral part of gluconeogenesis, and in adipose tissue it provides a supply of oxaloacetate in mitochondria for condensation with the acetyl-CoA to form citrate. Citrate, in turn, is the precursor of acetyl-CoA in the cytosol used for fatty-acid synthesis.

46   VITAMINS (BIOTIN)

Few other physiological processes of animals and microorganisms involve biotin-dependent carboxylases. β-Methylcrotonyl-CoA carboxylase catalyzes the conversion of β-methylcrotonyl-CoA to β-methylglutaconyl-CoA, a process involved in the catabolism of leucine. Biotin is also a cofactor for a *trans*-carboxylase, which effects the conversion of methylmalonyl-CoA and pyruvate to oxaloacetate and propionyl-CoA, a process of bacterial propionate metabolism. Biotin carboxylase also catalyzes formation of ATP from carbamyl phosphate and ADP in a biotin-promoted reaction (21).

It is well established that the biotin-dependent carboxylases utilize bicarbonate as the carboxyl donor. The role of the enzyme-biotin is accepting the CO_2 moiety and transferring it to the appropriate substrate. This process is represented as a composite of the following two reactions:

$$\text{enzyme-biotin} + HCO_3^- + ATP \underset{}{\overset{Mg^{2+}}{\rightleftharpoons}} \text{enzyme-biotin-}CO_2 + ADP + HPO_4^{2-}$$

$$\text{enzyme-biotin-}CO_2 + \text{substrate} \rightleftharpoons \text{substrate-}CO_2 + \text{enzyme-biotin}$$

The first of these reactions is common to all carboxylases, whereas the second depends on the substrate, which is different in each case. With *trans*-carboxylase, methylmalonyl-CoA replaces ATP and HCO_3^- as the carboxyl donor. During the formation of enzyme-biotin-CO_2, the carboxyl group attaches to the distal biotin nitrogen, and the biotin links covalently to the enzyme by means of a lysine residue.

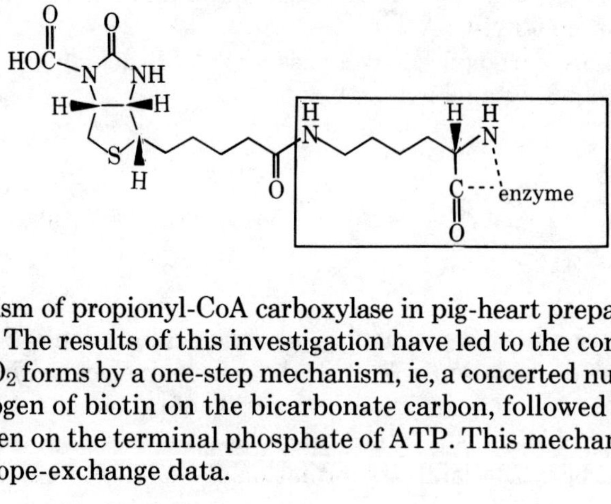

The mechanism of propionyl-CoA carboxylase in pig-heart preparations has been investigated (22). The results of this investigation have led to the conclusion that the enzyme-biotin-CO_2 forms by a one-step mechanism, ie, a concerted nucleophilic attack of the distal nitrogen of biotin on the bicarbonate carbon, followed by attack of the bicarbonate oxygen on the terminal phosphate of ATP. This mechanism agrees with the available isotope-exchange data.

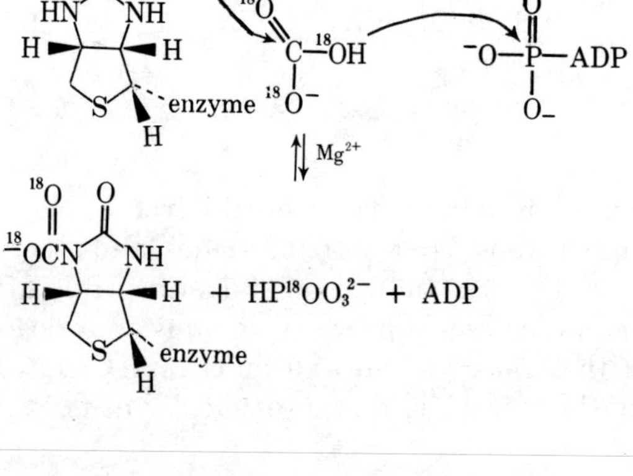

However, another mechanism involving an enzyme-biotin-PO_4^{2-} intermediate was proposed as early as 1960 and recently was given more support with a detailed mechanistic investigation (23–24).

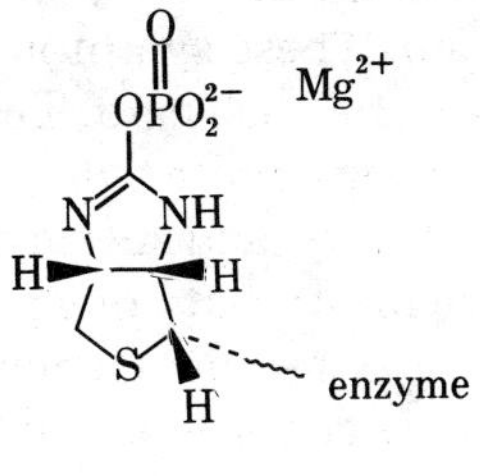

The transfer of CO_2 from enzyme-biotin-CO_2 to the substrate is believed to occur at a different site on the enzyme, with the mobile biotin prosthetic group transporting the active CO_2 from one site to the other (25). On the basis of model studies, the following mechanism, with acetyl-CoA carboxylase as an example, has been proposed (26):

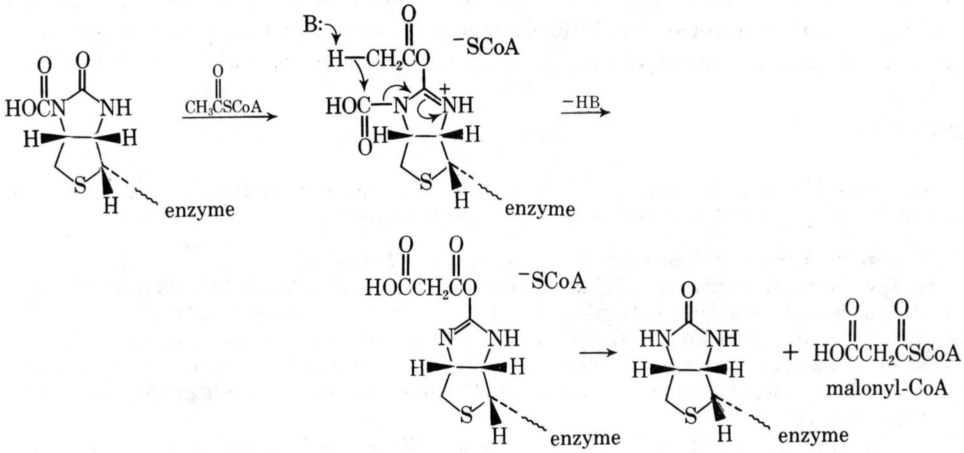

The preceding proposed mechanism cannot account for the previous findings in the studies of CO_2 transfer with propionyl-CoA carboxylase and pyruvate carboxylase that the replacement of a proton with CO_2 occurs with retention of configuration (27–29). The concerted mechanism, which accounts for the retention of configuration in both reactions of transcarboxylase and which involves proton transfer by means of enolbiotin, is described in refs. 30 and 31 and is based on proposals reported in refs. 32 and 33. This concerted mechanism does not require a base at the active site to facilitate proton transfer. As yet, there is no general agreement with respect to the mechanism of the CO_2 transfer from carboxybiotin to a substrate in biotin-dependent carboxylation reactions. However, there is some new evidence that favors the carbanion mechanism (34–35).

VITAMINS (BIOTIN)

Biotin Deficiency

Because of biosynthesis by intestinal flora, a deficiency of biotin seldom occurs in humans. In rare cases, a biotin deficiency is indicated by dermatitis, loss of appetite, nausea, vomiting, depigmentation, alopecia, weight loss, anemia, elevated blood cholesterol, and depression (36). These symptoms can be reversed by giving biotin at the level of adult requirement, 150–300 µg/d. The biotin deficiency can be produced artificially by a diet rich in raw egg white, which contains the glycoprotein avidin. Avidin binds biotin with high affinity, and such a complex is resistant to the digestive enzymes. More recently, a rare life-threatening genetic defect in biotin metabolism, that is, biotin-dependent-carboxylase deficiency, has been determined in a small number of young children. These conditions can be corrected by biotin administration.

In poultry, biotin is an essential vitamin for normal growth, feed conversion, and reproduction as well as healthy skin, feathers, and bones. Biotin deficiency in poultry causes reduced growth rate, impaired feed conversion, reduced feed intake, perosis, and other deformities causing leg weakness, poor feathering, and foot dermatitis. In broilers, a biotin deficiency causes breast blisters (when broilers rest on their breasts because of leg weakness), fatty liver and kidney syndrome, parrot beak, and death. Biotin deficiency also causes dramatic symptoms in swine, eg, reduced growth rate, impaired feed conversion, dermatitis, excessive hair loss, furry tongue, foot lesions, stiff-legged gait, squatness, and hind-leg spasms. These deficiencies are corrected by using biotin as a feed additive for poultry and swine (see Pet and livestock feeds).

BIBLIOGRAPHY

"Biotin" in *ECT* 1st ed., Vol. 2, pp. 519–525, by K. Hofmann, University of Pittsburgh; "Biotin" in *ECT* 2nd ed., Vol. 3, pp. 518–527, by J. D. Woodward, University of Reading.

1. F. Kogl, B. Tonnis, and Hoppe-Seyl, *Z. Physiol. Chem.* **242**, 43 (1936).
2. V. duVigneaud, K. Hofmann, D. B. Melville, and P. Gyorgy, *J. Biol. Chem.* **140**, 643 (1941).
3. D. B. Melville, K. Hofmann, E. Hague, and V. duVigneaud, *J. Biol. Chem.* **142**, 615 (1942).
4. V. duVigneaud, K. Hofmann, D. B. Melville, and J. R. Rachele, *J. Biol. Chem.* **140**, 763 (1941).
5. V. duVigneaud, D. B. Melville, K. Folkers, D. E. Wolf, R. Mozingo, J. C. Keresztesy, and S. A. Harris, *J. Biol. Chem.* **146**, 475 (1942); D. B. Melville, A. W. Moyer, K. Hofmann, and V. duVigneaud, *J. Biol. Chem.* **146**, 487 (1942).
6. S. A. Harris, D. E. Wolf, R. Mozingo, R. C. Anderson, G. E. Arth, N. R. Easton, D. Heyl, A. N. Wilson, and K. Folkers, *J. Am. Chem. Soc.* **66**, 1756 (1944); S. A. Harris, D. E. Wolf, R. Mozingo, G. E. Arth, R. C. Anderson, N. R. Easton, and K. Folkers, *J. Am. Chem. Soc.* **67**, 2096 (1945).
7. J. Trotter and Y. A. Hamilton, *Biochemistry* **5**, 713 (1966).
8. U.S. Pat. 2,489,234 (Nov. 22, 1949), M. W. Goldberg and L. H. Sternbach (to Hoffmann-La Roche, Inc.).
9. U.S. Pat. 3,876,656 (April 8, 1975), A. Hisao, A. Yasuhiko, O. Shigeru, and S. Hiroyuki (to Sumitomo Chemical Co., Ltd.).
10. P. Confalone, G. Pizzolato, E. Baggiolini, D. Lollar, and M. R. Uskoković, *J. Am. Chem. Soc.* **99**, 7020 (1977).
11. U.S. Pat. 4,284,557 (Aug. 8, 1981), E. G. Baggiolini, H. L. Lee, and M. R. Uskoković (to Hoffmann-La Roche, Inc.).
12. T. Ogawa, T. Kawano, and M. Matsui, *Carbohydr. Res.* **57**, C31 (1977); H. Ohrui, N. Sueda, and S. Emoto, *Agric. Biol. Chem.* **42**, 865 (1978); F. G. M. Vogel, J. Paust, and A. Nurrenbach, *Liebigs Ann. Chem.*, 1971 (1980).
13. P. Confalone, G. Pizzolato, D. Lollar, and M. R. Uskoković, *J. Am. Chem. Soc.* **102**, 1959 (1980); H. J. Monteiro, *Ann. Acad. Bras. Cienc.* **52**, 493 (1980).

14. A. Fliri and K. Hohenlohe-Oehringen, *Chem. Ber.* **113,** 607 (1980); R. A. Whitney, *Can. J. Chem.* **59,** 2650 (1981).
15. M. A. Eisenberg, *Adv. Enzymol.* **38,** 317 (1973).
16. R. J. Parry and M. G. Kunitani, *J. Am. Chem. Soc.* **98,** 4024 (1976).
17. F. Frappier, M. Jouany, A. Marquet, A. Olesker, and J.-C. Tabet, *J. Org. Chem.* **47,** 2257 (1982).
18. D. A. Trainer, R. J. Parry, and A. Gutterman, *J. Am. Chem. Soc.* **102,** 1467 (1980).
19. P. N. Achuta Murthy and S. P. Mistry, *Prog. Food Nutr. Sci.* **2,** 420 (1977).
20. S. J. Wakil, *J. Am. Chem. Soc.* **80,** 6465 (1958).
21. R. B. Guchait, S. E. Polakis, D. Hollis, C. Fenselau, and M. D. Lane, *J. Biol. Chem.* **249,** 6646 (1974).
22. R. Mazumder, T. Sasakawa, Y. Kaziro, and S. Ochoa, *J. Biol. Chem.* **237,** 3065 (1962).
23. D. R. Halenz and M. D. Lane, *J. Biol. Chem.* **235,** 878 (1960).
24. R. Kluger, P. P. Davis, and P. D. Adawadkar, *J. Am. Chem. Soc.* **101,** 5995 (1979).
25. J. Moss and M. D. Lane, *Adv. Enzymol.* **35,** 321 (1971); R. E. Borden, C. H. Fung, M. F. Utter, and M. C. Scrutton, *J. Biol. Chem.* **247,** 1323 (1972).
26. H. Kohn, *J. Am. Chem. Soc.* **98,** 3690 (1976).
27. D. Arigoni, F. Lynen, and J. Rétey, *Helv. Chim. Acta* **49,** 311 (1966).
28. D. J. Prescott and J. L. Rabinowitz, *J. Biol. Chem.* **243,** 1551 (1968).
29. I. A. Rose, *J. Biol. Chem.* **245,** 6052 (1970).
30. H. G. Wood and G. Zwolinski, *Crit. Rev. Biochem.* **4,** 47 (1976).
31. H. G. Wood and R. E. Barden, *Annu. Rev. Biochem.* **46,** 385 (1977).
32. I. A. Rose, E. L. O'Connell, and F. Solomon, *J. Biol. Chem.* **251,** 902 (1976).
33. J. Rétey and F. Lynen, *Biochem. Z.* **342,** 256 (1965).
34. J.-A. Stubbe, S. Fish, and R. H. Abeles, *J. Biol. Chem.* **255,** 236 (1980).
35. D. J. Kuo and I. A. Rose, *J. Am. Chem. Soc.* **104,** 3235 (1982).
36. J.-P. Bonjour, *World Rev. Nutr. Diet.* **38,** 1 (1981).

MILAN R. USKOKOVIĆ
Hoffmann-La Roche, Inc.

INOSITOL

The name inositol may be applied to nine possible isomers of 1,2,3,5-cis-1,2,3,4,5,6-hexahydroxycyclohexane, of which only one, *myo*-inositol [87-89-8], has been associated with vitaminlike properties. Accordingly, this review deals almost exclusively with the *myo* isomer unless otherwise indicated. The reader is referred to the nomenclature of the cyclitols as recommended by the IUPAC and IUB Commissions of Organic and Biochemical Nomenclature (1) as well as to refs. 2–3. Individual inositols are differentiated by italicized prefixes and hyphens; the locants or positional numbers are assigned according to recommended criteria (1). *myo*-Inositol is a ubiquitous cyclitol that occurs in plants and animals, as well as many lower forms of life, where it exists free and as covalently linked derivatives of various classes of compounds. Scherer is credited with the discovery of *myo*-inositol from muscle tissue, and he gave the substance the name *inosit* from the Greek for muscle (4). The conformational analysis of *myo*-inositol indicates that the most stable form should be the strain-reduced chair form (Fig. 1) which has a single axial hydroxyl group at position C-2 (5–6).

Occurrence

Plants. Free *myo*-inositol is widely distributed in many types of plants. However, it is most commonly present in higher plants as the hexakis-O-phosphate, phytic acid (see Food toxicants, naturally occurring). Phytic acid occurs to a great extent in seeds and cereal grains, accounting for as much as 80% of the total phosphorus content (7). It is isolated from plant tissue as phytin, the combined calcium–magnesium salt. In addition, cations of various essential metals, especially zinc, manganese, iron, and copper, are present; phytin represents a significant dietary source of these elements. The availability of the ions depends upon the gastrointestinal hydrolysis of the phosphoric esters by microflora or intestinal cellular phytase. These factors as well

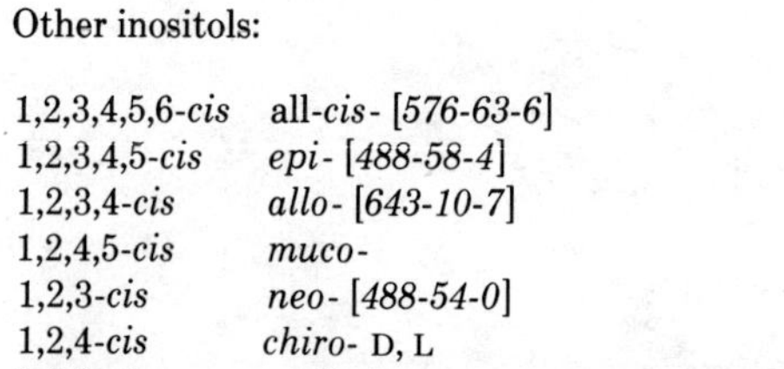

1,2,3,5-*cis*-4,6-*trans*-Hexahydroxycyclohexane

Other inositols:

1,2,3,4,5,6-*cis*	all-*cis*- [576-63-6]
1,2,3,4,5-*cis*	*epi*- [488-58-4]
1,2,3,4-*cis*	*allo*- [643-10-7]
1,2,4,5-*cis*	*muco*-
1,2,3-*cis*	*neo*- [488-54-0]
1,2,4-*cis*	*chiro*- D, L
1,3,5-*cis*	*scyllo*-

Figure 1. The structure of *myo*-inositol.

as food preparation help control the source of Ca^{2+} and phosphorus and the release of *myo*-inositol, which is otherwise unabsorbed across intestinal membranes as phytin. Lower phosphoric esters of *myo*-inositol exist as dephosphorylation products of phytic acid or as more stable metabolites, eg, tetra- and pentaphosphoric esters in mature cotton embryos (8). Tetra- and pentaphosphoric esters of *myo*-inositol are present in red blood cells in birds, where they apparently serve the same function as 2,3-diphosphoglyceric acid in human red blood cells, namely, to regulate oxygen binding to hemoglobin.

Phosphatidylinositol is a normal constituent of plant phospholipids, accounting for 10–12% of the phospholipid in mitochrondria and chloroplasts (9). The methyl ethers of *myo*- scyllo- [488-59-5], D-chiro- [643-12-9], L-chiro- [551-72-4], and *muco*-inositol [488-55-1] occur in higher plants where they may have a storage function (10). The galactoside of *myo*-inositol, galactinol (1-L-1-*O*-α-D-galactopyranosyl-*myo*-inositol), first isolated as a minor disaccharide of sugar-beet molasses, also is in many plants (11). Galactinol is a cofactor in the biosynthesis of galactosyl oligosaccharides, eg, raffinose, stachyose, and verbascose (12–13). An isomer of galactinol, a 4-*O*-β-D-galactopyranosyl-*myo*-inositol (formerly called 6β-galactinol), was isolated from rat mammary gland and from rat and human milk (14–15). A 4-*O*-β-D-mannopyranosyl-*myo*-inositol (formerly numbered 6-*O*) and a β-D-glucopyranosyl-*myo*-inositol of unknown linkage have been reported in yeast and *Phaseolus aureus*, respectively (16–18). Plants, but not animals, contain a variety of esters and ester glycosides of *myo*-inositol or its derivatives (19). In this class, *myo*-inositol can be esterified to the auxin hormone indole-3-acetic acid (IAA) and, in some instances, is glycosidically linked at position 5 to galactose or arabinose (20–21). A mechanism involving the synthesis and hydrolysis of these compounds for homeostatic regulation of the IAA levels during plant growth has been proposed (22) (see Plant-growth substances).

Animals. In its free state, *myo*-inositol occurs in significant amounts in brain, testis, semen, secretory tissues, and fetal blood, and the milk of rodents and humans contains appreciable levels of free *myo*-inositol (23). A correlation may exist between those species with higher levels of *myo*-inositol in their milk and the immaturity of their newborns; thus, *myo*-inositol may be required for rapid neonatal development (23–24). It is possible that the main function of free *myo*-inositol is as a precursor of the phosphoinositides of which phosphatidylinositol is the principal species in animal tissues. The rat-brain free-*myo*-inositol level is approximately equivalent to the K_m (Michaelis-Menten constant) for *myo*-inositol of cytidinediphosphodiglyceride: *myo*-inositol phosphatidyl transferase (EC 2.7.8.11) which is the enzyme responsible for the synthesis of phosphatidylinositol (25–26).

Despite the ability of most animal cells to synthesize *myo*-inositol from D-glucose 6-phosphate at measurable rates during embryogenesis, the cells of the adult show little activity, with the exception of testis and lactating mammary tissue (23). Accordingly, the *myo*-inositol of many tissues is believed to be regulated by the uptake of the vitamin from the blood against a concentration gradient (27). For example, plasma levels of adult rats are typically 50–60 μM, whereas the brain content is 4–5 $\mu mol/g$ fresh tissue (23–25). Alternatively, a very slow diffusion of biosynthesized *myo*-inositol may contribute to the significant intracellular-to-extracellular differential concentrations.

Bound forms of *myo*-inositol in animals are predominantly phospholipids consisting of phosphatidylinositol, phosphatidylinositol 4-phosphate (diphosphoinositide), and phosphatidylinositol 4,5-bisphosphate (triphosphoinositide). The long-chain acyl groups at the sn-1 and sn-2 positions of animal phosphoinositides are predominantly stearic and arachidonic acids, respectively. 4-O-β-D-Galactopyranosyl-myo-inositol occurs only in milk or mammary tissue (15). This is not surprising since it is the product of a transferase reaction between lactose and free myo-inositol catalyzed by β-galactosidase (28). Erythrocytes of avian species contain the myo-inositol 1,3,4,5,6-pentakisphosphate and a tetrakisphosphoric ester of myo-inositol occurs in the ostrich (29–30).

Found only in trace amounts, L-myo-inositol 1-phosphate appears in the biosynthetic pathway between D-glucose 6-phosphate and myo-inositol, and 1-D-myo-inositol 1-phosphate and cyclic myo-inositol 1,2-phosphate are catabolites of phosphatidylinositol.

Common Foods. In recent years, increased myo-inositol tissue levels have been associated with various diseased states. The dietary form of myo-inositol can affect its biological availability. The amount of total myo-inositol present in 487 foods has been determined by glc of the hexakis(trimethylsilyl) ether prepared after hydrolysis in 6 N HCl in sealed vials at 120°C for 40 h (31). The greatest amounts of myo-inositol were in fruits, beans, grains, and nuts. Also, various domestic and foreign wines have relatively high levels of myo-inositol (Table 1) (see also Wine).

Properties

myo-Inositol is a highly stable white crystalline powder. Its anhydrous molecular weight is 180.16, but crystals are commonly obtained as the dihydrate. Other chemical and physical properties are listed in Table 2.

Determination

Because they lack carbonyl carbons, the cyclitols were not as easily quantified as other polyhydroxy compounds, eg, glucose (33–35). Gas–liquid chromatographic analysis of inositols, when coupled with conventional techniques of sample preparation such as ion exchange (qv), lipid extraction, thin-layer chromatography, and paper electrophoresis, provides a powerful analytical capability for biochemical studies (see Analytical methods). Furthermore, when combined with mass spectrometry, a positive

Table 1. myo-Inositol Content of Selected Wines[a]

Name	µg/mL	mM
Rheinpfalz—1975	238	1.32
Michigan wines		
Aurora—1972, Sodus 35%	178	0.99
Lawton Delaware—1975	406	2.26
Seyval—1975	228	1.27
De Chaunac Leelanaw—1974	385	2.14
RY-1	595	3.31
Spartan Cellers—B	541	3.01

[a] myo-Inositol determination according to methods described in ref. 25.

Table 2. Physical Properties of *myo*-Inositol[a]

Property	Value
molecular formula	$C_6H_{12}O_6$
melting range, °C	225–227
density (anhydrous crystals at 15°C), g/cm^3	1.752
optical rotation	inactive
solubility, g/100 mL	
in water at 25°C	14
at 60°C	28
in ethanol	sl sol
in ether	insol

[a] Refs. 10, 32.

identification of the structure can be made (36). The mass spectrometer has been used to monitor the intensities of selective ions in the analysis of *myo*-inositol at levels of less than 50 pg.

Manufacture

The main source of *myo*-inositol for commercial production is from plant phytin by degradative processes (37). The heaviest concentration of phytin is in seeds and cereal grains, namely, corn, wheat, rye, oats, peas, beans, barley, rice, cottonseed, flaxseed, soybeans, and peanuts. Typically, the raw material is steeped in white aqueous sulfur dioxide, which facilitates the separation of hull, fiber, germ, gluten, and starch. Soluble proteins, sugars, gums, and soluble organic compounds (including phytic acid salts) are filtered in the steepwater. Phytate, which makes up ca 2% of the total solids, precipitates from the extract upon addition of milk of lime until pH 5–7 is reached. The filter cake is washed thoroughly with water; the product is commonly called phytin or calcium phytate although it contains traces of other metal cations.

Various processes have been developed to hydrolyze the phytate to *myo*-inositol and phosphate, including mild autoclaving temperatures and pressures, usually under alkaline (25 wt % calcium hydroxide slurry) or acid (60 wt % sulfuric acid) conditions (38–39). *myo*-Inositol can be obtained by heating phytate under pressure in water (39). The yield depends on the quality of the crude phytate and typically is 20–21% when highly purified calcium phytate is hydrolyzed but only 7–10% from crude steepwater phytate (39). Traces of *scyllo*-, *chiro*-, and *neo*-inositols accompany *myo*-inositol in the phytate hydrolysates. Adjusting the pH to more than pH 7 followed by filtration effects the removal of insoluble phosphate salts of calcium, etc. The filtrate and wash waters can be decolorized with activated charcoal. Crystallization of the *myo*-inositol can be accomplished from the concentrated liquors or by additions of ethanol to concentrated extracts. Recrystallization below 50°C yields a white crystalline powder as a dihydrate (mp 225–227°C).

Economic Aspects

Tsuno Rice Chemicals, Ltd., of Wakayama, Japan, is the leading *myo*-inositol producer, supplying 75–80% of the world market. The remaining sources are from the

People's Republic of China and Italy. A. E. Staley Manufacturing Company (Decatur, Illinois) was the previous leading supplier to the world market but stopped production in 1981. Staley's prices had fallen from $31/kg in the autumn of 1978 to $15/kg in 1981. Chugai International Corp., the exclusive East Coast distributor for Tsuno, reports that inositol production is still a profitable business and that there is a strong, steady demand. The annual production of the Tsuno plant is 17 metric tons, of which the United States imports 6 t. The 1981 price for *myo*-inositol from Tsuno was $19.25/kg in 50–450-kg lots.

Biological and Nutritional Significance

Most nutritional investigations regarding *myo*-inositol have been concerned with its role as an essential nutrient in microorganisms and mammals (40–42). *myo*-Inositol deficiency results in inadequate growth, alopecia, and eventual death in mice (43–44). Discrepancies in these findings were reported and it was suggested that there is a possible relationship between *myo*-inositol and pantothenate (45–47). Early studies on the dietary requirement of the rat for *myo*-inositol also gave contradictory results (48–53). Exogenous sources of *myo*-inositol, including impure diets and variable intestinal flora biosynthesis, contributed to the difficulty of interpreting these studies (47,53–54). In addition, endogenous sources, which include *in vivo* synthesis and possible tissue mobilization and exchange of *myo*-inositol, make difficult the study of its dietary role (54). Thus, the demonstration that *myo*-inositol is an independent lipotropic agent required use of carefully controlled diets that were not limited in B-complex vitamins, choline, and high quality protein and that contained adequate levels of unsaturated fatty acids (55–56). With such diets supplemented with 0.5 wt % phthalylsulfathiazole to prevent formation of *myo*-inositol by intestinal microorganisms, lactating rats developed severe fatty livers which could be completely prevented by the dietary supplementation of 0.05% *myo*-inositol or by weaning the rat pups (57). The significant conclusion from these studies was that the adult female rat requires a small amount of *myo*-inositol from whatever source to maintain normal hepatic phospholipid and lipoprotein metabolism. However, when the liver of a nursing *myo*-inositol-deficient female rat was stimulated to produce over 50% of the milk triglycerides, a fatty liver resulted (58). The mechanism of the lactation-induced fatty liver was explained by measuring the rate of liver triglyceride esterification from intravenously administered [1-^{14}C]palmitate; there were no differences between control rats and *myo*-inositol-deficient ones (59). In contrast, a significantly slower rate of appearance of ^{14}C-labeled serum triglyceride was observed in *myo*-inositol-deficient lactating dams compared with controls. It was concluded that the reduced release of hepatic triglycerides concurrent with normal delivery of serum-free fatty acids from depots to liver combined to cause the fatty liver condition. These studies demonstrate that decreased liver phosphatidylinositol levels correlate with the impaired release of lipoproteins, especially very low density lipoproteins (VLDL), high density lipoproteins (HDL), and intermediate density lipoproteins (IDL) (60). Phospholipids are essential to the association of lipids with apoHDL and the activity of plasma lecithin:cholesterol acyltransferase (61). In all aspects studied to date, the lipotropic function of *myo*-inositol in the lactating rat is analogous with that independently induced by the dietary deprivation of choline (62). Furthermore, dietary *myo*-inositol has been shown to prevent intestinal lipodystrophy in the female gerbil (63).

The fact that animal cells can synthesize *myo*-inositol from D-glucose 6-phosphate, especially in the embryonic or neonatal states and in adult testis at moderate rates, raises the question of the vitamin status of the cyclitol (23,64). It is known that *myo*-inositol is an essential growth factor for human cells grown in culture (42,65). It has been suggested that proliferating cells require greater phospholipid biosynthesis for increased membrane production. *De novo* synthesis of *myo*-inositol apparently cannot keep up with the requirement and, hence, growth retardation results. Rat pheochromocytoma cell death resulting from *myo*-inositol deficiency has been completely prevented by the administration of the B subunit of mouse-nerve growth factor (66).

Advances in the understanding of the nutritionally related aspects of *myo*-inositol have chiefly involved its lipotropic role and its precursor relationship to the phosphoinositides. However, significant studies have been reported on the biological function of these phospholipids in various systems. The main effect of external stimuli on selected target tissues is rapid and enhanced phosphatidylinositol metabolism (67). A number of agents stimulate phosphatidylinositol turnover in cells that utilize Ca^{2+} as their intracellular second messenger, and these responses are initiated by phosphatidylinositol breakdown into phosphorylinositol and 1,2-diacylglycerol (67). The latter may re-enter the phosphatidylinositol biosynthetic scheme, thus rapidly introducing ^{32}P from terminally labeled ATP into phosphatidic acid, cytidinediphosphodiglyceride, and phosphatidylinositol (67). It has been suggested that phosphatidylinositol in the brain might provide the arachidonic acid in the synthesis of prostaglandins and thromboxanes (68–69) (see Prostaglandins in Supplement Volume).

Various enzymes are stimulated by additions of phosphatidylinositol. For example, phosphatidylinositol is the endogenous activator of $(Na^+ + K^+)$–adenosine triphosphate in microsomes isolated from rabbit kidney (70). Tyrosine hydroxylase, the enzyme that catalyzes the rate-limiting step in the biosynthesis of dopamine and norepinephrine, is strongly affected by phosphatidylinositol (71). The phospholipid rapidly causes a reversible activation of the enzyme and a slow irreversible inactivation of the enzyme in the absence of the substrate. Acetyl CoA carboxylase from rat liver is a regulatory enzyme in fatty-acid biosynthesis that is activated by citrate; the process depends upon the presence of phosphatidylinositol (72). Various other membrane-bound enzymes, eg, alkaline phosphatase and 5'-nucleotidase, depend specifically upon phosphatidylinositol for their association with plasma membranes (73).

The minor phosphoinositides, phosphatidylinositol 4-phosphate and phosphatidylinositol 4,5-bisphosphate, are believed to play important roles in membrane function. These phospholipids bind Ca^{2+} (74). In stimulated nervous tissues, the 4- and 5-phosphate groups on the inositol ring undergo rapid turnover caused by a transient rise in intracellular Ca^{2+} levels, which is a response to the passage of the action potential (75). These phospholipids have also been involved with the regulation of Ca^{2+} binding to erythrocyte membrane and in the regulation of intracellular Ca^{2+} levels (76).

myo-Inositol Metabolism and Disease

A correlation has been observed between decreased peripheral motor and sensory nerve conduction velocities and decreased free *myo*-inositol in the nerves from experimentally diabetic rats (77). Dietary supplementation of 1 wt % *myo*-inositol increases plasma and nerve levels of free inositol and significantly improves motor nerve conduction velocities in these animals. However, when 3 wt % dietary *myo*-inositol is provided, a decreased motor conduction velocity occurs in both normal and diabetic rats. The observed decrease of free inositol in sciatic nerves of streptozotocin diabetic rats has been confirmed, and lower lipid-bound inositol has been shown to be in the nerves of the diabetic animals, suggesting another functional relationship between free and phospholipid *myo*-inositol levels (78). The effect of streptozotocin-induced diabetes on *myo*-inositol metabolism in rat testis has been studied (79). Over a two-week period, there was no apparent effect on the specific activity of inositol 1-phosphate synthetase, whereas significantly higher levels of free *myo*-inositol, glucose, and glucose 6-phosphate occurred. In studies of longer-term diabetes in streptozotocin-treated rats, the specific activity of inositol 1-phosphate synthetase is lower in the testis but not in the sciatic nerves of diabetic rats compared with controls (80). However, the specific activities of the kinases responsible for the synthesis of phosphatidylinositol 4-phosphate and phosphatidylinositol 4,5-bisphosphate are significantly lower in the sciatic nerves and brains of diabetic rats, suggesting that an altered metabolism of the inositol lipids in nerve tissue may contribute to the human diabetic neuropathy. It has been suggested that the inositoluria observed in human diabetes may arise from the inhibitory effect of glucose on renal tubular reabsorption of inositol (81). However, the activity of kidney inositol oxygenase, which catabolizes inositol, decreases markedly in experimental diabetes, which may explain the increased clearance of inositol (78).

The brains of galactosemic patients have elevated levels of galactitol and decreased free and lipid-bound *myo*-inositol, suggesting a possible correlation between *myo*-inositol metabolism and the neurotoxicity associated with this metabolism error (82–83). Further studies have shown that the phosphatidylinositol response to acetylcholine is impaired in synaptosomes from galactose-fed rats; thus, there may be a defect in the coupling of receptor–neurotransmitter interaction and phosphatidylinositol turnover (84).

A marked elevation in serum free-*myo*-inositol levels has been reported in patients with chronic renal failure (85–86). It has been suggested that kidney malfunction leads to abnormal glomerular clearance and oxidation of *myo*-inositol leading to the elevated levels of serum cyclitol (85,87). The question arises whether a deleterious effect may accompany the increased serum *myo*-inositol levels. Large doses of inositol increase phosphatidylinositol levels in the endoplasmic reticulum of rat liver but without adverse effects (88). There is a striking decrease in the sciatic nerve conduction velocity in normal rats fed a diet enriched in inositol for one week. These observations support the hypothesis that secondary hyperinositolemia may contribute to the pathogenesis of uremic polyneuropathy in subjects with chronic renal failure. A more extensive review of the nutritional significance, metabolism, and function of *myo*-inositol and phosphatidylinositol has recently been published (89).

BIBLIOGRAPHY

"Inositol" in *ECT* 1st ed., Vol. 7, pp. 877–881, by E. R. Weidlein, Jr., Mellon Institute; "Inositol" in *ECT* 2nd ed., Vol. 11, pp. 673–676, by E. R. Weidlein, Jr., Union Carbide Corporation.

1. *Eur. J. Biochem.* **57**, 1 (1975).
2. B. W. Agranoff, *Trends in Biochemical Sciences* **3**, N283 (1978).

3. D. J. Cosgrove in *Studies in Organic Chemistry 4. Inositol Phosphates—Their Chemistry, Biochemistry, and Physiology*, Elsevier, Amsterdam, 1980.
4. J. Scherer, *Ann.* **73,** 322 (1850).
5. L. Anderson in W. Pigman and D. Horton, eds., *The Carbohydrates*, 3rd ed., Academic Press, New York, 1972, p. 519.
6. S. J. Angyal in ref. 5, p. 195.
7. H. Mollgaard, K. Lorenzen, I. G. Hansen, and P. E. Christensen, *Biochem. J.* **40,** 589 (1946).
8. C. B. Sharma and J. W. Dieckert, *Phytochemistry* **13,** 2529 (1974).
9. M. Kates and M. O. Marshall, *Annu. Proc. Phytochem. Soc.* **12,** 115 (1975).
10. T. Posternak in *The Cyclitols*, Holden-Day, Inc., San Francisco, Calif., 1965, p. 127–143.
11. R. I. Brown and R. F. Serro, *J. Am. Chem. Soc.* **75,** 1040 (1953).
12. W. Tanner and O. Kandler, *Plant Physiol.* **41,** 1540 (1966).
13. W. Tanner, L. Lehle, and O. Kandler, *Biochem. Biophys. Res. Commun.* **29,** 166 (1967).
14. W. F. Naccarato and W. W. Wells, *Biochem. Biophys. Res. Commun.* **57,** 1026 (1974).
15. W. F. Naccarato, R. E. Ray, and W. W. Wells, *J. Biol. Chem.* **250,** 1872 (1975).
16. W. W. Wells, C.-H. Kuo, and W. F. Naccarato, *Biochem. Biophys. Res. Commun.* **61,** 644 (1974).
17. W. Tanner, *Z. Pflanzenphysiol.* **57,** 474 (1967).
18. J. Kemp and B. C. Longhman, *Biochem. J.* **142,** 153 (1974).
19. F. A. Loewus and M. W. Loewus in J. Preiss, ed., *The Biochemistry of Plants*, Vol. 3, Academic Press, New York, 1980, pp. 43–76.
20. J. Kopcewicz, A. Ehmann, and R. S. Bandurski, *Plant Physiol.* **54,** 846 (1974).
21. M. Ueda, R. S. Bandurski, *Phytochemistry* **13,** 243 (1974).
22. R. S. Bandurski and A. Schulze, *Plant Physiol.* **60,** 211 (1977).
23. L. E. Burton and W. W. Wells, *Dev. Biol.* **37,** 35 (1974).
24. R. Jenness, A. W. Erickson, and J. J. Craighead, *J. Mammal.* **53,** 34 (1972).
25. W. W. Wells, T. A. Pittman, and H. J. Wells, *Anal. Biochem.* **10,** 450 (1965).
26. J. A. Benjamins and B. W. Agranoff, *J. Neurochem.* **16,** 513 (1969).
27. G. Hauser, *Biochim. Biophys. Acta* **173,** 267 (1969).
28. C.-H. Kuo and W. W. Wells, *J. Biol. Chem.* **253,** 3550 (1978).
29. L. F. Johnson and M. E. Tate, *Can. J. Chem.* **47,** 63 (1969).
30. R. Isaacks, D. Harkness, R. Sampsell, J. Alder, S. Roth, C. Kim, and P. Goldman, *Eur. J. Biochem.* **77,** 567 (1977).
31. R. S. Clements, Jr., and B. Darnell, *Am. J. Clin. Nutr.* **33,** 1954 (1980).
32. E. Chargaff in W. H. Sebrell, Jr., and R. S. Harris, eds., *The Vitamins II*, Academic Press, Inc., New York, 1954, pp. 336–338.
33. C. C. Sweeley, R. Bentley, M. Makita, and W. W. Wells, *J. Am. Chem. Soc.* **85,** 2497 (1963).
34. D. W. Woolley in *Biological Symposia XII. Estimation of the Vitamins*, Jaques Cattell Press, Lancaster, Pa., 1947, p. 279.
35. S. J. Angyal and D. J. McHugh, *J. Chem. Soc.*, 1423 (1957).
36. W. R. Sherman, N. C. Eilers, and S. L. Goodwin, *Org. Mass Spectrom.* **3,** 829 (1970).
37. E. R. Weidlein, Jr., in ref. 32.
38. F. A. Hoglan and E. Bartow, *Ind. Eng. Chem.* **31,** 749 (1939).
39. E. Bartow and W. W. Walker, *Ind. Eng. Chem.* **30,** 300 (1938).
40. F. Kagl and N. van Hasselt, *Z. Physiol. Chem.* **242,** 74 (1936).
41. H. W. Buston and B. N. Pramanik, *Biochem. J.* **25,** 1656 (1931).
42. H. Eagle, V. I. Oyama, M. Levy, and A. Freeman, *Science* **123,** 845 (1956).
43. D. W. Woolley, *J. Biol. Chem.* **136,** 113 (1940).
44. *Ibid.*, **139,** 29 (1941).
45. G. J. Martin, *Science* **93,** 422 (1941).
46. D. W. Woolley, *Proc. Soc. Exp. Biol. Med.* **46,** 565 (1941).
47. D. W. Woolley, *J. Exp. Med.* **75,** 277 (1942).
48. P. L. Pavcek and H. L. Baum, *Science* **93,** 502 (1941).
49. E. Nielson and C. A. Elvehjem, *Proc. Soc. Exp. Biol. Med.* **54,** 89 (1943).
50. J. C. Forbes, *Proc. Soc. Exp. Biol. Med.* **54,** 89 (1943).
51. T. H. Jukes, *Proc. Soc. Exp. Biol. Med.* **45,** 625 (1940).
52. L. R. Richardson, A. G. Hogan, B. Long, and K. I. Itschner, *Proc. Soc. Exp. Biol. Med.* **46,** 530 (1941).
53. M. H. McCormick, P. N. Harris, and C. A. Anderson, *J. Nutr.* **52,** 337 (1954).

54. J. W. Halliday and L. Anderson, *J. Biol. Chem.* **217,** 797 (1955).
55. C. H. Best, J. H. Ridout, J. M. Patterson, and C. C. Lucas, *Biochem. J.* **48,** 448 (1951).
56. J. M. Beveridge and C. C. Lucas, *J. Biol. Chem.* **157,** 311 (1945).
57. L. E. Burton and W. W. Wells, *J. Nutr.* **106,** 1617 (1976).
58. R. Bickerstaffe in J. R. Falconer, ed., *Lactation*, Butterworth, London, 1971, p. 317.
59. L. E. Burton and W. W. Wells, *J. Nutr.* **109,** 1483 (1979).
60. *Ibid.*, **107,** 1871 (1977).
61. A. K. Sautar, C. W. Garner, H. N. Baker, J. T. Sparrow, R. L. Jackson, A. M. Gotto, and L. C. Smith, *Biochemistry* **14,** 3057 (1975).
62. B. Lombardi, P. Pani, and F. F. Schlunk, *J. Lipid Res.* **9,** 437 (1968).
63. D. M. Hegsted, K. C. Hayes, A. Gallagher, and H. Hanford, *J. Nutr.* **103,** 302 (1973).
64. F. Eisenberg, Jr., *J. Biol. Chem.* **242,** 1375 (1967).
65. H. Eagle, V. I. Oyama, M. Levy, and A. E. Freeman, *J. Biol. Chem.* **266,** 191 (1957).
66. J. Lakshmanan, R. Tarris, M. I. Lee, and D. A. Fisher, *Biochem. Biophys. Res. Commun.* **105,** 36 (1982).
67. R. H. Michell, *Biochim. Biophys. Acta* **415,** 81 (1975).
68. J. Marion and L. S. Wolfe, *Biochim. Biophys. Acta* **574,** 25 (1979).
69. S. M. Prescott and P. W. Majerus, *J. Biol. Chem.* **256,** 579 (1981).
70. J. G. Madersloot, B. Roelofsen, and J. DeGrier, *Biochim. Biophys. Acta* **508,** 478 (1978).
71. T. Lloyd, *J. Biol. Chem.* **254,** 7247 (1979).
72. H. W. Heger and H. W. Peter, *Int. J. Biochem.* **8,** 841 (1977).
73. M. G. Low and B. J. Finean, *Biochim. Biophys. Acta* **508,** 565 (1978).
74. H. S. Hendrickson and J. G. Fullington, *Biochemistry* **4,** 1599 (1965).
75. H. D. Griffin and J. N. Hawthorne, *Biochem. J.* **176,** 541 (1978).
76. J. T. Buckley and J. N. Hawthorne, *J. Biol. Chem.* **247,** 7218 (1972).
77. D. A. Greene, P. V. DeJesus, and A. I. Winegrad, *J. Clin. Invest.* **55,** 1326 (1975).
78. K. P. Palmano, P. H. Whiting, and J. N. Hawthorne, *Biochem. J.* **167,** 229 (1977).
79. T. P. Rancour and W. W. Wells, *Arch. Biochem. Biophys.* **202,** 150 (1980).
80. P. H. Whiting, K. P. Palmano, and J. N. Hawthorne, *Biochem. J.* **179,** 549 (1979).
81. R. S. Clements, Jr., and R. Reynertson, *Diabetes* **26,** 215 (1977).
82. W. W. Wells, T. A. Pittman, H. J. Wells, and T. J. Egan, *J. Biol. Chem.* **240,** 1002 (1965).
83. S. Segal in J. B. Stanbury, J. B. Wyngaarden, and D. S. Frederickson, eds., *The Metabolic Basis of Inherited Disease*, 3rd ed., McGraw Hill, New York, 1972, pp. 174–195.
84. A. S. Warfield and S. Segal, *Proc. Natl. Acad. Sci. U.S.A.* **75,** 4568 (1978).
85. R. S. Clements, Jr., P. V. DeJesus, Jr., and A. I. Winegrad, *Lancet* **i,** 1377 (1973).
86. L. M. Lewin, S. Melmed, and H. Bank, *Clin. Chim. Acta* **54,** 377 (1974).
87. E. Pitkanen, *Clin. Chim. Acta* **71,** 461 (1976).
88. K. Yagi and A. Kataki, *Ann. N.Y. Acad. Sci.* **165,** 710 (1969).
89. B. J. Holub in H. H. Draper, ed., *Advances in Nutritional Research*, Vol. 4, Plenum Publishing Corp., New York, 1982, pp. 107–141.

WILLIAM W. WELLS
Michigan State University

NICOTINAMIDE AND NICOTINIC ACID (B$_3$)

Nicotinamide [98-92-0] (1) is a member of the B-vitamin complex. Other commonly used names include niacinamide, nicotinic acid amide, nicotinamidum, vitamin B$_3$, vitamin PP (pellagra preventing), and 3-pyridinecarboxamide.

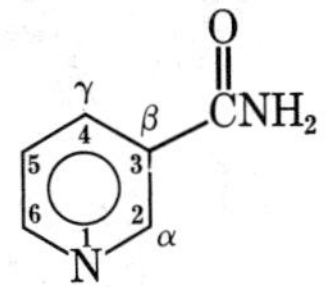

(1)
nicotinamide
3-pyridinecarboxamide

Both nicotinamide and nicotinic acid are building blocks for coenzyme I (Co I), nicotinamide–adenine dinucleotide (NAD), and coenzyme II (Co II), nicotinamide–adenine dinucleotide phosphate (NADP).

Lately, the American Institute of Nutrition (AIN) has suggested the designation niacin for both nicotinamide and nicotinic acid (1). Niacin, however, is usually understood to mean nicotinic acid. Moreover, the analytical method generally used in the past measured the total of nicotinic acid and nicotinamide and reported it as niacin. Thus, considerable confusion exists in the literature. The acceptance of the term vitamin B$_3$ for combined nicotinamide–nicotinic acid would be a desirable simplification. Vitamin B$_3$ has also occasionally been used as a name for pantothenic acid (2). At present, however, the term vitamin B$_5$ is the preferred name for pantothenic acid.

Before 1937, when the significance of nicotinamide as an antipellagra factor was recognized (3), pellagra was widespread in many parts of the world where corn was the main source of nutrition. A brief chronology of nicotinamide and nicotinic acid deficiency is given in Table 1. As late as the beginning of this century, hundreds of

Table 1. History

Year	Event
1735	Casal described pellagra as an endemic disease
1867	Huber obtained, via oxidation of nicotine with potassium dichromate, a substance, that
1873	Weidel named nicotinic acid
1894	nicotinamide was prepared by Engler
1912	nicotinic acid was isolated from rice bran by Suzuki and co-workers
1913	Funk isolated nicotinic acid from yeast
1914	Goldberger carried out investigations of the cause of pellagra
1935	Kuhn isolated nicotinamide from heart muscle
1936	Warburg and Euler simultaneously discovered that nicotinamide is a component of the hydrogen-transferring enzymes codehydrogenase I (= NAD) and codehydrogenase II (= NADP)
1937	Elvehjem and co-workers isolated nicotinamide from a liver extract and identified it as pellagra-preventing factor
1937	cure for pellagra using nicotinamide discovered by Spies

thousand cases of pellagra were documented in the United States (4). Deaths were not uncommon. A similar situation existed in the Mediterranean region. In 1938, after the relationships between pellagra, corn diet, and vitamin B_3 deficiency were clarified, Americans began to enrich cereal products (corn, wheat) with nicotinic acid or nicotinamide. The result was a dramatic decline in the number of pellagra deaths, as shown below (5). Today, acute avitaminoses are seen only rarely.

Year	1929	1935	1940	1945	1949	1956
Number of deaths in the United States	7358	3543	2133	914	321	70

The symptoms of vitamin B_3 deficiency are evident, above all, in the skin, digestive tract, and nervous system. Damage to the skin and mucous membranes occurs in cases of slight deficiency; severe cases lead to pellagra (*pellis agra* = diseased skin). The symptoms in humans are characterized by dermatitis and pigmentation, diarrhea, vomiting, and nervous disorders. Infections of the mouth's mucous membrane and tongue as well as of the stomach and intestinal mucous membrane also occur. Nervous disorders include insomnia, headaches, pains in the extremities, depression, hallucination, and neuroparalysis. A victim's electroencephalogram exhibits a characteristic alteration. In developed countries, pellagra symptoms are found principally in connection with chronic alcoholism, liver cirrhosis, and chronic diarrhea. The principal symptoms in dogs are infection of the mouth's mucous membrane and black coloration of the tongue (black-tongue disease). Swine exhibit raw skin and dermatitis in the ear, diarrhea, and weight loss. Poultry are afflicted with infection of the pharyngeal cavity, diarrhea, and disorders in growth and feathering. Laying hens lose weight and egg-output and hatching rates suffer; calves have severe diarrhea.

Occurrence

Nicotinamide and nicotinic acid occur in nature almost exclusively in bound form. Free nicotinic acid is more prevalent in plants, whereas in animals nicotinamide predominates. Almost all of the nicotinamide found in humans and animals is fixed as building blocks of NAD and NADP. The content of pyridine nucleotides in the organs is apparently genetically determined (see Table 2).

Pellagra and the nature of bound nicotinic acid in corn and wheat were the

Table 2. Pyridine Nucleotides in Rat Organs, mg/kg [a]

Organ	NAD	NADH	NADP	NADPH
liver	370	204	6	205
heart	299	184	4	33
kidney	223	212	3	54
brain	133	88	<2	8
thymus	116	35	<2	12
lung	108	52	9	18
pancreas	115	78	<2	12
testes	80	71	<2	6
blood	55	36	5	3

[a] Ref. 6.

subjects of intensive nutritional research (4–5,7–10). For wheat bran, two bound forms are described: a peptide with a molecular weight of 12,000–13,000 (11) and a carbohydrate complex that was named niacytin (12). Recent investigations show that nicotinic acid is bound to a macromolecule consisting of ca 60% polysaccharides and 40% peptides or glycopeptides. At present, it is believed that there are a number of bound forms of nicotinic acid in wheat, and that niacytin contains various peptides, hexoses, and pentoses.

A polysaccharide, consisting principally of arabinose, glucose, galactose, and xylose, was recently extracted from wheat bran (13). It contains 1.05% nicotinic acid in bound form. Partial hydrolysis yielded a fragment which was identified as β-3-O-(nicotinoyl)-D-glucose (**2**):

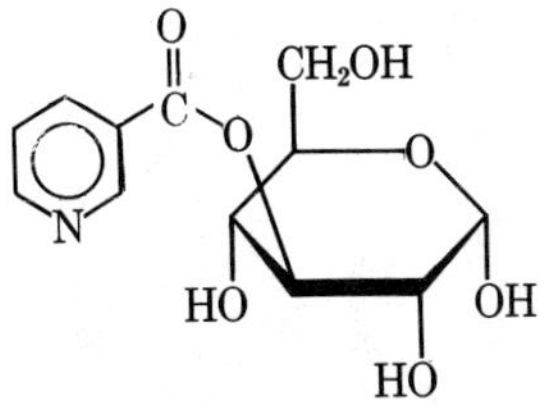

β-3-O-(nicotinoyl)-D-glucose

(**2**)

In the higher molecular fractions isolated from wheat bran, nicotinic acid is most probably attached primarily to the glucose moiety of the polysaccharide or glycoprotein. It now seems clear that the biological inactivity of bound nicotinic acid is caused by the hemicellulose-type bond which is resistant to digestive enzymes. In the case of corn, the bioavailability of nicotinic acid is considerably improved by pretreatment with alkali. Such a process is utilized in Mexico for the preparation of tortillas; the corn is pretreated with lime water (14).

Nicotinamide and nicotinic acid are commonly found in foodstuffs and feed (see Tables 3 and 4). They are especially concentrated in brewer's yeast, wheat germ, and liver; less so in grain and vegetables. Fruits and milk have very low concentrations. Tryptophan is considered to be a provitamin and assigned a vitamin B_3 equivalence of 1/60.

In cereals, vitamin B_3 is generally distributed in an inhomogeneous manner, occurring predominantly as bound nicotinic acid, as shown by the content of wheat, rye, and triticales (Table 5) (20).

The content of nicotinic acid in foodstuffs can be improved by technological means. Trigonelline (**3**) (*N*-methylnicotinate), an abundant component of green coffee beans, is demethylated to nicotinic acid during the roasting process. The ratio of trigonelline (**3**) to nicotinic acid plus nicotinamide is a criterion for the extent of roasting. A roasting number is defined as the ratio of trigonelline to nicotinic acid + nicotinamide (21).

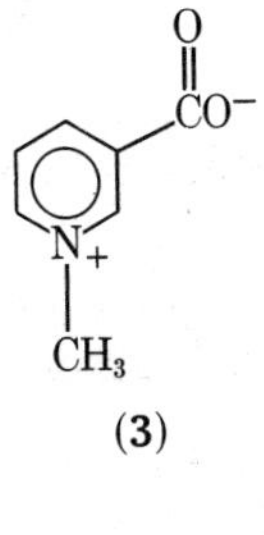

(**3**)

62   VITAMINS (NICOTINAMIDE AND NICOTINIC ACID)

Table 3. Nicotinic Acid and Nicotinamide (Vitamin B$_3$) Content of Foodstuffs, mg/kg [a]

Foodstuff	Vitamin B$_3$	Potential from tryptophan [b]
cow milk, fresh, whole, summer	0.8	7.8
cheese, cream cheese	0.8	7.4
egg, whole, raw	0.7	36.1
butter, salted	traces	traces
beef, lean, average, raw	52	43
beef liver, fresh	178[c]	
pork, lean, average, raw	62	38
pork liver, fresh	118[c]	
corn, whole	12[b]	
corn flour	traces	1
breakfast cereals, all bran	490	32
cornflakes		
fortified	210	9
unfortified	6	
wheat flour		
whole meal	56	25
white 72%, for bread		
fortified	20	23
unfortified	7	
brown 85%		
fortified	42	26
unfortified	12	
bread, white	14	16
macaroni, boiled	3	9
rice, polished, boiled	3	5
rice, bran	366–437[c]	
soybean flour, full fat	20	86
spaghetti, boiled	3	9
potatoes, raw	12	5
yeast, baker's, dry	257[c]	

[a] Ref. 15, unless otherwise noted.
[b] 60 mg tryptophan corresponds to 1 mg vitamin B$_3$.
[c] Ref. 16.

The average cup of coffee contains 1–2 mg nicotinic acid, and coffee thus supplies a considerable portion of vitamin B$_3$ requirement, as shown below.

Coffee beans	Vitamin B$_3$, mg/kg
green	20
medium-roasted	80–150
dark-roasted	up to 500

NICOTINAMIDE

Properties

Nicotinamide is a colorless, fine crystalline solid with a bitter taste. At room temperature, 1 g is soluble in ca 1 mL water, 1.5 mL 95% ethanol, or 10 mL glycerol. The compound is soluble in butanol, amyl alcohol, ethylene glycol, acetone, and chloroform, but only slightly soluble in ether or benzene. The physical properties are given in Table 6.

Table 4. Nicotinic Acid and Nicotinamide (Vitamin B$_3$) in Foodstuffs, mg/kg

Foodstuff	Ref. 17	Ref. 18	Ref. 19
corn, yellow	22	20	14–29
corn, gluten feed			50
rye	12		9–11
oats	15		9–12
milo	38		66–71
barley	61	50–60	70–99
wheat	55	50–60	34–65
wheat bran	200		86–334
tapioca flour	3		
potatoes	16		
soybean meal	22	30	18–19
rapeseed-extract meal	38		
cottonseed cake	40		29
palm kernel meal			11
groundnut cake	162	190	256
sunflower cake	202		
blood meal	58		38
meat and bone meal			17–65
fish meal		70	
skim milk, dried		11	
torula yeast	453	500	

Table 5. Nicotinic Acid and Nicotinamide (Vitamin B$_3$) Content of Wheat, Rye, and Triticales After Enzymatic Digestion, mg/kg

	Wheat	Winter triticales		Spring triticales		Rye
grain	48.3	17.9	16.3	16.0	15.6	15.3
flour	9.5	6.5	6.5	7.6	6.0	8.8
shorts	23.5	14.4	13.7	17.1	16.9	18.0
bran	171.4	58.2	58.0	57.3	58.4	27.4

Nicotinamide has been utilized in numerous reactions. Among the complex salts of nicotinamide, the bis(pyridine-3-carboxamide)–PdCl$_2$ complex can be used via a Mannich reaction for the synthesis of N-aminomethylated pyridine-3-carboxamides in good yields (22).

Important reactions of the ring nitrogen include quaternization (4) (23–28) and N-oxide (5) formation. The pyridinium ylides of nicotinamide are prepared from monochloromalonyl heterocyclic compounds (28):

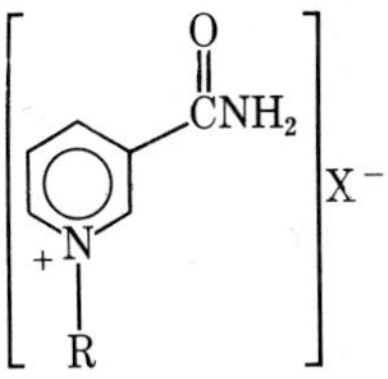

(4)
R = alkyl, aryl
X = halogen

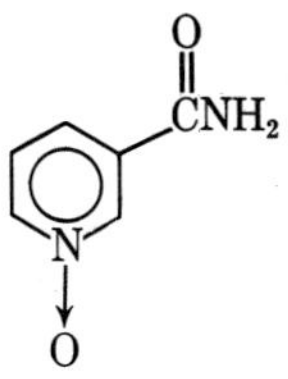

(5)
nicotinamide N-oxide

64   VITAMINS (NICOTINAMIDE AND NICOTINIC ACID)

Table 6. Physical Properties of Nicotinamide

Property	Value	Ref.
molecular weight	122.12	
melting point, °C		
stable modification	129–132	
unstable modifications		
V	116	29
IV	113	29
III	111	29
II	110	29
I	105	29
boiling range, °C (0.067 Pa)[a]	150–160	30
sublimation range, °C	80–100	29
density of crystals, g/cm^3	1.401	
true dissociation constants in H$_2$O at 20°C		
K_{b1}	2.24×10^{-11}	31
K_{b2}	3.16×10^{-14}	31
specific heat, kJ/(kg·K)[b]		
solid, 55°C	1.30	32
65°C	1.34	32
75°C	1.39	32
liquid, 135°C	2.18	32
heat of solution in H$_2$O, kJ/kg[b]	−148	32
heat of fusion, kJ/kg[b]	381	32
density of melt at 150°C, g/cm^3	1.19	32

[a] To convert Pa to mm Hg, multiply by 0.0075.
[b] To convert J to cal, divide by 4.184.

Special redox polymers, after reduction with Na$_2$S$_2$O$_4$, exhibit a reductive capacity of 0.94×10^{-3} equivalents per gram in relation to chloranil (33). Reactive quaternary nicotinamide monomers, containing p-vinylbenzyl, allyl, glycidyl, or vinyloxyethyl moieties, react with azobisisobutyronitrile or polyethyleneimine (34).

Reaction of 4-vinylpyridine-N-vinylpyrrolidone copolymer with N-(2-bromoethyl)-3-carbamoylpyridinium bromide and subsequent electrochemical or Na$_2$S$_2$O$_4$ reduction yields a polymer with reduced nicotinamide moieties which, in turn, may be used to reduce hemin (35). Nicotinamide-containing chloromethylstyrene polymers, after reduction with Na$_2$S$_2$O$_4$, can be utilized in the preparation of alloxantin from alloxan and hydrindantin from ninhydrin (36). N-Oxidation of nicotinamide with H$_2$O$_2$ in glacial acetic acid gives the N-oxide (5) in 82% yield (37). N-Chloronicotinamide is formed by treating an aqueous solution of nicotinamide with chlorine and sodium hydroxide (38).

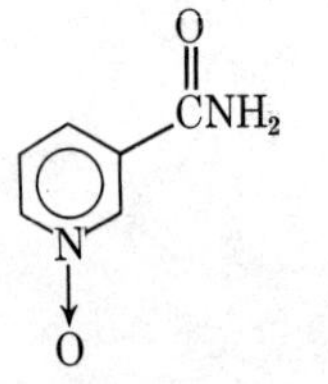

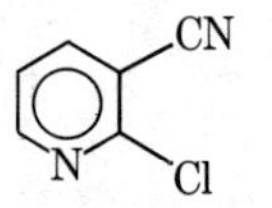

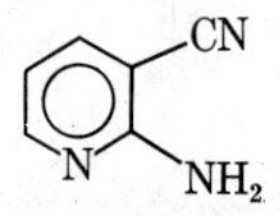

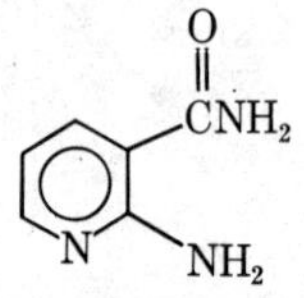

Treatment of (5) with PCl$_5$ in POCl$_3$ gives 2-chloronicotinonitrile (6) in 38% yield (39). Treatment with POCl$_3$ alone yields nicotinonitrile N-oxide, which may be converted with H$_2$S into nicotinic acid thioamide N-oxide (40). Reaction of (6) with NH$_3$ gives 2-aminonicotinonitrile (7), which may be converted to 2-aminonicotinamide (8) by treatment with H$_2$O$_2$ and KOH. Treatment of (6) with aqueous hydrochloric acid yields 2-hydroxynicotinic acid (41). Among the most interesting reactions of the aromatic ring are the Chichibabin reaction (formation of 2-aminonicotinamide (8) by treatment with NaNH$_2$ in N,N-dimethylaniline (42)) and electrochemical reduction to 3-aminomethylpiperidine (43). In the latter reaction, both the pyridine ring and the carboxamide group are hydrogenated. Nicotinamide may be converted to nicotinic acid esters (44), the nitrile (45), acylamidines (eg, (9)), and triazoles (eg, (10)) (46) by standard synthetic methods.

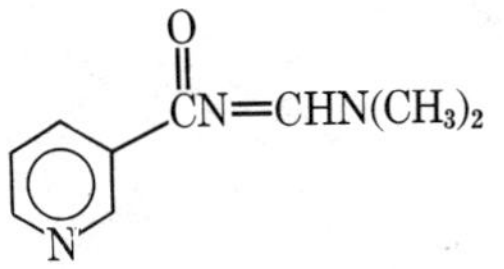

(9)

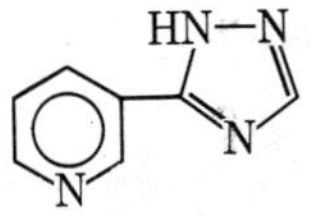

(10)

Preparation and Manufacture

The synthesis of nicotinamide using the nicotinonitrile route involves the following steps (Fig. 1): synthesis, from aliphatic starting materials, of an alkylpyridine, primarily 3-methylpyridine (11) or 2-methyl-5-ethylpyridine (MEP) (12); ammoxidation or ammonodehydrogenation of the alkylpyridine to nicotinonitrile (13) (in the case of 2-methyl-5-ethylpyridine, part of the carbon chain is oxidatively cleaved in this step); and hydrolysis of the nitrile to nicotinamide (1).

In an alternative method (see Fig. 1), nicotinic acid, made from 2-methyl-5-ethylpyridine or 3-methylpyridine, is treated with ammonia.

The synthesis of 3-methylpyridine (11) may be carried out via the gas-phase reaction of acetaldehyde, formaldehyde, and ammonia in the presence of a catalyst. However, the principal product is pyridine, which is produced at 40–50% yield. 3-Methylpyridine is obtained as a by-product at 20–30% yield (47–48).

Since large quantities of pyridine, which are required for herbicide production, are synthesized by this method, there is a sizable amount of 3-methylpyridine available. Most of the 3-methylpyridine required for the production of nicotinamide is obtained from this source.

To a lesser extent, acrolein serves as a starting material for 3-methylpyridine (49). In this reaction, 3-methylpyridine, the major product, is obtained at 40–50% yield. Pyridine, the minor product, is coproduced at 20–30% yield. By mixing propionaldehyde with the acrolein, it is possible to increase the yield of 3-methylpyridine to above 60% at the expense of the pyridine yield (50).

A completely different route starts with 2-methylglutaronitrile (14) (51), which is a by-product in the hydrocyanation of butadiene to adiponitrile. Cyclizing hydrogenation of (14) gives a 90% yield of 3-methylpiperidine (15). This is then dehydrogenated to give 3-methylpyridine (11) at 83.5% yield (52). This process has not yet been commercialized.

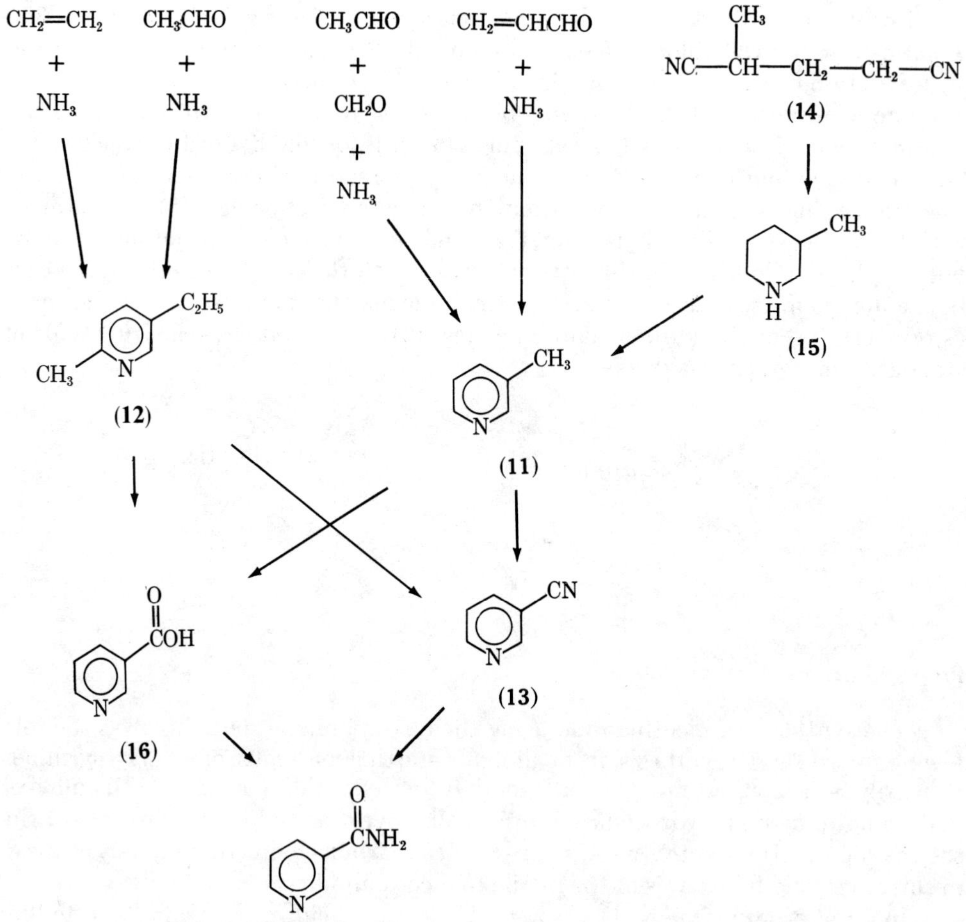

VITAMINS (NICOTINAMIDE AND NICOTINIC ACID)

Figure 1. Feedstocks and intermediates for the synthesis of nicotinamide.

2-Methyl-5-ethylpyridine (**12**) is made by a liquid-phase process in a 70% yield from acetaldehyde (53) or ethylene (80% yield) (54).

Ammoxidation of 3-methylpyridine is carried out on a commercial scale in the gas phase using heterogeneous catalysts (55–56); oxides of antimony, vanadium, and titanium (57), or antimony, vanadium, and uranium (58) are highly effective. With an antimony–vanadium–titanium catalyst, a reactor temperature of 360°C, and a molar feed ratio of 3-methylpyridine:NH$_3$:air:water vapor of 1:6:30:6, a 96% conversion and a 93% yield of nicotinonitrile were obtained (59). The ammoxidation of 2-methyl-5-ethylpyridine has only recently been sufficiently refined to be considered of commercial interest. A 75% yield of nicotinonitrile was obtained (60) with a catalyst consisting of oxides of vanadium, titanium, and zirconium; a molar feed ratio of 2-methyl-5-ethylpyridine:NH$_3$:air:water vapor of 1:46.7:48.3:71.7; and a reactor temperature of 385°C. A yield of 63% has been reported with a vanadium–titanium catalyst (61).

In ammonodehydrogenation, the alkylpyridine reacts with ammonia and an oxygen-carrier, such as V$_2$O$_5$, which must be reoxidized in a separate reactor (62). The ammonodehydrogenation of 3-methylpyridine and the aromatization of the 3-methylpiperidine precursor can be carried out in the same reactor (63–64).

Nicotinonitrile (**13**) is an important precursor of nicotinamide. Physical properties are given in Table 7.

The alkaline hydrolysis of nicotinonitrile (**13**) is one of the two commercial processes used to manufacture nicotinamide (**1**). This process offers the advantage that the rate K_1 of the principal reaction is fast compared to the rate K_2 of further saponification to nicotinic acid (**16**).

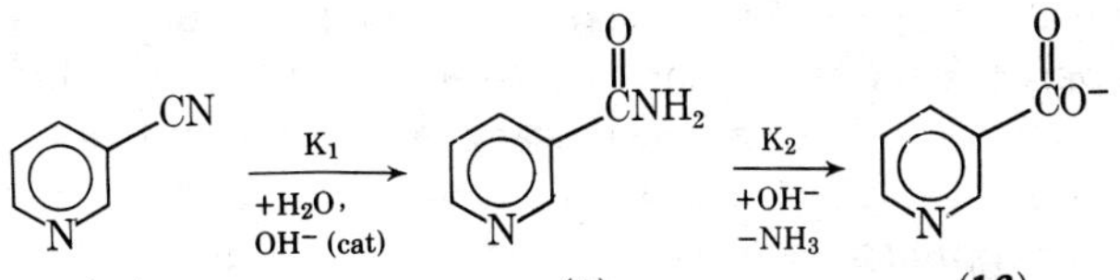

Thus, by maintaining a low concentration of hydroxide ion, it is possible to hydrolyze most of the nitrile while forming a quantity of nicotinic acid roughly corresponding to the quantity of hydroxide ion available (55,65). The hydrolysis is normally carried out with a catalytic amount of sodium hydroxide (66–68). Magnesium oxide (69), ammonia (66,70–71), manganese dioxide (72), complex chromium-containing metal oxides (73), and Raney copper (74) are also suitable. Higher temperatures shift the ratio of the reaction rates K_1 and K_2 strongly in favor of K_1 (67), permitting almost complete conversion of nicotinonitrile to nicotinamide. The process yields 100 kg of feed-grade nicotinamide from 85 kg of nicotinonitrile.

A maximum yield of 95.5–96.0% at a nitrile conversion of over 99% was obtained with NaOH (68) at a temperature of 160°C and the use of 2.95 mol % hydroxide, which however, gave a 3.5–4.0% nicotinic acid content in the product. Hydrolysis of nicotinonitrile with aqueous ammonia allows the production of either nicotinamide or nicotinic acid (70–71,79–80) depending on the temperature, pressure, and concentration.

As an alternative to the ammoxidation–hydrolysis route, a suitable alkylpyridine (3-picoline or 2-methyl-5-ethylpyridine) is first oxidized to nicotinic acid. The acid is then treated at 180–230°C, as a melt, with a large excess of gaseous ammonia at at-

Table 7. Physical Properties of Nicotinonitrile

Property	Value
molecular weight	104.10
melting point, °C	50
boiling point, °C	206.7
azeotrope with H$_2$O	
boiling point, °C	99.9
water content, g/L	963
solubility in water, g/L	
25.2°C	109
49.2°C	1000
density (g/cm^3)	0.870

mospheric pressure. The water formed in the reaction is continually removed (75). A metal salt catalyst accelerates the reaction (76).

Higher pressure has been investigated (77–78). Under proper reaction conditions, conversion is practically complete at 99% selectivity. On a commercial scale, the amidation is carried out at atmospheric pressure in a stirred-batch reactor. Excess ammonia can be recycled.

The reaction product is a nicotinamide melt containing small quantities of nicotinic acid and nicotinonitrile, the principal by-product of the amidation. The melt is dissolved directly in water or after solidification. Unreacted nicotinic acid is kept in solution by the addition of a weak base, eg, ammonia. The solution is decolorized with activated charcoal and nicotinamide is separated by crystallization. For pharmaceutical grade, recrystallization is necessary. A sidestream from the mother liquor is evaporated to dryness and recycled to the reactor. The yields in an industrial process for the preparation of pharmaceutical-grade nicotinamide are 90–95%, based on nicotinic acid.

The nicotinamide produced by any of these processes must be recrystallized before it can be utilized as a pharmaceutical product. Recrystallization can be performed in aqueous 2-methyl-1-propanol after adjusting the pH to 7.5–9.5 (81). Ion exchangers may be utilized (82). The crystallization can also be carried out in water in the presence of a strong shear-force field to improve filtration (83).

Economic Aspects

The producers of nicotinamide and nicotinic acid, production capacities, and raw materials are given in Table 8. The statistics published by the U.S. International Trade

Table 8. Production of Nicotinamide and Nicotinic Acid

Producer	Location	Product	Capacity[a], t/yr	Feedstock
Lonza	Visp, Switzerland	nicotinic acid	10,000	2-methyl-5-ethyl-pyridine (12)
		nicotinamide		3-picoline
Vitachem[b] Reilly Tar/ Degussa A.G.	Indianapolis, Indiana Antwerp, Belgium	nicotinamide	5,500	3-picoline
Nepera	Harriman, New York	nicotinic acid nicotinamide	2,000	3-picoline
Yuki Gosei	Japan	nicotinamide	2,000	3-picoline
Lisac	Spain	nicotinic acid	700	3-picoline
Craiova	Romania	nicotinic acid	120	2-methyl-5-ethyl-pyridine (12)
Givaudan	France	nicotinic acid	120	2-methyl-5-ethyl-pyridine (12)
Gema	Spain	nicotinic acid	500	3-picoline
Indian Drugs	India	nicotinic acid	1,000	2-methyl-5-ethyl-pyridine (12)
			21,940	

[a] Degussa estimate.
[b] Ref. 84.

Commission on the production, sales, and unit value of nicotinamide and nicotinic acid ceased with the year 1975 (see Table 9). At that time, Merck was the third largest U.S. producer, but discontinued production at the end of that year. The U.S. demand for nicotinamide and nicotinic acid is given in Table 10, imports and exports in Table 11, and the price history in Table 12.

Table 9. U.S. Production and Sales of Nicotinamide and Nicotinic Acid[a]

Year	Production, metric tons Nicotin-amide	Production, metric tons Nicotinic acid	Production, metric tons Total	Sales, metric tons Nicotin-amide	Sales, metric tons Nicotinic acid	Sales, metric tons Total	Unit value, $/kg Nicotin-amide	Unit value, $/kg Nicotinic acid
1966	474.4	1000.6	1475.0	431.8	810.6	1242.4	4.30	2.60
1969	459.9	892.7	1352.6	459.5	894.5	1354.0	4.08	2.23
1972			2483.4			2231.7		3.99
1975			2519.7			1662.9		4.45

[a] Ref. 85.

Nicotinamide and nicotinic acid are sold in three grades: pharmaceutical (USP, NF, Ph. Eur. (*European Pharmacopeia*)), food (USP/FCC (*Food Chemicals Codex*)), and feed–technical. Pharmaceutical and food grades are sold in drums (5–50 kg), and feed–technical grade in drums (50 kg) or bags (25 kg). All products should be stored in a dry place.

Analytical and Test Methods

Nicotinamide may be determined by either titration with perchloric acid (89) or uv (90–91). The latter method is also occasionally used to determine the nicotin-

Table 10. U.S. Market for Nicotinamide and Nicotinic Acid[a]

Year	Metric tons
1978	4800
1979	5030
1980	5300
1985[b]	6700

[a] Ref. 86.
[b] Estimated annual growth through 1985 is 4.9%.

Table 11. U.S. Imports and Exports of Nicotinamide and Nicotinic Acid[a]

Year	Imports, t Nicotinamide	Imports, t Nicotinic acid USP	Imports, t Nicotinic acid feed	Value, $1000	Exports, t Nicotinamide + nicotinic acid
1970		905.2[b]		2049	
1974	238.3	336.5	1253.1	ca 5040	
1978	341.2	932.1	2211.1		239.2
1980	297.1	655.5	1444.6		294.0

[a] Ref. 87.
[b] Sum of nicotinamide and nicotinic acid (USP + feed).

Table 12. Price History of Nicotinamide and Nicotinic Acid, 1970–1982, $/kg [a]

Year	Price Nicotinamide	Nicotinic acid
pharmaceutical grade		
1970	3.70	3.60
1974	5.30	4.60
1976	4.00	4.60
1978	6.60	5.25
1980	6.60	5.25
1982, March	8.00	7.50
feed–technical grade		
1980	6.10	6.10
1981	6.30	7.00
1982, March		7.00

[a] Ref. 88.

amide content of injection solutions of B-complex vitamins. Official standards of purity for the pharmaceutical grade are given in the USP (90) and Ph. Eur. (89) (Table 13).

The König reaction (92) is still often used to determine nicotinamide and nicotinic acid in vitamin mixtures or organic matrices. In this procedure, reaction with cyanogen bromide and an aromatic amine, eg, sulfanilic acid, causes opening of the pyridine ring. The postulated dialdehyde intermediate may react to form a yellow-to-orange-colored polymethine dye, which can be determined colorimetrically. The dye, however, is sensitive to pH and not very stable.

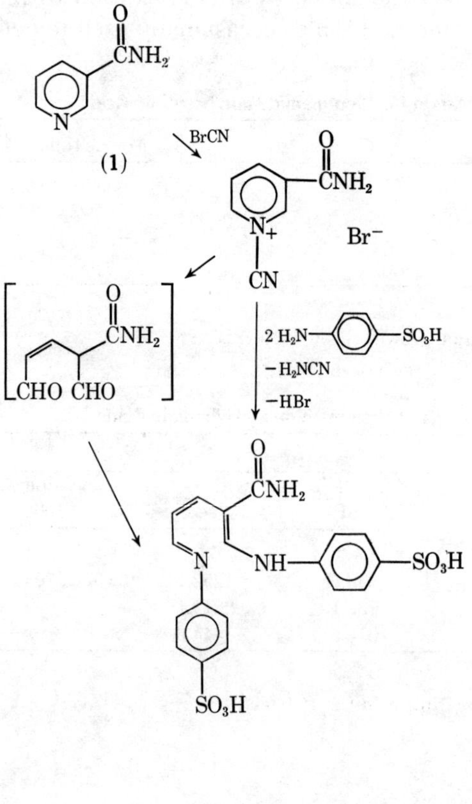

Numerous variations of this procedure have been suggested (93–94). In the AOAC procedure (95), the sample is extracted with KH_2PO_4 solution at pH 4.5. The extract is treated with BrCN and barbituric acid to form a colored compound which is then analyzed by uv. Samples of cereal products are first digested in an autoclave with $Ca(OH)_2$ solution; other samples are digested with $1\ N\ H_2SO_4$.

Chemical analysis generally measures only the sum of the nicotinamide and nicotinic acid contents, but with microbiological methods (96) the total biologically available vitamin B_3 content can be determined (nicotinamide, nicotinic acid, NAD, nicotinuric acid). *Lactobacillus plantarum* ATCC (American type culture collection) No. 8014 (96) or *Lactobacillus arabinosus* (97) serve as test bacteria.

Thin-layer chromatography can be used for simple mixtures such as multivitamin preparations (98–103). More powerful methods are under investigation.

Nicotinamide may be detected by gas chromatography, eg, as N-ethylnicotinamide (104), as nicotinonitrile (105), or as a silyl derivative (106).

The latest analytical developments have been in high pressure liquid chromatography, especially ion-paired, reversed-phase chromatography (107–112). With this method, nicotinamide and nicotinic acid may be simultaneously determined in multivitamin preparations and vitamin premixes. The method is also suitable for quantitatively determining nicotinamide metabolites in urine (113).

Biochemical Functions and Requirements

Nicotinamide is a building block of the hydrogen-transfer coenzymes NAD and NADP. These enzymes catalyze numerous hydrogenation and dehydrogenation reactions in intermediary metabolism. They are always bound to a substrate-specific apoenzyme in the form of freely dissociable complexes. These two coenzymes take part in most hydrogen-transfer reactions involved in metabolism. At present, at least 200 dehydrogenases that function with NAD or NADP are known. Some of these have

Table 13. Purity Standards for Pharmaceutical Nicotinamide

Specification or test	USP XX[a]	Ph. Eur. II[b]
identification	ir	melting point, ammonia odor when boiled with sodium hydroxide, yellow color with BrCN/aniline
	uv	
assay, %	98.5–101.0	>99.0%
melting range, °C	128–131	
loss on drying, %	<0.5	<0.5
residue on ignition, %	<0.1	
heavy metals, %	<0.003	<0.003
readily carbonizable substances	passes test	
appearance of solution		clear, color passes test
pH, 5% aqueous solution		6.0–8.0
chloride, ppm		<70
nucleus-nitrated impurities		passes test
sulfate ash, %		<0.1

[a] Ref. 90.
[b] Ref. 89.

VITAMINS (NICOTINAMIDE AND NICOTINIC ACID)

been isolated in crystalline form.

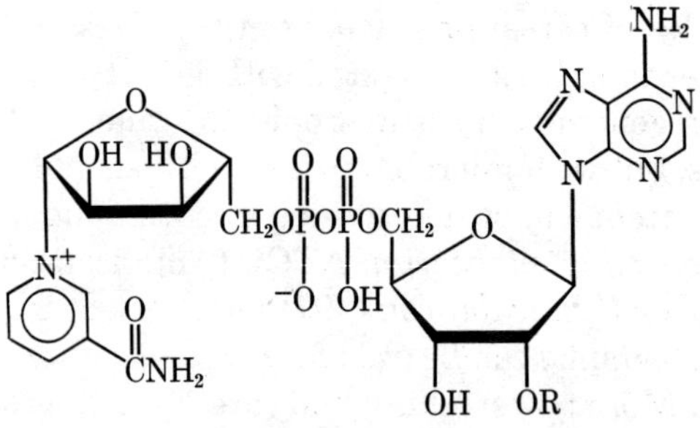

NAD: R = H
NADP: R = PO$_3$H$_2$

They participate in the synthesis and degradation of carbohydrates, fatty acids, and amino acids. In so doing, NAD accepts hydrogen from substrates XH$_2$ and transfers it as a hydride ion to other substrates.

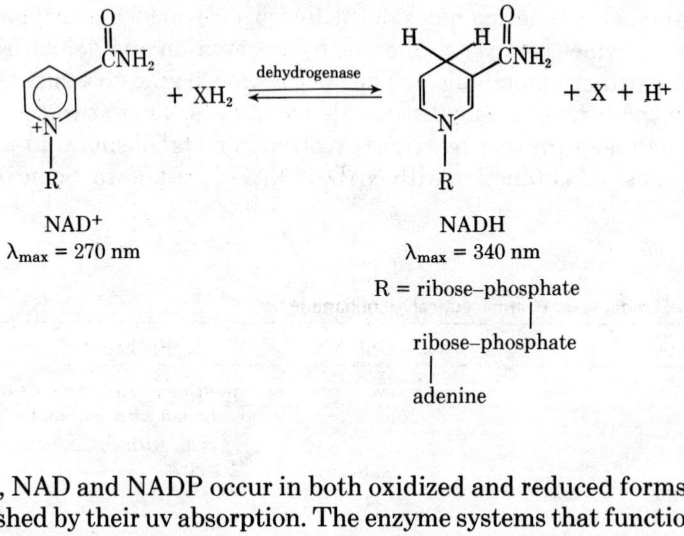

NAD$^+$
λ_{max} = 270 nm

+ XH$_2$ ⇌ (dehydrogenase) + X + H$^+$

NADH
λ_{max} = 340 nm

R = ribose–phosphate
 |
 ribose–phosphate
 |
 adenine

In cells, NAD and NADP occur in both oxidized and reduced forms, which may be distinguished by their uv absorption. The enzyme systems that function with NAD or NADP may thus be easily detected and quantitatively determined. In living cells, NAD occurs primarily in the oxidized form (NAD$^+$), whereas NADP is found mainly in the reduced form (NADPH).

In addition, NAD is essential for the final oxidative catabolism of substrates in the citrate cycle, ie, for energy production (see Citric acid). In this process, NAD, as the first member of a reaction series known as the respiratory chain, transfers hydrogen to the flavin enzymes. Other members of the respiratory chain include ubiquinone, cytochrome C, and cytochrome oxidase. At the end, oxygen is reduced to water. The energy thus released is utilized to synthesize three molecules of adenosine triphosphate

(ATP). A deficiency of vitamin B_3 gives rise to disturbances in the glycolysis function, the citric acid cycle, and the respiratory chain. Disturbances in synthetic processes, eg, fatty-acid synthesis, also occur. NAD can be endogenously synthesized from the amino acid L-tryptophan, provided the tryptophan requirement for protein synthesis is fulfilled. However, the tryptophan route is not very efficient, ie, in humans 60 mg tryptophan is equivalent to 1 mg vitamin B_3 (114–115). This ratio does not hold true in vitamin B_3-free diets or when tryptophan is limited (116) or during pregnancy. Almost all animal species except cats can utilize tryptophan in this manner. A vitamin B_3 equivalent of 1/45 mg has been determined for poultry (117) and 1/50 mg for rats. The small amounts of free nicotinamide or nicotinic acid present in animal and plant cells are utilized for NAD synthesis (118).

The biosynthesis of NAD in the mammalian liver starts from tryptophan

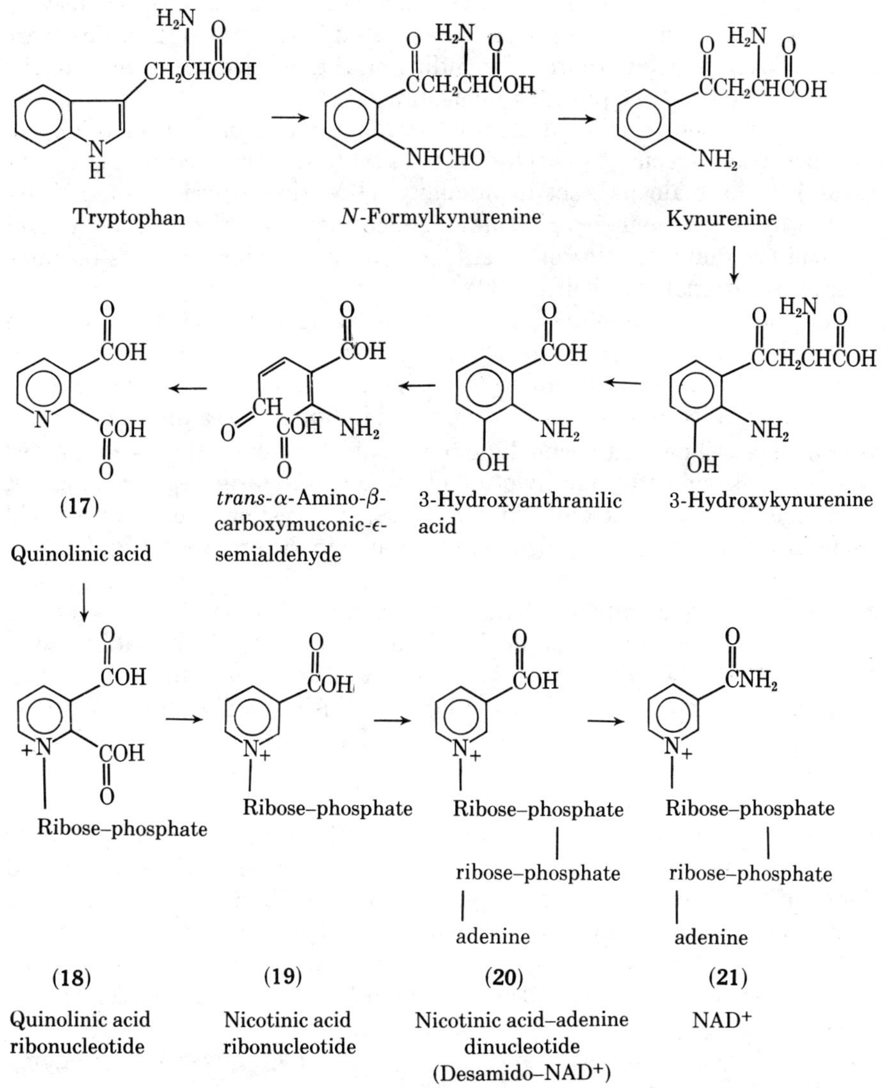

Figure 2. Biosynthesis of NAD from tryptophan.

(Fig. 2); neither free nicotinamide nor free nicotinic acid is involved. Nicotinamide, formed as a direct degradation product of NAD catabolism, is recycled to NAD by means of the intermediate nicotinamide mononucleotide in very efficient reactions. Radioactive labeling experiments in rat liver show that in mammals recycling via nicotinic acid (Preiss-Handler route) is insignificant (119–120). At abnormally high doses of nicotinamide, the metabolism seems to switch preferentially to this route, as demonstrated in mouse liver tissues (121). In microbial cells, recycling via nicotinic acid and, in some cases, an additional recycling route via nicotinic acid mononucleotide are utilized (122). Further investigations have confirmed that the rate constants, ie, the enzyme activities, determine which of the two recycling alternatives predominates.

Apparently, enzyme activities are different in different tissues (123–124). For rats, deamidase activity in the liver appears to be quite low (125). Human tissue cells do not have sufficient deamidase either. Nicotinamide can hydrolyze in the intestinal tract under the influence of the enzymes of the bacterial flora (126). NAD-cleaving enzymes of very high activity are reported to be present in the brain (127). There are three different NAD-recycling routes, depending on the relative importance of the species involved, known as the pyridine nucleotide cycle.

Figure 3 includes the principal urinary excretion products of NAD metabolism. The ratios vary from species to species (128–129). N^1-Methylnicotinamide (trigonellinamide) is the main product in humans. N^1-Methyl-2-pyridone-5-carboxamide, N^1-methyl-4-pyridone-5-carboxamide, nicotinamide, nicotinic acid, and nicotinuric acid are found to a lesser extent. N^1-Methylnicotinamide was recently identified as an additional metabolite (130).

The total amount of metabolites in humans is 3–30 mg per day (131). Should the pool of NAD and NADP be full, additional vitamin doses result in increased excretion of N^1-methyl-2-pyridone-5-carboxamide (132). Trigonelline itself does not appear to be a metabolite of nicotinamide or nicotinic acid in humans or animals (133). Investigations with rats show that trigonelline represents a fragment that is eliminated from bound nicotinic acid after methylation of these bound forms, eg, niacinogenes and niacytin (bound forms of nicotinamide in cereals) by digestion. Free nicotinic acid is not present in this process (134). Trigonelline is found in the urine of coffee drinkers (see Coffee).

Excretion of the two metabolites, trigonellinamide and N^1-methyl-2-pyridone-5-carboxamide, is the basis for the clinical evaluation of vitamin B_3 nutrition (134). The normal ratio of the two metabolites is 1.3–4.0. A ratio of less than 1.0 is an indication of potential vitamin B_3 deficiency (135). Humans normally excrete 7–10 mg per day of trigonellinamide. Interestingly, the excretion of nicotinamide metabolites increases considerably during pregnancy (136), when vitamin B_3 requirement is increased.

Requirements. The daily requirement of vitamin B_3 for humans is not precisely known. It depends on protein consumption, the leucine content of the protein, and the efficiency of the conversion of tryptophan to vitamin B_3. The best estimate of average requirement for an adult is 15–20 mg per day. The recommended dietary allowance (RDA) for adults, expressed as vitamin B_3 equivalents, is 6.6 mg per 4184 kJ (1000 food calories) and not less than 13 mg for an intake of less than 8368 kJ (2000 food calories) (137).

The recommended dietary allowances as published by the National Academy of Sciences (138–139) and the recommendation of the Deutsche Gesellschaft für Er-

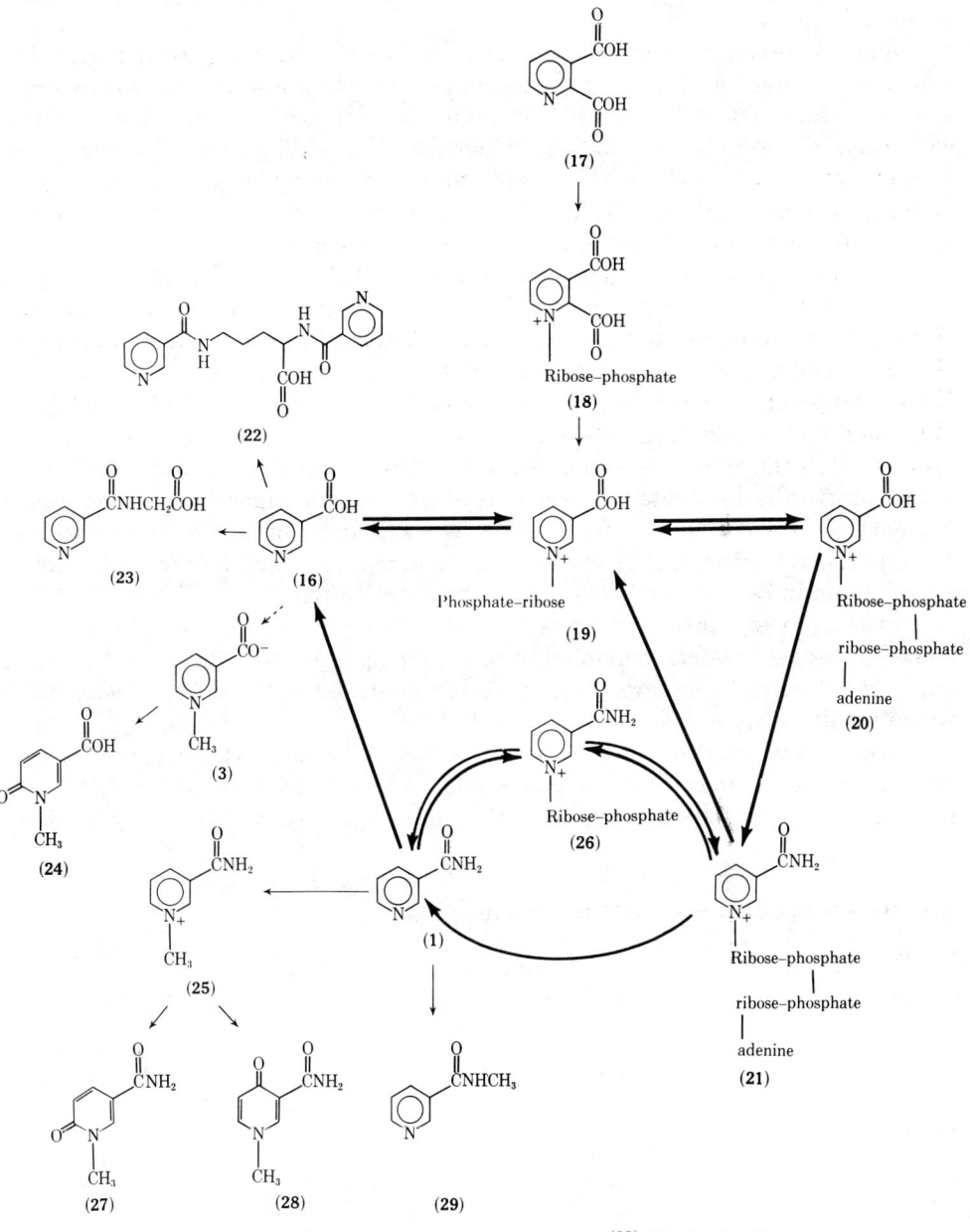

(1) Nicotinamide
(3) N^1-Methylnicotinamide (trigonelline)
(16) Nicotinic acid
(17) Quinolinic acid
(18) Quinolinic acid mononucleotide
(19) Nicotinic acid mononucleotide
(20) Nicotinic acid–adenine dinucleotide (desamido–NAD)
(21) Nicotinamide–adenine dinucleotide (NAD)
(22) Dinicotinoylornithine
(23) Nicotinuric acid
(24) N-Methyl-2-pyrindone-5-carboxylic acid
(25) N^1-Methylnicotinamide (trigonellinamide)
(26) Nicotinamide mononucleotide
(27) N^1-Methyl-2-pyridone-5-carboxamide
(28) N^1-Methyl-4-pyridone-3-carboxamide
(29) N^2-Methylnicotinamide

Figure 3. Metabolism of nicotinamide. Boldface pathways = pyridine nucleotide cycle.

nährung (DGE) are essentially the same (139). Recommended dietary allowances are given in Table 14.

A large excess of leucine in the diet causes a disturbance in tryptophan metabolism which results in reduced endogenous formation of NAD (140–141). Since corn is rich in leucine, this could be the cause for the frequent occurrence of pellagra in conjunction with a corn diet which is low in tryptophan. Addition of 10 g leucine per day to the human diet reduces NAD synthesis and induces the typical electroencephalogram alteration and psychotic manifestations of pellagra (142). In the presence of an amino acid imbalance, vitamin B_3 requirements increase sharply.

Vitamin B_3 is also essential to most animals (see Table 15). Its metabolic synthesis from tryptophan is of secondary importance, since most natural feedstuffs are relatively poor in this amino acid. Domestic animals are subject to constantly changing conditions and demands, and a vitamin B_3 deficiency directly influences productivity. No other B-group vitamin is required in such large quantities. Since the bioavailability of vitamin B_3 from the feedstuff is limited and often varies considerably (Table 4), synthetic B_3 is added to commercial feed or feed supplements, and a vitamin deficiency is thus practically eliminated. In commercial feed, the nicotinamide and amino acid content are correlated. For example, the presence of high concentrations of histidine, threonine, and leucine, together with a low tryptophan content, require an increased dose of vitamin B_3 (143–145) (see Pet and livestock feeds).

In some cases, considerably less vitamin B_3 is required than the recommended allowance because a safety margin is provided to compensate for conditions the animal might be subjected to, eg, stress, variations in climate, parasites, infection, etc, which would require a larger dose.

The recommendations for monogastric animals (swine, fowl, etc) cannot simply be extended to ruminants (cattle, sheep, goats), since ruminants have an additional fermentative preprocessing unit, the rumen. It is populated with a variety of micro-

Table 14. Vitamin B_3 Recommended Dietary Allowances, mg/d

Group	Age, yr	RDA[a]	DGE[b]
infants	0–0.5	5	4
	0.5–1.0	8	6
children	1–3	9	8
	4–6	12	14
	7–10	16	14
males	11–14	18	16
	15–18	20	16
	19–22	20	9–15
	23–50	18	9–15
	>50	16	9–15
females	11–14	16	16
	15–18	14	16
	19–22	14	9–15
	23–50	13	9–15
	>50	12	9–15
pregnant		+2	12
lactating		+4	16

[a] Ref. 137.
[b] Ref. 139.

Table 15. Requirements of Bioavailable Vitamin B$_3$ for Animals[a]

Animal	mg/kg feed
poultry	
broilers, growing	27–35
laying hens	10–26
turkeys, growing	20–50
ducks, growing	25–55
geese, growing	35–65
pheasants, growing	40
swine	
piglets, 1–10 kg	22
10–20 kg	18
pigs, 20–35 kg	14
35–60 kg	12
60–100 kg	10
others	
rats	20–30
dogs	12–25
cats	45–60
rabbits	180
trout	150–180

[a] Ref. 19.

organisms capable of synthesizing vitamins from simple raw materials. However, nicotinamide and nicotinic acid are often produced in insufficient quantities by the rumen flora of high performance ruminants such as modern dairy and beef cattle. Thus, vitamin B$_3$ not only becomes the limiting factor in hydrogen transfer in the rumen but also limits the growth of microorganisms. Consequently, the additional vitamin B$_3$ improves productivity in ruminants, such as higher milk production (146–147) and higher growth rates (148).

Toxicity

Toxicity after either single or repeated ingestion of nicotinamide (as well as nicotinic acid) is only observed at high concentrations or doses. The acute toxicity of nicotinamide and nicotinic acid is minimal. The LD$_{50}$ (oral) for rats was reported to be 3530 and 3540 mg nicotinamide per kg for male and female animals, respectively (149). In comparison, the corresponding LD$_{50}$ values for nicotinic acid, oral, for rats were 5210 and 4500 mg/kg (150). The LD$_{50}$ values of nicotinamide are somewhat lower than those of nicotinic acid, but these differences are toxicologically not relevant (151). The same applies to the acute toxicities of both substances after parenteral administration (see Table 16). The differences are comparable to the variations in LD$_{50}$ values reported for both substances by different authors.

Oral administration of nicotinamide to rats in concentrations of 1.0 and 2.0% in feed (1000 and 2000 mg/kg per day, 30 and 40 d) inhibited growth (152). Nicotinic acid under the same experimental conditions inhibited growth only in the 2% concentration. On further investigation, nicotinamide was administered for 8–12 wk to rats in concentrations of 0.1, 0.2, and 0.4% (corresponding to 100, 200, 400 mg/kg per day) in feed (153). Although 0.1% nicotinamide was tolerated with no recognizable side effects,

Table 16. Acute Toxicity, LD$_{50}$[a], of Nicotinamide and Nicotinic Acid, mg/kg Body Weight

Species	Application	Nicotinamide (1)	Nicotinic acid (16)
rat	oral	male, 3530	male, 5210
		female, 3540	female, 4500
	subcutaneous	2000–3000	4000–5000
mouse	oral	2500–3500	5000–7000
	intraperitoneal	1940[b]	3090[b]

[a] Unless otherwise stated.
[b] LD$_{25}$.

an influence on growth and development was observed at higher concentrations. The concentration at which there is no effect can be assumed to be 0.1% of feed. In contrast, weanling rabbits in another study were given 1.0 and 2.0% of nicotinamide in feed (corresponding to 300 and 600 mg/kg) for 20 d. No detrimental effects on growth were observed (154). Young guinea pigs, which were given 0.5, 1.0, and 2.0% of nicotinamide in feed (corresponding to 200, 400, and 800 mg/kg per day) also exhibited no growth inhibition (154). There are evidently species-specific and possibly age-specific differences in the reactions of animals to repeated administration of nicotinamide. No chronic toxicity tests or reproduction studies of nicotinamide are available.

In humans, a distinction should be made between single or repeated exposure, while handling nicotinamide at the workplace, and ingestion of nicotinamide, either daily in food (in which it may be either a natural component or added in an enrichment process) or therapeutically in medications. It has been reported that handling nicotinamide in the workplace does not cause the flush effect which is known to occur in handling nicotinic acid (155).

In the United States, nicotinamide is certified as a nutrient and dietary supplement. It has GRAS status in accordance with the CFR (151,156).

Uses

It is common practice in many parts of the world to enrich the human diet with nicotinamide or nicotinic acid (28) (see also Food additives). Deficiency diseases, which previously occurred in endemic form (pellagra), have thus been largely eliminated. A program of enriching bread and flour with nicotinamide, vitamin B$_1$, vitamin B$_2$, and iron was introduced in 1941 in the United States and was later expanded to include corn, corn products, rice, and other products. Standards of Identity have been specified for flour and farina, corn meals and grits, macaroni and noodle products, and rice (157) (Table 17).

Supplementation of foodstuffs with vitamins is the subject of numerous review articles (159–161), and its technology and problems have been described (162–164). Special attention has been paid to rice (165–166). To date, there are no official standards for ready-to-eat breakfast cereals, but recommendations have been published (167). Normal rates of supplementation are 10–30 mg/kg for wheat flour and 100–300 mg/kg for cornflakes.

Nicotinamide and nicotinic acid have a stabilizing effect on the pigmentation of cured meats (168–170). Moreover, they are reported to contribute to the formation of light-stable, red pigments (hemochrome) during the curing of meats and sausages (171) (see also Food processing).

Table 17. U.S. Standards of Identity for the Enrichment of Cereal Products with Vitamin B$_3$

Cereal	Required content, mg/kg[a] Min	Max
flour		52.9
farina	35.3	44.1
macaroni products	59.5	75.0
noodle products	59.5	75.0
corn meals	35.3	52.9
corn grits	35.3	52.9
rice	35.3	70.5

[a] Ref. 158.

Nicotinamide is administered to patients for prevention and treatment of hypovitaminosis. Severe deficiencies are treated with 50–500 mg/d. In addition to acting as vitamins, both nicotinamide and nicotinic acid exhibit special pharmacodynamic properties. In the simplest cases, hypovitaminoses can result from insufficient or unbalanced nutrition. Usually, the cause is an increase in normal nicotinamide requirement, eg, during pregnancy or in the case of gastrointestinal disturbances or alcoholism. At present, nicotinamide is widely used therapeutically as a vitamin (eg, Nicobion, Nicotamide, Aminicotin, Dipegyl, Nicotylamide, Benicot, etc), as a component of vitamin-B complexes, and in multivitamin preparations. Nicotinic acid, on the other hand, is rarely used in multivitamins. Such preparations are used to treat the liver, intestinal flora, dermatoses, and vitamin deficiency, and to increase the body's resistance to infectious diseases. Nicotinamide is especially useful in liver therapy in combination with other B vitamins and pharmaceuticals. Because of its importance in the normal functioning of many metabolic processes, it is also widely used in geriatric preparations. Furthermore, it is employed against anemia, as a synergist for the antituberculous agent Isoniazid, and to enhance the safety of chemotherapeutic agents.

In animal nutrition, particularly in modern, intensive livestock raising, nicotinamide, nicotinic acid, and other vitamins are added to feed as concentrates and premixes. A typical vitamin premix for chicks is given in Table 18. Supplementation rates for the enrichment of feed generally exceed the minimum requirements, and vitamin deficiency is prevented. In vitamin B$_3$ deficiency, symptoms include dermatitis, inflammation and ulceration of the mucous membrane (black-tongue disease), reflex disturbances within the nervous system (ataxia, epileptic attacks), and perosis. Among the less-specific, secondary symptoms are loss of appetite, inhibited growth, poor feed conversion, and lowered resistance to infections of the gastrointestinal tract.

Nicotinamide and nicotinic acid are equally effective for maintaining an optimal vitamin supply. However, nicotinamide is superior to nicotinic acid for treating an existing deficiency (172).

Nicotinamide promotes the growth of plants. It has been reported that spraying wheat seed with 0.1 and 0.01% nicotinamide solutions results in a higher content of auxins in the wheat sprouts during germination (173–174). Increased auxin content was also reported in corn plants (175), and dramatic effects were observed in the growth of rice. Treating seeds with an aqueous solution of 0.1 ppm of nicotinamide gave a 45% increase in growth (176) (see Plant-growth substances).

80 VITAMINS (NICOTINAMIDE AND NICOTINIC ACID)

Table 18. Vitamin Premix for Chicks

Vitamin	mg[a] per kg feed[b]
A	3000[c]
D_3	1500[d]
E	30[d]
K_3	3
B_1	3
B_2	8
B_3	50
B_6	7
B_{12}	0.030
C	150
biotin	0.15
choline	1500
folic acid	1.5
pantothenic acid	20

[a] Unless otherwise stated.
[b] Based on dry substances.
[c] RE, retinol equivalent is the new standard for vitamin A; 1 RE = 5 IU.
[d] International units (IU).

Nicotinamide is used as a brightener in electroplating baths at concentrations of 1–10 g/L. This has proved especially effective in the galvanization of zinc (177), cadmium (178), and palladium (179) (see Electroplating; Metallic coatings).

NICOTINIC ACID

Nicotinic acid (16) [59-67-6] (niacin, acidum nicotinum, vitamin B_3, vitamin PP, 3-pyridinecarboxylic acid) can be used instead of nicotinamide in some applications since it enters into the NAD biosynthesis quite effectively and is amidated (Preiss-Handler route).

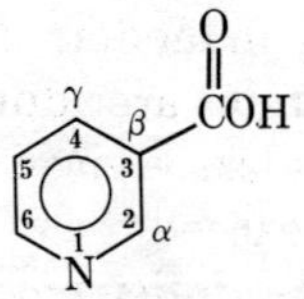

(16)
nicotinic acid
3-pyridinecarboxylic acid

Nicotinic acid is found principally in plants. In cereals, such as corn and wheat, it is bound to polysaccharides and peptides and is inactive as a vitamin.

Properties

Nicotinic acid is a fine crystalline compound with needle-shaped crystals. It has a sour taste and a weak odor. Its physical properties are given in Table 19.

Compared to other pyridinecarboxylic acids, nicotinic acid is stable. It is amphoteric and forms numerous salts with both acids and bases. Chemical reactions may involve the nitrogen atom, and carboxyl group, or the heterocyclic ring. The nitrogen

Table 19. Physical Properties of Nicotinic Acid

Property	Value	Ref.
molecular weight	123.11	
melting point, °C	236–237	
sublimation range, °C	≥150	180
density of crystals, g/cm³	1.473	
true dissociation constants in H₂O at 25°C		
K_a	1.50×10^{-5}	181
K_b	1.04×10^{-12}	181
isoelectric point in H₂O at 25°C, pH	3.42	181
pH of saturated aqueous solution	2.7	
solubility, g/L		
water		182
0°C	8.6	
38°C	24.7	
100°C	97.6	
ethanol, 96%		182
0°C	5.7	
78°C	76.0	
methanol		183
0°C	63.0	
62°C	345.0	
of sodium nicotinate hemihydrate		
water		182
0°C	95.0	
100°C	497.8	

atom is subject to quaternization, *N*-oxidation, and *N*-imine formation. Its reaction with methyl bromide is an example of quaternization which, when carried out with methyl nicotinate, yields 1-methyl-3-methoxycarbonylpyridinium bromide (**30**). This product may be converted to arecoline hydrobromide (**31**) (184) by treatment with KBH₄.

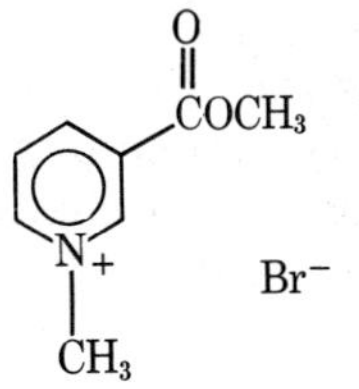

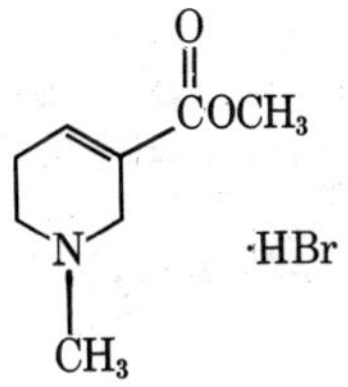

N-oxidation is effected in 80% yield by refluxing nicotinic acid with H₂O₂ in glacial acetic acid for two hours (185). The *N*-oxide (**32**) may be converted to 2-chloronicotinic acid (**33**) with PCl₅ in POCl₃ (186), or to 2-hydroxynicotinic acid (**34**) by heating in Nujol (187). The latter is also formed by the reaction of (**32**) with acetic anhydride. However, the main product of this reaction is 2-acetylnicotinic acid *N*-oxide (**35**). Small quantities of 6-hydroxynicotinic acid are also formed (187).

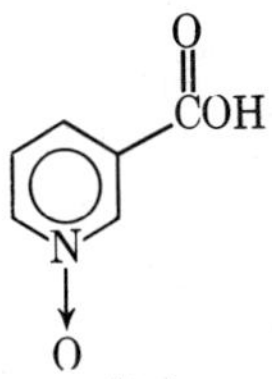

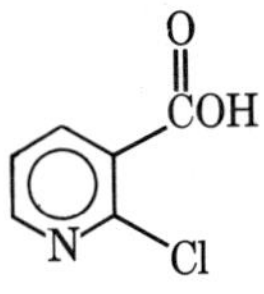

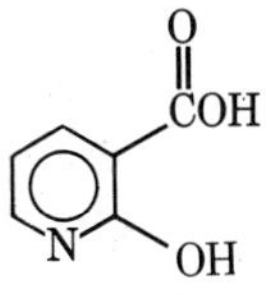

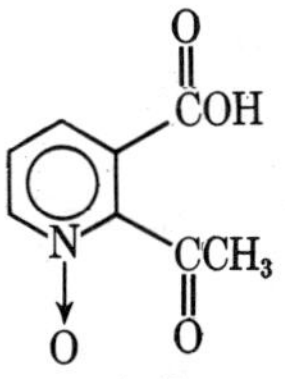

82   VITAMINS (NICOTINAMIDE AND NICOTINIC ACID)

Hydrogenation of nicotinic acid gives nipecotic acid at 88% yield (189). Depending upon reaction conditions, chlorination with thionyl chloride gives nicotinoyl chloride hydrochloride or acid chlorides of 5-chloro- and 5,6-dichloronicotinic acid (190). Iodination of a suspension of nicotinic acid sulfate in ethanol with iodine and H_2O_2 yields 3,5-diiodopyridine (191). The carboxyl group participates in reduction and decarboxylation and the formation of anhydrides, esters, thiolesters, oxazoles, imidazoles, thiazoles, and triazoles. Mixed anhydrides are intermediates in the synthesis of esters (192) and amides (193). Other synthetic routes to esters (194) and amides (195) employ the acid chloride, which can be made from nicotinic acid and thionyl chloride. Esters can also be prepared directly using catalysts such as arylsulfonic acids (196), arylsulfonyl chlorides (197), tertiary amines (198), 1-methyl-2-chloropyridinium iodide (199), and 2-chloropyridinium iodide (200). Cyclization of nicotinic acid with o-aminophenols in the presence of a polyphosphoric acid catalyst gives the corresponding oxazolenes and imidazoles (201). Imidazolines and oxazoles are formed with nicotinic acid and ethylenediamine (202) or aminophenols (201), imidazoles with o-phenylenediamine (203), and thiazoles with o-α-aminothiophenols (204). The triazolium compound (36) can be prepared at 98% yield from S-methyl-1,4-diphenylisothiosemicarbazide (37) and nicotinic acid with $POCl_3$ in pyridine and treating the reaction product with methanol and aqueous KI. Reduction with $NaBH_4$ (205) gives nicotinaldehyde (38) at 55% yield.

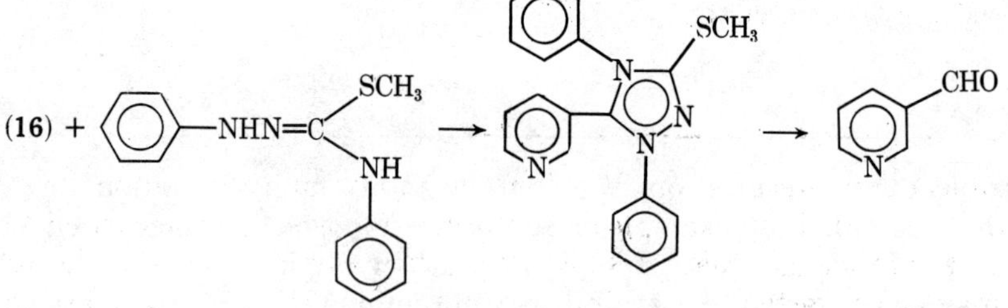

The aldehyde (38) can also be prepared directly from nicotinic acid and bis(4-methylpiperazinyl) aluminum hydride (206). The corresponding alcohol, 3-pyridinemethanol, is prepared in 81% yield by reducing methyl nicotinate with $NaBH_4$ in methanol (207). The decarboxylation of nicotinic acid at 138–250°C is described in ref. 208.

Preparation and Manufacture

Nicotinic acid is obtained by oxidizing 2-methyl-5-ethylpyridine (MEP) or 3-methylpyridine with nitric acid or air. The reaction may be carried out in either the liquid or gas phase; catalysts are sometimes employed.

Most nicotinic acid is produced by the liquid-phase oxidation of MEP (12); isocinchomeronic acid (39) is formed as an intermediate.

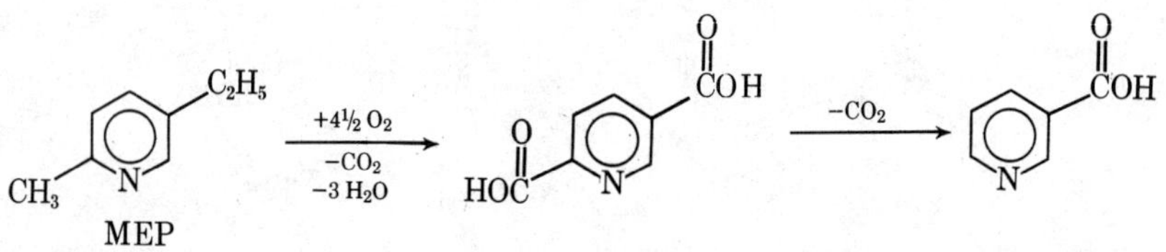

The following process is described in a Lonza A.G. patent (209). A mixture of 6.4% MEP and 33% aqueous nitric acid is heated in a titanium tube reactor to 330°C and 29 MPa (290 atm) for 5.5 s. The reaction mixture is concentrated, cooled, and centrifuged, and crystalline nicotinic acid nitrate is recovered. This product is dissolved in hot water, treated with activated charcoal, and acidified to pH 3.3. Nicotinic acid crystallizes upon cooling. Selectivity to nicotinic acid is ca 95%. In a variation of this process, MEP is heated at 220–240°C and 2.4–4.5 MPa (24–45 atm) for 5–60 min. Depending upon the conditions, the yield of nicotinic acid, based on MEP, is 84.5% (210) or 91.8% (211).

The oxidation of 3-methylpyridine (11) is stoichiometrically more efficient, since the entire carbon skeleton is retained.

However, oxidation of 3-methylpyridine proceeds much slower than that of MEP. Using nitric acid under analogous conditions, the oxidation of MEP is 29 times faster than that of 3-methylpyridine (212). Nevertheless, the growing number of publications is evidence of the interest in the methylpyridine route. In the liquid phase, the use of a catalyst containing cobalt acetate, manganese acetate, ammonium bromide, and zirconyl acetate gives a 97% conversion at a 85% selectivity to nicotinic acid (213). Oxidation with sodium dichromate in aqueous solution (pH 5.6), gives a 98.5% yield of sodium nicotinate. Nicotinic acid is precipitated by acidification and isolated in 95% yield (214). Air oxidation of aqueous 3-methylpyridine and sodium hydroxide gives a 76.6% yield of nicotinic acid (215). Oxidation with nitrogen dioxide in a solution of 1,2,4-trichlorobenzene, with a selenium dioxide catalyst, gives a 77% conversion at 85% selectivity (216). Hydrogen peroxide is a suitable oxidizing agent for the 3-methylpyridine–PdCl$_2$ complex (217). Finally, 3-methylpyridine can be oxidized biologically with mycobacteria (218).

The gas-phase oxidation of MEP or 3-methylpyridine with air or oxygen has not achieved commercial significance to date. According to one procedure, MEP, water vapor, and air are passed over a catalyst system consisting of oxides of vanadium, titanium, and potassium to yield 26% nicotinic acid and 18% reusable intermediates, such as 5-ethyl-2-pyridinealdehyde, 3-ethylpyridine, and 3-vinylpyridine. The conversion of MEP is 98%. With oxides of vanadium, iron, and zinc, a 60.6% yield is obtained (219). A catalyst containing oxides of vanadium, boron, and titanium gives a 99% conversion at 63% selectivity (220), whereas oxides of vanadium and potassium on silica give a 73% MEP conversion at 85% selectivity (221).

A selectivity of 51% at a 93% conversion of 3-methylpyridine is obtained with a vanadium–titanium–sodium catalyst (222). A 3-picoline–air mixture in the presence of a vanadium–titanium catalyst results in a conversion of 62.6% at a selectivity of 81.2% (223). With pure oxygen, water vapor, and a catalyst of silicon carbide and oxides of vanadium, chromium, antimony, and zinc, quantitative conversion of 3-picoline at 75.6% selectivity is achieved (224).

The synthesis of nicotinic acid from nicotinitrile (obtained by ammoxidation of 3-methylpyridine) via hydrolysis with aqueous ammonia is a variation of the hydrolysis of the nitrile to nicotinamide discussed previously.

Finally, nicotinic acid may be synthesized from pyridine. Bubbling CO_2 into a solution of pyridine and $LiAlH_4$ in dioxane gives an 80–90% yield of nicotinic acid (225).

Analytical and Test Methods

Assay is most readily determined by titration with sodium hydroxide (226) or by uv (227). Quality standards for pharmaceutical-grade nicotinic acid are published in the USP (227) and the Ph. Eur. (228) (Table 20). The König reaction, described above for nicotinamide, is also used for the determination of nicotinic acid (229). For microbiological determination, the same method is used as for nicotinamide (230). Nicotinic acid can be selectively detected by the bacterium *Leuconostoc mesenteroides*, ATCC No. 9135 (231).

Thin-layer chromatography is a semiquantitative method for determining nicotinic acid. For quantitative determination of derivatives, high pressure liquid chromatography is used. This permits the simple and accurate determination of the nicotinic acid content of vitamin premixes. As in the case of nicotinamide, reversed phases are preferred.

Biochemical Functions

Nicotinic acid is formed in animal metabolism by biosynthesis from tryptophan or by deamidation of nicotinamide. The converse, the direct amidation of nicotinic acid to nicotinamide, does not appear to be possible. Since deamidase activity in animal cells appears to be low, it has been postulated that the formation of nicotinic acid from nicotinamide occurs only when the pyridine–nucleotide cycle is overloaded, eg, by

Table 20. Purity Standards for Pharmaceutical-Grade Nicotinic Acid

Specification or test	USP XX[a]	Ph. Eur. III[b]
identification	ir, uv, red color with 2,4-dinitro-chlorobenzene/KOH, blue precipitate with NaOH/CuSO$_4$	yellow color with Br$_2$/NH$_4$SCN/ aniline, ammonia odor when boiled with NaOH, red-violet color with citric acid/acetic anhydride
assay, %	99.0–101.0	>99.5
loss on drying, %	<1.0	<1.0
residue on ignition, %	<0.1	
chloride, %	<0.02	<0.02
sulfate, %	<0.02	
heavy metals, %[a]	<0.002	<0.002
melting point, °C		234–237
nitro compounds		passes test
sulfate ash, %		<0.1

[a] Ref. 227.
[b] Ref. 228.

ingestion of massive, abnormally high doses of nicotinic acid or nicotinamide. In such cases, the detoxification mechanisms of the various species appear to operate at different efficiencies. This is suggested by the qualitative and quantitative differences found in the composition of metabolites in the urine (129). When consuming physiological dosages of niacin, humans excrete as metabolites 30–40% trigonellinamide (25) and 35–45% N^1-methyl-2-pyridinone-5-carboxamide (27) (232–233).

In contrast to animal cells, bacteria in the intestinal flora can easily amidate nicotinic acid or deamidate nicotinamide.

The metabolism of higher plants and some bacteria probably starts with the C_3 and C_4 building blocks, glyceraldehyde-3-phosphate and aspartic acid, in the synthesis of nicotinic acid and joins the normal NAD route at quinolinic acid (234) from which it proceeds as shown in Figure 2.

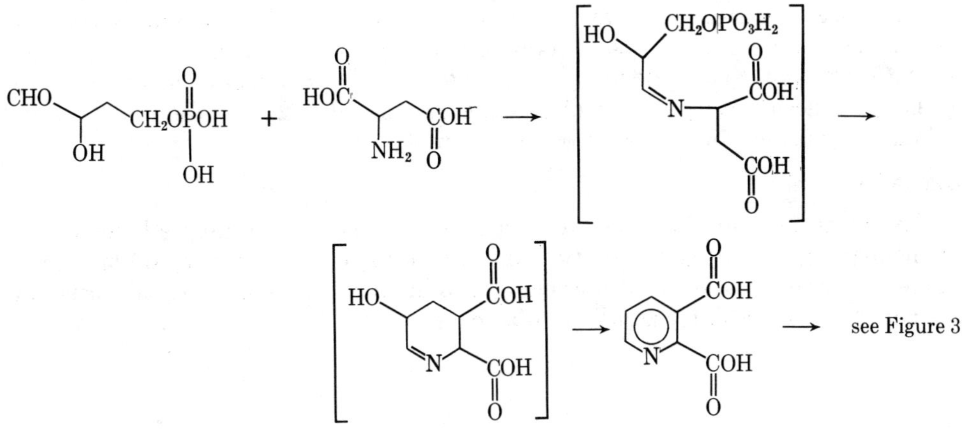

The synthetic origin of the glycosidically bound nicotinic acid in cereals is not clear.

Toxicity

The acute toxicity of nicotinic acid is low, as can be seen from the LD_{50} values which were determined for several rodents using different application methods (see Table 16) (151). In isochronic experiments (150), nicotinic acid (16) proved slightly less toxic than nicotinamide after single oral administrations. This difference, however, is not toxicologically relevant. Nicotinic acid and nicotinamide were administered to rats as a feed component in concentrations of 1.0 and 2.0% (1000 and 2000 mg/kg per day, 30 and 40 d) in isochronically conducted studies. The presence of 2% nicotinic acid in feed inhibited growth, whereas 1% nicotinic acid had no influence on the growth of the animals. Both concentrations resulted in formation of fatty livers (152).

In another investigation, the influence of nicotinic acid (150 mg/kg per day, 3 mo) on serum cholesterol level and atheroma was studied in the blood vessels of rabbits which were fed a cholesterol-containing diet (235). The control group, which only received the nicotinic acid (150 mg/kg per day, 3 mo), exhibited no reactions indicating incompatibility. These toxicity studies are an indication of the good compatibility of nicotinic acid in long-term oral administration. No chronic toxicity test or reproduction studies of nicotinic acid are available.

In the workplace, single or repeated contact with nicotinic acid can lead to a flush effect on the skin, including the face (155,236–237). The reddening is the result of increased peripheral blood circulation from dilation of capillaries at the skin surface. The flush effect is accompanied by sensations of itching, tingling, and warmness of the skin. It vanishes after the exposure has ended and is thus reversible. Nicotinic acid is certified in the United States as a nutrient and dietary supplement with GRAS status according to the CFR (151,238).

Uses

Nicotinic acid, like nicotinamide, is used for vitamin enrichment of cereal products (eg, wheat flour, bread, macaroni, corn products). Nicotinamide is preferred for supplementing special dietetic foodstuffs because of its better compatibility. Nicotinic acid is utilized as a meat additive. It is often used instead of nicotinamide as a vitamin supplement for animal nutrition. Its pharmacological side effects are insignificant for the levels at which it is added to feed. In the area of industrial applications, nicotinic acid can replace nicotinamide as a brightener for zinc (239–240), cadmium (241), and palladium (179) in electroplating baths.

Derivatives

Nicotinic acid and its ester and amide derivatives have widespread medical use as antihyperlipidemic agents (antilipemic agents) and peripheral vasodilators (see Table 21). In some cases, nicotinic acid is used in salt formation with basic drugs, eg, the peripheral vasodilator xanthinol niacinate.

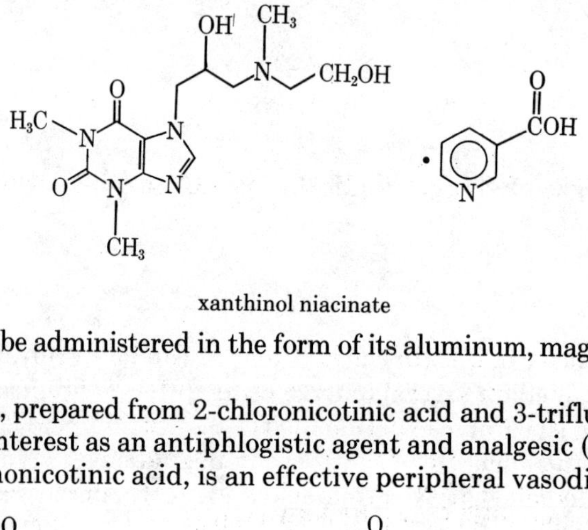

xanthinol niacinate

Nicotinic acid can be administered in the form of its aluminum, magnesium, or sodium salt.

Niflumic acid, prepared from 2-chloronicotinic acid and 3-trifluoromethylaniline, is of commercial interest as an antiphlogistic agent and analgesic (253). Nicergoline, an ester of 5-bromonicotinic acid, is an effective peripheral vasodilator (254).

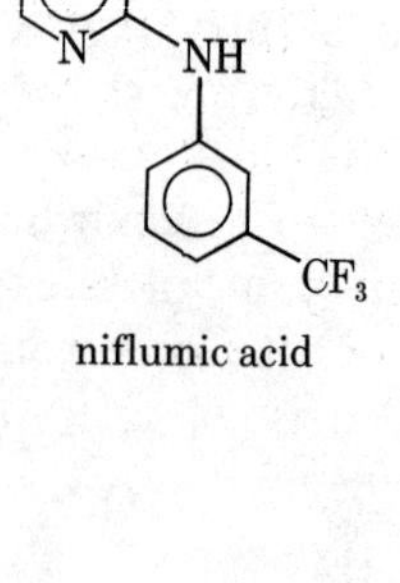

niflumic acid

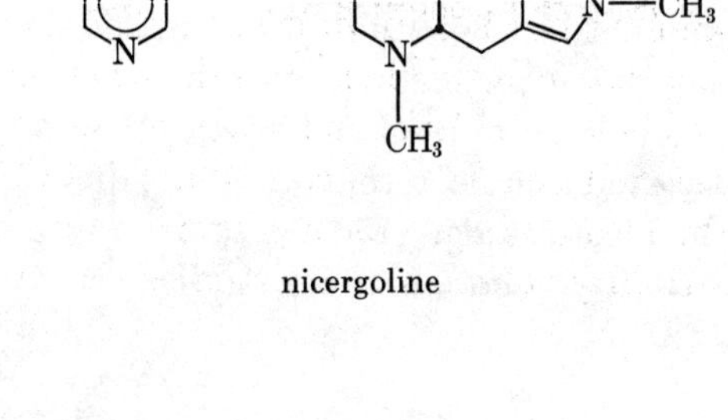

nicergoline

Table 21. Derivatives of Nicotinic Acid Used as Drugs

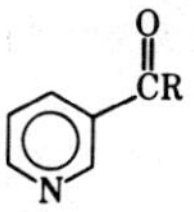

Name	CAS Registry No.	R	Medical use	Ref.
benzyl nicotinate	[94-44-0]	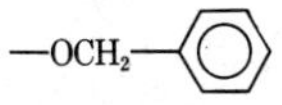 —OCH₂—C₆H₅	rubefacient	
etofibrate	[31637-97-5]	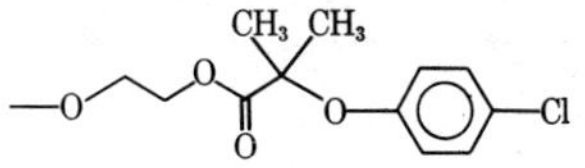 —OCH₂CH₂OC(CH₃)₂C(O)O—C₆H₄—Cl	antilipemic (lipid-reducing)	242
etofylline nicotinate	[13425-39-3]	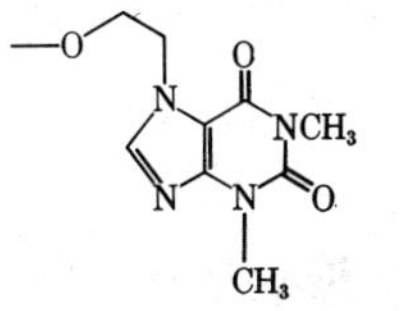 (theophylline-7-ethyl)	vasodilator	243
hepronicate	[7237-81-2]	 (—CH₂O)₃C–C₆H₁₃CH₃	vasodilator	244
inositol niacinate	[6556-11-2]	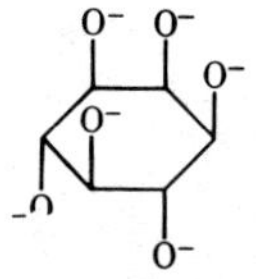 inositol hexaester	vasodilator	245
niaprazin	[27367-90-4]	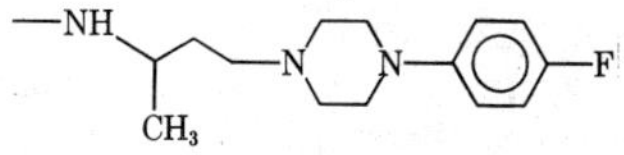 —NHCH(CH₃)CH₂–N(piperazine)–C₆H₄–F	antiallergic, sedative	246
nicametate dihydrogen citrate	[3099-52-3] [1641-74-3]	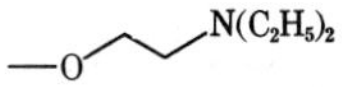 —OCH₂CH₂N(C₂H₅)₂	vasodilator	247
niceritrol	[5868-05-3]	—(OCH₂)₄C	antilipemic	248
nicoclonate	[10571-59-2]	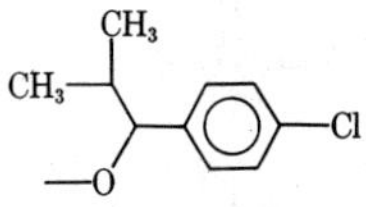 CH₃–C(CH₃)(O—)–C₆H₄–Cl	antilipemic	249
nifenazone	[2139-47-1]	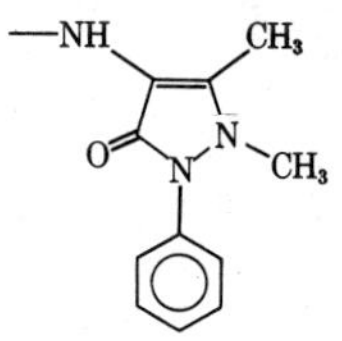 (4-aminoantipyrinyl)	analgesic, antipyretic	250
nikethamide	[59-26-7]	—N(C₂H₅)₂	respiratory stimulant	251
thurfylnicotinate	[70-19-9]	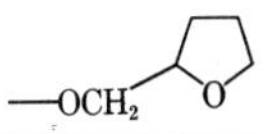 —OCH₂–(tetrahydrofuryl)	vasodilator (topical)	252

Nicotinyl alcohol (3-pyridylcarbinol, 3-pyridinemethanol) has application as an antilipemic and peripheral vasodilator. It may be prepared either by reduction of nicotinic acid esters with lithium aluminum hydride or, more economically, from nicotinonitrile (255):

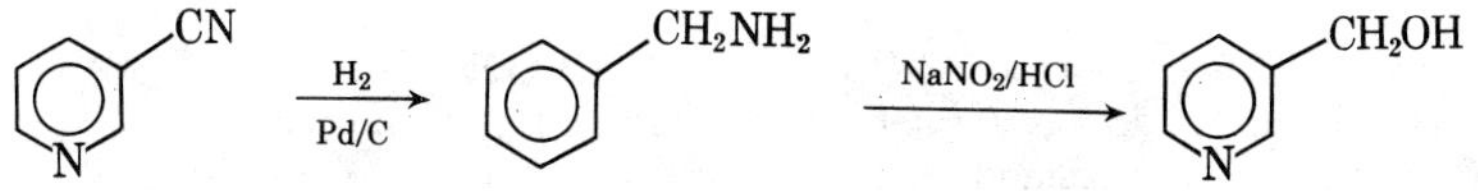

3-pyridyl–CN $\xrightarrow{H_2, Pd/C}$ 3-pyridyl–CH₂NH₂ $\xrightarrow{NaNO_2/HCl}$ 3-pyridyl–CH₂OH

Nicotinyl alcohol is often administered in the form of its tartrate.

BIBLIOGRAPHY

"Nicotinic Acid and Nicotinamide" in *ECT* 1st ed., Vol. 9, pp. 305–313, by A. P. Sachs, Consultant, and J. F. Couch, U.S. Department of Agriculture; "Vitamins, Nicotinic Acid" in *ECT* 2nd ed., Vol. 21, pp. 509–542, by C. del Río-Estrada, Universidad Nacional de Mexico, and H. W. Dougherty, Merck Sharp & Dohme.

1. *J. Nutr.* **105,** 134 (1975).
2. M. Windholz, ed., *The Merck Index*, 9th ed., Merck & Co., Rahway, N.J., 1976.
3. C. A. Elvehjem and co-workers, *J. Am. Chem. Soc.* **59,** 1767 (1937).
4. W. J. Darby, K. W. McNutt, and E. N. Todhunter, *Nutr. Rev.* **33,** 289 (1975).
5. E. Kodicek, *Nutr. Dieta* **4,** 109 (1962).
6. G. E. Glock and P. McLean, *Biochem. J.* **61,** 388 (1955).
7. E. Kodicek, *Br. J. Nutr.* **14,** 13 (1960).
8. B. D. A. Osifo, *Indian J. Nutr. Diet.* **8,** 17 (1971).
9. *Nutr. Rev.* **19,** 240 (1961).
10. *Nutr. Rev.* **32,** 124 (1974).
11. M. L. Das and B. C. Guha, *J. Biol. Chem.* **235,** 2971 (1960).
12. E. Kodicek and P. W. Wilson, *Biochem. J.* **76,** 27 P (1960).
13. R. Koetz, R. Amado, and H. Neukom, *Lebensm. Wiss. Technol.* **12,** 346 (1979).
14. R. O. Cravioto and co-workers, *J. Nutr.* **48,** 453 (1952).
15. E. M. McCance and R. A. Widdowson, *The Composition of Foods*, 4th ed., Elsevier, North-Holland Biomedical Press, Amsterdam, 1978.
16. C. A. Elvehjem, *Physiol. Rev.* **20,** 249 (1940).
17. A. Hennig, *Mineralstoffe, Vitamine, Ergotropica*, VEB Deutscher Landwirtschaftsverlag, Berlin, 1972, p. 350.
18. *Handbuch der Tierernährung*, Verlag Paul Parey, Hamburg, Berlin, 1969, p. 102.
19. H. Friesecke, *Niacin in Animal Nutrition*, Hoffmann-La Roche Technical Bulletin, Basel, Switzerland, 1980, p. 24.
20. P. Michela and K. Lorenz, *Cereal Chem.* **53,** 853 (1976).
21. H. Kwasny, *Lebensmittelchemie und Gerichtliche Chemie* **32,** 36 (1978).
22. S. M. Paraskewas, *Synthesis*, 47 (1978).
23. Ger. Offen. 2,557,397 (June 24, 1976); Brit. Pat. Appl. 74 55,213 (Dec. 20, 1974), J. Bradshow and co-workers.
24. R. Iff and M. Viscontini, *Helv. Chim. Acta* **59,** 2892 (1976).
25. H. Amal and co-workers, *Istanbul Univ. Eczacilik Fak. Mecm.* **11**(2), 158 (1975).
26. Ref. 25, p. 171.
27. T. Fujii and co-workers, *Chem. Pharm. Bull.* **25,** 3042 (1977).
28. G. Schindler and T. Kappe, *Synthesis*, 243 (1977).
29. L. Kofler and A. Kofler, *Ber. Dtsch. Chem. Ges.* **76,** 246 (1943).
30. R. Kuhn and H. Vetter, *Ber. Dtsch. Chem. Ges.* **68,** 2374 (1935).
31. W. Jellinek, *J. Phys. Chem.* **55,** 173, 179 (1951).
32. Degussa Company, Wolfgang, FRG, unpublished results.
33. Jpn. Kokai Tokkyo Koho 78 133,291 (Nov. 20, 1978); Jpn. Pat. Appl. 77 48,869 (April 26, 1977), H. Yamakita and K. Hayakawa.
34. Jpn. Kokai Tokkyo Koho 77 106,876 (Sept. 7, 1977); Jpn. Pat. Appl. 76 22,896 (March 3, 1976), E. Hasegawa and E. Tsuchida.
35. Jpn. Kokai Tokkyo Koho 77 105,988 (Sept. 6, 1977); Jpn. Pat. Appl. 76 22,897 (March 3, 1976), E. Hasegawa and E. Tsuchida.
36. T. Endo and M. Okawara, *J. Polym. Sci. Polym. Chem. Ed.* **17,** 3667 (1979).
37. E. C. Taylor, Jr., and A. J. Corvetti, *Org. Synth.* **37,** 63 (1957).
38. Jpn. Kokai Tokkyo Koho 78 31,666 (March 25, 1978); Jpn. Pat. Appl. 76 105,842 (Sept. 6, 1976), H. Miyazaki and S. Miyazaki.
39. Ref. 37, p. 12.
40. T. S. Garner, E. Wenis, and J. Lee, *J. Org. Chem.* **22,** 984 (1957).
41. E. C. Taylor, Jr., and A. J. Corvetti, *J. Org. Chem.* **19,** 1633 (1954).
42. W. T. Caldwell, F. T. Tyson, and L. Lauer, *J. Am. Chem. Soc.* **66,** 1479, 1484 (1944).

43. F. Šorm, *Collect. Czech. Chem. Commun.* **13,** 57 (1948).
44. Czech. Pat. 175,322 (Dec. 15, 1978); Czech. Pat. Appl. 75 8,645 (Dec. 17, 1975), J. Malek and co-workers.
45. East Ger. Pat. 30,872 (Sept. 15, 1965); East Ger. Pat. Appl. 84,085 (March 8, 1963), W. Kirsten and H. Schultz.
46. Y.-J. Lin and co-workers, *J. Org. Chem.* **44,** 4160 (1979).
47. Belg. Pat. 845,405 (Aug. 20, 1976); Brit. Pat. Appl. 75 34,971 (Aug. 22, 1978), J. I. Darragh (to Imperial Chemical Industries, Ltd.).
48. A. P. Ivanovskii, V. A. Sichanov, A. M. Kutin, and M. A. Korsunov, *Chimiceskaja Promyslennost (USSR)*, 26 (1972).
49. H. Beschke and H. Friedrich, *Chemiker Zeitung* **101,** 377 (1977).
50. U.S. Pat. 4,163,854 (Aug. 7, 1979); Appl. Ger. 2,703,070 (Jan. 26, 1977), H. Beschke and H. Friedrich (to Degussa Co.).
51. Ger. Offen. Appl. 2,514,004 (March 25, 1975), H. aus der Fünten and co-workers (to Dynamit Nobel).
52. Ger. Offen. Appl. 2,519,529 (May 2, 1975), G. Daum and H. Richtzenhain (to Dynamit Nobel).
53. A. Nenz and M. Pieroni, *Hydrocarbon Process.* **47**(11), 139 (1968).
54. Y. Kusunoki and H. Okazaki, *Hydrocarbon Process.* **53**(11), 129 (1974).
55. H. Beschke, H. Friedrich, H. Schaefer, and G. Schreyer, *Chem. Ztg.* **101,** 384 (1977).
56. H. Schaefer, *Chemie-Technik* **7,** 231 (1978).
57. Ger. Offen. Appl. 3,107,755 (Feb. 26, 1981), H. Beschke and H. Friedrich (to Degussa Co.).
58. Eur. Pat. Appl. EP 37,123 (Oct. 7, 1981); Jpn. Pat. Appl. 80 43,119 (April 1, 1980), O. Isatsugu, F. Itsuo, and M. Ichiro (to Takeda Chemical Industries, Ltd.).
59. Ger. Offen. Appl. 3,128,956 (July 22, 1981), H. Beschke and co-workers (to Degussa Co.).
60. Ger. Auslegeschrift 2,505,745 (May 3, 1979); Jpn. Pat. Appl. 74 28,597 (March 14, 1974), R. Ishioka and co-workers (to Showa Denko K.K.).
61. B. V. Suvorov and co-workers, *Kinet. Katal.* **18,** 957 (1977).
62. M. C. Sze and A. P. Gelbein, *Hydrocarbon Process.* **55**(2), 103 (1976).
63. Ger. Offen. 2,729,009 (Jan. 12, 1978); U.S. Pat. Appl. 701,694 (July 1, 1976), H. Richtzenhain and P. Janssen (to The Lummus Co.).
64. Ger. Offen. 2,729,072 (Jan. 12, 1978); U.S. Pat. Appl. 701,693 (July 1, 1976), A. Gelbein and co-workers (to The Lummus Co.).
65. *Swiss Chem.* **3**(5), 17 (1981).
66. U.S. Pat. 2,471,518 (May 31, 1949); Brit. Pat. Appl. 11357/55 (Aug. 28, 1942), B. F. Duesel and H. H. Friedman (to Pyridium Corp.).
67. Ger. Offen. Appl. 2,517,054 (April 17, 1975), H. Beschke, H. Friedrich, K.-P. Müller, and G. Schreyer (to Degussa Co.).
68. C. B. Rosas and G. B. Smith, *Chem. Eng. Sci.* **35,** 330 (1980).
69. Brit. Pat. 777,517 (March 28, 1956); Pat. Appl. (April 20, 1955), E. J. Gasson and D. J. Hadley (to The Distillers Company, Ltd.).
70. U.S. Pat. 4,008,241 (Feb. 15, 1977); U. S. Pat. Appl. 7,609,626 (Sept. 2, 1975), A. P. Gelbein and co-workers (to The Lummus Co.).
71. J. E. Paustian and co-workers, *Chemtech*, 174 (March 1981).
72. Ger. Offen. 2,131,813 (Oct. 19, 1972); U.S. Pat. Appl. 66,181 (Aug. 24, 1970), L. R. Haefele (to R. J. Reynolds Tobacco Co.).
73. Jpn. Pat. 74 127,976 (Dec. 7, 1974); Jpn. Pat. Appl. 73 43,624 (April 19, 1973), Y. Watabiki and co-workers (to Yuki Gosei).
74. Jpn. Pat. 73 22,710 (July 7, 1973); Jpn. Pat. Appl. 70 87,363 (Oct. 5, 1970), T. Takahashi and F. Matsuda (to Mitsui Toatsu Chemicals, Inc.).
75. U.S. Pat. 2,412,749 (Dec. 17, 1946), E. F. Pike and R. S. Shane (to Gelating Products Corp.).
76. S. K. Sana, *J. Indian Chem. Soc. Ind. News Ed.* **19,** 95 (1956).
77. U.S. Pat. 2,617,805 (Nov. 11, 1952), L. J. Wissow (to Merck & Co.).
78. A. Atsuaki, *J. Chem. Soc. Ind. Chem. Soc.* **57,** 495 (1954).
79. *Chem. Week*, 38 (Aug. 22, 1979).
80. U.S. Pat. 3,929,811 (Dec. 30, 1975); U.S. Pat. Appl. 463,883 (April 25, 1974), A. P. Gelbein, M. C. Sze, and J. E. Paustian (to The Lummus Co.).
81. Ger. Offen. Appl. 3,028,904 (July 30, 1980), H. Beschke and co-workers (to Degussa Co.).
82. Ger. Offen. Appl. 3,028,791 (July 30, 1980), H. Beschke and co-workers (to Degussa Co.).

83. U.S. Pat. 4,304,917 (Dec. 8, 1981); U.S. Pat. Appl. 160,684 (June 18, 1980), G. D. Suciu (to The Lummus Co.).
84. *Chem. Mark. Rep.*, 3, 19 (Feb. 2, 1981).
85. *Synthetic Organic Chemicals*, U.S. Production and Sales, U.S. Tariff Commission (1966, 1969, 1972) and U.S. International Trade Commission (1975), Washington, D.C.
86. *Chem. Ind.* **34**, 82 (Feb. 1982).
87. *Medicinal Chemicals*, Chemical Economics Handbook-SRI International (Feb. 1980–Dec. 1981), 565.9000X (Imports), 565.9001A (Exports).
88. *Chem. Mark. Rep.* 45 (March 8, 1982), and midyear issues for 1970, 1974, 1976, 1978, 1980–1981.
89. *Europäisches Arzneibuch*, Vol. II, Deutscher Apotheker-Verlag Stuttgart, Govi-Verlag G.m.b.H., Frankfurt, 1975, p. 298–299.
90. *United States Pharmacopeia*, Vol. XX, USP Convention, Inc., Rockville, Md., 1979, p. 548.
91. J. Kracmar, J. Kracmarova, and J. Zyka, *Pharmazie* **28**, 567 (1968).
92. W. J. König, *J. Prakt. Chem.* **69**, 105 (1904).
93. *Standard Methods of Chemical Analysis*, 6th ed., D. Van Nostrand Co., Inc., Princeton, N.J., 1963, p. 2366–2368.
94. H. D. Eilhauer, P. Fahrig, and G. Krautschik, *Z. Anal. Cnem.* **228**, 276 (1967).
95. *Official Methods of Analysis*, Association of Official Analytical Chemists, Washington, D.C., 1980, pp. 745–746.
96. *Ibid.*, pp. 759–761, 763–764.
97. *Methods of Vitamin Assay*, Interscience Publishers, a division of John Wiley & Sons, Inc., New York, 1966, pp. 172–176.
98. E. Nürnberg, *Dtsch. Apoth. Ztg.* **101**, 142 (1961).
99. M. M. Frodyma and Van T. Lieu, *Anal. Chem.* **39**, 815 (1967).
100. R. W. Frei and co-workers, *Anal. Chim. Acta* **49**, 527 (1970).
101. H.-C. Chiang, *J. Chromatogr.* **45**, 161 (1969).
102. R. T. Nuttall and B. Bush, *Analyst* **96**, 875 (1971).
103. S. E. Petrovic and co-workers, *Anal. Chem.* **40**, 1007 (1968).
104. A. J. Sheppard and A. R. Prosser, *Methods Enzymol.* **18**, 17 (1971).
105. J. Vessmann and S. Strömberg, *J. Pharm. Sci.* **64**, 311 (1975).
106. L. T. Sennello and C. J. Argoudelis, *Anal. Chem.* **41**, 171 (1969).
107. G. R. Scurray, *Food Chem.* **7**, 77 (1981).
108. R. L. Kirchmeier and R. P. Upton, *J. Pharm. Sci.* **67**, 1444 (1978).
109. S. P. Sood and co-workers, *J. Pharm. Sci.* **66**, 40 (1977).
110. M. C. Walker, B. E. Carpenter, and E. L. Cooper, *J. Pharm. Sci.* **70**, 99 (1981).
111. R. B. Wills, C. G. Shaw, and R. W. Day, *J. Chromatogr.* **15**, 262 (1977).
112. T. A. Tyler and R. R. Shrago, *J. Liquid Chromatogr.* **3**, 269 (1980).
113. J. S. Sandhu and D. R. Fraser, *Int. J. Vitam. Nutr. Res.* **51**, 138 (1981).
114. M. K. Horwitt and co-workers, *Am. J. Clin. Nutr.* **34**, 423 (1981).
115. W. A. Krehl, *Fed. Proc. Fed. Am. Soc. Exp. Biol.* **40**, 1527 (1981).
116. *Nutr. Rev.* **32**, 76 (1974).
117. D. H. Baker and co-workers, *J. Anim. Sci.* **36**, 299 (1973).
118. S. Chaykin, *Am. Rev. Biochem.* **36**, 146 (1967).
119. J. Keller, M. Liersch, and H. Grunicke, *Eur. J. Biochem.* **22**, 263 (1971).
120. J. Keller, M. Liersch, and H. Grunicke, *Verh. Dtsch. Ges. Inn. Med.* **78**, 1371 (1972).
121. C. Streffer and J. Beneš, *Eur. J. Biochem.* **21**, 357 (1971).
122. J. W. Foster and A. G. Moat, *Microbiol. Rev.* **44**, 83 (1980).
123. P. B. Collins and S. Chaykin, *J. Biol. Chem.* **247**, 778 (1972).
124. L.-F. H. Lin and L. M. Henderson, *J. Biol. Chem.* **247**, 8023 (1972).
125. J. Kirchner and co-workers, *J. Biol. Chem.* **241**, 953 (1966).
126. D. T. C. Ngo and V. C. Fullon, *Acta Med. Philipp.* **8**, 152 (1972).
127. K. Lang, *Biochemie der Ernährung*, Dr. D. Steinkopff Verlag, Darmstadt, 1979, p. 500.
128. E. Leifer, L. J. Roth, D. S. Hogness, and M. H. Corson, *J. Biol. Chem.* **190**, 595 (1951).
129. J. E. Mrochek and co-workers, *Clin. Chem.* **22**, 1821 (1976).
130. H. Holmen, H. Egsgaard, J. Funck, and E. Larssen, *Biomedical Mass Spectrometry* **8**, 122 (1981).
131. M. L. Blank and O. S. Privett, *J. Lipid Res.* **4**, 170 (1963).
132. P. R. Brown and co-workers, *J. Nutr.* **66**, 599 (1959).
133. J. B. Mason and E. Kodicek, *Biochem. J.* **120**, 515 (1970).

134. H. E. Sauberlich, R. P. Dowdy, and J. H. Skala, *Laboratory Tests for the Assessment of Nutritional Status*, CRC Press, Boca Raton, Fla., 1977, p. 70.
135. D. J. de Lange and C. P. Joubert, *Am. J. Clin. Nutr.* **15,** 169 (1964).
136. M. E. Loijkin, A. W. Wertz, and C. G. Dietz, *J. Nutr.* **46,** 335 (1952).
137. Recommended Dietary Allowances, 8th ed., Food and Nutrition Board, National Academy of Sciences, Washington, D.C., 1974.
138. *Nutr. Rev.* **33,** 147 (1975).
139. K. H. Bässler and K. Lang, *Vitamine*, 2nd ed., Dr. D. Steinkopff Verlag, Darmstadt, 1981, p. 54.
140. S. Ohguri, *J. Nutr. Sci. Vitaminol.* **26,** 141 (1980).
141. B. I. Magboul and D. A. Bender, *Proc. Nutr. Soc.* **40,** 16A (1981).
142. Ref. 139, p. 56.
143. H. Fisher, H. M. Scott, and B. C. Johnson, *Br. J. Nutr.* **9,** 340 (1955).
144. E. B. Patterson and co-workers, *Poult. Sci.* **35,** 499 (1956).
145. J. A. Manson, *Proc. Nutr. Soc.*, 120A (1975).
146. L. Kung, Jr., K. Gubert, and J. T. Huber, *J. Dairy Sci.* **63,** 2020 (1981).
147. D. O. Ridell, E. E. Bartley, and A. D. Dayton, *J. Dairy Sci.* **64,** 782 (1981).
148. F. M. Byers, *Anim. Nutr. Health* **6,** 20 (1979).
149. *Acute Oral Toxicity of Nicotinic Acid-Amide in Rats*, Central Institute for Nutrition and Food Research (TNO), Zeist, Netherlands, 1979, unpublished.
150. *Acute Oral Toxicity of Nicotinic Acid in Rats*, Central Institute for Nutrition and Food Research (TNO), Zeist, Netherlands, 1979, unpublished.
151. *Hearing Draft: Evaluation of the Health Aspects of Niacin and Niacinamide as Food Ingredients*, Contract No. FDA 223-75-2004/1979, Life Sciences Research Office, Bethesda, Md., 1978.
152. P. Handler and W. J. Dann, *J. Biol. Chem.* **146,** 357 (1942).
153. L. M. Horger and E. B. Gerheim, *Proc. Soc. Exp. Biol. Med.* **97,** 444 (1958).
154. P. Handler, *J. Biol. Chem.* **154,** 203 (1944).
155. F. A. Patty, *Industrial Hygiene and Toxicology*, Vol. II, John Wiley & Sons, Inc., New York, 1966, p. 1836.
156. 21 CFR 182.5535.
157. *Fed. Regist.* **42,** 14409 (March 15, 1977).
158. 21 CFR Parts 137 and 139.
159. R. S. J. Harris, *J. Agric. Food Chem.* **7,** 88 (1959).
160. C. L. Brooke, *J. Agric. Food Chem.* **16,** 163 (1968).
161. P. N. Ranum, *Cereal Chem.* **57,** 70 (1980).
162. Fr. Pat. Appl. 1,530,248 (July 5, 1967), (to Hoffmann-La Roche).
163. W. M. Cort, B. Borenstein, J. L. Harley, M. Osadca, and J. Scheiner, *Food Technol.* **30,** 52 (1976).
164. U.S. Pat. 3,767,824 (Oct. 23, 1973), W. L. Keyser and W. J. Zielinski (to The Quaker Oats Co.).
165. L. Vanossi, *Tec. Molitoria* **29,** 172 (1978).
166. C. L. Brooke, *Chem. Technol.*, 353 (1972).
167. R. H. Anderson, L. D. Maxwell, A. E. Mulley, and C. W. Fritsch, *Food Technol.* **30,** 110 (1976).
168. J. L. Kendrick and B. M. Watts, *J. Food Sci.* **34,** 292 (1969).
169. W. D. Brown and A. L. Tappel, *Food Res.* **22,** 214 (1957); A. L. Tappel, *Food Res.* **22,** 404 (1957).
170. S. J. Mulder and B. Krol, *Fleischwirtschaft* **55,** 1255 (1975).
171. M. E. Bailey and co-workers, *J. Agric. Food Chem.* **12,** 89 (1964).
172. D. H. Baker and co-workers, *Nutr. Rep. Int.* **14,** 115 (1976).
173. W. Sauermilch, *Beitr. Biol. Pflanz.* **36,** 91 (1961).
174. G. K. Daribaeva, *Izv. Akad. Nauk Kaz. SSR Ser Biol.* **10,** 20 (1972); *Chem. Abstr.* **77,** 57100b (1972).
175. R. Manteuffel and co-workers, *Biochem. Physiol. Pflanz.* **163,** 586 (1972).
176. Ger. Offen. 2,349,745 (April 25, 1974), S. Takeuchi and co-workers (to the Institute of Physical and Chemical Research).
177. Ger. Offen. 2,251,103 (Nov. 16, 1971), H. Yamagishi and T. Watanabe (to Japan Steel and Tube Corp.).
178. U.S. Pat. 4,177,131 (May 20, 1976), R. Merker and S. Lucca (to Metalux Corp.).
179. H. D. Hedrich and C. J. Raub, *Metalloberflaeche* **33,** 308 (1979).
180. O. de Conineck, *Bull. Soc. Chim. Fr.* **42**(2), 100 (1884).
181. R. F. Evans, E. F. G. Herington, and W. Kynaston, *Trans. Faraday Soc.* **49,** 1284 (1953).
182. Y. M. Slobodin and M. M. Goldmann, *Zh. Priklad. Khim. (Leningrad)* **21,** 859 (1948).

183. L. Hertelendi, *Z. Phys. Chem.* **192**, 379 (1943).
184. I. A. Kozello, A. Ya. Gasheva, and V. I. Khmelevskii, *Khim. Farm. Zh.* **10**(11), 90 (1976).
185. D. Dal Monte Casoni, *Boll. Sci. Fac. Chim. Ind. Bologna* **8**, 137 (1950).
186. E. C. Taylor, Jr., and A. J. Crovetti, *J. Org. Chem.* **19**, 1633 (1954).
187. P. Nantka-Namirski, *Acta Pol. Pharm.* **23**, 403 (1966).
188. B. M. Bain and J. E. Saxton, *J. Chem. Soc.*, 5216 (1961).
189. U.S. Pat. 3,159,639 (Dec. 1, 1964); U.S. Pat. Appl. 6,224,201 (Sept. 17, 1962), M. Freifelder (to Abbott Laboratories).
190. E. Späth and H. Spitzner, *Ber. Dtsch. Chem. Ges.* **59**, 1479 (1926).
191. J. Bartoszewski, *Acta Pol. Pharm.* **11**, 189, 192 (1954).
192. Jpn. Kokai Tokkyo Koho 77 148,082 (Dec. 8, 1977); Jpn. Pat. Appl. 76 65,908 (Jan. 4, 1976), I. Chibata and co-workers (to Tanabe Seiyaku Co., Ltd.).
193. Jpn. Kokai Tokkyo Koho 77 46,080 (April 12, 1977); Jpn. Pat. Appl. 75 121,121 (Oct. 6, 1975), S. Yamabe and co-workers (to Taiho Yakuhin Kogyo K.K.).
194. Belg. Pat. 867,515 (Sept. 18, 1978); Ital. Pat. Appl. 77 24,141 (May 30, 1977), M. Fazzini (Scharper S.p.A. per l'Industria Farmaceutical).
195. Ger. Offen. 2,525,024 (Dec. 30, 1976); Ger. Pat. Appl. (June 5, 1975), H. Reis and co-workers (to BASF).
196. Jpn. Kokai Tokkyo Koho 76 110,571 (Sept. 30, 1976); Jpn. Pat. Appl. 75 34,939 (March 25, 1975), Y. Kondo and co-workers (to Sanwa Chemical Laboratories).
197. Jpn. Kokai Tokkyo Koho 76 146,472 (Dec. 16, 1976); Jpn. Pat. Appl. 75 71,039 (June 11, 1975), M. Fujumoto (to Fujimoto Pharmaceutical Co., Ltd.).
198. Jpn. Kokai Tokkyo Koho 79 20,501 (July 27, 1979); Jpn. Pat. Appl. 71 23,403 (April 13, 1971), (to Donau-Pharmazie G.m.b.H.).
199. Jpn. Kokai Tokkyo Koho 79 66,682 (May 29, 1979); Jpn. Pat. Appl. 77 133,989 (Nov. 7, 1977), E. Sawai, K. Sumikawa, and H. Okada (to Sawai Pharmaceutical Co., Ltd.).
200. Jpn. Kokai Tokkyo Koho 78 116,830 (Oct. 11, 1978); Jpn. Pat. Appl. 77 30,075 (March 17, 1977), D. Fujimoto (to Fujimoto Pharmaceutical Co., Ltd.).
201. Y. D. Reddy and V. V. Somayajulu, *J. Indian Chem.* **56**, 511 (1979).
202. V. B. Piskov and co-workers, *Khim. Geterosikl. Soedin.*, 1112 (1976).
203. B. A. Porai-Koschits, O. F. Ginzburg, and L. S. Efros, *J. Gen. Chem. USSR* **17**, 1768 (1947).
204. U.S. Pat. 3,985,885 (Oct. 12, 1976); U.S. Pat. Appl. 7,615,027 (Sept. 19, 1975), V. L. Narayanan and R. G. Angel (to E. R. Squibb and Sons, Inc.).
205. G. Doleschall, *Tetrahedron* **32**, 2549 (1976).
206. Jpn. Kokai Tokkyo Koho 76 125,003 (Nov. 1, 1976); Jpn. Pat. Appl. 74 127,849 (Nov. 5, 1974), T. Mukaiyama, M. Muraki, and J. Himizu.
207. M. S. Brown and H. Rapoport, *J. Chem. Soc.*, 3261 (1963).
208. A. Bylicki, *Bull. Acad. Pol. Sci. Ser. Sci. Chim.* **6**, 639 (1958).
209. Ger. Offen. 2,046,556 (April 22, 1971); Swiss Pat. Appl. 14419-69 (Sept. 24, 1969), W. Schalk and co-workers (to Lonza A.G.).
210. Ger. Auslegeschrift 2,256,508 (Dec. 8, 1977); Jpn. Pat. Appl. 71 92,142 (Nov. 17, 1971), K. Masuda, H. Kizawa, and O. Niigata (to Nippon Soda Co., Ltd.).
211. U.S. Pat. 4,001,257 (Jan. 4, 1979); Pat. Appl. 603,133 (Aug. 8, 1975), K. Masuda, H. Kizawa, and Y. Otaki (to Nippon Soda Co., Ltd.).
212. T. S. Titova and co-workers, *Zh. Prikl. Khim.* (*Leningrad*) **42**, 910 (1969).
213. Ger. Auslegeschrift 2,165,035 (Jan. 26, 1978); Jpn. Pat. Appl. 70 122,247 (Dec. 30, 1970), M. Kubo and T. Horikawa (to Daicel, Ltd.).
214. Ger. Offen. 2,826,333 (Feb. 8, 1979); Ital. Pat. Appl. 25812 A-77 (July 18, 1977), G. Ghelli (to Luigi Stoppani S.p.A.).
215. E. Yu. Mednikova, A. N. Christyakov, and D. A. Sibarov, *Zh. Prikl. Khim.* (*Leningrad*) **51**, 2533 (1978).
216. Ger. Auslegeschrift 1,620,159 (Nov. 23, 1972); Ger. Pat. Appl. 40,947 (June 25, 1965), L. Rappen and O. Koch (to Rütgerswerke A.G.).
217. S. Paraskewas, *Synthesis*, 819 (1974).
218. USSR Pat. 228,688 (July 29, 1967), I. A. Golovleva.
219. Jpn. Pat. 92,409 (April 25, 1975); Jpn. Pat. Appl. 46,670 (Aug. 20, 1973), (to Shinnippon Seitetsu).
220. Jpn. Pat. 76 29,483 (March 12, 1976); Jpn. Pat. Appl. 100,952 (Sept. 4, 1974), T. Inoue and T. Hara (to Nippon Steel Chemical Co., Ltd.).

221. Ger. Auslegeschrift 1,097,991 (Jan. 26, 1961); Swiss Pat. Appl. 55 20,945 (June 16, 1955), W. Wettstein (to Ciba A.G.).
222. S. Järas and S. T. Lundin, *J. Appl. Chem. Biotechnol.* **27,** 499 (1977).
223. Jpn. Pat. 74 61,173 (June 14, 1974); Jpn. Pat. Appl. 72 102,566 (Oct. 12, 1972), K. Nakajima and T. Sato (to Japan Synthetic Chemical Industry Co., Ltd.).
224. Ger. Auslegeschrift 1,940,320 (Aug. 17, 1972); Ger. Pat. Appl. 1,940,320 (Aug. 7, 1969), R. M. Yokoyama and K. H. Sawada (to Teijin Chemicals, Ltd.).
225. Jpn. Kokai Tokkyo Koho 77 36,670 (March 23, 1977); Jpn. Pat. Appl. 75 110,875 (Sept. 16, 1975), M. Kawamata, S. Fujikake, and H. Tanabe (to Mitsui Toatsu Chemicals).
226. *Europäisches Arzneibuch*, Vol. III, Deutscher Apotheker-Verlag, Stuttgart, Govi-Verlag G.m.b.H., Frankfurt, 1978, p. 165.
227. Ref. 90, p. 547.
228. Ref. 226, pp. 164–165.
229. Ref. 95, pp. 743–745.
230. Ref. 95, pp. 759–761, 763–764.
231. B. C. Johnson, *J. Biol. Chem.* **159,** 227 (1945).
232. P. B. Perlzweig, W. S. Perlzweig, and F. Rosen, *Arch. Biochem.* **22,** 191 (1949).
233. P. B. Perlzweig, F. Rosen, and P. B. Pearson, *J. Nutr.* **40,** 453 (1950).
234. E. B. T. Leete, *Adv. Enzymol.* **32,** 295 (1969).
235. E. E. Cava and co-workers, *Proc. Staff Meeting Mayo Clinic* **34,** 502 (1959).
236. M. N. G. Dukes, ed., *Meyler's Side Effects of Drugs*, Excerpta Medica, Amsterdam, Oxford, Princeton, 1980, pp. 634, 734.
237. L. S. Goodman and A. Gilman, eds., *The Pharmacological Basis of Therapeutics*, 5th ed., McMillan Publishing Co., Inc., New York, 1975, pp. 736, 749.
238. 21 CFR 182.5530.
239. U.S. Pat. 3,969,219 (Aug. 6, 1975), J. C. Kosmos (to Sterling Drug, Inc.).
240. U.S. Pat. 4,081,336 (April 7, 1977), F. W. Eppensteiner and C. Steinecker (to Richardson Co.).
241. U.S. Pat. 4,045,305 (Sept. 14, 1975), J.-J. Fong and G. F. Hsu (to Minnesota Mining & Manufacturing Co.).
242. Ger. Offen. 1,941,217 (March 4, 1969), A. Scherm and D. Peteri (to Merz & Co.).
243. R. Fischbach and R. Haas, *Arzneim. Forsch.* **17,** 313 (1967).
244. U.S. Pat. 3,384,642 (May 21, 1968); Jpn. Pat. Appl. 65 62,859 (Oct. 12, 1965), N. Michio, K. Ryosuke, and A. Katsuo (to Yoshitomi Pharmaceutical Industries, Ltd.).
245. C. O. Badgett and C. F. Woodward, *J. Am. Chem. Soc.* **69,** 2907 (1947).
246. U.S. Pat. 3,712,893 (Jan. 23, 1973); Fr. Pat. Appl. 6,173,685 (Nov. 14, 1968), R. Y. Mauvernay and co-workers (to Centre Européen de Recherches Mauvernay).
247. F. F. Blicke and E. L. Jenner, *J. Am. Chem. Soc.* **64,** 1721 (1942).
248. Fr. Pat. M 2046 (Sept. 30, 1963), (to Société d'Etudes et de Recherches Pharmacotechniques).
249. U.S. Pat. 3,367,393 (Feb. 6, 1968); Fr. Pat. Appl. 970,539 (April 10, 1964), J. Nordmann and H. B. Swierkot (to Etablissement Kuhlmann).
250. Ger. Pat. 897,407 (Feb. 6, 1951), W. Heid.
251. Ger. Reichspatent 351,085 (Sept. 14, 1920), (to Ciba).
252. U.S. Pat. 2,485,152 (Oct. 18, 1949), M. Hartmann and E. Merz (to Ciba Pharmaceutical Products, Inc.).
253. U.S. Pat. 3,415,834 (Dec. 10, 1968); Brit. Pat. Appl. 63 50,274 (Dec. 19, 1963), C. Hoffmann and A. Faure (to UPSA).
254. L. Bernardi, *Arzneim. Forsch./Drug Res.* **29,** 1204 (1979).
255. U.S. Pat. 2,615,896 (Oct. 28, 1952), G. O. Chase (to Hoffmann-La Roche).

HERIBERT OFFERMANNS
AXEL KLEEMANN
HERBERT TANNER
HELMUT BESCHKE
HEINZ FRIEDRICH
Degussa Company

PYRIDOXINE (B₆)

Most of the significant developments in pyridoxine research occurred during the 1930s (1–3). In 1926, a U.S. physician, Joseph Goldberger, described the induction of a pellagralike dermatitis, later called rat acrodynia, in rats fed a diet lacking sources of the B complex (4). Paul György showed that the deficiency continued to exist even when B_1 and B_2 were included in the diet. This showed the existence of another growth factor, vitamin B_6, which was identified and named in 1934 (5). Several research groups undertook the isolation and characterization of vitamin B_6 and five different laboratories announced this achievement in 1938 (6). The chemical structure of pyridoxine, the isolated form of the vitamin, was elucidated and two independent syntheses were completed in the following year (7–8).

A report from Japan in 1931 described the isolation from rice bran of a substance having the empirical formula and melting point of pyridoxine hydrochloride, but the substance was not recognized as a vitamin (9). Subsequent investigations of vitamin B_6 by Esmond Snell and others established that it is not a single chemical entity but the group of six chemically related and biochemically interrelated substances shown below (10–11):

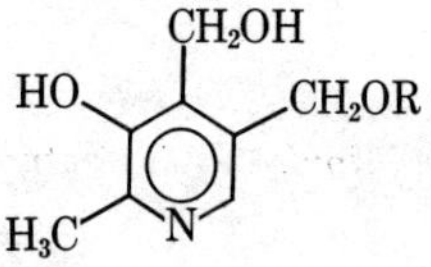

(1) R = H, pyridoxine
(4) R = OPO₃H₂, pyridoxine-5′-phosphate

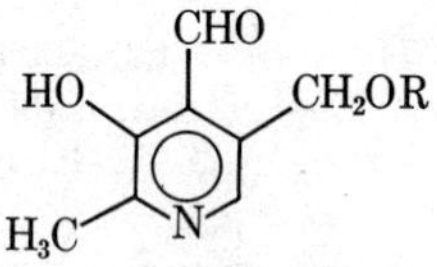

(2) R = H, pyridoxal
(5) R = OPO₃H₂, pyridoxal-5′-phosphate

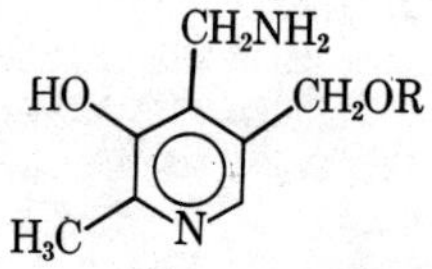

(3) R = H, pyridoxamine
(6) R = OPO₃H₂, pyridoxamine-5′-phosphate

An IUPAC recommendation made in 1960 proposed that the term pyridoxine be used in reference to all the naturally occurring pyridine derivatives with vitamin B_6 activity (12). More recently, the IUPAC–IUB Commission on Biological Nomenclature has recommended that only the term vitamin B_6 be used in this generic sense (13). Pyridoxine refers specifically to 5-hydroxy-6-methyl-3,4-pyridinedimethanol (1). The older name for compound (1), pyridoxol, is no longer used.

The six pyridine derivatives that exhibit significant vitamin B_6 activity are pyridoxine (1), pyridoxal (2), pyridoxamine (3), and their respective 5′-phosphates (4), (5), and (6). The vitamin B_6 group also includes minor components, including various isomers and metabolites.

Pyridoxine, pyridoxal, and pyridoxamine are about equally effective as forms of vitamin B_6 in animals. However, various microorganisms need specific forms, and it was the inability of *Lactobacillus casei* and *Streptococcus faecalis* to utilize pyridoxine that led to discovery of the other two (10).

In everyday usage, the term vitamin B_6 most frequently refers to pyridoxine hydrochloride, as this is the article of bulk manufacture that is used in commercial vitamin preparations and in vitamin enriched foods. However, pyridoxine comprises a small fraction of the total amount of vitamin B_6 in many foodstuffs.

Occurrence

Vitamin B_6 occurs in the tissues and body fluids of virtually all living organisms. This ubiquity of the vitamin reflects its essential role in amino acid biochemistry. Almost all foods in a typical human diet contain detectable amounts. Meat products, calf liver especially, are relatively rich sources; cereals (not corn) and vegetables are intermediate; whereas fruits, except bananas, are relatively meager sources (14). Having 9.75 µg/g, salmon is one of the richest dietary sources. Edible yeast contains 11 µg/g. Other yeast species have the highest vitamin B_6 contents of any natural source: up to 380 µg/g dry weight (15).

Technical fermentation processes have been developed for two of the B vitamins, riboflavin (vitamin B_2) and B_{12}, and overproduction of pyridoxine by bacterial organisms and by plant tissue cultures has been studied (16–17) (see also Fermentation). However, it appears unlikely that a fermentation process for pyridoxine will supplant its manufacture by chemical synthesis.

Properties

Physical properties of pyridoxine, pyridoxamine, pyridoxal, and their derivatives are listed in Table 1.

The chemical properties of pyridoxine that are related to its role as a vitamin are centered on the 4-hydroxymethyl group. Mild oxidation, for example with manganese dioxide, transforms pyridoxine into pyridoxal (11). Heating with aqueous ammonia results in conversion to pyridoxamine (11). Heating a neutral solution of pyridoxine in the absence of other nucleophilic species produces dimeric material (18). These last two reactions may proceed through the quinonemethane-type intermediate (7).

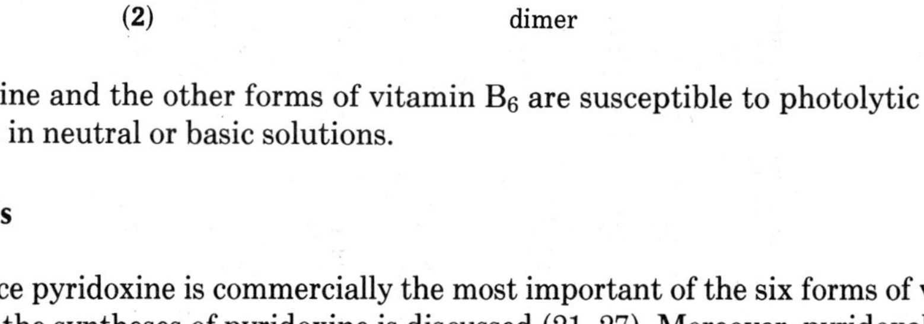

Pyridoxine and the other forms of vitamin B_6 are susceptible to photolytic decomposition in neutral or basic solutions.

Synthesis

Since pyridoxine is commercially the most important of the six forms of vitamin B_6, only the syntheses of pyridoxine is discussed (21–27). Moreover, pyridoxal is best prepared by oxidation of pyridoxine (11,19) and pyridoxamine, in turn, is obtained by catalytic reduction of pyridoxal oxime (20–21).

Condensation. Condensation reactions are the classical routes to pyridoxine. One of them was used in the original synthesis described in ref. 7, and they provided the basis for the early technical syntheses (21). Typically, they involve formation of the

Table 1. Properties of Pyridoxine, Pyridoxamine, Pyridoxal, and Their Derivatives

Substance	CAS Registry No.	Empirical formula	Mol wt	Form	Mp, °C	Solubility in water	pK values
pyridoxine (1)	[65-23-6]	$C_8H_{11}NO_3$	169.18	needles, acetone	160	sol	5.00, 8.96
pyridoxine hydrochloride	[58-56-0]	$C_8H_{12}ClNO_3$	205.64	colorless platelets or rods, ethanol/acetone	204–206 (dec)	22 g/100 mL	
pyridoxine-5′-phosphate (4)	[447-05-2]	$C_8H_{12}NO_6P$	249.16	colorless needles, water crystals	212–213 (dec)	sol	3.31–3.54, 7.90–8.21, 10.4–10.63
pyridoxamine (3)	[85-87-0]	$C_8H_{12}N_2O_2$	168.20		193	sol	
pyridoxamine dihydrochloride	[524-36-7]	$C_8H_{14}Cl_2N_2O_2$	241.12	deliquescent white platelets	226–227 (dec)	50 g/100 mL	
pyridoxamine-5′-phosphate (6)	[529-96-4]	$C_8H_{13}N_2O_5P$	248.18			sol	<2.5, 3.25–3.69, 5.76, 8.61
pyridoxamine-5′-phosphate dihydrate	[84878-64-8]	$C_8H_{13}N_2O_5P \cdot 2H_2O$	284.21	rhombic plates			
pyridoxal (2)	[66-72-8]	$C_8H_9NO_3$	167.16		oxime: 225–226 (dec)	50 g/100 mL	4.2, 8.68, 13
pyridoxal hydrochloride	[65-22-5]	$C_8H_{10}ClNO_3$	203.63	rhombic crystals	165 (dec)	50 g/100 mL	
pyridoxal-5′-phosphate (5)	[54-47-7]	$C_8H_{10}NO_6P$	247.14		oxime: 229–230 (dec)		
pyridoxal-5′-phosphate monohydrate	[41468-25-1]	$C_8H_{10}NO_6P \cdot H_2O$	265.16	needles		sol	<2.5, 4.14, 6.20, 8.69

pyridine ring by a Knoevenagel-type condensation of cyanoacetamide with a β-diketone, followed by a sequence of functional group modifications of substituents on the ring. The following outline of the Harris-Folkers route is given as an example. Some

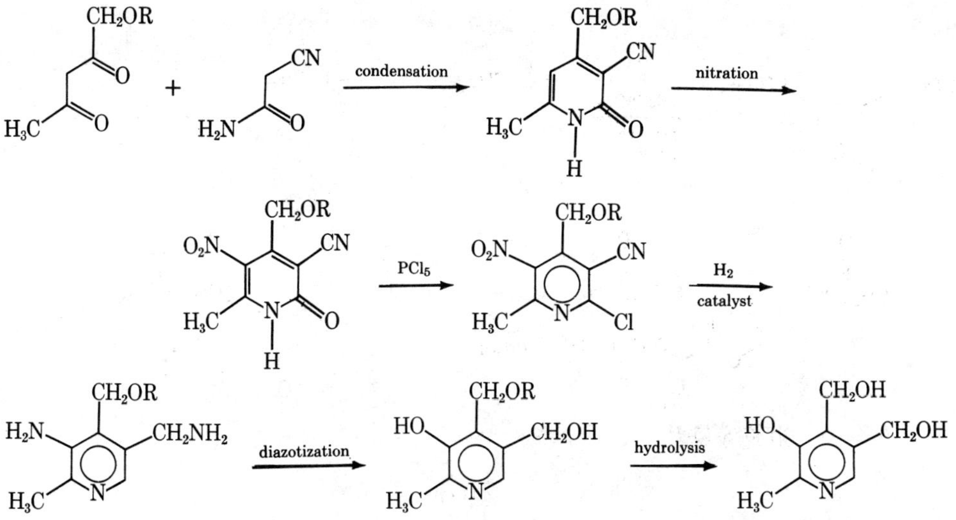

recent Japanese patents indicate that variations and refinements of this scheme are being studied, eg (28):

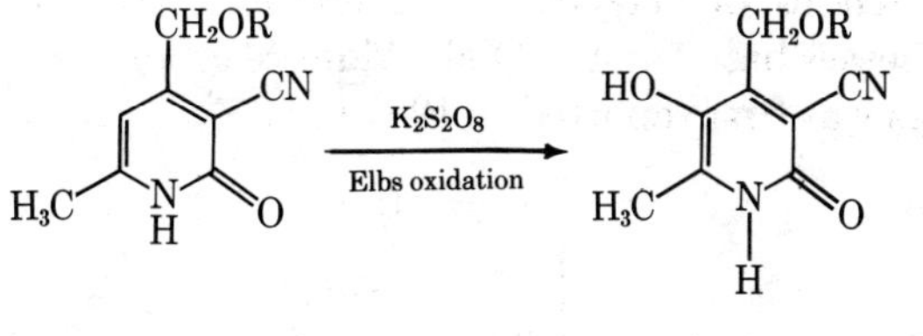

From 2-(α-Aminoethyl)Furans. 2-(α-Aminoalkyl)furans can be transformed into 3-hydroxypyridines (29). With appropriate substituents on the furan ring, the reaction sequence, which consists of oxidative addition of methanol to the ring followed by hydrolytic rearrangement, affords pyridoxine (30).

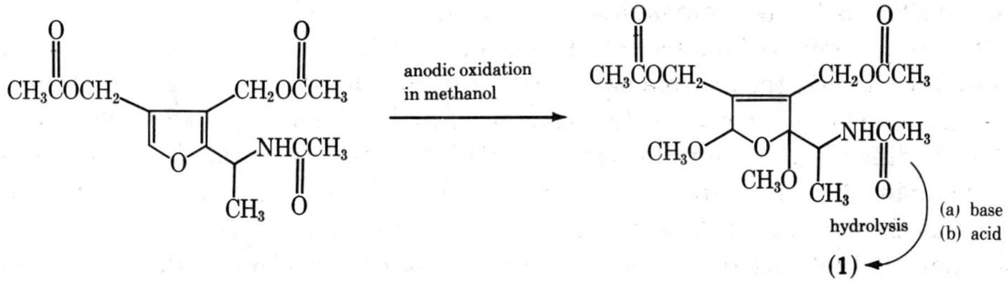

The process was never developed for technical production of pyridoxine because of the cumbersome reaction sequence required to make the necessary aminoethylfuran.

98   VITAMINS (B₆)

However, recent work in Japan showed that the α-aminoalkyl group can be introduced in a single step (31). This represents an important improvement in this category of vitamin B₆ synthesis. The starting material, 3,4-furandimethanol, is readily prepared by the Diels-Alder reaction of 2-butyne-1,4-diol and 4-methyloxazole (32).

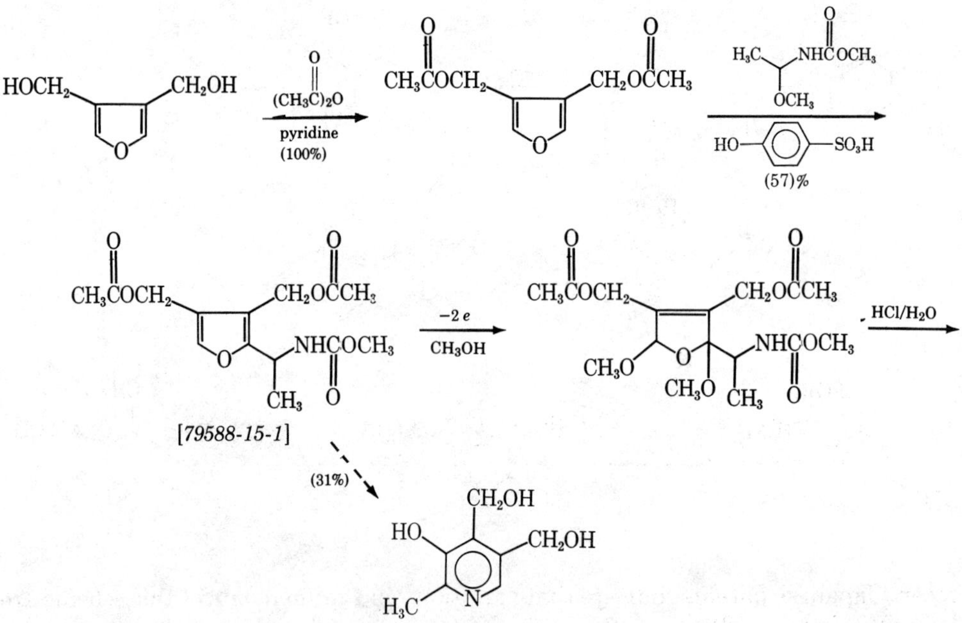

Diels-Alder Reactions of Oxazoles. Fundamental to vitamin B₆ production technology are processes based on the Diels-Alder reaction of oxazoles. This route to pyridine derivatives was first reported in 1957 (22,33).

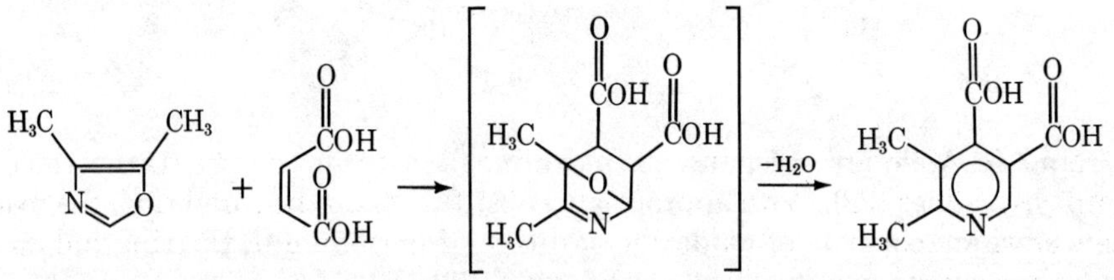

The chemistry was implemented into vitamin B₆ syntheses shortly thereafter by groups in the USSR, the United States (Merck), and the UK (Roche) (34–36).

Two requirements for a technically feasible synthesis emerged from these early studies. First is that the oxazole has a leaving group substituent at the 5-position to ensure that the bridging oxygen atom is retained in the collapse of the bicyclic adduct to the pyridine ring. Second, the substituents introduced with the dienophile must correspond in oxidation state to the hydroxymethyl groups. Otherwise, costly hydride reductions are necessary subsequent to the Diels-Alder step. These requirements are problematic in that such diene–dienophile combinations have attenuated Diels-Alder reactivity. There is an extensive patent literature regarding the development of the ideal combination of substituents and reactivity and the synthesis of the desired oxazoles.

The preferred substituents at the 5-position of the oxazole ring appear to be alkoxy and cyano (22–23,27). Ethoxy, propoxy, and trimethylsilyloxy are representative of those alkoxy groups that have been studied (37–39). The preferred dienophiles are derivatives of cis-2-butene-1,4-diol. These include the diacetate, ether derivatives including cyclic ethers, and cyclic ketals and acetals such as those formed with acetone and isobutyraldehyde (37,40–41). A typical reaction sequence is:

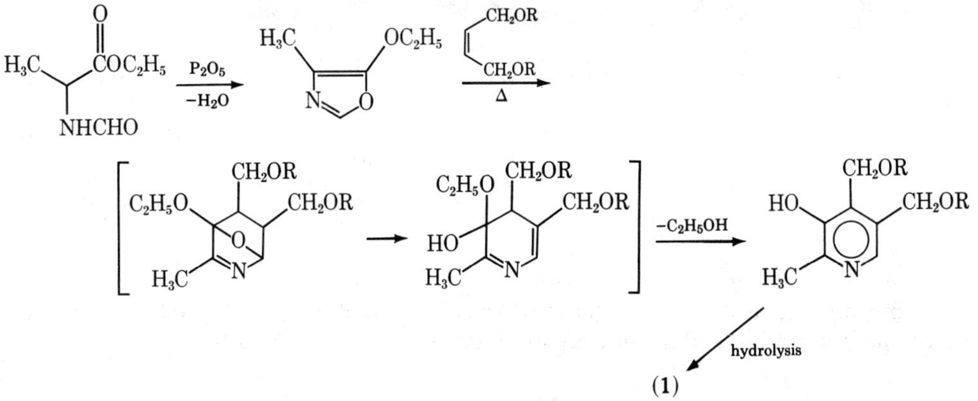

In the case where 4-methyl-5-cyanooxazole is the diene, the intermediate with a hemiacetal function is formulated as a cyanohydrin instead. Hydrogen cyanide is evolved instead of ethanol and appropriate precautions must be taken.

The cleavage of ether substituents at the 4'- and 5'-positions requires vigorous conditions. For example, one patented method involves heating at 126°C with 48 wt % hydrobromic acid. This gives the 4',5'-dibromo compound, in which the bromine atoms are subsequently displaced by acetate. Hydrolysis of the diacetate gives pyridoxine (42). The advantage of using more easily hydrolyzed derivatives of 1,4-butenediol is clear.

The solvent used for the Diels-Alder reaction must be high boiling because of the low reactivity of the 1,4-butenediol-based dienophiles. An excess of the dienophile is used as solvent in many procedures (43–44). N-Methylpyrrolidinone and sulfolane are superior solvents in the reaction with 5-alkoxyoxazoles (45). In most of the published examples, the temperatures used for the Diels-Alder reaction are 115–180°C. Reaction times of ≥ 20 h are used.

The reaction is said to be acid-catalyzed, and various organic acids including phenolic compounds have been used (44,46). In apparent contradiction to these reports, a Merck patent claims that yields of the Diels-Alder adduct are improved in the presence of acid-binding agents, eg, calcium and magnesium oxides (43).

A rather surprising catalytic effect is attributed to maleic anhydride in the following example (47):

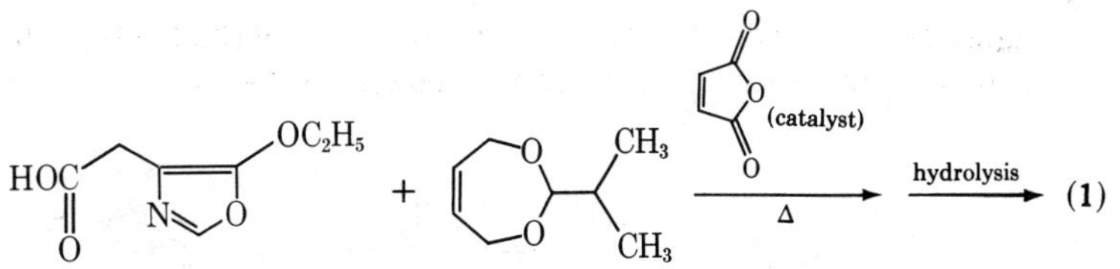

The redundant carboxylic acid group is lost by decarboxylation.

100 VITAMINS (B₆)

Variations of the general processes outlined above have been studied. In one such study, 4-methyloxazole without a leaving group in the 5-position was combined with a vinylsulfone dienophile (48). In this way, the Diels-Alder reactivity is enhanced but the oxidation state of the adduct and its substituents is not changed.

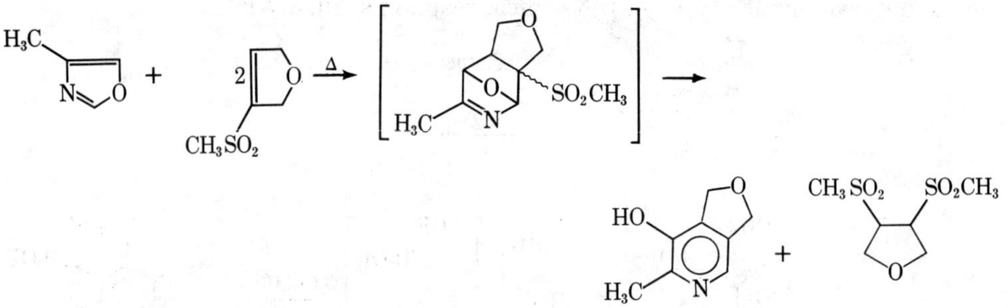

In other variations of the general process, oxazoles are not used explicitly as reactants but are instead generated and consumed *in situ*, eg (49):

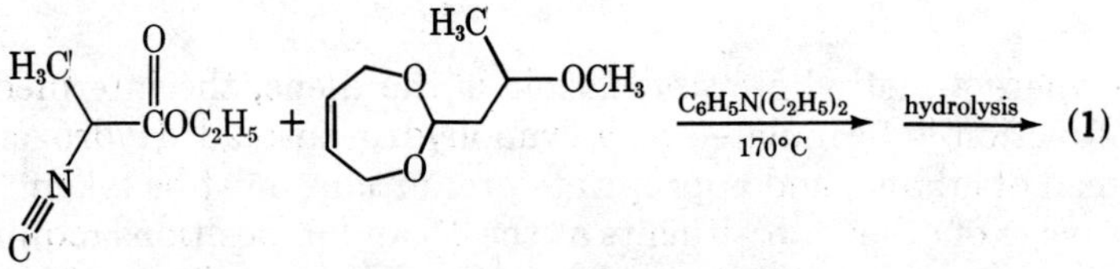

The formation of an oxazole by heating ethyl 2-isonitrilopropionate with tertiary amines is documented separately (50):

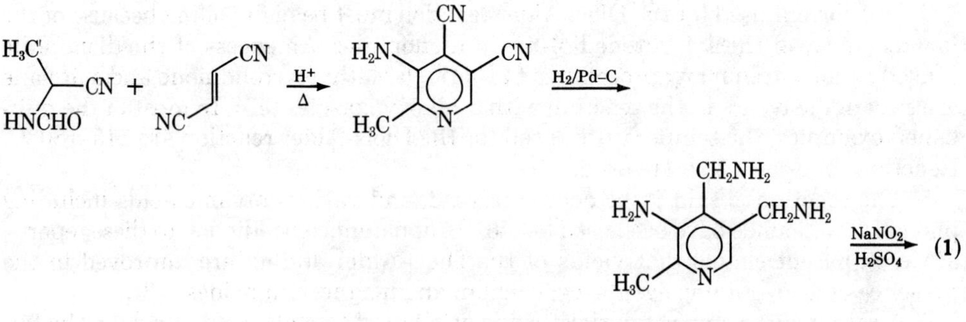

This latter example (51) appears to involve a 5-aminooxazole and, in this respect, is rather unique.

The raw materials required for pyridoxine production by the Diels-Alder reactions of oxazoles are those needed for the intermediates shown below:

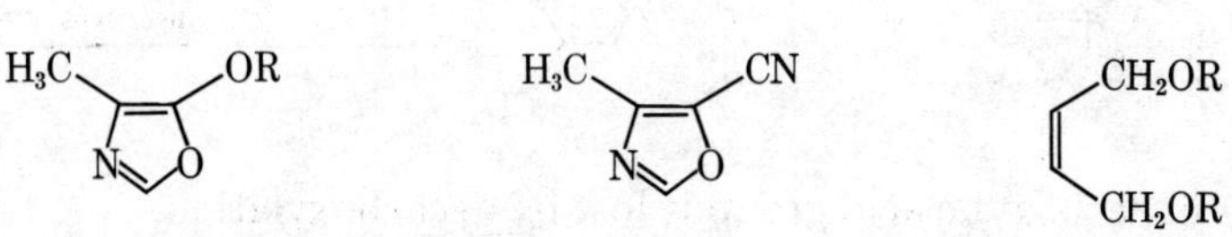

4-Methyl-5-alkoxyoxazoles are prepared from the esters or orthoesters of N-formylalanine (37). This is normally done with phosphorus pentoxide but creates a spent phosphate disposal problem. A patent for a vapor-phase catalytic process has been issued but the yield is low (52). The main raw-material requirement is for D,L-alanine, which costs (1982) $5.65/kg. The process described in refs. 41 and 47 involves the use of D,L-aspartic acid which is less expensive. 4-Methyl-5-cyanooxazole can be prepared from the corresponding ester and amide (53–54).

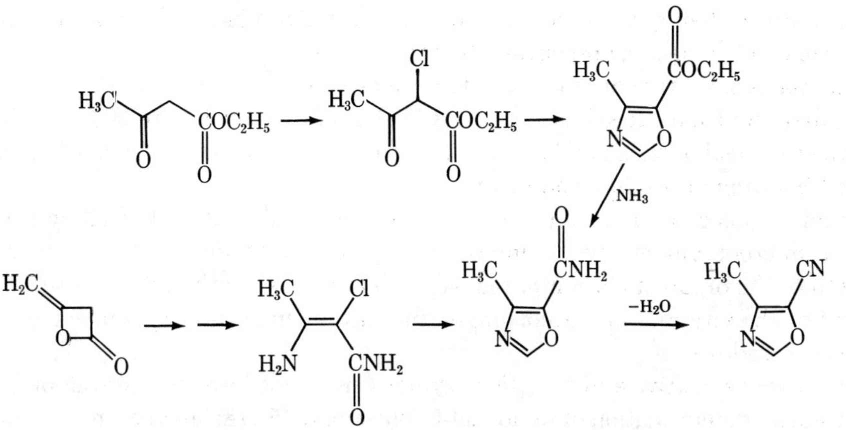

Ethyl acetoacetate ($2.12/kg, 1982) is the raw material used to make the ester in a procedure described in ref. 53. The ester is bypassed in an alternative route to the amide in which diketene ($1.35/kg, 1982) is the raw material (55).

Methods for dehydrating the amide to the nitrile are described in several patents. The earliest of these involves phosphorus pentoxide in quinoline; however, this procedure must also generate spent phosphate waste (56). Acetic anhydride is used as the dehydrating agent in another process and, in a third, the dehydration is effected catalytically in the vapor phase (57–58).

The dienophile components are derived from cis-2-butene-1,4-diol. This material is obtained from 2-butyne-1,4-diol, the condensation product of formaldehyde and acetylene (see Acetylene-derived chemicals).

Production and Economic Aspects

Pyridoxine hydrochloride is produced by four companies in the noncommunist world. They are Hoffmann-La Roche, BASF at its newly acquired Grindsted/Syntetic facility in Denmark, E. M. Merck in Darmstadt, FRG, and the Daiichi Company in Japan. The total annual production of these companies is ca 1700 metric tons, of which 400–500 t is consumed in the U.S. human and animal food and health markets.

In 1944, material produced by the Harris-Folkers-type processes was sold for $1000/kg. This price dropped by half during the next ten years, presumably reflecting competition and economy of scale in expanded production. During the next ten years, the price dropped by 90% to a low of $30/kg in 1967 as a result of the introduction of the oxazole processes. Since then, inflationary effects on labor, energy, and raw-material costs have caused a gradual upward drift in prices to the 1982 level of $47/kg.

Analytical and Test Methods

Determining the vitamin B_6 content of various foods, body fluids, etc, is very complicated for four reasons: the vitamin consists of six different chemical entities; each of them can exist in various ionic forms at different pH values; all of them are heat- and light-sensitive; and much of vitamin B_6 in a food or blood sample is bound, in part covalently, to proteins in the sample. Vitamin B_6 assay methods have been reviewed in depth in a recent monograph (1,59). The various methods that have been used, are in use, or being developed include animal, microbiological, enzymatic, fluorometric and colorimetric, and chromatographic assays.

In animal assays, rats or chicks are fed the test material with an otherwise B_6-deficient diet and their growth is compared to that of a control group. The method requires a great deal of time and large sample size, but avoids the problem of nondestructive liberation of protein-bound vitamin.

For microbiological assays, the test material is incubated with a vitamin-B_6-dependent microorganism, most commonly the yeast *Saccharomyces uvarum*, and the growth of the organism is compared with a blank. The method detects only the free forms of the vitamin and, accordingly, the sample must first be subjected to a hydrolytic procedure.

Several very sensitive and specific enzymatic methods have been developed for the quantitative determination of pyridoxal-5′-phosphate (5), the main coenzyme form of vitamin B_6. In a typical assay, the test material is incubated with a purified enzyme and a sample of its substrate. A product of the enzymic degradation is then determined quantitatively. For example, the amount of $^{14}CO_2$ evolved with L-tyrosine-1-^{14}C and L-tyrosine apodecarboxylase is a direct measure of (5) in the test sample (60).

Direct fluorometric determinations of the free forms of the vitamin are complicated by their photoinstability. Oxidative conversion to the stable pyridoxic acid (8) and its highly fluorescent lactone (9) provides the basis of a more reliable method.

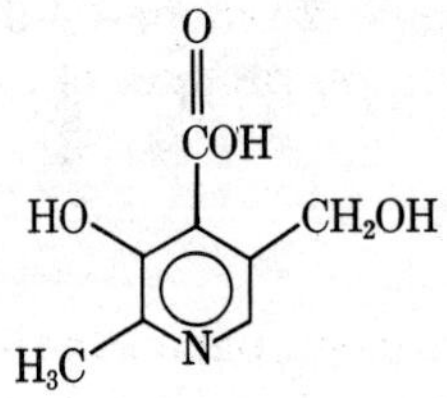

(8)
pyridoxic acid

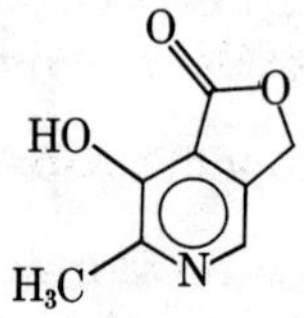

(9)

Colorimetric methods involve derivatization to stable, light-absorbing species, eg, phenylhydrazones of pyridoxal or the azo dyes formed with various diazotized aniline derivatives.

Chromatographic methods which have been used include thin-layer electrophoresis and column chromatography with ion-exchange resins and gradient-elution techniques. Development of hplc with either ion-exchange or reverse-phase column packings has resulted in the quickest assay methods available.

Nutrition

Human. *Sources and Requirements.* The vitamin B_6 content of typical U.S. diets is 1.1–3.6 mg/d (61). The approximate distribution of dietary sources of this amount is meats, 40.0%; vegetables, 22.2%; diary products, 11.6%; cereals, 10.2%; fruits, 8.2%; legumes, 5.4%; and eggs, 2.1% (62). The range for typical diets includes the recommended daily allowance of 2 mg; however, poor diets provide only 0.7–1.4 mg/d (61). Certain population groups have been identified for which vitamin B_6 requirements may be substantially higher. These include, for example, alcoholics, persons using oral contraceptives, and those with high protein intakes. Moreover, many drugs interfere with the functioning of vitamin B_6 (63).

The stability problems associated with various forms of vitamin B_6, particularly its light sensitivity, can result in a lowering of the B_6 content in foods during processing and storage. This aspect has received considerable study in relation to dairy products (64).

Biochemical Function. As enzyme cofactors, pyridoxal-5'-phosphate (**5**) and pyridoxamine-5'-phosphate (**6**) are required for several important biotransformations of amino acids (65). These include decarboxylation, transamination, deamination, racemization, and the metabolism of sulfur-containing amino acids. Pyridoxal (**2**) is also a cofactor for the phosphorylases of muscle tissue.

Decarboxylation and oxidative deamination involve the formation of a Schiff base between the amino acid and the pyridoxal moiety (see Enzymes). A di- or trivalent metal (M) also participates in nonenzymatic model reactions, but not in the corresponding enzymatic reactions:

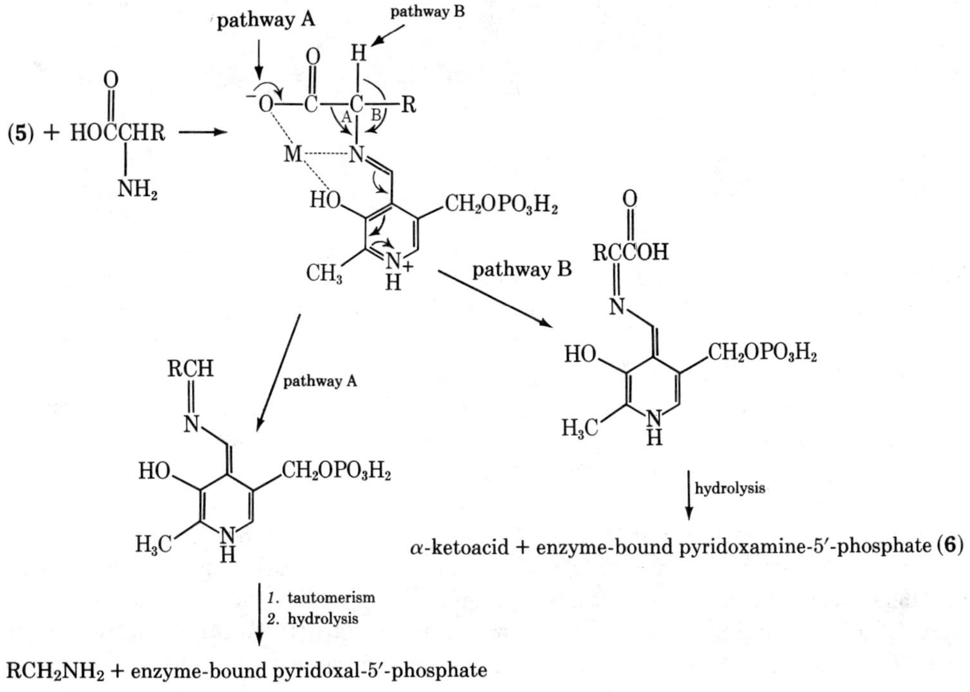

Pathway A, the decarboxylation pathway, is vital to any organism having a nervous system, since the neurotransmitters (biogenic amines) are produced in this way. These include dopamine, leading to norepinephrine and epinephrine, serotonin, tyramine, tryptamine, taurine, histamine, and GABA (γ-aminobutyric acid). For this reason, vitamin B_6 deficiency in humans is often manifested in dysfunction of the central and peripheral nervous systems.

The reversal of pathway B, the deamination pathway, indicates how both transamination and racemization occur. The latter process is important for bacteria that use D-amino acids in their cell-wall structures.

Deficiency-Related Diseases. Direct observation of the clinical consequences of vitamin B_6 deficiency in humans has not been easy for three reasons. Virtually all foods contain some B_6, and it may also be available from intestinal microflora (66). In addition, nutritionally compromised individuals may exhibit symptoms associated with deficiencies of any or all components of the vitamin B complex. However, a pattern of B_6 deficiency symptomology has emerged and the following clinical signs have been identified: eczema and seborrheic dermatitis in the ears, nose, and mouth; cheilosis, glossitis, and angular stomatitis; hypochromic and microcytic anemia; central nervous system changes, eg, hyperirritability, convulsive seizures, and hyperacusis; and abnormal electroencephalograms (67).

Possible vitamin B_6 deficiency in an individual can be assessed in various ways, of which the tryptophan load test is both commonly used and typical (68). The patient is fed tryptophan (50 mg/kg) and the level of urinary metabolites emanating from the non-B_6 mediated portion of the kynurenine metabolic pathway is interpreted as a measure of the B_6 levels.

Biosynthesis and Metabolism. The biosynthesis of pyridoxine in an *E. coli* strain has been studied (16,69). Three glycerol units make up the carbon skeleton; two are intact and a third is incorporated as acetaldehyde which forms via pyruvic acid.

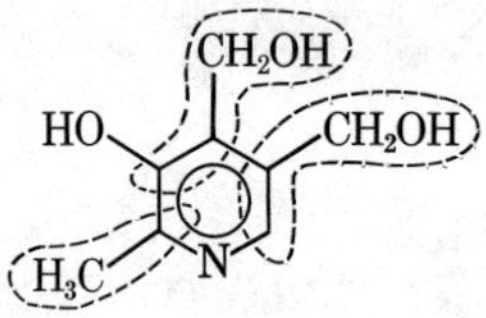

Subsequent studies provided further refinement of the biosynthetic pathway (70).

Humans are incapable of vitamin B_6 synthesis and must obtain it from dietary sources. Two important aspects of human B_6 metabolism are interconversion and clearance. Although humans are not biochemically equipped to produce any of the vitamin forms, human metabolism does carry out interconversions of the six forms. Thus, required levels of the enzyme cofactor forms (5) and (6) can be maintained by ingestion of, for example, only pyridoxine (1).

Clearance of vitamin B_6 is renal. Small amounts of forms (1)–(6) are excreted into the urine but the main urinary metabolite is 4-pyridoxic acid (8).

Animal Health. The importance of vitamin B_6 in the nutrition of farm animals and pets was reviewed in 1964 (71). Supplemental pyridoxine should be added to the feeds of all nonruminants to ensure their well being, growth, and productivity (71) (see Pet and livestock feeds). This conclusion was offered at a time when the price of pyridoxine (1) was dropping dramatically and, as a result, the addition of pyridoxine (hydrochloride) to commercial feeds became a common practice during the next ten years (72).

Toxicity

The toxicology of vitamin B_6 was reviewed in 1978 (2). The vitamin is relatively nontoxic and excess amounts in the body are readily removed by renal clearance. The acute toxicity of pyridoxine hydrochloride is 6000 mg/kg (oral) and 700 mg/kg (intravenous) in mice, and 3700 mg/kg (subcutaneous) in rats. No chronic toxicity was observed in dogs and rats at doses of 20 and 25 mg/kg per day, nor were teratogenic effects observed in rats at doses up to 80 mg/kg per day. Damage to the nervous system resulted in dogs fed 1000 mg/kg per day for several days; this is equivalent to 70,000 mg/d for a man. Vitamin B_6 administration in humans, either as a diet supplement or in therapeutic doses, rarely exceeds 50 mg/d.

BIBLIOGRAPHY

"Pyridoxine, Pyridoxal, and Pyridoxamine" in *ECT* 1st ed., Vol. 11, pp. 293–307, by S. A. Harris, G. E. Sita, and P. G. Stecher, Merck & Co., Inc.; "Pyridoxine" in *ECT* 2nd ed., Vol. 16, pp. 806–824, by S. A. Harris, E. E. Harris, and R. W. Burg, Merck & Co., Inc.

1. E. E. Snell in J. E. Leklem and R. D. Reynolds, eds., *Methods in Vitamin B-6 Nutrition, Analysis and Status Assessment*, Plenum Press, New York, London, 1981, pp. 1–19.
2. M. Brin in *Human Vitamin B_6 Requirements*, National Academy of Sciences, Printing and Publishing Office, Washington, D.C., 1978, pp. 1–20.
3. E. E. Snell in M. Florkin and E. H. Stotz, eds., *Comprehensive Biochemistry*, Vol. 11, Elsevier Publishing Co., New York, 1963, pp. 48–58.
4. J. Goldberger and R. D. Lillie, *U.S. Public Health Reports* **41**, 1025 (1926); *Chem. Abstr.* **20**, 2693 (1926).
5. P. György, *Nature (London)* **133**, 498 (1934); *Biochem. J.* **29**, 741 (1935).
6. S. Lepkovsky, *Science* **87**, 169 (1938); *J. Biol. Chem.* **124**, 125 (1938); J. C. Keresztesy and J. R. Stevens, *Proc. Soc. Exp. Biol. Med.* **38**, 64 (1938); *J. Am. Chem. Soc.* **60**, 1267 (1938); P. György, *J. Am. Chem. Soc.* **60**, 983 (1938); R. Kuhn and G. Wendt, *Chem. Ber.* **71**, 780 (1938); A. Ichiba and K. Michi, *Sci. Pap. Inst. Phys. Chem. Res. Jpn.* **34**, 623 (1938); *Chem. Abstr.* **32**, 7534 (1938).
7. S. A. Harris and K. Folkers, *J. Am. Chem. Soc.* **61**, 1245 (1939).
8. R. Kuhn, K. Westphal, G. Wendt, and O. Westphal, *Naturwissenschaften* **27**, 469 (1939); *Chem. Abstr.* **33**, 8201 (1939).
9. S. Otake, *J. Agric. Chem. Soc. Jpn.* **7**, 775 (1931); *Chem. Abstr.* **26**, 1323 (1932). The formula cited in the abstract is $C_8H_{10}NO_3 \cdot HCl$. $C_8H_{11}NO_3 \cdot HCl$ is correct.
10. E. E. Snell, B. M. Guirard, and R. J. Williams, *J. Biol. Chem.* **143**, 519 (1942); E. E. Snell, *J. Am. Chem. Soc.* **66**, 2082 (1944); E. E. Snell, *J. Biol. Chem.* **154**, 313 (1944); J. C. Rabinowitz and E. E. Snell, *J. Biol. Chem.* **169**, 631, 643 (1947); E. E. Snell and A. N. Rannefeld, *J. Biol. Chem.* **157**, 475 (1945); E. E. Snell, *J. Biol. Chem.* **157**, 491 (1945).
11. S. A. Harris, D. Heyl, and K. Folkers, *J. Biol. Chem.* **154**, 315 (1944); *J. Am. Chem. Soc.* **66**, 2088 (1944); D. Heyl, E. Luz, S. A. Harris, and K. Folkers, *J. Am. Chem. Soc.* **73**, 3430 (1951); E. A. Peterson and H. A. Sober, *J. Am. Chem. Soc.* **76**, 169 (1954); H. Wada, T. Morisue, Y. Nishimura, Y. Morino, Y. Sakamoto, and K. Ichihara, *Proc. Jpn. Acad.* **35**, 299 (1959); *Chem. Abstr.* **54**, 10089 (1960).
12. *J. Am. Chem. Soc.* **82**, 5575 (1960).
13. *Biochem. J.* **137**, 417 (1974).
14. H. Lieck and H. Sondergaard, *Int. Z. Vitaminforsch.* **29**, 68 (1958).
15. K. D. Lunan and C. A. West, *Arch. Biochem. Biophys.* **101**, 261 (1963); G. H. Scherr and M. E. Rafelson, *J. Appl. Bacteriol.* **25**, 187 (1962).

16. W. B. Dempsey in G. P. Tryfiates, ed., *Vitamin B$_6$ Metabolism and Role in Growth*, Food and Nutrition Press Inc., Westport, Conn., 1980, pp. 93–111.
17. Y. Yamada and K. Watanabe, *Agric. Biol. Chem.* **44,** 2683 (1980).
18. S. A. Harris, *J. Am. Chem. Soc.* **63,** 3363 (1941).
19. M. Iwata, *Bull. Chem. Soc. Jpn.* **54,** 2835 (1981), and literature cited therein.
20. P. Karrer, M. Viscontini, and O. Forster, *Helv. Chim. Acta* **31,** 1004 (1948).
21. J. M. Osbond in R. S. Harris, I. G. Wool, and J. A. Loraine, eds., *Vitamins and Hormones*, Vol. 22, Academic Press, New York, 1964, pp. 367–397.
22. M. Ya. Karpeiskii and V. L. Florent'ev, *Russ. Chem. Rev.* **38,** 540 (1969).
23. R. Lakhan and B. Termai in A. R. Katritzky and A. J. Boulton, eds., *Advances in Heterocyclic Chemistry*, Vol. 17, Academic Press, New York, 1974, p. 182.
24. I. J. Turchi and M. J. Dewar, *Chem. Rev.* **75,** 416 (1975).
25. H. König and W. Böll, *Chem. Ztg.* **100,** 105 (1976).
26. H. König in Z. Yoshida, ed., *New Synthetic Methodology and Biologically Active Substances, Studies in Organic Chemistry 6*, Kodansha, Ltd., Tokyo, Elsevier Science Publishing Co., New York, 1981, pp. 201–221.
27. C. S. Sodano, *Vitamins Synthesis, Production and Use, Advances Since 1970*, Noyes Data Corp., Park Ridge, N.J., 1978.
28. Jpn. Pats. 72 41,901 (Oct. 23, 1972); 73 24,395 (July 20, 1973); 74 07,157 (Feb. 19, 1974); 74 11,717 (March 19, 1974); 74 11,718 (March 19, 1974); 75 10,874 (April 24, 1975); 79 32,784 (1979); and 79 21,347 (1979), E. Matsumura.
29. N. Clauson-Kaas and co-workers, *Acta Chem. Scand.* **9,** 1, 9, 14, 17 (1955).
30. N. Elming and N. Clauson-Kaas, *Acta Chem. Scand.* **9,** 23 (1955).
31. T. Shono, Y. Matsumura, K. Tsubata, and J. Takata, *Chem. Lett.*, 1121 (1981).
32. Ger. Pat. 1,935,009 (Jan. 14, 1971), F. Graf and H. König (to BASF); *Chem. Abstr.* **74,** 64201 (1971).
33. G. Ya. Kondrat'eva, *Khim. Nauka Promyst.* **2,** 666 (1957); *Izv. Akad. Nauk SSSR, Otd. Khim. Nauk*, 484 (1959); *Chem. Abstr.* **53,** 21940 (1959).
34. C. H. Huang and G. Ya. Kondrat'eva, *Izv. Akad. Nauk SSSR, Otd. Khim. Nauk*, 525 (1962); *Chem. Abstr.* **57,** 15064 (1962).
35. E. E. Harris and co-workers, *J. Org. Chem.* **27,** 2705 (1962).
36. Belg. Pat. 626,620 (1962); Brit. Pats. 971,361 (1962) and 1,034,483 (1963); Fr. Pat. 1,343,270 (Nov. 15, 1963), (to Hoffmann-La Roche); *Chem. Abstr.* **60,** 11991 (1964).
37. Jpn. Pats. 73 15,949 (May 18, 1973); 73 30,636 (Sept. 21, 1973); 74 55,663 (May 30, 1974), (to Daiichi Seiyaku Co.); *Chem. Abstr.* **79,** 53287 (1973); **80,** 82700 (1974); **82,** 43393 (1975).
38. S. D. L'vova, Z. I. Itov, and V. I. Gunar, *Khim. Farm. Zh.* **12,** 102, 106 (1978); Z. I. Itov, S. D. L'vova, and V. I. Gunar, *Khim. Farm. Zh.* **13,** 59 (1979); *Chem. Abstr.* **92,** 110750 (1980).
39. H. Takagaki, N. Yasuda, M. Asaoka, and H. Takei, *Chem. Lett.*, 183 (1979); *Am. Chem. Soc. National Meeting*, Hawaii, 1979, Org. Abs. No. 28.
40. R. A. Firestone, E. E. Harris, and W. Reuter, *Tetrahedron* **23,** 943 (1967).
41. Jpn. Pat. 72 39,115 (1972), (to Takeda Chemical Industries, Ltd.); *Chem. Abstr.* **78,** 4293 (1973); T. Matsuo and T. Miki, *Chem. Pharm. Bull. Jpn.* **20,** 669, 806 (1972); U.S. Pat. 3,525,749 (Aug. 25, 1970), E. M. Chamberlin, E. E. Harris, and J. L. Zabriskie (to Merck & Co., Inc.); Fr. Pat. 1,384,099 (1965), W. Kimel and W. Leimgruber (to Hoffmann-La Roche); *Chem. Abstr.* **63,** 4263 (1965).
42. Jpn. Pat. 72 34,713 (1972), (to Daiichi Seiyaku Co.); *Chem. Abstr.* **78,** 43281 (1973); U.S. Pat. 4,144,239 (March 13, 1979), W. Böll and H. König (to BASF).
43. U.S. Pat. 3,822,274 (July 2, 1974), E. E. Harris and R. Currie (to Merck & Co., Inc.).
44. U.S. Pat. 3,250,778 (May 10, 1966), W. Kimel and W. Leimgruber (to Hoffmann-La Roche).
45. Ger. Pat. 2,218,739 (1972) and U.S. Pats. 3,796,720 (March 12, 1974), H. Miki and H. Saikawa (to Takeda Chemical Industries, Ltd.); *Chem Abstr.* **78,** 43288 (1973).
46. D. Szlompek-Nesteruk and co-workers, *Przem. Chem.* **54,** 238 (1975); Pol. Pat. 79,206 (Oct. 10, 1975); *Chem. Abstr.* **86,** 29644 (1977).
47. Jpn. Pat. 81 99,461 (Aug. 10, 1981), (to Takeda Chemical Industries, Ltd.); Ger. Pat. 3,100,502 (1981); *Chem. Abstr.* **96,** 20120 (1982).
48. W. Böll and H. König, *Liebigs Ann. Chem.*, 1657 (1979); Ger. Pat. 2,143,989 (1973) and U.S. Pat. 3,876,649 (April 8, 1975), W. Böll and H. König (to BASF); *Chem. Abstr.* **78,** 147815 (1973).
49. Jpn. Pat. 73 39,938 (Nov. 28, 1973), (to Daiichi Seiyaku Co.); *Chem. Abstr.* **80,** 120774 (1974).
50. Jpn. Pat. 79 20,493 (July 23, 1979), (to Daiichi Seiyaku Co.); *Chem. Abstr.* **91,** 157722 (1979).
51. Jpn. Pat. 81 128,761 (Oct. 8, 1981), (to Daiichi Seiyaku Co.); *Chem. Abstr.* **96,** 122530 (1982).

52. Jpn. Pat. 79 01,308 (Jan. 23, 1979), (to Daiichi Seiyaku Co.); *Chem. Abstr.* **90,** 204077 (1979).
53. J. W. Cornforth and R. H. Cornforth, *J. Chem. Soc.*, 93 (1953).
54. T. Rinderspacher and B. Prijs, *Helv. Chim. Acta* **43,** 1522 (1960).
55. U.S. Pat. 4,093,654 (June 6, 1978), D. L. Coffen (to Hoffmann-La Roche).
56. U.S. Pat. 3,222,374 (Dec. 7, 1965), G. O. Chase (to Hoffmann-La Roche).
57. U.S. Pat. 4,011,234 (March 8, 1977), H. Hoffmann-Paquotte (to Hoffmann-La Roche).
58. U.S. Pat. 4,026,901 (May 31, 1977), D. L. Coffen (to Hoffmann-La Roche).
59. B. E. Haskell in ref. 2, pp. 61–71.
60. H. N. Bhagavan, J. M. Koogler, and D. B. Coursin, *Int. J. Vit. Nutr. Res.* **46,** 160 (1976).
61. A. S. Mangay Chung, W. N. Pearson, W. J. Darby, O. N. Miller, and G. A. Goldsmith, *Am. J. Clin. Nutr.* **9,** 573 (1961).
62. H. E. Sauberlich in ref. 1, p. 214.
63. J. C. Bauernfeind and O. N. Miller in ref. 2, p. 94.
64. J. F. Gregory and J. R. Kirk in ref. 2, pp. 72–77.
65. E. E. Snell, P. M. Fasella, A. Braunstein, and A. Rossi Fanelli, eds., *Chemical and Biological Aspects of Pyridoxal Catalysis, International Union of Biochemistry Symposium 30*, The Macmillan Co., New York, 1963; see also M. Ebadi in ref. 2, pp. 129–161.
66. Ref. 2, p. 8.
67. Ref. 62, p. 213.
68. R. R. Brown in ref. 1, pp. 321–340.
69. R. E. Hill and I. D. Spenser, *Science* **169,** 773 (1970); R. E. Hill, I. Miura, and I. D. Spenser, *J. Am. Chem. Soc.* **99,** 4179 (1977).
70. G. J. Vella, R. E. Hill, B. S. Mootoo, and I. D. Spenser, *J. Biol. Chem.* **255,** 3042 (1980); G. J. Vella, R. E. Hill, and I. D. Spenser, *J. Biol. Chem.* **256,** 10469 (1981).
71. H. L. Fuller in ref. 21, pp. 659–676.
72. J. C. Bauernfeind, *Feedstuffs* **46,** 30 (1974).

DAVID L. COFFEN
Hoffmann-La Roche, Inc.

RIBOFLAVIN (B₂)

Riboflavin [83-88-5] (**1**) (vitamin B₂, vitamin G, lactoflavin, ovoflavin, lyochrome, hepatoflavin, uroflavin) has the chemical name 7,8-dimethyl-10-D-ribitylisoalloxazine, $C_{17}H_{20}N_4O_6$, mol wt 376.37.

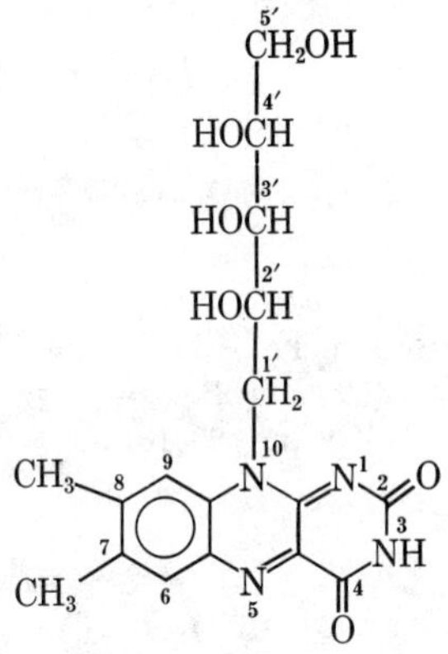

riboflavin
(**1**)

In 1933, Kuhn and co-workers first isolated riboflavin from eggs in a pure, crystalline state (1), named it ovoflavin, and determined its function as a vitamin (2). At the same time, impure crystalline preparations of riboflavin were isolated from whey and named lyochrome and, later, lactoflavin. Soon thereafter, Karrer and co-workers isolated riboflavin from a wide variety of animal organs and vegetable sources and named it hepatoflavin (3). Ovoflavin from eggs, lactoflavin from milk, and hepatoflavin from liver were all subsequently identified as riboflavin. The discovery of the yellow enzyme by Warburg and Christian in 1932 and their description of lumiflavin (4), a photochemical degradation product of riboflavin, was of great use for the elucidation of the chemical structure of riboflavin by Kuhn and co-workers (5). The structure was confirmed in 1935 by the synthesis by Karrer and co-workers (6) and Kuhn and Weygand (7). For therapeutic use, riboflavin is produced by chemical synthesis, whereas concentrates for poultry and livestock feeds are produced by fermentation using microorganisms, such as *Ashbya gossypii* and *Eremothecium ashbyii*, which have capacities to synthesize large quantities of riboflavin.

In the free form, riboflavin occurs in the retina of the eye, whey, and urine. Principally, however, riboflavin fulfills its metabolic function in a complex form. In general, riboflavin is converted into flavin mononucleotide (FMN, riboflavin-5′-phosphate) and flavin-adenine dinucleotide (FAD) (**2**), which serve as the prosthetic groups (coenzymes), ie, they combine with specific proteins (apoenzymes) to form flavoenzymes, in a series of oxidation–reduction catalysts widely distributed in nature. In several riboflavin coenzymes, the apoenzyme is covalently attached to C-8α through a linkage to the nitrogen of a histidine imidazolyl group or to the sulfur of a cysteine residue (see Amino acids). Riboflavin is not a nucleotide, since it is derived from D-ribitol rather than D-ribose, and therefore FMN and FAD are not truly nucleotides; yet this designation has been accepted overwhelmingly and continues to be used.

As a coenzyme component in tissue oxidation–reduction and respiration, riboflavin is distributed in some degree in virtually all naturally occurring foods. Liver, heart, kidney, milk, eggs, lean meats, and fresh leafy vegetables are particularly good sources of riboflavin (see Table 1). It does not seem to have long stability in food products (8).

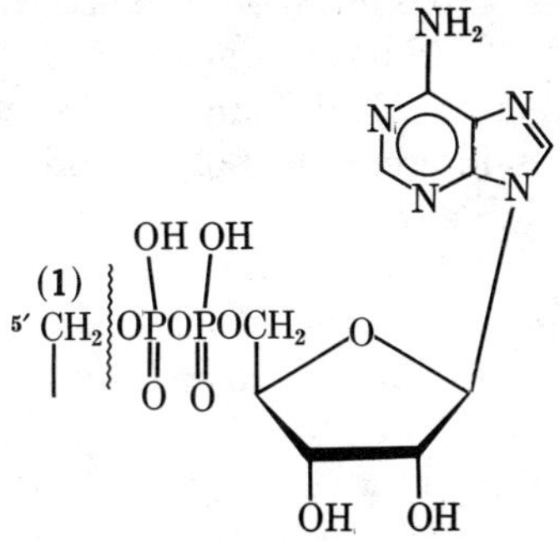

$$\text{FAD}\ (2)$$

Riboflavin is widely used in the pharmaceutical, food-enrichment, and feed-supplement industries. Riboflavin USP is administered orally in tablets or by injection

Table 1. Riboflavin Content of Various Foods[a]

Food	mg/100 g	Food	mg/100 g
fruits		fish	
apple, raw	0.01	cod, haddock, raw	0.17
banana, raw	0.04	salmon, canned	0.12
citrus, grapefruit, orange	0.03–0.04	tuna, canned	0.13
strawberry	0.03	whitefish, herring, halibut	0.17–0.29
vegetables		grain	
broccoli, raw	0.27	wheat, entire	0.10
cabbage, raw	0.05	wheat, germ	0.6–0.8
fresh green peas	0.14	rice, entire	0.06
mushroom	0.57	rye, entire	0.20
parsley	0.24	cereal products	
potato, raw	0.03	refined	
sweet corn	0.14	bread	0.07–0.10
sweet potato, raw	0.05	cereal	0.10
tomato, raw	0.03	soda cracker	0.10
meat		whole grain and enriched	
beef muscle	0.16–0.32	bread	0.12
pork muscle	0.19–0.33	cereal	0.20
chicken muscle	0.10–0.28	dairy products	
liver, beef, pork	3.00–3.60	cheese, cheddar	0.54
		eggs	0.48
		milk	0.15–0.17

[a] Averages from several sources, often of a wide range of analytical results; should be regarded as working estimates which vary with geography, season, and preparative method.

as an aqueous solution, which may contain nicotinamide or other solubilizers. As a supplement to animal feeds, riboflavin is usually added at concentrations of 2–8 mg/kg, depending on the species and age of the animal (see Pet and livestock feeds).

Properties

Riboflavin forms fine yellow to orange-yellow needles with a bitter taste. It melts with decomposition at 278–279°C (darkens at ca 240°C). The solubility of riboflavin in water is 10–13 mg/100 mL at 25–27.5°C, and in absolute ethanol 4.5 mg/100 mL at 27.5°C; it is slightly soluble in amyl alcohol, cyclohexanol, benzyl alcohol, amyl acetate, and phenol, but insoluble in ether, chloroform, acetone, and benzene. It is very soluble in dilute alkali, but these solutions are unstable. Various polymorphic crystalline forms of riboflavin exhibit variations in physical properties. In aqueous nicotinamide solution at pH 5, solubility increases from 0.1 to 2.5% as the nicotinamide concentration increases from 5 to 50% (9).

In aqueous solution, riboflavin has absorption maxima at ca 220–225, 266, 371, 444, and 475 nm. Neutral aqueous solutions of riboflavin have a greenish-yellow color and an intense yellowish-green fluorescence with a maximum at ca 530 nm and a quantum yield of $\Phi_f = 0.25$ at pH 2.6 (10). Fluorescence disappears upon the addition of acid or alkali. The fluorescence is used in quantitative determinations. The optical activity of riboflavin in neutral and acid solutions is $[\alpha]_D^{20} = +56.5-59.5°$ (0.5%, dil HCl). In an alkaline solution, it depends upon the concentration, eg, $[\alpha]_D^{25} = -112-122°$ (50 mg in 2 mL 0.1 N alcoholic NaOH diluted to 10 mL with water). Borate-containing solutions are strongly dextrorotatory, because borate complexes with the ribityl side chain of riboflavin; $[\alpha]_D^{20} = +340°$ (pH 12).

Photochemical decomposition of riboflavin in neutral or acid solution gives lumichrome (**3**), 7,8-dimethylalloxazine, which was synthesized and characterized by Karrer and co-workers in 1934 (11). In alkaline solution, the irradiation product is lumiflavin (**4**), 7,8,10-trimethylisoalloxazine; its uv–vis absorption spectrum resembles that of riboflavin. It was prepared and characterized in 1933 (5). Another photodecomposition product of riboflavin is 7,8-dimethyl-10-formylmethylisoalloxazine (12).

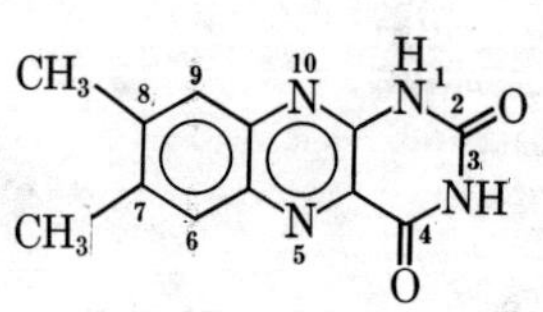

lumichrome
(**3**)

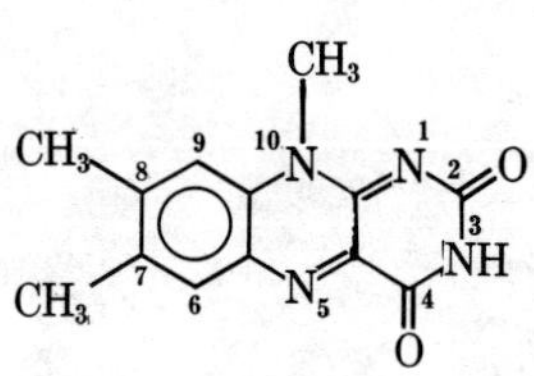

lumiflavin
(**4**)

Riboflavin is stable against acids, air, and common oxidizing agents such as bromine and nitrous acid (except chromic acid, $KMnO_4$, and potassium persulfate). Upon reduction by conventional agents such as sodium dithionite ($Na_2S_2O_4$), zinc in acidic solution, or catalytically activated hydrogen, riboflavin readily takes up two hydrogen atoms to form the almost colorless 1,5-dihydroriboflavin (**5**) (see Fig. 1), which is reoxidized by shaking with air. This oxidation–reduction system has con-

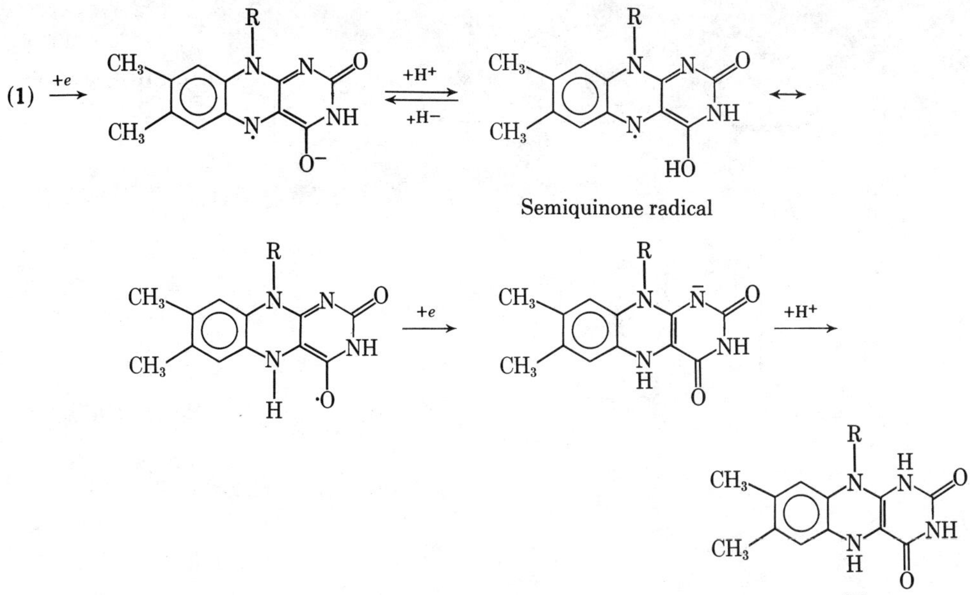

Figure 1. Formation of dihydroriboflavin. R = D-ribityl; (5) 1,5-Dihydroriboflavin.

siderable stability, a normal potential of −0.208 V (referred to as the normal hydrogen electrode), and is probably responsible for the physiological functions of riboflavin. In acidic solution (pH 1), flavins are reduced to dihydroflavins through intermediate semiquinone radicals (13–14). These radicals have been directly observed by esr (15–17).

Riboflavin forms a deep-red silver salt (1). The strong bathochromic shift of the spectra of riboflavin analogues occurring by interaction with Ag$^+$ can also be obtained with Cu$^+$ and Hg^{2+} complexes (18). These complexes contain the flavin and the metal ligand anion in a ratio of 1:1. Their color is probably due to a charge transfer between the metal and the flavin (19). The chelates with Fe(II/III), Mo(V/VI), Cu(I/II), and Ag(I/II) belong to this group; the last two are stable in the presence of water. Another group of metal complexes, radical chelates, are formed with Mn(II), Fe(II), Co(II), Ni(II), Zn(II), and Cd(II); in these cases, the radical character of the ligand is conserved (20).

Chemical Synthesis

In 1935, Karrer (6) and Kuhn (7) independently proved that riboflavin was 7,8-dimethyl-10-D-ribitylisoalloxazine by total synthesis (see Fig. 2). These syntheses are essentially the same and involve a condensation of 6-D-ribitylamino-3,4-xylidine (6) with alloxan (7) in acid solution. Boric acid as a catalyst increases the yield considerably (21). The intermediate (6) was prepared by a condensation of 6-nitro-3,4-

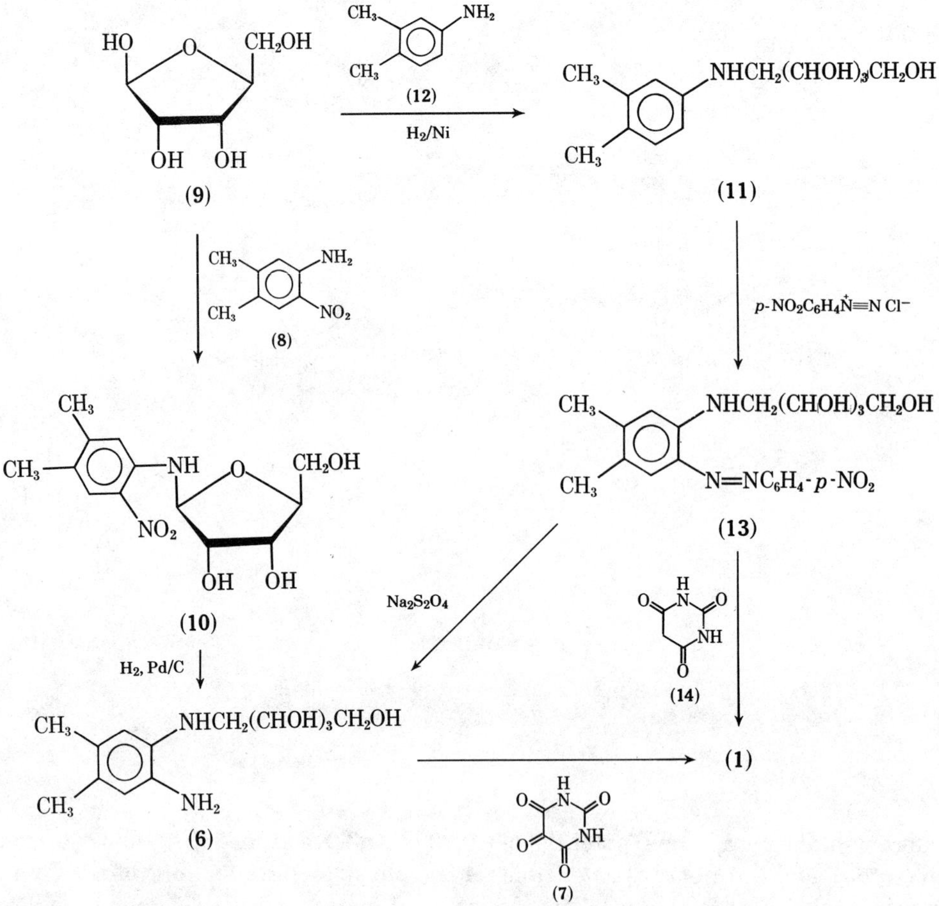

VITAMINS (B₂)

Figure 2. Syntheses of riboflavin.

xylidine (8) with D-ribose (9), followed by catalytic reduction of the riboside (10). The yield based on D-ribose was increased (22) by using N-D-ribityl-3,4-xylidine (11), which was prepared by the condensation of 3,4-xylidine (12) with D-ribose (9), and, followed by catalytic reduction, the reduced product was coupled with p-nitrophenyldiazonium salt to give 1-D-ribitylamino-2-p-nitrophenylazo-4,5-dimethylbenzene (13), which is reduced to (6) and treated with alloxan (23) to give riboflavin. Replacement of alloxan by 5,5-dichlorobarbituric acid (24), 5,5-dibromobarbituric acid, or 5-bromobarbituric acid (25) in the above syntheses yields riboflavin.

More conveniently, compound (13) was directly condensed with barbituric acid (14) in acetic acid (26). The same azo dye intermediate (13) and alloxantin give riboflavin in the presence of palladium on charcoal in alcoholic hydrochloric acid under nitrogen. This reaction may involve the reduction of the azo group to the o-phenylenediamine by the alloxantin which is dehydrogenated to alloxan (see Uric acid) (27).

Although not suitable for large-scale manufacture, the synthesis of riboflavin from lumazine derivatives is interesting with regard to the biosynthesis of riboflavin.

Thus, 5-amino-6-D-ribitylaminouracil (15) was condensed with a dimeric or trimeric aldol of biacetyl to give riboflavin (1) through the formation of intermediary 6,7-dimethyl-8-D-ribityllumazine (16) (28). A variation of the above synthesis involves the condensation of monomeric biacetyl and preformed (16) prepared by the condensation of (15) with biacetyl (29). The condensation of (15) with 4,5-dimethyl-1,2-benzoquinone (17) is another pathway to riboflavin, although in low yield (30).

Recently, a completely different and convenient synthesis of riboflavin and analogues was developed (31). It consists of the nitrosative cyclization of 6-(N-D-ribityl-3,4-xylidino)uracil (18) obtained from the condensation of N-D-ribityl-3,4-xylidine (11) and 6-chlorouracil (19) with excess sodium nitrite in acetic acid, or the cyclization of (18) with potassium nitrate in acetic acid in the presence of sulfuric acid, to give riboflavin-5-oxide (20) in high yield. Reduction with sodium dithionite gives (1). In another synthesis, 5-nitro-6-(N-D-ribityl-3,4-xylidino)uracil (21), prepared in situ from the condensation of 6-chloro-5-nitrouracil (22) with N-D-ribityl-3,4-xylidine (11), was hydrogenated over palladium on charcoal in acetic acid. The filtrate included 5-amino-6-(N-D-ribityl-3,4-xylidino)uracil (23) and was maintained at room temperature to precipitate (1) by autoxidation (32). These two pathways are suitable for the preparation of riboflavin analogues possessing several substituents.

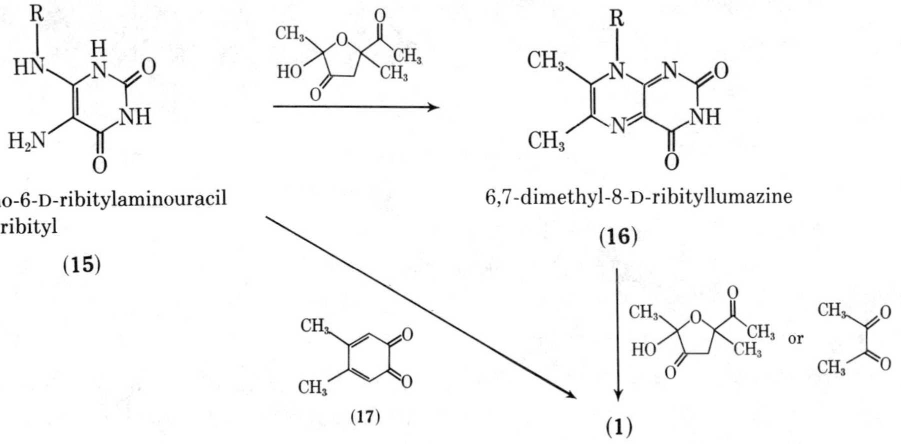

5-amino-6-D-ribitylaminouracil
R = D-ribityl
(15)

6,7-dimethyl-8-D-ribityllumazine
(16)

(17)

(1)

The chemistry of flavins, including several synthetic methods for the preparation of N-D-ribityl-3,4-xylidine (11) is reviewed in ref. 33.

Microbial Synthesis

Biosynthesis. Riboflavin is produced by many microorganisms, including

Ashbya gossypii
Asperigillus sp
Eremothecium ashbyii
Candida yeasts
Debaryomyces yeasts

Hansenula yeasts
Pichia yeasts
Azotobactor sp
Clostridium sp
Bacillus sp

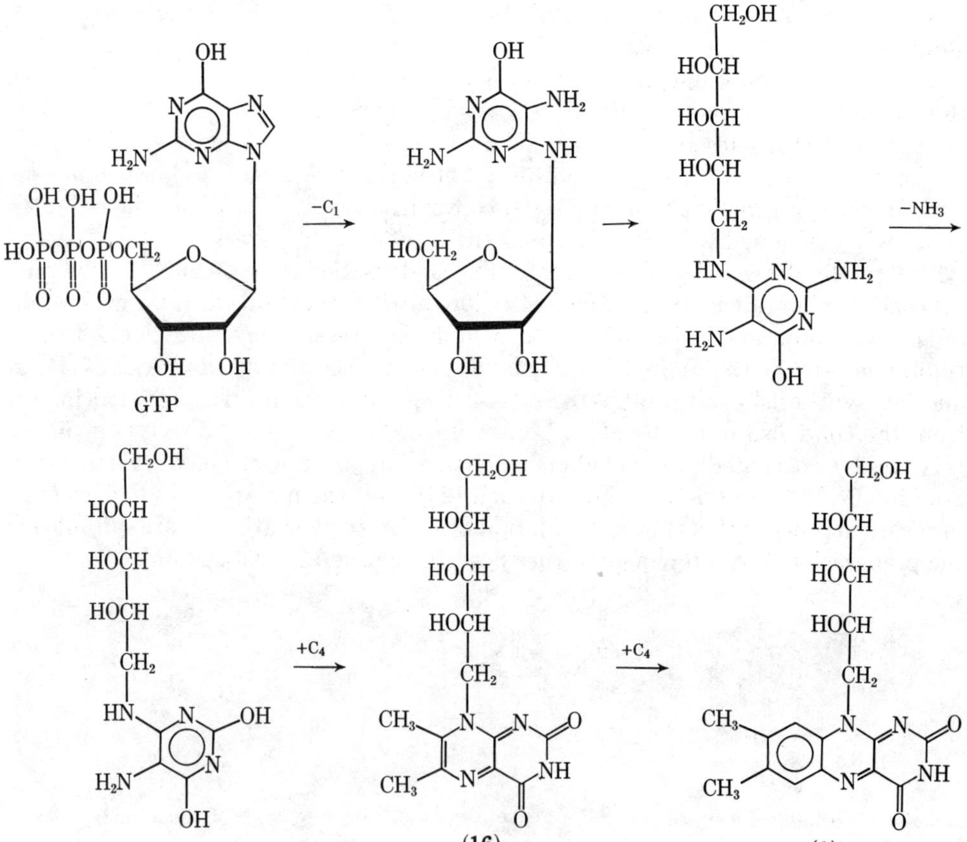

114   VITAMINS (B₂)

Figure 3. Biosynthesis pathway to riboflavin.

These organisms have been used frequently in the elucidation of the biosynthetic pathway (34). The mechanism of riboflavin biosynthesis shown in Figure 3 has been deduced from data derived from several experiments involving a variety of organisms. Included are the conversion of a purine such as guanosine triphosphate (GTP) to 6,7-dimethyl-8-D-ribityllumazine (16) (35), and the conversion of (16) to (1). This concept of the biochemical formation of riboflavin was verified *in vitro* under nonenzymatic conditions (36) (see also Microbial transformations).

Fermentation. Throughout the years, riboflavin yields obtained by fermentation have been improved to the point of commercial feasibility. Most of the riboflavin thus produced is consumed in the form of crude concentrates for the enrichment of animal feeds. Riboflavin was first produced by fermentation in 1940 from the residue of butanol–acetone fermentation. Several methods were developed for large-scale production (37). A suitable carbohydrate-containing mash is prepared and sterilized, and the pH adjusted to 6–7. The mash is buffered with calcium carbonate, inoculated with *Clostridium acetobutylicum*, and incubated at 37–40°C for 2–3 d. The yield is ca 70 mg riboflavin per liter (38) (see Fermentation).

Most varieties of *Candida* yeasts (qv) produce substantial amounts of riboflavin when glucose is the carbon source. Particularly, *Candida guilliermondia* and *Candida flaveri* produce high yields on a simple synthetic medium of low cost. Some modifications employing several *Candida* yeasts have been patented; eg, *C. intermedia var A*, a newly isolated microorganism assimilating lactose and ethanol, gave yields of 49.2 mg riboflavin/L from ethanol in the presence of biotin (39). *Candida T-3*, assimilating methanol, produces riboflavin by this method from 50 g methanol (40). *Candida* bacteria have the disadvantage of extremely low iron tolerance. Chelating agents, such as 2,2'-dipyridyl, are recommended to control the iron content (41).

Currently, most of the commercial riboflavin production by aerobic fermentation is obtained by biosynthesis with the yeastlike fungus *Eremothecium ashbyii*. Many variations for the production of riboflavin by *E. ashbyii* have been patented. Employing *E. ashbyii* grown on a yeast medium, riboflavin production on an industrial scale is said to have reached 11.6 g/kg from dried powder (42). Riboflavin is also obtained by fermentation with *E. ashbyii* preserved on Difco-treated millet seeds. The culture, kept for 7–8 d at 32°C on the medium containing crude collagen, corn extracts, unrefined plant oil, glucose, KH_3PO_4, trace elements, and H_2O at pH 7–8, gave 4.5–5.0 g/L (43). In another procedure, *E. ashbyii* was cultivated on a culture medium containing sources of assimilable nitrogen, essential minerals, and growth factors, along with sources of assimilable carbohydrates, unsaturated fatty acids, saccharides, and amino acids or their salts. Incubation at 29°C for 6 d on a rotary shaker with aeration gave average yields of 3.8 g/L (44). *Eremothecium ashbyii* grown in a culture medium containing the oil cake obtained after extractions of lipids from the biomass grown on hydrocarbons gives an even better yield (45). *Eremothecium ashbyii* cultured on a medium containing carbohydrates and vegetable oils also gave FAD (344 mg/L); the yield from a medium without vegetable oils was ca two-thirds less (46).

In operations similar to the *E. ashbyii* procedures, the closely related fungus *Ashbya gossypii* gave similar yields. Thus, a yield of 7.3 g/L was obtained with a lyophilized culture in a medium containing fat, leather glue, and corn extracts (47), and 6.420 g/L with bone or hide fat, alone or in a mixture with other plant or animal fats as the carbon source (48). The yield from immersed cultures of *A. gossypii* was increased to 6.93–7.20 g/L by use of waste fats or technical cod-liver oil (49).

Riboflavin is also made by aerobic culturing of *Pichia guilliermondii* on a medium containing n-C_{10}–C_{15} paraffins in a yield of 280.5 mg/L (50). A process employing *Pichia* yeasts, such as *P. miso*, *P. miso* Mogi, or *P. mogii*, in a medium containing a hydrocarbon as the carbon source, has been patented (51).

Processes employing *Torulopsis xylinus* (52), *Hansenula polymorpha* (53), *Brevibacterium ammoniagenes* (54), *Achromobactor butrii* (55), *Micrococcus lactis* (56), *Streptomyces testaceus* (57), and others have also been patented. These procedures yield, at most, several hundred mg of riboflavin per liter.

Analytical Methods

Riboflavin can be assayed by chemical, enzymatic, and microbiological methods. The most commonly used chemical method is fluorometry, which involves the measurement of intense yellow-green fluorescence with a maximum at 565 nm in neutral aqueous solutions. The fluorometric determinations of flavins can be carried out by measuring the intensity of either the natural fluorescence of flavins or the fluorescence of lumiflavin formed by the irradiation of flavin in alkaline solution (58). The recent development of a laser-fluorescence technique has extended the limits of detection for riboflavin by two orders of magnitude (59–60).

Polarography is applied in the presence of other vitamins, eg, in multivitamin tablets, without separation. The polarography of flavins is reviewed in ref. 61.

The microbial assay is based on the growth of *Lactobacillus casei* in the natural (62) or modified form. The lactic acid formed is titrated or, preferably, the turbidity measured photometrically. In a more sensitive assay, *Leuconostoc mesenteroides* is employed as the assay organism (63). It is 50 times more sensitive than *L. casei* for assaying riboflavin and its analogues (0.1 ng/mL vs 20 ng/mL for *L. casei*). A very useful method for measuring total riboflavin in body fluids and tissues is based on the riboflavin requirement of the protozoan ciliate *Tetrahymena pyriformis*, which is sensitive and specific for riboflavin. This method can be applied to large-scale nutrition studies.

Although riboflavin can be assayed more readily by chemical or microbiological methods than by animal methods, the latter are preferred for nutritional studies and as a basis of other techniques. Such assays depend upon a growth response; the rat or chick is the preferred experimental animal. This method is particularly useful for assaying riboflavin derivatives, since the substituents frequently reduce or eliminate the biological activity.

An enzymic method for assessing riboflavin deficiency in humans has been developed (64). It is based on the fact that NADPH-dependent glutathione reductase of red cells reflects riboflavin fluctuations.

Recently, high pressure liquid chromatography (hplc) has been extensively used for the riboflavin determinations. This method is usually automated and more rapid and sensitive than the microbial method. It has been used in combination with fluorometric detection for the riboflavin assay in foods (65), meat and meat products (66), and enriched and fortified foods (67), as well as in a simple assay for animal tissues (68). A rapid and efficient reversed-phase hplc method is described for the quantitative separation of flavin coenzymes and their structural analogues such as 5-deazaflavin [*19342-73-5*] (**24**) and 8-hydroxy-7-demethyl-5-deazariboflavin [*37333-48-5*] (**25**) (F_{420} chromophore) (69) (see under Derivatives).

Comprehensive reviews of the analytical methods for riboflavin are given in refs. 70–71.

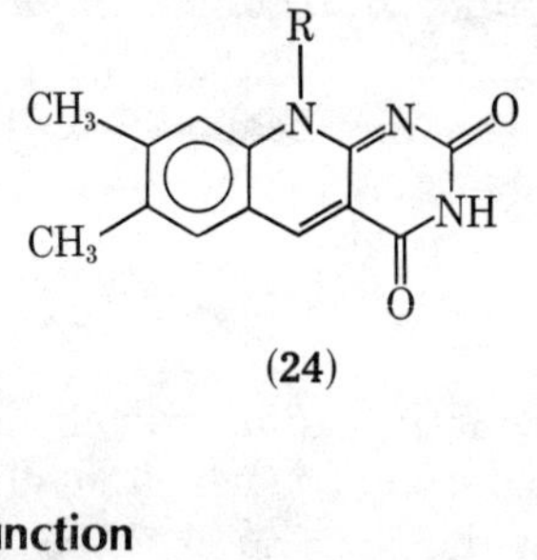

(24) 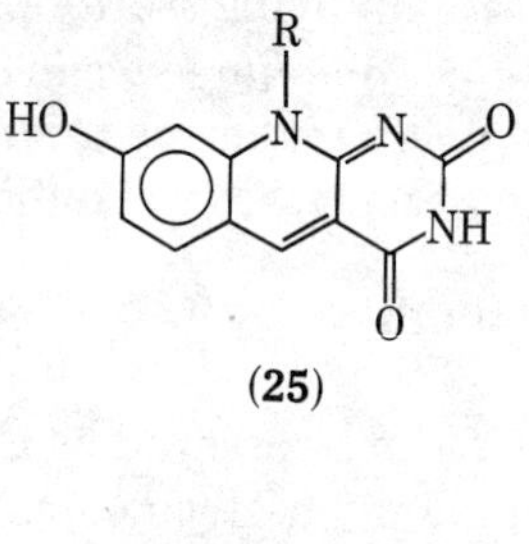 (25)

R = D-ribityl

Biological Function

In biological systems, riboflavin functions almost exclusively in the form of flavoproteins, in which the FMN or FAD is generally bound as prosthetic group or

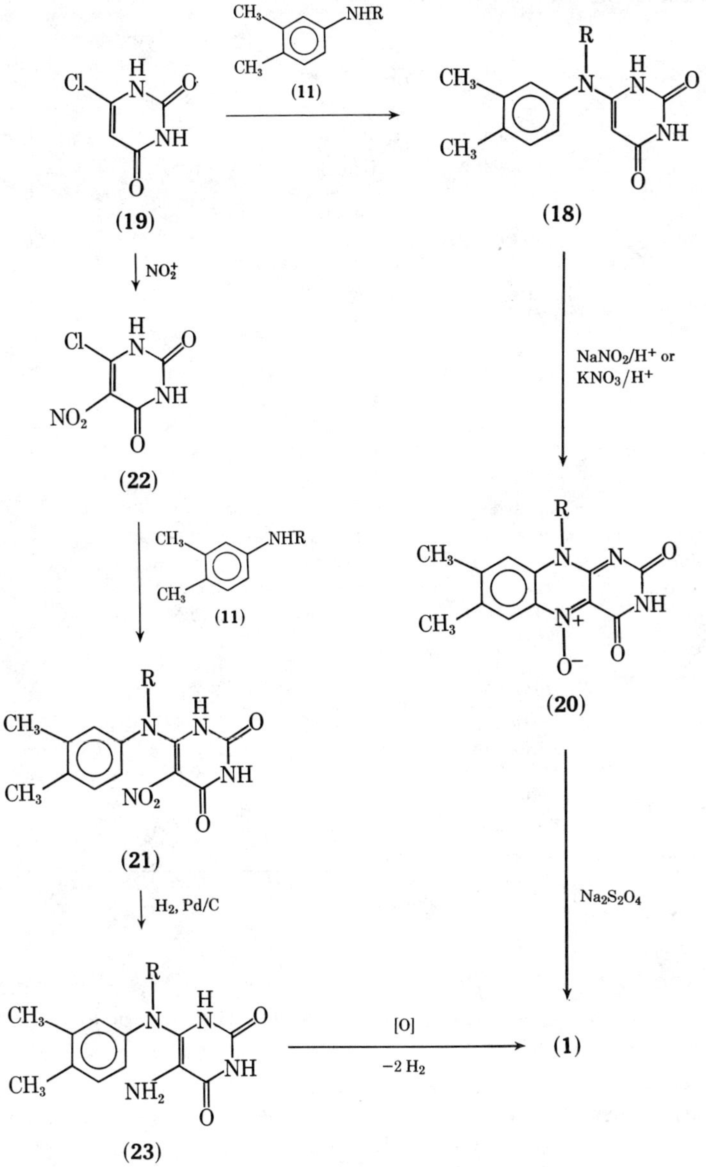

R = D-ribityl

coenzyme to specific proteins. These enzymes catalyze oxidation–reduction reactions (see Table 2). The flavin group of the oxidized coenzyme is reduced chemically or enzymatically to 1,5-dihydroflavin coenzyme, probably in two one-electron steps, each involving the addition of a single electron. Stable semiquinone radicals are formed as intermediates, because the unpaired electron is highly delocalized by the conjugated isoalloxazine structure.

In contrast to the nicotinamide nucleotide dehydrogenases, the prosthetic groups FMN and FAD are firmly associated with the proteins, and the flavin groups are

VITAMINS (B₂)

Table 2. Some Reactions Catalyzed by Flavoproteins[a]

Enzyme	Electron donor	Product	Coenzyme and other components	Electron acceptor
D-amino acid oxidase	D-amino acids	α-keto acids + NH_3	2 FAD	$O_2 \rightarrow H_2O_2$
L-amino acid oxidase (liver)	L-amino acids	α-keto acids + NH_3	2 FAD	$O_2 \rightarrow H_2O_2$
L-amino acid oxidase (kidney)	L-amino acids	α-keto acids + NH_3	2 FMN	$O_2 \rightarrow H_2O_2$
L(+)-lactate dehydrogenase (yeast)	lactate	pyruvate	1 FMN; 1 heme (cyt b_5)	respiratory chain
glycolic acid oxidase	glycolate	glyoxylate	FMN	$O_2 \rightarrow H_2O_2$
NAD^+-cytochrome c reductase	NADH	NAD^+	2 FAD, 2 Mo, NHI	cytochrome c_{ox}; respiratory chain
NAD^+-cytochrome b_5 reductase	NADH	NAD^+	FAD; Fe	cytochrome b_5
aldehyde oxidase (liver)	aldehydes	carboxylic acids	FAD; Fe, Mo	respiratory chain
α-glycerol phosphate dehydrogenase	glycerol 3-phosphate	dihydroxyacetone phosphate	FAD; Fe	respiratory chain
succinic dehydrogenase	succinate	fumarate	FAD; Fe, NIH	respiratory chain
acyl–CoA (C_6–C_{12}) dehydrogenase	acyl–CoA	enoyl–CoA	FAD	electron-transferring flavoprotein
nitrate reductase	NADPH	$NADP^+$	FAD; Mo, Fe	nitrate
nitrite reductase	NADPH	$NADP^+$	FAD; Mo, Fe	nitrite
xanthine oxidase	xanthine	uric acid	FAD; Mo, Fe	O_2
lipoyl dehydrogenase	reduced lipoic acid	oxidized lipoic acid	2 FAD	NAD^+
dihydroorotate dehydrogenase	dihydroorotic acid	orotic acid	2 FMN; 2 FAD, 4 Fe	

[a] Ref. 73.

usually only separated from the apoenzyme (protein) by acid treatment in water. However, in several covalently bound flavoproteins, the enzyme and flavin coenzymes are covalently affixed (see under Derivatives). In these cases, the flavin groups are isolated after the proteolytic digestion of the flavoproteins.

Many flavoproteins react directly with molecular oxygen to produce hydrogen peroxide. Some flavoproteins, such as the flavin-containing monooxygenase, give water instead of hydrogen peroxide. In this case, one atom of oxygen is introduced into a substrate to undergo hydroxylation, whereas the other oxygen atom is released as water. Several flavoproteins include metal complexes where these reactions take place. A number of reviews of the preparation, properties, and mechanism of action of these enzymes have been published (72).

Deficiency, Requirements, and Toxicity

Riboflavin is essential for mammalian cells. A lack in the human diet causes well-defined syndromes, such as angular stomatitis; glossitis (magenta tongue); reddened, shiny, and denuded lips; seborrhoeic follicular keratosis of the nasolabial folds, nose, and forehead; and dermatitis of the anogenital region (scrotum and vulva). An adult requires ca 1.5–3 mg riboflavin daily. Requirements are increased in pregnancy and lactation. The 1974 Recommended Dietary Allowances (RDA) of the Food and Nutrition Board, NAS–NRC, are 0.6 mg riboflavin per 1000 food calories, or 1.7 mg riboflavin for a sedentary person weighing 70 kg.

Riboflavin is essentially nontoxic. The LD$_{50}$ values in mice and rats by intraperitoneal injection are 340 mg/kg (74) and 560 mg/kg (75), respectively. The oral administration of 10 g/kg to rats or 2 g/kg to dogs showed no toxic effects (76).

Derivatives

Riboflavin-5'-Phosphate. Riboflavin-5'-phosphate [146-17-8] (vitamin B$_2$ phosphate, flavin mononucleotide, FMN, cytoflav), $C_{17}H_{21}N_4O_9P$, mol wt 456.5, is a microcrystalline yellow solid, mp 195°C, $[\alpha]_D^{28} = +44.5°$ (2% soln in conc HCl) with biological and enzymatic activity. It is prepared by phosphorylation of riboflavin with chlorophosphoric acid (77), pyrophosphoric acid (78), metaphosphoric acid (79), or catechol cyclic phosphate (80). It is soluble in water to the extent of 3 g/100 mL at 25°C as the sodium salt but tends to gel. Because of the high sensitivity of FMN to uv, it must be preserved in dark, tight containers.

Flavin mononucleotide was first isolated from the yellow enzyme in yeast by Warburg and Christian in 1932 (4). The yellow enzyme was split into the protein and the yellow prosthetic group (coenzyme) by dialysis under acidic conditions. Flavin mononucleotide was isolated as crystalline calcium salt and shown to be riboflavin-5'-phosphate; its structure was confirmed by chemical synthesis by Kuhn and Rudy (81). It is commercially available as monosodium salt dihydrate [6184-17-4], with a water solubility of more than 200 times that of riboflavin. It has wide application in multivitamin and B-complex solutions, where it does not require the solubilizers needed for riboflavin.

Riboflavin-5'-Adenosine Diphosphate. Riboflavin-5'-adenosine diphosphate [146-14-5] (flavin-adenine dinucleotide, FAD) (2), $C_{27}H_{33}N_9O_{15}P_2$, mol wt 785.56, was first isolated in 1938 from the D-amino acid oxidase as its prosthetic group by Warburg and Christian (82), who postulated it to be flavin-adenine dinucleotide. The structure was established by the first synthesis in Todd's laboratory (83); the monosilver salt of FMN was condensed with 2',3'-isopropylidene-adenosine-5'-benzyl phosphorchloridate, followed by removal of protective groups. It was also synthesized directly from FMN and adenosine-5'-monophosphate (AMP) with di-p-tolylcarbodiimide as the condensation agent (84). Another direct synthesis was achieved by dehydration between FMN and AMP with trifluoroacetic acid anhydride (85). A 40% yield was obtained by condensation of adenosine-5'-phosphoramidate and FMN using a mixture of pyridine and o-chlorophenol as the solvent (86). Condensation of AMP with FMN in ethoxyacetylene gives a 10–15% yield; by-products such as riboflavin-4',5'-cyclic phosphate are avoided (87). In addition to D-amino acid oxidase, FAD is the prosthetic group for the other flavoproteins including glucose oxidase, glycine oxidase, fumaric hydrogenase, histaminase, and xanthine oxidase.

Covalently Bound Flavins. The FAD prosthetic group in mammalian succinate dehydrogenase was covalently affixed to protein at the 8α-position through the linkage of 3-position of histidine (88–89). Since then, several covalently bound riboflavins (90) have been found successively from the enzymes listed in Table 3. The biosynthetic mechanism, however, has not been clarified.

120 VITAMINS (B$_2$)

Table 3. Covalently Bonded Flavins

Prosthetic group	CAS Registry No.	Formula	Enzyme
8α-(N^1-histidyl)flavin	[58525-92-1]	C$_{23}$H$_{27}$N$_7$O$_8$	thiamine dehydrogenase β-cyclopiazonate oxidocyclase L-galactonolactone oxidase cholesterol oxidase
8α-(N^3-histidyl)flavin	[37854-44-7]	C$_{23}$H$_{27}$N$_7$O$_8$	succinate dehydrogenase salcosine dehydrogenase D-hydroxynicotine oxidase choline oxidase
8α-S-cysteinylflavin	[35836-22-7]	C$_{20}$H$_{25}$N$_5$O$_8$S	monoamine oxidase (MAO) *Chlorobium* cytochrome C$_{552}$ *Chlomatium* cytochrome C$_{553}$
6-S-cysteinylflavin R = D-ribityl	[73647-60-6]	C$_{20}$H$_{25}$N$_5$O$_8$S	trimethylamine dehydrogenase

6-Hydroxyriboflavin. This compound [86120-61-8] (**26**) was isolated as a green coenzyme of the NADH dehydrogenase from *Peptostreptococcus eledenii*. It is not fluorescent, and its structure was established by synthesis (91). The 5′-monophosphate serves as a cofactor for glycolate oxidase from pig liver.

8-Nor-8-Hydroxyriboflavin. A new prosthetic group of red color has been isolated from NADH dehydrogenase of the electron-transferring flavoprotein of *Peptostreptococcus eledenii*. Its structure [52134-62-0] (**27**) has been established as the FAD derivative of 8-hydroxy-7-methylisoalloxazine. Proof has been obtained by the synthesis of 8-hydroxy-7-methylisoalloxazine models and stepwise degradation of the naturally occurring compound (92).

Roseoflavin. Roseoflavin [*51093-55-1*] (**28**), $C_{18}H_{23}N_5O_6$, mol wt 405.40, mp 274–297°C, $[\alpha]_D = -315°$ (0.1 M NaOH), was isolated from a culture medium of *Streptomyces davawensis* as dark, reddish-brown fine needles (from ethanol); the 8-methyl group of riboflavin is substituted by a dimethylamino group. This structure was confirmed by the synthesis. Roseoflavin shows antimicrobial activity against gram-positive bacteria (93).

R = D-ribityl

5-Deazariboflavin. In 5-deazariboflavin (**24**), the N-5 of riboflavin is replaced by CH; it serves as cofactor for several flavin-catalyzed reactions (94). It was first synthesized in 1970 (95); an improved synthesis was reported later (96).

A low potential electron carrier, the fluorescent factor F_{420} [*37333-48-5, 64885-97-8*] (**29**) (it absorbs maximally at 420 nm), possessing a 5-deazaflavin moiety, was isolated from methane-producing bacteria (97). F_{420} is an obligate intermediate for passage of an electron from H_2 to $NADP^+$ to generate NADPH. The structure of F_{420} was proposed as an 8-hydroxy-7-demethyl-5-deaza-FMN derivative. The chromophore is essentially identical to synthetic 8-hydroxy-7-demethyl-5-deazariboflavin (**25**) (98).

BIBLIOGRAPHY

"Riboflavin" in *ECT* 1st ed., Vol. 11, pp. 749–759, by G. E. Sita, Merck & Co.; "Riboflavine" in *ECT* 2nd ed., Vol. 17, pp. 445–458, by E. De Ritter, Hoffmann-La Roche, Inc.

1. R. Kuhn, P. György, and T. Wagner-Jauregg, *Ber. Dtsch. Chem. Ges.* **66,** 576 (1933).
2. *Ibid.*, 317 (1933).
3. P. Karrer and K. Schöpp, *Helv. Chim. Acta* **17,** 735, 771 (1934).
4. O. Warburg and W. Christian, *Biochem. Z.* **254,** 438 (1932).
5. R. Kuhn, H. Rudy, and T. Wagner-Jauregg, *Ber. Dtsch. Chem. Ges.* **66,** 1950 (1933).
6. P. Karrer, K. Schöpp, and F. Benz, *Helv. Chim. Acta* **18,** 426 (1935).
7. R. Kuhn, K. Reinemund, H. Kaltschmitt, K. Strobele, and H. Trischmann, *Naturwissenschaften* **23,** 260 (1935).
8. E. A. Woodcock and J. J. Warthesen, *J. Food Sci.* **47,** 545 (April 1982).
9. U.S. Pat. 2,407,412 (Sept. 10, 1946), D. V. Frost (to Abbott Laboratories).
10. A. W. Varnes, R. B. Dodson, and E. L. Wehry, *J. Am. Chem. Soc.* **94,** 946 (1972).
11. P. Karrer, H. Salmon, K. Schöpp, E. Schlittler, and H. Fritsche, *Helv. Chim. Acta* **17,** 1010 (1934).

12. E. C. Smith and D. Metzler, *J. Am. Chem. Soc.* **85**, 3285 (1963).
13. R. Kuhn and T. Wagner-Jauregg, *Ber. Dtsch. Chem. Ges.* **67**, 361 (1934).
14. L. Michaelis, M. P. Schubert, and C. V. Smythe, *J. Biol. Chem.* **116**, 587 (1936).
15. A. Ehrenberg, *Acta Chem. Scand.* **11**, 205 (1957).
16. B. Commoner and B. Lippincott, *Proc. Natl. Acad. Sci. U.S.A.* **44**, 1110 (1958).
17. F. Müller, P. Hemmerich, and A. Ehrenberg in H. Kamin, ed., *Flavins and Flavoproteins*, University Park Press, Baltimore, 1971, p. 107.
18. P. Hemmerich, *Experientia* **16**, 534 (1969).
19. P. Bamberg and P. Hemmerich, *Helv. Chim. Acta* **44**, 1001 (1961); K. H. Dubley, A. Ehrenberg, P. Hemmerich, and F. Müller, *Helv. Chim. Acta* **47**, 1354 (1964).
20. A. Ehrenberg and P. Hemmerich in T. P. Singer, ed., *Biological Oxidations*, Interscience Publishers, a division of John Wiley & Sons, Inc., New York, 1968, p. 722.
21. R. Kuhn and F. Weygand, *Ber. Dtsch. Chem. Ges.* **68**, 1282 (1935).
22. P. Karrer, *Helv. Chim. Acta* **18**, 1130 (1935).
23. *Ibid.*, **30**, 2101 (1947).
24. M. Tishler, J. W. Wellman, and K. Ladenburg, *J. Am. Chem. Soc.* **67**, 2165 (1945).
25. T. Matsukawa and K. Shirakawa, *Yakugaku Zasshi* **69**, 208 (1949).
26. M. Tishler, K. Pfister, R. D. Babson, K. Ladenburg, and A. J. Fleming, *J. Am. Chem. Soc.* **69**, 1487 (1947).
27. F. Bergel, A. Cohen, and J. W. Haworth, *J. Chem. Soc.*, 165 (1945).
28. R. M. Cresswell and H. C. S. Wood, *J. Chem. Soc.*, 4768 (1960).
29. Jpn. Pat. 10,031 (Nov. 13, 1959), A. Masuda (to Takeda Pharmaceutical Industries).
30. J. Davol and D. D. Evans, *J. Chem. Soc.*, 5041 (1960); see also R. M. Cresswell, T. Neilson, and H. C. S. Wood, *J. Chem. Soc.*, 477 (1961).
31. F. Yoneda, Y. Sakuma, M. Ichiba, and K. Shinomura, *J. Am. Chem. Soc.* **98**, 830 (1976).
32. F. Yoneda, Y. Sakuma, and K. Shinozuka, *J. Chem. Soc. Perkin Trans. 1*, 348 (1978).
33. T. Wagner-Jauregg in W. H. Sebrell, Jr., and R. E. Harris, eds., *The Vitamins*, Vol. V, Academic Press, Inc., New York, 1972, p. 19; J. P. Lambooy in R. C. Elderfield, ed., *Heterocyclic Compounds*, Vol. 9, John Wiley & Sons, Inc., New York, 1967, p. 118.
34. G. W. E. Plaut in M. Florkin and E. H. Stotz, eds., *Comprehensive Biochemistry*, Vol. 21, Elsevier Publishing Co., Amsterdam and New York, 1971, p. 11.
35. M. Mitsuda, K. Nakajima, and T. Nadamoto, *J. Nutr. Sci. Vitaminol.* **22**, 477 (1976); **23**, 71 (1977).
36. R. Rowan and H. C. S. Wood, *J. Chem. Soc.*, 452 (1968).
37. U.S. Pat. 2,202,161 (May 28, 1940), C. S. Miner (to Commercial Solvents Corp.).
38. U.S. Pat. 2,369,680 (Feb. 20, 1945), R. E. Maeda, H. L. Polland, and N. E. Rodgers (to Western Condensing Co.); U.S. Pat. 2,449,144 (Sept. 14, 1948), N. E. Rogers, H. L. Polland, and R. E. Maeda (to Western Condensing Co.).
39. Jpn. Pat. 73 19,958 (June 18, 1973), S. Sugawara and K. Sato (to Nippon Beet Sugar Manufacturing Co.).
40. Jpn. Pat. 76 19,187 (Feb. 16, 1976), Y. Ichihara, H. Abe, and A. Aoike (to Kuraray Co.).
41. U.S. Pat. 2,425,280 (Aug. 5, 1947), R. Hickey (to Commercial Solvents Corp.).
42. Ger. Pat. 1,936,238 (Jan. 28, 1971), E. M. Dikanskaya and A. A. Balabanova (to All-Union Scientific-Research Institute of Protein Biosynthesis).
43. Ger. Pat. 2,028,355 (Jan. 14, 1971), I. Nitelea and co-workers (to Romania, Antibiotic Plant).
44. Ger. Pat. 1,767,260 (Aug. 19, 1976), G. M. Miescher (to Commercial Solvents Corp.).
45. USSR Pat. 194,261 (June 15, 1974), E. M. Dikanskaya (to All-Union Scientific-Research Institute of Protein Biosynthesis).
46. Jpn. Pat. 74 22,711 (June 11, 1974), H. Koga, T. Sasaki, and T. Nagasaki (to Nisshin Flour Milling Co.).
47. Ger. Pat. 2,453,827 (May 15, 1975), T. Slave and co-workers (to Institutul de Cercetari Chimico Farmaceutice).
48. Pol. Pat. 66,611 (March 15, 1973), T. Szczesniak and co-workers (to Instytut Przemyslu Fermentacyjnego).
49. Pol. Pat. 76,481 (March 10, 1975), T. Szezesniak and co-workers (to Instytut Przemyslu Farmaceutycznego).
50. Ger. Pat. 2,037,905 (Feb. 3, 1972), T. Kamikubo and N. Hiroshima (to Kanegafuchi Chemical Industry Co.).

51. U.S. Pat. 3,433,707 (March 18, 1969), T. Matsubayashi and Y. Suzuki (to Dai Nippon Sugar Manufacturing Co. and Nitto Physico-Chemical Research Institute).
52. Jpn. Pats. 73 96,790 and 73 96,791 (Dec. 10, 1973), T. Fukukawa, T. Matsuyoshi, and J. Hiratsuka (to Mitsui Petrochemical Industries).
53. Jpn. Pat. 79 80,495 (June 27, 1979), S. Uragami (to Mitsubishi Gas Chemical Co.).
54. Jpn. Pat. 77 110,897 (Sept. 17, 1977), K. Nakayama, K. Araki, and S. Shimojo (to Kyowa Hakko Kogyo Co.).
55. Jpn. Pat. 77 54,094 (May 2, 1977), I. Chibata and co-workers (to Tanabe Seiyaku Co.).
56. USSR Pat. 511,742 (Sept. 25, 1976), T. E. Popova and co-workers (to All-Union Scientific-Research Institute of Protein Biosynthesis).
57. Jpn. Pat. 75 116,690 (Sept. 12, 1975), H. Umezawa and co-workers (to Sanraku-Ocean Co.).
58. J. Koziol in D. B. McCormick and L. D. Wright, eds., *Methods in Enzymology*, Vol. 43, Academic Press, Inc., New York, 1971, p. 253.
59. J. H. Richardson in D. B. McCormick and L. D. Wright, eds., *Methods in Enzymology*, Vol. 66, Academic Press, Inc., New York, 1980, p. 416.
60. N. Ishibashi, T. Ogawa, T. Imasaka, and M. Kunitake, *Anal. Chem.* **51**, 2096 (1979).
61. E. Knobloch in ref. 58, p. 305.
62. E. E. Snell and F. M. Strong, *Ind. Eng. Chem. Anal. Ed.* **11**, 346 (1939).
63. H. A. Kornberg, R. S. Langdon, and V. H. Cheldelin, *Anal. Chem.* **20**, 81 (1948).
64. J. A. Tillotson and E. M. Baker, *Am. J. Clin. Nutr.* **25**, 425 (1972).
65. P. J. Richardson, D. J. Favell, G. C. Gidley, and A. D. Jones, *Proc. Anal. Div. Chem. Soc.* **15**, 53 (1978).
66. C. Y. Wang and F. A. Moseley, *J. Agric. Food Chem.* **28**, 483 (1980).
67. J. F. Kamman, T. P. Labuza, and J. J. Warthesen, *J. Food Sci.* **45**, 1497 (1980).
68. K. Yagi and M. Sato, *Biochem. Int.* **2**, 327 (1981).
69. D. R. Light, C. Walsh, and M. A. Marletta, *Anal. Biochem.* **109**, 87 (1980).
70. W. N. Pearson in P. György and W. N. Pearson, eds., *The Vitamins*, Vol. VII, Academic Press, Inc., New York, 1967, p. 99.
71. H. Baker and O. Frank in R. S. Rivilin, ed., *Riboflavin*, Plenum Press, New York, 1975, p. 49.
72. K. M. Horowitt and L. A. Wittig in W. H. Sebrell, Jr., and R. E. Harris, eds., *The Vitamins*, Vol. V, Academic Press, Inc., New York, 1972, p. 53; C. A. Hamilton, *Prog. Bioorg. Chem.* **1**, 83 (1971); T. C. Bruice, *Prog. Bioorg. Chem.* **4**, 1 (1976); C. Walsh, *Annu. Rev. Biochem.* **47**, 881 (1978); V. Massey and P. Hemmerich in P. D. Boyer, ed., *The Enzymes*, Vol. 12, Academic Press, Inc., New York, 1976, p. 191.
73. E. E. Conn and P. K. Stumpf, *Outlines of Biochemistry*, 4th ed., John Wiley & Sons, Inc., New York, 1976, p. 207.
74. R. Kuhn and P. Boulanger, *Hoppe-Seyler's Z. Physiol. Chem.* **241**, 233 (1936).
75. K. Unna and J. G. Greslin, *J. Pharmacol. Exp. Ther.* **76**, 75 (1942).
76. V. Demole, *Z. Vitaminforsch.* **7**, 138 (1938).
77. U.S. Pats. 2,610,178 and 2,610,179 (Sept. 9, 1952), L. A. Flexser and W. G. Farkas (to Hoffmann-La Roche, Inc.).
78. U.S. Pat. 2,535,385 (Dec. 26, 1950), P. J. Breivogel (to White Laboratories).
79. M. Viscontini, C. Ebnother, and P. Karrer, *Helv. Chim. Acta* **35**, 457 (1952); M. Viscontini and co-workers, *Helv. Chim. Acta* **38**, 15 (1955).
80. T. Ukita and K. Nagasawa, *Chem. Pharm. Bull.* **7**, 465 (1959).
81. R. Kuhn and H. Rudy, *Ber. Dtsch. Chem. Ges.* **68**, 353 (1935).
82. O. Warburg and W. Christian, *Biochem. Z.* **296**, 294 (1938); **298**, 150 (1938); O. Warburg, W. Christian, and A. Griese, *Biochem. Z.* **297**, 417 (1938).
83. S. M. H. Christie, G. W. Kenner, and A. R. Todd, *Nature (London)* **170**, 924 (1952); *J. Chem. Soc.*, 46 (1954).
84. F. M. Huennekens and G. L. Kilgour, *J. Am. Chem. Soc.* **44**, 6716 (1955).
85. C. DeLuca and N. O. Kaplan, *J. Biol. Chem.* **223**, 569 (1956).
86. J. G. Moffatt and H. G. Khorana, *J. Am. Chem. Soc.* **80**, 3756 (1958).
87. H. Wassermann and D. Cohen, *Chem. Eng. News*, 47 (1962).
88. E. B. Kearney and T. P. Singer, *Biochem. Biophys. Acta* **17**, 596 (1955).
89. T. Y. Wang, C. L. Tsuo, and Y. L. Wang, *Sci. Sin.* **5**, 73 (1956).

90. T. P. Singer and D. E. Edmondson in ref. 59, p. 253 and references cited therein.
91. S. G. Mayhew, C. D. Whitefield, S. Ghisla, and M. Schuman-Jörns, *Eur. J. Biochem.* **44,** 579 (1974).
92. S. Ghisla and S. G. Mayhew, *Eur. J. Biochem.* **63,** 373 (1976).
93. S. Otani, M. Takatsu, M. Nakano, S. Kasai, R. Miura, and K. Matsui, *J. Antibiot.* **27,** 88 (1974); S. Kasai, R. Miura, and K. Matsui, *Bull. Chem. Soc. Jpn.* **48,** 2877 (1975).
94. P. Hemmerich, V. Massey, and H. Fenner, *FEBS Lett.* **84,** 5 (1977).
95. D. E. O'Brien, L. T. Weinstock, and C. C. Cheng, *J. Heterocycl. Chem.* **7,** 99 (1970).
96. F. Yoneda in ref. 59, p. 267.
97. D. Eirich, G. D. Vogels, and R. S. Wolfe, *Biochemistry* **17,** 4583 (1978).
98. W. T. Ashton, R. D. Brown, F. Jacobson, and C. Walsh, *J. Am. Chem. Soc.* **101,** 4419 (1979).

FUMIO YONEDA
Kumamoto University

THIAMINE (B₁)

Thiamine [59-43-8] (**1**), thiamin, vitamin B_1, aneurine, 3-[(4-amino-2-methyl-5-pyrimidinyl)methyl]-5-(2-hydroxyethyl)-4-methylthiazolium chloride, $C_{12}H_{17}N_4OSCl$, is a member of the vitamin-B complex group. The terms thiamine, thiamin, and vitamin B_1, usually denoting the hydrochloride [67-03-8] (**2**), are used interchangeably. The IUPAC–IUB (International Union of Biochemistry) name is thiamin, but thiamine is used in many official documents including the USP (1–2).

Thiamine is associated with carbohydrate metabolism. It is converted in the body to the pyrophosphate [154-87-0] (**3**) (cocarboxylase) which acts as a coenzyme in the decarboxylation of α-keto acids and several other reactions. A deficiency of thiamine in the diet leads to the beriberi syndrome characterized by deterioration of the nervous system. Thiamine is used for prophylactic and therapeutic purposes and as an enrichment of foodstuffs. The commercially available forms of thiamine are the chloride hydrochloride [67-03-8] (thiamine hydrochloride) (**2**) and the mononitrate [532-43-4] (**3**).

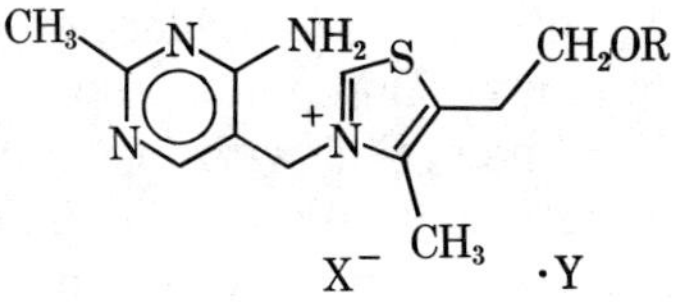

(1) R = H, X = Cl
(2) R = H, X = Cl, Y = HCl
(3) R = H, X = NO$_3$

(4) R = $-\overset{O}{\underset{OH}{\overset{\|}{P}}}-$OH

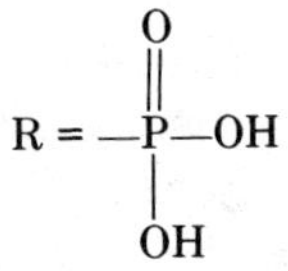

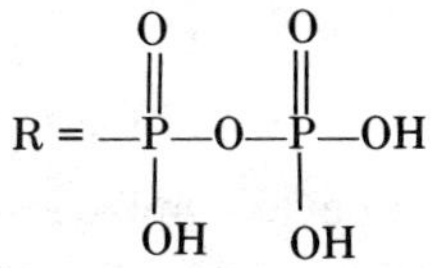

(5) R = $-\overset{O}{\underset{OH}{\overset{\|}{P}}}-O-\overset{O}{\underset{OH}{\overset{\|}{P}}}-$OH

(6) R = X$-\overset{O}{\underset{OH}{\overset{\|}{P}}}-O-\overset{O}{\underset{OH}{\overset{\|}{P}}}-O-\overset{O}{\underset{OH}{\overset{\|}{P}}}-O^-$

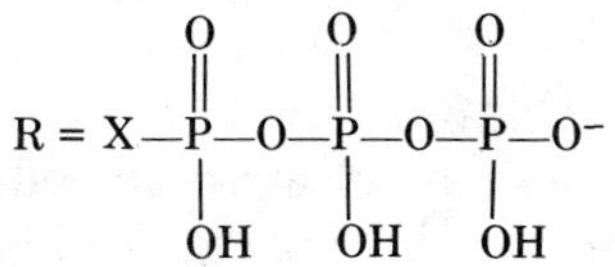

The world production of thiamine in 1977 was estimated at ca 2000 metric tons; the principal producers are Hoffmann LaRoche, Takeda Chemical Industries, Ltd., Tanabe Seiyaku Co. Ltd., and Synthetic Vitamin Factory. The price in 1981 was ca $35/kg.

In 1882, Takaki, during a nine months' cruise with the Japanese Navy, observed that more than half of the crew fed on polished rice suffered from beriberi. On his next cruise, the number of the patients was significantly reduced by changing to a western-style diet based on bread (3). In the 1890s in Jakarta, the first experimental beriberi was produced in fowl and the presence of an antianeuritic substance in rice bran was suggested (4). A few years later, it was clearly established that beriberi was caused by a dietary deficiency (5). In 1912, a substance was isolated from rice bran that effectively cured dietary polyneuritis in rats; it was named vitamine (amine essential for life) (6–7). At almost the same time, a similar substance was isolated independently from rice bran and yeast which was named orizanin and torulin, respectively. In accordance with the successive discovery of many other physiologically essential factors during that period, it was proposed that these should be designated as vitamins (e was omitted) A, B, C, D, E, etc (8). Vitamin B was later divided further into subclasses B_1, B_2, B_6, and B_{12}, as well as others. Pure crystals of thiamine hydrochloride were isolated by Jansen in 1926 and named aneurin (9–10). Elucidation of the chemical structure was achieved in 1936 independently by Williams (11) and Grewe (12), and the structure was confirmed by total synthesis (11,13–14). The name thiamine was first used by Williams.

Thiamine is found in the tissues of animals, plants, and microorganisms. It occurs in nature as a salt or as the monophosphate [523-40-1] (4), pyrophosphate [154-87-0] (cocarboxylase) (5), or triphosphate ester [3475-65-8] (6). The predominant form in animal tissue is the pyrophosphate, which exists largely as a protein complex bound to an enzyme. The most abundant form in plant tissue is free thiamine (1) (15).

In animals, including man, the organs with high thiamine concentration (μg/g of moist tissue) are the heart (2.8–7.9), kidney (2.4–5.8), liver (2.0–7.6), and brain (1.4–4.4); lesser amounts are found in the spleen, lung, adrenals, and muscle (16). Normal human blood contains ca 90 ng/mL, although the reported values vary considerably. A value below 40 ng/mL may be indicative of thiamine deficiency.

The occurrence of thiamine in foodstuffs is of nutritional importance. Concentration in individual foods varies widely (see Table 1). Considerable amounts are likely to be destroyed in cooking processes owing to diverse factors such as heat, metals, chlorine in the water, and reactive organic compounds present in the food. Therefore, naturally occurring thiamine in cereals is often supplemented with synthetic thiamine or its derivatives (see also Food additives).

Properties

Thiamine consists of two heterocyclic moieties, pyrimidine and thiazole, connected by a methylene group. It forms mono- or diacid salts.

Free thiamine base [70-16-8] (7), obtained by the neutralization of (1) with two equivalents of alkali, is unstable owing to the presence of a quaternary nitrogen atom (Fig. 1). It readily undergoes ring cleavage in water to give thiol-form thiamine [554-45-0] (12) by way of the pseudo-base (9). Addition of a third equivalent of alkali, sodium hydroxide, for instance, affords the sodium salt of the thiol [4710-54-7] (10). Actually, when an aqueous solution of (1) is made alkaline with sodium hydroxide, the solution turns yellow and then fades gradually. The yellow color is sustained longer when the neutralization is carried out in ethanol. This phenomenon is interpreted by the coexistence of another pathway (7) → (11) [80483-97-2] → (8) [79990-47-9] → (12) along with (7) → (9) → (12) and the yellow color is attributed to (8) or its salt. Neutralization of (10) with carbon dioxide yields a fat-soluble thiamine base [21682-72-4;

Table 1. Average Thiamine Concentration in Foodstuffs[a]

Foodstuff	μg/100 g
wheat germ	2050
dried brewer's yeast	1820
dried beans and peas	680
nut products	560
flower and fruit vegetables	70
leaf and stem vegetables	70
root and tuber vegetables	60
organ meats[b]	100
pork muscle	600–800
fish	50–90
egg	170

[a] Ref. 16.
[b] Liver, heart, kidney, and brain.

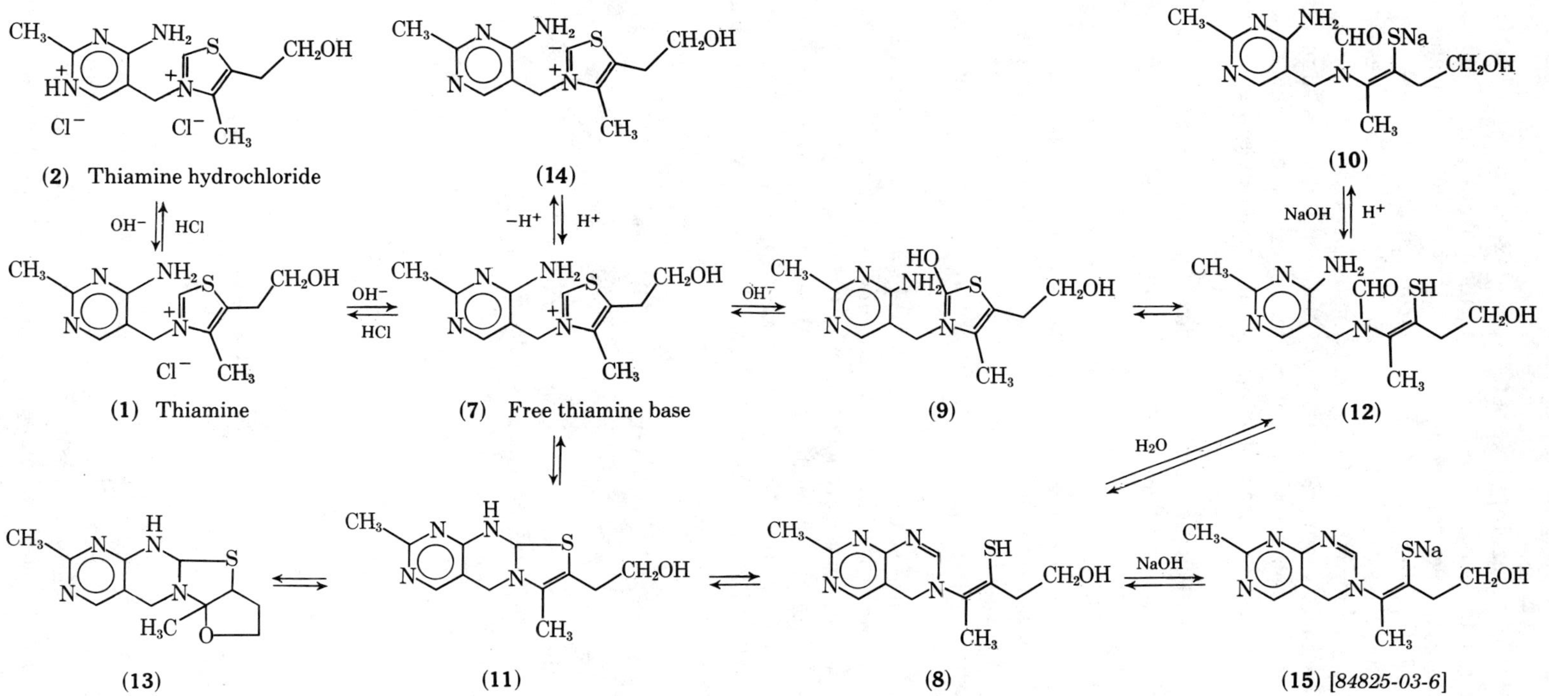

Figure 1. Structural changes of thiamine depending upon pH.

128   VITAMINS (B₁)

35922-43-1] **(13)**, presumably by the route: **(10)** → **(12)** → **(9)** → **(7)** → **(11)** → **(13)** (17). On the other hand, the hydrogen at the 2-position of the thiazolium ring is readily liberated owing to the electron-withdrawing effect of the adjacent quaternary nitrogen to form an ylid **(14)** [*84812-92-0*], which is of significance for the physiological function of thiamine. The pK_a of thiamine in an aqueous solution at 25°C is estimated as 12.6 for the C–H and 4.8 for NH₂ of the pyrimidine moiety (18).

Spectral data for thiamine, eg, the uv (19), the nmr (20), ^{13}C-nmr (21), ^{15}N-nmr (22), and the mass spectrum (23), as well as x-ray crystallographic data (24) are reported in the literature, as are molecular orbital calculations of the electron density (25).

Degradation. Although an aqueous solution of thiamine hydrochloride **(2)** is stable under normal conditions, it decomposes to 4-amino-2-methyl-5-pyrimidinemethanol [*73-67-6*] **(16)** and 4-methyl-5-thiazole-ethanol [*137-00-8*] **(17)** when heated in a sealed tube at 140°C (26). Heating **(2)** with 20% hydrochloric acid hydrolyzes the amino group to yield oxythiamine [*582-36-5*] **(18)**, an antagonist of thiamine (27). Concentrated solutions of free thiamine base in methanol or ethanol rapidly degrade at room temperature. The thiazole **(17)** is liberated, and the pyrimidine appears as an oligomer. The oligomer has one pyrimidine ring bonded at *N*-1 to the CH₂ side chain of another pyrimidine (28–29).

In the course of the structural elucidation of thiamine, it was readily cleaved into 4-amino-2-methyl-5-pyrimidinemethanesulfonic acid [*2908-73-8*] **(19)** and **(17)** by treatment with sulfite in weakly acidic solutions; this finding provided strong support for the present structure of thiamine (11). When this reaction is carried out in the presence of an aromatic base, pyridine for instance, the thiazolium moiety is replaced to give **(20)** [*54997-93-2*] (30). The complex mechanism by which these reactions with sulfite ion takes place has finally been established (31).

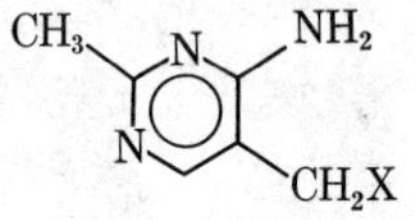

(16) X = OH
(19) X = SO₃H
(20) X = —N⁺(C₅H₅)

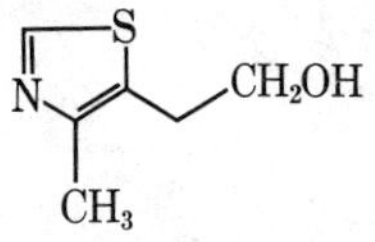

(17)

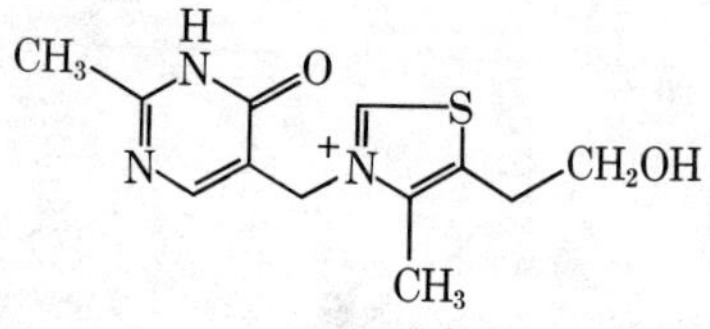

(18)

Similar reactions are observed in the degradation of thiamine by thiaminase (aneurinase), an enzyme found in certain clams, freshwater fish, and bacteria. Thiaminase is classified into thiaminase I and thiaminase II. The former promotes the replacement of the thiazolium moiety by another organic base, whereas the latter catalyzes the cleavage of thiamine into **(19)** and **(17)** (32).

On the other hand, the thiol form of thiamine undergoes a variety of degradations mainly by nucleophilic attack on either the *N*-formyl group of the enamine or the thiol group. Heating of an aqueous solution of (**10**) under nitrogen effects hydrolysis to 4-amino-2-methyl-5-pyrimidinylmethylamine [95-02-3] (**21**), 5-hydroxy-3-mercaptopentan-2-one (**22**), and formic acid (33). The ene-thiol group of thiamine (the chemical equivalent to a thioketone) reacts with a variety of nucleophilic reagents. Reactions of (**12**) with hydroxylamine, phenylhydrazine, and thiosemicarbazide liberate hydrogen sulfide to give the corresponding desulfurized derivatives (**23**)–(**27**) (34) (see Fig. 2). Similar reactions with ammonia and hydrazine yield the cyclized products (**28**) [305-25-9] and (**29**) [7199-84-0], respectively. The reaction with glycine affords dethiothiamine [13004-39-2] (**30**), presumably derived from the intermediate Schiff base by hydrolysis (35). A pyrimidodiazepine derivative [18641-53-7] (**31**), an intramolecularly desulfurized product, has been isolated from the reaction mixture when the solution of (**10**) is heated (36).

The thiol-form thiamine exists in a *Z*-conformation with respect to the double bond. Treatment of an aqueous solution of (**10**) with sulfur results in an equilibrium with the *E*-isomer [86291-33-0] (**32**) by way of the intermediate (**33**) [84812-91-9]. The *E*-isomer isolated as the benzyl and butyl disulfide derivatives, in contrast to the derivatives of the *Z*-isomer, showed practically no thiamine activity in rats (37).

Oxidation. The thiol form of thiamine is susceptible to oxidation. For example, heating an aqueous pH 8 solution of thiamine at 60°C for two hours gives thiamine disulfide [67-16-3] (**34**), thiothiamine [299-35-4] (**35**), a 2-thiazolone derivative [490-82-4] (**36**), and thiochrome [92-35-3] (**37**) (38). Thiamine is also obtained in good yield by the oxidation of (**10**) with iodine. Thiothiamine is an important intermediate in the manufacture of thiamine. Thiochrome is a yellow crystalline compound which, in solution, exhibits strong blue fluorescence. This property is used for the quantitative determination of thiamine. Thiamine disulfide is readily reduced to thiamine *in vivo* and is as biologically active as thiamine; the other three oxidation products possess no biological activity.

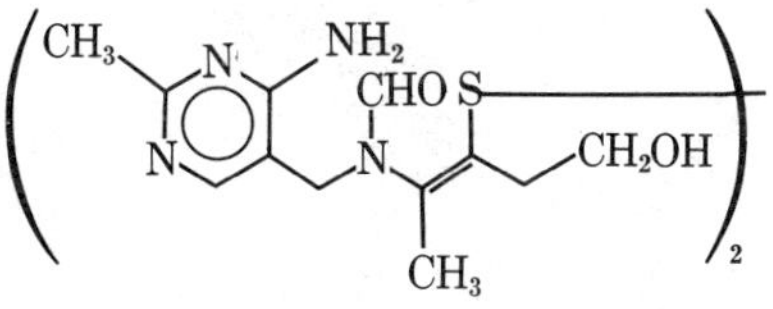

(**34**)

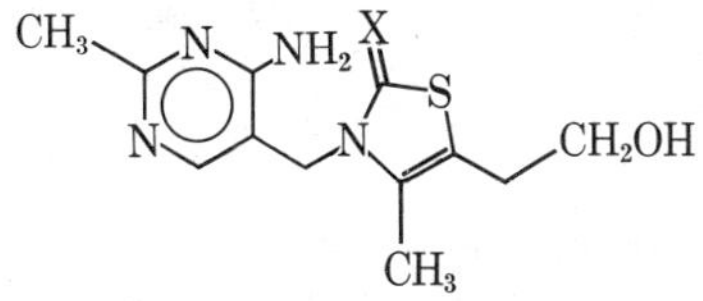

(**35**) X = S
(**36**) X = O

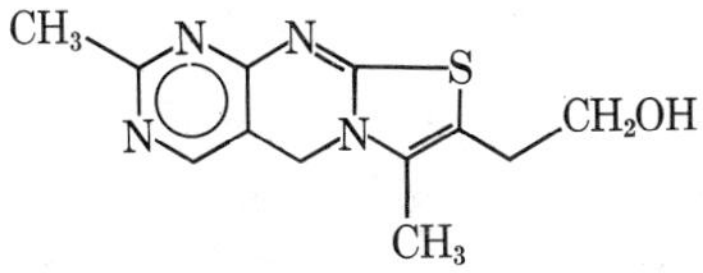

(**37**)

Figure 2. Degradation reactions of thiamine.

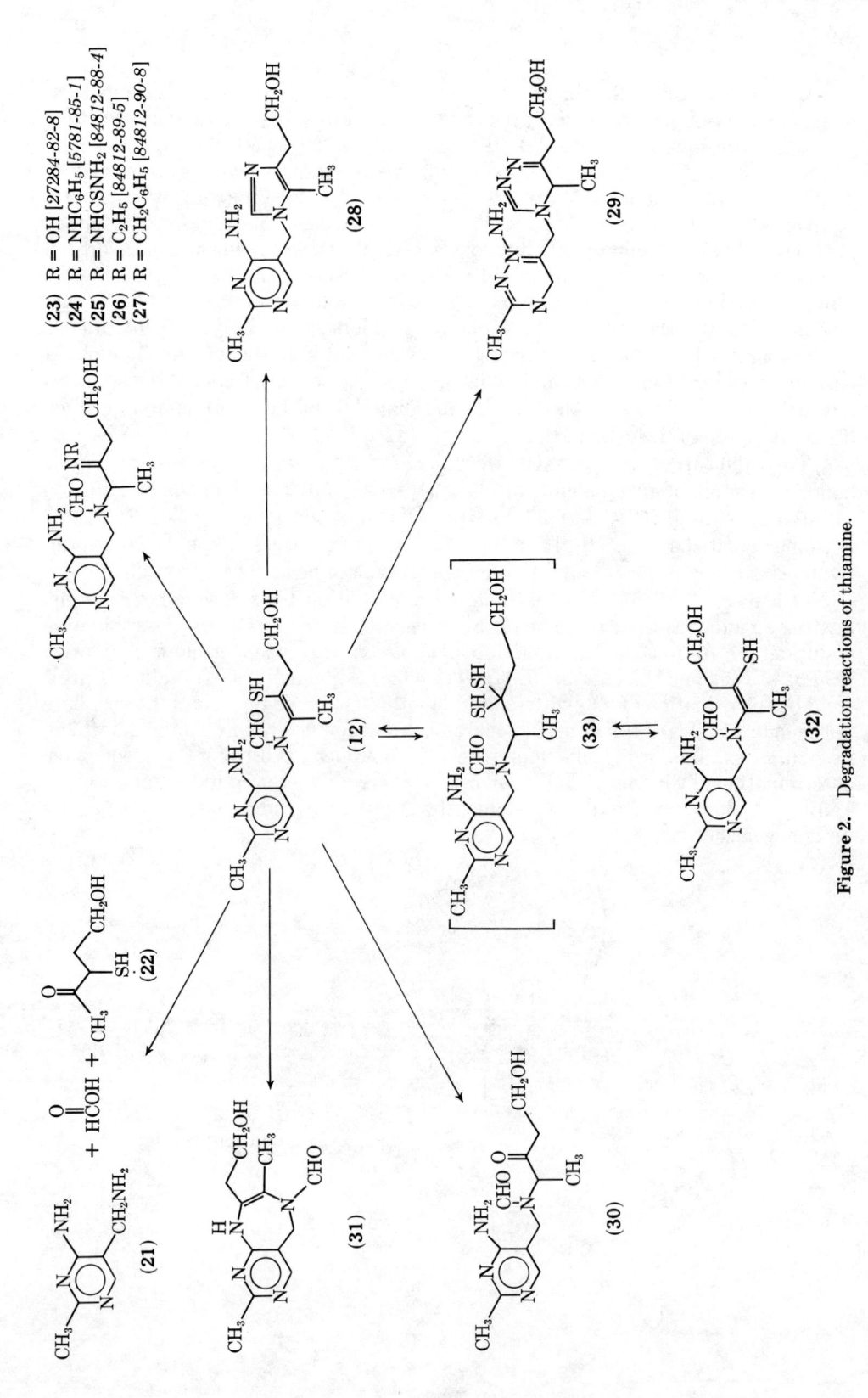

Reduction. Thiamine in the thiazolium form undergoes reduction of the thiazolium ring by lithium aluminum hydride, sodium borohydride, or sodium trimethoxyborohydride to afford tetrahydrothiamine [*15233-41-7*] (**38**) via the intermediate dihydrothiamine [*959-18-2*] (**39**) (39). This compound is synthesized by the one-step condensation of (**21**), (**22**), and formaldehyde. In the crystalline state, dihydrothiamine can exist as structure (**40**) [*297-94-9*] and two stereoisomers of (**41**) [*889-59-8*]; structure (**40**) is the most stable (40).

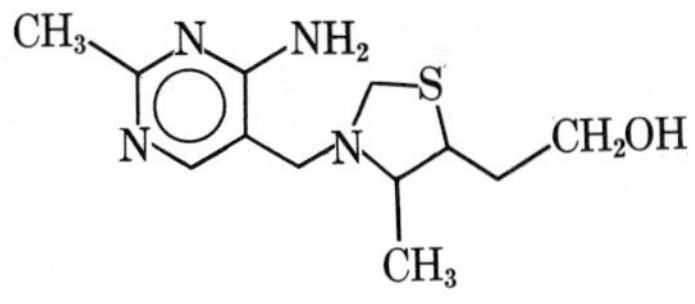

(38)

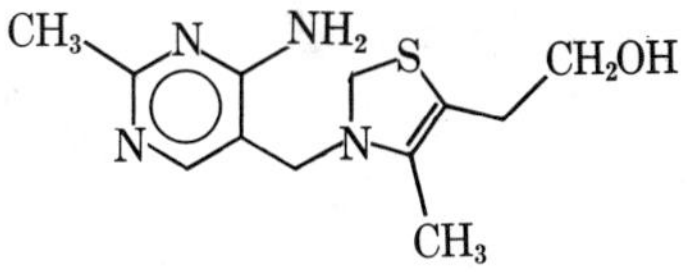

(39)

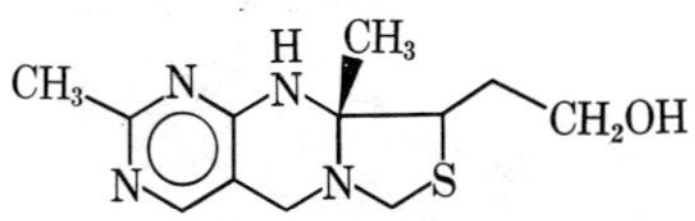

(40)

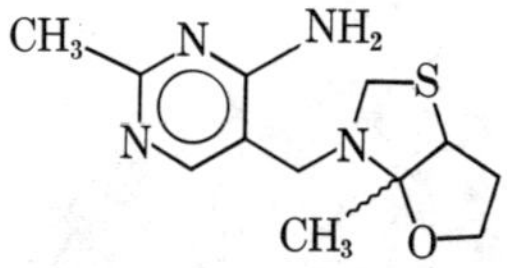

(41)

Although dihydrothiamine is chemically oxidized to thiamine (**41**), the possibility of it being a biogenetic precursor of thiamine has been ruled out (42).

Synthesis

The practical synthetic methods for thiamine are represented by several pathways. In the first, the pyrimidine and thiazole moieties are synthesized separately, followed by coupling (11,14). 4-Amino-5-bromomethyl-2-methylpyrimidine hydrobromide [*2908-71-6*] (**42**) and (**17**) give thiamine bromide hydrobromide [*4234-86-0*] (**43**), followed by the exchange of the bromide and chloride ions by treatment with silver chloride or ion-exchange resins (see Fig. 3).

The second method is characterized by the extension of the side chain at the 5-position of 4-amino-2-methyl-5-pyrimidinylmethylamine (**9**) followed by cyclization into the thiazole ring. 5-Acetoxy-3-chloropentan-2-one is condensed with *N*-(4-amino-2-methyl-5-pyrimidinyl) methyl thioformamide [*31375-20-9*] (**44**), prepared from (**9**) by formylation with formic acid and reaction with phosphorus pentasulfide. Hydrolysis of the resulting thiamine acetate with hydrochloric acid completes the synthesis (13).

In 1949, a 4-thiazoline-2-thione derivative, thiothiamine (**35**), was isolated from a reaction mixture obtained by the oxidation of thiamine (43). This compound is prepared in a good yield by the condensation of (**9**), carbon disulfide, and 3-chloro-5-hydroxypentan-2-one [*13045-13-1*] (**45**) (44). Based on these findings, an alternative industrial method, the thiothiamine method, was devised consisting of the transformations (**21**) → (**46**) [*1153-39-5*] → (**35**) → (**2**).

The syntheses of compounds (**17**), (**21**), and (**42**) are shown in Figure 4.

Figure 3. Synthetic methods for thiamine.

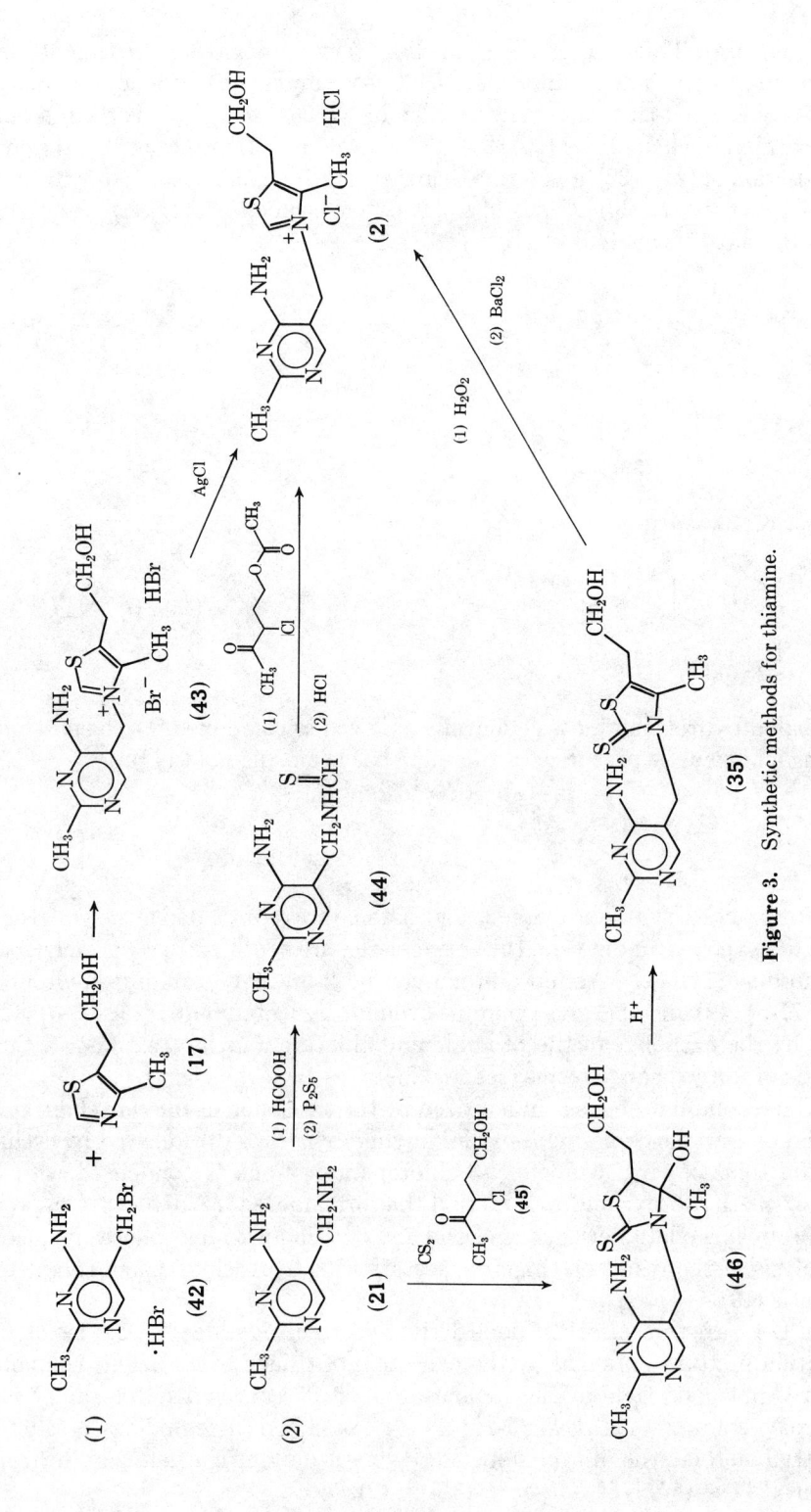

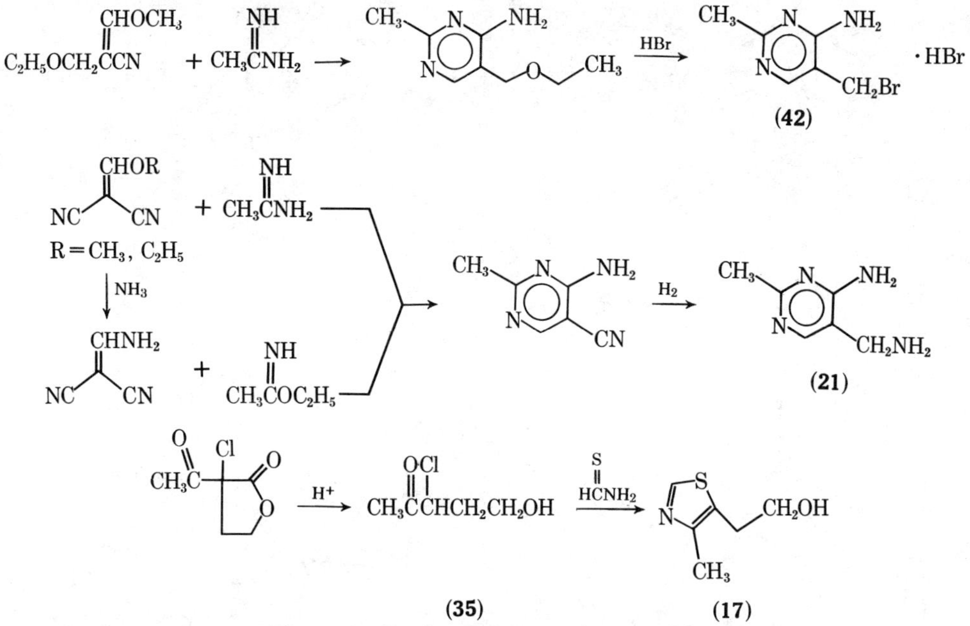

Figure 4. Syntheses of the starting materials.

Analytical Methods

Animal, microbiological, colorimetric, and fluorometric assay methods have been employed for the estimation of thiamine. At present, the most widely used is the fluorometric thiochrome method, where thiamine is oxidized to thiochrome (37) with potassium ferricyanide [USP method, (2)] or cyanogen bromide, and the fluorescence is compared with that of a standard thiamine solution.

Physiological Function

The biologically active form of thiamine is its pyrophosphate (5), cocarboxylase. It plays a vital role in carbohydrate metabolism as a coenzyme for the enzymes that catalyze the oxidative decarboxylation of α-keto acids and the formation or transfer of α-ketols. These biochemical reactions include the oxidative decarboxylation of pyruvic acid and 2-oxoglutaric acid, the formation of acetoin, and the reversible α-ketol transfer reactions catalyzed by transketolase (see also Citric acid; Enzymes).

The transketolase activity in the blood has been used clinically as an indicator of thiamine deficiency.

A number of hypotheses had been proposed concerning the active site in the thiamine molecule where the α-keto acids are attached; this controversy was settled in 1958 by the thiazolium ylid mechanism, ie, (14) (45). Thus, the above decarboxylation reactions are initiated by the addition of an α-keto acid to the 2-position of the thiazolium ylid (47) to form an adduct (48). The adduct, in turn, upon decarboxylation gives so-called active aldehyde (49), from which the above diverse biochemical reactions are derived. In the case of oxidative decarboxylation, for instance, the decarboxylated moiety is transferred into the citric acid cycle coupled with other four coenzymes,

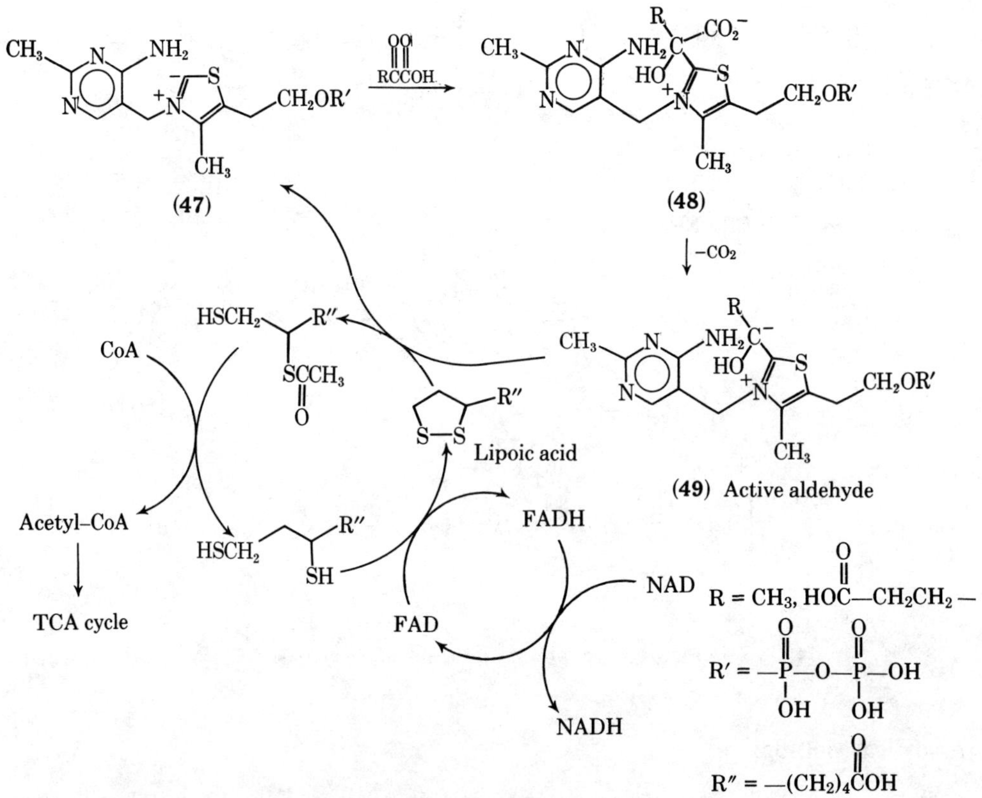

Figure 5. Mechanism of the oxidative decarboxylation of α-keto acids.

namely lipoic acid, coenzymes A (CoA), flavin adenine dinucleotide (FAD), and nicotinamide adenine dinucleotide (NAD) as shown in Figure 5 (45–46).

This mechanism was supported by the chemical synthesis of 2-(α-hydroxyethyl)thiamine [3670-41-5], a protonated species of (49) (46). More recently, the thiamine–pyruvate adduct as well as its pyrophosphate have been prepared (47).

Requirements, Deficiency, Toxicity. The maintenance requirements of thiamine for adults are in the range of 1.0–1.5 mg/d, depending on carbohydrate (food calorie) intake. The minimum requirement is substantially raised under certain circumstances such as pregnancy, hyperthyroidism, lactation, fever, diarrhea, liver disease, alcoholism, and physical exertion.

The pathological condition arising from thiamine deficiency is beriberi. It is characterized by degenerative changes in the nervous system, including a multiple peripheral neuritis, often accompanied by edema and serious effusions with or without cardiac hypertrophy and dilatation.

Thiamine deficiency is remedied by parenteral or oral administration of thiamine in doses up to 50 mg/d. Larger amounts up to 100 mg/d are given in cases of severe deficiency. High oral doses of thiamine are poorly absorbed from the digestive organs. Some derivatives of the thiol-form thiamine have been used in Japan as prodrugs for better absorption. The toxicity of thiamine is low with an oral LD_{50} of thiamine hydrochloride of 9000 mg/kg in mice (48).

Salts and Derivatives

Thiamine Hydrochloride. Thiamine chloride hydrochloride (2) (generally termed thiamine hydrochloride) forms colorless crystals melting at ca 248°C (dec). It is very soluble in water (1 g/mL), soluble in methanol, glycerol, and propylene glycol, sparingly soluble in ethanol, and practically insoluble in acetone, ether, chloroform, and benzene. A 1% solution in water has a pH of 2.7–3.3. The protonated salt has two positive charges, one associated with the thiazole ring and the other with the pyrimidine ring. The pyrimidine ring protonates at the 1-position (22).

Thiamine Mononitrate. Thiamine mononitrate (3), mp 196–200°C (dec), is more stable and less soluble in water (2.7 g/100 mL at 25°C) than the hydrochloride. Hence, the mononitrate is suitable for the dry pharmaceutical preparations and enrichment of food; however, only the hydrochloride is used for injection preparations.

Thiamine Pyrophosphoric Acid Ester. Thiamine pyrophosphoric acid ester (5), thiamine pyrophosphate, thiamine diphosphate, cocarboxylase, crystallizes as a hydrochloride monohydrate, mp 240–244°C (dec), from methanolic hydrogen chloride, and as a hydrochloride, mp 238–240°C, from absolute ethanol. The free ester is isolated as a tetrahydrate, mp 220–222°C (sinters at 130–140°C) (49). Thiamine pyrophosphate is freely soluble in water but essentially insoluble in organic solvents. It is very stable but the aqueous solutions are prone to decomposition into thiamine monophosphate and thiamine. It was first isolated from yeast in 1937 (50). Its synthesis is based upon the reaction of thiamine hydrochloride with a mixture of phosphoric acid and phosphoric anhydride followed by purification over ion-exchange resin (49). In an alternative method, thiamine disulfide monophosphate is phosphorylated, followed by reduction with cysteine (51).

Thiamine Triphosphoric Acid Ester. Thiamine triphosphoric acid ester, thiamine triphosphate (6), is another naturally occurring form of thiamine. Although it is suggested that thiamine triphosphate may be involved with nervous-system functioning (52), its biological significance still remains to be clarified. It crystallizes as the hygroscopic hydrochloride hemihydrate, mp 228–232°C (dec), or hydrochloride dihydrate, mp 203–213°C (dec); both are freely soluble in water but insoluble in common organic solvents. It is synthesized by the same procedures as the pyrophosphate (49,53–54).

Thiamine Disulfide. Thiamine disulfide (34), mp 177°C, was first prepared by the oxidation of thiol-form thiamine with iodine (55). It is readily reduced to thiamine by treatment with cysteine or various other reducing agents and is physiologically as active as thiamine *in vivo*.

Other Derivatives of Thiol-Form Thiamine. Treatment of thiamine with an extract of garlic gives a physiologically active fat-soluble derivative of thiamine, whose structure was determined as thiamine allyl disulfide, allithiamine [554-44-9] (50) (56–57). Its formation was interpreted by the reaction between the thiol form of thiamine and a constituent of garlic, allicine (CH_2=$CHCH_2SS(O)$-CH_2CH=CH_2). Allithiamine acts as a prodrug of thiamine which shows better absorption from the intestine and longer blood retention than thiamine (see Pharmacodynamics). Based on these findings, a number of thiol-form thiamine derivatives (51)–(58) have been developed and marketed, mostly in Japan (see Fig. 6).

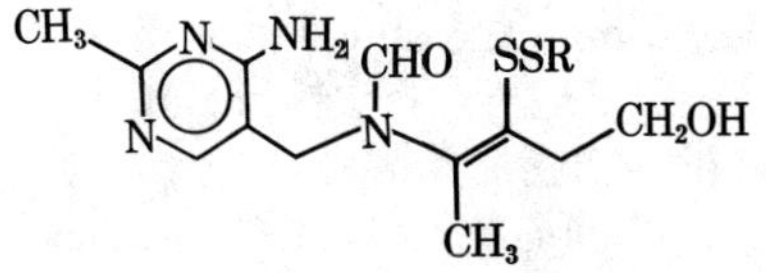

(50) R = CH$_2$CH=CH$_2$ Allithiamine
(51) R = CH$_2$CH$_2$CH$_3$ Prosultiamine [59-58-5]

(52) R = CH$_2$-⟨furan⟩   Fursultiamine [804-30-8]

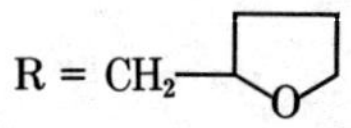

(53) R = CH$_2$CH$_2$CH(CH$_2$)$_4$COCH$_3$ Octotiamine [137-86-0]
          |
          SCCH$_3$
          ‖
          O

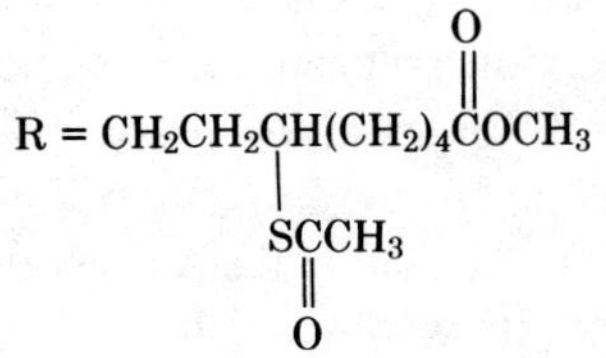

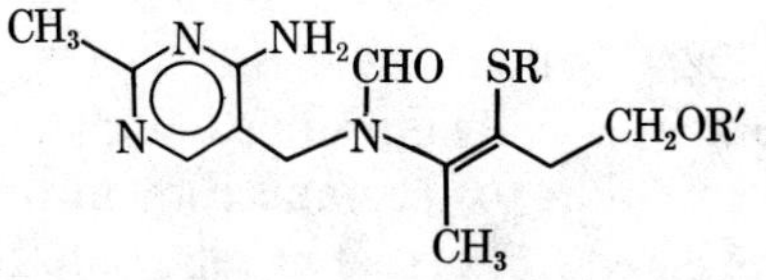

(54) R = R' = —C(=O)—C$_6$H$_5$ Dibenzoyl thiamine [299-88-7]

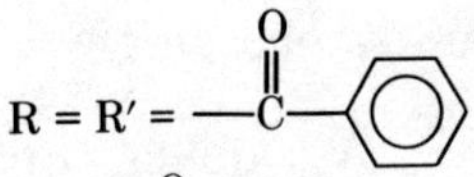

(55) R = —C(=O)—C$_6$H$_5$, R' = —P(=O)(OH)—OH   Benfotiamine [22457-89-2]

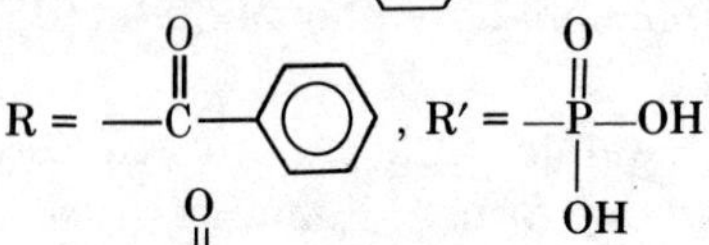

(56) R = R' = COC$_2$H$_5$ Cetotiamine [137-76-8]

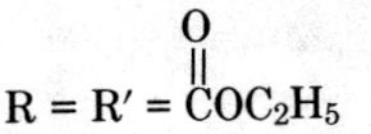

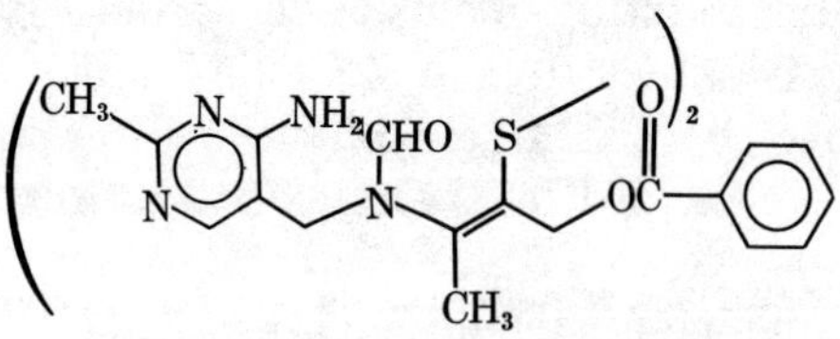

(57) Bisbentiamine [2667-89-2]

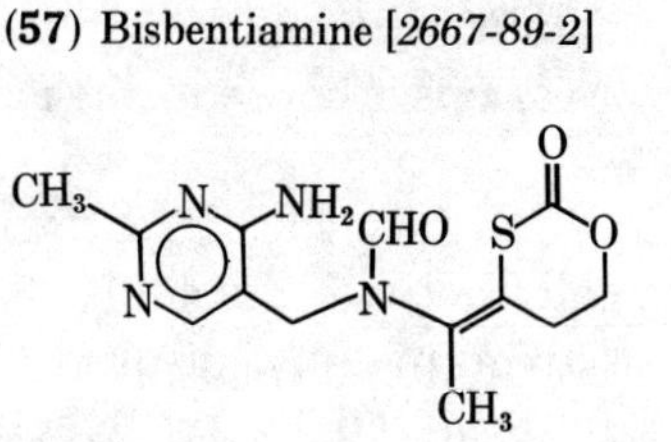

(58) Cyclothiamine [6092-18-8]

Figure 6. The thiol-form thiamine derivatives.

Analogues and Antimetabolites

In spite of extensive chemical modifications attempted since its structure was elucidated, no analogue has been found to exceed the biological activity of thiamine. However, several antagonists have been discovered in the course of these studies.

When the 2-methyl group in the pyrimidine moiety is replaced by hydrogen, other alkyl groups, or by benzyl, phenyl, alkoxy, or alkylthio groups, only the ethyl homologue (**59**) [*3505-34-8*] shows activity comparable to that of thiamine, whereas some of the other analogues are antagonistic to thiamine (58–62). Replacement of the 4-amino group by a hydrogen, alkyl amino, or mercapto group results in loss of activity (59,63–65), and the 4-hydroxy (4-oxo) analogue, oxythiamine (**18**), is known as a thiamine antagonist. Introduction of a methyl group into the 6-position also decreases the activity. Pyridine analogues (**60**) [*84825-02-5*] and (**61**) [*13860-65-6*], in which the 1- or 3-position of the pyrimidine ring is replaced by a carbon atom, have been synthesized. The pyrophosphate of (**61**) exhibits ca 13% of the coenzyme activity, whereas that of (**60**) is essentially inactive (66), indicating that the nitrogen atom at the 1-position is more important for the coenzyme activity than that at the 3-position. Extension or branching of the methylene group between the two rings leads to loss of activity (67).

In the thiazolium ring, hydrogen at the 2-position and 2-hydroxyethyl group at the 5-position are essential for activity in view of the mechanism of the biochemical reactions as the coenzyme. In fact, introduction of a methyl group at the 2-position (68–69) or modifications of the hydroxyethyl group (58,69–70) give virtually inactive compounds. Analogues in which the sulfur atom is replaced by oxygen (71), selenium (58), nitrogen (34,59), and a —CH=CH— group have been synthesized. None of these, however, shows thiamine activity, and the —CH=CH— analogue, pyrithiamine [*534-64-5*] (**62**) is the most potent thiamine antagonist known (69,72–75). On the other hand, some thiamine antagonists were found to be effective as a prophylactic agent for coccidosis of fowl, and a pyrimidinium derivative, amprolium [*121-25-5*] (**63**), has been marketed as a coccidiostat (76) (see Fig. 7) (see Chemotherapeutics, antiprotozoal).

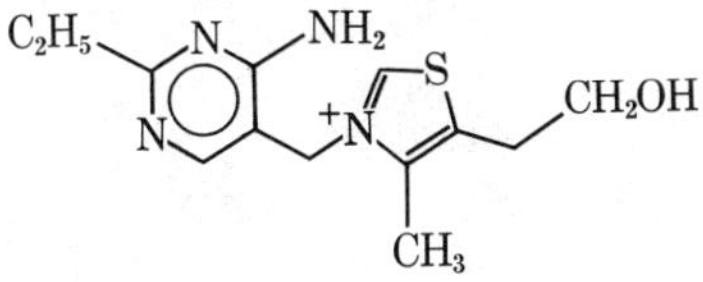

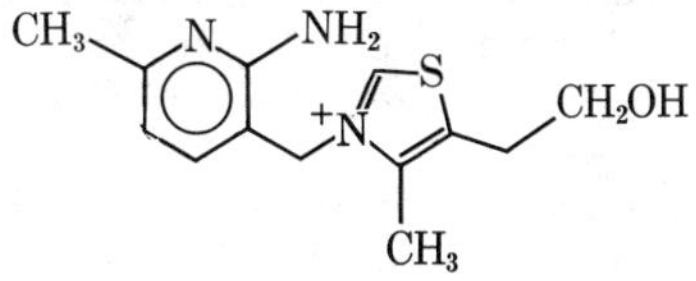

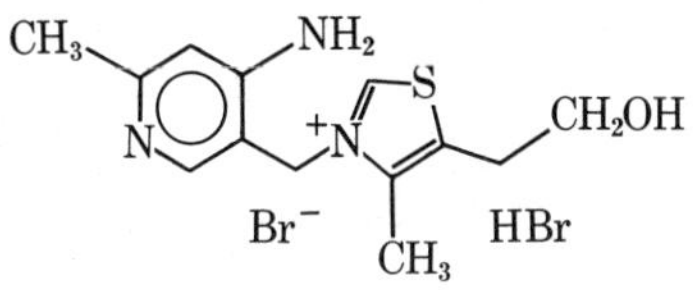

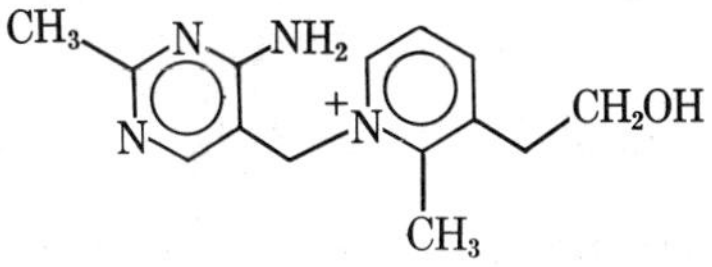

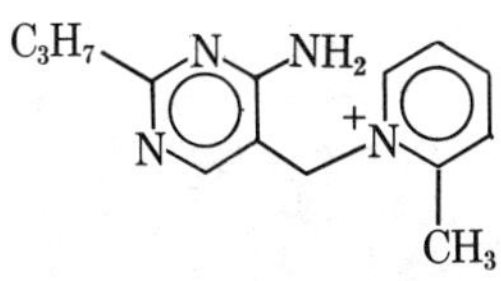

Figure 7. Thiamine analogues.

VITAMINS (B₁)

BIBLIOGRAPHY

"Thiamine" in *ECT* 1st ed., Vol. 14, pp. 38–48, by E. Pierson, Merck & Co., Inc.; "Thiamine" in *ECT* 2nd ed., Vol. 20, pp. 173–184, by P. I. Pollak, Merck & Co., Inc.

1. *Biochim. Biophys. Acta* **107,** 1 (1965).
2. *The United States Pharmacopeia*, XXth rev., United States Pharmacopeia Convention, Inc., Rockville, Md., 1980.
3. K. Takaki, *Transactions of Seiikai Suppl.* **4,** 29 (1885); *Lancet* **2,** 86, 189, 233 (1887).
4. C. Eijkman, *Geneeskd. Tijdschr. Ned. Indie* **36,** 214 (1896); *Arch. Hyg.* **58,** 150 (1906).
5. G. Grijns, *Geneeskd. Tijdschr. Ned. Indie* **41,** 3 (1901); **49,** 216 (1909).
6. E. A. Cooper and C. Funk, *Lancet* **ii,** 1266 (1911).
7. C. Funk, *J. Physiol. (London)* **43,** 395 (1911); *J. State Med.* **20,** 341 (1912).
8. J. C. Drummond, *Biochem. J.* **13,** 77 (1919); **14,** 660 (1920).
9. B. C. P. Jansen and W. F. Donath, *Proc. K. Akad. Wet. Amsterdam* **29,** 1390 (1926).
10. B. C. P. Jansen, *Nature (London)* **135,** 267 (1935).
11. R. R. Williams, R. E. Waterman, J. C. Keresztesky, and E. R. Buchman, *J. Am. Chem. Soc.* **57,** 536 (1935); R. R. Williams and J. K. Cline, *J. Am. Chem. Soc.* **58,** 1063, 1504 (1936).
12. R. Grewe, *Z. Physiol. Chem.* **242,** 89 (1936).
13. A. R. Todd and F. Bergel, *J. Chem. Soc.*, 364 (1937).
14. H. Andersag and V. Westphal, *Ber.* **70,** 2035 (1937).
15. G. Rindi and L. de Giuseppe, *Intern. Z. Vitaminforsch.* **31,** 321 (1961); *Biochem. J.* **78,** 602 (1961).
16. M. P. Lamden in W. H. Sebrell, Jr., and R. S. Harris, eds., *The Vitamins*, 2nd ed., Vol. V, Academic Press, Inc., New York, 1972, pp. 114–120.
17. A. Takamizawa, K. Hirai, T. Ishiha, and I. Makino, *Chem. Pharm. Bull.* **19,** 759 (1971).
18. R. Hopmann and G. P. Brugnoni, *Nature (London) New Biol.* **246,** 157 (1973).
19. G. D. Maier and D. E. Metzler, *J. Am. Chem. Soc.* **79,** 4386 (1957).
20. H. Z. Sable and J. E. Biaglow, *Proc. Natl. Acad. Sci. U.S.A.* **54,** 808 (1965).
21. R. E. Echols and G. C. Levy, *J. Org. Chem.* **39,** 1321 (1974).
22. A. H. Cain, G. R. Sullivan, and J. D. Roberts, *J. Am. Chem. Soc.* **99,** 6423 (1977).
23. M. Hesse, N. Bild, and H. Schmidt, *Helv. Chim. Acta* **50,** 808 (1967).
24. K. Kraut and H. J. Reed, *Acta Crystallogr.* **15,** 747 (1962).
25. B. Pullman and C. Spanjaad, *Biochim. Biophys. Acta* **46,** 576 (1961).
26. A. Watanabe, *Yakugaku Zasshi* **59,** 500 (1939); J. J. Windheuser and T. Higuchi, *J. Pharm. Sci.* **51,** 354 (1962).
27. T. Matsukawa and S. Yurugi, *Yakugaku Zasshi* **71,** 827 (1951).
28. N. Shimahara, N. Nakajima, and H. Hirano, *Chem. Pharm. Bull.* **22,** 2081 (1974); N. Shimahara, H. Asakawa, Y. Kawamatsu, and H. Hirano, *Chem. Pharm. Bull.* **22,** 2086 (1974).
29. J. A. Zoltewicz, G. Uray, and G. M. Kauffman, *J. Am. Chem. Soc.* **103,** 4900 (1981).
30. T. Matsukawa and S. Yurugi, *Yakugaku Zasshi* **71,** 1423, 1450 (1951); **72,** 33 (1952).
31. J. A. Zoltewicz and G. M. Kauffman, *J. Am. Chem. Soc.* **99,** 3134 (1977).
32. K. Murata, *Ann. N.Y. Acad. Sci.* **378,** 146 (1982).
33. A. Watanabe and Y. Asahi, *Yakugaku Zasshi* **77,** 153 (1957).
34. K. Masuda, *Yakugaku Zasshi* **81,** 533, 537, 541 (1961).
35. G. Kurata, T. Sakai, T. Miyahara, and H. Yokoyama, *Bitamin* **35,** 136 (1967); *Chem. Abstr.* **66,** 83320r (1967).
36. H. Hirano, *Yakugaku Zasshi* **77,** 1007 (1957).
37. M. Murakami, K. Takahashi, M. Iwanami, and H. Iwamoto, *Yakugaku Zasshi* **85,** 752 (1965).
38. T. Matsukawa and S. Yurugi, *Yakugaku Zasshi* **72,** 1599 (1952).
39. P. Karrer and H. Krishna, *Helv. Chim. Acta* **33,** 555 (1950); G. E. Bonovicino and D. J. Hennessy, *J. Am. Chem. Soc.* **79,** 6325 (1957).
40. H. Hirano, T. Iwatsu, and S. Yurugi, *Yakugaku Zasshi* **77,** 241, 244 (1957).
41. H. Hirano, H. Yonemoto, and Y. Hara, *Chem. Pharm. Bull.* **7,** 545 (1959).
42. C. Kawasaki, *Bitamin* **30,** 1 (1964); *Chem. Abstr.* **62,** 15240h (1965).
43. T. Matsukawa and T. Iwatsu, *Yakugaku Zasshi* **69,** 550 (1949); **70,** 28 (1950).
44. *Ibid.*, **71,** 720, 1215 (1951).
45. R. Breslow, *J. Am. Chem. Soc.* **80,** 3719 (1958); R. Breslow and E. McNelis, *J. Am. Chem. Soc.* **81,** 3080 (1959).

46. L. O. Krampitz, *Annu. Rev. Biochem.* **38**, 213 (1969); L. O. Krampitz, G. Greull, C. S. Miller, J. B. Bicking, H. R. Skeggs, and J. M. Sprague, *J. Am. Chem. Soc.* **80**, 5893 (1958).
47. R. Kluger, J. Chin, and T. Smyth, *J. Am. Chem. Soc.* **103**, 884 (1981); R. Kluger and T. Smyth, *J. Am. Chem. Soc.* **103**, 1214 (1981).
48. A. Aramaki, T. Kobayashi, K. Furuno, I. Ishikawa, Z. Suzuoki, and S. Shintani, *Bitamin* **16**, 240 (1959).
49. A. Wenz, G. Gottmann, and H. Koop, *Ann.* **618**, 210 (1958).
50. K. Lohmann and P. Schuster, *Naturwissenschaften* **25**, 26 (1937); *Biochem. Z.* **294**, 188 (1937).
51. S. Takei, H. Asakawa, and Y. Fukushima, *Takeda Kenkyusho Ho* **33**, 232 (1975).
52. Y. Itokawa and J. R. Cooper, *Biochem. Biophys. Acta* **196**, 274 (1970).
53. M. Viscontini, G. Bonetti, C. Ednother, and P. Karrer, *Helv. Chim. Acta* **34**, 1384 (1951).
54. T. Yusa, *J. Biochem.* **46**, 391 (1959).
55. O. Zima and R. R. Williams, *Ber.* **73**, 941 (1940).
56. M. Fujiwara and H. Watanabe, *Proc. Jpn. Acad.* **28**, 156 (1952).
57. T. Matsukawa, S. Yurugi, and T. Matsuoka, *Proc. Jpn. Acad.* **28**, 146 (1952); *Science* **118**, 325 (1952).
58. F. Schulz, *Z. Physiol. Chem.* **265**, 113 (1940).
59. H. Neef, K. D. Kohnert, and A. Schellenberger, *J. Prakt. Chem.* **315**, 701 (1973).
60. S. D. Verma and A. N. Day, *J. Indian Chem. Soc.* **40**, 283 (1963).
61. H. C. Koppel, B. H. Springer, R. K. Robins, and C. C. Chang, *J. Org. Chem.* **27**, 3614 (1962).
62. T. Okuda and C. C. Price, *J. Org. Chem.* **24**, 14 (1959).
63. A. Schellenberger, W. Rodel, and H. Rodel, *Z. Physiol. Chem.* **322**, 164 (1960).
64. A. Schellenberger and K. Winter, *Z. Physiol. Chem.* **322**, 164 (1960); **344**, 16 (1966).
65. A. Fumagalli, G. Checchi, and P. Pasotti, *Farmaco. Ed. Sci.* **26**, 736 (1971).
66. A. Schellenberger, K. Wendler, P. Creuzburg, and G. Hubner, *Z. Physiol. Chem.* **348**, 501 (1967).
67. J. Biggs and P. Sykes, *J. Chem. Soc.*, 1849 (1959).
68. F. Bergel and A. R. Todd, *J. Chem. Soc.*, 1504 (1937).
69. A. Schellenberger, K. Winter, G. Hübner, R. Schwaiberger, D. Helbig, S. Schumacher, R. Thieme, G. Bouillon, and K. P. Radler, *Z. Physiol. Chem.* **346**, 123 (1966).
70. A. Takamizawa, K. Hirai, and T. Ishiba, *Chem. Pharm. Bull.* **19**, 2222, 2229 (1971).
71. A. Dornow and H. Hell, *Chem. Ber.* **94**, 1248 (1961).
72. D. W. Wooley and A. G. C. White, *J. Biol. Chem.* **149**, 285 (1943).
73. A. Dornow and A. Schacht, *Chem. Ber.* **82**, 117 (1949).
74. R. F. Raffauf, *Helv. Chim. Acta* **33**, 102 (1950).
75. A. N. Wilson and S. A. Harris, *J. Am. Chem. Soc.* **73**, 2388 (1951).
76. E. F. Rogers, R. L. Clark, A. A. Pessolano, H. J. Becker, W. J. Leanza, L. H. Saret, A. C. Cuckler, E. McManus, M. Garzillo, C. Malange, W. H. Ott, A. M. Dickson, and A. Van Iderstine, *J. Am. Chem. Soc.* **82**, 2974 (1960).

General References

W. H. Sebrell and R. S. Harris, *The Vitamins*, 2nd ed., Vol. V, Academic Press, Inc., New York, 1972, pp. 98–133.
F. A. Robinson, *The Vitamin Co-Factors of Enzyme Systems*, Pergamon Press, Oxford, 1966, pp. 6–142.
A. F. Wagner and K. Folkers, *Vitamins and Coenzymes*, Interscience Publishers, a division of John Wiley & Sons, Inc., New York, 1964, pp. 17–45.
H. Z. Sable and C. J. Gubler, eds., *Ann. N.Y. Acad. Sci.* **378**, 1 (1982).

YOSHIKAZU OKA
Takeda Chemical Industries, Ltd.

VITAMIN A

The substance first designated vitamin A [11105-57-4], and subsequently vitamin A$_1$ [68-26-8] (1), is the primary isoprenoid polyene alcohol (all-E)-3,7-dimethyl-9-(2,6,6-trimethyl-1-cyclohexen-1-yl)-2,4,6,8-nonatetraen-1-ol. *Chemical Abstracts* prefers the base name retinol for (1) and adheres to a numbering system for the carotenoids to show the relationship between the two (see Terpenoids). As a consequence, in *Chemical Abstracts* vitamin A aldehyde is retinal (2); vitamin A acid, retinoic acid (3); vitamin A$_2$ [79-80-1] (4), 3,4-didehydrovitamin A, 3,4-didehydroretinol.

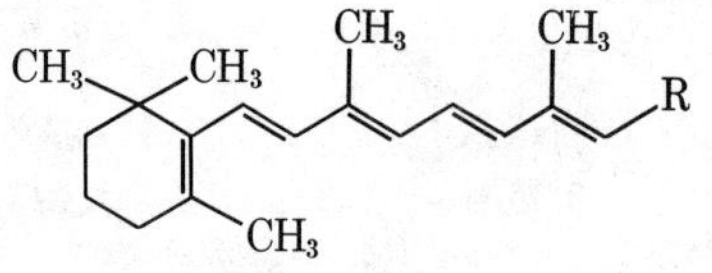

(1) R = CH$_2$OH
(2) R = CHO

$$\text{(3) R = } \overset{\overset{\text{O}}{\|}}{\text{C}}\text{OH}$$

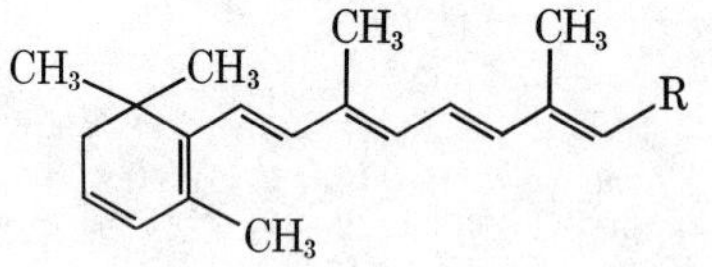

(4) R = CH$_2$OH

Vitamin A is indispensable for normal development and functioning of the ectoderm (ie, the cutaneous, mucosal, and epithelial tissues) and for normal growth, reproduction, and vision. Whereas its role in vision is largely understood (1–2), its other physiological modes of action remain unclear although progress has been made (3). The structural specificity required for a molecule to show vitamin A activity is very narrow. In general, any change in the molecule except at the terminal carbon atom (C-15) considerably reduces growth-promoting power. Vitamin A usually occurs in nature esterified as the palmitate. It is found in animal organisms only. The most abundant source is fish-liver oils. It is, in practically all cases, a metabolic product of carotenoids, notably β-carotene. For this reason, carotenoids that are a source of vitamin A in animals are called provitamin A. Vitamin A$_2$ (2) accompanies vitamin A in fish-liver oils. Physiologically, vitamin A$_2$ has only about 40% the biological activity of (1). As a vitamin A source, vitamin A$_2$ is important only for fish and lower animals. It is likely that in the course of evolution nature changed from the use of vitamin A$_2$ to vitamin A$_1$. Tadpoles, for instance, need vitamin A$_2$, but frogs after metamorphosis need vitamin A$_1$ for vision.

In 1909, a fat-soluble principle obtained from egg yolk which proved essential for life was described (4). Shortly afterwards, the same factor was found in butter fat, egg-yolk extract, and cod-liver oil. Over a period of 20 yr, vitamin A was differentiated from the other fat-soluble biologically active substances, namely the antirachitic vitamin D, the antioxidant vitamin E, and vitamin K which is essential for blood clotting. It became apparent that a positive correlation existed between growth-promoting activity and spectral extinction at 325–328 nm and the Carr-Price reaction (blue coloration at 620 nm with antimony trichloride in chloroform). This permitted a reasonably accurate estimation of the vitamin without resorting to laborious bioassays. In plant products, however, vitamin A activity increases with the intensity of the yellow carotene color. Crystalline β-carotene exerts high vitamin A activity and, in the rat, β-carotene is metabolized to vitamin A and stored as such in the liver. The structural elucidation of carotene and vitamin A in 1931 provided the simple chemical explanation that vitamin A has the structure of half of the β-carotene molecule with an added molecule of water in the end position (5). Today, practically all the vitamin A and β-carotene on the market is of synthetic origin.

Historical reviews of the chemical and biological aspects of vitamin A are given in refs. 6–9.

Biosynthesis

Practically all vitamin A found in nature is derived from carotenoids. The biosynthesis of carotenoids follows the usual pattern of terpenoid synthesis. Three units of acetyl–CoA give mevalonic acid, which is decarboxylated to the isoprenoid building block isopentenyl diphosphate. Dimerization leads to geranyl diphosphate, trimerization to farnesyl diphosphate, tetramerization to geranylgeranyl diphosphate. Head-to-head dimerization of this C_{20}-unit yields a colorless carotenoid phytoene, octahydrocarotene, which upon dehydration and cyclization is converted to various carotenoids. In this process, 24 acetyl groups (48 carbon atoms) are converted to a carotenoid molecule with 40 carbon atoms. In higher organisms, the carotenoid is then cleaved at the central 15,15′-double bond to yield two molecules of retinal. In some organisms, an oxidative mechanism beginning at one end of the polyene chain degrades the carotenoid through apo-carotenals to the C_{20}-compound retinal. Retinol, retinoic acid and its esters, and 11-cis-retinal are all derived then from retinal (Fig. 1). Comprehensive reviews of this subject have been published (10–11).

Acetyl–CoA → Isopentenyl diphosphate → Geranyl diphosphate →

Geranylgeranyl diphosphate —Dimerization→ Phytoene —Dehydrogenation/Cyclization→

β-Carotene (5) / Carotenoids —(O)→ Retinal (2) ⇌ (H)/(O) Retinol (1)

↘ (O) ↗ ↓ (O)

β-Apo-carotenoids   Retinoic acid (3)

(5)

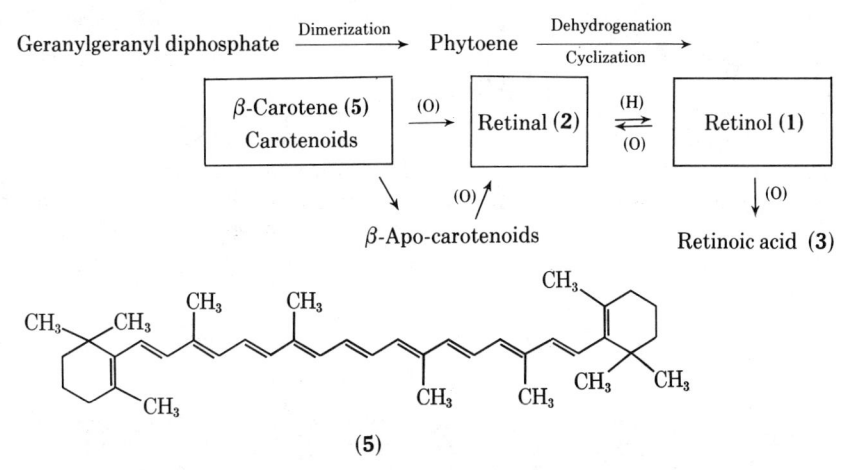

Figure 1. Biosynthetic route to vitamin A.

Stereochemical Aspects. The four double bonds in the vitamin A side chain theoretically permit 16 possible geometrical isomers. The double bonds in the polyene chain of carotenoids and vitamin A are of three types: those for which the adoption of a cis configuration involves very little steric hindrance between two hydrogen atoms (type a), and those for which a cis configuration leads to a serious clash between a methyl group and either a hydrogen atom (type b) or a methyl group (type c) (12):

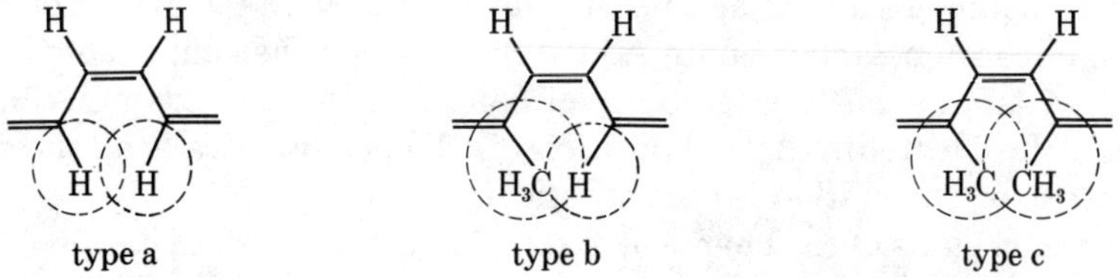

type a   type b   type c

Thus, at one time, only a limited number of isomers were believed capable of existence. Now, however, 13 of the 16 possible geometrical isomers of retinal and eight isomers of retinol have been prepared or shown to exist (13–17). All the unhindered isomers (9-cis, 13-cis, and 9,13-dicis) as well as one hindered isomer (11-cis) occur naturally. The analogous cis isomers of vitamin A_2 are of synthetic origin only. There is some evidence for the existence of 13-*cis*-vitamin A_2 in nature. The cis isomers of vitamin A_1 and A_2 can be obtained by stereoselective synthesis or cis–trans isomerizations. Isomerization is generally carried out by exposing the compound to heat or light, or by catalysis with iodine or acid (13).

The 7-cis isomers are stable only when kept in the dark (14–15). The 9-*cis*-vitamin A isomers are stable but may be isomerized through irradiation in the presence of some sensitizer (18) or by heating with a noble-metal catalyst (19). X-ray structural analyses have been reported for all-*trans*-vitamin A acetate (**6**) (20), all-*trans*-retinal (**2**) (21), and 11-*cis*-retinal (22).

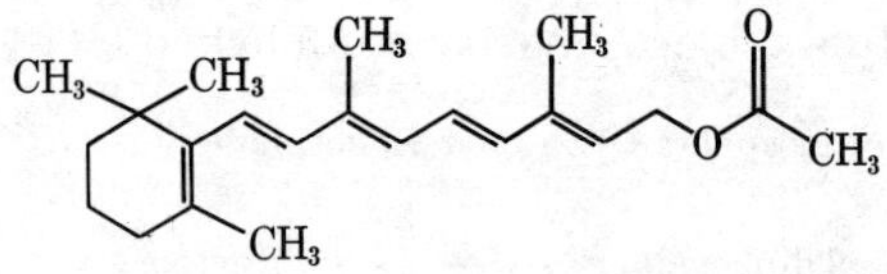

(6)

A saber-shaped bending of the polyene chain was found for all the trans compounds, an effect attributable to the nonbonded interactions between the side-chain methyl groups and the hydrogen atoms at C-11 and C-15. The cyclohexene ring double bond is not coplanar with the polyene chain in any of the compounds examined so far. The torsion angle between ring and chain is 58° for vitamin A acetate (**6**) and retinal and 40° for 11-*cis*-retinal.

Physical Properties

Vitamin A is a pale-yellow crystalline solid (mp 63–64°C) which can be recrystallized from ethyl formate or petroleum ether at low temperature. It is insoluble in water but dissolves readily in most organic solvents. The partition coefficient between petroleum ether and 83% ethanol is 0.82. Chromatographic purification and separation of vitamin A and its isomers is best carried out on nonacidic adsorbents such as alumina or magnesia. Separations by tlc, gc, and hplc have been successfully applied (23–25).

Vitamin A acetate (6) possesses a characteristic uv-absorption spectrum consisting of a single absorption band with a maximum at 325 nm in ethanol ($E_{1\,cm}^{1\%}$ 1830). Vitamin A (1) exhibits the same behavior with a slightly higher molar extinction. Vitamin A_2 (4) shows (in ethanol) a bathochromic shift of 25 nm. In addition, a minimum at 293 nm and two smaller maxima at 277 and 286 nm are observed.

The nmr technique is useful in vitamin A and carotenoid chemistry in the exact assignment and determination of stereochemistry and structure (26). Physical properties of various all-trans vitamin A compounds are listed in Table 1. Physical properties and vitamin A potencies of various stereoisomers of retinol (1) and retinal (2) are shown in Table 2.

Vitamin A acetate is the basis of definition of the international and USP units of biological activity. The content of vitamin A is often reported (since 1967) as retinol equivalent per liter (RE = 3.33 IU = 1 µg retinol or 6 µg β-carotene).

Chemical Properties

Vitamin A is oxidized readily by atmospheric oxygen, especially in the presence of light and heat. Its esters are less sensitive to oxidation. Addition of antioxidants such as the natural α-tocopherol, vitamin E, stabilizes vitamin A. Nevertheless, storage under exclusion of light and air in the cold is advisable. In high vacuum, vitamin A can be distilled. At 667 mPa (0.005 mm Hg), retinol (1) distills at 123°C, vitamin A acetate (6) at 132°C, and vitamin A palmitate at 213°C.

Typical reactions (and reagents) of vitamin A are shown in Figure 2. In all these reactions, the stereochemistry of the polyene chain remains unaffected if light and air are excluded. The same conversions can be carried out in the vitamin A_2 (4) series.

Concentrated hydrobromic acid converts retinol (1) at low temperature to retro-vitamin A [16729-22-9] (7), a yellow oil ($\lambda_{max}^{C_2H_5OH}$ 332, 348, 366 nm):

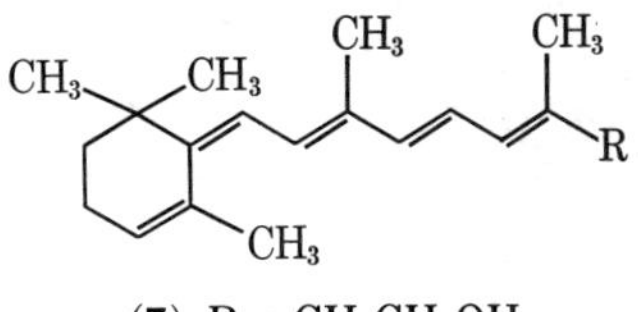

(7) R = CH_2CH_2OH
(8) R = CH=CH_2

Reisomerization of the double bonds back into the polyene chain is possible in retinoic acids only by conversion into their acid chlorides or, in the case of retinal (2), via its hydroquinone adduct (27).

The last two double bonds of the polyene chain participate readily in a Diels-Alder reaction with maleic anhydride (28). The product of vitamin A acetate (6) and maleic anhydride has a melting point of 96–96.6°C. The difference in reaction velocity of the

Table 1. Physical Properties of All-Trans Vitamin A Compounds

Compound	Structure no.	Molecular formula	Molecular weight	Mp, °C	Absorption in ethanol λ_{max}, nm	E_{max}[a]	Carr-Price reaction λ_{max}, nm	E_{max}[a]
vitamin A	(1)	$C_{20}H_{30}O$	286.4	62–64	325	1832	620	4800
vitamin A acetate	(6)	$C_{22}H_{32}O_2$	328.5	57–58	328	1550	620	4580
vitamin A palmitate		$C_{36}H_{60}O_2$	468.8	28–29	328	975	620	2535
vitamin A aldehyde	(2)	$C_{20}H_{28}O$	284.4	61–62	381	1530	664	3470
vitamin A acid	(3)	$C_{20}H_{28}O_2$	300.5	179–180	350	1510		
vitamin A acid, methyl ester		$C_{21}H_{30}O_2$	314.5	56/72–73	354	1415		
3,4-didehydroretinol	(4)	$C_{20}H_{28}O$	284.4	63–65	350	1455	693	4100
3,4-didehydroretinal		$C_{20}H_{26}O$	282.5	78–79	401	1470	741	4200
3,4-didehydroretinoic acid		$C_{20}H_{26}O_2$	298.5	183–184	370	1395		

[a] E_{max} expressed as the value of absorbance of a 1% ethanolic solution in a 1-cm cell.

Table 2. Physical Properties of Stereoisomers of Vitamin A

	Alcohol mp, °C	λ_{max}	$E_{max}{}^a$	Vitamin A potency	Aldehyde mp, °C	λ_{max}	$E_{max}{}^a$	Vitamin A potency
all-*trans*-vitamin A	62–64	325	1832	100%	57/65	381	1530	91%
13-*cis*-vitamin A	58–60	328	1686	75%	77	375	1250	93%
11-*cis*-vitamin A	oil	319	1220	23%	63.5–64.7	376.5	878	48%
11,13-di*cis*-vitamin A	86–88	311	1024	15%	oil	373	700	31%
9-*cis*-vitamin A	81.5–82.5	323	1477	22%	64	373	1270	19%
9,13-di*cis*-vitamin A	58–59	324	1379	24%	49/85	368	1140	17%

^{a} E_{max} expressed as the value of absorbance of a 1% ethanolic solution in a 1-cm cell.

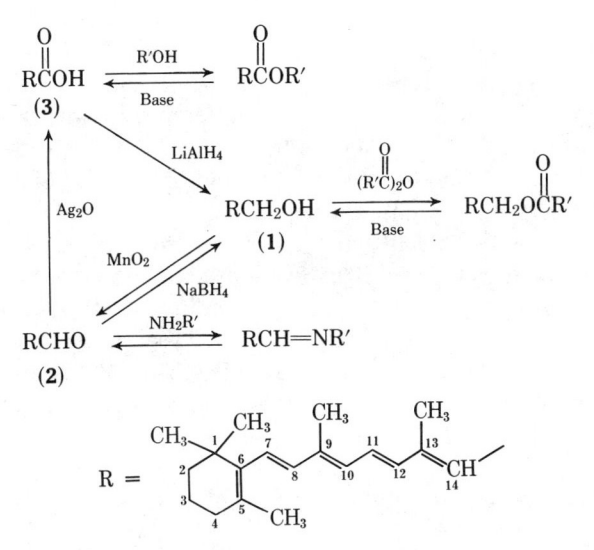

Figure 2. Typical reactions of vitamin A.

various cis–trans stereoisomers in this reaction has been used for their separation (29). Vigorous oxidation of retinol (1) with ozone leads to 2,2-dimethyl-6-oxoheptanoic acid; with potassium permanganate, acetic acid is obtained. Peroxy acids oxidize preferentially the cyclohexene double bond to give an oily 5,6-epoxy derivative.

N-Bromosuccinimide treatment of vitamin A compounds introduces a bromine into position 4 of the cyclohexene ring. Subsequent base-catalyzed elimination of HBr leads to vitamin A_2 (4) derivatives. On the other hand, substitution of the bromine by a hydroxyl group produces, after MnO_2-oxidation, 4-oxo-derivatives (30).

Vitamin A is not attacked by base; however, its allylic hydroxyl group and the polyene chain make it very sensitive to acid. Dilute acids cause dehydration with the formation of crystalline anhydrovitamin A (8), a compound with no significant vitamin A activity.

With strong Lewis acids in $CHCl_3$ solution, vitamin A gives characteristic colors. The blue color caused by $SbCl_3$ ($\lambda_{max}^{CHCl_3}$ 620 nm, $E_{1\,cm}^{1\%}$ 4800), the so-called Carr–Price reaction, is the basis of an analytical procedure (31).

Synthesis

Although it is possible to isolate vitamin A from natural sources, practically all the vitamin A used today is obtained by total synthesis. Of the numerous syntheses, only a few have been exploited commercially. Extensive reviews on this subject have been written (32–35).

The key intermediate in all industrial vitamin A syntheses is β-ionone (9), which until recently was synthesized from citral (10), which in turn had to be obtained from natural sources such as lemon grass or turpentine oil. Today, all β-ionone is manufactured from acetone (11). Figure 3 summarizes the various routes from acetone to (9). The industrial procedures leading from (9) to vitamin A acetate (6), together with the formulas of the key intermediates, are shown in Figure 4.

D.P.I. (Distillation Products Industries, Eastman-Kodak) and Glaxo proceed via β-propargyl-β-ionol (17) in a Grignard reaction with ketobutanal acetal to vitamin A aldehyde (36); A.E.C. (Alimentation Equilibré, Rhône-Poulenc) and Philips-Duphar convert the β-C_{15} aldehyde (14) via β-C_{18} ketone (16) into vitamin A aldehyde and acetate. Lengthening of β-C_{18} ketone by two carbons is effected by A.E.C. (37) with orthoformate and a Grignard reaction and by Philips-Duphar with an ethyl cyanoacetate condensation (38). Sumitomo condenses the β-C_{15} aldehyde (14) with ethyl senecioate to give a vitamin A acid ester (39). BASF rearranges β-ionol (15) to a C_{15} phosphonium salt which then in a Wittig reaction with 3-formylcrotonyl acetate yields vitamin A acetate (6) directly (40–41). Roche condenses the C_{14} aldehyde (12) with (Z)-3-methyl-2-penten-4-yn-1-ol (13) in a Grignard reaction. Partial hydrogenation and rearrangement leads to vitamin A acetate (42–43). Today, only the processes of Roche, BASF, and A.E.C. are used to produce vitamin A on a commercial scale.

New syntheses using 2,2,6-trimethylcyclohexanone as a key intermediate (44) or using carbon–carbon-bond and double-bond formation through activation with sulfones (45–47) have not been commercially exploited.

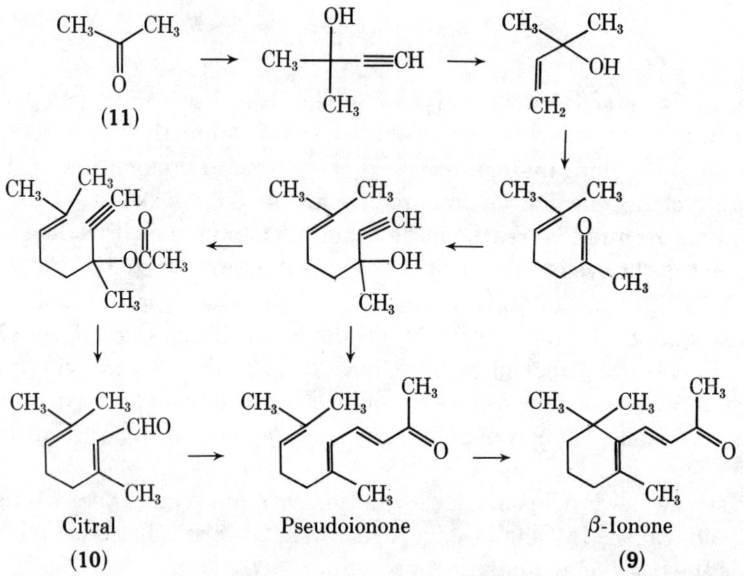

Figure 3. Routes from acetone to β-ionone.

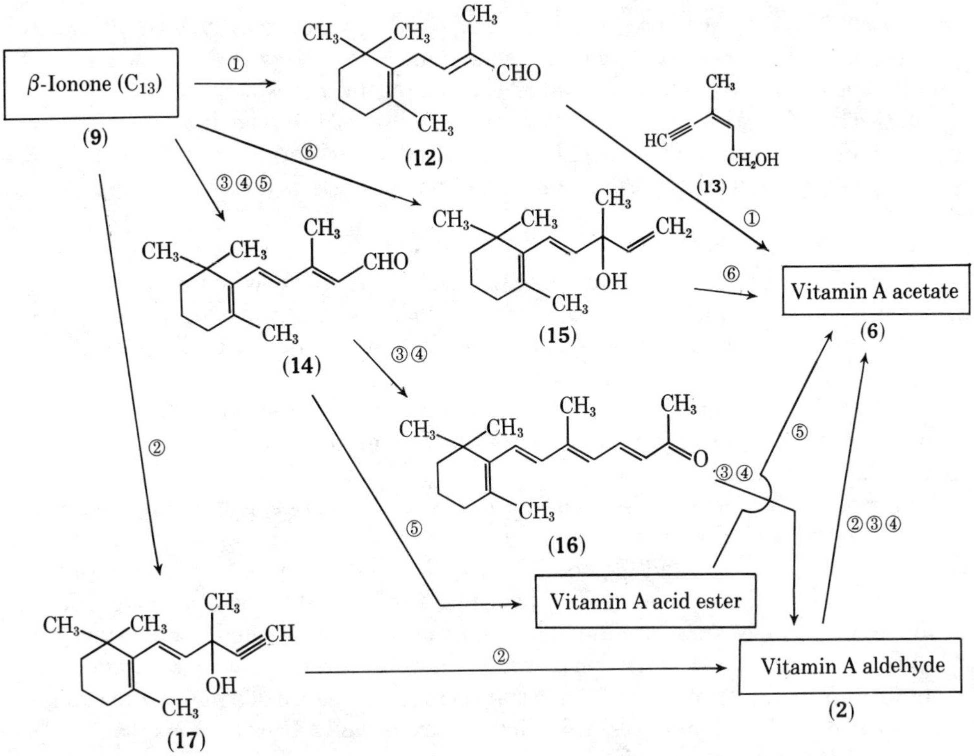

Figure 4. Industrial procedures leading from β-ionone to vitamin A acetate and formulas of key intermediates. ①, Roche; ②, D.P.I./Glaxo; ③, A.E.C.; ④, Philips; ⑤, Sumitomo; ⑥, BASF.

Vitamin A palmitate, an important commercial form of vitamin A, is obtained by transesterification of the acetate with methyl palmitate.

Synthesis of Labeled Compounds. Labeling of vitamin A and related compounds with ^{14}C, ^{13}C, ^{3}H, or ^{2}H is an indispensable tool for the investigation of their absorption, distribution, storage, and metabolism. For vitamin A, the various possibilities of labeling have been reviewed (48).

Cis Isomers of Vitamin A

Properties. Some physical properties of the stable *cis*-retinols and *cis*-retinals are shown in Table 2. The biological potency is, in all cases except the 13-cis compound, considerably less than that of (1).

13-cis-Vitamin A. 13-*cis*-Vitamin A [2052-63-3], 13-*cis*-retinol, formerly called neovitamin A, shows chemical behavior similar to that of retinol, but some of its reactions, such as the formation of anhydrovitamin A (8) and the Diels-Alder condensation with maleic anhydride, are much slower. In the Carr-Price test, it gives the same blue color observed with retinol. Separation of the two is easily accomplished by chromatography, the cis isomer being less adsorbed on alumina, magnesia, or sodium aluminum silicate. Isomerization to retinol (1) may be effected by catalytic quantities of iodine.

9-cis-Vitamin A. 9-*cis*-Vitamin A [22737-97-9] also shows the same color in the Carr-Price reaction and reacts rapidly with maleic anhydride. It is not isomerized in the presence of iodine or on heating. Isomerization to the all-trans form is, however, possible with catalytic amounts of noble-metal catalysts (19).

11-cis-Vitamin A. 11-*cis*-Vitamin A [*22737-96-8*] reacts slowly with maleic anhydride. Oxidation with active manganese dioxide gives 11-*cis*-retinal, the key compound in vision. Iodine catalysis completely inverts the 11-cis configuration.

9,13-dicis-Vitamin A. 9,13-di*cis*-Vitamin A [*29444-25-5*], because of its terminal cis configuration, reacts slowly with maleic anhydride. The Carr-Price color is the same as that observed with the other isomers. Iodine catalysis produces an equilibrium mixture with the 9-cis isomer.

11,13-dicis-Vitamin A. 11,13-di*cis*-Vitamin A [*17706-49-9*] can be isomerized by iodine catalysis to the all-trans compound. It reacts more slowly with maleic anhydride than the 11-cis compound. The other stereoisomers possessing a 7-cis double bond are notoriously unstable. They are usually characterized only by their spectral data and isomerize easily to the trans form.

Synthesis. There are two basically different ways of obtaining stereoisomers of vitamin A. One consists of isomerization of the all-trans compound through heating, sometimes in the presence of a catalyst, to an equilibrium mixture and then isolation of the various stereoforms. The other method uses syntheses specifically designed to give exclusively or predominantly one stereoisomer. The rapid progress in separation techniques, ie, hplc, tlc, and gc, combined with the development of sophisticated identification methods such as nmr and ms, make the first method attractive, at least for the preparation of small amounts. Thus, 13-*cis*-vitamin A can be separated from the all-trans compound by fractional crystallization. Similarly, 9-*cis*-vitamin A can be obtained from the all-trans by isomerization in the presence of a palladium catalyst (19) followed by fractional crystallization. Alternatively, 13-*cis*-, 9-*cis*-, and 9,13-*cis*-vitamin A may be prepared using a modified Sumitomo process (28,49). 9-*trans*- or 9-*cis*-ionylidene acetaldehyde was condensed with β-methylglutaconic ester or senecio ester in the presence of sodium or lithium amide in liquid ammonia. The product, 13-*cis*-retinoic acid or 9,13-di*cis*-retinoic acid, was then reduced with lithium aluminum hydride to the corresponding retinol. The 9-cis isomer was eventually obtained by isomerization of the 9,13-dicis compound. The hindered 11-*cis*- and 11,13-di*cis*-vitamin A was prepared using a modified Roche process (50). Isomers containing the 7-cis and 7,11-dicis geometry were made accessible for the first time by a photochemical isomerization that converts synthetic intermediates with trans stereochemistry into those with cis configuration. Standard reactions then furnish the highly unstable retinols or retinals (14–16). A thermally induced [1,5]-sigmatropic hydrogen shift of 9,10-allenic retinoids has been utilized for constructing the hindered 9,11,13-tri*cis*-retinol and retinal (**2**), as well as the 11,13-dicis and the 11-cis compounds (17). Although the stereoselectivity and yields are not high, the method appears to be specific for producing the elusive 11-cis isomers.

With just one exception, various cis isomers of vitamin A have no commercial significance. The exception is 13-*cis*-retinoic acid, which has been introduced for the treatment of cystic acne as isotretinoin, Roaccutane (Roche).

Vitamin A₂

Vitamin A_2, all-*trans*-3,4-didehydroretinol (**4**) (Table 1), shows reactions similar to those of vitamin A. Thus, it readily esterifies, oxidizes to the corresponding aldehyde, reacts with maleic anhydride, and gives an intense color in the Carr-Price test. Because of its additional double bond, it is more sensitive to atmospheric oxidation than is vitamin A. The unhindered cis isomers as well as the 11-cis and 11,13-dicis isomers have been prepared (51). Vitamin A_2 is of no importance commercially.

Retinoids

All the compounds related to either the natural forms or synthetic analogues of vitamin A have been named retinoids (52). These molecules are characterized by a polyene chain connected at one end to a cyclic end group, and at the other to a polar group such as a carboxyl, ester, amide, or alcohol group. The recognition of these compounds as a special type stems from their similar biological behavior. They play a role in controlling cell proliferation and differentiation and show inhibitory effects on the development of epithelial cancer (53–54).

All-*trans*-Retinoic acid (**3**) (Tretinoin, Airol, Roche) is being used for the treatment of acne. The aromatic retinoid (**18**), sold under the trade name Tasmaderm (Roche), is active against mild forms of acne. Another aromatic retinoid (**19**) is on the market (Tigason, Roche) for the treatment of severe forms of psoriasis and a series of other keratinizing dermatoses.

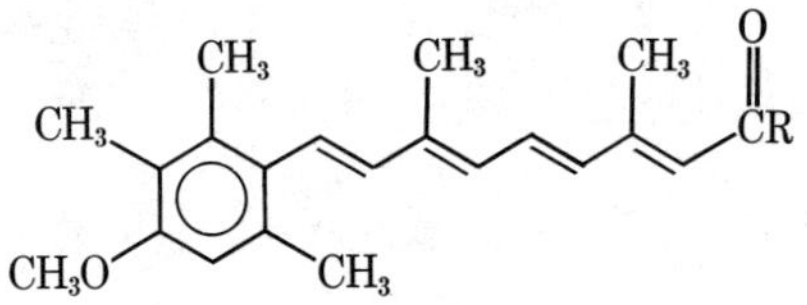

(**18**) R = NHC_2H_5
(**19**) R = OC_2H_5

Other even more active compounds are being evaluated. Already there is strong evidence that these compounds have a preventive effect on various forms of skin tumors and bladder carcinomas (54–56).

Synthesis of these compounds follows procedures well established in vitamin A synthesis. Preferentially, a Wittig reaction is used to join the cyclic end group with the polyene chain (see Phosphorus compounds). The various methods have been reviewed (53). Metabolic studies have been made (57–58), and the major metabolites have been synthesized (59).

The Provitamins A

In a broad sense, any compound that can be converted in the animal body to vitamin A should be called a provitamin A. Usage, however, has confined the term to those naturally occurring carotenoids that are transformed *in vivo* to vitamin A such as α-carotene, β-carotene, γ-carotene, and cryptoxanthin. Certain synthetic analogues also show vitamin A activity. Approximately 40 carotenoids are currently regarded as having vitamin A activity (11).

Provitamin A activity seems to require at least a cyclohexene or a 5,6-epoxycyclohexane ring and a polyene chain equivalent in length to that of vitamin A.

Carotenoids are yellow-to-red pigments occurring in all green tissues of plants as well as in many species of fungi, bacteria, and algae. Animals such as birds and fish acquire them by ingestion of vegetable products. Comprehensive reviews have been written (9–11). Carotenoids are usually composed of eight isoprene units linked in such a manner that the two methyl groups nearest the center of the molecule are separated by six carbon atoms and all the other methyl groups by five carbon atoms. A series of 11 conjugated double bonds constitutes the chromophoric system. This is illustrated by the structural formula of β-carotene (5), the carotenoid that gave the name to the whole class (see Fig. 5). More than 400 naturally occurring carotenoids are known (60); but only about 40 are provitamins A.

Properties. In Table 3, the properties of some common provitamins A are listed. β-Carotene (5) is by far the most important provitamin A. It is the most widely distributed carotenoid in nature and has been found in the leaves of practically all green plants.

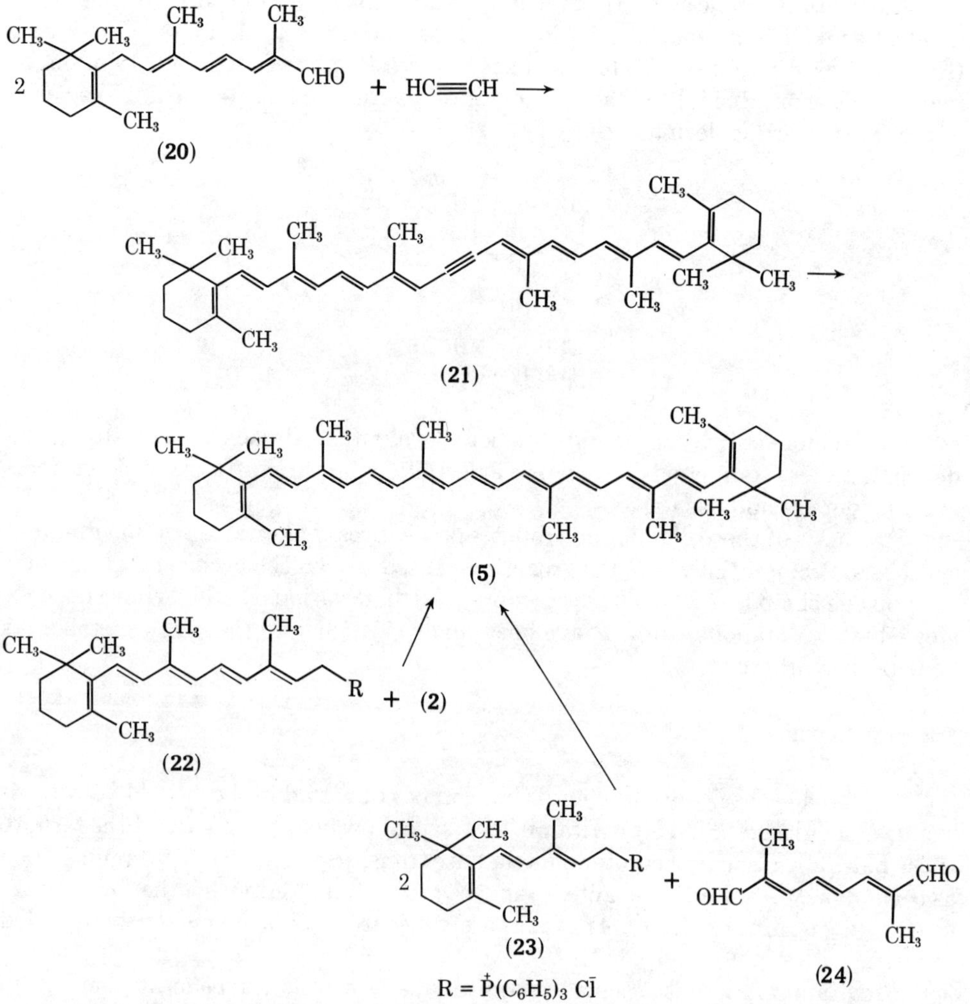

Figure 5. Synthetic routes to β-carotene. (20), β-Apo-14′-carotenal; (21), 15,15′-didehydro-β-carotene; (22), [3,7-dimethyl-9-(2,6,6-trimethyl-1-cyclohexen-1-yl)-2,4,6,8-nonatetraenyl] triphenylphosphonium chloride; (24), 2,7-Dimethyl-2,4-6-octatriendiol.

Table 3. Properties of Provitamin A Compounds

Provitamin A	Structure no.	Molecular formula	Mp, °C	Absorption in petroleum ether λ_{max}, nm	E_{max}[a]	Biological activity
β-carotene	(5)	$C_{40}H_{56}$	180	273	383	100%
				453	2592	= standard
				481	2268	
β-apo-8'-carotenic acid ethyl ester	(26)	$C_{32}H_{44}O_2$	137–138	444	2600	25%
β-apo-8'-carotenal	(25)	$C_{30}H_{40}O$	138–139	457	2640	40%
α-carotene		$C_{40}H_{56}$	162	422	1790	55%
				446	2720	
				473	2500	
γ-carotene		$C_{40}H_{56}$	152–154	437	2055	43%
				462	3100	
				494	2720	
cryptoxanthin		$C_{40}H_{56}O$	158–160	452	2370	57%
				480	1986	

[a] E_{max} expressed as the value of absorbance of a 1% ethanolic solution in a 1-cm cell.

The biological conversion of a carotenoid precursor to vitamin A can be accomplished by either symmetrical or asymmetrical fission or terminal oxidation. The metabolism of provitamins A takes place mainly during absorption in the intestinal mucosa (61).

Most carotenoids are unstable in the presence of oxygen and are converted to colorless products (see Fig. 6). The preferred site for oxygen attack, especially when the reagent is a peracid, is the cyclohexene double bond. Thus, β-carotene is converted by peracetic acid to 5,6:5',6'-β-carotene di-epoxide (62). Oxidative degradation with permanganate results in the production of β-apo-8'-carotenal (25) and β-apo-12'-carotenal (28). Iodine attacks primarily the 5–6 double bond of β-carotene to give crystalline 5,5',6,6'-tetraiodo-β-carotene. Subsequent treatment with sodium thiosulfate leads to 3,3'-didehydro-β-carotene (29) (63). In the presence of aluminum amalgam, hydrogen adds to the end of the polyene chain to give 7,7'-dihydro-β-carotene (30) (64). In a technical synthesis, β-carotene is converted with N-bromosuccinimide in acetic acid to 4,4'-diacetoxy-β-carotene, which upon saponification and oxidation yields canthaxanthin (65), an important food-coloring agent.

The provitamins A are insoluble in water and glycerol, poorly soluble in ethanol, and fairly soluble in chloroform and benzene. They are, like vitamin A, unstable when exposed to light, oxygen, or acids. Isolation is usually achieved by partitioning between solvents followed by chromatographic separation (66). In ethanol, (5) shows two maxima (452 and 481 nm) in its visible-absorption spectrum. The ir spectrum of β-carotene is not very characteristic. The Raman spectrum, on the other hand, exhibits intensive bands at 1150 and 1520 cm^{-1} which may serve to identify β-carotene even in a crude mixture. The interpretation of the nmr spectra in CDCl$_3$ is simplified by the symmetry of the β-carotene molecule. The mass spectrum shows three characteristic cleavage products at 444, 430, and 378 arising through elimination of toluene,

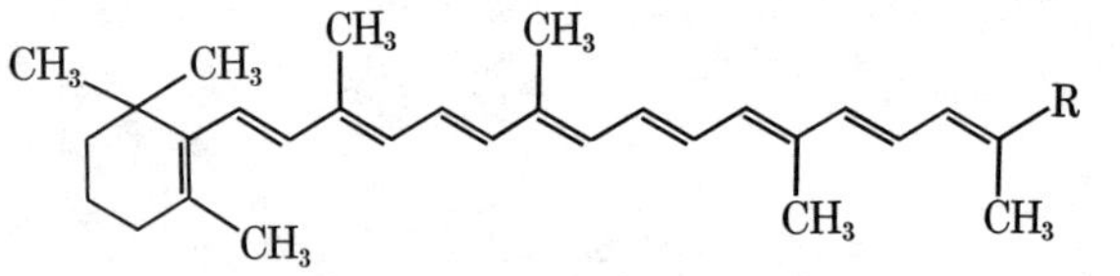

152 VITAMINS (A)

(25) R = CHO

(26) R = $\overset{\overset{\displaystyle O}{\|}}{C}OC_2H_5$

(27) R = CH=CH$\overset{\overset{\displaystyle O}{\|}}{C}CH_3$

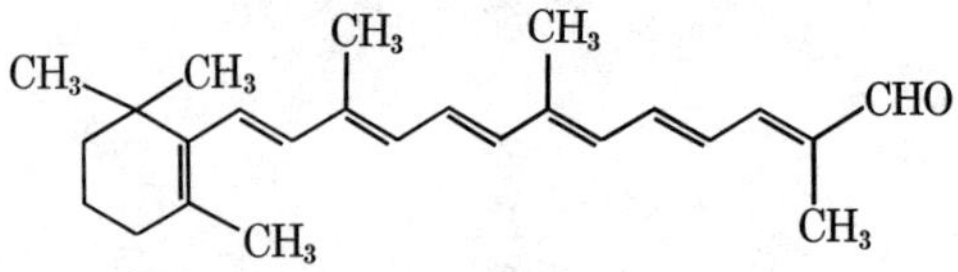

(28)

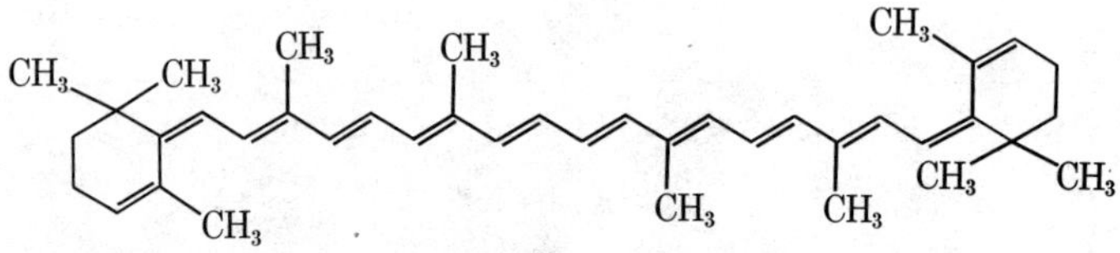

(29)

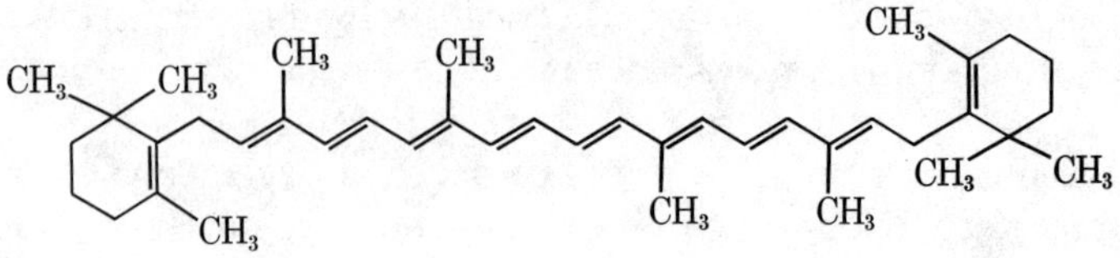

(30)

Figure 6. Carotenoids. (25), β-Apo-8'-carotenal; (26), β-apo-8'-carotenoic acid, ethyl ester; (27), citranaxanthin; (28), β-apo-12'-carotenal; (29), 3,3'-didehydro-β-carotene; (30), 7,7'-dihydro-β-carotene.

xylene, and 2,6-dimethylnaphthalene, which are formed from the polyene chain (26).

Syntheses. The syntheses of carotenoids have been reviewed (9,33–34). The most important provitamin A, β-carotene, is at present synthesized industrially in two different ways. Roche synthesizes it by condensation of two moles of the C_{19}-aldehyde (20) with one mole of acetylene. The intermediate 15,15'-didehydro-β-carotene (21) leads upon partial hydrogenation and isomerization to β-carotene (67). BASF makes use of Wittig condensations of either two moles of the C_{15} phosphonium salt (23) with a C_{10} dialdehyde (24) or of a C_{20} phosphonium salt (22) with retinal (2) (Fig. 5) (68–69).

Three other provitamins A, β-apo-8'-carotenal (25), β-apo-8'-carotenic acid ethyl ester (26) (70–71), and citranaxanthin (27) (68,72), are also produced on an industrial scale and sold mainly as food colorants (see Colorants for foods, drugs, and cosmetics).

Analytical Aspects and Standards

Analysis. Various bioassays have been developed for the determination of vitamin A activity. These methods have been based on such specific responses as night blindness, xerophthalmia, cornification of vaginal epithelium, blood levels, and liver storage as well as such nonspecific responses as growth during prophylactic or curative dosing or survival after a single dose (73–77). As a rule, however, vitamin A content is determined by chemical and physical methods. Four different analytical methods are being used for the exact determination of vitamin A content (78–79):

(1) The measurement of uv absorption is applicable for relatively pure samples.

(2) The Carr-Price reaction ($SbCl_3$ in $CHCl_3$) gives rise to an unstable but characteristic blue color with vitamin A and its esters. The same color develops with cis–trans isomers. Carotenoids disturb the determination.

(3) Retinol is converted in the presence of acids to anhydrovitamin A with a characteristic uv-absorption. This anhydrovitamin A method is more specific, but less sensitive, than the Carr-Price reaction.

(4) Measurement of fluorescence is a simple, sensitive, and very specific method. Prior chromatographic purification may, however, be necessary (80–81). Purification by hplc is the method of choice (82).

Analytical determination of vitamin A in food and feed necessitates first an alkaline hydrolysis of all vitamin A esters to retinol, followed by extraction and chromatographic purification. The product must be protected from light and air during all manipulations. Provitamins A also have to be purified by chromatographic methods prior to analyses, which are usually carried out by spectroscopic methods.

Standards. In 1934, the international unit of vitamin A was defined by the Permanent Commission of the Health Organization of the League of Nations as the vitamin A activity of 0.6 µg of β-carotene. When pure synthetic vitamin A had become available, the USP in 1948 established the USP unit of vitamin A as equal to 0.3 µg of vitamin A alcohol or 0.344 µg of vitamin A acetate. The same definition of an international unit was made in 1949 by the Committee on Biological Standardization of the World Health Organization, which also retained the definition that 1 IU of vitamin A equals 0.6 µg β-carotene.

Because the utilization of dietary provitamins is considerably poorer than that of retinol, the expression of the total vitamin A activity of a diet as IU has to be qualified by indicating the percentages of the activity coming from retinol and from the provitamins. Therefore, in 1967, a FAO–WHO Expert Committee decided to abandon the expression of vitamin A value of foods as IU and proposed that vitamin A activity be stated as the equivalent weight of retinol. The term "µg retinol equivalent" was introduced in 1969. By definition, one retinol equivalent (RE) is equal to 1 µg of retinol, or 6 µg of β-carotene, or 12 µg of other provitamin A carotenoids. In terms of IU, 1 RE is equal to 3.33 IU of retinol or 10 IU of β-carotene. These equivalencies are based primarily on rat bioassays (83).

Two reference standards are available for use in vitamin A bioassays. The USP Vitamin A Reference Solution is a solution of vitamin A acetate in cottonseed oil at a defined potency of 100,000 USP units per gram; it is supplied in the form of gelatin capsules containing 250 mg of this solution. The ANRC (American National Red Cross) Vitamin A Reference Standard is in the form of gelatin beadlets derived from the USP

Vitamin A Reference Solution and containing about 10,000 USP units of vitamin A per gram. The USP standard is suitable for direct oral dosing, but it is not stable when added to the test diet. The ANRC standard is stable in diets and suitable for use in bioassays.

The normal daily requirement of vitamin A for adults is about 5000 IU (or 1500 RE).

Biological Aspects

Although vitamin A is known to play a vital role in general metabolism, the only biological function in which its action is clearly understood is vision.

The first step after the uptake of vitamin A as such or after its formation from provitamin A in the intestine is esterification with long fatty acids. These esters are then transported via the lymph or blood to the liver where they are stored. The liver releases unesterified retinol slowly into the bloodstream at such a rate that its concentration remains roughly constant. Retinol-binding protein, a protein with a molecular weight of ca 21,000, functions as a carrier of vitamin A to a specific receptor site (84–86).

In the retina, vitamin A alcohol is enzymatically oxidized to all-*trans*-retinal (2) which in turn is isomerized, also enzymatically, to 11-*cis*-retinal. It is only this cis form that combines with the protein opsin to form the purple pigment rhodopsin, which is the photoreceptor in the rod cells of the retina. Rhodopsin is sensitive to light of low intensity.

In the cone cells of the retina, the photoreceptors are composed of 11-*cis*-retinal and different proteins. The pigment in this case is iodopsin, which is sensitive to light of high intensity and different colors.

In rhodopsin, 11-*cis*-retinal is bound to opsin via a Schiff's-base-type linkage. When exposed to light, both rhodopsin and iodopsin are bleached. The original protein and all-*trans*-retinal (2) are set free, and the energy released in this chemical break-up is imparted to the optic nerve, resulting in vision. The all-*trans*-retinal can then be reisomerized in the eye to the 11-cis form and the cycle repeated (Fig. 7) (87–92).

In freshwater fish, vitamin A₂ functions in an analogous manner forming the corresponding analogues of rhodopsin (porphyropsin). Other compounds related to vitamin A but not convertible to retinal *in vivo* cannot function in vision. However, some remain effective in promoting growth. Retinoic acid, for instance, when fed to rats maintains their growth although these rats go blind.

```
                         Rhodopsin
                        ↗         ↖ Light
   Opsin + 11-cis-retinal  ⇌  Opsin + all-trans-retinal (2)
                          Isomerase
         ↕                                    ↕
   Alcoholdehydrogenase              Alcoholdehydrogenase
        + NAD                              + NAD
         ↕                                    ↕
      11-cis-Retinol        ⇌        all-trans-Retinol (1)
         ↕                                    ↕
   11-cis-Retinylester      ⇌        all-trans-Retinylester
```

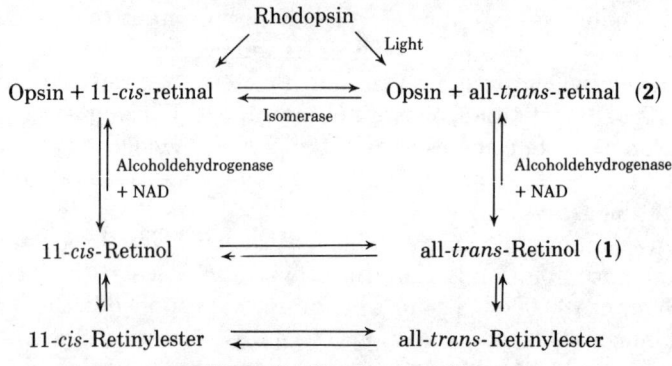

Figure 7. Rhodopsin cycle.

Vitamin A deficiency manifests itself in changes of the eye (xerophthalmia), resulting ultimately in blindness. It is estimated that more than 250,000 children around the world may go blind annually as a result of this nutritional deficiency. Other clinical manifestations associated with vitamin A deficiency are hyperkeratosis of the skin and conjunctional xerosis. In animals, lack of vitamin A causes cessation of skeletal growth, the appearance of xerophthalmia, and cornification of many epithelia, such as those of the respiratory, digestive, urinary, and vaginal tracts (93–94).

Typical symptoms of vitamin A overdosage are headache, lassitude, dizziness, nausea, and vomiting. Chronic hypervitaminosis A is characterized by loss of hair and typical cutaneous and mucosal changes such as desquamation of epithelial tissues, pruritus, and cracked lips. These symptoms are reversible and clear up on withdrawal of the vitamin A preparations.

Applications

The synthetic vitamin A acetate (6), propionate, and palmitate are the derivatives most generally used. They are offered in a number of application forms depending on the use in pharmaceuticals, food fortification, or animal feeds. Stabilizers (antioxidants) are commonly added but only those that are completely acceptable for food or pharmaceutical use. These include tocopherols (see Vitamin E), ascorbic acid (see Vitamin C) and derivatives, butylated hydroxyanisole, butylated hydroxytoluene, and propyl gallate (see Antioxidants and antiozonants).

In pharmaceuticals, oral or parenteral preparations are usually water-dispersed, oil-free vitamin A, utilizing a nonionic dispersing agent. Tablets and capsules for oral use generally contain finely dispersed vitamin A in gelatin or sugar. Stable gelatinized powder is also used. Margarine, cooking oil, and milk are fortified using oil concentrates of vitamin A. All of these forms are used in animal feeds, which consume most vitamin A (see Pet and livestock feeds).

The provitamin A, β-carotene, or apo-carotenoids are used when a yellow color is required or desirable, as in the case of margarine, butter, cheese, salad dressing, fruit juices, and carbonated beverages.

Economic Aspects

Vitamin A is sold but not necessarily manufactured by A.E.C./Rhône-Poulenc, BASF, Danochem, Duphar/Solvay, Glaxo, Merck (Darmstadt), and Roche. About three quarters of the production is for animal feed; the rest is for food fortification and pharmaceuticals. Trade names are Davitin, Dohyfral, Lutavit, Microvit, and Rovimix. Prices depend on application form and are $30–40 per 10^9 IU retinol in October 1982, ie, $100–133 per 10^9 RE.

BIBLIOGRAPHY

"Vitamins (Vitamin A)" in *ECT* 1st ed., Vol. 14, pp. 791–813, by W. Oroshnik, Ortho Research Foundation; "Vitamins (Vitamin A)" in *ECT* 2nd ed., Vol. 21, pp. 490–509, by W. Oroshnik, Chemo Dynamics.

1. G. Wald and R. Hubbard in R. A. Morton, ed., *International Encyclopedia for Food and Nutrition*, Vol. 9, Pergamon Press, Oxford, 1970, p. 267.
2. W. J. De Grip, S. L. Bonting, and F. J. M. Daemen, *Biochim. Biophys. Acta* **396,** 104 (1975).

3. L. M. De Luca, S. Adamo, P. V. Bhat, W. Sasak, C. S. Silverman-Jones, I. Akalovsky, J. P. Frot-Coutaz, T. R. Fletcher, and G. J. Chader, *Pure Appl. Chem.* **51**, 581 (1979).
4. W. Stepp, *Biochem. Ztschr.* **22**, 456 (1909).
5. P. Karrer and co-workers, *Helv. Chim. Acta* **14**, 1036, 1431 (1931).
6. W. H. Sebrell and R. S. Harris, ed., *The Vitamins*, 2nd ed., Vol. 1, Academic Press, New York, 1967.
7. *Vitam. Horm.* (*N.Y.*) **18**, 289 (1960).
8. U. Schwieter and O. Isler in W. Foerst, ed., *Ullmanns Encyclopaedia of Technical Chemistry*, 3rd ed., Vol. 18, Urban & Schwarzenberg, Munich, 1967.
9. O. Isler, ed., *Carotenoids*, Birkhäuser, Basel, Switzerland, 1971.
10. T. W. Goodwin, *The Biochemistry of the Carotenoids*, 2nd ed., Chapman & Hall, London, 1980.
11. J. C. Bauernfeind, ed., *Carotenoids as Colorants and Vitamin A Precursors*, Academic Press, New York, 1981.
12. B. C. L. Weedon in ref. 9.
13. L. Pauling, *Fortschr. Chem. Org. Naturst.* **3**, 203 (1939).
14. V. Ramamurthy and R. S. H. Liu, *Tetrahedron* **31**, 201 (1975).
15. A. E. Asato and R. S. H. Liu, *J. Am. Chem. Soc.* **97**, 4128 (1975).
16. V. Ramamurthy, M. Denny, and R. S. H. Liu, *Tetrahedron Lett.* **22**, 2463 (1981).
17. C. G. Knudsen, C. S. Carey, and W. H. Okamura, *J. Am. Chem. Soc.* **102**, 6355 (1980).
18. Ger. Pat. 2,210,800 (1972), M. Fischer, W. Wiersdorff, A. Nürrenbach, D. Horn, and F. Feichtmayr (to BASF).
19. Belg. Pat. 819,012 (1971), H. Stoller (to F. Hoffmann-LaRoche & Co.).
20. W. E. Oberhänsli, H. P. Wagner, and O. Isler, *Acta Crystallogr. Sect. B* **30**, 161 (1974).
21. T. Hamanaka, T. Mitsui, T. Ashida, and M. Kakudo, *Acta Crystallogr. Sect. B* **28**, 214 (1972).
22. R. Gilardi, I. L. Karle, J. Karle, and W. Sperling, *Nature* (*London*) **232**, 187 (1971).
23. C. D. Hickman, *Chem. Rev.* **34**, 51 (1944).
24. H. R. Bolliger and A. König in E. Stahl, ed., *Dünnschichtchromatography*, 2nd ed., Springer, Heidelberg, 1967.
25. M. Vecchi, J. Vesely, and G. Oesterhelt, *Chromatography* **83**, 447 (1973).
26. W. Vetter, G. Englert, N. Rigassi, and U. Schwieter in ref. 9.
27. H. O. Huisman, A. Smit, P. H. van Leeuven, and J. H. van Rij, *Recl. Trav. Chim. Pays-Bas* **75**, 977 (1956).
28. C. D. Robeson, J. D. Cawley, L. Weisler, M. H. Stern, C. C. Eddinger, and A. J. Chechak, *J. Am. Chem. Soc.* **77**, 4111 (1955).
29. H. atain and M. Debadard, *C. R. Hebd. Seances Acad. Sci.* **232**, 355 (1954).
30. P. Karrer and R. Entschel, *Helv. Chim. Acta* **43**, 94 (1960).
31. F. H. Carr and E. H. Price, *Biochemical J.* **20**, 497 (1926).
32. H. Mayer and O. Isler in ref. 9.
33. F. Kienzle and O. Isler in K. Venkataraman, ed., *The Chemistry of Synthetic Dyes*, Vol. 8, Academic Press, New York, 1978, p. 389.
34. F. Kienzle, *Pure Appl. Chem.* **47**, 183 (1976).
35. O. Isler, *Pure Appl. Chem.* **51**, 447 (1979).
36. U.S. Pats. 2,676,990; 2,676,992; 2,676,994 (1956), W. J. Humphlett, D. M. Burness, and C. D. Robeson (to Eastman-Kodak).
37. Fr. Pat. 1,243,824 (1960), G. J. M. Nicolaux, E. A. Gay, J. Martet, R. L. H. Mauge, C. M. T. Sandevoir, and A. J. A. Wasmer (to Alimentation Equilibré, Société de Chimie Organique et Biologique).
38. J. G. J. Kok and R. van Morselaar, *Chem. Weekbl.*, 48 (1973).
39. M. Matsui, S. Okano, K. Yamashita, M. Miyano, S. Kitamura, A. Kobayashi, T. Sato, and R. Mikami, *J. Vitaminol.* **4**, 178 (1958).
40. H. Pommer, *Angew. Chem.* **89**, 437 (1977).
41. *Ibid.*, **72**, 811 (1960).
42. O. Isler, W. Huber, A. Ronco, and M. Kofler, *Helv. Chim. Acta* **30**, 1911 (1947).
43. O. Isler, A. Ronco, W. Guex, N. C. Hindley, W. Huber, K. Dialer, and M. Kofler, *Helv. Chim. Acta* **32**, 489 (1949).
44. G. L. Olson, H. C. Cheung, K. D. Morgan, R. Borer, and G. Saucy, *Helv. Chim. Acta* **59**, 567 (1976).
45. M. Julia and D. Arnold, *Bull. Soc. Chim. Fr.*, 743, 746 (1973).
46. A. Fischli, H. Mayer, W. Simon, and H. J. Stoller, *Helv. Chim. Acta* **59**, 397 (1976).
47. G. L. Olson, H. C. Cheung, K. D. Morgan, C. Neukom, and G. Saucy, *J. Org. Chem.* **41**, 3287 (1976).
48. O. Isler, R. Rüegg, U. Schwieter, and J. Würsch, *Vitam. Horm.* (*N.Y.*) **18**, 295 (1960).

49. U. Schwieter, C. von Planta, R. Rüegg, and O. Isler, *Helv. Chim. Acta* **45**, 528 (1962).
50. W. Oroshnik, *J. Am. Chem. Soc.* **78**, 2651 (1956).
51. U. Schwieter and O. Isler in ref. 6.
52. M. B. Sporn, N. M. Dunlop, D. L. Newton, and J. M. Smith, *Fed. Proc. Fed. Am. Soc. Exp. Biol.* **35**, 1332 (1976).
53. H. Mayer, W. Bollag, R. Hänni, and R. Rüegg, *Experientia* **34**, 1105 (1978).
54. R. Lotan, *Biochim. Biophys. Acta* **605**, 33 (1980); R. K. Boutwell and A. K. Verna, *Pure Appl. Chem.* **51**, 857 (1979).
55. *Report on Diet, Nutrition, and Cancer*, National Research Council, National Academy of Sciences, Washington, D.C., 1982.
56. M. B. Sporn, R. A. Squire, C. C. Brown, J. M. Smith, M. L. Wenk, and S. Springer, *Science* **195**, 487 (1977).
57. R. Hänni, F. Bigler, W. Vetter, G. Englert, and P. Loeliger *Helv. Chim. Acta* **60**, 2309 (1977).
58. G. Englert, S. Weber, and M. Klaus, *Helv. Chim. Acta* **61**, 2697 (1978).
59. M. Rosenberger, *J. Org. Chem.* **47**, 1698 (1982); M. Rosenberger and C. Neukom, *J. Org. Chem.* **47**, 1779, 1782 (1982).
60. O. Straub, *Key to Carotenoids, List of Natural Carotenoids*, Birkhäuser, Basel, Switzerland, 1976.
61. H. Kläui and J. C. Bauernfeind in ref. 11.
62. P. Karrer and E. Jucker, *Helv. Chim. Acta* **28**, 427 (1945).
63. P. Karrer and G. Schwab, *Helv. Chim. Acta* **23**, 578 (1940).
64. P. Karrer and A. Rüegger, *Helv. Chim. Acta* **23**, 955 (1940).
65. P. Karrer and R. Entschel, *Helv. Chim. Acta* **41**, 402, 983 (1958).
66. S. Liaaen-Jensen in ref. 9.
67. O. Isler, H. Lindlar, M. Montavon, R. Rüegg, and P. Zeller, *Helv. Chim. Acta* **39**, 249 (1956).
68. H. Pommer, *Angew. Chem.* **89**, 440 (1977).
69. J. Paust, W. Reif, and H. Schumacher, *Liebigs Ann. Chem.*, 2194 (1976).
70. R. Rüegg, M. Montavon, G. Ryser, G. Saucy, U. Schwieter, and O. Isler, *Helv. Chim. Acta* **42**, 854 (1959).
71. O. Isler, W. Guex, R. Rüegg, G. Ryser, G. Saucy, U. Schwieter, M. Walter, and A. Winterstein, *Helv. Chim. Acta* **42**, 864 (1959).
72. H. Pommer and A. Nürrenbach, *Pure Appl. Chem.* **43**, 546 (1975).
73. C. I. Bliss and P. György in P. György, ed., *Vitamin Methods*, Vol. 2, Academic Press, New York, 1951, p. 41.
74. N. B. Guerrant in ref. 73, p. 1.
75. N. D. Embree, S. R. Ames, R. W. Lehman, and P. L. Harris, *Methods Biochem. Anal.* **4**, 43 (1957).
76. P. L. Harris, *Vitam. Horm.* (N.Y.) **18**, 341 (1960).
77. O. A. Roels in ref. 6.
78. *U.S. Pharmacopeia*, 20th ed., U.S. Pharmacopeial Convention, Inc., Rockville, Md., 1980, pp. 845, 933.
79. J. P. Vuilleumier, H. P. Probst, and G. Brubacher in J. Schormüller and W. Diemair, ed., *Handbuch der Lebensmittelchemie*, Vol. 2/2, Springer, New York, 1967, p. 686.
80. J. Kahan in S. P. Colowick and N. O. Kaplan, ed., *Methods in Enzymology*, Vol. 18C, Academic Press, New York, 1971, p. 574.
81. D. P. Parrish, *CRC, Crit. Rev. Food Sci. Nutr.* **9**, 375 (1977).
82. M. H. Bui and B. Blanc, *Experientia* **36**, 374 (1980).
83. *Committee on Dietary Allowances*, 9th revised ed., National Academy of Sciences, Washington, D.C., 1980; J. G. Bieri, *Am. J. Clin. Nutr.* **34**, 289 (1981).
84. L. M. De Luca in H. F. De Luca, ed., *Fat-Soluble Vitamins, Handbook of Lipid Research*, Vol. 2, Plenum Press, New York, 1978, p. 43.
85. F. Chytil and D. E. Ong, *Vitam. Horm.* (N.Y.) **36**, 1 (1978).
86. G. Venkalasmy, J. Glover and M. Colby, *Am. J. Clin. Nutr.* **30**, 1968 (1977).
87. G. Wald, *Angew. Chem.* **80**, 857 (1968).
88. R. A. Morton in H. J. A. Dartnall, ed., *Photochemistry of Vision*, Springer, New York, 1972.
89. E. L. Menger, *Acc. Chem. Res.* **8**, 81 (1975).
90. H. Langer, *Biochemistry and Physiology of Visual Pigments*, Springer, New York, 1973.
91. M. Sheves, K. Nakanishi, and B. Honig, *J. Am. Chem. Soc.* **101**, 7086 (1979).
92. V. Balogh-Nair and K. Nakanishi in C. Tamm, ed., *New Comprehensive Biochemistry Vol. 3, Stereochemistry*, Elsevier Biomedical Press, Amsterdam, 1982.

93. *WHO Tech. Rep. Ser.* **590,** 10 (1976).
94. G. Brubacher, J. P. Vuilleumier, G. Ritzel, R. Bruppacher, H. Stähelin, and M. Stransky, *Nutrition* **6,** 211 (1982).

OTTO ISLER
FRANK KIENZLE
F. Hoffmann LaRoche & Co.

VITAMIN B$_{12}$

Vitamin B$_{12}$ [68-19-9] is a red, crystalline cobalt complex synthesized by microorganisms. It belongs to a group of compounds named corrinoids. All corrinoids contain four reduced pyrrole rings, joined into a macrocyclic ring by links between the α positions; three of these links are formed by methylidene =C— groups and the fourth by a C–C bond. The corrinoids act as catalysts in certain carbon-skeleton rearrangements. Vitamin B$_{12}$ is essential for normal blood formation, certain fundamental metabolic processes, neural functions, and human, animal, and microbial growth and maintenance. Human requirements are extremely small, ca 1 μg daily.

Vitamin B$_{12}$ was discovered during the investigation of pernicious anemia (PA), a disease that for a long time resisted any form of therapy (1–10). In 1926, the first breakthrough was made with the discovery of the effectiveness of liver in the therapeutic control of the disease. This therapy was based on experiments on dogs in 1920. Liver in feed accelerated the regeneration of red cells in dogs made anemic by bleeding (11). Soon after, it was demonstrated that absorption of the antipernicious anemia principle required the combination of an intrinsic factor (IF), present in normal human gastric juice, with an extrinsic factor, present in animal protein (12). The extrinsic factor is vitamin B$_{12}$ and IF is a specific B$_{12}$-binding protein secreted by the stomach to enhance absorption of the vitamin. This discovery was followed by the extraction of the active factors from the liver, so-called purified liver extract (13). It was common practice to treat PA by injecting purified liver extract and testing the reticulocyte response and other improvements. Since 1928, various fractionation procedures have been used to obtain the antipernicious anemia factors from the liver, and finally, preparations were obtained that were active even in amounts of a few milligrams. In 1934, Whipple, Minot, and Murphy were awarded the Nobel Prize in medicine and physiology for liver therapy in anemia.

The fractionation of liver extracts was guided mostly by tests on PA patients. After the discovery in 1947 that *Lactobacillus lactis* Dorner requires liver extracts for growth (14), tests with this bacterium were performed. Shortly thereafter, the Merck team in the United States, headed by Folkers, announced the crystallization and partial chemical characterization of vitamin B$_{12}$ (15); this name was assigned to the antipernicious anemia factor because the number 12 was the first free number on the long list of B vitamins. Simultaneously and independently, vitamin B$_{12}$ was isolated at

Glaxo, UK, by Smith (16), using PA patients for bioassay. A bioassay was, however, not needed for the purified liver extracts whose red color was associated with clinical activity. For the preparation of 15 mg of pure crystalline vitamin B_{12}, ca 1000 kg of fresh liver was required. Isolation of crystalline vitamin B_{12} from a fermentation source utilizing *Streptomyces griseus* was also reported (17). In addition, vitamin B_{12} was obtained from the broth of *Streptomyces aureofaciens* (18) and neomycin fermentation (19).

Following the isolation of vitamin B_{12}, the teams of Folkers, Smith, and Todd obtained several hydrolytic fragments of the B_{12} molecule, including (R)-1-amino-2-propanol, α-ribazole-2′(3′)-phosphate (a phosphoriboside of 5,6-dimethylbenzimidazole), 5,6-dimethylbenzimidazole, cyanide, and cobalt (1,2). The cyanide could be removed by photolysis or reduction, giving vitamin B_{12a} with an H_2O or OH^- group in place of the cyanide. The term cobalamin (Cbl) was then suggested for the entire B_{12} molecule except for the CN^- group, and thus vitamin B_{12} is called cyanocobalamin (CNCbl) and vitamin B_{12a} aquacobalamin or hydroxocobalamin (aqCbl). The aqCbl is biologically fully active like CNCbl (1–2,9).

These degradation studies have given valuable information about the periphery of the molecule. It has proved, however, practically impossible to discover by chemical means only the nature of the structure surrounding the cobalt atom. The total structure of CNCbl was elucidated by x-ray crystallography in 1955 (20).

Vitamin B_{12}, like a number of other vitamins, occurs not singly but as a family of closely related compounds, most of which lack biological activity in animals. Soon after the first crystallization of vitamin B_{12}, analogues were obtained which differ fundamentally from the cobalamins, eg, pseudovitamin B_{12} (adeninylcobamide), containing a purine base in the place of 5,6-dimethylbenzimidazole, whereas cobinamide (**1**) lacks the nucleotide portion. Other analogues, eg, cobyric acid (**2**), lack both the nucleotide and 1-amino-2-propanol.

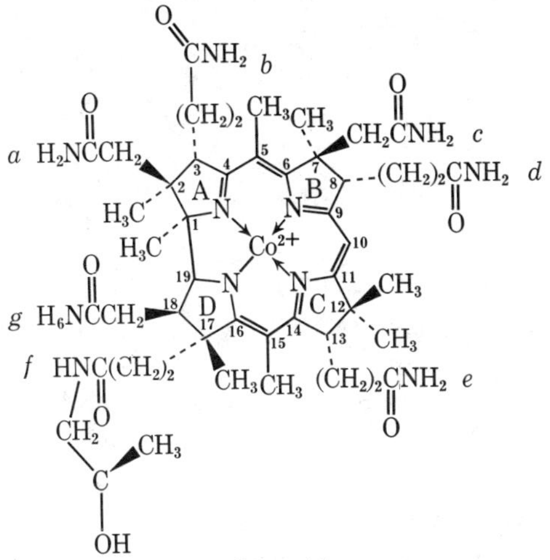

cobinamide
(axial ligands not shown)
(**1**)

The two coenzyme forms of vitamin B_{12} found in bacteria and animals are adenosylcobalamin (**3**) (AdoCbl), also named coenzyme B_{12} (21–22) and methylcobalamin

160 VITAMINS (B$_{12}$)

(MeCbl) (23). Animal sources contain the three main cobalamins: aqCbl, AdoCbl, and MeCbl. Cyanocobalamin occurs only sporadically in small amounts in biological systems. It is now established that CNCbl, isolated in early experiments, was an artifact made by the action of cyanide present in charcoal used as adsorbent or added as papain activator in proteolytic steps. Nevertheless, the name vitamin B$_{12}$ is still used sometimes as a synonym of CNCbl; in this article, vitamin B$_{12}$ mostly refers to the cobalamins and their biologically active analogues.

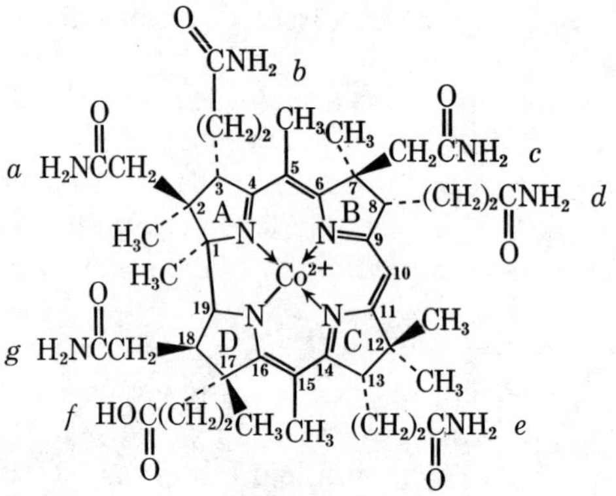

cobyric acid
(axial ligands not shown)
(2)

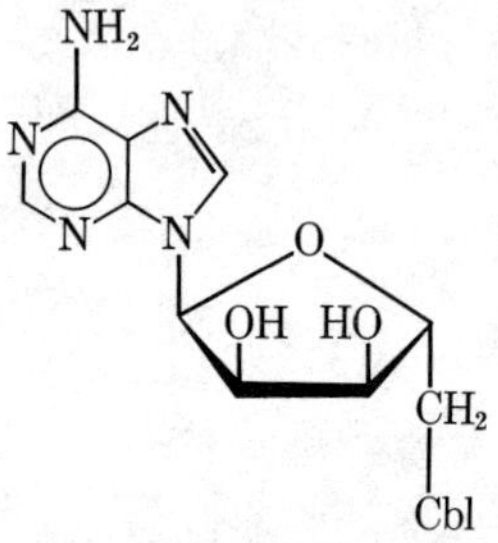

adenosylcobalamin (AdoCbl)
(3)

The structure of coenzyme B$_{12}$ (AdoCbl) was determined by x-ray crystallography (24). In 1964, Hodgkin was awarded the Nobel Prize in chemistry "for determining the structure of biochemical compounds essential in combating pernicious anemia." Further x-ray studies on vitamin B$_{12}$ were described later in a series of publications (25–32).

Vitamin B$_{12}$ is present in food in very small amounts. The main dietary sources are of animal origin. Plant material, in general, contains little or no B$_{12}$ (see Table 1).

Vitamin B$_{12}$ is sold under the trade names Bevatine-12, Berubigen, Betalin-12 crystalline, β-Twelv-Ora, Depinar, Dodecavite, Dodex, Endoglobin, Hepcovite, Normocytin, Poyamin, Rubramin PC, Sytobex, Vibalt, Vitron-C-Plus, Vi-Twel, and Tulag (35). With improvements of commercial yields, the price dropped from ca $800

Table 1. Vitamin B$_{12}$ Content of Foodstuffs[a]

Foodstuffs	B$_{12}$, µg/100 g
cows' milk	0.1–0.6
cheese	0.2–3.0
egg	0.4–1.25
beef, lean	2–3
ox kidney	30
ox liver	60
herring	14
mackerel	5
cod and shellfish	0.5–0.8
green leafy vegetables	0.01

[a] Refs. 33–34.

in 1954 to $18 per gram today (36). Manufacturers include Merck Sharpe & Dohme, United States; Farmitalia, Italy; Glaxo Laboratories, UK; Nippon Oil Company, Japan; and Richter Gedeon Vegyeszeti Gyar Rt., Hungary.

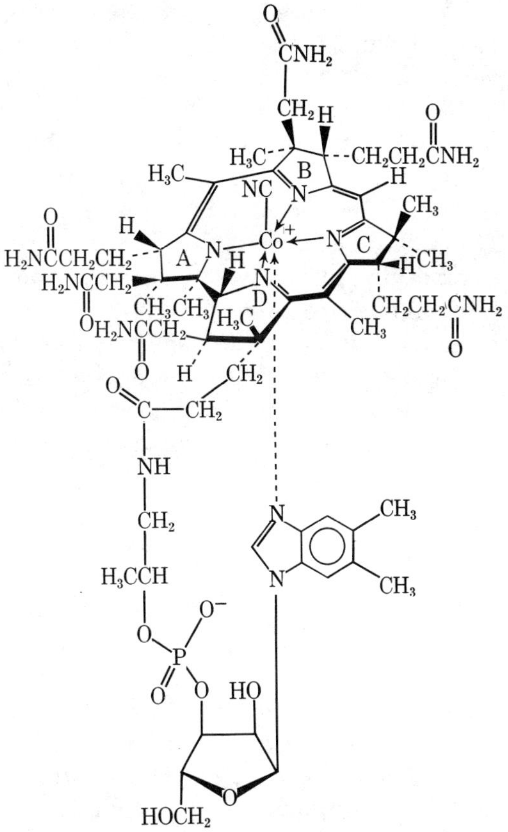

vitamin B$_{12}$ (CNCbl)

(4)

162   VITAMINS (B$_{12}$)

Structure

Vitamin B$_{12}$ (4) is a large, octahedral cobalt complex consisting of a porphyrinlike nucleus with cobalt in the middle, a benzimidazole (or a purine) nucleotide, and a cyano or another group attached to cobalt (37). Magnetic susceptibility measurements, which indicated the diamagnetic character of the vitamin, led to the assignment of the trivalent state of the cobalt ion. The nearly planar macroring system containing four reduced pyrrole rings joined by three bridge carbon atoms is designated corrin. The corrin nucleus differs from porphyrins by having a direct link between rings A and D, fewer double bonds, but more methyl groups and amidated side chains. The presence and position of the additional methyl groups fixes the corrin nucleus in a state corresponding to a reduced porphyrin and breaks the conjugation path between rings A and D. The corrin ring contains six double bonds arranged as a linearly conjugated system involving 12 of the 15 atoms comprising the inner margin of the macrocycle. The central cobalt atom has the capacity to coordinate up to six ligands. Of these coordination positions, four are occupied by the four nitrogens of the corrin ring. A fifth position, usually designated the α or lower position, is usually occupied either by the heterocyclic side chain or solvent; it may, however, be unoccupied. The group occupying the sixth position, the β or upper position, varies. Reactions involving this position are of fundamental importance for the chemistry and biology of vitamin B$_{12}$. In biological systems, the β ligand is usually one of the following groups: H$_2$O, OH$^-$ (for both groups the symbol aq is used), CH$_3$ (Me), or 5'-deoxyadenosyl (Ado); the last is bound to cobalt via C-5' (3). The CN group occurs rarely and is probably not of physiological origin. Like the α position, the β position may be unoccupied. The four cobalamins CNCbl, OHCbl, MeCbl, and AdoCbl are neutral complexes, and the ligands CN, OH, CH$_3$, and Ado are anionic. Because H$_2$O is a neutral ligand, H$_2$OCbl has a positive charge. The β (upper) ligand and one N atom in the corrin nucleus contribute usually two negative charges (in H$_2$OCbl, one) and satisfy two coordinate linkages. The remaining three N atoms of the corrin ring and N-3 of the benzimidazole nucleus contribute four neutral groups satisfying the remaining four coordination requirements. The phosphate ion contributes the third negative charge (in H$_2$OCbl, the second). The Cbl molecule has a very compact arrangement. The benzimidazole nucleus is approximately perpendicular to the corrin nucleus. The ribofuranose ring is nearly perpendicular to the benzimidazole nucleus and nearly parallel with the corrin nucleus.

The numbering in the corrin skeleton is the same as that of the porphyrin nucleus; number 20 is omitted in order to preserve the identity. Vitamin B$_{12}$ and many other important natural corrinoids have a regular pattern of substituents on the methylene carbon atoms of the reduced pyrrole rings; the substituents are four propionamide chains, three acetamide chains, and five methyl groups. In addition, there are three methyl groups on other positions (1, 5, and 15) of the ring. The amide (or carboxyl) groups are designated by the letters a to g as shown in (5), the structure of cobyrinic acid. The propionamide residues of vitamin B$_{12}$ (4) all lie on the α (lower) side, whereas the acetamide residues all lie on the β (upper) side of the corrin ring. Of the peripheral

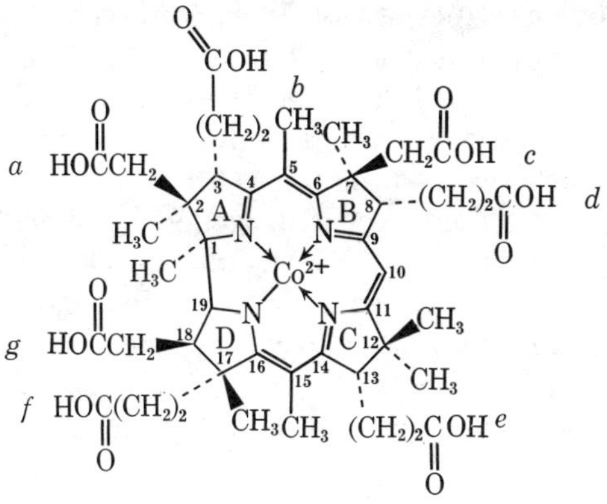

cobyrinic acid

(axial ligands not shown)

(5)

carboxamide residues, six are simple primary amides

$$(-\overset{O}{\underset{\|}{C}}NH_2)$$

The propionamide on the D ring, however, has a more complex structure. The carboxyl group of this substituent is amidated with (R)-1-amino-2-propanol, which in turn is attached by a phosphodiester bridge to the 3′-hydroxyl group of a ribonucleoside. In cobalamins, this ribonucleoside has the structure of α-D-ribofuranosyl-5,6-dimethylbenzimidazole (α-ribazole).

The heptacarboxylic acid (5) (with free carboxylic groups on all side chains of the corrin ring) is designated cobyrinic acid (5) and the a,b,c,d,e,g-hexaamide is cobyric acid (2). The nucleotide-free vitamin B_{12} derivative is designated cobinamide (1). Cobyrinic acid, cobyric acid, and cobinamide are natural precursors of vitamin B_{12}. The base-free vitamin B_{12} derivative is designated cobamide (Cba); it does not occur in nature, but the term cobamide is useful for nomenclature purposes. Using this term, Cbl is 5,6-dimethylbenzimidazolylcobamide, factor III is 5-hydroxybenzimidazolylcobamide, and pseudovitamin B_{12} is adeninylcobamide (1,7,38–41).

Many analogues of Cbl were isolated from natural sources. The two most prominent nucleotide-free analogues are cobyric acid (42) and cobinamide (43). Modifications of Cbl in which the 5,6-dimethylbenzimidazole moiety is replaced by another nucleotide base have been obtained from animal feces, sewage, activated sludge, and by biosynthesis. Modifications containing a purine base include adeninylcobamide (pseudovitamin B_{12}), 2-methyladeninylcobamide (factor A), 2-methylmercaptoadeninylcobamide, and guaninylcobamide. Although these analogues possess significant microbiological activity, they exhibit no activity in animals. The purine base is linked to ribose through N-7 by an α-glycosidic bond and to cobalt through N-9. Analogues of Cbl in which the 5,6-dimethylbenzimidazole moiety is replaced by other benzimidazole derivatives are also known. 5-Hydroxybenzimidazolylcobamide (factor III), found in sewage, is involved in the microbial synthesis of methane. 5-Methoxybenzimidazolylcobamide (factor IIIm) participates in the acetate synthesis by *Clostridium thermoaceticum*. Many analogues of Cbl were obtained by fermentation. When cobinamide and a suitable purine or benzimidazole base are incorporated in a medium for the growth of *Escherichia coli* 113-3, the corresponding cobalt complex is formed. Similar analogues of Cbl can be obtained from *Propionibacterium arabinosum* by incorporating the base into the medium (1,44–46).

Several cobalt-free corrinoids, eg, hydrogenocobyric acid, have been isolated, mostly in small amounts, from *Chromatium vinosum* (47) or some other photosynthetic bacteria. If *Chromatium* is grown in the presence of 5,6-dimethylbenzimidazole, the base is incorporated into the molecule and hydrogenocobalamin is the principal product. Metals such as Cu, Mn, Zn, Rh, Fe, and Co were inserted in isolated metal-free corrinoids. All cobalt-free corrinoids are without biological activity or are antagonists of Cbl (48–50).

Properties

Vitamin B_{12} crystallizes from water or water–acetone as red prisms (51). It darkens at 210–220°C, but does not melt below 300°C. The air-dried crystals contain 10–12% water of crystallization. Vitamin B_{12} dissolves only in highly polar solvents such as water (solubility ca 1.2%), methanol, lower aliphatic acids, phenols, dimethylformamide, dimethyl sulfoxide, and liquid ammonia. It is not soluble in acetone, ether, or benzene. The other cobalamins crystallize similarly and have similar thermal stability (52).

Action of Mineral Acids. Mild treatment of CNCbl with dilute mineral acids successively removes ammonia from the amide groups and gives a series of red, crystalline nucleotide-bearing acids. By electrophoresis at pH 6.5, followed by chromatography, three monocarboxylic acids, three dicarboxylic acids, and one tricarboxylic acid are separated. These experiments show that vitamin B_{12} contains at least three labile primary amide groups. A more rigorous acid degradation gives tetra-, penta-, hexa-, and heptacarboxylic acids and (R)-1-amino-2-propanol. Determination of the ammonia liberated after treating CNCbl with 20% HCl at 100°C gave a value of 5.74%, which corresponds to six mol NH_3 per mol CNCbl. Hydrolysis of CNCbl with 6 M HCl at 150°C for 20 h yields 5,6-dimethylbenzimidazole; milder conditions (6 M HCl at 100°C) yield α-ribazole. Hydrolysis with 1 M HCl at 100°C gives the nucleotide α-ribazole-2'(3')-phosphate (1,52–53). Hydrolysis with Ce(OH)$_3$ is generally used for the isolation of the nucleoside moiety of Cbl and of Cbl analogues; it gives very pure cobinamide (54). All attempts to remove cobalt reversibly from the corrin ring using hydrolytic or other methods were unsuccessful (49).

The corrin ring of the B_{12} molecule contains nine chiral (asymmetric) centers, three of which (at atoms 3, 8, and 13) are configurationally labile in strong acids. Treatment of CNCbl with concentrated $HClO_4$ leads to an equilibrium of two crystallizable cobalamins, the original and neo-Cbl, which is 13-epiCbl with the propionamide chain *e* on ring C located above the plane of the corrin ring; it is biologically inert. This reaction was studied on several corrinoids (55–57).

Exchange of Axial Ligands. Many ligand exchange reactions involve groups in which the coordination to the metal is through nitrogen (NH_3, N_3^-), oxygen (H_2O, OH^-), sulfur (SH^-, SO_3^{2-}), halogen, or carbon (CN^-, CH_3^-). Very important reactions involve the displacement of the heterocyclic base from the α coordination position by a solvent, usually water. This displacement occurs in acidic solutions and results from the protonation of the heterocyclic base. The protonation is associated with a characteristic change in the spectrum. The pK_a for the base-on $\rightleftharpoons$ base-off equilibrium depends on the nature of the β ligand (see Table 2). Displacement of H_2O, Ado^-, or CH_3^- from cobalt by cyanide was also studied. In the presence of cyanide ion, aqCbl and AdoCbl are converted to CNCbl. In contrast, MeCbl and other alkyl corrinoids are stable in the presence of 0.1 M cyanide in the dark (7,39,58–59).

Exchange reactions concerning both axial ligands were studied on a series of acidic nucleotide-free cyanoaquacorrinoids (eg, cobyric acid (2)) and on MeCbl. The CN^- ligand of cyanoaquacobyric acid slowly changes its position at room temperature to reach an equilibrium of two isomers, a (CN^- on lower side) and b (CN^- on upper side) (60–62). Chemical synthesis of MeCbl using Co(I)Cbl results in an equilibrium of two products; the minor product (ca 7%) is the isomer having CH_3^- at the α position. Heating at 80°C or exposure to light transforms the isomer with α-CH_3 to the more stable isomer with β-CH_3 (63–64). Isomeric pairs of various methylcorrinoids and of fluoroalkylcobalamins are described in refs. 65 and 66, respectively. The equilibrium $H_2OCbl^+ \rightleftharpoons HOCbl + H^+$ (pK_a = 6.9–7.8) is not a pure ligand exchange but only a ligand modification. The cobalt-bound water is acidic and reversibly loses a proton at neutral pH.

Degradation and Halogenation of the Corrin Ring. Most reactions of cobalamins that involve a change in the oxidation state of the cobalt ion are accompanied by the formation of biologically inactive by-products. In water solution, aqCbl is rapidly reduced and destroyed by ascorbic acid, whereas CNCbl, AdoCbl, and MeCbl are more stable. The reaction begins with formation of Co(II)Cbl; subsequent slower reactions yield yellow intermediates and eventually lead to release of cobalt (67). Treatment of CNCbl with ascorbic acid under aerobic conditions causes hydroxylation at C-5 and lactam formation between C-6 and C-7 (68). It was suggested that ascorbic acid destroys cobalamins in the human body (69–70), but others have found that in food, humans (67,71), and the rat, Cbl is not attacked by ascorbic acid (72). Chromic acid oxidation of B_{12} yields several pyrrolidine compounds derived from rings B and C, with the side chains intact. These compounds were the first indication of the presence of a pyrrole-like structure in B_{12} (73–74). Vitamin B_{12} can be readily halogenated on C-10 (75).

Methyl Transfer to Metals. The formation of trimethylarsine, methylmercury, and dimethylmercury in the environment are well-known examples of biological systems synthesizing highly toxic organic compounds from less-toxic inorganic substances. In nature, MeCbl is most probably the methylating agent not only for As and Hg salts, but also for salts of Pb, Pd, Pt, Sn, Te, Au, and Se. Under aerobic conditions, MeCbl reacts rapidly with a number of metal ions in aqueous media to give metal alkyls

Table 2. Displacement of 5,6-Dimethylbenzimidazole in Cobalamin by Water in Acidic Solution[a]

β Ligand	pK_a	β Ligand	pK_a
H_2O	−2.4	CH_3^-	2.7
CN^-	0.1	$CH_3CH_2CH_2^-$	3.84
$HC{\equiv}C^-$	0.7	$Ado^{-\,b}$	3.5
$H_2C{=}CH^-$	2.4		

[a] Ref. 58.

and aqCbl. The reactions involve predominantly carbanion (CH_3^-) transfer. The biosynthesis of methylmercury in sediments and by bacteria isolated from sediments is well established. The rate of methylmercury biosynthesis can be enhanced by adding B_{12} to certain bacterial cultures. The mechanism of alkyl transfer from Cbl to metal ions has been thoroughly studied (76–80).

Reduced Vitamin B_{12} as Catalyst. Vitamin B_{12} derivatives and B_{12} model compounds (81–82) catalyze the electrochemical reduction of alkyl halides and formation of C–C bonds (83–84), as well as the zinc–acetic acid-promoted reduction of nitriles (85), α,β-unsaturated nitriles (86), α,β-unsaturated carbonyl derivatives and esters (87–88), and olefins (89). It is assumed that these reactions proceed through intermediates containing a Co–C bond which is then reductively cleaved.

The reaction of Co(II)Cbl, generated by photolysis of MeCbl under a N_2 atmosphere, with DDT results in extensive dechlorination of the latter. In this radical-type reaction, chlorocobalamin, ClCbl, is formed as an intermediate which is readily hydrolyzed to aqCbl and Cl^- (90).

Redox Properties. The cobalt ion in Cbl exists in three oxidation states, each with different ligand-accepting abilities. In CNCbl, aqCbl, AdoCbl, and MeCbl, it is trivalent; aqCbl is readily reduced to Co(II)Cbl, even under mild conditions. At neutral pH, the two redox potentials of the aqCbl–Co(II)Cbl and the Co(II)Cbl–Co(I)Cbl couples are -0.04 and -0.85 V, respectively (67). For chemical reduction of aqCbl to Co(II)Cbl, thiols or carbon monoxide are used. Organocobalamins are reduced by photolysis to Co(II)Cbl which oxidizes slowly in the presence of air to aqCbl. Cobalt-(II)cobalamin contains a single unpaired electron in the $3d_{z^2}$ orbital of the cobalt ion; one of the two axial positions is unoccupied and Co(II)Cbl is regarded as five-coordinate. Chemically, Co(I)Cbl is obtained with strong reducing agents under anaerobic conditions, eg, $NaBH_4$ or zinc and NH_4Cl. It is a powerful reducing agent, and reacts rapidly with oxygen and reduces protons to hydrogen. It is therefore unstable in aqueous acid solution, less so in neutral or basic aqueous solution. It is frequently used for the preparation of organocobalamins, eg, AdoCbl and MeCbl. Cobalt(I)cobalamin contains two electrons in the $3d_{z^2}$ orbital. It has been postulated that the coordination number for the cobalt is four and that both axial ligands are vacant in Co(I)Cbl (7,39,91–94).

Photochemistry. Aquacobalamin is stable to light for long periods of time. Photolysis of CNCbl in air in aqueous solution gives aqCbl, but the reaction is reversed on standing in the dark. From the reversibility of the reaction, it can be concluded that no photoreduction but only photoaquation occurs. Organocorrinoids in solution are attacked by light. The photolytic decomposition of these compounds involves the homolysis of the Co–C bond with formation of Co(II)corrinoids and organic radicals. In the presence of oxygen, Co(II)corrinoids oxidize to the corresponding Co(III)corrinoids. Photolysis in the presence of cyanide gives CNCbl or $(CN)_2Cbl$. The photolysis of organocobalamins in the presence of oxygen rapidly produces aqCbl. The rate of photolysis of AdoCbl is about the same both in the presence and absence of oxygen, whereas that of MeCbl and EtCbl is very much increased by oxygen, which reacts with the generated radical, disturbing the reversibility of the homolysis. The action of light on AdoCbl under anaerobic conditions causes the formation of Co(II)Cbl and the adenosyl radical which rapidly cyclizes to the 5′,8-cyclic 5′-deoxyadenosine. In the presence of air, aqCbl and the 5′-aldehyde of adenosine are formed. The approximate quantum yields of the aerobic photolysis of the Co–C bond by visible light in water at neutral conditions are CNCbl, 0.0001; AdoCbl, 0.1; and MeCbl, 0.3. The quantum yields are almost independent of the irradiating wavelength (95–99).

Synthesis

The unsubstituted corrin ring has been synthesized using linear tetrapyrrolic compounds (98). Another method is described in ref. 100.

The total synthesis of cobyric acid (**2**) was achieved in 1960–1972 by the teams of Woodward (101–103) and Eschenmoser (104–109) in a collaborative effort (110–111). The configurationally labile chiral centers at the atoms 3, 8, and 13 were one of the many handicaps of this synthesis. In one of the two main variants of the synthesis of cobyric acid, schematically shown as

$$\begin{array}{c} A\!-\!B \\ || \\ D\!-\!C \end{array}$$

the two components, A–D (made by Woodward) and B–C (made by Eschenmoser), were linked to give the corrin macrocycle. During the synthesis of β-corrnorsterone, the masked form of the component A–D, the Woodward-Hoffmann rules of the conservation of orbital symmetry were developed in 1965 (112). In the newer variant, performed by Eschenmoser's team, the most interesting point is the photo-induced final macroring closure between A and D.

The path from natural cobyric acid to CNCbl and to several analogues was elucidated by the teams of Bernhauer and Friedrich (113–115). It was confirmed by Woodward and Wuonola (116). In this reaction, the mixed carbonic anhydride of cobyric acid (**2**) was combined with the (R)-1-amino-2-propanol ester of α-ribazole-2'(3')-phosphate. With analogues of (R)-1-amino-2-propanol, eg, 2-amino-2-methylpropanol, analogues of B_{12} were made.

The synthetic route to AdoCbl, MeCbl, and other organocorrinoids mostly used Co(I)Cbl (117–120); for reviews, see refs. 7, 39, 82, 98, 121–122. The most common synthetic routes are nucleophilic displacement reactions (eq. 1) or nucleophilic additions across polarized double bonds (eq. 2). Using equation 1, reactions with methyl halides (eg, CH_3I) give MeCbl, reaction with 5'-p-tosyladenosine gives AdoCbl:

$$Co(I) + RX \rightarrow Co(III)R + X^-; \quad RX = \text{alkyl halide or tosylate} \tag{1}$$

$$Co(I) + H^+ + CH_2\!\!=\!\!CH\overset{\overset{\displaystyle O}{\|}}{C}R \rightarrow Co(III)CH_2CH_2\overset{\overset{\displaystyle O}{\|}}{C}R; \quad R = \text{alkyl} \tag{2}$$

Biosynthesis

Vitamin B_{12} is produced nearly exclusively by bacteria. Corrins and porphyrins have common biosynthetic features (123–127). The best-known corrinoid intermediates on the way to B_{12} are cobyrinic acid (**5**), cobyric acid (**2**), and cobinamide (**1**). Biosynthetic experiments using ^{14}C and ^{13}C labeled precursors have shown that uroporphyrinogen III (uro'gen III) is an intermediate for B_{12} as well as for heme (125,128–131). The conversion of uro'gen III to cobyrinic acid (**5**) requires the introduction of seven methyl groups (derived from methionine), decarboxylation of the acetic acid side chain at C-12, ring contraction, and insertion of cobalt. The ring contraction occurs after methylations at positions 2, 7, and 20; it follows the elimination of C-20 and of the adjacent methyl group (132–139). Amidation on the carboxyls a, b, c, d, e, and g in structure (**5**) gives cobyric acid (**2**). Adenosylation on cobalt is performed in *Propionibacterium shermanii* probably after the first amidation (123). Insertion of (R)-1-amino-2-propanol (derived from L-threonine) into cobyric acid (**2**) leads to cobinamide (**1**). In the next steps, the nucleotide loop is formed in a reaction sequence that is analogous to the formation of phosphatidylcholine from 1,2-diacylglycerol. Both adenosine triphosphate (ATP) and guanosine triphosphate (GTP) activate the cobinamide (**1**); α-ribazole-5'-phosphate is also involved. The final step is the hydrolysis of Cbl-5'-phosphate to Cbl (123,127,140–141). The 5,6-dimethylbenzimidazole originates mostly from flavin mononucleotide (FMN), or via another precursor in some anaerobic microorganisms (142–144).

Aquacobalamin is converted to AdoCbl in microorganisms, and probably in mammalian cells, in three steps (eqs. 3–5) (145–147). Adenosylcobalamin is synthesized intramitochondrially and is the cofactor for methylmalonyl–CoA mutase (148):

$$\text{Co(III)Cbl} + e \xrightarrow{\text{Co(III)Cbl reductase}} \text{Co(II)Cbl} \qquad (3)$$

$$\text{Co(II)Cbl} + e \xrightarrow{\text{Co(II)Cbl reductase}} \text{Co(I)Cbl} \qquad (4)$$

$$\text{Co(I)Cbl} + \text{ATP} \xrightarrow{\text{adenosyltransferase}} \text{AdoCbl} + \text{PPP}_i \qquad (5)$$

where PPP_i = polyphosphoric acid. A series of human mutations affect the cellular uptake and utilization of Cbl (145,149). The conversion of aqCbl to MeCbl in cytosol takes place in the reaction catalyzed by methionine synthase. Most bacterial and animal cells produce mainly adenosylcorrinoids. However, *Streptomyces griseus* produces mainly MeCbl (150).

Microbial Production

Vitamin B_{12} is produced by bacteria and some blue–green algae (151–152) (see also Microbial transformations). Yeasts, such as *Candida utilis*, produce very little B_{12} (153). Since a variety of plants contain B_{12}-dependent enzymes, therefore they may have endogenous corrin compounds (153–154). Microorganisms producing B_{12} are nearly ubiquitous. Rumens have high concentrations of B_{12} (155–156). Because of the anterior position of the rumen, cobalamin is absorbed. In other mammals and in birds, the densest intestinal bacterial populations normally occur in the caecum and colon, and B_{12} produced in these areas is mostly excreted (151). However, the human small intestine often harbors considerable microflora, and at least two groups of bacteria, *Pseudomonas* and *Klebsiella*, may synthesize significant amounts of B_{12}. Free IF (intrinsic factor), often present in the small intestine, can bind the Cbl (157–158). Nevertheless, the role of intraluminal small intestinal bacteria in human B_{12} metabolism is complex, and probably depends on the type of luminal flora present. Various small-bowel disorders associated with stasis of intestinal contents and intraluminal bacterial overgrowth may cause megaloblastic anemia (158–159).

Another rich source of B_{12} is sewage (160), and attempts have been made to utilize sewage for commercial production of B_{12} (161–162). Such recovery is, however, complicated by the existence of Cbl analogues which are produced in various amounts (163–164) and would have to be separated by chromatography.

The B_{12}-producing microorganisms belong to the genera *Propionibacterium*, *Bacillus*, *Corynebacterium*, *Arthrobacter*, *Rhodopseudomonas*, *Protaminobacter*, *Streptomyces*, *Rhodospirillum*, *Actinomyces*, *Selenomonas*, and *Nocardia*. Some of these microorganisms utilize carbohydrates as carbon source, other microorganisms of the genus *Corynebacterium*, *Arthrobacter*, *Pseudomonas*, and *Nocardia* utilize hydrocarbons. Microorganisms of the genus *Pseudomonas* and *Protaminobacter* utilize methanol (165). Bacteria utilizing hydrocarbons or methanol have been described (166–170). Some *Mycobacteria* produce significant amounts of B_{12} (171). Many species belonging to the above groups of microorganisms have been utilized for commercial B_{12} production. Some microorganisms, especially bacteria of the genus *Propionibacterium*, produce >40 mg/L when grown on media containing sugars (166). Methane-producing bacteria synthesize B_{12} in excellent yields using methanol as carbon source.

Microbial production of B_{12} is generally carried out by submerged culture. The vitamin is mostly associated with the cellular material. Centrifugation yields a sludge, which, when dispersed in a minimum of water, water–alcohol, or water–acetone and heated, releases the vitamin into the solution. Addition of cyanide converts the cobalamins into CNCbl which is then extracted from the filtered solution. Many procedures have been reported for this step, including adsorption on charcoal (172), bentonite (173), ion-exchange resins (174–175), 2,4,5-trichlorophenol (176), aluminum oxide (177), or Amberlite XAD-2 (178). For elution, water, water–alcohol, organic bases, or hydrochloric acid are used, whereas extraction is carried out with phenols (179) or benzyl alcohol (180). Chromatography on aluminum oxide and crystallization from methanol–acetone or water–acetone complete the process (181–182). Cyanocobalamin is converted to aqCbl via sulfitocobalamin and nitritocobalamin (183) or with Pd(II) salts (184). Both CNCbl and aqCbl labeled with ^{57}Co, ^{58}Co, or ^{60}Co have been produced commercially (150).

Fermentation. Microorganisms of the genus *Propionibacterium* are widely used for B_{12} production. A typical fermentation medium is given in Table 3. The medium is sterilized in an autoclave at 121°C for 20 min. During fermentation, the pH is maintained at 6–7 by addition of NH_4OH or $CaCO_3$. For the preparation of Cbl, the species *P. shermanii*, *P. freudenreichii*, *P. technicum* may be used. In the initial anaerobic phase of fermentation (ca 3 d at 30°C), only cobinamide (1) is produced. In the second, aerobic phase (2–3 d), Cbl predominates. Maximum yields of Cbl depend

Table 3. Vitamin B_{12} Fermentation Medium

Constituent	Value
cornsteep liquor, mL[a]	80
glucose, g[b]	100
$CoCl_2 \cdot 6H_2O$, g	0.01
water	to make up 1 L

[a] Neutralized to pH 7 with NaOH.
[b] Sterilized separately as a 50% solution.

in part on the effectiveness of the oxygen exclusion (44). Under aerobic conditions, 5,6-dimethylbenzimidazole is synthesized, converting the cobinamide (1) to Cbl. The latter has a repressing effect on corrinoid synthesis. By exclusion of oxygen, neither 5,6-dimethylbenzimidazole nor Cbl are formed (142,185). With *P. freudenreichii*, the optimum pH is 7.0–7.5 at ca 30°C. Culture medium is given in Table 4. Glycine markedly improves the yield of Cbl (186). Maximum yields of Cbl by the above processes are given below:

mg/L	23	24	25	40
ref.	186	187	44	166

The medium for *Streptomyces* may contain soybean meal (3%), glucose (3%), $CaCO_3$, and $CoCl_2 \cdot 6H_2O$ (0.0025%). After aerobic incubation for 5–7 d, the cells are harvested by filtration or centrifugation. The Cbl may be released by acidification to pH 2, heating to 80–90°C, sonication, or by adding 2% NaCl. All cultures of *Streptomyces* form Cbl as grown in the above medium. The highest yields are ca 5 mg/L (44). Using *S. griseus* fermentation, [^{57}Co]–MeCbl and [^{57}Co]–AdoCbl have been obtained with high specific activities, 7.03–8.51 MBq/µg (190–230 µCi/µg) (150).

The medium for *Micromonospora purpurea* contains Bacto-beef extract, Bacto-tryptone, glucose, starch, yeast extract, and $CaCO_3$. After aerobic incubation for ca 2 d at 35°C and filtration, Cbl is adsorbed on charcoal and eluted with pyridine. The best yields are ca 6 mg/L (188).

A strain of *Nocardia*, named *N. rugosa*, synthesizes Cbl under submerged conditions. The optimal fermentation medium at pH 6.6 contains glucose (9%), enzymatic casein hydrolysate (0.5%), plant extract (2.5%), molasses (0.2%), $CoCl_2$ (0.001%), and mineral salts. The yields of Cbl are ca 4 mg/L after a fermentation period of 110 h (189–190).

In the aerobic fermentation using *Arthrobacter hyalinus*, a medium is used that contains low mol wt alcohols or ketones as carbon source. After cultivation for 60 h at 32°C, the separated cells are extracted with 2-propanol. Phenol extraction and column chromatography using triethylaminoethyl (TEAE) cellulose yield ca 1.1 mg Cbl/L (165). This yield is greatly improved (to ca 4 mg/L) if the medium is preincubated with *P. shermanii* and combined with a medium that was incubated with *A. hyalinus* (191).

The methanogenic bacteria *Methanobacterium arbophilicum*, *Methanobacterium formicicum*, *Methanobacterium ruminantium*, *Methanobacterium thermoautotrophicum*, and *Methanosarcina barkeri* produce corrinoids in good yields. The highest concentration of corrinoids (ca 4.1 nmol/mg cell, dry weight) were found in *M. barkeri* grown on methanol (192). Production of B_{12} by fermentation using a

Table 4. Culture Medium for Vitamin B_{12} Production with *Propionibacterium freudenreichii*

Constituent	Parts
yeast extract	20
glucose	25
glycine	0.2
$CoCl_2 \cdot 6H_2O$	0.008
water	1000

methane-producing mixed bacterium population of sewage origin under anaerobic septic conditions was also reported. Methanol, 0.5 vol %, is a good carbon source. Methane-producing bacteria accumulate 93–97% of the B_{12} produced in the cells. In this fermentation process, B_{12}-concentrations as high as 40 mg/L are obtained (193–194) (see also Fermentation).

Metal-free corrinoids have been isolated from the purple sulfur bacterium *C. vinosum* and the purple non-sulfur bacteria *Rhodospirillum rubrum* and *Rhodopseudomonas palustris*. The most important product of the phototrophic bacteria *C. vinosum* and *R. rubrum* is cobalt-free cobyric acid. The yields of corrinoids are ca 0.05 mg/L or 1 mg/100 g of wet bacteria. Certain phototrophic non-sulfur purple bacteria such as *Rhodopseudomonas spheroides* and *Rhodopseudomonas capsulata* form and secrete significant amounts (ca 3 mg/L) of metal-free corrinoids (48,50,195–197). Many metals can be incorporated in metal-free corrinoids (49,198–200).

Vitamin B_{12} Feed Supplements. Evaporation of the final fermentation broth gives low potency concentrates of B_{12} which are used as feed supplements. These final broths contain ca 3% solids and are first evaporated *in vacuo* to a solids content of 15–20%. The resulting syrups are drum-dried or spray-dried. These concentrates are reported to contain 20–60 mg/kg of Cbl. They may also contain other vitamins and antibiotics, depending on the source and method of production (201–204) (see also Pet and livestock feeds).

Stability. The cobalamins are stable when protected against light. Heating an aqueous solution at 94°C for 5 h does not affect the ^{13}C nmr spectra (67). In blood samples, Cbl concentrations show no change at room temperature during two weeks (205). Acids cause hydrolysis of the amide groups and inactivation. Treatment with 0.1 M NaOH at 100°C for 10 min in air gives inactive dehydrovitamin B_{12}, with a lactam ring fused to ring B. Radiation degrades CNCbl in solution, but freezing or benzyl alcohol stabilizes irradiated solutions. Vitamin B_{12} is not stable in the presence of thiamine and niacinamide. Solutions of CNCbl have been stabilized by various antioxidants, chelating agents, phosphate buffer (pH 4.6), and sodium chloride. Before formulation in vitamin–mineral products, CNCbl is stabilized by adsorption on Amberlite IRP-64 (methacrylic acid–divinylbenzene resin) (206).

Specifications, Standards, and Quality Control

Cyanocobalamin ($C_{63}H_{88}CoN_{14}O_{14}P$, mol wt 1355.39) contains, according to the USP (207), "not less than 96% and not more than 100.5% of $C_{63}H_{88}CoN_{14}O_{14}P$ calculated on the dry basis." Vitamin B_{12} is packaged and stored in tight, light-resistant containers; a USP Cyanocobalamin Reference Standard is available. Identification uses the absorption spectrum of a solution that exhibits uv–vis maxima at 278 ± 1 nm, 361 ± 1 nm, and 550 ± 2 nm; the ratio A_{361}/A_{278} is between 1.70 and 1.90 and the ratio A_{361}/A_{550} between 3.15 and 3.40 (207). Oral crystalline CNCbl is marketed by various manufacturers in 1000- and 500-μg strengths (208). Cyanocobalamin injection is a sterile solution of CNCbl in water or isotonic NaCl solution (207). Cobalamin concentrate is the dry, partially purified product resulting from the growth of a Cbl-producing microorganism. It contains no less than 0.5 mg Cbl per g (209). Radiolabeled cyanocobalamin, marketed as capsules or solution, contains [^{57}Co]–CNCbl or [^{60}Co]–CNCbl. The specific activity is no less than 18.5 kBq/μg (0.5 μCi/μg) of CNCbl (see Radioactive drugs). The expiration date is no later than six months after manufacture (209).

Analytical and Test Methods

Isolation and Identification. The isolation of Cbl compounds from natural sources begins with extraction with aqueous alcohols or acetone and evaporation of the solvent (1,210–212). The aqueous extract can be purified by adsorption and extraction. The nonpolar adsorbents Amberlite XAD-2 (178,210) or silanized (reversed-phase) silica gel, both available commercially (210), have replaced extraction with phenol.

Cobalamin mixtures can be separated by column chromatography on linters powder using aqueous n-butanol as eluant (213); column chromatography utilizing SP-Sephadex C-25 for the separation of MeCbl, AdoCbl, CNCbl, aqCbl, and sulfitocobalamin (214); high performance liquid chromatography (hplc) (215–216); affinity chromatography on columns of sepharose containing a covalently bound Cbl-binding protein (217–218); tlc using silanized silica gel (210), cellulose (219), or silica gel (211); ion-exchange chromatography on carboxymethyl cellulose (CMC) or diethylaminoethyl cellulose using water or aqueous buffer solutions as eluants (213); paper chromatography and paper electrophoresis (220). On tlc plates or paper chromatograms, cobalamins are visualized and identified by bioautography, eg, with a B_{12}-dependent *E. coli* mutant (211). For quantitative determination, absorption spectra and determination of cobalt (221) are suitable (see also Analytical methods; Chromatography, affinity).

Microbiological Assay. The microbiological assay (Table 5) is based on the fact that certain organisms grow slowly or not at all in the absence of B_{12} (222–225). The most common technique is the dilution assay using a series of tubes containing sterile medium. Graded doses of the test solution are added to some; to others, standard B_{12} solution. After inoculation and incubation, the microbial growth is measured.

Radioisotope-Dilution Assay. In most laboratories, cobalamin in serum (226), natural waters (216,227), foods (228), and many other natural sources, is measured by competitive ligand-binding radioassay (radioisotope dilution) (229). Commercial kits are available. For natural waters, the sensitivity of the radioassay ranges from 10.0 to ca 200 pg B_{12}/mL H_2O (216). A number of modifications have been developed (230–232). However, with all these modifications higher Cbl values were obtained than by the traditional microbiological methods (10,227,230,233–234). These elevated values occurred by radioassay only with crude IF samples contaminated by R protein (cobalophilin), a nonspecific B_{12} binder and did not occur with pure IF. It is postulated that Cbl analogues are present in human plasma which mask Cbl deficiency because current radioisotope dilution assays are not specific for true Cbl (235).

Table 5. Microbiological Assay of Cbl in Serum[a]

Assay microorganisms[b]	Microorganism responds to	Sensitivity, g Cbl/mL medium
Ochromonas malhamensis	Cbl and several other benzimidazole cobamides	10^{-12}
Lactobacillus leichmannii	Cbl and many clinically inactive Cbl analogues	10^{-11}
Euglena gracilis	Cbl and purine cobamides	10^{-12} to 10^{-15}
Escherichia coli	Cbl, purine cobamides, and cobinamide (1)	10^{-10} to 10^{-11}

[a] Refs. 222–225.
[b] Bioassay media sold by Difco Laboratories, Detroit, Mich.; the assay microorganisms by the American Type Culture Collection, Rockville, Md.

Upon the request of the FDA, the National Committee for Clinical Laboratory Standards (NCCLS) developed a standard for B_{12} radiodilution assays (236). The committee recognized three treatments of the problem of assay binder specifications.

1. Methods utilizing pure IF should not measure cobinamide (a Cbl analogue that is bound by R proteins but not by IF) up to 10 ng/mL in serum.
2. R protein can be blocked by the addition to the binder of, eg, cobinamide.
3. Cobalamin binders other than IF, eg, fish sera or TC (transcobalamin), must be validated by the direct demonstration that these binders do not measure analogues in human serum (237–238). Manufacturers of Cbl assay kits and many laboratories currently utilize one of these approaches to produce a specific binder (239–241). Comparison of commercial kits which were made in accord to the above approaches showed that the results with purified IF (method 1) and with crude IF plus cobinamide (method 2) were comparable to those obtained with the microbiological assay. Thus, these methods can be used to detect cobalamin deficiency (242–243). However, in other measurements using modified radioassay kits (244), it was discovered that normal Cbl levels could appear erroneously low.

The hypothesis of the presence of Cbl analogues in serum as the basis for the erroneous values obtained by some radioisotope-dilution assays (235) was rejected by several workers. No evidence for the existence of circulating Cbl analogues in humans could be found (226,245). The second hypothesis ascribing the erroneous assay values to a Cbl binder other than IF, which binds non-Cbl factors in serum (235), was also rejected. To test the second hypothesis, two assays were made, one using saliva (with R proteins) and the other using human gastric juice (with 99% IF) as Cbl-binding agents. In both assays, sera from patients with PA were differentiated from sera of control subjects (246).

Measurement of Cbl Absorption Using Radioactive Cbl. The patient's ability to absorb orally administered radioactive Cbl can be measured by determination of fecal, hepatic, urinary, plasma, or whole-body radioactivity (247). Oral doses of 0.5–2.0 µg are normally administered; the preferred tracer is ^{57}Co. Theoretically, whole-body counting should be the best method for measuring retention of radioactive Cbl (248). The urine radioactivity technique is, however, more convenient and is widely used (249). In the urinary excretion test, 1 mg of nonradioactive CNCbl is administered subcutaneously or intramuscularly 1–2 h after an oral dose of ca 1 µg of radioactive CNCbl. The urine voided in the 24 h after starting the test is collected and assayed; if the absorption of Cbl is clearly lower than in healthy controls, a second absorption test is recommended giving IF together with the oral Cbl dose (247). To overcome the disadvantages of administering radioactive Cbl with and without IF at a minimum 3-d interval, a dual-isotope technique involving the simultaneous administration of IF saturated with ^{60}Co-labeled Cbl and of free ^{57}Co-labeled Cbl was developed (250); a commercial kit is available (251).

Absorption Spectroscopy. Cobalamins are red or yellow-red compounds with distinct uv and visible spectra (39,82,252–255). The main bands in the absorption spectra above 300 nm result from π–π* transitions within the corrin ring. Intensity and position of the absorption bands are highly sensitive to the character of the axial ligands. For quantitative spectroscopy of corrinoids, the γ band (at 368 nm) of the dicyano form is very useful. Cobalamins and most other natural corrinoids are converted to the dicyano form by the action of cyanide and light. All natural dicyano-

Table 6. Absorption Bands of Cobalamins and Cobyric Acids in Water at Room Temperature [a]

Corrinoid	CAS Registry No.	Abbreviation	Wavelength, nm	Absorbance	pH
cyanocobalamin	[68-19-9]	CNCbl	278	15.5	2–11
			361	28.06	
			551	8.74	
dicyanocobalamin	[15041-09-5]	(CN)$_2$Cbl	279	14.4	12
			289	10.6	
			308	9.7	
			313	9.8	
			368	30.8	
			543	8.6	
			584	9.8	
aquacobalamin	[13422-52-1]	H$_2$OCbl	274	20.6	0–6.8
			317	6.1	
			351	26.5	
			499	8.1	
			527	8.5	
hydroxocobalamin	[13422-51-0]	HOCbl	279	19.0	8–11
			325	11.4	
			359	20.6	
			516	8.9	
			537	9.5	
adenosylcobalamin	[13870-90-1]	AdoCbl	260	34.7	7
			288	18.1	
			315	13.2	
			340	12.3	
			375	10.9	
			522	8.0	
methylcobalamin base-on	[13422-55-4]	MeCbl			
			266	19.1	7
			280	18.3	
			289	17.1	
			315	12.5	
			340	13.3	
			373	10.7	
			519	8.7	
base-off			264	24.8	weakly acidic
			278	19.1	
			286	20.4	
			304	22.4	
			378	8.7	
			461	8.9	
Co(II)Cbl (B$_{12r}$)	[14463-33-3]		288		7
			311	27.5	
			402	7.5	
			473	9.2	
Co(I)Cbl (B$_{12s}$)	[18534-66-2]		280	29.1	5–14
			288	29.4	
			386	28.0	
			455	2.5	
			545	2.8	
			680	1.7	
			800	1.4	
cyanoaquacobyric acid a	[18662-89-0]		274	11.1	7
			321	11.1	
			353	28.3	
			495	9.0	
			527	8.4	

Table 6 (continued)

Corrinoid	CAS Registry No.	Abbreviation	Wavelength, nm	Absorbance	pH
cyanoaquacobyric acid b			275	11.7	7
			322	11.6	
			355	27.4	
			496	8.7	
			528	8.2	

[a] Refs. 39, 253.

corrinoids have identical absorption spectra above 300 nm. For quantitative spectroscopy of CNCbl, the γ band at 361 nm (E_{max} = 207, at 1% soln and 1-cm pathlength) is mostly used. The spectral properties of cobyric acid (2) and several cobalamins are given in Table 6. The spectra of CNCbl, (CN)$_2$Cbl, H$_2$OCbl, and HOCbl are similar and characterized by the intense γ band in the 350–368 nm region. In organocorrinoids, the intensity of the γ band is considerably reduced. In dilute acid, organocorrinoids change from red to yellow; under these conditions, the base is protonated and no longer coordinated to the cobalt. The so-called base-on form is red, the base-off form, yellow.

Optical Rotary Dispersion and Circular Dichroism. Optical rotary dispersion (ord) (56,256–257) and circular dichroism (cd) (66,254,256,258–260) are sensitive methods for the detection of changes in the axial ligands or the corrin ring. For instance, the ord and cd spectra of 13-epicorrinoids are quite distinct from those of natural corrinoids (56). Likewise, the ord and cd spectra of the isomers a and b of methylcobyric acid are distinctly different (256). The cd spectra of the a and b isomers of cyanoaquacobyric acid are, however, only slightly different (261).

Electron Spin Resonance. Of the three oxidation states of corrinoids, only the Co(II)corrinoids are paramagnetic and thus suitable for esr spectroscopy. The esr spectrum is sensitive to the nature of the axial ligand. Thus, the highly resolved spectra of base-on Co(II)Cbl show seven hyperfine lines and superhyperfine splitting into three lines because of the interaction of the unpaired electron with the N nucleus of the axial base. The spectrum of base-off Co(II)Cbl (in acidic solution) does not show superhyperfine splitting (7,39). For the elucidation of the mechanisms of AdoCbl-dependent biological reactions, esr is a very useful tool. In these reactions, AdoCbl gives esr signals that support the hypothesis that radical formation may be important in the reactions of the Co–C bond. However, this is an area of some controversy, since some of the enzyme-catalyzed reactions dependent on AdoCbl do not show esr signals (7,259,262–264).

Proton Magnetic Resonance. Many ^{1}H nmr studies using deuterated solvents, such as D$_2$O, CF$_3$COOD, or (CD$_3$)$_2$SO, have been concerned with the assignment of the resonance positions of single protons or methyl groups and with the conformation of the corrin ring. The proton at C-10 of the corrin ring exchanges in acidic D$_2$O and CF$_3$COOD. It exhibits in cyanoaquacobinamide two resonances assigned to the axial ligand isomers a and b. The chemical shift of protons of coordinated water in cobinamide depends on the nature of the trans ligand (39,116,265–269).

Carbon-13 Magnetic Resonance. The ^{13}C-nmr spectroscopy is a very important tool for the study of corrinoids. Nearly all the lines of the ^{13}C-nmr spectra are well resolved single-carbon resonances (132,270–272). The ^{13}C-nmr spectroscopy has been used extensively in biosynthetic studies of porphyrins and vitamin B$_{12}$. It was shown that seven of the eight methyl groups of the corrin ring are enriched by feeding [^{13}CH$_3$]–methionine to *Propionibacterium shermanii*. The methyl group not enriched is one of those at C-12 (132–133,273–276).

Other Spectroscopic Techniques. Emission Mössbauer studies (277–278) and Field desorption mass spectrometry (279–280) are described.

Cobalamin-Transport Proteins

Specific extracellular, membrane-bound, and intracellular proteins are needed for absorption, storage, transport, and metabolism of cobalamins in blood and tissues (281–285). The extracellular transport proteins are the intrinsic factor (IF), R protein (cobalophilin, non-IF), and transcobalamin (TC). The membrane-bound transport proteins are the receptors for both the IF–Cbl complex, located in the ileum, and the TC–Cbl complex, located in the tissues (see Table 7). The intracellular Cbl binders are probably identical with the enzymes methionine synthase and methylmalonyl–CoA mutase.

The gastric glycoprotein IF binds dietary Cbl very specifically, whereas Cbl analogues are not bound. When bound to Cbl, IF forms a dimer. Most commercial kits for Cbl radioassay contain IF. Radioassay for IF is described in refs. 286–287. Uptake of Cbl by the ileal receptor protein is facilitated by IF. The Cbl subsequently appears in the portal blood where it is bound to TC, a plasma protein. The TC–Cbl complex binds to the tissue-cell receptors, which facilitate the uptake of the complex via pinocytosis. After proteolysis of TC in the lysosomes, Cbl is metabolized to MeCbl in the cytosol or to AdoCbl in the mitochondria.

In the inherited or acquired absence of one of these transport proteins, severe megaloblastic anemia may occur. An acquired absence of IF results in pernicious anemia. An absence of R protein, however, causes no anemia. Only inherited absence is known and the role of this protein is not fully understood. It binds not only Cbl but also Cbl analogues, eg, cobinamide.

The concentration of the Cbl-transport proteins in tissues and body fluids is extremely low, ca 15 µg TC/L serum. These proteins are isolated by affinity chromatography (288–290).

Table 7. Some Properties of the Cbl-Transport Proteins[a]

Protein	Mol wt	Glycosylation	Distribution
IF	44,200	yes	gastric parietal cells; gastric juice
IF–Cbl receptor	90,000 and 140,000	yes	ileal enterocyte
TC	38,000	no	plasma, enterocyte and other tissues
TC–Cbl receptor	50,000	yes	tissue–cell membranes
R protein	58,000	yes	plasma, saliva, milk, granulocytes

[a] Ref. 281.

Biochemical Functions

There are three mammalian essential enzymatic reactions requiring Cbl. The methyl analogue is involved in transmethylation. In the presence of methionine synthase, it transfers the cobalt-bound methyl group, accepted from 5-methyltetrahydrofolate, to homocysteine to give methionine. This reaction appears to be critical in the biosynthesis of methionine and nucleic acids and in the folic acid cycle for one-carbon metabolism. Methionine synthase is a soluble cytosolic enzyme. Lack of Cbl leads to failure of demethylation of tetrahydrofolate to tetrahydrofolic acid which is needed for the production of thymidylate and consequently DNA. The anesthetic gas N_2O inactivates the methionine synthase and leads to anemia (291) and myelin damage (292).

Another mammalian reaction requiring Cbl is the isomerization of (R)-methylmalonyl–CoA (MM–CoA) to succinyl–CoA in the presence of methylmalonyl–CoA mutase (eq. 6). This reaction is important in propionate metabolism. The extensive methylmalonic aciduria in Cbl deficiency is explained by the failure of isomerization of MM–CoA to succinyl–CoA.

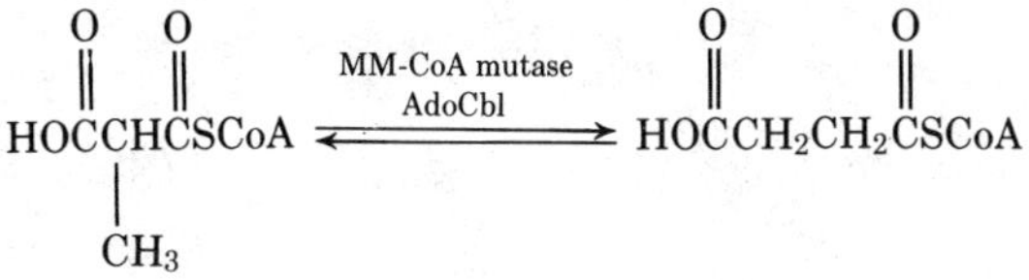

$$\underset{\underset{CH_3}{|}}{HOCCHCSCoA} \xleftrightarrow[AdoCbl]{MM\text{-}CoA\ mutase} HOCCH_2CH_2CSCoA \tag{6}$$

The third mammalian reaction requiring Cbl is the interconversion of α- and β-leucine (eq. 7):

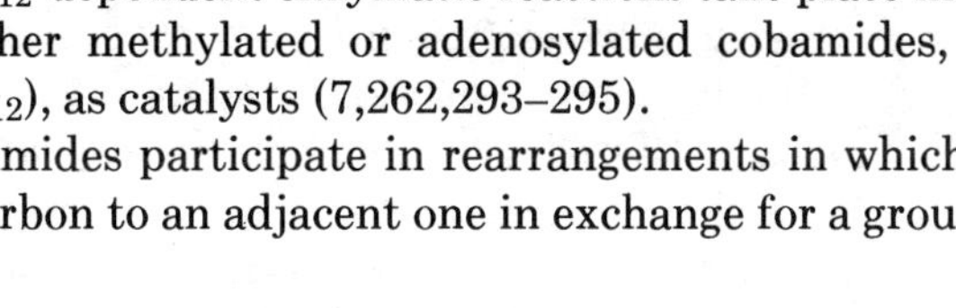

$$\text{L-}\alpha\text{-leucine} \xleftrightarrow{\text{leucine 2,3-aminomutase}} \beta\text{-leucine} \tag{7}$$

Many other B_{12}-dependent enzymatic reactions take place in microbial systems, utilizing many other methylated or adenosylated cobamides, eg, adeninyl–Cba (pseudovitamin B_{12}), as catalysts (7,262,293–295).

Adenosylcobamides participate in rearrangements in which a hydrogen atom moves from one carbon to an adjacent one in exchange for a group X (eq. 8):

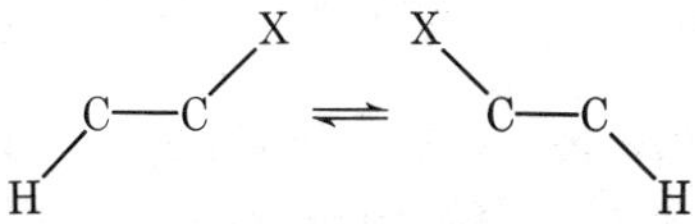

$$\underset{H}{\overset{X}{C-C}} \rightleftharpoons \underset{H}{\overset{X}{C-C}} \tag{8}$$

They also participate in the ribonucleotide reductase reaction in which the 2′-hydroxyl group of the ribose moiety of a ribonucleoside tri- or diphosphate is replaced by a hydrogen atom derived from the reducing agent. To date, ten rearrangements and one reduction are known. The structure of group X varies from reaction to reaction; it is, eg,

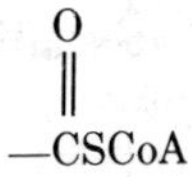

$$-\overset{\overset{\displaystyle O}{\|}}{C}SCoA$$

in equation 6 and —NH$_2$ in equation 7. In the other rearrangements, it may be —OH, —CH(NH$_2$)CO$_2$H, or —C(=CH$_2$)CO$_2$H. AdoCba is an intermediate hydrogen carrier. The migrating hydrogen is first transferred, from the substrate or from the reducing agent, to the 5′-carbon of AdoCba forming 5′-deoxyadenosine and then to the final product. The 5′-deoxyadenosine is an intermediate in all AdoCba-dependent hydrogen-transfer reactions. The catalysis involves, therefore, the reversible cleavage of the Co–C bond of AdoCba. This cleavage occurs most probably by homolysis where the transfer of hydrogen takes place by a free-radical mechanism (see eq. 9, where

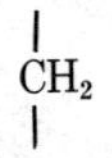

$$\overset{|}{\underset{|}{CH_2}}$$

and [Co^{3+}] symbolize the adenosyl and the corrin moieties, respectively, of Ado-Cba):

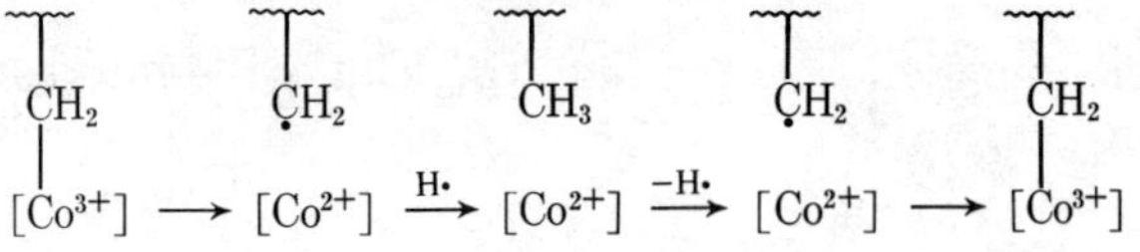

$$\underset{[Co^{3+}]}{\overset{\sim}{CH_2}} \longrightarrow \underset{[Co^{2+}]}{\overset{\sim}{\underset{\cdot}{CH_2}}} \overset{H\cdot}{\longrightarrow} \underset{[Co^{2+}]}{\overset{\sim}{CH_3}} \overset{-H\cdot}{\longrightarrow} \underset{[Co^{2+}]}{\overset{\sim}{\underset{\cdot}{CH_2}}} \longrightarrow \underset{[Co^{3+}]}{\overset{\sim}{CH_2}} \quad (9)$$

The mode of transfer of group X in the rearrangement in equation 8 is not known. In the absence of enzymological data, many workers use model reactions.

Methylcobamides are involved in the biosynthesis of methionine, methane, and acetate.

Medical Aspects. The amounts of Cbl stored in the body are relatively extensive, on the order of 2–5 mg, with ca 30–60% located in the liver and ca 30% in the muscle. The plasma Cbl concentration is 200–900 pg/mL. It takes several years of total Cbl deprivation before the body stores are depleted and symptoms of Cbl deficiency become manifest. True nutritional deficiency in humans is rare but may occur in vegetarians (10). Young animals show retarded growth and high mortality on a Cbl-free diet. Deficiency in humans has two effects: an interference with the normal maturation of dividing cells such as those of the bone marrow (megaloblastic bone marrow) and all types of mucosa; and morphological and functional abnormalities of the nervous system. The lesion in the dividing cells, caused by dietary Cbl deficiency, is named megaloblastic anemia. Pernicious anemia results from the atrophy of parietal cells of the stomach which produce IF. In the nervous system, B$_{12}$ deficiency may cause peripheral neuropathy, memory loss, or dementia (296–297).

Much attention has been devoted to determining human requirements for Cbl and formulating the recommended dietary allowance (RDA) (298–299). At present, the RDA is 3 µg/d for adults and 1.5 µg/d for infants. The average intake of Cbl in the United States is 5–15 µg/d depending on the income group (299). It was shown that the minimal daily requirement of dietary Cbl is about 0.5 µg and that an intake of 1.0 µg/d would cover the needs of most of the population (300).

Cobalamin deficiency is treated by the administration of enough Cbl to replenish the depleted stores. An intramuscular dose of 100 µg Cbl once per month is generally effective but some patients may need larger amounts more frequently. A monthly dose of 500 µg should serve patients with poor retention. Aquacobalamin is better retained than CNCbl, although antibodies against TC may develop after aqCbl administration (296). Oral Cbl is absorbed both with and without IF mediation. The former mechanism allows a Cbl absorption with an upper limit of ca 2 µg, whereas the latter mechanism allows an absorption of Cbl which is ca 1.2% of the dose. Using this mechanism, patients with PA are treated with oral doses of 500–1000 µg Cbl per day (301). Cobalamin-dependent forms of methylmalonic aciduria (congenital defects of Cbl metabolism) are treated with intramuscular injection of doses of Cbl, ca 1 mg/d (302).

Many preparations with a protracted Cbl effect upon injection (so-called depot preparations) have been made, including suspensions of Cbl in a vegetable oil, complexes of Cbl with tannin and zinc, and a complex of Cbl with tannin suspended in a solution of aluminum monostearate in a vegetable oil and isopropyl myristate (303).

Cobalamin is one of the least toxic therapeutic agents, and doses of 100 mg/kg (intraperitoneal) in rats and 10 mg orally in humans have shown no toxic effects.

BIBLIOGRAPHY

"Vitamin B$_{12}$" in *ECT* 1st ed., Vol. 14, pp. 813–828, by F. M. Robinson, Merck & Co., Inc.; "Vitamins (Vitamin B$_{12}$)" in *ECT* 2nd ed., Vol. 21, pp. 542–549, by S. B. Greenbaum, Diamond Shamrock Chemical Co.

1. A. F. Wagner and K. Folkers, *Vitamins and Coenzymes*, John Wiley & Sons, Inc., New York, 1964, p. 194ff.
2. E. Lester Smith, *Vitamin B$_{12}$*, Methuen, London, 1965, p. 1.
3. J. C. Krantz, *Historical Medical Classics Involving New Drugs*, The Williams and Wilkins Company, Baltimore, Md., 1974, p. 68ff.
4. K. Kunze and K. Leitenmaier in P. J. Vinken and G. W. Bruyn, eds., *Handbook of Clinical Neurology*, Vol. 28, North-Holland Publishing Co., Amsterdam, 1976, Part II, p. 141.
5. R. F. Schilling, *J. Lab. Clin. Med.* **91,** 893 (1978).
6. D. M. Matthews and J. C. Linnell, *Br. Med. J.* **2,** 533 (1979).
7. B. M. Babior and J. S. Krouwer, *CRC Critical Reviews in Biochemistry* **6,** 35 (1979).
8. P. Karlson, *Trends Biochem. Sci.* **4,** 286 (1979).
9. W. Friedrich, *Chem. Rundsch.* **32**(9), 1 (1979).
10. K. S. Lau, *Pathology* **13,** 189 (1981).
11. G. R. Minot and W. P. Murphy, *J. Am. Med. Assoc.* **87,** 470 (1926); G. H. Whipple, C. W. Hooper, and F. S. Robscheit, *Am. J. Physiol.* **53,** 236 (1920).
12. W. B. Castle, *Am. J. Med. Sci.* **178,** 748 (1929).
13. E. J. Cohn, G. R. Minot, G. A. Alles, and W. T. Salter, *J. Biol. Chem.* **77,** 325 (1928).
14. M. S. Shorb, *J. Biol. Chem.* **169,** 455 (1947).
15. E. L. Rickes, N. G. Brink, F. R. Koniuszy, T. R. Wood, and K. Folkers, *Science* **107,** 396 (1948).
16. E. Lester Smith and L. F. J. Parker, *Biochem. J.* **43,** VIII (1948).
17. E. L. Rickes, N. G. Brink, F. R. Koniuszy, T. R. Wood, and K. Folkers, *Science* **108,** 634 (1948).
18. J. V. Pierce, A. C. Page, E. L. R. Stokstad, and T. H. Jukes, *J. Am. Chem. Soc.* **71,** 2952 (1949).
19. W. G. Jackson, G. B. Whitfield, W. H. DeVries, H. A. Nelson, and J. S. Evans, *J. Am. Chem. Soc.* **73,** 337 (1951).

20. D. C. Hodgkin, J. Pickworth, J. H. Robertson, K. N. Trueblood, R. J. Prosen, and J. G. White, *Nature (London)* **176,** 325 (1955).
21. H. A. Barker, H. Weissbach, and R. D. Smyth, *Proc. Nat. Acad. Sci. U.S.A.* **44,** 1093 (1958).
22. H. Weissbach, J. Toohey, and H. A. Barker, *Proc. Nat. Acad. Sci. U.S.A.* **45,** 521 (1959).
23. K. Lindstrand, *Nature (London)* **204,** 188 (1964).
24. P. G. Lenhert and D. C. Hodgkin, *Nature (London)* **192,** 937 (1961).
25. D. C. Hodgkin, *Fed. Proc. Fed. Am. Soc. Exp. Biol.* **23,** 592 (1964).
26. D. C. Hodgkin, *Proc. R. Soc. London Ser. A* **288,** 294 (1965).
27. D. C. Hodgkin, *Angew. Chem.* **77,** 954 (1965).
28. D. C. Hodgkin in B. Zagalak and W. Friedrich, eds., *Vitamin B_{12}, Proceedings of the Third European Symposium on Vitamin B_{12}*, Zurich, 1979, Walter de Gruyter, Berlin, 1979, p. 19.
29. J. P. Glusker in G. Dodson, J. P. Glusker and D. Sayre, eds., *Structural Studies on Molecules of Biological Interest: a Volume in Honour of Professor Dorothy Hodgkin*, Clarendon Press, Oxford, 1981, p. 106.
30. K. Venkatesan, D. Dale, D. C. Hodgkin, C. E. Nockolds, F. H. Moore, and B. H. O'Connor, *Proc. R. Soc. London Ser. A* **323,** 455 (1971).
31. H. Stoeckli-Evans, E. Edmond, and D. C. Hodgkin, *J. Chem. Soc. Perkin Trans. 2*, 605 (1972).
32. J. Kopf, K. von Deuten, R. Bieganowski, and W. Friedrich, *Z. Naturforsch. Teil C* **36,** 506 (1981).
33. W. Friedrich, *Vitamin B_{12} und Verwandte Corrinoide*, Georg Thieme, Stuttgart, 1975, p. 170.
34. J. Marks, *A Guide to the Vitamins, Their Role in Health and Disease*, Medical and Technical Publishing Co., Lancaster, UK, 1975, p. 118.
35. J. Kirschbaum in K. Florey, ed., *Analytical Profiles of Drug Substances*, Vol. 10, Academic Press, New York, 1981, p. 185.
36. Ref. 35, p. 192.
37. Ref. 33, p. 3.
38. *Biochemistry* **13,** 1555 (1974).
39. H. P. C. Hogenkamp in B. M. Babior, ed., *Cobalamin—Biochemistry and Pathophysiology*, John Wiley & Sons, Inc., New York, 1975, p. 21.
40. N. H. Georgopapadakou and A. I. Scott, *J. Theor. Biol.* **69,** 381 (1977).
41. Ref. 33, p. 5.
42. K. Bernhauer, H. Dellweg, W. Friedrich, G. Gross, F. Wagner, and P. Zeller, *Helv. Chim. Acta* **43,** 693 (1960).
43. J. E. Ford, S. K. Kon, and J. W. G. Porter, *Biochem. J.* **50,** IX (1951).
44. D. Perlman in D. B. McCormick and L. D. Wright, eds., *Methods in Enzymology*, Vol. 18, Academic Press, New York, 1971, Part C, p. 75.
45. Ref. 2, p. 79.
46. Ref. 33, p. 176.
47. J. I. Toohey, *Proc. Nat. Acad. Sci. U.S.A.* **54,** 934 (1965).
48. B. Dresow, G. Schlingmann, L. Ernst, and V. B. Koppenhagen, *J. Biol. Chem.* **255,** 7637 (1980).
49. R. Bieganowski and W. Friedrich, *Z. Naturforsch. Teil C* **36,** 9 (1981).
50. V. B. Koppenhagen, E. Warmuth, G. Schlingmann, and B. Dresow in ref. 28, p. 635.
51. D. C. Hodgkin in L. Zechmeister, ed., *Progress in the Chemistry of Organic Natural Products*, Vol. 15, Springer, Wien, 1958, pp. 167–220.
52. Ref. 2, p. 30.
53. Ref. 33, p. 79.
54. W. Friedrich and K. Bernhauer, *Chem. Ber.* **89,** 2507 (1956).
55. Ref. 33, p. 85.
56. R. Bonnett, J. M. Godfrey, and V. B. Math, *J. Chem. Soc. C*, 3736 (1971).
57. R. Bonnett, J. M. Godfrey, V. B. Math, E. Edmond, H. Evans, and O. J. R. Hodder, *Nature (London)* **229,** 473 (1971).
58. Ref. 33, p. 119.
59. J. M. Pratt, *Inorganic Chemistry of Vitamin B_{12}*, Academic Press, London, 1972, p. 138.
60. W. Friedrich, *Biochem. Z.* **342,** 143 (1965).
61. Ref. 33, p. 130.
62. Ref. 59, p. 117.
63. W. Friedrich and M. Moskophidis, *Z. Naturforsch. Teil B* **25,** 979 (1970).
64. Ref. 33, p. 136.
65. M. Moskophidis, C. M. Klotz, and W. Friedrich, *Z. Naturforsch. Teil C* **31,** 255 (1976).

66. E. M. Tachkova, I. P. Rudakova, N. V. Myasishcheva, and A. M. Yurkevich, *Bioorganicheskaya Khimiya* **2,** 535 (1976).
67. H. P. C. Hogenkamp, *Am. J. Clin. Nutr.* **33,** 1 (1980).
68. B. Grüning and A. Gossauer in ref. 28, p. 141.
69. V. Herbert, E. Jacob, K. T. J. Wong, J. Scott, and R. D. Pfeffer, *Am. J. Clin. Nutr.* **31,** 253 (1978).
70. V. Herbert, L. Landau, R. Bash, S. Grosberg, and N. Colman in ref. 28, p. 1069.
71. M. Marcus, M. Prabhudesai, and S. Wassef, *Am. J. Clin. Nutr.* **33,** 137 (1980).
72. S. W. Thenen in ref. 28, p. 1065.
73. F. A. Kuehl, C. H. Shunk, and K. Folkers, *J. Am. Chem. Soc.* **77,** 251 (1955).
74. V. M. Clark, A. W. Johnson, I. O. Sutherland, and A. R. Todd, *J. Chem. Soc.*, 3283 (1958).
75. Ref. 2, p. 44.
76. J. M. Wood, A. Chen, L. J. Dizikes, W. P. Ridley, S. Rakow, and J. R. Lakowicz, *Fed. Proc. Fed. Am. Soc. Exp. Biol.* **37,** 16 (1978).
77. J. M. Wood and Y. T. Fanchiang in ref. 28, p. 539.
78. V. C. W. Chu and D. W. Gruenwedel, *Bioinorg. Chem.* **7,** 169 (1977).
79. A. M. Yurkevich, E. G. Chauser, and I. P. Rudakova, *Bioinorg. Chem.* **7,** 315 (1977).
80. W. P. Ridley, L. J. Dizikes, and J. M. Wood, *Science* **197,** 329 (1977).
81. G. N. Schrauzer, *Acc. Chem. Res.* **1,** 97 (1968).
82. D. Dodd and M. D. Johnson, *J. Organomet. Chem.* **52,** 1 (1973).
83. G. Rytz, L. Walder, and R. Scheffold in ref. 28, p. 173.
84. R. Scheffold, M. Dike, S. Dike, T. Herold, and L. Walder, *J. Am. Chem. Soc.* **102,** 3642 (1980).
85. A. Fischli, *Helv. Chim. Acta* **61,** 2560, 3028 (1978).
86. A. Fischli, *Helv. Chim. Acta* **62,** 882 (1979).
87. A. Fischli and D. Süss, *Helv. Chim. Acta* **62,** 48, 2361 (1979).
88. A. Fischli and J. J. Daly, *Helv. Chim. Acta* **63,** 1628 (1980).
89. A. Fischli and P. M. Müller, *Helv. Chim. Acta* **63,** 529, 1619 (1980).
90. M. C. M. Laranjeira, D. W. Armstrong, and F. Nome, *Bioorg. Chem.* **9,** 313 (1980).
91. D. Lexa, J. M. Savéant, and J. Zickler, *J. Am. Chem. Soc.* **102,** 2654, 4851 (1980).
92. S. Bachman, Z. Gasyna, and A. Furmaniak-Suwalska, *Stud. Biophys.* **71,** 209 (1978).
93. J. M. Savéant, N. de Tacconi, D. Lexa, and J. Zickler in ref. 28, p. 203.
94. Ref. 33, p. 99.
95. Ref. 33, p. 111.
96. Ref. 59, p. 256.
97. A. Vogler, R. Hirschmann, H. Otto, and H. Kunkely, *Ber. Bunsenges. Phys. Chem.* **80,** 420 (1976).
98. A. W. Johnson, *Chem. Soc. Rev.* **9,** 125 (1980).
99. D. J. Lowe, K. N. Joblin, and D. J. Cardin, *Biochim. Biophys. Acta* **539,** 398 (1978).
100. R. V. Stevens in ref. 28, p. 119.
101. R. B. Woodward, *Pure Appl. Chem.* **17,** 519 (1968).
102. *Ibid.*, **25,** 283 (1971).
103. *Ibid.*, **33,** 145 (1973).
104. A. Eschenmoser, *Q. Rev. Chem. Soc. London* **24,** 366 (1970).
105. A. Eschenmoser, *Pure Appl. Chem. Suppl.* **2,** 69 (1971).
106. A. Eschenmoser, *Naturwissenschaften* **61,** 513 (1974).
107. A. Eschenmoser, *Chem. Soc. Rev.* **5,** 377 (1976).
108. A. Eschenmoser and C. E. Wintner, *Science* **196,** 1410 (1977).
109. A. Eschenmoser in ref. 28, p. 89.
110. J. H. Krieger, *Chem. Eng. News*, 16 (1973).
111. T. H. Maugh, *Science* **179,** 266 (1973).
112. R. B. Woodward, *Chem. Soc. Spec. Publ.* **21,** 217 (1967).
113. W. Friedrich, G. Gross, K. Bernhauer, and P. Zeller, *Helv. Chim. Acta* **43,** 704 (1960).
114. W. Friedrich in H. C. Heinrich, ed., *Vitamin B$_{12}$ und Intrinsic Factor, 2. Europäisches Symposium, Hamburg, 1961*, Ferdinand Enke, Stuttgart, 1962, p. 8.
115. W. Friedrich, *Z. Naturforsch. Teil B* **18,** 455 (1963).
116. R. B. Woodward in ref. 28, p. 37.
117. E. Lester Smith, L. Mervyn, A. W. Johnson, and N. Shaw, *Nature (London)* **194,** 1175 (1962).
118. A. W. Johnson, L. Mervyn, N. Shaw, and E. Lester Smith, *J. Chem. Soc.*, 4146 (1963).
119. K. Bernhauer, O. Müller, and G. Müller, *Biochem. Z.* **336,** 102 (1962).
120. D. Autissier, P. Barthelemy, and L. Penasse, *Bull. Soc. Chim. Fr. Pt. II*, 192 (1980).

121. Ref. 33, p. 106.
122. Ref. 2, p. 58.
123. H. C. Friedmann in ref. 39, p. 75.
124. Ref. 33, p. 153.
125. A. I. Scott, *Tetrahedron* **31,** 2639 (1975).
126. A. I. Scott, *Philos. Trans. R. Soc. London Ser. B* **273,** 303 (1976).
127. H. C. Friedmann and L. M. Cagen, *Annu. Rev. Microbiol.* **24,** 159 (1970).
128. J. S. Seehra and P. M. Jordan, *J. Am. Chem. Soc.* **102,** 6841 (1980).
129. A. R. Battersby, C. J. R. Fookes, G. W. J. Matcham, and E. McDonald, *Nature (London)* **285,** 17 (1980).
130. A. I. Scott, C. A. Townsend, K. Okada, and M. Kajiwara, *Trans. N.Y. Acad. Sci.* **35,** 72 (1973).
131. A. I. Scott, C. A. Townsend, K. Okada, M. Kajiwara, and R. J. Cushley, *J. Am. Chem. Soc.* **94,** 8269 (1972).
132. A. I. Scott in ref. 28, p. 247.
133. A. R. Battersby in ref. 28, p. 217.
134. A. R. Battersby and E. McDonald, *Bioorg. Chem.* **7,** 161 (1978).
135. N. G. Lewis, R. Neier, G. W. J. Matcham, E. McDonald, and A. R. Battersby, *J. Chem. Soc. Chem. Commun.*, 541 (1979).
136. G. Müller, R. Deeg, K. D. Gneuss, G. Gunzer, and H. P. Kriemler in ref. 28, p. 279.
137. G. Müller, K. D. Gneuss, H. P. Kriemler, A. I. Scott, and A. J. Irwin, *J. Am. Chem. Soc.* **101,** 3655 (1979).
138. V. Y. Bykhovsky in ref. 28, p. 293.
139. A. R. Battersby and K. Frobel, *Chem. Unserer Zeit* **16,** 124 (1982).
140. Ref. 33, p. 162.
141. H. C. Friedmann in ref. 28, p. 331.
142. P. Renz, J. Hörig, and R. Wurm in ref. 28, p. 317.
143. L. Lamm, J. A. Hörig, P. Renz, and G. Heckmann, *Eur. J. Biochem.* **109,** 115 (1980).
144. J. A. Hörig and P. Renz, *Eur. J. Biochem.* **105,** 587 (1980).
145. W. A. Fenton and L. E. Rosenberg, *Annu. Rev. Genet.* **12,** 223 (1978).
146. H. Ohta and W. S. Beck, *Arch. Biochem. Biophys.* **174,** 713 (1976).
147. W. A. Fenton and L. E. Rosenberg, *Biochem. Biophys. Res. Commun.* **98,** 283 (1981).
148. W. A. Fenton and L. E. Rosenberg, *Arch. Biochem. Biophys.* **189,** 441 (1978).
149. I. Mellman, H. F. Willard, P. Youngdahl-Turner, and L. E. Rosenberg, *J. Biol. Chem.* **254,** 11847 (1979).
150. E. V. Quadros, A. Hamilton, D. M. Matthews, and J. C. Linnell, *J. Chromatogr.* **160,** 101 (1978).
151. P. F. Uphill, F. Jacob, and P. Lall, *J. Appl. Bacteriol.* **43,** 333 (1977).
152. E. Grieco and R. Desrochers, *Can. J. Microbiol.* **24,** 1562 (1978).
153. J. M. Poston and B. A. Hemmings, *J. Bacteriol.* **140,** 1013 (1979).
154. J. M. Poston, *Phytochemistry* **17,** 401 (1978).
155. G. W. Bigger, J. M. Elliott, and T. R. Rickard, *J. Anim. Sci.* **43,** 1077 (1976).
156. A. L. Sutton and J. M. Elliott, *J. Nutr.* **102,** 1341 (1972).
157. *Nutr. Rev.* **38,** 274 (1980).
158. M. J. Albert, V. I. Mathan, and S. J. Baker, *Nature (London)* **283,** 781 (1980).
159. L. J. Brandt, L. H. Bernstein, and A. Wagle, *Ann. Intern. Med.* **87,** 546 (1977).
160. Ref. 33, p. 174.
161. Ger. Pat. 922,126 (Feb. 21, 1955), K. Bernhauer and W. Friedrich (to Aschaffenburger Zellstoffwerke).
162. Ger. Pat. 941,150 (April 5, 1956), K. Bernhauer and W. Friedrich (to Aschaffenburger Zellstoffwerke).
163. W. Friedrich and K. Bernhauer, *Angew. Chem.* **71,** 284 (1959).
164. *Ibid.*, **65,** 627 (1953).
165. U.S. Pat. 4,119,492 (Oct. 10, 1978), I. Kojima, H. Sato, and Y. Fujiwara (to Nippon Oil Company, Ltd.).
166. T. Kamikubo, M. Hayashi, N. Nishio, and S. Nagai, *Appl. Environ. Microbiol.* **35,** 971 (1978).
167. T. Toraya, B. Yongsmith, A. Tanaka, and S. Fukui, *Appl. Microbiology* **30,** 477 (1975).
168. K. Sato, S. Ueda, and S. Shimizu, *Appl. Environ. Microbiol.* **33,** 515 (1977).
169. G. Dumenil, A. Cremieux, R. Couderc, J. Chevalier, H. Guiraud, and D. Ballerini, *Biotechnol. Lett.* **1,** 371 (1979).

170. G. Dumenil, A. Cremieux, J. Chevalier, and H. Guiraud, *Biotechnol. Lett.* **3,** 285 (1981).
171. V. Karasseva, J. G. Weiszfeiler, and Z. Lengyel, *Zentralbl. Bakteriol. Parasitenk. Infektionskr. Abt. 1: Orig. Reihe A* **239,** 514 (1977).
172. U.S. Pat. 2,505,053 (April 25, 1950), F. A. Kuehl and L. Chaiet (to Merck & Co., Inc.).
173. U.S. Pat. 2,626,888 (Jan. 27, 1953), S. Kutosh, G. B. Hughey and R. Malcolmson (to Merck & Co., Inc.).
174. U.S. Pat. 2,628,186 (Feb. 10, 1953), W. Shive (to Research Corp.).
175. Ger. Pat. 953,643 (June 14, 1952), H. M. Shafer and A. J. Holland (to Merck & Co., Inc.).
176. Ger. Pat. 964,090 (Jan. 17, 1953), R. G. Denkewalter and co-workers (to Merck & Co., Inc.).
177. Ger. Pat. 1,037,066 (Feb. 12, 1959), K. Bernhauer and W. Friedrich (to Aschaffenburger Zellstoffwerke).
178. H. Vogelmann and F. Wagner, *J. Chromatogr.* **76,** 359 (1973).
179. W. Friedrich and K. Bernhauer, *Z. Naturforsch. Teil B* **9,** 755 (1954).
180. G. O. Rudkin and R. J. Taylor, *Anal. Chem.* **24,** 1155 (1952).
181. U.S. Pat. 2,563,794 (Aug. 7, 1951), E. L. Rickes and T. R. Wood (to Merck & Co., Inc.).
182. U.S. Pat. 2,582,589 (Jan. 15, 1952), H. H. Fricke (to Abbott Laboratories).
183. U.S. Pat. 3,167,539 (Jan. 26, 1965), E. Lester Smith (to Glaxo Laboratories).
184. R. Bieganowski and G. Klar, *Chem. Ztg.* **106,** 235 (1982).
185. N. M. Datsyuk, R. V. Kucheras, V. D. Rurik, S. P. Gudz, M. F. Kostruba, and S. Y. Kremenetskaya, *Microbiology Engl. Transl.* **47,** 365 (1978), translated from *Mikrobiologiya* **47,** 451 (1978).
186. U.S. Pat. 3,411,991 (Nov. 19, 1968), P. G. Lim (to Hercules, Inc.).
187. P. Rapp, thesis, University of Stuttgart, 1968.
188. U.S. Pat. 3,169,100 (Feb. 9, 1965), M. J. Weinstein (to Schering Co.).
189. Ger. Pat. Document 1,046,258 (April 24, 1956), A. Di Marco and C. Spalla (to Farmitalia).
190. A. Di Marco, G. Boretti and C. Spalla, *Sci. Repts. Ist. Super. Sanità* **1,** 355 (1961).
191. U.S. Pat. 4,210,720 (July 1, 1980), I. Kojima, H. Sato, and Y. Fujiwara (to Nippon Oil Co.).
192. J. Krzycki and J. G. Zeikus, *Curr. Microbiol.* **3,** 243 (1980).
193. U.S. Pat. 3,964,971 (June 22, 1976), B. Johan and co-workers (to Richter Gedeon Vegyeszeti Gyar Rt.).
194. U.S. Pat. 3,979,259 (Sept. 7, 1976), B. Johan and co-workers (to Richter Gedeon Vegyeszeti Gyar Rt.).
195. T. Kamikubo, K. Sasaki, and M. Hayashi, *J. Nutr. Sci. Vitaminol.* **23,** 179 (1977).
196. G. Schlingmann, B. Dresow, L. Ernst, and V. B. Koppenhagen, *Liebigs Ann. Chem.*, 2061 (1981).
197. Ger. Pat. Document 2,908,769 (March 6, 1979), V. Koppenhagen and co-workers (to Gesellschaft für Biotechnologische Forschung).
198. V. B. Koppenhagen and J. J. Pfiffner, *J. Biol. Chem.* **246,** 3075 (1971).
199. E. Warmuth, thesis, University of Braunschweig, 1976.
200. B. Elsenhans, thesis, University of Braunschweig, 1974.
201. A. S. Hester and G. E. Ward, *Ind. Eng. Chem.* **46,** 238 (1954).
202. V. F. Pfeiffer, C. Vojnovich, and E. H. Heger, *Ind. Eng. Chem.* **46,** 843 (1954).
203. J. A. Garibaldi, K. Ijechi, N. S. Snell, and J. C. Lewis, *Ind. Eng. Chem.* **45,** 838 (1953).
204. U.S. Pat. 2,619,420 (Nov. 25, 1952), T. H. Jukes (to American Cyanamid Co.).
205. N. P. Kubasik, M. Graham, and H. E. Sine, *Clin. Chim. Acta* **95,** 147 (1979).
206. Ref. 35, p. 250.
207. *The United States Pharmacopeia XX*, United States Pharmacopeial Convention, Inc., Rockville, Md., 1980, p. 182.
208. V. Lacroce and W. H. Crosby, *Arch. Intern. Med.* **141,** 1558 (1981).
209. Ref. 207, p. 161.
210. W. A. Fenton and L. E. Rosenberg, *Anal. Biochem.* **90,** 119 (1978).
211. J. Farquharson and J. F. Adams, *Br. J. Nutrition* **36,** 127 (1976).
212. J. Farquharson and J. F. Adams, *Am. J. Clin. Nutr.* **30,** 1617 (1977).
213. Ref. 33, p. 11.
214. J. A. Begley and C. A. Hall, *J. Chromatogr.* **177,** 360 (1979).
215. E. P. Frenkel, R. L. Kitchens, and R. Prough, *J. Chromatogr.* **174,** 393 (1979).
216. R. A. Beck, *Anal. Chem.* **50,** 200 (1978).
217. J. F. Kolhouse and R. H. Allen, *Anal. Biochem.* **84,** 486 (1978).
218. H. Kondo, J. F. Kolhouse, and R. H. Allen, *Proc. Nat. Acad. Sci. U.S.A.* **77,** 817 (1980).
219. R. B. Silverman and D. Dolphin, *J. Chromatogr.* **194,** 273 (1980).

220. Ref. 33, p. 145.
221. P. Bruno, *Anal. Lett.* **14**, 1493 (1981).
222. R. Strohecker and H. M. Henning, *Vitamin Assay, Tested Methods*, Verlag Chemie, Weinheim, 1966, p. 164.
223. Ref. 33, p. 190.
224. Ref. 207, p. 903.
225. Ref. 2, p. 106.
226. J. M. England and J. C. Linnell, *Lancet* **2**, 1072 (1980).
227. G. M. Sharma, H. R. DuBois, A. T. Pastore, and S. F. Bruno, *Anal. Chem.* **51**, 196 (1979).
228. P. J. Casey, K. R. Speckman, F. J. Ebert, and W. E. Hobbs, *J. Assoc. Off. Anal. Chem.* **65**, 85 (1982).
229. K. S. Lau, C. W. Gottlieb, L. R. Wasserman, and V. Herbert, *Blood* **26**, 202 (1965).
230. S. P. Rothenberg, *Am. J. Clin. Pathol.* **75**, 75 (1981).
231. A. Castro, A. Cid, P. A. Buschbaum, and L. Clark, *Res. Commun. Chem. Pathol. Pharmacol.* **24**, 583 (1979).
232. J. Lindemans, J. Vankapel, and J. Abels, *Clin. Chim. Acta* **95**, 29 (1979).
233. D. L. Mollin, A. V. Hoffbrand, P. G. Ward, and S. M. Lewis, *J. Clin. Pathol.* **33**, 243 (1980).
234. K. L. Cohen and R. M. Donaldson, *J. Am. Med. Assoc.* **244**, 1942 (1980).
235. J. F. Kolhouse, H. Kondo, N. C. Allen, E. Podell, and R. H. Allen, *N. Engl. J. Med.* **299**, 785 (1978).
236. *NCCLS Proposed Standard: PSLA-12. Guidelines for evaluating a B_{12} (cobalamin) assay*, National Committee for Clinical Laboratory Standards, Villanova, Pa., March 1980.
237. L. R. Witherspoon, *J. Nucl. Med.* **22**, 474 (1981).
238. R. B. Gilbert and E. A. Mailhot, *J. Am. Med. Assoc.* **246**, 734 (1981).
239. H. M. Waters, J. A. Thornton, R. F. Stevens, A. H. Gowenlock, J. E. Maciver, and I. W. Delamore, *J. Clin. Pathol.* **34**, 972 (1981).
240. G. Reynoso, R. Tuggey, H. Hansen, P. A. Fontelo, S. Konopka, and J. T. Miller, *Am. J. Clin. Pathol.* **75**, 786 (1981).
241. E. Nexø and P. Gimsig, *Scand. J. Clin. Lab. Invest.* **41**, 465 (1981).
242. N. P. Kubasik, M. Ricotta, and H. E. Sine, *Clin. Chem.* **26**, 598 (1980).
243. I. W. Chen, E. B. Silberstein, H. R. Maxon, M. Sperling, and E. Barnes, *J. Nucl. Med.* **22**, 447 (1981).
244. R. J. Lefebvre, A. S. Virji, and B. F. Mertens, *Am. J. Clin. Pathol.* **74**, 209 (1980).
245. J. A. Begley and C. A. Hall, *J. Clin. Pathol.* **34**, 630 (1981).
246. R. Zacharakis, M. Muir, and I. Chanarin, *J. Clin. Pathol.* **34**, 357 (1981).
247. N. I. Berlin, K. Boddy, J. D. Cook, R. A. Dudley, R. Gräsbeck, P. A. McIntyre, S. M. Lewis, Y. Najean, J. E. Pettit, and R. F. Schilling, *J. Nucl. Med.* **22**, 1091 (1981).
248. T. Smith and R. Hesp, *Br. J. Radiol.* **52**, 832 (1979).
249. R. F. Schilling, *J. Lab. Clin. Med.* **42**, 860 (1953).
250. A. Celada, V. Herreros, and A. Donath, *Blut* **42**, 87 (1981).
251. P. A. Domstad, Y. C. Choy, E. E. Kim, and F. H. DeLand, *Am. J. Clin. Pathol.* **75**, 723 (1981).
252. Ref. 59, p. 44.
253. Ref. 33, p. 46.
254. E. Nexø and H. Olesen, *Biochim. Biophys. Acta* **446**, 143 (1976).
255. M. Moskophidis, *Z. Naturforsch. Teil C* **34**, 689 (1979).
256. *Ibid.*, **36**, 497 (1981).
257. G. L. Eichhorn, *Tetrahedron* **13**, 208 (1961).
258. I. P. Rudakova, T. A. Pospelova, E. M. Tachkova, and A. M. Yurkevich in ref. 28, p. 183.
259. Ref. 33, p. 70.
260. R. Fugate, C. A. Chin, and P. S. Song, *Biochim. Biophys. Acta* **421**, 1 (1976).
261. W. Friedrich, *Z. Naturforsch. Teil B* **21**, 595 (1966).
262. B. M. Babior in ref. 39, p. 141.
263. J. R. Pilbrow in ref. 28, p. 505.
264. V. D. Ghanekar, R. J. Lin, R. E. Coffman, and R. L. Blakley, *Biochem. Biophys. Res. Commun.* **101**, 215 (1981).
265. J. D. Brodie and M. Poe, *Biochemistry* **10**, 914 (1971).
266. *Ibid.*, **11**, 2534 (1972).

267. J. M. Wood and D. G. Brown in J. D. Dunitz and co-workers, eds., *Structure and Bonding*, Vol. 11, Springer, Berlin, 1972, p. 47.
268. P. Y. Law, D. G. Brown, E. L. Lien, B. M. Babior, and J. M. Wood, *Biochemistry* **10,** 3428 (1971).
269. A. Cheung, R. Parry, and R. H. Abeles, *J. Am. Chem. Soc.* **102,** 384 (1980).
270. D. Doddrell and A. Allerhand, *Proc. Nat. Acad. Sci. U.S.A.* **68,** 1083 (1971).
271. D. Doddrell and A. Allerhand, *Chem. Commun.*, 728 (1971).
272. L. Ernst, *Liebigs Ann. Chem.*, 376 (1981).
273. A. R. Battersby, M. Ihara, E. McDonald, J. R. Stephenson, and B. T. Golding, *J. Chem. Soc. Chem. Commun.*, 404 (1973).
274. A. I. Scott, C. A. Townsend, and R. J. Cushley, *J. Am. Chem. Soc.* **95,** 5759 (1973).
275. C. E. Brown, D. Shemin, and J. J. Katz, *J. Biol. Chem.* **248,** 8015 (1973).
276. T. J. Simpson, *Chem. Soc. Rev.* **4,** 497 (1975).
277. K. Inoue and A. Nath, *Bioinorg. Chem.* **7,** 159 (1977).
278. S. Tyagi, K. Inoue, and A. Nath, *Biochim. Biophys. Acta* **539,** 125 (1978).
279. H. R. Schulten and H. M. Schiebel, *Naturwissenschaften* **65,** 223 (1978).
280. H. M. Schiebel and H. R. Schulten, *Tetrahedron* **35,** 1191 (1979).
281. C. Sennett, L. E. Rosenberg, and I. S. Mellman, *Annu. Rev. Biochem.* **50,** 1053 (1981).
282. E. Jacob, S. J. Baker, and V. Herbert, *Physiol. Rev.* **60,** 918 (1980).
283. R. Gräsbeck in ref. 28, p. 743.
284. C. A. Hall, *J. Lab. Clin. Med.* **94,** 811 (1979).
285. B. Seetharam and D. H. Alpers, *Annu. Rev. Nutr.* **2,** 343 (1982).
286. K. J. Andersen, *Scand. J. Clin. Lab. Invest.* **39,** 685 (1979).
287. J. A. Begley and A. Trachtenberg, *Blood* **53,** 788 (1979).
288. D. W. Jacobsen, Y. D. Montejano, and F. M. Huennekens, *Anal. Biochem.* **113,** 164 (1981).
289. T. Toraya and S. Fukui, *J. Biol. Chem.* **255,** 3520 (1980).
290. K. Sato, E. Hiei, S. Shimizu, and R. H. Abeles, *FEBS Lett.* **85,** 73 (1978).
291. R. Deacon, M. Lumb, J. Perry, I. Chanarin, B. Minty, M. Halsey, and J. Nunn, *Eur. J. Biochem.* **104,** 419 (1980).
292. D. H. Small and P. R. Carnegie, *Trends in Neurosciences* **4,** X (1981).
293. J. M. Poston and T. C. Stadtman in ref. 39, p. 111.
294. B. Zagalak, *Naturwissenschaften* **69,** 63 (1982).
295. J. M. Poston, *J. Biol. Chem.* **255,** 10067 (1980).
296. C. A. Hall in H. L. Ioachim, ed., *Pathobiology Annual*, Vol. 9, Raven Press, New York, 1979, p. 257.
297. D. M. Matthews and J. C. Linnell, *Eur. J. Pediat.* **138,** 6 (1982).
298. *Recommended Dietary Allowances*, 9th ed., Food and Nutrition Board, National Academy of Sciences, National Research Council, Washington, D.C., 1980.
299. D. S. McLaren, *Am. J. Clin. Nutr.* **34,** 1611 (1981).
300. S. J. Baker and V. I. Mathan, *Am. J. Clin. Nutr.* **34,** 2423 (1981).
301. H. Berlin, R. Berlin, and G. Brante, *Acta Med. Scand.* **184,** 247 (1968).
302. D. Leupold, *Klin. Wochenschr.* **55,** 57 (1977).
303. U.S. Pat. 3,219,532 (Nov. 23, 1965), K. M. Kristensen (to A/S Dumex (Dumex, Ltd.)).

General References

D. Dolphin, ed., *Vitamin B*$_{12}$. *Vol. I, Chemistry; Vol. II, Biochemistry and Medicine*, John Wiley & Sons, Inc., New York, 1982.
D. B. McCormick and L. D. Wright, eds., *Methods in Enzymology*, Vol. 67, *Vitamins and Coenzymes*, Academic Press, New York, 1980, Part F, pp. 1–108.
K. B. G. Torssell, *Natural Product Chemistry*, John Wiley & Sons, New York, 1983.

WILHELM FRIEDRICH
Universität Hamburg

VITAMIN D

Vitamin D developed in the twentieth century as a dietary supplement to treat and prevent rickets, a disease in which the organic matrix of new bone is not mineralized. Natural vitamin D is made in the skin of animals during exposure to sunlight, and a lack of exposure caused a need for vitamin D as a dietary additive to supplement the natural vitamin D production in humans and animals (see Pet and livestock feeds).

Vitamin D referred to a substance that possessed antirachitic activity. Research during the 1970s revealed that vitamin D is better defined as those natural or synthetic substances that are converted by animals into metabolites that control calcium and phosphorus homeostasis in a hormonal-like manner. Vitamin D_2 and vitamin D_3 are the two economically important forms. The other D vitamins have relatively little biological activity and are only of historical interest. Vitamin D_2 (ergocalciferol) is active in humans, other mammals, such as cattle, swine, and dogs but is inactive in poultry. It is prepared by the uv irradiation of ergosterol (1), a plant sterol. Vitamin D_3 (cholecalciferol) is active and occurs naturally in all animals. It is produced by the irradiation of 7-dehydrocholesterol (2). The D vitamins are fat-soluble.

(1) ergosterol
(24-methyl-cholesta-5,7,22-triene-3β-ol;
provitamin D_2)

(2) 7-dehydrocholesterol
(cholesta-5,7-diene-3β-ol;
provitamin D_3)

The early development of vitamin D technology closely followed the development of the bone disease rickets in children and a similar condition in adults known as osteomalacia. Rickets is characterized by the body's inability to calcify the collagen matrix of growing bone which results in wide epiphyseal plates and large areas of uncalcified bone called osteoid. The resultant lack of rigidity of bones leads to the ends becoming twisted and bent, particularly in long bones. The ribs develop a bumpy and uneven texture known as rosary ribs and the legs become bowed. Also, the cranium becomes soft and misshapen. In adults, a similar problem occurs. No long-bone growth occurs, but new bone which is being continually remodeled activates cells to resorb bone followed by osteoblast-mediated bone-growth replacement (1–2).

The discovery, growth, and treatment of these diseases paralleled the history of the industrial revolution (3). The advent of the use of soft coal, the migration to cities, and the tendency of people and animals to spend less time in sunshine caused a decline in the ability of populations to synthesize sufficient quantities of natural vitamin D_3. This led to the increased incidence of rickets beginning around the middle of the 1600s. Important developments in vitamin D history are given in refs. 4–68.

Nomenclature

The vitamin D compounds are steroidal materials and thus are named according to the IUPAC–IUB rules for nomenclature (69) (see Table 1). The common name, vitamin D, is used throughout industry for simplicity. The trivial name calciferol has also been used extensively with the prefix ergo- and chole-, which indicate vitamin D$_2$ (3) and vitamin D$_3$ (4), respectively (see Steroids). Historically, a number of substances were referred to as vitamin D and were distinguished from one another by a subscript numeral, eg, vitamin D$_2$, vitamin D$_3$, etc. Vitamin D$_1$ [520-91-2] is a mixture of vitamin D$_2$ and lumisterol.

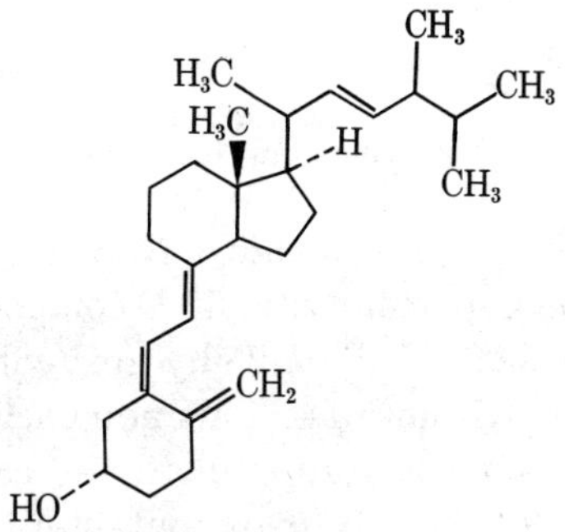

(3) vitamin D$_2$
(ergocalciferol; (5Z,7E,22E-(3S))-9,10-seco-5,7,10(19)-22-ergostatetraen-3-ol; ercalciol)

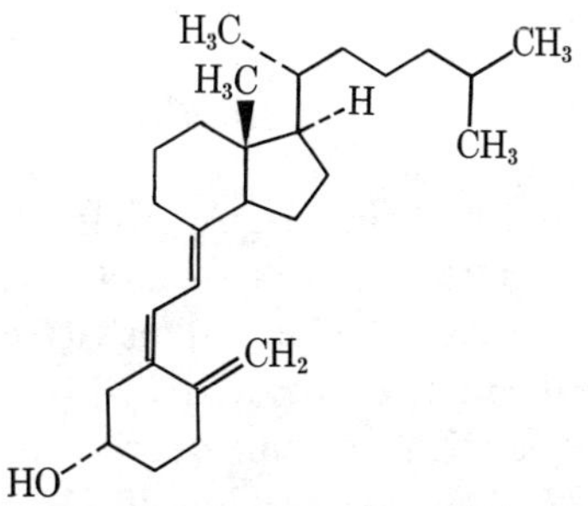

(4) vitamin D$_3$
(cholecalciferol; (5Z,7E)-(3S)-9,10-seco-5,7,10(19)-cholestatrien-3-ol; calciol)

Vitamin D and its isomers are characterized as 9,10-secosteroids, which are steroid

Table 1. Vitamin D Substances

Common name	CAS Reg. No.	Provitamin	Structure	Trivial name	IUPAC–IUB name
vitamin D$_2$	[50-14-6]	ergosterol	(3)	ergocalciferol	9,10-seco-5,7,10(19),22-ergostatetraen-3β-ol
vitamin D$_3$	[67-97-0]	7-dehydrocholesterol	(4)	cholecalciferol	9,10-seco-5,7,10(19)-cholestatrien-3β-ol
vitamin D$_4$	[511-28-4]	22,23-dihydroergosterol			24-methyl-9,10-seco-5,7,10(19)-cholestatrien-3β-ol
vitamin D$_5$	[71761-06-3]	7-dehydrositosterol		sitocalciferol	24-ethyl-9,10-seco-5,7,10(19)-cholestatrien-3β-ol
vitamin D$_6$	[481-19-6]	7-dehydrostigmasterol			24-ethyl-9,10-seco-5,7,10(19),22-ergostatetraen-3β-ol
vitamin D$_7$	[20304-51-2]	7-dehydrocampesterol			24-methyl-9,10-seco-5,7,10(19)-cholestatrien-3β-ol

molecules with a severed 9,10 bond of the steroid B-ring. The relationship between the provitamin steroid (perhydro-1,2-cyclopentanophenanthrene ring system) and the 9,10-secosteroid nucleus is shown in structures (5) and (6).

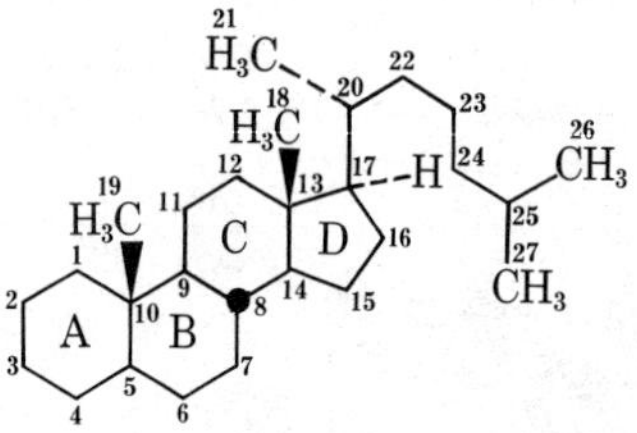

(5) cholestane 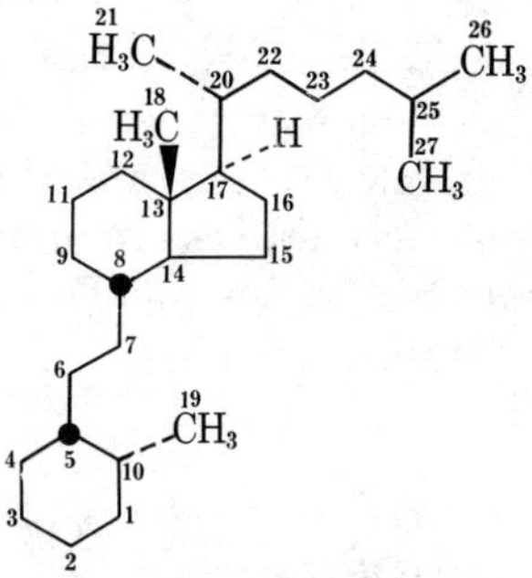   (6) 9,10-secocholestane (calcitane)

In 1981, the IUPAC–IUB Joint Commission on Biochemical Nomenclature proposed that there be a set of trivial names for the important vitamin D compounds, including calciol [67-97-0] for vitamin D, calcidiol [19356-17-3] for 25-hydroxy-vitamin D, and calcitriol [32222-06-3] for 1α,25-dihydroxy-vitamin D_3. This nomenclature has met with varying degrees of acceptance as has the proposal to use calcine [69662-75-5] (deoxy-vitamin D_2) and ercalcine [68323-40-0] (deoxy-vitamin D_3) to name the triene hydrocarbon carbon structure for 9,10-seco-5,7,10(19)-cholestatriene and 9,10-seco-5,7,10(19),22-ergostatetraen, respectively. In systematic nomenclature, calcitane would be used for the basic 27-carbon skeleton instead of 9,10-secocholestane.

The synthesis of the vitamin D usually involves the photochemical conversion of the provitamin steroid, which contains a 5,7-conjugated diene. Fission of the 9,10 bond occurs with concomitant formation of a 9,10-secotriene. The two main active vitamin D structures are shown above.

Occurrence

The amounts of provitamins D_2 and D_3 in various plants and animals are listed in Table 2.

Fish-liver oil, liver, milk, and eggs are good natural sources of the D vitamin. Most milk sold in the United States is fortified with vitamin D. Fish oil is the only commercial source of natural vitamin D, and the content of vitamins D varies according to species as well as geographically. Atlantic cod contain 100 IU/g, whereas oriental tuna (Percomorpli) contain 45,000 IU/g of oil.

Vitamin D_3 rarely occurs in plants. However, *Solanum malacoxylon*, *Cestrum diurnum*, and *Trinetum flavescens* have recently been shown to contain 1,25-dihydroxy-vitamin D activity (70–75). The vitamin D content in various plant and animal materials is shown in Table 3.

Physical Properties

The physical properties of the provitamins and vitamins D_2 and D_3 are listed in Table 4. The values are listed for the pure substances, although these are not usually isolated in normal production.

Table 2. Occurrence of the Provitamins D in Selected Plants and Animals, Parts per Thousand of Total Sterol

cottonseed oil	28	gallstones, man	0.25
rye grass	15	*Mytilus edulis*, sea mussel	100
wheat germ oil	10	*Modiolvs demissus*, ribbed mussel	370
carrot	1.7	*Ostrea virginica*, oyster	80
cabbage	0.5	*Ostrea edulis*, oyster	34
skin, pig	46	*Asterias rubens*, starfish	3.8
skin, chicken feet	25	common sponges	20
skin, rat	19	common coral	10
skin, mouse	9	*Aspergillus niger*, mold	1000
skin, calf	7	*Cortinellus shiitake*, mushroom	1000
skin, human infant	1.5	*Claviceps purpurea*, ergot	900
skin, human adult	4.2	*Saccharomyces cerevisiae*, yeast	800
liver, Japanese tuna	11	*Penicillium puberculum*, mold	280
liver, Atlantic cod	4.4	*Fucus vesiculosus*, alga, seaweed	0.8
liver, shark	1.0	*Tubifex* sp, waterworm	210
liver, halibut	0.6	*Lumbricus terrestris*, earthworm	170
liver, tuna	1.0	*Tenebrio molitor*, mealworm	120
eggs, Chinese duck	60	*Gyronomus* sp, goat	61
eggs, cod (roe)	5.5	*Cancer pagurus*, common crab	15
eggs, hen	1.6	*Daphnia* sp, water flea	7.5
wool fat, sheep	3.9	*Musca domestica*, housefly	7.0
milk, cow	2.3	*Crangon vulgaris*, shrimp	3.8
pancreas, beef	1.8	*Homarus vulgaris*, lobster	2.5
spinal cord, beef	1.2	*Helix pomatia*, edible snail	97
blood serum, cow	1.5	*Arion empiricorum*, slug, red road snail	220
herring oil	0.5	*Littorina littorea*, periwinkle	170
heart, calf	0.32	*Sepia* sp, cuttlefish	12

Chemical Properties

Provitamins. 3β-Hydroxy steroids, which contain the 5,7-diene system and can be activated with uv light to produce vitamin D compounds, are called provitamins. The two most important provitamins are ergosterol (**1**) and 7-dehydrocholesterol (**2**). These are produced in plants and animals, respectively, and 7-dehydrocholesterol is produced synthetically on a commercial scale. Small amounts of hydroxylated derivatives of the provitamins have been synthesized in efforts to prepare the metabolites of vitamin D, but these products do not occur naturally. The provitamins do not possess physiological activities with the exception that provitamin D_3 acts as a precursor to vitamin D_3. Like other 3β-hydroxy steroids, the provitamins precipitate upon treatment with digitonin (**7**). This property is not exhibited by the vitamin D substances.

Provitamin D_2. Ergosterol is isolated exclusively from biological sources. The commercial product is ca 90–100% pure and often contains up to 5 wt % 5,6-dihydroergosterol. Usually, the isolation of provitamin D_2 from natural sources involves the isolation of the total sterols and the separation of the provitamins from other sterols. The isolation of the total sterols involves extraction of the total fat, its saponification, and then reextraction of the unsaponifiable portion with an ether. The sterols are in the unsaponifiable portion. Another method is the saponification of the total material

Table 3. Distribution of Vitamin D Activity[a]

Sample	Amount	Refs.
phytoplankton	0	77–78
sargassum (a gulfweed)	some activity	79
clover hay		
sun-cured	slight	27
dark-cured	0	27
mushrooms (Agaricus campestris)	0.21 IU/g	80
milk		
winter (bovine)	5.3 IU/L	81
summer (bovine)	53 IU/L	81
milk (human)	63 IU/L	82
milk colostrum (human)	315–635 IU/L	82
egg yolk	150–400 IU/g	83
butter	4–8 IU/g	84
fish-liver oils	50–45,000 IU/g	85

[a] Ref. 76.

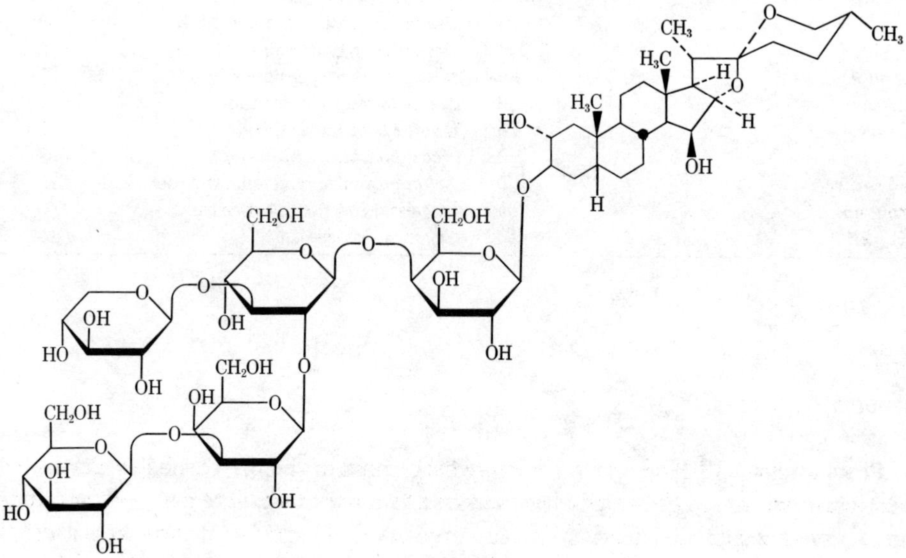

(7) digitonin

followed by isolation of the nonsaponifiable fraction. Separation of the sterols from the unsaponifiable fraction is by crystallization from a suitable solvent, eg, acetone or alcohol. Ethylene dichloride alone or mixed with methanol has been used commercially for recrystallization. In the case of yeasts, it is particularly difficult to remove the ergosterol by simple extraction, since this gives only ca 25% of the ergosterol. Industrially, therefore, the ergosterol is obtained by preliminary digestion with hot alkalies or with amines (87–92). Variations of the isolation procedure have been developed. For example, after saponification, the fatty acids may precipitate as calcium salts, which tend to adsorb the sterols. The latter are then recovered from the dried precipitate by solvent extraction.

Provitamin D_3. Provitamin D_3 is made from cholesterol, and its production begins with the isolation of cholesterol from one of its natural sources. Cholesterol occurs in

Table 4. Physical Properties of Provitamins and Vitamins D_2 and D_3

Properties	7-Dehydro-cholesterol	Ergosterol	Vitamin D_2	Vitamin D_3
melting point, °C	150–151	165	115–118	84–85
color and form	solvated plates from ether–methanol	hydrated plates from alcohol; needles from acetone	colorless prisms from acetone	fine colorless needles from dilute acetone
optical rotation (α_D^{20}), °				
acetone			82.6	83.3
ethanol			103	
chloroform	−113.6	−135	52	51.9
ether			91.2	
petroleum ether			33.3	
benzene	−127.1			
coefficient of rotation per °C in alcohol			0.515	
uv max, nm	282	281.5	264.5	264.5
specific absorption, E_{max} (at 1% conc)	308		458.9 ± 7.5	473.2 ± 7.8
potency[a], IU/g			40 × 10⁶	40 × 10⁶
biological activity			in mammals	in mammals and birds
chicken efficacy, %			8–10[b]	100
solubility, g/100 mL				
acetone at 7°C			7	
acetone at 26°C			25	
absolute ethanol at 26°C	sl sol	0.15	28	
ethyl acetate at 26°C			31	
water	insol	insol	insol	insol

[a] The international standard for vitamin D is an oil solution of activated 7-dehydrocholesterol (3). The IU is the biological activity of 0.025 µg of pure cholecalciferol.
[b] Recent studies have claimed an efficacy as high as 10% (86).

many animals, but commercially, it is extracted from wool grease. Wool grease is obtained by washing wool after it is sheared from sheep. This grease is a mixture of fatty-acid esters which contain ca 15 wt % cholesterol. The alcohol fraction is obtained after saponification, and the cholesterol is separated usually with zinc chloride. Cholesterol is also extracted from the spinal cords and brains of animals, especially cattle, and from fish oils.

The cholesterol (8) is converted to 7-dehydrocholesterol (2) (see Fig. 1). This process usually involves the Ziegler allylic bromination of the 7 position followed by dehydrobromination (93). Esterification of the cholesterol is necessary to prevent oxidation by the brominating agent. Allylic bromination may be accomplished with a variety of brominating agents, eg, N-bromosuccinimide, N-bromophthalimide, or 5,5-dimethyl-1,3-dibromohydantoin (94). Bromine in carbon disulfide can be used if the free-radical bromination is photocatalyzed (95). A mixture of 7α- and 7β-bromo cholesteryl esters is obtained and is treated with an appropriate base to dehydrohalogenate the molecule and give the 7-dehydrocholesteryl ester (11) (96). The proper conditions for this reaction are necessary to generate a high yield of the desired 7-dehydro product instead of the undesired cholesta-4,6-dien-3β-ol ester (12), which forms as a by-product. Various reagents can be used to perform the dehydrohalogenation. Trimethyl phosphite or pyridine bases, particularly trimethylpyridine, have

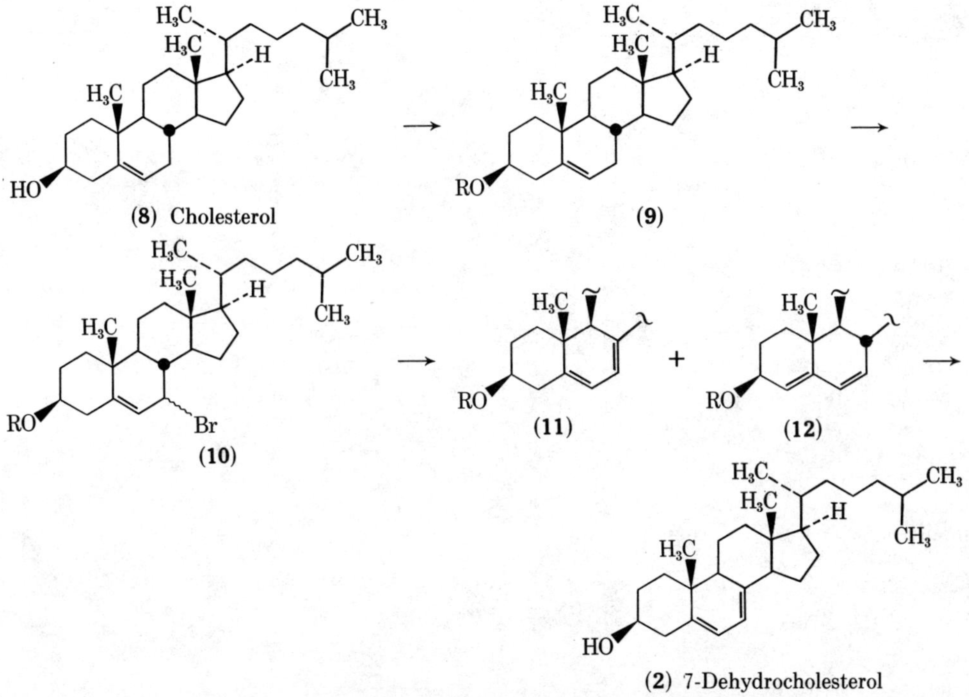

192 VITAMINS (D)

Figure 1. Conversion of cholesterol to 7-dehydrocholesterol.

been used; the symmetrical collidine is the reagent of choice (94,97). Recently, *t*-butylammonium fluoride has been used to improve the yield of high quality 5,7-diene (98).

The 7α-bromo steroid (10) can also be treated with sodium phenyl selenolate (99). The resultant 7β-phenyl selenide (13) compound can be oxidized and the corresponding phenyl selenoxide (14) eliminated to form the 7-dehydrocholesteryl ester.

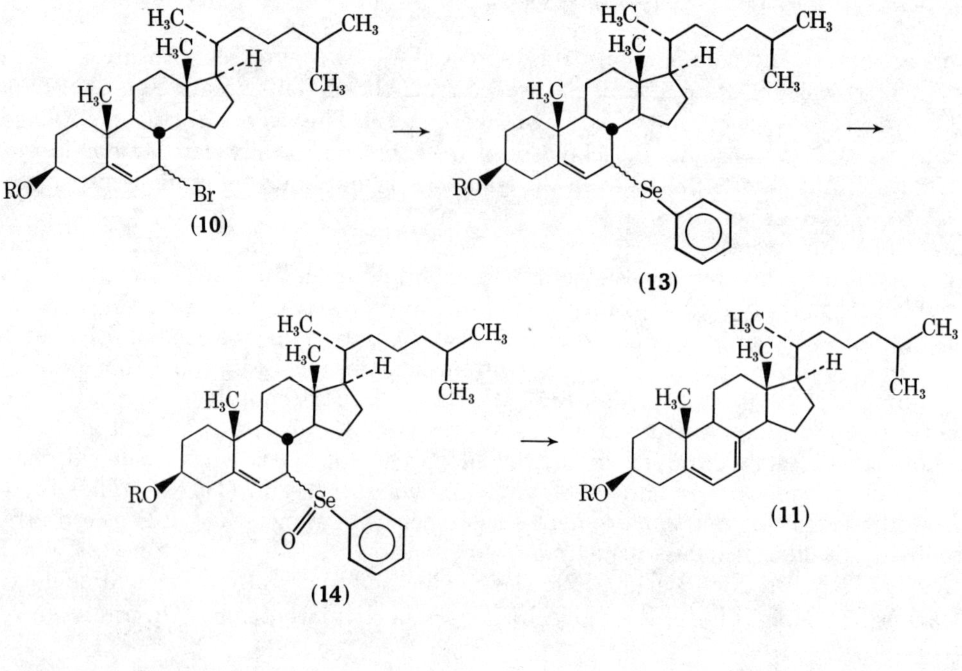

7-Dehydrocholesterol has also been made by the Windaus procedure (Fig. 2); modifications of it yield the 7-benzoate (**17**), which is deesterified thermally (42,100–101). However, the yields are substantially lower than those achieved by the bromination–dehydrohalogenation method.

The 7-tosylhydrazone and 7-phenyl sulfoxide have also been introduced and eliminated to prepare the 5,7-diene in other types of elimination reactions (102–103). The methods of choice are the allylic bromination–dehydrohalogenation procedures, and the commercial yields, in converting cholesterol to 7-dehydrocholesterol are 35–50%.

Vitamin D. The irradiation of the provitamins to produce vitamin D as well as several isomeric substances was first studied with ergosterol. The chemistry is identical for the D_3 series and yields analogous isomers, except that the side chains of the steroid nuclei are slightly different in the two series. In 1932, a scheme for the irradiation of ergosterol leading to vitamin D_2 was proposed (102). Twenty years later, the mechanism of the irradiation of the provitamins to vitamin D and its photoisomers was further elucidated (103–104). A number of products are associated with the irradiation process; these are shown in Figure 3, in which the current understanding of the interrelationships of these isomers are diagrammed. The geometry and electronic characteristics of these molecules have been well established by x-ray analysis and valence force-field calculations (105–106).

The irradiation process, which converts 7-dehydrocholesterol to vitamin D, occurs in the skin of animals if sufficient sunlight is available. Figure 3 shows the various

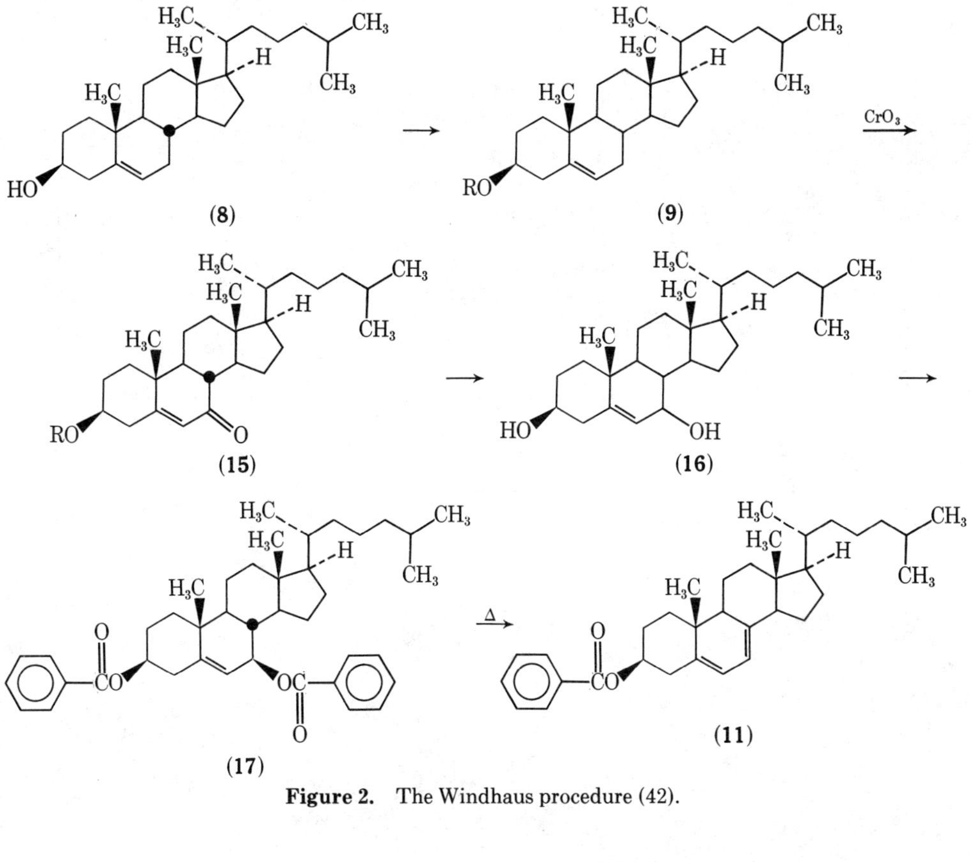

Figure 2. The Windhaus procedure (42).

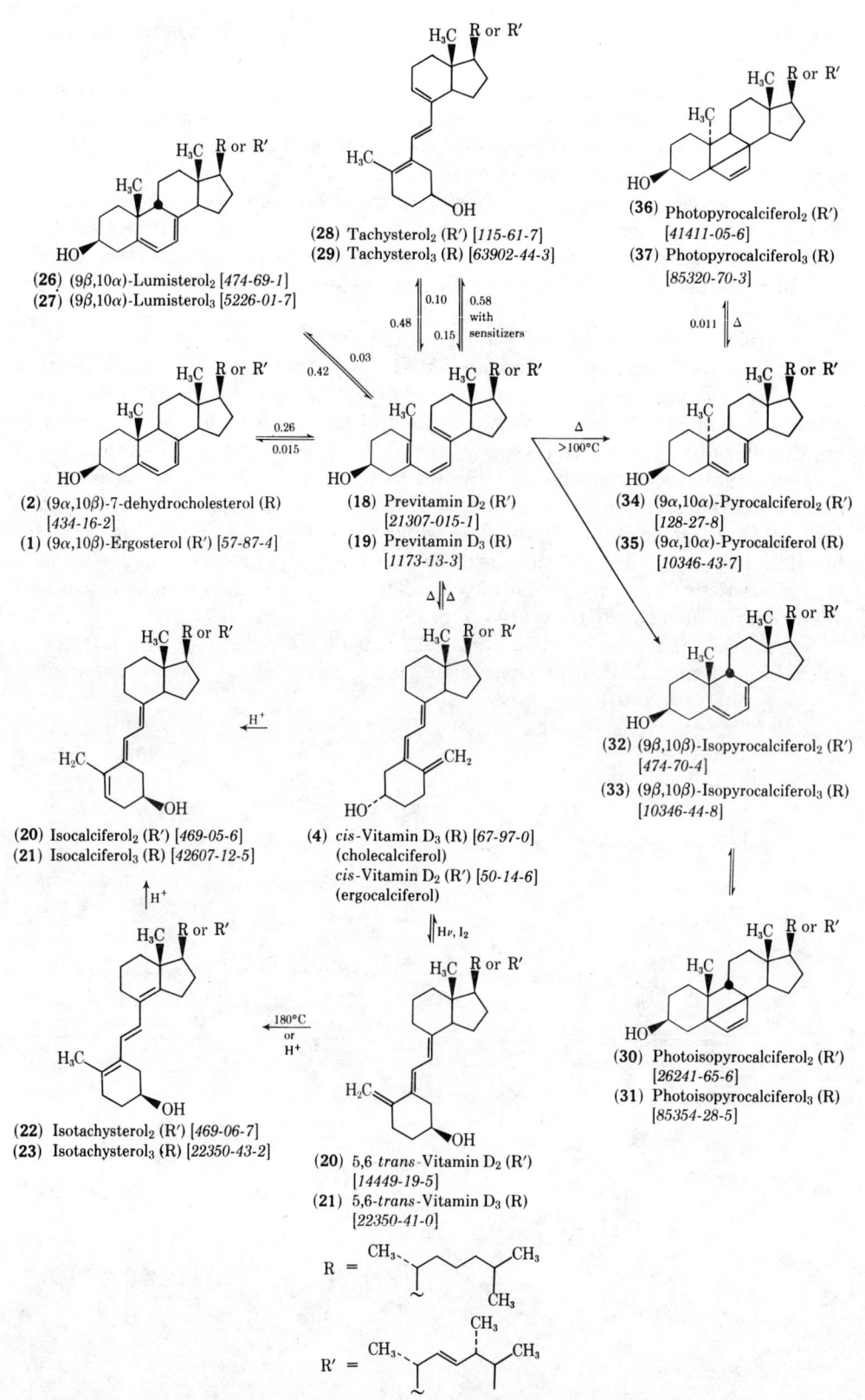

Figure 3. Photochemical and thermal isomerization products of vitamin D manufacture. The quantum yields of the reactions are listed beside the arrows for the given reactions (94).

photochemical and thermal isomerizations, which occur during the generation of vitamin D *in vivo* as well as during its synthetic photochemical preparation (107). The initial step involves ring opening of the B-ring of the sterol by ultraviolet activation of the conjugated diene. The absorbance of uv energy activates the molecule, and the $\pi \rightarrow \pi^*$ excitation (absorption, 250–310 nm; λ_{max} = 291 nm, ϵ = 12,000) results in the opening of the 9,10 bond and the formation of the (Z)-hexadiene, previtamin D (**18**) or (**19**). Previtamin D undergoes thermal equilibration to vitamin D (**3**) or (**4**). Additionally, the $\pi \rightarrow \pi^*$ excitation of previtamin D can result in ring closure to the provitamin (**1**) or (**2**) or to the other 9,10-antiisomer, lumisterol (**26**) or (**27**), which has a 9β,10α configuration. It can also exhibit (Z) $\rightleftharpoons$ (E) photoisomerization to the 6,7-(E)-isomer, tachysterol (**28**) or (**29**) (108).

The photoinduced cyclization reactions occur by conrotatory bond formation to give the 9,10-antiisomers, whereas thermal cyclization (at >100°C) leads to the two 9,10-syn isomers, (9α,10α)-pyrocalciferol (**34**) or (**35**) and (9β,10β)-isopyrocalciferol (**36**) or (**37**), by a disrotatory bond formation mechanism (81). Ultraviolet over-irradiation leads to photopyro- (**36**) or (**37**) and photoisopyrocalciferol (**30**) or (**31**), respectively. Normal irradiation conditions for the production of vitamin D include sufficiently low temperatures so that these products do not form.

The conversion of previtamin D (**18**) or (**19**) at lower temperatures (at ≤80°C) by thermal isomerization to give the cis vitamin (calciferol (**3**) or (**4**)) involves an equilibrium which is established between the pre- and *cis*-vitamin D and is described in Table 5 (109).

The equilibrium composition is normally ca 80% *cis*-vitamin D and 20% previtamin D. This reaction is an intramolecular {1–7}H sigmatropic shift and occurs through a rigid cyclic transition state (110). The overirradiation of *cis*-vitamin D leads to the formation of suprasterols of the type shown in structure (**38**) (111–112) (see Fig. 4). Prolonged irradiation of the mixture of isomers can also lead to toxisterols of the type shown in structures (**39**) and (**40**) (113–114).

More than 20 members of this type of substance have been identified. There is little evidence that these materials are toxic; however, their nomenclature leads to

Table 5. Interconversion Time of Previtamin D$_3$ and Vitamin D$_3$ at −20°C, 20°C, and 40°C [a]

Formation, %	Vitamin D$_3$ from previtamin D$_3$			Previtamin D$_3$ from vitamin D$_3$	
	at −20°C, d	at 20°C, d	at 40°C, h	at 20°C, d	at 40°C, h
2	27	0.2	0.4	2.2	3.7
5	68	0.4	1.1	8.2	11.2
7	96	0.5	1.5		18.6
10	140	0.8	2.3		43.3
20	269	1.6	4.8		
30	474	2.6	7.8		
40	681	3.8	11.3		
50	926	5.2	15.6		
60	1230	7.0	21.2		
70	1628	9.5	29.2		
80	2204	13.3	43.4		
90	2910	23.3			

[a] Ref. 109.

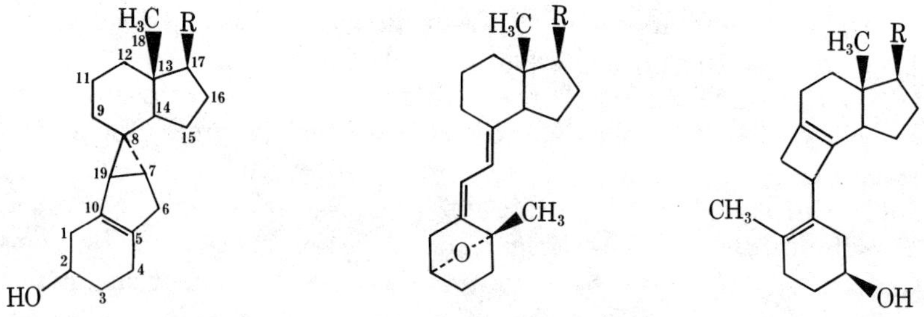

196 VITAMINS (D)

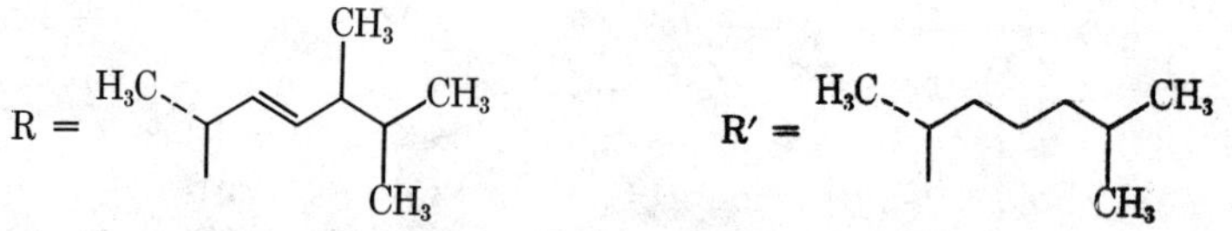

Figure 4. Overirradiation products of vitamin D.

(38) Suprasterol₂ II   (39) Toxisterol₂-E   (40) Toxisterol₃-E₁

misunderstanding. These compounds generally show little if any biological activity.

The irradiation of calciferol in the presence of iodine leads to the formation of 5,6-*trans*-vitamin D (**20**) or (**21**) (115–116). 5,6-*trans*-Vitamin D as well as *cis*-vitamin D (**3**) or (**4**) is converted to isovitamin D on treatment with mineral or Lewis acids (117–118). Isocalciferol also forms upon heating of 5,6-*trans*-vitamin D. Isotachysterol forms from isovitamin D or vitamin D upon treatment with acid, and its production appears to be the result of sequential formation of *trans*- and isovitamin D from calciferol. These reactions are the basis of the antimony trichloride test for vitamin D (119–120).

The irradiation of the 5,7-diene provitamin to make vitamin D must be performed so as to optimize the production of previtamin D and its conversion to *cis*-vitamin D under conditions that avoid the development of unwanted isomers. The optimization is achieved by controlling the extent of irradiation as well as the wavelength of the light source. The best frequency for the irradiation to form previtamin D is 295 nm (121–123). The conversion of previtamin to the tachysterol is favored when 254 nm light is used with a quantum efficiency of 0.48, compared to 0.1 for the reverse reaction. Sensitized irradiation has been used to favor the reverse reaction; fluorenone sensitizes the triplet-state conversion of tachysterol to previtamin D (124–125).

The molecular extinction coefficients of the four main components of the irradiation mixture favor the absorption of light above 300 nm by tachysterol (see Table

Table 6. Molecular Extinction Coefficients of Irradiation Products[a]

λ, nm	7-Dehydro-cholesterol	Previtamin D₃	Tachysterol₃	Lumisterol₃
254	4,500	725	11,450	4,130
300	1,250	930	11,250	1,320
330	25	105	2,940	30
340	20	40	242	25
350	10	25	100	20

[a] Ref. 126.

6). A yield of 83% of the preisomer at 95% conversion of 7-dehydrocholesterol is obtained by irradiation first at 254 nm followed by reirradiation at 350 nm with an yttrium aluminum garnet (YAG) laser to convert tachysterol to previtamin D. A similar approach with laser irradiation at 248 nm (KrF) and 337 nm (N$_2$) has been described (127). However, the use of laser light must be made more economical before it can be used in commercial manufacture of vitamin D.

The irradiation of the provitamin has been done on the acetate and benzoate esters, although the free alcohol form of the provitamin is usually used (128). The uv irradiation of 7-dehydrocholesterol or ergosterol results in the steady diminution in concentration of the provitamin, initially giving rise to predominantly previtamin D. The pre- levels reach a maximum as the provitamin level drops below ca 10%. The concentration of the previtamin then falls as it is converted to tachysterol and lumisterol, which increase in concentration with continued irradiation (see Fig. 5). Temperature, frequency of light, time of irradiation, and concentration of substrate all affect the ratio of products.

Biochemistry. Vitamin D is introduced into the bloodstream either from the skin after natural synthesis by the irradiation of 7-dehydrocholesterol in the epidermis or by ingestion and absorption through the gut wall of vitamin D$_2$ or vitamin D$_3$ (129). Sixty to eighty percent of the vitamin introduced in the blood is taken up by the liver, where it is hydroxylated at C-25 (130–135). This hydroxylation occurs in the endoplasmic reticulum and requires NADPH, Mg^{2+}, and O$_2$. 25-Hydroxylation also occurs in intestinal homogenates of chicks (136). However, 25-hydroxylation outside the liver appears not to occur in mammals (133). 25-Hydroxycholecalciferol is the main circulating form of vitamin D$_3$, and its normal concentration levels are 15–35 mg/mL (137–139). It is transported on a specific α-globulin to the kidney, where further hydroxylation takes place at C-1 or C-24 in response to low calcium levels. Low phosphate levels also stimulate 1,25-dihydroxycholecalciferol production, which in turn stimulates intestinal calcium absorption. To produce biological response in bone or intestine, C-1-hydroxylation must take place to give 1,25-dihydroxycholecalciferol (140). This occurs exclusively in the kidney mitochondria and is catalyzed by a mixed-function monooxygenase with a specific cytochrome P-450 (54,141–144). 1α,25-Dihydroxycholecalciferol is responsible for intestinal calcium transport as well as bone calcium mobilization (145–146).

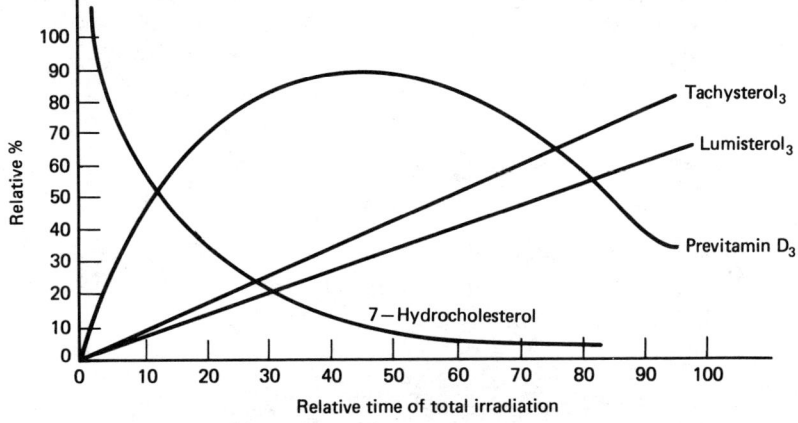

Figure 5. Reaction profile of irradiation of 7-dehydrocholesterol.

198   VITAMINS (D)

Further side-chain oxidation may be necessary for phosphate transport (147–148). 24,25-Dihydroxycholecalciferol is produced by kidney mitochondria but can be produced elsewhere (149–150). 1α,24,25-Trihydroxy-vitamin D_3 stimulates intestinal calcium mobilization to the extent of ca 60% of the 1,25 activity and 10% of the 1,25 activity on bone resorption (151). It is less active in chicks and is excreted. The function of this naturally occurring substance is not fully understood. Other forms of the hydroxylated vitamin D have also been isolated, characterized, and synthesized, and their function is the subject of continuing research.

Regulation of vitamin D_3 metabolism appears to be controlled by serum calcium, serum phosphorus, and parathyroid hormone. Normal calcium levels are 9.6 mg/mL. When calcium serum levels fall below 9.5 mg/mL, 1,25-dihydroxy-vitamin D_3 is made; at and above this level, 24,25-dihydroxycholecalciferol is made, and 1α-hydroxylase activity discontinues.

The calcium homeostasis mechanism involves a hypocalcemic stimulus, which induces the secretion of parathyroid hormone. This causes phosphate diuresis in the kidney, which stimulates the 1α-hydroxylase activity and causes the hydroxylation of 25-hydroxy-vitamin D_3 to 1α,25-dihydroxycholecalciferol. Parathyroid hormone and 1,25-dihydroxycholecalciferol act at the bone site cooperatively to stimulate calcium mobilization from bone (see Hormones).

Interaction of vitamin D and its metabolites with sex hormones has been demonstrated, particularly in birds where egg-laying functions combine calcium needs and reproductive activity. The metabolites of vitamin D behave as hormones. As such, they play an active role in the endocrine system to maintain harmony, with the other hormones, of the various body functions. The metabolism of vitamin D_2 follows a pathway similar to that described for vitamin D_3.

Synthesis

The total synthesis of vitamin D involves many steps and is not an economical route to the production of commercial material. The syntheses are useful for the preparation of molecules containing isotopes of various atoms for radioactive tracer work or for making derivatives of vitamin D. The early synthesis involved oxidative degradation of the vitamin D molecule to obtain the C- and D-ring portion with the intact side chain. Recombination of this molecule with an appropriate structure containing the A-ring was then carried out by a Wittig type condensation. In 1959, vitamin D_3 was synthesized by a total synthesis from 3-methyl-2-(2-carboxyethyl)-2-cyclohexenone (41), and the Wittig reaction was used extensively (152).

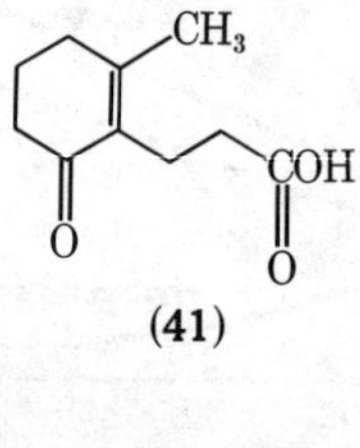

(41)

The 5,6-*trans*-vitamin D_3 was prepared, followed by photochemical isomerization into vitamin D.

Direct preparation of the previtamin and vitamins D is described in refs. 153–154.

Also, a vinyl allene synthesis has been used to prepare vitamin D materials, and the total synthesis of hydroxy-vitamin D with d-carvone has been described (155–157). The synthesis and chemical characterization of most of the natural vitamin D_3 compounds as well as several synthetic analogues have been described, eg, 25-hydroxy, 1,25-dihydroxy vitamin D_3, 24,25-dihydroxy [40013-87-4], 1,24,25-trihydroxy [50648-94-7], 25,26-dihydroxy [29261-12-9], 1-hydroxycalcitroic acid [71204-89-2], 25-hydroxy-26,23-lactone [71203-34-6], and homocalcitroic acid.

The chemical synthesis of 25-hydroxy-vitamin D_3 has been achieved by several groups (158–164). It depends on the availability of the precursor 25-oxo-27-norcholesterol. Grignard reaction followed by introduction of the 7-dehydro function and irradiation allows 25-hydroxy-vitamin D_3 to be isolated. Fucosterol (stigmasta-5,24(28)-dien-3β-ol) as well as bile acids, pregnenolone, and desmosterol have been used as starting materials (161).

The product yield of the first chemical synthesis of 1,25-dihydroxy-vitamin D_3 was <0.005% (165). The principal analogue that was synthesized was the 1α-hydroxy-vitamin D_3 (166–174). The 1α-hydroxy group was introduced by epoxidation of Δ^1, which was obtained from cholesterol, and subsequent reduction of the oxido group. Separation of the 1β-hydroxy or oxido group was done in a variety of ways to achieve the stereospecific product. A key intermediate compound is 1,25-dihydroxycholesterol (161,175–179).

One efficient process permits direct C-1 hydroxylation of a variety of vitamin D compounds (180). The method involves conversion of vitamin D tosylates to 3,4-cyclovitamin D derivatives, allylic oxidation with SeO_2, followed by acid-catalyzed solvolysis to yield 1α-hydroxy derivatives. This procedure gave 10–15% yields of 1,25-dihydroxy-vitamin D_3 or 1α-hydroxy-vitamin D_3.

24,25-Dihydroxy-vitamin D_3 has been isolated and chemically characterized (181). The biosynthesized material migrates exclusively with the synthetic 24(R) epimer (182–183). The first synthesis of 24,25-dihydroxy-vitamin D_3 was of a racemic mixture, and it was followed by the stereospecific syntheses of the two epimers (184–191).

The chemical syntheses of 1,24(R),25-trihydroxy-vitamin D_3 [56142-94-0] and 1,24(S),25-trihydroxy-vitamin D [56142-95-1] have been reported (192–193).

The chemical synthesis of 25,26-dihydroxy-vitamin D_3 has been described, and it has been determined that the biologically occurring epimer is 25(R),26-dihydroxy-vitamin D_3 (185,194–196). The 23,25-dihydroxy-24-oxo metabolite has also been isolated (197).

Preliminary data concerning 1α-hydroxycalcitroic acid (1-hydroxy-24-nor-9,10-secochola-5,7,10(19)-trien-23-oic acid), 25-hydroxy-26,23-lactone vitamin D_3 and homocalcitroic acid (9,10-secochola-5,7,10(19)-trien-24-oic acid) have been reported, and the biological activity of these metabolites is being studied (198). The synthetic chemistry associated with these and other metabolites of vitamin D_2 and vitamin D_3 is described in refs. 45, 129, and 199.

Manufacture

Most of the vitamin D produced in the world is made by the photochemical conversion of 7-dehydrocholesterol. Ergosterol is not used as extensively as it once was, because it offers no real price advantage and, upon irradiation, it gives vitamin D_2 which is inactive in poultry. Vitamin D_2 is used primarily in formulations developed early in the history of vitamin D chemistry for human and animal uses. Irradiation of 7-dehydrocholesterol or ergosterol is carried out by dissolving the steroid in an appropriate solvent, eg, peroxide-free diethyl ether. Solvents such as methanol, cyclohexane, and dioxane have also been used. The solution is pumped through uv-transparent quartz reactors which permit the light from high pressure mercury lamps to impinge upon the solution, which is recycled, until the desired degree of irradiation has been achieved. This results in a mixture of unreacted 7-dehydrosterol, previtamin D, vitamin D, and irradiation by-products.

The solvent is then evaporated, and unconverted sterol precipitates upon treatment with an appropriate solvent, eg, alcohol. Recovered sterol is reused in subsequent irradiations. The chilled slurry is filtered to recover the 7-dehydrocholesterol and the solvent is evaporated to yield vitamin D resin. The resin is a pale yellow-to-amber oil that flows freely when hot and becomes a brittle glass when cold; the activity of the resin is $20-30 \times 10^6$ IU/g. The active cis-vitamin D can be crystallized to give the USP product from a mixture of hydrocarbon solvent and aliphatic nitrile, eg, benzene or acetonitrile, or from methyl formate (200–201). The resin is formulated without further purification in most uses. Chemical complexation has also been used for purification.

In 1938, it was estimated that 7.5×10^{13} quanta of light were required to convert ergosterol to 1 USP unit of vitamin D_2 (202). The value was later determined to be 9.3×10^{13} quanta. Among the light sources used for irradiation are carbon arcs, metal-corded carbon rod, magnesium arcs, and mercury-vapor lamps; the high pressure mercury lamp is most widely used. Higher yields and more favorable isomer distribution can be achieved if the frequency of light is kept at 275–300 nm.

Several arrangements of the components of the simple reactor shown in Figure 6 are possible and have been used. An important feature is a cooling jacket which controls the high temperature (ca 800°C) of the mercury-vapor lamps. Water solutions for cooling may contain salts for screening frequencies of light. Light below 275 nm can be filtered by aromatic compounds as well as by a 5 wt % lead acetate or another inorganic salt solution. Glass filters can also be used as screens for frequencies outside the chemical range. Photosensitizers, eg, eosin, erythrosin, dibromodinitrofluorescein, and others, have been suggested to limit light frequencies and improve the isomer distribution. The vitamin D resin is stabilized against oxidation by the addition of ≤1 wt % butylated hydroxyanisole or butylated hydroxytoluene.

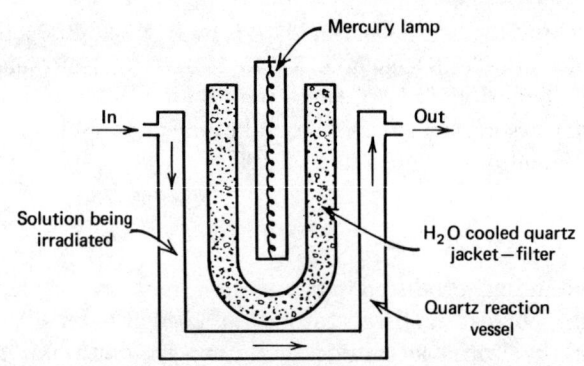

Figure 6. Ultraviolet-light-transparent quartz reactor for sterol irradiation.

Shipping and Handling

Vitamin D and its products are sensitive to uv light, heat, and air. Its sensitivity to these conditions is exaggerated by the presence of heavy-metal ions, eg, iron. Care should be taken to store and ship vitamin D and its various product forms so that exposure to these conditions are minimized. If it is minerally stabilized and microencapsulated, it is protected and stable. It should also be stored in a cool dry place under nitrogen. Vitamin D is generally recognized as safe when used in accordance with good manufacturing or feeding practices (203).

The provitamins are also unstable to heat and light. They generally should be stored in a dark, cool, dry place. The product is more stable if shipped with 10–15 wt % methanol rather than in a dry form.

Economic Aspects

Vitamin D is available in a variety of forms. Cod-liver oil and percomorph-liver oil, which are relatively low and high potency products, respectively, historically were good sources of vitamin D. Recent cost increases of these materials have caused a decline in their market position. Cod-liver oil sold for $0.40–0.45/L in 1970 and as high as $1.45/L in 1979. The prices of the fish and cod-liver oils and of vitamin D_2 and ergosterol from 1955 to 1979 are shown in Table 7.

Vitamin D_3 is produced as a resin (20–30 × 10^6 IU/g) which sold in 1980 for ca $1.02/10^6$ IU. It is usually formulated into more usable forms at a variety of dosages, ie, 200,000 IU/g, 400,000 IU/g, or 1,000,000 IU/g. The 1979 prices of vitamin D were $24.25/10^9$ IU ($9.70/kg of product at 400,000 IU/g) (204). In the United States in 1979, $3.6 × 10^6 worth of vitamin D was sold.

U.S. production and sales values for vitamin D from 1955 to 1978 are listed in Table 8. These numbers appear to be low, and estimations of world demand have been as high as 1300 × 10^{12} IU of vitamin D_3 in 1979. This is divided into thirds for Europe, the United States, and the rest of the world. Of this demand, 90% is estimated for animal-feed fortification and 10% for food and pharmaceutical uses. It is estimated that this demand will grow by 50% to 1750 × 10^{12} IU by 1985. A substantial proportion of the vitamin D is imported, and with all uses included, it is estimated that >80% of the sales are of vitamin D_3.

Table 7. Price History of Vitamin D Sources and of Ergocalciferol and Ergosterol

Year	Sources Fish-liver oil, $/$10^6$ IU	Sources Cod-liver oil, $/L	Vitamin D Ergocalciferol (crystalline vitamin D_2), $/g	Ergosterol, (provitamin D_2), $/$10^6$ IU
1955	0.12	0.40	0.60	0.025
1960	0.12	0.37	0.54	0.025
1965	0.12	0.40	0.48	0.015
1970	0.12	0.42	0.48	0.015
1975		1.39	0.63	
1979		1.45	0.63	0.065

Table 8. U.S. Production and Sales of Vitamin D [a]

Year	Production, metric tons Vitamin D_2	Vitamin D_3	Sales, metric tons Vitamin D_2	Vitamin D_3	Unit sales value, $/kg Vitamin D_2	Vitamin D_3
1955	0.91	0.45	0.45	0.45	939.16	798.06
1960	0.40	0.99	0.21	0.34		522.49
1965	0.51		0.40	0.74	396.83	573.20
1970	5.90[b]		3.63[b]		311.11[b]	
1972	5.90[b]		4.08[b]		343.92[b]	
1974	7.70[b,c]		4.99[b]		673.40[b]	
1976		5.90		4.08		674.61
1978		11.51		3.44		617.66

[a] Ref. 204.
[b] Combined value for vitamins D_2 and D_3.
[c] Includes data for provitamin D_3.

Analytical and Test Methods

Pure vitamin D_3 was adopted in 1949 as the international standard; USP also uses cholecalciferol as a standard (205). Samples of reference standard may be purchased from USP Reference Standards, New York. One international unit of vitamin D activity is that activity demonstrated by 0.025 µg of pure vitamin D_3. One gram of vitamin D_3 is equivalent to 40×10^6 IU or USP units. The international chick unit (ICU) is identical to the USP unit.

The standard chemical and biological methods of analysis are those accepted by the *United States Pharmacopeia XX* as well as the one accepted by the AOAC (206–209). The USP method involves saponification of the sample, whether dry concentrate, premix, powder, capsule, tablet, or aqueous suspension, with aqueous alcohol KOH; solvent extraction; solvent removal; chromatographic separation of vitamin D from extraneous ingredients; and colorimetric determination with antimony trichloride and comparison with a solution of USP cholecalciferol reference standard.

The AOAC recognizes a similar procedure except that the unsaponifiable material is treated with maleic anhydride to remove the trans-isomer (208). The antimony trichloride colorimetric assay is performed on the trans-isomer-free material. This procedure cannot be used to distinguish certain inactive isomers, eg, isotachysterol; if present, these are included in the result, giving rise to a false high analysis. A test must be performed to check for the presence of isotachysterol.

High pressure liquid chromatography (hplc) is used to separate the pre and cis isomers of vitamin D_3 from other isomers and allows their analysis by comparison with the chromatograph of a sample of pure reference *cis*-vitamin D_3, which is equilibrated to a mixture of pre and cis isomers (207,209–210). This method is more sensitive and provides information on isomer distribution as well as the active pre and cis isomer content of a vitamin D sample. It is applicable to most forms of vitamin D, including the more dilute formulations, ie, multivitamin preparations containing at least 1 IU/g (211). The practical problem of isolation of the vitamin material from interfering and extraneous components is the limiting factor in the assay of low level formulations.

A number of methods have been developed for the chromatographic separation of vitamin D and related substances. Some of these methods are listed in Table 9.

Table 9. Methods for Chromatographic Separation of Vitamin D and Related Steroids[a]

Method	Comment	Refs.
paper chromatography		
quinoline-impregnated paper		213–214
reversed-phase paper		215–218
column chromatography		
alumina		219
floridin		220
celite		221–224
	useful for multivitamin tablets	225
silicic acid	useful for metabolites	226–228
factice	can resolve vitamin D_2 and vitamin D_3	229
Sephadex LH-20	highly useful for metabolites	230
high pressure liquid chromatography		
Zorbax-Sil support	useful for vitamin D_2 in capsules	231
	separates 24(R),25-dihydroxy-vitamin D_3 from 24(S),25-dihydroxy-vitamin D_3	232
	useful for metabolites	233
	assay of 25-hydroxy-vitamin D_3	234–235
ODS[b]-Permaphase	useful for vitamin D metabolites	236
silica gel	commercial vitamin D_3 assay	209–210, 237
thin-layer chromatography		
silicic acid	effective for irradiation mixtures	226
silica gel G	two-dimensional	238–240
gas chromatography		241
	separates trimethylsilyl ethers of vitamin D_3	242
	cyclizes vitamin D_3	243–244
	useful for 25-hydroxy-vitamin D_3	245
	useful for multivitamin tablets	246
	useful for vitamin D_2 in milk	247

[a] Ref. 212.
[b] ODS = octadecyl (C_{18}) silane.

Biological Assay. The generally recognized biological method is the rat line test. Rachitic rats are fed diets containing the vitamin D sample. This test measures bone growth on the proximal end of the tibia or distal end of the ulna, which is stained with silver nitrate. This test is not applicable to products offered for poultry feeding. The AOAC recognizes another procedure which measures the vitamin D sample activity in increasing bone ash of growing chicks compared to the activity of USP cholecalciferol reference standard.

Physical Methods. Vitamins D_2 and D_3 exhibit uv absorption curves that have a maximum at 264 nm and an E (absorbance) of 450–490 at 1%. The various isomers of vitamin D exhibit characteristically different uv absorption curves. Mixtures of the isomers are difficult to distinguish. However, when chromatographically separated by hplc, the peaks can be identified by stop-flow techniques based on uv absorption scanning. The combination of elution time and characteristic uv absorption curves can be used to identify the isomers present in a sample of vitamin D.

Infrared and nmr spectroscopy have been used to help distinguish between vitamins D_2 and D_3 (248–250). X-ray crystallographic techniques are used to determine the vitamin D structure, and gas chromatography also is a method for assaying vitamin D (259) (see Table 9).

204 VITAMINS (D)

Provitamin D. The molecular extinction coefficient of 7-dehydrocholesterol at 282 nm is 11,300 and is used as a measure of 7-dehydro isomer content of the provitamin (252–253). High pressure liquid chromatography can also be used to analyze the provitamins.

There is a variety of chemicals that show characteristic colors when applied to the provitamins. Some of these are listed in Table 10.

Health and Safety Factors

Disease States. Although rickets is the most common disease associated with vitamin D deficiency, many other diseases can occur and are related to vitamin D. These can involve a lack of the vitamin, deficient synthesis of the metabolites from the vitamin, deficient control mechanisms, or defective organ receptors. The control of calcium and phosphorus is essential in the maintenance of normal cellular biochemistry, eg, muscle contraction, nerve conduction, and enzyme function. The interaction of the metabolites and the various target organs is depicted in Figure 7 with some vitamin D-associated disease states.

In the treatment of diseases where the metabolites are not being delivered to the system, synthetic metabolites or active analogues have been successfully administered.

Table 10. Chemical Test Methods for Provitamin D

Name of reaction	Components	Results	Interpretation
revised Salkowski reaction	$CHCl_3$ + H_2SO_4 (conc)	deep red acid layer	differentiates from sterols lacking conjugated diene (red color in $CHCl_3$ layer); acid gives green fluorescence
Lieberman-Burchard reaction	$CHCl_3$; acetic acid–H_2SO_4 added dropwise	red color develops and changes to blue-violet to green	can be quantitative; cholesterol acts similarly, but red color lasts longer
Tortelli-Jaffé reaction	acetic acid + 2 wt % Br_2 in $CHCl_3$	green	sterols with ditertiary double bonds; vitamin D and compounds that give similar bonds upon isomerization or reaction
Rosenheim reaction	$CHCl_3$ + trichloroacetic acid in H_2O	red color develops and changes to light blue	
Rosenheim reaction	$CHCl_3$ + lead tetraacetate in CH_3COOH is added; then trichloroacetic acid is added	green fluorescence	not given by esters of provitamin D; can be used to distinguish between provitamin and provitamin ester; sensitive to 0.1 µg and is quantitative
chloral hydrate	mixture of crystalline provitamins and chloral hydrate heated slowly; melts at 50°C	color develops and changes red to green to deep blue	other sterols, eg, cholesterol, do not react to give color
antimony trichloride reaction	$CHCl_3$ + $SbCl_3$	red color	
Chugaev reaction	glacial acetic acid plus acetyl chloride and zinc chloride heated to boiling	eosin-red greenish yellow fluorescence	1:80,000 sensitivity

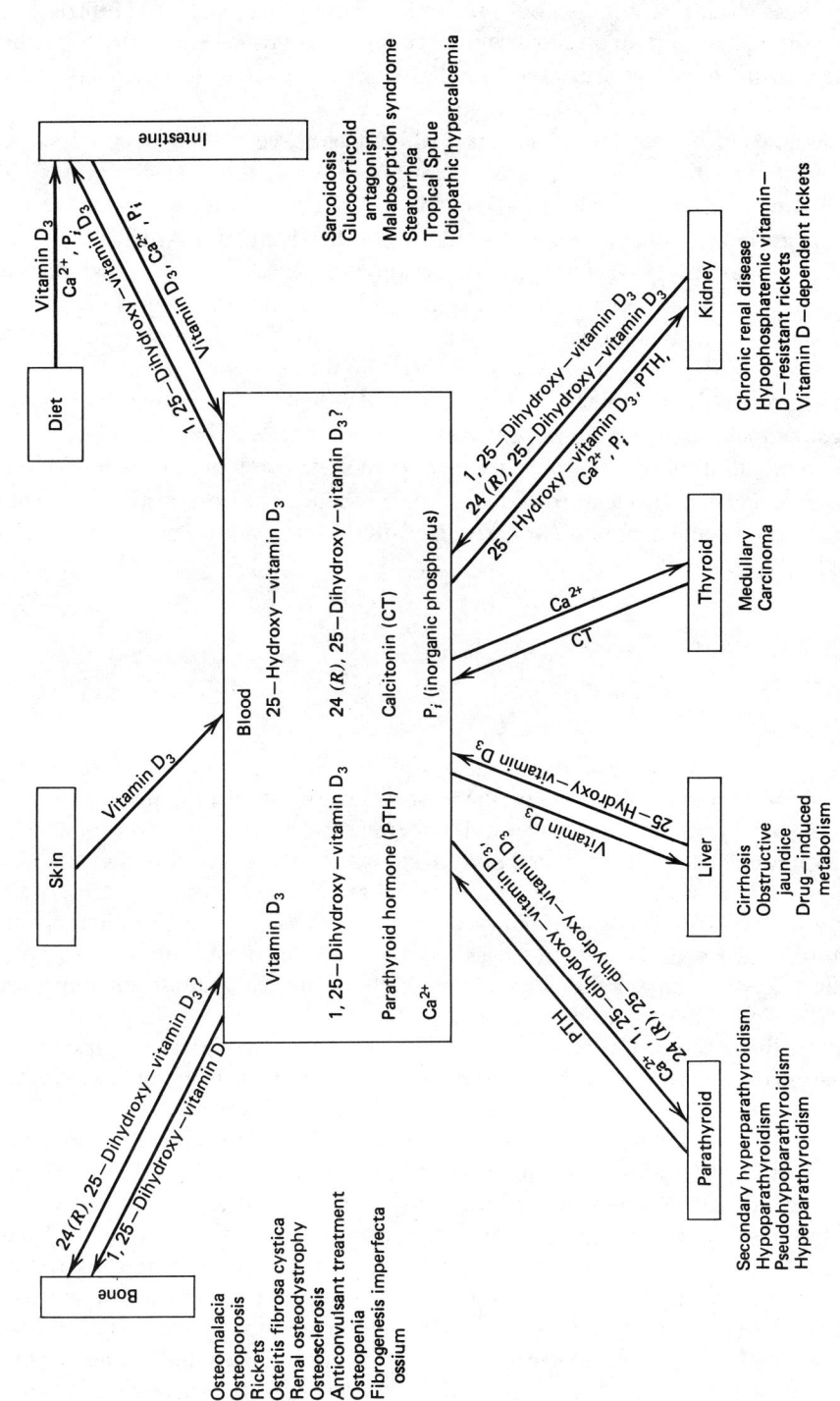

Figure 7. Human disease states related to vitamin D (45).

Many of these clinical studies are outlined in references 45 and 129. Useful treatment of many animal and human diseases can be achieved; ie, vitamin D_3 metabolites have been successfully used for treatment of milk fever in cattle, turkey leg weakness, and osteoporosis and renal osteodystrophy in humans.

Toxicology. The vitamins D are toxic when doses exceed 1000–3000 IU/kg body weight per day over extended periods of time. Vitamin D mobilizes bound calcium from the bone, and toxic levels cause considerable increases in plasma calcium and urinary excretion of phosphate and calcium. Mobilized calcium is absorbed by soft tissues, eg, kidneys and vessel media. Loss of appetite, gastrointestinal disturbances, head and joint pain, and muscle weakness are typical clinical symptoms. Death from hypervitaminosis D is usually caused by renal failure. The poisoning symptoms are usually reversed when ingestion of the material is stopped (254).

The lowest published lethal doses of ergocalciferol in dogs are as follows: oral, 4 mg/kg; intraperitoneal, 10 mg/kg; intravenous, 5 mg/kg; and intramuscular, 5 mg/kg (255). The metabolites of vitamin D are usually more toxic than the vitamin because the feedback mechanisms that regulate vitamin D concentrations are circumvented. 25-Hydroxycholecalciferol has a one-hundred-fold increase in toxicity over vitamin D when fed to chicks (256).

Uses

Most of the vitamin D sold is synthetic. Vitamin D_2 as a concentrate or in microcrystalline forms is used in many pharmaceutical preparations, although vitamin D_3 is preferred by many. Vitamin D_2 in the form of irradiated yeast has been used as a feed supplement for cattle, swine, and dogs, but its use has declined over the last several years. The swine and poultry industries are the largest markets for vitamin D_3. Swine accounts for 25.6% and poultry for 43.3% of vitamin D consumption in animal agriculture. The beef and dairy industries account for 30.3% of vitamin D consumption, of which 22% is by dairy calves. The remaining 0.7% of total animal consumption of vitamins is largely in prepared pet foods.

Crystalline vitamin D is used for medicinal preparations and formulations in the pharmaceutical industry as well as for the fortification of fresh and evaporated milk and nonfat dry milks. Preparations based on the use of the resin are less expensive than crystalline D and are more useful. Essentially all milk produced in the United States is fortified with vitamin D. Cereals, margarine, and animal feeds are also fortified with vitamin D. The resin must be diluted to a proper dosage form, and many supplement preparations are marketed, eg, tonics, drops, capsules, and tablets; oil-based injectables are also available. Combination formulations are used widely, particularly with vitamin A for humans and with vitamins A and E for animals. Preemulsified products are used for increased bioavailability. Water-miscible formulations have also been developed (257). Solutions of vitamin D in oil or in oil on dry carriers, eg, corn or flour, are used in animal feeds. These diets contain high mineral contents, and the vitamin D formulation is not stable long-term unless protected. Several stable forms are patented and sold commercially; they include beadlets of dry suspensions in gelatin, carbohydrates, wax, and cellulose derivatives (258–262).

Dietary Requirements. Vitamin D is tremendously important for growth and maintenance of good health. According to the NRC, the vitamin D requirement for optimum health is ca 400 IU/d, regardless of age. This amount of vitamin D gives ample protection from rickets, provided a sufficient amount of the other essential nutrients, including calcium and phosphorus, is supplied (263). Recommended amounts of vitamin D per kilogram of feed are as follows: starting and growing chicks, 200 IU; laying and breeding hens, 500 IU; turkeys, 400 IU; ducks, 200 IU; and swine 125–200 IU. Calves require 600 IU per 100 kg of body weight (264–266). Practical feeding levels are somewhat higher to assure delivery of the vitamin to the animals.

Animals exposed to sunlight for extended lengths of time do not require substantial dietary vitamin D. Current livestock management practices place an emphasis on high productivity, and most feed manufacturers recommend vitamin D supplementation of diets. Recommendations for practical levels of vitamin D in feeds for various animals, as recommended by some manufacturers, are listed below, in units of 10^6 IU/t:

Poultry
  chickens
    broilers   2–4
    replacement birds   1–3
    layers   1–3
    breeding hens   2–4
  turkeys
    starting   3–5
    growing   2–4
    breeding   3–5
  ducks
    market   1–3
    breeding   2–4
Swine
  prestarter (to 10 kg)   2–3
  starter (10–35 kg)   1–2
  growing–finishing (35 kg to   1–2
    market)
  gestation   1–2
  lactation   2–3
  boars   2–3
Fish
  trout   1–2
  catfish (channel)   1–2

Dairy cattle
  calf starter   1–2
  calf milk replacer   2–4
  replacement heifers   1–2
  dry cows   1–2
  lactating cows   2–4
  bulls   2–4
Beef cattle
  calf starter   1–2
  replacement heifers   1–2
  feedlot   2–3
  dry pregnant cows   1–2
  lactating cows   1–2
  bulls   1–2
Sheep
  fattening lambs   2–3
  breeding   1–2
Dogs   5–1
Cats   1–2
Horses   1–2

BIBLIOGRAPHY

"Vitamins (Vitamin D)" in *ECT* 1st ed., Vol. 14, pp. 828–849, by H. R. Rosenberg, E. I. du Pont de Nemours & Co., Inc.; "Vitamins (Vitamin D)" in *ECT* 2nd ed., Vol. 21, pp. 549–573, by Sheldon B. Greenbaum, Diamond Shamrock Chemical Co.

1. H. M. Frost, *Bone Dynamics in Osteoporosis and Osteomalacia*, Surgery Monograph Series, Charles C Thomas, Publisher, Springfield, Ill., 1966.

2. H. F. Deluca in R. B. Alfin-Slater and D. Kratchevsky, eds., *Nutrition and the Adult: Micronutrients*, Plenum Press, New York, 1980, p. 207.
3. W. F. Loomis, *Sci. Am.* **223**(6), 77 (1970).
4. G. T. Smerdon, *J. Hist. Med.* **5**, 397 (1950).
5. J. H. Bennett, *Treatise on the Oleum Jecoris Aselli or Cod Liver Oil*, Edinburgh, UK, 1848.
6. W. Mozolowski, *Nature (London)* **143**, 121 (1939).
7. M. Kassowitz, *Die normale Ossifikation und die Erkrankungen des Knochensystems bei Rachitis und Hereditaerer Syphylis*, Braumueller, Wien, Austria, 1881.
8. G. Pommer, *Untersuchungen Uber Osteomalacia und Rachitis*, Vogel, Leipzig, Germany, 1885.
9. C. Tanret, *Ann. Chim. Phys. Sect. 5* **17**, 493 (1879).
10. C. Tanret, *Compt. Rend.* **108**, 98 (1889).
11. T. A. Palm, *Practitioner* **45**, 270 (1890).
12. H. Fehling, *Z. Geburtshilfe Gynaekol.* **30**, 471 (1894).
13. C. Funk, *Die Vitamine*, Bergmann, Wiesbaden, FRG, 1914.
14. L. Findlay and co-workers, *Lancet* **1**, 825 (1922).
15. E. Mollanby, *J. Physiol. (London)* **52**, L 111 (1919); *Lancet* **196**, 407 (1919).
16. K. Huldschinski, *Dtsch. Med. Wochenschr.* **45**, 712 (1919).
17. E. V. McCollum, N. Simmonds, T. H. Parsons, P. G. Shipley, and E. H. Park, *J. Biol. Chem.* **45**, 332 (1921).
18. E. V. McCollum, N. Simmonds, J. E. Becker, and P. G. Shipley, *Bull. Johns Hopkins Hosp.* **33**, 229 (1922).
19. A. F. Hess and L. J. Unger, *J. Am. Med. Assoc.* **74**, 216 (1920).
20. T. F. Zucker, A. M. Pappenheimer, and M. Barnett, *Proc. Soc. Exp. Biol. Med.* **19**, 167 (1922).
21. E. V. McCollum, N. Simmonds, J. E. Becker, and P. G. Shipley, *J. Biol. Chem.* **53**, 293 (1922).
22. A. F. Hess and M. G. Gutman, *J. Am. Med. Assoc.* **78**, 29 (1922).
23. H. Goldblatt and K. N. Soames, *Biochem. J.* **17**, 294 (1923).
24. H. Steenbock, *Science* **60**, 224 (1924).
25. A. F. Hess and M. Weinstock, *J. Biol. Chem.* **62**, 301 (1924).
26. A. F. Hess, M. Weinstock, and F. D. Helman, *J. Biol. Chem.* **63**, 305 (1925).
27. H. Steenbock and A. Black, *J. Biol. Chem.* **64**, 263 (1925).
28. O. Rosenheim and T. A. Webster, *Lancet* **1**, 1025 (1925).
29. E. V. McCollum, N. Simmonds, J. E. Becker, and P. G. Shipley, *J. Biol. Chem.* **65**, 97 (1925).
30. A. Windhaus and H. Hess, *Hachr. Ges. Wiss. Guttingen Math. Physik. Kl.*, 175 (1927).
31. O. Rosenheim and T. A. Webster, *Biochem. J.* **21**, 389 (1927).
32. E. H. Reerrink and A. Van Wijk, *Biochem. J.* **23**, 1294 (1929).
33. A. F. Hess, *Rickets, Osteomalacia and Tetany*, Lea & Febiger, Philadelphia, Pa., 1929, p. 90.
34. O. N. Massengale and M. Nussmeier, *J. Biol. Chem.* **87**, 423 (1930).
35. A. Windaus and co-workers, *Justus Liebigs Ann. Chem.* **492**, 226 (1932).
36. A. Windaus and R. Langer, *Justus Liebigs Ann. Chem.* **508**, 105 (1934).
37. J. Waddell, *J. Biol. Chem.* **105**, 711 (1934).
38. A. Windaus, H. Lettie, and F. Schenck, *Justus Liebigs Ann. Chem.* **520**, 98 (1935).
39. U.S. Pat. 2,226,674 (Dec. 16, 1941), A. G. Baer, E. H. Reerrink, A. Van Wijk, and J. Van Niekerk (to Hartford National Bank & Trust Co.).
40. A. Windaus and F. Bock, *Z. Physiol. Chem. Hoppe-Seglers* **245**, 168 (1937).
41. H. Brockmann and co-workers, *Z. Physiol. Chem. Hoppe-Seglers* **241**, 104 (1936); **245**, 96 (1937).
42. U.S. Pat. 2,098,984 (Nov. 16, 1937), A. Windaus and F. Schenck (to Winthrop Chem. Co., Inc.).
43. D. Crowfoot and J. Dunitz, *Nature (London)* **162**, 608 (1948).
44. D. Crowfoot-Hodgkin, B. M. Rummer, J. D. Dunitz, and K. Trueblood, *J. Chem. Soc.*, 4945 (1963).
45. A. W. Norman, *Vitamin D—The Calcium Homostatic Steroid Hormone*, Academic Press, Inc., New York, 1979, p. 19.
46. A. W. Norman, *Am. J. Physiol.* **211**, 829 (1966).
47. J. E. Zull, E. Czarnowski-Misztal, and H. F. DeLuca, *Science* **149**, 182 (1965).
48. J. W. Blunt, H. F. DeLuca, and H. K. Schnoes, *Chem. Commun.*, 801 (1968); J. W. Blunt, H. F. DeLuca, and V. Tanaka, *Proc. Natl. Acad. Sci. U.S.A.* **61**, 1503 (1968).
49. M. R. Haussler, J. Myrtle, and A. W. Norman, *J. Biol. Chem.* **243**, 4055 (1968).
50. D. E. M. Lawson, P. W. Wilson, and E. Kodicek, *Biochem. J.* **115**, 269 (1969).
51. E. Kodicek, D. E. M. Lawson, and P. W. Wilson, *Nature (London)* **228**, 763 (1970).
52. J. F. Myrtle and A. W. Norman, *Science* **171**, 78 (1971).

53. T. Suda, H. F. DeLuca, H. K. Schnoes, Y. Tanaka, and M. F. Holick, *Biochemistry* **9,** 4776 (1970).
54. D. R. Fraser and E. Kodicek, *Nature (London)* **228,** 764 (1970).
55. A. W. Norman, R. J. Midgett, J. F. Myrtle, and H. G. Nowicki, *Biochem. Biophys. Res. Commun.* **42,** 1082 (1971).
56. R. Gray, I. Boyle, and H. F. DeLuca, *Science* **172,** 1232 (1971).
57. A. W. Norman and co-workers, *Biochem. Biophys. Res. Commun.* **42,** 1082 (1971).
58. R. J. Midgett, A. M. Spielvogel, J. M. Coburn, and A. W. Norman, *J. Clin. Endocrinol. Metab.* **36,** 1153 (1973).
59. D. E. M. Lawson, D. M. Fraser, E. Kodicek, H. R. Morris, and D. H. Williams, *Nature (London)* **230,** 228 (1971).
60. A. W. Norman, J. F. Myrtle, R. J. Midgett, H. G. Nowicki, V. Williams, and G. Popjak, *Science* **173,** 51 (1971).
61. M. F. Holick, H. K. Schnoes, H. F. DeLuca, T. Suda, and R. J. Cousins, *Biochemistry* **10,** 2799 (1971).
62. E. J. Semmler, M. F. Holick, H. K. Schnoes, and H. F. DeLuca, *Tetrahedron Lett.* **40,** 4147 (1972).
63. A. S. Brickman, J. W. Coburn, A. W. Norman, *N. Engl. J. Med.* **287,** 891 (1972).
64. D. R. Fraser, S. W. Kooh, H. P. Kind, M. F. Holick, Y. Tanaka, and H. F. DeLuca, *N. Engl. J. Med.* **289,** 817 (1973).
65. H. C. Tsai and A. W. Norman, *Biochem. Biophys. Res. Commun.* **54,** 622 (1973).
66. R. M. Wing, W. H. Okamura, M. R. Pirio, S. Sine, and A. W. Norman, *Science* **186,** 939 (1974).
67. P. F. Brumbaugh and M. R. Haussler, *J. Biol. Chem.* **249,** 1251 (1974).
68. J. G. Haddad and J. Walgate, *J. Biol. Chem.* **251,** 4803 (1976).
69. *Eur. J. Biochem.* **124,** 223 (1982).
70. R. H. Wasserman, J. D. Henion, M. R. Haussler, and T. A. McCain, *Science* **194,** 853 (1976).
71. M. Peterlik, K. Bursac, M. R. Haussler, M. R. Hughes, and R. H. Wasserman, *Biochem. Biophys. Res. Commun.* **70,** 797 (1976).
72. J. P. Simonit, K. M. L. Morris, and J. C. Collins, *J. Endocrinol.* **68,** 18 (1976).
73. H. Zucker and W. A. Rambeck, *Centralbl. Veterinaermed.* **28,** 436 (1981).
74. W. A. Rambeck and co-workers, *Z. Pflanzenphysiol.* **104,** 9 (1981).
75. R. H. Wasserman and co-workers, *Nutr. Rev.* **33,** 1 (1975); *J. Nutr.* **106,** 457 (1976); *Biochem. Biophys. Res. Commun.* **62,** 85 (1975).
76. Ref. 45, p. 49.
77. J. C. Drummond and E. R. Gunther, *Nature (London)* **126,** 398 (1930).
78. J. C. Drummond and E. R. Gunther, *J. Exp. Biol.* **11,** 203 (1934).
79. H. H. Darby and H. T. Clarke, *Science* **85,** 318 (1937).
80. A. Scheunert, M. Schieblich, and J. Reschke, *Hoppe-Segler's Z. Physiol. Chem.* **235,** 91 (1935).
81. J. E. Campion, K. M. Henry, S. K. Skon, and J. Mackintosh, *Biochem. J.* **31,** 81 (1937).
82. J. C. Drummond, C. H. May, and N. E. G. Richardson, *Br. Med. J.* **2,** 757 (1939).
83. H. J. Heinz, H. J. Heinz Co., Pittsburgh, Pa., 1950.
84. K. H. Coward and B. G. E. Morgan, *Br. Med. J.* **2,** 1041 (1935).
85. C. E. Bells in W. H. Sebrell and R. S. Harris, eds., *The Vitamins*, 1st ed., vol. 2, Academic Press, New York, 1954, p 132.
86. P. S. Chen and H. B. Bosman, *J. Nutr.* **83,** 133 (1964).
87. U.S. Pat. 2,395,115 (Feb. 19, 1946), K. J. Goering (to Anheuser-Busch, Inc.).
88. U.S. Pat. 2,874,171 (Feb. 17, 1959), H. A. Nelson (to Upjohn Co.).
89. U.S. Pat. 3,006,932 (Oct. 31, 1961), J. Green, S. A. Price, and E. E. Edwin (to Vitamins, Ltd.).
90. U.S. Pat. 2,794,035 (May 28, 1957), O. Hummel (to Zellstaff-Fabrik Waldhof).
91. U.S. Pat. 2,865,934 (Dec. 23, 1958), R. A. Fisher (to Bioferm. Corp.).
92. Ger. Pat. 1,252,674 (Oct. 26, 1967), K. Petzoldt, K. Kliesich, and H. J. Koch (to Schering A.G.).
93. K. Ziegler, *Ann.* **551,** 80 (1942).
94. S. Bernstein, L. J. Benovi, L. Dorfman, K. J. Sax, and Y. Subbarow, *J. Org. Chem.* **14,** 433 (1949); U.S. Pat. 2,498,390 (Feb. 21, 1950), S. Bernstein and K. J. Sax (to American Cyanamid Co.).
95. U.S. Pat. 2,446,091 (May 4, 1968), J. Van der Vliet and W. Stevens (to Hartford National Bank and Trust Co.).
96. H. Schaltagger, *Helv. Chim. Acta* **33,** 2101 (1950).
97. F. Hunziker and F. X. Mullner, *Helv. Chim. Acta* **41,** 70 (1958).
98. M. Rappoldt, J. Hoogendoorn, and L. F. Pauli in A. W. Norman, K. Schaefer, and D. V. Herrath, eds.,

Proceedings of the Fifth Workshop on Vitamin D, Vitamin D—Chemical, Biochemical and Clinical Endocrinology of Calcium Metabolism, Williamsburg, Va., Feb. 1982, Walter de Gruyter, Berlin, 1982.
99. W. G. Salmond, M. A. Serta, A. M. Cain, and M. C. Sobala, *Tetrahedron Lett.* (20), 1683 (1977).
100. L. E. Fieser, *J. Am. Chem. Soc.* **75,** 4394 (1953).
101. U.S. Pat. 2,505,646 (Apr. 25, 1950), W. E. Meuly (to E. I. du Pont de Nemours & Co., Inc.).
102. A. Windaus, F. VonWerder, A. Luttringhaus, and E. Fernholz, *Ann.* **499,** 188 (1932).
103. L. Velluz, G. Amiard, and B. Goffinet, *Bull. Soc. Chim. France* **22,** 1341 (1955).
104. E. Havinga, R. J. DeKoch, and M. Rappoldt, *Tetrahedron* **11,** 276 (1960).
105. P. B. Braun, J. Hornstra, C. Knobles, E. W. M. Rutten, and C. Romer, *Acta Crystallogr.* **B29,** 463 (1973).
106. E. Havinga, *Experimentia* **29,** 1181 (1973); citation of calculations by C. Altona in a private communication.
107. M. Holick and co-workers in ref. 98, p. 1151.
108. A. L. Hoevoet, A. Verloop, and E. Havinga, *Rec. Trav. Chim.* **74,** 788 (1955).
109. K. H. Hanewald, M. P. Rappoldt, and J. R. Roborgh, *Rec. Trav. Chim. Pays-Bas* **80,** 1003 (1961).
110. A. Verloop, A. L. Koevoet, and E. Havinga, *Rec. Trav. Chim. Pays-Bas* **76,** 689 (1957).
111. W. G. Dauben, I. Bell, T. W. Hutton, G. F. Laws, A. Rheiner, and H. Urscheler, *J. Am. Chem. Soc.* **80,** 4117 (1958).
112. W. H. Okamura, M. L. Hammond, A. J. C. Jacobs, and J. Von Thiegil, *Tetrahedron Lett.* (52), 4807 (1976).
113. E. Havinga, *Chimia* **30,** 27 (1976); D. H. R. Barton and co-workers, *Chem. Comm.*, 65 (1976).
114. F. Boosman, H. J. C. Jacobs, E. Havinga, and A. Van den Gen, *Tetrahedron Lett.* (7), 427 (1975).
115. A. Verloop, A. L. Koevoet, R. Van Moorselase, and E. Havinga, *Rec. Trav. Chim. Pays-Bas* **78,** 1004 (1959).
116. A. Verloop, A. L. Koevoet, and E. Havinga, *Rec. Trav. Chim. Pays-Bas* **74,** 1125 (1955).
117. T. Kobagashi, *J. Vitaminol. (Jpn).* **13,** 255 (1967).
118. H. H. Inhoffen, G. Quinkert, J. H. Hess, and H. M. Erdman, *Chem. Ber.* **89,** 2273 (1956).
119. C. H. Nields, W. C. Russell, and A. Zimmerli, *J. Biol. Chem.* **136,** 73 (1940).
120. J. B. Wilkie, S. W. Jones, and O. L. Kline, *J. Amer. Pharm. Assoc. Sci. Ed.* **47,** 395 (1958).
121. S. Kobayshi and M. Yasumura, *J. Nutr. Sci. Vitaminol.* **19,** 123 (1973); S. Kobayshi and co-workers, *J. Nutr. Sci. Vitaminol.* **26,** 545 (1980).
122. D. H. R. Barton, R. H. Hesse, M. M. Pecket, and E. Rizzardo, *J. Am. Chem. Soc.* **95,** 2748 (1973).
123. K. Pfoertner and J. D. Weber, *Helv. Chim. Acta* **55,** 921, 937 (1972).
124. S. C. Eyley and D. H. Williams, *J. Chem. Soc. Chem. Comm.*, 858 (1975).
125. E. C. Snoeren, M. R. Daha, J. Lugtenburg, and E. Havinga, *Rec. Trav. Chim. Pays-Bas* **89,** 261 (1970).
126. W. G. Dauben and R. B. Phillips, *J. Am. Chem. Soc.* **104,** 355 (1982).
127. V. Malatesta, C. Willis, and P. A. Hackett, *J. Am. Chem. Soc.* **103,** 6781 (1981).
128. U.S. Pat. 3,661,939 (May 9, 1972), M. Toyoda, Y. Tawara (to Nisshin Flour Milling Co., Ltd., Japan).
129. A. W. Norman, K. Schaefer, D. V. Herrath, H. G. Grigoleit, J. W. Coburn, H. F. DeLuca, E. B. Mower, and T. Suda, *Vitamin D Basic Research and Its Clinical Application*, Walter de Gruyter, Berlin, 1979.
130. E. Kodicek, *Ciba Found. Symp.*, 161 (1956).
131. A. W. Norman and H. F. DeLuca, *Biochemistry* **2,** 1160 (1963).
132. P. F. Neville and H. F. DeLuca, *Biochemistry* **5,** 2201 (1966).
133. E. B. Olson, Jr., J. C. Knutson, M. H. Bhattacharyya, and H. F. DeLuca, *J. Clin. Invest.* **57,** 1213 (1976).
134. G. Ponchon, A. L. Kennan, and H. F. DeLuca, *J. Clin. Invest.* **48,** 2032 (1969).
135. M. Horsting and H. F. DeLuca, *Biochem. Biophys. Res. Commun.* **36,** 251 (1969).
136. S. A. Holick, M. F. Holick, T. E. Tavela, H. K. Schnoes, and H. F. DeLuca, *J. Biol. Chem.* **251,** 397 (1976).
137. J. G. Haddad and T. J. Hahn, *Nature* **244,** 515 (1973).
138. R. Belsey, M. B. Clark, M. Bernat, J. Glowacki, M. F. Holick, H. F. DeLuca, and J. T. Potts, Jr., *Am. J. Med.* **57,** 50 (1974).
139. J. A. Eisman, R. M. Shepard, and H. F. DeLuca, *Anal. Biochem.* **80,** 198 (1977).

140. M. F. Holick, H. K. Schnoes, H. F. DeLuca, T. Suda, and R. J. Cousins, *Biochemistry* **10**, 2799 (1971).
141. R. Gray, I. Boyle, and H. F. DeLuca, *Science* **172**, 1232 (1971).
142. J. G. Ghazarian, H. K. Schnoes, and H. F. DeLuca, *Biochemistry* **12**, 2555 (1973).
143. J. G. Ghazarian, C. R. Jefcoate, J. C. Knutson, W. H. Orme-Johnson, and H. F. DeLuca, *J. Biol. Chem.* **249**, 3026 (1974).
144. J. I. Pedersen, J. G. Ghazarian, N. R. Orme-Johnson, and H. F. DeLuca, *J. Biol. Chem.* **251**, 3933 (1976).
145. I. T. Boyle, L. Miravet, R. W. Gray, M. F. Holick, and H. F. DeLuca, *Endocrinology* **90**, 605 (1972a).
146. T. C. Chen, L. Castillo, M. Korycka-Dahl, and H. F. DeLuca, *J. Nutr.* **104**, 1056 (1974).
147. R. Kumar, D. Harnden, and H. F. DeLuca, *Biochemistry* **15**, 2420 (1976).
148. D. Harnden, R. Kumar, M. F. Holick, and H. F. DeLuca, *Science* **193**, 493 (1976).
149. M. F. Holick, H. K. Schnoes, H. F. DeLuca, R. W. Gray, I. T. Boyle, and T. Suda, *Biochemistry* **11**, 4251 (1972).
150. Y. Tanaka, L. Castillo, H. F. DeLuca, and N. Ikekawa, *J. Biol. Chem.* **252**, 1421 (1977).
151. M. F. Holick, A. Kleiner-Bossaller, H. K. Schnoes, P. M. Kasten, I. T. Boyle, and H. F. DeLuca, *J. Biol. Chem.* **248**, 6691 (1973).
152. H. H. Inhoffen, K. Irmscher, H. Hirschfeld, U. Stache, and A. Kreutzer, *Chem. Ber.* **91**, 2309 (1958); H. H. Inhoffen and co-workers, *Chem. Ber.* **92**, 1564 (1959).
153. T. M. Dawson, J. Dixon, P. S. Littlewood, B. Lythgoe, and A. K. Saksena, *J. Chem. Soc. C*, 2960 (1971).
154. B. Lythgoe, M. E. N. Nambudiry, and J. Tideswell, *Tetrahedron Lett.* (31), 3685 (1977).
155. M. L. Hammond and co-workers, *J. Am. Chem. Soc.* **100**, 4907 (1978).
156. P. Condran and co-workers, *J. Am. Chem. Soc.* **102**, 6259 (1980).
157. E. G. Baggiolini, J. A. Jacobelli, B. M. Hennessy, and M. R. Uskokovic in ref. 98.
158. J. A. Campbell, D. M. Squires, and J. C. Babcock, *Steroids* **13**, 567 (1969).
159. S. J. Halkes and N. P. VanVliet, *Rec. Trav. Chim. Pays-Bas* **88**, (1969).
160. J. W. Blunt and H. F. DeLuca, *Biochemistry* **8**, 671 (1969).
161. M. Morisaki, J. Rubio-Lightbourn, and N. Ikekawa, *Chem. Pharm. Bull.* **21**, 457 (1973).
162. J. J. Partridge, S. Faber, and M. R. Uskokovic, *Helv. Chim. Acta* **57**, 764 (1974).
163. U.S. Pat. 4,172,076 (Apr. 4, 1977), A. Hirsch and J. Pikl (to Diamond Shamrock, subsequently assigned to A. L. Labs, Jan. 1, 1982).
164. U.S. Pat. 4,226,770 (Oct. 7, 1980), E. Kaiser.
165. E. J. Semmler, M. F. Holick, H. K. Schnoes, and H. F. DeLuca, *Tetrahedron Lett.*, 4147 (1972).
166. R. G. Harrison, B. Lythgoe, and P. W. Wright, *Tetrahedron Lett.*, 3649 (1973).
167. R. G. Harrison, B. Lythgoe, and P. W. Wright, *J. Chem. Soc.*, 2654 (1974).
168. C. Keneko, S. Yamada, A. Sugimoto, Y. Eguchi, M. Ishikawa, T. Suda, M. Suzuki, S. Kabuta, and S. Sasaki, *Steroids* **23**, 75 (1973).
169. D. H. R. Barton, R. H. Hesse, M. Pechet, and E. Rizzardo, *J. Am. Chem. Soc.* **95**, 2748 (1973).
170. A. Furst, L. Labler, W. Meier, and K. H. Pfoertner, *Helv. Chim. Acta* **56**, 1708 (1973).
171. M. F. Holick, S. A. Holick, T. Tavela, B. Gallagher, H. K. Schnoes, and H. F. DeLuca, *Science* **190**, 576 (1975).
172. W. H. Okamura, M. L. Hammond, M. R. Pirio, R. M. Wing, A. Rego, M. N. Mitra, and A. W. Norman in A. W. Norman and co-eds., *Proceedings of the Second Workshop on Vitamin D and Problems Related to Uremic Bone Disease*, Walter de Gruyter, Berlin, 1975, p. 1.
173. U.S. Pat. 4,199,518 (Apr. 22, 1980), W. R. Glave, R. L. Johnson, and A. L. Hirsch (to Diamond Shamrock, subsequently assigned to A. L. Labs, Jan. 1, 1982).
174. U.S. Pat. 4,287,129 (Sept. 1, 1981), W. H. Klausmeier, R. L. Johnson, and A. L. Hirsch (to Diamond Shamrock, subsequently assigned to A. L. Labs, Jan. 1, 1982).
175. K. Ochi and co-workers, *J. Chem. Soc. Perkin I*, 165 (1979).
176. T. A. Narwid, J. F. Blunt, J. A. Iacobelli, and M. R. Uskokovic, *Helv. Chim. Acta* **57**, 781 (1974).
177. J. Rubio-Lightbourn, M. Morisaki, and N. Ikekawa, *Chem. Pharm. Bull.* **21**, 1854 (1973).
178. M. Morisaki, K. Bannai, and N. Ikekawa, *Chem. Pharm. Bull.* **21**, 1853 (1973).
179. M. Morisaki, J. Rubio-Lightbourn, N. Ikekawa, and T. Takeshita, *Chem. Pharm. Bull.* **21**, 2568 (1973).
180. H. E. Paaren, D. E. Hamer, H. K. Schnoes, and H. F. DeLuca, *Proc. Natl. Acad. Sci. U.S.A.* **75**, 2080 (1978).

181. M. J. Holick, H. K. Schnoes, H. F. DeLuca, R. W. Gray, I. T. Boyle, and T. Suda, *Biochemistry* **11**, 4251 (1972).
182. Y. Tanaka, H. Frank, H. F. DeLuca, N. Koizumi, and N. Ikekawa, *Biochemistry* **14**, 3293 (1975).
183. Y. Tanaka, H. F. DeLuca, N. Ikekawa, M. Morisaki, and N. Koizumi, *Arch. Biochem. Biophys.* **170**, 620 (1975).
184. H.-Y. Lam, H. K. Schnoes, H. F. DeLuca, and T. C. Chen, *Biochemistry* **12**, 4851 (1973).
185. M. Seki, J. Rubio-Lightbourn, J. Morisaki, and N. Ikekawa, *Chem. Pharm. Bull.* **21**, 2783 (1973).
186. J. Redel, P. Bell, F. Delbarre, and E. Kodicek, *C.R. Acad. Sci.* **278**, 529 (1974).
187. J. Redel, N. Bazely, Y. Calando, F. Delbarre, P. A. Bell, and E. Kodicek, *J. Steroid Biochem.* **6**, 117 (1975).
188. N. Ikekawa, M. Morisaki, N. Koizumi, Y. Kato, and T. Takeshita, *Chem. Pharm. Bull.* **23**, 695 (1975).
189. G. Milhaud, M.-L. Labat, J. Redel, *C. R. Acad. Sci.* **279**, 827 (1974).
190. M. Seki, N. Koizumi, M. Morisaki, and N. Ikekawa, *Tetrahedron Lett.*, 15 (1975).
191. J. J. Partridge, V. Toome, and M. R. Uskokovic, *J. Am. Chem. Soc.* **98**, 3739 (1976).
192. N. Ikekawa, M. Morisaki, N. Koizumi, M. Sawamura, Y. Tanaka, and H. F. DeLuca, *Biochem. Biophys. Res. Commun.* **62**, 485 (1975).
193. J. J. Partridge, S. J. Shivey, E. G. Baggiolini, B. Hennessy, and M. Uskokovic in A. W. Norman and co-eds., *Vitamin D: Biochemical, Chemical and Clinical Aspects Related to Calcium Metabolism*, Walter de Gruyter, Berlin, 1977, p. 47.
194. H.-Y. Lam, H. K. Schnoes, and H. F. DeLuca, *Steroids* **25**, 247 (1975).
195. J. Redel, P. Bell, F. Delbarre, and E. Kodicek, *C. R. Acad. Sci.* **276**, 2907 (1973).
196. J. Redel, P. A. Bell, N. Bazely, Y. Calando, F. Delbarre, and E. Kodicek, *Steroids* **24**, 463 (1974); J. Redel, N. Bazely, Y. Tanaka, and H. F. DeLuca, *FEBS Lett.* **94**, 228 (1978).
197. A. W. Norman and co-workers, *Biochemistry* **22**, 1798 (1983).
198. H. F. DeLuca and H. K. Schnoes in A. W. Norman and co-eds., *Vitamin D: Recent Basic Advances and Their Clinical Application*, Walter de Gruyter, Elmsford, N.Y., 1979.
199. D. E. M. Lawson, *Vitamin D*, Academic Press, Inc., New York, 1978.
200. U.S. Pat. 3,334,118 (Aug. 1, 1967), K. Schaff, S. Schmuhler, and H. C. Klein (to Nopco Chemical Co.).
201. U.S. Pat. 3,665,020 (May 23, 1972), R. Marbet (to Hoffmann-La Roche, Inc.).
202. R. S. Harris, J. W. M. Bumker, and L. M. Moser, *J. Am. Chem. Soc.* **60**, 2579 (1938).
203. 21 CFR § 592.5993.
204. *Chemical Economics Handbook*, Stanford Research Institute, Menlo Park, Calif., 1980.
205. *Chron. WHO* **3**, 747 (1965).
206. *The United States Pharmacopeia XX (USP XX–NF XV)*, The United States Pharmacopeial Convention, Inc., Rockville, Md., 1980; *The National Formulary XV*, The American Pharmaceutical Association, Washington, D.C., 1980.
207. W. Horwitz, ed., *Official Methods of Analysis of the Association of Official Analytical Chemists*, 13th ed., Washington, D.C., 1980.
208. F. Mulder, *J. Assoc. Off. Anal. Chem.* **60**, 989 (1977).
209. H. Hofsass, N. J. Alicino, A. L. Hirsch, L. Amieka, and L. D. Smith, *J. Assoc. Off. Anal. Chem.* **61**, 774 (1978).
210. F. Mulder and co-workers, *J. Assoc. Off. Anal. Chem.* **64**, 61 (1981).
211. *J. Assoc. Off. Anal. Chem.* **64**, 524 (1981).
212. Ref. 45, p. 64.
213. R. B. Davis, J. M. McMahon, and G. Kalnitsky, *J. Am. Chem. Soc.* **74**, 4483 (1952).
214. D. Kritchevsky and M. R. Kirk, *J. Am. Chem. Soc.* **74**, 4484 (1952).
215. E. Kodicek and D. R. Ashby, *Biochem. J.* **57**, XII (1954).
216. A. R. Terepka, P. S. Chen, and B. Jorgensen, *Endocrinology* **68**, 996 (1961).
217. C. Michalec and J. Strasek, *J. Chromatogr.* **4**, 254 (1960).
218. P. S. Chen, A. R. Terepka, and N. Remsen, *Anal. Chem.* **35**, 2030 (1963).
219. G. N. Festenstein and R. A. Morton, *Biochem. J.* **60**, 22 (1955).
220. J. Green, *Biochem. J.* **49**, 45 (1951).
221. J. B. DeWitt and M. X. Sullivan, *Anal. Chem.* **18**, 117 (1946).
222. D. T. Ewing, T. D. Schlabach, M. J. Powell, J. W. Vaitkus, and O. D. Bird, *Anal. Chem.* **26**, 1406 (1954).
223. A. Fujita and K. Numata, *J. Vitaminol.* **4**, 299 (1958).
224. S. W. Jones, W. W. Morris, and J. B. Wilkie, *J. Assoc. Agric. Chem.* **42**, 180 (1959).

225. J. G. Theiragt and D. J. Campbell, *Anal. Chem.* **31,** 1375 (1959).
226. A. W. Norman and H. F. DeLuca, *Anal. Chem.* **35,** 1247 (1963).
227. G. A. Fischer and J. J. Kabara, *Anal. Biochem.* **9,** 303 (1964).
228. E. B. Mawer and J. Backhouse, *Biochem. J.* **112,** 255 (1969).
229. H. H. A. Dollwet and A. W. Norman, *Anal. Biochem.* **25,** 297 (1968).
230. M. F. Holick and H. F. DeLuca, *J. Lipid Res.* **12,** 460 (1971).
231. D. F. Tomkins and R. J. Tscherne, *Anal. Chem.* **46,** 1602 (1974).
232. N. Ikekawa and N. Koizumi, *J. Chromatogr.* **119,** 227 (1976).
233. G. Jones and H. F. DeLuca, *J. Lipid Res.* **16,** 448 (1975).
234. K. T. Koshy and A. L. Vanderslik, *Anal. Biochem.* **74,** 282 (1976).
235. I. Bjorkhem and I. Holmberg, *Clin. Chem. Acta* **68,** 215 (1976).
236. E. W. Matthews, P. G. H. Byfield, K. W. Colston, K. M. A. Evans, L. S. Galante, and I. MacIntyre, *FEBS Lett.* **48,** 122 (1974).
237. H. Hofsass, A. Grant, N. J. Alicino, and S. B. Greenbaum, *J. Assoc. Agric. Chem.* **59,** 251 (1976).
238. D. R. Fraser and E. Kodicek, *Biochem. J.* **96,** 59P (1965).
239. A. L. Fisher, A. M. Parfitt, and H. M. Lloyd, *J. Chromatogr.* **65,** 493 (1972).
240. C. K. Parekh and R. H. Wasserman, *J. Chromatogr.* **17,** 261 (1965).
241. H. Ziffer, W. J. VendenHeuvel, E. O. Haahti, and E. C. Horning, *J. Am. Chem. Soc.* **82,** 6411 (1960).
242. J. R. Evans, *Clin. Chim. Acta* **42,** 167 (1972).
243. T. K. Murray, K. C. Day, and E. Kodicek, *Biochem. J.* **98,** 29P (1966).
244. L. V. Aviolo and S. W. Lee, *Anal. Biochem.* **16,** 193 (1966).
245. D. Sklan, P. Budowski, and M. Katz, *Anal. Biochem.* **56,** 606 (1973).
246. D. O. Edlund and F. A. Filippini, *J. Assoc. Agric. Chem.* **56,** 1374 (1973).
247. J. G. Bell and A. A. Christie, *Analyst* **99,** 385 (1974).
248. J. Carol, *J. Pharm. Sci.* **50,** 451 (1961).
249. W. W. Morris, Jr., J. B. Wilkie, S. W. Jones, and L. Friedman, *Anal. Chem.* **34,** 381 (1962).
250. R. Strohecker and H. M. Henning, *Vitamin Assay-Tested Methods*, CRC Press, Cleveland, Ohio, 1966, p. 281.
251. Trink-Toan, H. F. DeLuca, and L. F. Dahl, *J. Org. Chem.* **41,** 3477 (1976).
252. T. G. Hogness, A. E. Sidwell, Jr., and F. P. Zscheile, Jr., *J. Biol. Chem.* **120,** 239 (1937).
253. W. Huber, G. W. Ewing, and J. Kriger, *J. Am. Chem. Soc.* **67,** 609 (1945).
254. K. Diem and C. Lentner, eds., *Scientific Tables*, Ciba Geigy, Ltd., Basle, Switz., 1971, p. 464.
255. *Z. Gesamte Exp. Med.* **116,** 138 (1950).
256. R. L. Morrissey, R. M. Cohn, R. N. Empson, Jr., H. L. Greene, O. D. Taunton, and Z. Z. Ziporin, *J. Nutr.* **107,** 1027 (1977).
257. U.S. Pat. 2,417,299 (Mar. 11, 1947), L. Freedman and E. Green (to U.S. Vitamin Corp.).
258. U.S. Pat. 2,702,262 (Feb. 15, 1955),. Burley and A. E. Timrech (to Charles Pfizer and Co., Inc.).
259. U.S. Pats. 2,777,797 and 2,777,798 (Jan. 15, 1977), M. Hochberg and M. L. MacMillan (to Nopco Chemical Co.).
260. U.S. Pat. 2,827,452 (Mar. 18, 1978), H. Schlenk, D. M. Sand, and J. A. Tillotson (to University of Minnesota).
261. U.S. Pat. 3,067,104 (Dec. 4, 1962), M. Hochberg and C. Ely (to Nopco Chemical Co.).
262. U.S. Pat. 3,143,475 (Aug. 4, 1964), A. Koff and R. F. Widmer (to Hoffmann-LaRoche, Inc.).
263. *Recommended Dietary Allowances*, 7th ed., National Academy of Science, National Research Council, Washington, D.C., 1968, p. 1964.
264. *Nutritional Requirements of Poultry*, 5th ed., National Academy of Science, National Research Council, Washington, D.C., 1966, p. 1345.
265. *Nutritional Requirements of Swine*, 6th ed., National Academy of Science, National Research Council, Washington, D.C., 1968, p. 1599.
266. *Nutritional Requirements of Dairy Cattle*, 3rd ed., National Academy of Science, National Research Council, Washington, D.C., 1966, p. 1349.

General References

Refs. 2, 45, 85, 98, 129, 172, 193, and 199 are general references.

ARNOLD L. HIRSCH
A. L. Laboratories, Inc.

VITAMIN E

The term vitamin E [59-02-9] was used originally to denote a partially characterized material in vegetable oils that was essential to maintain rat fertility. Discovery of this fat-soluble vitamin in 1922 and early pioneer investigations of its natural distribution, isolation, and identification were owing primarily to Evans, Burr, and Emerson (University of California) and Mattill and Olcott (University of Iowa). Other investigators subsequently found more than one naturally occurring substance and several synthetic organic compounds that acted like or had a sparing effect on vitamin E in the body. Furthermore, the symptoms of vitamin E deprivation were discovered to vary according to the animal species. Consequently, vitamin E activity for many years designated a general type of physiological activity that could cure or prevent any symptom of vitamin E deficiency in animals.

Four naturally occurring compounds possessing vitamin E activity have been isolated and identified and designated tocopherols (1)–(4) (Fig. 1). Tocopherol is derived from the Greek meaning "to bring forth offspring." All are methyl derivatives of tocol, 2-methyl-2-(4',8',12'-trimethyltridecyl)-6-chromanol. About 25 years later, four additional compounds analogous to the tocopherols were characterized (1). These compounds are methyl derivatives of tocotrienol, 2-methyl-2-(4',8',12'-trimethyltrideca-3',7',11'-trienyl)-6-chromanol and contain three unsaturated bonds in the side chain. α-Tocotrienol (5) is shown in Figure 1.

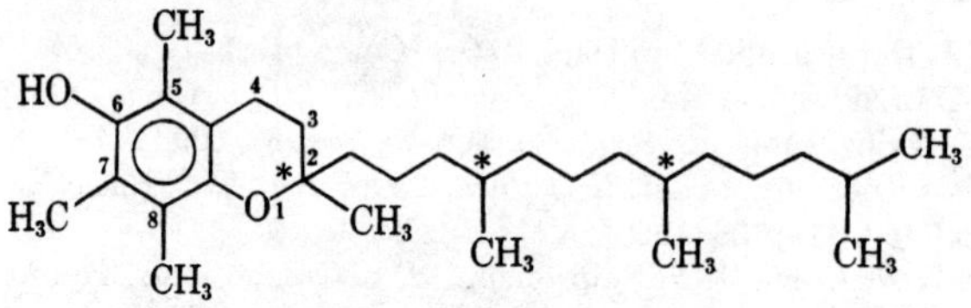

(1) α-Tocopherol (5,7,8-trimethyltocol)
(RRR) [59-02-9], all-rac [2074-53-5]

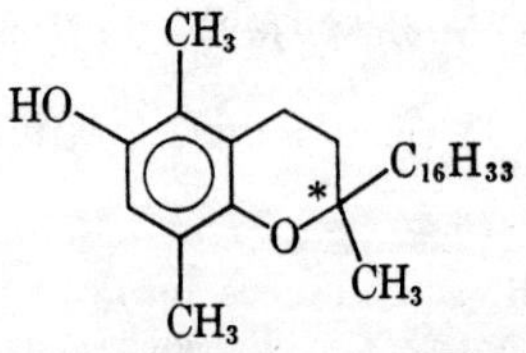

(2) β-Tocopherol [148-03-8]
(5,8-dimethyltocol)

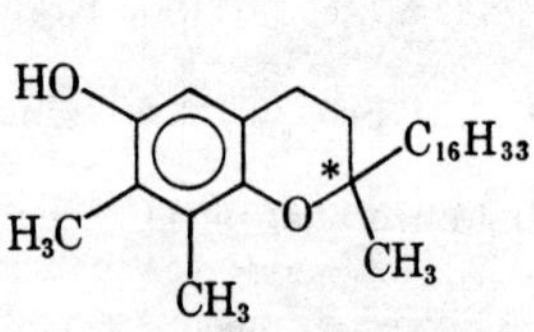

(3) γ-Tocopherol [54-28-4]
(7,8-dimethyltocol)

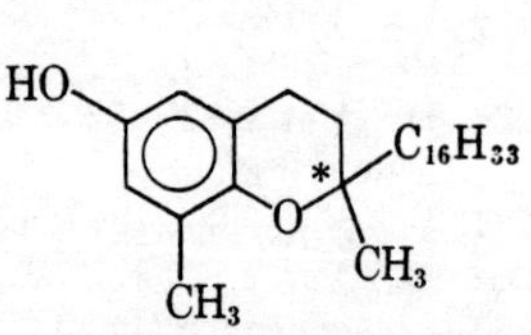

(4) δ-Tocopherol [119-13-1]
(8-methyltocol)

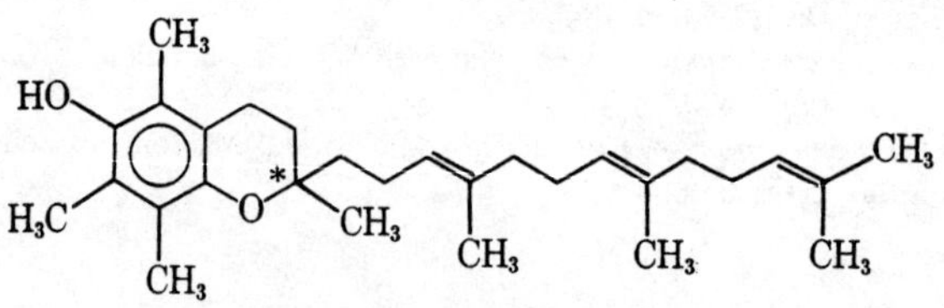

(5) α-Tocotrienol [1721-51-3, 6829-55-6]
(5,7,8-trimethyltocotrienol)

Figure 1. The four naturally occurring tocopherols and α-tocotrienol. Asterisks denote asymmetric centers.

Although the tocopherols and tocotrienols are closely related chemically, they have widely varying degrees of biological effectiveness. α-Tocopherol [59-02-9] (1) ($C_{29}H_{50}O_2$, formula weight 430.72), with a completely methylated aromatic ring and a saturated side chain, has the highest activity. The natural stereoisomer, $2R,4'R,8'$-R-α-tocopherol [59-02-9], can be called (RRR)-α-tocopherol. Similarly, $2S,4'R,8'$-R-α-tocopherol can be called 2-*epi*-α-tocopherol, a mixture of (RRR)- and (SRR)-α-tocopherols can be called 2-*ambo*-α-tocopherol, and α-tocopherol prepared from synthetic phytol or isophytol should be called *all-rac*-α-tocopherol [2074-53-5] (2). The *all-rac*-α-tocopherol and its esters are less potent on a weight basis than the corresponding (RRR) forms. Because of the high potency of α-tocopherol and because it is the predominant form in animal tissues, the term α-tocopherol is now used widely instead of vitamin E. For precision, the specific isomer form of α-tocopherol should always be indicated. The term vitamin E should be reserved as the generic descriptor for all tocol and tocotrienol derivatives having qualitatively the biological activity of α-tocopherol (2).

Both α-tocopherol and α-tocopheryl acetate [58-95-7] are clear, odorless, viscous, slightly yellow oils. α-Tocopheryl acetate is the principal commercial form of vitamin E in food fortification, dietary supplements, and medicinals, and for domestic animals as a source of vitamin E activity. The tocopherols are used as dietary supplements and in food technology as antioxidants (qv) to retard the development of rancidity in fatty materials (see also Food additives).

Tocopherols are widely distributed in foods in an unesterified form and occur in the highest concentration in cereal-grain oils. For example, crude corn and wheat oils may contain 200 mg tocopherol/100 g. On the other hand, certain vegetable oils, such as coconut and olive, are practically devoid of tocopherol. The proportion of the most active form, α-tocopherol, also varies widely. Whereas 90% of the tocopherol in safflower oil is α-tocopherol, only 20% of corn and soybean oil tocopherols is α-tocopherol. The γ form predominates in corn oil, whereas the γ and δ forms predominate in soybean oil. Wheat-oil tocopherols contain ca 65% of the β form.

The per capita daily intake of α-tocopherol in the United States has been estimated from data on total purchases of food to be ca 15 mg. Most is derived from margarine, salad oils, and shortening, with other foods supplying relatively little of the α-tocopherol ingested. Assays of food as eaten, however, indicate that average diets may supply daily α-tocopherol intakes of only 7–9 mg in the United States and Canada (3–5) and even lesser amounts in other countries. Farm animals receive most of their α-tocopherol from fresh grains, alfalfa-leaf meal, and fresh grasses, if available. The amount of α-tocopherol in eggs, milk, or meat depends on the amount of α-tocopherol received by the animal. Thus, the content of this vitamin in dairy products and meats is seasonal and highly variable.

Appreciable losses of vitamin E may occur during the processing of foods, including cooking and baking. The amount of tocopherol left in refined salad oil depends on the severity of the refining process. Losses during the processing of cereal grains depend partially on the amount of germ that is retained (6). The α-tocopherol content of bleached wheat flour is very low. Proper cooking of vegetables does not cause much loss of α-tocopherol, but the portions of the plant with the highest content of α-tocopherol are frequently discarded as inedible (7). Frozen and fresh vegetables have similar vitamin E contents, which, however, are substantially reduced by canning. Even freezing does not always prevent loss of tocopherols (3). Loss of α-tocopherol during dehydration, irradiation, or canning of meats may range up to 50% (8) (see also Food processing).

Another factor influenced both by distribution and by processing loss is the cumulative biological value of the tocopherols other than α-tocopherol. The Food and Nutrition Board of the NRC has concluded that the tocopherols other than α-tocopherol in a mixed diet contribute vitamin E activity equivalent to ca 20% of the indicated α-tocopherol content (4,9). To amalgamate the various sources of activity, the Board also recommends that the total activity be expressed as milligrams of α-tocopherol equivalents (2), where 1 mg (RRR)-α-tocopherol = 1 α-tocopherol equivalent.

Properties

The melting range of (RRR)-α-tocopherol is 2.5–3.5°C. A characteristic physical property is fat solubility. Both α-tocopherol and α-tocopheryl acetate are freely soluble in glyceride oils, ethanol, chloroform, and acetone, but are insoluble in water.

The optical activity of natural (RRR)-α-tocopherol is very slight. The specific rotation of (RRR)-α-tocopherol in 2,2,4-trimethylpentane is only +0.16°. A much greater specific rotation, $[\alpha]_D^{25} = +31.5°$, is shown by a compound obtained from (RRR)-α-tocopherol by oxidation with alkaline potassium ferricyanide (10). Because there is no specific rotation for the corresponding oxidation product from all-rac-α-tocopherol, a method is provided for distinguishing between the (RRR) and all-rac forms.

The uv absorption spectrum for α-tocopherol in ethanol shows λ_{max} at 292 nm ($A_{1\,cm}^{1\%}$ = 75.8) and λ_{min} at 256 nm. Acetylation of the phenolic hydroxy group shifts λ_{max} to 284 nm ($A_{1\,cm}^{1\%}$ = 43.6) and λ_{min} to 255 nm. In the infrared region, a band at ca 8.6 µm is characteristic of the chromane structure of tocopherols. The magnetic susceptibility of α-tocopherol has been determined as $\chi = -9.714 \times 10^{-6}$ m^3/mol (-0.773×10^{-6} cgs), and the presence or absence of characteristic bands for methyl groups or aromatic hydrocarbons enables differentiation among mono-, di-, and trimethyl-substituted tocols by nmr. Mass spectral data show a cleavage of the heterocyclic ring followed by loss of the isoprenoid side chain (11). Physical properties of commercial vitamin E products are given in Table 1.

Oxidation of α-tocopherol is catalyzed by light and accelerated by unsaturated fatty acids, metal salts, or alkalies. Oxidation proceeds according to reactions shown in Figure 2. Further oxidation (chromic oxide) yields dimethylmaleic anhydride, 2,3-butanedione, acetone, a C_{16} acid, a C_{18} ketone, and a C_{21} lactone. From these products, Fernholz (12) in 1938 correctly postulated the structure of α-tocopherol. Examples of other reaction products of the tocopherols are given in refs. 13–14.

The metabolism of α-tocopherol is limited. Evidence for the appearance of the products in Figure 2 is sound only for α-tocopherylquinone (8); two other products formed *in vivo* are α-tocopheronic acid (12) and its lactone (13) (15).

(**12**) α-tocopheronic acid [*1948-76-1*]   (**13**) α-tocopheronolactone [*3121-68-4*]

Table 1. Some Physical Properties of Commercial Vitamin E Products

Product	CAS Reg. No.	Mp, °C	Sp gr$_{25°C}^{25°C}$	n_D^{20}	$A_{1\,cm}^{1\%}$ in ethanol
(RRR)-α-tocopheryl acetate	[58-95-7]	25	0.950–0.964	1.4940–1.4985	40–44 at 284 nm
(RRR)-α-tocopheryl acetate concentrate[a]		liq	0.920–0.950	[b]	[b]
(RRR)-α-tocopheryl hydrogen succinate; white, crystalline	[4345-03-3]	73–78			35–40 at 284 nm
mixed tocopherols concentrate[c]		liq	0.920–0.950	[b]	[b]
all-rac-α-tocopherol	[2074-53-5]	liq	0.947–0.958	1.5030–1.5070	71–76 at 292 nm
all-rac-α-tocopheryl acetate	[7695-91-2]	liq	0.950–0.964	1.4940–1.4985	40–44 at 284 nm

[a] Contains ≥40% (RRR)-tocopheryl acetate.
[b] Varies with potency of concentrate.
[c] Contains ≥50% total tocopherol, at least half of which is (RRR)-α-tocopherol.

In addition, a dimer and a trimer have been recovered from liver following administration of α-tocopherol. Structure (11) (Fig. 2) was proposed for the dimer obtained by oxidation of α-tocopherol with alkaline potassium ferricyanide (10).

The ready reversibility of α-tocopherol with certain of its oxidation products suggests that vitamin E could act in the body as an integral part of some oxidation–reduction system or enzyme. The existence of a free semiquinone radical of α-tocopherol has been demonstrated by uv irradiation at low temperatures (16). A free-radical formation has been postulated (17).

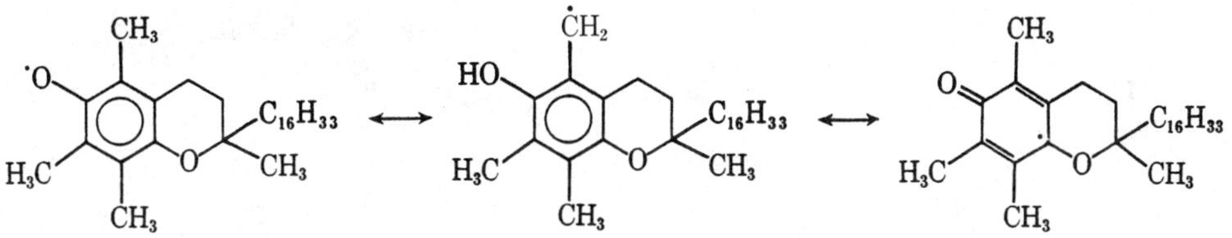

Other studies with singlet oxygen, eg, ref. 18, suggest that formation of epoxides, eg, α-tocoquinone-2,3-oxide (10) (Fig. 2), may be a main pathway of tocopherol oxidation. Intermediates such as these might well be the critical factors in one or more oxidation–reduction reactions *in vivo*.

α-Tocopherylhydroquinone (α-tocohydroquinone) and α-tocopherethoxide show activity in preventing and curing muscular dystrophy in hamsters, whereas α-tocopherylquinone and α-tocopherylhydroquinone are effective in rabbits. Rats can utilize all three compounds to a slight degree for preserving reproductive function, but other oxidized products, including the lactone and dimer, are ineffective.

Esterification of α-tocopherol with aliphatic acids ordinarily does not markedly affect biological activity, but etherification almost completely eliminates biopotency. The tocopheramines and N-methyltocopheramines show varying degrees of activity; some are as active as all-rac-α-tocopherol (19). α-Tocopherothiol shows activity in the erythrocyte hemolysis test. Other derivatives of α-tocopherol in which ring methyl

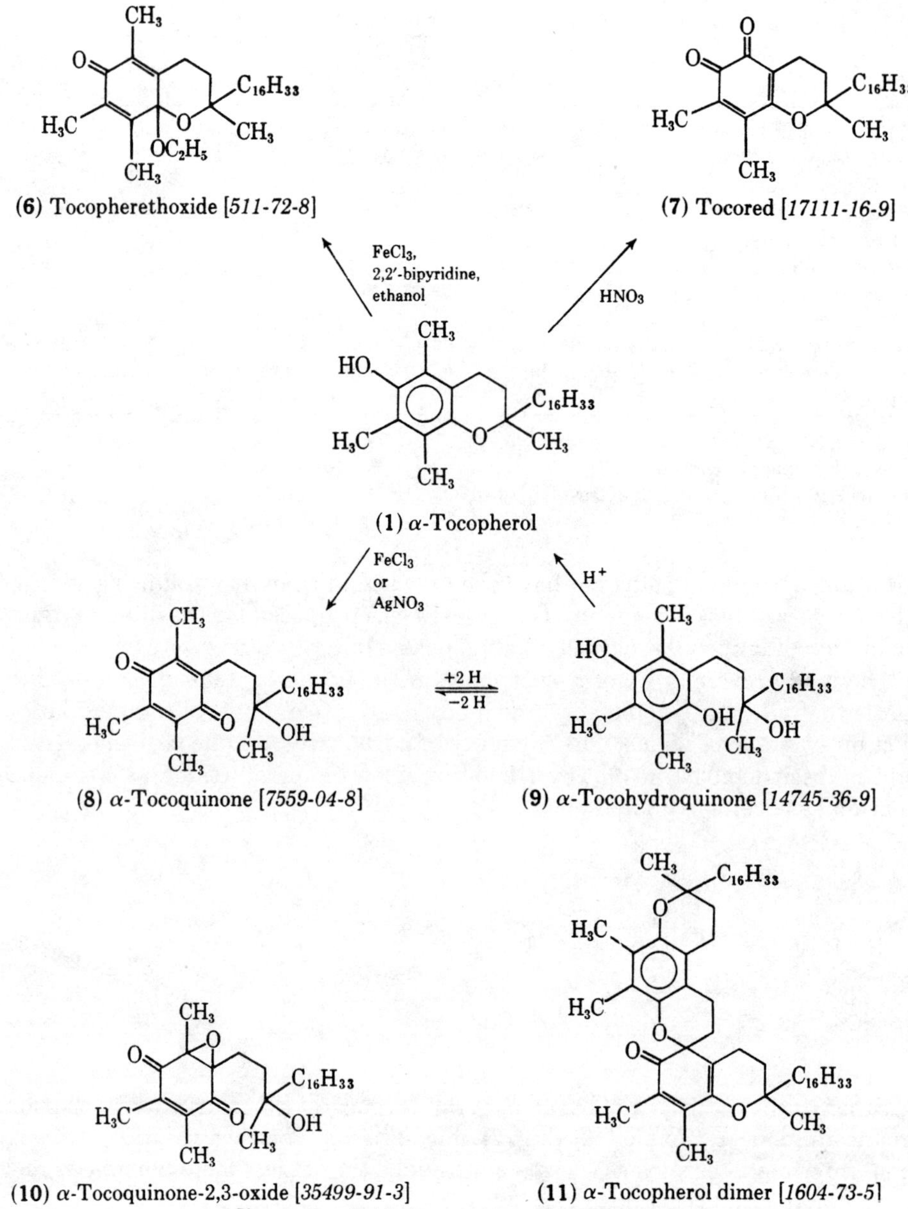
(6) Tocopherethoxide [511-72-8]
(7) Tocored [17111-16-9]
FeCl3, 2,2′-bipyridine, ethanol
HNO3
(1) α-Tocopherol
FeCl3 or AgNO3
H+
+2 H
−2 H
(8) α-Tocoquinone [7559-04-8]
(9) α-Tocohydroquinone [14745-36-9]
(10) α-Tocoquinone-2,3-oxide [35499-91-3]
(11) α-Tocopherol dimer [1604-73-5]

Figure 2. Oxidation products of α-tocopherol.

groups are substituted or in which the phytyl side chain is modified are less potent than α-tocopherol itself.

Manufacture and Processing

The naturally occurring tocopherols are obtained from vegetable oil sources, particularly the distillate from the deodorization step during the refining process. An important step, technically, is concentration of the tocopherols by molecular distillation (20). As one example, alkali-refined soybean oil containing 0.19% of a mixture of α-, γ-, and δ-tocopherols is distilled in a centrifugal-type molecular still; the tocopherol fraction, which distills below 240°C under 0.53 Pa (0.004 mm Hg) pressure, is collected. These conditions minimize loss of heat-sensitive materials. After as much as possible of the sterols and other substances in the fraction have been removed by crystallization from acetone at −10°C and the glycerides have been removed by saponification, the tocopherols in the unsaponifiable matter are further concentrated by a second molecular distillation. Thus, a fraction containing at least 60% mixed tocopherols is obtained (21). The residual sterols and glycerides are sold.

Concentrates of mixed tocopherols can be obtained from vegetable oil sources by one or more of the following treatments (22): esterification, saponification, fractional extraction, ion exchange, and precipitation of sterols with alkaline earth halides. The tocopherols other than α-tocopherol can be converted to the more biologically active (*RRR*)-α-tocopherol by methylation of the aromatic ring (23).

The four natural tocopherols have been synthesized in their all-rac forms. Only one, *all-rac-*α-tocopheryl acetate, is commercially significant. Although the commercial methods have not been described in detail, the processes are probably based on the procedures reported almost simultaneously in 1938 by several research teams (24–26). These workers condensed 2,3,5-trimethylhydroquinone (**14**) directly with phytol (25), phytyl halide (24,26), or phytadiene (26). Later, synthetic isophytol (**15**) was used, which today appears to be the preferred intermediate. Either phytol or isophytol is synthesized readily from acetone by one of several possible routes. The condensation with trimethylhydroquinone is carried out in acetic acid or in an inert solvent, such as benzene, with an acidic catalyst, eg, zinc chloride, formic acid, or boron trifluoride ethyl etherate (23).

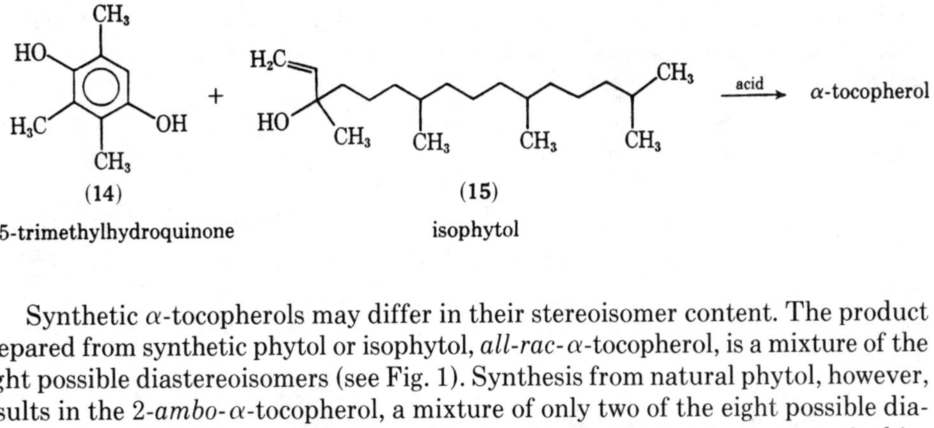

(**14**)   (**15**)

2,3,5-trimethylhydroquinone   isophytol

Synthetic α-tocopherols may differ in their stereoisomer content. The product prepared from synthetic phytol or isophytol, *all-rac-*α-tocopherol, is a mixture of the eight possible diastereoisomers (see Fig. 1). Synthesis from natural phytol, however, results in the 2-*ambo-*α-tocopherol, a mixture of only two of the eight possible diastereoisomers (27). The configuration of the 2 carbon is of prime importance for biological activity; the (2*S*) epimer is only 21% as active as the (2*R*) epimer (28). The (2*R*) and (2*S*) epimers can be separated by fractional crystallization (22). However, the (*RRR*)-α-tocopherol can now be prepared by direct synthesis (29–30), and the total synthesis of all eight diastereoisomers of α-tocopheryl acetate is described in ref. 31.

The all-rac forms of β-, γ-, and δ-tocopherols can be synthesized by the same procedure as *all-rac-*α-tocopherol. Instead of trimethylhydroquinone, the appropriate dimethylhydroquinone or monomethylhydroquinone is used. Thus, with synthetic phytol or isophytol, 2,5-dimethylhydroquinone yields *all-rac-β*-tocopherol, 2,3-dimethylhydroquinone yields *all-rac-γ*-tocopherol, and methylhydroquinone yields *all-rac-δ*-tocopherol and other monomethyl tocols. The yields of these compounds are usually lower than that obtained in the synthesis of *all-rac-α*-tocopherol because the hydroquinones can condense with more than one molecule of phytol. Monoesters, such as the monobenzoates, of the hydroquinones give better yields.

A number of other procedures have been used for the synthesis of tocol homologues and analogues (32). Esters of the tocopherols are prepared readily. Typical laboratory procedures for the acetate and the succinate are given in references 33 and 34, respectively.

Vitamin E products are shipped in several sizes of plastic-lined metal containers ranging from 1-kg cans to 190-kg drums. Dry products intended for feed uses are usually shipped in multiwalled bags or in fiber drums containing up to 50 kg.

Economic Aspects

The 1980 world production of both natural and synthetic preparations of vitamin E is estimated at 5300 metric tons; growth is expected to keep pace with increasing recognition of need. The 1980 U.S. production was 3297 t (35), augmented by maximum imports of 331 t mainly from France and Japan (36). The principal U.S. producers of natural vitamin E are Eastman Kodak and Henkel and of synthetic vitamin E are Hoffmann-LaRoche and BASF Wyandotte. International producers include BASF, Eisai, AEC (Rhone-Poulenc SA), Nisshin-Badische, and Molecular Distillations.

The price of vitamin E has decreased steadily, whereas usage has increased. Prices for pharmaceutical-grade α-tocopheryl acetate are given in Table 2 (37). The prices for the all-rac form have been lower than those for the (*RRR*) form since 1954, reflecting the higher potency of the (*RRR*) form and influenced by the availability of suitable raw material for production of the (*RRR*) form.

Specifications

Specifications and standards for various vitamin E preparations intended for pharmaceutical use are described in *The United States Pharmacopeia XX* (38). All products should contain not less than 96.0% and not more than 102.0% of the appropriate form. They should be labeled to indicate both the chemical and stereoisomeric forms.

Label claims for tocopherol preparations are based on the equivalents listed below. Because the original International Unit of vitamin E has been discontinued, the Pharmacopeia has adopted the USP unit in its place. The two units are numerically equal, however, and the International Unit continues to be used in labeling some vitamin E products.

Table 2. Bulk Prices of Pharmaceutical-Grade α-Tocopheryl Acetate, $/kg

	1948	1951	1954	1960	1970	1982
(*RRR*)-α-tocopheryl acetate	250	250	185	122	68	68
all-rac-α-tocopheryl acetate	750	350	136	90	50	27

1 mg all-rac-α-tocopheryl acetate	= 1.0 IU	= 1.0 USP Unit
1 mg all-rac-α-tocopheryl hydrogen succinate	= 0.89 IU	= 0.89 USP Unit
1 mg all-rac-α-tocopherol	= 1.1 IU	= 1.1 USP Unit
1 mg (RRR)-α-tocopheryl acetate	= 1.36 IU	= 1.36 USP Unit
1 mg (RRR)-α-tocopheryl hydrogen succinate	= 1.21 IU	= 1.21 USP Unit
1 mg (RRR)-α-tocopherol	= 1.49 IU	= 1.49 USP Unit

Some revision of these equivalents to reflect a greater relative activity for the (RRR) forms has been suggested, as well as adoption of (RRR)-α-tocopheryl acetate as a new international standard (39).

International Units are still used by the Food Chemicals Codex (40). Specifications described by this organization are particularly important for food fortification and dietary supplements. Vitamin E preparations include both (RRR)- and all-rac-α-tocopherols and α-tocopheryl acetates, mixed tocopherol concentrates, and (RRR)-α-tocopheryl hydrogen succinate. All preparations must show low acidity, not more than 0.004% heavy metals (as Pb), and not more than 10 ppm Pb. Except for the various concentrates, all preparations must contain not less than 96.0% and not more than 102.0% of the appropriate product. In addition, preparations comprising the natural α-form must show a specific rotation $[\alpha]_D^{25}$ of not less than +24° for the oxidation product with alkaline potassium ferricyanide.

Analytical and Test Methods

Chemical and Physical Methods. The methods of analysis for pure or highly concentrated free α-tocopherol preparations are rather simple. Reaction with the Emmerie-Engel reagent (2,2'-bipyridine (α,α'-dipyridyl) and ferric chloride) gives a red color from combination of the bipyridine with the ferrous ions resulting from reduction of the ferric ions; this color is directly proportional to the amount of tocopherol present which thus can be determined quantitatively. 4,7-Diphenyl-1,10-phenanthroline (bathophenanthroline) instead of 2,2'-bipyridine provides greater sensitivity. Alternatively, pure α-tocopherol in 0.5 N alcoholic sulfuric acid is titrated with 0.01 N ceric sulfate to a blue end point with diphenylamine as indicator. Spectrophotometry is sometimes used to measure the specific absorbancy, which is proportional to the α-tocopherol concentration.

Analysis of low potency concentrates, original vegetable oils, or foods and feeds is considerably more difficult. The Emmerie-Engel color reaction is nonspecific and occurs to varying degrees with all the tocopherols and tocotrienols and other reducing materials. Hence, the method of assay must provide for isolation of the α-tocopherol from interfering compounds. Determination of the other tocopherols and tocotrienols is often a secondary goal.

A variety of approaches to vitamin E assay enable the analyst to choose the technique to fit a particular problem (41–42). Tocopherols and tocotrienols can be separated by column chromatography on one of several adsorbents. Paper chromatography (43) or tlc (42) are suitable for small amounts. Gas liquid chromatography (glc), currently used for several official methods (38,40), has been most valuable for separating all the forms of vitamin E activity (44) and has even been reported to separate the four pairs of diastereoisomers in all-rac-α-tocopherol (45). The rapid development of high pressure liquid chromatography (hplc) as an analytical tool has also brought its application to the analysis of tocopherols (46). The compounds isolated by these methods can be measured by colorimetry, ultraviolet spectrometry, or spectrofluorometry.

The α-tocopherol esters commonly used in vitamin preparations and enriched foods and feeds must be hydrolyzed before analysis. This step is critical because tocopherols are sensitive to oxygen under alkaline conditions. However, if oxygen is excluded from the system and an effective antioxidant is added (43), the esters may be saponified without loss of tocopherol. Hydrolysis of the esters before any other treatment permits analysis for total tocopherol content. Hydrolysis subsequent to oxidative destruction of the unesterified tocopherols permits analysis only for the added α-tocopherol ester. These two approaches have been combined into an official method (47). In addition, the difference in optical rotations of the alkaline potassium ferricyanide oxidation products of (*RRR*)- and *all-rac-α*-tocopherols provides a distinguishing assay for pharmaceuticals, food supplements, or feed supplements (48).

Bioassay Methods. The classical procedure for determining α-tocopherol activity is a modification of the Evans resorption–gestation method (49). Female rats are raised on a diet free of vitamin E and mated with normal males. Conception and the first half of pregnancy are normal. However, unless a vitamin E supplement is administered within the first 10–12 d of pregnancy, the embryos die and are resorbed. The mother rat is apparently unharmed and may be used repeatedly. If the dose of α-tocopherol administered as a test supplement is above a certain critical amount, pregnancy proceeds and terminates normally. The critical dose of α-tocopherol varies somewhat between assays (0.3–1.0 mg), but the response is an all-or-none (alive-or-dead) type, a criterion that is easily determined. Consequently, the bioassay properly controlled and carefully conducted is specific and reasonably accurate, but is used infrequently because it is time-consuming.

Another biological procedure is the hemolytic test in rats (50). This test is based on the fact that adult rats receiving adequate vitamin E in their diet have red blood cells that are resistant to hemolysis induced by chemical action of such compounds as dialuric acid and hydrogen peroxide. Changing the animals' diet to one deficient in vitamin E causes the red blood cells to become susceptible to hemolysis. A satisfactory bioassay constitutes the administration of various amounts of α-tocopherol (both standard and test substance), either prophylactically or curatively, to the rats during the depletion period and measurement of the change in degree of resulting red-cell hemolysis. However, this test is not specific for α-tocopherol and may respond to compounds that do not prevent other physiological manifestations of vitamin E deficiency.

Among other biological methods based on muscular dystrophy, tissue storage, growth, chick encephalomalacia, or testicular degeneration, a recent description of prevention of muscle myopathy shows promise (51). In general, preparation of animals for a vitamin E bioassay requires a lengthy depletion period.

Biochemical Function, Clinical Evaluation, and Toxicology

The exact mechanism whereby α-tocopherol functions in the body as vitamin E is not known. It probably acts as an antioxidant controlling redox reactions in a variety of tissues and organs, particularly in protecting the cellular membrane from free radicals generated during peroxidation of unsaturated lipids. That this is the exclusive role has been challenged (52), however, and the high isomeric specificity of the α-tocopherol structure in physiological reactions suggests other roles as an integral part of the biological machinery.

Experiments with animals deficient in vitamin E have revealed perhaps the largest variety of disorders associated with the nutritional lack of any single vitamin. Vitamin E deficiency affects the reproductive system (testicular degeneration, defective development of the embryo); the muscular system (skeletal muscular dystrophy, cardiac necrosis and fibrosis, ceroid pigment in smooth muscle); the circulatory system (exudative diathesis, erythrocyte hemolysis, anemias); the skeletal system (incisor depigmentation); and the nervous system (encephalomalacia, deposition of lipofuscin pigment). In addition, vitamin E deficiency may be manifested by a number of other conditions, such as discolored adipose tissue, liver necrosis, lung hemorrhage, kidney nephrosis and postmortem autolysis, and creatinuria. The gross and microscopic pathology of various species resulting from vitamin E deficiency is described in refs. 53–54.

Vitamin E ameliorates the adverse effects of various toxic agents (eg, lead and chlorinated solvents), drugs (eg, adriamycin), and a number of suboptimal dietary (eg, low protein) and environmental (eg, high oxygen, ozone, or nitrogen dioxide) conditions. Supplementation of well-balanced diets with pharmacological amounts of vitamin E enhances humoral immunity and increases phagocytosis in animals. Vitamin E also interacts metabolically with selenium. This relationship has been clarified recently by the discovery that selenium is a constituent of glutathione peroxidase, an enzyme that destroys peroxides.

Signs of vitamin E deficiency have been positively identified in humans, in children first (55) but confirmed subsequently in adults with experimental nutritional deprivation (56). The populations most likely to be naturally deficient are premature infants, patients with cystic fibrosis (57) or β-thalassemia (58), and patients with fat-malabsorption syndromes. Although the incidence of deficiency in man appears relatively low, the signs are sufficiently numerous and varied to establish a positive correlation between vitamin E deficiency in man and other animals.

Recommended Dietary Allowances for vitamin E range from 3 mg α-tocopherol equivalents for infants to 10 mg equivalents for adult men (9).

Various investigators have concluded that a need for supplementary or therapeutic amounts of vitamin E in humans exists for several conditions: fat-malabsorption syndromes, excessive intakes of polyunsaturated fats, intermittent claudication, and certain nutritional and hemolytic anemias (59–60). Other potential uses have been reviewed (61). Tocopherol administration also deserves further evaluation in stasis ulcers and habitual abortion. Subnormal reproductive functions in both males and females were among the clinical problems first treated with α-tocopherol, but results were conflicting. The broad group of diseases of collagen tissue, the so-called collagenoses, have been treated empirically with vitamin E; such treatment may have value in lupus erythematosus, primary fibrositis, Dupuytren's contracture, and Peyronie's disease.

Infants, particularly those born prematurely, deserve special attention. They are poorly supplied with vitamin E because the placenta limits the passage of α-tocopherol from the mother to the fetus. Conversely, the colostrum and early milk are very rich in tocopherol and apparently designed to supply infants with relatively large amounts of vitamin E during the first few days of life. Formulas based on cow's milk supply relatively little α-tocopherol, and therefore, most infant-feeding preparations are now supplemented with vitamin E. The importance of adequate vitamin E nutriture in the young has been emphasized by its value in reducing excessive aggregation of platelets, correcting the hemolytic anemia of low birth-weight infants, and preventing or lessening the severity of retrolental fibroplasia or bronchopulmonary dysplasia during prolonged oxygen administration (62).

α-Tocopherol has been administered orally in large doses to a number of species and is generally well tolerated. Occasional adverse effects, eg, ref. 63, appear to result from inhibition of absorption of other nutrients. Isolated but inconsistent case reports have suggested several side effects in humans consuming 0.4–1.0 g daily, but most adults appear to tolerate these doses (9). The likelihood of serious adverse effects from consuming 100–1000 units daily appears to be very low (64), and adult humans have been given levels as high as 500–600 mg/kg body weight daily for five months. Parenteral administration of α-tocopherol as an oil concentrate is not recommended. Local reactions at the site of the injection are common, and utilization of the tocopherol is very poor. On the other hand, injectable preparations in water-dispersible emulsifiers are well utilized.

Uses

Both α-tocopherol and its esters are constituents of multivitamin and single-dose nutrient capsules or other liquid dietary supplements. Crystalline α-tocopheryl hydrogen succinate or oily forms embedded in materials such as gelatin are particularly useful in the formulation of multivitamin tablets and dry-fill capsules. α-Tocopherol or its esters dispersed or solubilized in hydrophilic media are used for malabsorption conditions and as injectable preparations (65). (RRR)-α-Tocopheryl polyethylene glycol (1000) succinate (a waxy solid) forms clear water solutions at concentrations up to 20%, thus permitting the use of vitamin E in aqueous formulations without requiring a solubilizer. Supplements for human use range from a few milligrams in multivitamin preparations to 500–1000 mg in certain dietary supplements. Specialty items, such as ointments, salves, and suppositories containing vitamin E, provide other outlets for α-tocopherol.

Owing to the absence of widely recognized signs of vitamin E deficiency in humans, foods are generally not fortified. However, most infant formulas are supplemented with α-tocopheryl acetate to furnish 5–10 units per reconstituted quart. Nonfat dry milk solids are unquestionably deficient (66). Fortification of some cereals restores losses due to processing (6). Both chicken and, particularly, turkey meats benefit from dietary incorporation of vitamin E to prevent rancidity during refrigerated or frozen storage. Tocopherol may also be a useful adjunct to inhibit nitrosamine formation and oxidation in bacon and other foods (see N-Nitrosamines) (67).

Animal feeds, primarily in the poultry industry, consume ca 40% of the commercial production (see also Pet and livestock feed). Increasing amounts are being used, however, in supplements and concentrates for calves, lambs, swine, and beef and dairy cattle, as well as for fortifying rations of fur-bearing animals, eg, mink, and pets, particularly cats. The vitamin E naturally present in feedstuffs is relatively unstable in stored, mixed feeds. To replace tocopherol that might have been destroyed and to ensure that the animal's requirements are met, stable α-tocopheryl acetate is added at concentrations ranging from 2.5 to 20 units per pound (5.5 to 44 units per kilogram) of feed. It is usually added as a dry, granular, free-flowing, nondusting powder containing from 20,000 to 125,000 units per pound (44,000 to 276,000 units per kilogram) or as a high potency oil concentrate.

Unesterified tocopherols have been used as antioxidants to some extent by food technologists and by pharmaceutical formulators. They are particularly effective in essential oils, rendered animal fats, and mineral oil. In pharmaceutical preparations, tocopherols stabilize other substances, eg, vitamin A and unsaturated lipids, during shelf life and also in the gastrointestinal tract. Although not the best antioxidant *per se*, α-tocopherol is uniquely suited for this purpose because it is absorbed and deposited in body tissues, whereas other antioxidants are destroyed or excreted.

BIBLIOGRAPHY

"Vitamin E" in *ECT* 1st ed., Vol. 14, pp. 849–858, by P. L. Harris and N. D. Embree, Eastman Kodak Co.; "Vitamins (Vitamin E)" in *ECT* 2nd ed., Vol. 21, pp. 574–585, by D. C. Herting, Tennessee Eastman Co.

1. J. F. Pennock, F. W. Hemming, and J. D. Kerr, *Biochem. Biophys. Res. Commun.* **17**, 542 (1964).
2. International Union of Nutritional Sciences, *Nutr. Abstr. Rev.* **48**, 831 (1978).
3. R. H. Bunnell, J. Keating, A. Quaresimo, and G. K. Parman, *Am. J. Clin. Nutr.* **17**, 1 (1965).
4. J. G. Bieri and R. P. Evarts, *J. Am. Diet. Assoc.* **62**, 147 (1973).
5. J. N. Thompson, J. L. Beare-Rogers, P. Erdödy, and D. C. Smith, *Am. J. Clin. Nutr.* **26**, 1349 (1973).
6. D. C. Herting and E. E. Drury, *J. Agric. Food Chem.* **17**, 785 (1969).
7. V. H. Booth and M. P. Bradford, *Int. Z. Vitaminforsch.* **33**, 276 (1963).
8. M. H. Thomas and D. H. Calloway, *J. Am. Diet. Assoc.* **39**, 105 (1961).
9. Food and Nutrition Board, National Research Council, *Recommended Dietary Allowances*, 9th ed., National Academy of Sciences, Washington, D.C., 1980, p. 63.
10. D. R. Nelan and C. D. Robeson, *J. Am. Chem. Soc.* **84**, 2963 (1962).
11. S. E. Scheppele, R. K. Mitchum, C. J. Rudolph, Jr., K. F. Kinneberg, and G. V. Odell, *Lipids* **7**, 297 (1972).
12. E. Fernholz, *J. Am. Chem. Soc.* **60**, 700 (1938).
13. J. L. G. Nilsson, *Acta Pharm. Suecica* **6**, 1 (1969).
14. W. A. Skinner, R. M. Parkhurst, J. Scholler, and K. Schwarz, *J. Med. Chem.* **12**, 64 (1969).
15. E. J. Simon, A. Eisengart, L. Sundheim, and A. T. Milhorat, *J. Biol. Chem.* **221**, 807 (1956).
16. L. Michaelis and S. H. Wollman, *Biochim. Biophys. Acta* **4**, 156 (1950).
17. W. A. Skinner, *Biochem. Biophys. Res. Commun.* **15**, 469 (1964).
18. G. W. Grams, *Tetrahedron Lett.* **50**, 4823 (1971).
19. J. G. Bieri and E. L. Prival, *Biochem.* **6**, 2153 (1967).
20. N. D. Embree, *Chem. Rev.* **29**, 317 (1941).
21. M. H. Stern, C. D. Robeson, L. Weisler, and J. G. Baxter, *J. Am. Chem. Soc.* **69**, 869 (1947).
22. T. Rubel, *Vitamin E Manufacture*, Chemical Process Review No. 39, Noyes Development Corp., Park Ridge, N.J., 1969.
23. S. Kasparek in L. J. Machlin, ed., *Vitamin E: A Comprehensive Treatise*, Marcel Dekker, Inc., New York, 1980, Chapt. 2.
24. P. Karrer, H. Fritzsche, B. H. Ringier, and H. Salomon, *Helv. Chim. Acta* **21**, 520 (1938).
25. F. Bergel, A. Jacob, A. R. Todd, and T. S. Work, *Nature* **142**, 36 (1938).
26. L. I. Smith, H. E. Ungnade, and W. W. Prichard, *Science (Washington, D.C.)* **88**, 37 (1938).
27. H. Mayer, P. Schudel, R. Rüegg, and O. Isler, *Helv. Chim. Acta* **46**, 650 (1963).
28. S. R. Ames, M. I. Ludwig, D. R. Nelan, and C. D. Robeson, *Biochem.* **2**, 188 (1963).
29. J. W. Scott, F. T. Bizzarro, D. R. Parrish, and G. Saucy, *Helv. Chim. Acta* **59**, 290 (1976).
30. C. Fuganti and P. Grasselli, *J. Chem. Soc., Chem. Commun.* (22), 995 (1979).
31. N. Cohen, C. G. Scott, C. Neukom, R. L. Lopresti, G. Weber, and G. Saucy, *Helv. Chim. Acta* **64**, 1158 (1981).

32. O. Isler, P. Schudel, H. Mayer, J. Würsch, and R. Rüegg in R. S. Harris and I. G. Wool, eds., *Vitamins and Hormones*, Vol. 20, Academic Press, Inc., New York, 1962, pp. 389–405.
33. C. D. Robeson, *J. Am. Chem. Soc.* **64**, 1487 (1942).
34. L. I. Smith, W. B. Renfrow, and J. W. Opie, *J. Am. Chem. Soc.* **64**, 1084 (1942).
35. *Synthetic Organic Chemicals: United States Production and Sales, 1980*, USITC Publication 1183, U.S. Government Printing Office, Washington, D.C., 1981, p. 118.
36. *Imports of Benzenoid Chemicals and Products, 1980*, USITC Publication 1163, U.S. Government Printing Office, Washington, D.C., 1981, pp. 86, 97.
37. *Chem. Mark. Rep.* **221**(6), 47 (1982).
38. *The United States Pharmacopeia XX (USP XX–NF XV)*, The United States Pharmacopeial Convention, Inc., Rockville, Md., 1980, p. 846.
39. S. R. Ames, *J. Nutr.* **109**, 2198 (1979).
40. Food and Nutrition Board, National Research Council, *Food Chemicals Codex*, 3rd ed., National Academy Press, Washington, D.C., 1981, p. 330.
41. R. H. Bunnell in P. György and W. N. Pearson, eds., *The Vitamins*, 2nd ed., Vol. 6, Academic Press, Inc., New York, 1967, Chapt. 6.
42. J. G. Bieri in G. V. Marinetti, ed., *Lipid Chromatographic Analysis*, Vol. 2, Marcel Dekker, Inc., New York, 1969, Chapt. 8.
43. Vitamin E Panel, Analytical Methods Committee, *Analyst (London)* **84**, 356 (1959).
44. H. T. Slover, J. Lehmann, and R. J. Valis, *J. Am. Oil Chem. Soc.* **46**, 417 (1969).
45. C. G. Scott, N. Cohen, P. R. Riggio, and G. Weber, *Lipids* **17**, 97 (1982).
46. P. Taylor and P. Barnes, *Chem. Ind. (London)* (20), 722 (1981).
47. S. R. Ames, *J. Assoc. Off. Anal. Chem.* **54**, 1 (1971).
48. S. R. Ames and E. E. Drury, *J. Assoc. Off. Anal. Chem.* **58**, 585 (1975).
49. K. E. Mason and P. L. Harris, *Biol. Symp.* **12**, 459 (1947).
50. C. I. Bliss and P. György in ref. 41, Chapt. 6.
51. L. J. Machlin, E. Gabriel, and M. Brin, *J. Nutr.* **112**, 1437 (1982).
52. J. Green and J. Bunyan, *Nutr. Abstr. Rev.* **39**, 321 (1969).
53. K. E. Mason in R. S. Harris and K. V. Thimann, eds., *Vitamins and Hormones*, Vol. 2, Academic Press, Inc., New York, 1944, pp. 107–153.
54. K. E. Mason, P. L. Harris, R. S. Harris, and H. A. Mattill in W. H. Sebrell, Jr. and R. S. Harris, eds., *The Vitamins*, Vol. 3, Academic Press, New York, 1954, Chapt. 17.
55. H. H. Gordon and J. P. deMetry, *Proc. Soc. Exp. Biol. Med.* **79**, 446 (1952).
56. M. K. Horwitt in ref. 32, pp. 541–558.
57. P. M. Farrell, J. G. Bieri, J. F. Fratantoni, R. E. Wood, and P. A. diSant'Agnese, *J. Clin. Invest.* **60**, 233 (1977).
58. E. A. Rachmilewitz, A. Shifter, and I. Kahane, *Am. J. Clin. Nutr.* **32**, 1850 (1979).
59. P. R. Dallman in D. G. Nathan and F. A. Oski, eds., *Hematology of Infancy and Childhood*, W. B. Saunders Co., Philadelphia, Pa., 1974, pp. 97–150.
60. L. Corash and co-workers, *N. Engl. J. Med.* **303**, 416 (1980).
61. M. K. Horwitt, *Nutr. Rev.* **38**, 105 (1980).
62. F. A. Oski, *N. Engl. J. Med.* **303**, 454 (1980).
63. T. P. Murphy, K. E. Wright, and W. J. Pudelkiewicz, *Poult. Sci.* **60**, 1873 (1981).
64. P. M. Farrell in ref. 23, Chapt. 11A.
65. U.S. Pat. 3,639,587 (Feb. 1, 1972), S. R. Ames (to Eastman Kodak Company).
66. D. C. Herting and E. E. Drury, *Am. J. Clin. Nutr.* **22**, 147 (1969).
67. W. Fiddler and co-workers, *J. Agric. Food. Chem.* **26**, 653 (1978).

General References

Ref. 23 is a general reference.
P. P. Nair and H. J. Kayden, eds., "Vitamin E and its Role in Cellular Metabolism," *Ann. N.Y. Acad. Sci.* **203** (1972).
W. H. Sebrell, Jr. and R. S. Harris, eds., *The Vitamins*, Vol. 5, Academic Press, Inc., New York, 1972, Chapt. 16.
C. deDuve and O. Hayaishi, eds., *Tocopherol, Oxygen and Biomembranes*, Elsevier/North Holland Biomedical Press, Amsterdam, The Netherlands, 1978, 374 pp.

B. Lubin and L. Machlen, eds., "Vitamin E: Biological, Hematological, and Clinical Aspects," *Ann. N.Y. Acad. Sci.* **393** (1982).

DAVID C. HERTING
Eastman Kodak Company

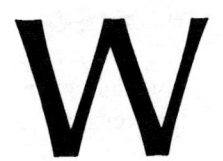

W

WALLBOARD. See Laminated and reinforced wood.

WAR GASES. See Chemicals in war.

WARFARIN. See Poisons, economic.

WASTES, INDUSTRIAL

Industrial wastes contribute at least two thirds of the organic matter entering the watercourses in the United States and, in order to improve this situation, a thorough investigation of the problems and production processes is needed. The EPA is currently making progress in this area. Each industry was studied by a separate contractor in order to determine treatments and acceptable levels of residual wasteloads at acceptable costs. The EPA established minimum wasteloads of key contaminants for each main wet industry. State and local governments may require more treatment in cases where local water quality is not maintained by the industrial treatment and effluent guidelines. Therefore, both effluent and stream standards are of importance in the planning of industrial waste treatment.

Some solid wastes produced by industry may be similar to municipal refuse, eg, from paper mills, plastic plants, and food-processing plants. Such industrial waste comprises ca 0.5 kg per capita per day of all municipal solid wastes collected (ca 25% of that generated). It is rapidly increasing in quantity as industrial production expands (1).

Airborne industrial effluents consist mostly of carbon monoxide, particulates (suspended visible matter), sulfur oxides (SO_x), hydrocarbons, and nitrogen oxides (NO_x). These contaminants originate from several sources including transportation, stationary sources of fuel combustion, industrial processes, solid-wastes disposal, and other miscellaneous sources of operation (see Air pollution).

In the United States, nine waste-treatment firms handle more than 50% of all commercial waste disposal and treatment: Browning-Ferris Industries, CECOS International, Chem-Clear, Chemical Waste Management, Conversion Systems, IT Corporation, Rollins Environmental Services, SCA Chemical Services, and US Ecology. Data on the various methods are given in Table 1.

The cost of waste treatment depends upon many factors such as plant location, environmental-control regulations, and type of wastes produced. It usually ranges from 1 to 10% of production costs. Overall costs can be minimized by utilizing the professional services of a highly qualified and experienced engineering firm. Information about such establishments can be obtained from the EPA in Washington, D.C., the local agency in charge of environmental control, or the American Association of Air Pollution Engineers and the American Academy of Environmental Engineers. Membership lists of these organizations are published in journals such as the *Journal of the Water Pollution Control Federation*, which also contains information on manufacturers and distributors of equipment and supplies (3).

Although capital and operating costs of waste treatment have been variously estimated at $220–660 per kilogram of BOD per day or 2–8¢/1000 L, methods of treatment and effectiveness vary greatly from industry to industry. The cost of treatment increases, of course, with increasing efficiency.

Both industry and regulatory agencies must develop a dependable method to determine the economics of waste treatment. These costs must eventually be passed on to the consumer. If the price increase of the product is too high, industry may then have to choose among several alternatives: cease certain production, curtail some production, seek government subsidy, change type of production, merge with a competitor in order to be able to afford waste treatment, relocate, or try to ease the regulations.

Anerobic wastewater treatment uses less energy and generates less sludge than aerobic treatment. Today, when energy and sludge disposal are so costly, anaerobic treatment is increasingly applied (4).

Classification

The sources and characteristics of the principal industrial pollutants are given in Table 2.

Table 1. Waste-Disposal Methods[a]

Waste management	Capacity, wet, 1000 t 1981	1980	Change, %	Volume received, wet, 1000 t 1981	1980	Change, %
land fill	37,372	25,672	46	1,965	2,182	−10
land treatment and solar evaporation	1,400	1,447	−3	282	300	−6
chemical treatment	1,305	1,105	18	734	544	35
deep-well injection	1,095	1,095	0	475	475	0
resource recovery	341	341	0	83	83	0
incineration	102	102	0	80	85	−6
Total	41,615	29,762	40	3,610	3,669	−1

[a] Ref. 2.

Table 2. Sources, Characteristics, and Treatment of Industrial Wastes[a]

Waste-producing industry and materials	Origin	Characteristics	Treatment and disposal methods
agriculture	chemicals, irrigation return flows, crop residues, and liquid and solid animal wastes	highly organic and BOD detergent cleaning solutions	biological oxidation basins; some composting and anaerobic digestion; land application
fertilizer	spills, cooling waters, washing of products, boiler blowdowns	sulfuric, phosphoric, and nitric acids; mineral elements, P, S, N, K, Al, NH_3, NO_3, etc, Fl, some suspended solids	neutralization, detain for reuse, sedimentation, air stripping of NH_3, lime precipitation
pesticides	washing and purification products such as 2,4-D and DDT	high organic matter, benzene-ring structure, toxic to bacteria and fish, acid	dilution, storage, activated-carbon adsorption, alkaline chlorination
asbestos	cleaning and crushing ore	suspended asbestos and mineral solids	detention in ponds, neutralization and land filling
acids	dilute wash waters; many varied dilute acids	low pH, low organic content	upflow or straight neutralization, burning when some organic matter is present
candles	wax spills, stearic acid condensates	organic (fatty) acids	anaerobic digestion
cement	fine and finish grinding of cement, dust-leaching collection, dust control	heated cooling water, suspended solids, some inorganic salts	segregation of dust-contact streams, neutralization and sedimentation
coal processing	cleaning and classification of coal, leaching of sulfur strata with water	high suspended solids, mainly coal; low pH, high H_2SO_4 and $FeSO_4$	settling, froth flotation, drainage control, and sealing of mines
energy	scrubber power plant wastes, scrubbing of gaseous combustion products by liquid water	particulates, SO_2, impure absorbents or NH_3, NaOH, etc	solids removal usually by settling, pH adjustment, and reuse
nuclear power	processing ores; laundering of contaminated clothes; research-laboratory wastes; processing of fuel; power-plant cooling waters	radioactive elements, can be very acidic and "hot"	concentration and containing, or dilution and dispersion
steam power	cooling water, boiler blow-down, coal drainage	hot, high volume, high inorganic and dissolved solids	cooling by aeration, storage of ashes, neutralization of excess acid wastes
explosives	washing TNT and guncotton for purification, washing and pickling of cartridges	TNT, colored, acidic, odorous, and contains organic acids and alcohols from powder and cotton, metals, acids, oils, and soaps	flotation, chemical precipitation, biological treatment, aeration, chlorination of TNT, neutralization, adsorption

Table 2 (*continued*)

Waste-producing industry and materials	Origin	Characteristics	Treatment and disposal methods
formaldehyde	residues from manufacturing synthetic resins and from dyeing synthetic fibers	normally high BOD and HCHO, toxic to bacteria in high concentrations	trickling filtration, adsorption on activated charcoal
food			
bakeries	washing and greasing of pans	high BOD, grease, floor washings, sugars, flour, detergents	amenable to biological oxidation
beet sugar	transfer, screening, and juicing waters; draining from lime sludge; condensates after evaporator; juice and extracted sugar	high in dissolved and suspended organic matter, containing sugar and protein	reuse of wastes, coagulation, and lagooning
brewed and distilled beverages	steeping and pressing of grain; residue from distillation of alcohol; condensate from stillage evaporation	high in dissolved organic solids, containing nitrogen and fermented starches or their products	recovery, concentration by centrifugation and evaporation, trickling filtration; use in feeds digestion of slops
cane sugar	spillage from extraction, clarification; evaporator entrainment in cooling and condenser waters	variable pH, soluble organic matter with relatively high BOD of carbonaceous nature	neutralization, recirculation, chemical treatment, some selected aerobic oxidation
canned goods	trimming, culling, juicing, and blanching of fruits and vegetables	high in suspended solids, colloidal and dissolved organic matter	screening, lagooning, soil absorption of spray irrigation
coffee	pulping and fermenting of coffee bean	high BOD and suspended solids	screening, settling, and trickling filtration
cornstarch	evaporator condensate or bottoms when not reused or recovered, syrup from final washes, wastes from bottling process	high BOD and dissolved organic matter; mainly starch and related material	equalization, biological filtration, anaerobic digestion
dairy products	dilutions of whole milk, buttermilk, and whey	high in dissolved organic matter, mainly protein, fat, and lactose	biological treatment aeration, trickling filtration, activated sludge
fish	rejects from centrifuge; pressed fish; evaporator and other wash water wastes	very high BOD, total organic solids, and odor	evaporation of total waste; barge remainder to sea
meat and poultry products	stockyards; slaughtering of animals; rendering of bones and fats; residues in condensates; grease and wash water; picking chickens	high in dissolved and suspended organic matter, blood, other proteins, and fats	screening, settling or flotation, trickling filtration

Table 2 (*continued*)

Waste-producing industry and materials	Origin	Characteristics	Treatment and disposal methods
rice	soaking, cooking, and washing of rice	high BOD, total and suspended solids (mainly starch)	lime coagulation, digestion
soft drinks	bottle washing; syrup storage-tank drains	high pH, suspended solids, and BOD	screening, plus discharge to municipal sewer
yeast	residue from yeast filtration	high in solids (mainly organic) and BOD	anaerobic digestion, trickling filtration
fuel oil	spills	high in emulsified and dissolved oils	leak and spill prevention, flotation
glass	many sources	red color, alkaline nonsettleable suspended solids	calcium chloride precipitation
glue	lime and acid washes	high COD, BOD, pH	to aerobic biological pond
hospital	washing, sterilization of facilities, used solutions, spills	bacteria, various chemicals, radioactive materials	discharge to municipal sewers; holding and biological aeration in large facilities
iron-foundry products	sand from hydraulic discharge	high suspended solids, mainly sand; some clay and coal	selective screening, drying of reclaimed sand
laundry	washing of fabrics with soap or detergents	high turbidity, alkalinity, and organic solids	screening, chemical precipitation, flotation and adsorption
detergents	washing and purifying soaps and detergents	high in BOD and saponified soaps	flotation and skimming, precipitation with $CaCl_2$
leather	unhairing, soaking, deliming, and bating of hides	high total solids, hardness, salt, sulfides, chromium, pH precipitated lime, and BOD	equalization, sedimentation, and biological treatment
metal container	cutting and lubricating metals, cleaning can surfaces	metal fines, lubrication oils, variable pH, surfactants, dissolved metals	oil separation, chemical precipitation, collection and reuse, lagoon storage; final carbon absorption
metal-plated products	stripping of oxides, cleaning and plating of metals	acid, metals, toxic, low volume, mainly mineral matter	alkaline chlorination of cyanide; reduction and precipitation of chromium; lime precipitation of other metals
mortuary	body fluids, wash waters, spills	blood salt, formaldehydes, high BOD, infectious diseases	discharge to municipal sewer holding and chlorination
naval stores	washing of stumps, drop solution, solvent recovery, and oil-recovery water	acid, high BOD	by-product recovery, equalization, recirculation, and reuse, trickling filtration
oil fields and refineries	drilling muds, salt, oil, and some natural gas; acid sludges and miscellaneous oils from refining	high dissolved salts from field; high BOD, odor, phenol, and sulfur compounds from refinery	diversion, recovery, injection of salts; acidification and burning of alkaline sludges

Table 2 (*continued*)

Waste-producing industry and materials	Origin	Characteristics	Treatment and disposal methods
paints and inks	solvent-based rejected materials scrubbers for paint vapors; refining or removing inks	contain organic solids from dyes, resins, oils, solvents, etc	settling ponds for detention of paints, lime coagulation of printing inks
petrochemicals	contaminated water from chemical production and transportation of second-generation oil compounds	high COD, TDS[b], metals, COD/BOD ratio, and compounds inhibitory to biol. action	recovery and reuse, equalization and neutralization, chemical coagulation, settling or flotation, biological oxidation
pharmaceutical products	mycelium, spent filtrate, and wash waters	high in suspended and dissolved organic matter, including vitamins	evaporation and drying; feeds
phosphate and phosphorus	washing, screening, floating rock, condenser bleed-off from phosphate reduction plant	clays, slimes and tall oils, low pH, high suspended solids, phosphorus, silica and fluoride	lagooning, mechanical clarification, coagulation and settling of refined waste
photographic products	spent solutions of developer and fixer	alkaline, containing various organic and inorganic reducing agents	recovery of silver; discharge of wastes into municipal sewer
plastics and resins	unit operations from polymer preparation and use; spills and equipment washdowns	acids, caustic, dissolved organic matter such as phenols, formaldehyde, etc	discharge to municipal sewer, reuse, controlled discharge
plywood	glue washings	high BOD, pH, phenols, potential toxicity	settling ponds, incineration
pulp and paper	cooking, refining, washing of fibers, screening of paper pulp	high or low pH, color, high suspended, colloidal, and dissolved solids, inorganic fillers	settling, lagooning, biological treatment, aeration, recovery of by-products
rubber	washing of latex, coagulated rubber, exuded impurities from crude rubber	high BOD and odor, high suspended solids, variable pH, high chlorides	aeration, chlorination, sulfonation, biological treatment
steel	coking of coal, scrubbing of blast-furnace flue gases, and pickling of steel	low pH, acids, cyanogen, phenol, ore, coke, limestone, alkalies, oils, mill scale, and fine suspended solids	neutralization, recovery, and reuse, chemical coagulation
textiles	cooking of fibers; desizing of fabric	highly alkaline, colored, high BOD and temperature, high suspended solids	neutralization, chemical precipitation, biological treatment, aeration or trickling filtration
toxic chemicals	leaks, accidental spills, and refining of chemicals	various toxic dissolved elements and compounds such as Hg and PCBs	retention and reuse, change in production
water production	filter backwash; lime-soda sludge; brine; alum sludge	minerals and suspended solids	direct discharge to streams or indirectly through holding lagoons
wood furniture	wet spray booths and laundries	organics from staining and sealing wood products	evaporation or burning

Table 2 (continued)

Waste-producing industry and materials	Origin	Characteristics	Treatment and disposal methods
wood preserving	steam condensates	high in COD, BOD, solids, phenols	chemical coagulation; oxidation pond and other aerobic biological treatment

[a] Ref. 5.
[b] TDS = total dissolved solids.

Gaseous Wastes. Gaseous wastes can be classified into several categories: pure gases or vapors, combinations of gases and solids, combinations of gases and liquds, and combinations of gases, liquids, and solids. The last three are usually considered gaseous wastes because the gas is the carrier for the solid or liquid phase.

Typical pure gases that might be emitted from an industrial operation are hydrogen, hydrogen sulfide, carbon monoxide, carbon dioxide, etc. A pure gas may be considered an air pollutant if it is either toxic or obnoxious in odor or creates a visible plume. Table 3 shows the exposure limits of various types of gases in these categories, as recommended by the ACGIH.

In a gas containing solid particles, generally, the solid particles are neither toxic nor offensive in odor and the carrier gas is usually air or combustion products. The solid or particulate matter in the gas must be of such a size that the effluent gas can

Table 3. Recommended TLV[a] **for Airborne Contaminants**[b]

Compound	ppm	Compound	ppm
ammonia	25	hydrogen cyanide	10
benzene	10	hydrogen bromide	3
biphenyl	0.2	hydrogen fluoride	3
bromine	0.1	hydrogen sulfide	10
carbon dioxide	5000	iodine	0.1
carbon disulfide	20	isopropyl alcohol	400
carbon monoxide	50	maleic anhydride	0.25
carbon tetrachloride	10	methyl chloride	100
chlorine	1	naphthalene	10
chloroform	10	nitric acid	2
ethanolamine	3	nitric oxide	25
ethyl acetate	400	nitrogen dioxide	5
ethyl alcohol	1000	ozone	0.1
ethyl ether	400	phenol	5
ethylene oxide	50	phosgene	0.1
fluorine	1	phthalic anhydride	1
formaldehyde	2	sulfur dioxide	5
formic acid	5	toluene	100
hexane	100	turpentine	100
hydrogen chloride	5	xylene	100

[a] Exposure during 8-h period.
[b] Ref. 6.

carry these particles into the atmosphere. An example is the effluent from grinding or milling applications such as lime, cement-dust silica, or asbestos dust. Some solid particles are larger than 1 μm, some smaller. They are easily airborne and, when inhaled by humans, can cause respiratory ailments and, if toxic, may have serious consequences.

In a combination of gases and liquids, the liquids are in the form of small droplets and are carried by the effluent gas stream to adjacent areas. Such liquids may or may not be toxic, but can cause difficulties similar to particulate matter. Large quantities of water vapor or organic liquids entrained in air are examples of this type of gaseous effluent.

Liquid Wastes. The liquid wastes discharged from industrial processes are so varied as to defy classification. They may contain dissolved gases or solid particles. Gases, if deemed undesirable, are usually separated and may be treated as a gaseous waste. Solids may be separated by filtration (qv), centrifugation (qv), or sedimentation (qv).

Pure liquid waste is either aqueous or nonaqueous. Aqueous wastes contain a high concentration of water with small amounts of dissolved inorganic or organic materials. If the dissolved material is primarily organic and biodegradable, biological methods can be used for separation. If the waste is primarily water with dissolved inorganic solids, chemical or physical treatment, ion exchange, or distillation may be more practical.

Liquid waste can be further classified by its combustibility. Pure organic liquids can be burned in air but may thereby present an air-pollution problem.

Solid Wastes. Solid waste may be suitable for a sanitary land fill. Waste suitable for land fill is stable organic or inorganic material which either remains in the soil or is biologically degraded. However, potential toxicity of the degradation products must be considered. Finally, solid waste should be classified into waste that may or may not be compacted or reduced to uniform size by shredding, and waste that cannot. Compaction and size reduction greatly facilitate landfill operations.

The segregation of wastes into their various classifications is an important step in waste management, especially if recycling is possible. Frequently, segregation is the first step. In general, a variety of different wastes should not be combined in a single disposal area; segregation at the source usually facilitates handling. Many plants channel all of their liquid waste into a single lagoon, whereas segregation at the source would permit simpler methods to be employed. For example, combustible organic waste should not be combined with a highly aqueous waste before treatment, since the combustible waste may be burned easily in an incinerator but, if combined with aqueous waste, requires a more expensive disposal method. Furthermore, an organic waste that gives off a toxic vapor when burned (eg, chlorinated hydrocarbons) should not be combined with organic waste which could be burned to carbon dioxide, nitrogen, and water vapor. Similarly, a combustible gaseous waste should not be combined with an air stream containing particulate matter before treatment. The latter could be treated with a separation device such as a cyclone or bag filter, but the addition of a combustible fume could create an explosion hazard. The former should be treated by incineration, and the latter by a separation method.

The possibility of recycling material should always be considered before disposal. Many waste streams contain products that can be recycled into the process or saved as a valuable commodity. In the past, many industries have disregarded such considerations but today, with pollution regulations requiring each industry to take a careful look at its waste products, recovery is more attractive. For example, recovery by activated carbon of valuable solvents from printing or coating operations reduces air pollution as well as cost (see also Recycling).

Air Contaminants

The industrial air contaminants, as estimated by the EPA in 1970, are given in Table 4. Although concentrations have changed little, actual quantities have increased.

Treatment of waste gas is usually not complicated from an engineering standpoint, and a number of methods are available (7). However, because of the large volume of gas emitted from various industrial processes, large equipment is needed and the cost is high. First, particulates or solid matter are removed (see Gas cleaning; Air-pollution control methods).

After the waste has been treated with the methods discussed below, it is either acceptable for disposal in the atmosphere or must be treated further in an additional step called designed effluent disposal, eg, by mixing with sufficient air, the concentration of the waste gas can be reduced to an acceptable level.

Condensation of Vapors. A gas emitted from a process may be a pure vapor, which is easily condensed at temperatures below its normal boiling point. A condenser then serves for recovery. However, most waste gases are mixtures of air or inert gas and solvent vapors. The solvent content might be so small that condensation may not be practical. Thus, the temperature of the gas must be reduced to the dew point of the solvent concentration and, as liquid is removed from the gas, the dew point continues to decrease, making 100% recovery possible. Therefore, it is practical in some cases to remove a portion of some vapors present in air by supercooling if the dew point of the vapor is high enough. This generally requires refrigeration and heat exchangers (see also Heat-exchange technology).

Solvent Extraction. The use of activated carbon for the removal of solvent vapors from air solves a pollution problem, while at the same time recovering the solvent. Recovery varies from 95 to 99% (see Adsorption separation; Solvent recovery).

Odor Control. Although the presence of highly toxic gases in minute quantities in the air is difficult to detect, even extremely small amounts of certain malodorous materials can pollute the air; for example, downwind from a fat-rendering plant, a sickening odor permeates the atmosphere. A number of treatment methods are available.

Table 4. Estimated U.S. Emissions of Air Pollutants, 10^6 t/yr[a]

Source	CO	Particulates	SO_x	Hydrocarbons	NO_x
transportation	110.0	0.7	1.0	19.5	11.7
fuel combustion in stationary sources	0.8	6.8	26.5	0.6	10.0
industrial processes	11.4	13.1	6.0	5.5	0.2
solid-waste disposal	7.2	1.4	0.1	2.0	0.4
miscellaneous	16.8	3.4	0.3	7.1	0.4
Total	146.2	25.4	33.9	34.7	22.7

[a] Ref. 1.

Incineration. A waste gas can be incinerated if it contains organic compounds that rapidly oxidize at high temperatures (see Incinerators).

Direct-flame incineration is applicable if a combustible substance is present above its lower combustible limit when mixed with air, ie, if the waste gas burns with a visible flame when mixed with air and has a heating value >3.72 MJ/m^3 (100 Btu/ft^3).

Gases with a lower heating value may become combustible when preheated with air to ca 343°C. Blast-furnace gas is a typical example of a fuel with a low heating value which sustains combustion. Typical waste gases which may be burned in this manner are hydrogen cyanide, hydrogen sulfide, carbon monoxide, hydrogen, etc. Hydrogen sulfide produces a toxic product, sulfur dioxide, which is removed by further scrubbing.

The contaminant in the waste gas may serve as a part or all of the fuel in the system. Direct-flame combustion should be employed only where the contaminant supplies at least 50% of the fuel value of the mixture. The equipment for direct-flame incineration can be either a conventional industrial burner or combustor utilizing either forced or induced draft. Petroleum refineries and chemical plants employ flare-type burners. The ground flare is used at or near grade level and sufficient space must be provided for safety purposes. It is normally used in oil fields, gas fields, or other open areas. The tower flare is utilized in refineries and chemical plants to keep the flame well above the process equipment, protecting against fire hazards. Tower flares have a pilot burner for continuous automatic ignition. They consist of a pipe with a special flame-holding device at one end discharging the gas directly into the atmosphere. The waste gas is ignited and burned at this point with the flame-holding or retaining device ensuring stable combustion.

Flares frequently require steam injection to prevent smoking when burning waste hydrocarbon gases. Steam injection is used for hydrocarbons with a carbon-to-hydrogen weight ratio $>3:1$, eg, ethane or propane.

In general, flares are a practical method of direct-flame combustion but they are affected by atmospheric conditions, especially high winds, and are not a method of complete waste disposal.

Thermal incineration is utilized for waste-gas streams containing small amounts of combustible organic material. Instead of injecting waste gas directly through a burner along with auxiliary fuel, the burner is fed exclusively with the auxiliary fuel and is used to heat the waste gas to 93–149°C above the autoignition temperature of the organic waste present, usually 537–815°C. Most industrial effluents of this type emanate from ovens, paint-spraying operations, etc, and involve organic materials carried in air supplying sufficient oxygen for combustion.

Thermal incineration is used for waste gases containing organic materials at or below 25% of the lower explosive limit for the particular solvent present. In the incineration of wastes containing higher percentages of organic materials, flame propagation from the incinerator back to the source of the waste gas must be prevented.

The design of the incinerator depends upon time, temperature, and turbulence. The operating temperature must be high enough to oxidize the organic contaminants, residence time within the incinerator must be long enough to complete the reaction, and the turbulence sufficient to mix the air and the waste material.

In the system shown in Figure 1, the burner is actually installed in the waste duct, whereas a tunnel-type burner is fired from the outside. Refractories are required at the temperatures involved; however, stainless steel may be satisfactory in certain cases.

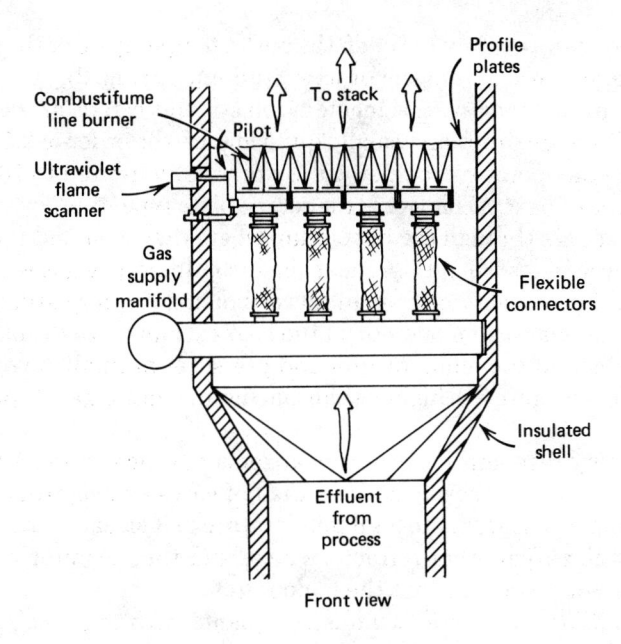

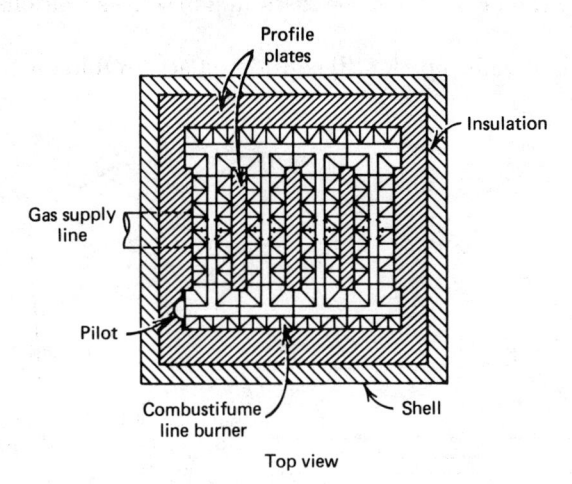

Figure 1. Duct-type fume burner. Courtesy Maxon Premix Burner Co.

Catalytic incineration is used for waste containing low concentrations of combustibles in air. The catalyst is generally a noble metal, such as platinum or palladium, dispersed on the surface of a catalyst support, eg, a silica honeycomb or a screen of nichrome wire. Use of a catalyst permits a low operating temperature which reduces the auxiliary fuel cost. The catalyst, however, may be expensive, and it may be poisoned or blanketed which reduces its activity (see Exhaust control, industrial).

Any waste-gas incineration system offers the possibility of heat recovery. Generally, the incinerator has an exit temperature of 426–871°C. The gases may be discharged directly to the atmosphere, but they can be used to reduce the auxiliary fuel input. The waste heat from the combustion reaction can be used to preheat the incoming fumes to a temperature between the outlet temperature of the incinerator and the incoming temperature from the process equipment (Fig. 2).

Heat exchangers for waste-gas incineration systems can be of the shell-and-tube or plate types. Pressure drops are usually quite low, on the order of 2.5 kPa (18.75 mm Hg), and maximum operating temperatures are usually limited to 760°C.

Another type of heat exchanger is the rotary-plate type. It utilizes revolving metal plates which first pass through the hot stream where they heat and then into the cold stream where heat is desorbed to preheat the incoming waste gases. These units are commonly used for recovery of heat in power boilers, refinery stills, and steel-mill furnaces. Some gas leaks from one side of the heat exchanger to the other because the seals do not withstand the temperatures and pressures normally involved, but when pressure on the exit gases is higher than on the incoming gases, pollution can be avoided.

The refractory regenerative heat exchanger is a cyclic system. A large refractory mass or similar device absorbs heat from the hot gases exiting from the incinerator and when this mass is heated, the high temperature exit stream is reversed by means of a switching valve to another refractory mass and the incoming cold waste gas is passed over the hot refractory into the incinerator.

Dispersion. For many years, industrial plants with a gaseous waste disposal problem vented the gas into a stack or chimney, regardless of the contaminants or pollutants. With present emphasis on air-pollution abatement, stacks are not always the solution. They can be used in some cases, and in some situations they are the only means available.

A short stack or vent pipe is still considered acceptable for certain waste gases

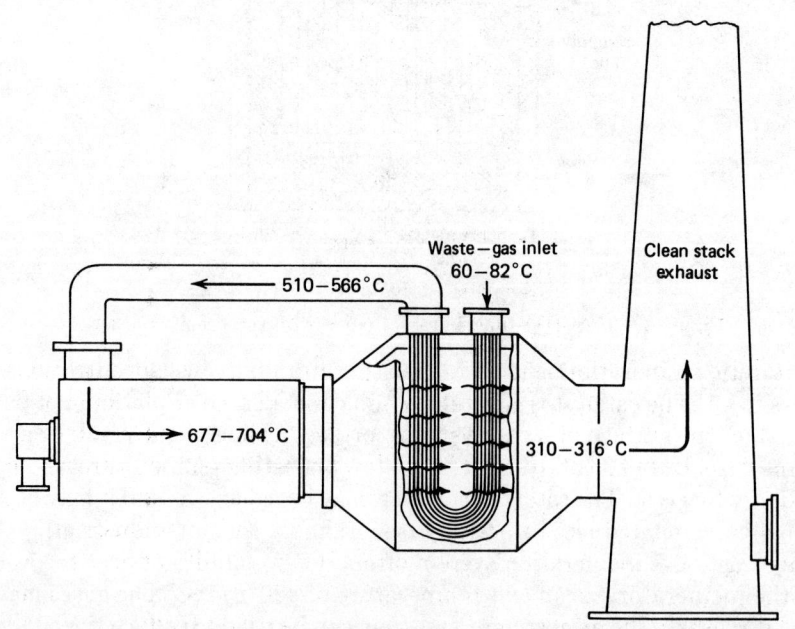

Figure 2. Waste-gas heat-recovery and incineration unit. Courtesy Thermal Research & Engineering Corp.

that do not contain pollutants, but for some waste gases containing small amounts of pollutants the tall stack or chimney may be utilized safely and to economic advantage. A high stack conveys waste gases containing toxic or particulate matter to a point high above the ground level; after normal dispersion into the atmosphere, the ground-level concentration of these materials will be well below permissible levels. This applies to concentrations at the base of the stack, and at a distance. Today, tall stacks or chimneys promote low concentrations of pollutants above the inversion layer and, therefore, these pollutants are not trapped in or under the inversion layer to cause high ground-level concentrations.

In most areas, stacks must be 100–135-m high and, in certain areas, even 304–335 m. Many factors influence the acceptability of a stack for the dispersion of pollutants, eg, the vicinity of residential areas and buildings; the prevailing wind direction and velocity; humidity and rainfall; topography; and the amount of pollutant exiting from the top of the stack and its exit velocity.

The permissible ground-level concentration of particular gaseous wastes is generally available from literature, but it may vary from locality to locality, depending on regulations.

The plume of smoke or waste gas exiting the stack usually has a conical shape. As it expands, it touches the ground at some distance from the base of the stack. The ground-level concentration of contaminant reaches a maximum at a distance from the stack which is usually about ten times the effective stack height, that is, the height of the stack plus the plume resulting from the discharge velocity and buoyancy of the gas.

A stack should be ca 2.5 times the height of any nearby buildings. Ejection velocities for the stack should be >182 m/s to allow the stack gases to escape a turbulent gas wake. Exit velocities on the order of 304 m/s are desirable. There is a critical wind velocity for every stack exit velocity above which there is no corresponding rise of the waste gas caused by that exit velocity. Above this critical wind velocity, gas temperatures and flow no longer affect the ground-level concentration.

With stacks of diameters <1.5 m and height <61 m, the plume may touch the ground leading to excessive ground concentrations. For such small stacks, the design basis is unreliable and results are unpredictable. When stack gases have been subjected to atmospheric diffusion, and turbulence induced from surrounding buildings is not a factor, the ground-level concentration should be on the order of 0.001–1% of the stack concentration, assuming that the stack is properly designed.

In many industrial complexes, multiple stacks within a short distance of one another cause problems of ground concentration of various contaminants, because their plumes combine at some point.

Some sulfur dioxide emanating from burning fossil fuels and certain industrial operations, such as smelting, reaches the atmosphere as a pollutant. Most of it is diffused from stacks, but, with the present emphasis on pollution control, several methods have been developed for its removal and recovery of the sulfur as sulfur, hydrogen sulfide, or sulfuric acid (see Sulfur recovery). These methods are usually expensive. However, at present, the high sulfur-bearing fuel is replaced with low sulfur fuel. This reduces the capital expense of scrubbers, but may result in an increase in operating costs resulting from a higher fuel cost.

Liquid Wastes

Effects on Streams. Streams and rivers assimilate a certain quantity of waste contaminants before reaching a polluted state (see Water, water pollution). A polluted stream contains an excessive amount of specific contaminants (see Table 5).

Effects on Sewage Plants. Industry often presumes that its waste can be disposed of in the public sewer system, where it is accepted by the municipality. However, waste discharge should not be permitted into the sewer system without disclosure of its contents, the system's ability to handle them, and the effects upon the system. A sewer ordinance, restricting the types and concentrations of waste admitted, is a protection (see Water, sewage).

To remove contaminants, a sewage-treatment plant must have treatment capability and sufficient capacity. In theory, a sewage-treatment plant should be able to handle any type of industrial waste, but many plants fall short of this aim. Joint treatment of municipal and compatible industrial wastewaters may offer great advantages, but economic factors are usually crucial.

Characteristics of municipal and industrial sewage are compared in Table 6.

The BOD of industrial waste affects mainly the biological or secondary treatment units of a sewage-treatment plant. If these units are large enough and the industrial waste can be biodegraded as rapidly as domestic sewage, it can be handled by the plant. Suspended solids, readily settable, can be disposed of along with domestic sludge as long as the industrial sludge can be drained and decomposed and extra sludge-treatment unit capacity is provided. Special process design and operations disintegrate

Table 5. Industrial Contaminants of Water[a]

Contaminant	Source	Effects
inorganic salts	oil refineries, desalination plants, munitions manufacture, pickle curing	interference with industrial usage, municipal water, and agriculture
acids, alkalies	chemical manufacturing	corrosion of pipelines and equipment; destruction of aquatic life
organic matter	tanneries, canneries, textile mills, etc	food for bacteria; oxygen depletion
suspended matter	paper mills, canneries, etc	suffocation of fish eggs; stream deterioration
floating solids and liquids	slaughterhouses, refinery oil	objectionable appearance, odor; interference of oxygen transfer
heated water	cooling waters from most industries and power plants	acceleration of bacterial action, reduction of oxygen saturation
color	textile, metal-finishing, and chemical plants	objectionable appearance
toxic chemicals	munitions manufacture, metal plating, steel mills, etc	alteration of stream biota and animal and human predators
microorganism	tanneries, municipal-industrial sewage plants	unsafe for drinking
radioactive particles	nuclear power plants, chemical laboratories	concentration in fish; unsafe for drinking
foam producers	glue manufacture, slaughterhouses, detergent manufacture	objectionable appearance

[a] Ref. 5.

Table 6. Municipal Sewage and Industrial Wastes

Waste	BOD, ppm	Suspended solids, ppm
municipal sewage	250	300
pulp and paper mills		
no chemical pulping	150	350
with chemical pulping	5000	2500
tannery	1200	2000
cannery	150–1000	250–1500

floating or colored material and other harmful constituents. Excess industrial volume curtails retention time. Thus, most units are enlarged when detention time is a significant factor in overall plant effectiveness.

Stream Protection

The quality of receiving waters can be preserved by stream standards or effluent standards. A polluter might be required to discharge only a certain quantity of contaminant or a certain concentration with a certain volume of wastewater. Many state agencies determine the best usage of a stream and assign certain quality standards to that water use. Any polluter violating these standards is cited and must abate the pollution. More recently, some agencies establishing the receiving-water quality desired and attempted to maintain this quality by controlling each waste discharge to the minimum contaminant units per unit of production or per capita. The latter is generally determined on an industry-wide basis from an analysis of effective treatment potential on an economically feasible basis. The stream-standards system used in New York State is shown in Table 7, the effluent standard for pulp and paper mills for Pennsylvania in Table 8.

Organic Content Permitted in Receiving Waters. If a regulatory agency decides to allow an industry to use a portion of the assimilating capacity of the receiving water, an acceptable organic level must be established. The decision is usually based upon preservation of a certain minimum concentration which is determined by many factors, including the nature of the organic matter, the amount of receiving water as a diluent as well as the amount of oxygen present, and the physical, biological, and chemical nature of the stream.

The industry must determine the ultimate (20 d) BOD of its waste and the quantity discharged per day. The stream deoxygenation rate k_1 and reaeration rate k_2, as well as the travel time of the waste downstream, must be computed. The dissolved-oxygen sag curve, shown in Figure 3, can be constructed from the Streeter-Phelps sag-curve formula:

$$D_t = \frac{k_1}{k_2 - k_1} L_a (10^{-k_1 t} - 10^{-k_2 t}) + D_a \cdot 10^{-k_2 t}$$

where D_t = dissolved oxygen deficit at waste travel time t, downstream, and L_a = ultimate BOD in the receiving water after the industrial waste has mixed with the stream water and D_a = its corresponding oxygen deficit.

Several simplifications of the above formula have been proposed as well as methods for arriving at true values of the stream reaction rates k_1 and k_2. Variations

Table 7. New York State Classes and Standards for Fresh Surface Water[a]

Class[b]	Use	Minimum dissolved oxygen, mL/L	Coliform bacteria median, no./100 mL	pH	Water standards[c] Toxic wastes, deleterious substances, colored wastes, heated liquids, odor-producing substances[d]	Floating solids, settleable solids, oil, and sludge deposits
AA	source of unfiltered public water supply and any other usage	5.0[e] 4.0[f]	not to exceed 50	6.5–8.5	none in sufficient amounts or at such temperatures as to be injurious to fish life or make the waters unsafe or unsuitable	none attributable to sewage, industrial wastes or other wastes
A	source of filtered public water supply and any other usage	5.0[e] 4.0[f]	not to exceed 5000	6.5–8.5		
B	bathing and any other usages except as a source of public water supply	5.0[e] 4.0[f]	not to exceed 2400	6.5–8.5		none readily visible and attributable to sewage, industrial wastes or other wastes
C	fishing and any other usages except public water supply and bathing	5.0[e] 4.0[f]	not applicable	6.5–8.5	none in sufficient amounts or at such temperatures as to be injurious to fish life or impair the waters for any other best usage	

| D | 3.0 | not applicable | 6.0–9.5 | none in sufficient amounts or at such amounts or at such temperatures as to prevent fish survival or impair the waters for agricultural purposes or any other best usage | natural drainage, agriculture, and industrial water supply |

[a] Ref. 8.
[b] Class B and C waters and marine waters shall be substantially free of pollutants that unduly affect the physical or chemical nature of the bottom; and interfere with the propagation of fish. Class D and SD (marine) are assigned only where a higher water-use class cannot be attained after all appropriate waste-treatment methods are utilized. Any water falling below the standards of quality for a given class shall be considered unsatisfactory for the uses indicated for that class. Waters falling below the standards of quality for Class D, or SD (marine), shall be Class E, or SE (marine), respectively, and considered to be in a nuisance condition.
[c] These Standards do not apply to conditions brought about by natural causes. Waste effluents discharging into public water supply and recreation waters must be effectively disinfected. All sewage-treatment plant effluents shall receive disinfection before discharge to a watercourse or coastal and marine waters. The degree of treatment and disinfection shall be as required by the state pollution-control agency. The minimum average daily flow for seven consecutive days that can be expected to occur once in ten years shall be the minimum flow to which the standards apply.
[d] Phenolic compounds cannot exceed 0.005 mg/L; no odor-producing substances that cause the threshold-odor number to exceed 8 are permitted; radioactivity limits are to be approved by the appropriate state agency, with consideration of possible adverse effects in downstream waters from discharge of radioactive wastes, and limits in a particular watershed are to be resolved when necessary after consultation between states involved.
[e] Trout surface water.
[f] Not a trout surface water.

Table 8. Pennsylvania Raw-Waste Standards for Pulp and Paper Mills[a]

Type of product or process	Kilograms of suspended solids per metric ton of product 3-d average[c]	8-h average	Population equivalent[b] per metric ton of product based on 5-d BOD 3-d average	8-h average
group A				
tissue paper	82.5	880	16.53	20.66
glassine paper	27.5	33	6.20	8.26
parchment paper	44.0	49.5	8.26	12.40
miscellaneous papers	27.5	33	2.07	4.13
flax papers–condenser	412.5	456.5	123.97	144.63
group B (specialty group)				
fiber paper	880	935	82.64	97.11
asbestos paper	137.5	203.5	119.83	144.63
felt paper	231	253	24.79	26.86
insulating paper	2475	2750	134.30	144.63
specialty papers	1100	1320	55.79	66.12
group C (coarse paper)	99	132	14.46	20.66
group D (integrated mills)				
wood preparation	88	110	16.53	20.66
pulp, sulfite	3300	3850	14.46	16.53
pulp, alkaline	330	385	8.26	14.46
pulp, groundwood	126.5	143	33.06	35.12
pulp, deinked unfilled stock	550	715	154.96	206.61
pulp, deinked filled stock	440	550	247.93	330.59
pulp, rag cooking	1540	1705	196.28	206.61
bleaching				
long-fiber stock, multisingle-stage bleaching and short-fiber stock, single-stage bleaching	66	77	1.24	2.48
short-fiber stock, multistage bleaching	181.5	203.5	12.40	14.46
papermaking	110	137.5	30.99	35.12

[a] Ref. 5.
[b] Number of people producing the same amount of waste.
[c] All averages are for periods of consecutive operation.

in values of the reaction rate can result in vastly different oxygen sag curves and correspondingly different degrees of waste treatment required. Considerable cost differences derive from these situations. Industry often decides to overtreat its waste (at greater costs) to protect itself against the unpredictable nature of the stream and conservative decisions of the regulatory agencies.

Stream Sampling. Receiving streams should be sampled under the most critical conditions likely to be encountered, ie, at the lowest streamflow, highest river temperature, and maximum flow of industrial waste. However, it is not possible to sample under these three extreme conditions simultaneously, and the stream is usually sampled under the most critical conditions existing, generally during the hot, dry summer period. The findings are then projected to critical design conditions.

A receiving water should be sampled as follows: as frequently as possible in consecutive stretches below the waste entrance; at least every 400 m and less if the stream character changes suddenly; below and above every main source of contamination or

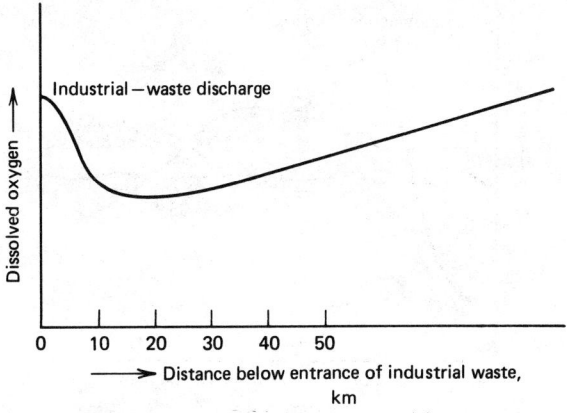

Figure 3. Oxygen sag curve (5).

dilution; at appropriate sections across the entire stream; at every significant pool or riffle area; and just before daylight and at midday when algae effects in the stream are suspected. Analyses for dissolved oxygen, temperature, BOD, and streamflow should be made at each sampling point. Biological analyses for microscopic algae, protozoa, fish, etc, often yield information on pollution not readily obtained from chemical data.

Volume Reduction. Although reduction of waste volume *per se* may not result in a reduction of contaminants in wastewater, it will reduce the capital expense. Contaminants can be reduced by incorporating more waste matter in the final product, or by dry-collecting waste material rather than washing it into drains. This procedure, however, increases the solid-waste disposal problem and creates, for example, by burning, an air-pollution problem. All environmental impacts must be considered.

Reduction of Waste Concentration. Reduction of pollution in a wastewater effluent stream can be accomplished by changing mechanics or operation of the process, eliminating a step in (or an entire) process, or substituting for the raw material or a process chemical. Segregating small volumes of concentrated wastes creates a large volume of relatively weak waste. Both may be treated differently and sometimes more economically and effectively than when combined. However, this step is only recommended when either production costs or total pollution is lowered.

Neutralization. The adjustment of pH may be required to protect sewer lines and plant structures from corrosion, reduce the demand for coagulation chemicals, provide optimum bacterial activity in biological treatment, prevent odors, and protect receiving waters from detrimental effects of highly alkaline or acid wastes. The pH can be adjusted by adding acid (usually sulfuric) or base (usually sodium hydroxide or lime), mixing acid wastes with alkaline wastes, acidifying with flue gas or compressed CO_2, acidifying with submerged combustion (burning of fuel under water), and passing acid wastes through limestone beds.

Neutralization should be applied where it is justified economically, is required by receiving-water quality standards, or eliminates a nuisance to the environment.

Equalization and Proportioning. When industrial wastes are introduced into a municipal (domestic) sewage-treatment system in a uniform volume and strength in accordance with sewage flows, more effective combined treatment results. The industrial wastes are held or retained for a period of time sufficient to produce at the

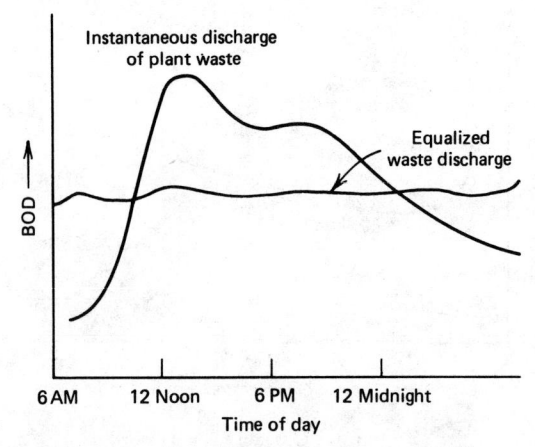

Figure 4. Equalization of waste streams (5).

discharge end of the basin a waste more uniform in quality (see Fig. 4). The optimum retention time depends upon the various industrial manufacturing processes. For example, the process shown in Figure 4 resulting in a high BOD waste at 12 noon and once every 24 h may require a 24-h holding period and mixing during retention. This may be accomplished by proper distribution and baffling, mechanical agitation, aeration, or a combination of all three.

The plant waste may also be discharged in proportion to the sewage flow at the treatment plant or to the river flow into which the waste flows. Proportioning wastes assures that neither treatment plants nor receiving streams are overloaded with slug loads of industrial wastes. Proportioning can be accomplished by pumping or discharging industrial plant wastes at previously planned or instantly signaled rates. The system requires flow measurement, signal transmission, and valve responses at waste pumping devices (see Fig. 5).

Proportioning of plant wastes can also prevent overloading of a particular contaminant in the system at any one time. This may require reduction of industrial plant waste flows when municipal flows and contaminants are high.

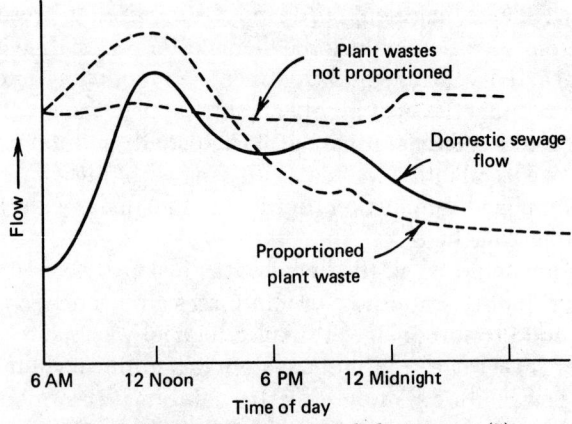

Figure 5. Flow of proportional plant wastes (5).

Removal of Suspended Solids. Many industrial wastes, such as those from cannery, pulp and paper, and tannery industries contain suspended matter. In general, these solids are removed by differential sedimentation, flotation (qv), or fine screening. Since suspended matter in wastes is generally more dense than water, it can be separated by providing ample detention time for settling. In some instances, these solids contain sufficient surface area relative to their density that they may stratify, be buoyed up, and subsequently floated to the surface and skimmed. Air dissolved in wastewater under presssure and then released at atmospheric pressures is generally used to effect flotation of suspended solids. Suspended solids are normally present at 1–20 vol %.

Sedimentation or flotation produces a sludge with 2–12 wt % solids. These two methods generally reduce the solids by ca 60–90%. Retention time ranges from 1 to 3 h and may even be as long as 24 h or more, whereas only 20–30 min are necessary when removing solids by flotation. A typical sedimentation-removal curve is shown in Figure 6.

Screening is an economical and effective means of rapid separation of relatively large-size suspended solids from the remaining waste.

Removal of Colloidal Solids. These solids are 1–200 μm and are small enough to exhibit stability by virtue of the slight residual electrical charge (generally negative), but large enough to interfere with the passage of light and therefore cause turbidity. They do not settle unless destabilized, coagulated, and flocculated into larger masses with sufficiently higher densities than water. The coagulants are normally electrolytes with a strong positive charge when dissolved in water, eg, iron or aluminum salts. The coagulants appear to react simultaneously with the negative hydroxyl ions and the negative colloidal impurities in the wastewaters. Both charges are reduced substantially by the reaction. In many instances, colloidal solids may represent from one third to one half of the total oxygen demand of an industrial wastewater. Little benefit is gained by employing chemical coagulation to remove solids that are not truly colloidal; larger ones could be removed by settling whereas smaller ones are not affected (see Flocculating agents).

Removal of Dissolved Inorganic Material. Salts of nitrates, phosphates, carbonates, chlorides, or sulfates are often contained in industrial water. They are seldom removed before discharge to the receiving waters because it is expensive, until recently it has not been considered essential, and it is the downstream user's responsibility to remove them. However, unless the downstream user finds it absolutely essential to remove all inorganic salts from the water, he or she fails to do so. Consequently, in general,

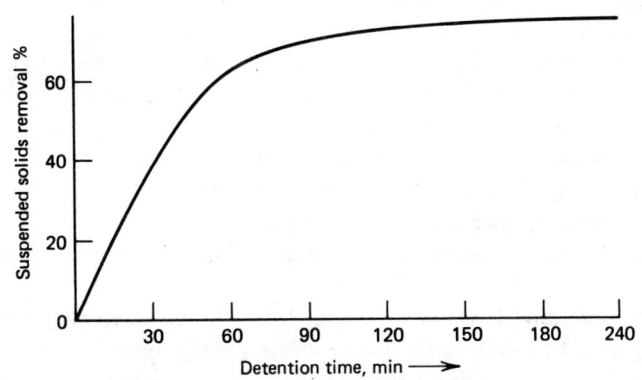

Figure 6. Sedimentation-removal curve (5).

each subsequent downstream use tends to add more of these salts to the streams. Inorganic salts can cause diseases and discomforts, affect an industrial product, cause scale in pipelines and other equipment, accelerate algal growth, increase the hardness of water, and enhance metal corrosion. Inorganic salts are removed by ion exchange, selective membranes, evaporation, and certain biological treatments.

Removal of Dissolved Organic Material. Manufacture of textiles, paper, canned goods, and leather, as well as food products such as milk, cheese, and meat, yields a certain amount of dissolved organic matter, resulting in a reduction of dissolved oxygen in the watercourses (see Table 2). Because these solids are amenable to bacterial degradation, waste-treatment methods employ biological means. Activated sludge, trickling filtration, rotating biological contactors, and oxidation pond treatment are used. Activated-sludge treatment consists of aerating biological flocculent growths within the industrial wastes. Surface for biological oxidation is created on the flocculent growths. With sufficient detention time for biological reaction, ample supply of oxygen from pure oxygen or air, and a readily degradable organic matter to serve as bacterial food, activated-sludge treatment is very effective. In trickling filtration, a solid medium such as rock, plastic, or glass supplies the surface for biological oxidation. The surface must be resistant to deterioration and relatively lightweight, and allow for sufficient pore or void space for large volumes of industrial waste. Oxygen for biological oxidation derives from the void area in the medium. As the growth builds up on the medium, some sloughs off and is removed by clarification. After filtration, some seed material is usually recycled to keep the filter bed active. Both activated sludge and trickling filtration can remove 80–90% BOD. Rotating circular contactors, half immersed in wastewater at all times, build biological growths on their surfaces, oxidize organic matter, and slough off stabilized sludge. Oxidation pond treatment is less effective; it requires a large land area and a long detention time, and emanates odors unless properly designed or aerated. It is, however, not expensive to build and operate. Some ponds develop an algal growth, especially in shallow basins, which assists in providing oxygen to the bacteria as well as in removing inorganic nutrients. A reduction of 50–75% BOD can usually be obtained with these waters with a 1–3-d detention time.

Disposal of Sludge Solids. Screening, settling, flotation, biological oxidation, and coagulation all produce various quantities and types of sludges, in addition to an improved effluent. Often the immediate and continual removal and subsequent effective ultimate disposal of these sludges controls the efficiency of the entire treatment process. If these sludges tend to build up in treatment basins and are not removed, they decompose, rise, and find their way into the treatment plant effluent.

Sludge usually still contains 90–95% H_2O, which has to be removed. Filtration and sand drying beds are popular methods for concentration, although centrifugation and lagooning are also used. These processes give sludges with 15–35 wt % solids. In this form, they can be readily disposed.

Ultimate disposal can be by incineration, land fill, high temperature high pressure oxidation, fertilization of farmlands, or barging or pumping to the sea. These methods, however, often cause further pollution problems.

Sludge solids are reused in a few industries, such as in paper and tannery plants.

Joint Treatment with Municipal Sewage. It is often advantageous for an industrial manufacturing plant to deliver its entire wastewater, under control, to a municipal sewage system. Certain state and Federal legislation encourages industries to combine their wastes with those of nearby municipalities. Incentives are the following advantages to the industries:

Responsibility is placed with one owner, while at the same time the cooperative spirit between industry and municipality is increased, particularly if sharing of costs is mutually satisfactory.

Only one chief operator is required, whose sole obligation is the management of the treatment plant; he or she does not have to perform the miscellaneous duties often given to the industrial employee in charge of waste disposal, and the chances of mismanagement and neglect which may result if industrial production people operate waste-treatment plants are eliminated.

Since the operators of such a large treatment plant usually receive higher pay than the operators of municipal plants, better trained people are usually available.

Construction costs are less for a single plant than for two or more. Furthermore, municipalities can apply for state and Federal aid, but private industry cannot.

The land required for plant construction and for disposal of waste products is obtained more easily by the municipality.

Operating costs are lower, since more waste is treated at a lower rate per unit of volume.

Some wastes may add valuable nutrients for biological activity to counteract other industrial wastes that are nutrient deficient. Thus, bacteria in the sewage are added to organic industrial wastes as seeding material. These microorganisms are vital to biological treatment when the necessary BOD reduction exceeds ca 70%. Similarly, acids from one industry may help neutralize alkaline wastes from another industry.

The treatment of all wastewater generated by the community in a municipal plant or plants enables the municipality to assure a uniform treatment which could be the best obtainable with modern techniques.

Joint treatment gives the municipality full control of the river's resources and permits it to use the capacity of the river to the best advantage for the public at large with close monitoring of effluent quality.

Among the many problems arising from combined treatment, the most important is the character of the industrial wastewater reaching the disposal plant. Equalization and regulation of discharge of industrial wastes are sometimes necessary to prevent rapid change in the environmental conditions of the bacteria and other organisms which act as purifying agents, to ensure ample chemical dosage in coagulating basins and adequate chlorination to kill harmful bacteria before the effluent is discharged to a stream.

Since most sewage plants employ some form of biological treatment, it is essential for satisfactory operation that the waste mixture be homogenous in composition, uniform in flow rate, and free from sudden dumpings (shock loads) of deleterious industrial wastes; not highly loaded with suspended matter; free of excessive acidity or alkalinity and low in content of chemicals that precipitate on neutralization or oxidation; practically free of antiseptic materials and traces of toxic metals; low in potential sources of high BOD, such as carbohydrates, sugar, hydrolyzed starch, and cellulose; and low in oil and grease content.

Industrial contaminants, their limiting concentrations, and pretreatments are given in Table 9.

Municipal ordinances protect the treatment plant from violation, and sewer-rental charges defray the costs of construction and operation resulting from acceptance of the industrial wastes.

Table 9. Industrial Contaminants and Their Discharge into Municipal Sewage Systems[a]

Contaminating factor	Concentrations accepted in municipal sewer systems	Reason for limitation	Pretreatment, if contaminant is excessive
BOD, 20°C	300 ppm	exerts a disproportionately high percentage of oxygen-demanding organic matter to municipal wastewater	change in industrial manufacturing process, equalization, biological pretreatment
color	visible in dilutions of 4 parts sewage to 1 part industrial waste	color is normally not removed by domestic sewage-treatment plants, appears in the combined, treated effluent, readily detected and visually undesirable	change in industrial manufacturing process, chemical pretreatment to remove color, equalization or proportioning
suspended solids	350 ppm	overloads disproportionately normal domestic sewage-treatment plants	change in industrial manufacturing process, equalization, sedimentation pretreatment
pH	5.5–9	corrosion of sewers and treatment plant equipment causes a diminution or a malfunctioning of biological treatment units	equalization, neutralization, change in industrial manufacturing process
grease	100 ppm	interferes with plant operating equipment, overloads sludge-handling treatment units	change in industrial manufacturing process, installation of grease traps

heavy metals Cu, Cr Ni, Zn	1 ppm 5 ppm	equalization, chemical and sedimentation pretreatment
nonorganic matter and toxic chemicals	none to be toxic to bacteria serving the treatment plant or humans or animals near sewage plant	change of industrial plant process, advanced pretreatment techniques
flammable liquids, foaming agents, various solids	none in quantities that cause a hazard to the environment or a nuisance to plant operation	removal by process change or physical means, such as screening
temperature	65.5°C	change in industrial process, use of cooling water systems
turbulence	none resulting from direct connections or faulty sewer construction	separate sewer
refractory organic matter	none	change in industrial manufacturing process, carbon adsorption pretreatment
refractory mineral matter	boron, 0.7 ppm NaCl, 1000 ppm	change in industrial manufacturing process, membrane separation or distillation of wastewater

	inhibits biological action in municipal sewer units	
	possible toxicity, interference with operations	
	nuisance and interference with operation	
	hastens corrosion, drives out dissolved oxygen, volatilizes hazardous gases such as H_2S	
	lowers capacity of sewers	
	contaminates sewage-plant effluent for downstream reuse	
	contaminates sewage-plant effluent for irrigation waters	

[a] Ref. 5.

Although a combination of municipal and industrial waste treatment is most desirable, it is difficult to arrange because of frequent lack of cooperation and compatibility between the parties. Systems engineering combines political, scientific, and cost factors toward optimum solutions to treatment problems. Accessibility to sewers and proximity to treatment plants and receiving rivers is essential for industrial use of public sewers.

If, in the past, a sewer system has accepted all types of industrial wastes regardless of volume or character, this practice may be expected to continue. On the other hand, if industrial wastes and municipal sewage have been treated separately, it might be difficult to change this situation. Any change in policy requires considerable effort, education, and good will of all parties. The engineer must recognize that even with proper justification, it may be impossible to overcome long-time customs. Good relationship between industry and municipal officials is essential.

Sewer-Service Charge. Municipalities are governed by elected officials with different views about industrial patronage or subsidy for local industry. Some prefer to accept wastes at a minimum charge, a flat fee, or a water-use basis. Others believe that industry should either solve its own environmental problems or pay a sewer-service charge which covers capital and operating costs. In general, industry accepts a reasonable charge, but often objects to comprehensive and complicated charges, which may lead to conflicts.

Sewage Treatment. A secondary biological-treatment plant of adequate size can be utilized to treat a readily decomposable industrial waste containing organic matter from, eg, dairies, canneries, slaughterhouses, and tanneries. All these wastes may contain contaminants which can interfere with effective treatment when combined with domestic sewage. For example, the pH of dairy wastes can affect biological oxidation, cannery effluents contain strongly alkaline waste which interferes with biological oxidation, slaughterhouse wastes contain grease and blood, and tannery wastes chromium, sulfides, and lime, which are incompatible with sewage treatment. Proper pretreatment and plant operation, however, can remedy these problems.

The treatment of the wastes from a paper mill requires equipment that removes finely divided suspended solids. Such waste could not be treated in a municipal high rate trickling-filter plant designed primarily for BOD removal. Metal-plating waste creates similar problems. A complete analysis of the industrial waste must be carried out to ascertain its compatibility for various treatments.

Receiving Stream Water Quality. A stream that must be maintained at a high water quality requires the maximum onshore waste treatment, ie, a minimum of secondary treatment. Conventional biological treatment systems often do not adequately remove sufficient amounts of contaminants, and a specific treatment such as chemical precipitation followed by adsorption on activated carbon may be more efficient than a secondary-type trickling-filter plant. Occasionally, drastic measures are needed, such as a change in raw material.

Sometimes a legal requirement demands a so-called complete waste treatment even if the receiving stream shows little deterioration. In this situation, industry should negotiate with the regulatory agency in order to operate at optimum economic efficiency.

Volume Ratio of Industrial to Municipal Waste. A small volume of industrial waste can usually be assimilated in a municipal sewage-treatment system regardless of its contaminants. In general, municipal-plant operators tend to accept small volumes of any industrial wastes. If the industrial waste volumes are large, however, a separate plant is needed.

Costs. Comparison of system costs on the basis of total capital expenditures and lowest capital outlay can be misleading and even erroneous over the long term since they overlook the annual cost of operation and maintenance. Economic alternatives must be based upon net annual costs which include operation, maintenance, and capital amortization. In a study for the Department of the Interior, some of the principal benefits of pollution abatement were evaluated, including recreational facilities, land use, water withdrawal, waste treatment, and in-place water uses. The overall stream water quality can be assessed with a Stream-Pollution Index.

Solid Wastes

The quantities involved are overwhelming, and the problem of solid-waste disposal is most serious (9). Only a few methods are available; they are discussed below. In recent years, reclamation of materials has been considered as one means of solving the pollution problem.

Incineration. Municipal and industrial incinerators account for 30–50% of the total trash disposal within the United States (10). Incineration provides complete combustion of the feed which is made up of a wide spectrum of materials. Waste materials range from wet garbage with heating values up to 4.65 MJ/kg (2000 Btu/lb) to plastics such as polystyrene with heating values up to 44 MJ/kg (19,000 Btu/lb). It is difficult to provide the correct amount of air for combustion of such waste mixtures in a large incinerator.

Incinerators should be designed to operate with excess air, usually on the order of 100–150% above calculated requirements. Underfire air should be kept to a minimum to exclude particulate matter from the waste-gas stream, whereas overfire air provides ample oxygen and turbulence in the combustion space above the fuel. The temperature should be 760–982°C in order to keep smoke formation and odor to a minimum. Sufficient combustion volume provides residence time for the burn-out of all flying particulate matter. The average heat release is 18.6 MJ/(m^3·h) [30 Btu/(ft^3·min)]. A secondary chamber or zone should be provided in every incinerator and is required by most municipal and state codes. The gas residence time in the incinerator should be 1–2 s. If the incinerator has a grate, low loading rates per square meter of grate surface should be observed, even in forced-draft incinerators (10). The rate should not exceed 293 kg/(m^3·h) (see also Incinerators; Furnaces).

Composting. Composting is a means for reducing the volume of solid waste and returning a partly usable product to the industrial plant or community (11–12).

Composting is accelerated biological degradation of organic waste resulting in a mature compost containing mainly nitrogen, phosphorus, and potassium. It has limited nutritive value for vegetation, but is a conditioner for layers of soil and replaces certain essential nutrients.

Composting plants have been built throughout the world but only very few have been installed in the United States because land is expensive. Furthermore, the odor associated with the degradation of waste creates a problem, and the market for the products is limited. In Europe, however, there is a tradition of waste recovery, whether it be heat, water, or solids. A Danish corporation has been building composting plants with capacities of 5–300 t/d for over 30 yr and has installed these throughout the world except in the Western Hemisphere. In this system, a rotary cylinder similar to a cement kiln is fed waste material at one end and pathogenically pure compost is delivered at the other.

Glass and metal particles are removed from municipal or industrial trash by special techniques and pretreatment of the dry waste. Shredding or granulation ensures homogeneity and controls moisture content. Pretreatment accelerates the decomposition rate at temperatures where pathogenic organisms are destroyed. Once the waste material has been added to the kiln, air is added at low pressure and in controlled amounts over the length of the cylinder. In this manner, an environment is created where the action of aerobic microorganisms ensures rapid decomposition of the waste. Normal operating temperatures are ca 60°C and are spontaneously developed without additional heat. The weight reduction from the feed to the final compost is 30–40%. The resulting material can be used as fertilizer or soil conditioner and is usually applied during the autumn and winter months to the ground at rates of 12–49 t/ha.

Sanitary Landfill. Perhaps the most widely used solid-waste disposal method today is the "sanitary landfill," a misnomer (13–14). Although landfill operations do not fill land and are not sanitary, they can be beneficial. Swampy areas near industrial complexes can be filled in and leveled using municipal trash, thus recovering valuable land for industrial use. The earth is removed with dredges or bulldozers and trash trucks dump the refuse into the cavity. The waste is distributed over the area, compacted, and covered with a layer of soil ca 15 cm thick; several layers can be deposited.

Compacted materials resist erosion and thus provide a better landfill than non-compacted materials. Most landfill is heterogenous, although homogenous waste is more desirable. Wastes that are not biodegradable, such as glass, plastics, and metals, do not decompose and will always remain in their original form.

Water-soluble components of waste that can be leached by rain or surface water will ultimately find their way into nearby springs and streams. Toxic contamination presents a serious problem. Another serious problem is the possibility of harboring rodents and vermin. This can be prevented by immediately covering the waste with soil. Shredding before filling also prevents rodent infestation and odors.

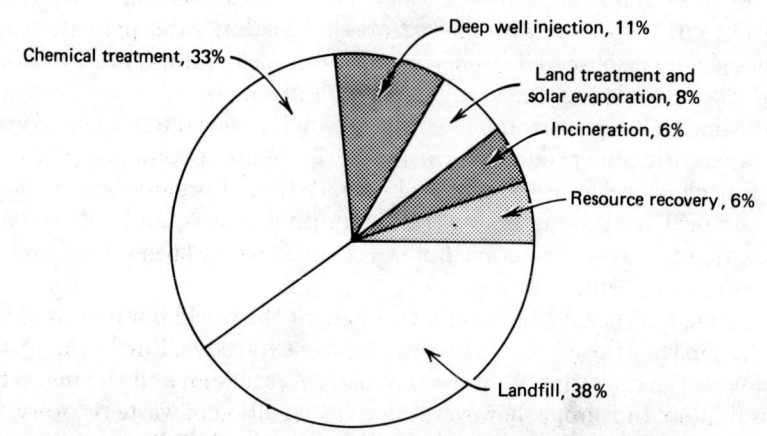

Figure 7. Waste handled by commercial hazardous-waste service companies, percentage according to treatment and disposal methods (1980) (2).

Table 10. Hazardous Waste Management[a]

Type of waste management	Waste	Price 1981, $
land fill	drum	35–50/208-L drum[b]
	bulk	55–83/t
land treatment	all types	5.29–23.8/m^3 [c]
incineration	relatively clean liquids, high fuel value	13.2–52.9/m^3
	liquids	52.8–238/m^3
	solids, highly toxic liquids	396–793/m^3
chemical treatment	acids and alkalies	2.11–925/m^3
	cyanides, heavy metals, highly toxic wastes	66–793/m^3
resource recovery	all types	66–264/m^3
deep-well injection	oily wastewaters	15.9–39.6/m^3
	toxic rinse waters	132–264/m^3
transportation		10.3/(t·km)

[a] Ref. 2.
[b] 208 L (55 gal) drum.
[c] To convert m^3 to gal, multiply by 264.

Landfill can be efficient and economical when proper engineering techniques are applied. The cost of sanitary land fill in 1978 was $2–5 per metric ton for small operations and $0.75–2.50 per metric ton for large operations. Since then, costs have been fluctuating.

Recycling of Products and Energy. Systems involving recovery of resources from solid wastes depend primarily upon economics and local situations such as characteristics of refuse and land available for filling (15) (see also Fuels from waste). The following overall systems are used: incineration with recovery of materials from the ash residue, with heat recovery to produce industrial steam, or both; incineration with heat recovery used to generate electric power; pyrolysis, with recovery of oil, char, and inorganic materials; composting producing humus and inorganic materials; material recovery (paper, aluminum, ferrous metals, and glass) by separation of mixed refuse into marketable components; and recovery of organic material in public utility boiler furnaces as supplemental fuel.

In addition, pilot-plant data indicate that classification to separate and recover glass and metals followed by anaerobic digestion to generate methane as energy source and ultimate disposal of the residual solids may be feasible.

Hazardous Wastes

After passage of the Resource Conservation and Recovery Act (RCRA) in 1976, an effort was made to identify hazardous chemicals and devise a method for their safe disposal. The Act provides that a priority list of chemicals be established by the EPA, including concentration limits. The present list contains 123 chemicals; limits have been established for 65. Since most industries producing these wastes discharge relatively small quantities, they contract for treatment and disposal (2,16–17). The treatment depends upon the type of chemical and the system selected by the commercial service company (see Fig. 7). Prices are given in Table 10.

BIBLIOGRAPHY

"Wastes, Industrial" in *ECT* 1st ed., Vol. 14, pp. 896–914, by H. E. Orford, W. B. Snow, and W. A. Parsons, Rutgers University; "Wastes, Industrial" in *ECT* 2nd ed., Vol. 21, pp. 625–651, by R. D. Ross, Thermal Research & Engineering Corp.

1. H. C. Perkins, *Air Pollution*, McGraw-Hill Corp., New York, 1974.
2. L. D. Helsing, *Protecting the Environment*, Institute of Chemical Waste Management, Booz, Allen, and Hamilton, Inc., Bethesda, Md.
3. *J. Water Pollut. Control Fed.* **54,** 89 (March 1982).
4. *Chem. Eng.*, 121 (Nov. 15, 1982).
5. N. L. Nemerow, *Industrial Water Pollution*, Addison-Wesley Corp., Reading, Mass., 1978.
6. N. I. Sax, *Dangerous Properties of Industrial Materials*, 5th ed., Van Nostrand Reinhold Co., New York, 1979, Section I, pp. 15–24.
7. A. C. Stern, ed., *Air Pollution*, Vols. I–II, Academic Press, New York, 1962.
8. I. Grossman, *J. Sanit. Eng. Div. Am. Soc. Civ. Eng.* **94,** 13 (1968).
9. G. Tchobanoglous, *Solid Wastes*, McGraw-Hill Corp., New York, 1977.
10. N. L. Nemerow, *Industrial Solid Wastes*, Ballinger Publishing Co., Cambridge, Mass., 1983.
11. *Composting—Theory and Practice for City, Industry, and Farm*, The J. G. Press, Emmaus, Pa., 1981.
12. C. G. Golueke, *Composting*, Rodale Press, Inc., Emmaus, Pa., 1972.
13. *Land Cultivation of Industrial and Municipal Solid Wastes*, EPA 600/2-78-1402, U.S. Environmental Agency, Washington, D.C., 1978.
14. *Sanitary Landfills*, National Center for Resource Recovery, Inc., Lexicon Books, D. C. Heath and Co., Lexington, Mass., 1974.
15. P. A. Vesiland and A. E. Rimer, *Unit Operations in Resource Recovery Engineering*, Prentice Hall Co., Englewood, N.J., 1981.
16. T. Bonner and co-workers, *Hazardous Waste Incineration Engineering*, Noyes Data Corp., Park Ridge, N.J., 1981.
17. A. C. Scurlock and co-workers, eds., *Incineration in Hazardous Waste Management*, U.S. Environmental Protection Agency, Washington, D.C., 1975.

General References

A. G. Calley, C. F. Forster, and D. A. Stafford, eds., *Treatment of Industrial Effluents*, John Wiley & Sons, Inc., New York, 1976.
R. A. Conway and R. D. Ross, *Handbook of Industrial Waste Disposal*, Van Nostrand Reinhold Co., New York, 1980.
C. E. Adams and W. W. Eckenfelder, eds., *Process Design Techniques for Industrial Waste Treatment*, Environmental Press, Nashville, Tenn., 1974.
B. Koziorowski and J. Kucharski, *Industrial Waste Disposal*, Pergamon Press, New York, 1972.
E. J. Middlebrooks, *Industrial Pollution Control. Vol. I. Agro-Industries*, John Wiley & Sons, Inc., New York, 1979.
Report on the Council on Environmental Quality on Ocean Dumping, U.S. Government Printing Office, Washington, D.C., 1970.
C. Warren, *Biology and Water Pollution*, 1971.
C. G. Wilber, *The Biological Aspects of Water Pollution*, Charles C Thomas Publishing Co., Springfield, Ill., 1969.

NELSON L. NEMEROW
University of Miami

WATER

    Sources and utilization, 257
    Properties, 276
    Pollution, 295
    Analysis, 315
    Supply and desalination, 327
    Industrial water treatment, 367
    Municipal water treatment, 385
    Sewage, 407
    Reuse, 420
    Treatment of swimming pools, spas, and hot tubs, 427

SOURCES AND UTILIZATION

Development of water resources and exploration for new sources of supply represent man's efforts to accommodate progressively enlarging domestic, agricultural, and industrial demands for water. The world's water continually circulates among the oceans, atmosphere, and land masses. The upper limits of supply obtainable from these three environments are governed by the nature of occurrence and by the movement of water within and among them. The quality of water is also governed by various chemical and physical interactions within and among these domains.

Sources

Hydrologic Cycle. Hydrology is the science that deals with the earth's water, its occurrence in nature, and its chemical reactions and physical relations with other substances of the earth, including mineral, animal, and plant matter. A fundamental perception of hydrology is the continuing movement of water from the oceans to the atmosphere, to the land, and back to the oceans. The heat of the sun and the forces of gravity provide the energy for this circulation. Water evaporated from the surfaces of lakes, reservoirs, oceans, and other water bodies, from wet ground, and from plants rises into the atmosphere as water vapor. When the airborne vapor condenses, it reverts from gas to liquid and falls as rain or as snow or some other form of ice.

Most rain or meltwater reaching the land surface flows overland as surface water to rills and streamlets, which direct the water to streams. The latter conduct it to lakes, bays, and the oceans (Fig. 1). Part of the water falling on the land surface, which is rarely >40%, infiltrates into the unsaturated zone or zone of aeration, which consists of the soil and underlying shallow rocks lying above the water table. Rates of infiltration vary temporally at a given locality because of changing antecedent moisture conditions and irregular rainfall frequency and amounts; rates vary geographically as a result of terrain differences, eg, soil texture and land slope, and uneven distribution of rainfall. Water reaching the unsaturated zone may evaporate back to the atmosphere directly or through plants by transpiration. Evaporation and transpiration from vegetated areas commonly occur together and are termed evapotranspiration. A portion of downward percolating moisture adds to underlying groundwater in the zone of saturation. The water then moves slowly downgradient and laterally with the mass of groundwater, discharging into streams or other outlets. Another portion may be dis-

258   WATER (SOURCES AND UTILIZATION)

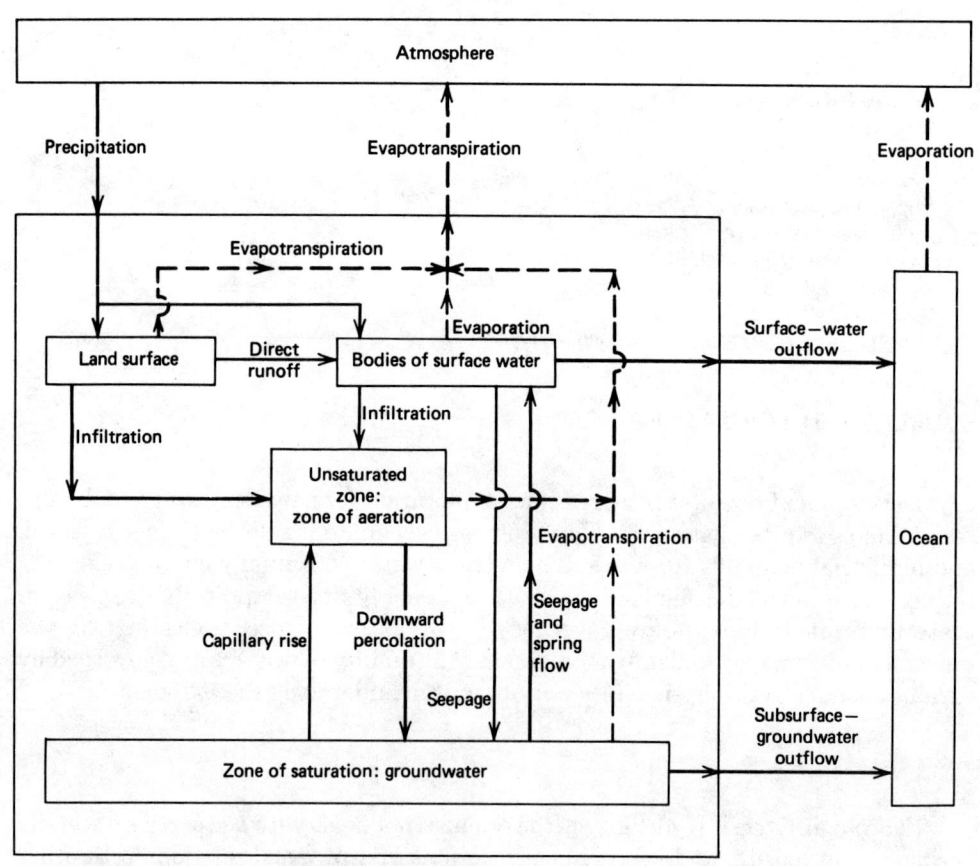

Figure 1. Components of the hydrologic cycle and directions of flow of water under natural conditions. Heavy lines represent principal flow paths; thin lines, minor flow paths; solid lines, flow of liquid water; and dashed lines, flow of gaseous water (1).

charged to the atmosphere by evapotranspiration where deep-rooted plants tap the zone of saturation.

In these ways, transfer of water proceeds endlessly among the earth's environments, although nonuniformly in space and time, and this orderly process is termed the hydrologic cycle. Figure 1 shows the components of moisture storage and flow within the cycle (1). The diagram assumes natural conditions of water flow, ie, those unaffected by development or other artificial disturbances. Movement of water in nature is much more complicated than implied by Figure 1, since it is affected, for example, by human interferences.

World's Water Supply. From a practical standpoint, the total amount of water on earth is fixed. Most of the water that is used is recycled or reused. However, regional and local water problems, eg, inadequate supply and deteriorating quality, are growing. Internationally since the early 1950s, the acquisition of water information and assistance in water management has been under the aegis of various organizations, including principally the United Nations, the International Monetary Fund, the World Bank, and programs conducted by individual countries, such as the Agency for International Development (AID) of the United States. These organizations address fundamental water and sanitation requirements and provide assistance in technical training, institutional design, and water-resources inventory, development, and management.

Principal water environments of the earth are listed in Table 1 (2–3). The table includes estimated volumes of water in storage and in transit within each environment and the estimated average time of residence of water during its passage through each environment. These average values oversimplify actual residence-time conditions, because many different water storage and flow phenomena operate in each environment and rates of migration of water in nature vary greatly. Residence-time estimates are gross values and indicate only the approximate time that water remains in each segment of the hydrologic cycle.

A large amount of water falls as precipitation each year (ca 5.17×10^5 km^3 (1.24×10^5 mi^3) worldwide). The atmosphere contains only ca 1.3×10^4 (3120 mi^3) of water (4). Although the average residence time is brief, atmospheric water is a primary link in the water cycle and the only continuous source of freshwater supply.

Precipitation is measured by networks of gauging stations and by remote sensing from aircraft and earth satellites. There is more data for land areas than for oceans, although ca 80% of the precipitation falls on the oceans. Synoptic tracking and measurement of atmospheric moisture distribution and air-mass migration, made possible by new satellite technology, had aided in the extrapolation of data on rainfall and other weather phenomena. Worldwide mean annual precipitation is ca 86 cm, but the annual and seasonal rates vary greatly from place to place (5). Some humid land areas receive >10 m of precipitation annually; some arid regions receive only a few millimeters a year. In the conterminous United States, ca 1 m falls each year, which is equal to ca 9600 km^3 (2304 mi^3) of water.

Most of the world's water supplies are derived from rivers. Many nations have installed stream-flow-gauging networks to measure the flow of principal streams, but for much of the world, the network of stations is sparse. The estimated total flow or runoff from all rivers, gauged and ungauged, is ca 3.83×10^4 km^3 yearly or 105 km^3 daily (9200 mi^3 yearly or 25.2 mi^3 daily) (6). The Amazon river contributes nearly 20% of the world's river discharge to oceans. The total amount of water stored in all river channels is only ca 1280 km^3 (308 mi^3). Total channel storage worldwide is equal to ca 2 wk of the river flow worldwide. The brief residence time in channels explains the

Table 1. Estimated Volumes of Water in Storage and Average Time of Residence in the Earth's Environments[a]

Environment	Volume, km^3 [b]	Average residence time
atmospheric water	1.30×10^4	8–10 d
oceans and open seas	1.37×10^9	>4000 yr
freshwater lakes and reservoirs	1.25×10^5	from days to years
saline lakes and inland seas	1.04×10^5	
river channels	1700	2 wk
swamps and marshes	3600	years
biological water	700	1 wk
moisture in soil and unsaturated zone (zone of aeration)	6.5×10^4	2 wk to 1 yr
groundwater	$(4–60) \times 10^6$	from days to tens of thousands of years
frozen water (glaciers and ice caps)	3.0×10^7	tens to thousands of years

[a] Refs. 2–3.
[b] To convert km^3 to mi^3, divide by 4.16.

need for dams and reservoirs which expand storage capacity, extend residence time, and improve continuity of flow. These engineering works in conjunction with generally perennial groundwater inflow to stream channels and intermittent replenishment by surface runoff make possible the use of streams as sources of continuous, principal water supplies.

Swamps and marshes worldwide contain only a small volume of the world's water, ie, ca 3600 km^3 (864 mi^3). Movement of water in swamps and marshes commonly is sluggish. Though of small significance as a direct source of human water supply, swamps and marshes are key ecological settings for plant and animal life, which indirectly bears on human food supply.

Water stored in and passing through plants and animals (biological water) amounts to only ca 700 km^3 (168 mi^3) worldwide. Although the total quantity of water contained in plant and animal tissues is small, the amount discharged worldwide from vegetation by transpiration and evaporation to the atmosphere is enormous (see Table 2).

Water stored in the unsaturated zone or zone of aeration, ie, the shallow subsurface zone above the water table in which the soil and rocks are only partially saturated with water, may amount to 6.5×10^4 km^3 (1.56×10^4 mi^3). Overall, water content of the unsaturated zone generally remains more or less constant compared to river-channel storage, but the water content of shallow soil may fluctuate widely with varying atmospheric conditions, eg, temperature, wind, moisture gradient, precipitation rates, etc, that influence soil moisture levels. Depleted soil moisture is characteristic of drought.

Groundwater is a large and important component of most hydrologic systems. It is the sustaining source of stream flow and an ubiquitous source of water for wells and springs. Estimates of the total amount of groundwater in storage range widely: $(4–60) \times 10^6$ km^3 (96×10^4 and 1400×10^6 mi^3). The great breadth of estimates is attributable to different assumptions as to the average porosity of earth materials and the thickness of water-bearing zones in the earth's crust. Even the most conservative computation yields an amount of water exceeding the combined storage in rivers, lakes, reservoirs, swamps, marshes, and the atmosphere. For much of the world, present levels of groundwater development are small compared to potential supplies. However, concentrated pumpage has generated economic, physical, and chemical problems of groundwater development and use in parts of some regions.

Table 2. Estimated Amount of Irrigation Water Applied and Consumed in 1970, km^3 (mi^3) [a]

Continent	Estimated amount of water applied	Estimated consumptive loss through evaporation — Loss	Percent of amount applied
Asia	1400 (336)	1100 (264)	79
North America	208 (50)	129 (31)	62
Europe	125 (30)	83 (20)	67
Africa	108 (26)	92 (22)	82
South America	54 (13)	46 (11)	82
Australia and Oceania	13 (3.1)	10 (2.4)	77
Total	*1909 (458)*	*1460 (350)*	*76*

[a] Ref. 7.

Ordinarily, the residence time for water migrating through groundwater reservoirs is extremely long. Although rapidly moving shallow groundwater may remain underground only a few minutes, hours, or days before discharging to nearby surface outlets, most groundwater moves slowly, ie, at rates of a few meters to a few hundred meters per year, and may remain in groundwater reservoirs for centuries or millenia before discharging to streams or other outlets. Distribution and movement of groundwater are controlled by geologic conditions.

Oceans and open seas, containing ca 1.32×10^9 km^3 (3.17×10^8 mi^3) of saline water, constitute the principal water reservoir of the earth (4). The ocean basins have an average depth of ca 3.65 km and contain ca 97% of all the water. Owing to the extensive interface between ocean surfaces and the atmosphere (the surface of the oceans amounts to 80% of the earth's surface), total annual evaporation to the atmosphere is ca 4×10^5 km^3 (9.6×10^4 mi^3). By comparison, evaporation from land areas and inland water bodies totals only ca 7×10^4 km^3 (1.68×10^4 mi^3). The oceans also serve as natural sinks for the storage and distribution of heat energy and are important to the maintenance of the environment of the biosphere.

Freshwater lakes contain ca 1.25×10^5 (3×10^4 mi^3) of water (4,8). Saline lakes contain slightly less. Their combined volume is ca 1.04×10^5 km^3 (2.5×10^4 mi^3). Large lakes have active roles in regional hydrologic systems through water capture, storage, flow regulation, and evaporation of water from their surfaces to the atmosphere. The many small lakes and ponds collectively hold only a minor amount of the world's total water supply. Lake Baikal in the USSR is the world's largest and deepest body of freshwater. It contains 2.63×10^4 km^3 (6300 mi^3) of water, nearly 1250 km^3 (300 mi^3) more water than the five Great Lakes combined. Although the combined surface area of the Great Lakes is larger, their average depth is considerably less than that of Lake Baikal. The Caspian Sea is the largest saline lake, containing ca 8×10^4 km^3 (1.9×10^4 mi^3) of water, equal to 75% of the total volume of all saline lakes. The Great Salt Lake of the western United States is a shallow lake that contains only 47 km^3 (11.3 mi^3) of saline water.

Reservoirs, once filled, function similarly to lakes. Residence time for water flowing through and stored in lakes and reservoirs is highly variable, being subject to natural and man-induced influences, including precipitation, evaporation, natural and regulated stream flow and lake levels, and withdrawals for water supply. The total gross capacity of the world's main reservoirs, including those with dams greater than 150 m in height or with a total volume exceeding 18.3×10^6 m^3 (20×10^6 yd^3), is ca 2500 km^3 (600 mi^3) (9). These include the 116 largest dams in the world. The volume of water stored in reservoirs in the United States is ca 450 km^3 (108 mi^3). Worldwide statistics on dams and reservoirs are compiled in ref. 10.

The liquid equivalent of the world's glaciers and ice caps is ca 3×10^7 km^3 (7.2×10^6 mi^3), which is about equal to the estimated volume of groundwater in storage. Extensive bodies of ice in the polar regions influence heat-energy distribution and other important climatic and geophysical conditions worldwide. Though many ice caps are small and thin, the volume of the Greenland ice cap is ca 4.2×10^6 km^3 (1×10^6 mi^3). Antarctica is the largest mass of frozen water in the world and has a volume of ca $(40-47) \times 10^6$ km^3 (($9.6-11.2) \times 10^6$ mi^3) (11). It represents ca 85% of all ice on the earth and the frozen equivalent of ca 64% of all water excluding the oceans.

Mountain glaciers contain only a minor fraction (ca 3.3×10^5 km^3 (8×10^4 mi^3))

of the world's water, but they are significant components of local hydrology and water resources. These mountain and highland snow accumulations during winter months constitute important local sources of snowmelt runoff for municipal, irrigation, and industrial water supplies.

For further information on the world's water supply, see refs. 12–13.

Utilization

Withdrawal uses involve the diversion or withdrawal of water from its source, and in nonwithdrawal uses, the water is used within its natural setting (14). Withdrawal uses require removal of groundwater or surface water from its place of occurrence to a place of use. It includes water intake, which refers to water required by industrial processes, and the comparable agricultural term water requirement, which refers to moisture needed for optimum crop production. Common examples of withdrawal usage include water derived from wells, springs, lakes, and rivers for municipal and industrial supply or for agricultural irrigation. Generation of hydroelectric power, which requires great quantities of river water, is a withdrawal use because the water is diverted through power plants.

Diverted or withdrawn water used more than once by an industrial plant, farm, or other user is termed recycled. Withdrawn water that is not consumed in the course of its use may be returned to streams, lakes, groundwater reservoirs, or other natural sources, and thereby be available for subsequent repeat withdrawal. The sum of successive withdrawals of water in this manner by one or a series of users is the cumulative withdrawal use. Inasmuch as multiple use of water is common, comparison of the magnitude of the total supply available with the total volume utilized has limited significance. Except for water consumed, the opportunity for cyclic reuse within a plant or successive downstream usage is limitless.

The magnitude of consumptive usage of a water source is statistical information of vital management importance because evaporation of water to the atmosphere during its use (cooling-water systems and irrigation are the best examples) or incorporation into a product, eg, canned food, represents water that is not available for recycling. Consumptive usage may constitute a sizable fraction of total usage (see Tables 2 and 3). Evaporation rates and, hence, consumptive rates are typically greatest in hot, arid regions where water supplies commonly are limited.

Nonwithdrawal use refers to in-place use of water in its natural setting for purposes such as navigation, seafood production, fishing, disposition and dilution of wastes and heat, and recreation. Measurement of the amount and value of water used in this manner is subjective and inexact, and data on the magnitude of many of these uses are, therefore, lacking or of limited reliability.

Water Planning and Management

Most decisions relating to water apportionment and management are intricate, involving social, political, technical, and economic issues (16). Modern construction technology and earth-moving capacity enable massive alterations of natural terrain and water conditions: marshes and swamps can be drained, valleys inundated, rivers redirected, and water conveyed over great distances. In addition to familiar water concerns that include adequate supply, satisfactory quality, and dependability, comprehensive water management involves such issues as land use and zoning, recreational and aesthetic elements, aquatic and wildlife habitats, health and property hazards, and many other public interests.

Computer-supported methods of systems analysis and simulation modeling have been devised to supplement traditional empirical and analytical methods for evaluation and resolution of developmental and management questions (17–19). Methods for coping with uncertainties inherent in the available information, in synthesis and projection of data, and in hydrologic and management techniques and methodologies have been incorporated into the analytical process (20).

Extending Water Supplies. *Interbasin Transfer of Water.* Long-distance transfer of water from a river basin with excess water to one with a low supply can alleviate water-supply shortages. However, relatively few large water-transport projects have been undertaken, mainly for economic and political reasons. Also, regions with excess water supply are reluctant to export water that later may be needed internally. The best known water-transfer plans for North America include the Pacific Southwest Water Plan, the Texas Water Plan, and the North American Water and Power Alliance. These three projects would be capable of relocating ca 136×10^9 m^3 (36×10^{12} gal) of water per year from areas of presently excess supply to areas of deficient supply.

Desalination. Desalination involves the extraction of mineral matter from saline or brackish water to produce water low in dissolved solids (21). Although desalination technology is improving in efficiency and cost, the continued relatively high cost of desalinated water has retarded applications. Saltwater sources are used where the need for freshwater is pressing and alternative less expensive sources are unavailable. Disposal of the brine generated by the conversion processes may pose economic and environmental problems. The production capacities of desalting plants in arid regions range from a few cubic meters of output per day to as much as 15×10^4 m^3/d (40×10^6 gpd). Their combined output totals ca 2×10^6 m^3/d (5.25×10^8 gpd). Except for those few localities where energy prices are very low, water produced by these plants costs from ca 22¢/m^3 (85¢/1000 gal) to several times that amount. In the United States, ca 330 plants produce almost 3.8×10^5 m^3/d (10^8 gpd) of desalinated water, 70% of which is used for industrial purposes.

Recycling. Recycling of wastewater is practiced only on a moderate scale, but it has considerable potential (22). Industrial recycling of process and cooling water is increasing. Recirculation of irrigation tail water is common practice. Reclamation of sewage water for irrigation and industrial uses is a well-established practice also, as indicated below (23–27).

Year	Amount of sewage water reclaimed for irrigation and industrial use, in 10^6 m^3/d (10^8 gpd)
1960	2.38 (6.3)
1965	2.54 (6.7)
1970	1.97 (5.2)
1975	2.01 (5.3)
1980	1.78 (4.7)

However, certain factors have limited the use of recycled sewage water, and the preceding figures indicate generally declining usage since 1960. Costs of treatment and recirculation of the water often makes it noncompetitive with conventional sources of water. Psychological resistance inhibits use of treated waste and sewage water for public consumption, even in regions of serious water shortage (see Water, sewage).

Weather Modification. Techniques for influencing rates of precipitation and its geographic distribution have been developed over the last half century, but with mixed results (28). Local success under favorable meteorological conditions has been reported (29). In the United States, attempts to induce or to modify precipitation by seeding clouds with silver iodide, frozen carbon monoxide, and other artificial nuclei have been the subject of Federally sponsored research and development. Seeding of winter storms in local areas of the western United States increases snow and water precipitation by ca 5–25%. Stimulating precipitation from summer cumulus clouds is considerably more complex, and the results have been less conclusive (30–31).

Artificial Recharging of Groundwater. Subject to certain technical and geological requirements, groundwater can be artificially replenished or recharged by surface-spreading techniques, infiltration ponds, injection wells, and induced infiltration of surface water from pumped wells (32). Artificial recharge can be used to relieve heavy draft on groundwater sources, to control seawater intrusion into groundwater reservoirs, and to improve groundwater quality. Temporary surpluses of surface water, eg, during periods of high stream stage or flooding, can be used to replenish groundwater supplies. Storm runoff is routed to recharge basins where the water infiltrates into the ground. Though not widely practical, wastewater, treated and untreated, can be stored underground for later withdrawal and utilization (33).

Community and Rural Water Supplies. Trends in community and rural water-supply and sanitary-system development are indicators of cultural and economic progress. The magnitude of domestic water use reflects living standards since modern kitchen and sanitary conveniences and ornamental vegetation require water. Per-capita rates of water use in large cities of the world are ca 0.25–0.60 m^3/d (66–160 gal/d): Moscow and New York, 0.60 m^3/d (160 gal/d); Paris, 0.50 m^3/d (130 gal/d); London, 0.263 m^3/d (69 gal/d) (34). Since 1900, per-capita rates in the United States have risen four- or fivefold; in the USSR, they have increased about fourfold; and in most nations of Europe, they have at least doubled. It is expected that the increase in world population will continue and that most of the growth will occur in nations of lowest economic status (35). Substantial rises in per-capita usage of water are anticipated only for economically and industrially advancing nations.

Total community and rural water use throughout the world in 1970 is estimated to have been 120 km^3 (32 × 10^{12} gal) (34). Compared to most other categories of water use, this is a small quantity. In most nations, water supply for domestic purposes is the smallest of the primary uses, ie, industrial, agricultural, recreational, and domestic. Nevertheless, it is generally given highest developmental and protective priority since drinking water is a requirement for existence.

Estimated annual domestic water use by continent is as follows: Asia, 40 km^3 (10.6 × 10^{12} gal); North America, 40 km^3 (10.6 × 10^{12} gal); Europe, 30 km^3 (7.9 × 10^{12} gal); Africa, 4 km^3 (1.1 × 10^{12} gal); South America, 4 km^3 (1.1 × 10^{12} gal); Australia, 1 km^3 (2.6 × 10^{11} gal); and Oceania, 1 km^3 (2.6 × 10^{11} gal). Approximately 17% of the total domestic usage supply is consumed (34).

Industrial Water Supplies. Virtually all industrial processes are dependent on water for cooling and condensing, cleansing, conveyance of wastes (including heat), drinking and sanitary purposes, incorporation into products, and numerous other purposes.

Manufacturing industries in the United States alone withdrew ca 80 km^3 (21.1 × 10^{12} gal) of water in 1978 and discharged ca 72 km^3 (19 × 10^{12} gal) (24). The balance, ca 9 km^3 (2.4 × 10^{12} gal), dissipated by evaporation or leakage or was incorporated into products. These figures exclude those for electric-power generating plants and other plants that used <76 × 10^3 m^3/yr (20 × 10^6 gal/yr) of water.

Industrial water usage is commonly expressed volumetrically in terms of the amount of water required to produce a unit amount of a product, eg, a ton of steel, a barrel of petroleum, a loaf of bread, a ton of cotton, or 1000 MW of energy. However, water requirements for production of a given product are not fixed; they may change with a change in production processes. Where the cost of water is a significant factor in the production process, reduction of the magnitude of water required is an economic benefit. Identification of the nature of the production process should accompany specification of water requirements. Chemicals, primary metals, and paper account for about three fourths of the water used in manufacturing (see Table 3).

Approximately one half to two thirds of the water used in mostly industrially developed nations is for power generation and cooling. Water systems of power-generating plants range from direct-flow systems with massive flowthrough of water but little evaporative loss to multiple recycling systems that demand considerably less water but lose a large percentage through evaporation. Water for cooling processes in nuclear-energy plants is a relatively new and sizable use (36).

Industries must dispose of or treat water contaminated with wastes or heat. Reduction in the amount of liquid effluents by internal recycling and reuse lowers the unit cost of water and the volume of effluent requiring treatment or disposal. However, significant evaporative water loss accompanies most recycling processes.

Agricultural Water Supply. Irrigation of commercial crops is the predominant use of water in agriculture. Additional agricultural applications include stock watering, gardening, cleaning, and other farm purposes. Worldwide, most farmland is in moderately humid to semiarid regions. Although only a little more than 15% of the world's

Table 3. Water Intake and Discharge by U.S. Manufacturing Industry Groups in 1978, km^3/d (10^9 gal/d (gpd))[a]

Manufacturing industry group	Water intake	Percent of total water intake of all groups	Water recirculated and reused	Water discharged
chemicals and allied products	16,374 (4,326)	33.3	33,047 (8,731)	14,799 (3,910)
primary metal industries	12,839 (3,392)	26.1	13,149 (3,474)	11,855 (3,132)
paper and allied products	7,430 (1,963)	15.1	34,883 (9,216)	6,681 (1,765)
petroleum and coal products	4,440 (1,173)	9.0	26,908 (7,109)	3,649 (964)
food and related products	2,801 (740)	5.7	3,883 (1,026)	2,456 (649)
other	5,291 (1,398)	10.8	17,574 (4,643)	4,777 (1,262)
Total	*49,175* *(12,992)*	*100.0*	*129,444* *(34,199)*	*44,217* *(11,682)*

[a] Ref. 15.

cultivated land is irrigated, that land yields more than half the monetary value of total agricultural production. Increasing worldwide food requirements has resulted in expansion of the total irrigated area from 3.9×10^7 ha (1.5×10^5 mi^2) at the start of the present century to ca 2.36×10^8 ha (9.1×10^5 mi^2) in 1970 (34).

The FAO is the primary source of international statistical information on agriculture (37–38). Farmland in the People's Republic of China, India, the United States, and the USSR accounts for ca 60% of the world's irrigated land. Based on FAO projections, total irrigated land worldwide may amount to ca 4.14×10^8 ha (1.6×10^6 mi^2) by the year 2000. In India, ca 310 km^3 (81.9×10^{12} gal) of water was applied in 1970. Applications in the United States amounted to 165 km^3 (43.6×10^{12} gal), and in the USSR, 136 km^3 (35.9×10^{12} gal).

Consumptive loss of water by evaporation and transpiration generally is extremely high in irrigated fields. Total water application for irrigation throughout the world is ca 1900 km^3 (5.02×10^{14} gal), of which ca 79% is consumed by these processes (see Table 2). Technological improvements in the efficient use of irrigation water include reduced seepage losses and consumptive processes, strategically timed applications at controlled rates, sprinkler and drip applications, and increased plant uptake.

Utilization in the United States. The gross water resources available within the United States greatly exceed demands for them. The average annual precipitation of ca 0.77 m, of which ca 4.5 km^3/d (1.2×10^{12} gal/d) is shed as natural runoff, exceeds nationwide withdrawal usage ca threefold. However, precipitation is not evenly distributed over the nation, nor are the other natural conditions that determine water-supply availability. In addition, development of water geographically is not uniform, and levels of demand are uneven. Water usage in the United States in 1980 is given in Table 4 (27). Hydroelectric power generation is the largest use of water in the United States and far exceeds all other withdrawal uses combined. There is a negligible loss by evaporation from hydroelectric power generation.

Adequacy of Sources. Water shortages are most commonly associated with dry regions of the West but occur in wetter parts of the Nation as well. Many humid eastern localities experience periodic water-supply deficiencies resulting from drought and inadequate storage systems. For uses that require water of good quality, deterioration of quality represents, in effect, loss of water supply.

Table 5 indicates the adequacy of water sources in the 21 water-resources regions in the United States. In the Lower Colorado region, an arid area, both water withdrawals and consumptive use exceed the natural supply originating within the basin. Withdrawals in excess of available natural supply is possible in the poorly watered Colorado basin only because water is imported, groundwater is mined, and available water is reused. In the Colorado region and other arid regions, only limited increase is possible and would require expanded use of effective conservation measures, importation of additional water, and other management measures that extend or supplement present supplies.

Trends and Projections. The most recent national projection of future water use is given in ref. 39 and includes evaluations of historic and projected water withdrawal, consumptive usage, and adequacy for the period 1955–2000 for all principal categories of water use. Groundwater resources and their role in national water supply are considered more fully than in previous assessments. The assessment report predicts a reduction of ca 14% in withdrawal rates and an increase of 2% in the rate of consumption during 1975–2000. These conclusions are based on anticipated expansion

Table 4. Summary of Water Withdrawals, Sources, and Use in the United States from 1950 to 1980, 10^6 m^3/d[a,b]

Withdrawal	1950	1955	1960	1965	1970	1975	1980
Use							
public supplies	53	64	79	91	102	110	129
rural domestic and livestock	13.6	13.6	13.6	15.1	17.0	18.5	21.2
irrigation	337	416	416	454	492	530	568
industrial							
thermoelectric power	151	273	379	492	643	757	795
other industrial uses	140	148	144	174	178	170	170
Total withdrawals (exclusive of hydroelectric power)	695	915	1,032	1,226	1,432	1,586	1,683
hydroelectric power	4,160	5,680	7,570	8,710	10,600	12,500	12,500
Source							
groundwater							
fresh	129	178	189	227	257	310	333
saline	na	2.3	1.5	1.9	3.8	3.8	3.4
surface water							
fresh	530	681	719	795	946	984	1,098
saline	38	68	117	163	201	261	269
reclaimed sewage	na	0.76	2.3	2.6	1.9	1.9	1.9
consumption, freshwater	na	na	231	291	329	363	379

[a] Ref. 27.
[b] To convert m^3 to gal, multiply by 264.

of water-recycling practices in manufacturing and electric-generating plants in efforts to meet effluent standards and to minimize attendant costs (see Regulatory agencies). However, assumptions, conclusions, and projections contained in the assessment are not universally accepted. It is apparent from the many disparate predictions that assumptions, methodologies, and results vary widely (40).

Water Quality

Water quality and quantity are interdependent, interacting elements of water systems (41–42). Pumping of wells, for example, may induce distant groundwater of inferior quality to migrate toward the locality of pumping, or the pumping may induce surface water of inferior quality to flow into the groundwater reservoir. The interrelations of flow conditions and chemistry are relevant in the case of a stream receiving waste-liquid effluents: both the stream-flow characteristics and the chemical nature of the stream water govern the effectiveness of stream processes in the dilution and assimilation of the waste liquid.

Natural Water. The exposure of water to earth materials and its participation in natural biological, hydrological, and geochemical processes alters its chemical and physical character. Solutes contained in water are derived from many sources, including products of biological activity, of the erosion and weathering of soil and rocks, and of the solution and precipitation reactions occurring during the flow of water on and through earth materials. The chemical characteristics of natural water are described in ref. 43.

Although rainwater initially evaporates to the atmosphere as distilled water, it

Table 5. U.S. Water Runoff and Withdrawals by Region in 1975, 10^6 m^3/d [a,b]

Region	Average runoff	Estimated dependable supply for 1980	Withdrawals	Freshwater consumed	Annual flow exceeded in 90% of years reported	Fresh surface water withdrawn
New England	254	83.3	53.0	1.7	185	16.7
Mid-Atlantic	318	136	197	6.1	257	83.3
South Atlantic-Gulf	746	284	163	14.0	488	90.8
Great Lakes	284	261	136	4.2	204	132
Ohio	473	182	136	4.5	284	129
Tennessee	155	53.0	41.6	1.1	106	37.8
Upper Mississippi	246	117	71.9	3.0	136	60.6
Lower Mississippi	299	94.6	60.6	20.8	144	41.6
Souris-Red-Rainy	23.5	11.4	1.5	0.3	7.6	1.1
Missouri Basin	204	114	132	56.8	110	94.6
Arkansas-White-Red	276	75.7	56.8	34.0	136	23.5
Texas-Gulf	121	64.3	83.3	30.3	41.6	36.7
Rio Grande	18.9	11.4	20.4	13.2	7.6	11.4
Upper Colorado	49.2	49.2	15.5	6.4	30.3	14.8
Lower Colorado	12.1	7.6	32.2	23.8	3.8	13.2
Great Basin	28.4	34	26.1	13.6	11.4	20.4
Pacific Northwest	795	265	125	41.6	560	98.4
California	235	106	193	87.0	114	83.3
Totals, contiguous United States	4538	1950	1545	367	2826	989

[a] Refs. 26–27.
[b] To convert m^3/d to gal/d, multiply by 264.

acquires dissolved salts and particulate matter from dust and other material in the atmosphere. In addition, rain contains dissolved gases, eg, carbon dioxide and sulfur dioxide, and aerosols that influence the chemistry of rainwater and snow. Although quantities of mineral matter in atmospheric moisture are generally minute, the repeated precipitation of salts and other airborne chemicals with rain and snow totals many tons per square kilometer in a year. Chemical constituents of precipitation generally are detectable in stream water (44).

After reaching the land surface, rainwater and snowmelt water move over and through the earth's crust, causing the dissolution and precipitation of matter during movement toward rivers or other places of discharge. Rivers of North America carry an average load of 33 metric tons of dissolved minerals in solution from each square kilometer of land area (45). Mean composition values for seawater and river water are given in Tables 6 and 7, respectively. Seawater composition does not differ greatly worldwide, except in the nature and proportions of minor constituents (48). The dominant constituents everywhere are sodium and chlorine (see Water, analysis).

Streams, on the other hand, may differ widely in chemical character as the result of differences imposed by geographic variations in the chemical character of precipitation, climate and weather, vegetation, organisms, soil and rocks, and introduced pollutants. During periods of low flow when stream drainage is derived mainly from groundwater discharge, the chemical component of stream water commonly reflects the chemical character of shallow groundwater. For these reasons, the gross values given in Table 7 have only general significance.

Table 6. Average Concentration of the Thirteen Most Prevalent Chemical Constituents of Seawater[a]

Constituent	Concentration, mg/L	Constituent	Concentration, mg/L
Cl^-	19	Br^-	0.065
Na^+	10.5	Sr^{2+}	0.008
SO_4^{2-}	2.7	SiO_2	0.0064
Mg^{2+}	1.35	B	0.0046
Ca^{2+}	0.40	F^-	0.0013
K^+	0.38	N	0.0005
HCO_3^-	0.142		

[a] Ref. 46.

Table 7. Mean Chemical Composition of River Water in the World[a]

Constituent	Concentration mg/L	meq/L
silica (SiO_2)	13	
iron (Fe)	0.07	
calcium ion (Ca^{2+})	15	0.750
magnesium ion (Mg^{2+})	4.1	0.342
sodium ion (Na^+)	6.3	0.274
potassium ion (K^+)	2.3	0.059
bicarbonate ion (HCO_3^-)	58	0.958
sulfate ion (SO_4^{2-})	11	0.233
chloride ion (Cl^-)	7.8	0.220
nitrate ion (NO_3^-)	1	0.017
dissolved solids	90	
hardness as $CaCO_3$	55	
noncarbonate	7	

[a] Ref. 47.

The chemical character of groundwater is generally more uniform than stream water, and under natural conditions, the chemistry of groundwater changes at imperceptibly slow rates. Hydrogeologically similar conditions controlling the chemistry of groundwater generally prevail over relatively extensive land areas. However, vertical variations in chemistry among aquifers hydraulically insulated from one another by impermeable rock layers may be large even in the same locality. For example, tight layers of rock separate freshwater and saline aquifers in close, verticle proximity to each other.

The chemical nature of rocks through which groundwater moves generally influences the chemistry of the water (49–50). The chemical analyses of groundwater from various common rock types listed in Table 8 show features of both uniformity and irregularity of water chemistry within each rock type. Igneous and metamorphic rocks, composed mostly of silicate minerals of low solubility, typically yield water low in dissolved minerals. Sedimentary rocks, on the other hand, consist of greater varieties of rock materials accumulated from various sources, and they yield water of more assorted solute content. Sedimentary rocks are derived from chemical precipitates, evaporites, detrital accumulations, and products of rock weathering.

Practical management of water quality limits the term water pollution to chemical

Table 8. Chemical Analyses of Groundwater from Various Kinds of Rock, mg/L[a]

Constituent	Igneous rocks			Metamorphic rocks			Sedimentary rocks		
	Rhyolite, west of Los Alamos, N.M.	Granite, West Warwick, R.I.	Gabbro, Laurel, Md.	Basalt, Camas, Wash.	Mica schist, Wilkesboro, N.C.	Willimantic gneiss, Willimantic, Conn.	Quartzite, Cliffs Shaft Mine, Mich.	Catahoula sandstone, Collins, Miss.	Navajo sandstone, east of Mexican Water, Ariz.	Ocala limestone, Lake City, Fla.
SiO_2	55	20	21	49	26	13	7.6	25	11	25
Al	0.1	0	0.3			0.1	0	0.2	0.2	0.1
Fe	0.08	0.19	1.2	0.04	2.6	0.09	0.21	0.41	0.50	0.17
Mn	0	0	0.16		0	0	0.20	0	0	0
Cu		0				0			0	0
Zn		0.07				0.06				0.02
Ca	4.4	6.5	16	13	10	19	32	2.4	0.8	39
Mg	1.4	2.6	10	9.0	1.6	5.1	16	0.5	0.9	15
Na	11	5.9	15	6.6	5.5	4.4	8.5	2.6	144	7.5
K	1.2	0.8	1.5	2.8	1.0	3.2	3.1	2.0	0.8	1.3
HCO_3	42	38	20	88	45	39	144	18	217	106
CO_3	0	0	0		0	0	0	0	33	0
SO_4	1.9	0.9	59	4.9	3.0	30	39	1.4	64	1.8
Cl	2.0	5.0	25	6.9	2.5	5.8	8.0	2.5	16	9.8
F	0.6	0.5	0.1	0.1	0.1	0.7	0.2	0.1	0.3	0.5
NO_2	0.3	1.5	2.2	8.4	1.4	15	2.7	0	0.5	0.8
PO_4		0	0			0	0			0.1
specific conductance at 25% conc, µS	80	76	259	181	92	178	323	38	630	327
pH	7.2	7.6	5.6	7.7	6.9	6.9	7.9	6.2	9.2	8.0

[a] Ref. 50.

or physical changes of sufficient magnitude to hamper utility of the water for designated purposes and to changes caused only by human activities. For management and regulation of water resources, localized sources of pollutants or point sources must be distinguished from sources of more widespread occurrence or nonpoint sources (51–52). Polluting substances are chemical, biological, or physical (see Water, water pollution; Water, industrial water treatment; Water, municipal water treatment).

Criteria and Regulatory Standards. The term water quality refers to the level of suitability of water for specified purposes. Criteria and standards for water vary widely among nations. In 1978, the EPA published guideline criteria for 52 of the most common constituents and characteristics of water influencing its quality (53). Substances considered to be most significant to human health protection and to freshwater and marine aquatic environments, including natural and polluting chemicals, are discussed. The criteria are not intended for regulatory use, but rather for guidance in the development of water-quality management programs.

Hardness of water affects its utility for some purposes, but because it has not been proven to affect health, few governmental criteria have been advanced for it. It is caused by the presence of polyvalent metallic ions. In freshwater, these are principally calcium and magnesium. Other metals, eg, iron, strontium, and manganese, contribute hardness when present in relatively high concentrations. As indicated by Table 9, limits of hardness permissible for industrial uses range widely and are determined by the nature of products and production processes. Table 9 lists the maximum hardness concentration for raw-water sources acceptable to industry for the six principal categories of industrial water usage in the United States (54). In 1980, the EPA issued criteria for 64 additional pollutants or pollutant groups determined to be toxicants (55). A 65th toxicant, dieldrin, was added later.

National drinking-water regulations issued by the EPA in 1977 are for use in the governmental control and management of drinking-water quality (56). In contrast with water-quality criteria, these specified permissible levels of concentration of chemicals constitute legal standards. The standards are listed in Table 10 (57).

Economic Aspects

Water is a necessary commodity for a nation's economic strength and growth. The influence of water resources on the economies of regions or nations appears to decrease with higher levels of economic status. In the United States, for example, large water-development projects continue to play a role in economic activity and growth,

Table 9. Industrially Accepted Maximum Hardness Levels of Raw Water[a]

Industry	Maximum concentrations, mg/L as $CaCO_3$
electric utilities	5000
textile	120
pulp and paper	475
chemical	1000
petroleum	900
primary metals	1000

[a] Ref. 54.

Table 10. Maximum Contaminant Levels for Drinking Water in the United States[a]

Constituent	Maximum concentration, mg/L unless otherwise specified
inorganic chemicals	0.05
arsenic	0.05
barium	1
cadmium	0.010
chromium	0.05
lead	0.05
mercury	0.002
nitrate (as N)	10
selenium	0.01
silver	0.05
fluoride	1.4–2.4
organic chemicals turbidity, Tu (turbidity unit)	1–5
coliform bacteria (mean value), μL	10
endrin	0.002
lindane	0.004
methoxychlor	0.1
toxaphene	0.005
2,4-D((2,4-dichlorophenoxy)acetic acid)	0.1
2,4,5-TP Silvex (2-(2,4,5-trichlorophenoxy)propionic acid)	0.01
radionuclides	
radium-226 and radium-228, Bq/L[b]	0.19
gross alpha-particle activity, Bq/L[b]	0.56
gross beta-particle activity, μSv/yr[c]	40

[a] Ref. 57.
[b] To convert Bq to Ci, divide by 3.7×10^{10}.
[c] To convert Sv to rem, multiply by 100.

population distribution, and employment opportunity, but water projects generally are no longer the primary factor they were in the developing years of the nation (58).

The value placed on water continues to be low for most uses, but competition among water uses is increasing. As supply sources become more fully appropriated, pollution takes its economic toll and the costs of water development, storage, and distribution rise. Fortunately, water is a migratory resource that can be used and reused in successive processes and by successive parties; the exception is usage that consumes the water. Each use changes the quantity, quality, and location of the water, and in these ways influences its value for subsequent uses. Thus, economic planning for water necessarily entails conjunctive evaluation of water-supply sources, contemplated uses for the water, and the economic and environmental impacts of its usage and subsequent release. Further, water is one facet of comprehensive land and natural-resources management, and it should be evaluated in this context.

Market prices of water may not be reliable measures of its economic value, and determination of cost is difficult and subjective. Subsidies, taxes, regulatory measures, and monopolistic influences must be considered as must environmental and water-quality concerns (59–61). The monetary value of water is governed more by its importance to production processes than by its worth to water consumers. Among advanced national economies, industry, irrigation agriculture, and power generation account for the bulk of water use; all of these uses produce goods and services. Water used directly by consumers commonly is only a minor fraction of total water usage. In the United States, for example, the amount of water used for public and rural domestic supplies is only ca 1% of the total (see Table 4).

The value of water as used for irrigation varies with climatic and other environmental conditions, the nature of crops grown (vegetables and fruits are more costly than forage crops and grain), the stage of crop maturity at the time of irrigation, the efficiency of application, and the effects of governmental economic support measures and other regulatory influences. Irrigation water values in the United States are ($12–32)/1000 m^3 (($45–120)/10^6 gal) at the farm headgate, with most estimates being ca $16/1000 m^3 ($60/10^6 gal) (62).

The value of water for municipal and domestic purposes are the highest among all uses, eg, ca $81/1000 m^3 ($300/10^6 gal) for in-house uses, and ca $52/1000 m^3 ($198/10^6 gal) for lawn and garden irrigation in the West, but only ca $13/1000 m^3 ($48/10^6 gal) in the East. Potable water is produced in relatively small quantities and is costly to treat and distribute; therefore, it commands the highest prices among water uses.

Water supply is not generally a principal economic factor in industrial production, since water-supply costs often are only a few percent or less of production costs (63). In water-deficient areas, where water costs are highest, recycling technology is used extensively. The cost of recycling water for industrial cooling, the largest industrial use, is low. Recycle cooling systems, in which heat is expelled to the atmosphere through a cooling tower, operate in the United States at costs of ($2.00–3.40)/1000 m^3 (($7.50–13.00)/10^6 gal) (64). The value of water for once-through cooling, in which the heated water is discharged to the environment after circulating through a cooling system only once, cannot exceed the recycling values (see also Heat-exchange technology).

Process water acquires foreign matter, eg, dissolved minerals, sediment, etc, during its use and therefore is more expensive to recycle than cooling water and often requires treatment. Thus, the cost is several times greater than the expense of recycling cooling water, which ordinarily does not lessen in quality. The mean value of recycling industrial process water in the United States ranges from <$4 to ca $21 per 1000 m^3 (<$15 to ca $78 per 10^6 gal), the average being $10.30/1000 m^3 ($39/10^6 gal) (64).

BIBLIOGRAPHY

"Water, Sources and Utilization" in *ECT* 2nd ed., Vol. 21, pp. 652–668, by R. G. Dressler, Trinity University.

1. O. L. Franke and N. E. McClymonds, *U.S. Geol. Surv. Prof. Pap.* **627-F,** 39 (1972).
2. R. L. Nace, ed., *Scientific Framework of World Water Balance*, No. 7 of *Technical Papers in Hydrology*, United Nations Educational, Scientific, and Cultural Organization, New York, 1971, Table 2.
3. R. L. Nace, *Water of the World*, U.S. Government Printing Office, Washington, D.C., 1980, pp. 9–11.
4. *Ibid.*, 20 pp.
5. H. J. Critchfield, *General Climatology*, 3rd ed., Prentice-Hall, Inc., Englewood Cliffs, N.J., 1974, pp. 61–66.
6. Ref. 3, p. 8.

7. United Nations Educational, Scientific, and Cultural Organization, *World Water Balance and Water Resources of the Earth*, The UNESCO Press, Paris, France, 1978, p. 600.
8. J. Nemec in J. C. Rodda, ed., *Facets of Hydrology*, John Wiley & Sons, Ltd., Chichester, England, 1976, pp. 331–362.
9. F. van der Leeden, *Water Resources of the World, Selected Statistics*, Water Information Center, Inc., Port Washington, New York, 1975, Table 7-15, pp. 466–468.
10. International Commission on Large Dams, *World Register of Dams*, Paris, France, 1973, 998 pp.; *World Register of Dams, Supplement 1*, Paris, France, 1973, 297 pp.
11. C. R. Bentley and co-workers, *Physical Characteristics of the Antarctic Ice Sheet*, Folio 2 of *Antarctic Map Folio Series*, American Geographical Society, New York, 1964.
12. V. I. Korzoun, A. A. Sokolov, and co-eds., *Atlas of World Water Balance*, The UNESCO Press, Paris, 1977, 36 pp., 65 maps.
13. R. L. Nace, *The Physical Basis of Hydrology, Part A, Chapter 1, The Science of Hydrology*, manuscript in review for *Technical Papers in Hydrology*, United Nations Educational, Scientific, and Cultural Organization, New York, 1981.
14. C. R. Murray and E. B. Reeves, *U.S. Geol. Surv. Circ.* **765**, 3 (1977).
15. Bureau of the Census, *1977 Census of Manufactures, Water Use in Manufacturing*, U.S. Department of Commerce, Washington, D.C., 1981, p. 8-6.
16. A. Wiener, *The Role of Water in Development, An Analysis of Principles of Comprehensive Planning*, McGraw-Hill Book Co., New York, 1972, 483 pp.
17. N. Buras, *Scientific Allocation of Water Resources*, American Elsevier, New York, 1972, 208 pp.
18. G. Fleming, *Computer Simulation Techniques in Hydrology*, Environmental Science Series, American Elsevier, New York, 1975, 333 pp.
19. A. K. Biswas, ed., *Systems Approach to Water Resources*, McGraw-Hill, Inc., New York, 1976, 429 pp.
20. J. L. Cohon, *J. Water Resources Res.* **18**(1), 1 (Feb. 1982).
21. *The A-B-C of Desalting*, PB-243 556, U.S. Department of the Interior, Office of Water Research and Technology, Washington, D.C., 1977, 30 pp.
22. L. K. Cecil, ed., *Complete Water Reuse, Industry's Opportunity*, American Institute of Chemical Engineering, New York, 1973, 728 pp.
23. K. A. MacKichan and J. C. Kammerer, *U.S. Geological Surv. Circ.* **456**, (1961).
24. C. R. Murray, *U.S. Geol. Surv. Circ.* **556**, (1969).
25. C. R. Murray and E. B. Reeves, *U.S. Geol. Surv. Circ.* **676**, (1972).
26. *Ibid.*, **765**, (1977).
27. W. B. Solley, E. B. Chase, and W. B. Mann IV, *U.S. Geol. Surv. Circ.* **1001**, (1982).
28. W. N. Hess, ed., *Weather and Climate Modification*, John Wiley & Sons, Inc., New York, 1974, pp. v–vi.
29. Ref. 5, pp. 384–405.
30. U.S. Water Resources Council, *The Nation's Water Resources, 1975–2000, Second National Water Assessment*, Vol. 1, Washington, D.C., 1978, p. 22.
31. A. J. Large, *Wall Street J.* **33**, 23, 29 (June 14, 1982).
32. C. J. Schmidt and co-workers, *J. Am. Water Works Assoc.* **70**, 140 (1978).
33. J. Braunstein, ed., *Underground Waste Management and Artificial Recharge*, Vol. 2, American Association of Petroleum Geology, New Orleans, La., 1973, 931 pp.
34. Ref. 7, 663 pp.
35. Council on Environmental Quality and U.S. Department of State, *The Global 2000 Report to the President: Entering the Twenty-First Century, The Technical Report*, U.S. Government Printing Office, Washington, D.C., 1980, pp. 11–12.
36. E. V. Giusti and E. L. Meyer, *U.S. Geol. Surv. Circ.* **745**, (1977).
37. Food and Agriculture Organization of the United Nations, *Irrigation, Drainage and Salinity: An International Source Book*, FAO and UNESCO, Paris, France, 1973, 510 pp.
38. Food and Agriculture Organization of the United Nations, *Production Yearbook, 1971*, Vol. 25, Rome, Italy, 1972, 829 pp.
39. Ref. 30, Vols. 1 and 2.
40. W. Viessman, Jr., *State and National Water Use Trends to the Year 2000*, Serial No. 96-12, Congressional Research Service, The Library of Congress, Washington, D.C., 1980, 297 pp.
41. W. Stumm and J. J. Morgan, *Aquatic Chemistry*, 2nd ed., Wiley-Interscience, New York, 1981, 780 pp.

42. E. A. Jenne, ed., *Chemical Modeling in Aqueous Systems*, No. 93 of *ACS Symposium Series*, American Chemical Society, Washington, D.C., 1977, 914 pp.
43. J. D. Hem, *U.S. Geol. Surv. Water-Supply Pap.* **1473,** (1970).
44. E. Gorham, *Geol. Soc. Am. Bull.* **72,** 795 (1961).
45. D. A. Livingstone, *U.S. Geol. Surv. Prof. Pap.* **440-G,** 28 (1963).
46. Ref. 43, p. 11.
47. Ref. 45, p. 41.
48. U.S. Naval Oceanographic Office, *Handbook of Oceanographic Tables*, Washington, D.C., 1966, pp. 3–25.
49. Ref. 43, pp. 287–309.
50. D. E. White, J. D. Hem, and G. A. Waring, *U.S. Geol. Surv. Prof. Pap.* **440-F,** (1963).
51. U.S. Environmental Protection Agency, *National Water Quality Inventory—1977 Report to Congress*, EPA-440/4-78-001, U.S. Government Printing Office, Washington, D.C., 1978, 138 pp.
52. U.S. Comptroller General, *National Water Quality Goals Cannot Be Attained Without More Attention to Pollution from Diffused or Nonpoint Sources*, CED-78-6, General Accounting Office, Washington, D.C., 1977, 47 pp.
53. *Quality Criteria for Water*, U.S. Environmental Protection Agency, Washington, D.C., July 1976, 256 pp.
54. *Ibid.*, p. 76.
55. U.S. Environmental Protection Agency, *Fed. Regist.* **45**(231), 79317 (Nov. 28, 1980).
56. *National Interim Primary Drinking Water Regulations*, U.S. Environmental Protection Agency, Washington, D.C., 1977, 159 pp.
57. U.S. Council on Environmental Quality, *Environmental Quality—1979, The Tenth Annual Report of the Council*, U.S. Government Printing Office, Washington, D.C., 1980, p. 156.
58. National Water Commission, *Water Policies for the Future, Final Report to the President and to the Congress of the United States*, U.S. Government Printing Office, Washington, D.C., June 1973, 579 pp.; P. Rogers, "The Future of Water," *The Atlantic Monthly* **252,** 80 (July 1983).
59. R. A. Young and S. L. Gray, *Economic Value of Water: Concepts and Empirical Estimates*, Accession No. PB 210 356, National Water Commission, National Technical Information Service, Springfield, Va., 1972, 337 pp.
60. A. V. Kneese and co-workers, *Economics and the Environment: A Materials Balance Approach*, Resources for the Future, Inc., Washington, D.C., 1970.
61. J. L. Wilson, R. L. Lenton, and J. Porras, *Groundwater Pollution: Technology, Economics, and Management*, Report No. TR 208, Massachusetts Institute of Technology, Cambridge, Mass., 1976, 311 pp.
62. Ref. 58, pp. 42–43.
63. Ref. 59, p. 159.
64. Ref. 58, p. 43.

GERALD MEYER
U.S. Geological Survey

PROPERTIES

The most striking feature of the earth, and one lacking from the neighboring planets, is the extensive hydrosphere. Water is the solvent and transport medium, participant, and catalyst in nearly all chemical reactions occurring in the environment. It is a necessary condition for life and represents a necessary resource for humans. It is an extraordinary complex substance. Structural models of liquid water depend on concepts of the electronic structure of the water molecule and the structure of ice. Hydrogen bonding between H_2O molecules has an effect on almost every physical property of liquid water.

Natural water systems contain numerous minerals and often a gas phase. They include a portion of the biosphere and organisms, and their abiotic environments are interrelated and interact with each other. The distribution of chemical species in waters is strongly influenced by an interaction of mixing cycles and biological cycles.

Human civilization interferes more and more with the cycles that connect land, water, and atmosphere, and pollution seriously affects water quality. In order to assess the stresses caused to aquatic ecosystems by chemical perturbation, the distribution of pollutants and their fate in the environment must be investigated (see also Air pollution).

Structure

The Water Molecule. Bond formation has little effect on the $1s^2$ electrons of the oxygen, but the remaining eight electrons of the water molecule form four hybrid orbitals, two of which contain bonding-pair electrons and are directed along the O—H bond axes. The other two are nonbonding, ie, they contain lone-pair electrons. They are symmetrically located above and below the H—O—H molecular plane and form roughly tetrahedral angles with the bond hybrids (Fig. 1). These lone-pair electrons are responsible for the molecule's large induced dipole moment.

Hydrogen bonding is the specific association of the hydrogen of one molecule with the lone-pair electrons of a neighboring molecule, a situation prevalent in liquid water

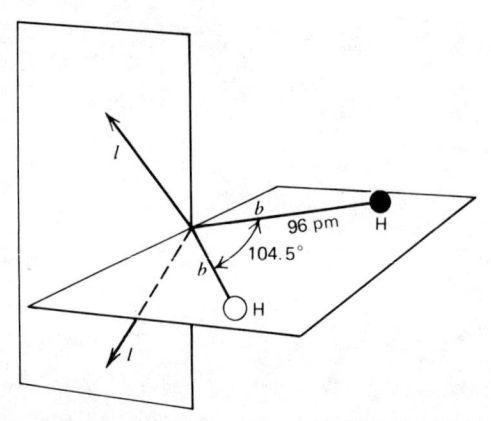

Figure 1. The structure of the isolated water molecule. b, Hybrid orbital with bonding-pair electrons; l, hybrid orbital with lone-pair electrons (1).

and ice. It is a useful concept in describing some of the extraordinary physical properties of water (eg, high melting and boiling points, low density) and their dependence on temperature and pressure.

The planar H—O—H bond angle in the isolated water molecule is 104.5°C. The O—H bond length in the water-vapor molecule is 96 pm (1). These values are slightly dependent on the vibrational and rotational states of the molecule.

Ice. Ordinary hexagonal ice (ice I) is one of at least 13 distinct forms of ice, but it is the only one stable at a pressure of ca 101.3 kPa (1 atm). Its structure is well established. The O—H distance, ca 101 pm, and the H—O—H angle, ca 104.5°, are not much greater than in the isolated molecule. Every water molecule is associated with its four nearest neighbors by intermolecular hydrogen bonds. In the ice-I structure, every oxygen atom is at the center of a tetrahedron formed by four other oxygen atoms at a distance of 276 pm (Fig. 2). Between adjacent oxygen atoms, a hydrogen nucleus has the choice of two positions, 101 pm from either of the two oxygen atoms (2).

The tetrahedral arrangement leads to an open lattice consisting of puckered layers (Fig. 3). These layers contain hexagonal rings of water molecules, similar in structure to the chair structure of cyclohexane. This open-crystal structure of ice produced by vacant shafts is an important characteristic and responsible for the low density of ice compared to that of liquid water.

Liquid Water. The structure of liquid water is not known. Various theories, however, account for many of its properties. Models for the structure of water in the liquid state are based on either the continuum theory or the mixture theory, but both make use of the structure of the solid (4–8).

In the mixture model, liquid water is represented as a mixture of distinguishable states, with one type of water bound (hydrogen-bonded polymers or clusters) into an icelike lattice and the other type free or monomeric (broken hydrogen bonds).

In the continuum theory, the water is homogeneous with, at least at low tem-

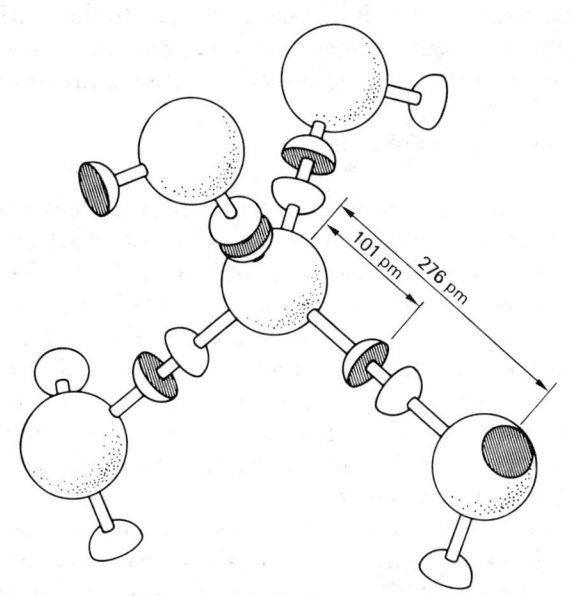

Figure 2. Water molecule in ice I and its four closest neighbors; tetrahedral arrangement of the oxygen atoms (large spheres) and the approximate dimensions; the hydrogen atom may occupy one of the two positions (shaded half-spheres) between the oxygen atoms (1).

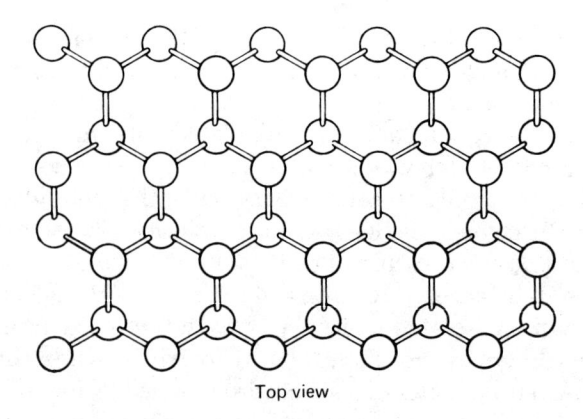

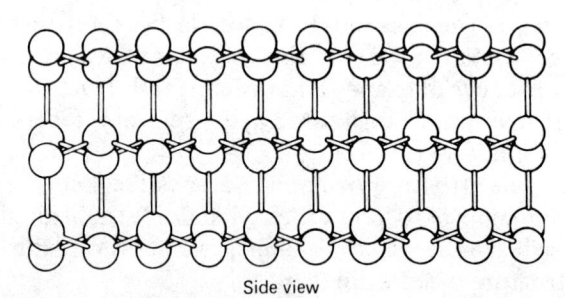

Figure 3. Arrangement of oxygen atoms in the ice-I lattice. The open shafts are characteristic and responsible for the low density of ice (3). Courtesy of American Institute of Physics.

peratures, essentially complete hydrogen bonding. The hydrogen bonds may be regarded as distorted rather than as intact or broken (7). In the continuum model, the average bond energy varies with temperature and pressure because of differences in distances and angle distortions while preserving tetrahedral coordination.

Thermodynamics and Physical Properties

The temperature dependence of the isopiestic heat capacity at constant pressure of 101.3 kPa (1 atm) and the thermodynamic functions derived from the heat capacity are shown in Figure 4. The heat capacity of liquid water is ca twice that of ice at 0°C or steam at 100°C. This thermodynamic quantity of the liquid is nearly constant but exhibits a slight minimum near 35°C. The changes of thermodynamic properties at melting and boiling points and for sublimation are given in Table 1. Properties of water vapor, liquid water, and ice I are given in Tables 2–4. The extremely high heat of vaporization, relatively low heat of fusion, and unusual values of other thermodynamic properties, including melting point, boiling point, and heat capacity can be explained by the presence of hydrogen bonding.

The density of pure liquid water and ice at atmospheric pressure as a function of temperature is shown in Figure 5. The liquid shows a maximum in density (or minimum in volume) $\rho = 0.999973$ g/cm^3 at 3.98°C, ie, above the melting point. This anomalous property plays an important part in controlling temperature distribution and vertical circulation in lakes and oceans (see Solar energy). It may be explained

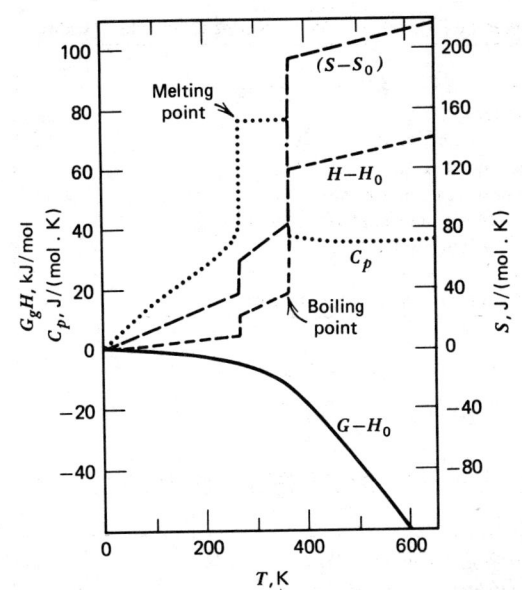

Figure 4. Enthalpy, entropy, free energy, and isopiestic heat capacity of H_2O at 101.3 kPa (1 atm) pressure, calculated from heat capacity measurements:

$$H_T - H_0 = \int_0^T C_p dT + \Delta H_{pc}; \quad S_T - S_0 = \int_0^T \frac{C_p}{T} dT + \Delta S_{pc}; \quad G_T - G_0 = H_T - H_0 - TS_T$$

The subscript pc indicates phase change. Thermodynamic constants for phase change are given in Table 1 (1,9). To convert J to cal, divide by 4.184.

Table 1. Thermodynamic Constants for Phase Changes of H_2O [a]

Property	Fusion[b]	Vaporization[b]	Sublimation[c]
temperature, K	273.15	373.15	273.16
isopiestic heat capacity change ΔC_p, J/(mol·°C)[d]	37.28	−41.93	
enthalpy change ΔH, kJ/mol[d]	6.01	40.66	51.06
entropy change ΔS, J/(mol·°C)[d]	22.00	108.95	186.92
volume change ΔV, cm³/mol	−1.621	3.01×10^4	
internal energy change ΔE, kJ/mol[d]	6.01	37.61	48.79

[a] Ref. 1.
[b] At 101.3 kPa (1.013 bar, 1 atm).
[c] At ice I liquid–vapor triple point.
[d] To convert J to cal, divide by 4.184.

by two competing effects, a configurational and a vibrational contribution to the coefficient of thermal expansion. As temperature is increased, the open four-coordinated structure of water is further broken down, increasing angular deformation and thus reducing the volume; furthermore, the amplitudes of anharmonic intermolecular vibrations (of the first-neighbor distances) increase, thus enlarging the volume. The first effect predominates below 4°C where thermal expansion is negative (1).

These effects explain another peculiar feature of water, namely, its minimum compressibility, which is not seen in other liquids. The compressibility at 101.3 kPa (1 atm) decreases from 0 to 46°C and goes through a minimum near 46.5°C (Fig. 6).

Table 2. Selected Properties of Water Vapor at 101.3 kPa (1 atm)[a]

Property	Value
molecular weight	18.015
heat of formation, kJ/mol[b] at 100°C	242.49
viscosity, mPa·s (= cP) at 20°C	96 × 10⁻⁶
velocity of sound, m/s at 100°C	405
diffusion coefficient, cm²/s at 100°C in air	0.380
specific volume, cm³/g at 100°C	1729.6
specific heat, J/(g·K)[b] at 100°C	2.078
thermal conductivity, W/(cm·K) at 110°C	2.44 × 10⁻⁴

[a] Refs. 5 and 9.
[b] To convert J to cal, divide by 4.184.

Table 3. Selected Properties of Liquid Water at 101.3 kPa (1 atm)[a]

Property	Value
heat of formation, kJ/mol[b] at 25°C	285.890
ionic dissociation constant, M^{-1} at 25°C	10^{-14}
heat of ionization, kJ/mol[b] at 25°C	55.71
apparent dipole moment, C·m[c]	6.24 × 10⁻³⁰
viscosity, mPa·s (= cP), at 25°C	0.8949
velocity of sound, m/s at 25°C	1496.3
density, g/cm³	
at 25°C	0.9979751
at 0°C	0.99987
freezing point, °C	0.0
boiling point, °C	100.0
isothermal compressibility, nPa⁻¹ at 25°C over the range of 0.1–1 MPa[d]	0.45
specific heat at constant volume, J/(g·K)[b] at 25°C	4.17856
thermal conductivity, W/(cm·K) at 20°C	0.00598
temperature of maximum density, °C	3.98
dielectric constant at 17°C and 60 MHz	81.0
electrical conductivity, S/cm at 25°C[e]	<10⁻⁸

[a] Refs. 5 and 9.
[b] To convert J to cal, divide by 4.184.
[c] To convert C·m to D, divide by 3.336 × 10⁻³⁰.
[d] To convert MPa to atm, divide by 0.101.
[e] S (siemens) = 1/ohm.

The minimum in the compressibility becomes less pronounced as pressure is applied and disappears near 300 MPa (3000 atm).

The pressure–temperature relationship for water is given in Figure 7. The triple point, at which ice, liquid water, and water vapor coexist at equilibrium, is at 0.01°C and 611 Pa (4.58 mm Hg).

The critical values of temperature, pressure, and volume are 374.15°C, 22.135 MPa (218 atm), and 59.1 cm³/mol, respectively. Above the critical point, the distinction between liquid and vapor properties vanishes.

For a pure shear flow τ (force per unit area) in the x direction with a velocity gradient dv_x/dz in the z direction, the shear viscosity η is the proportionality factor

Table 4. Selected Properties of Ice at 101.3 kPa (1 atm)[a]

Property	Value
heat of formation, kg/mol at 0°C	292.72
Young's modulus of elasticity, MPa[b] at −10°C	967
density, g/cm^3 at 0°C	0.9168
coefficient of cubical thermal expansion, cm^3/(g·°C) at 0°C	120×10^{-6}
coefficient of linear thermal expansion, °C^{-1} at 0°C	52.7×10^{-6}
isothermal compressibility, nPa^{-1} at 0°C	0.12
specific heat, J/(g·K)[c] at 0°C	2.06
thermal conductivity, W/(m·K)	210
dielectric constant at −1°C and 3 kHz	79

[a] Refs. 5 and 9.
[b] To convert MPa to psi, multiply by 145.

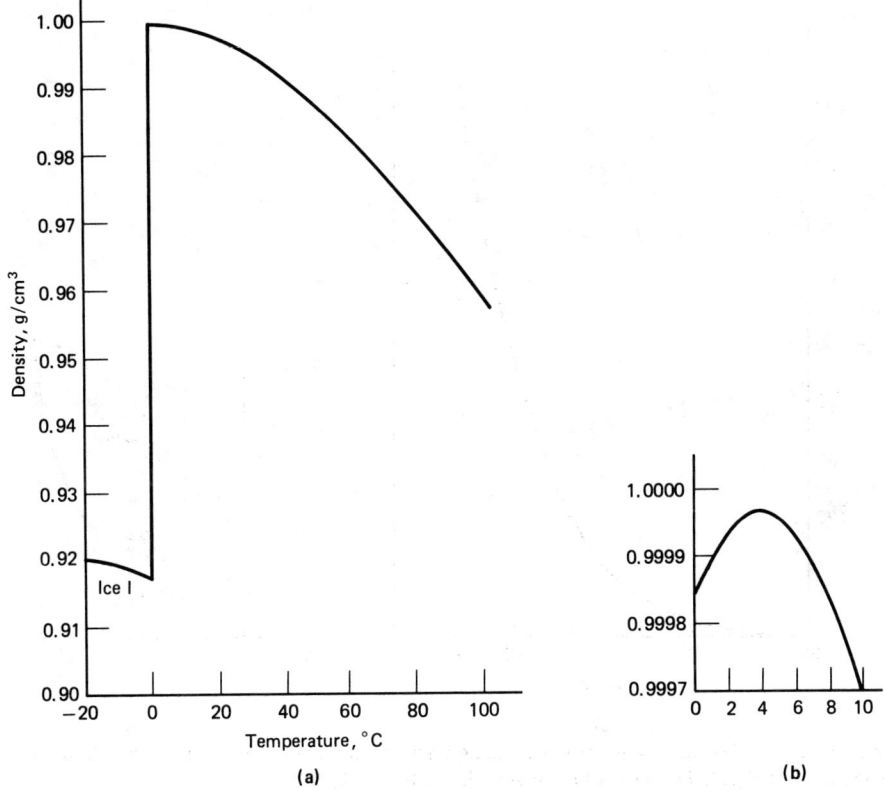

Figure 5. (a) The density of ice and liquid water at 101.3 kPa (1 atm) as a function of temperature; (b) the density in the domain of its maximum (1,10–11).

defined by the following equation:

$$\tau = \eta \cdot \frac{dv_x}{dz}$$

The shear viscosity of water and its dependence on temperature and pressure are given in Figure 8. With the exception of water, liquids become more viscous under pressure.

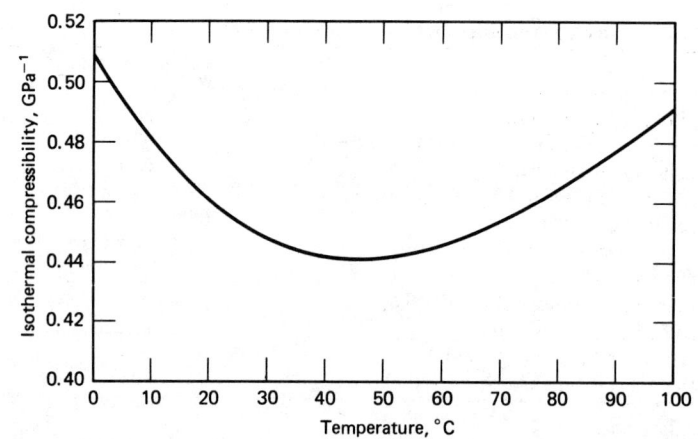

Figure 6. The coefficient of isothermal compressibility for liquid water at 101.3 kPa (1 atm) pressure (1,11). To convert GPa to psi, multiply by 145,000.

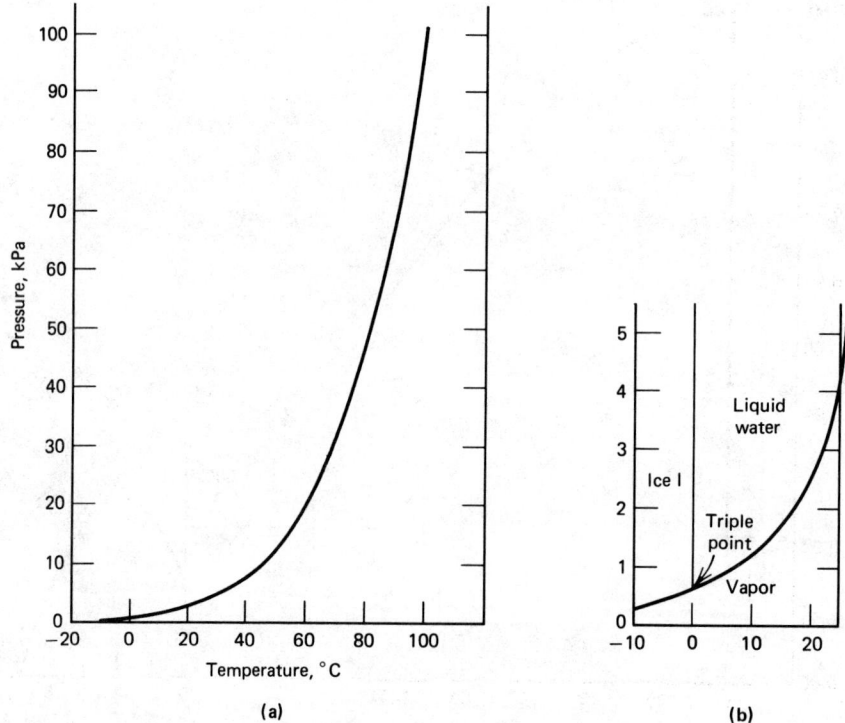

Figure 7. (a) The vapor pressure of ice and liquid water as a function of temperature; (b) the vicinity of the triple point (1,12). To convert kPa to atm, divide by 101.3.

Like most properties of water, the viscosity at temperatures <30°C exhibits some anomalous behavior. With increasing pressure the viscosity decreases, passes through a minimum, and finally increases normally.

The velocity of sound in a medium depends on the density and compressibility and thus on temperature and pressure. The speed of sound in pure water as a function of temperature in the range of 0–100°C is shown in Figure 9. Sound signals are subject to two types of degradation: dispersion, scattering, and reflection represent changes

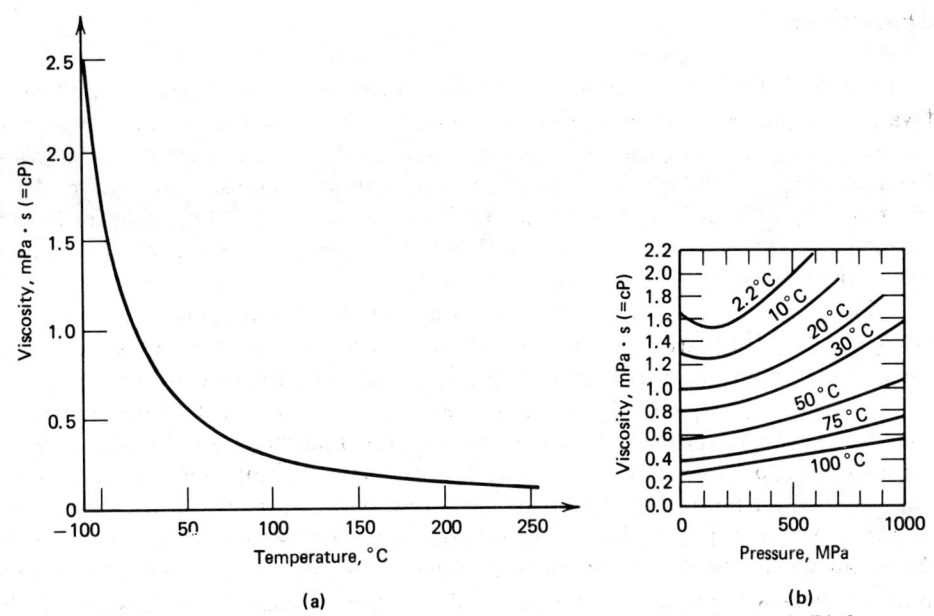

Figure 8. (a) The shear viscosity of water as a function of temperature (1,13–14); (b) the pressure dependence of the shear viscosity of water (15). To convert MPa to psi, multiply by 145.

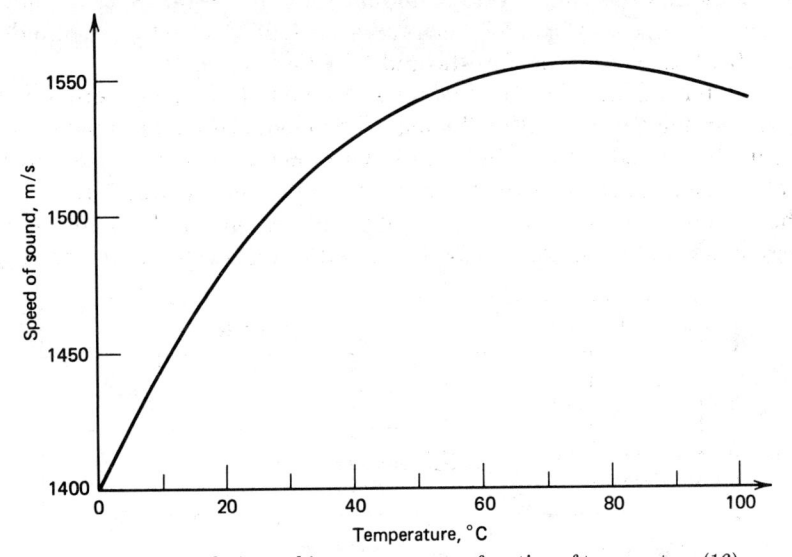

Figure 9. Speed of sound in pure water as a function of temperature (16).

in direction; and absorption, together with unavoidable shrinkage due to spherical spreading, represents a loss of energy. In seawater, the salinity, in addition to influencing the density and compressibility, determines the velocity of sound (5). An anomalous absorption of sound, found in seawater but not in freshwater, is owing to the strong absorption by electrolytes, especially by a $MgSO_4$ complex, an important constituent of seawater. The effect of this hydrated complex on the viscosity of water is very pronounced. Sound is propagated through complex paths in the ocean because of strong gradients in temperature, pressure, and salinity (see also Insulation, acoustic).

Natural Waters

Hydrological Cycle and Water Reservoirs. In hydrological studies, the transfer of water between reservoirs is of primary interest. The magnitude of the main reservoirs and fluxes are given in Figure 10. The oceans hold ca 76% of all the earth's water. Most of the remainder (21%) is contained in pores of sediments and sedimentary rocks. A little more than 1% (or 73% of freshwater) is locked up in ice (17). The other freshwater reservoir of significant size is groundwater. Lakes, rivers, and the atmosphere hold a surprisingly small fraction of the earth's water.

The oceans are subdivided into surface ocean (100–1000 m) and deep ocean. The zone separating the warmer surface water from the lower, cooler layer (oceanic thermocline) is characterized by a density gradient that prevents mixing (18,20).

The exchange of water between the various reservoirs includes ocean mixing, evaporation, precipitation, runoff, and percolation. The exchange of water between surface and deep ocean (mixing by upwelling and downwelling) and subsequent circulation of water masses are the most important water-transport processes. About 20 times more water is added to the surface ocean by mixing than by river runoff (18). The ocean–atmosphere–continents system may be looked at as a great heat engine; its water cycle is driven by the sun (see Solar energy). Evaporation uses 23% of the energy of solar radiation. The oceans provide 86% and the continents 14% of the annual total atmospheric water. The average annual amount of worldwide precipitation is 0.7 m, ranging from 0 to 0.25 m for desert areas, to 0.25–0.40 m for grasslands or open woodland, 0.40–1.25 m for dry forests, and >1.125 m for wet forest.

The mean residence time for a water molecule in the atmosphere is ca 10 d.

Rocks and the Course of Weathering. The chemical characteristics of natural waters must be viewed in the light of the environmental history of the water and the chemical reactions of the rock–water–atmosphere systems (17,20–24).

The sedimentary rocks, oceans, and atmosphere have probably been formed by the reaction of primary igneous rocks (bases) with excess volatiles (acids) originating

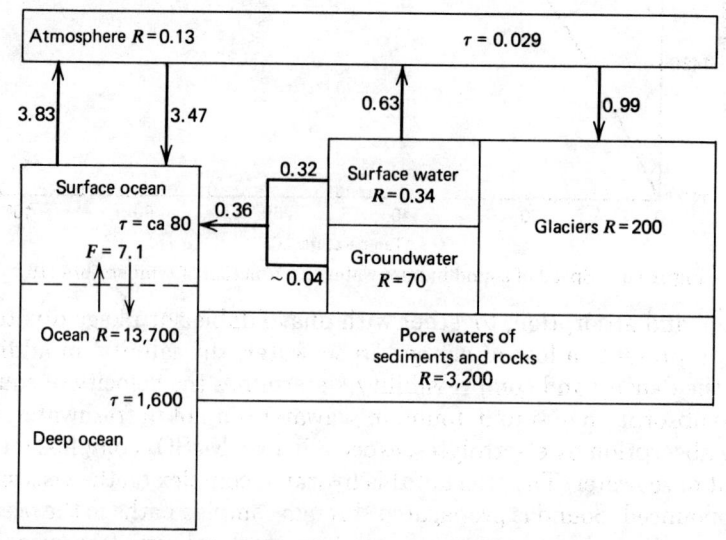

Figure 10. The principal reservoirs in the hydrological cycle. R, Reservoirs in units of 10^{14} metric tons; F, fluxes in units of 10^{14} t/yr; τ, residence time, yr. R/F = volume/input–output.

from the earth's interior and concentrating at the surface by distillation. Seawater is one product of these reactions.

An imaginary equilibrium model has been proposed for such a system (22). Since the compositions of igneous rocks or sediments have hardly changed in 6×10^8 yr, the composition of seawater has probably remained fairly constant (17,21). Current models for seawater composition emphasize the importance of steady-state balances between the various inputs (rivers) and outputs (sedimentation, atmospheric recycling to land). In the hydrochemical cycle, water acts as transport medium of weathered-rock material in suspended and dissolved form and as chemical reactant, converting this material into soil sediments, sedimentary rocks, and solutes (Fig. 11). In this cycle, the sea occupies a central position, whereas the atmosphere provides acids (CO_2) and oxidants. Typical weathering reactions are given below (25).

Congruent dissolution reactions

$$SiO_2(s) + 2 H_2O \rightarrow H_4SiO_4$$
quartz

$$CaCO_3(s) + H_2CO_3 \rightarrow Ca^{2+} + 2 HCO_3^-$$
calcite

$$Al_2O_3 \cdot 3H_2O(s) + 2 H_2O \rightarrow 2 Al(OH)_4^- + 2 H^+$$
gibbsite

$$Ca_5(PO_4)_3(OH)(s) + 4 H_2CO_3 \rightarrow 5 Ca^{2+} + 3 HPO_4^{2-} + 4 HCO_3^- + H_2O$$
apatite

$$Mg_2SiO_4(s) + 4 H_2CO_3 \rightarrow 2 Mg^{2+} + 4 HCO_3^- + H_4SiO_4$$
forsterite

Incongruent dissolution reactions

$$2 NaAlSi_3O_8(s) + 2 H_2CO_3 + 9 H_2O \rightarrow 2 Na^+ + 2 HCO_3^- + 4 H_4SiO_4 + Al_2Si_2O_5(OH)_4(s)$$
albite (Na-feldspar)   kaolinite

$$7 NaAlSi_3O_8(s) + 6 H_2CO_3 + 20 H_2O \rightarrow 6 Na^+ + 6 HCO_3^-$$
$$+ 10 H_4SiO_4 + 3 Na_{0.33}Al_{2.33}Si_{3.67}O_{10}(OH)_2(s)$$
Na-montmorillonite

$$Al_2Si_2O_5(OH)_4(s) + 5 H_2O \rightarrow 2 H_4SiO_4 + Al_2O_3 \cdot 3H_2O$$
kaolinite   gibbsite

$$CaMg(CO_3)_2(s) + Ca^{2+} \rightarrow Mg^{2+} + 2CaCO_3(s)$$
dolomite   calcite

Redox reactions

$$3 Fe_2O_3(s) + H_2O + 2 e \rightarrow 2 Fe_3O_4(s) + 2 OH^-$$

$$2 OH^- \rightarrow H_2O + \tfrac{1}{2} O_2 + 2 e$$

$$3 Fe_2O_3(s) \rightarrow 2 Fe_3O_4(s) + \tfrac{1}{2} O_2$$
hematite   magnetite

$$FeS_2(s) + 1\tfrac{1}{2} H_2O + 4\tfrac{3}{4} O_2 + 4 e \rightarrow Fe(OH)_3(s) + 2 SO_4^{2-}$$

$$2 H_2O \rightarrow O_2 + 4 H^+ + 4 e$$

$$FeS_2(s) + 3\tfrac{1}{2} H_2O + 3\tfrac{3}{4} O_2 \rightarrow Fe(OH)_3(s) + 4 H^+ + 2 SO_4^{2-}$$
pyrite

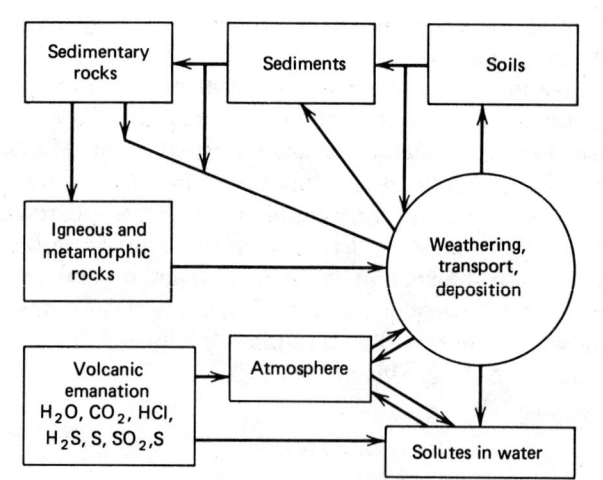

Figure 11. Interaction of the cycle of rocks with that of water (20).

Similarly, the composition of freshwater may be interpreted as a consequence of the interaction of water with atmospheric gases (mostly carbon dioxide) and mineral rocks.

Carbonate minerals ($CaCO_3$) and some silicates (ie, forsterite) dissolve congruently. Most other weathering processes are incongruent dissolutions; ie, part of the solid dissolves, leaving behind a solid phase different in composition from the original. The conversion of feldspar to kaolinite, an important weathering reaction, can be regarded as a hydrolysis process in which the silicate becomes replaced by OH^- in a different structural configuration. The reaction may go through various Al–silicate intermediates (weathering sequence).

The dissolution processes are influenced by the environment, including pH, complex formation, and changes in the oxidation state. Naturally occurring organic matter, such as water-soluble anions of low molecular weight or low molecular weight organic acids, may form chelate complexes, whereas natural redox processes may control the pH. For example, the oxidation of pyrite yields ferric oxide and acid which may react with the rocks; on the other hand, hematite (Fe_2O_3) may, upon reduction, produce large quantities of OH^-. Physical and biological weathering may accelerate the relatively slow chemical weathering process by further comminuting inorganic material, thereby increasing the specific surface area susceptible to weathering.

Biochemical Cycle and Oxidation–Reduction Processes. The ecological system may be considered as a unit of the environment that contains a biological organization made up of all the organisms interacting reciprocally with the chemical and physical environment. The maintenance of life in aquatic ecosystems results directly or indirectly from the steady impact of solar energy (qv). Photosynthesis is carried out mostly by algae and water plants; it may be conceived as a disproportionation of water into an oxygen reservoir and hydrogen, which forms high energy bonds with C, N, S, and P compounds that are incorporated as organic matter in the biomass (20). Various organisms catalytically decompose the unstable products of photosynthesis through energy-yielding electron-transfer reactions (reduction–oxidation processes, respiration). These organisms use this source of energy for their metabolic needs.

The composition of natural waters is strongly influenced by the growth, distribution, and decay of algae and other organisms. Organisms regulate the oceanic and lacustrine composition and its variation with depth. Dissolved constituents are taken up by organisms. Their remains sink under the influence of gravity and are gradually destroyed by oxidation. The superposition of this particular cycle upon the ordinary mixing cycle in the ocean or in lakes accounts for the variation in depth and distribution of chemical properties.

Chemical Composition. The concentrations of the principal elements in marine and freshwaters are given in Figure 12. The concentrations of biologically regulated components (ie, C, N, P, Si) vary with depth and are markedly influenced by the growth, distribution, and decay of phytoplankton and other organisms. The concentrations of other constituents, especially the salts (Cl^-, SO_4^{2-}, Mg^{2+}, and Na^+), are remarkably constant and are different from those in fresh waters. For most elements,

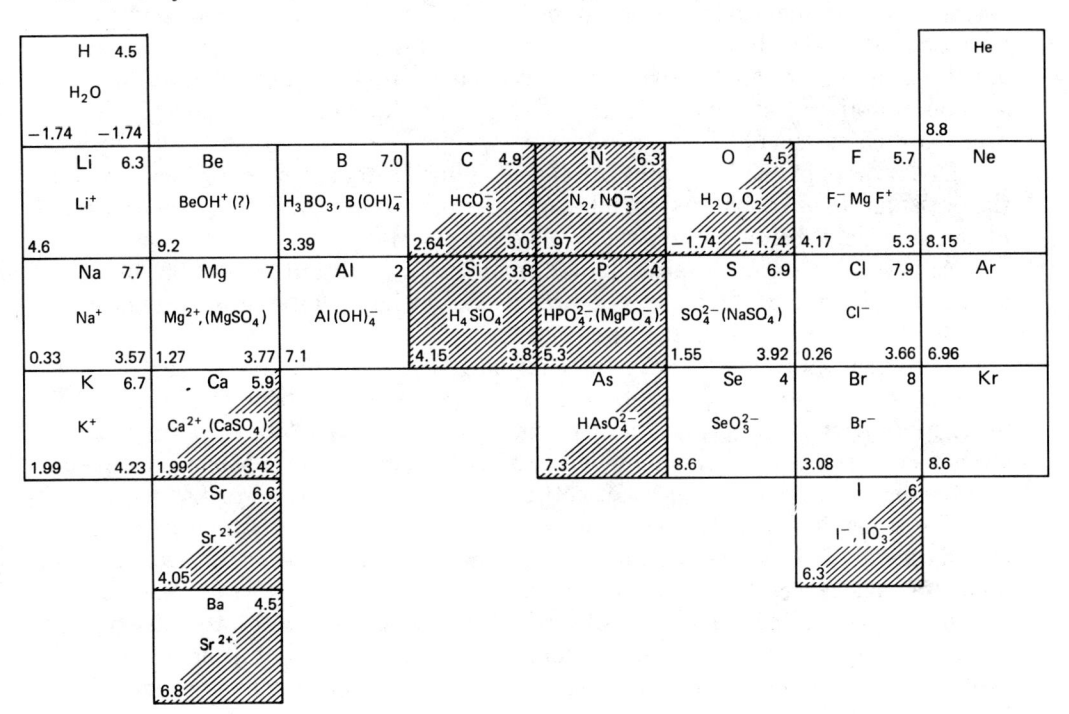

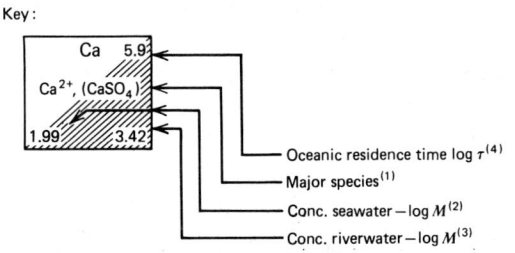

Figure 12. Elements in natural waters, their form of occurrence, and concentration. Elements whose distribution is significantly affected by biota are shaded; P, N, and Si (fully shaded) are often depleted in surface waters. Species in parentheses are the main ion pairs in seawater. Concentrations (M = mol/L) in seawater are taken from ref. 26, and concentrations in river water from ref. 27. Data are uncorrected for atmospheric recycling of marine salts (20,28).

a mass balance appears to exist between input into the sea (mostly by rivers) and its removal (mostly by sedimentation). Thus, the oceans are often assumed to be at steady state. The mean residence time τ (yr) of an element or constituent is defined as average time a given element spends in the ocean before being removed by sedimentation and other processes:

$$\tau = \frac{\text{total mass}}{\text{input or output}} = \frac{M}{Q}$$

where M is the concentration of a particular constituent in seawater times the volume of the ocean water, and Q is the product of the concentration of the particular element in average river water and the annual flux of river water to the ocean or the quantity of this element being removed from the water by sedimentation (see also Mass transfer). Some sea salts, especially NaCl, are recycled through the atmosphere (as aerosols from droplets of sea spray) to the land masses where they are washed out by rain. Chloride and sodium, originating from the atmosphere, for example, contribute about one third of the annual river flux of these elements. The other main constituents are mainly derived from the weathering of rocks, and the fraction for atmospherically cycled salts is small (17,28).

The concept of residence time can also be applied to lakes where the flow through the outlet has to be considered. In lakes it is often convenient to define a relative residence time, ie, a residence time relative to that of water.

At present, it is not certain whether the flow of hydrothermal solutions into the oceans and the chemical reactions occurring there by high temperature alteration of basalts are of significance in the mass balance.

The observed residence times range from 100 yr for aluminum to ca 10^8 yr for chlorine (Figs. 12 and 13), reflecting the great variation in geochemical reactivity; unreactive elements simply accumulate in the ocean and have large τ values. Figure 13 shows the strong correlation between the mean oceanic residence time of an element and the distribution of that element between seawater and crystal rocks.

The transition between rivers and oceans occurs in estuaries, which are semienclosed coastal bodies of water within which seawater is diluted with freshwater from land drainage. Several types of water circulation may occur (vertically mixed or stratified), depending on topography, tidal currents, and freshwater discharge. Estuaries are of great environmental concern. Many marine organisms require the estuarine environment for part of their life cycle; thus, pollution in an estuary may have long-range ecological consequences.

Estuaries with their salinity gradients and tidal movements represent gigantic natural coagulation tanks. In coagulation, colloidal particles agglomerate and are subsequently removed by sedimentation. Clay minerals, organic matter, colloidal hydroxides of iron(III), and heavy metals tend to coagulate and accumulate in the estuary. The trapping of organic matter may impart anaerobic conditions in the lower part of estuarine water.

Chemical Variety. The term species refers to the actual form in which a molecule or ion is present in solution (20,30–31). The various forms in which metal ions may occur in seawater are shown in Figure 14. It is difficult to distinguish between dissolved and particulate forms. The definition by pore size of membrane filter (0.45 μm) is considered questionable, and dissolved species may often include some or most of the colloidal forms of an element. The inorganic fraction of suspended particulate matter

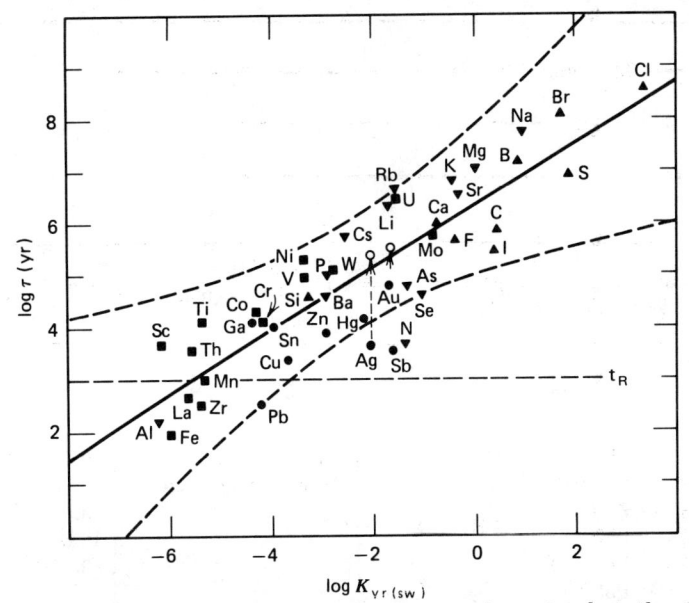

Figure 13. The relationship between the mean oceanic residence time [τ (yr)] and the seawater-crustal rock partition function [$K_{yr(SW)}$] of the elements. ▼, pretransition metals; ■, transition metals; ●, B-metals; ▲, nonmetals. Open symbols indicate [τ (yr)]-values estimated from sedimentation rates. The solid line indicates the linear regression fit and the dashed lines the Working-Hotelling confidence band at the 0.1% significance level. The horizontal broken line indicates the time required for one stirring revolution of the ocean (τ_R) (29).

consists mainly of minerals formed by weathering of terrestrial rocks. The organic particulate matter is primarily composed of living organisms, their decay, and metabolic products. Inorganic trace elements may be sorbed or bonded onto them.

Dissolved inorganic species present in seawater are principally electrolytes, uncharged species, and dissolved gases. The species patterns for 58 cations in model seawater (pH 8.2) and freshwater (pH 6 and 9) have been derived from calculations based on stability constants of complexes (32). The principal species of the more important elements occurring in natural waters are given in Figure 12.

Interactions between cations and anions may range from electrostatic attraction (ion pairs) to covalent bonding, ie, unidentate or multidentate (chelate) complexes. Some cations form their strongest complexes with electron donors from the first row of the periodic table (hard ligands, F, O, N) and bond mainly by means of electrostatic interactions. Others form their most stable complexes with the ligand atom in the second, third, or fourth row of the periodic table (soft ligands, S, P, I, Cl, N). Strong bonds are produced mainly by covalent interactions.

Four general groups summarize cation species in natural waters (31–32).

The first group includes cations that are so weakly complexed in both freshwater and seawater that they are dominated by the free cation.

In the second group, the species are dominated by hydrolysis (ligands having oxygen). Other anions (preferentially, fluorine ions) compete with the hydroxyl ion, and their complexes are formed depending on the stability constant. For example, the stability increases rapidly with an increase in charge on the metal ion, and those ions with the smallest radii form the most stable complexes.

The third group includes cations whose species are dominated by complexation

290 WATER (PROPERTIES)

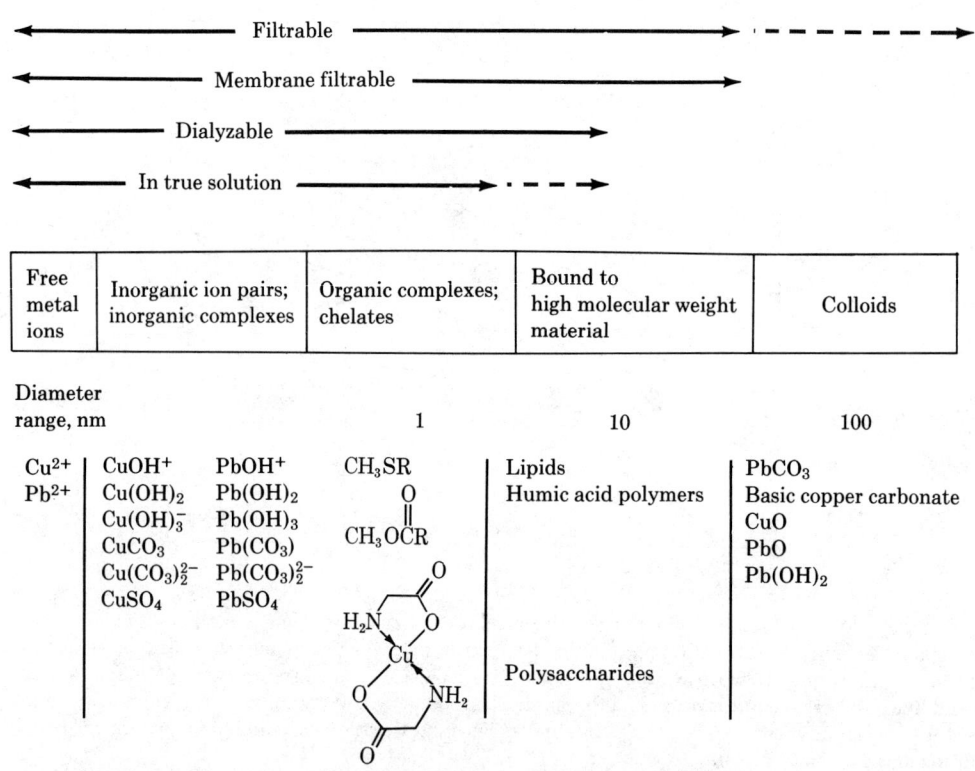

|←——————— Filtrable ———————→| · – – – – →
|←——— Membrane filtrable ———→|
|←——— Dialyzable ———→|
|←—— In true solution ——→| · – – →

Free metal ions	Inorganic ion pairs; inorganic complexes	Organic complexes; chelates	Bound to high molecular weight material	Colloids

Diameter range, nm   1   10   100

Cu^{2+}	$CuOH^+$	$PbOH^+$	CH_3SR	Lipids	$PbCO_3$
Pb^{2+}	$Cu(OH)_2$	$Pb(OH)_2$	CH_3OCR (with C=O)	Humic acid polymers	Basic copper carbonate
	$Cu(OH)_3^-$	$Pb(OH)_3$			CuO
	$CuCO_3$	$Pb(CO_3)$			PbO
	$Cu(CO_3)_2^{2-}$	$Pb(CO_3)_2^{2-}$			$Pb(OH)_2$
	$CuSO_4$	$PbSO_4$			
			(Cu-glycinate chelate structure)	Polysaccharides	

Figure 14. Metal species occurring in seawater (20).

with chloride. The metal cations in this class as well as the transition metal cations form insoluble sulfides and soluble complexes with S^{2-} and HS^-. The tendency toward complex formation increases with the capability of the cation to take up electrons and with increasing electronegativity of the ligand (ie, increasing tendency of the ligand to donate electrons). In the series F, O, N, Cl, Br, I, S, the electronegativity decreases from left to right, whereas the stability of complexes dominated by chloride increases (20).

The transition-metal cations constitute a fourth group whose species in natural water vary widely. For these cations, the stability of complexes increases in the series $Mn^{2+} < Fe^{2+} < Co^{2+} < Ni^{2+} < Cu^{2+} > Zn^{2+}$ according to a well-established rule on the sequence of complex stability (20).

Pollution

The products of human activities find their way into the environment and disturb ecosystems. Pollution has altered the surroundings to the detriment of humanity. In the last decades, the pollutional load increased, and its character has changed (20). Aquatic pollutants consist of salts (Cl^-, SO_4^{2-}, Na^+, Mg^{2+}, K^+, Ca^{2+}), nutrients (compounds intimately involved with the life cycle), and trace constituents (heavy metals, synthetic organic compounds, xenobiotic substances) (18).

Salts and minerals do not have a measurable impact on the sea; however, they may contaminate fresh and coastal waters by causing a local change in the geochemical composition (see Water pollution).

The balance between photosynthesis and respiration in a receiving water (lakes, estuaries, and coastal basins) may become perturbed by pollution with inorganic plant (algal) nutrients (eg, phosphorus and nitrogen compounds). Excessive production of algae in surface waters is followed, subsequent to the settling of the biota and its debris into the deeper waters, by the respiration (oxidative mineralization) of this organic material by oxygen creating an oxygen deficit or even anaerobic conditions in the deeper waters. Many trace constituents (eg, mercury, pesticides, phenols) are harmful to specific aquatic life and other organisms.

The fate of a pollutant in an aquatic system may be expressed as follows:

$$\text{accumulation} = \text{input} - \begin{pmatrix} \text{interphase} \\ + \\ \text{intraphase} \\ \text{mass transfer} \end{pmatrix} - \begin{pmatrix} \text{chemical} \\ + \\ \text{biological} \\ \text{reaction} \end{pmatrix} - \text{output}$$

By evaluating the strength of anthropogenic and natural emission sources, identifying the relevant pathways, and characterizing the various processes (dilution, dispersion, transport, partitioning, adsorption, volatilization, chemical and biological transformation, biodegradation) that govern the behavior of chemicals in the environment, present and future fluxes can be predicted, as well as the distribution, fate, and residual activities of pollutants and their effects on ecological systems and on human health (20,33–39).

Pollutant Distribution. Of particular importance for the aquatic ecosystem are the distribution of volatile substances (gases, volatile organic compounds) between the atmosphere and water, and the sorption of compounds at solid surfaces (settling suspended matter, biological particles, sediments, soils).

Under equilibrium or near equilibrium conditions, the distribution of volatile species between gas and water phases can be described in terms of Henry's law. The rate of transfer of a compound across the water–gas phase boundary can be characterized by a mass-transfer coefficient and the activity gradient at the air–water interface. In addition, these substance-specific coefficients depend on the turbulence, interfacial area, and other conditions of the aquatic systems. They may be related to the exchange constant of oxygen as a reference substance for a system-independent parameter; reaeration coefficients are often known for individual rivers and lakes (40–42).

The solid-solution interface is of greatest importance in regulating the concentration of aquatic solutes. Suspended inorganic and organic particles and biomass, sediments, soils, and minerals (eg, in aquifers and infiltration systems) act as adsorbents. The reactions occurring at interfaces can be described with the help of surface-chemical theories. The adsorption of polar substances, eg, metal cations, M, anions, A, and weak acids, HA, on hydrous oxide or clay or organically coated surfaces may be described in terms of surface-coordination reactions:

$$ROH + M^+ \rightleftharpoons ROM + H^+$$

$$ROH + HA \rightleftharpoons RA + H_2O$$

With the help of equilibrium constants, the extent of adsorption can be predicted as a function of pH and solution variables (43–44).

Based on this model, the partitioning of most inorganic elements between seawater

WATER (PROPERTIES)

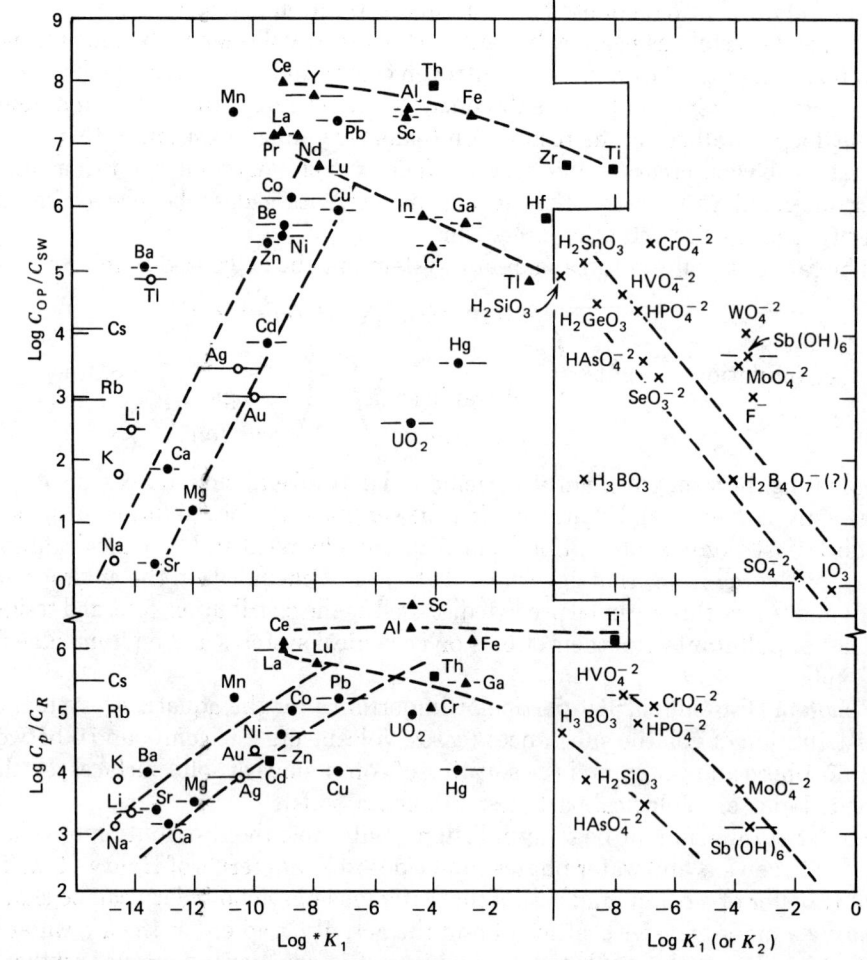

Figure 15. Adsorption of polar substances. Plots of the distribution coefficient between solid and liquid phases in the ocean (C_{OP}/C_{SW}) and in river water (C_p/C_R) as a function of log $*K_1$ (the first hydrolysis constant) as a measure of the bond strength between cations and the oxygen donor group of the hydrous oxide surface, and log K_1 or log K_2 (first and second dissociation constants of acids); the strength of the bond between Fe or Mn and the O of the oxyanion is inversely related to log K_2 or log K_1. Acids are indicated by x on the graph. Valence of cations: ○, 1; ●, 2; ▲, 3; ■, 4 (45).

and pelagic clays and between river water and particles suspended in rivers, can be explained. The partitioning of various cations between solid and liquid phases in the ocean or in rivers is presented in Figure 15. It can be correlated with the bond strength between cations (metals) and the oxygen of the hydrous oxide surface, ie, with $*K_1$. The distribution of oxyanions between the solid and liquid phases in oceans and rivers is related to the corresponding ligand-exchange surface-equilibrium constants or inversely to the acidity constants of the corresponding acids, K_1 or K_2. The higher log K_1 or K_2, the weaker is the bond between Fe or H and the ligands (dissociated acids), and therefore, the lower is the ratio of concentrations in the solid and liquid phases (45).

Nonpolar organic compounds are hydrophobic, ie, they are readily soluble in nonpolar solvents but only sparingly soluble in water. The tendency of a compound to dissolve in a nonpolar solvent is often measured by the octanol–water distribution coefficient K_{ow}, which is considered an indicator for the lipophilicity of a substance (tendency to become dissolved in the lipids). Such hydrophobic nonpolar substances tend to reduce the contact with water, seek relatively nonpolar environments, and tend to accumulate at solid surfaces of an aqueous solution. Because the nonpolar part of the solid surface has an affinity for the lipophilic solute, the organic material present in natural solids enhances the sorption of these substances.

The extent of sorption of nonpolar organic trace compounds (ie, halogenated hydrocarbons) on natural solid materials (river and lake sediments, biota, activated sewage sludge) is described by linear isotherms or by a partition coefficient K_p. For different natural sorbents containing >0.1% organic carbon (f_{oc} >0.001), the partition coefficient of a particular hydrophobic compound K_p can be related to its octanol–water distribution coefficient K_{ow} and the organic carbon content f_{oc} of the solid material (12), as expressed by the equation below

$$K_p = a \cdot K_{ow}^n \cdot f_{oc}$$

where $a = 0.49$ and $n = 0.72$.

Lipophilic substances tend to become concentrated in organisms, and the biomagnification resulting from the accumulation of ecologically or toxicologically harmful substances in the food chain is important in ecotoxicology.

BIBLIOGRAPHY

"Water Properties" in *ECT* 2nd ed., Vol. 21, pp. 668–688, by Ralph A. Horne, Woods Hole Oceanographic Institution.

1. D. Eisenberg and W. Kauzmann, *The Structure and Properties of Water*, Oxford University Press, London, 1969.
2. L. Pauling, *J. Am. Chem. Soc.* **57**, 2680 (1935).
3. C. M. Davis and T. A. Litovitz, *J. Chem. Phys.* **42**, 2563 (1965).
4. D. T. Hawkins, *J. Solution Chem.* **4**(8), 623 (1975).
5. R. A. Horne, *Marine Chemistry*, Wiley-Interscience, a division of John Wiley & Sons, Inc., New York, 1969.
6. R. A. Horne, ed., *Water and Aqueous Solutions. Structure, Thermodynamics, and Transport Processes*, Wiley-Interscience, a division of John Wiley & Sons, Inc., New York, 1972.
7. J. A. Pople, *Proc. R. Soc. London Ser. A* **205**, 163 (1951).
8. E. W. Lang and H. D. Lüdemann, *Angew. Chem.* **94**, 351 (1982).
9. N. E. Dorsey, *Properties of Ordinary Water-Substance in All Its Phases: Water Vapor, Water and All the Ices*, American Chemical Society, Reinhold Publishing Corporation, New York, 1940.
10. J. Westall and W. Stumm in A. O. Hutzinger, ed., *The Handbook of Environmental Chemistry*, Vol. 1, Pt. A, Springer Verlag, Berlin, 1980.
11. G. S. Kell, *J. Chem. Eng. Data* **12**, 66 (1967).
12. R. P. Schwarzenbach and J. Westall, *J. Environ. Sci. Technol.* **15**, 1360 (1981).
13. L. D. Eicher and B. J. Zwolinski, *J. Phys. Chem.* **75**, 2016 (1971).
14. L. Korson, W. Drost-Hansen, and F. J. Millero, *J. Phys. Chem.* **73**(1), 34 (1969).
15. K. E. Bett and J. B. Cappi, *Nature* **207**, 620 (1965).
16. V. A. Del Grosso and C. W. Mader, *J. Acoust. Soc. Am.* **52**(5), 1442 (1972).
17. R. M. Garrels and F. T. Mackenzie, *Evolution of Sedimentary Rocks*, W. W. Norton & Company, Inc., New York, 1971.
18. W. S. Broecker, *Chemical Oceanography*, Harcourt Brace Jovanovich, Inc., New York, 1974.
19. R. J. Chorley, *Water, Earth, and Man*, Methuen & Co. Ltd., London, 1969.
20. W. Stumm and J. J. Morgan, *Aquatic Chemistry, An Introduction Emphasizing Chemical Equilibria*

in Natural Waters, 2nd ed., Wiley-Interscience, a division of John Wiley & Sons, Inc., New York, 1981.
21. F. T. Mackenzie in J. P. Riley and G. Skirrow, eds., *Chemical Oceanography*, 2nd ed., Vol. 1, Academic Press, Inc., New York, 1975, pp. 309–364.
22. L. G. Sillén in M. Sears, ed., *Oceanography*, American Association for the Advancement of Science (AAAS), Washington, D.C., 1961, pp. 549–581.
23. R. E. McDuff and F. M. M. Morel, *Environ. Sci. Technol.* **14**(10), 1182 (1980).
24. G. D. Nicholls in J. P. Riley and R. Chester, eds., *Chemical Oceanography*, 2nd ed., Vol. 5, Academic Press, Inc., New York, 1976, pp. 81–101.
25. Ref. 20, p. 531.
26. P. G. Brewer in ref. 21, pp. 415–496.
27. D. A. Livingstone, *Chemical Composition of Rivers and Lakes*, U.S. Geological Survey Professional Paper 440-G, Washington, D.C., 1963.
28. H. D. Holland, *The Chemistry of the Atmosphere and Oceans*, Wiley-Interscience, a division of John Wiley & Sons, Inc., New York, 1978.
29. M. Whitfield and D. R. Turner, *Nature* **278**, 132 (1979).
30. J. P. Riley and R. Chester, eds., *Introduction to Marine Chemistry*, Academic Press, Inc., New York, 1971.
31. W. Stumm and P. A. Brauner in ref. 21, pp. 173–235.
32. D. R. Turner, M. Whitfield, and A. G. Dickson, *Geochim. Cosmochim. Acta* **45**, 855 (1981).
33. L. J. Thibodeaux, *Chemodynamics, Environmental Movement of Chemicals in Air, Water, and Soil*, Wiley-Interscience, a division of John Wiley & Sons, Inc., New York, 1979.
34. I. J. Tinsley, *Chemical Concepts in Pollutant Behavior*, Wiley-Interscience, a division of John Wiley & Sons, Inc., New York, 1979.
35. G. L. Baughman and L. A. Burns in O. Hutzinger, ed., *Environmental Chemistry*, Vol. 2, Pt. A, Springer Heidelberg, FRG, 1980, pp. 1–17.
36. W. B. Neely, *Chemicals in the Environment. Distribution·Transport·Fate·Analysis*, Marcel Dekker, Inc., New York, 1980.
37. *Water-Related Environmental Fate of 129 Priority Pollutants*, Vols. I and II, EPA-440/4-79-029a, Environmental Protection Agency, Washington, D.C., 1979.
38. J. F. Pankow and J. J. Morgan, *Environ. Sci. Technol.* **15**, 1155 (1981).
39. D. Mackay and S. Paterson, *Environ. Sci. Technol.* **15**, 1006 (1981).
40. D. Mackay and T. K. Yuen, *Proceedings of 14th Canadian Symposium, 1979: Water Pollution Research*, Canada.
41. E. C. Tsivoglou and L. A. Neal, *J. Water Pollut. Control Fed.* **48**(12), 2669 (1976).
42. C. Matter-Müller, W. Gujer, and W. Giger, *Water Res.* **15**, 1271 (1981).
43. P. W. Schindler, *Thalassia Jugosl.* **11**, 101 (1975).
44. W. Stumm, R. Kummert, and L. Sigg, *Croat. Chem. Acta* **53**(2), 291 (1980).
45. Y. H. Li, *Geochim. Cosmochim. Acta* **45**, 1659 (1981).

CHRISTINA MATTER-MÜLLER
WERNER STUMM
Swiss Federal Institute of Technology

POLLUTION

Water is omnipresent on the earth. Constant circulation of water from the ocean to the atmosphere (evaporation) and from the atmosphere to land and the oceans (precipitation, runoff, etc) is generally known as the hydrologic cycle (see Fig. 1) (1–2). Within the hydrologic cyclic, there are several minor and local subcycles where water is used and returned to the environment.

The volume of the freshwater amounts to only one thirtieth of the 1.25×10^9 km^3 (300×10^6 mi^3) of the water in salty oceans. Approximately one third of the freshwater exists permanently as snow and ice (3). A large portion of the remaining freshwater has infiltrated too far undergound or is partially polluted with minerals and chemicals and therefore is not readily usable. The entire life system on the earth depends on the remaining freshwater sources; therefore, it is essential to protect the quality of the available waters.

The principal hydrologic parameters involved in the storage and transport of freshwater were studied as early as 1894 and are summarized in Table 1 (2).

Freshwater is withdrawn from various sources (rivers, lakes, groundwater, etc) and used many times before its discharge to the ocean. Water uses can generally be classified as follows: public water supply (domestic); industrial; commercial and institutional, eg, restaurants, schools; agricultural; and livestock.

These applications require withdrawal of water from a source and subsequent treatment and conveyance to the point of use. Water is also used without being withdrawn from a source, eg, for navigation, recreation, wild and aquatic life propagation, hydroelectric-power generation, and waste assimilation and transport. The principal

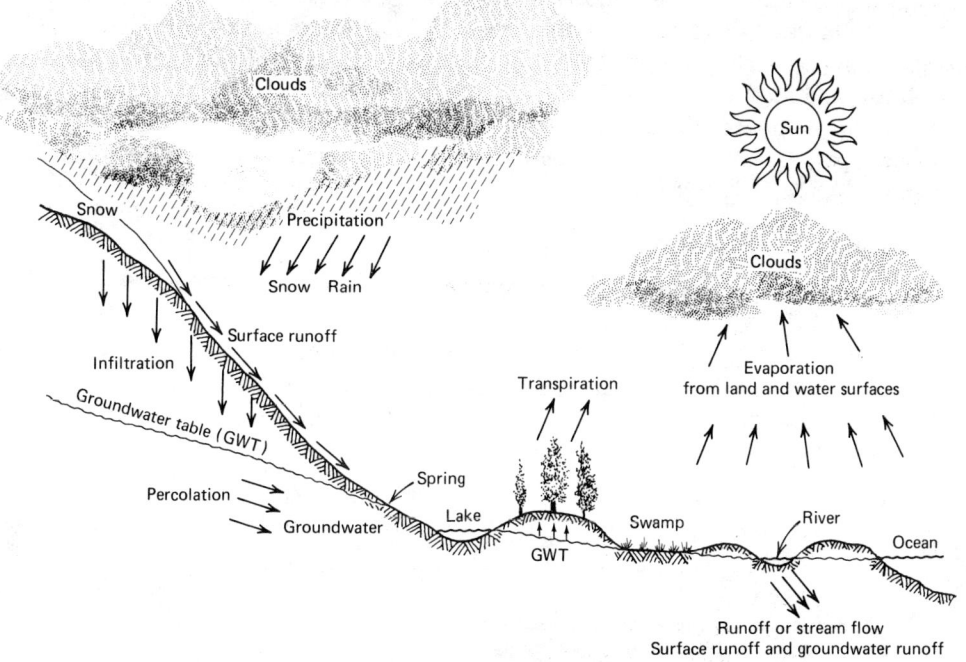

Figure 1. Schematic diagram of the hydrologic cycle.

Table 1. Principal Hydrologic Parameters[a]

Storage	Transport
atmospheric water	evaporation from land surface; oceans, lakes, and other water bodies; and snow and ice
oceans and seas	precipitation
rivers	runoff
lakes and reservoirs	infiltration
swamps	
biological water	
soil moisture and groundwater	
frozen water (ice and snow)	

[a] Ref. 3.

types of withdrawal uses and their average rates are given in Table 2. Some of these withdrawal rates represent multiple uses of the same water along main rivers in metropolitan and industrialized areas.

Water-use data and withdrawal rates for public water-supply systems are well documented by municipalities. The U.S. Public Health Service (USPHS) (5), American Water Works Association (AWWA) (6–10), and Federal Housing Administration (FHA) (11) have compiled statistics at regular intervals for >23,000 municipalities in the United States.

Quantitative information on industrial water use is less complete, and water use within industrial groups varies significantly. A complete survey is required to accurately estimate the total industrial water-use requirements. Comprehensive information on industrial water requirements is given in refs. 12–31.

The largest consumers of water in the United States are thermal power plants (eg, steam and nuclear power plants) and the iron and steel, pulp and paper, petroleum refining, and food-processing industries. They consume >60% of the total industrial water requirements (see also Power generation; Food processing; Pulp; Petroleum, refinery processes). The requirements for selected industrial categories are given in Table 3 (4,32).

Table 2. Average Daily Water Requirements for Categories of Withdrawal Use in the United States[a]

Use	Average daily requirement[b], 10^6 m^3/d (mgd)[c]
public[d]	102.2 (27,000)
rural	
domestic	9.84 (2,600)
livestock	7.19 (1,900)
agriculture	492.10 (130,000)
industrial	794.90 (210,000)

[a] Ref. 4.
[b] Estimated 206 × 10^6 people served.
[c] mgd = million (10^6) gallons per day.
[d] Residential and municipal (domestic).

Table 3. Water-Use Requirements and Effluent Discharge of Selected Industries[a]

Industry	Total water intake[b], m³/s	Total effluent discharge, m³/s
food	97.3	90.6
meat products	12.6	12.0
dairy products	6.7	6.4
canned, cured, and frozen foods	15.5	14.4
grain mill products	9.0	8.1
sugar processing	29.2	27.9
beverages	13.4	11.7
textiles	18.5	16.3
lumber and wood	14.1	11.2
paper and pulp	270.3	249.4
chemicals	537.2	501.1
petroleum and coal products	172.2	146.1
rubber and plastics	16.2	15.4
leather and leather products	1.9	1.8
stone, clay, and glass products	30.1	26.1
primary metals	600.7	563.5
fabricated metals	8.1	7.8
machinery, except electrical	22.7	21.7
electrical equipment and supplies	15.2	14.1
transportation equipment	37.6	35.2
miscellaneous industries	6.2	5.9

[a] Most water-use data reflect 1968 information on manufacturing industries (4,32–33).
[b] To convert m³/s to mgd, multiply by 4.381×10^{-2}.

Water-Quality Management

Because of the crucial importance of water, its quality must be managed and preserved. The traditional practice of discharging partially treated spent waters (wastewater) into receiving streams for eventual cleanup and restoration is no longer adequate because of the increases in population density and industrial activities. In addition, new industrial chemicals and synthetic products significantly altered the characteristics of the pollutants that are being discharged. Thus, the water pollution problem may reach crisis proportions.

Quality Criteria. Whether the water of a given source is suitable for a specific use depends on the criteria or standards for that use. Earlier standards were primarily intended to protect the public health and were administered by USPHS.

The first water-quality criteria were adopted by the Treasury Department on October 21, 1914, based on the recommendations developed by the USPHS (34). These standards specified limits for bacteriological quality. Subsequent revisions between 1925 and 1942 included physical and chemical standards and defined the minimum analytical requirements based on the size of the community being served.

These standards were further revised between 1956 and 1962 and provided the basis for the European and international standards developed by the WHO. In 1968, the American Water Works Association (AWWA) further refined these standards by adopting the Quality Goals for Potable Water, which were more restrictive than the USPHS standards.

The USPHS drinking-water standards were primarily developed for interstate waters that are subject to Federal quarantine regulations. Most departments of health, however, have adopted these standards as general standards for all public water supplies.

Even though the USPHS standards provided protection for public water-supply systems, the control and discharge of pollutants were primarily monitored through state and local governments. The original Federal Water Pollution Control Act (FWPCA) was not passed until 1948. It gave temporary responsibility to the Federal government for interstate water pollution. The primary control and enforcement authority remained with local governments, however. Amendments passed in 1956 and 1966 primarily involved the transfer of responsibility between the Departments of the Interior and Health, Education, and Welfare. The FWPCA amendments in 1965 extended Federal authority to all navigable waters in the United States (interstate and intrastate waters).

The formation of the EPA in 1970 initiated specific water pollution control programs for restoring national water quality. Amendments in 1973 (PL 92-500) established specific goals and targets to make the nation's waters suitable for fishing and swimming and achieve a zero discharge of pollutants by 1985 (35). In addition, the National Pollutant Discharge Elimination System (NPDES) was established to provide the basic regulatory mechanism. It empowers the states to issue discharge permits, based on specific pollutants, to all owners of point-source discharges. The limitations are governed by the desired quality of the receiving stream. Each state was required to develop and adopt specific water-quality standards that meet or exceed the following water-quality criteria established by the U.S. EPA:

All waters must be free from substances attributable to wastewater or other discharges that settle to form objectionable deposits; float as debris, scum, oil, or other matters to form nuisances; produce objectionable color, odor, taste, or turbidity; injure, are toxic, or produce adverse physiological responses in humans, animals, or plants; and produce undesirable or nuisance aquatic life. Other data are given in Tables 4 and 5.

Table 4. Water Uses and Quality Control[a]

Classification	Quality-control parameters
water supply, recreation	coliform bacteria, color, turbidity, pH, dissolved oxygen, toxic materials, taste, odor, temperature
swimming, aquatic life[b], recreation	coliform bacteria, pH, dissolved oxygen, toxic materials, color, turbidity, temperature
industrial, agricultural, navigation, aquatic life	dissolved oxygen, pH, floating and settleable solids, temperature
navigation, cooling water	nuisance-free conditions, floating materials, pH

[a] Ref. 36.
[b] Fish.

Table 5. Water Quality Criteria, U.S. EPA (1976)[a]

Water quality characteristic or pollutant	Domestic water supply	Aquatic life Freshwater	Aquatic life Marine water	Swimming	Irrigation of crops
alkalinity as $CaCO_3$	[b]	≥20 mg/L	[b]	[b]	[b]
ammonia		0.02 mg/L			
arsenic	50 µg/L				100 µg/L
beryllium		11–1100	µg/L[c]		100 µg/L[d]
boron					750 µg/L
cadmium	10 µg/L	0.4–12 µg/L	5.0 µg/L		
chromium	50 µg/L	100 µg/L			
chlorine		2–10 µg/L	10 µg/L		
fecal coliform bacteria[e]	200/100 mL			200/100 mL	
color[f], Pt–Co scale	75 units				
copper	1.0 mg/L	0.1 times 96-h LC_{50}			
cyanide		5.0 µg/L	5.0 µg/L		
total dissolved gases		not to exceed 110% saturation value			
iron	0.3 mg/L	1.0 mg/L	LC_{50}		
lead	50 µg/L	0.01 times 96-h LC_{50}			
manganese	50 µg/L		100 µg/L (mollusks)		
mercury	2.0 µg/L	0.05 µg/L	0.1 µg/L		
nickel		0.01 times 96-h LC_{50}			
nitrates, nitrites	10 mg/L				
oil and grease	none	0.01 times 96-h LC_{50}			
dissolved oxygen		5.0 mg/L, min			
aldrin-dieldrin		0.003 µg/L	0.003 µg/L		
chlordane		0.01 µg/L	0.004 µg/L		
chlorophenoxy herbicides					
2,4-D	100 µg/L				
2,4,5-TP	10 µg/L				
DDT		0.001 µg/L	0.001 µg/L		
demeton		0.1 µg/L	0.1 µg/L		
endosulfan		0.003 µg/L	0.001 µg/L		
endrin	0.2 µg/L	0.004 µg/L	0.004 µg/L		
Guthion		0.01 µg/L	0.01 µg/L		
heptachlor		0.001 µg/L	0.001 µg/L		
lindane	4.0 µg/L	0.01 µg/L	0.004 µg/L		
malathion		0.1 µg/L	0.1 µg/L		
methoxychlor	100 µg/L	0.03 µg/L	0.03 µg/L		
mirex		0.001 µg/L	0.001 µg/L		
parathion		0.04 µg/L	0.04 µg/L		
toxaphene	5 µg/L	0.005 µg/L	0.005 µg/L		
pH	5–9	6.5–9.0	6.5–8.5		
phenol	1 µg/L	1.0 µg/L	1.0 µg/L		
phosphorus, elemental		0.1 µg/L	0.1 µg/L		
phthalate esters		3.0 µg/L			
polychlorinated biphenyls		0.001 µg/L	0.001 µg/L		
selenium	10 µg/L	0.01 times 96-h LC_{50}			
silver	50 µg/L	0.01 times 96-h LC_{50}			
dissolved solids	(250 mg/L for Cl^- and SO_4^{2-})				
sulfide (H_2S)		2 µg/L	2 µg/L		

[a] Ref. 35.   [b] No limitations specified for this use classification.
[c] Ranges in value represent the criteria for soft and hard freshwaters, respectively.
[d] For irrigation on neutral to alkaline fine-textured soils the criterion is 500 µg/L.
[e] 14/100 mL for shellfish harvesting.
[f] Increased color may not reduce the depth of compensation point for photosynthetic activity by more than 10%.

Pollution Control

In past years, municipal wastewaters were treated to improve their appearance and bacteriological safety. Treatment included reduction in biochemical oxygen demand (BOD), suspended solids, pathogens, and inorganic dissolved solids. Before 1960, the activated-sludge process, developed in the UK in 1914, was used (37). It was based on the capabilities of microorganisms to assimilate organic compounds through oxidative and respiratory mechanisms. The organisms involved could be flocculated and settled in conventional sedimentation vessels (clarifiers).

The effluent was relatively clear and required only disinfection before discharge to receiving waters. Preliminary treatment removed grit, sand, and floating debris. It was followed by primary clarification for the removal of suspended solids, grease, and scum. The effluent from the primary clarifiers, containing colloidal and dissolved organic matter, was subjected to activated-sludge treatment and disinfection (eg, chlorination) before discharge. A typical process flow sheet is given in Figure 2 (38); the process components are shown in Figure 3.

When industrial wastewaters are mixed with municipal wastes, as in many urban systems, toxic and inhibitory materials are removed in the pretreatment system where nutrient chemicals, eg, nitrogen and phosphorus, are added. The estimated construction costs for the treatment of municipal wastewaters are given in Figures 4–6 (39).

Biological processes (eg, activated sludge, trickling filters, etc) generate primary and biological sludges. These by-products require further treatment and processing

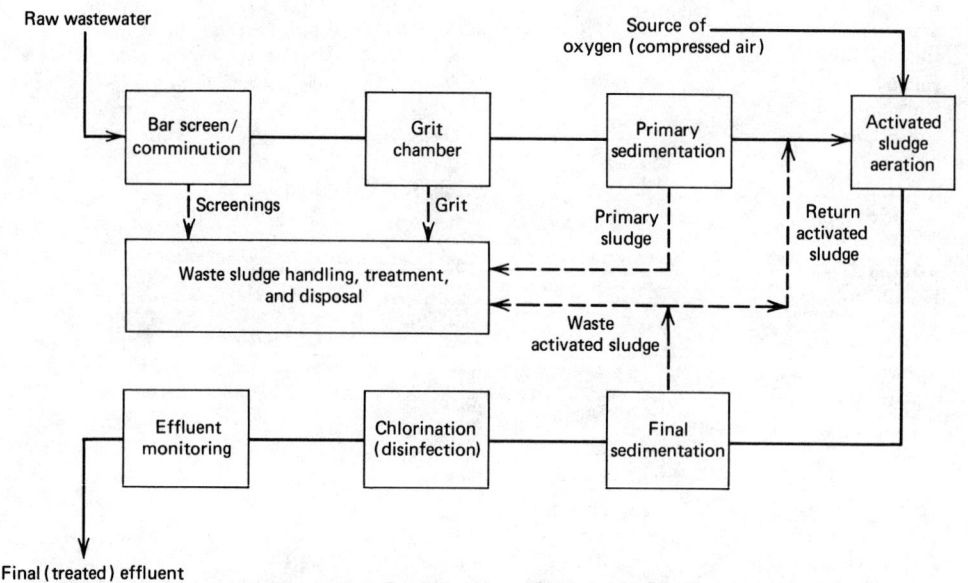

Figure 2. Typical flow sheet for a domestic wastewater treatment plant utilizing the activated-sludge process.

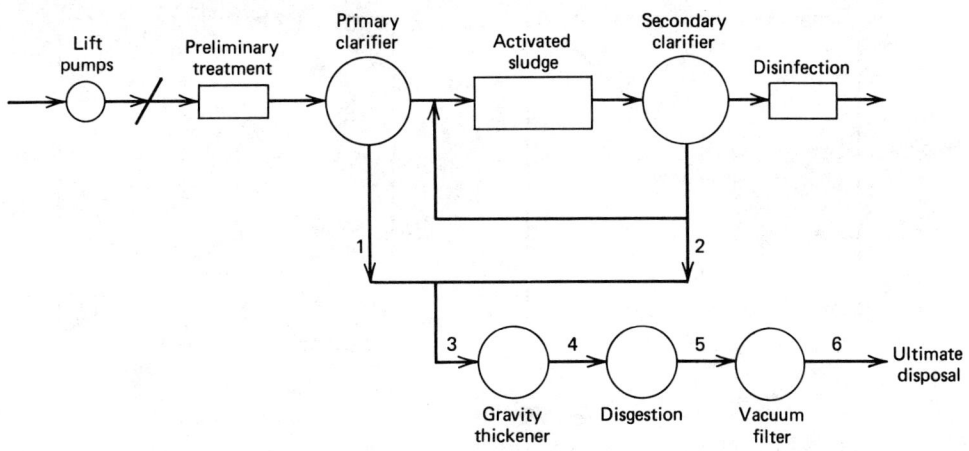

Figure 3. Process components for the activated-sludge process, with aerobic digestion at plants <473.1 m³/h and two-stage anaerobic digestion at plants >473.1 m³/h (39–40). To convert m³/h to mgd, divide by 157.7.

Point no.	Sludge, kg/m³	Concentration, %
1	129.4	4
2	98.3	0.8
3	227.7	2.6
4	227.7	8
5	113.8	5
6	113.8	20

before they are suitable for ultimate disposal. In general, the waste sludge is thickened, stabilized by anaerobic or aerobic digestion, and dewatered before land application or land disposal (39–40). The design parameters for the selection and sizing of various process units are given in Table 6 and Figure 3. The capital and annual operation and

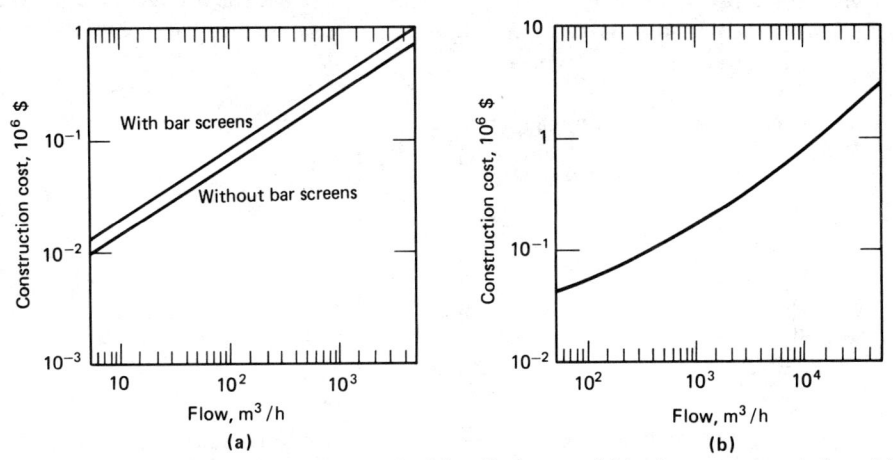

Figure 4. Installed capital cost estimates for (**a**) preliminary and (**b**) primary treatment of municipal wastewaters (ENR (*Eng. News Record*) Index: 2475, Sept. 1976). To convert m³/h to mgd, divide by 157.7.

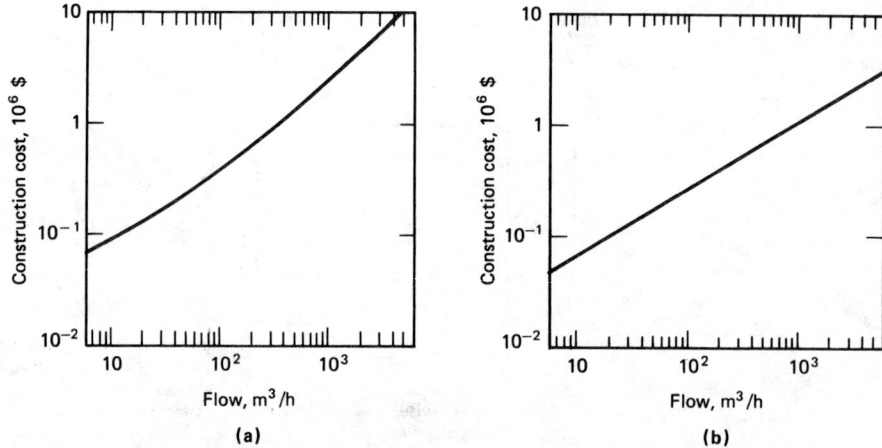

Figure 5. Installed capital cost estimates for (a) conventional activated-sludge and (b) clarification process units (ENR (*Eng News Record*) Index: 2475, Sept. 1976). To convert m³/h to mgd, divide by 157.7.

maintenance cost estimates for the treatment of municipal wastewaters, utilizing the activated-sludge process, are presented in Figure 7.

Before 1960, research and development work was directed toward the improvement of aerobic biological treatment systems. Fixed media were utilized on which microorganisms could attach and grow, eg, stone or plastic trickling filters, rotating biological contactors, etc (see Fig. 8) (41–42). More recent developments in the design of attached-growth biological systems include the use of synthetic and redwood media, as well as granular media (see Figs. 9 and 10).

The post-World War II growth in industrial activity has significantly altered the composition of wastewaters in urban treatment facilities. Pollutants that are resistant to biological oxidation have become predominant (eg, synthetic detergents, petrochemicals, synthetic rubber, etc), requiring the development of new nonbiological processes and approaches to water-pollution control. Today, the industrial-wastewater engineer must be familiar with the manufacturing process and the chemistry of the raw materials, products, and by-products.

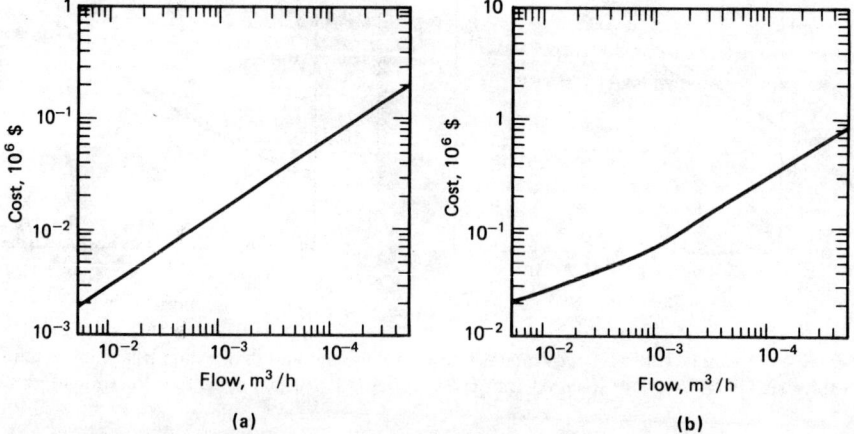

Figure 6. Installed capital cost estimates for waste (a) sludge pumping and (b) chlorination (ENR (*Eng. News Record*) Index: 2475, Sept. 1976). To convert m³/h to mgd, divide by 157.7.

Table 6. Design Parameters for the Activated-Sludge Process[a]

Parameter, mg/L	Influent[b]	Effluent
BOD_5[c]	210	20
COD[d]	400	45
TSS[e]	230	20
total-P	11	7
NH_3–N	0	0
UOD[f]	406	107

[a] Refs. 39–40.
[b] Aerobic digestion at plants <473.1 m³/h (3 mgd) and two-stage anaerobic digestion at plants >473.1 m³/h.
[c] 5-d Biochemical oxygen demand.
[d] Chemical oxygen demand.
[e] Total suspended solids.
[f] Ultimate oxygen demand.

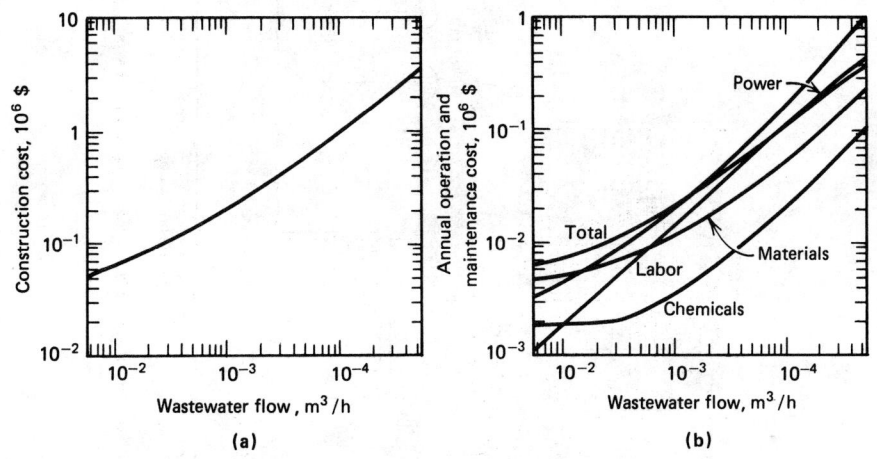

Figure 7. (a) Installed capital and (b) annual operation and maintenance costs for the treatment of municipal wastewaters utilizing the activated-sludge process (ENR (*Eng. News Record*) Index: 2475, Sept. 1976). To convert m³/h to mgd, divide by 157.7.

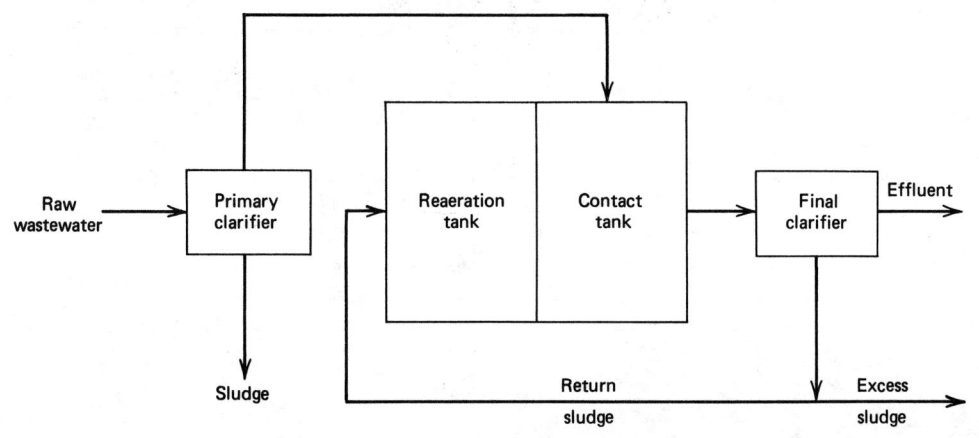

Figure 8. Contact stabilization plant. Conventional activated-sludge process. The reaeration and contact tanks can be replaced by an aeration tank (41).

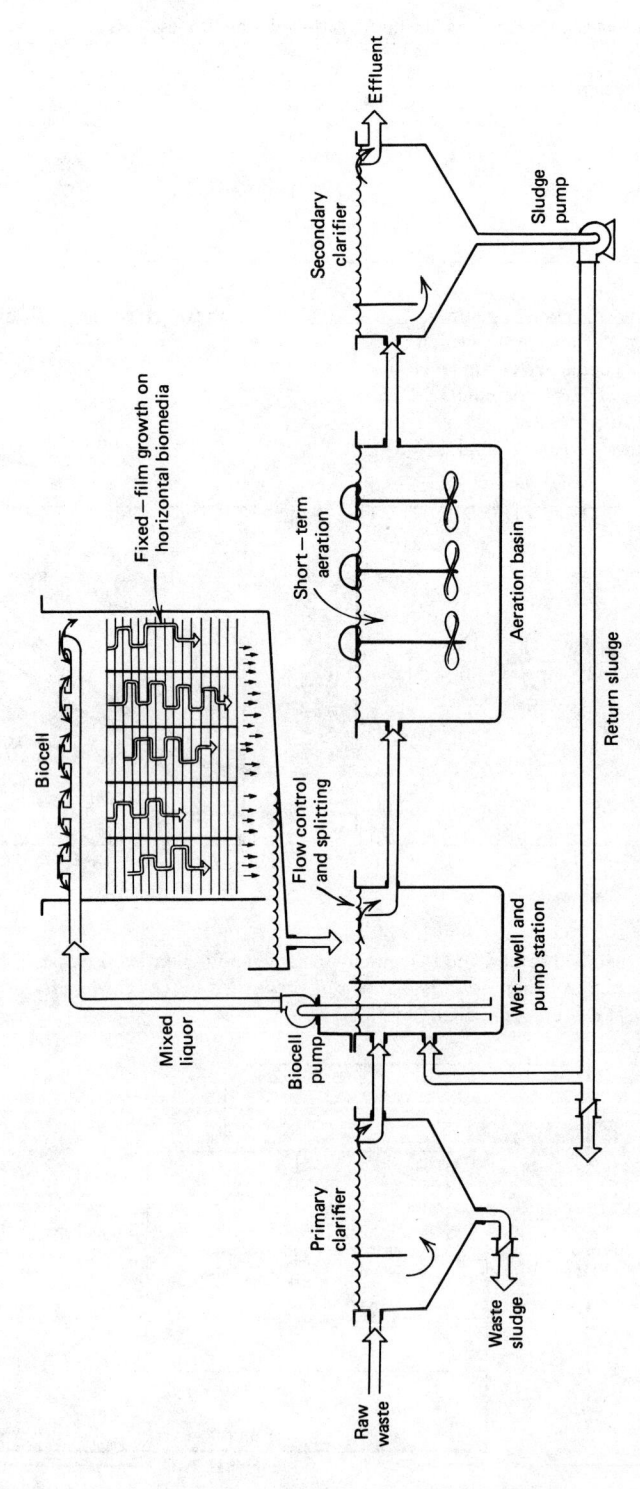

Figure 9. ABF (activated bio filter) process flow diagram (redwood medium) (42–43).

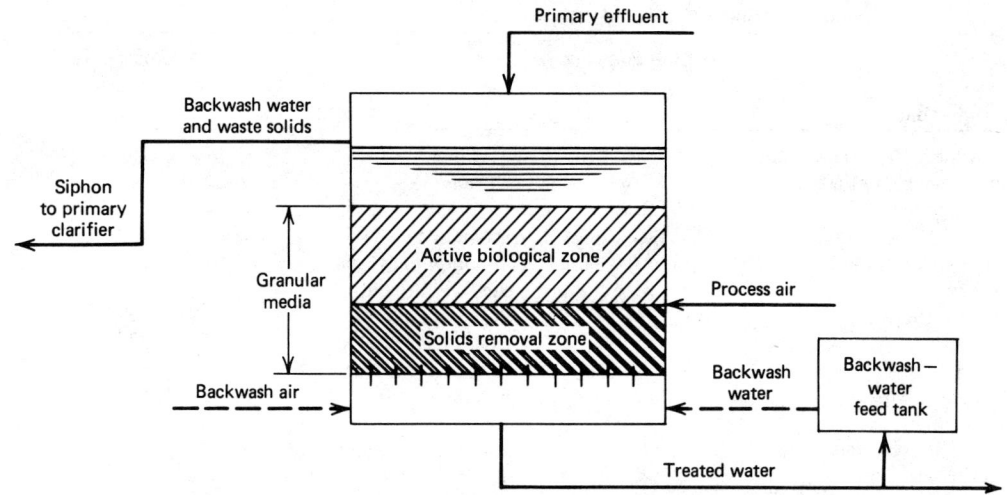

Figure 10. Biological aerated filter. Courtesy of Envirotech, Inc.

Primary Investigation. The basic systems engineering approach is the most suitable method for developing a solution to industrial wastewater management. The preliminary investigation includes a plant survey and the characterization of the wastewater source.

A wastewater survey provides the facts and data necessary to complete a wastewater management plan which comprises the following steps: segregation of clean water for potential recovery and reuse; isolation and segregation of noncompatible waste streams for separate pretreatment (eg, ion-exchange-regeneration wastes, inorganic and organic waste streams, waste streams that contain potentially toxic compounds, etc); isolation and segregation of concentrated waste streams and streams containing solvents for possible recovery or separate treatment and disposal (ie, thermal decomposition with heat recovery); and identification and characterization of batch discharges and spills in order to incorporate protective systems (eg, equalization) for the treatment facility.

In-Plant Waste Control. Pollution can be reduced or eliminated by process modification, chemical and raw materials substitution, or recovery of by-products. In ad-

Table 7. Membrane Rejection Rates

Compound	Percent
lindane	84
DDT[a]	99.5
DDD[b]	99.9
2,4-D[c]	99.9
chlorinated hydrocarbons	98.95–100
organophosphorus pesticides	98.05–98.88
miscellaneous pesticides	72–100

[a] Dichlorodiphenyltrichloroethane.
[b] 1,1-Dichloro-2,2-bis(p-chlorophenyl)ethane.
[c] 2,4-Dichlorophenoxy acetic acid.

Table 8. Adsorption Capacities of Activated Carbon

Compound	Adsorption capacity, mg/g	Compound	Adsorption capacity, mg/g
bis(2-ethylhexyl) phthalate	11,300	guanine	120
butyl benzyl phthalate	1,520	styrene	120
heptachlor	1,220	1,3-dichlorobenzene	118
heptachlor epoxide	1,038	acenaphthylene	115
endosulfan sulfate	686	4-chlorophenyl phenyl ether	111
endrin	666	diethyl phthalate	110
fluoranthene	664	2-nitrophenol	99
aldrin	651	dimethyl phthalate	97
PCB-1232	630	hexachloroethane	97
β-endosulfan	615	chlorobenzene	91
dieldrin	606	p-xylene	85
hexachlorobenzene	450	2,4-dimethylphenol	78
anthracene	376	4-nitrophenol	76
4-nitrobiphenyl	370	acetophenone	74
fluorene	330	1,2,3,4-tetrahydronaphthalene	74
DDT	322	adenine	71
2-acetylaminofluorene	318	dibenzo[a,h]anthracene	69
α-BHC	303	nitrobenzene	68
anethole	300	3,4-benzofluoranthene	57
3,3-dichlorobenzidine	300	1,2-dibromo-3-chloropropane	53
2-chloronaphthalene	280	ethylbenzene	53
phenylmercuric acetate	270	2-chlorophenol	51
hexachlorobutadiene	258	tetrachloroethylene	51
γ-BHC (lindane)	256	o-anisidine	50
p-nonylphenol	250	5-bromouracil	44
4-dimethylaminoazobenzene	249	benzo[a]pyrene	34
chlordane	245	2,4-dinitrophenol	33
PCB-1221	242	isophorone	32
1,1-dichloro-2,2-bis(p-chlorophenyl)ethane (DDE)	232	trichloroethylene	28
		thymine	27
acridine yellow	230	toluene	26
benzidine dihydrochloride	220	5-chlorouracil	25
β-benzenehexachloride	220	N-nitrosodi-n-propylamine	24
n-butyl phthalate	220	bis(2-chloroisopropyl) ether	24
N-nitrosodiphenylamine	220	phenol	21
phenanthrene	215	bromoform	20
dimethylphenylcarbinol	210	carbon tetrachloride	11
4-aminobiphenyl	200	bis(2-chloroethoxy)methane	11
2-naphthol	200	uracil	11
α-endosulfan	194	benzo[ghi]perylene	11
acenaphthene	190	1,1,2,2-tetrachloroethane	11
4,4'-methylenebis(2-chloroaniline)	190	1,2-dichloropropene	8.2
		dichlorobromomethane	7.9
benzo[k]fluoranthene	181	cyclohexanone	6.2
acridine orange	180	1,2-dichloropropane	5.9
1-naphthol	180	1,1,2-trichloroethane	5.8
4,6-dinitro-o-cresol	169	trichlorofluoromethane	5.6
1-naphthylamine	160	5-fluorouracil	5.5
2,4-dichlorophenol	157	1,1-dichloroethylene	4.9
1,2,4-trichlorobenzene	157	dibromochloromethane	4.8
2,4,6-trichlorophenol	155	2-chloroethyl vinyl ether	3.9

Table 8. (continued)

Compound	Adsorption capacity, mg/g	Compound	Adsorption capacity, mg/g
2-naphthylamine	150	1,2-dichloroethane	3.6
pentachlorophenol	150	1,2-*trans*-dichloroethylene	3.1
2,4-dinitrotoluene	146	chloroform	2.6
2,6-dinitrotoluene	145	1,1,1-trichloroethane	2.5
4-bromophenyl phenyl ether	144	1,1-dichloroethane	1.8
p-nitroaniline	140	acrylonitrile	1.4
1,1-diphenylhydrazine	135	methylene chloride	1.3
naphthalene	132	acrolein	1.2
1-chloro-2-nitrobenzene	130	cytosine	1.1
1,2-dichlorobenzene	129	benzene	1.0
p-chloro-*m*-cresol	124	EDTA	0.86
1,4-dichlorobenzene	121	benzoic acid	0.76
benzothiazole	120	chloroethane	0.59
diphenylamine	120	N,N-dimethylnitrosamine	6.8×10^{-5}

dition, process modification generally increases product yield by incorporating control devices.

Substitution of a calcium by a magnesium base in the paper industry has significantly reduced the discharge of pollutants. The iron and steel industry has replaced sulfuric acid by hydrochloric acid in the pickling process, and the hydrochloric acid can be recovered.

In-plant control permits the recovery of metals from electroplating wastewater. Silver and nickel are generally recovered with ion exchange or electrodialysis. The recycling of water itself is widely practiced, eg, as steam condensates or cooling waters.

Wastewater Treatment. The end-of-pipe treatment and discharge criteria should be determined before selecting a suitable treatment process. It is an objective of the Clean Water Act to establish industry-specific limitations or standards for the discharge of conventional and nonconventional pollutants. The conventional pollutants include BOD_5 (5-d BOD), COD, suspended solids, fecal coliforms, temperature, pH, oil, grease, and phosphates. The nonconventional pollutants include a variety of inorganic and organic materials, which are termed priority pollutants (35). The industry-specific effluent standard plus the local water-quality criteria established for the receiving water body provide the basis for wastewater-treatment selection.

Treatment technologies utilize physical, chemical, and biological methods. Selection depends primarily on the type of pollutants and the effluent-discharge criteria.

The technologies developed for the treatment of municipal wastewaters generally apply to the removal of conventional pollutants, eg, dissolved organics, suspended and floating materials (see Fig. 11). A fixed or a suspended-growth biological system generally provides the basis of the treatment process combined with physical separation units (clarification–sedimentation) for the removal of grit, oil, grease, and biological solids (45). These process systems can remove >90% of BOD_5 and suspended solids.

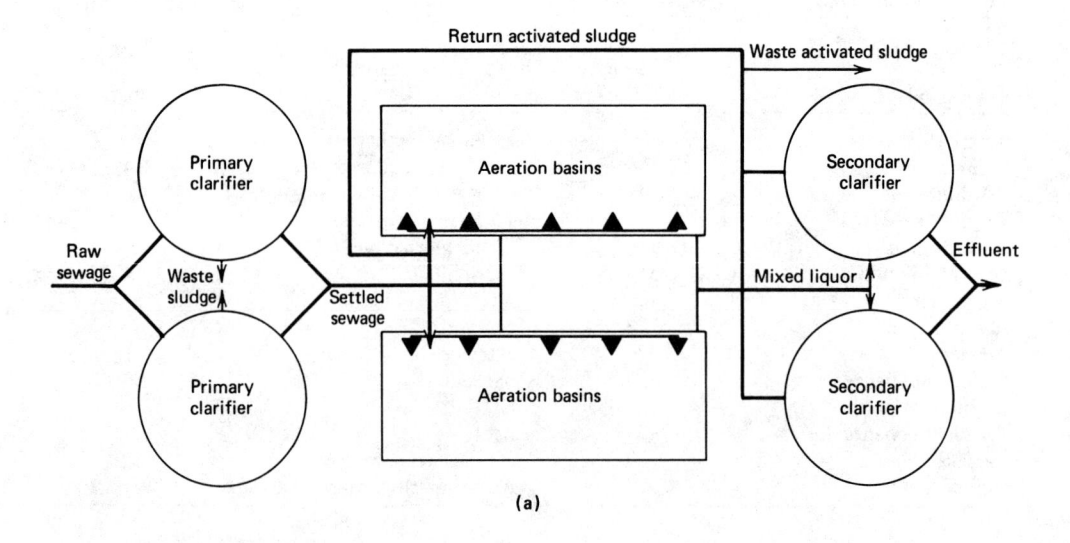

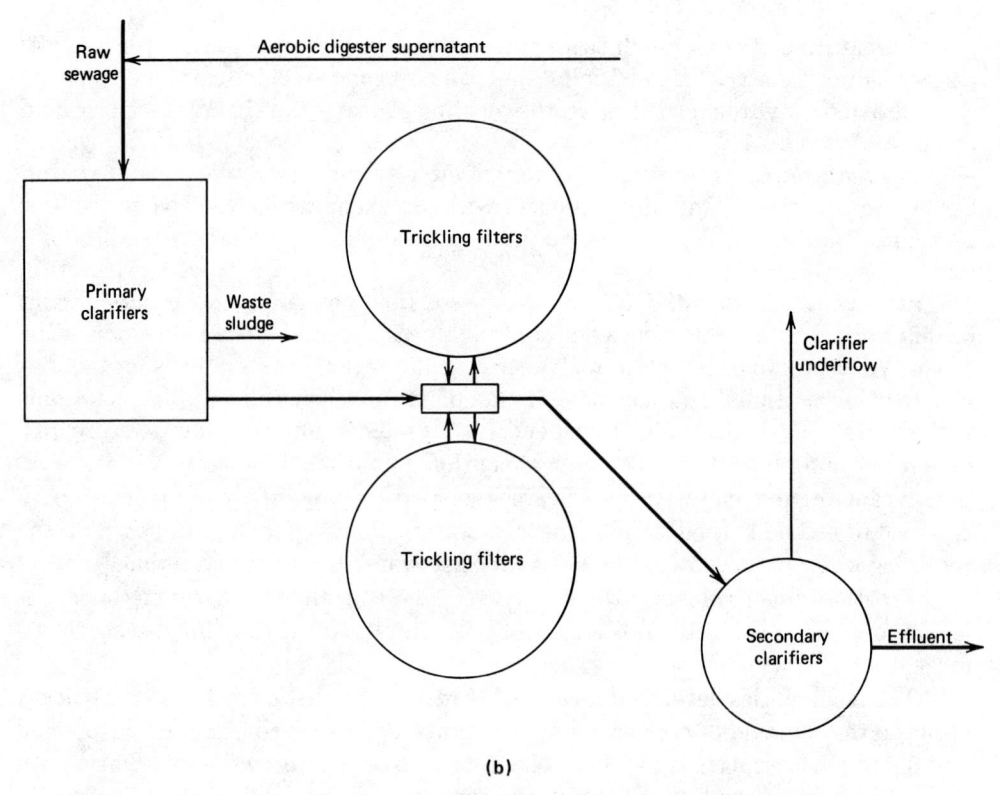

Figure 11. Typical process flow sheets for the treatment of conventional pollutants (44).

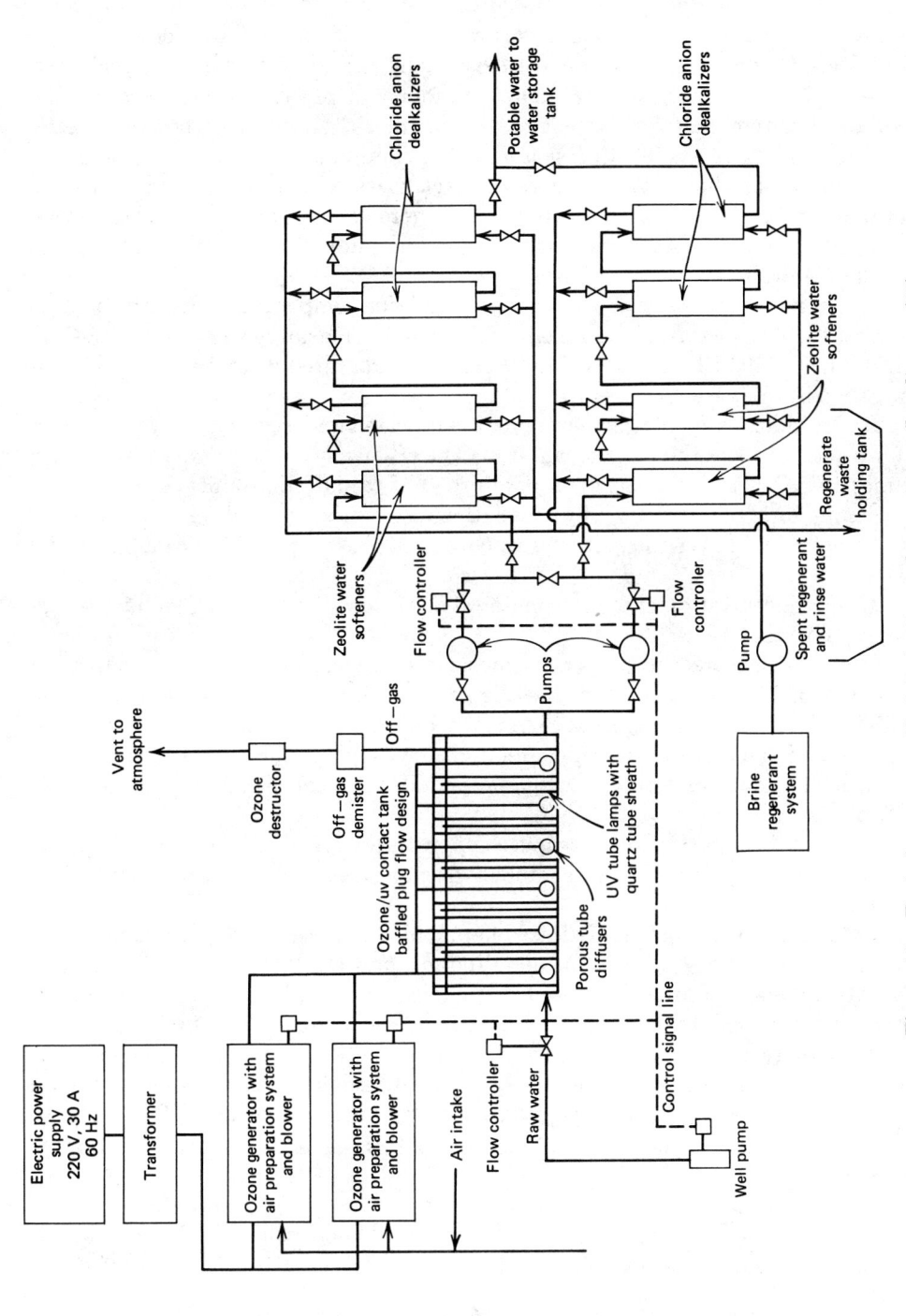

Figure 12. Typical ozone–uv oxidation plus ion-exchange process flow diagram (53).

Toxic Organic Materials. The term toxic organics includes synthetic organic compounds such as pesticides, herbicides, PCBs, and chlorinated hydrocarbons, usually produced by the manufacturers and formulators of these products.

Because these compounds persist over a long period of time in a natural environment, the most effective treatment technology at present is incineration. The recommended temperatures for incineration of chlorinated hydrocarbons and pesticides range from 982 to 1482°C. A sustained high temperature prevents the emission of degradation products. The incinerator stack gases generally contain HCl vapors that require the installation of scrubbers. In general, a vortex burner provides satisfactory performance. The mixture is preheated, the liquid vaporized, and the gases heated to ignition temperature (46–47) (see Incinerators).

The vortex burner maintains stable combustion temperature when the organic concentration in the waste is sufficiently high and has a heating value of ca 10.5–12.6 MJ/kg (4500–5400 Btu/lb). Auxiliary fuel may be required when the chloride concentration in the waste exceeds 70% (48).

In wet-air oxidation, the aqueous mixture is heated under pressure in the presence of air, which oxidizes the organic material. The efficiency of the oxidation process is a function of reaction time and temperature. The oxidation products are generally less complex and can be treated by conventional biological methods (49). The reactor usually operates between 177 and 321°C with pressures of 2.52–20.8 MPa (350–3000 psig).

If the concentration of organic material is too low, the following technique discussed below may be used.

Membrane Separation. Reverse osmosis (qv) or ultrafiltration (qv) can be used to concentrate toxic organic substances, depending on the type of compounds and the stability of the membrane against chemical attack (see also Membrane technology). In general, high molecular weight compounds have higher rejection rates than those obtained with lower molecular weight compounds. Typical rejection rates are given in Table 7.

Chemical Treatment. Some organic compounds are attacked by chemical reagents such as potassium permanganate, sodium hydroxide, calcium hypochlorite, and ozone (47–48).

Potassium permanganate oxidizes heptachlor with ca 80–90% efficiency (48). Sodium hydroxide degrades malathion, lindane, and DDT (48,50). Ozone oxidizes dissolved organic compounds, including toxic substances, because of an oxidation potential higher than that of permanganate, hydrogen peroxide, or hypochlorite. The ozonation system must be designed to carry the reaction to completion in order to prevent the generation of toxic intermediates. Ultraviolet radiation in conjunction with ozone is a highly effective degradation technique for malathion, DDT, pentachlorophenol, dichlorobutane, dichlorobenzene, PCBs, and chloroform (51–52) (see Fig. 12).

Ozone alone oxidizes phenolic compounds to carbon dioxide and water. However, at concentrations of 1.5–2.5 parts ozone per part of phenol, the phenolic compounds can be converted to less toxic and biodegradable intermediates in a cost-effective manner (54).

Adsorption. Organic compounds are adsorbed on activated carbon and synthetic resins (eg, XAD-2 and XAD-4, Rohm and Haas Co.). This technique depends on the properties of the compound being removed and the regenerative capability of the

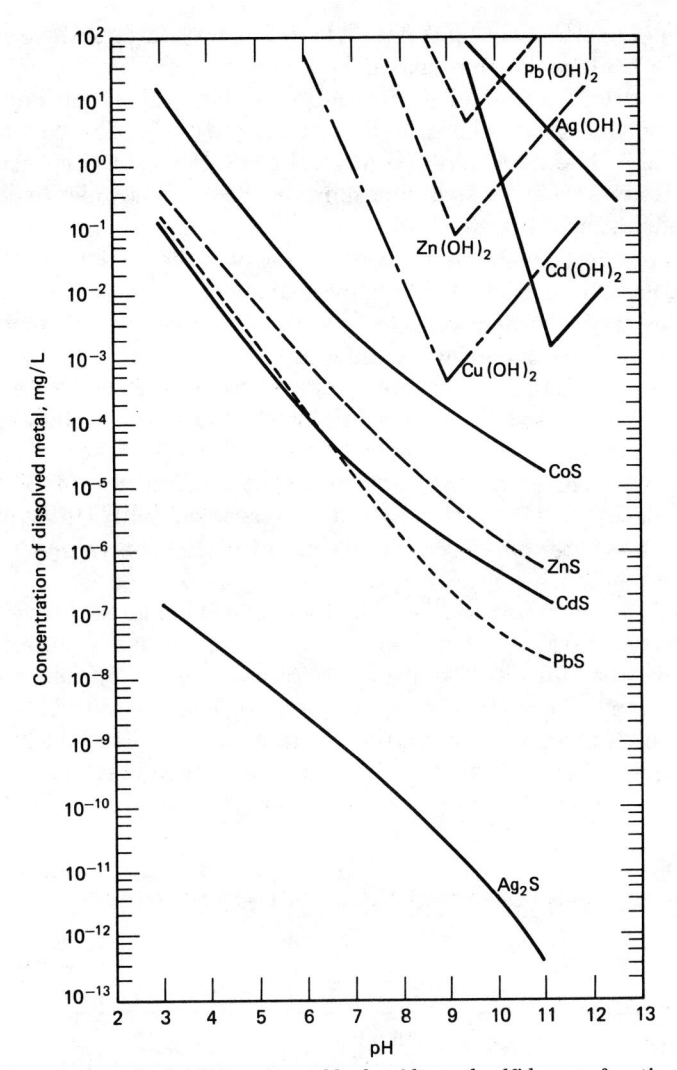

Figure 13. Theoretical solubilities of metal hydroxides and sulfides as a function of pH (56).

Table 9. Adsorption Capacity of Amberlite XAD-4 Polymeric Resin[a]

Compounds	Maximum solubility in water, g/L	Influent concentration, mg/L	Adsorption capacity, kg/m³
phenol	82	250	12.5
2-chlorophenol	26	350	38.5
2,4-dichlorophenol	4.5	430	81.6
2,4,6-trichlorophenol	0.9	510	192.3

[a] Ref. 55.

adsorbent. The EPA has developed carbon-adsorption isotherms for various toxic organic compounds, and the results are shown in Table 8 (55). The following compounds are not adsorbed on activated carbon: acetone cyanohydrin, butylamine, choline chloride, cyclohexylamine, diethylene glycol, ethylenediamine, hexamethylenediamine, morpholine, and triethanolamine.

Synthetic polymeric adsorbents have a high porosity, large surface area, and an inert hydrophobic surface. These resins can be regenerated chemically, which produces a concentrated waste stream requiring further treatment or disposal. The adsorption capacity of the polymeric adsorbent XAD-4 for a group of chlorinated hydrocarbons is given in Table 9. The EPA may recommend a combination of air stripping and carbon adsorption wherein air stripping removes most of the volatile organics and adsorption removes the rest (56).

Heavy Metals. Heavy metals of particular concern in the treatment of wastewaters include copper, chromium, zinc, cadmium, mercury, lead, and nickel. They are usually present in the form of organic complexes, especially in wastewaters generated from textiles finishing and dye chemicals manufacture.

Inorganic heavy metals are usually removed from aqueous waste streams by chemical precipitation in various forms (carbonates, hydroxides, sulfide) at different pH values. The solubility curves for various metal hydroxides, when they are present alone, are shown in Figure 13. The presence of other metals and complexing agents (ammonia, citric acid, EDTA, etc) strongly affects these solubility curves and requires careful evaluation to determine the residual concentration values after treatment (see Table 10) (57–58).

Other methods, including activated carbon, ion exchange, and reverse osmosis, can be used to concentrate waste streams and remove the heavy metals. Activated carbon is effective in reducing hexavalent chromium, mercury, and many metals complexed by organic liquids. Similarly, various ion-exchange resins have been found to be effective in reducing metal ions from solution. The spent resins and activated carbon may require chemical regeneration, however, which may produce a concentrated waste stream that again requires treatment.

Table 10. Typical Residual Concentrations after Chemical Precipitation[a]

Metal	Agent	Residual concentration, mg/L
cadmium	soda ash	0.3
chromium (hexavalent)	sodium bisulfite and lime	0.05
chromium (total)	caustic, lime	0.5
copper	caustic, lime	0.5
nickel	soda ash	0.5
zinc	caustic, lime	0.5

[a] Ref. 57.

BIBLIOGRAPHY

1. R. K. Linsley and J. B. Franzini, *Water Resources Engineering*, McGraw-Hill Book Company, New York, 1979.
2. G. M. Fair, J. C. Geyer, and D. A. Okun, *Water and Wastewater Engineering*, John Wiley & Sons, Inc., New York, 1966.
3. D. K. Todd, *Groundwater Hydrology*, John Wiley & Sons, Inc., New York, 1980.
4. H. W. Gehm and J. I. Bregman, *Handbook of Water Resources and Pollution Control*, Van Nostrand Reinhold Company, New York, 1976.

5. J. H. Feth, *Water Facts and Figures for Planners and Managers*, Publication No. 24, U.S. Public Health Service, Washington, D.C., 1962.
6. American Water Works Association, *A Survey of Operating Data for Water Works in 1960*, AWWA staff report, New York, 1964.
7. American Water Works Association, *Supplement to a Survey of Operating Data for Water Works in 1960*, AWWA staff report, New York, 1964.
8. B. R. Beattie and H. S. Forster, Jr., *Proceedings of the 1979 Annual AWWA Conference*, American Water Works Association, New York, 1979.
9. D. W. Browne, F. Minton, and C. Barnhill, *J. Am. Water Works Assoc.* **72**(9), 506 (1980).
10. American Water Works Association, *AWWA Committee Report on Water Use*, New York, 1972.
11. F. P. Lineweaver, Jr., J. C. Geyer, and J. B. Wolff, *A Study of Residential Water Use*, Federal Housing Administration, U.S. Department of Housing and Urban Development, Washington, D.C., 1967.
12. U.S. Bureau of the Census, *Water Use in Manufacturing*, 1967 Census of Manufacturers, Publication No. MC67(1)-7, 1971.
13. National Association of Manufacturers, *Water in Industry*, Washington, D.C., 1965.
14. United Nations Department of Economic and Social Affairs, *Water for Industrial Use*, U.N. Publication No. E/3058 ST/ECA/50, New York, 1958.
15. F. B. Walling and L. E. Otts, Jr., *Water Requirements of the Iron and Steel Industry*, USGS water supply paper 1330-H, Reston, Va., 1967.
16. L. E. Otts, Jr., *Water Requirements of the Petroleum Refining Industry*, USGS water supply paper 1330-G, Reston, Va., 1963.
17. O. D. Mussey, *Water Requirements of the Pulp and Paper Industry*, USGS water supply paper 1330-A, Reston, Va., 1961.
18. O. D. Mussey, *Water Requirements of the Copper Industry*, USGS water supply paper 1330-E, Reston, Va., 1961.
19. O. D. Mussey, *Water Requirements of the Rayon and Acetate Fiber Industry*, USGS water supply paper 1330-D, Reston, Va., 1957.
20. C. N. Durfor, *Water Requirements of the Carbon Black Industry*, USGS water supply paper 1330-B, Reston, Va., 1956.
21. H. L. Conklin, *Water Requirements of the Aluminum Industry*, USGS water supply paper 1330-C, Reston, Va., 1956.
22. C. R. Murray, *Estimated Use of Water in the United States, 1965*, USGS circular no. 556, 1968.
23. J. H. Feth, *Water Facts and Figures for Planners*, USGS circular no. 601-I, 1973.
24. C. R. Murray, *Estimated Use of Water in the United States, 1970*, USGS circular no. 676, 1972.
25. *The Nation's Water Resources—the First National Assessment*, Water Resources Council, Washington, D.C., 1968.
26. U.S. Environmental Protection Agency, *Development Document for Effluent Limitations Guidelines for the Iron and Steel Manufacturing Point Source Category (Proposed)*, EPA 440/1-80/024b, Washington, D.C., 1980.
27. U.S. Environmental Protection Agency, *Development Document for Effluent Limitations Guidelines and Standards for the Metal Finishing Point Source Category (Proposed)*, EPA 440/1-82/091b, Washington, D.C., 1982.
28. U.S. Environmental Protection Agency, *Contractor's Engineering Report for the Development of Effluent Limitations Guidelines and Standards for the Pharmaceutical Manufacturing Point Source Category*, EPA 440/1-80/084a, Washington, D.C., June 1980.
29. U.S. Environmental Protection Agency, *Development Document for Effluent Limitations and Standards for the Inorganic Chemicals Manufacturing Point Source Category (Proposed)*, EPA 440/1-79/007, Washington, D.C., 1979.
30. U.S. Environmental Protection Agency, *Development Document for Effluent Limitations and Standards for the Pulp, Paper, and Paperboard and the Builders' Paper and Board Mills Point Source Categories*, EPA 440/1-80/025b, Washington, D.C., 1980.
31. U.S. Environmental Protection Agency, *Development Document for Effluent Limitations Guidelines and Standards for the Petroleum Refining Point Source Category*, EPA 440/1-79/014b, Washington, D.C., 1979.
32. U.S. Army Corps of Engineers, *Northeastern United States Water Supply Study*, U.S. Army COE, North Atlantic Division, New York, 1977.
33. U.S. Department of Commerce, *Water Use in Manufacturing*, Department of Census Publications, Washington, D.C., 1973.

34. United States Department of Health, Education and Welfare, *Public Health Service Drinking Water Standards, 1962*, U.S. Public Health Service, Washington, D.C., 1962.
35. U.S. Environmental Protection Agency, *A Handbook of Key Federal Regulations and Criteria for Multimedia Environmental Control*, Industrial Environmental Research Laboratory, Research Triangle Park, N.C., EPA-600/7-79-175, 1979.
36. W. W. Eckenfelder, Jr., *Principles of Water Quality Management*, CBI Publishing Company, Inc., Boston, Mass., 1980.
37. H. W. Van Gills, *Bacteriology of Activated Sludge*, report no. 32-1G-TNO, Research Institute for Public Health Engineering, the Hague, The Netherlands, 1964.
38. Metcalf, Eddy, Inc., *Wastewater Engineering: Collection, Treatment, and Disposal*, McGraw-Hill Book Company, New York, 1972.
39. *Wastewater Treatment Systems, Performance and Cost*, Roy F. Weston, Inc., West Chester, Pa., 1977.
40. U.S. Environmental Protection Agency, *Innovative and Alternative Technology Assessment Manual*, Office of Water Program Operations, Washington, D.C., EPA-430/9-78-009, 1978.
41. Roy F. Weston, Inc., *Process Design Manual for Upgrading Existing Wastewater Treatment Plants*, prepared for U.S. EPA, technology transfer, program no. 17090 GNQ, 1971.
42. M. Ramanathan and W. E. Verley, *Evaluation, Design and Startup of an Innovative and Cost-effective Wastewater Treatment Plant at Concord, New Hampshire*, paper presented at the 36th Annual Meeting, Virginia Water Pollution Control Association, Inc., Charlottesville, Va., 1982.
43. *Activated Biofiltration*, bulletin no. KL 42330-2, Neptune Microfloc, Inc., Corvallis, Ore., 1980.
44. H. D. Stensel, *Biological Aerated Filter*, presented at the I/A Technologies Seminar sponsored by the U.S. Environmental Protection Agency, 1982.
45. U.S. Environmental Protection Agency, *Process Control Manual for Aerobic Biological Wastewater Treatment Facilities*, Office of Water Program Operations, Washington, D.C., EPA-430/9-77-006, 1977.
46. Y. H. Kiang, *Chem. Eng. Progr.* **72,** 71 (1976).
47. J. J. Santoleri, *Chem. Eng. Progr.* **69,** 68 (1973).
48. G. M. Leigh, *J. Water Pollut. Control Fed.* **41**(11), R450 (1969).
49. *Technology and Application—Summary*, Zimpro Environmental Systems, Inc., Rothschild, Wisc., 1978.
50. M. V. Kennedy, B. J. Stojanovic, and F. L. Shuman, Jr., *J. Agric. Food Chem.* **20**(2), (1972).
51. H. W. Prengle, Jr., *Proceedings of the Seminar on Wastewater Treatment and Disinfection with Ozone*, International Ozone Association, Vienna, Va., 1977.
52. R. C. Rice, *AIChE Symp. Ser., Water 1980*, **77**(209), (1981).
53. Project design files, confidential, Roy F. Weston, Inc., West Chester, Pa., 1982.
54. K. H. Lanouette, *Chem. Eng.* **48**(22), (Oct. 1977).
55. R. A. Dobbs, R. J. Middendorf, and J. M. Cohen, *Carbon Adsorption Isotherms for Toxic Organics*, Municipal Environmental Research Laboratory, U.S. Environmental Protection Agency, Cincinnati, Ohio, EPA-600/8-80-023, 1980.
56. *Chem. Week*, 44 (June 22, 1983).
57. The Permutit Company, *Proceedings of Seminar on Metal Wastes Treatment Featuring the Sulfex Process*, The Permutit Co., Paramus, N.J., 1978.
58. K. H. Lanoutte, *Chem. Eng.* **48**(22), (Oct. 1977).

MUTHIAH RAMANATHAN
Roy F. Weston, Inc.

ANALYSIS

Since 1970, new analytical techniques, eg, ion chromatography, have been developed, and others, eg, atomic absorption and emission, have been improved (1–5). Detection limits for many chemicals have been dramatically lowered. Many wet chemical methods have been automated and are controlled by microprocessors which allow greater data output in a shorter time. Perhaps the best known continuous-flow analyzer for water analysis is the Autoanalyzer system manufactured by Technicon Instruments Corp. (Tarrytown, N.Y.) (6). Isolation of samples is maintained by pumping air bubbles into the flow line. Recently, flow-injection analysis has also become popular, and a theoretical comparison of it with the segmented flow analyzer has been made (7–9). Several books are available regarding automated chemical analysis (10–11) (see Analytical methods; also Biomedical automated instrumentation).

Although simple analytical tests often provide the needed information regarding a water sample, recent discoveries, such as the formation and presence of chloroform and other organohalides in drinking water, require some very specialized methods of analysis. The separation of trace metals into total and uncomplexed species also requires special sample handling and analysis (12).

A list of all water analyses would be extremely long since, under some conditions and with enough time, water can solubilize everything to some extent. Fortunately, a great deal can be learned about a water supply by carrying out a few physical and chemical tests. These simple tests might be all that are needed to characterize a water supply for many purposes, and it is usually the purpose for which the water is to be used that determines the type and extent of testing. The methods described in this review are intended primarily for freshwater analysis and may not be suitable for the analysis of saline water. In addition, there are several books, manuals, and annual review articles available that detail water analysis for different substances and uses (13–21).

Physical Properties

Temperature. Water temperature is an important parameter in calculations of oxygen solubility, calcium carbonate saturation and stability, and various forms of alkalinity, as well as in determining basic hydrobiological characteristics. The temperature should be taken *in situ* for accuracy, and a standard mercury thermometer with readings to the nearest 0.1°C should be used. It should be calibrated against a precision thermometer certified by the National Bureau of Standards. A thermistor is preferable when attempting to measure temperature at different depths and for automated monitoring and surveillance, and should be similarly calibrated (see Temperature measurement).

Specific Conductance. The specific conductance depends on the total concentration of the dissolved ionized substances, ie, the ionic strength of a water sample. It is an expression of the ability of the water to conduct an electric current. Freshly distilled water has a conductance of 0.5–2 μS/cm, whereas that of potable water generally is 50–1500 μS/cm. The conductivity of a water sample is measured by means of an a-c Wheatstone-bridge circuit with a null indicator and a conductance cell. Each cell has an associated constant which, when multiplied by the conductance, yields the specific conductance.

The concentration of dissolved ionic substances can be roughly estimated by multiplying the specific conductance by an empirical factor of 0.55–0.9, depending on temperature and soluble components. Since specific conductance is temperature dependent, all samples should be measured at the same temperature. Alternatively, an appropriate temperature-correction factor obtained by comparisons with known concentrations of potassium chloride may be used. Instruments are available that automatically correct conductance measurements for different temperatures.

Color. Many water samples have a yellow to brownish-yellow color which is caused by natural substances, eg, leaves, bark, humus, and peat material. Turbidity in a sample can make the measurement of color uncertain and is usually removed by centrifugation prior to analysis. The color is usually measured by comparison of the sample with known concentrations of colored solutions. A platinum–cobalt solution is used as the standard, and the unit of color is that produced by 1 mg/L platinum as chloroplatinate ion. The standard is prepared from potassium chloroplatinate (K_2PtCl_6) and cobalt chloride ($CoCl_2 \cdot 6H_2O$). The sample may also be compared to suitably calibrated special glass color disks.

Three special tristimulus light filters are available which, when combined with a specific light source and a filter photometer, can be used to obtain color data (20). Although this method provides added precision and accuracy, it is seldom worth the extra effort required (see Color).

Turbidity. Turbidity in natural water is caused by the presence of suspended matter which scatters light. The suspended material is usually clay or silt, finely divided organic and inorganic material, or microorganisms. It is measured either visually or by a nephelometric method. The former is based on the light path through a sample that just causes the image of the flame of a standard candle to disappear. Longer light paths are indicative of lower turbidities. Suspensions of clay are used as standards and results are reported in Jackson turbidity units (JTU). The lowest turbidity value that can be measured using this method is 25 JTU.

The nephelometric method involves illuminating the sample in a turbidimeter and measuring the amount of light scattered at 90° to the incident beam. The higher the intensity of scattered light, the greater the sample turbidity. A formazan polymer suspension is used as the standard and results are reported in nephelometric turbidity units (NTU). The greater precision, sensitivity, and wider applicability of this method make it preferable to the visual method.

Taste and Odor. The measurement of taste and odor is somewhat subjective and depends on the personal judgements of individuals. Panels of not less than five observers, and preferably more than ten, are used. The sample is diluted with odor-free water until a ratio at which the odor is just perceptible is determined; this ratio is called the threshold odor number (TON). A similar method is used to detect a distinct taste in water (see Flavor characterization).

Dissolved Solids. Dissolved solids are materials that pass through a glass-fiber filter and remain after evaporation and drying at 180°C.

Suspended Solids. Suspended solids are determined by filtering a known volume of water through a glass-fiber filter and weighing the filter before and after filtration. The filter is dried at 105–110°C, and the weight difference is equal to the suspended solids.

pH. The pH of most natural waters is 4–9 and should be measured *in situ* since it is subject to change once a sample has been isolated. It can be measured either colorimetrically or potentiometrically. The former method relies on color-indicator papers which are impregnated with pH-sensitive dyes. The potentiometric method requires the use of a pH-sensitive glass electrode, which is generally interference-free at pH 4–9. The electrode should be calibrated frequently against standard buffer solutions of known pH (see Hydrogen-ion concentration; Ion-selective electrodes).

Principal Mineral Constituents and Gases

Alkalinity. The alkalinity of a water sample is its acid-neutralizing capacity. Bicarbonate and carbonate ions are the predominant contributors to alkalinity in most waters, and their chemical equilibria generally maintain the pH of 5–9. The presence of enough hydroxide ion to affect the alkalinity determination in natural waters is rare. Silica, borate, or phosphate do contribute to the overall alkalinity if present in large enough quantities.

The alkalinity is determined by titration of the sample with a standard acid (sulfuric or hydrochloric) to a definite pH. If the initial sample pH is >8.3, the titration curve has two inflection points reflecting the conversion of carbonate ion to bicarbonate ion and finally to carbonic acid (H_2CO_3). A sample with an initial pH <8.3 only exhibits one inflection point corresponding to conversion of bicarbonate to carbonic acid. Since most natural-water alkalinity is governed by the carbonate–bicarbonate ion equilibria, the alkalinity titration is often used to estimate their concentrations.

Acidity. Acidity is the base-neutralizing capacity of a sample of water. It is determined by titration of the sample with standard base to pH 8.3 (phenolphthalein end point). Generally, a sample is not reported as acidic unless its initial pH is <4.5. The acidity may be the result of free acids, eg, HCl or H_2SO_4, or the hydrolysis of certain metal cations.

The free-acid content is determined by titration of a cold solution to pH 4.5. The total acidity is determined by titration to pH 8.3 in a boiling solution. Some natural-water samples might be complex, and the best determination of acidity results from visual inspection of the plotted titration curve.

Hardness (Calcium and Magnesium). The term hardness has its origins in the household laundry use of water. Some waters were considered harder to use for laundering because they required more soap to produce suds. The hardness of water was then taken to be a measure of the capacity of the water to cause the precipitation of soap. This is actually a result of the reaction of soap with calcium or magnesium ions. Thus, water hardness is generally a measure of the total concentration of calcium and magnesium. The portion of hardness that disappears with boiling is temporary or carbonate hardness and is primarily caused by calcium and magnesium bicarbonates. They precipitate as their carbonates by the loss of carbon dioxide during boiling. The hardness remaining after boiling is the permanent or noncarbonate hardness. Other cations, eg, strontium or barium, also contribute to hardness, but their concentration is usually insignificant (see Dispersants; Drycleaning and laundering).

Calcium and magnesium can be titrated readily with disodium ethylenediaminetetraacetate, with Eriochrome Black T as the indicator. The solution is buffered at pH 10.0. Certain metal ions interfere with this procedure by causing fading or indistinct end points. Cyanide, sulfide, or hydroxylamine can be used to eliminate or minimize the interferences.

Hardness can also be calculated by summation of the individually determined alkaline earths by means of atomic absorption analysis. Basic samples must be acidified, and lanthanum chloride must be added to minimize interferences from phosphate, sulfate, and aluminum. An ion-selective electrode that utilizes a liquid ion exchanger is also available for hardness measurement; however, this electrode is susceptible to interferences from other dissolved metal ions.

Magnesium, calcium, barium, and strontium can also be determined by ion chromatography with m-phenylenediamine in perchloric acid as the eluent. Ion chromatography by conductimetric detection has been described, and applications to environmental waters have been discussed (1,22–23).

Sodium and Potassium. Sodium and potassium can be determined by either atomic emission or absorption. Large concentrations of sodium can interfere with the potassium determination in either of these methods. Excess sodium can be added to both the potassium standards and samples to minimize any variations in the samples. Proper positioning of the flame helps reduce sodium interference in atomic absorption.

Chloride. Chloride is common in freshwater because almost all chloride salts are very soluble in water. Its concentration is generally 10^{-4} to 10^{-3} M. Chloride can be titrated with mercuric nitrate. Diphenylcarbazone, which forms a purple complex with the excess mercuric ions at pH 2.3–2.8, is used as the indicator. The pH should be controlled to ± 0.1 pH unit. Bromide and iodide are the principal interferences, whereas chromate, ferric, and sulfite ions interfere at levels greater than 10 mg/L. Chloride can also be determined by a colorimetric method based on the displacement of thiocyanate ion from mercuric thiocyanate by chloride ion. The liberated SCN^- reacts with ferric ion to form the colored complex of ferric thiocyanate. The method is suitable for chloride concentrations from 10^{-6} to 10^{-3} M.

Ion chromatography can be used to determine chloride concentrations of 2–1000 ppb with a carbonate–bicarbonate eluent (23). Fluoride, nitrite, phosphate, bromide, nitrate, and sulfate do not interfere and can be measured simultaneously with a total analysis time of <30 min.

An ion-selective electrode is available for chloride analysis; chloride can be measured potentiometrically at 10^{-6}–1 M. Iodide and sulfide are the principal interferences.

Sulfate. Mine drainage may contribute to high sulfate concentrations resulting from pyrite oxidation. High concentrations may exhibit a cathartic action. Sulfate concentrations of 10^{-5} to 10^{-3} M can be titrated in an alcoholic solution with standard barium chloride and using Thorin as an indicator. Barium reacts with Thorin to form a deep red complex. The sample should be kept at pH 2–5, and the sample must be passed through a strong cation-exchange column prior to the titration. This is done to remove any multivalent cations which would also form intensely colored complexes with Thorin. High concentrations of sulfate can be determined by gravimetric analysis. The sulfate precipitates as barium sulfate. The ion-chromatographic response to sulfate is linear from 2 to 10^4 ppb (23).

Nitrate and Nitrite. Nitrate is usually present in trace quantities in surface waters but occasionally occurs in high concentrations in some groundwaters. If present in excessive amounts, it can contribute to the illness infant methemoglobinemia. Nitrate is an essential nutrient for many photosynthetic autotrophs. Nitrite is an intermediate in the reduction of nitrate as well as in the oxidation of ammonia; it is also used as a corrosion inhibitor in some industrial processes.

Nitrite can be determined by reaction with sulphanilamide to form the diazo compound, which couples with N-(1-naphthyl)ethylenediamine dihydrochloride to form an intensely colored red azo dye. Nitrate can be determined in a similar manner after reduction to nitrite. Suitable reducing agents are cadmium filings or hydrazine. This method is useful at a nitrogen concentration of 10^{-7}–10^{-4} M.

Nitrate can also be measured potentiometrically with an ion-selective electrode at 10^{-5}–10^{-1} M (24–25). This method is suggested as a screening method for determining the approximate nitrate concentration (20). Ion chromatography can be used for nitrate concentrations of 2 to 10^4 ppb (23).

Fluoride. A fluoride concentration of ca 1 mg/L is helpful in preventing dental caries. Fluoride is determined potentiometrically with an ion-selective electrode. A buffer solution of high total ionic strength is added to the solution to eliminate variations in sample ionic strength and to maintain the sample at pH 5–8, the optimum range for measurement. (Cyclohexylenedinitrilo)tetraacetic acid (CDTA) is usually added to the buffer solution to complex aluminum and thereby prevent its interference. If fluoroborate ion is present, the sample should be distilled from a concentrated sulfuric acid solution to hydrolyze the fluoroborate to free fluoride prior to the electrode measurement (26–27).

Several colorimetric procedures for fluoride are available, but it is usually desirable to distill the sample from concentrated sulfuric acid prior to analysis to eliminate interferences. One method is based upon bleaching a dye formed by the reaction of zirconium and sodium 2-(p-sulfophenylazo)-1,8-dihydroxy-3,6-naphthalenedisulfonate (SPADNS reagent) (28).

Phosphate. Phosphorus occurs in water primarily as a result of natural weathering, municipal sewage, and agricultural runoff. The most common form in water is the phosphate ion. A sample containing phosphate can react with ammonium molybdate to form molybdophosphoric acid ($H_3P(Mo_3O_{10})_4$). This compound is reduced with stannous chloride in sulfuric acid to form a colored molybdenum-blue complex, which can be measured colorimetrically. Silica and arsenic are the chief interferences.

Some water samples contain phosphorus forms other than phosphate, eg, polyphosphate, hexametaphosphate, and organic phosphates. These forms can be hydrolyzed to phosphate in hot sulfuric acid solution and determined by the preceding method. The more refractory organic phosphates require digestion in a sulfuric acid–ammonium persulfate solution. Ion chromatography can also be used to measure PO_4^{3-} at 2 to 10^4 ppb (21).

Boron and Borates. Boron is an essential element for plant growth; however, concentrations >2 mg/L are harmful to some plants. Natural-water concentrations of boron are usually well below this value, although higher concentrations can occur as a result of industrial waste effluents or cleaning agents.

Two colorimetric methods are recommended for boron analysis. One is the curcumin method, where the sample is acidified and evaporated after addition of curcumin reagent. A red product called rosocyanine remains; it is dissolved in 95 wt % ethanol and measured photometrically. Nitrate concentrations >20 mg/L interfere with this method. Another colorimetric method is based upon the reaction between boron and carminic acid in concentrated sulfuric acid to form a bluish-red or blue product. Boron concentrations can also be determined by atomic absorption spectroscopy with a nitrous oxide–acetylene flame or graphite furnace. Atomic emission with an argon plasma source can also be used for boron measurement.

Silica. The silica content of natural waters is usually 10^{-5} to (5×10^{-4}) M. Its presence is considered undesirable for some industrial purposes because of the formation of silica and silicate scales. The heteropoly-blue method is used for the measurement of silica. The sample reacts with ammonium molybdate at pH 1.2, and oxalic acid is added to reduce any molybdophosphoric acid produced. The yellow molybdosilicic acid is then reduced with 1-amino-2-naphthol-4-sulfonic acid and sodium sulfite to heteropoly blue. Color, turbidity, sulfide, and large amounts of iron are possible interferences. A digestion step involving $NaHCO_3$ can be used to convert any molybdate-unreactive silica to the reactive form. Silica can also be determined by atomic absorption with a nitrous oxide–acetylene flame or by atomic emission involving either a d-c or inductively coupled argon plasma source.

Oxygen. The solubility of atmospheric oxygen in water depends primarily on pressure, temperature, and salt content. The most reliable results for oxygen measurement are obtained from fresh samples. The longer the time lag between sampling and measurement, the greater the chance that the dissolved oxygen concentration diminishes because of chemical or biological activity in the sample. The Winkler titration has been the preferred method for dissolved oxygen determination for many years. Several modifications of the basic method have been made to minimize interferences (20). The basis of this analysis is the quantitative oxidation of alkaline manganous hydroxide by the oxygen in the sample. Upon acidification in the presence of excess iodide, an amount of iodine equivalent to the dissolved oxygen is released. The iodine can then be titrated with standard sodium thiosulfate.

Considerable care is required in the collection of samples for dissolved-oxygen analysis. The errors associated with sampling can be minimized by an *in situ* method. Membrane-covered dissolved-oxygen electrodes are ideally suited to this purpose. These electrodes can be either polarographic or galvanic. The electrodes are protected by an oxygen-permeable polymeric membrane, which provides a rigorously defined diffusion barrier against solution impurities. For the polarographic type, an external voltage must be applied to the working or sensing electrode and must be sufficient to reduce molecular oxygen. With the galvanic type, two solid metal electrodes are used, such that the reduction of oxygen occurs spontaneously and no external voltage source is necessary. In both cases and under steady-state conditions, the resulting current is proportional to the dissolved-oxygen concentration in solution. The use and operation of dissolved-oxygen electrodes are thoroughly described in several publications (29–30).

Minor Mineral Constituents and Gases

Metals. The method of choice for metal analysis generally depends upon the concentration as well as the number of metals to be analyzed per sample and, to a lesser degree, the number of samples. When there is a large number of metals per sample, an emission spectroscopy method is preferred because of its simultaneous multielement capabilities. This technique has become much more popular for water analysis because of the development of a new emission source, the argon plasma. This source combines the advantages of the flame with the high temperature of an electric arc or spark. The plasma can be generated either through an induction process or by a direct current. When coupled with a polychromator in a direct-reading fashion, this method can be used to measure up to 60 elements in approximately one minute (31–33). Limits of detection for many elements are comparable to or better than flame atomic absorption and, in some cases, to graphite-furnace atomic absorption.

Atomic absorption spectroscopy is more suited to samples where the number of metals is small, because it is essentially a single-element technique. The conventional air–acetylene flame is used for most metals; however, elements that form refractory compounds, eg, Al, Si, V, etc, require the hotter nitrous oxide–acetylene flame. The use of a graphite furnace provides detection limits much lower than either of the flames. A cold-vapor-generation technique combined with atomic absorption is considered the most suitable method for mercury analysis (34).

Nonmetals. *Arsenic.* Total arsenic concentration can be determined by reduction of all forms to arsine (AsH_3) and collection of the arsine in a pyridine solution of silver diethyldithiocarbamate. Organoarsenides must be digested in acidic potassium persulfate prior to reduction. The complex that forms is deep red, and this color can be measured spectrophotometrically. Reduction is carried out in an acidic solution of $KI-SnCl_2$, and AsH_3 is generated by addition of zinc.

Atomic absorption spectroscopy is an alternative to the colorimetric method. Arsine is still generated but is purged into a heated open-end tube furnace or an argon–hydrogen flame for atomization of the arsenic and measurement. Arsenic can also be measured by direct sample injection into the graphite furnace. The detection limit with the air–acetylene flame is too high to be useful for most water analysis.

Selenium. Selenium is determined by atomic absorption after the organoselenides are broken down with acidic persulfate and all forms of selenium have been converted to H_2Se. The reduction is brought about in acidic solution of $KI-SnCl_2$ or borohydride, and H_2Se is generated by addition of zinc. The dihydrogen selenide is purged into an open-end tube furnace or argon–hydrogen flame for atomization and measurement. Selenium can also be determined by direct sample injection into the graphite furnace.

Cyanide. Industrial processes frequently discharge significant concentrations of cyanide, which can be extremely toxic at very low levels. Cyanide compounds are classified as either simple or complex. It is usually necessary to decompose complex cyanides by an acid reflux. The cyanide is then distilled into sodium hydroxide to remove compounds that would interfere in analysis. Extreme care should be taken during the distillation as toxic hydrogen cyanide is generated. The cyanide in the alkaline distillate can then be measured potentiometrically with an ion-selective electrode. Alternatively, the cyanide can be determined colorimetrically. It is converted to cyanogen chloride by reaction with chloramine-T at pH <8. The CNCl then reacts with a pyridinebarbituric acid reagent to form a red-blue dye.

Bromide and Iodide. The spectrophotometric determination of trace bromide concentration is based on the bromide catalysis of iodine oxidation to iodate by permanganate in acidic solution. Iodide can also be measured spectrophotometrically by selective oxidation to iodine by potassium peroxymonosulfate ($KHSO_5$). The iodine reacts with colorless leucocrystal violet to produce the highly colored leucocrystal violet dye. Greater than 200 mg/L of chloride interferes with the color development. Trace concentrations of iodide are determined by its ability to catalyze ceric ion reduction by arsenous acid. The reduction reaction is stopped at a specific time by the addition of ferrous ammonium sulfate. The ferrous ion is oxidized to ferric ion, which then reacts with thiocyanate to produce a deep red complex.

Both of these halides can also be determined potentiometrically with an appropriate ion-selective electrode. Sulfide and cyanide both interfere with the electrode response.

Gases. *Hydrogen Sulfide.* Sulfide ion from 10^{-7} to $1\ M$ can be measured potentiometrically with an ion-selective electrode. Mercuric ion interferes at concentrations $>10^{-7}\ M$. The concentration of hydrogen sulfide can be calculated knowing the sample pH and the pK_a for H_2S.

Ammonia. The most reliable results for ammonia are obtained from fresh samples. Storage of acidified samples at 4°C is the best way to minimize losses if prompt analysis is impossible. The sample acidity is neutralized prior to analysis. Ammonia concentrations of 10^{-6}–$0.5\ M$ can be determined potentiometrically with the gas-sensing, ion-selective electrode. Volatile amines are the only known interferents.

The most common colorimetric technique involves a reaction between ammonia and a reagent containing mercuric iodide in potassium iodide (Nessler reagent) to form a reddish-brown complex. Turbidity, color, and hardness are possible interferences that can be removed by preliminary distillation at pH 9.5.

Organic Materials

Nonspecific Organics. *Biochemical Oxygen Demand.* The biochemical oxygen demand (BOD) test is an empirical determination of the oxygen requirement of a sample. It is most often applied to wastewaters, industrial effluents, and polluted waters. The decrease in the dissolved oxygen concentration resulting primarily from biological action is measured after storage for 5 d at 20°C.

If the sample has very low initial dissolved oxygen, it must be aerated; if it has a very high BOD, it may be necessary to dilute the sample. The dilution water is a phosphate buffer containing magnesium sulfate, calcium chloride, and ferric chloride. The percentage dilution must usually be determined by trial and error, although rough estimates may be made depending on sample type (20).

The dissolved oxygen concentrations are determined immediately and after five days. The method for dissolved measurement involves either a modified Winkler titration or a membrane-covered oxygen electrode. The difference between initial and final dissolved oxygen multiplied by the dilution factor is the BOD value.

Chemical Oxygen Demand. The chemical oxygen demand (COD) test measures the oxygen equivalent of the organic matter in a sample that is susceptible to oxidation by a strong oxidant (20). The sample is refluxed with a known amount of potassium dichromate in sulfuric acid for two hours. Silver sulfate is added to catalyze the oxidation of straight-chain compounds, and mercuric sulfate is added to the sample prior to refluxing to react with chloride ion and prevent its oxidation. The amount of unreacted dichromate is then titrated with standard ferrous ammonium sulfate.

Organic Carbon. The total organic carbon (TOC) in a water sample is determined by injecting a microliter sample into a heated, packed tube in a stream of oxygen. The water is vaporized and carbon is converted to carbon dioxide, which is detected with a nondispersive infrared analyzer. Nitrogen is bubbled through the acidified sample prior to injection to remove inorganic carbon and other volatiles.

Detergents. The most widely used surfactant in synthetic detergents is the readily biodegradable linear alkyl sulfonate (LAS). Since the detergent industry began using this ingredient, the occurrence of foaming incidents has all but disappeared (see Surfactants and detersive systems). The methylene-blue method is used to measure

LAS in addition to alkyl sulfates and alkyl poly(ethoxyl sulfate)s. The blue salt that forms is extracted with chloroform and measured colorimetrically. There are many organic and inorganic interferences associated with this method and, since a large number of surfactants react with methylene blue, the test is designated as one for methylene-blue-active substances. If this test shows a significant amount of surfactant, confirmation of the LAS should be made by other methods, eg, thin-layer chromatography followed by infrared spectroscopy (35).

Oil and Grease. Industrial processes contribute most of the grease and oil found in water. Oil and grease can cause problems in wastewater-treatment processes as well as prevent the use of the sludge as fertilizer. The amount of oil and grease can be determined by extraction of an acidified sample with an organic solvent followed by evaporation of the solvent and weighing the residue. During the last part of the solvent evaporation, the sample must not be overheated to prevent the loss of low boiling oils.

Specific Organics. Large quantities of specific organic materials are used annually in industrial and agricultural applications, and these compounds or their degradation products are present in surface and groundwaters. The volume of publications dealing with analytical methodology for specific organics is enormous. The development of glass capillary columns has improved the application of gas–liquid chromatographic (glc) separations. In addition, the application of the mass spectrometer (ms) as a detector for gas–liquid chromatography has made the positive identification of peaks possible. High performance liquid chromatography (hplc), which involves various detectors, can be used to measure hydrophilic and hydrophobic organic compounds in water.

Various methods for the glc monitoring of EPA Consent Decree Priority Pollutants in water have been described (36) (see Regulatory agencies). The determination of organic pollutants in water by glc and ms methods has also been detailed (37–38). Nonvolatile organic compounds in drinking water have been determined by hplc (39) (see Water, water pollution).

Pesticides. Chlorinated hydrocarbon insecticides are determined with an electron-capture detector following extraction with an organic solvent (20). If polychlorinated biphenyls (PCBs) are known to be present or if the extract contains so many pesticides that separation by glc is difficult, the extract should be passed through a Florisil column. Elution of the column with different solvents allows certain group separations of the pesticides. The following organochlorinated pesticides and PCBs have been determined by the method: lindane, heptachlor, heptachlor epoxide, aldrin, dieldrin, p,p'-1,1-dichloro-2,2-bis(p-chlorophenyl)ethane (p,p'-TDE), p,p'-1,1,1-trichloro-2,2-bis(p-chlorophenyl)ethane (p,p'-DDT), o,p'-DDT, p,p'-1,1-dichloro-2,2-bis(p-chlorophenyl)ethylene (p,p'-DDE), endrin, p,p'-methoxychlor, α-endosulfan, β-endosulfan, cis-chlordane, trans-chlordane, Aroclor 1248, Aroclor 1254, and Aroclor 1260. Quantitation is by comparison of chromatograms with standard concentrations of pure compounds treated in an identical manner. The phenoxy acid herbicides (2,4-dichlorophenoxy)acetic acid (2,4-D), silvex, and (2,4,5-trichlorophenoxy)acetic acid (2,4,5-T) can be determined by electron-capture detection after extraction and conversion to the methyl esters with BF_3-methanol. The water sample must be acidified to pH ≤ 2 prior to extraction with chloroform.

Identification of the pesticides is based on retention time on at least two dissimilar glc columns. A nonpolar and a relatively polar packing are generally used, for example, OV-17 and a mixture of QF-1 and DC-200.

Organophosphorus pesticides are determined with a flame photometric detector, which is sensitive only to compounds containing phosphorus or sulfur (20). The water sample is extracted with benzene, and the solvent is concentrated to a small volume prior to glc analysis. The following organophosphates are determined by this method: ethyl guthion, guthion, trithion, ruelene, diazinon, di-syston, ethion, imidan, malathion, methyl parathion, methyl trithion, parathion, thimet, and trolene (see Insect control technology).

High performance liquid chromatography with electrochemical detection has been used to determine 2–7 ppb of carbamate pesticides in water (40). The investigated pesticides were aminocarb, asulam, *sec*-butyl phenylmethylcarbamate (BPMC), carbaryl, carbendazim, chlorpropham, desmedipham, and phenmedipham.

Trihalomethanes. Wherever chlorine is used as a disinfectant in drinking-water treatment, trihalomethanes (THMs) generally are present in the finished water. The THMs usually formed are trichloromethane (chloroform), bromodichloromethane, dibromochloromethane, and tribromomethane (bromoform). There are four main techniques for the analysis of THMs: headspace, liquid–liquid extraction (lle), adsorption–elution (purge–trap), and direct aqueous injection. The final step in each technique involves separation by gas–liquid chromatography with a 2 mm ID coiled glass column containing 10 wt % squalene on chromosorb-W-AW (149–177 μm (80–100 mesh)) with detection generally by electron capture.

The purge–trap method is the most widely accepted method for THMs as well as for other purgeable organohalides (41). The purgeable organics are stripped from the water sample with a stream of inert gas and are adsorbed in a porous polymer trap. The compounds are thermally desorbed from the trap into a gas chromatograph. All of the lle methods involve extraction of a small volume of water with a much smaller volume organic solvent (42). An aliquot of the extract is then gas-chromatographed. Different lle techniques have been evaluated to be as sensitive and accurate as the purge–trap methods (42–43).

Radioactive Materials

Radioactivity in environmental waters can originate from both natural and artificial sources. The natural or background radioactivity usually amounts to $\leq$100 mBq/L. The development of the nuclear power industry as well as other industrial and medical uses of radioisotopes (qv) necessitates the determination of gross alpha and beta activity of some water samples. These measurements are relatively inexpensive and are useful for screening samples. The gross alpha or beta activity of an acidified sample is determined after an appropriate volume is evaporated to near dryness, transferred to a flat sample-mounting dish, and evaporated to dryness in an oven at 103–105°C. The amount of original sample taken depends on the amount of residue needed to provide measurable alpha or beta activity.

Alpha counting is done with an internal proportional counter or a scintillation counter. Beta counting is carried out with an internal or external proportional gas-flow chamber or an end-window Geiger-Mueller tube. The operating principles and descriptions of various counting instruments are available, as are techniques for determining various radioelements in aqueous solution (20,44). A laboratory manual of radiochemical procedures has been compiled for analysis of specific radionuclides in drinking water (45). Detector efficiency should be determined with commercially available sources of known activity.

Bacteria

A bacteriological examination of water is primarily carried out to determine the possible presence of harmful microorganisms. Testing is actually done to detect relatively harmless bacteria called *colon bacilli*, commonly called the coliform group, which are present in the intestinal tract of man and animals. If these organisms are present in a water in sufficient number, then this is taken to be evidence that other harmful pathogenic bacteria may also be present.

One standard test used to determine the presence of the coliform group is called the multiple-tube fermentation technique (sometimes called the presumptive test). If this test indicates the presence of these bacteria, then a confirmed test must be done. If only negative colonies or no colonies develop during this test, it is considered negative; otherwise, a completed test must be undertaken. Positive results obtained in the completed test are evidence for the presence of coliform bacteria. Testing methods have been given by the APHA, and the detailed procedures contained therein should be consulted (20).

Another standard test, which is much simpler and more convenient, is the membrane filter technique. A suitable volume of sample is filtered through a sterile, 0.45-μm membrane filter. The filter is placed in a petri dish containing a specific growth medium (M-Endo nutrient broth, M-Endo medium) and incubated for 24 h at 35°C. If after this time the colonies show the characteristic green sheen, this is taken as positive evidence for the presence of the coliform group (see Water, sewage).

Fecal Coliforms. Fecal coliforms are those originating from the intestines of warm-blooded animals. Fecal coliforms can be determined by a multiple-tube procedure, which must be applied to a positive presumptive test for optimum recovery of fecal coliforms (20). Incubation must be at 44.5 ± 0.2°C for 24 ± 2 h. Gas production during incubation is positive evidence of fecal coliform pollution.

A membrane filter technique can also be used to determine the presence of fecal coliforms, and this procedure is said to be 93% accurate (20). A sample is passed through a membrane filter, and this filter is placed in a petri dish containing an enriched lactose medium. The dishes are incubated at 44.5 ± 0.2°C for 24 h. Following the incubation period, the fecal coliform colonies appear blue.

Both multiple-tube and membrane-filter methods are also available for testing for the fecal streptococcal group (20). These assays can be used to provide supplementary data regarding the bacteriological quality of water. Other fecal indicators should also be used concurrently because of the survival characteristics of the fecal streptococci.

BIBLIOGRAPHY

"Water (Municipal)" in *ECT* 1st ed., Vol. 13, pp. 946–962, by H. O. Halvorson, University of Illinois; "Water (Analysis)" in *ECT* 2nd ed., Vol. 21, pp. 688–707, by Marvin W. Skougstad, U.S. Geological Survey.

1. F. C. Smith and R. C. Chang, *Crit. Rev. Anal. Chem.* **9,** 197 (1980).
2. J. D. Mulik and E. Sawicki, eds., *Ion Chromatographic Analysis of Environmental Pollutants*, Vol. 2, Ann Arbor Science Publishers, Inc., Ann Arbor, Mich., 1979.
3. W. B. Robbins and J. A. Caruso, *Anal. Chem.* **51,** 889A (1979).
4. B. V. L'vov, *Spectrochim. Acta Pt. B* **33B,** 153 (1973).
5. R. K. Skogerboe, *Toxicol. Environ. Chem. Rev.* **2,** 209 (1978).
6. L. T. Skeggs, *Am. J. Clin. Pathol.* **28,** 311 (1957).

7. J. Ruzicka and E. H. Hansen, *Anal. Chim. Acta* **78,** 145 (1975).
8. *Ibid.*, **99,** 37 (1976).
9. L. R. Snyder, *Anal. Chim. Acta* **144,** 3 (1980).
10. J. K. Foreman and P. B. Stockwell, eds., *Topics in Automatic Chemical Analysis*, Vol. 1, Ellis Horwood Ltd., W. Sussex, UK, 1979.
11. J. K. Foreman and P. B. Stockwell, eds., *Automatic Chemical Analysis*, Ellis Horwood Ltd., W. Sussex, UK, 1975.
12. T. M. Florence and G. E. Batley, *Crit. Rev. Anal. Chem.* **9,** 219 (1980).
13. M. J. Fishman, D. E. Erdmann, and T. R. Steinheimer, *Anal. Chem.* **53,** 182R (1981).
14. D. S. Polcyn, *J. Water Pollut. Control Fed.* **53,** 620 (1981).
15. F. B. DeWalle, D. Norman, J. Sung, E. S. K. Chian, and M. Giabbai, *J. Water Pollut. Control Fed.* **53,** 659 (1981).
16. C. E. Hamilton, ed., *Manual on Water*, 5th ed., ASTM Special Technical Pub. No. 422A, American Society for Testing and Materials, Philadelphia, Pa., 1978.
17. M. A. Forbes, ed., *Analytical Methods Manual*, Inland Waters Directorate, Environment Canada, Ottawa, Ontario, Can., 1979.
18. J. K. Kopp and G. D. McKee, eds., *Methods for Chemical Analysis of Water and Wastes*, 3rd ed., EPA/600/4-79-020, Washington, D.C., 1979.
19. H. S. Hertz, W. E. May, S. A. Wise, and S. N. Chesler, *Anal. Chem.* **50,** 428A (1978).
20. *Standard Methods for the Examination of Water and Wastewater*, 15th ed., American Public Health Association, Washington, D.C., 1981.
21. *Instrumentation for Environmental Monitoring*, rev., Vol. II, Water LBL-1, Technical Information Department, Lawrence Berkeley Laboratory, University of California, Berkeley, Calif., 1980.
22. H. Small, T. S. Stevens, and W. C. Bauman, *Anal. Chem.* **47,** 1801 (1974).
23. R. A. Wetzel, C. L. Anderson, H. Schleicher, and G. D. Crook, *Anal. Chem.* **51,** 1532 (1979).
24. D. Langmuir and R. I. Jacobson, *Environ. Sci. Technol.* **4,** 835 (1970).
25. D. R. Keeney, K. H. Byrnes, and J. J. Green, *Analyst* **95,** 383 (1970).
26. M. S. Frant and J. W. Ross, Jr., *Anal. Chem.* **40,** 1169 (1968).
27. J. E. Harwood, *Water Res.* **3,** 273 (1969).
28. E. Bellack and P. J. Schoubof, *Anal. Chem.* **30,** 2032 (1968).
29. K. H. Mancy and T. Jaffe, *Analysis of Dissolved Oxygen in Natural and Wastewater*, U.S. Public Health Service Publication No. 999-WP-37, Washington, D.C., 1966.
30. M. L. Hitchman, *Measurement of Dissolved Oxygen*, John Wiley & Sons, Inc., New York, 1978.
31. A. F. Ward, *Am. Lab.* **10,** 79 (1978).
32. Barnes, R. M., *Application of Plasma Emission Spectrochemistry*, Heyden & Sons, Inc., Philadelphia, Pa., 1979.
33. G. W. Johnson, *Spectrochim. Acta Pt. B*, **34B,** 197 (1978).
34. G. E. Millward and A. LeBihan, *Water Res.* **12,** 179 (1978).
35. H. H. Hellman, *Fresenius Z. Anal. Chem.* **293,** 359 (1978).
36. L. H. Keith, K. W. Lee, L. P. Provost, and D. L. Present, *ASTM Special Technical Publication STP 686*, American Society for Testing and Materials, Philadelphia, Pa., 1979, pp. 85–197.
37. D. Beggs, *NBS Spec. Publ. (US)* **519,** 169 (1979).
38. W. E. Pereira and B. A. Hughes, *J. Am. Water Works Assoc.* **72,** 220 (1980).
39. B. Crathorne, C. D. Watts, and M. Fielding, *J. Chromatogr.* **185,** 671 (1979).
40. J. L. Anderson and D. J. Chesney, *Anal. Chem.* **52,** 2156 (1980).
41. T. A. Bellar and J. J. Lichtenberg, *J. Am. Water Works Assoc.* **23,** 234 (1974).
42. R. C. Dressman, A. A. Stevens, J. Fair, and B. Smith, *J. Am. Water Works Assoc.* **71,** 392 (1979).
43. M. M. Varma, M. R. Siddique, K. T. Doty, and A. Machis, *J. Am. Water Works Assoc.* **71,** 389 (1979).
44. K. Haberer and U. Stuerzer, *Fresenius Z. Anal. Chem.* **299,** 177 (1979).
45. H. L. Krieger, *U.S. NTIS PB Report No. 253258*, National Technical Information Service, Washington, D.C., 1976, 61 pp.

R. B. SMART
West Virginia University

K. H. MANCY
University of Michigan

SUPPLY AND DESALINATION

Of the surface of the earth, 71% (3.60×10^8 km^2 or 1.39×10^8 mi^2) is covered by oceans; their average depth is 6 km and their volume is 8.54×10^8 km^3 (ca 2×10^8 mi^3). Unfortunately, this huge quantity of water is unsuitable for many human uses. Water with over 1000 ppm salt is usually considered unfit for human consumption, but in some parts of the world, people and land animals are forced to survive with much higher concentrations of salts. It has been recorded that when the native population in parts of Mexico was given desalinated potable water, the people refused at first to drink it as its taste was strange (1).

Freshwater is generally considered to be potable water with less than 500 ppm or 0.05% dissolved solids. Rain is the source of freshwater, and its precipitation of over 1.3×10^{15} L/d (3.4×10^{14} gal/d) over the earth's surface averages about 1.05 m/yr. Extremes range from almost zero in North Chile's desert bordering the Pacific Coast to >25.4 m in some tropical forests and on some high slopes where the high, cold mountains condense floods from the clouds.

Even rain is not pure water. Reports from the U.S. Geological Survey show it contains 2.3–4.6 ppm of solids, or a yearly precipitation of 2.5–5 t/km^2 (6.5–13 t/mi^2). Recent work in the United States and Europe has concentrated on the rather dangerous results of technological progress, the SO_2 and NO_x emissions that end up in the rain, lowering its pH from the slightly acidic pH of 5.6 for uncontaminated rain to $\geq$pH 4 for acid rains. These rains are having serious effects on surface waters and their aquatic population as well as potentially effecting changes in surface water composition (2). About 85% of the 32×10^6 t of SO_2 that is emitted annually in North America originates in the United States and the balance is from Canada (3).

About 60% of the land area of the earth is arid or semiarid, not generally considered habitable. Mountainous areas and the polar areas covered by ice must also be subtracted from the land available for human habitation and agriculture. Of the small fraction remaining, the preferred land areas support the expanding world population more generously; so increasing amounts of better lands are used for living—for urban areas, industries, roads, airfields, etc. Areas for living and industry usually require much more water per unit than do farm lands. Thus, agricultural areas with adequate water become smaller as population and industry increase, and the demands for water multiply.

The oceans hold about 97% of the earth's water. More than 2% of the total water and over 75% of the freshwater of the world is locked up as ice in the polar caps. Of the remaining 1% of total water that is both liquid and fresh, some is groundwater at depths of >300 m and therefore impractical to obtain, and only the very small difference, possibly 0.06% of the total water of this planet, is available for human use as it cycles from sea to atmosphere to land to sea. Only recently have humans been able to regulate that cycle to their advantage, and then only infinitesimally in some few isolated places.

Wells produce groundwater, stored from previous rains. However, the fact that in recent years wells have had to be made deeper and deeper to reach water shows that groundwater is being used faster than it is being replenished. Water lying in deep strata for millions (10^6) of years is being mined like other minerals, never to be replaced. In Libya, recent oil drilling found a lake 100 m below the dry sand, hundreds of square kilometers in area, and 750 m deep (4). It has been estimated that this lake will supply irrigation of 800 km^2 (309 mi^2) for 300 yr, but the pumping of this water is as final an act as the pumping of Libya's petroleum, which probably dates from the same lush geological era. Once pumped, neither can be replenished.

Transport of Freshwater

For centuries, containers have been used to carry freshwater, usually for longer distances than would be practical for conduits. Trucks and railways have used tanks. The bridge-tunnel over and under the Chesapeake Bay has a 380,000-L (1.4 × 10^6-gal) tank that is filled by a tank truck which hauls water 19 km. Ships have carried and still do carry water halfway around the world in ballast tanks; the tankers otherwise returning empty from oil deliveries may make the return voyage filled with freshwater as precious to the oil-rich, water-poor country as the oil is to its market.

Systems of dams, canals, and aqueducts were developed to carry freshwater considerable distances to growing cities, and to irrigate agricultural lands. Records of ancient Rome indicate that 14 aqueducts, averaging over 144 km in length, carried 1.175 × 10^6 m^3 H$_2$O/d (3.1 × 10^8 gal/d) from the surrounding highlands by gravity. The Romans depended on gravity flow in open channels and had no concept of friction losses in pipes, even though Pliny lists standard lead pipe in circumferences up to ca 2.5 m. (This represented the widths of strips of lead formed around a mandrel and "burned" or welded in a single seam). The aqueducts of Istanbul are even more dramatic. The engineers' competence developed with increasing needs for water, and conduits became longer.

Initiated ca 120 yr ago, New York City's water system is truly an engineering marvel. Farsighted action in the late nineteenth century also gave the city extensive upstate watershed rights. This system, with a storage capacity of 2 × 10^9 m^3 (5.3 × 10^{11} gal), can safely furnish the city with an average of 5 × 10^6 m^3/d (1.3 × 10^9 gal/d). The system is not adequate today, but it has served well except in years of serious drought. Water enters the city via two tunnels, one built in 1917 and the other in 1937. The total cost of a third tunnel is now estimated at $3.5 × 10^9 in 1981 dollars, but could be as high as $11 × 10^9 by the year 2009, the date scheduled for its completion (5).

The 1974 Safe Drinking Water Act put all public water supplies in the United States under Federal supervision, and the 94th Congress authorized a six-state water study, the High Plains Study Council, to develop plans for increasing water supply in the area. If its recommendations are accepted, the government will move huge amounts of water from the Missouri and Arkansas Rivers to the high plains (6). It could take as long as 9 yr to design, 25 yr to build, and is estimated by the Army Corps of Engineers to cost $(6–25) × 10^9.

The problem of bringing water to southern California continues to be one of much controversy (7). The project to build the so-called Peripheral Canal, a 67-km long, 120-m wide channel to carry water from the Sacramento River delta to an existing aqueduct and then to southern California, was passed by California's legislature in 1980.

In some places and under certain conditions, freshwater can be obtained by desalination of seawater more cheaply than by transporting water. This is true when all costs of the tremendous investments in dams, reservoirs, conduits, and pumps to move the water are considered. Before the rapid escalation of fuel costs from 1973 through 1980, the cost of desalination of seawater to definitely supply southern California would have been less than transport to the Peripheral Canal. This would have been the case even if there were an unlimited supply of water in the mountains of northern California, a condition that does not appear to exist. It has been shown that before 1973 a seacoast town could have been supplied with $(7-12) \times 10^4$ m^3 $((1.8-3.2) \times 10^7$ gal) of freshwater more cheaply by desalination than by damming and piping water a distance of ≥ 160 km (8). Furthermore, desalination could at that time have become less costly with improved technology, whereas transporting water is not likely to decrease substantially in cost. Desalination increases the total amount of freshwater; transportation never does. The reduction of water costs, if possible at all, will be limited. Desalination is an energy-intensive process, and no scenarios conceivable today predict much reduction of energy costs. The cost savings obtained as consequences of advances in desalination technology have been completely nullified by the increase in energy costs since the oil embargo of 1973.

The Water Problem

Many cities of the world do not levy a separate tax on water distributed, and even in those places where water is in shortest supply, a minimal ration may be free to everyone. The problem that wasted water and unmetered water add to the overall water demand is not new. In ancient Rome, fountains were connected to the public water by privately installed and owned lead pipelines, many of which were unrecorded, illegal, and hence untaxed. Frontius, the Water Commissioner of Emperor Nerva of Rome in 96 AD, developed crude meters to increase revenue and cut demand.

Today in the United States, three times as much water is used per capita than in 1900; with inclusion of all industrial and agricultural uses, this quantity is probably ten times as great as at the turn of the century. Individual usage in some southern cities, with swimming pools, lawns, air conditioners, and other local demands, can be as much as twice the national average. Population increase multiplies the total withdrawals, particularly in cities, where it may be as much as 1060 L/d (280 gal/d) per person. In New York City, for example, per capita consumption grew from 690 L/d (182 gal/d) in 1970 to 760 L/d (200 gal/d) in 1981.

An analysis of water use in the United States and an estimate for the next 50 yr predicts withdrawals will increase by 400% and consumptive uses by >100% (9). Most of this increase must be met by reuse; ie, in manufacturing, recycling must rise from 1.3 times reuse to 6.3 times reuse by the year 2020.

In some areas of the United States, twice as much water is pumped from the ground as soaks into it. Although the amount of groundwater within 0.8 km of the surface is estimated at 3.8×10^{16} m^3 (10^9 gal), in some places the water table has dropped by 1–5 m for each year of the present generation, thus exhausting a historical treasure. Often this withdrawal causes sinking of the ground level, as it has near Houston, Texas, Mexico City, and Florida. Las Vegas, Nevada, is growing rapidly upon a ground level that has sunk about a meter in recent years as a result of greatly increased mining of prehistoric water. This water supplies many hectares of swimming pools, many thousands of "tons of refrigeration" for air conditioning, and other water uses. It is water from wells in a desert where only 7.5–10 cm of rain falls each year.

Southern California is suffering from seawater intrusion in certain aquifers because of overpumping. To minimize the problem, the Orange County Water District in 1979 began a pioneering effort of injecting sewage purified by reverse osmosis (RO) into the aquifer with the objective of reducing, if not stopping, seawater intrusion (10) (see Reverse osmosis).

Water is far from evenly distributed in the United States, with major shortages in some very populous areas. California and the southwest have always been water short, but in the first half of the 1960s and the late 1970s, the northeast, too, experienced water shortages. For the United States as a whole, the demand for water will become equal to the total supply before 1985. Some areas will be desperately short much sooner. Water shortages are acute in years of low rainfall, as in 1957 when over 1000 communities in 47 of the 48 contiguous states restricted the use of water. Restrictions of water use in one or more states have become virtually annual occurrences.

Two out of five U.S. cities have inadequate water supplies, and at least a quarter of the U.S. population faces serious water shortages. Yet, half the states—with two thirds of the industry and over half of the population—have direct access to as much as they can draw of the approximately 3.3×10^8 km^3 (8.8×10^{10} gal) of seawater. The solids content of this water, mainly salts, varies from 25,000 ppm (2.5%) in the Baltic Sea to over 40,000 ppm (4%) in some of the more confined gulfs of the Indian Ocean. The waters of the wide oceans are almost constant at 35,000 ppm (3.5%). Also, many inland areas have access to large quantities of water too brackish to drink.

Some of the most attractive areas of the world, particularly islands and beaches, are almost devoid of freshwater. This living space, and space for resort hotels, is lost. The biggest and one of the fastest growing of the industries of the world is tourism. It is a particularly attractive industry to developing countries, and in some of these it may be almost the only nonagricultural industry. Tourism accounts for a substantial use of the available freshwater, and it may be stifled or entirely prevented in otherwise attractive places if there is insufficient freshwater. To conserve available water supplies, some hotels might use double water systems, ie, seawater for flushing, but they still must provide on the order of 400–600 L/d (106–159 gal/d) per tourist to assure a comfort to guests unaccustomed to water shortages. Production of these quantities of water by desalination techniques has become an important expense to the hotels. Today, the cost of production of desalinated water can be as high as $4.00–5.25 per thousand liters ($15–20/1000 gal), and in some locations even higher.

The island of Bermuda derives the largest percentage of its income of any country from tourism (over 150,000 visitors per year). By law, each house there must have a roof area of 11 m^2 for each occupant to catch the average annual rainfall of 148 cm and drain it to a basement reservoir. The U.S. Virgin Islands have similar regulations. Vegetation is cleared from hill slopes above the natural limestone and coral which is plastered to give drains and catchments for heavy seasonal rains. But even with all this, some water has to be imported at ca $4/1000 L ($15/1000 gal). Today, much water is desalinated locally at equally high or higher costs.

The newly acquired wealth of the countries in the Middle East has given an unexpected boost in desalination for the region. In the last decade, Saudi Arabia has added over 2.3×10^6 m^3/d (6×10^8 gal/d) of capacity including a 15,000 m^3 (4×10^6 gal) seawater RO (reverse osmosis) plant, one of the largest seawater RO plants ever built, and a single-site 946,000 m^3 (2.5×10^8 gal) multistage flash (MSF) plant, similarly the largest distillation complex ever built, to be completed in 1984.

In many of the developing nations, lack of water hampers the exploitation of profitable material resources (11). Proven mines on Egypt's Red Sea Coast cannot be operated; others on the same coast are operable only with desalination water for processing phosphate ores and for mining lead and zinc. The workers are rationed 15–35 L (4–9.25 gal) water per day. Fishing industries on South America's arid Pacific Coast cannot be expanded for lack of water to process the haul. These represent great losses in the world's supplies of minerals and foods. There are many other examples. Not all freshwater shortages are in the torrid tropics; a substantial iron-ore deposit in a small waterless island off the coast of Iceland needs water for workers. Needed is a desalination plant or some means of utilizing as liquid water the heavy fogs that prevail there. Off-shore oil-drilling rigs around the world similarly need water. Technological progress in the last decade has made readily available packaged small $[(10–400) \times 10^3$ L/d or $(2{,}500–106{,}000)$ gal/d] seawater RO and vapor-compression systems. In fact, virtually all off-shore drilling rigs today have their own seawater-conversion plants. Desalination facilities, and in turn desalinated water, become readily available contingent only on the availability of funds.

Cities on the lower reaches of a large river, with many cities above, use water that has been through sewers upstream many times. On the lower Mississippi, a water inventory indicates such reuse averages 14 times, with biochemical oxidation of the wastes during the flow between cities. The Rhine River, in passing through several countries, all of which drink and dump into its waters, causes international problems of pollution. Many so-called hard contaminants present in sewage increase almost proportionally with reuse of the water. Such materials do not ferment or oxidize under ordinary sewage treatments, and some are known carcinogens which present serious health problems to potable waters drawn from these bodies.

The U.S. government spent ca 150×10^6 in fiscal 1969 on water-resources research relating to artificial rainmaking, soil conservation, waste treatment, desalting, public health, and planning research. This level of annual expenditure, large compared to the rest of the world, controls with a need estimated at $>\$100 \times 10^9$ for water facilities worldwide within 10–15 yr (12). All of the reports and results of research on water are available from the Office of Saline Water (OSW).

Saline Water for Municipal Distribution. Only a very small amount of potable water is actually taken internally, and it is quite uneconomical to desalinate all municipally piped water, although all distributed water must be clear and free of harmful bacteria. Most of the water piped to cities and industry is used for little more than to carry off extremely small amounts of waste materials or waste heat. In many locations, seawater can be used for most of this service. If chlorination is required, it can be accomplished by direct electrolysis of the dissolved salt (13). Against the obvious advantage of economy, there are also several disadvantages. Use of seawater requires different detergents; sewage-treatment plants must be modified; the usual metal pipes, pumps, condensers, coolers, meters, and other equipment corrode more readily; chlorination could cause environmental pollution; and dual water systems must be built and maintained.

Pipes, valves, fittings, and almost all other components of small equipment are now available in plastic or ceramics, which do not corrode in salt water and are less expensive than the metals now used. Synthetic detergents are now available for use with seawater, although a final rinse with freshwater may be desired. Saltwater sewage can be treated successfully. Dual water systems using freshwater and seawater, are already in use on shipboard and in many island resort hotels. Many of these also have seawater systems for fire fighting. This trend will grow.

Some inland municipalities now distribute water with salt content exceeding 1000 ppm, water so brackish as to be unpleasant to the taste, even though it is distributed as potable water. Each home may produce or have delivered the very small requirement of freshwater for drinking and cooking. Small membrane desalinators are becoming available; these produce the relatively few liters of potable water required (see Membrane technology; Ion exchange). They can be purchased from and periodically serviced by service companies in some communities. The cost per liter of potable water produced in the home is several times as great as the cost in a central desalination plant, but the amount of desalinated water needed is only a small fraction of the total supply.

Home desalinators are possible only for industrialized countries with a central service organization. They will eventually become available on a rental-service contract basis, as is standard practice for water softeners in many communities. Although rental of water softeners is common in the United States, no membrane-system desalter rental is known to exist at present.

Alternatively, small amounts of potable water may be delivered by truck to distribution centers or to tanks on house roofs. This system exists in Kuwait, which has many filling stations from which tank-truck operators buy water at $1.00/kL ($3.88/1000 gal) for distribution at $2.40–3.00 ($9.00–11.00/1000 gal). Although much water is directly piped to residences in Kuwait today, that system has replaced only part of the filling-station distribution system.

Water in Industry. Freshwater for industry can often be replaced by saline or brackish water, usually after sedimentation, filtration, and chlorination (electrical or chemical) or other treatments (14). Such treatment is not necessary for the largest user of water, the electric power industry, which in the United States will be passing through its heat exchangers about one quarter of the total supply of surface water by 1985 and about one half by 1995 (15). Single stations of 1000 MW may heat as much as 12 GL/d (3.2×10^9 gal/d) as much as 10–15°C.

Cooling towers circulate the warmed water downward against a rising stream of air which removes heat by evaporating a part of the water to cool the balance. In the United States, cooling towers are not so common as in Europe because of the large bodies of water available. Towers in the United States are usually rectangular, with many gridworks of redwood for the water to descend over in films. In Europe, they are more often the shape of a Venturi (corseted) waist, and virtually all new cooling towers in the United States are also of the Venturi variety.

Construction of new power plants in the coal region of the western United States presents serious problems to the power industry in states whose laws dictate zero effluent. In these plants, cooling-tower water withdrawn from rivers cannot be returned to the river. In these situations, cooling-tower effluent is purified by distillation (vapor-compression plants have predominated) and by a combination of distillation and membrane technology. The converted water then is used as boiler feedwater; the plant blowdown (effluent) is evaporated from open-air lined pools, and pool sediment is periodically buried back in the coal mine with the flue ashes.

Although 2.4×10^5 L (ca 63,000 gal) water is used to make a metric ton of steel, a half ton of gasoline, or 360 kg of acetate fiber, little if any is required chemically in any of these processes. Recycling can reduce industrial requirements by a factor of

10–50. Much of this water, particularly that for cooling, and often that for washing, can be saline. Some petroleum refiners have used salt water to remove heat (water's principal role in gasoline production) and some have actually produced table salt by evaporation in cooling towers.

The pulp and paper industry has tried for many years to use salt water for some of the 4×10^5 L (ca 10^5 gal) required to make a metric ton of paper (see Pulp; Paper). Here, however, salt is disadvantageous to the chemical processes, either in pulping the lignocellulose or in the recovery of values from the black liquor after pulping, and can corrode expensive papermaking machinery. The possibility of recovering and reusing at least part of this water after membrane processing is under study.

The textile industry, also a large consumer of water, must always be located in areas with abundant water supply (see Textiles, survey). Before long, advances in purification of textile industry effluents (16) are expected to free the industry of this limitation and enable textile plants to locate in virtually any area.

The great fluid reservoir of heat, the alternative to water, fresh or salt, is the atmosphere. Increasingly, industry is using air coolers to dissipate the large quantities of heat from power plants, petroleum distillation, and other process use. In 1969, $50 $\times 10^6$ was spent on new air coolers and this amount is expected to increase by 20% each year in the near future to reduce correspondingly the use of cooling water (17).

Air coolers almost invariably add considerably to plant cost, but they are competitive in operating cost based on direct once-through use of water that requires no treatment. If the alternative to air coolers is the use of water that requires substantial treatment or pumping costs, the air coolers will cost less to operate.

Water for Agriculture. Two liters (ca 0.5 gal) of water in some form is the daily requirement of the average human, depending on many personal and external conditions. However, at least several hundred liters per day is required for the growing of vegetables, fruit, and grain that make up the absolute minimal daily food ration for a vegetarian. About 480 L (ca 130 gal) is required to produce an egg, and 31,000 L (8200 gal) to produce a kilogram of beef, based on the cereals required by the animals. The quality of water (in particular, its salinity) is of considerable importance to agriculture. There are crops that can grow in water of relatively high salinity, but most crops cannot tolerate salinity exceeding 1000 ppm. This limit led to the treaty between the United States and Mexico in which the United States agreed to limit the salinity of the lower Colorado River reaching Mexico to 1000 ppm salt, a requirement to be met by constructing a 400 ML (100×10^6-gal) RO desalination plant in Yuma, Arizona. Money problems appear to have decreased the size of the actual plant to be built, and for a diversity of reasons the project is much behind schedule.

A loaf of bread contains little of the more than a metric ton of water necessary to grow the wheat therein. The water content of vegetables comes from the 1000–2000 times their weight needed to grow them. Some of the main losses in agriculture may be reduced by agronomists and plant physiologists. Studies are being made to determine whether it is possible to supply less water to the roots, with better absorption there and less losses by transpiration through the leaves. Certainly, plant structures of an entirely different type will be necessary to prevent, for example, the loss by transpiration in a field of corn on an August day, an amount that may be equivalent to more than a centimeter of rainfall. However, the recently developed high yield grains, which use much less water than the conventional varieties, are a great aid to the developing nations.

It is possible to breed plants that have more efficient systems for utilization of water, and agricultural technology can help existing crop plants by spraying impervious coatings on them. Extremely small amounts of long-chain, fatty alcohols reduce evaporation losses from quiet lakes or reservoirs to less than 5% of the normal surface evaporation.

Waterproofing sandy soils to prevent drain-through has been successful in increasing crops as much as 400% with the same rainfall. A special plow lifts the soil to allow melted asphalt to be layered in overlapping impermeable strips 82 cm wide and 50 cm below the surface (18). Waterproofing the surface in Israel by compacting with chemicals increases runoff to basins or fields on slopes below. In many places, barren slopes have been coated with asphalt or concrete in layers as thin as 0.3 cm to catch rain, which is conducted to catch basins for irrigation or other uses. In St. Thomas, U.S. Virgin Islands, such catchment basins have, however, now been abandoned in favor of desalination.

The one seventh of the world's crop lands that are irrigated produce one quarter of the world's crops. Irrigation's main losses result directly from seepage and evaporation from the open water-carrying channels and the soil. Only a small fraction of the water withdrawn from the irrigation ditch or pipe is absorbed by the plants. Plastic films, as ground covers through which the plants protrude, prevent some losses but at great expense for film and labor. Cheaper systems are necessary to assure better water utilization by plants. Other possible goals would be food plants with membranes capable of separating freshwater from brackish water, to give a nonsalty crop. Progress has been made in both of these directions, and some plants have been developed that accumulate salt from the ground.

Vegetables and fruits such as melons are profitably grown in four or five crops per year by hydroponics, the growth of plants in large, shallow concrete tanks containing no soil but gravel and water with added nutrients. Such installations can even be found in areas known for their water scarcity such as Aruba and St. Croix (U.S. Virgin Islands) where both desalinated seawater and brackish water are used for the purpose. Much water is still necessary per unit weight of crop, but the largest losses of ordinary irrigation are prevented, as indeed they must be because of the comparatively high cost of the water. Such concentrated agriculture is very expensive in preparation of land area, but economical for water and labor requirements. Production is high in the tropics, and hydroponics offers a major opportunity to many developing countries.

Other locations where desalinated water of either brackish or sea origin is used for agriculture are the Channel Island of Guernsey, Israel, Libya, and countries in the Arabian Gulf including Saudi Arabia. The cost of such agricultural use is at best marginal and at worst exorbitant depending on amounts of water required.

Economic Aspects

According to figures compiled in 1978 by the National Wildlife Federation, residential water rates per 28.3 m^3 (1000 ft^3) were $3.99 in El Paso, Texas; $4.48 in Albuquerque, New Mexico; and $4.55 in Los Angeles, California (19). By contrast, residents of water-abundant states in the northeast were apparently paying considerably more: $7.79 in Newark, New Jersey; $8.90 in Boston, Massachusetts; and $8.13 in Philadelphia, Pennsylvania. Water costs have continued to escalate. This was painfully demonstrated in New Jersey during a drought in 1981 when conservation caused substantial shrinking of demand and private water companies had to double their water prices. The continued increase of demand and reduction of supply spell real and relentless water-cost increases for the future in every part of the United States. One of the possible ways to assure at least adequate supplies, and possibly moderate these cost increases, is water reuse.

Desalination Development Programs and Associations

Office of Saline Water (OSW)–Office of Water Research and Technology (OWRT).

The U.S. Department of the Interior, long active in many areas of water resources and management, was authorized by Congress in 1952 to organize the OSW with a 2×10^6 budget for 5 yr to develop economical processes for desalination. This budget was repeatedly expanded to a total of almost 318×10^6 authorized through fiscal 1980 for research and development efforts including the construction of demonstration plants to prove the technical and economic feasibility of desalination processes. One fundamental principle guiding these efforts and enunciated by one president of the United States is that all scientific and engineering knowledge and skills developed in the wide reaches of this program are to be immediately and freely available to every nation and every person in the world as a generous contribution to the betterment of the lives of all.

This has come to pass, for desalination technology developed in the United States has been of tremendous benefit to the world at large, even though the U.S. desalination industry has not always shared in the profits from the exploitation of this technology. The U.S. need for desalting technology projected by the OWRT is given in Figure 1. In its effort, OSW and its successor, OWRT, supported the complete gamut of activities

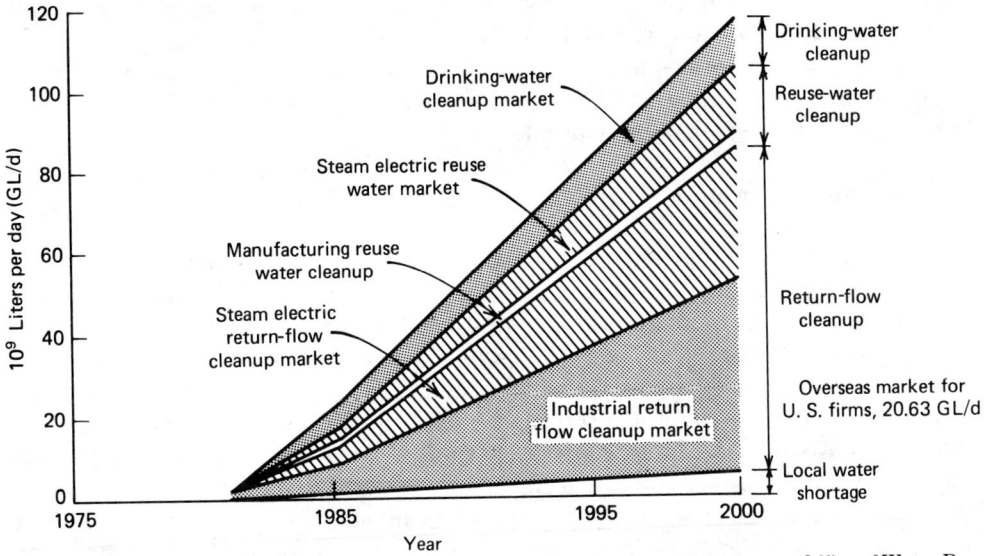

Figure 1. Projected U.S. need for desalting technology. Courtesy of the former Office of Water Research and Technology (OWRT). The OWRT was closed August 24, 1982. The unit GL/d = 264.2×10^6 gpd (gal/d).

from fundamental through demonstration plants and covered all desalination technologies. In membrane technology, it was responsible for a number of landmarks including the discovery of the desalination properties of cellulose acetate RO membranes, seawater conversion with membranes, construction of spiral-wound modules, composite-film seawater membranes, and very large membrane modules. The early demonstration distillation plants expounding diverse distillation techniques have also made their mark (see Hollow-fiber membranes).

The training of many of today's industrial leaders in desalination technology was supported by these agencies.

United Nations. The Resources and Transportation Division of the United Nations has sponsored many desalination projects with various developing nations. Some of these have been financed by the U.N. Special Fund. The Resources and Transportation Division sponsored the Interregional Seminar on the Economic Application of Water Desalination in September 1965, in which 34 countries participated; the related economic water desalination study was published as *Proposals for a Costing Procedure and Related Technical and Economic Considerations* (20). Also sponsored was a detailed survey of the water needs of 43 developing nations (21). The report examined the current and future needs in those water-short areas in relation to their economics and the technical and economic prospects for desalinating sea or brackish water at attractive prices under the prevailing conditions (11).

European Federation of Chemical Engineering. The European Federation of Chemical Engineering has a working party for freshwater from the sea consisting of representatives from 13 countries. It has held seven symposia in different parts of the world, and its published proceedings contain much valuable material (22).

International Desalination and Environmental Association (IDEA). This association is an outgrowth of a regional association in the Caribbean under the name Association of Caribbean Desalination Plant Owners and Operators. It is specifically devoted to desalination and, beginning in 1976, it began organizing congresses in desalination and water reuse. The proceedings of four of its conferences have been published (23). It also publishes a bimonthly newsletter (24).

Water Supply Improvement Association. This association is devoted to water-supply improvement and considers desalination among other techniques. The association publishes a semiannual journal.

Manufactured Freshwater

The possibility of producing freshwater from seawater or brackish water by separation of the salts opens a new dimension in the supply of freshwater. Areas bordering the sea would have an available raw material without limit or cost of transportation to the water facility. The successful realization of desalination by the combined effort of chemists, chemical engineers, mechanical engineers, and metallurgists, as opposed to the search for and transport of existing freshwater, gives new hope for adequate water in many, but not all, cases. Table 1 shows large plant sales in the last two decades, and Table 2 shows a growth in use of desalted water.

Cost remains the main stumbling block to desalination procedures, but freshwater is now one of the materials produced as it is needed. Water has considerable differences from other materials: the very low value of the commodity and the very large amount of the product needed. Thus, plant costs might be expected to be high compared to

Table 1. Worldwide Sales of Land-Based Desalting Plants with Capabilities of at Least 3.785 ML/d (10^6 gal/d), 1957–1980

Year	Capacity, ML/d Municipal and discharge water	Industrial and power	Total, including demonstration, military, and recreation
1957–1959	19		19
1960–1962	8	15	23
1963–1965	30	4	34
1966–1968	225	115	360
1969–1971	225	170	400
1972–1975	530	170	700
1976–1978	1890	190	2080
1979–1980	2025	320	2400

Table 2. Use of Desalted Water and Volume of Desalting Capacity Designated for Consumption by General Classifications, 1980

Classification	Capacity, ML/d (mgd)
municipal	4920 (ca 1300)
industrial	1440 (ca 380)
power	420 (ca 110)
discharge	365 (ca 97)
military	80 (ca 21)

unit sales value of production, and the capital charges for the production equipment do make up a very large percentage of costs in comparison to other production industries.

Obstacles attend this new solution of the freshwater problem. These obstacles are enlargements of those familiar to the chemical engineer in purifying other cheap or worthless raw materials into valuable products by treatment with chemicals or thermal or electrical energy. They are quite different from the previous main problem of water supply—the happenstance of finding a river or lake nearby or of making a fortunate geological strike.

In many places, desalination is even more urgent than the production of food, which is limited by water shortages. These shortages exist both in the petroleum rich countries and in many of the least-developed and poorest countries of the world. With the former, the improved water supply obtained through desalination has already had stunning effects; little hope for progress is on the horizon for the poor countries.

Freshwater may be obtained from saline solution by many processes, but all fall into one of two classes: freshwater is separated from seawater, which becomes more concentrated (water from saline solution); or salt or more concentrated brine is separated from seawater, which becomes less concentrated until an acceptable level is reached (salt from water). Usually the first type of process, water from saline solution, is the easier.

Unfortunately, the mechanism of the separation always requires much more energy than that of breaking the molecules of water loose from the solution: evaporation requires the latent heat of evaporation to make steam simultaneously with the separation, and making ice requires that the latent heat of freezing be removed. The theoretical thermodynamic energy requirement of separating water from brine, because of the heat of solution of the salt, is quite small. Starting from seawater having 35,000 ppm salts, the minimum energy requirement would be 38.9 kJ/L (155 Btu/gal) to produce dry salt. This minimum energy requirement is greatly reduced if evaporation is not carried so far: starting with 8000 L (ca 2100 gal) seawater, to produce 4000 L freshwater and 4000 L (ca 1060 gal) brine of twice the original concentration would require only 5.50 kJ/L but twice the volume of seawater would have to be handled. The energy requirement could be reduced to a theoretical minimum of 1.89 kJ/L (4 Btu/gal) by using an indefinitely large volume of seawater, whose concentration would be changed only very slightly. However, the pumping, deaerating, and possible chemical treatment necessary to accomplish the separation become correspondingly infinitely expensive. These costs fall and the thermodynamic separating costs rise as the amount of seawater processed decreases and the percent yield of product freshwater increases. Also, because other energy costs, eg, heat accompanying the change of phase of water, usually do not change substantially with the strength of the solution, the total energy required does not vary as much proportionally as does the much smaller heat of solution.

Water, sodium and magnesium, chlorides, iodine, bromine, and magnesium metal are the primary products derived from seawater today (see Sodium compounds; Magnesium compounds; Iodine; Bromine; Magnesium and magnesium alloys). Many other elements are certain to be economically obtained from the ocean as technology for the recovery improves.

Materials for Desalination Plants. The water produced by all of the desalination plants operational around the world is rapidly approaching 8 GL/d (2.1×10^9 gal/d). The price of this water as reported in the most widely used term, price per 4000 L (1000 gal), varies from ca \$0.60 to \$0.80 for large brackish-water-conversion reverse-osmosis plants in the continental United States to about \$15.00–25.00 for the seawater converted by either RO or distillation in the Caribbean to perhaps twice that or more in some seawater-conversion plants in the Arabian Peninsula. Referring to cost has its pitfalls because of the different ways in which accountants may deal with the variety of factors that are conventionally taken into consideration in deriving cost figures. Energy is one of the most important factors in the sum of parameters affecting desalination costs and since 1973 it has become the single overriding component. The fact that energy will continue to have a principal role in determining the cost of desalinated water speaks ill of any potential of important price reduction, particularly for converted seawaters. Capital, interest, amortization, depreciation, and maintenance costs come next in the ranking of costs.

Often in optimization studies of several processes, as energy costs decrease, equipment costs rise and vice versa. The most economical balance is always the goal. Unfortunately, in the desalination industry, the quest for lowest possible cost has had some rather devastating consequences. In distillation plants, which in the past were accepted after meeting performance characteristics for a month or two on stream, has had tragic consequences. The use of inadequate materials of construction in many locations combined with poor operation by virtually untrained hands led to rapid deterioration and failure of plants long before their estimated design life. This situation is slowly changing. Reference 25 gives the more important recommendations for use of metals and alloys for acid-dosed distillation plants based upon OSW studies. In addition to metals and alloys, a wide variety of plastic and polymeric materials and different concrete compositions have become an integral part of materials in construction of desalination plants. The plastics and polymers are used in pipes, sheets, membranes, ion-exchange resins, gaskets, coating and linings, etc. Other materials are also becoming part of the operations, among them various solvents, heat-transfer reagents, refrigerants, and similar thermodynamics reagents. Chemicals that are strictly expendable are those used for treating raw seawater and brackish water such as sulfuric acid, lime, algicides, antifoams, and antiscalants of the conventional type, eg, polyphosphates or the new high temperature polymeric variety such as Belgard (CIBA-GEIGY). Smaller amounts of other chemicals for treating product water are also used, such as chlorine, ozone, activated-carbon or higher alcohols to stop evaporation in storage basins.

These materials are used in large plants weighing thousands of metric tons and processing many millions (10^6) of liters of water flowing through kilometers of tubes and hundreds of square meters of sheets of plastic. Very small desalination units are used in private homes, just as they are now used at condominiums and resort hotels. These plants are available, but because of the relatively small number in which they are produced, they are still too expensive to permit popular use. Today, desalination plants span the range from membrane plants producing a few liters per day to membrane plants producing millions (10^6) of liters per day. In the world's largest complex of its type, to be completed in 1984, there will be a $1 \times$ GL/d (250 mgd) MSF (multistage flash) capacity in a single location (Al Jobail, SWCC, Saudi Arabia).

Not all phases of the world desalination industry have been profitable. Profitability has varied in different countries. Credit market and government guarantees play a very important part in the ability of corporations to get business in this extremely competitive market. The desalination industry will continue to grow worldwide, fueled by increase of population, the need to protect the available finite water supply from pollution, and the increased demand for water in less-developed countries.

Evaporation Processes for Desalination

Evaporation with Surface for Boiling and for Condensing. Sailors have used simple evaporation apparatus to make drinking water for almost 400 yr, and steamships have traditionally used evaporators, often multiple-effect, to prepare water for the boilers and potable use. In such processes, heat is transferred from prime steam coming from a boiler in single-purpose plants, or more often, as is the case of dual-purpose plants, from the exhaust or some intermediary stage of a turbine. The steam is condensed on one side of the first or heated metal surface, the evaporating surface, to boil the seawater contained in a suitable vessel on the other side of the tubular surface. Obviously, the required area of heat-transfer surface (hence the amount of metal required for a given duty) is reduced by increasing the coefficient of heat transfer through the surface. For standard tubes, the heat-transfer coefficient, reported in $W/(m^2 \cdot K)$ [$Btu/(ft^2 \cdot h \cdot °F)$], is usually considerably below 5.7 $kW/(m^2 \cdot K)$ [1000 $Btu/(ft^2 \cdot h \cdot °F)$] but values up to 14.2 (2500) have been reached by Westinghouse, General Electric, and others. If these high transfer rates can be obtained, it will be possible to reduce the heat-transfer area, and hence the material cost, provided no other considerations intervene.

The vapors from the evaporation pass to a second metal surface where they are condensed to give freshwater on the tubular condensing surface. This condensing surface is cooled, for example, by circulating cooling water on the other side.

If the evaporation is violent, a small amount of mist from the boiling seawater is carried over, giving 10–100 ppm or more of salt in the freshwater, depending on the design of the evaporator and the violence of ebullition. This entrainment may be reduced to 1–5 ppm (or even less) by proper design of the evaporation surface in relation to the evaporator vessel or by passing the vapors through a demister or entrainment separator, which changes the direction of vapor flow in passing through a pad of metallic mesh, glass fibers, or other materials. Because of their greatest mass and inertia compared to that of the vapors, the droplets of mist strike the metal or other parts of the demister instead of passing in the vapor streams around them and they are coalesced into films that drain back to the evaporating liquid.

In a once-through or single-effect evaporator, the thermal energy consumption is high since all heat transferred through the evaporating surface is lost to the condensing surface. But, in multiple-effect evaporators, several units or effects are operated at successively lower pressures. Vapors from the evaporation at the higher pressure condense and give up heat to the heating surface of the next effect; the heat of condensation evaporates about the same amount of water in each successive effect. Allowing for heating of the liquid, losses, etc, a triple-effect evaporator may give a gain ratio of about 2 kg evaporation per kilogram of steam used, and a quadruple-effect, about three.

Multiple effects have been used for many years in the salt, sugar, and other process industries. The usual maximum number is 4–6 effects, to balance the lower steam costs against the increased capital costs that accompany a large number of effects, to minimize the total cost per unit of evaporation. In one demonstration unit of the OSW at Freeport, Texas, the number of effects in producing freshwater from seawater has been increased to 12. This unit gives a net evaporation or gain ratio of about 100 kg freshwater produced per kilogram of boiler steam used. In such plants, large numbers of tubular surfaces for evaporating and condensing are required for recovery of heat from the streams of salt water and freshwater to and from the many effects. In this 12-effects, 4 ML/d (1-mgd) unit, more than 30 separate bundles of tubes are used. Use of the vertical-tube evaporator (VTE) has not been a great success, and relatively few plants of this type are to be found in the field.

Another multiple-effect concept that has been developed and exploited successfully is the low temperature multi-effect distillation (LTMED) (26). This development, which operates at an upper limit of 70°C, is illustrated schematically in Figure 2. Figure 3 shows two 6 ML/d (1.6-mgd) LTMED plants installed in St. Thomas, U.S. Virgin Islands. The virtues of this approach appear to be lower fuel consumption estimated by the manufacturer at 2.4–2.8 kg fuel per 1000 m^3 (0.15–0.175 lb/1000 ft^3), and lower capital cost because the whole plant is built of aluminum.

In the vapor-compression evaporator (VCE), the vapors from boiling water inside the tubes are compressed mechanically to give a steam temperature high enough to allow condensation on the outside of the tubes. The single tubular surface is used now for both evaporating and condensing. Boiling continues as long as does the compression of vapors formed. Mechanical power from an internal-combustion engine may drive the compressor, and the waste heat from the engine block and the exhaust gas may be used to preheat the seawater. This system is also used where electric power is available, or where power is generated by boiler steam which after going through the turbine is utilized for another purpose.

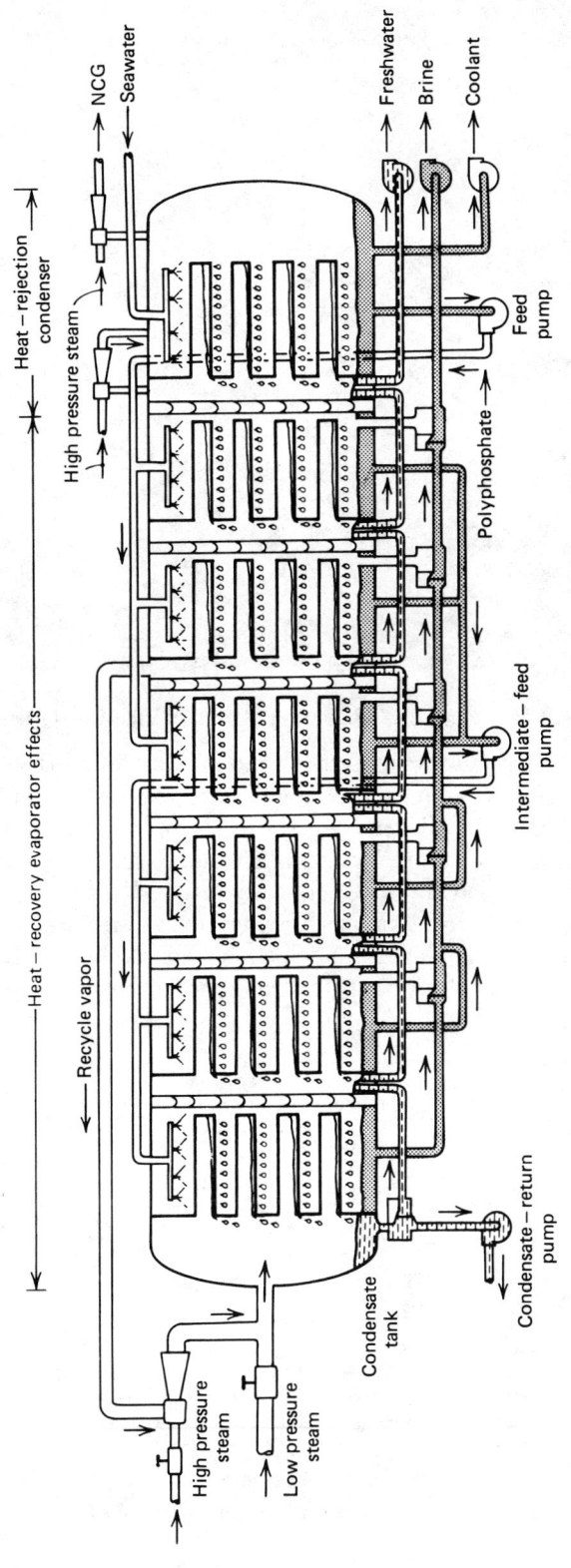

Figure 2. Low temperature multi-effect distillation (LTMED). Courtesy Israel Desalination Engineering.

Figure 3. Two 4.73 ML/d (1.25 mgd) LTMED plants. Courtesy Israel Desalination Engineering.

The largest VCE ever built was a demonstration plant for 4 ML/d (1 mgd) by the OSW in Roswell, New Mexico, for evaporating alkaline brackish water having a mixed salt content of ca 15,000 ppm. The compressor was to operate across the two effects, to increase the amount of evaporation, and the energy required was 57 kJ/L (227 Btu/gal) freshwater produced and plant capacity was 4 ML/d.

Today, at least four different vapor-compression systems are commercially utilized. The spray film vapor compression, which is extensively used for hotels, small industrial plants, and power stations, is illustrated in Figure 4. Its capacity is usually 10,000–120,000 L/d [(2.6–32) × 1000 gpd]. The vertical-tube flow vapor-compression

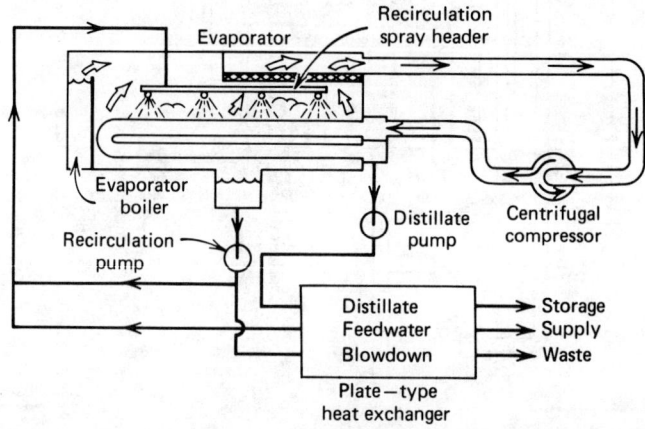

Figure 4. Basic process flow diagram of spray film vapor-compression distiller. Courtesy Aqua Chem.

unit has a similar range. The steam jet ejector vapor-compression unit operates in sizes up to 1.4×10^5 L/d (3.7×10^4 gpd) and the low temperature all-aluminum vapor-compressor unit operates at temperatures below 55°C (see Fig. 5). These units appear to have been particularly successful because their lower temperature operation removes many of the potential corrosion and scaling problems and permits the use of many polymer materials as well as aluminum. The vertical-tube vapor-compression unit has been used as a brine concentrator for cooling-towers blowdown in the southwestern United States. These units have been built with capacity of up to 1.4×10^6 L/d (3.7×10^5 gal/d) and have a potential of achieving $\geq 4 \times 10^6$ L/d (≥ 1 mgd). The construction materials have been exclusively stainless 316L and use titanium heat-exchange tubing, and the units are sold with 30-yr guarantees (Fig. 6). Another feature of this system is the use of seeding techniques for the control of scale.

At Roswell, New Mexico, in connection with the OSW VC program, a lined 0.3 km² (3.3 ft²) pond was built which allowed no seepage of the concentrate from the evaporator, but provided surface evaporation for the water from this brine. This lined-pond approach for drying the concentrated brine effluent of the unsuccessful OSW VC systems is also used with all the newly built coal-fired power plants of the western United States. After drying, the sediment is periodically buried together with some of the flue ashes.

Conventional Multiflash Evaporation. Flash evaporation is the vaporization, often violent, of a part of the water in a stream of hot seawater as it passes into a low pressure chamber. The vapors are formed adiabatically. The sensible heat given up by the cooling liquid as it comes to equilibrium with the lower vapor pressure is equated to the latent heat of the vapors formed. A high ratio of flash temperatures and pressures,

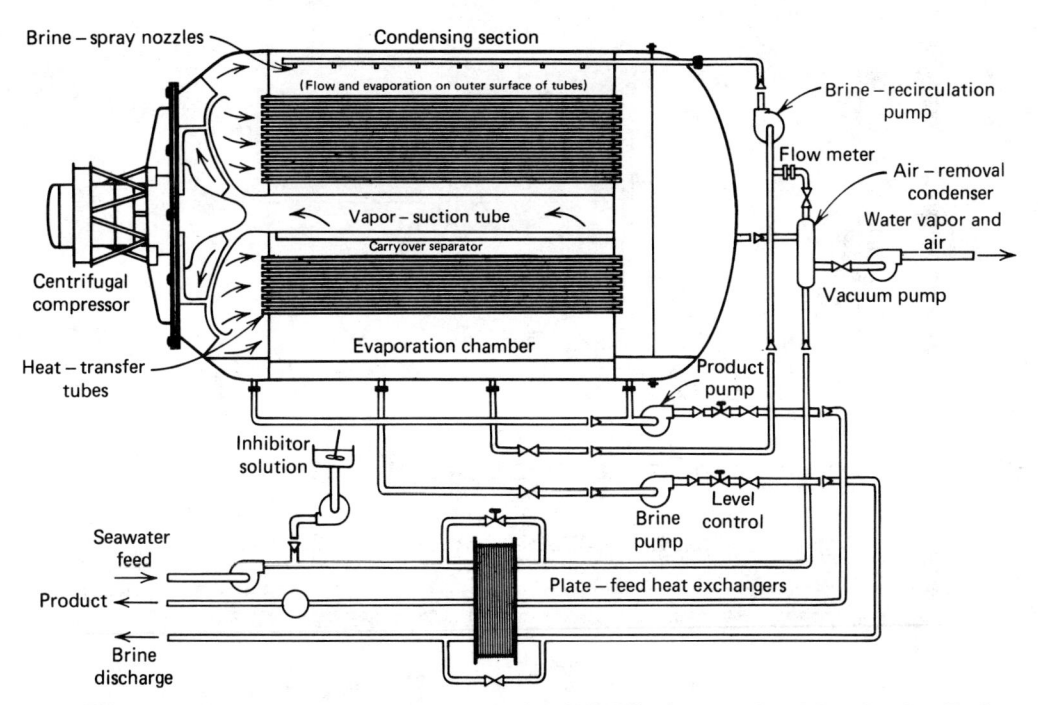

Figure 5. Low temperature vapor compression (LT-VC). Courtesy Israel Desalination Engineering.

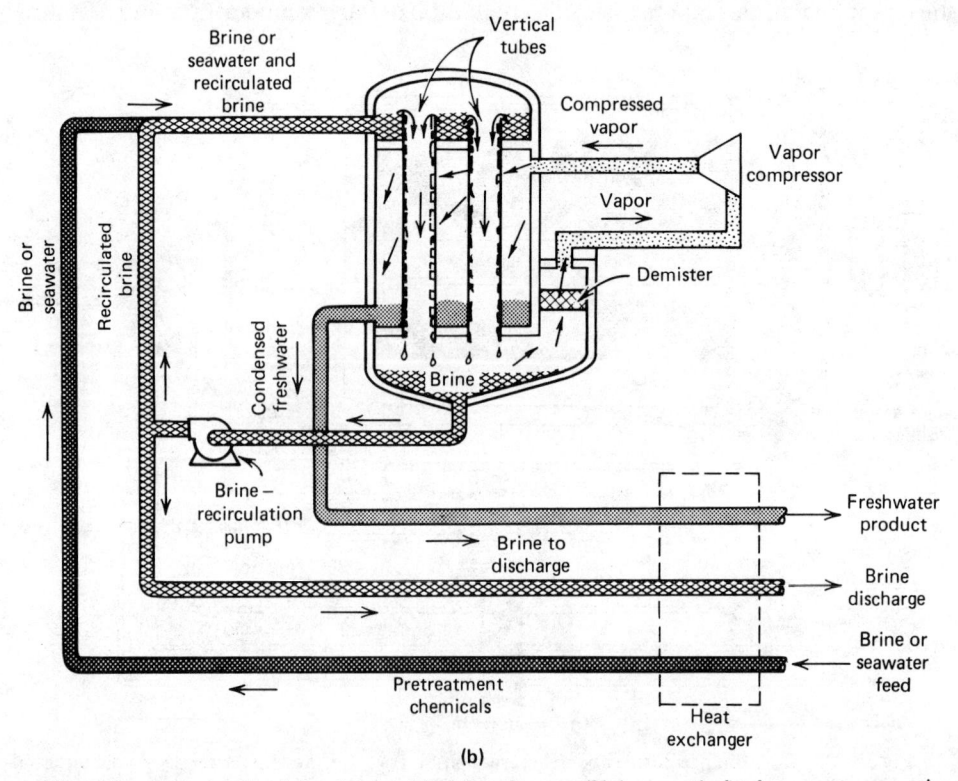

Figure 6. Pictorial view (a) and a simplified flow diagram (b) for a vertical-tube vapor-compression process. Courtesy Resources Conservation Company.

as in the discharge of condensate from a stream trap to atmospheric pressure, causes an approach to equilibrium with almost explosive violence. The vapors formed are withdrawn and condense as they preheat the raw seawater passing the system of condensing tubes.

A large number of stages of such flash evaporations at successively lower pressures give what is called a multistage flash (MSF) evaporation. The condensing of the vapors to heat the feed operates countercurrently to the MSF to give a cooling-evaporation and heating–condensation heat interchange. There is no evaporation heat-transfer surface, but there is a condensation heat-transfer surface.

After passing from the last of the series of condenser preheaters, the seawater is finally heated in the brine heater to its highest temperature before the first or highest pressure of the MSF. About 85% of the present world capacity for desalination of seawater now uses the MSF system. Multistage flash will probably remain the most important system (in terms of water produced) for making freshwater from the sea at least until the year 2000.

Principles of the MSF. The substance of the flash process is illustrated in the single-stage schematic of Figure 7. Here the seawater is heated in the brine heater to just below saturation temperature t_{max} at P_{max}. The water then enters the boiler at reduced pressure P_1. This pressure reduction causes the feed to flash or to attain equilibrium with the stage-saturation conditions dictated by the pressure P_1. The rejected brine is discharged and fresh seawater brought into the system. The vapor formed in the flashing process condenses on the condenser or heat-recovery tubes, raising the temperature of the incoming feed to temperature t_{in}. The brine heater supplies whatever additional energy is needed to bring the feed to t_{max} before flashing begins.

The fraction of brine that is evaporated at this first stage is given by

$$\frac{C(t_{max} - T_1 + \Delta\text{BPE})}{L}$$

where BPE = boiling point elevation and L = 2.324 MJ/kg (1000 Btu/lb).

In this equation, the assumption is made that equilibrium has been attained in the stage, a condition that in practice is certainly not achieved. This single-stage concept can be extended in a multiple-stage arrangement. Here, the pressure is de-

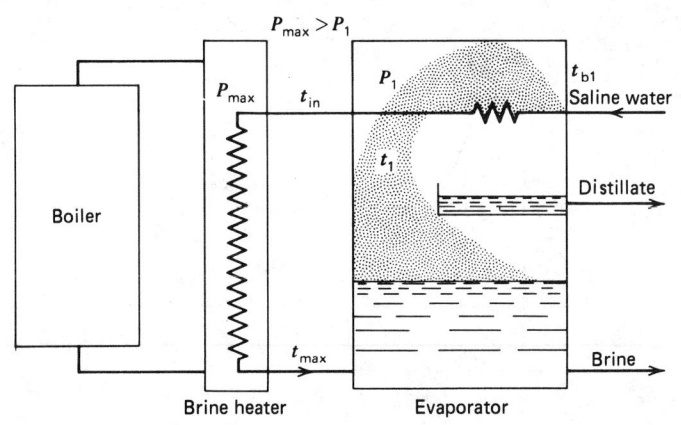

Figure 7. Flash distillation (brine heater and first stage) (27). Courtesy of Longman Group Ltd., London.

creased in each successive stage until vapor volume equilibrium and heat-rejection considerations fix the conditions for the last stage. In most cases, the temperature in the last stage is of the order of 30°C. The upper temperature t_{max} is fixed by the precipitation of $CaSO_4$ and for most plants is 122°C.

The relationship of capital and energy costs is shown in Figure 8.

Table 3 compares the virtues and shortcomings of the MSF approaches. All distillation plants are rated by a performance ratio and the performance ratio is thus defined as the number of kilograms of product produced for a kilojoule of steam.

Scale Control. Obtaining maximum performance from a seawater distillation unit requires minimizing the detrimental effects of scale formation. The term scale describes deposits of calcium carbonate, magnesium hydroxide, or calcium sulfate which can form in the brine heater and the heat-recovery condensers. The carbonates and the hydroxide are conventionally called alkaline scales, and the sulfate, nonalkaline scale. The presence of bicarbonate, carbonate, and hydroxide ions, the total concentration of which is referred to as the alkalinity of the seawater, leads to the alkaline-scale formation. In seawater, the bicarbonate ions decompose to carbonate and hydroxide ions giving most of the alkalinity.

$$2\ HCO_3^- \xrightleftharpoons{heat} CO_3^{2-} + CO_2 + H_2O$$

$$CO_3^{2-} + H_2O \xrightleftharpoons{heat} 2\ OH^- + CO_2$$

The kinetics of the formation of the magnesium hydroxide and calcium carbonate are functions of the concentration of the bicarbonate ions, the temperature, and the

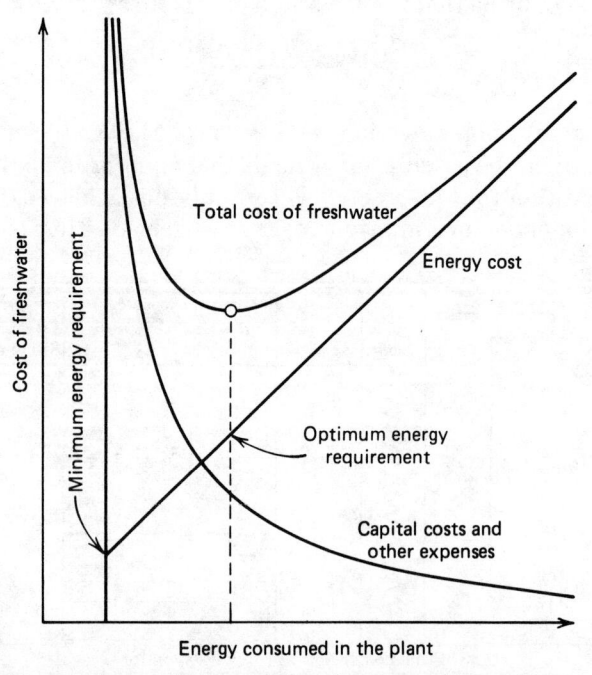

Figure 8. Qualitative relationship between the minimum and the optimum energy requirements of a seawater-conversion plant.

Table 3. Advantages and Disadvantages of Multistage Flash Processes

Advantages	Disadvantages
Brine recirculation process	
small seawater make-up rate permits use of continuous acid treatment for scale control or high temperature antiscale additives, Belgard type	control of proper acid dosage is critical
	acidic attack of metals is a high risk if brine pH is not properly controlled
allows process to operate up to 120°C without alkaline scale	higher operating skill requires daily laboratory analysis and close oxygen meter instrument
deaerates seawater before it is heated and introduced into evaporator	more susceptible to $CaSO_4$ scaling during plant upsets or abnormal plant operations
higher process temperature reduces heat-transfer surface and evaporator shell size	plant start-up requires considerable operator attention
not sensitive to raw-seawater temperatures	uses dangerous acids for scale control
	corrosion products from the shell enter recirculation brine and may cause fouling and corrosion of tubes
Once-through process	
simple operation	seawater deaeration takes place in first evaporating stage
skilled operators not required	
higher on-stream factor	more susceptible to foaming when organics present
negligible seawater concentration, low boiling-point rise	evaporator tubes exposed to nondeaerated seawater
uses nonacidic scale control: polyphosphate, polyacrylates, Belgard type high temperature antiscale additives	limited to maximum top temperature of 100°C
	limited plant economy 8:1
small seawater retention time, better success of scale control	sensitive to raw-seawater temperature
scale chemical dosage rate not critical	
equipment less expensive for capacities of up to 577 m³/d (1.5 × 10⁶ gal/d)	
instrumentation not complex	
capacity modulated by seawater flow change	
cleanest and lowest salinity seawater in contact with tubes	

rate of release of CO_2 from the solution. At temperatures up to 82°C, $CaCO_3$ predominates, but as the temperature exceeds 93°C, $Mg(OH)_2$ becomes the principal scale. Thus, in seawater, there is considerable tendency for surfaces to scale with increase in temperature.

The interrelationship of nonalkaline scales ($CaSO_4$, $CaSO_4 \cdot \frac{1}{2}H_2O$, $CaSO_4 \cdot 2H_2O$) depends on temperature and the concentration of $CaSO_4$. In order to assure that no hemihydrate scale forms, MSF operators must run their plants in such a manner as to assure that the concentration of the total dissolved solids does not exceed 70,000 ppm at temperatures of 120°C. With average salinity seawater, plants can operate at a concentration factor of 2, but in the Middle East where water salinity can be as high as 50,000 ppm, the concentration factor should not exceed 1.4. Under no circumstances should the total dissolved solids exceed 70,000 ppm, ie, twice the concentration of normal seawater at 120°C. A number of options for controlling scale formation are used in plant operations around the world.

The first is use of mechanical means including thermal shock. Although rare today, this practice can be found in use with the few obsolete submerged-tube evaporators.

The second approach, the so-called seeding technique, provides preferential sites for the nucleation of scale which permits the heat-transfer surfaces to remain clean of scale. Extensive studies of this technique have been conducted, and field use was reported in the USSR as early as the mid-1960s is used there to this very day (28). Today, all of the Resources Conservation Company's vapor-compression units use the seeding technique.

The use of ion-exchange methods is the third possible approach. For calcium, the exchange can be represented as

$$2\,RNa + Ca^{2+} \rightarrow R_2Ca + 2\,Na^+$$

Magnesium can be exchanged in a similar fashion. To date, cost considerations have prevented the use of ion exchange for scale control.

The fourth approach is the use of polyphosphate-based blends including proprietary chemicals. The exact mechanism of the observed effect is not clearly understood. In the polyphosphate mode of operation, the polyphosphate is dosed in quantities of 2–5 ppm; periodically, sludges resulting from phosphate treatment are removed by acid cleaning (see Dispersants).

In the fifth option, the hydrogen ion from added acid decomposes the bicarbonate ions.

$$HCO_3^- + H^+ \rightleftharpoons CO_2\uparrow + H_2O$$

About 120 ppm of sulfuric acid must be provided for normal seawater. Control of acid dosing is critical; the amount of acid must be stoichiometric to the alkalinity expressed as $CaCO_3$. In conjunction with acid dosing, the CO_2 formed must be removed and some sodium hydroxide added to maintain ca pH 8 in the system. Alternatively, less-than-stoichiometric amounts of acid can be added to retain some alkalinity in the untreated feed. In either case, CO_2 removal is done with packed columns. Acid-dosed feed is passed through a column with air flow that sweeps the CO_2 from the feed saturated with carbon dioxide. This is usually followed with a deaeration during which both the air and CO_2 are reduced to the levels needed to minimize, if not eliminate, corrosion. Although acid dosing does permit higher operating temperatures, it has had a devastating effect on plant life.

The sixth option involves use of a new family of polymers, the so-called high temperature scale-control chemicals. These are compounds that, added in 3–8 ppm, lead to lattice distortion and the formation of a nonadhering scale. Belgard (CIBA-GEIGY) was the first compound exemplifying this type of MSF operation, which is now steadily displacing acid in the operation of MSF plants around the world with important contribution to plant life (29) (see Dispersants).

The different distillation-related processes discussed briefly here are powered by either thermal or mechanical energy. Thermal energy can be provided by steam generated specifically for the plant. Plants so energized are referred to as single-purpose plants. This approach is typical of the smaller distillation plants. Although it is not the most economical solution, it is the only one available for small self-contained plants at remote locations. In bigger installations, the steam providing the thermal energy is usually a by-product of power generation; it may be, for example, the low pressure steam from a power plant. This dual-purpose plant is the prevailing type for virtually all multimillion (10^6) liter installations around the world.

Freezing. The concept of the freezing desalination process is based upon thermodynamics and the phase diagram of an NaCl–H_2O solution. As the temperature of the solution is reduced to the freezing point of water, pure water solidifies, leaving a more concentrated, lower freezing brine. A freezing desalination process must have a freezer, a means of washing the ice crystals formed from brine on their surface, and a melter to process the ice formed. Schematically, these components are given in Figure 9. There are two important advantages that make such freeze-desalting, in principle, most desirable and seemingly competitive. Specifically, the heat of solidification is much less than that of evaporation, offering energy savings. Second, the freeze process operates at relatively low temperature, reducing the corrosion problem and, in turn, cost of materials. These two features of freezing-based desalination have provided considerable incentive in process development. A number of systems and pilot plants have resulted (30–37), but no commercial freezing desalination installation is yet in service and no firm plans to build any. The inability to develop sound engineering and economically competitive systems after 30 yr and considerable financial investment is perhaps a classic example of a conceptually simple process that is suppressed by the engineering complexity needed to realize it.

Even with an engineering breakthrough, freeze-desalting is unlikely to emerge as an economic competitor for seawater conversion for some time. It could, however, have important use in the processing of complex industrial-waste fluids where resource recovery could enhance the economics.

Reverse Osmosis (RO). Reverse osmosis is a process that consists of letting the solution flow under pressure through an appropriate porous membrane and withdrawing membrane-permeated product, generally at atmospheric pressure and ambient temperature. In this process, the permeate or product is enriched in one or more of the constituents of the mixture, and the solution on the high pressure side of the membrane is left with higher or lower concentration. The reverse-osmosis membrane need not be treated, and no phase change is involved in the process.

Reverse osmosis is a process of wide-ranging and continuously growing applications. Its application to the desalination of sea and brackish waters is the single largest use of this emerging technology (38–41) (see Reverse osmosis).

Today, RO is established as a reliable water-conversion technique for brackish water. Since its commercialization in 1975, seawater RO has been gradually accepted worldwide. Today, more than 1.2×10^9 L/d (3.2×10^8 gal/d) is converted by RO, of which ca 10% is seawater. An operational RO water-conversion system, whether it be

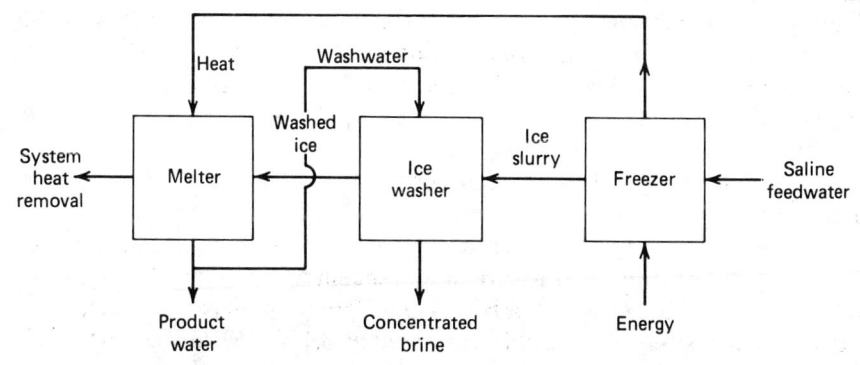

Figure 9. Basic components in the freezing desalination process.

for brackish water or for seawater, has four functions: pretreating, pressurizing, demineralizing, and stabilizing the water.

The first function, the pretreatment, must assure that the water, as it arrives for demineralization, is of a quality compatible with the membranes used. Pretreatment consists essentially of removal of suspended solids, pH modification, and possibly some addition of antiscalants. Depending on the water to be processed and the specific membrane requirements, the first step here can vary from a simple cartridge filtration through a complex in-line coagulation and filtration process. Again, depending on the nature of the membrane, the water to be demineralized must meet particulate-matter requirements. These are most reliably monitored today by the silting density index (SDI) (42). Some membranes can tolerate SDIs as high as 6. As a rule, hollow-fiber membranes (qv) are much less tolerant to particulate matter than are the spiral-wound types.

Once water is brought to the proper quality in terms of SDIs, it is ready to be pressurized to a level required for the conversion. Pressures of 1.7–3.0 MPa (17–30 atm) are used in systems for brackish-water conversion and 6.0–6.8 MPa (60–68 atm) for seawater-conversion systems. This is accomplished with a high pressure pump directed toward the membrane-holding container or the permeator, the heart of the system where demineralization is achieved. Permeators in use today are of four basic designs: plate-and-frame, tubular, spiral-wound, and hollow-fiber. These few configurations can be of a variety of membrane materials, ranging from improved versions of cellulose acetate (43) through polyamide and the most recent composite film membranes (44).

Membranes are of two types, fixed and dynamic, but only the former are economically significant for this application. Both types basically consist of two layers. The first is the functional layer, which performs the demineralization; the second is the support layer, which supports the functional layer and permits the water to flow through it. In the fixed membranes, these two-layered structures are produced either by modifying a surface layer of the porous support material (cellulose acetate and polyamide) or by applying the functional level to the surface of the porous support film. In the case of dynamic membranes, the functional layer is built *in situ* in pores of a supporting material by a chemical added to the fluid to be passed through the system. No dynamic membranes have been used in desalination. From the different module concepts mentioned above, only the hollow-fiber, spiral-wound, and, to an insignificant extent, frame-and-plate membranes are part of the world inventory of operating desalination plants.

The type of membrane and specific configuration of the membrane-conversion device are rated by their characteristics. The most important of these are flux, salt rejection, and recovery, followed by the relative stability of the membranes in respect to time, the resistance of the membrane to fouling, chemically induced deterioration, and the functional life of the membrane.

Flux. Flux is the quantity of water that can flow through the membrane. The water-transport coefficient is determined by a multitude of factors including the type and thickness of the membrane. The flux is proportional to the applied pressure. It should be noted that osmotic pressure of a brackish water is in the range of 69–345 kPa (10–50 psi) and that of seawater is 0.24–3.0 MPa (35–440 psi). Thus, for any kind of fluxes, brackish systems should be operated at 6.1–6.9 MPa (900–1000 psi). Improving the RO technology requires among other things, membranes with more efficient transport coefficients and higher fluxes. It should also be noted that the membrane configuration plays an important role in the flux. Spiral membranes have fluxes of 615–1230 L/(m²·d) [15–30 gal/(ft²·d)] with membrane areas per unit volume of 1053 m²/m³ (43 ft²/gal); the hollow-fiber configuration has fluxes of 0.8–1.2 L/(m²·d) [0.02–0.03 gal/(ft²·d)] and membrane areas of 16,400 m²/m³ (668 ft²/gal).

Salt Rejection. Salt rejection is the second key membrane parameter, as the purpose of the membrane is to demineralize.

Depending on the particular situation, four approaches are possible: single stage with a single membrane assembly, single stage with parallel membrane assemblies (Fig. 10), multistage with product staging, and multistage with reject staging (Fig. 11). Pretreatment is an integral part of any membrane plant. In the long run, the quality of the water arriving at the membrane surface is the cardinal factor in determining the life of the membrane. At present, membranes are warranted for brackish waters by the manufacturer for 4 yr, but life up to 6 yr has been recorded. Seawater membranes, on the other hand, are only warranted for 2 yr. The scheme shown in Figure 10 is generally used in brackish-water conversion; that in Figure 11 is utilized in seawater-conversion plants.

Most plant failures are caused by precipitation of the sparingly soluble salts from the water on the feed side of the membrane. This happens when the solubility limit for salts such as $CaCO_3$ or $CaSO_4$ is exceeded in the immediate vicinity of the membrane surface. These concentrations are usually higher than the concentration of the feed. Once precipitation begins, the situation is likely to worsen spontaneously and may often lead to membrane failure. The concentration of such foulants can be controlled by the percent of incoming water converted. For brackish-water conversion, this can be 40–95%; in the case of seawater, it is 20–35%, and the upper limit is seldom, if ever, exceeded. The percent conversion affects both the cost of production and the amount of product water obtained. Many water bodies contain some colloidal particles (ie, particles smaller than 1 µm) of a variety of organic compounds, heavy metals, and silica (45). Some surface waters are notably rich in particulate matter; in addition to

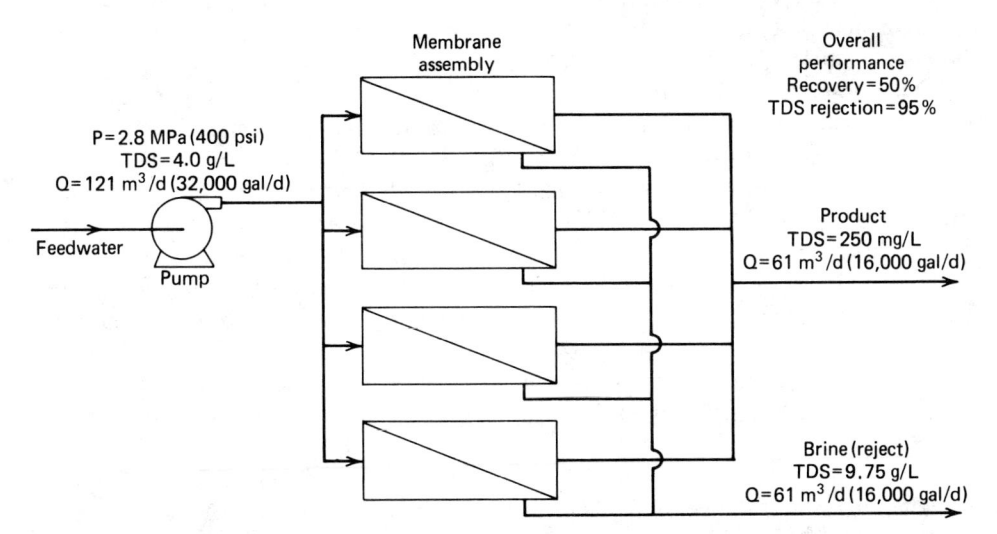

Figure 10. Single-stage RO-plant configuration with parallel membrane assemblies. Courtesy U.S. Agency for International Development.

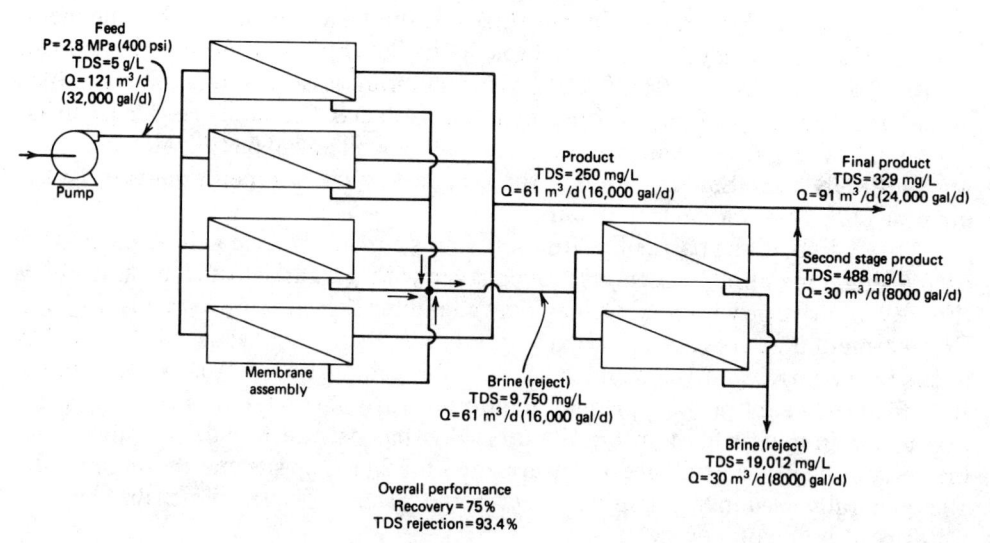

Figure 11. Multistage RO-plant configuration with reject staging. This configuration is occasionally referred to as a cascade, tapered, or pyramidal configuration. This diagram illustrates a two-stage unit. Three-stage units (with recoveries of 85–90%) are also used. Courtesy U.S. Agency for International Development.

clays and silts, they carry a diversity of algae. Iron and manganese oxides which also can be found in some waters have been known to impair the function of permeators. Last but not least, bacteria that cannot pass through the membranes can build up on the feed side and also impair membrane performance. Some bacteria can also attack cellulose acetate membranes. Although less frequent as a cause of failure, greases, oils, and various organics can also lead to permeator failure. Fouled permeators can sometimes be cleaned with detergents, but often the damage is irreversible (46). Some membrane material, particularly the polyamides, are strongly chlorine sensitive. These membranes can only operate on water from which chlorine has been removed. This is usually accomplished with bisulfite injection in an amount in excess of chlorine in the feed. Failure to dechlorinate can lead to rapid membrane failure.

Membrane technology, in comparison with alternative water-conversion processes, offers hopes for energy conservation. Reverse-osmosis conversion of brackish water uses 6.8–11.4 kJ/L (25–41 Btu/gal); 28.5–38 kJ/L (102–136 Btu/gal) is required for seawater conversion. Much effort has been invested in devising energy-recovery schemes especially for seawater plants where attempts are made to recover part of the 414–469 kPa (60–68 psi) discharged at ca 140 kPa (20 psi) below the pressure of the feedwater stream to the membranes. Much of the discharge pressure could be recovered in seawater conversion. Table 4 shows energy recovery data for seawater and brackish water from a recent study (47). At present, there appear to be no satisfactory and proven energy-recovery systems in operation. The few experimental systems have been plagued with major mechanical and maintenance problems (48). Much work is done to resolve the problems.

Tables 5 and 6 show operations and cost data of the 12 ML/d (3.2×10^6-gal/d) brackish-water RO-conversion plant at Cape Coral, Florida (49). These data are representative of brackish-water-conversion plants in the 4–20 ML [(1–5.3) $\times 10^6$-gal/d] size run in the contiguous United States. Operations data are similar for con-

Table 4. Energy Recovery for Reverse Osmosis[a]

Plant size, 1000 m³/d	Product water recovery, %	Brine flow rate, m³/d[b]	Available pressure, MPa[c]	Available[d] power, kW Ideal[e]	Turbine Shaft[f]	Equivalent electric motor
Brackish-water reverse osmosis[g]:						
4	50	9.93	2.5	109	96	101
	70	4.26	2.4	46	40	42
	80	2.49	2.3	26	23	24
20	50	49.65	2.5	545	480	505
	70	21.28	2.4	230	200	210
	80	12.45	2.3	130	115	120
Seawater reverse osmosis[h]:						
4	20	39.77	5.9	1027	900	950
	25	29.76	5.9	770	675	710
	30	23.78	5.8	590	520	550
20	20	198.87	5.9	5133	4500	4750
	25	148.80	5.9	3840	3375	3550
	30	115.89	5.8	2950	2600	2750

[a] Ref. 47.
[b] To convert m³/d to gpm, multiply by 4.403.
[c] To convert MPa to psi, multiply by 145.
[d] Recoverable.
[e] Ideal power at 100% efficiency.
[f] Shaft power at 88% efficiency.
[g] Feedwater pressure at 689.4 kPa (6.8 atm).
[h] Feedwater pressure at 6205 kPa (61.25 atm), TDS = 35,000 mg/L and temperature = 25°C.

version of similar-type water elsewhere in the world, but cost varies greatly from location to location. The largest brackish-water plant today in Riyadh, Saudi Arabia, converts 160 ML/d (42.3 × 10⁶ gal/d). One of the largest operational seawater plants is in Jeddah, also in Saudi Arabia, and has 14 ML/d (3.7 × 10⁶ gal/d) capacity. It uses double-pass to achieve the needed product water specification. By contrast, a somewhat smaller plant, the 12 × ML/d (3.2 × 10⁶-gal/d) plant at Key West, Florida, is the largest single-pass seawater-conversion plant in the world. The newest seawater-RO plant opened in June 1983 in Malta. It has a 23 ML/d (6 × 10⁶ gal/d) capacity and is now the largest seawater-RO plant in the world. At present, as much as 200 ML (52.8 × 10⁶-gal) seawater-RO-conversion plant capacity is expected to become operational in the next three years, most of it in the oil-rich Middle East.

Reverse osmosis has also been successful in wastewater recovery with the pioneering effort of Orange County, California (50). In this ambitious project, 20 ML (5.3 × 10⁶-gal) wastewater is demineralized by RO and injected in underground aquifers to prevent seawater encroachment. Another ambitious project planned for the near future will be the largest RO plant ever built, a 300 ML/d (79.3 × 10⁶-gal/d) plant designed to assure that the salinity of the brackish water of the lower Colorado River will not exceed 1000 ppm as it enters Mexico. This plant will be built as one of the requirements of a treaty between the United States and Mexico (51).

Electrodialysis. Dialysis is membrane permeation by molecules or ions. In electrodialysis (ED), the diffusion of ions is facilitated by an electric current (see Electrodialysis). The process is most widely used for the processing of brackish water.

354 WATER (SUPPLY AND DESALINATION)

Table 5. Summary of Operating Conditions for the Cape Coral Reverse-Osmosis Plant, Jan. 1978–May 1979[a]

Property	Value[b]
total water produced: product plus blend	4.23×10^6 m^3
average daily water production	6930 m^3/d
average raw-water blend (22%)	1510 m^3/d
potential water production at 17,000 m^3/d (4.5 mgd)	9.21×10^6 m^3
overall plant utilization	46%
feedwater pH range	5.8–6.0
sulfuric acid dosage	125 mg/L
SHMP dosage[c]	3.4 mg/L
caustic dosage	21 mg/L
chlorine dosage	4.4 mg/L
free chlorine residual	1.5–2.0 mg/L
feedwater pressure range	2.0–2.2 MPa (20–21.4 atm)
membrane cleaning	none
cartridge-filter replacement	once

[a] Ref. 49.
[b] To convert m^3 to gal, multiply by 264.2.
[c] SHMP = sodium hexametaphosphate.

Table 6. Actual Cost Data for the Cape Coral Reverse-Osmosis Plant, April 1977–May 1979[a]

Property	Value
total water produced	5.18×10^6 m^3 [b]
total plant capability for same period	
product alone, 11,400 m^3/d[b]	8.98×10^6 m^3 [b]
product plus blend, 17,000 m^3/d[b]	13.50×10^6 m^3 [b]
item, ¢/m^3	
chemicals (sulfuric acid, SHMP, chlorine, and caustic)	1.5
electrical (entire plant plus well field)	7.7
labor (salary plus fringes)	3.1
membrane replacement[c]	3.3
equipment maintenance, 11,400 m^3/d flow[d]	1.1
Total operating cost	16.7
amortized capital, $4,495,355 at 7% for 30 yr[e]	14.6
Total	31.3

[a] Ref. 49.
[b] To convert m^3 to gal, multiply by 264.2.
[c] Assumes that all membranes will be replaced after 4 yr with a flow of 11,400 m^3/d (3×10^6 gpd) at a total cost of $550,000.
[d] Maintenance estimated at 1% of capital cost/yr.
[e] All components of plant, excluding permeators and high pressure pumps and including cartridge filtration, chemical-feed systems, degassifiers, transfer and high service pumps, ground storage, chlorination system, and building, are sized to handle a product water flow of 18,900 m^3 (5×10^6 gal).

Worldwide, over 300 ML/d (79.3×10^6-gal/d) are converted in about 300 plants. Electrodialysis utilizes two different types of specially developed polymeric membranes, one permeable to anions and the other to cations. The electric energy required is proportional to the concentration of salt in the saline water. The ionic movements and resulting demineralization in the electrodialysis process, as diagrammed in Figure

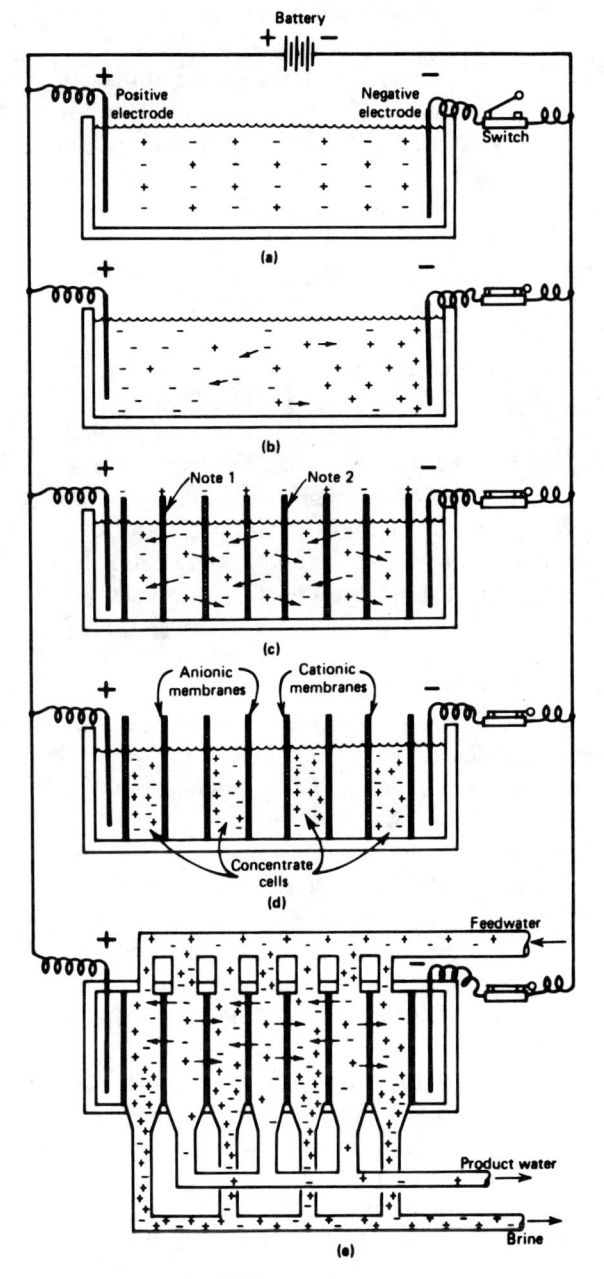

Figure 12. Ion movements in the electrodialysis process. Courtesy U.S. Agency for International Development.

(a) Many of the substances which make up the total dissolved solids in brackish water are strong electrolytes. When dissolved in water, they ionize; ie, the compounds dissociate into ions which carry an electric charge. Typical of the ions in brackish water are Cl^-, Na^+, HCO_3^-, Mg^{2+}, SO_4^{2-}, and Ca^{2+}. These ions tend to attract the dipolar water molecules and to be diffused, in time, fairly evenly throughout a solution.

(b) If two electrodes are placed in a solution of ions, and energized by a battery or other direct-current source, the current is carried through the solution by the charged particles and the ions tend to migrate to the electrode of the opposite charge.

(c) If alternatively fixed-charge membranes, which are selectively permeable to ions of the opposite charge, are placed in the path of the migrating ions, the ions are trapped between the alternate cells formed. A positively fixed-charge (anionic) membrane allows negative ions to pass but repels positive ions. A negatively fixed-charge (cationic) membrane allows positive ions to pass, but repels negative ions.

(d) If this continues, almost all the ions become trapped in the alternate cells, which lack ions, have a lower level of dissolved constituents, and have a high resistance to current flow.

(e) The phenomenon illustrated above is used in electrodialysis to remove ions from incoming saline water on a continuous basis. Feedwater enters both the concentrate and product cells. Up to about half of the ions in the product cells migrate and are trapped in the concentrate cells. Two streams emerge from the device: one of concentrated brine and the other with a much lower concentration of TDS (product water).

12, lead to the production of two streams, the brine, which carries the concentrated ions, and the product water. Hundreds of cell pairs of the type illustrated are found in a typical network of stacks of a commercial unit so that considerable amounts of current can be carried. When an electrodialysis unit is in operation, feedwater passes in parallel paths through all of the cells providing a continuous flow of product and brine.

A commercial electrodialysis unit (Fig. 13) consists of a d-c power supply; a membrane stack; a circulation pump; and related hardware. Depending on the water to be processed, pretreatment may also be needed for trouble-free operation. As electrodialysis uses direct current, the main element of the power-supply system is a rectifier converting ac to the dc applied to the electrodes on the membrane stack. The membrane stack itself consists of the cathode and anion-permeable membranes. Figure 14 shows special spacers and requisite plumbing to transport water to and from the stack. The membranes are flat polymer sheets formed on fabric backing to provide strength. The membranes contain ion-transfer sites so designed as to permit the selective passing of cations, anions, or both. In commercial installations, two approaches are used in the mode of incorporation of the ion-transfer sites. The leading U.S. manufacturer, Ionics, locates transfer sites uniformly throughout the membrane; one of the leading Japanese manufacturers, Mitsubishi, distributes the ion-transfer sites as discrete points within the membrane. The size, thickness, and specific composition of the membranes varies depending on the application and the manufacturer. The thinner the membrane, the lower its resistance. The water content of the membrane is also important as it affects ion passage. The higher the water content, the lower the membrane resistance and the easier the ion flow.

All these parameters can be and are varied by the manufacturer in an effort to offer the best system for a particular water chemistry. There are two main types of membranes in commercial ED systems. The Ionics anion-transfer membranes are made of cross-linked copolymers of vinyl monomers and contain quaternary ammonium anion-exchange groups. In the electrodialysis reversal (EDR) plants, these are backed with modacrylic fiber, ie, a copolymer of vinyl chloride and acrylonitrile (see Acrylic and modacrylic fibers). The cation-transfer membranes (52), on the other hand, consist of sulfonated copolymers of vinyl compounds. They are also cast on the same synthetic reinforcing fabrics used for the anion membranes.

Mitsubishi's membranes have powdered anion- and cation-exchange resin uniformly dispersed throughout the polypropylene films that make up the membranes. The film is specially treated to create micro cracks (53) which permit the ion transport; the ion-exchange resins provide the needed selectivity.

The spacers together with the membranes allow the water to flow and as such are of great importance. A cell usually consists of two membranes with a spacer between them; of course, hundreds of such cells constitute a stack. The current is brought into the system with a pair of electrodes of platinized titanium or niobium.

The pump in a dialysis system circulates the water through the stack. The circulation varies based on specific designs of flow through the stack with pumping pressures of 340–520 kPa (50–75 psi). These low pressures permit use of a great deal of plastic tubing throughout the system, reducing the incidence of corrosion and simplifying manufacture.

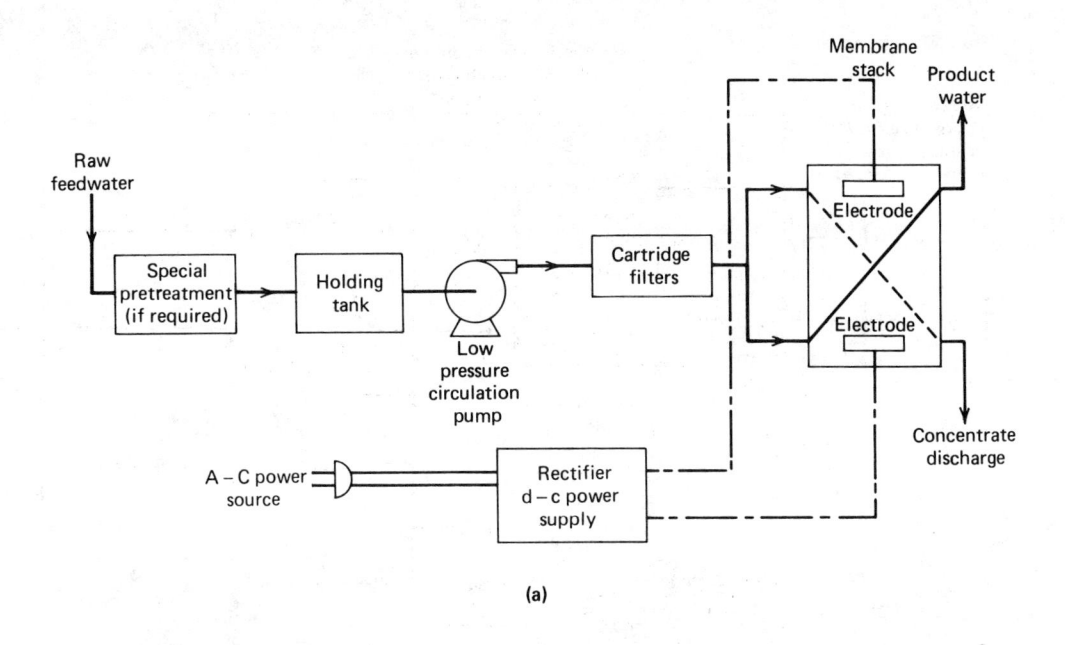

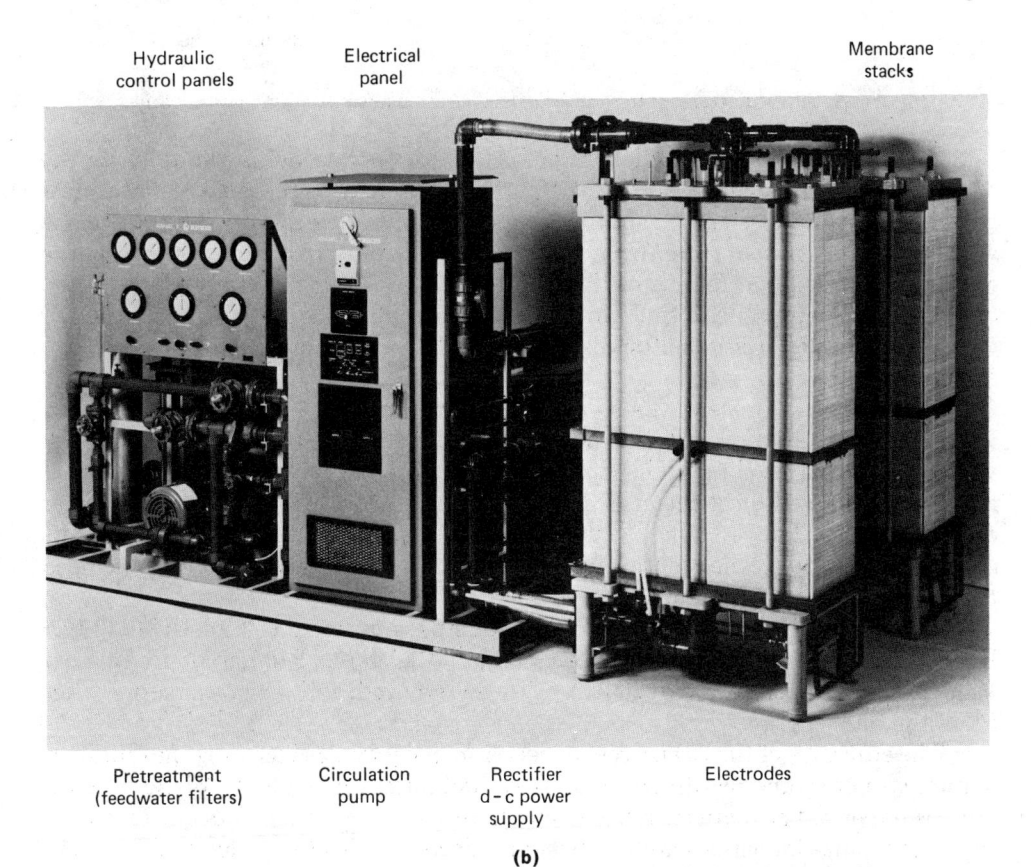

Figure 13. Schematic (**a**) and basic components (**b**) of an electrodialysis unit. Courtesy Ionics and U.S. Agency for International Development and Ionics.

357

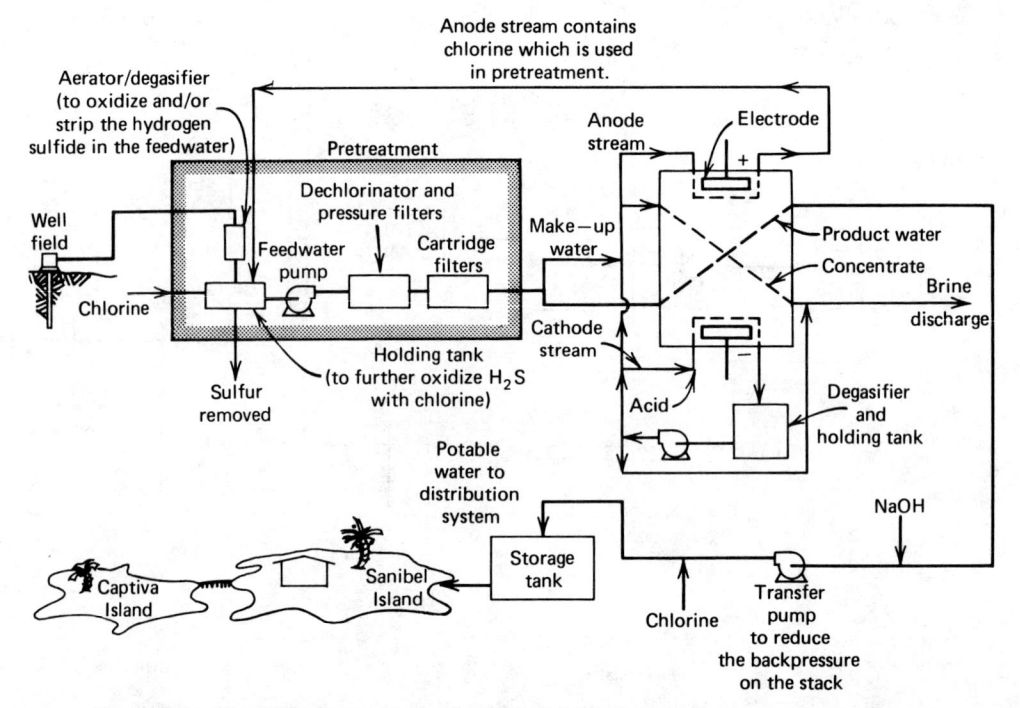

Figure 14. Flow diagram for the Sanibel Island electrodialysis facility. Courtesy Ionics.

Today, two basic electrodialysis processes are operational in the field. In the original electrodialysis process (Fig. 13), water is pumped to the stacks where, after passing through the cells, it is split into product and brine columns. This process is widely used, but it does suffer from the precipitation of sparsely soluble salts in the stacks and consequent fouling of the membranes. This occurs despite the addition of acid or polyphosphate to the brine stream. The fouling requires periodic cleaning of the stacks either *in situ* or after disassembly of the system, leading to increased cost of operations.

In the second process, the so-called EDR or reversal process, both cells are identical in construction and, 3–4 times per hour, the polarity of the electrodes is reversed with special valves. Following the reversal of polarity and flow, the product water is discharged until the cells and lines are flushed out and the desired water quality restored. The 2–3 min that this reversal takes, scale, slime, and other deposits are broken up and flushed out of the cells. This cleaning action eliminates the need to add either acid or polyphosphate.

Of the power used in ED plants, the pumps consume 1.9–3.8 kJ/L (6.8–13.6 Btu/gal) of product; the membrane stack uses 2.37 kJ/L (8.5 Btu/gal) of product per 1000 mg of TDS (temperature, depth, salinity) removed; and the power supply uses 5% of the total energy.

The efficiency of the ED process depends on many factors including the efficiency of pumps, motors and rectifiers, stack losses, and water temperature. Water temperature is important as it affects conductance in the cells. As the temperature increases, the conductance increases and the resistance decreases, leading to lower power consumption. As the other energy-use factors are not sensitive to temperature, its effect is felt only in large, high TDS plants.

Proper pretreatment is requisite for trouble-free operations. Depending on specific waters to be processed, the pretreatment can be minimal or consequential. However, in the presence of transition metals, Fe in excess of 0.3 mg/L, Mn over 0.10 mg/L, or H_2S over 0.3 ml/L, and turbidity over 0.21 Tu (turbidity units), pretreatment is requisite.

Table 7 gives the operating data (54) for the 8×10^6-L/d (2.1×10^6-gal/d) electrodialysis plant in Sanibel, Florida; Figure 14 shows a flow sheet of the plant.

Electrodialysis is a widely accepted technology in the conversion of brackish water with TDS in the range of 1000–4000 ppm. At present, the largest ED plant built is located in Corfu, Greece (55–56), with capacity of 15 ML/d (4×10^6 gal/d); information on the feed and blowdown for this installation is given in Tables 8 and 9. There are also a number of plants with capacities in the order of 4×10^6 L/d (10^6 gal/d) in operation. The EDR process has brought great improvement in the technology, and there has been an addition of 80×10^6 L/d (2.1×10^6 gal/d) capacity in ca 140 new plants during 1977–1980. The entry of this technology in seawater conversion is likely, though at present only a handful of seawater ED plants are operational, all of them in Japan. Research in ED is concentrated toward developing low resistance membranes with antifouling characteristics, skin membranes, and improved spacers (57–58).

Desalination by Renewable Energy Sources. The sun, ocean, geothermal energy, waves, and direct labor are all renewable energy sources that have been tapped for desalination projects. Although wood and biomass can also be sources of energy, these two sources are not considered in this section, because there is no work with these to power desalination processes. The exploration of desalination using renewable energy

Table 7. Operating Data for the Sanibel Island Electrodialysis Facility [a]

Property	1974	1975	1976	1977	1978	1979, 6 mo
production						
design capacity, m³/d [b]	4563	6813/7948	7948	7948	7948	7948
annual production, 1000 m³ [b]	1044	1269	1484	1460	1109	727
load factor, %	63	53	51	49	42	50
recovery ratio, %	82	74	78	76	85	89
quality						
feedwater TDS, mg/L	2470	2570	2440	2700	2460	2500
product water TDS, mg/L	517	445	479	489	479	482
consumables per unit of product						
power kW·h/10³ L	2.50	2.83	2.44	2.63	2.27	2.35
acid, kg/10³ L	0.93	0.77	0.80		0.83	0.55
chlorine, kg/10³ L	0.17	0.20	0.12	0.07	0.07	0.07
soda ash, kg/10³ L	0.26	0.24	0.17	0.38	0.38	0.20
costs (¢/10³ L)						
well field and pretreatment	3.22	3.38	3.37	8.17	5.20	5.7
power	7.13	18.21	7.92	33.88	8.63	9.40
chemicals	5.01	1.75	1.32	7.72	1.84	1.34
membranes	1.0	2.43	15.34	1.77	2.82	2.14
parts (stack)	1.8	0.37	9.51	1.58	0.92	1.08
labor	7.62	1.37	1.03	2.85	19.15	10.82
Total	14.55	17.30	18.78	22.88	24.49	22.58

[a] Ref. 54.
[b] To convert m³ to gal, multiply by 264.2.

Table 8. Feedwater Sources for the Corfu Electrodialysis Plant [a,b]

Quantity	Source			
	Gardiki	Nickolas	Karterion	Chrysiis
feedwater, m^3/d[c]	7,460	984	1,480	11,400
constituents, mg/L				
sodium	18	55	21	15
calcium	188	90	180	476
magnesium	51	31	58	70
chloride	36	35	36	36
bicarbonate	281	207	268	329
sulfate	422	244	446	1,147
TDS	996	662	1,009	2,073

[a] Refs. 55–56.
[b] With a feedwater flow of ca 31,200 m^3/d[c], there is a recovery of ca 70%, with 14,800 m^3/d[c] of product water and a concentrate discharge of ca 6430 m^3/d[c]. Table 9 shows the characteristics of the feedwater, product, and concentrate streams. The salt rejection is ca 64%.
[c] To convert m^3 to gal, multiply by 264.2.

Table 9. Characteristics of the Feed, Product, and Concentrate Blowdown at the Corfu Facility [a]

Constituent	Feedwater, mg/L	Product water, mg/L	Concentrate blowdown, mg/L
sodium	19	7	44
calcium	295	105	873
magnesium	58	21	153
chloride	36	13	89
bicarbonate	294	105	757
sulfate	699	249	2082
TDS	1401	500	3998

[a] Refs. 55–56.

sources has been prompted by the desire to conserve the standard fuels, the difficulties of obtaining standard fuels in remote regions, and especially, their high cost. Basic solar stills have been in use at least since the seventeenth century, and recently there have been serious studies of a diversity of concepts for advanced solar-powered desalination systems (see Solar energy). The exploration of the other sources, although of notable potential, has seen less than either serious or consequential investment of effort to effect their practical realization, and even solar distillation remains almost unused. The actual amount of solar energy received depends, of course, on time of year, time of day, and obstruction to sun rays. In some areas, for calculation purposes, it is assumed that radiation is received for about 8 h/d with variation depending on locality and atmospheric conditions. Charts showing average radiation distribution for many areas of the world are available.

Alternatively, photovoltaics can be used to convert the solar energy directly to d-c electrical energy, which in turn becomes the energy for desalination. The photovoltaic cells (qv) can be either of flat-panel or ribbon configuration; the latter is capable of significant increase in output when exposed to focused solar-energy exposure.

Because incident solar radiation fluctuates, any installation that is to be fed by solar energy must have suitable energy-storage schemes, and most such systems are expensive.

Solar stills use solar energy to increase the energy in a confined area and thus distill the feedwater without boiling. The device that can now be referred to as a classic solar still consists of a water-filled basin covered by glass or another medium that is transparent to the solar radiation, which heats the water to be desalinated. The vapor produced condenses on contact with a cooler surface and is collected as product.

There are two basic types of solar stills (58–59). Passive stills, which are the more common in the field, are built in a variety of forms (Fig. 15). Figure 16 shows the passive solar still at Puerto Chale in Mexico. Nonpassive solar stills use other sources of power as well to improve operations. All presently operational solar stills are relatively small capacity and only truly suitable for supplying water to families or small communities in underdeveloped nations. They produce up to about 5000 L/d (1320 gal/d). Table 10 lists the areas and capacities of four operating solar stills.

The latest trend in solar-powered desalination installations is to abandon both the passive and nonpassive stills in favor of much more complex solar systems. Figure 17 shows a flow diagram for a solar reverse-osmosis system built by the French Atomic Energy Commission at Cadarach (60). A similar system has also been installed in Jeddah, Saudi Arabia. It is powered by an 8-kW array of Mobil-Tyco photovoltaic modules (61).

Summary

In summary, desalination today is making a considerable contribution to the world's potable water supply as well as water for industry, ranging from boiler feedwater for the power industry to ultrapure water for the electronics industry. The process industries are also looking at desalination technology either for better process water, or as means of conservation through recycling of wastewaters. In some industries, it is considered a technique for recovering some process chemicals from water and lowering the cost of purification of polluting discharges. Some desalinated water is also used in agriculture but to a limited extent because of cost. If and when it is completed, the brackish-water RO plant to process the lower Colorado River as it exits from the Imperial Valley of California will be the principal desalination project with important implications for agriculture.

Though desalination technologies are diverse, MSF has been for some time, and will remain well into the next century, the main process for desalination of seawater. Inroads are being made by the multi-effect processes and, in particular, by the low temperature ME processes. Despite some notable advantages, including lower energy consumption, they will continue to lag materially behind MSF.

Important advances in the membrane field were responsible for the commercialization of seawater RO. This technique both saves energy and offers much-reduced plant construction time. Compared with distillation, RO treatment of seawater has had a meteoric success in the last five years and further growth of almost 20% is expected. Considering brackish-water RO and EDR as well, the membrane processes have become a integral and growing part of the desalination scene. In fact, their combined growth appears to exceed that of distillation.

The steady growth in water consumption around the world without any real

University of California solar still no.1

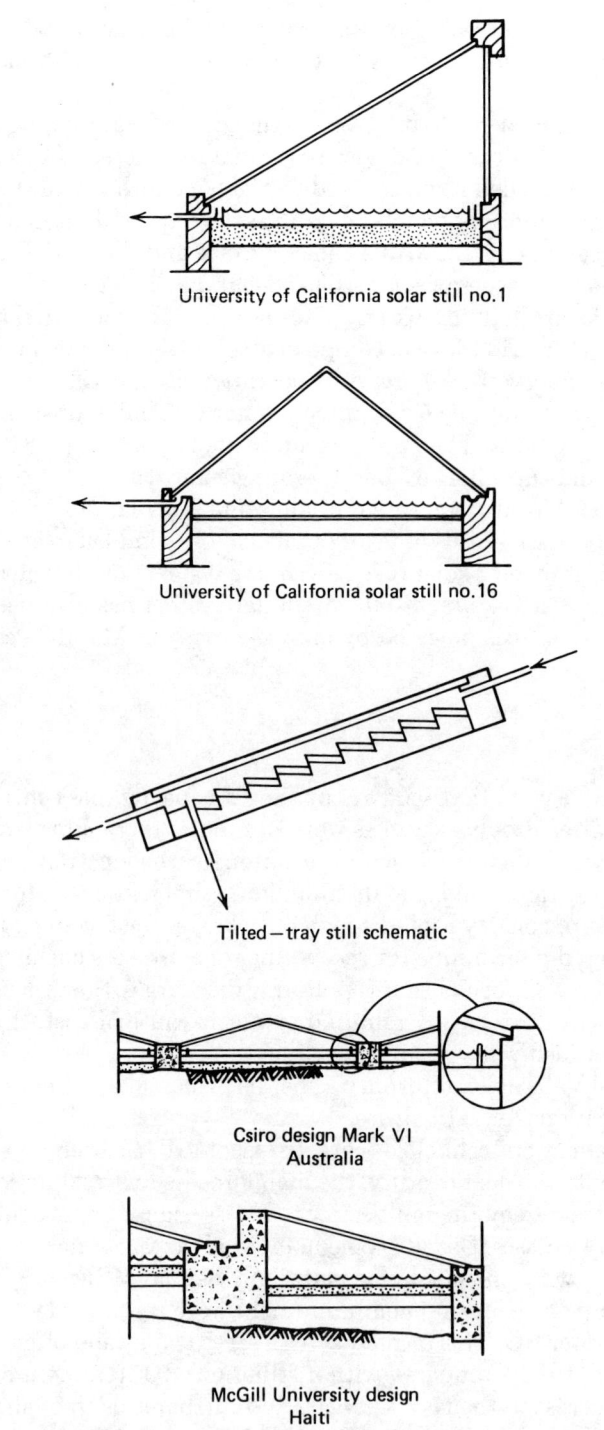

University of California solar still no.16

Tilted—tray still schematic

Csiro design Mark VI
Australia

McGill University design
Haiti

Figure 15. Cross sections of some common types of passive solar stills. Courtesy U.S. Agency for International Development.

Figure 16. The Puerto Chale solar still in Mexico. Courtesy Digas.

Table 10. Solar Stills

Location	Capacity, L/d[a]	Surface area, m^2 [b]
Aeania, India	4920	185.8
La Gonare, Haiti	946	na
Puerto Chale, Mexico	975	211.6
Satmos Island, Greece	3785	169.3

[a] To convert L to gal, divide by 3.785.
[b] To convert m^2 to ft^2, multiply by 10.76.

growth of available water resources is focusing increasing attention on water reuse. Desalination technology is expected to be an integral part of all water-reuse schemes. Water reuse is already receiving serious consideration today, and within the next ten years there will be water-reuse schemes in virtually all areas of water use–not only in water-scarce regions, but also in water-abundant regions such as the industrial northeast of the United States. Municipal authorities will become intimately involved in conversion of their wastewater in the future.

Much progress has been effected in desalination technology, but all desalination processes remain energy-intensive. Though energy costs have prompted the consideration of renewable energy resources as sources for desalination, these sources are unlikely to become important energy sources for desalination, except in certain limited circumstances, unless there are significant unforeseen breakthroughs in either energy conversion or desalination technologies.

Despite the cost of desalination technology, it has made water available in places

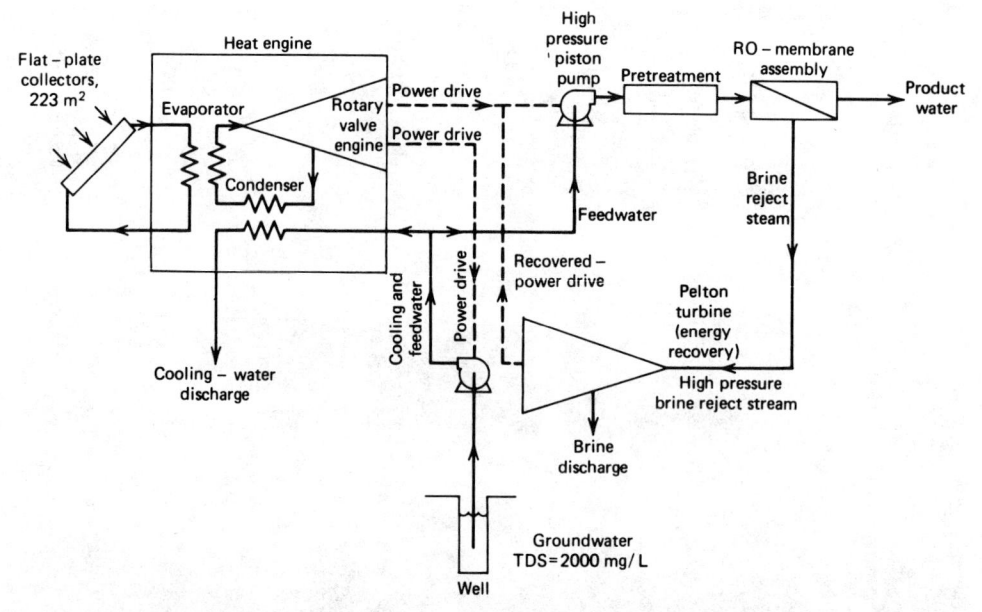

Figure 17. Flow diagram of the experimental solar RO unit at Cadarache, France. Feedwater flow = 1.38 L/s, product water flow = 0.69 L/s, energy consumption without recovery = 0.89 kW·h/m^3, energy recovery = 0.21 kW·h/m^3, energy consumption with recovery = 0.68 kW·h/m^3 (60). Courtesy CEA (Commissariat d'Energie Atomique), France.

where it was not before. Not only has water become available in these places, but the quantities available have also opened prospects for industrial development. This has led to important improvements in the standard of living with prospects for even further improvements in countries fortunate enough to afford the technology.

Abbreviations

BPE	= boiling point elevation
ED	= electrodialysis
EDR	= electrodialysis reversal
LTMED	= low temperature multi-effect distillation
LT-VC	= low temperature vapor compression
mgd	= million (10^6) gallons per day
MSE	= mechanical support equipment
MSF	= multistage flash
NCG	= noncondensable gases
OSW	= Office of Saline Water
OWRT	= Office of Water Research and Technology
RO	= reverse osmosis
SDI	= silting density index
TDS	= temperature, depth, salinity
VC	= vapor compression
VCE	= vapor-compression evaporation
VTE	= vertical-tube evaporation
VT-VC	= vertical-tube vapor compression

BIBLIOGRAPHY

"Water, Demineralization" in *ECT* 1st ed., Suppl. 1, pp. 908–930, by E. R. Gilliland, Massachusetts Institute of Technology; "Water Supply and Desalination" in *ECT* 2nd ed., Vol. 22, pp. 1–65, by D. F. Othmer, Polytechnic Institute of Brooklyn.

1. R. V. Salazar, private communication, 1976.
2. A. LaBastille and T. Spiegal, *National Geographic Magazine*, 652 (Nov. 1981).
3. J. Urghart, *Wall Street Journal* (Feb. 24, 1982).
4. *Bus. Week* (Feb. 22, 1969).
5. A. O. Sulzberger, *New York Times* (Aug. 10, 1981).
6. W. Schmidt, *New York Times* (Aug. 11, 1981).
7. P. Shabekov, *New York Times* (Feb. 27, 1980).
8. J. Koenig, *J. Am. Water Works Assoc.* **51,** 845 (1959).
9. *Environ. Sci. Technol.* **5,** 115 (1969).
10. D. G. Argo, *Proceedings Second Annual Conference*, National Water Supply Improvement Association, 1974.
11. *Water Desalination in Developing Countries*, U.N. Publication 64 11 B 5-1964, United Nations, 1964.
12. S. F. Singer, *Environ. Sci. Technol.* **3,** 197 (1969).
13. *Chem. Eng.*, 23 (May 3, 1982).
14. W. M. Wagner and D. R. Finnegan, *Chem. Eng.*, 71 (Feb. 7, 1983).
15. *Chem. Eng. News*, 50 (Feb. 24, 1969).
16. C. A. Brandon, J. L. Gaddis, D. A. Jernigan, and H. G. Spencer in R. Bakish, ed., *Proceedings, Congress on Desalination and Water Reuse*, International Desalination and Environmental Association, Teaneck, N.J., 1981.
17. *Chem. Week*, 22 (July 19, 1968).
18. *Chem. Eng. News* (Aug. 12, 1968).
19. R. Renhold, *New York Times* (Aug. 8, 1981).
20. *Proposals for a Costing Procedure and Related Technical and Economic Considerations*, UN Publication, United Nations.
21. *Water Needs of Developing Nations*, UN Publication, United Nations.
22. A. Delyanis and E. Delyanis, eds., *Proceedings, Congress by Working Party on Fresh Water from the Sea of EFCE*, 1962, 1967, 1970, 1976, 1978, 1980.
23. *Proceedings, International Congress on Desalination and Water Reuse, Mexico*, Elsevier Publishing Co., Amsterdam, 1976; *Proceedings, International Congress on Desalination and Water Reuse, Tokyo*, International Desalination and Environmental Association, Teaneck, N.J., 1977; R. Bakish, ed., *Proceedings, International Congress on Desalination and Water Reuse, Nice*, International Desalination and Environmental Association, Teaneck, N.J., 1979; R. Bakish, ed., *Proceedings, International Congress on Desalination and Water Reuse, Bahrain*, International Desalination and Environmental Association, Teaneck, N.J., 1981.
24. *Pure Water Publication*, International Desalination and Environmental Association, Englewood, N.J., Oct. 1972.
25. *Desalination Materials Manual*, Office of Water Research and Technology, U.S. Department of the Interior, Washington, D.C.
26. D. Hoffman, *National Water Supply Improvement Association Journal* (July 1981).
27. A. Porteus, *Saline Water Distillation Processes*, Longman Group Ltd., London, 1975, p. 30.
28. V. B. Cherrozubov and co-workers, *Proceedings of the 1st International Symposium on Water Desalination*, Washington, D.C., Oct. 1965, p. 139.
29. *The Application of Belgard EV and Belgard EVN in Sea Water Evaporators*, CIBA-GEIGY pamphlet DB 21, CIBA-GEIGY, Ardsley, N.J., 1980.
30. O. K. Buros, *Desalination Manual*, U.S. Agency for International Development, International Desalination and Environmental Association, Teaneck, N.J., 1981.
31. R. B. Cox, *Sea Desalination Science and Technology*, April 1969.
32. W. Johnson, *State of the Art, Survey and Economic Comparison of Freezing Processes*, Office of Water Research and Technology, U.S. Department of the Interior, Washington, D.C., 1976.
33. W. Johnson, *Desalination* **31,** 417 (1979).

34. *10,000 GPD Secondary Refrigerant Freeze Desalination Plant*, National Technical Information Service Publication PB 1251906, U.S. Department of Commerce, Washington, D.C., 1973.
35. *Economics of a Freeze Desalting Process Using Cold Sea Water Effluent of Liquid Natural Gas Plant*, Office of Water Research and Technology Report 52, U.S. Department of the Interior, National Technical Information Service No. PB 259-272, U.S. Department of Commerce, Washington, D.C., 1976.
36. *Desalting Plans and Progress, An Evaluation of the State of the Art and Future Research and Development Requirements*, 2nd ed., National Technical Information Service No. PB 290 75, U.S. Department of the Interior, Washington, D.C., 1978.
37. R. S. Robertson, *Developmental Testing of a Secondary Refrigerant Freezing Desalination Pilot Plant at the Wrightsville Beach Test Facility*, Office of Water Research and Technology, U.S. Department of the Interior, Washington, D.C., 1978.
38. S. Sourirajan, ed., *Reverse Osmosis and Synthetic Membranes*, National Research Council, Canada, 1977.
39. S. Sourirajan in A. Turbak, ed., *Synthetic Membranes*, American Chemical Society, 1981.
40. L. Dresner and J. S. Johnson, Jr., in K. Spiegler and A. D. Laird, eds., *In Principles of Desalination*, Academic Press, New York, 1980.
41. A. Turbak, ed., *Synthetic Membranes*, Vols. 1 and 2, American Chemical Society, Washington, D.C., 1981.
42. *SDI Procedure Sec.*, DuPont Technical Bulletin No. 401, E. I. du Pont de Nemours & Co., Inc., Wilmington, Del., Sept. 1, 1977, p. 1.
43. S. Loeb and S. Sourirajan, *Sea Water Demineralization by the Surface Skinning Process*, Sea Water Research Quarterly Progress Report 59-3, Department of Engineering, University of California at Los Angeles, Los Angeles, Calif., June–Aug. 1958.
44. J. E. Cadotte and R. Peterson in A. F. Turbak, ed., *Thin Film Composite Reverse Osmosis Membrane*, American Chemical Society Symposium Services, Washington, D.C., 1981.
45. M. C. Mickley and G. Coury, *Natl. Water Supply Improvement Assoc. J.* **9,** 11 (Jan. 1982).
46. D. B. Guy and R. Singh, *Natl. Water Supply Improvement Assoc. J.* **9,** 35 (Jan. 1982).
47. R. Singh and S. Cabibo, *Burns and Roe Study*, Burns & Roe, Paramus, N.J., 1980.
48. C. T. Sackinger, *The Energy Requirements of Sea Water Desalination Reverse Osmosis*, VS MSF MEW & S, April–May, 1980.
49. E. Shannon, *Design and Start Up of the 3 MGD RO Plant at Cape Coral, Florida*, OWRT Technology Transfer Shop, July 19–20, 1979, Virginia Beach, Va.
50. D. G. Argo, *5th Annual Conference Proceedings*, National Water Supply Improvement Association, July 1977.
51. D. L. Krull in ref. 50; F. Engstrom, I. G. Taylor, and L. A. Haus in ref. 50.
52. Bulletins N-CR61.1E, March 1982, and CRCI.OE, May 1982, Ionics, Inc., Watertown, Mass.
53. M. Kishim, S. Serizawa, and M. Nakano in *Proceedings, International Congress on Desalination and Water Reuse*, International Desalination and Environmental Association, Teaneck, N.J., 1977.
54. I. Watson and R. DeRowitsch in *Proceedings of the 7th Annual Conference*, National Water Supply Improvement Association, Ipswich, Mass., 1979.
55. G. Andreadis and J. W. Arnold in *Proceedings of the 6th Annual Conference*, National Water Supply Improvement Association, Ipswich, Mass., 1978.
56. J. W. Arnold in R. Bakish, ed., *Proceedings, International Congress on Desalination and Water Reuse, Tokyo*, International Desalination and Environmental Association, Teaneck, N.J., 1979.
57. N. L. Goldstein in ref. 56.
58. M. E. Mattson and J. E. Lundstrom in *Technical Proceedings of the 7th Annual Conference of WSIA*, 1979.
59. D. C. Dunham, *Fresh Water from the Sun*, U.S. Agency for International Development, Washington, D.C., 1979.
60. A. Maurel in ref. 56.
61. W. Boesch, *Desalination* **41,** (1982).

ROBERT BAKISH
Bakish Materials Corporation

INDUSTRIAL WATER TREATMENT

In addition to the water required for drinking and sanitary use by employees and for fire protection, most industrial establishments require substantially larger amounts of water for cooling, boiler feed, and process use. For these industrial uses, the quality of the water required for specific applications is as important as the availability of a sufficient quantity. This is particularly true for such industries as steam-electric power generation, chemical processing, petroleum refining, the manufacturing of electronics components, food processing (qv), and the preparation of alcoholic and nonalcoholic beverages (see Power generation; Petroleum, refinery processes; Beverage spirits, distilled). Industrial water treatment involves the production from available water supplies of boiler feedwaters, cooling waters, or process waters that have the composition or properties required for each water-using system or process. A number of the processes discussed in this section are used for the treatment of waters discharged from industrial water systems or from municipal or industrial wastewater treatment plants in order to make these waters reusable for industrial applications (see Water, municipal water treatment; Water, reuse).

Industrial Uses of Water

Cooling is the largest industrial use for water (see Table 1). The general quality criteria for cooling waters are that they should neither permit the formation of deposits, which reduce heat transfer or increase resistance to the flow of water, nor permit significant deterioration of the materials of construction in the cooling system (2). Specific cooling-water quality requirements are highly individual and vary with the composition of the water supply, type of cooling system, materials of construction and other design aspects, and such operating characteristics as temperatures, flow rates, uniformity of operations, and quality of operating personnel. Waters used as make-up for industrial cooling systems include seawater, which is widely used for once-through cooling systems and for some large recirculating evaporative cooling systems in coastal locations, and deionized water or steam condensate, which is supplied to a limited number of recirculating evaporative cooling systems (see Table 2).

Process water is the second largest use for industrial water. Here, too, quality requirements are highly individualized according to the needs of the process in which the water is being used and the characteristics of the individual plant. The requirements range from virtually no restrictions on water quality for coal washing and other mineral-industry uses to quality similar to that of conductivity water for use in processing electronic components.

Boiler feed is the third largest industrial use of water. Although the quality of the feed and boiler water needed depends largely on the operating pressure of the boiler, it is also influenced by the type of boiler and auxiliaries, the heat flux, and the uses of the steam. Table 3 indicates the range of water-quality requirements for boilers as given in the consensus guidelines for water-tube boilers developed by the American Society of Mechanical Engineers Committee on Water in Thermal Power Systems (4).

368 WATER (INDUSTRIAL TREATMENT)

Table 1. Industrial Plant Water Intake, Recirculation, and Reuse by Large Users in 1978[a,b]

Industrial group	Condensing	Process	Feed	Other	Total
food and food products					
intake	1,321	984	136	360	2,801
recirculation	2,487	1,113	140	418	4,158
paper and paper products					
intake	2,385	4,493	329	223	7,430
recirculation	2,407	31,927	526	23	34,883
chemicals and chemical products					
intake	12,536	2,854	538	447	16,375
recirculation	29,266	1,893	1,745	144	33,048
petroleum and coal products					
intake	3,460	231	621	129	4,441
recirculation	26,518	182	204	4	26,908
primary metal industry					
intake	8,679	3,649	193	318	12,839
recirculation	8,239	4,690	34	133	13,096
transportation equipment					
intake	310	276	30	269	885
recirculation	4,432	2,850	1,389	235	8,906
Total, all industries[d]					
intake	*28,691*	*12,487*	*1,847*	*1,746*	*44,771*
recirculation	*73,349*	*42,655*	*4,038*	*957*	*120,999*

Water use, 10^6 m^3/yr[c]

[a] Ref. 1.
[b] Manufacturing industries using more than 3.8×10^6 m^3/yr (10^9 gal/yr).
[c] To convert m^3 to gal, multiply by 264.
[d] Manufacturing industries only. The thermal-electric generation industry uses more than twice as much water as the combined manufacturing industries; more than 99% of it is used for condensing and cooling.

Table 2. Sources of Industrial Water Withdrawals in the United States in 1980[a]

Type of industrial use	Fresh water, 10^6 m^3/yr[b]	Saline and brackish water, 10^6 m^3/yr[b]	Total
commercial	22.0		22.0
manufacturing	135.5	45.0	180.5
steam-electric utilities	351.2	215.8	567.0
Total	*508.7*	*260.8*	*769.5*

[a] Ref. 3.
[b] To convert m^3 to gal, multiply by 264.

Water-Caused Problems

Water-caused problems in industry affect production costs and cause unscheduled shutdowns, which can result in extremely costly loss of production. Water-caused problems may reduce heat transfer, reduce water flow, cause premature equipment deterioration or failure, and reduce product quality or yield.

Table 3. Guidelines for Water Quality in Modern Industrial Water-Tube Boilers for Reliable Continuous Operations[a]

Drum pressure, MPa[b]	Boiler feedwater			Boiler water		
	Iron, ppm Fe	Copper, ppm Cu	Total hardness, ppm CaCO$_3$	Silica, ppm SiO$_2$	Total alkalinity[c], ppm CaCO$_3$	Specific conductance, μS/cm
0.1–2.2	0.100	0.050	0.300	150	700[d]	7000
2.2–3.2	0.050	0.025	0.300	90	600[d]	6000
3.2–4.2	0.030	0.020	0.200	40	500[d]	5000
4.2–5.3	0.025	0.020	0.200	30	400[d]	4000
5.3–6.3	0.020	0.015	0.100	20	300[d]	3000
6.3–7.0	0.020	0.015	0.050	8	200[d]	2000
7.0–10.4	0.010	0.010	0.000	2	0[e]	150
10.4–13.9	0.010	0.010	0.000	1	0[e]	100

[a] Ref. 3.
[b] To convert MPa to psi, multiply by 145.
[c] Minimum level of OH alkalinity in boilers <6.9 MPa (1000 psi) must be individually specified with regard to silica solubility and other components of internal treatment.
[d] Alkalinity not to exceed 10% of specific conductance.
[e] Zero in these cases refers to free sodium or potassium hydroxide alkalinity. Some small variable amount of total alkalinity is present and measurable with the assumed congruent control or volatile treatment used at these high pressures.

Reduced heat transfer develops from the formation of deposits on cooling or heating surfaces and results in increased energy costs, reduced output, or both. Deposits may consist of sedimentary materials, eg, sand, silt, or clay; corrosion products, eg, rust, formed in place or transported from another part of the system; crystalline scales formed *in situ*; bacterial or fungal slimes; or other materials. Usually, water-system deposit consist of mixtures of several types of materials. Reduced water flow results from the same types of deposits but usually in larger accumulations and at locations other than heat-transfer surfaces. Premature equipment deterioration or failure may be caused by contact with the waters used in industrial systems. Corrosion of metallic equipment and components is the most common manifestation, but nonmetallic materials, eg, wood, concrete, plastics, and asbestos cement, also deteriorate as a result of contact with industrial waters. Equipment deterioration may also be caused by erosion or abrasion by solids suspended in the water; equipment may fail catastrophically as a result of water-caused problems, as when formation of an insulating scale causes boiler tubes to overheat and blow out. Reduction of product quality can be caused by impurities present in the water entering a system or by those which develop as a result of corrosion, bacterial growth, or other problems within the system.

Types of Industrial Water Systems

Whether scale or other deposits, corrosion, or biological growths develop in a water system depends largely upon the water composition. Because of the low concentrations of constituents in most industrial waters, their quantities are expressed in mg/L or μg/L or parts per million (10^6; ppm) or parts per billion (10^9; ppb), respectively. The metric terms are equal to the common terms unless the water density is greater than

1.0, in which case ppm or ppb must be multiplied by the density (g/cm^3) for conversion to mg/L or µg/L, respectively. Other terms used for reporting water-analysis data include milliequivalents per liter (meq/L), particularly for ion-exchange calculations. Table 4 shows the relationships of the various water-analysis units.

Two of the most frequently used water-analysis terms, ie, total hardness and alkalinity, are always expressed as $CaCO_3$. Total hardness is normally the sum of the calcium and magnesium concentrations, both expressed as $CaCO_3$ (5). Alkalinity is the quantitative capacity of a compound, mixture, or solution to react with a strong acid to a designated pH (6), usually 4.5 for total or methyl orange or M alkalinity and 8.3 for phenolphthalein or P alkalinity.

It is not only the composition of the make-up water entering a system that controls the development of waterside problems. Changes in water composition develop in most types of water systems during operation. With the increase in recirculation and reuse of industrial waters (Table 1), the composition of the water within the system is increasingly important. The causes of water-related problems in cooling-water systems, boilers, and process-water systems can be better understood by considering the basic types of industrial water systems and the changes in water chemistry that take place during their operation.

Once-through systems are the simplest. These include plant water-supply systems or sanitary water systems; cooling-water systems drawing water from rivers, lakes, or the sea and returning it directly thereto after passage through heat-exchange equipment; and automatic sprinkler and fire-protection systems (see Heat-exchange technology; Plant safety). The water problems that develop in once-through systems depend upon the composition of the influent water since there is little, if any, change in water composition during its passage through these systems.

Nonevaporative, closed circulating systems include hot-water heating, high temperature water, chilled water, combined chilled and hot water, solar heating, snow melting, and brine systems. These systems frequently involve a cyclic change in temperature during operation and minor changes in composition develop primarily as a result of a reaction of the water with system components. In theory, such systems require no make-up water because they are closed and, therefore, water-caused problems should be substantially nil. Unfortunately, these systems are rarely as closed as their designers would like to believe, and the constant influx of make-up water leads to problems such as corrosion caused by the oxygen in the water, scale in heated systems supplied with hard make-up waters, and bacterial or fungal growths (7).

Table 4. Units Used for Expressing Water Composition

Unit	Abbreviation	To convert to mg/L, multiply by
milligrams per liter	mg/L	1.0
parts per million (10^6)	ppm	1.0[a]
micrograms per liter	µg/L	0.001
parts per billion (10^9)	ppb	0.001[a]
milliequivalents per liter	meq/L	50.0[b]
equivalents per million (10^6)	epm	50.0[b]
pounds per 1000 gallons	lb/1000 gal	120.0

[a] For brackish or saline waters whose density exceeds 1.0, ppm or ppb must also be multiplied by the water density in g/mL for conversion to mg/L or µg/L.
[b] mg/L as $CaCO_3$.

Evaporative, closed circulating systems include low pressure steam-heating boilers and high pressure power-generation boilers. In these, steam is generated and condensed, and the condensate is almost completely recycled as boiler feedwater. The boiler water pH rises as a result of loss of dissolved carbon dioxide and thermal breakdown of bicarbonate to carbonate and hydroxide. If there are significant losses of steam or condensate, ie, in-leakage from steam condensers, the dissolved solids increase and scale formation may develop. Air or carbon dioxide dissolving in cooled condensate make the water corrosive. In addition to attacking metallic surfaces with which it comes into contact, the cooled condensate carries corrosion products back to the boiler where they form heat-insulating deposits (see also Steam).

Condensing systems, in addition to those associated with steam-boiler systems, include dehumidifiers in air-conditioning systems. The condensate produced in these systems is essentially distilled water, which dissolves oxygen and carbon dioxide. Small amounts of the latter produce quite low pH values in unbuffered waters. Both of these gases promote corrosion.

Evaporative, noncirculating systems include process steam boilers, which operate at a substantial make-up rate; automatic icemakers; and nonspray humidifiers. Such systems are characterized by a rapid rate of increase in the dissolved-solids concentration of the water in the equipment, a change which often leads to serious scale formation even when the make-up water is fairly soft.

Evaporative, open circulating systems include spray-type humidifiers and cooling-water systems, eg, cooling towers, evaporative condensers, or evaporative coolers. The evaporation increases the dissolved solids, leading to scale formation. In addition, gases and solids are scrubbed from the air through which the water is sprayed. Oxygen and sometimes sulfur oxides from stack gases are dissolved, increasing the corrosivity of the water. Particulates scrubbed from the air abrade moving parts, form heat-insulating deposits, promote localized corrosion, or result in undesirable growths of microorganisms.

In most operating water systems, there are temperature increases which promote scale formation and corrosion. Dissolved-solids concentrations increase in evaporative systems and thereby promote scale formation. Condensation and dilution promote corrosion. Any contact with air does the same, but spraying of the water through air aggravates other problems by providing an opportunity to add suspended solids and microorganisms to the water. Finally, these effects are frequently combined in various ways which increase damage to the water system or increase interference with its operation. Table 5 indicates the changes in water composition resulting from the use of water from a single public water supply without chemical treatment in a variety of water systems. The values are from analyses of waters taken from operating systems.

Treatments

Ideally, the water-quality and treatment requirements for each industrial plant and water use should be evaluated beginning with early design to anticipate possible water-caused operating problems and to provide the most cost-effective means for preventing or minimizing these problems during the life of the plant. The evaluation should take into consideration the quantity and quality of available water sources, including their ranges of variations; quantitative and qualitative requirements for

Table 5. Water Composition Changes in Untreated Operating Systems, mg/L[a]

Property	Make-up supply[b]	Steam-heating boiler	Steam condensate	Cooling tower, No air pollution	Cooling tower, Flue gas absorption	Chilled water
total hardness as $CaCO_3$	20	78	0	392	450	21
M alkalinity as $CaCO_3$	10	35	1	152	0	11
P alkalinity as $CaCO_3$	0	15	0	0	0	2
free carbon dioxide	2.5	0	5	5	[c]	0
free acid as $CaCO_3$	0	0	0	0	44	0
chloride	3	15	0	72	72	4
sulfate	9	43	0	216	520	8
iron	0	5	2	0	3	2
copper	0	0	0	0	3	1
total dissolved solids	31	156	2	642	977	33
pH	6.9	10.1	5.6	8.0	3.2	9.3

[a] Ref. 8.
[b] New York City's Catskill water supply.
[c] Not determined.

each plant water use; possible cascading or sequential use of water for multiple-plant applications; in-plant recycling of water, alternative treatment processes capable of meeting the various water-quality requirements; space requirements for treatment processes; applicable effluent-discharge regulations; and quality and quantity of operating personnel.

In external water-treatment processes, water is passed through treatment equipment, in which some desired change in water composition is produced either by physical or chemical means. External treatment processes are selected for the removal from water of specific impurities or classes of impurities, eg, suspended and colloidal solids, dissolved solids, or dissolved gases. Examples are the removal of suspended matter in a filter, of dissolved oxygen in a deaerator, and of hardness in a cation-exchange-resin softener.

In internal-treatment processes, chemicals are added to water in an operating system or near the point of introduction of the water into the system in order to produce a desirable change in the water composition or to minimize an undesirable effect. Internal treatments may also be selected for the removal of a specific impurity, as with the reduction of dissolved oxygen by the addition of a sulfite or hydrazine. More often, internal treatments are not selected for bringing about a specific change in water composition, but for minimizing such water-caused problems as deposit or scale formation, corrosion, or the accumulation of algae or biological slimes. Examples of such internal treatments are the addition of polymer dispersants to prevent sediment from settling and forming deposits, acid to adjust the pH to a range less favorable to calcium carbonate scale formation, crystal modifiers to prevent the formation of scales, corrosion inhibitors to form protective films on metals, and microbiocides to prevent the development of organic growths (see Corrosion and corrosion inhibitors; Dispersants; Coatings, marine).

Some treatment objectives can only be attained by external treatment, as in the reduction or removal of solids. Other treatment objectives can only be achieved by internal treatment, as in the formation of corrosion-inhibiting films on metal surfaces.

Some treatment objectives can be attained either by external or internal treatment. Thus, in water-supply systems, scale formation may be avoided by lime or ion-exchange softening of the water or by the addition of a polyphosphate to the water. Still other treatments require the use of both external and internal treatment. An example of this is the removal of dissolved oxygen from feedwater for high pressure boilers. The bulk of the oxygen is removed most economically by passing the water through a deaerator, but this equipment cannot remove the last traces of oxygen. Therefore, it is also necessary to add a chemical oxygen scavenger to the deaerator effluent in order to produce oxygen-free feedwater.

External. External treatment processes are used for the reduction or removal of suspended or colloidal solids, dissolved solids, or dissolved gases. They may be applied to total plant water supplies, to water entering a subsystem, eg, boiler feedwater, or to water being recirculated through a subsystem, eg, steam condensate or cooling water. All or only a portion of the recirculating water may be passed through treatment equipment. A part of a recirculating water flow can be treated with side-stream filtration in which a few percent of the flow of recirculating cooling-tower water is filtered and returned to the cycle in order to limit the concentration of suspended solids in the system.

Suspended or colloidal solids can be reduced or removed by rough screening, sedimentation (qv), centrifugal separation (qv), straining, filtration (qv), coagulation, flocculation, magnetic separation (qv), or combinations of these processes (see Flocculating agents). Clarification frequently describes a combination of these processes. The selection of the appropriate method or methods for each case depends upon the size, specific gravity, and concentration of suspended particles in the raw water and the maximum concentration acceptable for the final use of the water. Particle size is an important consideration for the use of screening, straining, filtration, and coagulation. Particle specific gravity must be considered for sedimentation and centrifugal separation. Particle concentration influences the selection of clarification processes and is of particular importance for straining and polishing. The magnet responsiveness of the particles is critical for magnetic separation.

Rough screening is used for the removal of large objects, eg, logs, fish, masses of water weeds or algae, and other floating debris. Depending upon the screen opening, it may be used for the removal of particles as small as ca 5 mm. Equipment for rough screening ranges from trash racks with bars spaced several inches apart to screens with openings as small as 5 mm.

Sedimentation is used for the removal of larger particles with densities greater than 1.0 g/cm^3 and those that settle within a reasonable time; sedimentation can be used to remove particles as small as 0.05 mm. In accordance with Stoke's Law, the settling rate of a particle in water is directly proportional to the particle size and to the difference between the specific gravities of the particle and water and indirectly proportional to the viscosity of the water. For most water uses, sedimentation alone does not effect an adequate reduction of suspended solids. Therefore, sedimentation is commonly used for preliminary treatment of highly turbid waters, ie, for removing the readily settleable solids. Sedimentation basins are usually quite large in area and may be equipped with mechanical means for periodic or continuous removal of the settled solids. The capacity and efficiency of sedimentation equipment can be enhanced by including inclined-plate or tube separators, which substantially reduce the vertical distance through which particles must settle before they can be separated from the body of the water.

Centrifugal separation is like sedimentation but centrifugal force is applied for improved results. The particles to be removed must have a specific gravity greater than that of water and those as small as ca 0.1 mm can be removed. Some uses of hydrocyclones are for centrifugal separation of sand from well-water supplies and mill scale particles from pump-seal lubricant water.

Straining and filtration are used for the removal of all types of particulate matter as small as ca 0.01 mm. Woven mesh is available in a broad range of openings and a variety of weaves and materials. Granular media for filtration include sand and gravel, anthracite coal, and garnet. Mixed granular-media filters provide greater in-depth loading and, therefore, longer runs between backwashing than do single-media filters. Granular-media filters are available for gravity or pressurized operation, upflow or downflow, and manual or automatic backwash. They can be used for waters containing up to ca 500 mg/L of suspended matter, whereas woven-media filters or strainers are practical for waters with about one tenth this concentration. Woven-media filters precoated with filter aids, eg, diatomaceous earth, are used to polish waters containing ca 10 mg/L or less of suspended solids, and remove particles as small as 0.001 mm (see Diatomite). Membrane filters are used for ultrafiltration (qv) to remove even smaller particles.

Flocculation and coagulation of smaller and lighter particles of suspended matter accelerate settling and improve filtration. Salts of aluminum or ferric ion or high molecular weight polyelectrolytes are added to the water as coagulating and flocculating agents. Direct filtration of coagulant-treated waters through deep-bed granular media filters effect $\geq 90\%$ removal of suspended particles from waters containing 150–200 mg/L suspended matter. Better results are obtained if flocculation and coagulation are carried out in a clarifier, which is comprised of a high velocity mixing zone with short residence time for rapid dispersal of the coagulant throughout the water, a slower mixing zone with longer residence time for agglomeration of the initial fragile particles of floc, and a zone for settling the floc and decanting the clarified water which is usually filtered to remove fine, slowly settleable floc particles. The positive charges of the coagulant cations or molecules neutralize the negative charges on the surfaces of the suspended particles; the particles then combine as the larger, more rapidly settleable floc. Optimum results require careful adjustment of pH and coagulant dosage. The raw-water quality also affects the coagulation process. Low suspended solids provide limited floc-forming nuclei. Various organic constituents tend to stabilize the colloidal suspension, thereby making coagulation more difficult.

Clarification may be carried out in horizontal-flow, sludge-blanket, or solids-contact clarifiers. Horizontal-flow clarifiers are made up of a rapid mixing zone, a slow mixing zone, and a rectangular sedimentation basin which is usually equipped with hydraulic or mechanical means for intermittent or continuous removal of the settled sludge which accumulates in the bottom of the basin. The other two clarifiers are vertical-flow units, which are usually cylindrical. In sludge-blanket clarifiers, the raw water and coagulant chemicals are added in a central cylinder. The initial floc particles settle toward the bottom while being slowly stirred. They then pass beneath the edge of the central cylinder and flow upward at a much slower vertical velocity, which permits the floc to form a loose but reasonably stable sludge blanket. The clarified water passes upward through this blanket, eventually overflowing to a collection ring at the top of the unit. The solids-contact clarifier is essentially a modification of the sludge-blanket clarifier, in which the sludge is mechanically recirculated through the central tube so that the entering mixture of raw water and coagulant chemicals comes into contact with a high concentration of floc particles. This accelerates floc development and forms a denser floc. Clarification may be used alone, but it is usually followed by filtration in granular media equipment. Clarification is an essential pretreatment of waters containing suspended or colloidal matter prior to membrane processes, eg, ultrafiltration or reverse osmosis (qv).

The removal of dissolved solids from water ranges from partial to virtually complete removal of a single constituent, eg, silica, or a limited group of constituents, eg, hardness, to the complete removal of all dissolved substances resulting in the production of an ultrapure water. A variety of external treatment processes is available for reducing and removing dissolved solids, including chemical precipitation, ion exchange (qv), evaporation (qv), reverse osmosis, and electrodialysis (qv).

The equipment used for chemical precipitation resembles that used for clarification. Precipitants are mixed with the water in clarification equipment, in which the chemical reactions and agglomeration of the resulting precipitates into readily settleable and filterable form occur. Lime softening is probably the most widely used chemical precipitation for water treatment. Bicarbonate or temporary hardness can be reduced by treatment with lime:

$$Ca(HCO_3)_2 + Ca(OH)_2 \rightarrow 2\ CaCO_3\downarrow + 2\ H_2O \qquad (1)$$

$$Mg(HCO_3)_2 + Ca(OH)_2 \rightarrow CaCO_3\downarrow + MgCO_3 + 2\ H_2O \qquad (2)$$

$$MgCO_3 + Ca(OH)_2 \rightarrow CaCO_3\downarrow + Mg(OH)_2\downarrow \qquad (3)$$

Reduction of noncarbonate or permanent hardness requires treatment with soda ash in addition to lime:

$$CaCl_2 + Na_2CO_3 \rightarrow CaCO_3\downarrow + 2\ NaCl \qquad (4)$$

$$MgSO_4 + Ca(OH)_2 \rightarrow Mg(OH)_2\downarrow + CaSO_4 \qquad (5)$$

$$CaSO_4 + Na_2CO_3 \rightarrow CaCO_3\downarrow + Na_2SO_4 \qquad (6)$$

At equilibrium, the lime-soda effluent contains ca 35 mg/L calcium and 30 mg/L magnesium, both as $CaCO_3$; but the concentration of the effluent produced in practice depends upon the chemical dosages, pH, time of reaction, and water temperature. Lime softening at ambient temperature, or cold-lime softening, is used for plant water supplies. Side-stream treatment of recirculating, cooling-tower water proceeds at the somewhat higher temperatures of water returning from condensers and other heat exchangers. Hot-lime softening is used for boiler-feedwater treatment. This water must be heated in any case, and lime softening is more effective at higher temperatures. The hot-lime effluent is usually passed through a sodium-cycle cation exchanger to complete the removal of hardness. Lime treatment also reduces bicarbonate alkalinity, phosphate, iron, and manganese by precipitation and silica by adsorption on the freshly precipitated magnesium hydroxide.

A variety of ion-exchange cycles is used for the partial or complete removal of some or all dissolved salts from water (9). The most widely used ion-exchange process is zeolite water softening, which involves the replacement of the calcium and magnesium in the water with sodium by passage of the water through salt-regenerated cation exchangers.

$$Ca^{2+} + Na_2R \rightleftharpoons 2\ Na^+ + CaR \qquad (7)$$

Although the calcium and magnesium and any other multivalent cations present are substantially removed by this process, the total dissolved solids may actually increase because the equivalent weight of sodium is higher than the weights of calcium and magnesium (see also Molecular sieves).

Total dissolved solids can be reduced by passage of water through acid-regenerated cation-exchange resins, followed by any one of various processes used for neutralization of the acidic effluent. Aeration or vacuum degasification removes most of the carbonic acid produced and the residual hydrochloric, sulfuric, and other mineral acids can then be neutralized by the addition of alkali. The resulting water is completely soft and has a total dissolved-solids content equal to the sodium salts of the mineral acids. Essentially the same results can be achieved by blending the acidic cation-exchange effluent with some of the same water that has been softened by passage through a salt-regenerated cation exchanger. The ratio of the two waters is selected so that the sodium bicarbonate content of the softened water is sufficient to neutralize the mineral acid in the acidic water. If a completely soft water is not required, the acidic effluent can be blended with enough raw water so that the acidity is neutralized by the alkalinity of the supply water.

For complete removal of the dissolved solids, the acidic cation-exchange effluent can be passed through an alkali-regenerated anion exchanger. The degree of removal of the dissolved solids depends upon the types of cation- and anion-exchange resins used and the process techniques selected. Acid-regenerated carboxylic acid cation-exchange resins convert most of the alkalinity to carbonic acid but do not exchange the cations equivalent to the salts of stronger acids, as acid-regenerated sulfonic acid cation-exchange resins do.

The alkalinity of water can be reduced by passing softened water through a salt-regenerated anion-exchange resin. This process is used to prepare make-up water for low pressure boilers from supplies that are relatively alkaline. Acid is not required for regeneration and the effluent produced does not create the serious condensate line corrosion problems caused by the carbon dioxide in the steam which is characteristic of high alkalinity make-up waters.

Membrane-separation processes are based on synthetic organic membranes that have pore sizes small enough to prevent or impede the passage of colloidal particles or ions. Electrodialysis, reverse osmosis, and ultrafiltration are membrane processes used for reduction or removal of dissolved solids from industrial waters (10). Ion-exchange membranes, which are alternate layers of cation- and anion-permeable membranes, operate under an applied electric current and can be used to deionize water by electrodialysis. The process is used to treat brackish waters and produces a continuous flow of low solids product and enriched-solids waste. In reverse osmosis, pressure is applied to the supply water on one side of a semipermeable membrane to reverse the normal flow of water from dilute to concentrated solution. Suitable membranes are available in sheet or hollow-fiber form and in a variety of mechanical modules. Depending upon the solids concentration in the supply water and the operating conditions, salt removals of 85–99% are obtainable by reverse osmosis.

Evaporation was the earliest process used for reducing the dissolved-solids content of waters. It is still widely used, although the increased cost of fossil fuels makes it less and less attractive economically. Simple distillation suffices for small-volume applications, eg, in laboratories, but larger-volume uses, particularly desalting of seawater and brackish waters, require more efficient processes, eg, vapor-compression distillation and multiple-effect evaporation. Modifications include long vertical tube units and flash-distillation units.

Dissolved gases can be removed by either physical (external) or chemical (internal) methods. Among the former, thermal or vacuum degasifiers, commonly called deaerators, are used for the removal of all gases dissolved in water in accordance with Henry's Law. They are used most frequently for reducing dissolved oxygen and carbon dioxide, but the removal is never complete. Commercial, thermal, ie, steam-heated, deaerators are guaranteed to reduce residual oxygen to ca 7 mg/L (27 mg/gal). Vacuum deaerators can reduce residual oxygen to ca 200 mg/L (760 mg/gal) and dissolved carbon dioxide to less than 10 mg/L, depending upon the original gas concentration, the water temperature, and the deaerator design.

Aeration is widely used for the removal of dissolved gases other than oxygen, particularly carbon dioxide, hydrogen sulfide, and methane. The equipment is called an aerator, degasifier, or decarbonator and usually resembles a cooling tower in principle. The entering water cascades downward through a column containing splash bars or other fill which breaks up the water to expose the maximum surface to an upward flow of air. As with vacuum degasifiers, the gas removal depends upon the original composition of the water, the water temperature, and the aerator design characteristics, but the treated water is saturated with oxygen.

The broad range of changes in water composition attainable by the use of different external water-treatment processes is implied in Table 6, which shows the effluent water analyses produced through the treatment of a single water supply by ten alternative external treatments.

Internal. In contrast to external treatment, most internal water-treatment processes are directed toward modification of the water quality within a system in order to minimize deposit formation, corrosion, and biological growths rather than toward the reduction or removal of constituents that cause these problems. One of the few exceptions is the treatment of boiler feedwater with sulfite or hydrazine for removal of residual dissolved oxygen by chemical reduction. Although the following sections separately discuss treatments for the control of deposits, corrosion, and biological growths, in practice these are not handled separately. Internal treatment formulations for most industrial water systems usually include chemicals for both deposit and corrosion control, and these chemicals must be compatible with those added for biological control when this is required.

Internal treatment for deposit control is accomplished by means of a variety of chemical additives, which act as precipitants, chelating agents (qv), dispersants, or scale inhibitors. Precipitants and chelating agents are used usually with dispersants for internal boiler-water treatment (12). The precipitation of calcium as the readily dispersable hydroxyapatite ($3Ca_3(PO_4)_2.Ca(OH)_2$) is probably the most widely used internal treatment for boilers. Other boiler-water treatments prevent scale formation by the chelation of calcium and magnesium with sodium salts of ethylenediaminetetraacetic acid (EDTA) or nitrilotriacetic acid (NTA). In cooling and process waters, deposits are usually controlled with dispersants and scale inhibitors (13). Among the most widely used dispersants for both boiler- and cooling-water treatment are a number of polymers and copolymers, eg, poly(acrylic acid) and its salts, acrylamide–acrylic acid copolymers, poly(maleic acid), sulfonated polymers, and many others. In general, polymers with molecular weights below ca 20,000 are used as dispersants, whereas those with molecular weights of 10^6 or more are used as flocculants. Natural products and chemically modified natural products, eg, tannins, lignosulfonates, and carboxymethyl cellulose, are also used as dispersants; however, these are being replaced

Table 6. Changes in Composition Produced Through Treatment of a Single Sample of Raw Water by Ten Alternative External-Treatment Processes, mg/L[a,b]

Constituent	Raw, untreated water	Cationic polymer and filtration	Alum clarification and filtration	Sodium-cycle cation exchange[c]	Weakly acidic cation exchange[c]	Deionization with individual exchanger beds[c]	Deionization with mixed-bed polisher[c]	Cold lime	Warm lime, 49°C	Hot lime-soda	Hot lime-zeolite
P[d] alkalinity, as CaCO$_3$	0	0	0	0	0	0	0	35	35	40	40
M[e] alkalinity, as CaCO$_3$	120	120	93	93	0	4	3	70	70	75	75
chloride	48	51	51	51	51	0	0	51	51	51	51
calcium, as CaCO$_3$	135	135	135	0	49	0	0	50	30	10	0
magnesium, as CaCO$_3$	53	53	53	0	19	0	0	48	20	10	0
hardness, as CaCO$_3$	188	188	188	<0.1	68	0	0	98	51	20	<0.1
phosphate	1	1	1	1	1	0	0	0.5	0.1	0.1	0.1
sodium	29	29	29	116	29	0.4	0.01	29	113	113	122
silica	7	7	7	7	7	0.02	0.005	7	2	1	1
sulfate	84	84	97	84	84	0	0	84	84	84	84
nitrate	13	13	13	13	13	0	0	13	13	13	13
total dissolved solids	307	307	315	307	187	2	0.03	267	214	204	204
suspended solids	50	0.5	0.5	0.5	0.5	0.1	0.1	0.5	0.5	0.5	0.2
pH (at 25°C)	7.5	7.5	6.5	7.5	4.3	9.0	7.2	10.1	10.1	10.1	10.1

[a] Ref. 11.
[b] To convert mg/L to mg/gal, multiply by 3.785.
[c] After pretreatment by polymer and filtration.
[d] P = phenolphthalein (measures [OH$^-$]).
[e] M = methyl orange (measures [HCO$_3^-$] as well as [OH$^-$]).

by polymers that can perform the same job at much lower concentrations or can be tailored for the dispersion of specific foulants.

Some polymers also act as inhibitors for crystalline scales, eg, calcium carbonate; these include the polyphosphates, phosphonates, and phosphate esters. Many of these compounds are more effective inhibiting one type of crystalline scale than other types. The [nitrilotris(methylene)]trisphosphonates are more effective in preventing calcium carbonate scales, whereas [1,2-ethanediylbis[nitrilobis(methylene)]]tetrakisphosphonates are more effective in the prevention of calcium sulfate scales (see Phosphorus compounds).

Calcium carbonate, the most common scale in water systems, can also be minimized in evaporative recirculating cooling systems by the carefully controlled addition of an acid, usually sulfuric acid, which reduces the alkalinity of the circulating water. In evaporative systems, eg, boilers and cooling-water systems, the prevention of scale formation requires limiting the total dissolved solids of the water in the system as well as adding a dispersant and an acid. The solids are limited to a predetermined value by controlled continuous bleeding or intermittent blowdown of a small portion of the circulating water. The maximum dissolved-solids concentration to be maintained in such a system is estimated from the make-up water composition, temperatures of heat-exchange surfaces, and other factors by means of the Langelier Saturation Index (14–15).

Corrosion control by internal treatment involves one or more of the following techniques: chemical reduction of dissolved oxygen, pH adjustment, protective film formation with corrosion inhibitors, and dispersion of suspended solids. Chemical oxygen scavengers are used primarily in boiler-water treatment. For the most part, either sodium sulfite or hydrazine is used. Because of their expense, they are used only for removal of the residual oxygen remaining in boiler feedwater after mechanical deaeration. To a small extent, these oxygen scavengers are also used for controlling corrosion in closed, recirculating water systems.

Increased pH reduces the overall corrosion rates of most metals in industrial water, but pH adjustment by itself is almost never sufficient to attain acceptably low corrosion rates. An exception to this is the treatment of boiler waters with volatile amines, eg, cyclohexylamine, morpholine, or diethylaminoethanol. These distill from the boiler with the steam and dissolve in the steam condensate, where they neutralize carbon dioxide and increase the pH sufficiently to minimize acidic corrosion of condensate piping (16). With many corrosion-inhibitor systems, the pH must be reduced and controlled within a narrow range to achieve optimum corrosion control.

Treatment with corrosion inhibitors is widely practiced for controlling corrosion in cooling-water systems and closed recirculating systems. Corrosion inhibitors may function by interfering with the anodic reaction in a corrosion cell, the cathodic reaction, or both. Historically, the chromates have been the most widely used and effective corrosion inhibitors for industrial water systems. They protect all metals used in these systems, are effective over a broad pH range, provide some microbicidal activity, and permit rapid restoration of damaged protective films. Unfortunately, chromates are environmentally objectionable, so that their use is limited. Other corrosion inhibitors used in industrial water systems include polyphosphates, orthophosphates, zinc compounds, nitrites, phosphate esters, phosphonates, azoles, molybdates, and silicates.

Many attempts have been made to apply the theoretically and environmentally attractive approach of adjusting the water composition in accordance with the Langelier Saturation Index so as to deposit an egg-shell layer of calcium carbonate, which would protect the underlying metal from corrosion without interfering excessively with heat transfer or water flow. These attempts have been generally unsuccessful for two reasons. First, most industrial water systems operate at different temperatures at different parts of the system, so that adjustment of the water composition to protect the metal at one water temperature leaves other sections unprotected or subject to heavy scale deposition. Second, surface and water conditions often lead to the deposition of a porous rather than an egg-shell layer of calcium carbonate, which significantly reduces heat transfer and permits corrosion under the deposit.

Chemical treatment for inhibiting corrosion in industrial water systems usually includes several components. Two or more different corrosion inhibitors may be included for their synergistic effect in reducing the corrosion rates of mild steel as well as requiring low inhibitor concentrations. Among the widely used inhibitor mixtures for recirculating cooling waters are chromate with polyphosphate, zinc, or both; zinc with polyphosphate, phosphonate, molybdate, or lignin derivatives; and polyphosphate with phosphonate or silicate. Many of the combinations require pH control within 1–1.5 pH units (usually 6.5–7.5 pH) to obtain the desired minimum corrosion rate or to prevent precipitation of zinc. Most formulations include a polymeric dispersant, eg, a polyacrylate, which keeps sediment, scales, corrosion products, and organic particles in suspension, thereby minimizing the risk of pitting, corrosion under the deposit, or other forms of localized corrosion. With these dispersants present, the protection of nonferrous metals sometimes decreases and specific corrosion inhibitors for these metals are then included. Such inhibitors include benzotriazole, tolyltriazole, and sodium 2-mercaptobenzothiazole.

Corrosion inhibitors are widely used for the protection of low pressure steam-heating boilers and hot-water heating systems. Borax-buffered sodium nitrite mixtures, which usually include one of the above inhibitors for nonferrous metal applications, are widely used. Much corrosion damage to steam-heating boilers occurs during seasonal disuse. Steam-condensate piping is protected against oxygen corrosion by the addition of a long-chain aliphatic amine, usually octadecylamine, to the boiler water or steam. The amine is carried over with the steam and, on condensation, forms a thin hydrophobic film which restricts access of the corrosive condensate to the metal surface.

Biological growths are usually controlled in industrial water systems by treatment of the water with oxidizing or nonoxidizing biocides, although several mechanical procedures are used, particularly for cooling-water systems (17–18). The latter procedures include automated, direct cleaning of condenser or heat-exchanger tubes for removal of microbiological slimes, installation of opaque covers on head basins of cooling towers to prevent algae growth, and periodic reversal or discontinuance of water flow to raise water temperatures high enough to kill organisms. Although such microorganisms as bacteria, fungi, and algae are the most widespread causes of biological problems, macroorganisms cause severe problems in some systems. Of particular importance are barnacles, mussels, and bryozoa in seawater cooling systems, and asiatic clams (*corbicula*) in once-through freshwater cooling systems.

Chlorine, an oxidizing biocide, is by far the most widely used water-system biocide. Depending upon the type, size, and location of the system, it may be added as chlorine gas, sodium or calcium hypochlorite, or various chlorine-yielding N-chloro compounds, eg, sodium dichloroisocyanurate. The frequency and dosage of chlorination varies, depending upon the type of water system, chlorine demand of the water, and other technical and regulatory site-specific factors.

Concerns about the environmental effects of chlorine have led to regulatory restrictions on the chlorination of some industrial waters. Other, more expensive oxidizing biocides have been substituted for chlorine. Although bromine is too difficult to handle for these applications, several bromine-yielding products are used. These include bromine chloride, 1-bromo-3-chloro-5,5-dimethylhydantoin, and 2,2-dibromo-3-nitrilopropionamide. Chlorine dioxide, despite its much higher cost, has replaced chlorine in many food-processing and cooling-water systems because its microbiocidal activity is virtually unaffected by pH, it does not react with ammonia, it does not create off-tastes by reaction with phenolic compounds, it does not form trihalomethanes by reaction with lignin derivatives, and it forms fewer chlorinated organic compounds (see Chlorine oxygen acids and salts). Ozone (qv) has also been used as a microbiocide in industrial waters. Nonoxidizing biocides for industrial waters include a wide variety of organic compounds. Environmental considerations have ruled out the use of such inorganics as mercury and arsenic compounds.

Quaternary ammonium compounds (qv) are the largest class of nonoxidizing microbiocides used in water systems. Most of them used for this purpose have 8–25 carbon chains with aromatic rings present in some. The surfactant properties of the quaternaries provide an advantage over other microbiocides, in that they tend to loosen and penetrate existing slime accumulations, thereby permitting more rapid and effective access of the biocide to encapsulated microorganisms.

A number of sulfur-containing microbiocides are used. Methylenebisthiocyanate is widely used in cooling systems to control various fungi, algae, and bacteria. It is particularly effective against sulfate-reducing bacteria (*desulfovibrio*), which seriously corrode some of these systems. Other sulfur-containing microbiocides include bis-(trichloromethyl) sulfone, tetrahydro-3,5-dimethyl-2H-1,3,5-thiadiazine-2-thione, disodium ethylenebisdithiocarbamate, sodium dimethyldithiocarbamate, 5-chloro-2-methyl-4-isothiazolin-3-one, and 2-methyl-4-isothiazolin-3-one. Bis(tri-n-butyltin) oxide is frequently used in mixtures with quaternary ammonium compounds. Sodium tri- and pentachlorophenates and acrolein, once widely used as cooling-water microbiocides, have largely been replaced because of their undesirable environmental effects.

Most commercially formulated products include several microbiocides combined with dispersants and surfactants which aid in breaking up accumulations of slimes or algae. In contrast to chemicals used for deposit or corrosion control, microbiocides, particularly the nonoxidizing type, are added intermittently. This is done in part for economic reasons and in part to minimize any tendency for organisms to become resistant to the chemicals. For the latter reason, it is common practice to alternate two different types of biocides in industrial water systems.

Programs. Water treatment for an industrial plant frequently includes various external and internal treatment processes. The total plant water supply may be clarified and filtered. A part of this treated water may be used without further treatment for general plant uses. Another part may be used as make-up water to systems, eg, cooling systems, within which it is treated internally for the control of deposit formation, corrosion, and biological growths. Still other portions may be softened, deionized, or otherwise treated, externally or internally, to meet the quality or performance requirements for use as boiler feed or in specific processes.

The development of an effective treatment program for an industrial water system requires evaluation of the potential water-caused problems in the case of a new system or of the problems that have occurred in an existing system. As complete information as possible is needed about the design, construction, and operation of all water-using equipment; about the composition and range of composition variation of the water supplied to the system; and about the changes in composition likely to occur within the system during operation. For process waters, information is needed on established water-quality requirements for the process and the nature of the products that come into contact with the water.

A complete water-treatment program includes equipment, instrumentation, chemicals, and instructions. It should provide for feeding treatment chemicals, testing the make-up and treated waters, controlling the treatment process, and monitoring the effectiveness of the treatment. In addition to treatment for control of deposits, corrosion, and biological growths, chemicals are usually required for cleaning new or fouled systems, preconditioning of metal surfaces, and water testing. Written instructions and training of operators and supervisors should be provided with regard to the safe handling of treatment chemicals; startup, operation, maintenance, and shutdown of equipment and instruments; sampling and testing of water; maintaining records of treatment and operations; controlling treatment, including recognizing and responding to deviations, upsets, and emergencies; and monitoring the effectiveness of the program.

Economic Aspects

Shipments of chemicals for the treatment of the 333×10^9 m^3 (88×10^{12} gal) of water used for industrial and municipal applications in the United States in 1980 totaled $>10 \times 10^6$ t with a value of $\$1.8 \times 10^9$ (3). Table 7 summarizes the consumption and value of the principal classes of chemicals used for water treatment in the United States during 1980.

Health and Safety Factors

Almost every chemical used for water treatment can be harmful under some circumstances. Therefore, operating personnel who handle the full-strength treatment chemicals should be fully informed of potential hazards, provided with necessary personal protective equipment, and trained in the safe handling of each chemical used. Physical facilities for safe handling and use of treatment chemicals should also be provided. Information on potential hazards and safe-handling procedures is available from product labels, manufacturers' material safety data sheets, and other sources (19–20).

Many chemicals added for internal treatment remain in the water when the latter is ultimately discharged from the system and from the industrial plant in which it is used. Therefore, consideration must be given to the effects of the chemicals of choice on the operation of the plant's wastewater-treatment system and their environmental acceptability in the plant's wastewater discharge. The wastewater discharge must meet all national, state, and local regulatory requirements. In the United States, wastewater-discharges permits issued by the EPA or the individual states define the limitations on the total wastewater leaving a point source, regardless of the in-plant systems from

Table 7. Water-Treatment Chemical Use in the United States During 1980[a]

Chemical	Amount used, t	Cost 10^6 $	$/kg
Based on total water use[b]			
acids	413,000	32.8	0.08
activated carbon	47,600	56.7	1.19
alkalies	3,180,000	168.0	0.05
biocides			
chlorine	590,000	95.0	0.16
other inorganic chlorine compounds	125,000	160.0	1.28
other biocides	93,000	144.0	1.55
coagulants	1,160,000	240.0	0.20
filter media			
diatonite and perlite	127,000	19.6	0.15
others	218,000	26.4	0.12
ion-exchange resins	43,000	123.5	2.87
Based on industrial water use[c]			
boiler-treatment compounds	144,000	172.0	1.19
coagulants	371,000	76.7	0.20
cooling-water treatment compounds	170,000	290.0	1.71
process-water treatment compounds	128,000	182.0	1.42

[a] Ref. 3.
[b] 333×10^9 m^3 (88×10^{12} gal).
[c] 281×10^9 m^3 (74×10^{12} gal).

which the water came. An exception is for the steam-electric utility industry, for which composition standards apply individually to discharges of cooling water (once-through or cooling-tower blowdown), boiler blowdown, ash-transport water, metal-cleaning wastes, and other water uses regardless of whether or not they are combined in the plant before discharge (21).

BIBLIOGRAPHY

"Water (Industrial)" in *ECT* 1st ed., Vol. 14, pp. 926–946, by E. P. Partridge, Hall Laboratories, Inc.; "Water, Industrial Water Treatment" in *ECT* 2nd ed., Vol. 22, pp. 65–82, by J. K. Rice and D. E. Simon II, Cyrus Wm. Rice Division, NUS Corporation.

1. *1977 Census of Manufacturers*, Vol. 1, *Subject Statistics, Water Use in Manufacturing*, Bureau of the Census, U.S. Department of Commerce, Washington, D.C., 1977, pp. 8-36–8-41, 8-58–8-63.
2. *Cooling Tower Manual, Water Chemistry*, Cooling Tower Institute, Houston, Texas, Jan. 1981, Chapt. 6, pp. 6.18–6.22.
3. A. C. Gross, *Industry Study 244—Water Treatment Chemicals*, Predicasts, Inc., Cleveland, Ohio, Aug. 1981, 177 pp.
4. D. E. Simon, *Proceedings of the 36th International Water Conference*, Engineers' Society of Western Pennsylvania, Pittsburgh, Pa., Nov. 1975, pp. 65–69.
5. *Standard Methods for the Examination of Water and Wastewater*, 15th ed., American Public Health Association, Washington, D.C., p. 193.
6. Ref. 5, p. 253.
7. S. Sussman, *Heat./Piping/Air Cond.*, 103 (April 1965).
8. S. Sussman, *7th Annual Liberty Bell Corrosion Course*, Philadelphia, Pa., Sept. 1969, p. IV–11.
9. S. B. Applebaum, *Demineralization by Ion Exchange*, Academic Press, New York, 1968, 389 pp.

10. R. E. Lacey, *Chem. Eng.*, 56 (Sept. 4, 1972).
11. A. S. Krisher, *Chem. Eng.*, 79 (Aug. 28, 1978).
12. J. W. McCoy, *The Chemical Treatment of Boiler Water*, Chemical Publishing Co., New York, 1981, 301 pp.
13. J. W. McCoy, *The Chemical Treatment of Cooling Water*, Chemical Publishing Co., New York, 1974, 237 pp.
14. F. N. Kemmer, ed., *The Nalco Water Handbook*, McGraw-Hill Book Co., New York, 1979, pp. 4-13–4-17.
15. *Betz Handbook of Industrial Water Conditioning*, 8th ed., Betz Laboratories, Inc., Trevose, Pa., 1980, pp. 177–179.
16. J. S. Beecher, *Power*, 74 (July 1981).
17. L. D. Jenson, ed., *Biofouling Control Procedures*, Marcel Dekker, Inc., New York, 1977, 113 pp.
18. D. S. Richardson, *Chem. Eng.*, 103 (Dec. 13, 1982).
19. F. K. Kinoshita, *Toxicology of Cooling Water Treatment Chemicals*, Paper No. TP207A, Cooling Tower Institute, Houston, Texas, 1979, 15 pp.
20. *Handling Water Treatment Chemicals Safely*, Publication WTP-129, Cooling Tower Institute, Houston, Texas, 1982, 1 p.
21. 40 CRF 423.

General References

References 2, 5, 9, 12, 13, 14, and 15 are general references.
Drew Principles of Industrial Water Treatment, 1st ed., Drew Chemical Corp., Boonton, N.J., 1977, 310 pp.
P. Hamer, J. Jackson, and E. F. Thurston, *Industrial Water Treatment Practice*, Butterworths, London, 1961, 514 pp.
C. E. Hamilton, ed., *Manual on Water*, ASTM Special Publication 442A, American Society for Testing and Materials, Philadelphia, Pa., 1978, 472 pp.
J. W. McCoy, *Microbiology of Cooling Water*, Chemical Publishing Co., New York, 1980, 249 pp.
A. G. Ostroff, *Introduction to Oilfield Water Technology*, 2nd ed., National Association of Corrosion Engineers, Houston, Texas, 1979, 394 pp.; broader coverage of industrial water treatment than suggested by title.

S. SUSSMAN
Olin Water Services

MUNICIPAL WATER TREATMENT

Although the human physiological need for water is less than two liters per day (0.5 gal/d), the average production and distribution of potable water through public water supplies in the United States is ca 600 liters per capita per day (158 gal/d) (1). This amount is used for household and sanitary needs as well as urban, industrial, and municipal requirements (see Water, industrial water treatment). The average per-capita cost of water in the United States is ca $0.26/m^3 ($0.98/1000 gal).

Water Quality

The principal sources of water are surface supplies, eg, streams, rivers, lakes, reservoirs, and groundwater supplies, ie, those obtained from deep wells. The general properties of the two sources are listed in Table 1. Selection of a supply source is based on the specific properties of the supplies in question and local availability.

Until the passage of the Federal Safe Drinking Water Act (SDWA) (Public Law 93-523), the states regulated the quality of the water furnished to water utility customers. The SDWA required that the EPA promulgate interim water-quality regulations, which would be subject to revision upon recommendation of the NAS. The Act requires the EPA to establish recommended maximum contaminant levels (MCLs) at which no known or anticipated adverse effects on the health of persons occur and which allow an adequate margin of safety. The interim regulations were to protect health by means of generally available technology and were based partly on cost and other factors.

The National Interim Primary Drinking Water Regulations (NIPDWR) were promulgated by the EPA and became effective nationwide in June 1977 (2). Additional regulations on radioactivity in water were set forth in 1976 and, in November 1979, trihalomethane (THM) content was regulated (3–4). Table 2 shows a summary of these water standards. States were required to adopt regulations no less stringent than the NIPDWR in order to retain regulatory control over the quality of water provided by the public water supplies in their state.

Secondary regulations were promulgated in July 1979 and relate to the effect of water quality on the public welfare, but not health, as provided in the NIPDWR (6)

Table 1. General Properties of Natural Water Supplies

Property	Surface	Groundwater
mineralization	low	high
dissolved oxygen	usually saturated	very low
H$_2$S	absent, unless polluted	may be present
pollution	common	uncommon
color	common	uncommon
turbidity	common	uncommon
Fe and Mn	uncommon	common
quantity	variable unless impounded	constant
organics	variable	variable

386 WATER (MUNICIPAL TREATMENT)

Table 2. National Interim Primary Drinking-Water Regulations[a]

Parameter	Maximum contaminant level
inorganics, mg/L[b]	
arsenic	0.05
barium	1.0
cadmium	0.010
chromium	0.05
lead	0.05
mercury	0.002
nitrate (as N)	10.0
selenium	0.01
fluoride	1.4–2.4[c]
organics (pesticides), mg/L[b]	
endrin	0.0002
lindane	0.004
methoxychlor	0.1
toxaphene	0.005
(2,4-dichlorophenoxy)acetic acid	0.1
(2,4,5-dichlorophenoxy)acetic acid	0.01
turbidity, ntu[d]	1–5[d]
microbiological[e]	
radioactivity, Bq/L[f]	
gross alpha	0.56
^{226}Ra and ^{228}Ra	0.19
gross beta	1.85
trihalomethanes, mg/L[b]	0.10[g]

[a] Refs. 2–4.

[b] To convert L to gal, divide by 3.785; to convert mg/L to lb/10^6 gal, multiply by 8.34.

[c] A function of the yearly average of the maximum daily air temperature.

[d] ntu = nephelometric turbidity unit; an arbitrary unit based on the standard instrument used (see Turbidity). The turbidity is measured at an entry point to the distribution system; the allowed turbidity level is a function of the effect of turbidity on disinfection. See also ref. 5.

[e] Levels depend on the analytical method used, and the frequency of sampling depends upon the size of utility.

[f] To convert Bq/L to pCi/gal, multiply by 102.3.

[g] Applicable to utilities serving over 10,000 people.

(see Table 3). These were required to be adopted by the states, but enforcement was not required.

Several changes were instituted in all water-quality regulations by the requirements of the Safe Drinking Water Act. Among these were

1. The quality had to be measured at the free-flowing tap of the consumer.

2. Turbidity became a health-related parameter because of its interference with disinfection or with the maintenance of a disinfectant residual.

3. Violations of the NIPDWR had to be reported to the public.

4. Many smaller systems were regulated as public water supplies. Included were all supplies that served an average of over 25 persons or 15 connections for more than 60 d/yr.

Many water suppliers were required to improve their treatment processes in order to satisfy the regulations. Those mainly affected were the smaller systems serving less than 100 consumers. Thus, the economic impact was greatest on these utilities.

Table 3. Secondary Regulations Promulgated in 1979 by the EPA[a]

Parameter	Secondary maximum contaminant level
chloride, mg/L	250
color, units	15
copper, mg/L	1
corrosivity	noncorrosive
foaming agents, mg/L	0.5
iron, mg/L	0.3
manganese, mg/L	0.05
pH	6.5–8.5
odor, threshold odor number	3
sulfate, mg/L	250
total dissolved solids, mg/L	500
zinc, mg/L	5

[a] Ref. 6.

In 1982, regulations were developed to control the levels of volatile organic compounds in potable water supplies (7). This resulted from the unanticipated finding that many groundwater supplies in the United States were contaminated with industrial solvents that could affect human health. The main contaminants were the chlorinated hydrocarbons, eg, carbon tetrachloride, trichloroethylene, the dichloroethylenes, tri- and dichloroethanes, and vinyl chloride.

Despite the widespread contamination noted above, there has been an increasing trend toward the use of available groundwaters, primarily because of their lower pollutant content and the assurance of constant supply. The main problem is the low rate at which supplies are replenished. Historically, surface supplies have been favored as they are more readily available. The ratio of treatment plants that use surface water to those that use groundwater is ca 2:5 for the 60,000 community water supplies in the United States. However, it is the larger plants that generally use surface-water sources.

Treatment Methods

The choice of which alternative sources of water to use for municipal consumption must be based on such factors as available quantity, continuity of availability, treatment and capital costs, and assurance that the chosen treatment process will provide water that is biologically safe, noncorrosive and nondamaging to plumbing fixtures and water heaters and containers, and aesthetically acceptable (8). The choice of treatment is a function of the raw-water quality. Most surface waters must be treated for the removal of suspended solids. These solids include clay, protozoan cysts, and organic debris resulting from partial breakdown of leaves and other plant materials. Some surface waters contain organic color, which is usually removed for aesthetic reasons. In addition, many utilities must either remove the natural organic compounds, which are precursors for trihalomethanes (THMs); use a disinfectant other than chlorine; remove the THMs formed; or modify their treatment scheme to reduce THM formation. Some groundwater supplies also contain such precursors. Waters obtained from groundwaters or impounded supplies may contain reduced iron and manganese, which must be removed to prevent precipitation of the oxidized forms in the distribution system or at the point of use.

A complete sequence of treatment for a municipal plant includes application of copper sulfate to a reservoir or lake for algae control; addition of activated carbon and chlorine at the head of the plant for taste, odor, and bacteria control; addition of a coagulant in a rapid mix tank for turbidity removal; passage through a flocculation tank to promote floc growth; passage through a horizontal or vertical sedimentation tank; pH adjustment; and gravity filtration (see Flocculating agents; Filtration). The water from the filters, after the addition of chlorine, is usually stored in a clearwell until it is pumped to a service or storage tank. Some or all of the preceding methods may be used in any given case for a turbid surface water. If the water is also hard or is hard rather than turbid, it can be softened by lime, lime-soda, or an ion-exchange process in addition to or instead of the coagulation–flocculation–sedimentation steps prior to filtration, chlorination, and distribution (see Ion exchange). The types of treatment used by the 100 largest U.S. cities are summarized in Table 4 (9).

Since the SDWA was passed in 1974, emphasis in treatment has been on the control of trace organics in public water supplies. The discovery that chloroform and the other bromo and chloro haloforms form as results of disinfection of water with chlorine led to consideration of treatment-process modification, alternative disinfectants, increased emphasis on the removal of natural organics, and removal of the haloforms after they form. Various of these alternatives have been included in treatment schemes.

Disinfection. As shown in Table 4, most of the largest plants in the United States practice disinfection and usually by chlorination (see Disinfectants and antiseptics). Disinfection is the destruction of pathogenic organisms. It is not uncommon for small plants to distribute untreated and undisinfected water from uncontaminated deep wells. This practice can lead to serious consequences in the event of a cross-connection or other type of contamination of the well or distribution system, as occurred in 1965 in Riverside, California, resulting in 18,000 cases of gastroenteritis (10). The water was not disinfected because it was from deep wells and was considered bacteriologically safe. The absence of a residual disinfectant led to distribution of contaminated water. In 1973, a water supply in a labor camp in Homestead, Florida, became contaminated, leading to 210 cases of typhoid fever. Inadequate disinfection and use of a shallow well

Table 4. Treatment Processes Used by Municipal Plants in the 100 Largest U.S. Cities[a]

Process	Number of cities	Percentage of total population[b]
chlorination	98	98.8
sedimentation and coagulation	68	63.8
filtration	76	56.8
softening	28	18.1
iron and manganese removal	10	5.6
no treatment	1	0.1
fluoridation[c]	34	35.0

[a] Ref. 9.

[b] The total population as determined by the 1960 census was 60,000,000.

[c] At the end of 1979, the use of fluoridation had increased appreciably as several large cities started fluoride addition, including New York City, Atlanta, and Seattle.

were identified as the causes (11). Disinfection of a public water supply was first practiced in the United States in 1908 (12). The agent used was a solution of calcium hypochlorite.

One of the main advantages of chlorination over such disinfectants as ozone, radiation, gamma rays, etc, is that free, available chlorine residual, ie, HOCl and OCl$^-$, can be maintained throughout the system. Consequently, the residual can inactivate any contamination that may be introduced after the initial disinfection and the presence of a residual indicates that such contamination has not occurred. When chlorine gas, which is the most common disinfectant in all but the smallest plants, is added to water, the following reactions take place:

$$Cl_2 + H_2O \rightleftharpoons H^+ + Cl^- + HOCl \tag{1}$$

$$HOCl \rightleftharpoons H^+ + OCl^- \tag{2}$$

The equilibrium constant for equation 1 is 4.5×10^{-4} at 25°C and the K_a for equation 2 is 2.5×10^{-8} at 25°C. Figure 1 shows the distribution of chlorine species as a function of pH based on these constants. The bactericidal-active forms are Cl_2 and HOCl; thus, at pH values above 7.6, less than 50% of the total chlorine is available in an effective form. Since many potable waters are treated and distributed at pH values greater than 7.6, it is often necessary that a high free residual be maintained in the water to provide adequate protection. There must be a least 0.2 mg/L (0.76 mg/gal) of free available chlorine residual in the distribution system. The total residual chlorine includes free and combined residuals; the latter results principally from chlorine that has reacted with ammonia or amines to produce chloramines. The reactions may be illustrated by ammonia:

$$HOCl + NH_3 \rightleftharpoons NH_2Cl + H_2O \tag{3}$$
$$\text{monochloramine}$$

$$NH_2Cl + HOCl \rightleftharpoons NHCl_2 + H_2O \tag{4}$$
$$\text{dichloramine}$$

$$NHCl_2 + HOCl \rightleftharpoons NCl_3 + H_2O \tag{5}$$
$$\text{trichloramine}$$
$$\text{(nitrogen trichloride)}$$

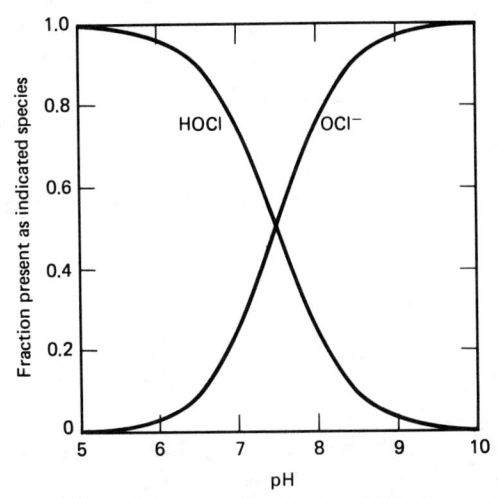

Figure 1. Effect of pH on the distribution of chlorine species (13).

Chloramines, owing to their reduced oxidation potential, are much less effective disinfectants than hypochlorous acid. They are frequently used as disinfectants, however, in systems where it is difficult or impossible to maintain a free available chlorine residual because of associated taste and odor problems (13). In these cases, ammonia or an ammonium salt is added to the plant effluent after disinfection with Cl_2 to provide a combined residual in the system. The reduced effectiveness requires that there be a higher residual content, ie, up to 3 mg/L (11.4 mg/gal) (see Chloramines and bromamines).

Another advantage to the use of chloramines is that they do not react with the natural organic compounds to form THMs. Many treatment plants that use a strong oxidant to reduce color or a strong disinfectant to assure bacteriological safety have used a free chlorine residual for a limited time followed by the addition of ammonia or an ammonium salt, which reacts with the chlorine converting it to monochloramine.

Chlorine may also serve as an oxidant for H_2S and Fe(II) as well as for various organic impurities. These reactions result in a chlorine demand which must be satisfied before effective disinfection can be accomplished. Figure 2 shows the effect of adding chlorine to a sample containing reducing agents. The first horizontal portion of the curve represents the chlorine demand of the water, ie, that amount of chlorine that reacts with reducing agents. The next portion of the curve represents chlorine reaction with ammonia and its derivatives to yield a combined residual. This is followed by further oxidation to ineffective dichloroamine and nitrogen trichloride until the breakpoint is reached, after which every additional part of chlorine added results in free residual.

In some cases, it is necessary to use large amounts of chlorine to oxidize organic materials that cause tastes and odors or high levels of bacteria. This is referred to as superchlorination. The large excess must be removed before distribution because of the objectionable taste and odor of the chlorine. Dechlorination is accomplished by adding a reducing agent, eg, sulfur dioxide, sodium sulfite, or sodium bisulfite. Chlorine can also be absorbed by passage through activated carbon.

Prechlorination is the addition of chlorine to the raw water prior to any other treatment, and postchlorination is the addition at various points through the plant and possibly again after all other treatments. It is sometimes necessary to add chlorine at locations in the distribution system to maintain a residual throughout the system.

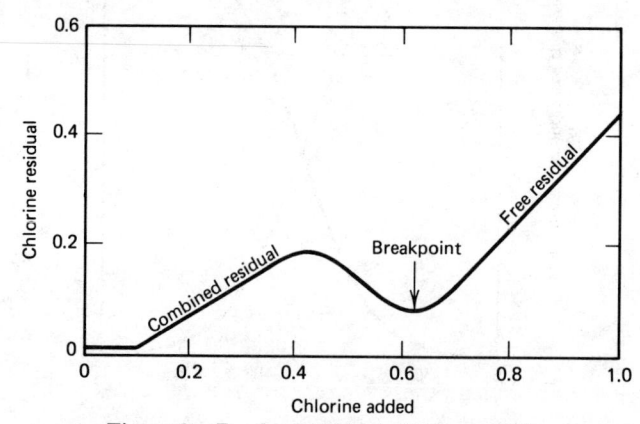

Figure 2. Breakpoint chlorination curve (13).

Chlorine, which is present as either HOCl or OCl$^-$, reacts with some natural organic compounds to produce THMs. The concentrations of the THM precursors correlate approximately with the color and with the total organic carbon content.

Other disinfectants that have been used or proposed include bromine, iodine, ozone, and chlorine dioxide. Ozone (qv) has been used in Europe principally because of consumers' objections to the taste and odor of chlorine residuals (14). However, ozone has not been used to any significant extent in the United States because of its very high oxidation potential, which results in the inability of water to carry any residual. Ozone is produced by the corona discharge of high voltage electricity in air. Since the efficiency is generally less than 1% and the process is energy-intensive, the cost of ozone is high compared to chlorine.

Bromine has been used little, but iodine has been studied extensively and used in some systems (15). It has a lower oxidation potential than chlorine, thus allowing more residual in the system. In addition, it is a good viricide and cysticide and, since it is solid, it is easier to handle than gaseous chlorine, particularly for small plants. Iodine was used extensively for emergency disinfection by the U.S. armed forces (13).

Chlorine dioxide is a strong oxidizing agent and can be produced from sodium chlorite by reaction with chlorine.

$$2\ NaClO_2 + Cl_2 \rightarrow 2\ NaCl + 2\ ClO_2 \qquad (6)$$

Chlorine dioxide has a higher oxidation potential than chlorine at high pH values at which Cl$_2$ is relatively ineffective. Chlorine dioxide has been used, particularly in Europe, for many years to reduce the offensive tastes and odors resulting from chlorine and, in the United States, for the oxidation of Mn^{2+} or Fe^{2+} and the reduction of chlorine-associated tastes and odors (13) (see Chlorine oxygen acids and salts).

The health authorities have been concerned about the residual by-products of chlorine dioxide, ie, the chlorite and chlorate ions. Their physiological effect, primarily methomoglobinemia in infants, has limited the use of ClO$_2$ in the United States (16).

Sedimentation and Filtration

Since most surface waters contain varying amounts of suspended solids, including silt, clay, bacteria, viruses, etc, it is necessary to remove them prior to distribution to the domestic or industrial consumer. They not only affect the acceptability of the water but interfere with disinfection. The principal treatment processes are sedimentation (qv) and filtration (qv). Sedimentation alone is rarely adequate for the clarification of turbid waters and is of little or no value for the removal of such very fine particles as clay, bacteria, etc. Table 5 shows the effect of particle size on the sedimentation rate of a solid having a specific gravity of 2.65 in water at 20°C.

In many plants that treat surface waters, there is a presedimentation reservoir ahead of the treatment units. The reservoir allows the larger particles to settle as well as to provide a volume buffer against changes in water quality. Table 5 shows that the only particles that settle within a reasonable period of time are silt, sand, and possibly some larger bacteria. The listed values are for an undisturbed system and the times would be much longer for actual cases where currents tend to upset the settling. Further treatment, which is necessary to produce potable water, may involve only filtration

Table 5. Sedimentation Rate as a Function of Particle Diameter[a]

Equivalent spherical radius	Approximate size	Sedimentation rate (time to settle 30 cm)
10 mm	gravel	0.3 s
1 mm	coarse sand	3 s
100 μm	fine sand	38 s
10 μm	silt	33 min
1 μm	bacteria	55 h
100 nm	colloid	230 h
10 nm	colloid	6.3 yr
1 nm	colloid	63 yr

[a] Ref. 17.

through sand or multiple-medium filters or may require considerable pretreatment, eg, coagulation and flocculation before filtration.

The use of sand filtration as a method of clarifying water dates to the small sand beds designed for Paisley, Scotland, in 1804 and, later, to the larger sand filters used by the Chelsea Water Company of London in 1828 to filter Thames River water. The first use of sand filters in the United States was in Poughkeepsie, New York, in 1872 (12). These were slow sand filters, ie, the filter rates were low compared to current filter rates. Table 6 compares the properties of slow sand filters with rapid sand filters and high rate filters. Slow sand filters have large surface areas because of the low rates that must be used. Higher rate filters were developed because of the very large surface areas that would be required for the filtration of water for the larger cities. There are slow sand filters still in use in the United States and in Europe (19).

The first patent relating to rapid sand filters was issued in 1884 for the use of ferric chloride as a coagulant ahead of the filters (20). The coagulant duplicates the slimy, gelatinous layer of bacteria, algae, and fungi that built up on the surface of slow sand filters and played an important part in the filtering action. This layer is called the *schmutzdecke* (German, dirty layer). The first large, municipal, rapid sand filter was built at Little Falls, New Jersey, in 1909 (12).

Dual-medium filters for municipal and industrial application are described in refs. 21–22. One of the first designs involved an inverted-medium loading, in which very fine, dense garnet sand was used at the bottom of the filter; this was covered by less dense but coarser silica sand, and then by granulated carbon of even lower density and larger particle size. This type of construction provides larger pore sizes for the initial contact with the water to be filtered and, thereby, allows deeper penetration of smaller particles into the bed and the consequent use of more of the filter medium for filtration. Backwashing returns the filter to the original particle-size distribution because of the density differences of the filter media.

The size of particles removed by such filters is less than the size of the passages. The mechanism of removal includes adsorption of the impurities at the interface between the media and the water either by specific chemical or van der Waals attractions or by electrostatic interaction when the medium particles have surface charges opposite to those on the impurities to be removed.

Neither rapid sand nor mixed-media filters remove appreciable quantities of colloidal particles without adequate pretreatment. Even though it is widely believed

Table 6. Typical Properties of Filter Types[a]

Property	Slow sand	Rapid sand	High rate
design filtration rate, L/(s·m^2)[b]	0.081–0.162	1.36	2.72–6.83
bed size needed to filter 1 m^3 of water per square meter of filter medium per minute, m^2 [c]	143	1.15	0.39
type of medium	ungraded	carefully graded sand with larger sizes at bottom	several types but usually carefully graded sand and carbon
bed depth	30 cm gravel beneath 90–110 cm sand	30–45 cm gravel beneath 60–75 cm sand	30–45 cm sand beneath 45–60 cm granular carbon
period of filter use between cleanings	20–60 d	24–72 h	12–72 h
cleaning procedure	removal of surface layer for cleaning or washing surface	backwash by forcing water up through filter; can use surface jets or submerged jets to fluidize bed with water or air	same as rapid sand
amount of water used for cleaning, % of filtered water	0.2–0.6	1–6	1–3
depth of filtration	surface only	first few cm	in depth
pretreatment	none	coagulation and sedimentation	coagulation and sedimentation

[a] Ref. 18.
[b] To convert L/(s·m^2) to gal/(s·ft^2), divide by 40.74.
[c] To convert m^2 to ft^2, multiply by 10.76.

that filters are an effective barrier against unsafe water, the effluent may be as colored, as turbid, or as bacteriologically unsafe as the water applied. In contrast, slow sand filters require no pretreatment, as the slow passage through the bed allows the particles to contact and attach to the *schmutzdecke*.

The two steps in the removal of a particle from the liquid phase by the filter medium are the transport of the suspended particle to the surface of the medium and interaction with the surface to form a bond strong enough to withstand the hydraulic stresses imposed on it by the passage of water over the surface. The transport step is influenced by such physical factors as concentration of the suspension, medium particle size, medium particle-size distribution, temperature, flow rate, and flow time. These parameters have been considered in various empirical relationships that help predict filter performance based on physical factors only (23–24). Most recent attention has been on the interaction between the particles and the filter surface, and the mechanisms that have been postulated are based on adsorption or specific chemical interactions (25).

The goal of filtration in the modern municipal treatment plant is a maximum of 0.1 ntu (nephelometric turbidity unit), which ensures a sparkling, clear water (8). Freedom from disease organisms is associated with freedom from turbidity, and complete freedom from taste and odor requires no less than such clarity. The NIPDWR requires that the maximum contaminant level for turbidity at the point of entry into the distribution system be 1.0 ntu unless it can be shown that levels up to 5 ntu do not interfere with disinfection, interfere with the maintenance of a chlorine residual in the distribution system, nor interfere with bacteriological analyses.

Coagulation and Flocculation. The removal of colloidal particles, eg, turbidity, color, and bacteria, requires agglomeration of these particles prior to filtration. Agglomeration can be carried out chemically, that is by interaction of the colloidal particles with materials added as coagulants, or physicochemically by neutralizing the charge on the particles or by interparticle bridging. The term coagulation is applied to the addition of any material that causes agglomeration of the colloids, and flocculation refers to the process of gentle agitation which builds floc particles large enough to settle rapidly. A physicochemical mechanism has been proposed for coagulation and is based on charge neutralization by double-layer compression, which allows the particles to approach closely enough for short-range van der Waals forces to cause agglomeration (26). Flocculation is the growth of a three-dimensional structure by interparticle bridging (see also Size enlargement).

The repulsive forces that act to stabilize the suspension are the charge on the particles (like particles have like charges and therefore repel one another) and hydration. The attractive or destabilizing forces are Brownian movement, van der Waals forces, and gravity. As shown in Table 5, the effect of gravity is insignificant in the sedimentation of colloidal particles. Thus, it is possible to destabilize colloids by increasing the attractive forces or by decreasing the repulsive forces.

Among the various factors that result in charges on colloidal particles are ionization of surface groups, eg, acid groups on organic color particles and saltlike bonds on the surface of clay particles that result in ion-exchange reactions with the solution. In general, the colloidal particles in natural waters are negatively charged. Clay, for example, has a net negative charge on the large surface with positive charges on the edges. Color particles and bacteria are also negatively charged. This charge cannot be measured directly, as the particles are surrounded by a sheath of counterions in a relatively fixed layer; see Figure 3. From this fixed layer, the charge is gradually neutralized by the higher concentration of counterions in the diffuse double layer. The measurements that can be made of this charge are limited by the impossibility of getting the particle independent of the counterions. The rate at which the particle and some of its counterions move under an impressed potential difference can be measured by various techniques, eg, microelectrophoresis. However, it is only the charge at the slipping plane or shear plane where the net force between the particle and the bulk solution is small enough so that the particle moves independently. The charge at this plane is referred to as the zeta potential and is shown in Figure 3 as ψ_ζ. The charge on the particle is ψ_0.

The other important stabilizing force is hydration, since hydrophilic colloids are involved. The waters of hydration modify the exterior of the particle, so that it approaches the properties of the bulk water. Owing to this, the particles show no tendency to approach or to coalesce. The destabilizing effect of the Brownian movement brings the particles into contact simply because of their random wanderings through the solution. If two particles happen to collide or penetrate each other's repulsive sphere, they may be held together by van der Waals forces. These forces are proportional to the reciprocal of the distance between the particle to the fifth power. This means that

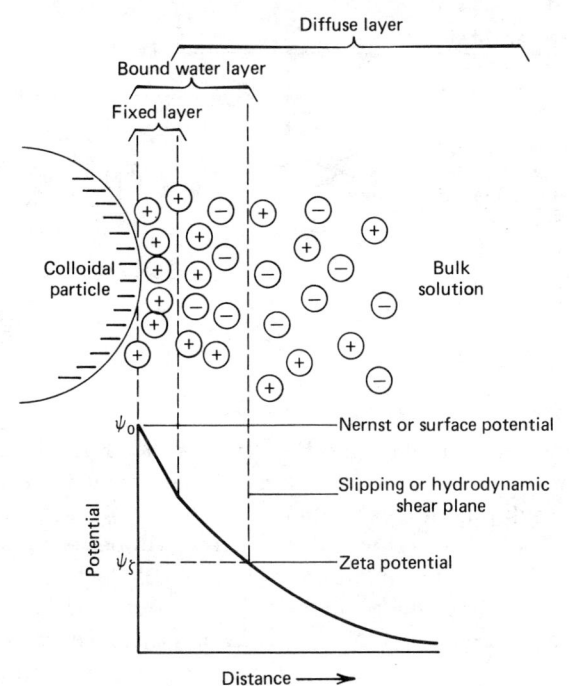

Figure 3. Double-layer model of colloidal particle.

attractive forces generally act over a much shorter distance than do repulsive forces.

It can be seen from these two factors, ie, particle charge and van der Waals forces, that the charge must be reduced or the double layer must be compressed to allow the particles to approach each other closely enough so that the van der Waals forces can hold them together. There are two approaches to the accomplishment of this goal: reaction of the charged surface sites with an opposite charge on an insoluble material and neutralization of the charge by opposite charges concentrated in the fixed layer or the immediate environment. It is difficult to distinguish between these two as they result in the same effect on the particle, insofar as this effect can be measured. Since the particles are negatively charged, they can be coagulated by organic or inorganic cationic polyelectrolytes. A particle that consists of an organic chain that is negatively charged or uncharged may still be adsorbed, but this does not result in reduction of the particle charge.

Many metal ions react with water to produce hydrolysis products that are multiply charged inorganic polymers. These may react specifically with negative sites on the colloidal particles to form relatively strong chemical bonds, or they may be adsorbed at the interface. In either case, the charge on the particle is reduced.

There are two classes of coagulants used in municipal water treatment: inorganic metal salts, containing iron and aluminum, and organic polymers. The metal salts have been used since the early days of water treatment as the primary coagulants, although they were first utilized as pretreatment for filtration. The metal salts hydrolyze when added to water, according to the typical series of reactions shown in Table 7. From the values in Table 7, it can be seen that the predominant equilibrium species for the hydrolysis of aluminum and iron(III) ions over the pH range of interest in water

Table 7. Hydrolytic Reactions of Metal Ions

Reaction	pK	Ref.
$Al^{3+} + H_2O \rightleftharpoons AlOH^{2+} + H^+$	5.03	27
$2\,Al^{3+} + 4\,H_2O \rightleftharpoons Al_2(OH)_4^{+} + 4\,H^+$	6.27	27
$Al^{3+} + 3\,H_2O \rightleftharpoons Al(OH)_3 + 3\,H^+$	9.1	28
$Al^{3+} + 4\,H_2O \rightleftharpoons Al(OH)_4^{-} + 4\,H^+$	21.84	28
$Fe^{3+} + H_2O \rightleftharpoons FeOH^{2+} + H^+$	3.05	23
$Fe^{3+} + 2\,H_2O \rightleftharpoons Fe(OH)_2^{+} + 2\,H^+$	6.31	23
$Fe^{3+} + 3\,H_2O \rightleftharpoons Fe(OH)_3 + 3\,H^+$	−0.5[a]	
$Fe^{3+} + 4\,H_2O \rightleftharpoons Fe(OH)_4^{-} + 3\,H^+$	18.0	24

[a] Calculated from $pK_w = 14$, $pK_{sp} = 42.5$.

treatment, ie, pH 5–9, are $Al(OH)_3$ and $Fe(OH)_3$, as shown in Figures 4 and 5, which express the concentration of the various species as a function of pH. These equilibrium conditions may occur during the coagulation, flocculation, and sedimentation processes in treatment plants. It has been shown that the reactive species may not be those that predominate under equilibrium conditions but rather the less-hydrolyzed species that form in the first few seconds or minutes of the reaction with water (29–30). These species are positively charged and effectively interact with the negative colloidal impurities.

The metal salts reduce the alkalinity in the water; therefore, it may be necessary to add base in the form of lime or soda ash. One part of technical aluminum sulfate ($Al_2(SO_4)_3 \cdot 14H_2O$) reduces the alkalinity as $CaCO_3$ by 0.55 parts and one part of technical ferric sulfate, $Fe_2(SO_4)_3 \cdot H_2O$, by 0.68 parts. The reaction is

$$Fe^{3+} + 3\,HCO_3^{-} \rightleftharpoons Fe(OH)_3 + 3\,CO_2 \qquad (7)$$

The selection of a particular metal salt is based on such factors as local availability, convenience, economics, and effectiveness for the specific treatment problem.

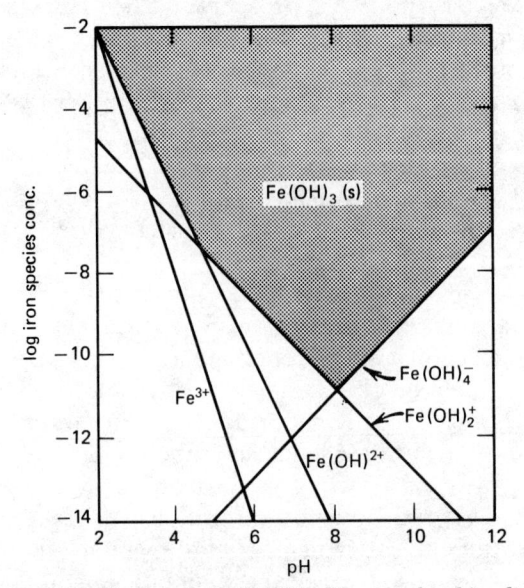

Figure 4. Equilibrium solubility domain for $Fe(OH)_3$. The cross-hatching defines the area of stability of solid $Fe(OH)_3$.

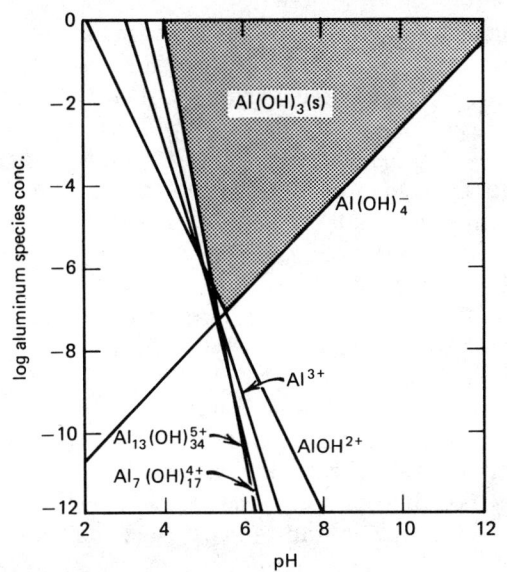

Figure 5. Equilibrium solubility domain for Al(OH)$_3$. The cross-hatching defines the area of stability of solid Al(OH)$_3$.

Aluminum sulfate (commercial product Al$_2$(SO$_4$)$_3$.14H$_2$O) is available as either a granular solid containing 17 wt % Al$_2$O$_3$ or a solution containing 8.3 wt % Al$_2$O$_3$. This is commonly referred to as alum or filter alum. The higher cost of transporting the solution may be more than compensated by the ease in handling and savings in labor and equipment required for feeding and dissolving the solid (31).

Ferric sulfate (commercial product Fe$_2$(SO$_4$)$_3$.H$_2$O) contains a minimum of 20 wt % iron(III). It is available only as a solid, which must be dissolved immediately before use. The solution must be kept concentrated to avoid premature hydrolysis and precipitation of Fe(OH)$_3$. Such concentrated solutions have low pH values and, thus, prevent hydrolysis but are very corrosive. Containers must therefore be coated with or be constructed of corrosion-proof materials (see Corrosion and corrosion inhibitors).

Ferrous sulfate (commercial product FeSO$_4$.7H$_2$O) is referred to as copperas. This may be the least expensive iron salt, particularly in localities where it may be available as a by-product of the manufacture of titanium dioxide or the pickling of steel (qv). Its high solubility as the hydroxide (ca 4 mg/L as iron) precludes its use as a coagulant; therefore, it must be oxidized to the +3 state. This can be accomplished readily above pH 6 by oxygen or chlorine. Each part of FeSO$_4$.7H$_2$O requires 0.03 parts of oxygen or 0.126 parts of chlorine for oxidation to the ferric state.

Sodium aluminate (NaAlO$_2$) normally contains an excess of sodium hydroxide or soda ash to maintain a sufficiently high pH to prevent Al(OH)$_3$ precipitation prior to its addition to the water as a coagulant.

The organic coagulants in use may be either natural or synthetic. Among the useful natural polymers are starches, gums, and gelatin. There are four classes of organic polymeric coagulants: cationic, anionic, ampholytic (frequently referred to as polyelectrolytes), and nonionic. The cationic class comprises quaternary ammonium compounds (qv), eg, poly(dimethyldiallylammonium chloride). Poly(sodium methacrylate) is an example of an anionic polyelectrolyte. Poly(lysine–glutamic acid), having both —NH$_3^+$ and —COO$^-$ groups, is an example of an ampholytic polyelectrolyte.

Nonionic polymers, eg, poly(ethylene oxide), have no ionizable sites but do have sites that can be made acidic or basic by distortion of the electronic distribution. Polyelectrolytes can act as charge-neutralizing agents for oppositely charged species. However, only cationic polyelectrolytes are significant in this respect because of the dominance of negatively charged natural colloids. Flocculation of colloids by polymers predominantly results from adsorption of the polymer at the interface between the colloid and the solution. This is followed by polymer–polymer cross-linking to build a three-dimensional floc particle (32–33) (see Flocculating agents). The apparent anomaly of coagulation of a negative colloid by an anionic polyelectrolyte or by a nonionic polymer can be explained in terms of the reduced free energy of the system resulting from hydrogen bonding or van der Waals forces between the colloid and the polymer chain despite the apparent reduction in entropy.

The removal of turbidity by the coagulation–flocculation–sedimentation–filtration process is the principal chemical treatment, other than chlorination, used in the United States, since most surface waters contain colloidally suspended solids, which cannot be removed by filtration alone. In a conventional or horizontal-flow plant, the coagulant is usually added in a rapid-mix tank or is mixed by passage through a baffled chamber. The water, with its very fine microfloc, is passed into a flocculation basin, where it is gently agitated to bring the microflocs into contact with one another, thereby forming flocs that are large enough to settle during the transit of the water through the next unit, a settling basin. The transit time through the settling basin may be 4–12 h, depending upon the design and the demand. The settled water then passes to the filters. Prior to filtration, the pH may be adjusted and the water may be chlorinated. After filtration, the water may be postchlorinated before distribution. In contrast, up-flow, solids-contact units contain all of the coagulation–flocculation–sedimentation steps in a single tank. The residence time in such a unit is 2–4 h. The time is greatly reduced because water is passed up through a blanket of sludge which increases the possibility of attachment of the microflocs to the heavier, previously precipitated floc. The coagulant is added to the raw water at the bottom of the vertical unit; flocculation can be accomplished in a section prior to passage up through the sludge. After overflowing of the unit, the water is treated in the same manner as effluent from a horizontal unit. The rate of water flow in a typical up-flow unit is 3–6 cm/min (0.75–1.5 gal/(min·ft^2). A unit of 9.8 m dia and ca 4.9 m deep produces ca 5700 m^3/d (1.5 × 10^6 gal/d), which corresponds roughly to the requirements of a city of 10,000 inhabitants.

In cases where both the turbidity and color are low, the flocculation, the sedimentation steps, or both, can be eliminated. The dosage of the coagulant used, either metal salt or polymer, or both, is much lower than in conventional cases where the goal is to produce a rapidly settling floc. Rather, the plant is designed to produce a floc that filters well with a minimum of clogging of the filter. These processes are called either direct filtration, where a flocculation time of 5–30 min is used ahead of filtration, or contact flocculation, where no flocculation precedes filtration.

The optimum pH range for the use of alum as the coagulant for turbidity removal is 7–9. The dosage required, depending on the concentration of the colloid, is 5–50 mg/L (19–190 mg/gal). The optimum pH range for the use of ferric salts is 6–10, and the concentrations are similar to those used for alum. The pH may affect the structure of and the charge on the polymer, but otherwise the polymers are essentially insensitive to pH. The dosage of the polymers may be as low as 0.03–0.1 mg/L (0.1–0.4 mg/gal).

Color can be removed effectively and economically with either alum or ferric sulfate at pH values of 5–6 and 3–4, respectively. The reaction is stoichiometric and is a specific reaction of the coagulant with the color to form an insoluble compound (34). The dosage required may be as high as 100–150 mg/L (380–570 mg/gal). Raw-water colors may be as high as 450–500 units on the APHA color scale. The secondary MCL (maximum contaminant level) for color in the finished water is 15 units, although most municipal treatment plants produce water that seldom exceeds 5 units.

The treatment units used for color removal are the same as those used for turbidity removal. However, the pH must be increased prior to filtration so that the metal hydroxides are removed by the filters. At low pH values, metal ions or their soluble complexes readily pass through the filters and form insoluble species in storage tanks and in the distribution system. For iron salts, it is important that the pH be greater than 6 as the oxidation of iron(II) to iron(III) occurs rapidly above this pH in the presence of dissolved oxygen or other strong oxidants (35).

Softening. A water is classified as hard if it contains more than 120 mg of divalent ions per liter (454 mg/L), usually calcium and magnesium, expressed as $CaCO_3$. Hard water is problematic as cleaning in it requires a greater amount of soap than is needed in soft water to produce lather. Also, the calcium and magnesium salts that cause hardness have negative solubility coefficients and precipitate with an increase in temperature. A deposit of scale forms on heat-transfer surfaces, leading to localized overheating or clogging of service lines.

The American Water Works Association Water Quality Goals recommend a maximum total hardness of 80 ppm for municipal purposes (8). Municipal softening plants, however, distribute waters containing 70–150 ppm; the final quality is established based on such factors as public demand and economics. The hardness of waters provided in the 100 largest U.S. cities in 1960, treated or untreated, was 0–738 ppm with a median value of 90 ppm. Only 13 cities distributed water with hardnesses exceeding 180 ppm. Twenty-eight of the 100 cities soften their water supplies.

The two principal methods of softening water for municipal purposes are addition of lime or lime-soda and ion exchange. The choice method depends upon such factors as the raw-water quality, the local cost of the softening chemicals, and means of disposing of waste streams.

Lime and Lime-Soda Processes. The first softening plant in the early 1900s used the lime softening process with fill and draw units. Later, continuous-treatment units, which greatly increased the amount of water that could be treated in a facility of given size, were developed. Today, more than 1000 municipalities soften water; most of these are in the Midwest and in Florida. However, concern for the adverse effect of soft water on cardiovascular disease (CVD) may limit the number of plants that introduce softening (40).

The lime or lime-soda process results in the precipitation of calcium as calcium carbonate and magnesium as magnesium hydroxide. The solubilities of these compounds are shown in Figure 6 as functions of pH. When lime is used alone, only the carbonate hardness is reduced (the carbonate hardness is present as calcium or magnesium bicarbonate). The additional use of soda ash can reduce the noncarbonate hardness by providing additional carbonate ion. The reactions involved in the various steps of the process are listed below:

The reactions of any free CO_2 with the added lime:

$$CO_2 + Ca(OH)_2 \rightarrow CaCO_3 + H_2O \tag{8}$$

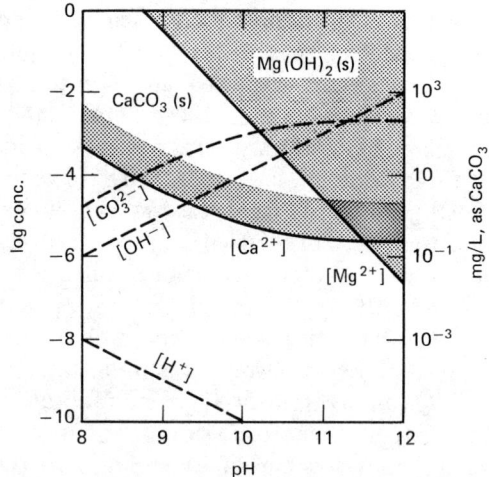

Figure 6. Equilibrium solubility domains for $CaCO_3$ and $Mg(OH)_2$ at a total carbonic species concentration of 5×10^{-3} M. The shaded areas above the Mg^{2+} and Ca^{2+} curves define the areas of stability of solid $Mg(OH)_2$ and $CaCO_3$, respectively.

This reaction provides no softening but occurs preferentially, since the CO_2 is the strongest acid present in the system.

The reaction of calcium carbonate hardness with lime, when the calcium can be represented as the bicarbonate at the usual pH values:

$$Ca(HCO_3)_2 + Ca(OH)_2 \rightarrow 2\ CaCO_3 + 2\ H_2O \tag{9}$$

The reaction of magnesium carbonate hardness with lime:

$$Mg(HCO_3)_2 + 2\ Ca(OH)_2 \rightarrow 2\ CaCO_3 + 2\ H_2O + Mg(OH)_2 \tag{10}$$

The reaction of calcium noncarbonate hardness with soda ash:

$$CaSO_4 + Na_2CO_3 \rightleftharpoons CaCO_3 + Na_2SO_4 \tag{11}$$

It is necessary to use soda ash to provide the CO_3^{2-}, as the available CO_3^{2-} would be consumed in reactions 9 and 10. The noncarbonate hardness may be represented as sulfate, although any anion except carbonate or bicarbonate could be present.

The reaction of magnesium noncarbonate hardness with lime and with soda ash is a two step reaction since reaction 12 produces a reasonably soluble calcium salt that must react with CO_3^{2-} in order to cause calcium precipitation:

$$MgSO_4 + Ca(OH)_2 \rightarrow Mg(OH)_2 + CaSO_4 \tag{12}$$

$$CaSO_4 + Na_2CO_3 \rightleftharpoons CaCO_3 + Na_2SO_4 \tag{13}$$

Reactions 11 and 13 are the same but represent two different sources of $CaSO_4$.

From reactions 8–13, it can be seen that the addition of lime always serves three purposes and may serve a fourth. It removes, in order, CO_2, calcium carbonate hardness, and magnesium carbonate hardness (reactions 8, 9, and 10, respectively). Where magnesium noncarbonate hardness must be removed, the lime converts it to calcium noncarbonate hardness (reaction 12). Soda ash, then, removes noncarbonate hardness according to reaction 11 or 13.

The needed amounts of lime and soda ash can be calculated from the stoichiometry of the reactions. The effluent quality is a function of the solubilities of calcium carbonate and magnesium hydroxide and of the quantities of softening chemicals added. The acceptable level of total hardness can be decided and usually is 70–120 mg/L (265–454 mg/gal), expressed as $CaCO_3$. The sum of the solubilities of calcium carbonate and magnesium hydroxide is ca 50–70 mg/L (190–265 mg/gal), depending upon the pH. The sum of the concentrations of the carbonic species $HCO_3^- + CO_3^{2-}$, expressed as $CaCO_3$ or carbonate hardness, can be as low as 20 mg/L (76 mg/gal) but is usually 30–50 mg/L (114–190 mg/gal) because of $CaCO_3$ supersaturation. It is desirable to reduce the carbonate hardness to the lowest practical value that allows the maximum concentration of noncarbonate hardness to remain. This decreases the cost of treatment, as soda ash costs at least twice as much as lime. Additionally, ca 190% more soda ash than lime is needed to remove the same amount of hardness. It is economically more favorable to minimize the amount of calcium, since twice as much lime is required to remove magnesium as to remove calcium. Unfortunately, all of the hardness left cannot be present as magnesium, since magnesium hardness above 40 ppm causes scaling in hot-water heaters at 16°C, which is the normal temperature setting. For this reason, the magnesium concentration is usually reduced to 40 ppm.

Since the effluent from a softening unit is usually supersaturated with calcium carbonate at the usually high pH values, it is necessary to reduce the pH to a value that allows the solution to be exactly saturated for the calcium-ion and carbonate-ion concentrations present. The relationship is

$$CaCO_3(s) + H^+ \rightleftharpoons Ca^{2+} + HCO_3^- \tag{14}$$

The equilibrium expression is

$$K = \frac{(Ca^{2+})(HCO_3^-)}{(H^+)} = K_{sp}/K_2 \tag{15}$$

or

$$pH_s = pK_2 - pK_{sp} + pCa + p(alk) \tag{16}$$

where pH_s is the pH of saturation, pK_2 is the second ionization constant for carbonic acid, pK_{sp} is the solubility product constant of $CaCO_3$, pCa is the negative logarithm of the molar concentration of calcium ion, and p(alk) is the negative logarithm of the bicarbonate ion concentration, which is measured analytically as the total titratable bases present at pH 6–9. The reduced pH is usually attained by recarbonation, ie, by the addition of CO_2. The carbon dioxide is obtained from the combustion of fuel oil or from liquid CO_2.

Modifications of the basic process are undersoftening, split recarbonation, and split treatment. In undersoftening, the pH is raised to 8.5–8.7 to remove only calcium. No recarbonation is required. Split recarbonation involves the use of two units in series. In the first or primary unit, the required lime and soda ash are added and the water is allowed to settle and is recarbonated just to pH 10.3, which is the minimum pH at which the carbonic species are present principally as the carbonate ion. The primary effluent then enters the second or secondary unit, where it contacts recycled sludge from the secondary unit resulting in the precipitation of almost pure calcium carbonate. The effluent settles, is recarbonated to the pH of saturation, and is filtered. The advantages over conventional treatment are reductions in lime, soda ash, and CO_2 requirements; very low alkalinities; and reduced maintenance costs because of the stability of the effluent. The main disadvantages are the necessity for very careful pH control and the requirement for twice the normal plant capacity.

The principal use for split treatment is for water with high magnesium content. This process also requires two units in series, which doubles the size of the plant. The ratio of the fraction of raw water in the primary unit to the fraction treated in the secondary unit is such that, when all of the magnesium is removed in the primary unit and none in the secondary, the mixed effluents contain 10 ppm magnesium as magnesium or 41 ppm as $CaCO_3$. For example, 75% of raw water containing 40 ppm magnesium would be treated in the primary unit and 25% in the secondary unit. Sufficient lime is added ahead of the primary unit to remove all of the magnesium or at least to an amount less than 1 ppm. This requires a pH of 11.1–11.3 with an excess of ca 70 ppm of caustic alkalinity. The effluent from the primary unit is mixed with the bypassed raw water in the secondary unit with no further addition of chemicals. The excess lime in the primary effluent is used to remove the calcium hardness from the bypassed raw water. In this process, the excess lime required for magnesium removal is used to soften raw water, whereas in conventional treatment this excess is wasted through recarbonation. Another advantage is the very low alkalinity [18–20 mg/L (68–76 mg/gal)] of the finished water. This, of course, reduces the necessary amount of soda ash.

One of the main problems associated with lime or lime-soda softening is the disposal of the sludge. Depending upon the ratio of calcium to magnesium removed and upon the amount of soda ash used, the sludge produced is 2.8–3.6 times the weight of the lime added. The principal methods of disposal have been lagooning, discharge into the nearest water course, or discharge to the sanitary sewer system. The latter two are no longer acceptable methods of disposal because of the resultant pollution load.

A method of disposal for large plants is recalcination, ie, regeneration of lime from the $CaCO_3$ by heating the $CaCO_3$ in a kiln to remove the carbon dioxide. The lime can be reused in the plant, the CO_2 used for recarbonation, and the excess lime sold. In cases where $Mg(OH)_2$ has precipitated with $CaCO_3$, it is necessary to remove the magnesium prior to recalcination. This is accomplished by selectively dissolving the magnesium with CO_2 from the kiln at ca pH 10.3. At lower pH values, $CaCO_3$ dissolves and, at higher values, magnesium remains in the sludge. In some plants, the sludge is vacuum-filtered to ca 40–50% solids and then used for agricultural purposes or disposed of in a land fill.

Ion Exchange. For waters with high noncarbonate hardness or high magnesium content, or both, the use of lime or lime-soda may be very expensive. In such cases, ion-exchange softening may be more economical, particularly if brines are locally available at little cost for regeneration of the resins. Ion-exchange softening involves exchanging sodium or hydrogen ions for the calcium or magnesium ions in the raw water. The exchange reactions are equilibria that favor the formation of polyvalent ion–resin complexes over monovalent ion–resin complexes. The raw water to be softened is passed through the resins that are initially saturated with sodium or hydrogen ions. The exchange is referred to as the exhaustion reaction and may be shown as $Ca^+ + 2 NaR \rightarrow R_2 + 2 Na^+$, where R is the resin matrix. The regeneration step is simply the reverse of the exhaustion step with the concentration of the regenerant ion increased to 5–25% in order to reverse the reactions. For sodium-ion regeneration, sodium chloride is used and for hydrogen-ion regeneration, either sulfuric or hydrochloric acid is used. The theoretical requirement for 98 wt % NaCl to regenerate the resin is 453 g for each 686 g of hardness removed (expressed as $CaCO_3$). The actual requirements vary from 544 to 1088 g, depending upon the specific resin used and the regeneration conditions. The water passing through the exchanger is reduced to 0–2 mg/L (0–7.6 mg/gal) hardness. Since this is usually much lower than the final acceptable hardness, only a portion of the water is passed through the exchanger. Enough is bypassed to give the desired final hardness.

The resins used are highly cross-linked organic polymers with acidic functional groups. The most common of the resins used are sulfonated copolymers of styrene and divinylbenzene.

There are several modifications of the ion-exchange process. For example, water with high total solids can be treated partially by sodium-cycle exchange followed by aeration to remove CO_2 and, consequently, part of the anions. However, sodium-cycle exchange increases the soluble-solids content as two sodiums (at wt $2 \times 23.0 = 46.0$) are substituted for one calcium (40.0) or magnesium (24.3). Hydrogen-cycle exchange replaces one calcium or one magnesium with two hydrogens and, thus, reduces the total solids content. The disadvantage is the reduction in pH. The resultant water is very corrosive and the pH must be raised before distribution.

Another possible modification is the use of seawater as the regenerant. Even though it contains calcium and magnesium, but only 2.7 wt % NaCl, it sometimes can be purified by coagulation, filtration, and chlorination less expensively than salt can be purchased. The lower concentration reduces the regeneration efficiency by 40–50%.

The two types of installations used are fixed-bed and continuous-regeneration units. The continuous units consist of a closed circuit containing two sections, one for softening and one for regeneration; the resin is circulated countercurrent to the raw-water flow through the softening tank. The resin then is pulsed periodically into the regeneration tank, where it is regenerated and rinsed; then it is pulsed back to the softening tank. Thus, there is continuous softening without the downtime customary with fixed-bed units.

Taste and Odor Control. Tastes and odors in surface waters result from the action of biological organisms, eg, algae, or from pollution by industry, domestic seepage, or agriculture. Groundwaters may have taste and odor if they are polluted or if they contain gases, eg, H_2S or CH_4; the latter always contains associated impurities that have taste and odor. Removal of these can be accomplished by adsorption with activated carbon; oxidation with chlorine, potassium permanganate, or ozone; or aeration.

Organic materials are generally removed by addition of powdered activated carbon. The carbon may be added at any point in the plant, although it is advantageous to have as much contact as possible. The adsorption reaction is slow at room temperature, since it is diffusion-controlled. Oxidation with chlorine, potassium permanganate, or ozone may destroy tastes and odors or it may intensify them, depending upon the particular compounds involved. For example, chlorination of phenolic compounds leads to greatly increased tastes and odors. For this reason, the system must be studied in the laboratory prior to water treatment.

Hydrogen sulfide and methane can be removed by aeration, although the largest reduction in hydrogen sulfide may result from oxidation by the dissolved oxygen introduced during the aeration. At low pH values, the product is sulfate, whereas at high pH values, the product is free sulfur.

Control of Organic Compounds. There are two groups of organic compounds that have been of recent concern: the trihalomethanes, which result from the use of chlorine as a disinfectant, and volatile organic solvents, which percolate through the soil thereby contaminating groundwaters. Trihalomethanes (THMs) result from the reaction of chlorine with natural organic precursors according to the general equation

$$HOCl + precursors \rightarrow THMs \qquad (17)$$

The THMs of concern include chloroform, dichlorobromomethane, chlorodibromomethane, and bromoform (see Chlorocarbons and chlorohydrocarbons). Control is based on the use of an alternative disinfectant, removal of precursors prior to chlorination, or removal of the THMs after formation. Alternative disinfectants are the same as those used for disinfection with the same advantages and disadvantages. There have been varying degrees of success with this approach. The removal of precursors has been accomplished by optimizing coagulation, using activated carbon, or oxidizing with ozone, $KMnO_4$, or ClO_2. The removal of THMs after formation can be accomplished by air stripping or adsorption. In some cases, a combination of these processes must be used to reduce the THM concentration to less than the MCL of 0.10 mg/L (0.4 mg/gal). Some plants using highly colored surface sources have had average concentrations as high as 2 mg/L (7.6 mg/gal).

Volatile organic contaminants occur primarily in groundwaters as a result of the disposal of industrial solvents on the ground or in soakage pits. The removal of these compounds has best been accomplished by the use of air stripping or adsorption on activated carbon.

Iron and Manganese Removal. Groundwaters or water withdrawn from the depths of reservoirs may contain iron and manganese in the +2 or reduced state, in which they are soluble. Either one in equilibrium with dissolved oxygen exists in an oxidized and, therefore, insoluble state ($Fe(OH)_3$ and MnO_2). If the reduced metal ions are allowed to remain in the finished water and then come into contact with the atmosphere, the oxidized forms precipitate upon domestic fixtures or clothes, yielding a reddish-brown stain from the iron and a dark-brown-to-black stain from the manganese.

Both ions can be removed by oxidation and subsequent filtration. Aeration is adequate for iron(II) oxidation at above pH 6, but the oxidation of manganese(II) is much too slow, even at higher pH values, for effective removal. Potassium permanganate or chlorine dioxide are frequently used for the oxidation of manganese; however, their use must be followed by coagulation prior to filtration because of the formation of colloidal MnO_2.

Fluoridation. The practice of adding fluoride ion to domestic water supplies has been increasing since the first experiments in the city of Grand Rapids, Michigan, in 1945. It has been shown that dental decay is reduced significantly by maintaining a fluoride residual of ca 1 mg/L (3.8 mg/gal) in the public water supplies (36). The concentration recommended by the U.S. Public Health Service is 0.7–1.2 depending upon the annual average maximum daily air temperature, since more water is ingested at higher temperatures (37).

The early programs were designed as demonstration studies and were followed closely by periodic dental examinations. When the U.S. Public Health Service formally endorsed fluoridation in 1950, many cities started the practice which has grown tremendously. In 1969, there were over 2550 water systems serving fluoridated water to over 4500 communities and 84×10^6 people in the United States. Over 7×10^6 people in Canada and over 38×10^6 people in 30 other countries were supplied with fluoridated water (38). In 1975, it was estimated that over 105×10^6 people in the United States consumed drinking water with a fluoride concentration equal to or greater than 0.7 mg/L (2.7 mg/gal) (39).

The sources of the fluoride ion vary, but the most common are sodium fluoride, sodium fluorosilicate (Na_2SiF_6), and fluorosilicic acid (H_2SiF_6). The solids NaF, Na_2SiF_6, and CaF_2 are added to the water with accurate dry feeders. Sodium fluoride can also be fed from a constant-strength, saturated solution containing ca 4 wt % NaF. Solutions of H_2SiF_6 are also fed with liquid feeders. The accuracy of such feeders has been estimated to be ca 99% (38). Careful control is possible because of the development of a specific-ion electrode for F^- analysis.

The first state to require fluoridation by law was Connecticut in 1965. This has been followed by Michigan, Delaware, Illinois, South Dakota, and Ohio. Australia, Brazil, Chile, and Ireland have compulsory fluoridation. The practice has been approved and recommended by the American Dental Association, the American Medical Association, the U.S. Public Health Service, and the American Water Works Association.

Economic Aspects

The cost of producing a potable water varies considerably depending upon the raw-water quality, the quality required by regulations extant, the treatment required, the size of the system, and the local costs of such items as labor, power, and chemicals.

In the United States, the costs have been shown to be much greater for small systems than for large systems treating comparable raw water to the same final quality (40). There were estimated to be 53,200 community water systems serving populations of 25–10,000 (21% of the population) and 8000 community systems serving populations greater than 10,000 (79% of the population). The total population served by community systems was estimated to be 213×10^6. In addition, it was estimated that over 200,000 non-community systems existed, each serving less than 25 persons (2). The total unit cost, including amortization of capital, was shown to vary from U.S. $0.11/kL ($0.42/1000 gal) for large systems to U.S. $0.46/kL ($1.74/1000 gal) for the smallest systems (40). The cost to the consumer, of course, varies even more widely depending on distribution costs as well as political and social pressures.

The imposition of more stringent water-quality regulations has a much greater impact on smaller systems in that economics of scale mitigate the impact on the larger systems. This was demonstrated by estimating the impact of the NIPDWR on such systems. The costs were increased by as much as a factor of 8 for the smallest system with essentially no impact on larger systems (40).

BIBLIOGRAPHY

"Water (Municipal)" in *ECT* 1st ed., Vol. 14, pp. 946–962, by H. O. Halvorson, University of Illinois; "Water, Municipal Water Treatment" in *ECT* 2nd ed., Vol. 22, pp. 82–104, by J. E. Singley, University of Florida.

1. *Drinking Water and Health*, National Academy of Health, Washington, D.C., 1977, Part I, p. I-3.
2. *Fed. Regist.* **40,** 59566 (Dec. 24, 1975).

3. *Fed. Regist.* **41,** 28402 (July 8, 1976).
4. *Fed. Regist.* **44,** 68624 (Nov. 29, 1979).
5. *Standard Methods for the Examination of Water and Wastewater*, 15th ed., American Public Health Association-American Water Works Association-Water Pollution Control Federation, Washington, D.C., 1980.
6. *Fed. Regist.* **44,** 4218 (July 19, 1979).
7. *Fed. Regist.* **40,** 9350 (March 4, 1982).
8. *J. Am. Water Works Assoc.* **60,** 1317 (1968).
9. C. N. Dufor and E. J. Becker, *J. Am. Water Works Assoc.* **56,** 23 (1964).
10. E. C. Ross, K. W. Campbell, and H. J. Ongerth, *J. Am. Water Works Assoc.* **58,** 165 (1966).
11. K. R. Pfeifer, *J. Am. Water Works Assoc.* **65,** 803 (1973).
12. M. N. Baker, *The Quest for Pure Water*, American Water Works Association, New York, 1949.
13. G. C. White, *Handbook of Chlorination*, Van Nostrand Reinhold, New York, 1972.
14. C. Gomella, *J. Am. Water Works Assoc.* **64,** 39 (1972).
15. A. P. Black, W. C. Thomas, R. N. Kinman, W. P. Bonner, M. A. Keirn, J. J. Smith, and A. A. Jabero, *J. Am. Water Works Assoc.* **60,** 69 (1968).
16. A. E. Greenberg, *Water Disinfection with Ozone, Chloramines, or Chlorine Dioxide*, American Water Works Association, Denver, Colo., 1980, p. 1.
17. S. T. Powell, *Water Conditioning for Industry*, McGraw-Hill Book Co., New York, 1954.
18. J. Arboleda-Valencia, *Teoria, Diseno y Control de los Procesos de Clarificacion del Agua*, Pan American Center for Sanitary Engineering and Environmental Sciences, Pan American Health Organization, Lima, Peru, 1973.
19. G. S. Logsdon and E. C. Lippy, *J. Am. Water Works Assoc.* **74,** 649 (1982).
20. U.S. Pat. 293,740 (Feb. 19, 1884), I. S. Hyatt.
21. T. R. Camp, *J. Am. Water Works Assoc.* **53,** 1478 (1961).
22. W. R. Conley, *J. Am. Water Works Assoc.* **53,** 1473 (1961).
23. G. Biederman quoted by K. Schlyter, *Trans. R. Inst. Technol. Stockholm*, 196 (1962).
24. H. Leugweiler, W. Buser, and W. Feitknecht, *Helv. Chim. Acta* **44,** 805 (1961).
25. J. L. Cleasby, *Proceedings, American Water Works Association Seminar*, No. 20153, American Water Works Association, Denver, Colo., 1980.
26. V. K. LeMar and T. W. Healy, *J. Phys. Chem.* **67,** 2417 (1963).
27. H. Kubota, thesis, University of Wisconsin, Madison, Wisc., 1956.
28. Z. G. Szabo, L. J. Czanyi, and M. Kaval, *Z. Anal. Chem.* **147,** 401 (1955).
29. J. E. Singley and A. P. Black, *J. Am. Water Works Assoc.* **59,** 1549 (1967).
30. J. H. Sullivan and J. E. Singley, *J. Am. Water Works Assoc.* **60,** 1280 (1968).
31. R. W. Ockershavsen, *J. Am. Water Works Assoc.* **57,** 309 (1965).
32. A. P. Black and M. R. Vilaret, *J. Am. Water Works Assoc.* **61,** 209 (1969).
33. F. B. Birkner and J. J. Morgan, *J. Am. Water Works Assoc.* **60,** 175 (1968).
34. A. P. Black, J. E. Singley, G. P. Whittle, and J. S. Maulding, *J. Am. Water Works Assoc.* **55,** 1347 (1963).
35. W. Stumm and G. F. Lee, *Ind. Eng. Chem.* **53,** 143 (1961).
36. H. V. Churchill, *J. Am. Water Works Assoc.* **29,** 1399 (1931).
37. *Public Health Service Drinking Water Standards*, Public Health Service Publication No. 956, U.S. Public Health Service, Washington, D.C., 1962.
38. F. J. Maier, *J. Am. Water Works Assoc.* **62,** 3 (1970).
39. *Fluoridation Census 1975*, U.S. Public Health Service, Center for Disease Control, Atlanta, Ga., 1977.
40. R. G. Stevie and R. M. Clark, *J. Am. Water Works Assoc.* **74,** 13 (1982).

J. E. SINGLEY
Environmental Science & Engineering, Inc.

SEWAGE

Sewage is the spent water supply of a community. Because of infiltration of groundwater into loose-jointed sewer pipes, the total amount of water treated may exceed the amount consumed. Sewage contains about 99.95% water and ca 0.05% waste material (1). The higher the per capita consumption of water, the weaker (more dilute) is the sewage, which may also be affected by industrial wastes. Sewage flow is greater during daylight hours and varies little in large cities, whereas in many small communities, the late-night flow is almost all groundwater.

Per capita flow varies from <378 L/d (2) for a residential community to >1134 L/d for highly industrialized areas. The concept of population equivalent is used for evaluating industrial waste contributions to sewage. It is applied when planning for hydraulic and BOD loadings.

Because of large equipment and land requirements, capital costs for wastewater-treatment plants are high. A collection system that conveys both sanitary and storm flows must be designed to deal with high peak flows at the treatment plant; detention basins are usually provided in order to smooth the flow into the plant and reduce the sudden peak flow. In the absence of such precautions, it may be necessary to by-pass a portion of the flow.

Some materials, such as gasoline and heavy metals, may damage the collection system, cause explosions, or upset the treatment processes. Therefore, they cannot be discharged directly into the system. These prohibitions are governed by sewer ordinances.

Strength of sewage is expressed in terms of BOD, total solids, suspended solids, fixed solids, volatile solids and filterable solids. The BOD is a measure of the load placed on the oxygen resources of the receiving waters. Treatment efficiency is usually evaluated on the basis of BOD removal by the plant. Unless otherwise stated, BOD signifies the biochemical oxygen demand for five days at 20°C (BOD_5). The significant parameters of wastewater are evaluated according to the analytical methods set forth by the U.S. Public Health Association (3).

Sewage Treatment and Disposal Systems

Sewage works were originally constructed for reasons based primarily on public health concepts. However, prevention of disease is not the sole purpose of modern water sanitation practice; it is also a means to protect the oxygen resources of the receiving water (2). An effluent that does not significantly reduce the oxygen concentration of the receiving water into which it is discharged can be expected to have a low concentration of food for microorganisms that deplete the dissolved oxygen. Thus, protection of the oxygen resources of the receiving water prevents the spreading of disease and satisfies aesthetic considerations.

Private and Rural Disposal Systems. In areas not served by sewers, human and other water-carried wastes are disposed of in primitive privies, cesspools, or septic tanks (4). In the more developed areas privies have almost disappeared. Cesspools are simply pits in the ground into which wastewater is allowed to flow, and in many parts of North America they are not permitted. Cesspool water seeps into the ground, leaving the solid matter in the pit, thus groundwater in the area might become contaminated. Septic tanks are widely used in smaller communities and outlying suburbs of larger communities. The tank is kept full of waste and functions as both a sedimentation tank and anaerobic digester. Sanitary and kitchen wastes flow into the tank; grease and light material rise to the top. Heavier materials settle to the bottom and decompose anaerobically. Baffles are placed at the inlet and outlet and a grease trap is usually provided in front of the tank, which has to be cleaned periodically of solids. Good practice requires a minimum volume of 5670 L (1500 gal). The effluent flows to a tile field where it is disposed of in the soil. The tile field is composed of perforated field tile fed by a manifold and underlain with granular material, usually gravel. Clogging of the soil under the tile field must be prevented. As more areas are served by municipal systems, septic tanks are becoming less common. Malfunctioning septic tanks cause odor problems and present a public health hazard. Furthermore, septic tanks and cesspools may form a closed system in which the waste discharged may return in the water supply. In many developing nations, human waste, called night soil, is not discarded, instead, it is used as crop fertilizer and for biogas production (5). Although it is not good public health practice, it is utilization of a valuable resource (see Fuels from biomass; Fuels from waste).

The Imhoff tank is similar in many respects to the septic tank, operating as a sedimentation tank and digester. It is composed of two chambers, one above the other and the shape may be square, circular, or rectangular. The depth ranges from 7.7 to 10.6 m (6). Sewage flows slowly into the upper chamber at ca 0.3 m/s and solids settle out and slide through a slot into the lower chamber, where the flowing wastewater is detained for ca 2 h. Solids held in the lower, or digestion, chamber have an initial water content of ca 95%. After a digestion time of 30–60 d the water content of the digesting sludge is reduced to ca 80% which greatly reduces the volume. The sludge is withdrawn at intervals for further dewatering and disposal. Gases produced during digestion are allowed to escape to the atmosphere through vents located at the tank sides. Solids buoyed up by gas are prevented from escaping to the upper chamber by deflector plates.

Small Communities. Small communities and recent subdivision additions to larger communities, which have not yet been connected to municipal collection systems, must have a means of waste disposal. Septic tanks are a possibility, but require periodic servicing and cleaning. Furthermore, the soil is not always suitable for accepting the effluent. An alternative is the package plant. These units are commercially produced to serve small areas. They furnish primary treatment and some secondary treatment, and require only minimal operating supervision. Capacity can be varied as needs dictate. In general, public health authorities prefer such installations instead of septic tanks.

City Disposal Systems. The systems described above are totally insufficient for population centers larger than small and rural communities. The operations necessary for treating sewage are outlined in Figure 1. In primary treatment large particles are removed by screening and sedimentation; 15 to 60% of the BOD can be removed in this manner. Colloidal and dissolved substances are removed by secondary, or biological treatment where the microorganisms of the process utilize the waste material for food. A commonly accepted definition of tertiary treatment is the use of any process or operation for further removals (2).

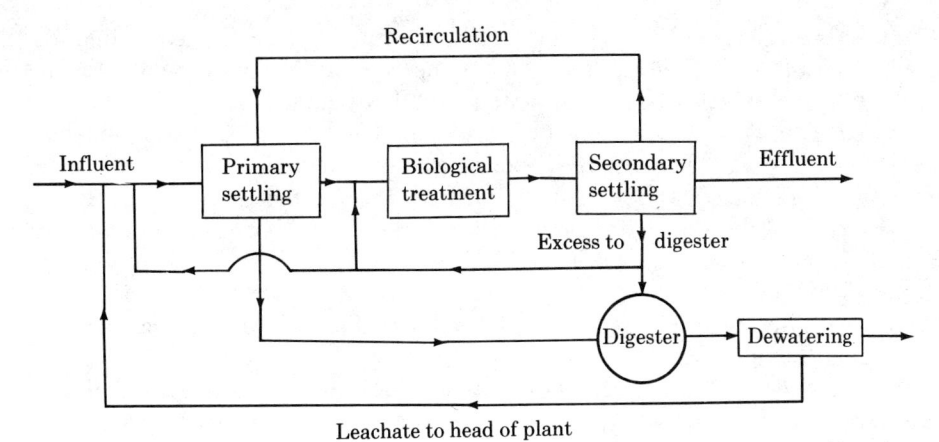

```
                         Recirculation
                    ┌──────────────────────┐
                    │                      │
Influent    ┌─────────┐   ┌──────────┐   ┌──────────┐   Effluent
──────────▶ │ Primary │──▶│Biological│──▶│Secondary │──────────▶
            │settling │   │treatment │   │ settling │
            └─────────┘   └──────────┘   └──────────┘
                                        Excess to ▼ digester
                                          ┌─────────┐   ┌──────────┐
                                          │ Digester│──▶│Dewatering│──▶
                                          └─────────┘   └──────────┘
              Leachate to head of plant
```

Figure 1. Sewage-treatment operations.

Operation and Maintenance. Engineering and construction firms can be contracted to take charge of treatment-plant operation and maintenance and manage other aspects necessary to meet performance goals, such as hiring and cost controls (7). Costs to the municipalities are reported to be competitive with the more traditional approach of management by municipal employees (8).

Primary Treatment

Collection. It is desirable, whenever possible, to collect sewage by gravity flow (9). Otherwise pumping is necessary and flow velocities must be maintained above 0.6 m/s in order to prevent solids from settling out in the pipes. Low flow velocities, long detention times, and relatively high temperatures cause operating problems in many areas. Sewage is best treated if it arrives at the treatment plant in a fresh condition. The plant usually has multiple treatment units and treatment continues without interruption during periods of routine maintenance or repairs.

The pumps must be protected against large objects (wood, etc) in the flowing sewage. Coarse racks with clear openings >5 cm are placed in advance of grit chambers. Settling tanks should have clear openings of 2.5–5 cm which must be cleaned frequently (generally by hand) to prevent excessive hydraulic head loss. Rakings are usually buried, incinerated, digested with sludge, or composted.

Mechanically cleaned racks allow smaller clear openings because head loss does not become so high. Mechanical cleaning can be intermittent or continuous. Intermittent cleaning is cycled by float-operated switches controlled by a float in the influent channel.

Comminutors macerate floating material into particles too small (0.5–1 cm) to clog pumps. Comminutors have almost completely replaced racks and screens with small openings.

Grit Chambers. These are placed ahead of sedimentation chambers and are designed to remove inorganic suspended matter of ca 0.02 cm diameter, but not organic matter. Removal of grit provides protection against excessive wear of pumps and loss of volumetric capacity of sludge digesters. Large digesters in plants serving low-lying sandy areas can lose as much as one third of their capacity in a few years (see Dewatering in Supplement Volume). Organic matter in the flowing wastewater is deposited when the flow velocity is below 23 cm/s. The mixture of grit and organic material is called detritus and is unsuitable for landfilling. Velocities above 38 cm/s cause the smaller particles of grit to be resuspended (10). Thus, a fairly narrow velocity range is required. The velocity is controlled by means of a Parshall flume placed immediately after the grit chamber. These flumes, sometimes called open-channel venturis, operate on the principle of critical flow (11) and measure the volumetric flow rate. However, they also control the velocity of flow in the grit chamber. Grit chambers often have a parabolic shape because flow velocity in the flume follows a parabolic function with water depth. The amount of grit removed per 3.78×10^3 m^3 (10^6 gal) is 29–339 L (7.7–89.6 gal). In smaller plants, grit is removed periodically by hand, whereas in larger plants an automatic conveyor system is used.

Sedimentation. In the sedimentation (settling) tanks, solids settle and oil and grease rise to the surface where they can be skimmed and transferred to the sludge digester. Tanks may be circular or rectangular and depths are 2.1–4.5 m. Current practice favors the circular configuration. The bottom is sloped at ca 1% in rectangular tanks and ca 8% in circular tanks to facilitate sludge removal. Detention periods are 2–6 h. Design is on the basis of application rate, usually 0.472 L/(m^2·s) [1000 gal/(yd^2·d)] for primary settling tanks and 0.283–0.377 L/(m^2·s) [600–800 gal/(yd^2·d)] for secondary (humus) tanks.

Secondary Treatment

The two main processes utilized for secondary treatment are the trickling filter and activated sludge (12). Existing biological-treatment methods are logical developments from sewage farms (irrigation areas) to intermittent sand filters to contact (fill and draw) beds. Numerous modifications of the basic processes have evolved over the years but the underlying principles remain unchanged (13).

Modifications have been developed in response to specific operating needs. The basis of biological treatment is the formation of an environment in which the microorganisms can thrive under controlled conditions (14). The microorganisms originate in the sewage. A suitable environment is rich in food and maintained in an aerobic state. It has been said that a sewage-treatment plant is a river in miniature (14).

Irrigation by sewage (sewage farms) provides water for irrigation and some sewage stabilization. This disposal method is still applied in some arid regions but is not desirable from a public health viewpoint. It cannot be used when the applied sewage comes in contact with crops. It may be used for orchards but with a certain risk. A notable case involved sewage contamination of ripening olives; odors cannot be avoided.

Intermittent sand filters resemble slow sand filters used for the treatment of potable water. The sewage is applied to a sandy area and allowed to flow slowly downward. Raw sewage can be applied at rates as high as 8.4 L/(ha·s) [0.9 gal/(acre·s)] and as high as 84.4 L/(ha·s) [9 gal/(acre·s)] after biological treatment. The latter would be considered a tertiary treatment step. Surface accumulations are periodically removed. This method may not be applied to areas on top of extensive limestone formations because of the danger to active wells. Biological films that form on the sand grains undergo continuous stabilization. Beds must be rested between dosings.

Fill-and-draw beds operate as the name indicates. A tank is filled with sewage and allowed to remain full for a while. It is then drained and allowed to rest. Air is drawn into the bed during emptying. Bed loadings are ca 21.1 L/(ha·s) [2.25 gal/(acre·s)]. This method is used more for industrial than for municipal wastes.

Trickling Filters. The so-called trickling filter is not a filter but a bed of stones or other coarse material (packing) over which the sewage flows. In terms of the total number of installations, it is the most widely used biological treatment process. However, the greatest total volume of waste is treated by the activated-sludge process (12).

Settled sewage is distributed by slowly rotating arms equipped with nozzles and deflectors and is allowed to flow downward over the packing. Air is drawn into the filter by the temperature differential between the inside and outside of the filter, thus maintaining a supply of oxygen. Until recently, stones with diameters of 2.5–10 cm were used as filter medium. These stones permitted sufficiently loose packing to allow free flow of water and air and left sufficient openings to prevent clogging by biological slimes. Stone filter depth is 0.9–4.2 m, and is usually 1.8 m. Plastic filter media are being used more and more (15). They require filters that are much deeper than those using stones and pumping costs can be expected to be high over the life of the plant. However, higher removal efficiencies are reported.

Sewage flows slowly downward over the filter medium and the effluent is collected in vitrified tile underdrains which collect the filter effluent and circulate air into the filter. The underdrains discharge into a main collection channel which, in turn, discharges into the final settling tank.

The importance of the final settling, or humus, tank can be seen by an examination of what occurs in the trickling filter itself. A new filter is broken in by applying settled sewage as in the normal operation. After a period of time the microbial, or zoogleal, mass forms on the filter medium and stabilizes the waste. Waste material is first adsorbed, and then assimilated by the microorganisms.

Much of the waste material is transferred to the microbial cells and is utilized for cell synthesis and energy (16). Dead and dying organisms are separated from the zoogleal mass and collected in settling tanks to prevent the clogging of the filter. A rate of application lower than that of the primary tanks is necessary because the organisms to be removed are considerably lighter than the material removed in primary tanks; the application rate is usually 28 L/(m^2·s) [18.6 gal/(yd^2·s)]. Trickling filters may be classified on the bases of hydraulic loading per unit area per unit time and the amount of BOD per cubic meter per day.

Low Rate. The low rate filter is most commonly used in small plants. The loadings are 0.22–0.43 m^3/(ha·s) [23.5–46 gal/(acre·s)] and 0.16–0.32 kg BOD/(m^3·s) [0.6–1.2 g BOD/(gal·s)] (7,17). With proper operation, 80–85% of BOD is removed. Generally, appreciable amounts of nitrite and nitrate are found in the effluents, indicating a high degree of organic stabilization.

A portion of the effluent is recirculated, in order to smooth out flow, keep the food concentration constant, lower film thickness and control psychoda flies, and reseed the applied sewage with acclimatized organisms. The psychoda, or filter fly is a very small insect that breeds in thick trickling-filter slimes. It does not bite, but can be a nuisance. Its radius of flight is small, but it can be carried great distances by the wind. The fly can be controlled in the development phase by occasional flooding of the filter or chlorination of the applied sewage.

Intermediate Rate. At 0.43–1.08 m^3/(ha·s) [46–115 gal/(acre·s)], certain operational difficulties were frequently encountered and therefore this range was avoided for many years. It now appears that these difficulties stemmed primarily from the inability of the hydraulic load to keep the filter slimes from becoming too thick and thereby clogging the filter. The use of relatively large filter stones solves this problem.

High Rate. By increasing the hydraulic loading of sewage to 1.08 m^3/(ha·s) [115 gal/(acre·s)], BOD loadings per unit volume of filter were increased, which, in turn, increased the organic material in the effluent. High recirculation ratios maintained the high hydraulic loading, and at the same time reduced the organic concentration in the effluent. Hydraulic loadings are 1.08–4.32 m^3/(ha·s) [115–462 gal/(acre·s)] and up to 1.44 kg BOD/(m^3·s) [5.5 g BOD/(gal·s)], but removal efficiency is only 65–75%.

Superrate. High BOD removal efficiencies (97%) at hydraulic loadings of 10.8 m^3/(ha·s) [1150 gal/(acre·d)] were obtained in experimental plants with plastic media. In these plants, much of the microbial mass remains in the recirculated effluent. This process is, in effect, a modification of activated sludge. Organic loadings are ca 1.60 kg BOD/(m^3·s) [6.1 g BOD/(gal·s)].

Modifications. Modifications of the trickling-filter included improvement in media, air circulation, loading, and means of bringing the waste into contact with the zoogleal mass. In one modification, sewage was introduced at various levels of a very deep filter in order to distribute the load more evenly. This can be compared to the step–aeration modification of the activated sludge process. Hydraulic loadings of this modification are as high as 54 m^3/(ha·s) [5800 gal/(acre·s)].

Rotating Biological Contactors. Rotating biological contactors (RBC) (18) are composed of multiple plastic or stainless steel disks which rotate slowly through incoming settled wastewater; ca 40% of the disk area is submerged. A growth similar to trickling-filter slime develops on the contactor surface and oxygen is supplied from the air when the disk is exposed. Recirculation of effluent is not necessary. The principal design parameter is the wastewater flow rate per unit disk area, ca 155 L/(m^2·d) [103 gal/(yd^2·d)]. Peripheral speeds are ca 18 m/min. Using four stages it is possible to obtain an overall BOD reduction of 90%. The RBCs, compared to activated sludge, have low energy requirements, resist shock loads, and produce a highly nitrified effluent.

Activated Sludge. In this process, in contrast to trickling filters, activated sludge floc is suspended in the moving wastewater stream. It originated in efforts to purify sewage by blowing air into it. It was observed that flocs composed of voraciously feeding organisms developed after prolonged aeration. After aeration was stopped, the floc settled. Addition of fresh sewage to the tank containing the sludge produced high purification in a reasonable time, and therefore, the floc was called activated sludge. As in the case of the trickling filter, it was first operated as a fill-and-draw bed. Further research proved that continuous operation could be practiced.

Some activated sludge is returned to the aeration tank influent and the excess sludge discharged to digestion. The sludge–sewage mixture is then purified by aeration, and the aeration tank effluent is settled to remove floc from the plant effluent. The last step removes the organic material transferred from the flowing wastewater to the cell mass. The floc is formed in sewage by aerobic growth of unicellular and filamentous bacteria, eg, protozoa and other organisms. These organisms gain food and energy by feeding upon the sewage.

Aeration tanks are usually rectangular in cross section, 3–4.5 m deep and 9 m wide. The ratio of surface length to width should be >5:1 to ensure thorough mixing. Detention periods in the aeration tanks are 4–8 h. This is a strictly aerobic process and air requirements are high. Satisfaction of oxygen requirements is difficult because oxygen is sparingly soluble in water (10 mg/L or 38 mg/gal).

For aeration, diffused-air or mechanical units are used. Air diffusers are commonly used in North America, but some mechanical aeration units are installed in plants with capacities >3,800 m^3/d (10^6 gal/d). Aeration transfers oxygen to the sewage and maintains aerobic conditions, mixes sewage and floc, and keeps the floc in suspension.

The air introduced from the diffusers creates a spiral flow pattern, thus promoting the mixing of floc and sewage and eliminating dead spaces. Rotating brushes, surface impellers, or submerged vertical turbine aerators provide mechanical aeration. Rotating brushes and mechanical surface aerators help the transfer of oxygen from the air at the air–water interface. Draft tubes are required for deep tanks. Slow speed turbines, located ca 0.5 m above the bottom of the aeration tank, distribute compressed air throughout the tank. They can be used with deep, narrow tanks, but compressed air is required.

The floc returned to the aeration tank (10–50%) has the same function as trickling-filter slime. However, the floc concentration can be varied as operational needs dictate. Usually the so-called mixed liquor–suspended solids (MLSS) are 600–4000 mg/L (2.3–15 g/gal).

A very important parameter in routine control of the process is the ratio of the MLSS volume to the dry weight, called the Sludge Volume Index (SVI). A well operating plant has an SVI of 50 to 100, but operational difficulties occur when the SVI approaches 150. At this point the sludge settles poorly.

The activated-sludge process is microbiological in nature and factors that promote or inhibit growth are important, including the pH, temperature, and oxidation–reduction potential. The pH determines which organisms predominate in the system. Bacteria are the dominant organisms at pH 6.5–9, fungi below pH 6.5. In the presence of metabolic products, buffering capacity must be adequate.

Modifications are usually the result of particular operating needs. Future improvements can be expected by process development rather than changes in hardware (19).

High Rate Activated Sludge. A low MLSS concentration of 200–500 mg/L (0.8–1.9 g/gal) is maintained (15), resulting in a high ratio of food to microbial mass, which keeps the floc in an active growth phase; excess food is discharged in the effluent. Although the BOD is not removed at a high rate, there are areas in which this technique is adequate. New York City uses this method because the sewage is weak and temperatures are low. Poor results were obtained in Philadelphia and Los Angeles because the sewage is strong, temperatures are higher, and pipe detention times are long (16).

Completely Mixed Activated Sludge. Intimate mixing of returned sludge and settled wastewater, homogenizes the contents of the aeration tank, including the organic load. The wastewater is thus subjected immediately to attack by fully developed organisms; biological stability is enhanced by this design (20).

The net growth rate of the microbial population approaches zero by increasing the aeration time in order to extend the period in which the activated sludge is in the endogenous-growth phase. Although it is not possible to operate an extended aeration system without sludge accumulation, the resulting sludge is inert to activated-sludge organisms.

Tapered and Step Aeration. In New York City, plants scattered throughout the five boroughs are treating more than 3.8×10^6 m^3 (10^9 gal) of sewage each day. In conventional plants, sewage is introduced at one end of the aeration tank and flows straight through. The oxygen demand at the influent end exceeds that at points further along the flow path. In order to satisfy the oxygen requirements, the number of diffusers was increased at the beginning and decreased toward the effluent end. This arrangement is known as tapered aeration and is now common in conventional plants. When all of the floc of sewage is added at the beginning of the aeration tank, the oxygen demand is high because of the large supply of microbial food. In step aeration, the sewage (food) is fed at various intermediate points along the aeration tank. The microbiological food is thus applied evenly and oxygen requirements are lower. This modification has increased the capacity of existing overloaded plants.

Biosorption. This process makes use of a phenomenon observed but ignored by many investigators. If activated sludge floc and raw sewage are mixed and aerated, there is first a significant reduction in BOD of the effluent. It is followed by a notable BOD rise and then by a decrease. Originally, the initial decrease had been attributed to experimental error; however, it was shown to be due to adsorption of colloidal material onto the activated sludge floc. Plants serving Austin, Texas, and Bergen County, New Jersey, were converted from being overloaded to underloaded by changing from conventional activated sludge to biosorption.

Kraus Process. The municipal wastewater-treatment plant at Peoria, Illinois, received a waste high in carbohydrates, causing a nitrogen and oxygen deficiency in the activated sludge which was light and settled poorly. Addition of a sludge-digester nitrogenous supernatant to the plant influent corrected the imbalance and the resulting floc settled well. Removal of organic material from the flowing wastewater was improved (16).

Sludge Digestion and Disposal

Putrescible material collected from primary and secondary tanks must be disposed of cheaply and efficiently because it is a nuisance source. It can be stabilized by biological means. Raw sludge is ca 95% water and the 5% solids cannot be easily separated. As sludge is broken down, the water content is lowered and the volume is greatly reduced. Water content of sludge reduced from 95 to 90% reduces the sludge volume by one-half and the reduction to 85% reduces the volume to one-third of the original. Fresh sludge has a gray color and can be easily pumped. It has a most disagreeable odor, due primarily to thiols.

Digestion. Digestion reduces the volume and the pathogenic organisms. Digested sludge is black and granular and has a tarry color. Sludge is withdrawn periodically from the settling tanks and allowed to flow by gravity to a collection well. From there it is pumped to a digester where it is thoroughly mixed.

Sludge digestion can be aerobic or anaerobic. The equipment used for aerobic digestion closely resembles that used for the activated-sludge process (18). Aerobic digestion is mostly applied to waste-activated sludge. Most of these digesters are for small-to-medium plants and most package plants have aerobic digesters. Aerobic digesters are operated as either batch or continuous-flow reactors; the latter type is more common.

Sludge is destroyed by microorganisms and the kinetics of their life processes is temperature dependent. Short anaerobic digestion detention times are obtained at 35°C. Even shorter detention times are possible at 52–54°C, but detention in this range is costly. An increase in detention time occurs at 35–43°C and then a progressive decrease takes place until 52–54°C. This variation is caused by a change in character of the dominant process organisms.

In older installations, the digesting sludge was heated by hot-water coils in the tank periphery. It was thought that adequate mixing resulted from turbulence induced by gases generated, however, this was not the case. Heat was transferred only in the immediate vicinity of the coils, and the remainder of the tank was not heated well. External heat, fueled by sludge digestion gases, results in better mixing and more uniform temperatures. External heat exchangers have almost completely replaced hot-water coils in new plant construction.

Sludge gas consists of 65–70% methane and 25–30% carbon dioxide. Anaerobic sludge digestion also produces traces of H_2S and hydrogen. Generally, the gases generated in sludge digestion are sufficient to provide for the immediate fuel needs of the plant in order to maintain the digester temperature, provide hot water, and fuel for incineration of sludge (if practiced) and generators. Electricity constitutes the principal operating cost of a treatment plant, but significant savings may be possible with on-site generation. In any case, stand-by generation must be provided in case of power failure.

Volatile acids, reported as acetic acid, are the most important operational parameter. In a well-operating digestion process, the value should be <1 g/L (3.8 g/gal). A value >6 g/L (23 g/gal) indicates malfunctioning; optimum pH is 6.8–7.2, and a pH <6.8 indicates excessive volatile acid production. Formerly, lime was added to the digester contents if the pH showed an undesirable drop. However, the reduction in pH indicated a change in organism that could not be remedied with lime (2).

Aerobic digestion causes fewer maintenance problems than anaerobic digestion; capital costs are lower, and supernatant BOD is lower. The advantages of anaerobic digestion are lower energy requirements because mixing and aeration are not necessary and the production of methane is a useful by-product. Furthermore, the solids content of the digested sludge is higher and, therefore, the volume of sludge to be handled significantly lower. Addition of a thickener-clarifier before aerobic digestion greatly reduces this difference.

Disposal. Digested sludge is reasonably inert but has a high water content. It can be dewatered by filtration or by drying on open or covered sand beds. Open sand beds are exposed to the atmosphere and the sludge dries by downward drainage and evaporation. A covered bed resembles a greenhouse with high temperatures and humidity. In both cases, sludge is allowed to flow by gravity from the digester to the sand beds and then permitted to stand for the required time. The dried sludge is scraped from the beds and taken to ultimate disposal. Sludge may also be dewatered by vacuum filtration.

The dried sludge may be incinerated or used for landfill or fertilizer. Sewage sludge does not have wide application as a fertilizer (qv), partly for aesthetic considerations and partly because of the low nutritive values. Refusal by the public to accept sewage sludge as fertilizer came long before recognition of the danger of toxin accumulation.

Many substances that are separated from the flowing wastewater may appear essentially unchanged in the final digested sludge. Concentrations of undesirable materials may be greatly increased in the sludge. The ultimate disposal site must be chosen with great care and contained in order to prevent release of toxic substances to the surrounding environment. Toxic materials may concentrate through biomagnification as they proceed up the food chain.

Ocean Disposal. Disposal of raw or treated sludge by barging to sea was practiced for many years by some coastal cities, but today is highly controversial, and it appears that this method will be no longer economically feasible. New Federal regulations require that sludge be taken to disposal sites about 160 km from the coast, whereas formerly, disposal sites were permitted within 20 km offshore. Transportation costs are expected to be so high that ocean disposal will be discontinued. New methods of sludge stabilization and volume reduction must be developed.

Chlorination of Effluent

It is common practice to chlorinate wastewater effluents for bacterial control. Regulations vary from state to state, but all require chlorination to specified residual concentrations. Regulations change with the season and the most stringent are in effect during the swimming season. A phenomenon not fully understood is aftergrowth. Immediately after discharge to the receiving water, the bacterial count is low, but then suddenly rises. Questions have been raised concerning the benefit of effluent chlorination. It has been found that some unusual organic industrial chemicals pass unchanged through conventional treatment plants. Chlorination of these substances gives products that are suspected of being carcinogenic. Drinking such waters may be dangerous to health. These considerations have brought about renewed interest in ozone, other halogens, and ultraviolet irradiation as means of bacterial control.

Storm-water overflow is chlorinated, and in some plants the influent may be chlorinated for odor control.

Other Disposal Problems

Watercraft-Waste Disposal. The popularity of recreational boating has brought with it the problem of disposing of wastes generated by the boat users. In many rivers and lakes, this has become a serious problem. A large number of pleasure craft can place, in a weekend, a pollutional load equivalent to a small community on a medium-sized water body. As legislation was passed by the states, it became apparent that this problem would require a solution at the national level. Various technologies are available, but no agreement could be reached on which to use. Units available include the holding tank, macerator–chlorinator, and on-board incinerators. The Coast Guard was given the task of recommending a solution but became mired in bureaucratic tugs-of-war. It was agreed that holding tanks would keep the waste from reaching the receiving waters directly from the boat. However, availability of facilities to receive the retained waste material ashore was limited. In many cases, the pumped-out waste was discharged to the receiving water as soon as the sun had set. After several years of indecision, rules were promulgated that allowed the use of the macerator–chlorinator for a period, followed by total conversion to holding tanks. Macerator–chlorinators chop the waste into fine particles for adequate chlorine contact and are regarded as potential sources of bacterial contamination. Incinerators did not achieve the same acceptance on pleasure craft as they did on railroads.

Surfactants. Long-chain soaps degrade using biological treatment and digestion (see Soap). In the early 1950s, synthetic detergents (syndets) came into wide use; they are mainly based on alkylbenzenesulfonates (ABS) with phosphates as builders (see Surfactants and detersive systems). The early syndets were resistant to biodegradation and passed unchanged through wastewater-treatment processes. Excessive foam formed in activated-sludge aeration tanks, points of high turbulence, and receiving waters. Suds production was not essential for cleaning efficiency, but was a merchandizing asset. Producers were reluctant to replace high suds detergents with a low suds product. In Europe, sewer ordinances were enacted, which forbid discharge of biologically hard detergents after a grace period. In North America the approach was that of gentle persuasion. In most countries, substitution of high foaming by biodegradable detergents was accepted as the best solution.

Phosphates. Phosphates are used as builders in detergents and function primarily as water softeners. A critical NO_3^- (N):PO_4^{3-} (P) ratio of 1:15 is necessary before algal blooms occur, which are essential for eutrophication (see Water pollution). It is necessary to reduce the amount of phosphate entering a treatment plant, which phosphates pass unchanged. The total phosphate entering the sewer systems has been greatly reduced by introduction of low phosphate detergents. In the period of controversy over lowering phosphates it was also suggested that highly nitrified effluents might be undesirable.

Health and Safety Factors

Wastewater-treatment plants have numerous hazards to be expected in a chemical-process plant. Worker safety is covered by applicable OSHA, state and local standards, but two hazards require special notice (9); the potential for infection by pathogenic organisms is always present, and plant workers require innoculation against the common waterborne diseases. In addition, since wastewater treatment plants utilize deep water-filled tanks, provision must be made against drowning. Proper railings should be installed around the tanks and life-saving equipment should be available for immediate use. A special hazard is the biomass concentration in aeration tanks. Ingestion of this floc has caused a number of fatalities.

Government Regulations

The modern approach to wastewater treatment, protection of the oxygen resources of the receiving waters, requires that all aspects of the problem be addressed, ie, the systems approach. The Ohio River Sanitation Commission (ORSANCO) is an excellent example of basin-wide management dealing with situations that involve several political entities. This approach has been adopted in several other regions.

A succession of Federal agencies and administrations has been charged with dealing with wastewater. At present this responsibility resides with the EPA. It has been proposed that most of the EPA functions be turned back to the states. It remains to be seen if the states, with conflicting needs and priorities, will be able to deal successfully with these problems.

The National Pollutant Discharge Elimination System (NPDES) is a cornerstone of the Federal efforts to control water pollution (21). It determines what can be discharged to a publicly owned treatment plant. Indirect discharges may not be required to obtain an NPDES permit but must meet pretreatment effluent limitations and conditions of the NPDES permit of the treatment plant cannot be exceeded.

Recent Developments

In recent years, some wastewater-treatment plants have utilized pure oxygen in place of air. This permits higher oxygen concentration gradients within the liquid phase and thus allows higher biomass concentrations (22). The use of pure oxygen requires less energy, but the cost of pure oxygen must be taken into account. The aeration tank is under slight pressure and three stages are usually provided. Mixed liquor-suspended solids are 3–10 g/L (11–38 g/gal). Significant increases in volumetric loading rates, reduction in sludge production, and lowered treatment costs are reported.

High pressure application is under investigation (23). The vertical tube reactor (VTR) uses an extended vertical U-tube with concentric tubes for achieving elevated pressures and temperatures and longer retention time for wet chemical reaction. The tubes may extend as much as 1.5 km into the earth. The waste is pumped into the inner tube along with air. As the waste stream and air flow downward the stream undergoes pressurization due to hydrostatic head. Heat is generated by the exothermic reactions of the chemical oxygen demand (COD). The process is reported to produce a stabilized effluent in the outer tube and spares land area.

The fluidized-bed bioreactor (FBBT) (24) increases the capacity of existing plants. Primary effluent is passed upward through the columnar reactor filled with sand or carbon with sufficient velocity to fluidize the bed. An attached biomass develops on the bed particles. Intimate contact between the biomass and waste is provided and improved removals are reported. Oxygen is provided by a deep U-tube reactor. No biomass recirculation is required and a secondary clarifier is not necessary.

Another recently introduced process for sludge stabilization and volume reduction operates on the principle of superoxidation. In the oxyozosynthesis process, small amounts of ozone are applied to acidified sludge in a pressurized vessel. The reactor contents are maintained in a highly turbulent state. The product is reported to have the consistency of coarse paper and be biologically stable. Though promising, no explanations have been advanced for process dynamics and no cost figures have been given (25).

BIBLIOGRAPHY

"Sewage" in *ECT* 1st ed., Vol. 12, pp. 191–207, A. M. Buswell, Illinois State Water Survey Division; "Water, Sewage" in *ECT* 2nd ed., Vol. 22, pp. 104–115, by J. R. Pfafflin, University of Windsor, Ontario, Canada.

1. G. M. Fair, J. C. Geyer, and D. A. Okun, *Water and Wastewater Engineering*, Vols. 1 and 2, John Wiley & Sons, Inc., New York, 1968.
2. C. A. MacInnis, "Municipal Wastewater," in *The Encyclopedia of Environmental Science and Engineering*, Vol. 1, Gordon and Breach Science Publishers, New York, 1976, pp. 587–600.
3. *Standard Methods for the Examination of Water and Wastewater*, 15th ed., American Public Health Association, New York, 1981.
4. J. Salvato, *Environmental Sanitation*, 2nd ed., John Wiley & Sons, Inc., New York, 1972.
5. D. Théry, *Resource Management and Optimization*, Vol. 1, No. 4, Harwood Academic Publishers, New York, 1981, pp. 289–314.
6. K. Imhoff, *Taschenbuch der Stadt Entwässerung*, R. Oldenburg Verlag, Munich, 1960.
7. N. Basta, *Chem. Eng.*, 79 (June 14, 1982).

8. J. M. Lynch and F. Burde, *N.J. Effluents*, 14–15 (Aug. 1982).
9. *Wastewater Treatment Plant Design, Manual of Practice No. 8*, Water Pollution Control Federation, Washington, D.C., 1977.
10. T. Camp, *Trans. ASCE* **111,** 895 (1942).
11. J. Stevens, ed., *Hydrographic Data Book*, Lupold and Stevens, Inc., Portland, Ore., 1969.
12. K. C. Flynn, *J. Water Pollut. Control Fed.* **54**(4), 328 (1982).
13. K. Imhoff and G. Fair, *Sewage Treatment*, 2nd ed., John Wiley & Sons, Inc., New York, 1956.
14. E. B. Phelps, *Stream Sanitation*, John Wiley & Sons, Inc., New York, 1944.
15. J. C. Germain, *J. Water Pollut. Control Fed.* **38**(2), 192 (1966).
16. R. E. McKinney, *Microbiology for Sanitary Engineers*, McGraw Hill Book Co., New York, 1962.
17. *Recommended Standards for Sewage Works*, Health Education Service, Albany, N.Y., 1977.
18. T. D. Reynolds, *Unit Operations and Processes in Environmental Engineering*, Wadsworth, Inc., Belmont, Calif., 1981.
19. J. K. Bewtra, "Biological Treatment of Wastewater," in *The Encyclopedia of Environmental Science and Engineering*, Vol. 1, Gordon and Breach Science Publishers, New York, 1976, pp. 109–126.
20. D. F. Othmer, *Catalyst*, Vol. 2, No. 1, Catalyst Publishing Co., New York, 1973, pp. 9–14.
21. J. R. McWhirter, *The Use of High Purity Oxygen in the Activated Sludge Process*, Vol. 1, CRC Press, Inc., West Palm Beach, Fla., 1978.
22. W. Goldfarb, "Federal Water Pollution Control Law Since 1972," in *Advances in Environmental Science and Engineering*, Vol. 2, Gordon and Breach Science Publishers, New York, 1979, pp. 3–36.
23. W. F. Nolte, "Effects of Elevated Pressure on Secondary Treatment of Wastewater," in *Advances in Environmental Science and Engineering*, Vol. 4, Gordon and Breach Science Publishers, New York, 1981, pp. 161–169.
24. *Meeting the Challenge of the 80's*, U.S. Environmental Protection Agency, Washington, D.C., 1980.
25. C. Newkumet, *Sludge*, 16 (July/Aug. 1980).

JAMES R. PFAFFLIN
Consultant

REUSE

An essentially constant supply of water is available for utilization by all living creatures; however, it is necessary to manage the water resources of a region in such a way that the available water meets all legitimate needs.

Use dictates the quality required. Potable water must be bacteriologically safe, and toxic substances must be present at levels that are accepted as safe (1–3) (see Table 1). In addition, the water must be aesthetically acceptable. Water that is suitable for drinking may be inadequate for many industrial processes. On the other hand, many industrial processes can use water that is not pure enough to drink.

Water consumption is increasing. In 1950, U.S. industry consumed 290×10^6 m^3 H_2O/d (7.7×10^7 gal/d). By 1970, consumption had increased to 624×10^6 m^3/d (1.65×10^8 gal/d), and in 1980, to 964×10^6 m^3/d (2.55×10^8 gal/d) (4).

Formerly, water was accepted by a second user for reuse while it was still under control of the first user (5). Today, the used water is treated in such a manner that it can be used again before ultimate disposal. Furthermore, a distinction can be made between direct reuse, where the water is reclaimed without dilution or natural purification, and indirect use, where treated used water is returned to the environment for subsequent utilization as a raw water supply.

A notable example of controlled water reuse was utilization of secondary sewage effluent from the Back River Wastewater Treatment Plant in Baltimore by the Sparrows Point Works of Bethlehem Steel (6). The Sparrows Point plant was supplied primarily by wells located near the brackish waters of Baltimore harbor. Increased draft on the wells had led to saltwater intrusion. Water with chloride concentration as high as 10 mg/L is unsuitable for many steelmaking operations. Rollers, for example, are pitted by such waters. However, treated effluent from the Back River Plant can be used for some operations, such as coke quenching, and $>4 \times 10^5$ m^3/d (10^8 gal/d) are piped 13 km to Sparrows Point. This arrangement has proved economical to both parties for >40 yr.

Water can seldom be reused directly. The treatment required depends on the intended second use. Disposal costs of the wastewater must be included in any economic analysis, and additional treatment for reuse may be justified when this expense is included. Costs of reclamation depend on the location, water scarcity, availability of public water supplies, and the intended reuse.

Owing to increased public awareness of water pollution, stringent waste-disposal regulations have been introduced, and best available technology (BAT) must be applied to industrial wastewater treatment by the mid-1980s (7). The cost–benefit analyses in BAT applications are controversial. Whereas three decades ago waste disposal was indicated only by an arrow in the process flow diagram, today it is a significant part of the operating costs. Pure water is expensive, and high quality sources are becoming more scarce. It is frequently necessary to accept lower quality water and subject it to costly treatment. Used water as a source of raw water supply has long been permitted by public health authorities in water-poor areas. However, in regions with an ample supply of fresh drinking water, there is resistance to the inclusion of reused water. A Gallup Poll indicated a 50% opposition to the use of reclaimed water in drinking water (8).

Table 1. Maximum Allowable Concentrations of Significant Constituents in Municipal (Domestic) Water Supplies[a]

Constituent	µg/L	Constituent	µg/L
arsenic	50	manganese	50
barium	1,000	mercury	2
cadmium	10	methoxychlor	100
chromium	50	nitrate (nitrogen)	10,000
copper	1,000	phenol	1
endrin	0.2	selenium	10
iron	300	silver	50
lead	50	toxaphene	5
lindane	4		

[a] Ref. 2.

Conventional secondary wastewater treatment does not produce an effluent suitable for direct reuse, and a tertiary treatment step is required. Secondary treatment reduces BOD, COD (chemical oxygen demand), and ≥90% of suspended solids. Microbial concentrations are greatly reduced, and remaining organic material in the effluent is further decomposed by natural processes. Some nitrogen and phosphorus compounds are not removed and may cause eutrophication in the receiving waters. Contaminants that cannot be removed by conventional means are termed refractory, and they range from simple inorganic salts to complex organic substances such as pesticides, herbicides, and surfactants. Recycling of effluents from wastewater treatment plants in an essentially closed system normally would result in unacceptably high concentrations of many undesirable substances. The components of typical tap water are compared with those of a secondary effluent in Table 2 (9).

Continued recycling of effluent would soon given an unusable product, and tertiary treatment, although expensive, is essential. In some instances, considerable amounts of water can be saved by careful housekeeping and process change.

In many cases, the quality of a stream or another water source can be adequately improved by removing more BOD or suspended solids. In other instances, the effluent is prepared for groundwater recharge which may require only the removal of nutrient. A classification of wastewater treatment processes is given in Table 3.

Tertiary Treatment

Chlorination. Chlorination kills bacteria and is routinely included in secondary treatment. For some special uses, it gives a water quality that is acceptable for blending with other water in storage reservoirs.

Chemical Precipitation. When applied after conventional secondary treatment, chemical precipitation removes heavy metals and gives a high quality effluent. In this process, an insoluble compound is formed and the resulting precipitate allowed to settle. Charges on colloidal particles are neutralized, and particles agglomerate. The resulting floc sweeps other material, including bacteria, from suspension. Phosphates are removed by addition of soluble aluminum or iron salts. Insoluble aluminum phosphate or ferric phosphate are formed, but the removal may require adsorption on hydroxide floc. Addition of lime is the cheapest and most easily controlled method. At high pH, lime, in addition to removing hardness, forms a precipitate of hydroxy-

Table 2. Increase of Tap Water Components in Wastewater

Component	Tap water, mg/L	Secondary wastewater, mg/L	Difference, mg/L
BOD	2	20	18
COD	12	100	88
alkylbenzenesulfonates	0.4	6.8	6.4
cations			
Na^+	50	125	75
K^+	3	13	10
NH_4^+	0.1	16	15.9
Ca^{2+}	42	60	18
Mg^{2+}	16	19	3
anions			
Cl^-	66	143	77
NO_3^-	5	12	7
NO_2^-	0.15	1.5	1.3
HCO_3^-	198	296	98
CO_3^{2-}	0.1	1	1
SO_4^{2-}	56	84	28
SiO_3^{2-}	29	43	14
PO_4^{3-}, total	8.1	25	17
$PO_4^{3-\,b}$	0.3	25	25
hardness[c]	158	235	77
alkalinity[c]	164	242	78
total dissolved solids	382	700	318
pH	8.0	7.4	−0.6

[a] Ref. 9.
[b] Other than orthophosphates.
[c] As $CaCO_3$.

apatite, $Ca_5(OH)(PO_4)_3$. Lime also assists in removing sulfides and fluorides. The sludge is putrescible and a nuisance source, and disposal is difficult.

Filtration. Filtration is usually a misnomer for tertiary processes that remove particulate matter. Small particles are removed by adsorption rather than by physical straining. If secondary effluents contain a high concentration of solids, filter beds clog and binding occurs at the bed surface. Energy losses become high, and short circuiting passage of dirty water. Sand, mixed media, and diatomaceous earth are the most common filter materials.

Rapid sand filters have a limited use in tertiary wastewater treatment. Suspended solids accumulate in the filter bed, and BOD in the water is reduced. Adsorbed foreign matter is removed by backwashing; frequency depends on the surface application rate and concentration of suspended material. For backwashing, water may be used alone at a rate of 100–120 mL/(cm²·min) (25–30 gal/(ft²·min)), or a combination of air and water may be used, with water applied at a rate of 40–60 mL/(cm²·min) (10–15 gal/(ft²·min)) and air at ca 120 mL/(cm²·min) (30 gal/(ft²·min)). Air serves to thoroughly agitate the filter. Particle range of the medium is 0.5–0.8 mm, bed depth is ca 60 cm, and the application rate of raw water is 8–10 mL/(cm²·min) (2–2.5 gal/(ft²·min)). Most of the particles are removed in the upper part of the filter, and bed porosity in this region decreases rapidly as the filtration progresses, excessive head loss results, and

Table 3. Wastewater Treatments[a]

Process	Substance treated or removed
Biological	
conventional secondary treatment	suspended solids, BOD, bacteria
conventional process modifications	suspended solids, BOD, bacteria, nutrients
anaerobic denitrification	nitrates
algae harvesting	nitrates and phosphates
Chemical	
ammonia stripping	ammonia nitrogen
ion exchange	nutrients
electrodialysis	salts
chemical precipitation	suspended solids and phosphates
Mechanical	
activated carbon adsorption	organic matter and suspended solids
sedimentation	suspended solids and bacteria
filtration	suspended solids and bacteria
microstrainers	very small particles
reverse osmosis	salts
distillation	salts
foam separation	detergents
vapor-compression evaporation	concentration of wastewater
wasteheat evaporation	concentration of wastewater
steam stripping	hydrogen sulfide and ammonia

[a] Ref. 7.

the filter clogs. Effective filter depth can be increased by mixed media, with the lightest and coarsest material at the top and the heaviest and least porous at the bottom. This is the opposite of the case where the bed contains material of constant density. In a mixed-media filter, three or more materials are used, including coal, sand, and garnet, with specific gravities of 1.65, and 2.65, and 4.00, respectively. Coal has the greatest porosity and garnet the least. After backwashing, where the filter acts as a fluidized bed, the filter materials stratify and the coal is on top. The resulting particles are graduated from ca 1 mm at the top to 0.15 mm at the bottom, and the entire filter bed is utilized for particulate removal. The garnet at the bottom forces the water to pass a much finer barrier than that provided by sand alone, whereas the coal layer does not clog as readily. Filter depths are 60–75 cm, and application rates of 4–120 mL/(cm$^2\cdot$min) (1–30 gal/(ft$^2\cdot$min)) are common.

In diatomaceous-earth filtration, the powdered filter aid is built upon a relatively loose septum to screen out suspended solids. The filter becomes clogged, and pressure losses become excessive; backwashing is then necessary. The smallest removable particle is 0.5–1 μm (see Diatomite).

Both vacuum and pressure filters are used. Turbidity is more easily removed by vacuum filters, usually at 85% efficiency. Flow rates are low, ca 4 mL/(cm$^2\cdot$min) [1 gal/(ft$^2\cdot$min)]; these filters are not practical for treating large volumes.

Microstrainers. Microstrainers are rotating steel screens with extremely fine stainless steel mesh (85–170 perforations per square centimeter (13–26/in.^{2})). The flowing liquid enters the open end of the drum and passes through the mesh to the effluent end. The mesh traps solid impurities and rotates with the drum. A wash-water spray washes the trapped solids into a hopper for final disposal. The mesh is washed with filtered effluent discharged from jets fitted into the drum and then exposed to uv radiation to inhibit microbial growth. The mesh is washed with chlorine water at intervals of seven to twenty-eight days in order to control slime growth; removal efficiencies are 30–55% of the applied BOD and 40–60% of suspended solids.

Effluent Polishing. The term polishing is sometimes applied to the preparation of effluents of exceptional clarity. Plant effluent is collected in a sump and pumped through a tube-settling unit, followed by mixed-media filtration. Filter effluent is collected in a storage tank, which also serves as a reservoir for backwash water. The tank further acts as a chlorine-contact chamber. The tube settler is a supplemental solids-separation device which allows the filter to operate efficiently during severe operational upsets, such as excessive loading of organic matter or increased hydraulic loading. In such instances, the bulk of the suspended solids in the plant effluent is removed in the tube settler and the filter is not overloaded. These installations produce an effluent low in BOD and suspended solids. The polishing units continue to function well when suspended solids are loaded at a rate of 2 g/L (0.12 oz/gal). The tube settler and mixed media can be used for phosphate removal after suitable treatment for coagulation and flocculation.

Foam Separation. Conventional secondary treatment cannot remove detergents that resist biological treatment such as alkylbenzenesulfonates. Although ABS has been almost completely replaced by linear alkylbenzenesulfonates in household detergents, foam (qv) continues to be a problem. Foam-separation techniques have been applied for removing refractory matter such as organic hydrates and nitrogen compounds (10). A sparger disperses small bubbles of gas, usually air, throughout the waste. The rising gas collects suspended solids and surface-active substances while the foam collects at the liquid–liquid interface. It is removed and collapsed to yield a waste concentrate.

For small installations, column foam separators are more suitable. Waste flows downward in the column whereas gas spargers, located at the bottom, give countercurrent flow. The foam generated is carried upward to a foam breaker and collector.

Trough-type separators are more effective for larger municipal and industrial treatment facilities. The operating principles are the same, but flow is through a trough rather than a column.

Activated-Carbon Adsorption. Treatment of waste by activated-carbon adsorption shows promise of becoming an important tertiary treatment process. It has long been applied for the removal of tastes, odor, and color but has not been suitable for treatment of effluents containing high concentrations of organic matter because of rapid decrease of surface area. As the surface of the carbon becomes coated with adsorbed material, less surface area is available for further adsorption. Activated carbon is available either as fine powder or granules. Use of the latter is preferred for wastewaters. In a packed-bed installation, water is passed through a column filled with activated-carbon particles, and the organic content of the water is reduced along the path of flow. However, the carbon at the influent end becomes exhausted and must be regenerated. A fluidized bed gives more uniform concentration gradient throughout the bed. Packed beds accumulate solids more rapidly than fluidized beds. Activated-carbon adsorption can remove 80% of the COD and 70% of BOD. Regeneration of carbon is accomplished in multistage hearth furnaces in a steam–air mixture at 76°C. Approximately 5% of the carbon is lost in regeneration.

Ion Exchange. Ion exchange (qv) is a method of softening hard water. Calcium and magnesium are exchanged for a cation, usually sodium. Ion-exchange beds can remove 95% of phosphate, 85% of nitrates, 100% of sulfates, and 45% of COD. Color and organic matter are efficiently removed by cationic and anionic mixed beds, but the latter tend to foul the beds. Cost and frequency of regeneration are principal disadvantages of ion exchange.

Oxidation Ponds. Oxidation ponds, or sewage lagoons, long popular in Europe, now also find wide application in North America. Careful design is necessary for successful operation. Although oxidation ponds are, in essence, large open ponds into which waste is led, careful planning allows natural purification processes to proceed in an orderly manner giving a stable and high quality effluent. The main disadvantage is production of odors offensive to nearby residents. In addition, relatively large areas are required.

Many facilities utilize air diffusers or surface aerators for supplying additional oxygen. The capacity of an oxidation pond is mostly a function of the oxygen transferred from the atmosphere to the pond water but also depends on other physical and biological variables. These ponds, usually designed for 2.2–3.3 g $BOD/(m^2 \cdot d)$ (6.5–9.7 oz/(acre·d)), are ca 1.2 m deep. Mixing is very important. If waste material were uniformly distributed throughout the pond, the oxygen demand would be constant throughout the pond. This is, however, not the case, and oxygen requirements vary from area to area. The processes involved in waste stabilization are mostly aerobic, but facultative organisms are dominant at the bottom of the pond. Most of the oxygen is transferred from the surface, but additional oxygen may be added by aerators and diffusers; most surface aerators operate intermittently (11). Natural surface agitation is important, and prevailing wind direction must be considered for surface aeration as well as deflection of odors affecting downwind residents. The pond must be deep enough to avoid weed problems but shallow enough to allow sunlight to penetrate. Algae maintain aerobic conditions. The pond functions as a sedimentation basin, and it may be necessary to periodically remove deposited solids. Current practice is to feed influent at the pond center and withdraw effluent at the periphery. Multiple ponds are used in series (7–8).

Water discharged from an oxidation pond may be used directly for low quality industrial purposes or mixed with other waters (12).

Reverse Osmosis. In reverse osmosis (qv), a solution or suspension flows under pressure through a membrane; the product is withdrawn on the other side. This process can treat dissolved solids concentrations ranging from 1 mg/L to 35 g/L (13). The principal constraint is the requirement that the waste material be relatively nonfouling. Recent advances have been mostly in membrane development, and pilot studies are required (14). Energy costs can be significant, and it is frequently necessary to pretreat influent in order to minimize fouling. Reverse osmosis can deal with particles <1 to 600 nm in size.

Electrodialysis. In reverse osmosis pressure achieves the mass transfer. In electrodialysis (qv), dc is applied to a series of alternating cationic and anionic membranes. Anions pass through the anion-permeable membranes but are prevented from migrating by the cationic permeable membranes. Only ionic species are separated by this method, whereas reverse osmosis can deal with nonionic species. The advantages and disadvantages of reverse osmosis are shared by electrodialysis.

Vapor-Compression Evaporation and Waste Heat Evaporation. Both of these processes remove water from contaminants rather than contaminants from water. They are better suited for industrial installations where excess energy is available. The water thus produced is of high quality and can be used directly. An important advantage is the concentration of waste-residue volume with attendant economies of handling and transportation (see Evaporation).

The saturation temperature of a vapor rises when it is mechanically compressed and its latent heat is available at a higher temperature. Application of this heat to an aqueous stream evaporates part of the water, producing a distillate of pure water. Application of vapor compression has grown significantly since 1960.

Cooling loads can be transferred from process heat exchangers to a wastewater-evaporation system, thus reducing cooling-water requirements and total wastewater volume.

In both processes, care must be taken in dealing with organic compounds that either steam distill or form azeotropes.

BIBLIOGRAPHY

"Water Reuse" in *ECT* 2nd ed., Vol. 22, pp. 116–124, by James R. Pfafflin, University of Windsor, Ontario, Canada.

1. *Quality Criteria for Water*, Environmental Protection Agency, Washington, D.C., 1976.
2. *Fed. Regist.* **4,** 57332 (Aug. 27, 1980).
3. *Guidelines for Water Reuse*, EPA 600/8-80-036, Environmental Protection Agency, Washington, D.C., 1980.
4. W. Hales, *Chemtech*, 532 (Sept. 1982).
5. R. Kasperson, *Public Acceptance of Water Reuse*, University Press of New England, Hanover, N.H., 1977.
6. A. Wolman, *Sewage Works J.* **20**(1), 1 (1948).
7. A. D. Holiday, *Chem. Eng.*, 118 (Apr. 19, 1982).
8. *J. Am. Water Works Assoc.* **65**(8), 513 (1973).
9. *Studies of Wastewater Reclamation and Utilization*, pub. no. 9, State Water Pollution Control Board, Sacramento, Calif., 1954.
10. R. B. Grieves, C. J. Randall, and R. K. Wood, *Air Water Pollut. Int. J.* **8**(8/9), 501 (1964).
11. T. D. Reynolds, *Unit Operations and Processes in Environmental Engineering*, Wadsworth, Inc., Belmont, Calif., 1982.
12. W. W. Eckenfelder, Jr., *Principles of Water Quality Management*, CBI Publishing, Boston, Mass., 1980.
13. S. Sourirajan, "Reverse Osmosis—Process Fundamentals" in *The Encyclopedia of Environmental Science and Engineering*, Gordon and Breach Science Publishers, Inc., New York, 1976, p. 738.
14. E. C. Kaup, *Chem. Eng.*, 46 (Apr. 2, 1973).

General Reference

W. J. Cooper, ed., *Chemistry in Water Reuse*, Ann Arbor Science, Woburn, Mass., 1981.

JAMES R. PFAFFLIN
Consultant

TREATMENT OF SWIMMING POOLS, SPAS, AND HOT TUBS

Swimming pools have experienced rapid growth since World War II and now number about 4×10^6 in the United States. Most numerous (90%) are the residential pools which are about equally divided between above-ground and in-ground types, but the latter are growing at a faster rate. In addition, hundreds of cities operate public pools, and many schools, clubs, and camps have outdoor and indoor pools. Most motels and many hotels, condominiums, and apartment complexes also have pools on their premises. Although commercial pools account for only about 10% of the total, they represent about half of the water that must be treated. The average sizes of residential and commercial pools are ca 64.6 and 363.4 m^3 (17,000 and 96,000 gal), respectively. The number of spas and hot tubs has also grown rapidly since 1970, with an estimated 400,000 spas and hot tubs in the United States in a growing market.

Swimming Pools

Most swimming pools are of the recirculating type in which the water is changed only infrequently. Through continuous treatment by mechanical filtration and chemical additions, the same water can be recycled for many years before the pool need be drained and refilled. Sanitizing chemicals must be added regularly to kill and control disease-carrying bacteria introduced by swimmers and dirt entering the water. It is also necessary to destroy algae whose spores are carried into the water by wind and rain. Unchecked algal growth results in discolored water, unsightly growth on the walls and bottom of the pool, clogging of filters, and provides breeding ground for bacteria. The pH of pool water must be maintained within the proper range for swimmer comfort and optimal effectiveness of chlorine sanitizers. In order to control the corrosive or scaling tendencies of pool water, it is also necessary to maintain a proper balance between pH, alkalinity, and hardness. Undesirable trace metals such as iron, manganese, or copper are sometimes found in feedwater or formed by corrosion of pool equipment. Unless removed, these metals discolor pool water and cause stains, especially damaging to plaster pool surfaces. Filtration of pool water is necessary for removal of suspended solids which otherwise cloud water and interfere with the disinfection process. Good circulation of water is important for proper filtration and dispersal of sanitizers and other chemicals added to pool water.

Sanitizers. Chlorine and its compounds have been and continue to be the foremost chemicals for disinfecting swimming-pool water (1–2) (see also Disinfectants). Chlorine gas, sodium hypochlorite solution, calcium hypochlorite, and chloroisocyanurates provide free available chlorine (FAC), ie, HOCl + ClO$^-$, and are the most commonly used swimming-pool sanitizers (Table 1). Available chlorine is a measure of the oxidizing power of the active chlorine (Cl$^+$) in a compound expressed in terms of elemental chlorine. The recommended free available chlorine concentration is 1.0 ppm in an unstabilized pool and 1.0–1.5 ppm in a stabilized pool. Because of field experience and other work, the industry is moving toward higher recommended FAC levels. Free available chlorine and combined available chlorine (CAC, ie, chloramines) in swimming-pool water are determined colorimetrically by the Palin DPD test using N,N-diethyl-p-phenylenediamine. The biocidal properties of chlorine are due principally to the formation of hypochlorous acid which kills bacteria, algae, and other microorganisms. The rate of disinfection is a function of the free available chlorine concentration, the type of organism, contact time, pH, temperature, pool-water recirculation rate (ie, mixing), and presence of interfering substances such as turbidity,

Table 1. Chlorine Sanitizers Used in Swimming Pools

Compound	Formula	Typical form	FAC, %
chlorine	Cl$_2$	liquefied gas	100
trichloroisocyanuric acid	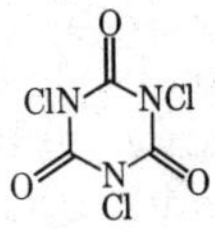 (structure)	tablets, sticks	89–91
calcium hypochlorite	Ca(OCl)$_2$	granules, tablets, briquettes	65 or 70
sodium dichloroisocyanurate	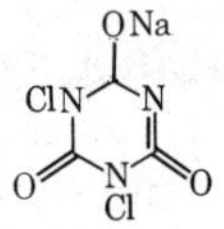 (structure)	granules	62–63
sodium dichloroisocyanurate dihydrate	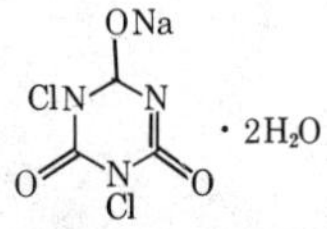 (structure) · 2H$_2$O	granules	55–56
lithium hypochlorite	LiOCl	granules	35
sodium hypochlorite	NaOCl	solution	10–15

organic matter, ammonia, and urea (3). The efficiency of disinfection is determined bacteriologically by the standard plate-count (SPC) technique and by examination for fecal coliform bacteria (4). Other sanitizing agents such as bromine, iodine, chlorine dioxide, and ozone have limited use in swimming-pool sanitation because of high cost or other disadvantages (see Chlorine oxygen acids and salts).

Chlorine. Chlorine gas is used as a biocidal agent exclusively in large commercial pools requiring a high feed rate to maintain the desired residual. Some state and local health codes specify chlorine gas as the required or preferred sanitizer for public pools above a certain size. When chlorine dissolves in water, it hydrolyzes to produce hypochlorous acid, hypochlorite ion, and hydrochloric acid:

$$Cl_2 + H_2O \rightleftharpoons HOCl + H^+ + Cl^-$$

$$HOCl \rightleftharpoons H^+ + ClO^-$$

The dissociation of hypochlorous acid depends upon pH and, to a much lesser extent, temperature, with almost 100% HOCl present at pH 5 and almost 100% ClO$^-$ at pH 10. Because chlorine gas is acidic, soda ash (Na$_2$CO$_3$) or caustic soda solution (NaOH) is necessary to maintain the proper pH.

Chlorine is supplied as a liquefied gas in steel cylinders of 45–68 kg capacity. It requires special metering and feeding equipment and experienced supervision. Because of its toxicity, elaborate precautions against leakage are needed. Although Cl$_2$ gas is not expensive, initial equipment costs, cylinder demurrage, and potential hazards make this method of disinfection impractical for all but very large, heavily used pools. In new, large public pools, hypochlorites or chloroisocyanurates are used increasingly because of the health hazards of chlorine gas. In the UK, chlorine usage is being curtailed for this reason (see Alkali and chlorine products).

Sodium Hypochlorite Solution. Sodium hypochlorite (liquid bleach), usually bottled as a 10–15% available Cl$_2$ solution, is a widely used disinfectant for swimming pools. In pool water, it produces hypochlorite ion and hypochlorous acid:

$$NaOCl \rightleftharpoons Na^+ + ClO^-$$

$$ClO^- + H^+ \rightleftharpoons HOCl$$

Sodium hypochlorite is widely used in California and Florida, the two states with the largest number of in-ground pools, where it is a popular sanitizer with the home-pool owner because of low cost per unit package. It is also widely used for small public pools and for semicommercial pools such as at motels and camps. Disadvantages of sodium hypochlorite are low storage stability and bulkiness which requires a large storage area near large pools.

Calcium Hypochlorite. This chemical, marketed since 1928, is one of the most widely used swimming-pool water disinfectants. Calcium hypochlorite is a crystalline material. It is a convenient source of available chlorine and is sold in granular or tablet form for use in home, semiprivate, and commercial pools. When dissolved in water, $Ca(OCl)_2$ forms hypochlorous acid and hypochlorite ion similar to NaOCl. Because of the presence of small amounts of stabilizing alkalinity in $Ca(OCl)_2$ and liquid bleach, use of these products requires periodic addition of acid to adjust the pH. Calcium hypochlorite has superior storage stability and higher available Cl_2 concentration than liquid bleach, which reduces storage requirements and purchasing frequency.

Chlorinated Isocyanurates. The cyanuric-acid-based sanitizers are crystalline, stable compounds with moderate-to-high available Cl_2 contents. They were introduced for pool use in 1958, and had wide acceptance (5). The sodium salt, sold in granular form, dissolves rapidly, whereas the trichloro compound dissolves very slowly and is widely used in the form of tablets or sticks in feeders, floating devices, or in the pool skimmer.

Other Sanitizing Agents. Specialty sanitizing agents include LiOCl, 1,3-dichlorohydantoin (6), 1,3-dichloro-5,5-dimethylhydantoin (7), 3-chloro-4,4-dimethyl-2-oxazolidinone (8), 1-chloro-3-bromo-5,5-dimethylhydantoin (9), Br_2 (10–12), I_2 (13–15), O_3 (16–17), ClO_2 (16,18), Ag^+ as silver nitrate (19–20), ultraviolet light (21), and NaOCl-producing electrolyzers. Some of these compounds or sanitation methods do not oxidize organic matter or destroy nitrogen compounds and therefore require regular superchlorination or shock treatment to maintain water quality and prevent algal blooms. A polymeric sanitizer, poly(hexamethylenebiguanide) hydrochloride, requires addition of H_2O_2 for oxidizing organic matter. Ozone and uv do not leave residuals and therefore can only serve as secondary sanitizers.

Stabilizers. Cyanuric acid has been successfully used to stabilize chlorine derived from chlorine gas, and hypochlorites or chloroisocyanurates against decomposition by sunlight. It has replaced other less-effective stabilizers such as chloramines (1) and sulfamic acid (22). Cyanuric acid and its chlorinated derivatives form a complex ionic and hydrolytic equilibrium system consisting of ten isocyanurate species. The 12 isocyanurate equilibrium constants have been determined by potentiometric and spectrophotometric techniques (23–24). Other measurements of two of the equilibrium constants important in swimming-pool water report significantly different and less-precise results than the above study (8,25–27).

In the presence of excess cyanuric acid, the predominant chlorinated species is the monochloroisocyanurate ion, $HClCy^-$ (where Cy represents the triisocyanurate anion). Therefore, the only significant equilibria in pool water are

$$H_3Cy \rightleftharpoons H^+ + H_2Cy^-$$

$$HClCy^- + H_2O \rightleftharpoons H_2Cy^- + HOCl$$

$$HOCl \rightleftharpoons H^+ + ClO^-$$

$$Cy = \begin{bmatrix} O \\ NN \\ ONO \end{bmatrix}^{3-}$$

At 25°C, pH 7.5, 1.5 ppm FAC, and 50 ppm cyanuric acid, the extents of the above reactions in the forward direction are ca 81, 2, and 48%, respectively (23). Although the monochloroisocyanurate ion hydrolyzes to only a small degree, it serves as a reservoir of HOCl because of its rapid hydrolysis. Indeed, this reaction is so fast that HClCy$^-$ behaves like FAC in all wet methods of analysis. Furthermore, since HClCy$^-$ absorbs uv only below 250 nm, which is filtered out of solar radiation by the earth's atmosphere, it is more resistant to decomposition than ClO$^-$, which absorbs sunlight at 250–350 nm and represents the principal mode of chlorine loss in unstabilized pools (28). Depending on concentration, Br$^-$ and I$^-$ reduce or cancel the stabilization of FAC by cyanuric acid (29) (see also Cyanuric and isocyanuric acids).

Based on the above equilibria, the concentration of HOCl in the normal pH range varies inversely with the total concentration of cyanurate. Increased concentration of cyanuric acid, therefore, results in decreased effectiveness of FAC. This has been confirmed by laboratory studies in buffered distilled water which showed 99% kill times of *S. faecalis* at 25°C increasing linearly with cyanuric acid concentration at pH 7 and 9 (5). Other studies in distilled water have found a similar effect of cyanuric acid on kill times of bacteria (30–32). By contrast, studies in pool water showed greater kill times than in distilled water even in the absence of cyanuric acid, indicating that unknown variables, possibly ammonia and organic nitrogen compounds, in swimming-pool water apparently have a greater effect on kill time than cyanuric acid or available chlorine under some conditions (33). Ammonia greatly increases the kill time of *S. faecalis* by chlorine whether cyanuric acid is present or not (29).

With a sanitizer such as Cl$_2$, NaOCl, Ca(OCl)$_2$, or LiOCl, the recommended initial cyanuric acid concentration to be used is 50 ppm (34). Since chloroisocyanurates increase the cyanuric acid concentration, the initial concentration is reduced to 20 ppm. No significant increase in stabilization occurs beyond 50–100 ppm, and since an excessively high cyanuric acid concentration may slow down the rate of disinfection, it is recommended that atypically high concentrations be reduced to below 100 ppm by partial drainage of the pool water and refilling with fresh water. Cyanuric acid is determined turbidimetrically after precipitation as melamine cyanurate.

Superchlorination. Superchlorination or shock treatment of pool water is necessary since accumulation of organic matter and nitrogen compounds consumes free available chlorine and impedes the process of disinfection. Reaction of chlorine with constituents of urine or perspiration (NH$_3$, urea, and amino acids) produces chloramines (N–Cl compounds) which are less effective disinfectants than chlorine but in some test procedures (eg, with *o*-toluidine) may be measured as free available chlorine. Monochloramine (NH$_2$Cl), eg, is only $1/280$ as effective as HOCl against *E. coli* (35). Chloramines (primarily NHCl$_2$ and NCl$_3$) are usually responsible for complaints of eye or mucous membrane irritation and objectional odors. Swimmers may blame this condition on too much chlorine, but the problem is caused by insufficient chlorine.

During superchlorination or shock treatment, ammonia is oxidized to nitrogen by breakpoint chlorination.

$$2\ NH_3 + 2\ HOCl \rightarrow 2\ NH_2Cl + 2\ H_2O \tag{1}$$

$$2\ NH_2Cl + 2\ HOCl \rightarrow 2\ NHCl_2 + 2\ H_2O \tag{2}$$

$$2\ NHCl_2 + H_2O \rightarrow N_2 + HOCl + 3\ HCl \tag{3}$$

$$2\ NH_3 + 3\ HOCl \rightarrow N_2 + 3\ HCl + 3\ H_2O \tag{4}$$

Conversion of NH$_3$ to monochloramine (eq. 1) is rapid and causes an essentially linear increase in CAC with chlorine dosage. Further addition of chlorine results in formation of unstable dichloramine (eq. 2) which decomposes to N_2 according to equation 3 causing a reduction in CAC (36). At breakpoint, the process is essentially complete, and further addition of chlorine causes an equivalent linear increase in free available chlorine. Small concentrations of combined chlorine remaining beyond breakpoint are due primarily to organic chloramines. Breakpoint occurs slightly above the theoretical Cl:N ratio (1.5) because of competitive oxidation of NH_3 to nitrate (37–38). Organic matter consumes chlorine and its oxidation also increases the breakpoint chlorine demand. Cyanuric acid does not interfere with breakpoint chlorination (33). Amino acids are also decomposed during superchlorination or shock treatment, with formation of N_2 and other oxidation products; however, the process is much slower than with ammonia (39) (see Chloramines and bromamines).

It is recommended that stabilized outdoor pools be superchlorinated (ie, treated with 5 ppm FAC) every other week when the average afternoon temperature is below 27°C and weekly when the temperature is above 27°C. The FAC should be allowed to decrease to ≤3.0 ppm before swimmers enter the water. When shock treatment is required to eliminate algae or objectionable odors, the chlorine dosage should be increased to 10 ppm (34). Calcium hypochlorite, because of its convenience, is widely used for superchlorination and shock treatment. Sodium hypochlorite, LiOCl, or chlorine gas are also used. Chloroisocyanurates would result in excessive cyanuric acid concentrations, and are therefore not recommended.

Algicides. Algal growth in pools is unsightly, a safety hazard to swimmers, and usually a result of poor pool maintenance. It can cause slipperiness, development of odors, cloudy and discolored water, chloramine formation, increased chlorine demand, bacterial growth, and stubborn stains. Low FAC, high temperatures, sunlight, and certain mineral nutrients promote algae growth. Such growth can be prevented by continuous pumping to ensure efficient sanitizer dispersal and by maintaining the proper pH range and free available chlorine content, supplemented by periodic superchlorination or shock treatment.

The three forms of algae commonly found in swimming pools are the fast-growing green algae, slow-growing blue-green algae (sometimes called black algae), and mustard or yellow algae. The first type is relatively easy to control because it remains suspended in the water; the latter two are more resistant to treatment because they grow on the pool wall where they become firmly attached by penetrating cracks, crevices, and tile grouting. As with bacteria, studies have shown that algae (particularly mustard algae) are more difficult to control in stabilized pools. If algae become entrenched, superchlorination or shock treatment of the water with a hypochlorite (repeated as necessary) in conjunction with mechanical treatment remove most algae. Patches on pool surfaces are removed by trichloroisocyanuric acid or $Ca(OCl)_2$ contained in a porous bag suspended in the affected area.

Algicides other than chlorine are used to a small extent, principally in the hotter southern and western regions of the United States. They serve as a backup to chlorine primarily as a preventive or corrective measure against unbalanced pool conditions. The most widely used algicides are the quaternary ammonium compounds (quats), of which the mixed *n*-alkyldimethylbenzylammonium chloride type (with C_{12}–C_{18} alkyl groups) are the most common (see Quaternary ammonium compounds). Formulation with other algicides improves their effectiveness, eg, *n*-alkyl-dimethyldichlorobenzylammonium chloride and copper bis[2,2′,2″-nitrilotris(ethanol)-*N,O,O′*]

(copper triethanolamine complex) (40). Quats are absorbed on filter media (41–43) and consequently require high initial doses and frequent replenishments. The quats are surfactants and therefore can cause foaming at sufficiently high concentrations. A polymeric quaternary, poly(oxyethylene(dimethylimino)ethylene(dimethylimino)ethylene dichloride), is reportedly non-foaming. Other algicides used to a smaller extent contain $AgNO_3$, copper compounds ($CuSO_4$ and copper chelates with citric acid, gluconic acid, triethanolamine, EDTA, etc), sodium dimethyldithiocarbamate, or 2-chloro-4,6-bis(ethylamino)-s-triazine. The effectiveness of various chemicals against algae has been evaluated in numerous studies under laboratory conditions (40–42,44–46). The concentrations necessary to kill algae (algicidal doses) are generally appreciably higher than those required to prevent or control the growth of algae (algistatic doses) and vary from species to species. Some algicides have chlorine demands which must be taken into account in maintaining the proper FAC (42).

pH Control. The optimal range for bather comfort and efficiency of disinfection of chlorine (Cl^+) sanitizers is pH 7.2–7.6 where the biocidal agent HOCl represents 47–69% of FAC. At pH 7.0, this percentage increases to 78%, but eyes and mucous membranes become irritated and metallic parts of the pool, pump, filter, and ladder corrode more rapidly. At higher pH, the HOCl fraction decreases (eg, to 26% at pH 8.0) and the ClO^- fraction correspondingly increases leading to greater photochemical decomposition and reduced rate of disinfection. Hypochlorite ion is only $1/80$ as effective as HOCl against *E. coli* (35). High pH lowers the buffer capacity and increases carbonate ion concentration, which promotes scale formation which is not only unsightly but can clog pipes and reduce the efficiency of pool heaters and filters (Fig. 1). Sodium dichloroisocyanurate has little effect on pH, whereas the hypochlorites (Ca, Li, and Na) raise it, and Cl_2 gas and trichloroisocyanuric acid lower it. However, the acidity of Cl_2 gas and trichloroisocyanuric acid is partially offset by the alkalinity of the make-up water. The pH of pool water is readily controlled with inexpensive chemicals. Hydrochloric acid solution or sodium bisulfate lowers it, whereas sodium carbonate raises it. The pH is measured colorimetrically with phenol red indicator. High FAC

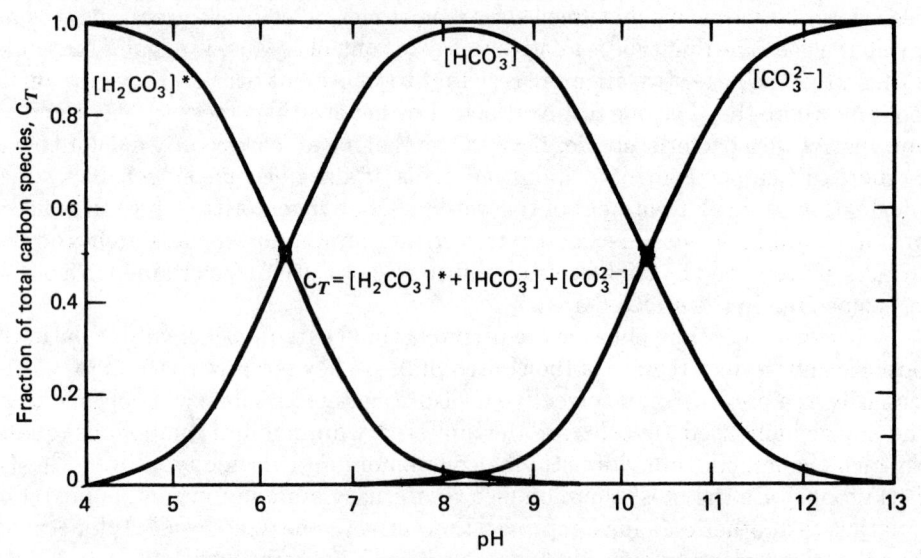

Figure 1. Distribution of carbonate species as a function of pH.

causes lower pH readings due to bleaching of the indicator and resultant HCl formation. Dechlorination with thiosulfate causing higher readings due to formation of hydroxyl ion, eg,

$$HOCl + 2\,S_2O_3^{2-} \rightarrow Cl^- + S_4O_6^{2-} + OH^-.$$

Alkalinity. In swimming-pool water at its normal pH range, the so-called carbonate alkalinity is due primarily to bicarbonate with a very small contribution from carbonate. It is expressed in terms of ppm of equivalent $CaCO_3$. Because of its buffering capacity, alkalinity resists changes in pH when sanitizing chemicals are added to pool water. Since cyanuric acid is 77% neutralized at pH 7.4, cyanurate ion also contributes to alkalinity. Therefore, in stabilized pools the alkalinity determination is corrected for 0.3 × ppm cyanuric acid in the normal pH range (7.2–7.6). Sodium bicarbonate is generally added to increase alkalinity, muriatic acid (HCl) or sodium bisulfate ($NaHSO_4$) to reduce it. In general, with acidic sanitizers such as chlorine gas or trichloroisocyanuric acid, alkalinity should be in the 100–120 ppm range, whereas, with basic products such as calcium, lithium, or sodium hypochlorite, a lower alkalinity of 50–80 ppm is recommended. Alkalinity is determined by titration with standard acid using a mixed bromcresol green–methyl red indicator after dechlorination of the sample with thiosulfate. Because of formation of hydroxyl ion in the latter reaction, the alkalinity requires correction. This correction, approximately equal to the FAC in the pH range 7.2–7.6, however, is significant only at high FAC, eg, during superchlorination or shock treatment.

Hardness. Hardness (qv) is defined as the soap-consuming capacity of water. It is due principally to cations such as Ca^{2+} and Mg^{2+} and to lesser extent (because of lower concentrations) to Sr^{2+}, Ba^{2+}, Al^{3+}, Fe^{2+}, Mn^{2+}, etc. It is expressed in terms of ppm of equivalent $CaCO_3$. In swimming-pool water, the hardness is caused primarily by Ca^{2+}. The concentration of Ca^{2+} must be at or near the $CaCO_3$ saturation value in order to prevent deterioration of the pool and its equipment. At 7.4 pH, 2000 ppm total dissolved solids, 85 ppm alkalinity and 25°C, the saturation Ca hardness is 474 ppm. Values at other pool conditions can be calculated with the Langelier saturation index discussed below. Hardness is raised with $CaCl_2$ and is lowered by draining some of the pool water and adding water of lower hardness or by passing the pool water through a softener or demineralizer. Calcium hardness is determined by titration with EDTA at pH 12–13 using Eriochrome Blue Black R, Eriochrome Black T, calmagite, hydroxy naphthol blue, or other suitable indicator.

Balanced Water. Materials of construction used in pools are subject to the corrosive effects of water, eg, iron and copper equipment can corrode due to presence of dissolved O_2 and anions such as Cl^- and SO_4^{2-}, whereas concrete and plaster can undergo simple dissolution. Maintaining proper water balance minimizes and may even prevent such corrosion by deposition of a protective layer of crystalline $CaCO_3$ (calcite) using the natural Ca and alkalinity in pool water.

$$Ca^{2+} + CO_3^{2-} \rightleftharpoons CaCO_3(s)$$

The ability of water to deposit a protective film can be determined by the Langelier saturation index (LSI) which is the logarithm of the degree of $CaCO_3$ saturation, ie, $[Ca^{2+}][CO_3^{2-}]/K_s$ (47).

$$LSI = pH_a - pH_s = pH_a + pCa^{2+} + pAlk - p(K_2/K_s) - 9.3$$

where concentrations are ppm, K_2 is the second ionization constant of H_2CO_3, K_s is the solubility-product constant of $CaCO_3$, $p(K_2/K_s) = 7.372 - 156.61/T - 0.01656\,T + (2.5\sqrt{I} + 3.63 \cdot I)/(1 + 3.3\sqrt{I} + 2.6 \cdot I)$ (62), T = K, I = ionic strength = 2.5×10^{-5} × ppm total dissolved solids, pH_a is the measured pH, and pH_s is the pH that the water should have to be at $CaCO_3$ saturation (ie, water that does not deposit or dissolve $CaCO_3$).

The numerical value of the LSI for water describes its deviation from equilibrium, ie, >0 = oversaturated, 0 = saturated, and <0 = undersaturated. The extent of saturation of a given water can also be determined experimentally by means of the marble test (48). Addition of finely ground $CaCO_3$ causes a pH decrease or increase, depending on whether the water is oversaturated or undersaturated, respectively. The higher the index, the greater the tendency for either $CaCO_3$ precipitation (LSI > 0) or corrosion (LSI < 0). Water with a sufficiently negative index is not only corrosive to metal or concrete, but it can also dissolve an existing protective film of $CaCO_3$. The extent and rate of $CaCO_3$ precipitation is a function of the saturation index (ie, the degree of supersaturation), the presence of impurities, and the buffer intensity (49) which in turn depends on the alkalinity and pH of the water. Elevated temperature (eg, in heaters) can cause precipitation of $CaCO_3$:

$$Ca^{2+} + 2\,HCO_3^- \xrightarrow{\Delta} CaCO_3 + CO_2 + H_2O.$$

Excessive deposition of $CaCO_3$ (scaling) is not desirable, because it can lead to flow restriction in piping and tubing, reduced heat transfer in heaters, plugging of filters, and even cloudiness. Therefore, the LSI should be maintained within an acceptable range, ie, ±0.5, but preferably as close to zero as possible. Water can be brought to a state of approximate equilibrium (balance) by suitable adjustments in pH, alkalinity, and hardness. Regular addition of chemicals and make-up water and introduction of contaminants by bathers, wind, and rain, causes the LSI to drift in one direction or the other. Therefore, to keep water in balance for a maximum time, the target for balancing must be on the opposite side of zero from that usually found (50).

Other Chemical Treatments. Pool water may occasionally contain metallic impurities such as copper, iron, or manganese which enter the pool with the make-up water or by corrosion of metallic parts in the circulation system. These dissolved metals can impart an unattractive color, cause stains, and catalyze the decomposition of available chlorine. Superchlorination oxidizes soluble Fe^{2+} and Mn^{2+} to the highly insoluble $Fe(OH)_3$ and MnO_2 which can be removed by filtration or allowed to settle to the bottom of the pool from where they are removed by vacuuming. For dispersed colloids, coagulation may be necessary. The water is adjusted to pH 7.2–7.6 and sprinkled with alum, $Al_2(SO_4)_3 \cdot 14H_2O$, 15.2 g/m³ (2 oz/1000 gal). The trace metals, entrapped in the floc of aluminum hydroxide that forms, are removed by filtration or vacuuming after settling. The hydrolysis of the alum consumes a small amount of alkalinity (ca 8 ppm) which can be replenished with $NaHCO_3$.

Water-soluble, high molecular weight polymers can be used as flocculating agents (qv) to increase the settling rate or strength of a chemical floc, thereby acting as a filter aid to control the depth of floc penetration (51). These polymers can also serve as coagulating agents for removal of precipitated trace metals or other undesirable turbidity. The polymers may be anionic, cationic, or nonionic and are called polyelectrolytes when they have ionizable chains, eg, polyacrylates, poly(styrene sulfonate), poly(diallyldimethylammonium chloride), poly(ethylene oxide), etc (52).

Antiscaling agents such as polyacrylates (low molecular weight) or organic phosphonates, eg, (1-hydroxyethylidene)diphosphonic acid, prevent or control precipitation of $CaCO_3$, $Fe(OH)_3$, etc. They function as chelating agents (qv) or dispersants (qv) to prevent excessive formation of hard scale by promoting crystal distortion.

Compatibility with chlorine must be considered in selecting ancillary chemicals since many of these additives have chlorine demands and therefore may interfere with pool disinfection.

Test Kits. Proper pool management requires routine analysis for free and combined chlorine and pH and, less frequently, alkalinity, hardness, and cyanuric acid. These analyses can conveniently be carried out at the poolside with simple test kits which are available at moderate cost. The more elaborate kits with photometers for color intensity measurement and meters for pH measurement are more accurate and more expensive. Test strips for chlorine and pH determinations are not as accurate as tests employing solutions. Water clarity (ie, turbidity) can be measured by the test kit or visually. Digital titrators are available for alkalinity and hardness (Ca and Mg) which read directly in ppm. Test kits are also available for N and Fe determinations. Other analyses such as total dissolved solids, trace metals (eg, Mn, Cu, etc) are carried out as needed by analytical laboratories. Bacteriological analysis of pool water can be obtained through local health departments or commercial laboratories. Detailed procedures for chemical, bacteriological, and biological analysis of water are given in ref. 4.

Filtration. Efficient filtration for removal of suspended particles is essential for pool water. Filters may be of the fixed-bed, precoat, or cartridge type and operate under vacuum or pressure. The most common filter media are sand, diatomaceous earth (DE), anthracite, and paper or cloth cartridges. The cartridges remove particles as small as 20 μm, sand to 8 μm, and DE to 2 μm. Since the visual limit is 30–35 μm, most properly operating filters do a satisfactory job (53). Sand and DE filters are backwashed, whereas cartridge filters are manually cleaned by rinsing in water. Backwashing removes accumulated insoluble matter which would eventually increase resistance to flow (head) and reduce filtration rates. Diatomaceous earth filters require a precoat of DE after each filter cycle (see Diatomite). Use of compressed air (bumping) to dislodge the DE reduces water usage and allows longer filter cycles. The DE is sometimes mixed continuously with unfiltered water before entering the filter. Alum improves the efficiency of sand filters (54). The pump serves not only to circulate pool water through the filter and chlorinator but also to mix the pool water, thereby diluting and dispersing sanitizers and other additives whether added manually or through feeders. Acceptable water quality requires 3–4 turnovers per day which results in 95–98% removal of suspended matter (at equilibrium) (54). Vacuum cleaners, which derive their suction from the filter pump, are used to remove solids from the walls and bottom of the pool. Some automatic pool cleaners moving around the bottom of the pool are efficient enough to eliminate the need for hand vacuuming.

Feeders (Dispensers). Feeders dispense the chemicals in gaseous, liquid, and solid (both granular and compacted) forms. Many health departments require that public pools have approved feeding devices for the daily application of all chemicals, including sanitizing agents. A slurry feeder for DE on diatomite filter installations may also be required.

Chlorine gas is usually fed from a chlorine cylinder equipped with a pressure gauge, reducing valve, regulating valve, feed-rate indicator, and aspirator-type injector for dissolving the chlorine gas in water. The feeder can be manually, or more desirably automatically, controlled utilizing continuous amperometric or potentiometric measurement of the free chlorine residual. The chlorine solution is normally introduced into the return line to the filter.

Liquid feeders employ positive-displacement metering pumps for adding aqueous solutions of sodium or calcium hypochlorite. The feed solutions are typically stored in polyethylene tanks of various capacities up to about 0.19 m^3 (50 gal).

Erosion-type feeders are commonly used for solid sanitizers, generally in a compacted form (typically tablets of calcium hypochlorite or trichloroisocyanuric acid) contained in a cartridge which is immersed into recirculating pool water flowing through a chamber. The dissolving rate of the sanitizer is controlled by varying the flow rate of water and the depth of immersion of the tablets. A simple flow-through feeder for Ca(OCl)$_2$ tablets or briquettes consists of a vertical cylinder with inlet and outlet valves and is based on feeding a saturated solution of Ca(OCl)$_2$. Another device for solid sanitizers is a floater. A popular type consists of a cartridge containing tablets or sticks, which is inserted to a variable depth into a hollow plastic flotation ring. Several units may be used which float freely around the pool continuously dispensing sanitizer. A very simple method of pool disinfection is to add tablets or sticks of sanitizing agent (usually trichloroisocyanuric acid) to the skimmer basket located in the filter return line. The filter pump must remain on until complete dissolution of the sanitizer to avoid excessively high local concentrations of available chlorine which could cause corrosion. A skimmer device for continuous feeding of Ca(OCl)$_2$ uses tablets or briquettes in a plastic cylindrical container with a variable opening in the top.

Feeders for controlling pH when using chlorine gas sanitation are typically based on pumping liquid soda ash, but dry chemical feeders can also be used, such as a type based on adding granular soda ash from a hopper by means of a reciprocating screw feeder to a solution chamber fed by pool water with the overflow mixing with the recirculating pool water. This type can also be used with granular sanitizing agents such as calcium or lithium hypochlorite, or sodium dichloroisocyanurate.

Spas and Hot Tubs

The basic principles of treating swimming-pool water apply also to spas and hot tubs. However, spas and tubs are not miniature swimming pools but are unique in treatment requirements because of use patterns and a high ratio of bather to water. For example, four people in a 1.9-m^3 (500-gal) spa or tub have a sanitizer demand equal to 160 people in a 75.7-m^3 (20,000-gal) swimming pool. Spa or hot-tub water is an excellent environment for the growth of microbes because of optimal temperatures and the introduction of nutrients from bathers. Analysis of unsanitized spa water has shown that a load of 10^3–10^4 microorganisms per mL in a 1.1-m^3 (300-gal) spa can result from a 15-min use by a single bather. Currently, there are no national standards governing sanitation of spas and hot tubs. Standards proposed by the National Spa and Pool Institute (NSPI) are like those for swimming pools (55). Public spas and tubs are under the jurisdiction of local health departments. From the microbiological standpoint, the ideal recommendations exclude bacteria and visible algae. However, local health codes may allow a maximum limit of 200 bacteria per mL as determined by the SPC procedure, and no positive results in a most probable number (MPN) determination for coliforms, ie, organisms that indicate fecal contamination (4).

Spa and hot-tub sanitation is dominated by chlorine- and bromine-based disinfectants. Public spas and tubs usually employ automatic feeders, eg, Cl_2 gas feeders, to maintain a disinfectant residual. Private or residential spas and tubs can use automatic chemical feeding or generating devices, or they can be sanitized manually with granular or liquid products. The most widely used products for private spa and tub sanitation are sodium dichloroisocyanurate and 1-chloro-3-bromo-5,5-dimethylhydantoin. Granular products are normally added before and after use, whereas solids, eg, stick-bromine, are placed in skimmers or feeders. The NSPI recommends a residual of 1.0–3.0 ppm FAC for chlorine sanitizers with CAC not exceeding 0.2 ppm (55). For bromine sanitizers, the maintenance of a 0.8–3.0 ppm residual is required.

Treatment-regime tests with $Ca(OCl)_2$ and chloroisocyanurates killed all pathogenic bacteria, including the *Pseudomonas* that cause skin infections, within 30 s by 2 ppm FAC at pH 7.4. In a series of tests under stabilized conditions (50–100 ppm cyanuric acid), an initial concentration of 3 ppm FAC (the maximum chlorine concentration allowed in swimming pools for bather exposure by the EPA) could not provide water that met the NSPI standards for bacteria or the maintenance of at least a 1 ppm FAC residual. Increasing the initial FAC to 4 ppm resulted in the bacterial criteria being met 100% of the time even when the FAC residual fell below 1 ppm. This demonstrated the necessity of having an initial FAC ≥4 ppm for heavy loads. Statistical analysis of the data showed that cyanuric acid stabilization did not affect the loss of chlorine during the use periods, that the main loss of FAC was caused by bathers, and that the rate of FAC loss was a linear function of the number of bathers. However, the use of chloroisocyanurates for both sanitation and shock treatment resulted in a build-up of cyanuric acid to above the 150 ppm NSPI limit within 15 user periods (ie, four bathers with 15 min of exposure at 40°C), necessitating frequent changing of the water. Furthermore, these studies showed combined chlorine levels to range from 0.8 to 1.6 ppm after a four-bather exposure exceeding the NSPI limit of 0.2 ppm CAC. Shock treatment with ten times the CAC content after each user period was found satisfactory in reducing these CAC levels to 0.2 ppm or less. Based on these studies, it was concluded that a spa or tub should have a 4–5 ppm FAC before bather entry and a 1–3 ppm FAC while the spa or tub is in use. The spa or tub water should be shock treated after each use, and the water replaced every 30–90 d, or more often under heavy use (56).

In addition to replacing spa or tub water because of cyanuric acid build-up from chloroisocyanurate sanitizers, the NSPI recommends that the water be replaced at least monthly or more frequently when often used because bathers contribute microorganisms, perspiration, body oils, lotions, etc, which affect water chemistry, total dissolved-solids concentration, water surface tension (resulting in foaming), and sanitizer demand (55). High bather density causes these compounds to accumulate to relatively high concentrations as compared to a swimming pool that has a lower bather-to-water ratio.

Chemicals such as defoamers, sequestrants, and flocculating agents are used in spas and tubs. In addition, water tints and fragrances are added, further complicating the water chemistry. These chemicals are of concern in spa and tub sanitation since they might reduce the bacterial kill efficiency by lowering the FAC because of chlorine demand or by interfering in the disinfection process. Representative types of commercial ancillary chemicals such as defoamers (aqueous silicone solutions), flocculants (polyelectrolytes), and sequestrants (organic phosphonates) exert a chlorine demand of 0.2–0.3 ppm (see Defoamers). The sanitizer demand of these chemicals must be compensated for to ensure the maintenance of the residual concentrations stated on the product's label.

In wooden tubs, the maintenance of a sanitizer residual becomes complicated due to the leaching of tannins and other organic matter from the wood into the water. The sanitizer demand of these substances must be overcome in order to maintain proper residual concentrations. As the tub ages, the leaching of these materials decreases, but bleaching of the wood may occur as the lignin (qv) in the wood reacts with sanitizers.

In contrast to the closed system of a swimming pool, a spa or tub is equilibrated with atmospheric CO_2 because of the violent aeration which removes CO_2 from the water. At equilibrium, the alkalinity is defined by the partial pressure of CO_2 in the air (49,57–58) and the temperature and pH of the water. Therefore, at a given pH and temperature, the alkalinity in an open-system spa or tub is fixed and is generally substantially lower than that in a closed-system swimming pool which exhibits variable alkalinity. Consequently, higher calcium concentrations are required to maintain a balanced water condition. As in swimming pools, the efficiency of chlorine sanitizers in spas and tubs is dependent on pH, which should be maintained in the range 7.2–7.6.

Spas and hot tubs are manufactured in various shapes and sizes and are of Gunite or fiberglass construction with the latter having either a gel coat or acrylic finish. An average spa has a depth of 0.74 m, a diameter of 2.13 m, and a water capacity of 2.08 m^3 (550 gal). Typical hot tubs are 1.52–1.83 m in diameter with a 1.22 m depth and are constructed of redwood, oak, cedar, cypress, or mahogany. The hot-water support systems of spas and hot tubs are basically the same, consisting of a pump, filter, heating system, and air blower. The heated water is violently mixed by an integrated forced-air system and venturi-type jets. The filter systems are usually cartridge type and the heating systems either electric or gas.

Economic Aspects

The distribution of U.S. sanitizer consumption for swimming pools, spas, and hot tubs is given in Table 2. Spas and hot tubs account for ca 4% of the total. This market is growing at an average of ca 4%/yr with chloroisocyanurates showing a higher growth rate than the inorganic sanitizers. The consumption of algicides in 1978 was ca 1650 metric tons. Quats account for about 78% of the market which is growing at about 6%/yr. The value of ancilliary chemicals used in residential swimming pools, spas, and hot tubs in 1981 was ca $33 × 10^6$.

Table 2. U.S. Sanitizer Consumption for Swimming Pools, Spas, and Hot Tubs

Compound	Percent
sodium hypochlorite	34
calcium hypochlorite	25
chloroisocyanurates	22
chlorine gas	17
others	2

Health and Safety Factors. In spas and tubs, diseases can be transmitted by contact with and ingestion of the water, inhalation of microorganisms contained in aerosols produced by the aerated water, and infection through close personal contact. Skin infections may be enhanced by the high water temperatures which open pores and remove protective oils. To date, the only confirmed disease outbreak associated with hot-water systems (whirlpools) has been nonpruritic rashes caused by *Pseudomonas aeruginosa* (59–62).

The use pattern of a spa or tub is dictated by personal preference. The water temperature of spas and tubs is normally maintained at 36.7–40.0°C. Although information from spa dealers suggests 15 min at 40°C, and 30 min at 36.7°C, individuals are known to spend longer periods. The NSPI recommends a reasonable time limit, followed by a shower and cool down, and, if desired, another brief stay. Long exposure at these temperatures may result in nausea, dizziness, or fainting. Spas and tubs should never be used while under the influence of alcohol (55). Elderly persons and those suffering from heart disease, diabetes, or high or low blood pressure should not use a spa or hot tub without prior consultation with a physician (55). Most important, the water temperature should not exceed 40°C.

Chlorine-based swimming-pool and spa and hot-tub sanitizers irritate eyes, skin, and mucous membranes and must be handled with extreme care. The toxicities are as follows: for chlorine gas, TLV = 1 ppm; acute inhalation LC_{50} = 137 ppm for 1 h (mouse) (63). The acute oral LD_{50} (rats) for the liquid and solid chlorine sanitizers are NaOCl (100% basis) 8.9 g/kg (64), 65% $Ca(OCl)_2$ 850 mg/kg, sodium dichloroisocyanurate dihydrate 735 mg/kg, and trichloroisocyanuric acid 490 mg/kg. Cyanuric acid is essentially nontoxic (oral $LD_{50} \geq$ 20 g/kg in rabbits), a mild eye irritant, and not a skin irritant.

Swimming-pool sanitizers should be kept in closed containers in a cool, dry area, segregated from other nonpool materials (such as paints, solvents, etc), and should never be mixed with each other or with other materials. Explosive NCl_3 can be generated on mixing hypochlorites and isocyanurates together (especially in the presence of water) or with nitrogen compounds (eg, ammonia, urea, quats, etc), mixing chloroisocyanurates with alkaline materials (eg, soda ash solution), or storing moist or wet chloroisocyanurates in closed or unvented containers. Mixing hypochlorites and chloroisocyanurates with pool acid (HCl) generates toxic chlorine gas. At elevated temperatures, ie, in a fire, swimming-pool chemicals release toxic or hazardous gases, eg, $Ca(OCl)_2$ gives O_2 and Cl_2; sodium dichloroisocyanurate gives CO_2, N_2, ClCN, $(CN)_2$, Cl_2, CO, NCl_3, and N_2O; trichloroisocyanuric acid gives Cl_2, N_2, ClC(O)NCO, and trichloroisocyanuric acid vapor; and cyanuric acid gives isocyanic acid.

BIBLIOGRAPHY

"Water, Treatment of Swimming Pools" in *ECT* 2nd ed., Vol. 22, pp. 124–134, by J. P. Faust and A. H. Gower, Chemical Division, Olin Corporation.

1. G. C. White, *Handbook of Chlorination*, Van Nostrand Reinhold Co., New York, 1972, Chapt. 8.
2. *Chem. Week*, 34 (March 2, 1977).
3. G. M. Fair, G. C. Geyer, and D. A. Okun, *Water and Wastewater Engineering*, Vol. 2, John Wiley & Sons, Inc., New York, 1967, pp. 31–36.
4. *Standard Methods for the Examination of Water and Wastewater*, 15th ed., American Public Health Association, American Water Works Association, Activation Project Control Plan, Washington, D.C., 1980.

5. J. R. Anderson, *Swimming Pool Data and Reference Annual* **32,** 86 (1965).
6. B. N. Israel, Ph.D. thesis, University of Wisconsin, Madison, Wisc., 1962.
7. L. O. Petterson and U. Grzeskowiak, *J. Org. Chem.* **24,** 1414 (1959); A. P. Brady, K. M. Sancier and G. Sirine, *J. Am. Chem. Soc.* **85,** 3101 (1963).
8. *Chem. Br.* **19**(6), 468 (1983).
9. J. R. Brown, D. M. McLean, and M. C. Nixon, *Can. J. Public Health* **54,** 267 (1963).
10. H. Farkas-Himsley in A. E. Jolles, ed., *Bromine and Its Compounds*, Academic Press, New York, 1966, p. 554.
11. T. F. LePointe, G. Inman, and J. D. Johnson in J. D. Johnson, ed., *Disinfection, Water and Wastewater*, Ann Arbor Publishers, Inc., Ann Arbor, Mich., 1975, p. 301.
12. H. A. Galal, Ph.D. thesis, Radcliffe College, Cambridge, Mass., June 1961.
13. H. C. Marks and F. B. Strandskov, *N.Y. Acad. Sci.* **53,** 163 (1950).
14. E. V. Putnam, *Parks and Recreation* (April 1961).
15. A. P. Black and co-workers, *J. Am. Water Works Assoc.* **60,** 69 (1968).
16. R. G. Rice and J. A. Cotruvo, eds., *Ozone/Chlorine Dioxide Oxidation Products of Organic Materials*, Ozone Press International, Cleveland, Ohio, 1978.
17. F. L. Evans, ed., *Ozone In Water and Wastewater Treatment*, Ann Arbor Science Publishers, Inc., Ann Arbor, Mich., 1972.
18. W. J. Masschelein in R. G. Rice, ed., *Chlorine Dioxide*, Ann Arbor, Publishers, Inc., Ann Arbor, Mich., 1979, p. 156.
19. J. C. Morris, *The Sanitarian* **16,** 221 (1954).
20. G. C. White in ref. 1, p. 697.
21. G. C. White in ref. 1, p. 703.
22. J. C. Morris in S. D. Faust and J. V. Hunter, eds., *Principles and Applications of Water Chemistry*, John Wiley & Sons, Inc., New York, 1967, p. 23.
23. J. E. O'Brien, J. C. Morris, and J. N. Butler in A. J. Rubin, ed., *Chemistry of Water Supply and Treatment*, Ann Arbor, Publishers, Inc., Ann Arbor, Mich., 1975, p. 333.
24. J. E. O'Brien, Ph.D. thesis, Harvard University, Cambridge, Mass., 1972.
25. H.-C. Hu, *J. Am. Water Assoc.* **73,** 150 (1981).
26. M. Pinsky and H.-C. Hu, *Environ. Sci. Technol.* **15,** 423 (1981).
27. J. Gardiner, *Water Res.* **7,** 823 (1973).
28. J. J. Bishop, J. P. Faust, and S. I. Trotz, *Comparative Light Stability of Aqueous Chlorine, Bromine, and Iodine*, National Swimming Pool Institute Paper, 12th Annual Convention, New Orleans, La., Jan. 1969.
29. G. D. Nelson, *Swimming Pool Disinfection with Chlorinated Isocyanurates*, Special Report 6862, Monsanto Co., St. Louis, Mo., revised May 1975.
30. L. S. Stuart and L. F. Ortenzio, *Soap Chem. Spec.* **40,** 79 (1964).
31. F. E. Swatek, H. Raj, and R. E. Kalbus, *National Swimming Pool Institute Convention*, Las Vegas, Nev., Jan. 21, 1967.
32. G. P. Fitzgerald and M. E. DerVartanian, *Appl. Microbiol.* **15,** 504 (1967).
33. R. W. Linda and R. C. Hollenback, *J. Environ. Health* **40,** 308 (1978).
34. *The HTH New Look Water Book*, Olin Corporation, Stamford, Conn., 1981.
35. N. A. Clarke, G. Berg, P. W. Kabler, and S. L. Chang, *International Conference on Water Pollution Research*, Pergamon Press, London, Sept. 1962.
36. I. W. Wei and J. C. Morris in ref. 23, p. 297.
37. A. T. Palin, *Water Water Eng.* **54,** 151 (1950).
38. Ref. 37, p. 189.
39. Ref. 37, p. 248.
40. R. P. Adamson, M.S. thesis, Arizona State University, Tempe, Ariz., 1978.
41. G. P. Fitzgerald, *Appl. Microbiol.* **8,** 269 (1960).
42. G. P. Fitzgerald, *Water and Sewage Works* **115,** 65 (1968).
43. W. W. Cochran and J. B. Hatlen, *Stud. Med.* **10,** 493 (1962).
44. G. P. Fitzgerald, *Algicides*, No. PB 198130, National Technical Information Service, U.S. Department of Commerce, Springfield, Va., 1971.
45. G. P. Fitzgerald, *Water and Sewage Works* **121,** 85 (1974).
46. C. M. Palmer and T. E. Maloney, *Ohio J. Sci.* **55,** 1 (1955).
47. W. F. Langelier, *J. Am. Water Works Assoc.* **38,** 169 (1946).
48. W. W. Auttman, *J. Am. Water Works Assoc.* **31,** 640 (1939).

49. W. Stumm and J. J. Morgan, *Aquatic Chemistry*, John Wiley & Sons, Inc., New York, 1981.
50. J. P. Brennan, J. P. Faust, and J. J. Tepas, *Parks Maintenance* **31,** 21 (1978).
51. G. L. Culp and R. L. Culp, *New Concepts in Water Purification*, Van Nostrand Reinhold Co., New York, 1974, p. 84.
52. S. Gutcho, *Waste Treatment With Polyelectrolytes*, Noyes Data Corp., Park Ridge, N.J., 1972.
53. *Pool News*, 10 (Feb. 22, 1971).
54. D. G. Thomas, *Swimming Pool Operators Handbook*, National Swimming Pool Foundation, Washington, D.C., 1972, p. 9.
55. *Suggested Health and Safety Guidelines for Public Spas and Hot Tubs*, U.S. Department of Health and Human Services Publication No. 99-960, Centers for Disease Control, Atlanta, Ga., 1981.
56. *PACE Easy Care Guide for Spas and Hot Tubs*, Olin Corporation, Stamford, Conn., 1981.
57. J. M. Edmond and J. M. Gieskes, *Geochim. Cosmochim. Acta* **34,** 1261 (1970).
58. J. F. Thomas and R. R. Trussel, *J. Am. Water Works Assoc.* **66,** 185 (1970).
59. *Morbidity and Mortality Weekly Report* **28,** 182 (1979).
60. J. Washburn and co-workers, *J. Am. Med. Assoc.* **235,** 2205 (1976).
61. W. F. Sausher and co-workers, *J. Am. Med. Assoc.* **239,** 2362 (1978).
62. *Morbidity and Mortality Weekly Reports* **30,** 329 (1981).
63. *Registry of Toxic Effects of Chemical Substances*, Vol. 2, U.S. Department of Health and Human Services, Washington, D.C., 1980 ed.
64. Industrial Biotest Labs, Northbrook, Ill., 1970.

General References

References 1 and 34 are also general references.
Pool & Spa News, Liesure Publications, Inc., Los Angeles, Calif., published semi-monthly.
Spa & Sauna, Hester Communications, Inc., Santa Anna, Calif., published monthly.
Swimming Pool Weekly and Swimming Pool Age, Hoffman Publications, Inc., Fort Lauderdale, Fla., published monthly.
Swimming Pool Data and Reference Annual, Hoffman Publications, Inc., Fort Lauderdale, Fla., published each year in September.
Swimming Pool and Natural Bathing Places, an annotated bibliography 1957–1966, U.S. Department of Health, Education and Welfare, Public Health Service, Bureau of Disease Prevention and Environmental Control, National Center for Urban and Industrial Health, Washington, D.C.
Swimming Pools, Hoffman Publications, Inc., Fort Lauderdale, Fla., revised periodically.
G. C. White, *Handbook of Chlorination*, Van Nostrand Reinhold Company, New York, 1972.
The National Spa and Pool Institute, Washington, D.C., publishes guides and manuals for public and private pools and spas.
Public Swimming Pools, American Public Health Association, Washington, D.C., 1981.

JOHN A. WOJTOWICZ
J. PHILIP FAUST
FRANK A. BRIGANO
Olin Corporation

WATER GLASS. See Silicon compounds.

WATERPROOFING AND WATER/OIL REPELLENCY

Early techniques of waterproofing textiles, paper (qv), leather (qv), and masonry led to the coating of masonry with pitch and asphalt (qv) and of textiles with wax or plastic which made them impermeable to water and air (see Textiles). However, in the early 1900s, textiles were treated to repel water but remain permeable to air; this resulted in increasingly comfortable rainwear. In the mid-1900s, textiles were treated to repel water, oil, and soil and remain permeable to air. Chemicals that are the same as or closely related to the repellents for textiles are now useful on paper, leather, and masonry.

More than 1500 articles and patents on textile water repellents are reported in *Chemical Abstracts* for the period 1970–1981. Many patents describe commercial products; however, commercially important patents are difficult to identify. Therefore, the patent literature is useful chiefly as a source of general information on chemical classes and for details of specific patent coverage. A thorough discussion of water repellency to 1963, including the theoretical background, is given in ref. 1.

Principles of Repellency

When a drop of liquid rests on a solid surface, the shape of the drop depends on the equilibrium among three forces, as shown in Figure 1. The contact angle θ depends on three surface tensions: the liquid–air interface γ_{LA}; the solid–air interface γ_{SA}; and the solid–liquid interface γ_{SL}. Wetting of the solid by the liquid results from reduction of the contact angle so that the liquid spreads easily. A surface is made repellent by raising the contact angle, so that spreading and migration into capillaries do not occur.

The spreading coefficient S_{SL} for a liquid on a solid is

$$S_{SL} = \gamma_{SA} - (\gamma_{LA} + \gamma_{SL})$$

The term γ_{SA} is the same as the free energy of the solid surface, and γ_{SL} is the free energy per unit area of the solid–liquid interface. If the spreading coefficient is positive, spreading can occur spontaneously and the liquid should spread to form a thin film on the solid. A liquid spreads on a surface because spreading causes a decrease in free surface energy.

Conversely, spreading does not occur, ie, the solid repels the liquid, if the spreading coefficient is negative. This is because γ_{SA} is low, ie, the surface is a low energy solid or has been treated with a chemical that converts it to a low energy solid. Surfaces that contain hydrocarbon, siloxane, or fluorocarbon groups are low energy surfaces. These repel a liquid unless the surface tension of the liquid γ_{LA} and the interfacial tension of the solid–liquid interface γ_{SL} are sufficiently low to make the spreading coefficient positive.

The measurement of contact angles is a convenient means for judging the degree of wetting or repellency for a liquid on a plane-solid surface. Measurements of advancing or receding contact angles, which form when the liquid moves outward or contracts on the surface, are more reproducible than measurements of static contact

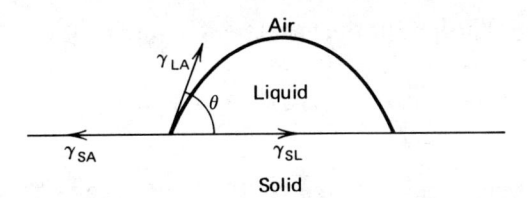

Figure 1. Contact angle for a liquid on a surface, $\cos\theta = \dfrac{\gamma_{SA} - \gamma_{SL}}{\gamma_{LA}}$.

angles (2). The advancing angle most nearly represents the angle for a surface completely covered with a finish.

The critical surface tension (CST) of a surface is a measure of the repellent properties of that surface. It is the maximum surface tension for a liquid that has a contact angle θ equal to 0°. Advancing contact angles on low energy solids depend approximately on only the surface tensions of the wetting liquids (3). The CST for a plane surface can be estimated by measuring the contact angles of a series of liquids, eg, alkanes, on the surface. A plot of cosines of contact angles against surface tensions of the liquids, as shown in Figure 2, gives approximately a straight line. Extrapolating this line to $\cos\theta = 1$ (contact angle = 0°) gives the surface tension required for a liquid to wet the solid. The principle applies also to surfactant solutions on low energy solids (4). Unfortunately, the CST cannot be measured directly on textile yarns, eg, cotton, that are complex bundles of twisted fibers. The CST for yarns may approximate that for chemically related plane surfaces.

A repellent functions by lowering the CST to a value below the surface tensions of liquids that are to be repelled. For example, a fluorochemical finish on a nylon fabric can lower the CST from ca 43 mN/m (= dyn/cm) for the untreated fabric to <10 mN/m for the treated fabric. Such treated fabric would repel water with a surface tension of 72 mN/m and oils or solvents with surface tensions down to ca 20 mN/m.

Most of the surfaces that require repellent treatments are not smooth but contain capillaries into which a liquid can migrate, even though the contact angle of the liquid

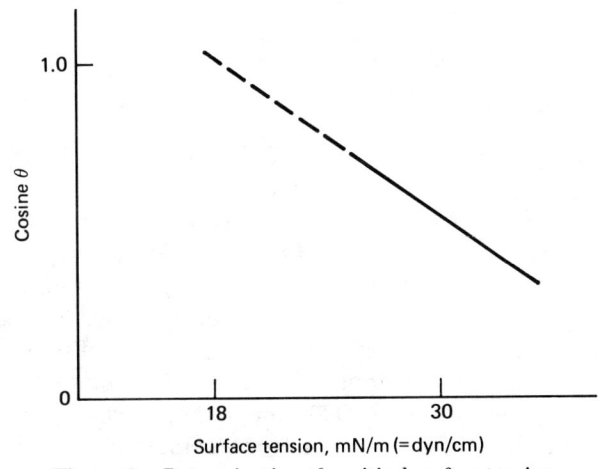

Figure 2. Determination of a critical surface tension.

on the surface is >0°. The law for the movement of liquids into an idealized capillary is given by the equations

$$\Delta P = \frac{2\gamma_{LA} \cos\theta}{r} = \frac{2(\gamma_{SA} - \gamma_{SL})}{r}$$

where ΔP is the pressure required to force the liquid back out of the capillary, and r is the diameter of the capillary. ΔP is positive, ie, the liquid rises in the capillary, if the liquid wets the capillary with a contact angle $\theta < 90°$ ($\cos\theta > 0$). If $\theta = 90°$, the liquid in the capillary is the same height as the liquid outside. If θ is >90°, $\cos\theta$ is negative, making ΔP negative. In the latter case, the solid repels the liquid, and pressure is necessary to force the liquid into the capillary. The second equation for ΔP shows that, for a negative value of ΔP, γ_{SA} (the free energy or surface tension of the solid in air) must be less than γ_{SL} (the interfacial tension of the solid in the liquid).

The limiting contact angle for repellency of a solid surface for a liquid depends on the physical form of the solid. The capillaries in textiles, paper, leather, or masonry are not ideal, smooth, and round. In a textile or in paper, the wicking of liquids can be lateral, ie, through the sides of fibers, or longitudinal, ie, along the length of fibers (5–7). In masonry, liquids can migrate through an array of irregular capillaries.

In a bundle of fibers, as in a textile yarn, fabric, or paper, the fibers may repel a liquid even though the contact angle is <90° (5). Pressure on the liquid that contacts the fibers, however, can force the liquid into the capillaries or interstices between fibers, thereby effecting wicking. The pressure required for wicking into repellent-treated fibers depends upon the critical surface tension of the treated fibers and the geometry of the system. Wicking occurs even without pressure on the liquid if the advancing contact angle is <70°; wicking occurs because the liquid surface can contact fibers beneath the surface.

The combination of lateral and longitudinal wicking into a fiber assembly illustrates the gradual wicking of aqueous fluids, solvents, or oils into fabrics, paper, or nonwoven products (see Nonwoven textiles). The gradual advance of a moving liquid front along capillaries can also explain the migration of liquids through masonry and concrete.

Textiles

The history of waterproofing textiles includes a long use of impermeable coatings followed by treatments that make the individual fibers of fabrics repellent without loss of permeability to air and water vapor. Waterproof, nonporous fabrics in antiquity were coated with natural fats and oils, waxes (qv), pitch, and asphalt. Eventually, vulcanized natural rubber became an important waterproofing material (see Rubber, natural). Owing to superior physical properties, coatings based on synthetic polymers replaced natural polymers and waxes. Fabric treatments that provided water repellency without loss of fabric permeability to air and water vapor were first introduced in the mid-1800s. In the early to mid-1930s, the first repellent finishes that combined repellency with durability to washing and dry cleaning were introduced.

Several publications describe the early development of waterproofing and water repellency (8–10). Reviews of the patent literature summarize the technology of water repellents and soil retardants for textiles from the early 1950s to 1979 (11–13).

According to the American Association for Textile Technology, Inc., the textile industry uses the following definitions for fabric treatments that repel water (14).

Water-repellent fabrics resist wetting or repel waterborne stains; they pass AATCC Test Method 22.

Shower-resistant fabrics protect against water penetration during a light or brief shower and pass AATCC Test Methods 22 and 42 (shower).

Rain-resistant fabrics protect against water penetration during a rain of moderate intensity and pass AATCC Test Methods 22 and 35 (rain).

Storm-resistant fabrics protect against water penetration during a heavy rain and pass AATCC Test Methods 22 and 35 (storm).

Waterproof is normally used to describe plastic, plastic-coated, or nonbreathable fabrics. However, the term has been used for some chemically coated fabrics that allow penetration of air.

Durable finishes maintain a high level of performance after laundering, dry cleaning, or both.

The levels of water repellency required to pass the preceding tests depend upon the specifications for particular fabrics. For example, a water-repellent fabric suitable for rainwear may have to show a spray rating (AATCC Test Method 22) of 80.

Oil-repellent fabrics resist wetting with oil and repel oilborne stains. The level of performance of such fabrics is judged by AATCC Test Method 118.

Waterproof Finishes. Waterproofing results from coating a fabric and filling the pores with film-forming material, eg, varnishes, rubber, nitrocellulose, waxes, tar, and plastics. The continuity of the film provides water resistance. Waterproof fabrics were long used for rainwear, tents, tarpaulins, and covers. Until the 1930s, the discomfort from impermeable rainwear, eg, slickers, was generally an acceptable price for staying dry. Except for tents, tarpaulins, and covers, coated fabrics have been largely replaced by plastics and by fabrics treated with water and oil repellents that do not reduce permeability to air and water vapor. The coating materials used for waterproofing are applied to fabrics as hot melts, eg, wax or some synthetic polymers; solvent solutions; or aqueous latexes.

Repellent Finishes. The following are six classes of repellent textile finishes in order of increasing use by weight: pyridinium compounds, organometallic complexes, waxes and wax–metal emulsions, resin-based finishes, silicones, and fluorochemicals. From 1975 to 1980, fluorochemicals constituted ca 50% of the weight volume and 90% of the dollar volume of all repellents (15). Fluorochemical finishes have achieved their preeminent position because they alone repel oils and oilborne stains as well as water and waterborne stains with excellent durability and minimum effect on other fabric properties. These benefits plus excellent dry-soil repellency have also made fluorochemicals important finishes for carpets. Table 1 summarizes the durabilities of the commercial classes of repellents (16).

Pyridinium Compounds. Most pyridinium compounds have been removed from the market because of the toxicities of the intermediates used in the manufacture of the compounds and the superiority of fluorochemical repellents. Other disadvantages of pyridinium products are their odors and the need for scouring fabrics after pyridinium treatment. The most important pyridinium compound used for water repellency is stearamidomethylpyridinium chloride

$$\underset{\substack{|\\CH_2-\overset{+}{N}\bigcirc}}{C_{17}H_{35}\overset{\overset{O}{\|}}{C}NH} \quad Cl^-$$

Table 1. Durabilities of Hydrophobic Finishes

Class	Durability Washing	Dry cleaning
pyridinium compounds	excellent	good
organometallic complexes	fair	fair
waxes and wax–metal emulsions	fair	poor
resin-based finishes	good	good
silicones	good	good
fluorochemicals	excellent	excellent

which forms from the reaction of stearamide with paraformaldehyde and pyridine hydrochloride. Application of the product to fabric in the presence of sodium acetate as a buffer, followed by drying and curing at 135–205°C, leads to the formation of methylenedistearamide (N,N'-methylenebisstearamide).

Important pyridinium water repellents include Velan PF (Imperial Chemical Industries) and Zelan AP paste (DuPont).

Organometallic Complexes. Werner-type complexes of chromium and long-chain carboxylic acids, eg, stearic acid, are water repellents for fabrics of natural and synthetic fibers (see Clathration). Commercial repellents include Quilon C, M, and S (DuPont) and Phobotex CR (CIBA-GEIGY). These complexes have a small market in the textile industry. They are neutralized with an amine, eg, hexamethylenetetramine, or sodium hydroxide, are padded onto the textile, and comprise 0.5–2 wt % of the fabric. Maximum repellency results from curing at 150–170°C. Chrome complexes have fair-to-good durability to laundering and dry cleaning (see Drycleaning and laundering). They are sensitive to alkali and some detergents and soaps. The green color of the products may produce an objectionable, although slight, change in the color of fabric. Aluminum complexes of long-chain acids are colorless but are generally less effective than chromium complexes.

Waxes and Wax–Metal Emulsions. Waxes and wax–metal emulsions are the lowest priced, widely used repellents. In addition to their low price, they also have the advantage of being applicable by padding or exhausting without requiring curing. Their market share has shrunk from a peak in the 1950s, however, because the finishes have only poor-to-fair durability to washing and dry cleaning. In addition, they tend to show streaks from abrasion.

Several types of wax and wax–metal emulsions are repellents (10,17). Among these are wax dispersions without metal salts and wax dispersions containing aluminum or zirconium salts. The products that do not contain metal salts are anionic emulsions of wax, used alone or in combination with durable-press resins. Specific compositions are proprietary. Their chief use is on nylon, polyester, and acetate fabrics.

Wax–metal emulsions are useful on a variety of fabrics, alone or in combination with resins. Wax–aluminum emulsions, which have largely lost their market share to wax–zirconium emulsions, usually contain paraffin wax, aluminum acetate or formate, and a dispersing agent and protective colloid, eg, glue. They tend to develop amine odors when used with thermosetting resins.

Wax–zirconium salt emulsions are useful on a wide variety of natural and synthetic fibers. Some finishes have considerable durability to laundering. The properties and use of such repellents are reviewed in ref. 18.

Resin-Based Finishes. Resin-based water repellents are durable finishes that are intermediate in price and superior to pyridinium compounds for ease of application. These resins are melamine or glyoxal derivatives or polyacrylates, and they are generally blended with waxes. In some cases, the resin helps disperse the wax in the repellent formulation. The resin provides water repellency and binds the wax on the fabric.

Resin-based repellents may be used alone or in combination with durable-press resins. They are widely used as extenders for fluorochemical repellents. When used alone, several of the resin-based finishes require an acid catalyst and curing at temperatures to 175°C for maximum repellency and durability. When coapplied with durable-press finishes, which themselves require a catalyst such as magnesium chloride or zinc nitrate, the catalyst and curing conditions for the durable-press finish provide the necessary conditions for the repellent.

Resin-based finishes are generally applied to fabrics by padding, although exhaust application is practical for some products. The concentration of repellent solids required for good initial repellency and good durability is 1–4% based on the weight of fabric. Important commercial products include Aerotex 96 and Permel Resin B (American Cyanamid), Nalan W and Nalan GN (DuPont), Phobotex FTC (Ciba-Geigy), and Norane 16 (Sun). The compositions of these products are proprietary.

Silicones. The most widely used silicones are polymers of methyl(hydrogen)siloxane

$$\left[OSiH \right]_n \quad \text{with CH}_3 \text{ substituent}$$

and of dimethylsiloxane

$$\left[OSi \right]_n \quad \text{with two CH}_3 \text{ substituents}$$

Silicones are second only to fluorochemicals in terms of the volume used as textile repellents. They provide better repellency than any other commercial repellent except fluorochemicals and are widely used on cellulosic and synthetic fiber fabrics. Other advantages of silicones are that they provide stain resistance; good durability to washing and dry cleaning; improved tear strength; a soft, slick hand; and improved fabric sewability (19). Their disadvantages are their relatively high prices, the need for curing on fabrics, and a tendency to cause yarn slippage in woven fabrics.

According to a review of silicones as used in the textile industry, the two fundamental reactions are hydrolysis and condensation (20); for example,

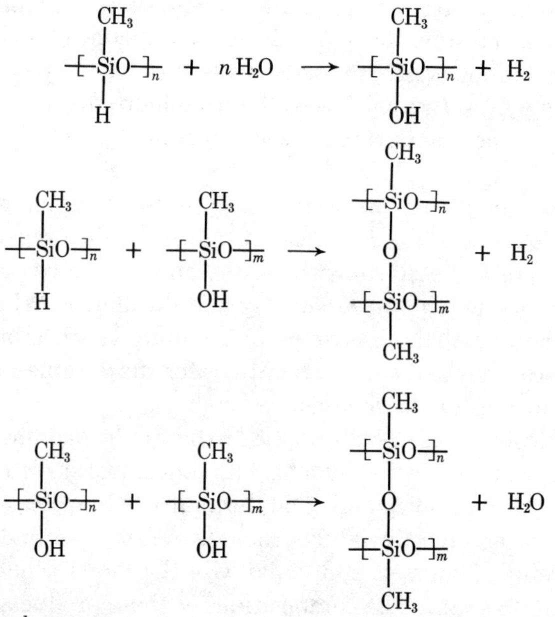

$$\begin{array}{c}CH_3\\|\\-[SiO]_n-\\|\\H\end{array} + n\,H_2O \longrightarrow \begin{array}{c}CH_3\\|\\-[SiO]_n-\\|\\OH\end{array} + H_2$$

$$\begin{array}{c}CH_3\\|\\-[SiO]_n-\\|\\H\end{array} + \begin{array}{c}CH_3\\|\\-[SiO]_m-\\|\\OH\end{array} \longrightarrow \begin{array}{c}CH_3\\|\\-[SiO]_n-\\|\\O\\|\\-[SiO]_m-\\|\\CH_3\end{array} + H_2$$

$$\begin{array}{c}CH_3\\|\\-[SiO]_n-\\|\\OH\end{array} + \begin{array}{c}CH_3\\|\\-[SiO]_m-\\|\\OH\end{array} \longrightarrow \begin{array}{c}CH_3\\|\\-[SiO]_n-\\|\\O\\|\\-[SiO]_m-\\|\\CH_3\end{array} + H_2O$$

where $m \geq n$.

Catalysts for these reactions include inorganic metal salts, eg, tin chloride; organic metallic compounds, eg, zinc salts of carboxylic acids; titanium esters; and epoxy resins.

Polydimethylsiloxane is the basic polymer used in silicone repellents. If the polymer is terminated with methyl groups, it is inert; however, if terminated with hydroxyl groups, it can be cross-linked. Continuous, durable coatings result from the use of curable blends of polydimethylsiloxane and polymethyl(hydrogen)siloxane. The silicone finish encapsulates individual fibers.

Silicones are supplied as aqueous emulsions or as solvent solutions; Dow Corning Corp. and Union Carbide Corp. are suppliers. Emulsions are usually applied to fabrics by padding or exhaustion. Solvent solutions can be applied by spraying. With either type of product, coapplication of a catalyst is necessary. The level of silicone solids on the weight of fabric should be 0.5–1.5%. Most of the silicone emulsions can be coapplied with durable-press resins. Curing occurs at 150°C.

The water repellency of silicone finishes results from the low critical surface tension (ca 22 mN/m or dyn/cm) produced by the methyl groups in the silicone that are oriented away from the fiber surface. The CST is lower than produced by any compounds except fluorochemicals (see Silicon compounds, silicones).

Fluorochemicals. Fluorochemicals are the main class of repellents for textiles. They are the only repellents that provide repellency to water, waterborne stains, oil, oilborne stains, and (especially for carpets) dry soil. Many fluorochemicals are commercially available from several suppliers. They differ in composition because of differences in their chemical intermediates and in manufacturers' technology. Specific products also provide special properties, eg, repellency and durability on certain fabrics and carpets. The first company to market fluorochemical repellents was Minnesota

Mining and Manufacturing Company in the 1950s (trade name Scotchgard). DuPont entered the market soon thereafter with the products Zepel and Teflon. Several other companies, including CIBA-GEIGY, Asahi, and Imperial Chemical Industries, also supply such finishes. The specific compositions are proprietary.

Fluorochemical finishes are used on rainwear, upholstery, drapery, automotive roofing material, and carpeting. Textile fibers used for these markets include all the natural and synthetic fibers. The benefits from fluorochemical finishes on these textiles compensate for the relatively high prices of the products.

The performance of fluorochemical repellents depends upon the presence of perfluorocarbon chains, $CF_3CF_2CF_2\cdots$. The products are polymeric and nonpolymeric. The polymeric types, which are used primarily for textiles, are copolymers of fluoroalkyl acrylates or methacrylates:

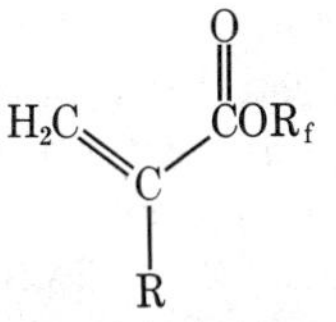

$$H_2C=C(R)COR_f$$

where R_f is the perfluoroalkyl chain and R is H or CH_3. The length of the perfluoroalkyl group is at least C_4 and for many products, it is C_8 or a mixture of chain lengths averaging C_6–C_8.

The comonomers in the polymeric products are esters of acrylic or methacrylic acid containing alkyl groups of several carbon atoms, alkylamide groups, or polyether groups. The comonomers modify the physical properties of the polymers (eg, control the hand of the finished fabric) or improve the performance of the polymers (eg, provide soil release) and reduce the cost of the product. The nonpolymeric fluorochemical repellents may be esters which, in some cases, are formulated with nonfluorinated polymers. For use on carpets, the formulations are designed to make the treated carpet surface sufficiently hard to minimize dry soiling. Soft finishes on nylon carpet may retain dry soil.

Fluorochemicals repel both water and oil because they produce an extremely low energy surface (21–29). They reduce the critical surface tensions of surfaces to such low levels that even oils or solvents of low surface tension are repelled. The effectiveness of the fluorochemicals depends upon orientation of the molecules on the fiber surface so that the perfluoroalkyl chains are directed away from the surface. The result is a CST as low as 5–10 mN/m (= dyn/cm). By contrast, the CST of polytetrafluoroethylene is ca 18 mN/m. Fluorochemical finishes are formulated with nonfluorinated extenders whenever possible. The extenders may be resin-based water repellents. The extenders not only reduce the cost of the finish but may also improve water and oil repellency and durability (30–31). Fluorochemicals that have soil-releasing as well as repellent properties cannot be used with extenders, which would destroy the soil-releasing property (10).

Quarpel is an important combination of fluorochemical finish and resin-based extender developed by the U.S. Army Natick Laboratories for military use. This finish contains 4–6 wt % commercial fluorochemical emulsion, 4–6 wt % resin-based repellent emulsion, 0.1 wt % acetic acid, and 5 wt % isopropyl alcohol. If necessary, the formulation includes a catalyst to cross-link the resin-based component. Quarpel provides excellent initial water and oil repellency and excellent durability to washing and dry cleaning.

Fluorochemical repellents are available as emulsions or solvent solutions. The most widely used emulsions for fabrics and carpets are cationic. Anionic and nonionic emulsions comprise a smaller portion of the products. Solvent solutions are used in some applications, eg, for upholstery or carpets (10). The emulsifier in a fluorochemical formulation affects the repellency and durability of the product (31). The emulsifier provides product stability to aging and to application conditions. A mixture of emulsifiers is sometimes needed for optimum performance. Surfactants used in formulations or added to finish baths for improved bath stability must be nonrewetting and must have a minimum adverse effect on oil repellency.

The choice of application conditions for fluorochemical repellents depends on the textile and the repellent formulation. Suppliers provide detailed recommendations in their product bulletins. In applications on home-furnishing fabrics, where a high level of repellency may not be required, the repellent can be applied by spray or foam finishing or by padding. Where a high level of repellency is required, as in rainwear, pad application is the most effective method. Foam finishing, the newest application technology, may become equally as useful as padding, even for a high level of repellency (32–35). Spray treatments are common for carpets. Carpet treatments often include the use of carpet antistats (see Antistatic agents). The usual concentration of repellent solids on any textile is 0.2–0.5% of the fiber weight, but repellency is reported to improve with up to 1.5 wt % solids (31).

There should be no silicones on fabrics or carpets prior to treatment with fluorochemicals, because traces of silicone lubricants can eliminate the oil repellency of the fluorochemical. The effect may result from poor wettability of the silicone-treated fiber with the fluorochemical repellent. Also, traces of silicone can dissolve in oils and reduce surface tensions, so that the oils migrate into the fabric or carpet.

Many fluorochemical finishes for fabrics require curing up to 175°C. Curing allows melt-spreading of the fluorochemical to ensure maximum leveling of the finish on the fibers and to promote optimum orientation of the chemicals on the fibers. Curing is not generally required for nonpolymeric repellents, eg, those used on carpets or for on-location application to any textile.

Nonwoven products often require fluorochemical finishes (36). Hospital uses for nonwovens often demand repellency to water, oil, and alcohol; the coapplication of a fluorochemical and a resin-based nonfluorochemical repellent is designed to meet such rigid use specifications. Treatments generally involve padding of 0.1–0.3 wt % fluorochemical solids and 0.5–1.5 wt % resin-based repellent, followed by curing at up to 175°C.

Several investigators have studied mechanisms of stain repellency and stain release of fabrics (37–42). Fluorochemicals repel oily as well as other stains, eg, dried blood and certain pigments. Some fluorochemicals have been designed specifically to be hydrophobic in air and hydrophilic in water (38,40).

Fluoroalkyl-substituted siloxanes are oil and water repellents. In a study of ten experimental products, the best repellents had perfluorinated, straight-chain substituents with seven or more carbon atoms (43). Two examples are

$$CF_3(CF_2)_8\overset{\overset{O}{\|}}{C}NH(CH_2)_3Si(OC_2H_5)_3$$

and

$$CF_3(CF_2)_6CH_2O(CH_2)_3Si(OC_2H_5)_3$$

Amides are better repellents than ethers. Mixtures of fluorinated and nonfluorinated silanes and resins provide good water- and oil-repellent properties that are durable to washing and dry cleaning (see Fluorine compounds, organic).

Fabric Construction for Water Repellency. Fabric construction affects the performance of water repellents. Waterproof finishes, which coat fibers to provide impermeable films, are applicable to a wide variety of fabrics. Close weaves are little better than open weaves for a cohesive film. Hydrophobic finishes, which make individual fibers repellent without altering fabric porosity, must be applied to fabrics whose pores are close together. The high contact angle of water or oil on such treated fibers prevents the passage of the fluid, but wide pores allow the passage of liquid even if the fibers are highly repellent (44).

The relation of rainwear fabric construction to the performance of repellents is reviewed in ref. 9. The twist, ply, and coarseness of yarn affect performance. The tightness and the nature of the weave, eg, twill vs poplin, have some effect on repellency, with tight weaves and poplins being more repellent. Some reports indicate that differences in fabric roughness affect repellency (31,44). Roughness reduces repellency, especially with low-to-moderate concentrations of repellent. Mechanical action on fabrics, even after treatment, can reduce repellency if the action increases fiber roughness or exposes fibers that have little repellent.

Modification of Textile Fibers. The reaction of hydrophobic chemicals with textile fibers offers the possibility of permanent repellency without alteration of the other physical properties of fibers. The disadvantage is the difficulty of carrying out chemical reactions on fibers in commercial textile-mill operations. Most studies of fiber modification for repellency have involved attempts to achieve permanent water repellency of cotton (qv). The etherification and esterification of cellulose have been most effective in terms of achieving durable water repellency (45). Etherification of cellulose produces bonds with good resistance to hydrolysis during washing with alkaline detergents. However, etherifications can require complex processes. Some investigators speculated that stearamidomethylpyridinium chloride reacts with cellulose to produce an ether of the structure

$$C_{17}H_{35}\overset{\overset{O}{\|}}{C}NHCH_2OR,$$

where R is a cellulose residue. The best evidence is that stearamidomethyl cellulose contributes to repellency but forms only to the extent of 1.0–1.5 wt % of the cellulose. Water repellency results from the effects of this ether and methylenedistearamide

$$(C_{17}H_{35}\overset{\overset{O}{\|}}{C}NH)_2CH_2$$

(46).

Water-repellent cellulose has been made by a low degree of substitution of cellulose with a fluorinated ether with the general structure (47):

$$C_nF_{2n+1}CH_2OCH_2\overset{}{C}H\!\!-\!\!CH_2$$

Treatment of cotton with aqueous sodium hydroxide and the fluorobutanol or fluorooctanol ether produces a fabric that repels oil and water. The effect is durable on the fabric even to boiling in alkaline soap solution.

Esterifications of cellulose are simpler than etherifications, although the reactions are generally not commercially attractive. Esterifying cellulose with aliphatic monobasic acid chloride in dimethylformamide at 30–85°C with pyridine as an acid scavenger produces water-repellent cotton (48). The degree of substitution required for moderate-to-good water repellency is only 0.02–0.36 ester groups per monomer unit. Water repellency generally improves with increasing chain length and reaction temperature. Apparently, esterification occurs on the surfaces of the fibers without significant change in the fiber's internal structure and tensile properties.

Direct acylation with isopropenyl esters of long-chain fatty acids also makes cellulose water-repellent (49). Only minor changes in tensile strength of the fabric occur. Acylation results from application of the ester to the fabric and heating so as to substitute one out of 1000 anhydroglucose units. The reaction is

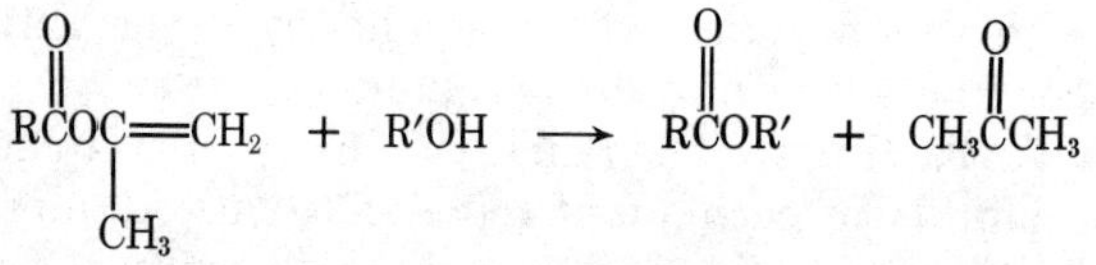

$$\underset{\underset{CH_3}{|}}{RCOC}=CH_2 + R'OH \longrightarrow RCOR' + CH_3CCH_3$$

where R is a fatty acid chain and R' is an anhydroglucose unit.

The addition of nonpolymeric fluorochemicals to synthetic fibers during spinning can produce water and oil repellency (50). The addition of

$$[(CF_3)_2CFO(CF_2)_3CNHCH_2CH_2]_2NC(CH_2)_3COH$$

to melts of nylon-6 and nylon-6,6 produces fibers that are impervious to oil-based stains and are only slightly wetted by water-based stains. The modified fibers can be dyed despite their water repellency. The additives migrate to the surfaces of the fibers so that 1 wt % produces fibers with low surface energy. A subsequent study shows that compatible fluorochemical additives can be applied to fabric so as to penetrate the fiber surfaces but concentrate near the surface (51). The water and oil repellency of the treated fabrics is durable to laundering, dry cleaning, and abrasion.

Radiation grafting of reactive repellents onto fibers has been studied as a potential commercial process (52–53). Methyl hydrogen siloxanes have been grafted onto Vinylon poly(vinyl alcohol) fiber and Tetron polyester fiber (53). Exposure of the siloxane on the fiber to a radiation dosage of 7–46 kGy (70–460 rad) causes the formation of radicals that react with the fiber to produce excellent water repellency. The repellency is highly durable to dry cleaning and abrasion.

Commercial Treatment of Carpet Fibers. The treatment of nylon fibers with fluorochemical finishes to provide durable soil, water, and oil repellency is an important part of carpet-fiber manufacture. The threadline application of fluorochemicals during fiber manufacture allows the carpet manufacturer to purchase treated fiber and makes topical treatments of carpets unnecessary. Fiber treatments are more durable to wear and cleaning than topical carpet treatments. Carpets containing such treated fibers benefit from a finish that covers the entire length of the fibers, in contrast with spray-applied fluorochemicals on carpets, which are deposited only on the upper portions of the fibers.

The first company to market threadline-treated nylon carpet fiber was Allied Corporation (Anso IV carpet fiber). Several other fiber manufacturers followed, including DuPont, Dow-Badische, Hoechst, and American Enka. Several patents assigned to Allied describe fluorochemicals useful for application to nylon or polyester fibers (54–58). Among the chemicals described in these patents are phthalic and pyromellitic acid esters, including compounds of the general formula

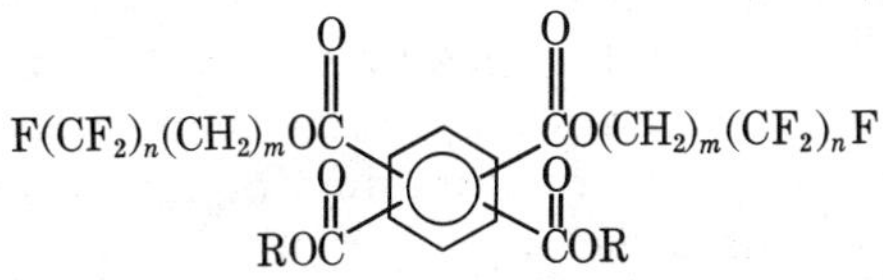

$$F(CF_2)_n(CH_2)_mOC\text{-Ar-}CO(CH_2)_m(CF_2)_nF$$
$$ROC\qquad\qquad COR$$

where m and n are 2–20 and R is a halogenated hydroxyalkyl group.

Fluorochemical repellents are applied to nylon carpet fiber in combination with lubricants and other finish ingredients. The fiber may be filament or staple. The finish allows normal fiber processing, carpet manufacture, and dyeing.

Economic Aspects. The U.S. manufacturers of water and oil repellents for textiles and a few trade names in parentheses of the most widely used repellents are, alphabetically (59): American Cyanamid Co. (Aerotex); Anscott Chemical Industries, Inc.; Apex Chemical Corp.; Arkansas Co., Inc. (Aquarol); Arol Chemical Products Co.; Astro Industries, Inc.; Auralux Corp.; Bercen Inc.; Callaway Chemical Corp.; Chem-Mark Inc.; Chemonic Industries, Inc.; CIBA-GEIGY Corp. (Phobotex); CNC Chemical Corp. (Aridry, Aquafilm); Continental Chemical Co.; Crest Chemical Corp.; Crown Metro Inc. (Cravenette); Diamond Shamrock, Process Chemicals Div.; Dow Corning Corp. (Dow Corning emulsion); DuPont (Zepel, Teflon, Nalan); Eastern Color & Chemical Co.; Emkay Chemical Co.; General Electric Co.; Glo-Tex Chemicals, Inc.; Hart Products Corp.; Hydrolabs, Inc.; International Minerals & Chemical Corp.; Jersey State Chemical Co. (Extendopel); Leatex Chemical Co.; Lutex Chemical Corp.; Lyndal Chemical Co., 3M Co. (Scotchgard); Moore & Munger Marketing, Inc.; Moretex Chemical Products, Inc.; Niacet Corp.; Organic Chemical Corp.; Original Bradford Soap Works, Inc.; Piedmont Chemical Industries, Inc.; Raffi & Swanson, Inc.; Raytex Chemical Corp.; Reliance Chemical Products Co.; Sandoz Colors & Chemicals Co.; Scher Chemicals Inc.; Soluol Chemical Co., Inc.; Southern U.S. Chemical Co., Inc.; Sidney Springer Co., Inc.; Star Chemicals Inc.; Sun Chemical Corp. (Norane); SWS Silicones Corp.; Synthron Inc. (Synthopel); Chas. S. Tanner Co.; Tritex Chemical Corp.; Union Carbide Corp.; U.S. Oil Co.; Valchem; Vikon Chemical Co.; and Witco Chemical Corp.

Data for manufacturers' capacities are proprietary, and product prices are not widely known. However, the total U.S. sales of repellents in 1982 were probably ca \$70 $\times$ 10^6, including nearly \$63 $\times$ 10^6 for fluorochemical finishes. The main U.S. manufacturers include 3M Co., DuPont, American Cyanamid Co., CIBA-GEIGY Corp., Dow Corning Corp., and Union Carbide Corp. Principal foreign manufacturers include Asahi and Imperial Chemical Industries. The prices of repellents change frequently.

In 1982, they ranged from <$2.20/kg for emulsions of wax, wax–metal, or resins to >$6.60/kg for fluorochemical emulsions. Solids contents in the emulsions are generally 15–30 wt %.

Test Methods. Many tests are useful for testing textiles for water and oil repellency and for testing the durability of finishes to washing and dry cleaning (60). The most widely used tests in the United States are described in refs. 61–63, eg, spray and drop tests, hydrostatic pressure tests, immersion tests, and repellency tests for carpets. The selection of tests depends upon the textile being tested, ie, whether it is a fiber, fabric, or nonwoven product. The selection also depends on the nature of the finish, ie, whether it is waterproof, water-repellent, water- and oil-repellent, or solvent-repellent. Special tests may be necessary for fabrics or garments that must meet specific requirements, such as repellency of shell and lining for a lined garment (64). Several supplementary tests are important, depending on the textile. These include air and water-vapor permeability, repellency to specific chemicals or vapors, fabric stiffness or softness, color, effects on dyes, and durability of physical properties and color to exposure to light or atmospheric gases.

In the following descriptions of some widely used repellency tests, fabric designates any woven, knitted, or nonwoven textile.

The spray test is one of the most commonly used tests for fabrics and nonwoven products. The test material is held tightly on an embroidery hoop and mounted at a 45° angle to the horizontal and 15 cm below a spray nozzle. The fabric is sprayed with 250 mL of water. The degree of repellency is rated by comparing the sprayed fabric or nonwoven with pictures on a standard chart.

The rain test simulates the effects of rainfall; the hydrostatic head on the spray controls the intensity of spraying. The repellency is rated by the weight of water that penetrates the fabric and is absorbed by a blotter mounted behind the fabric at a specific intensity of spraying.

Hydrocarbon resistance measures a fabric's resistance to wetting by a selected series of liquid hydrocarbons with a range of surface tensions. Drops of numbered test liquids are placed on the fabric surface and observed for wetting. The fabric rating is based on the liquid that does not wet the fabric surface in a specified time. The time for AATCC Test Method 118 is 30 s. Suppliers of fluorochemical repellents may recommend different times or a different series of hydrocarbons in the evaluation of treated fabrics.

Alcohol holdout involves placing drops of aqueous ethyl alcohol solutions of concentrations 10, 20, . . . 100 wt % on a nonwoven fabric. The rating for the fabric is based on the most concentrated solution that does not penetrate the fabric in five minutes. A second alcohol holdout test determines the degree of wetting of a nonwoven product by drops of a 50 wt % aqueous solution of isopropyl alcohol. Resistance to wetting of the top side of the fabric is rated from 5 for no wetting to 0 for complete wetting, based on pictures on a standard chart.

The hydrostatic-pressure test is performed on fabric mounted under the orifice of a conical well. The fabric is subjected to increasing water pressure at a constant rate until leakage occurs at three points on the fabric's undersurface. The rating is the height of the water head in centimeters above the fabric.

The static absorption test measures the water repellency of fibers and yarns in a fabric, not the repellency of the total fabric surface. A weighed portion of fiber, yarn, or fabric is immersed in water under a hydrostatic head of 8.9 cm for 20 min; it is then removed and reweighed to determine the percentage of water absorbed.

The specifications for water, oil, solvent, or soil repellency vary with the use of the textile, the nature of the textile, and the nature of the repellent (30–31,36,38,41). Suppliers of repellents may specify minimum levels of repellency for the approved use of a repellent. Nonwoven products for hospital use are generally required to meet critical specifications.

Tests for carpets treated with fluorochemical soil repellents include tests for water-drop resistance, hydrocarbon resistance, carpet soiling, and dry soiling. One wet-soiling test involves soiling with a muddy shoe and drying and judging the ease of removal of the dried soil. The most important test is generally the resistance of the carpets to soiling under practical conditions of soiling and maintenance.

Health and Safety Factors. The toxicological properties of repellents vary widely. Only the suppliers of the products can provide more than general information. Most products have low levels of toxicity, but suppliers urge caution in use. Many suppliers assure the safety of their repellents, not only as shipped but also as used on textiles and perhaps next to human skin. The products as supplied should be assumed to be skin and eye irritants. Those handling the chemicals should always avoid contact of the products with skin and eyes and avoid exposure to vapors. Operators should use gloves, goggles or spectacles, and clothing sufficient to minimize physical contact.

Paper and Paperboard

Paper that repels water may be designated waterproof or water-resistant. Waterproof paper is paper or paperboard that is resistant to both water and water vapor. Chemical modifications of paper fibers include processes that develop water repellency (65). However, apparently none has been commercialized. Water resistant or water-repellent paper or paperboard is of one of two kinds. One kind is analogous to water-repellent textiles; that is, it repels water because of a treatment of individual fibers without formation of a film impermeable to water and water vapor. The second kind is pigment-coated paper or paperboard (see Paper; Papermaking additives).

Waterproof. Waterproof paper or paperboard is coated or laminated with a molten thermoplastic material or coated with an aqueous latex or solvent solution of a polymer. Among the commercially important thermoplastic products are asphalt, wax, polyethylene, poly(ethylene-co-methacrylic acid), ethylene–vinyl acetate copolymers, poly(ethylene terephthalate), poly(vinylidene chloride), and combinations of thermoplastic synthetic resins and wax. The resulting paper or paperboard is useful for boxes, bags, drum or case liners, roofing material, and covers for machinery and goods in storage and transport (see Packaging materials, industrial products). The excellent barrier properties of some resins to water, oil, water vapor, oxygen, and some other gases make paper and paperboard coated with these resins especially valuable for flexible packaging and boxes. Details of resin properties and recommendations for application are available in the suppliers' technical bulletins.

In the extrusion-coating process, the hot melt is extruded as a semiliquid film through a slit die. The thin polymer film is drawn to its final gauge in the air gap between the die and the point of contact with the paper or board. Bonding to the base sheet is by mechanical means as the coated sheet passes between a pressure roll and a chill roll. With a single extruder and die, it is possible to make a wide range of coating gauges by varying the take-off speed of the coater and the screw speed of the extruder. The important variables and resulting properties of extrusion coatings for ethylene–

vinyl acetate copolymers are described in ref. 66. The conditions for extruding coatings of Eastman's Tenite 6N5DP polyterephthalate onto paper and paperboard to provide tough coatings with excellent water and grease resistance are described in ref. 67.

Poly(vinylidene chloride)-coated papers and boards are useful for many kinds of packages. Latex application, as on air-knife coaters, may require two or three coatings for adequate coverage. The resulting paper or board has excellent resistance to water, greases, water vapor, oxygen, and other gases (68).

Methods of making both the flat surfaces and cut edges of paperboard resistant to water and water vapor for use in shipping containers are reviewed in ref. 69. The best treatments involve the use of low molecular weight precondensates of phenol–formaldehyde resins or melamine or hydrocarbon resins, but embrittlement of the board by the resin is a problem.

Water-Resistant. Developing the resistance of paper or paperboard that is not pigment-coated to the penetration of liquids, especially water and aqueous solutions, is called sizing. A sizing agent is a chemical that makes cellulose-fiber surfaces hydrophobic, so that the contact angle with water or aqueous solutions is high (70). The sizing agent must be held on the fibers to resist removal by a liquid. Internal sizing is achieved by the addition of repellent chemicals to the aqueous suspension of pulp fibers before sheet formation, ie, at the wet end of a paper machine. The process is also called beater or engine sizing. Surface or tub sizing is achieved by the application of repellent chemicals to the surface of the sheet, often before the sheet has been dried. Internal or surface sizing may also be designed for repellency against nonaqueous fluids, eg, oils, greases, or alcohols. Repellents against water or aqueous solutions are hydrocarbon derivatives or siloxanes. However, repellents against oils, greases, and alcohols are fluorochemicals or mixtures of fluorochemicals and hydrocarbon derivatives.

Internal Sizing. Over 70% of all paper and paperboard produced in the United States is sized to some extent (71). Among the main grades are wrapping and bag papers, bond paper, offset and other printing papers, building paper and paperboard, boxboard, and food-container board. The most important internal sizes for water holdout include rosin size, emulsified wax, fortified sizes, bituminous emulsions, and silicones (72).

The chemistry and application of rosin size and the factors that affect sizing are described in refs. 70–73. Rosin is the most widely used internal size. Partially or completely saponified gum rosin, wood rosin, or tall-oil rosin is added to a dilute pulp slurry and precipitates upon addition of papermaker's alum (aluminum sulfate, $Al_2(SO_4)_3.14.5H_2O$) (see Tall oil). Rosin is a complex mixture of compounds and consists primarily of monocarboxylic acids with alkylated hydrophenanthrene structures (74). A main constituent is abietic acid. Many modifications of rosin size are available that are superior to unmodified rosin size for stability, ease of handling, color, and sizing efficiency.

Alum functions as a precipitant and retention aid and is a critical part of the sizing agent (70,73). The amount of alum required depends on several factors, eg, the amount of size, the nature of the fiber, and water conditions. The best utilization of size and alum is at pH 4–5. The use of sodium aluminate with alum allows sizing at a higher pH (75). The retention of the size precipitate by paper fibers depends at least partly on the electrostatic attraction of the negatively charged cellulose fibers for positively charged aluminum salts on the surface of the precipitate particles (76). Size retention apparently also results from a simple collision process between the precipitate particles and cellulose fibers, followed by adsorption (77). Size precipitate deposits as particles or clumps of particles along the fiber surfaces (78). Sizing efficiency depends on many factors described in the literature (79–80).

Several sizes are important in addition to rosin and alum (81). Two are fortified size and wax emulsions. Fortified or chemically modified rosin size is widely used. The size most used is a product of abietic acid and maleic anhydride. The presence of three carboxylic acid groups on the size molecule, instead of only one, may reduce the tendency of the aluminum salt to form large particles before adsorption by the cellulose fibers. Anionic emulsions of wax, especially paraffin or microcrystalline wax, are important. They are insolubilized with alum or sodium aluminate (82). The amount of wax used in sizing is ≤0.5 wt % of the fiber weight; the amount is limited because of the tendency of the wax to reduce sheet strength and make the sheet slippery.

Synthetic sizes have become increasingly important since 1960, particularly for papers that have special sizing requirements, eg, sizing in the presence of calcium carbonate as a filler or sizing at neutral-to-alkaline pH. One of the most commonly used synthetic sizes is hexadecylketene dimer (Aquapel or Hercon, Hercules Co.) (see Ketenes and related substances). The ketene dimer is believed to react with cellulose to form an ester linkage (83). With 0.05–0.3 wt % of the chemical, a monomolecular film on the fibers apparently reacts with the cellulose during drying. The sized paper is repellent to both acid and alkaline liquids. Other synthetic internal sizes include silicones; acrylic polymers; copolymers of maleic anhydride with vinyl monomers, eg, styrene or an olefin; and alkenyl succinic anhydrides, eg, Fibran sizing agent (National Starch & Chemical Corp.) (84–86).

Surface Sizing. Surface sizing involves the application of chemicals to the surface of paper, generally at the size press of a paper machine. However, it also can be done at the calender stack (for paperboard), at a coating station on the paper machine, at a spraying location on the machine, or in an off-machine converting operation. Surface sizing is also a process for modifying other surface properties of paper or paperboard, eg, printability or smoothness, as by the application of a film-forming material that has little effect on the holdout of water or hydrocarbons. Surface-sizing chemicals that produce repellency to liquids without the formation of an impermeable film are of three types: hydrocarbon derivatives, silicones, and fluorochemicals (87). The first two repel water and aqueous solutions; fluorochemicals repel oils and greases and may also repel water.

Hydrocarbon derivatives. The most widely used hydrocarbon derivatives in surface sizing for water repellency are wax emulsions, chromium complexes, and ketene dimers. Wax emulsions may be anionic, cationic, or nonionic. In addition to water repellency, they may impart an improved finish to the paper and increased scuff resistance. Paper and paperboard grades that may be surface-sized with wax emulsions are meat wrap, bag paper, cup stock, and folding boxboard (88).

Chromium complexes of long-chain fatty acids, eg, stearic acid and myristic acid, are excellent water repellents and release agents for paper and paperboard (see Abherents). The chemicals are Werner-type complexes, eg, Quilon chrome complex (DuPont), that react with the hydroxyl groups of cellulose to bond permanently to the fiber surface (89). The fatty-acid groups are oriented away from the surface to produce both water repellency and release effects. The chrome complexes are supplied as dark green solutions in water–isopropyl alcohol mixtures. For application to paper, they are diluted with water, adjusted to pH 3.0–4.0 by the addition of a weak base, and applied at 0.5–2.4 g/m² (1.6–7.9 oz/1000 ft²) depending on the paper or board. Curing at 80–105°C is essential for developing maximum performance. Chrome complexes are used for numerous grades of paper and paperboard, including paper bags for ice cubes, wet vegetables, meat, or refuse disposal; material wraps for storage and shipment; and paperboard boxes for wet vegetables, meat, or overseas shipment.

Ketene dimers (Aquapel or Hercon, Hercules) are excellent water repellents on paper and paperboard and are used at 0.1–3.2 kg/t (0.2–6.4 pounds per short ton) of paper (90). As in internal sizing, the ketene dimers probably react with cellulose to form an ester linkage (83).

Silicones are effective water repellents and release agents even when used at low concentrations. They are supplied by Dow Corning Corp. and Union Carbide Corp. A novel technique for imparting water repellency, wet strength, and release properties to paper by reaction with gaseous trichloromethylsilane is described in ref. 91. In a dwell time of 0.5 s, paper with low moisture content reacts with the silane in a chamber containing the silane and dehydrated air.

Fluorochemicals. Fluorochemicals have had increasing use as oil and grease repellents since the preparation of chromium complexes of perfluorinated carboxylic acids in 1953 (92). Fluoroalkyl phosphates have gradually displaced the chrome complexes because of FDA approval of the phosphates for use on paper and paperboard in contact with food. The phosphates are

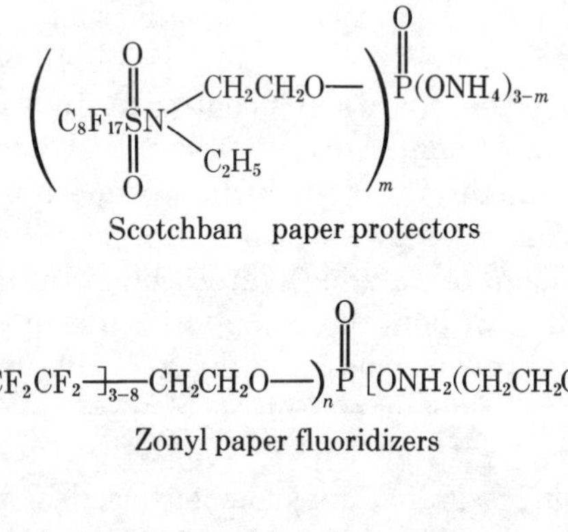

$$\left(C_8F_{17}SN\underset{C_2H_5}{\overset{CH_2CH_2O-}{\diagdown}}\right)_m P(ONH_4)_{3-m}$$

Scotchban paper protectors

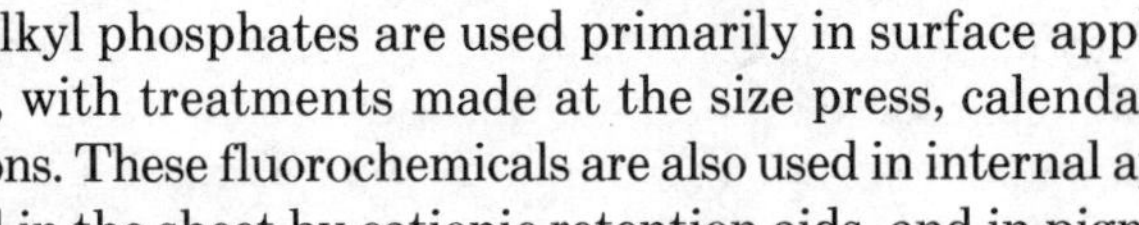

$$\left(F{+}CF_2CF_2{+}_{3-8}CH_2CH_2O-\right)_n P[ONH_2(CH_2CH_2OH)_2]_{3-n}$$

Zonyl paper fluoridizers

where *m* and *n* each are 2 or are mixtures of 1 and 2.

The fluoroalkyl phosphates are used primarily in surface applications to paper and paperboard, with treatments made at the size press, calendar stack, or in off-machine operations. These fluorochemicals are also used in internal applications, where they are retained in the sheet by cationic retention aids, and in pigment coatings. For surface applications, the commercial aqueous alcoholic solutions are diluted with water and applied at 0.10–0.30% fluorochemical solids based on the weight of paper. The products may be applied alone or with starch or other film-forming materials. The applications and performance of fluorochemicals are described in ref. 93. Technical bulletins supplied by DuPont and 3M Company give detailed information on use of the products.

Pigment Coating. Grades of pigment-coated paper and paperboard that may contact moisture during use must have water resistance or wet-rub resistance. The papers include coated tags, labels, wallpapers, decorative papers, offset printing paper, and some publication papers. The paperboards are primarily those coated grades used for packaging. The choice of adhesive used in a pigment coating determines the water resistance of coated paper or paperboard (94). Casein, soybean protein, and synthetic resins produce the highest water resistance. The addition of formaldehyde increases the water resistance when protein is the adhesive. Starch–pigment combinations yield coatings that are water-sensitive.

Test Methods. The most widely used tests for sizing or water repellency are standard and suggested tests and routine control (RC) tests of the Technical Association of the Pulp and Paper Industry; these are described in refs. 65 and 95, respectively. The tests are also described elsewhere in the literature (93,96–98).

Leather

Water-resistant treatments are useful on leather garments, shoes, and boots to reduce the tendency of leather to become stiff and uncomfortable after wetting and drying. Before the introduction of synthetic water repellents, the best way to protect leather was to apply large concentrations (up to 50% of the leather weight) of oil, tallow, or grease. The treated leather was uncomfortable because the pores of the leather were filled and the water-vapor permeability was low. Today, available water repellents are useful at low concentrations and make the leather surface repellent without reducing permeability to water vapor. Fluorochemical repellents also provide repellency to oils and greases, so that the treated leather resists staining. The theoretical and practical aspects of water repellents for leather are reviewed in ref. 99. Water should bead on the surface of water-repellent leather because of treatment of the individual fibers, both on the surface and in the center of the leather. The fibers should not become wet even with prolonged soaking, and the leather should be permeable to water vapor for comfort.

Commonly used repellents for leather are silicones, chrome complexes of long-chain fatty acids, and fluorochemicals. A water repellent may also be a hydrophobic chemical insolubilized in the leather. A simple water-repellent treatment consists of forming an aluminum soap in leather by the two-step process of applying a soap and then applying an aluminum salt (99).

The silicones, as supplied by Dow Corning Corp. and Union Carbide Corp., are an important class of repellents (99–101). Numerous patents describe the use of silicones for leather (102–104). Silicones are linear polymers, eg, dimethylsiloxane polymers, that bond to the leather. They can be applied as aqueous dispersions in the presence of glutaraldehyde during fat-liquoring, whereby the silicone is probably bonded to the leather with glutaraldehyde acting as a bridge between the silicone and amide groups of the leather (100). Silicones can also be applied to leather as solvent solutions in a post-tanning operation. The concentration of silicone used is 4.5–15 wt % to provide repellency that resists extended flexing (100–101).

Organic titanates can be used to cross-link silicones on leather (105–107). Tetrabutyltitanate and tetrakis(β-aminoethoxy)titanium cross-link silicones, eg, poly(dimethyl siloxane), to increase both water repellency and durability. For example, good repellency results from the impregnation of leather with a solution of 10 wt %

tetrabutyltitanate in butyl acetate, followed by impregnation with a 9:1 mixture of silicones and tetrabutyltitanate (106).

Chrome complexes of stearic and myristic acids, eg, Quilon chrome complexes (DuPont), provide water and aqueous stain resistance, dimensional stability, and lubricity. The products may also enhance the appearance and durability of leather. According to the DuPont product bulletin, the chrome complex should be used at 10–40 wt % of the fat-liquor composition (108). In postfat-liquoring, the concentration should be 6–10% of the leather weight. The chrome complex reacts with the leather molecules to form a permanent bond.

Fluorochemical repellents may be nonpolymeric or polymeric. A chrome complex and fluoropolymers that are used most widely are Scotchgard leather chemicals (3M Co.) (109). The compositions are proprietary. Fluorochemicals provide a high degree of water repellency as well as repellency to aqueous stains, oils, grease, and oilborne stains. The products may be applied by exhaustion, padding, brushing, or spraying so that 1.5–3% solids is deposited on the leather. A nonpolymeric fluoroester has been applied to sheepskin to produce water and oil repellency (110) (see Leather).

Test Methods. Tests for measuring the water and oil repellency of leather include ASTM tests and tests recommended by chemical suppliers:

Spray test (ASTM D 1913). Water is sprayed on taut surface, and the rating is based on comparison with a standard chart (111).

Water absorption, static (ASTM D 1815). The weight of water absorbed during 30 min, 2 h, or 4 h, with all surfaces of leather exposed to the water, is measured (111).

Dynamic water resistance by Dow Corning tester (ASTM D 2098). Water penetration is measured during flexing in Dow Corning Tester (111).

Dynamic water resistance by Maeser tester (ASTM D 2099). Water penetration is measured during flexing in Maeser tester (111).

Oil repellency (AATCC Test Method 118). Oil repellency is measured by the kit test or oleophobic rating, as described for paper in ref. 61.

Concrete and Masonry

Water resistance is an important factor in concrete and masonry construction for the safety, health, and comfort of building occupants (see Cement). Several texts on concrete construction describe the methods for achieving water resistance (112–115). The term waterproof describes concrete and masonry that is completely impervious to water and water vapor, whether or not the water is under pressure. Waterproof construction implies the use of a membrane or a film that provides a barrier and covers all surface pores or capillaries. The term water repellent describes concrete or masonry that repels water without significantly reduced permeability of the structure to the passage of water vapor. In the following discussion, concrete and masonry are used synonymously.

Several problems related to moisture movement in concrete are expansion, shrinkage, and cracking; efflorescence, staining, and mildew; lowered resistance to freezing and thawing; increased thermal and electrical conductivity; chemical attack; damage to contents from leakage; damage to finished walls; corrosion of reinforcing steel; and damage to structures from settlement. Care in choosing aggregate and in selecting a water–cement ratio leads to a watertight concrete that may not require additional waterproofing or water repellency (116).

Waterproof. Waterproofing may be achieved by the application of membranes to the surface of concrete structures. The sheet materials include such different products as roofing felt, poly(vinyl chloride), polyethylene, butyl rubber, neoprene, and sheet lead. Alternatively, some membranes may be applied as liquids that form elastomeric membranes on the concrete surface. Examples are liquid polysulfide polymers and polyurethanes (117). The selection of membrane materials has been described in the literature (112–114).

Waterproofing may also be achieved by the application of sealants (qv), eg, tar, asphalt, some solvent-based or latex paints, or mastics. Such coatings may be applied by brush, roller, spray, or trowel. Two coats are usually required. In some cases, such vapor barriers are unsuitable because they trap moisture in a wall. The proper selection and application of coatings depend on the nature of the concrete structure, ie, whether it is made up of exterior or interior walls above or below grade, floors, or roofs (118).

Water-Repellent. Three techniques used for water repellency are modification of cement by the addition of waterproofers, use of repellent additives to the concrete mix, and surface treatment of concrete structures with repellents. The modification of portland cement by the addition of calcium stearate, aluminum stearate, or gypsum treated with tannic acid and oils yields proprietary cements (119). The manufacturers claim waterproof properties. Considerable controversy exists, however, whether these cements produce concrete that is superior to carefully mixed concrete without such additives.

Admixtures are sometimes used to decrease capillary flow and the permeability of liquids in concrete (120–122). These additives may be pore-filling materials, eg, chalk, Fuller's earth, or talc. They may also be repellents, eg, mineral oil, asphalt emulsions, wax emulsions, or salts of fatty acids, especially stearates. Stearate damp-proofing admixtures form calcium stearate, which lines the pores of cells in the concrete. Such an admixture may also reduce the chance of freezing of any water that penetrates the concrete to help prevent cracking and spalling.

The third class of water repellents consists of materials applied to the surface of concrete for above-grade structures, where water pressure on the concrete is small or negligible. Repellents that may be used are oils, waxes, soaps, resins, and silicones (123). They may be applied to masonry surfaces by brush, roller, conventional air spray, or airless spray. Oils provide short-term benefits on surfaces that are not damaged by discoloration from the oil. For example, bridge decks may be treated with linseed oil to prevent spalling from salts and freeze–thaw cycles. Waxes are highly resistant to water. They partially fill pores as well as produce water repellency. They have the disadvantage of changing the appearance of masonry, migrating to the surface at high temperatures, and showing poor abrasion resistance. Among soaps, the most commonly used are calcium stearate and aluminum stearate. For surface treatments, they are generally supplied as solutions in organic solvents. Their primary use is on natural stone surfaces. Many kinds of synthetic resins are marketed as water repellents for concrete. The resins include acrylates, epoxies, styrene–butadiene rubber, chlorinated rubber, and urethanes. Their compositions are generally proprietary, and they are supplied in organic solvents. Some products have the disadvantage of changing the appearance and color of masonry.

Concrete can also be made water-repellent by the polymerization of vinyl monomers on the surface. A treatment that may be practical for highway bridge decks is the application of methyl methacrylate, isodecyl methacrylate, or isobutyl methacrylate (124). Benzoyl peroxide is the polymerization catalyst, and trimethylolpropane trimethacrylate is the cross-linking agent. External heat polymerizes the monomer to a depth of 1.4–3.8 cm.

The most widely used repellents are silicones. One application to masonry penetrates at least 0.3 cm without a change in the masonry appearance. The pores of the masonry are not sealed, and moisture can escape. Such treatments have been effective >5 yr (125). Silicones are available as sodium methyl siliconate in water or silicone polymers of proprietary compositions in organic solvents. Dow Corning Corp. and Union Carbide Corp. are suppliers. Silicone films may bridge minor cracks in structures. They are excellent repellents, except against water under hydrostatic pressure. Silicone films are capable of considerable elongation; thus, they are unaffected by small deformations (126). Silicones in organic solvents are useful for brick, concrete, stucco, or terrazzo surfaces. Water-based silicone compositions are used on natural stone for a good bond (127).

Test Methods. Tests for water repellency of concrete and masonry have been described by several authors (112,128). The NBS evaluated 55 clear-water repellents for masonry surfaces (129). Performance tests included water absorption, water-vapor transmission, resistance to efflorescence, change in appearance, and durability to accelerated and outdoor weathering. The authors give recommendations for the application of repellents, discuss the theory of water flow through concrete, and summarize a survey of field experiences with waterproofing.

BIBLIOGRAPHY

"Waterproofing and Water Repellency" in *ECT* 1st ed., Vol. 14, pp. 862–980, by J. David Reid and Ruby K. Worner, U.S. Dept. of Agriculture; "Waterproofing and Water Repellency" in *ECT* 2nd ed., Vol. 22, pp. 135–156, by J. D. Reid and W. J. Connick, Jr., U.S. Dept. of Agriculture.

1. J. L. Moillet, ed., *Waterproofing and Water-Repellency*, Elsevier Publishing Co., Amsterdam, The Netherlands, 1963.
2. R. E. Johnson and R. H. Dettre, *J. Phys. Chem.* **69,** 1507 (1965).
3. H. W. Fox and W. A. Zisman, *J. Colloid Sci.* **7,** 428 (1952).
4. M. K. Bernett and W. A. Zisman, *J. Phys. Chem.* **63,** 1241, 1911 (1959).
5. A. B. D. Cassie, *Disc. Faraday Soc.* **3,** 239 (1948).
6. F. M. Fowkes, *J. Phys. Chem.* **57,** 98 (1953).
7. F. W. Minor, A. M. Schwartz, E. A. Wulkow, and L. C. Buckles, *Text. Res. J.* **29,** 931 (1959).
8. J. W. Rowen and D. Gagliardi, *Am. Dyest. Rep.* **36,** 533 (1947).
9. C. A. Davis, *Am. Dyest. Rep.* **56,** 555 (1967).
10. J. M. May, *Am. Dyest. Rep.* **58**(20), 15 (1969).
11. M. W. Ranney, *Waterproofing Textiles 1970*, Noyes Data Corp., Park Ridge, N.J., 1970.
12. M. W. Ranney, *Soil Resistant Textiles 1970*, Noyes Data Corp., Park Ridge, N.J., 1970.
13. C. S. Sodano, *Water and Soil Repellents for Fabrics*, Noyes Data Corp., Park Ridge, N.J., 1979.
14. American Association for Textile Technology, Inc., *A Review of Water Repellent Finishes and Coatings*, Monograph 105, New York, 1970, p. 6.
15. E. G. Hochberg and Associates, *Textile Finishing Agents, A Marketing Technical Study*, Chester, N.J., 1976, pp. 111–127 (private issue for clients).
16. *Ibid.*, p. 111.
17. Ref. 1, pp. 52–62.
18. W. B. Blumental, *Ind. Eng. Chem.* **42,** 640 (1950).
19. Ref. 15, p. 117.
20. J. K. Campbell, *Text. Chem. Color.* **1**(8), 370 (1969).

21. W. A. Zisman in R. F. Gould, ed., *Contact Angle, Wettability, and Adhesion*, No. 43 in *Advances in Chemistry*, American Chemical Society, New York, 1964, pp. 1–51.
22. F. Schulman and W. A. Zisman, *J. Am. Chem. Soc.* **74**, 2123 (1952).
23. F. Schulman and W. A. Zisman, *J. Colloid Sci.* **7**, 465 (1952).
24. A. H. Ellison, H. W. Fox, and W. A. Zisman, *J. Phys. Chem.* **57**, 622 (1953).
25. E. F. Hare, E. G. Shafrin, and W. A. Zisman, *J. Phys. Chem.* **58**, 236 (1954).
26. E. G. Shafrin and W. A. Zisman, *J. Phys. Chem.* **64**, 519 (1960).
27. M. K. Bernett and W. A. Zisman, *J. Phys. Chem.* **67**, 1534 (1963).
28. R. E. Johnson, Jr., and R. H. Dettre in E. Matijevic, ed., *Surface and Colloid Science*, Vol. 2, Wiley-Interscience, New York, 1969, pp. 85–153.
29. R. E. Johnson, Jr. and R. H. Dettre, *J. Colloid Interface Sci.* **21**, 610 (1966).
30. P. B. Roth, *Am. Dyest. Rep.* **60**(7), 34 (1971).
31. E. J. Grajeck and W. H. Peterson, *Text. Res. J.* **32**, 320 (1962).
32. G. F. Clifford, *Am. Dyest. Rep.* **68**(1), 32 (1979).
33. J. D. Turner, *Text. Chem. Color.* **12**(3), 19 (1980).
34. G. F. Clifford, *Am. Dyest. Rep.* **69**(4), 19 (1980).
35. R. J. Lyons, *Am. Dyest. Rep.* **69**(4), 22 (1980).
36. J. F. Colvert, *Tappi* **59**, 129 (1976).
37. T. Fort, Jr., H. R. Billica, and T. H. Grindstaff, *J. Am. Oil Chem. Soc.* **45**, 354 (1968).
38. S. Smith and P. O. Sherman, *Text. Chem. Color.* **1**, 20 (1969).
39. L. Benisek and G. H. Cranshaw, *Text. Res. J.* **41**, 425 (1971).
40. A. G. Pitman, J. N. Roitman, and D. Sharp, *Text. Chem. Color.* **3**, 175 (1971).
41. C. J. Bierbrauer, K. D. Goebel, and D. P. Landucci, *Am. Dyest. Rep.* **68**(6), 19 (1979).
42. B. M. Latta and S. B. Sello, *Text. Res. J.* **51**, 579 (1981).
43. J. W. Bovenkamp and V. B. Lacroix, *Ind. Eng. Chem. Prod. Res. Dev.* **20**, 130 (1981).
44. O. C. M. Dorsett, *Text. Manuf.*, 112 (1970).
45. H. A. Schuyten, J. D. Reid, J. W. Weaver, and J. G. Frick, Jr., *Text. Res. J.* **18**, 396 (1948).
46. H. A. Schuyten, J. W. Weaver, J. G. Frick, Jr., and J. D. Reid, *Text. Res. J.* **22**, 424 (1952).
47. R. J. Berni, R. R. Benerito, and F. J. Philips, *Text. Res. J.* **30**, 576 (1960).
48. R. D. Deanin and D. C. Patel, *Text. Res. J.* **40**, 970 (1970).
49. L. S. Silbert, S. Serota, G. Maerker, W. E. Palm, and J. G. Phillips, *Text. Res. J.* **48**, 422 (1978).
50. F. Mares and B. C. Oxenrider, *Text. Res. J.* **47**, 551 (1977).
51. Ibid., **48**, 218 (1978).
52. U.S. Pat. 4,063,885 (Dec. 20, 1977), T. Mares, J. C. Arthur, Jr., and J. A. Harris (to United States).
53. N. Nishide and H. Shimizu, *Text. Res. J.* **45**, 591 (1975).
54. U.S. Pat. 4,193,880 (Mar. 13, 1979), R. M. Marshall (to Allied Chemical Corp.).
55. U.S. Pat. 4,195,105 (Mar. 25, 1980), F. Mares, B. C. Oxenrider, and C. Woolf (to Allied Chemical Corp.).
56. U.S. Pat. 4,209,610 (June 24, 1980), F. Mares, B. C. Oxenrider, and C. Woolf (to Allied Chemical Corp.).
57. U.S. Pat. 4,219,625 (Aug. 26, 1980), F. Mares, B. C. Oxenrider, and C. Woolf (to Allied Chemical Corp.).
58. U.S. Pat. 4,283,292 (Aug. 11, 1981), R. M. Marshall and K. C. Dardoufas (to Allied Chemical Corp.).
59. *Textile Chemist and Colorist Buyer's Guide*, American Association of Textile Chemists and Colorists, Research Triangle Park, N.C., 1981, pp. 110, 125–222.
60. C. A. Norris in ref. 1, pp. 265–296.
61. *AATCC Technical Manual*, American Association of Textile Chemists and Colorists, Research Triangle Park, N.C., 1982.
62. *Textile Test Methods, Federal Test Method Standard No. 191*, General Services Administration, Washington, D.C., July 20, 1978.
63. *Disposables Association Recommended Tests*, International Nonwovens and Disposables Association, New York.
64. Ref. 44, pp. 7–9.
65. K. Ward, Jr., *Chemical Modification of Papermaking Fibers*, Marcel Dekker, Inc., New York, 1973.
66. R. T. Van Ness, *Tappi* **59**, 113 (1976).
67. M. L. Carroll, Jr., *Tappi* **57**(8), 64 (1974).

68. F. C. Caruso, *Tappi* **57**(12), 139 (1974).
69. D. J. Fahey, *Indian Pulp Pap.* **22**(6), 361 (1967).
70. R. W. Davison and H. M. Spurlin in K. W. Britt, ed., *Handbook of Pulp and Paper Technology*, 2nd ed., Van Nostrand Reinhold Co., New York, 1970, pp. 355–367.
71. E. M. Engel in C. E. Libby, ed., *Pulp and Paper Science and Technology*, Vol. II, McGraw-Hill Book Co., New York, 1962, pp. 40–59.
72. *Ibid.*, p. 49.
73. J. J. Keavney and R. J. Kulick in J. P. Casey, ed., *Pulp and Paper Chemistry and Chemical Technology*, 3rd ed., Vol. III, John Wiley & Sons, Inc., New York, 1981, pp. 1547–1592.
74. C. A. Genge, *Anal. Chem.* **31**, 1750 (1959).
75. H. E. Berg, *Tappi* **39**, 153A (1956).
76. E. J. Vandenberg and H. M. Spurlin, *Tappi* **50**, 209 (1967).
77. E. Strazdins, *Tappi* **48**, 157 (1965).
78. R. W. Davison, *Tappi* **47**, 609 (1964).
79. Ref. 71, p. 55.
80. Ref. 73, pp. 1566, 1571–1573.
81. M. G. Halpern, *Paper Manufacture*, Noyes Data Corp., Park Ridge, N.J., 1975, pp. 126–185.
82. R. B. Porter and J. Behnke, *Tappi* **42**, 104A (1959).
83. J. W. Davis, W. H. Roberson, and C. A. Weisgerber, *Tappi* **39**, 21 (1956).
84. Ref. 70, p. 365.
85. Ref. 71, pp. 52–53.
86. U.S. Pat. 3,102,064 (Aug. 27, 1963), O. B. Wurzburg and E. D. Mazzarella (to National Starch and Chemical Corp.).
87. M. L. Cushing in ref. 73, pp. 1667–1714.
88. *Ibid.*, pp. 1703–1705.
89. J. W. Trebilcock, *Sizing, Short Course Notes*, Technical Association of the Pulp and Paper Industry, Atlanta, Ga., 1981, pp. 85–88.
90. J. Cochran, *Tappi* **42**(1), 113A (1959).
91. C. F. Robbart, *Tappi* **59**(10), 68 (1976).
92. U.S. Pat. 2,662,835 (Dec. 15, 1953), T. S. Reid (to Minnesota Mining & Manufacturing Co.).
93. C. A. Schwartz, *Sizing, Short Course Notes*, Technical Association of the Pulp and Paper Industry, Atlanta, Ga., 1981, pp. 71–75.
94. J. P. Casey, *Pulp and Paper Chemistry and Chemical Technology*, 2nd ed., Vol. III, Interscience Publishers, New York, 1961, p. 1716.
95. *TAPPI Routine Control Methods*, Technical Association of the Pulp and Paper Industry, New York, 1966.
96. Ref. 73, pp. 1550–1557.
97. Ref. 71, pp. 57–58.
98. *Zonyl RP Paper Fluoridizer*, technical bulletin, E. I. du Pont de Nemours & Co., Wilmington, Del., 1974, p. 18.
99. T. C. Thorstensen, *J. Am. Leather Chem. Assoc.* **73**(5), 196 (1978).
100. R. J. Heit, *J. Am. Leather Chem. Assoc.* **64**(11), 553 (1969).
101. F. P. Luvisi, W. J. Hopkins, E. M. Filachione, and J. Naghski, *J. Am. Leather Chem. Assoc.* **64**(12), 614 (1969).
102. U.S. Pat. 2,834,393 (Apr. 28, 1959), J. W. Gilkey (to Dow Corning Corp.).
103. U.S. Pat. 2,894,967 (July 14, 1959), J. W. Gilkey (to Dow Corning Corp.).
104. U.S. Pat. 3,002,949 (Oct. 3, 1961), S. Nitzsche and E. Pirson (to Wacker Chemie GmbH).
105. K. A. Andrinaov, V. I. Sidorov, P. I. Levenko, and A. I. Anpilogova, *Kozh. Obuv. Prom.* **16**(3), 27 (1974).
106. K. M. Zurabjan, V. V. Nepomnina, and L. V. Slobodskich, *Kozarstvi* **30**(10), 296 (1980); *Chem. Abstr.* **94**, 123128 (1981).
107. N. V. Vakhrameeva, K. M. Zurabyan, R. G. Drits, V. V. Nepomnina, K. S. Pushchevaya, and L. V. Slobodskich, *Kozh. Obuv. Promst.* **23**(6), 43 (1981).
108. *Quilon Chrome Complexes*, product information bulletin, E. I. du Pont de Nemours & Co., Wilmington, Del., 1979.
109. *Leather Chemical FC-233*, Pts. A and B, product information bulletin, Minnesota Mining & Manufacturing Co., St. Paul, Minn., Oct. 24, 1972.
110. P. E. Ingham and J. DeVeer, *Rep. Wool Res. Org. N.Z.*, 77 (1980); *Chem. Abstr.* **94**, 158212 (1981).

111. *1977 Annual Book of ASTM Standards*, Part 21, American Society for Testing and Materials, Philadelphia, Pa.
112. P. Maslow, *Chemical Materials for Construction*, Structures Publishing Co., Farmington, Mich., 1974, pp. 31–34, 433–471.
113. W. H. Taylor, *Concrete Technology and Practice*, McGraw-Hill Book Co., New York, 1977.
114. I. Biczok, *Concrete Corrosion and Concrete Protection*, Chemical Publishing Co., Inc., New York, 1967.
115. D. F. Orchard, *Concrete Technology*, 3rd ed., Vol. I, John Wiley & Sons, Inc., New York, 1973.
116. Ref. 113, p. 234.
117. Ref. 112, pp. 463–470.
118. Ref. 112, pp. 441–445.
119. Ref. 115, p. 61.
120. Ref. 112, pp. 31–34.
121. Ref. 113, pp. 234–235.
122. Ref. 115, pp. 73–74.
123. Ref. 112, pp. 449–454.
124. D. W. Fowler, J. T. Houston, and D. R. Paul in *Polymers in Concrete*, publication SP-40, American Concrete Institute, Detroit, Mich., 1973, pp. 93–117.
125. Ref. 112, p. 450.
126. Ref. 114, p. 491.
127. Ref. 112, p. 452.
128. V. M. Malhorta, *Testing Hardened Concrete: Nondestructive Methods*, Iowa State University Press, Ames, Iowa, and American Concrete Institute, Detroit, Mich., 1976, pp. 136–137.
129. E. J. Clark, P. G. Cambell, and G. Frohnsdorff, *National Bureau of Standards Technical Note 883*, National Bureau of Standards, Washington, D.C., Oct. 1975, 79 pp.

MASON HAYEK
E. I. du Pont de Nemours & Co., Inc.

WATTLE BARL. See Leather.

WAXES

Wax usually refers to a substance that is a plastic solid at ambient temperature and, on being subjected to moderately elevated temperatures, becomes a low viscosity liquid. Because it is plastic, wax usually deforms under pressure without the application of heat. The chemical composition of waxes is complex: they usually contain a broad variety of molecular weight species and reactive functional groups, although some classes of mineral and synthetic waxes are totally hydrocarbon compounds.

For centuries, the honeycomb of bees, ie, beeswax, was the material commonly referred to as wax. Substances having typical wax characteristics have traditionally come from insect, eg, beeswax, vegetable, eg, carnauba wax, and animal, eg, spermaceti, origins (1). Waxes from mineral and synthetic sources have been developed and surpass in tonnage and commercial importance those waxes from traditional sources. A tabulation of U.S. wax imports for 1971, 1976, and 1981 and their approximate values are given in Table 1.

It is important to recognize that unlike some other natural products used industrially in substantial volume wherein the plant producing the product is cultivated, eg, rubber trees or cotton (qv), the harvesting of vegetable and insect waxes is from wild, unmanaged sources (see Rubber, natural). Weather conditions and natural disasters severely affect the stability of supply and price. For example, Brazil, once a principal exporter of beeswax, has reduced production because of the development of a vicious strain of bee which hampers harvesting of the wax.

Insect and Animal Waxes

Beeswax. White [8012-89-3] and yellow [8006-40-4] beeswax have been known for over 2000 years, especially as used in the fine arts (3). References to wax prior to the nineteenth century are most probably to beeswax. Beeswax is secreted by bees, eg, *Apis mellifera*, *A. dorsata*, *A. flores*, and *A. Indica*, and used to construct the combs in which bees store their honey. The wax is harvested by removing the honey and melting the comb in boiling water; the melted product is then filtered and cast into cakes. The yellow beeswax cakes can be bleached with oxidizing agents, eg, peroxide or sunlight, to white beeswax, a product much favored in the cosmetic industry. Generally beeswax produced in the United States is not suitable for bleaching.

The composition of beeswax varies depending on its geographical origin. One analysis for yellow beeswax is given in Table 2. The structures of the components of beeswax are discussed in refs. 5–6 (see also Carboxylic acids).

Beeswax typically has a melting point of 64°C, a penetration (hardness) of 2.0 mm at 25°C and 7.6 mm at 43.3°C (ASTM D 1321), a viscosity of 1470 mm^2/s (= cSt = 67 SUs) at 98.9°C, an acid number of 20, and a saponification number of 84.

Beeswax is used widely in applications associated with the human body because it is safe to ingest in limited quantities and to apply to the skin (7). The FDA (U.S. Food & Drug Administration) affirmed the status of beeswax as GRAS in 1978 (8–10). Commercial specifications are given in ref. 11.

Demand for beeswax in the United States is ca 4500 metric tons per year and is supplied in part by domestic production. In 1981, the United States imported ca 1100 t

Table 1. U.S. Imports of Waxes, 1971, 1976, and 1981[a]

	1971			1976			1981		
Property	t	Fas[b], $1000	Fas[b], $/kg	t	Fas[b], $1000	Fas[b], $/kg	t	Fas[b], $1000	Fas[b], $/kg
beeswax (unbleached)	1,368	2,008	1.48	1,412	3,591	2.53	1,088	4,856	4.47
spermaceti	226	117	0.52	[c]	[c]	[c]	[c]	[c]	[c]
candelilla	1,727	1,680	0.97	1,192	1,493	1.26	379	855	2.25
carnauba	4,235	3,457	0.86	3,053	5,864	1.92	2,733	5,094	1.87
Japan	46	66	1.43	36	90	2.67	49	433	8.93
ouricury	78	77	0.97	14	40	2.84	6	28	4.74
montan	2,157	961	0.44	2,233	1,766	0.79	1,946	1,359	0.71
petroleum waxes	12,176	505	0.02	20,858	5,212	0.24	81,379	41,250	0.51
mineral waxes[d]									
FRG	2,114	2,006	0.95	3,912	4,117	1.06	2,662	5,938	2.23
Republic of South Africa	6,029	1,583	0.26	10,668	4,125	0.40	8,997	8,756	0.97

[a] Ref. 2.
[b] Fas = freight along side: value at port of export.
[c] The U.S. Endangered Species Act terminated the importation of any product produced from the sperm whale.
[d] Mineral waxes imported from the FRG are probably chemically modified montan waxes and the imports from the Republic of South Africa are Fischer-Tropsch waxes.

of unbleached beeswax: 23 t from Canada, 190 t from Brazil, 190 t from the Dominican Republic, and 122 t from Chile. Approximately 230 t of bleached beeswax was imported from the Netherlands in 1981.

The largest markets for beeswax are in the manufacture of candles and cosmetics (qv) and smaller quantities are used in such diverse applications as polishes (qv), confections, medicinals, and in the preservation of art (12). Except in Australia, the use of beeswax as a desensitizer in explosives based on cyclotrimethylenetrinitramine has largely been replaced by microcrystalline waxes (13) (see Explosives and propellants). The industrial markets, except cosmetics, are limited because of the high price of beeswax. Adulteration of pure beeswax with less-costly petroleum-derived paraffin or microcrystalline waxes is possible and detection of such adulteration has been addressed in the literature and by industry (14–15).

Spermaceti. Spermaceti [8002-23-1] is a wax that precipitates on cooling of the head oil from the sperm whale, *Physeter macrocephalus*, which is a protected species. Spermaceti has not been an article of commerce in the United States since the production and importation of sperm whale products was banned in 1976. This product has been replaced by blends of other natural and synthetic products.

Vegetable Waxes

The aerial surfaces of almost all multicellular plants are covered by a layer of wax (16). With the advent of more sensitive analytical tools, eg, gas–liquid chromatography and mass spectrometry, investigations of the character of epicuticular wax of many species has been undertaken (17–20). However, only a very few species growing, primarily those in semiarid climates, produce wax in such quantities that commercial recovery is economically feasible.

468   WAXES

Table 2. Analysis of Yellow Beeswax[a]

Component	Value, wt %
esters of wax acids	
myricyl palmitate, $C_{15}H_{31}COOC_{30}H_{61}$	23
lacceryl palmitate, $C_{15}H_{31}COOC_{32}H_{65}$	2
myricyl cerotate, $C_{25}H_{51}COOC_{30}H_{61}$	12
myricyl hypogaete, $C_{13}H_{27}CH=CHCOOC_{30}H_{61}$	12
ceryl 2-hydroxypalmitate, $C_{14}H_{29}CH(OH)COOC_{26}H_{51}$	8–9
acid esters	4–4.5
diesters	6–6.5
acid diesters, triesters, hydroxy diesters	3–3.5
Total	70–71.5
cholesteryl esters of fatty acids	1
coloring matter: 3-hydroxyflavone[b]	0.3
lactone: ω-myristolactone, $O(CH_2)_{13}CO$	0.6
free alcohols: C_{34}–C_{36}	1–1.25
free wax acids	
lignoceric acid, $C_{23}H_{47}COOH$	1–1.5
cerotic acid, $C_{26}H_{51}COOH$	3.8–4.4
melissic acid, $C_{24}H_{59}COOH$	2
psyllic acid, $C_{31}H_{63}COOH$	1.3–1.5
hypogaeic acid, $C_{13}H_{27}CH=CHCOOH$	1.5
Total	9.6–10.9
hydrocarbons	
pentacosane, $C_{25}H_{52}$	0.3
heptacosane, $C_{27}H_{56}$	0.3
nonacosane, $C_{29}H_{60}$	1–3
hentriacosane, $C_{31}H_{64}$	8–9
melene, $C_{28}H_{57}CH=CH_2$	2.5
Total	12.1–15.1
moisture and mineral impurities	1–2

[a] Ref. 4.

[b] 3-Hydroxyflavone =

Candelilla. Candelilla wax [8006-44-8] is harvested from the shrubs *Euphorbiea antisiphilitica*, *Euphorbiea cerifera*, and *Pedilanthus pavonis* in the Mexican states of Coahuila and Chihuahua and, to a very small degree, in the Big Bend region of Texas in the United States (21). The entire mature plant is uprooted and immersed in boiling water acidified with sulfuric acid. The wax floats to the surface and is filtered. The approximate chemical composition of candelilla is

Components	wt %
wax esters	28–29
alcohols, sterols, and neutral resins	12–14
hydrocarbons	49–50
free acids	7–9
moisture	2–3
inorganic residue	1

Typically, candelilla wax has a melting point of 70°C, a penetration or hardness of 0.3 mm at 25°C, an acid number of 14, and a saponification number of 55.

The marketing of candelilla wax is controlled by the Mexican government through the Banco Nacional de Credito Rural S. A., which establishes pricing and availability. The harvesting of candelilla wax is labor-intensive and brings much-needed revenue to an economically depressed area. Candelilla production in Mexico is now ca 1000 t/yr, down from ca 2700 t/yr. Approximately half of the production is exported to the United States. The March 1982 price was $4.23/kg.

The principal market for candelilla wax used to be as a component in chewing-gum base, but its use as a cosmetics raw material, eg, in lipsticks, is now the main market. Unlike most other waxes, candelilla exhibits little change in volume on phase change from solid to liquid or liquid to solid. Thus, the wax is desirable as a component of investment casting waxes and candelilla can be used as a protective coating for certain citrus fruits (22).

A specification for candelilla wax is given in ref. 23. Candelilla wax meets the requirements established by the FDA, 21 CFR 172.615 (chewing gum), 175.105 [adhesives (qv)], 175.320 (resinous and polymeric coatings for polyolefin films), and 176.180 [components of paper (qv) and paperboard in contact with dry food].

Carnauba. The source of carnauba wax [8015-86-9] is the palm tree, *Copernicia prunicia*, whose wax-producing stands grow almost exclusively in the semiarid northeast section of Brazil and principally in the states of Ceara and Piani. Carnauba wax forms on the fronds of the palm. The wax is removed by cutting the fronds, drying them, and mechanically removing the wax. Impurities are removed from the wax by melting and filtering or centrifuging; the former method produces the cleaner wax, although dirt left after centrifuging is acceptable to certain markets, eg, carbon-paper ink in which any dirt is ground with carbon black.

The chemical composition of carnauba wax is approximately as follows:

Component	wt %
aliphatic and aromatic esters	84–85
aliphatic esters	40
ω-hydroxy esters	13
cinnamic aliphatic diesters	8
free fatty acids	3–3.5
alcohols	2–3
lactides	2–3
hydrocarbons	1.5–3.0
resins	4–6
moisture and inorganic residue	0.5–1.0

Carnauba wax is a very hard (penetration of 0.2 mm at 25°C and 0.3 mm at 43.3°C), high melting (mp 82–86°C) material with a viscosity of 3960 mm^2/s (= cSt = 165 SUs) at 98.9°C, an acid number of 8, and a saponification number of 80. Its wide combination of desirable properties make it an attractive raw material in a variety of uses. Carnauba is outstanding in its ability to gel high concentrations of organic solvents or oils, a characteristic taken advantage of in the formulation of paste polishes and carbon inks. Carnauba polishes to a high gloss and thus is widely used as a polishing agent for such items as candies or pills. It also serves as a base for wax emulsions applied to certain perishable products, notably citrus fruits, to retard moisture loss. Among the many additional uses for carnauba wax are in the formulation of investment casting waxes, polystyrene phonograph records, leather finishes, and lipsticks. Wide fluctuations in price and availability have caused markets once served by carnauba to turn to synthetic waxes.

Carnauba wax is marketed in four basic grades: yellow, light fatty gray, fatty gray, and chalky. The grade in large part is determined by the age of the frond from which the wax is removed; the lighter color, ie, yellow, is the highest grade. Carnauba wax is harvested twice a year.

The marketing of carnauba wax is controlled by the Brazilian government through the Comite do Financiamento do Producao. Harvesting of the fronds helps maintain employment in the economically depressed northeast, and the Comite purchases, stores, and markets the harvest depending on the country's need for exchange and more favorable prices. To that extent, the price of carnauba is not a free-market price. Although the annual production of carnauba wax can vary substantially, on the average ca 12,000 t is produced, of which ca 10,000 t is available for export.

A specification for carnauba wax is given in ref. 24. The wax is on the GRAS list of the FDA.

Japan Wax. Japan wax [8001-39-6] is a fat and is derived from the berries of the species *Rhus succedanea*, a small tree native to Japan and China which is cultivated for its wax. Japan wax is composed of triglycerides, primarily tripalmitin. Typical properties of Japan wax are

mp, °C	52.8
acid number	18
saponification number	217
specific gravity at 25°C	0.98

Japan wax is used in the formulation of candles, polishes, lubricants, and as an additive to thermoplastic resins (see Lubrication and lubricants). It also is used in certain food-related applications (9,25).

Ouricury Wax. Ouricury wax [68917-70-4] is a brown wax obtained from the fronds of the palm *Syagrus coronata*, which grows in Bahia in northeast Brazil. Ouricury is difficult to harvest as it does not flake off the frond as does carnauba wax; rather, it must be scraped free. Ouricury is sometimes used as a replacement for carnauba in applications that do not require a light-colored wax.

Douglas-Fir Bark Wax. Douglas-fir (*Pseudotsuga menziesii*) bark extracted with a hot organic mixture of aliphatic and aromatic hydrocarbons yields a light-to-dark-green wax [8050-89-3] having a melting point of 59°C, an acid number of 80, a saponification number of 145, a penetration of 1.0–1.5 mm at 25°C (ASTM D 1321), and a viscosity of ca 4285 mm^2/s (= cSt = 200 SUs) at 99.8°C. An analysis of the wax is as follows:

Component	wt %
monocarboxylic acids	37.0
dicarboxylic acids	6.3
hydroxy acids	16.5
aliphatic alcohols	26.1
phenolic acids	14.1

Although a commercial plant was built to produce ca 910 t/yr of this wax in 1974 in the northwest United States, the venture was not successful because no profitable market for the wax developed (26).

Rice-Bran Wax. Rice-bran wax [8016-60-2] is extracted from crude rice-bran oil. It can be degummed, the fatty acid content reduced by solvent extraction, and bleached. The wax is primarily composed of esters of lignoceric acid (43 wt %), behenic acid (16 wt %) and C_{22}–C_{36} alcohols (28 wt %). The FDA specifications are (27)

mp, °C	75–80
free fatty acids (max), wt %	10
iodine number, max	20
saponification number	75–120

One commercial grade of rice-bran wax has a melting point of 79–83°C, acid number of 13 (max), saponification number of 75–88, and an iodine number of 5 (max) (28).

Rice-bran wax has been approved as a release agent in processing plastic packaging materials intended for food-contact applications and is used as a base for lipsticks (29) (see Abherents).

Jojoba. The cultivation of the jojoba bean, *Simmondsia chinensis*, occurs primarily in semiarid areas of Costa Rica, Israel, Mexico, and the United States (30). The oil, which is referred to as a liquid wax as it contains little or no glycerides and rather long-chain esters, can be hydrogenated to form a solid, jojoba wax (31). Jojoba oil's close resemblance chemically and physically to sperm whale oil, which is no longer available in the United States, and the interest in the oil as a cosmetic raw material are important factors in the oil's gaining broader use.

The oil is made up of ca 80 wt % of esters of eicos-11-enoic and docos-13-enoic acids, and eicos-11-en-1-ol, and docos-13-en-1-ol, ca 17 wt % of other liquid esters, and the balance being free alcohols, free acids, and steroids. The March 1982 price of jojoba oil was ca $22/kg.

Jojoba oil is used primarily in the formulation of cosmetics and derivatives are being studied for various applications (32). Hydrogenated jojoba oil, which is a true wax, is being used in the manufacture of candles, but until the price of the oil decreases, only low volume specialty applications are being considered.

Castor Wax. Castor wax [8001-78-3] is catalytically hydrogenated castor bean oil. The wax has a melting point of 85–88°C, acid number of 2, saponification number of ca 179, and an iodine number of 4.

Bayberry Wax. Bayberry wax [8038-77-5] is removed from the surface of the berry of the bayberry (myrtle) shrub by boiling the berries in water and skimming the wax from the surface of the water. The wax is green and made up of primarily lauric, myristic, and palmitic acid esters. The melting point is 42–48°C, the acid number is ca 15, the saponification number is ca 220, and the iodine number is ca 6. The wax has an aromatic odor and is used in the manufacture of candles and other products where the distinctive odor is desirable. Colombia is the main supplier of this wax.

Mineral Waxes

Montan Wax. Montan wax [8002-53-7] is derived by solvent extraction of lignite (qv). The earliest production on a commercial scale was in Germany during the last half of the nineteenth century, and the GDR at Amsdorf continues to produce ca 85% of the world's supply of montan wax, ie, 27,300–50,000 t/yr. The montan wax production at Amsdorf is part of a massive coal-mining operation from a continuous vein and raw material is expected to last for decades. Montan wax is produced in the United

States in Ione, California, as well as in Treysa, FRG. Montan wax is reportedly available from the USSR and the People's Republic of China, but no product from these sources has been available recently in world markets.

The chemistry and properties of a montan wax depend to a great degree on the material from which it is extracted, but all montan wax contains varying amounts of resin and asphalt, depending on the source (33) (see Table 3).

The wax component of montan is a mixture of long-chain (C_{24}–C_{30}) esters (62–68 wt %), long-chain acids (22–26 wt %), and long-chain alcohols, ketones, and hydrocarbons (7–15 wt %). The resin portion is ca 70 wt % terpenes and polyterpenes and 30 wt % resinic acid and oxyresinic acid; the asphalt portion is believed to be polymerized esters of oxyresinic acid.

The economics of the commercial production of montan wax has been reviewed in terms of production in the United States and the GDR (34). In 1979, the United States consumed ca 3409 t of montan wax, of which 44% was domestically produced.

By far the largest use for montan wax in the United States is as a component in one-time hot-melt carbon-paper inks. Montan wax is also useful in the formulation of various types of polishes. In addition to one-time carbon ink, a second large market for montan wax is as a raw material in the production of chemically modified montan waxes. Hoechst A.G., at its plant at Gersthofen, FRG, converts montan wax to the wax acid and removes the resin and asphalt portions in the process. The product, a light-colored wax acid, in turn reacts with alcohols to form esters and other chemicals and waxes useful in the manufacture of polishes and plastics lubricants.

Peat Waxes. Peat waxes are much like montan waxes in that they contain three main components: a wax fraction, a resin fraction, and an asphalt fraction. The amount of asphalt in the total yield is influenced strongly by the solvent used in the extraction. Montan waxes contain ca 50 wt % more of the wax fraction than peat waxes and correspondingly lower percentages of the resin and asphalt fractions. The wax fraction in peat wax is chemically similar to that of the wax fraction in montan wax.

Ozokerite and Ceresin Waxes. Ozokerite wax [8001-75-0] was a product of Poland, Austria, and the USSR where it was mined. True ozokerite no longer seems to be an article of commerce, at least in the Western Hemisphere, where it has been replaced with blends of petroleum-derived paraffin and microcrystalline waxes. These blends are designed to meet the specific physical properties required by the application involved.

Ceresin wax [8001-75-0] originally was a refined and bleached ozokerite wax but now is a paraffin wax of very narrow molecular weight distribution or blend of petroleum waxes.

Table 3. Properties of Montan Waxes from Varying Sources

Property	Czechoslovakia	GDR	United States	USSR
melting point, °C	82	86	87	86
viscosity (at 90°C), mPa·s (= cPs)	350	70	200	60
acid number	35	32	48	26
saponification number	97	92	112	88
resin, wt %	31	13	11	14
asphalt, wt %	8	5	12	2

Petroleum Waxes. Waxes derived from petroleum are hydrocarbons of three types: paraffin [64742-43-4] (clay-treated), semimicrocrystalline (intermediate), and microcrystalline [64742-42-3] (clay-treated); semimicrocrystalline waxes are not generally marketed as such (35) (see also [64742-26-3], acid-treated; [64742-33-2], chemically neutralized; [64742-60-5] and [64742-51-4], hydrotreated; and [8002-74-2], paraffin and hydrocarbon waxes, untreated). The quality and quantity of the wax separated from the crude oil depends on the source of the crude oil and the degree of refining to which it has been subjected prior to wax separation; however, some crude oils contain little or no wax. Generally, crudes containing high concentrations of wax are in the Appalachian (Pennsylvania) and midcontinent areas in the United States as well as in Venezuela, Romania, the USSR, Burma, and Sumatra.

A paraffin wax is a petroleum wax consisting principally of normal alkanes. Paraffin, microcrystalline, and semicrystalline waxes may be differentiated using the refractive index of the wax and its congealing point as determined by ASTM D 938 (36). Semimicrocrystalline and microcrystalline waxes are petroleum waxes containing substantial proportions of hydrocarbons other than normal alkanes. The former has a kinematic viscosity at 98.9°C < 10 mm^2/s (= cSt). A microcrystalline wax has a kinematic viscosity at 98.9°C of ≥10 mm^2/s (= cSt).

Refractive indexes of paraffin waxes are shown in Figure 1 (in shaded area).

Wax	n_D
microcrystalline wax	
midcontinent origin, 76.7°C	1.4381
midcontinent origin, amber, 87.8°C	1.4389
midcontinent origin, brown, 87.7°C	1.4390
Pennsylvania origin, 100°C	1.4426

Paraffin wax is macrocrystalline, brittle, and it is composed of 40–90 wt % normal paraffins and the remainder is C_{18}–C_{36} isoalkanes and cycloalkanes. Paraffin wax has little affinity for oil content; fully refined, below 1 wt %; crude scale, 1–2 wt %; and slack [64742-61-6], above 2 wt % and a melting point above 43.3°C. Within these classes, the melting point of the wax determines the actual grade and it varies from ca 46.1°C to 71.1°C. Typical physical properties of petroleum waxes are listed in Table 4.

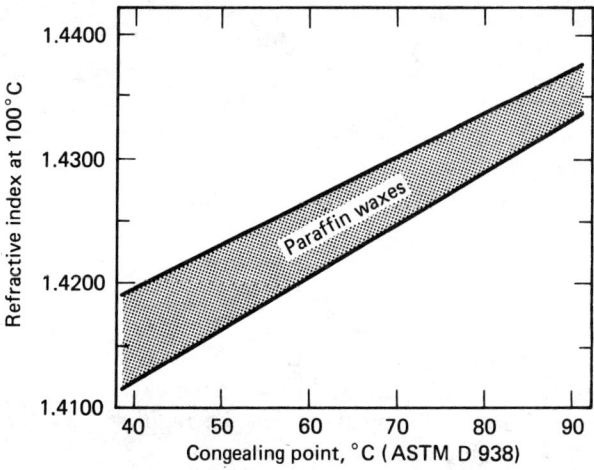

Figure 1. Congealing point–refractive index relationship.

Table 4. Typical Physical Properties of Petroleum Waxes

Property	Paraffin	Microcrystalline
flash point, °C	204, min	260, min
viscosity at 98.9°C, mm^2/s (SUs)	4.2–7.4 (40–50)	10.2–25 (60–120)
melting range, °C	46–68	60–93
refractive index at 98.9°C	1.430–1.433	1.435–1.445
average mol wt	350–420	600–800
carbon atoms per molecule	20–36	30–75
other physical aspects	friable to crystalline	ductile-plastic to tough-brittle

The separation of paraffin wax from crude oil occurs during distillation, as shown in Figure 2. The distillate is processed to remove oil to the degree desired through solvent extraction. It is then decolorized, usually by hydrogenation, but percolation through bauxite is also used. Microcrystalline wax is produced either from the residual fraction of crude-oil distillation or from crude-oil tank bottoms (37). After deasphalting of the residual fraction, heavy lubricating oil is removed by solvent extraction. The degree of solvent extraction is dictated by the economics of the lubricating-oil market. The filtrate is crude petrolatum, a dark-colored, unctuous material containing oil and microcrystalline wax in varying percentages but usually ca 40 wt % wax and 60 wt % oil.

This material is then solvent-extracted for the wax. Because microcrystalline wax has great affinity for oil, the oil content of the wax is 1–4 wt %, depending on the grade of wax. Unlike paraffin wax, oil is held tightly in the crystal lattice and does not migrate to the surface.

The high molecular weight fractions in crude oil precipitate from solution in large storage tanks and form a sludge that is removed periodically. These crude-oil tank bottoms are essentially crude oil with very high wax contents and are processed as indicated in Figure 2.

Two broad classes of microcrystalline waxes are produced: plastic grades, those with a penetration greater than 11 which are usually produced from crude petrolatum, and hard grades, which have a penetration of 11 or less, and usually are made from crude-oil tank bottoms.

Petroleum waxes are produced in massive quantities throughout the world. Subject to the wax content in the crude, paraffin and, to a substantially lesser degree, the microcrystalline wax are produced in almost all countries of the world that refine crude oil. Thirty-three refineries in the United States and five in Canada are capable of producing ca 1.36×10^6 t/yr of petroleum waxes, and production in the United States is estimated to be ca 9.09×10^5 t/yr of fully refined paraffin, crude-scale, and microcrystalline waxes. Tonnage and dollar values of petroleum wax exports from the United States are listed in Table 5.

The FDA has established basic regulations (39) for petroleum wax. The principal requirement is the maximum uv absorbtivity, ie, 280–400 nm, which an extract of the wax must not exceed. The FRG has established a food-use regulation for microcrystalline and hard paraffin (40). Many other nations' regulations refer to the FDA law. A specification for petroleum wax is given in ref. 41.

The Japanese authorities consider petroleum wax to be a natural product and

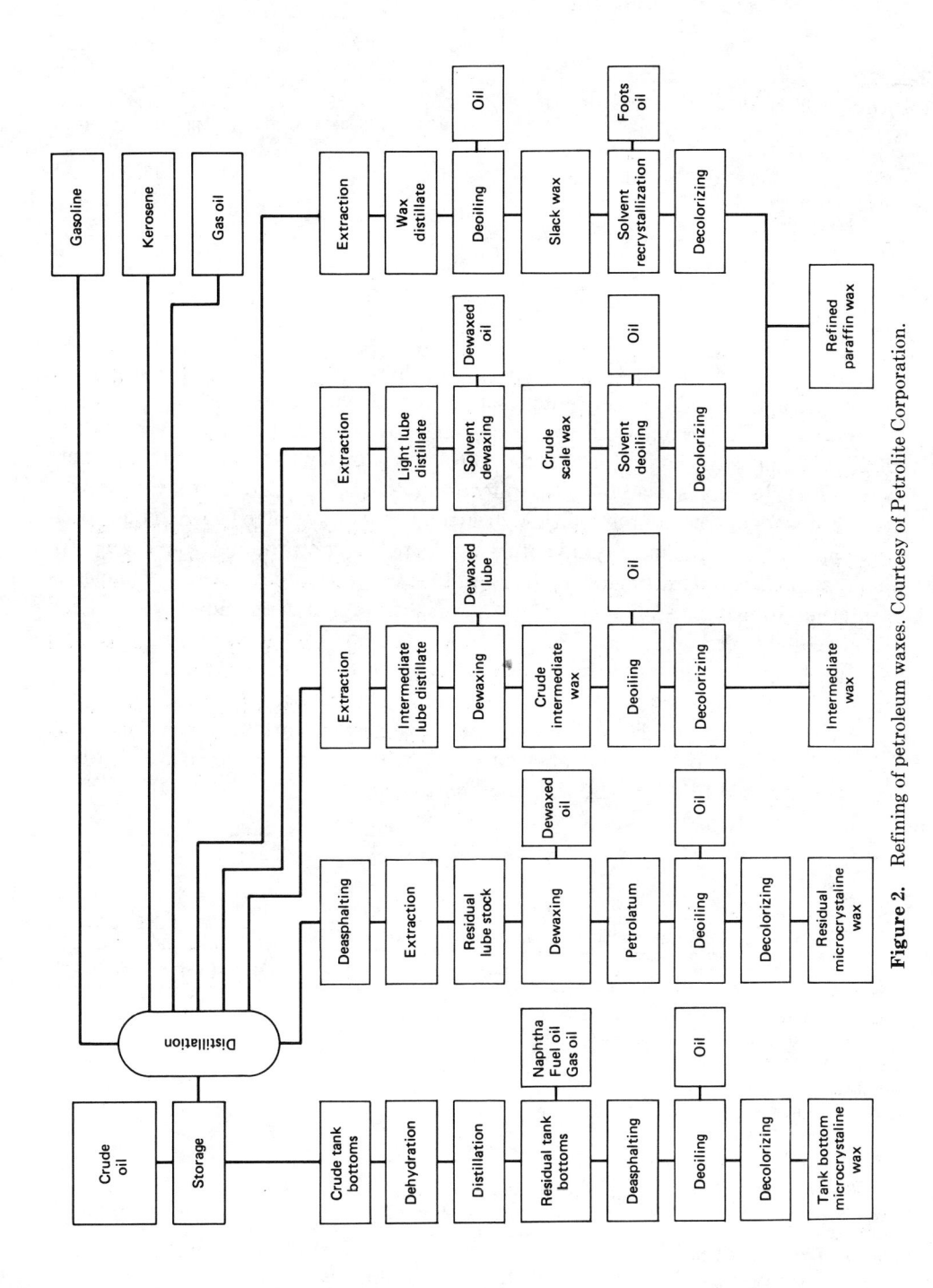

Figure 2. Refining of petroleum waxes. Courtesy of Petrolite Corporation.

Table 5. U.S. Exports of Petroleum Waxes [a]

Wax	1971 t	1971 $1000	1971 $/kg	1976 t	1976 $1000	1976 $/kg	1981 t	1981 $1000	1981 $/kg
microcrystalline wax	30,196	10,231	0.34	17,109	14,749	0.86	19,520	26,663	1.37
crystalline paraffin wax, fully refined	77,045	16,093	0.21	24,663	9,886	0.40	14,742	14,412	0.98
crystalline paraffin wax, not including the fully refined product	100,833	9,693	0.10	35,115	6,362	0.18	41,292	16,377	0.40

[a] Ref. 38.

allow its use in such products as chewing gum, whereas a low molecular weight polyethylene is considered synthetic and its use is not allowed in such applications.

Petroleum wax is outstanding as a cost-effective moisture and gas barrier; thus, food packaging is the largest market for petroleum waxes in the United States, where over 50% of the wax is used in this application (42–44). Much of the petroleum wax produced is food-grade quality, although such quality may well be used in non-food-grade applications to simplify inventorying. Depending on the food-packaging application, blends of paraffin and microcrystalline wax may be used as well as in blends with other additives, eg, higher molecular weight polyethylene and ethylene–vinyl acetate copolymers.

In 1981, in the United States, ca 7.95×10^5 t of fully refined paraffin, scale, and microcrystalline waxes was consumed in the markets listed in Table 6 (45).

The three main uses for slack waxes and crude petrolatum (the raw material from which microcrystalline waxes are made) are fireplace logs (1.14×10^5 t/yr), cable filling compounds (4.5×10^4 t/yr), and wallboard particle board (7.3×10^4 t/yr) (59–61) (see Insulation, electric; Laminated wood-based composites).

Table 6. Markets for Fully Refined Paraffin, Scale, and Microcrystalline Waxes

Use	1000 t	Refs.
coating and impregnating corrugated board	154	
coatings for cups and other containers	131	
candles	102	
coatings and laminates for flexible packages	100	46
ethylene–vinyl acetate-based hot-melt adhesives	59	47
antiozonants for rubber	45	48
raw material for chlorination	30	
processing aids for thermoplastic resins	20	
crayons	19	
emulsions	19	49
coatings for folding cartons	18	
miscellaneous uses, eg, carbon-paper ink, lubricants, floor polishes, dental waxes, investment casting waxes, sealants, electrical insulation, paper size, munitions, cosmetics, chewing-gum base, industrial coatings, thermal storage, cheese coatings, and mold-release agents	99	50–58

Synthetic Waxes

Polyethylene Waxes. Low molecular weight (less than ca 10,000) polyethylenes [9002-88-4] having waxlike properties are used in conjunction with petroleum waxes in food-packaging applications. These polyethylenes are made either by high pressure polymerization, low pressure (Zeigler-type catalysts) polymerization, or controlled thermal degradation of high molecular weight polyethylene (62–64). Because of the substantial plant investment required, the number of plants producing low molecular weight polyethylenes has been limited. These polyethylenes are produced in the United States, Japan, the GDR, and the FRG (see Olefin polymers).

The FDA allows the use of polyethylene in indirect food-additive applications under the olefins regulation (65). This regulation couples a lower molecular weight limit with certain solubility tests. Some low molecular weight polyethylenes, although meeting the molecular weight test, fail the solubility tests but are not of the structure nor meet the molecular weight requirements of the FDA synthetic wax regulations (66), ie, the polyethylenes must have molecular weights of 500–1200. Therefore, at least in the United States, caution is recommended when considering the use of polyethylenes with molecular weights of 1200–2000 in food-related applications. A specification for polyethylene and synthetic petroleum wax is given in ref. 11.

Fischer-Tropsch Waxes. Polymethylene wax [68649-50-3] production is based on the Fischer-Tropsch synthesis, which is basically the polymerization of carbon monoxide under high pressure and over iron catalysts to produce hydrocarbons (67) (see Fuels, synthetic, liquid fuels). The South African Coal, Oil, & Gas Corporation, Ltd. (SASOL) has the capacity to produce annually ca 27,000 t hard and 11,000 t soft polymethylene waxes (68). The United States probably imports ca 13×10^3 t of polymethylene waxes, some classified as petroleum wax but most classified as mineral wax (see Table 1). These synthetic waxes have melting points from ca 45°C to ca 106°C. Except for their use in adhesives, FDA regulations (69) covering these waxes limit their food-related uses to those waxes with a minimum congealing point of 93.3°C.

Chemically Modified Hydrocarbon Waxes. Hydrocarbon waxes of the microcrystalline, polyethylene, and polymethylene classes are chemically modified to meet specific market needs. In the vast majority of cases, the first step is air oxidation of the wax with or without catalysts (70). The product has an acid number usually no higher than 30 and a saponification number usually no lower than 25. An alternative step is the reaction of the wax with a polycarboxylic acid, eg, maleic, at high temperature (71). Through its carboxyl groups, the oxidized wax can be further modified in such reactions as saponification or esterification. Oxidized wax is easily emulsified in water through the use of surfactants or simple soaps, and films resulting therefrom are used widely in floor polishes and coatings (72–73).

Approximately 4500 t of low molecular weight polyethylene was produced in the United States in 1981, mainly by Allied Chemical Corporation and Eastman Chemical Products Co. Probably the largest single use is in hot-melt adhesives. Other large markets are plastics-processing aids, antislip and antimar printing-ink additives, polishes, and cosmetics (74). Oxidized polyethylene wax and its derivatives probably represent ca 15% of the unmodified polymer produced in the United States.

Low molecular weight polyethylenes are available with a variety of properties (see in Table 7). In addition to homopolymers of ethylene, copolymers of ethylene, propylene, butadiene, and acrylic acid with waxlike properties are also produced.

WAXES

Table 7. Typical Properties of Selected Synthetic Hydrocarbon Waxes

Wax	Mp[a], °C	Ring-and-ball[b], °C	Penetration[c], 0.1 mm 25°C	Penetration[c], 0.1 mm 60°C	Viscosity at 149°C, mPa·s (= cP)	Density at 23°C, g/cm³
Allied A-C 6 polyethylene	106	106	4	20	220	0.92
BASF A polyethylene	108	108	3	15	450	0.92
Ciech WP-2 polyethylene	110	101	3	15	200	0.93
Epolene N-12 polyethylene	110	113	1	9	220	0.94
Hoechst PA 130 polyethylene	125	126	1	4	320	0.93
Leuna LE 114 polyethylene	115	116	2	7	260	0.93
Paraflint H-1 polymethylene	108	108	2	15	7	0.94
Polywax 500 polyethylene	86	86	7	61	3	0.94
Polywax 2000 polyethylene	125	125	1	2	50	0.96
Veba A227 polyethylene	108	108	2	10	260	0.93

[a] ASTM D 127.
[b] ASTM D 36 (MOD).
[c] ASTM D 1321.
[d] ASTM D 3236.

Substituted Amide Waxes. The product of fatty acid amidation has unique waxlike properties (75). Probably the most widely produced material is N,N'-distearoylethylenediamine [110-30-5], which has a melting point of ca 140°C, an acid number of ca 7, and a low melt viscosity. Because of its unusually high melting point and unique functionality, it is used in additive quantities to raise the apparent melting point of thermoplastic resins and asphalts, as an internal–external lubricant in the compounding of a variety of thermoplastic resins, and as a processing aid for elastomers.

Analytical Methods

Most of the routine test procedures on waxes are for the measurement of the physical properties of waxes and are used to compare the properties of waxes within a class. One cannot directly substitute a wax from one class with a wax of another class, even though their physical and some chemical properties are the same (76). The vast differences in the chemistry of the waxes makes such substitution to achieve equal performance almost impossible. The ASTM, TAPPI, and the American Wax Importers & Refiners Association (AMERWAX) have developed widely used wax-testing procedures.

Melting and Congealing Points. A wax's reported melting point greatly depends on the method used and the characteristics of the wax (77). Drop melting point (ASTM D 127) is suitable for amorphous waxes, eg, microcrystallines, but is not reliable for higher viscosity synthetic waxes, for which ring-and-ball softening point (ASTM D 36) should be used. Waxes whose time–temperature cooling curves exhibit plateaus, eg, paraffin wax, may be evaluated by ASTM D 87. The congealing point of a wax (ASTM D 938) is the temperature at which a wax, once melted, ceases to flow. The melting point of many natural waxes is determined with either open or closed capillary tubes. Differential scanning calorimetry is being more widely used on all types of waxes and gives the investigator a quantitative melting-point profile (78–79).

Hardness (Penetration). The standard test for the hardness of waxes in industry is the penetration test (ASTM D 1321). This test measures the depth in tenths of a millimeter that a needle of a certain configuration under a given weight penetrates the surface of a wax at a given temperature. A series of penetrations made at different temperatures, rather than at a single temperature, is preferred.

Color. On solidification of a wax and depending on factors such as the rate of cooling, the amount of occluded air, and surface finish, the color of solidified samples of the same wax may be different. For this reason, the color of many waxes is judged only when they are molten, although some commercial standards for certain waxes, eg, carnauba, are based on the color of the solid wax.

The accurate measurement of color in light-colored, ie, amber to off-white to white, waxes is difficult but very important because of the additional processing costs required to achieve the light color. The two most widely used color standards providing numerical measurement are ASTM D 1500, which is used to measure dark-brown to off-white color, and ASTM D 156, which is used to measure off-white to pure white.

Oil Content. The production of petroleum waxes involves the removal of oil; therefore, the oil content or the percentage of low molecular weight fractions is one indication of the quality of the wax. Oil content is determined (ASTM D 721) as that percentage of the wax soluble in methyl ethyl ketone at $-31.7°C$.

Viscosity. Although traditionally of little importance in the evaluation of vegetable and insect waxes, viscosity is an important test for mineral and synthetic waxes. One of the most frequently used tests, ASTM D 88, is used to measure the time in seconds required for a specified quantity of wax at a specified temperature to flow by gravity through an orifice of specified dimensions. This viscosity is expressed in Saybolt Universal seconds (SUs) at the temperature of the test. The SI unit for kinematic viscosity is mm^2/s ($=$ cSt).

BIBLIOGRAPHY

"Waxes" in *ECT* 1st ed., Vol. 15, pp. 1–17, by C. J. Marsel, New York University; "Waxes" in *ECT* 2nd ed., Vol. 22, pp. 156–173, by E. S. McLoud, Consultant.

1. P. E. Kolattukudy, *Chemistry and Biochemistry of Natural Waxes*, Elsevier Scientific Publishing Co., Amsterdam, 1976.
2. *U.S. Imports for Consumption*, Reports IM145 and IM146, U.S. Department of Commerce, Washington, D.C., 1971, 1976, 1981.
3. R. White, *Stud. Conserv.* **23,** 57 (1978).
4. A. H. Worth, *The Chemistry and Technology of Waxes*, 2nd ed., Reinhold Publishing Co., New York, p. 92.

5. D. T. Downing and co-workers, *Aust. J. Chem.* **14,** 253 (1961).
6. G. J. Bloomquist and L. L. Jackson, *Prog. Lipid Res.* **17,** 319 (1979).
7. *Evaluation of the Health Aspects of Beeswax (yellow and white) as Food Ingredients*, PB 262-656, U.S. Department of Commerce, National Technical Information Service, Washington, D.C., 1975.
8. *Fed. Regist.* **43**(68), 14643 (April 7, 1978).
9. *Scientific Literature Review on Generally Recognized as Safe (GRAS) Food Ingredients—Beeswax and Japan Wax*, PB 223-854, National Technical Information Service, U.S. Department of Commerce, Washington, D.C., May 1973.
10. 21 CFR 184.1973.
11. *Food Chemicals Codex*, 3rd ed., National Academy of Sciences–National Research Council, Washington, D.C., 1980, p. 34.
12. E. C. G. Packard, *Museum News/Technical Supplement*, No. 19, The Walters Art Gallery, Baltimore, Md., Oct. 1967.
13. J. Eadie and W. C. Wilson, *Esters of 2,4,6-Trinitrobenzoic Acid and the Interfacial Tension Between Wax and TNT*, Report 637, Department of Defense, Australian Defense Scientific Service Materials Research Laboratory, Maribyrnong, Victoria, Australia, Aug. 1975.
14. C. A. Blair, *Proc. Soc. for Anal. Chem.*, 118 (May 1974).
15. A. P. Tulloch, *J. Am. Oil Chem. Soc.* **50,** 269 (1973).
16. S. Hamilton and R. J. Hamilton, *Top. Lipid Chem.* **3,** 199 (1972).
17. A. P. Tulloch, *J. Am. Oil Chem. Soc.* **50,** 367 (1973).
18. K. E. Murray and H. R. Schulten, *Chem. Phys. Lipids* **29,** 11 (1981).
19. S. Nishimoto, *J. Sci. Hiroshima Univ. Ser. A* **38**(1), 151 (April 1974).
20. A. P. Tulloch and co-workers, *Can. J. Bot.* **58,** 2602 (1980).
21. *Monograph on Candelilla Wax*, PB 287-762, National Technical Information Service, U.S. Department of Commerce, Washington, D.C., Aug. 10, 1978.
22. O. Paredes-Lopez and co-workers, *J. Sci. Food Agric.* **25,** 1207 (1974).
23. Ref. 11, p. 34
24. Ref. 11, p. 73.
25. 21 CFR 73.1(b)(2); 175.105; 175.170(a)(5); and 182.70.
26. D. G. Good and F. S. Trocino, *Chem. Eng.*, 70 (May 27, 1974).
27. 21 CFR 172.890.
28. Noda Wax Co., Ltd., Kanagawa-Ken, Japan, 1982.
29. *Fed. Regist.* **44,** 69649 (Dec. 4, 1979); 21 CFR 178.3860.
30. *Chem. Week*, 31 (Feb. 7, 1979).
31. T. K. Miwa, ed., *Jojoba*, Vol. 1, Jojoba Plantation Products, Inc., Los Angeles, Calif., 1980.
32. V. K. Bhatia and I. B. Gulati, *J. Sci. Ind. Res.* **40,** 45 (Jan. 1981).
33. D. Foedisch, *Am. Inkmaker*, 30 (Oct. 1972).
34. *Fed. Regist.* **46,** 45223 (Sept. 10, 1981); 38555 (July 28, 1981); 16287 (March 12, 1981).
35. W. M. Mazee, *Modern Petroleum Technology*, 4th ed., Applied Science Publishers, Ltd., Barking, UK, pp. 782–803.
36. *Petroleum Waxes—Characterization, Performance and Additives*, STAP NO. 2, Technical Association of the Pulp and Paper Industry, Atlanta, Ga., 1963, p. 16.
37. F. J. Ludwig, *Anal. Chem.* **37,** 1732 (Dec. 1965).
38. *U.S. Exports, Schedule E, Commodity by Country*, Report FT410, U.S. Department of Commerce, Washington, D.C., 1971, 1976, 1981.
39. 21 CFR 172.886 and 178.3710.
40. *Bundesgesundheitblatt* **22,** 377 (1979).
41. Ref. 11, p. 222.
42. L. Roth and J. Weiner, *Waxes, Waxing and Wax Modifiers*, Bibliography Series 198, The Institute of Paper Chemistry, Appleton, Wisc., 1961.
43. *Ibid.*, Supplement I, 1967.
44. J. Weiner and V. Pollack, *Waxes, Waxing and Wax Modifiers*, Bibliography Series 198, Supplement II, The Institute of Paper Chemistry, Appleton, Wisc., 1973.
45. F. V. Snyder, *Chem. Purchasing* **16,** 75 (May 1980).
46. Eur. Pat. 0-034-682-A1 (Feb 9, 1981), L. Guy (to Mobil Oil Corp.).
47. U.S. Pat. 4,299,745 (Nov. 10, 1981), D. A. Godfrey (to Eastman Kodak Co.).
48. F. Jowett, *The Prevention of Ozone Attack on Rubber Surfaces by the Use of Petroleum Waxes*, American Chemical Society, Rubber Division, Washington, D.C., Spring Meeting, March 27, 1979.

49. A. K. Plitt, *Sizing, Short Course Notes*, No. 01063381, Technical Association of the Pulp and Paper Industry, Atlanta, Ga., April 1981, pp. 89–96.
50. U.S. Pat. 4,088,671 (May 9, 1978), D. Skalla, W. C. Weisenhelder, and H. Alsberg (to The Richardson Company).
51. W. J. Hackett, *Maintenance Chemical Specialties*, Chemical Publishing Co., Inc., New York, 1972.
52. *Waxing, Agriculture Handbook, No. 66*, U.S. Department of Agriculture, Washington, D.C., revised 1968.
53. Brit. Pat. 1,322,137 (July 4, 1973), (to Robert Bosch G.m.b.H.).
54. R. B. Porter, *TAPPI Monogr. Ser.* **33**, 127 (1971).
55. U.S. Pat. 3,998,676 (Dec. 21, 1976), R. A. Plauson and W. McBride (to United States of America).
56. Brit. Pat. 1,347,502 (Feb. 20, 1974), F. Jowett and J. P. Plows (to Middlesex Oil and Chemical Works, Ltd.).
57. N. A. Mancini in C. den Ouden, ed., *Thermal Storage of Solar Energy*, TNO and Martinus Nijhoff Publishers, The Hague, The Netherlands.
58. T. E. H. Downes, *S. Afr. J. Dairy Technol.* **4**(2), 101.
59. U.S. Pat. 4,260,395 (April 7, 1981), T. J. Anderson.
60. Can. Pat. 1,033,504 (June 27, 1978), V. T. D'Orazio and J. E. Cullen (to S. C. Johnson & Son, Inc.).
61. U.S. Pat. 4,246,435 (Jan. 20, 1981), P. F. Johnson (to General Cable Corp.).
62. U.S. Pat. 2,504,400 (April 18, 1950), M. Erchak, Jr. (to Allied Chemical Corp.).
63. G. Baum, J. Boelter, and I. Y. Strauss, *Proceedings of Mid-Year Meeting, May 1979*, Vol. 57, Chemical Specialties Manufacturers Association, Washington, D.C., 1979, pp. 160–164.
64. Brit. Pat. 1,450,285 (Sept. 22, 1976), (to Henry Moutroy Co., Ltd.).
65. 21 CFR 177.1520.
66. 21 CFR 172.888 and 178.3720.
67. M. E. Dry, *Energiespectrum* **1**, 298 (Oct. 1977).
68. J. H. LeRoux and co-workers, *Int. Symp. Anal. Chem. Explor. Min. Process Mater.*, 164 (1977).
69. 21 CFR 175.250.
70. U.S. Pat. 3,060,163 (Oct. 23, 1962), M. Erchak, Jr. (to Allied Chemical Corp.).
71. U.S. Pat. 2,766,214 (Oct. 9, 1956), M. Erchak, Jr. (to Allied Chemical Corp.).
72. W. F. Baxter and C. M. Davis, *U.S. Defensive Publication Serial No. 248918*, Tennessee Eastman Company, Kingsport, Tenn., July 10, 1973.
73. *Emulsion Brochure*, Release No. 102, Petrolite Corp., Tulsa, Okla., 1978.
74. C. D. Holt, *Rigid Vinyls: A Review and What's New*, Society of Plastics Engineers, Brookfield Center, Conn., March 1977, pp. 61–74.
75. E. S. Pattison, ed., *Fatty Acids and Their Industrial Applications*, Marcel Dekker, Inc., New York, 1968, p. 85.
76. *Wax Technology*, Chemical Specialties Manufacturers Association, Inc., Washington, D.C., 1973.
77. A. W. Marshall, *Anal. Chem.* **22**, 842 (June 1950).
78. B. Flaherty, *J. Appl. Chem. Biotechnol.* **21**, 144 (May 1971).
79. C. Giavarini and F. Pochetti, *J. Thermal Anal.* **5**, 83 (1973).

General References

Reference 4 is also a general reference.
H. Bennett, *Industrial Waxes*, Chemical Publishing Co., Inc., New York, 1975.
H. Bennett, *Commercial Waxes*, Chemical Publishing Co., Inc., New York, 1956.

C. SCOTT LETCHER
Petrolite Corporation

WEED KILLERS. See Herbicides.

WEIGHING AND PROPORTIONING

Weighing is the operation of determining the weight of material in one or more objects or in a definite quantity of bulk material. The terms weight and weighing are used somewhat loosely in this connection, since what is desired is the mass of the material. The weight is a measure of the gravitational force on a mass and thus is dependent upon the location of the weighing station and upon its altitude. Since it is common practice to calibrate scales against dead weights at the location of the installation, for practical purposes the weighing operation actually determines the mass of material. Measuring by weight rather than by volume eliminates variations owing to changes in specific gravity of liquids with temperature and variations in density of solids due to voids.

Proportioning is the weighing and controlling of two or more materials to a specific formula to make a definite blend of the materials for a mixed product or for a chemical process.

The four most common applications of weighing in industry are the measuring of incoming material by weighing received shipments; controlling ingredients to the proper proportions; putting the product into packages of uniform weight (either by packaging directly on a scale or by using a scale to check the performance of other filling equipment); and measuring outgoing shipments for billing and transportation-charge purposes. In addition, for interdepartmental accounting, materials are frequently measured by weight when being transferred from one department to another. Also, in processing operations, the rate of material flow to various pieces of equipment, such as grinders or kilns, is very important to secure best efficiency of operation (see also Conveying in Supplement Volume).

Types of Equipment

A scale is a device or machine used to perform a weighing operation. There are many types of scales and several principles of operation used (1). The type of scale to be used depends upon the conditions of the operation and the performance required.

In mechanical scales, the load is measured either by comparing it with a known weight or by directly measuring the distortion of a spring caused by the load. With the even-balance type, the load is measured by direct comparison with known weights. Chemical scales consist of three basic elements: load-receiving element, steel yard rod, and mechanical indicator, ie, dial or beam. Electronic scales consist of a load-receiving element, an electrical transducer, ie, a load cell, and a digital indicator. Electronic scales have existed since 1965 (2). Electronic scales have brought a new capability to check weighters, enabling them to keep track of individual and total weights as well as weight deviations with greater accuracy than previously possible (3).

In beam-type scales such as shown in Figure 1, the force due to the load, reduced by the multiplication of the lever system, is measured by the position of a known weight (the poise) on a graduated beam. The poise is moved by hand until the beam is in balance. Many beam scales also use fixed weights at the tip of the beam to offset definite amounts of load. For instance, a 500-kg scale could have a beam graduated to 100 kg by one thousand 0.1 kg and four 100-kg tip weights. With this scale, any load

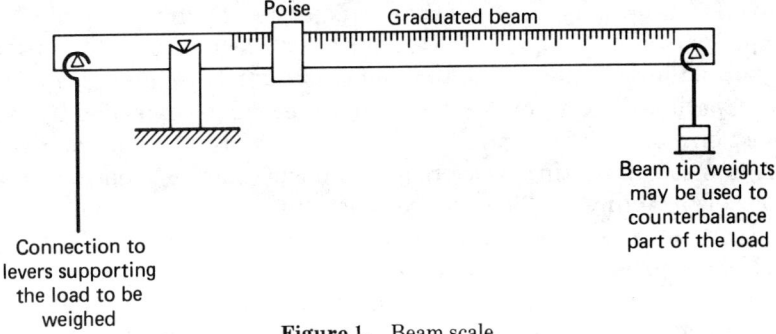

Figure 1. Beam scale.

to 500 kg can be weighed with 0.1-kg graduations. With the poise at 100 kg, tip weights would be added until the beam moved down. Then the poise would be moved back until the beam balanced. Both even-balance and beam-type scales are characterized by high sensitivity in relation to capacity, since motion of the scale is required only for the measurement of the final small percentage of the total weight.

Automatic indicating scales eliminate the manual operation of placing weights on a platter for an even-balance scale, or positioning a poise on a graduated beam for a beam-type scale. In pendulum-type dial or automatic indicating scales, as shown in Figure 2, the known weight (the pendulum) is moved by rotation of a sector or cam until it counterbalances the force from the load, and the amount of movement is measured and indicated by an indicator which is driven by a rack and pinion. The load is placed on a platform or in a hopper supported by levers which reduce the force due to the load by a factor called the multiplication of the lever system. Since it is frequently

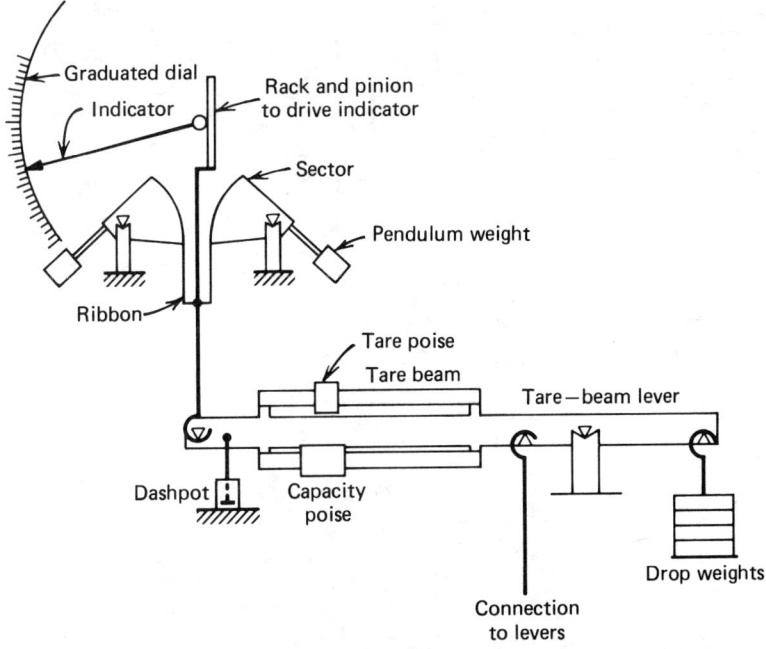

Figure 2. Pendulum scale.

desired to offset the empty weight of the container, most automatic indicating scales are equipped with a tare beam which is mounted on a lever. A poise on this beam can be moved until the empty weight of the container is offset so the scale indicator is at zero. The capacity of the automatic indicating scale can be extended by use of a capacity poise on a beam, or this additional poise can be used to offset additional tare weight. Automatic indicating scales require a device such as a dashpot to dampen oscillations of the indicator. The dashpot normally consists of a plunger moving in a cylinder containing oil. Frequently, drop weights are added to the scale to extend its range. For instance, the dial might have a capacity of 500 kg by 1 kg. By adding four drop weights, each equivalent to 500 kg, the capacity of the scale can be extended to 2500 kg. Since each drop weight is equivalent to one dial, the drop weights are frequently called unit weights.

In spring-type automatic indicating scales, such as shown in Figure 3, the deflection of a spring is measured and indicated by an indicator. A dashpot is required to dampen the oscillations of this scale also. In some cases, the load is supported directly from the springs. Many spring scales, however, are connected to levers and may use a tare-beam lever with poises and unit weights the same as a pendulum-type automatic indicating scale.

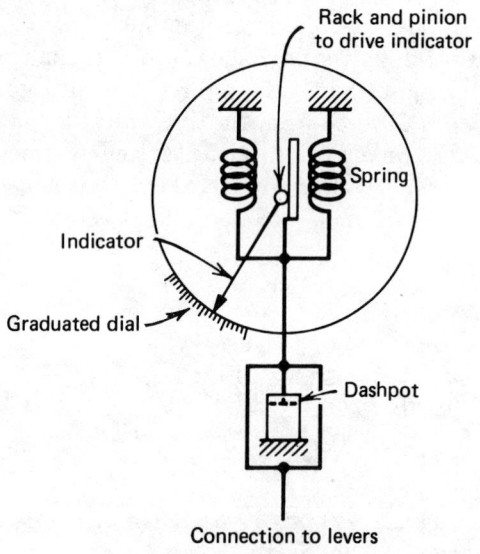

Figure 3. Spring scale.

A hydraulic scale measures weight by supporting the load on a piston in a hydraulic capsule, as shown in Figure 4, and measuring the hydraulic pressure developed

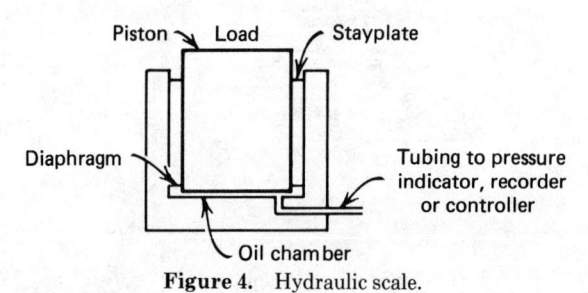

Figure 4. Hydraulic scale.

by the use of a Bourdon tube or similar pressure indicating device (see Pressure measurement). Since in many weighing operations it is required that the load be supported at three or more points, the individual hydraulic pressures developed by the various capsules are added by a hydraulic summing device by means of a mechanical lever, where the total force can be measured by a mechanical scale, or by using suitable electrical sensing elements and adding the electrical signals. The hydraulic equipment thus performs the same function as a mechanical lever system in reducing the load to an easily measured force. Compressed air can be used to measure weight by measuring the air pressure required to lift a load that is supported by a diaphragm.

Weighing can also be done by electrical means. One method is to measure the change in length of a steel column, as shown in Figure 5, or the bending of a steel beam, as shown in Figure 6, by measuring the change in resistance of grids of wire or foil bonded to the steel member. These grids are called strain gauges. Four strain gauges are connected in the form of a Wheatstone bridge as shown in Figure 7. When the load is applied as shown in Figures 5 and 6, gauges A_1 and A_2 are shortened so their resistance decreases and gauges B_1 and B_2 are lengthened so their resistance increases. Referring to Figure 7, when a voltage such as 15 V at 60 Hz is applied across the bridge, the output from the bridge would be 30 mV for a load equivalent to the full rating of the load cell. This output voltage is applied to a resistor and slide-wire potentiometer circuit which is connected to a servo amplifier. Since there is normally some initial load on the load cell, the RI is adjusted so that it balances the output of the cell with the initial load applied. The span-adjusting resistor RS is adjusted until the full output from the potentiometer balances the output from the load cell for the desired capacity of the scale. When a load within the capacity of the scale is applied to the load cell,

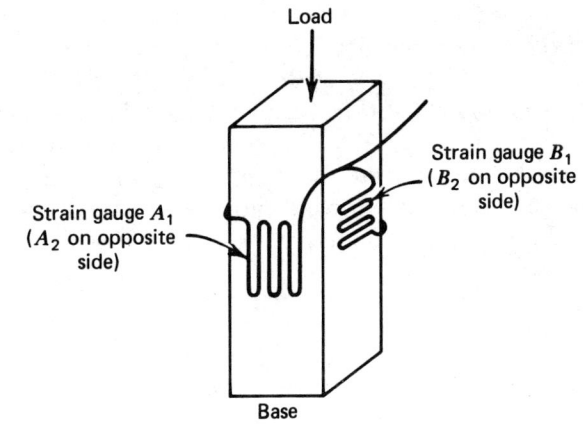

Figure 5. Strain-gauge load cell, column type.

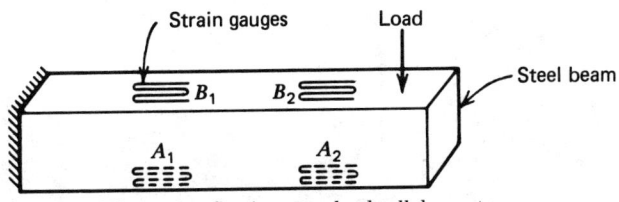

Figure 6. Strain-gauge load cell, beam type.

486 WEIGHING AND PROPORTIONING

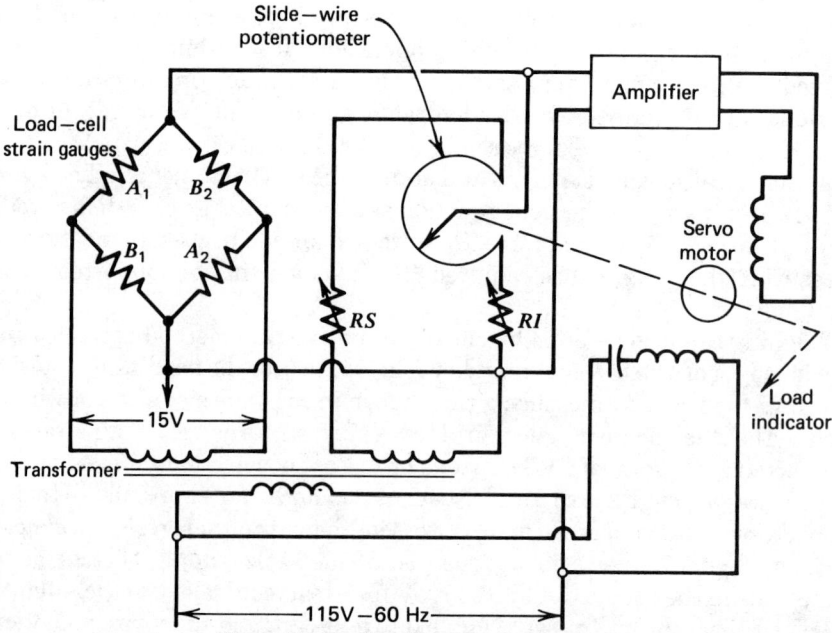

Figure 7. Load-cell scale, analogue type.

an output signal is applied to the potentiometer circuit. If the position of the sliding contact of the potentiometer does not correspond to the output from the load cell, a signal is amplified and applied to the servo motor to drive the potentiometer contact until the potentiometer balances the output from the load cell. An indicator, also driven by the servo motor, indicates the corresponding load on a dial. Figure 8 illustrates how a digital indication of a load can be obtained. By exciting the Wheatstone bridge with direct current, a direct-current signal is developed by the load cell. The RI is adjusted to offset the initial load, and the RS is adjusted to provide the proper voltage, depending upon the capacity of the scale, to apply to a digital d-c voltmeter or a voltage-to-frequency converter with a frequency meter to provide the digital indication. Figure 9 shows how the output from four load cells, which might be supporting a platform or a tank, can be combined to indicate the total load on the load cells, and also shows how range steps can be added to increase the resolution of the weight in-

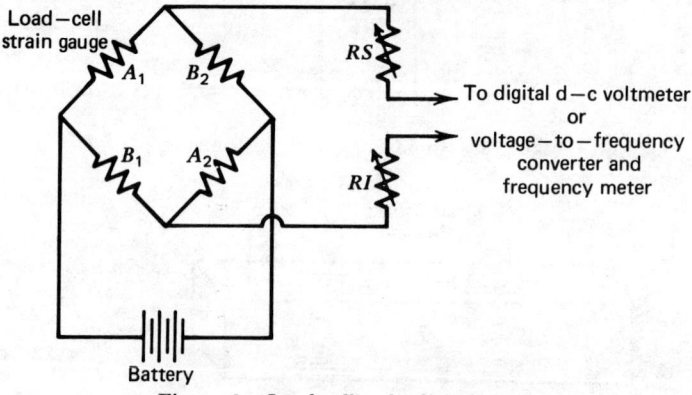

Figure 8. Load-cell scale, digital type.

Figure 9. Load-cell scale; 4 load cells, 5 ranges.

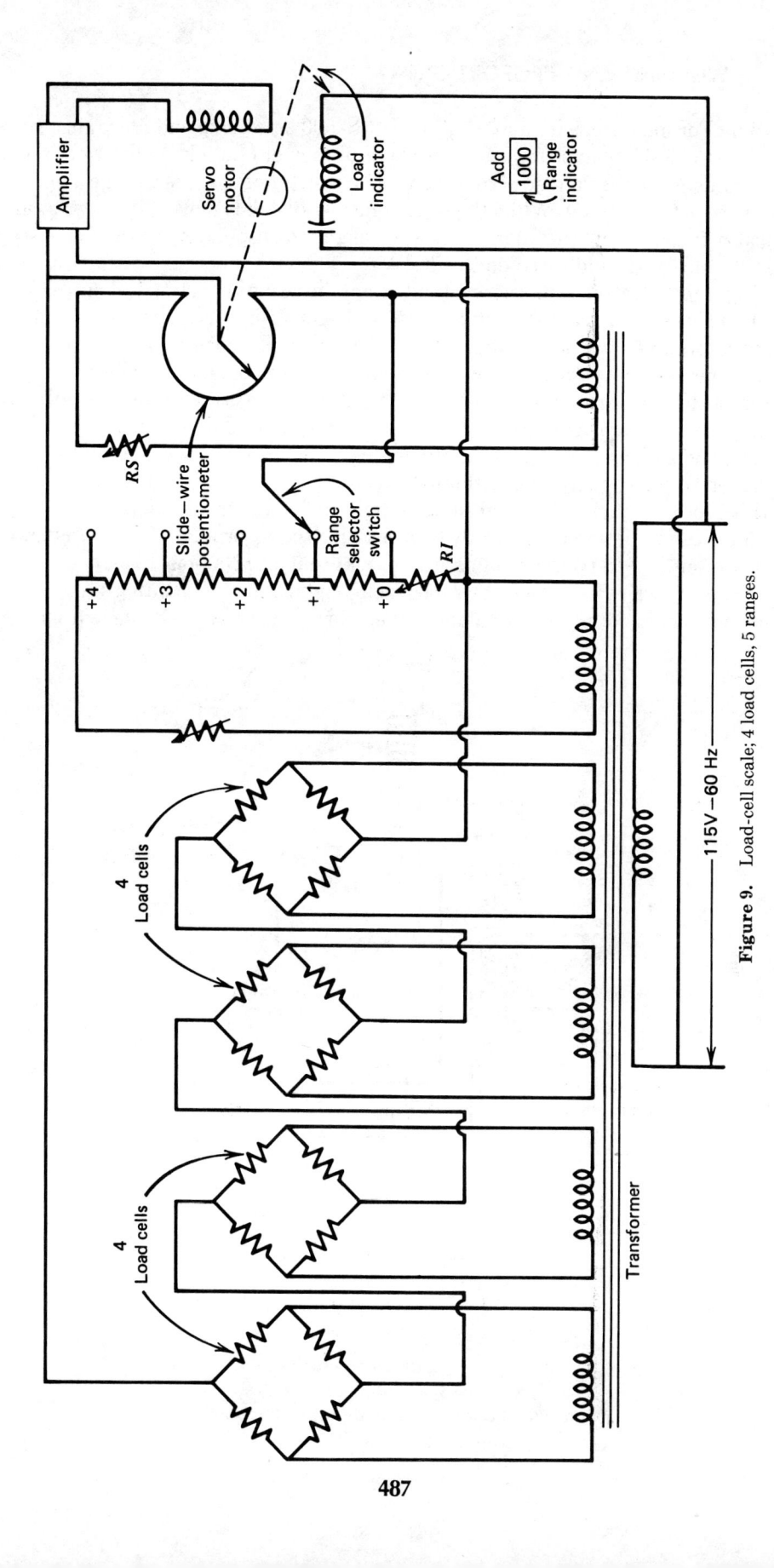

dication. For instance, with a dial equivalent to 500 kg by 1 kg and four range steps, a load up to 2500 kg can be measured with a resolution of 1 kg. The calibration resistor RC is adjusted so the voltage across each range resistor equals the voltage across the slide-wire potentiometer. When the range selector switch is moved, it moves a range indicator flag advising the operator to add weight corresponding to the range steps in use to the weight indicated on the dial.

Figure 10 illustrates another method of obtaining a direct electrical output from a scale. This uses a linear differential transformer consisting of a primary winding and two equal secondary windings spaced so that, when a soft iron core is centered in the transformer, the outputs of the two secondary windings balance each other and the output of the transformer is zero. When the soft iron core is moved up, the output of the upper secondary increases, and when the core is moved down, the output of the lower secondary increases. By proper design this output can be made linear. The output can be read directly on a meter or through an amplifier. The most common application for this type of scale is "over-and-under weighing," as for checkweighers.

Mass can be determined directly by measuring the absorption of γ or β rays owing to the presence of the unknown quantity of material (see Radioisotopes).

Counting scales have been developed from digital scales. A counting scale, usually a top dash loading or platform-balance type, allows multiples of the same item to be

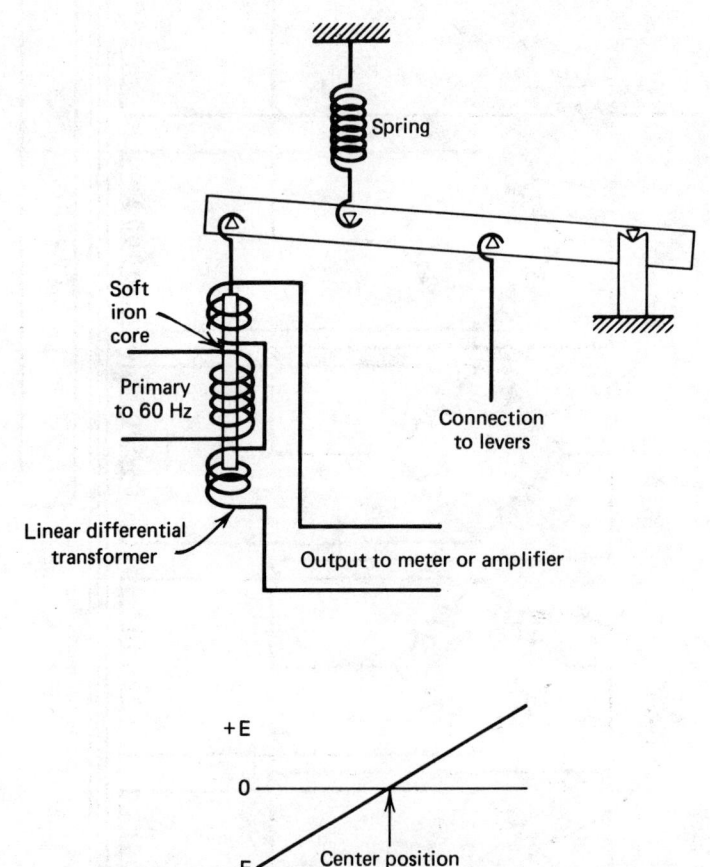

Figure 10. Scale using linear differential transformer.

weighed. The result is placed in the scale memory and displayed digitally. The scale can then be used for counting unknown quantities. The digital display always shows the total count.

Batch weighing is widely used in the chemical and food industries. The weighing system provides automatic weighing of a single ingredient at the start of the weighing process. There can be weighing of sequential additions of several ingredients as well as time control for processes such as heating and stirring.

With continuous-process weighing, the weighing and controlling is done while the ingredients are moving. For most processes, materials are transported by moving belts. The weight of the material on the belt is detected by a sensor (4).

Factors Affecting Weighing

Variations in the Force of Gravity. Weight is a measure of the force with which a mass is attracted to the earth. The basic force equation is

$$F = ma$$

where F is force, in newtons (poundals); m is mass, in kg; and a is acceleration, in cm/s² (ft/s²). For weight rather than force,

$$W = mg$$

where g is the acceleration due to gravity (5). The value of g is not the same everywhere on the earth. The earth is rotating on its axis, and thus a body at the equator is subject to small centrifugal force, whereas a body at the poles is not. Also, the earth is not spherical, but bulges at the equator, so that a body at one of the poles is actually nearer to the center of the earth. For both of these reasons, the acceleration due to gravity and the weight of a given mass increase with increasing latitude from the equator to the poles. In addition, the acceleration due to gravity decreases with increasing altitude. The effect of these factors is shown in Helmert's equation for g:

$$g = 980.616 - 2.5928 \cos 2\phi + 0.0069 \cos^2 2\phi - 3.086 \times 10^{-6} H$$

where ϕ is the latitude in degrees and H is the altitude in cm. Values from this equation can be used to compare the weight of a definite mass at different locations, such as the location where a scale was manufactured and originally calibrated as compared to a location where it will be installed:

$$W = \frac{mg}{g_c}$$

where W is scale reading, in kg; m is mass, in kg; g is g at point of installation, in cm/s²; and g_c is g at a point of calibration, in cm/s². Weighing equipment that measures force, such as spring scales including the popular strain-gauge load cell and hydraulic or pneumatic scales, may have to be recalibrated at the point of installation. As an example, Table 1 shows g for various locations.

Conventional balances, beam scales, and pendulum-type automatic indicating scales measure an unknown mass by comparing it with a known mass. Since a variation in the acceleration due to gravity has the same effect on the known mass as on the unknown, this factor does not affect the operation of these scales.

490   WEIGHING AND PROPORTIONING

Table 1. Readings for g at Various Locations

Location	Acceleration due to gravity, g, cm/s²	Location	Acceleration due to gravity, g, cm/s²
United States		*Other countries and areas*	
St. Michael, Alaska	982.192	Arctic Red River, Canada	982.434
Quiet Harbor, Alaska	981.624	Vancouver, Canada	980.949
Los Angeles	979.595	Canal Zone, Panama	978.243
San Francisco	979.965	Greenwich, UK	981.188
Denver	979.609	Königsberg, FRG	981.477
Hartford, Conn.	980.336	Rome, Italy	980.367
Washington, D.C.	980.095	Monrovia, Liberia	978.165
Key West	978.970	Bergen, Norway	981.922
Miami	979.053	Stockholm, Sweden	981.843
Honolulu	978.591	Balia, Brazil	978.331
St. Louis	980.001	Kingston, Jamaica	978.591
New York	980.267	North or South Pole, sea level	983.217
El Paso	979.124	equator, sea level	978.039
Seattle	980.733	equator, 5000 m above sea level	976.496

Buoyant Effect of the Air. Since most weighing operations are performed in air, a variation in density of the air between the point of calibration of a force-measuring scale and its point of use will affect the calibration (1,5). This effect is so small it can be ignored except for the most precise measurements. When weighing materials of very low density, there can be a measurable difference owing to variation in air density between the point of origin and the destination.

Moisture Content. Materials subject to variable moisture content are frequently treated on a dry weight basis or on some standardized moisture-coated basis. In such cases, the actual moisture content of the material being weighed must be determined, and the actual weight converted to the desired basis.

$$W_D = W(1 - m_c/100)$$

where W_D is dry weight; W is actual wet weight; and m_c is moisture content as percent of wet weight; or

$$W_D = \frac{W}{1 + m_c/100}$$

where W_D is dry weight; W is actual wet weight; and m_c is moisture content as percent of dry weight.

Commercial Requirements

When the weight measurement is used as the basis of paying for material or for labor or services, the weighing equipment is subject to state or local weights and measures inspection and must meet certain specifications and tolerances (6). This requirement does not apply to equipment used only for control of processes or to measure yield or inventory.

Accessories are available for most types of scales. Printing attachments record the weight as well as time, date, and other data on a ticket, sheet, or strip. Circular or strip recorders provide a continuous record, plotting weight vs time. Remote indication duplicates the scale indication at a remote point. Analogue-to-digital converters are used for direct digital control or to transmit weight information to remote printers or to digital computers for record or control purposes. Control devices can be coupled directly to the scale. A paddle on the indicator that cuts a light beam to a photocell (in electrical circuitry) or that cuts a low pressure air jet (in pneumatic circuitry) can be used for simple or complex control functions. A potentiometer coupled to the indicator shaft or a pneumatic transducer coupled to the lever system provides analogue electrical or pneumatic outputs for control. A good scale is a very accurate sensor, but extreme care must be taken to make sure the coupling of the control device to the scale does not reduce that accuracy by introducing friction or other variable forces (see also Instrumentation and control).

Choice of Scale

Since the error in a scale, as in many other instruments, is approximately the same throughout its indicating capacity, for best accuracy a scale should be chosen so that most of the weighings to be made on it are in the upper range of its indicating capacity. The capacity of a scale can be extended by the use of drop weights for a mechanical scale or range steps for a load-cell scale. This permits greater accuracy in weighing through a wide range of loads. For example, a 500-kg dial scale with four 500-kg drop weights can be expected to weigh to within 0.5 kg for the first 500 kg and within 2.5 kg at its capacity of 2500 kg. A 2500-kg dial scale would be expected to weigh within 2.5 kg through the 2500-kg range.

As a guide to the selection of the most suitable type of scale, the relative advantages and disadvantages of the various types of weighing equipment are discussed below.

Mechanical Scales. Advantages: high accuracy; relatively inexpensive, particularly in smaller capacities; comparatively simple maintenance because of mechanical construction; no electricity required; lends itself to simple control equipment when a simple contact such as a magnetically operated mercury switch is used for the control element; very accurate primary element for use of other controls such as pneumatic or electronic controllers; and capable of high overload capacity with relatively little sacrifice of accuracy.

Disadvantages: relatively complicated and expensive installation for large sizes; damage from such causes as corrosive fumes or liquids; remote indication adds to cost; excessive vibration may cause damage and difficulty of reading; and motion of the platform or weighing container is relatively high, particularly for small-capacity scales.

Hydraulic Load-Cell Scales. Advantages: compact and self-contained unit when a single load cell suffices, such as on a crane or hanging scale; inexpensive installation; relatively easily protected from corrosion; effects of vibration can be eliminated by damping; and motion of platform or weighing container is relatively small.

Disadvantages: additional equipment, either mechanical or electrical, is required if it is necessary to totalize the load on several load cells, such as for a large platform scale; and expensive electrical equipment must be added to perform control operations.

Pneumatic Load Cell. Advantages: compact unit; inexpensive installation; no electricity required; readily adaptable to pneumatic control equipment; simple construction lending to ease of maintenance; vibration is no serious problem; and motion of platform or weighing container is relatively small.

Disadvantages: limited to single-cell application so that a large platform, hopper, or tank cannot be supported on several cells and the total forces added; a supply of compressed air is required; and speed of response is relatively low.

Electronic Load-Cell Scales. Advantages: easily protected from corrosion; inexpensive installation; relatively easy remote indication and use of one remote unit to read from several scales; easily adapted to controls; and motion of platform or weighing container is very small.

Disadvantages: relatively expensive, particularly small-capacity units; somewhat slower recovery from sudden changes in ambient temperature; load-cell mountings must be designed to minimize side forces; and electricity is required.

Nuclear Scales. Nuclear scales measure mass directly by absorption of γ or β rays (7).

Advantages: a simple installation; not affected by extraneous forces on container such as connections to tank or hopper and tension or stiffness of belt conveyor; no contact with material is required; no moving parts subject to wear or corrosion; and electrical output permits direct readout and adaptation to controls.

Disadvantages: variations in geometry, as well as in the density of the material being measured, will affect accuracy; proper shielding and care are required to avoid exposure of personnel to radiation; frequent calibration is required to compensate for loss of source emission with time; and electricity is required.

Forms of Scales

Scales of various types are available in different forms to facilitate the weighing operation (8). Bench scales are small units, generally with a platform, on which small quantities can be placed by hand or moved across the scale on a roller conveyor installed on the platform. Portable scales generally have a platform near the floor, and the scale is on wheels so it can be moved about from place to place. Bench scales can be placed on a wheeled stand for similar use. Floor scales are platform scales installed in the floor. Motor-truck scales for weighing highway vehicles may be installed in a pit or may be self-contained so that they can be moved from one site to another.

Railroad-track scales are installed in a pit and support the proper length of rail to weigh a railroad car. For static weighing of cars, the rails are generally long enough to support the complete car, although with two-draft weighing, the car can be weighed one end at a time. Some railroad weighing is done with the cars coupled and in motion. In this case the rails of the scale may be long enough to weigh the pair of axles at each end of the car, or short, weighing a single axle at a time. An adding machine or calculator is used to obtain total car or train weight.

Where a product is conveyed by roller hangers from a monorail, scales weigh a section of the overhead rail. Hopper or tank scales support a hopper for dry materials or tank for liquids. Generally, the dead weight of the hopper or tank is offset on the scale so the scale reads the net weight of material.

Solid bulk materials, from fine powders to coarse quarried lumps, are carried in screw, bucket, vibrator, and other types of conveyors. The belt conveyor can best be used with on-the-fly weighing systems. A measurement of the belt speed is fed into the totalizing device that provides the total weight of material that has passed over the belt between two consecutive readings. Increased automation of bulk solids handling has increased the use of belt conveyors and increased demands for belt weighing systems (9). Controls can be provided to stop the flow after a predetermined amount of material has been delivered or to control the rate of feed. Because of the effect of the stiffness of the belt, such equipment is generally limited in accuracy. Under ideal conditions, accuracy in which the error is ca 0.5% can be obtained.

Weighing of definite objects is accomplished by placing the object on a suitable scale. Objects that are accounted for by count rather than by weight can be measured either by dividing the net weight by the known average weight of one object or by the use of scales equipped with special counting attachments. One type of attachment is the fixed-ratio scoop. This scale is equipped with one or more scoops mounted so that a piece in a scoop counterbalances a definite number of pieces on the platform. Thus, by counting relatively few pieces placed in the scoops, a total count is easily determined. Another type is the variable-ratio scoop. Here the scoop is moved along the lever connected to the scale. With 1, 2, 5, or 10 pieces in the scoop, it is moved along the lever until the scale balances. The count is indicated on a graduated bar.

The continuous-process-control systems have evolved mainly from pneumatic analogue controllers. Because of the microprocessor, there has been implementation of digital techniques to perform analogue control in continuous-process controllers (10). The nature of digital-control circuitry can be used for batch-control applications.

Measuring Definite Quantities

For the measurement of bulk materials, which may be handled in anything from small packages to railroad cars, the weighing operation can consist of measuring the gross weight of material and container and subtracting the tare weight of the container. When the tare weight is known, its weight can be offset on the scale so that the net weight is indicated. The weighing operation can be performed by any of the following three methods.

The material is brought to the scale, which is fixed in position, and placed on it. Bench scales are used for weighing small packages, floor scales for weighing larger loads, and motor truck or railroad track scales for weighing materials in trucks and cars. The scale should be located where it is handy for the weighing operation but does not get unnecessary abuse from excessive traffic that is not weighed.

The scale is brought to the load; scales of various types and capacities are available in portable form so that they can be moved about and temporarily located where the weighing operation is to take place.

Scales can be built into such material-handling devices as cranes, lift trucks, and conveyor belts so that the weighing operation can take place while the material is being transported. A hopper or tank scale can be installed so that it accumulates, weighs, and discharges the material as it is being transported. The quantity to be weighed can be measured in one draft or in multiple drafts. A hopper scale used for multiple draft weighing of material in transit is commonly called a bulk weigher, as shown in Figure 11. An accumulating or surge hopper with a feed gate is installed above the scale hopper, which has a discharge gate. A surge hopper is also installed below the scale

494 WEIGHING AND PROPORTIONING

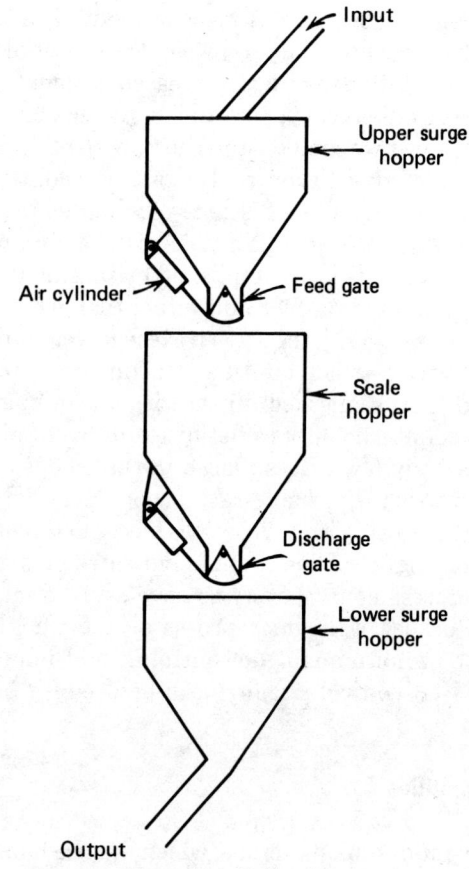

Figure 11. Bulk weigher.

hopper. The scale is generally equipped with an adjustable cutoff device to interrupt the flow of materials to the scale by closing the feed gate above the scale hopper, a signal device to close the scale discharge gate when the scale is almost empty, and a readout device to operate an adding machine. In operation, material flows into the scale until the feed gate is closed by the cutoff device. The actual load on the scale is transmitted to the adding machine, the discharge gate is opened, the scale discharges to a predetermined point, and the discharge gate is closed. The actual weight of material left on the scale is also transmitted to the adding machine which subtracts the two to arrive at the net weight of that draft. This operation is repeated until the quantity to be measured has passed through the scale. Since actual weights of repeated drafts are used, with this system it is possible to weigh a definite amount, such as contained in a railroad car or ship, with an error of <0.1% of net weight. With the input from storage and the output to a car or ship, the device is a shipping scale. With the input from a car or ship and the output to storage, the device is a receiving scale.

Since net weight of materials is desired when measuring definite quantities, various approaches can be taken. For example, any one of the following methods of weighing the material in a railroad car can be used. The methods are arranged in approximate increasing order of accuracy: weighing the loaded car on a coupled-in-motion track scale with the net weight determined by subtracting the stenciled weight of the car; weighing the material, either as it is put into the car or removed from the car, on a belt-conveyor scale; weighing the gross weight of the car on a static railroad track scale and determining the actual empty weight of the car, either before the loading operation or after the unloading operation; weighing the complete load either delivered to or received from the car as a single draft in a hopper or tank scale; and using a bulk weigher as described above to measure the material either being placed in or removed from the car.

Controlling Quantity

The most common application for the controlling of quantities of materials is to fill containers to predetermined weights. One method is the net-weighing process, in which material is fed into a fixed container such as a hopper or tank permanently attached to the scale. The dead weight of this container can be counterbalanced, so that the scale indicates the weight of material to be added; or the weight of container plus the desired weight of material can be counterbalanced, so that the scale indicates the deviation from the desired weight. The filling operation may be performed manually by an operator observing the scale and stopping the flow of material at the proper time, or automatically by equipping the scale with a suitable control device to stop the feeding equipment at the desired weight. The hopper or tank is then discharged into the shipping container. In the gross-weighing method, the shipping container is placed directly on the scale and filled. The tare weight of the individual container is offset on the scale, and if this weight varies, the tare weight must be adjusted for each container to maintain a uniform net weight. In manual operation, the tare is adjusted by the operator and the filling operation is controlled by the operator. Equipment for performing such operations automatically is available. The tare can be set and the feeding equipment can be controlled automatically.

Continuous Weighing and Controlling

Sometimes it is required to determine the weight per centimeter or meter of material in sheet or strip form as it is being transported or in the process of manufacture. This weight information, with suitable controls, can be used to control moisture content, addition of sizing or other materials, or an extrusion process. The material can pass over a roll which is weighed or over a power-driven belt conveyor mounted on a scale. Direct mass measurement with a nuclear scale can also be used for this application.

Weighing and controlling bulk material that is continuously conveyed is often necessary for optimizing the performance of such devices as grinders or pulverizers, for adding materials to a process at a continuous rate, or for controlling additives, such as to water supplies. A belt-conveyor scale can be installed in a belt conveyor, or a short belt feeder can be mounted on a scale. If a belt-conveyor scale is mounted in the conveyor, the scale can be equipped with controls to maintain the feed rate within limits by controlling the operation of the device feeding the material to the conveyor. A short belt feeder mounted on a scale can have controls to regulate either the rate of feed to the scale or the speed of the belt on the scale to maintain a uniform flow. Direct mass measurement with a nuclear scale can also be used to measure and control such a continuous stream of material. Uniform continuous flow of solid or liquid materials can also be achieved by the loss-in-weight system.

496   WEIGHING AND PROPORTIONING

Proportioning

There are several methods of proportioning ingredients to obtain the desired formulation. Selection of the proper method depends upon the processing equipment and also upon the accuracy required. Where a definite quantity of a mixture is to be injected into a process, such as being placed in a batch mixer, or where the greatest accuracy is required, the batch system is generally used. In this system, definite amounts of each of the ingredients are weighed and delivered to the process. The various ingredients can be weighed consecutively in a single hopper scale, or each ingredient can be weighed on a separate scale. The latter method provides the greater accuracy because each scale can be of the proper capacity for its specific ingredient, and also, any error in the weighing of one ingredient does not affect the accuracy of weighing of another. As a compromise between the two methods described, a sequence method of weighing can be used wherein one ingredient is weighed and then discharged and succeeding ingredients are weighed and discharged in the same manner all on one scale. If all ingredients weigh about the same amount, accuracies of quite high order can be obtained. A point to be considered, however, is the possibility of contamination in case the same equipment is used for different ingredients for different formulas. Another factor is the time involved. With a scale for each material, all ingredients are weighed simultaneously. The scale operation can be manual, wherein an operator controls the feed and discharge, or the operation can be automatic in varying degrees. With proper control, automatic equipment can function in conjunction with material handling and processing equipment so that the full process cycle proceeds automatically.

The following factors should be considered in the choice of equipment for a particular batching operation: accuracy to which each ingredient must be weighed; frequency in changing the formula or proportion of ingredients; number of formulas or products for which equipment is to be used; importance of avoiding any contamination of ingredients; speed of operation required; and volume of storage of raw materials required.

Two examples illustrate different approaches to different problems. The first example is the operating station for weighing one to five of a possible ten ingredients on one scale (contamination is not a problem, large capacity storage is not required, and speed and accuracy requirements can be met). Because the formula is changed quite frequently, controls that permit selecting the ingredients to be used and setting the amount of each at a point remote from the scale are chosen, and these controls are furnished in duplicate. Thus, as a series of formula A are being weighed, the operator can be setting up the next ingredients on the formula-B controls. The second example is the operating station for a system where a large number of scales is used to weigh the ingredients (large volume storage is required, and high accuracy in combination with high speed is necessary). A number of different formulas are used, but the amount of each ingredient in a formula is rarely changed. The controls for each scale are located at the corresponding scale, and the interlocking controls for the scales, as well as the control for delivering the batches to the two mixers and to the ultimate destination, are incorporated in the station.

When the manufacturing process is continuous, consideration can be given to special forms of weighing equipment that deliver a continuous controlled stream of material. Equipment to perform this operation may be of the continuous-feeder type or of the loss-in-weight type. In the case of a continuous feeder for dry materials, a suitable belt conveyor is installed on a scale. The belt runs at a definite controlled rate and material is delivered to the belt by means of a feeder. The rate of delivery of material from the feeder is controlled by the scale. The scale is preset to the desired weight of material to be delivered per unit of time, and corrections to the feed rate are made when the feed rate deviates from the desired rate. Accuracies with ca 1% error for short intervals of time are possible with equipment of this type.

For the loss-in-weight system of weighing, a hopper or tank containing a supply of the material is suspended from a scale. This hopper or tank is discharged by a controllable feeder. If the scale is of the beam type, a poise is moved automatically along the beam at a rate in accordance with the flow rate desired for the material. In the case of an automatic indicating scale, a flow-rate controller is used. The scale then controls the discharge feeder so as to deliver material at the preset rate. Since correction is continually made back to the weight in the hopper, this system can be said to have a closed-loop control as compared to the open-loop control of the continuous feeder. Accuracies within ca 0.5% can be achieved. The system has the disadvantage that two units must be used for each material to maintain continuous flow. One unit is being refilled while the other is discharging.

Both the continuous feeder and the loss-in-weight feeder can be equipped with remote controls, which can be connected to equipment for measuring an uncontrolled flow so the controlled stream or streams can be definitely proportioned to the uncontrolled stream.

Scales and Computers

Scales can be modified to perform some types of computing directly, eg, counting scales. A strain-gauge load-cell scale can be built with variable gain to act as a direct-reading counting scale by adjusting the gain to read the correct count for a definite number of pieces and then putting the unknown quantity on the scale to read actual count. By providing an adjustable-ratio lever system or an adjustable-gain load-cell scale, material with a known moisture content can be caused to read on a dry weight or a definite percentage moisture basis. Scales equipped with analogue- or digital-readout devices can feed weight information directly into analogue or digital computers (11–13). Small special-purpose computers can be used for counting by dividing the lot weight by the weight per piece, for determining dry weight or weight on a specified moisture-content basis, and for certain proportioning, optimizing, and other applications. Weight information fed directly into a central computer can be used for inventory control, proportioning with corrections for actual weights delivered, optimizing production, and other requirements. With proper interface equipment, weight information can be transmitted to distant computers. A microprocessor-based control system can also be used (14).

Feeding Equipment Suitable for Weighing Operations. For best performance, careful consideration must be given to the choice of proper equipment for the feeding of materials to scales for such operations as filling containers or for proportioning ingredients to a definite formula. The feeder should deliver material at as uniform a rate as possible and should respond quickly to control signals from the scale. When the feeder stops, there is a column of material in the air that falls into the weigh hopper. The cutoff point can be preset to allow for the average weight of this column, but deviations from the average weight will result in errors. For best accuracy, the weight in the column after cutoff should be small in proportion to the total weight. When the weighing operation is to be performed relatively fast, such as in filling the container in less than a minute, or when the rate of flow changes from such causes as a variation in head of material above the feeder, the feeder should be capable of being slowed down before the desired weight is in the scale. For this operation, the bulk of the material is added to the scale at a high rate of feed and the speed of flow is decreased shortly before the desired weight is in the scale. When the control element on the scale signals the completion of the weighing operation, the feeder should stop quickly. This means that motor-operated feeders may require brakes or plugging controls or the use of an auxiliary gate to stop the flow instantly.

The following are recommended types of feeders for various materials:

For free-flowing materials that do not tend to arch or "rat hole" and flood, electric vibrating feeders, belt feeders, screw feeders, and power-operated gates are used.

When the material is predominantly in large lumps, recommended types include electric vibrating feeders, belt or slat feeders, and screw feeders of proper design if lump size is not too large.

For finely pulverized materials that tend to flood, recommended feeders are as follows: *1.* Screw feeders of sufficient length, preferably double-flight or half-pitch. An auxiliary gate may be required. *2.* Rotary vane or star feeders. The compartment or pocket size should be small enough so the discharge of one pocket will not cause an unacceptable error. For some applications, it may be desirable to use a rotary vane feeder in conjunction with a vibrating feeder. With this combination, flooding is prevented, and the desired smooth control of flow is achieved with the vibrating feeder. *3.* Air-slide feeders. If sufficient length cannot be provided to prevent the possibility of flooding, a rotary vane feeder may be desirable to deliver material to the air slide.

Control of the weighing of liquids is usually accomplished by power-operated valves. The valve must close quickly and should be of proper design and size for the material to be handled at the rate of flow desired. Pneumatically operated diaphragm valves are frequently used for larger sizes and solenoid operated valves for small sizes. Valves of special construction to provide two rates of feed are available, or two valves in tandem with an adjustable bypass as shown in Figure 12 can be used. Both valves are open for the fast rate of feed. Valve *A* closes at the first cutoff when near the desired weight. Valve *B* closes at the desired weight for final cutoff. The hand valve in the bypass is adjusted to give the desired slow or dribble rate of feed for the maximum accuracy (see also Conveying in Supplement Volume).

Discharging the Scale. When material is weighed into a hopper or tank on a scale, as in the net-weight method of filling containers or in a batch system, the hopper or tank can be emptied completely by means of a gate or valve. When a gate is used, very little control can be exercised over the rate of discharge. In some cases, particularly when the scales of a batching system discharge onto a conveyor belt, it is very desirable to control the rate of discharge. This can easily be done by using a suitable feeding device in place of the gate. By proper adjustment of the discharge rate, a certain amount of premixing can be done on the belt. Also, with a controllable feeder as the discharge device, it is not necessary to empty the hopper or tank completely. By proper setting of the discharge control point, a definite amount of material can be left in the scale hopper or tank. This means the material is weighed out of, as well as into, the

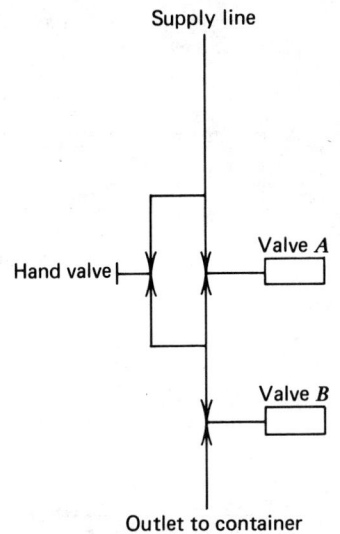

Figure 12. Two-speed liquid feeding arrangement.

scale. With materials that cling to the sides of a hopper, this has a definite advantage in that every bit of material does not have to be removed. In the case of liquids, more rapid discharge of a definite amount can be accomplished because the last material does not have to be discharged with zero head. Of course, this scheme cannot be used when different materials are accumulated in the scale hopper, because the composition of the material remaining in the hopper would vary.

Factors Affecting Scale Performance. Since weighing scales are highly accurate instruments, they should be protected from unnecessary abuse. Corrosive atmospheres and excessive dust will cause deterioration of weighing equipment unless it is properly protected. Some forms of load cells for electric, hydraulic, or pneumatic scales can be adequately protected by proper protective coatings. Exposed parts of mechanical scales can be coated with protective material if it does not interfere with the weighing operation. If mechanical-scale level systems are enclosed, they can be quite effectively protected from corrosion and dust by purging the enclosure with clean air. Indicating elements of scales can be protected by locating away from the damaging atmospheres or by purging with clean air.

Certain precautions must be observed in the installation of weighing equipment and in the connection of other devices, such as feeding equipment to such elements of the scale as the hopper or tank. Pipes conveying liquid to or from a weight tank should be equipped with connections that are suitably flexible for the capacity of the scale and the movement of the weigh tank. Precautions must be observed to see that any exterior forces such as might be caused by changes in temperature of the piping are applied to the scale in a horizontal rather than a vertical direction. When weighing dusty materials, it is very desirable to keep the dust out of the surrounding air. This is done by using dust seals or flexible connections where the material flows into and out of the scale hopper. Since the filling of a hopper or tank displaces air in the hopper or tank, it is essential to make certain that either the displaced air is allowed to flow out through a filter or through a vent pipe, or that the increased pressure does not exert a vertical force on the scale. This can be accomplished by designing the weigh hopper or tank so that material is fed in through the side of the hopper with the flexible con-

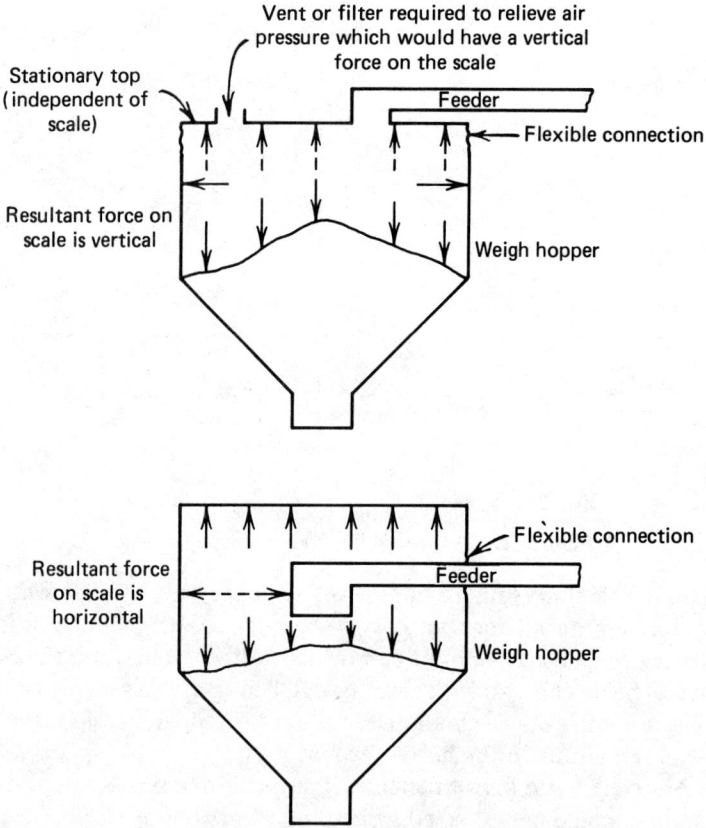

Figure 13. Effect on scale of air pressure in weigh hopper. Arrows indicate direction of air pressure. Solid arrows show force against scale; dotted arrows show force against stationary area.

nection in the vertical plane as shown in Figure 13. Horizontal forces applied to a weigh hopper can be overcome by suitable check links or stay rods without affecting the weighing accuracy of the scale. Vertical forces of unknown magnitude or of magnitudes that cannot be definitely controlled may have a serious effect on the accuracy of any weighing equipment.

BIBLIOGRAPHY

"Weighing and Proportioning" in *ECT* 1st ed., Vol. 15, pp. 25–34, by Donald B. Kendall, Toledo Scale Company; "Weighing and Proportioning" in *ECT* 2nd ed., Vol. 22, pp. 220–241, by Donald B. Kendall, Toledo Scale Company.

1. D. B. Kendall in D. M. Considine, ed., *Handbook of Applied Instrumentation*, McGraw-Hill Book Co., Inc., New York, 1964, pp. 5-41 to 5-54.
2. G. Wilson, *Weigh. Meas.* **61**(10), 9 (1977).
3. *Mod. Mater. Handl.*, 66 (Feb. 5, 1982).
4. J. Hall, *Instrum. Control Syst.*, 29 (Aug. 1978).
5. G. B. Anderson and R. C. Raybold, *Natl. Bur. Stand. U.S. Tech. Note* **436** (1969).
6. *Natl. Bur. Stand. Handb.* **44** (1983).
7. S. Rowe, *Instrum. Control Syst.* **40**, 85 (1967).
8. *Terms and Definitions for the Weighing Industry*, Scale Manufacturers' Association, Inc., Washington, D.C., 1964.

9. E. de S. Snead, *Pit Quarry*, 63 (Dec. 1976); H. van den Berge, H. A. Klasens, and A. Kopmels, *Control Eng.*, 52 (Sept. 1977).
10. H. M. Morris, *Control Eng.*, 61 (July 1982).
11. C. W. Hibscher, *Mod. Mater. Handl.* **23,** 82 (1968).
12. C. W. Hibscher and R. J. Phillips, *Unique Computer Controlled Batching*, ASME paper no. 61-WA-183, New York, 1961.
13. E. S. Savas, *Computer Control of Industrial Processes*, McGraw-Hill Book Co., Inc., New York, 1965.
14. G. Dale-Smith and P. Whiting, *Electron. Eng.*, 67 (July 1977).

General References

Scale Men's Handbook of Metrology, National Scale Men's Association, Libertyville, Ill., 1980.
H. A. Nielsen, Jr., *Scale Calibration Electronically, A Proposal for a Dual Sealing Standard for Large Capacity Commercial Scales*, paper presented at the 15th Annual Western States and Weights and Measures Conference, Boise, Idaho, Aug. 24, 1972.
G. E. Wood and D. P. Ivacheff, *Iron Steel Eng.*, 65 (Jan. 1977).
H. Colijn, *Weighing and Proportioning of Bulk Solids*, Trans Tech Publications, Aedermannsdorf, Switz., 1975.
H. M. Morris, *Control Eng.*, 78 (Sept. 1978).
S. Arvola, *Rock Prod.*, 62 (Nov. 1975).
"Control System in a Small Box Has a Big Impact on the Operating Efficiency of Jumbo Air Freighters," *Mater. Handl. Eng.*, 64 (Feb. 1979).

Trade Magazines

Feed & Grain Times, Johnson Hill Press, Fort Atkinson, Wisc.
Grain Age, Eden Prairie, Minn.
Industrial Equipment News, New York, New York
Material Handling Engineering, Cleveland, Ohio
Modern Materials Handling, Boston, Mass.
Production & Inventory Management Review, Hollywood, Fla.
Weighing & Measurement, Rockford, Ill.

WELDING

Welding comprises a group of processes whereby the localized coalescence of materials is achieved through application of heat and/or pressure (1–3). It represents a fabrication technology of tremendous scope; welding principles are used in nearly every product fabricated from metals and, increasingly, nonmetals. Obvious examples are bridges, buildings, ships, space vehicles, automobiles, offshore platforms, pipelines, pressure vessels, and consumer appliances, and less evident examples include the enormous number of miniature welds used in microcircuits. This field of great technical complexity utilizes robots, computer-controlled machines, and high concentrations of energy in the form of plasma arcs, lasers, and electron beams; very rapid processes that involve heat transfer, chemical reactions, and metallurgical reactions and requires complex methods of analysis to control stress, distortion, and fracture (see also Metal surface treatments; Metal treatments; Solders and brazing alloys).

The principal welding processes are of comparatively recent origin. Forge welding, soldering, and brazing can be traced to ancient times, but the modern processes of arc welding, resistance welding, and gas welding (as well as thermit welding) were discovered in the 1880s. The use of welding in production and repair increased slowly through the early 1900s and received a boost by successes in emergency ship repair during World War I. Welding had increased use in building construction during the 1930s, and new welding processes, eg, submerged arc welding, were discovered. World War II provided a tremendous impetus, with spectacular uses in all-welded ships and for military tanks, and with new processes developed for welding aluminum. Since the 1940s many new welding techniques have appeared, including electron beam, laser, and ultrasonic processes (4).

Thus, since the origins of modern welding in the 1880s, many processes have evolved, and today the American Welding Society recognizes over sixty methods of welding and more than twenty allied processes employing thermal cutting, thermal spraying, and adhesive bonding (1), some of which are identified in Table 1.

Arc Welding Processes

In arc welding processes, the coalescence of metals is achieved through the intense heat of an electric arc which is established between the base metal and an electrode. The six arc processes listed in Table 1 are differentiated by various means of shielding the arc from the atmosphere (1–3).

In arc processes, either d-c or a-c current is used to establish and maintain an arc between the base metal and an electrode. The electrode itself may be consumed by melting and thus become part of the weld, acting as a filler metal, or it may be a nonconsumable material (tungsten). In the latter case, the heat of the arc may simply be used to fuse adjacent base metal (autogenous welding), or a separate filler metal may be added. It is essential that the molten pool of material under the arc, as well as the adjacent solidified but still high temperature metal, be protected from oxygen, nitrogen, and other elements of the atmosphere, since these react with the metal to form oxides and other products that reduce the strength and toughness of a weld. Conse-

Table 1. Welding Processes[a]

Process	Abbreviation	Process	Abbreviation
Arc welding		*Resistance welding*	
shielded-metal arc welding	SMAW	resistance spot welding	RSW
gas–tungsten arc welding	GTAW	projection welding	RPW
plasma arc welding	PAW	resistance seam welding	RSEW
gas–metal arc welding	GMAW	flash welding	FW
flux-cored arc welding	FCAW	upset welding	US
submerged-arc welding	SAW	*Solid-state welding*	
Oxyfuel-gas welding	OFW	diffusion welding	DFW
oxyacetylene welding	OAW	explosion welding	EXW
oxyhydrogen welding	OHW	friction welding	FRW
pressure–gas welding	PGW	ultrasonic welding	USW
Soldering and brazing		*Other processes*	
dip soldering	DS	electron-beam welding	EBW
wave soldering	WS	electroslag welding	ESW
resistance brazing	RB	laser-beam welding	LBW
furnace brazing	FB	thermit welding	TW

[a] Ref. 1.

quently, various forms of shielding are provided around the arc in the different processes.

Shielded-Metal Arc Welding. The essential features of the SMAW process (commonly called stick-electrode welding) are shown in Figure 1. The arc acts between the consumable electrode wire and the base metal, and droplets of molten filler metal are transferred to the weld pool. The unique feature of this widely used process is the role of the electrode coating. The heat of the arc vaporizes the coating which provides the necessary shielding atmosphere for the arc. The products of the coating also provide a slag covering over the weld metal, thus providing further protection from the atmosphere.

In addition to these functions, the coating introduces fluxing agents to the weld pool, assists in establishing the electrical characteristics of the arc, and can be used to provide additional filler metal to the weld; other, more subtle characteristics are possible in various electrode types. There is wide variety in electrode-coating compositions, depending on desired welding characteristics. Common ingredients are cellulose to provide shielding gas, titanium dioxide to provide slag, and sodium silicate to act as a binder and fluxing agent.

Gas–Tungsten Arc Welding. The GTAW process (often called the tungsten–inert-gas or TIG process) is shown in Figure 1(**b**). Here, the electrode is nonconsumable and shielding is provided by the flow of inert gas through the welding-torch nozzle. Argon and helium are commonly used, as well as argon–helium mixtures; traces of oxygen or hydrogen are sometimes present. The arc may be used alone to fuse base metals together or, as shown in Figure 1(**b**), separately added filler rod can be fed into the weld pool.

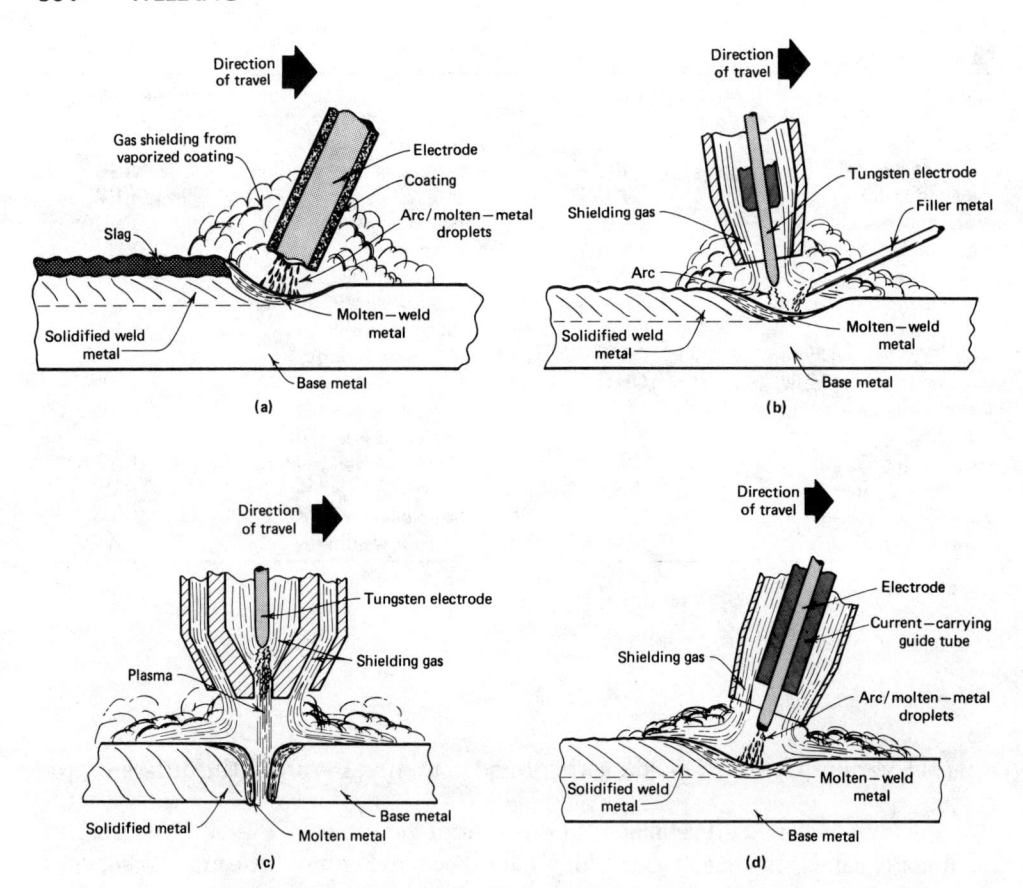

Figure 1. (a) The shielded-metal arc-welding (SMAW) process. (b) The gas–tungsten arc-welding (GTAW) process. (c) The plasma arc welding (PAW) process. (d) The gas–metal arc-welding (GMAW) process.

Plasma Arc Welding. In the PAW process, shown in Figure 1(c), the arc is between a nonconsumable electrode and the base metal, similar to the GTAW process. The unique feature is the flow of inert gas through a constricted orifice, resulting in a highly concentrated arc. There is also an outer sheath of shielding gas. In another version, the arc is between the electrode and the constricting orifice (nontransferred arc mode). The intensely concentrated plasma penetrates deeply into the base metal (see also Plasma technology in Supplement Volume).

Gas–Metal Arc Welding. The features of the GMAW process (also called the metal–inert-gas or MIG process) are shown in Figure 1(d). A consumable bare-wire electrode is fed continuously through the welding torch; a flow of inert gas through the nozzle provides shielding. In addition to argon and helium, carbon dioxide is also used as shielding gas, as well as various mixtures of these gases.

Flux-Cored Arc Welding. The FCAW process is illustrated in Figure 2(a). The characteristic feature is the hollow, flux-filled, consumable electrode which is fed continuously through the welding torch. The vaporization of the flux provides both arc shielding and a slag blanket over the weld. In a variation of the process, a shielding gas flows through the welding nozzle to provide additional protection of the arc.

Submerged-Arc Welding. In the SAW process, shown in Figure 2(**b**), a loose flux is blanketed over the region of the arc. A consumable bare-wire electrode is fed continuously through the torch into the weld. The molten- and granular-flux blankets provide the necessary arc shielding and slag cover for the solidified weld.

Welding Systems

The various welding processes result in systems of varying complexity. They include at least the electrode and a device for holding or feeding it, the work piece, the power source, and heavy-duty cabling to provide a complete electrical circuit. Provisions for supply and control of gas and control of wire feed and movement of the electrode assembly are required, depending on process type and degree of automation.

Welding systems are generally classified as manual, semiautomatic, and automatic. In manual welding, the operator must maintain the arc, feed in filler metal, and provide travel and guidance along the joint. In semiautomatic welding, the welding machine maintains the arc and feeds filler metal, while the operator controls joint travel

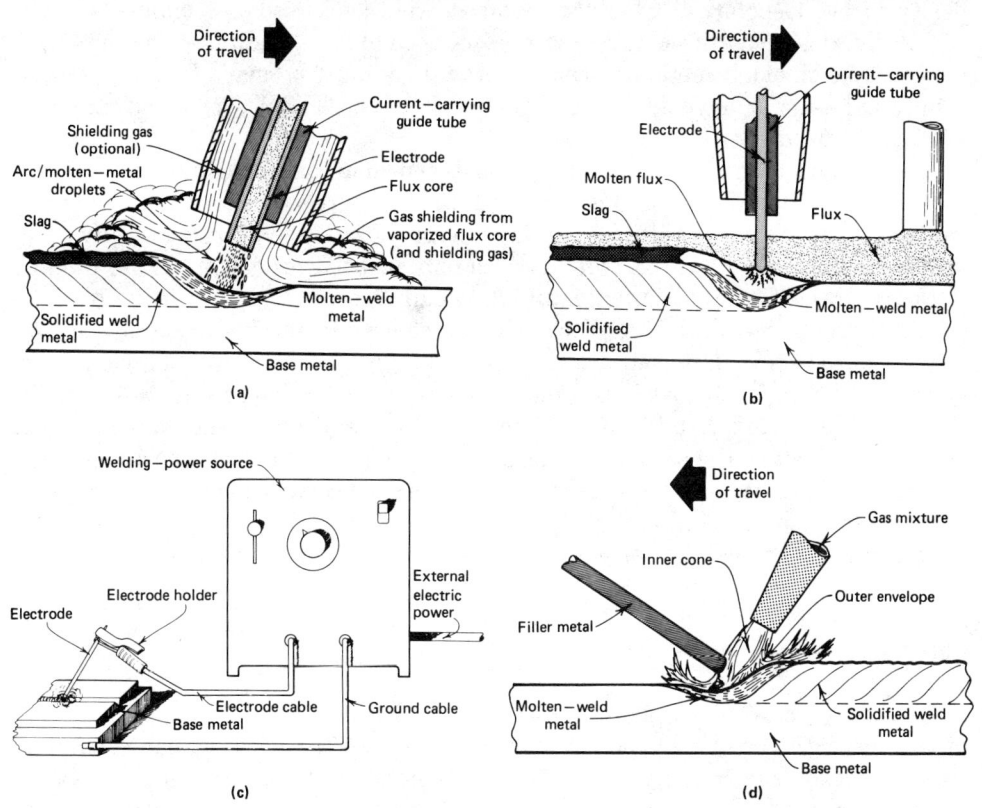

Figure 2. (**a**) The flux-cored arc-welding (FCAW) process. (**b**) The submerged arc-welding (SAW) process. (**c**) The shielded-metal arc system. (**d**) The oxyacetylene welding process.

and guidance. In automatic welding, the machine assumes all of the preceding functions. Automatic processes may be further classified, depending on the degree of feedback control used in controlling the welding variables. Automatic systems may be dedicated to a specific type of production, or they may be flexible, programmable robot systems.

The power source is the core of any welding system. Electrical power is provided by direct-line power or a generator; the latter is driven by an engine or an electric motor. In the case of welding with a-c current, the welding power source may be in the form of a transformer working from line power, or may be an engine driven a-c generator. Welding with d-c power may require a transformer–rectifier working from single- or three-phase line current, or a motor-driven generator. Welding power sources are further differentiated by controls imposed on the output current or voltage. One type of power source, used primarily for manual welding, is capable of providing essentially constant current. Another type, used for semiautomatic and automatic processes, is capable of providing essentially constant voltage.

The system for shielded-metal arc welding, shown in Figure 2(c), is the simplest system. It consists of the power source, electrode and holder, the base metal, and the electrical cables or leads. When the arc is struck, a complete electrical circuit is provided. With d-c welding, the electrode may be either negative (straight polarity) or positive (reverse polarity). Shielded-metal arc welding is used only manually.

A gas–tungsten arc welding system is more complex. In addition to the components of the shielded-metal arc system, provisions must be made for the inert-gas supply and water or air cooling of the welding torch. The GTAW system may be either manual or automated.

A semiautomatic flux-cored arc system is shown in Figure 3. Controls and drive motor are required to continuously feed the welding wire to the torch. If a shielding gas is used, provisions for control of this gas supply are needed and the torch configuration is different. The flux cored system may be used for gas metal arc welding (GMAW) by using the shielding gas and the welding torch configuration for use with gas. Both the FCAW and GMAW systems can be used in automatic operations.

An automatic submerged-arc welding system requires a power source, control system, and wire-feed provisions similar to those shown in Figure 3, as well as provisions for adding flux. In addition, controls are necessary to automatically move the welding torch along the seam of the base metal. This could be done by a motorized carriage moving along a straight beam, or by a positioner capable of placing the welding torch at any position over a base metal that is itself continually changing position. The submerged-arc process may also be used semiautomatically.

Other Major Processes

Oxyfuel-Gas Welding. This process, commonly called gas welding, uses the heat of combusting gases to melt and coalesce base metals. Although several different fuel gases, eg, propylene, hydrogen, or methane, can be added to the oxygen, the oxyacetylene flame is the most widely used because its high flame temperature (3100°C) is needed to weld steel. The essential features of the process are shown in Figure 2(d). The heat of the inner cone of the flame melts both the base and filler metal. The overall system is simple; it consists of high pressure tanks of the gases, gas regulators, hoses,

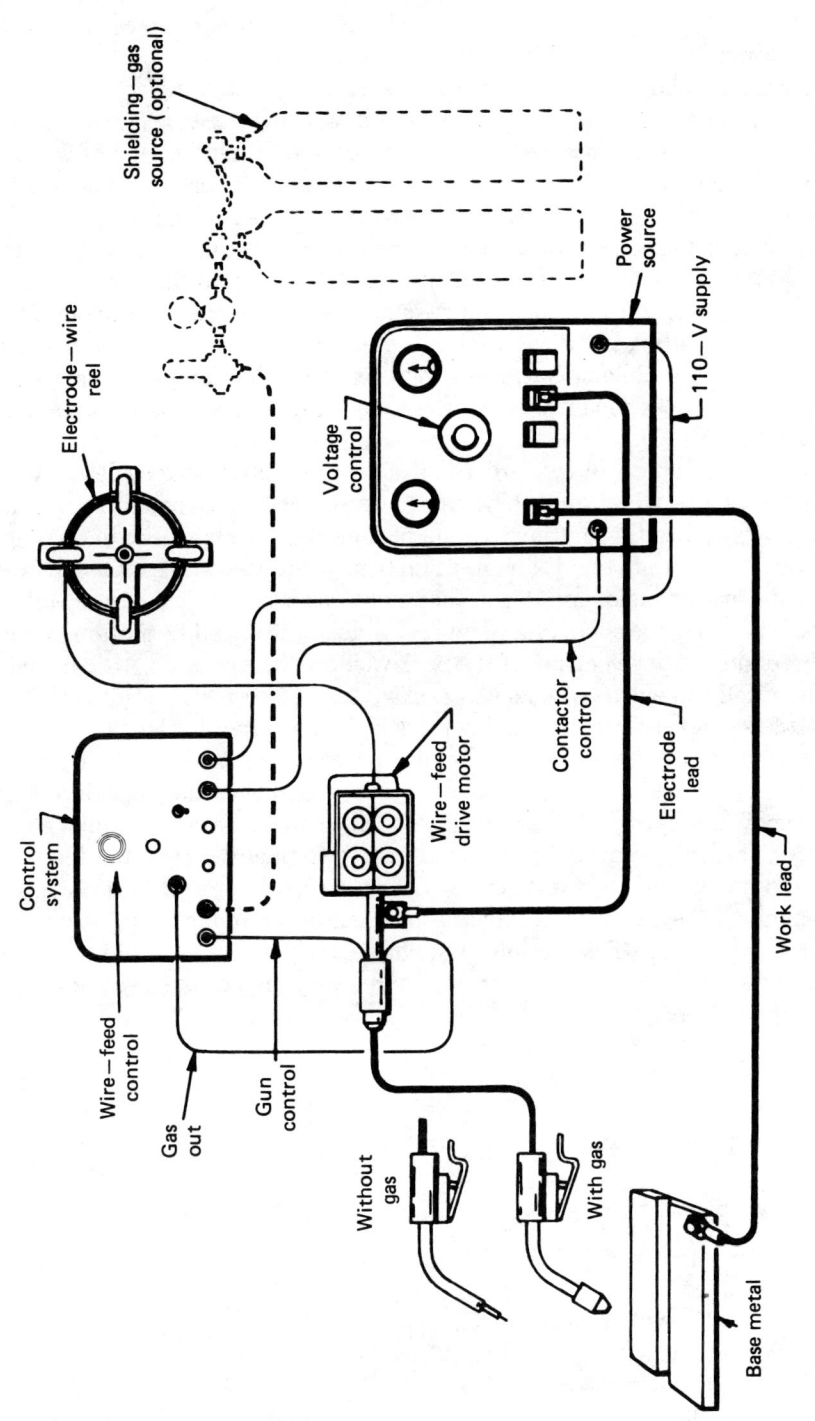

Figure 3. The flux-cored arc-welding system (2).

and the welding torch. Gas welding, once the most widely used welding method, has been supplanted by arc processes in commercial manufacture. However, it is still used in small welding shops, for repairs, and by the home craftsman.

Resistance Welding. As noted in Table 1, resistance welding comprises several processes; the most widely used is resistance spot welding (RSW). The principles are quite different from the processes previously described (see Fig. 4). The workpieces are firmly clamped between copper electrodes and an electric current is passed through the assembly. Heat is generated by the electrical resistance of the components, with maximum heat at the interface between the workpieces. A nugget of metal at the interface region is melted, at which moment the current is shut off and the clamping force on the electrodes released. The entire sequence typically requires about a second. Resistance spot welding is widely used in the production of thin gauge sheet-metal assemblies, eg, automobile bodies. It is a high speed production process and suitable for automation. Although mainly used for welding steels, it is also used with aluminum and other materials (5).

Electroslag Welding. In this process, the heat of molten slag coalesces the base and filler metals. Electric current flows from a consumable electrode through a molten metallurgical slag into the molten weld metal. The electric resistance of the slag provides the heat to maintain the slag molten and to melt the weld and base metal. In order to contain the molten metal and slag in the gap between the vertical base-metal plates, water-cooled copper shoes are placed on each side of the gap. The molten metal progressively solidifies at the bottom of the pool, welding the plates together in one pass. Although initially an electric arc melts the slag, steady electroslag welding is a nonarc process. It has extensive use in welding thick sections of metal. Deposition rates are very high. It is an automatic process that can be used only in the vertical direction.

Electron-Beam Welding. This welding process achieves the heat necessary for coalescence by bombarding the base metals with a concentrated stream of electrons (see Fig. 5). Thus, electrons emitted by the heated filament are accelerated between cathode and anode and then focused into a narrow beam onto the base metal. The kinetic energy of the electrons is converted to heat upon impact. The electron beam permits the welding of thick sections, but with a very narrow zone of molten metal. Electron-beam welding is usually performed in a very high or hard vacuum. Modifi-

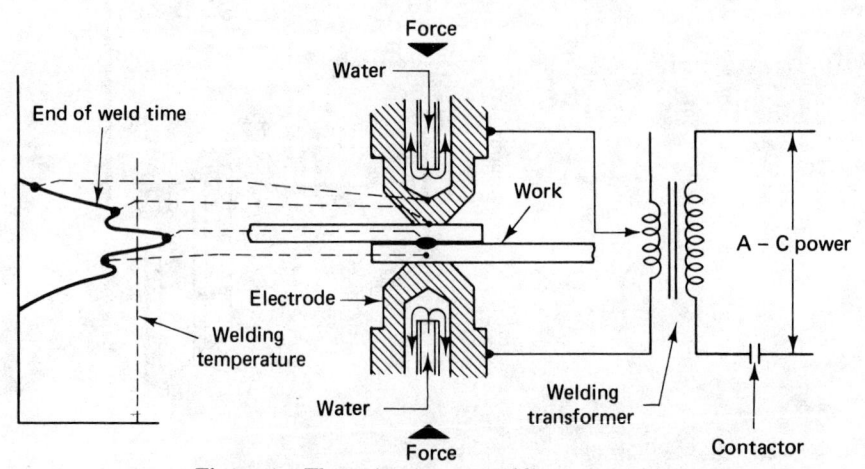

Figure 4. The resistance spot welding process (2).

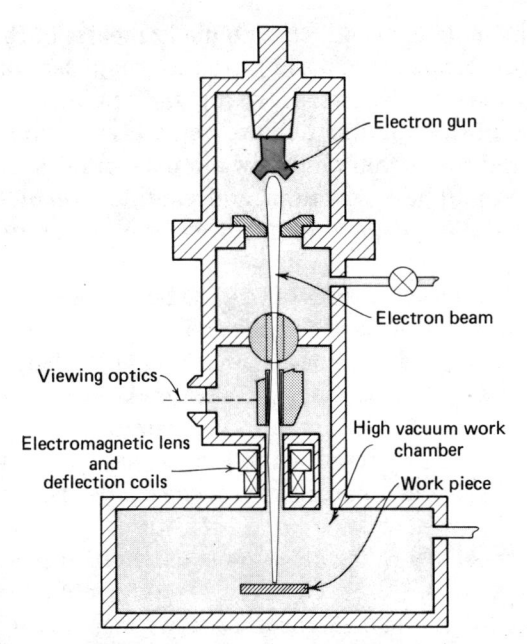

Figure 5. Electron-beam welding system.

cations have led to soft and out-of-vacuum welding, although penetration capability is sacrificed (6).

Laser-Beam Welding. The heat of coalescence is produced by focusing the beam of light (photons) from a high power laser on the base metal (see Lasers). Laser welding is considered a high energy (ca 10^6 W/cm^2) welding process, with weld characteristics similar to the electron beam. Thus, at low energy, surface melting occurs, but at high energy, a deep penetration of the base metal is possible. Unlike the electron beam, the laser does not require a vacuum. Since the laser beam is essentially a light beam, it is easily controlled by mirrors, beam splitters, and other optical devices, which leads to great flexibility in its use (6).

Brazing and Soldering. In brazing and soldering processes, a molten filler metal flows into the joint of the base metal at a temperature below the melting point of the base metal. Brazing and soldering processes are largely differentiated by the means of heating the parts to be joined. A common feature is that the braze or solder metal flows by capillary action into the closely fitted joints of the base metal. Bonding is through formation of intermetallic compounds at the interface between base and filler metals. Although these are the common features of brazing and soldering, they are differentiated by the melting temperature of the filler metal. In brazing processes, the filler metal is liquid >450°C, whereas in soldering, the melting point of the filler metal is <450°C (7).

Physics and Metallurgy

In most welding processes, the local regions of adjoining base metals and any added filler metal are melted and resolidified. These features are present, for example, in the arc, gas, electroslag, electron beam, and laser welding processes. The analogy of this action to the casting process is often made, where the molten base and filler metals are cast in the mold formed by the nonmolten parts of the base metals. This analogy is of some use because a range of metallurgical phase transformations is involved as the alloy system solidifies. However, welding differs from conventional casting by the speed of the solidification process. Only a local region of material is melted in welding, and the surrounding, low temperature base metal acts as a large heat sink, producing rapid heat flow from, and solidification of, the weld zone. This produces complex metallurgical microstructure and physical properties not typical of casting.

Heat Flow. Certain basic features of welding heat flow are illustrated by Figure 6(a), which represents a welding heat source moving along the joint between two base metals. The central region, under the heat source, is molten. The surrounding elliptical curves are temperature isotherms which are clustered toward the direction of travel of the heat source and show the temperature distribution at a given instant during welding. Different welding processes, plate thicknesses, and welding speeds modify the details of the clustering and magnitudes of the isotherms, without changing the principles of the behavior shown.

An important aspect of welding heat flow is the thermal cycle at a given location in the material. The nature of the cycle depends on the intensity of the heat source, the speed of welding, the thermal characteristics of the material, and the location in the material. The general behavior is shown in Figure 6(b). Curve 1, representing point 1 of Figure 6(a), is adjacent to the weld and actually reaches the melting point as the arc passes. At points 2 through 5, two characteristics are evident. First, peak temperatures that occur along the line n-n' in Figure 6(a) are progressively lower because the heat spreads over larger regions. Second, there is a progressively greater time delay in reaching peak temperature because of the time required for heat conduction. The weld thermal cycle, involving the peak temperatures achieved and the speed of heating and cooling, and possibly preheating and postheating, accounts for many of the subsequent complexities of welding metallurgy.

Solidification. The heat of the electric arc melts a portion of the base metal and any added filler metal. The force of the arc produces localized flows within the weld pool, thus providing a stirring effect which mixes the filler metal and that portion of the melted base metal into a fairly homogeneous weld metal. The pool of molten weld metal begins to solidify as soon as the welding heat source moves past a given location. There is a very rapid transfer of heat away from the weld to the adjacent, low temperature base metal, and solidification begins nearly instantaneously. The stages of solidification of the weld pool are shown in Figure 7. Thus, the initial crystal dendrites form at the fusion boundary of the weld-pool cavity (Fig. 7(a)); these grow to large columnar grains around the periphery of the weld (Fig. 7(b)), while the interior is still molten. Solidification then progresses to the weld surface, as shown in Figure 7(c).

However, the thermal cycle of the welding and solidification process have caused additional complex effects to occur. The differing peak temperatures reached at different locations from the weld significantly affect the microstructure of the base metal in the region surrounding the weld. This region, known as the heat-affected zone, contains various types of recrystallization. The most pronounced is at the fusion boundary, where large grain sizes characteristic of the adjacent weld metal are formed. The effects diminish away from the weld until, at some distance, there is no discernable change in the base metal.

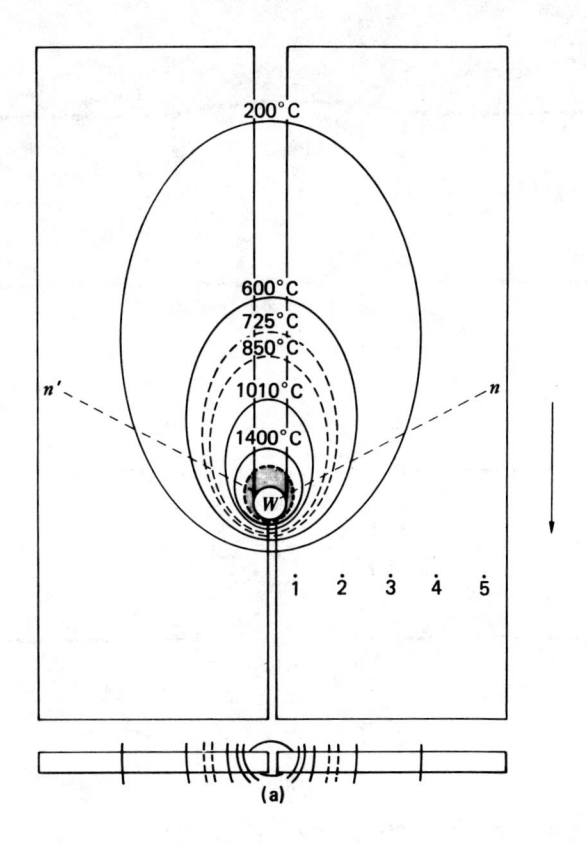

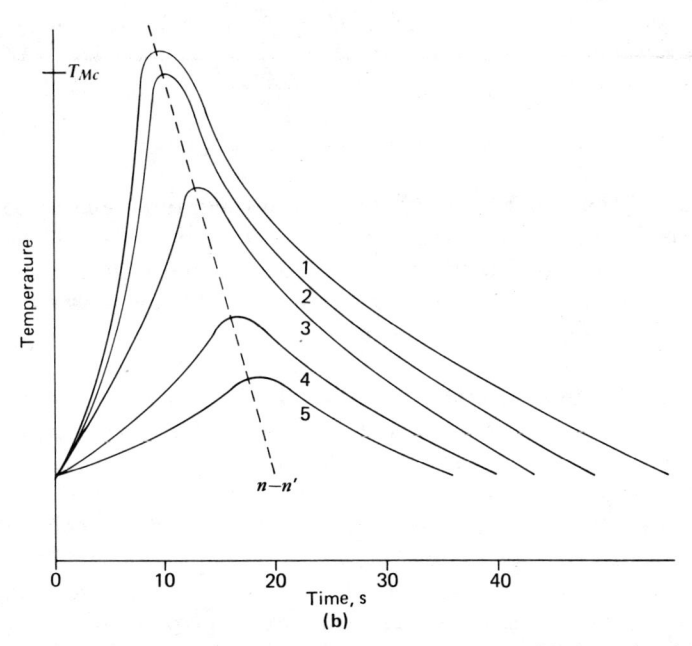

Figure 6. (a) Temperatures in the region of a moving welding arc; (b) thermal cycles at various locations in a plate (8).

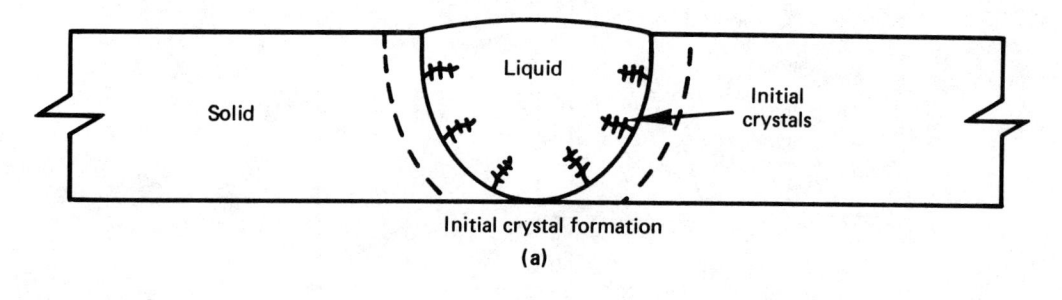

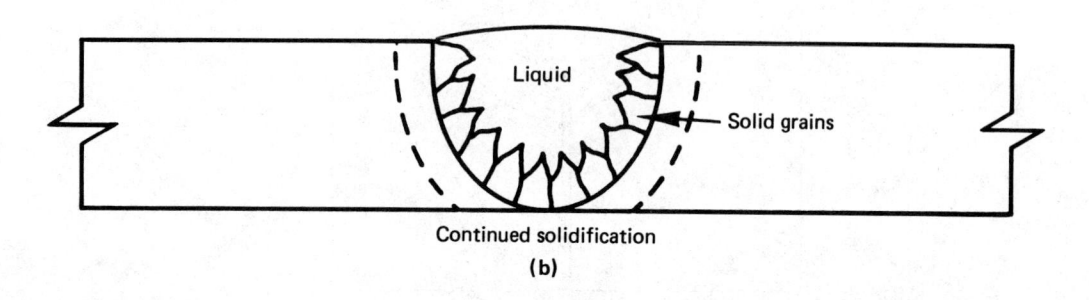

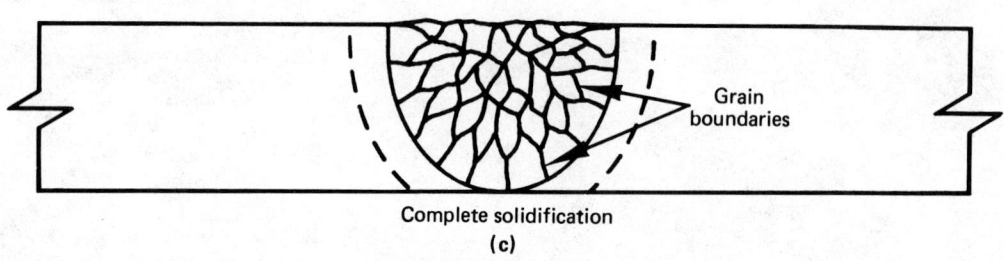

Figure 7. Stages of solidification at a weld cross section (8).

Metallurgy. Welding metallurgy deals with the interactions of base and filler metals and the interactions of these with various chemicals injected into the weld via gases, electrode coverings, fluxing and slagging agents, and surface contaminants. For example, a number of gas–metal reactions are possible while the weld is still at high temperature. Oxygen, nitrogen, water vapor, and carbon dioxide are gases that react with ferrous metals to yield products harmful to the metallurgical properties of a weld. The nature of slag–metal reactions that occur in the molten state strongly depends on the composition of the flux or the electrode coating. Flux chemistry may be altered to control removal of specific weld-metal impurities, such as addition of manganese or silicon to provide strong deoxidizing action, or additions to enhance slag removal, with all of these influencing the final metallurgical characteristics of the weld.

The microstructure of a weld is the overall arrangement of grains, grain boundaries, and phases that exist once solidification occurs. An important metallurgical tool for understanding weld microstructure is the phase diagram, which, for a given alloy composition, relates material phase and temperature. For example, the iron–carbon (Fe–C) phase diagram shows that a steel of 0.25% carbon melts above

1520°C; at a slightly lower temperature, it consists of a mixture of molten metal and delta iron; below 1500°C, it is transformed to liquid metal and austenite; at 1480°C, it solidifies completely to austenite; and below 815°C, a mixture of austenite and ferrite exists, which is transformed to ferrite and pearlite at 727°C (see Steel).

However, phase diagrams represent equilibrium conditions which do not prevail in welding. These nonequilibrium conditions result in changes in the temperatures of phase transformation and in microstructures that have solidified before attaining equilibrium. For example, a very rapid cooling of 0.25% carbon steel results in a martensitic microstructure, a material of greater hardness than achieved under slow cooling. Cooling rates as well as composition are essential to interpreting the microstructures of deposited weld metal and the surrounding heat-affected zone.

Thus, the metallurgy of welds, comprising the weld metal and surrounding heat-affected zone, is not only influenced by the chemistry of the materials involved, but also by the welding process, the specific procedures for applying the process, and the heat-transfer characteristics (determined by material, mass, and geometry) of the welded joint (8–10).

Base and Filler Metals

There is hardly a metal that cannot, or has not, been joined by some welding process. From a practical standpoint, however, the range of alloy systems that may be welded is more restricted.

Properties. The properties of materials are ultimately determined by the physics of their microstructure. For engineering applications, however, materials are characterized by various macroscopic physical and mechanical properties. Among the former, the thermal properties of materials are particularly important in welding, including melting temperature, thermal conductivity, specific heat, and coefficient of thermal expansion. The latter property greatly influences structural distortion that may occur in welding. The electrical conductivity of a material is important in any welding process where base or filler metal is a part of the welding electrical circuit. Material density is also of interest.

The response of materials to force is characterized by mechanical properties, eg, elastic modulus, yield stress, tensile strength, ductility, hardness, and impact or fracture strength. The fatigue characteristics are also of great importance. The various mechanical properties should be known over a temperature range that covers the expected service temperatures. The fracture resistance, for example, of most materials is temperature sensitive.

Almost all metals can be welded, but their welding varies greatly in difficulty and success. The term *weldability* specifies the capacity of a metal, or combination of metals, to be welded under fabrication conditions into a suitable structure which provides satisfactory service. It is not a precisely defined concept, but encompasses a range of conditions, eg, base- and filler-metal combinations, type of process, procedures, surface conditions, and joint geometries of the base metals (11). A number of tests have been devised to measure weldability. These generally are intended to determine the susceptibility of welds to cracking.

Base Metals. *Carbon Steels.* These steels do not contain alloying elements but, in addition to iron, only carbon and manganese in appreciable quantities. They are mainly used for structural purposes at ordinary temperatures, eg, beams and columns, storage tanks where pressure and strength are important, and piping at temperatures where corrosion is not a factor. Carbon steels are the most easily welded metals. The heat of welding has a metallurgical effect on the base metal according to its composition. This effect may reduce the strength or the corrosion resistance of a metal or otherwise change its properties, but this effect is lower in carbon steels than in other steels. Steels containing up to 0.30% carbon and 1.00% manganese can be readily welded in thicknesses up to ca 5 cm without special techniques. All welding processes discussed above are used to weld carbon steels.

Low Alloy Steels. These are carbon steels to which other elements have been deliberately added to impart a particular property. Common alloying elements include nickel, chromium, molybdenum, vanadium, and silicon. Nickel improves the mechanical properties of a steel at low temperatures, whereas chromium, molybdenum, and vanadium or combinations of these elements improve steel properties at elevated temperatures, and silicon improves the mechanical properties at ordinary temperatures. Low alloy steels are not used where corrosion is a prime factor and are usually considered separately from stainless steels. Low alloy steels are used for pressure piping systems operating at high temperatures, for storing liquid nitrogen or oxygen and similar materials at low temperatures, or where strength and low weight are required at ambient temperatures. Although less easily welded than plain carbon steels, low alloy steels are readily welded using preheating and suitable filler metal. In some cases, it may be necessary to subject the welded joint or the entire weldment to heat treatment in order to restore its properties. All of the welding processes previously mentioned may be used for low alloy steels. The filler metal is usually also low alloy steel.

Stainless Steels. The stainless steels are either austenitic chromium–nickel or straight chromium steels. They are widely used because of their excellent corrosion resistance. There are about twenty different austenitic chromium–nickel steels, which contain varying amounts of chromium and nickel. Other alloying elements are niobium, molybdenum, manganese, and silicon. Upon welding, these steels undergo a metallurgical change, known as carbide precipitation, which reduces the corrosion resistance of the weld and heat-affected zone. Stabilized stainless-steel grades, to which niobium or titanium has been added, and the extra-low carbon grades, may be welded without encountering carbide precipitation. Alternatively, the unstabilized grades may be welded and the weldment subjected to special heat treatment, which dissolves the precipitated carbides and restores the original corrosion resistance. The stainless steels have higher mechanical properties than the carbon steels and are readily welded by all arc welding processes, with a filler metal selected to provide a weld deposit of the same general composition as the base metal.

The chromium stainless steels contain only chromium in amounts of 3 to $\geq 27\%$. The metallurgy of these steels is rather complex. They harden to different degrees; generally, hardenability increases with increasing chromium content. They may be arc welded, using carefully planned procedures including pre- and postheat treatment. The filler metal is either chromium or chromium–nickel steel.

Aluminum and Aluminum Alloys. These are used wherever lightness or atmospheric corrosion resistance are required, or where mildly corrosive fluids are involved. Typical applications include railroad tank cars, pressure vessels, and tanks for storing chemicals and dairy products. Aluminum alloys are classified as heat-treatable or nonheat-treatable. Strengths of commercially pure aluminum and nonheat-treatable alloys are developed by strain hardening and by alloying elements of which magnesium, manganese, and silicon are typical examples. The alloying elements in heat-treatable aluminum alloys are dissolved in the pure aluminum at a high temperature by a process known as solution heat treatment. Specific aluminum alloys are available in different tempers from hard to soft.

The welding heat produces an annealing effect in aluminum and its alloys at or near the weld, thus reducing the strength. In the heat-treated alloys, properties can be restored by a postwelding heat treatment. Gas–metal arc, gas–tungsten arc, and resistance welding are most widely used to join aluminum and its alloys. Filler metal is chosen to match the composition of the base metal.

Magnesium Alloys. In its pure state, magnesium (qv) does not have sufficient mechanical strength for structural purposes and must be alloyed with other elements, such as aluminum, zinc, manganese, and zirconium; rare earths and thorium are used for high temperature applications. Magnesium alloys may be divided into room-temperature and elevated-temperature service groups. Certain magnesium alloys are subject to stress corrosion. Weldments subjected to corrosive attack over a period of time may crack adjacent to the weld seams if the residual stresses are not removed. Gas–tungsten arc welding and gas–metal arc welding are recommended for joining magnesium, the former for thinner materials and the latter for thicker materials.

Nickel and Nickel Alloys. This group of metals includes commercially pure nickel and a variety of nickel alloys such as Monel, Inconel, Incoloy, and Hastelloy. Commercially pure nickel has a nominal nickel content of 99.4%. It has a high corrosion resistance and is magnetic. A low-carbon version of commercially pure nickel may be used where maximum ductility is required at temperatures >315°C. It has the same corrosion resistance as commercially pure nickel. The Monel alloys are nickel–copper alloys with a nominal composition of 67% nickel and 30% copper. They have good strength, toughness, and resistance to corrosion and are slightly magnetic. The Inconel alloys are nickel–chromium alloys having a nominal composition of 77% nickel and 16% chromium. These alloys offer a high resistance to scaling and are used at elevated temperatures. The Incoloy alloys are nickel–iron–chromium alloys of ca 32% nickel, 21% chromium, and 45% iron. These alloys have high oxidation and heat resistance. The Hastelloy alloys are nickel–molybdenum–iron alloys that are resistant to corrosive mineral acids and possess good mechanical properties. All of these nickel alloys can be welded with the shielded-metal arc, submerged-arc, gas–metal arc, or gas–tungsten arc process.

Copper and Copper Alloys. The coppers are divided into oxygen-bearing and oxygen-free coppers. Numerous copper alloys are of commercial importance, including those with zinc (brasses), with tin (phosphor bronzes), and with aluminum (aluminum bronzes); all are weldable. When welding copper itself, it must be free of oxygen if the joint strength is required to be equal to that of the base metal. Copper alloys can be welded with the shielded metal arc, gas–metal arc and gas–tungsten arc processes.

Reactive and Refractory Metals. The reactive and refractory metals, originally used in the aerospace industry, are now welded for many applications. Reactive metals, which are zirconium, titanium, and beryllium, have a strong affinity for oxygen and nitrogen at elevated temperatures and form very stable compounds. At lower temperatures, they are highly resistant to corrosion. The refractory metals, tungsten, molybdenum, tantalum, and niobium, exhibit some of the same characteristics (see also Refractories). They all have extremely high melting points, relatively high density, and high thermal conductivity. The reactive metals have lower melting points and densities, and, except for zirconium, have higher coefficients of thermal expansion. The metals of both groups are difficult to weld. Their high affinity for oxygen and other gases at elevated temperatures excludes processes that utilize fluxes, or where heated metal is exposed to the atmosphere. Small amounts of impurities cause brittleness. The surfaces must be well prepared and very clean to maintain a contamination-free environment during welding and cooling. Beryllium, because of its toxicity, requires special precautions. Gas–tungsten arc welding may be used for all metals, and gas–metal arc welding for some. Electron-beam and laser welding may also be used.

Filler Metals. Filler metals are added to a weld by melting a consumable electrode or a separate wire fed into the weld pool. In the first category, the filler metal is part of the welding electrical circuit and may be in the form of short lengths of covered wire, as in shielded-metal arc welding, or in the form of continuous reels of wire used in semiautomatic and automatic welding processes. Solid wire is used in the gas–metal and submerged-arc welding processes, whereas a hollow, flux-filled wire is used in flux-cored arc welding. More filler metal in the form of iron powder is sometimes added to the electrode coating or the flux used in submerged arc welding. In the second category, the filler metal may be in the form of short lengths of bare, solid wire, as used in gas welding or manual gas–tungsten arc welding, or in continuous reel form, used in automatic gas–tungsten arc welding.

Filler metals are manufactured in many special forms for welding the commercial alloy systems previously described. The American Welding Society (AWS) has issued specifications covering the various filler-metal systems and processes (2). Thus, AWS A5.1 covers carbon steel covered arc welding electrodes, AWS A5.16 is for titanium and titanium alloy bare welding rod and electrodes, and AWS A5.28 is for low alloy steel filler metals for gas-shielded arc welding. A typical specification covers classification of relevant filler metals, chemical composition, mechanical properties, testing procedures, and matters related to manufacture, eg, packaging, identification, and dimensional tolerances. New specifications are issued occasionally, in addition to ca 30 established specifications. Filler-metal specifications are also issued by the ASME and the DOD. These are usually similar to AWS specifications, but should be specifically consulted when they are referenced.

Design

Welded Joints. The weld joint is the geometric arrangement between two pieces of base metal brought together for purposes of welding (12). There are only five recognized welding-joint configurations: butt, corner, tee, lap, and edge joints (see Fig. 8). Thus, a butt joint is located between two members in approximately the same plane, whereas a corner joint brings the edges together with an included angle between the planes of the two parts. A tee joint brings the edge of one part onto the planar surface of the second part with a 90° angle between the planes. In a lap joint, one member overlaps another, and an edge joint brings the edges of two parts together without an included angle.

Weld Types. A weld may be applied to the various joints in different ways, as shown in Figure 9. The fillet weld joins corners and tees. In the plug weld, the two pieces are joined by weld metal deposited in a prepared hole in the overlying piece. For a spot weld, heat is applied to the overlying plate, creating fusion at the interface. Resistance

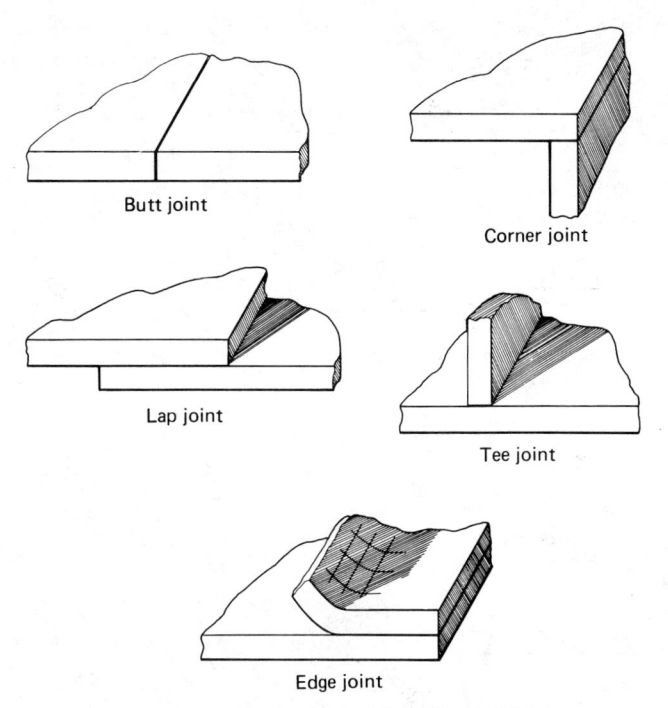

Figure 8. The five basic types of welded joint.

welding is generally used for the spot weld and the seam weld. The groove weld joins base metals in a butt joint. For the backing weld, the root of the original weld is first removed by chipping or gouging. Surface welds are used to build up the surface of parts with special materials or to replace worn material. Flange welds are applied to edge joints.

The edges of the joint to be welded must be prepared. The simplest joint is a square butt joint welded from one or both sides. However, such welds can only be used on thin-gauge material, since it is difficult to penetrate the joint. Various edge preparations permit fusion across the thickness of the material. Bevel and V grooves are typically prepared by flame cutting, whereas J and U grooves require machining. The selection of single or double grooves is dictated by a variety of practical considerations. Thus, for access to only one side of a surface, a single groove is needed. Double grooves require additional preparation, but their costs are usually offset by a net reduction in weld cross-sectional area, with savings in metal and welding time.

Another design factor is the position in which the weld must be made, which identifies the general orientation of the surface or of the axis of the weld (2). The positions are referred to as flat, horizontal, vertical and overhead. In the flat position, welding is performed from the upper side of the joint, and the surface is horizontal. In a horizontal weld, the axis is horizontal, but the surface is inclined at an angle. Most processes are applied to the flat position, which is the most convenient and gives the least defects. The most difficult position is the overhead position, and those welds are more prone to defects. Manual or semiautomatic processes require higher skill for welding in these positions. Special fixtures and positioning equipment are often used to permit welding in the flat or horizontal position.

518 WELDING

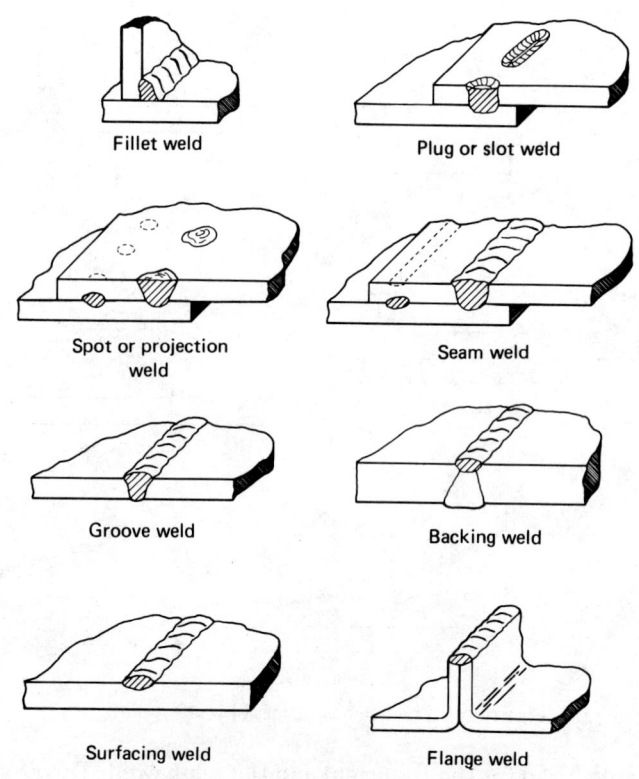

Figure 9. The eight types of weld.

Stress and Distortion. The forces acting on a structure are transmitted through the welded joints; that is, the joint is subjected to simple tension (or compression), bending, shear, or torsional stresses, or to combinations of these stresses owing to combined loading situations. Weldments must be of proper size, length, and location to withstand the loads imposed during service.

The magnitude and nature of the load are considered in design. The load may be essentially quasistatic, cyclic, or impact. Many structural failures, for example, have been caused by supposedly innocuous structural details welded in place without consideration of their effects on fatigue properties. The service temperatures are also important, since they affect the fracture resistance of a material.

The welding process itself may induce significant stress and distortion, primarily because the molten filler metal is at maximum volume. With cooling and solidification, the material contracts and thereby exerts stresses on the surrounding base metal. Without structural restraint, these stresses cause distortion until they are relieved or minimized. Distortions that occur in simple fillet and butt welds are shown in Figure 10. However, constraint imposed by the surrounding structure may prevent distortion. In this case, significant residual stress may be present, such as tensile stress along the weld axis in a plate. The residual stresses owing to welding may be quite high, approaching or reaching the yield stress of the material. These stresses, in combination with metallurgical conditions, may cause cracking in the weld.

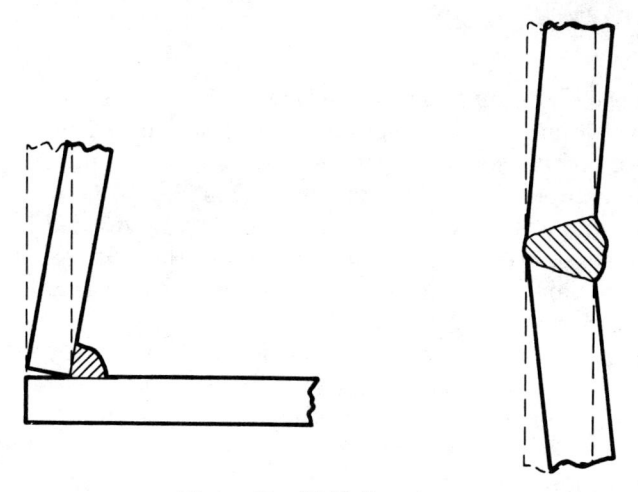

Figure 10. Weld distortion.

Testing

The integrity of welded structures depends on the integrity of the welds, and much attention is given to testing methods, such as destructive tests, nondestructive tests, and general weld inspection. An objective of many tests is to determine whether welds contain specific defects, such as porosity, slag inclusions, cracks, or lack of fusion (13–14).

Destructive tests destroy the specimens or a portion of the production under examination but provide direct information on properties such as tensile strength, impact strength, ductility, and corrosion resistance.

Standard destructive tests include determination of chemical composition of base or weld metal; corrosion-resistance tests; metallographic tests with microscopic examination of polished and etched specimens; and hardness tests across the base metal, heat-affected zone, fusion boundary, and weld metal. Mechanical tests include tensile, impact, and guided-bend tests. Destructive tests may be used to evaluate the suitability of welding processes or specific welding procedures for a given application. Simple bend or break tests are used for welder qualification. Test specifications are available.

Nondestructive evaluation (also termed nondestructive testing or nondestructive inspection) has extensive use in weld testing (13). Nondestructive tests do not impair the serviceability of the material or component under test (see Nondestructive testing). The most widely used tests for evaluation of welds are liquid penetrant, magnetic particle, ultrasonics, and radiography. Acoustic-emission tests are increasingly used. Nondestructive tests detect and characterize, in terms of size, shape, and location, the various types of weld discontinuities that can occur.

Weld inspection involves the duties of personnel responsible for judging the quality of welding with regard to specifications (14). Some of these duties involve the visual inspection of welds to determine if they are of the proper size, location, and type and are free of defects. Specifications of materials used must be checked, as must equipment and procedures.

Economic Aspects

More than 850 U.S. companies are involved in the direct manufacture of products associated with welding, such as power sources and electrodes as well as a wide range of accessories, eg, grinders, shears, presses, heat treating equipment, safety items, positioners and manipulators (15). An extensive distributor network is involved in sales of welding products. Sales of welding equipment are divided into arc welding, gas welding, and resistance welding equipment, accessories, and consumable material, ie, electrodes, wires, and gases. These sales have continued to grow over many years; the 1981 figure is close to 2×10^9 (16).

The Standard Industrial Classification (SIC) system of the United States Government shows eleven industrial groups utilizing 90% of the persons employed as welders and flamecutters in the following occupations: fabricated metal products and nonelectrical machinery and construction involve >40% of the welders and flamecutters (2); other industries, such as motor vehicles, primary metal, repair services and aircraft, ships, and railroad equipment accounting for 5–6% each. In 1978, the U.S. Bureau of Labor Statistics listed 650,000 persons employed as welders and flamecutters, with continued increases projected (17). The increased use of automated and robotic welding systems will affect the growth of these occupations. *Moody's Industrial Manual* indicates that goods using welding in some stage of fabrication account for at least 30% of the gross national product (18).

Health and Safety Factors

Welding is carried out in a wide range of industrial environments, such as field construction sites, factory production floors, and small shops. Welders, therefore, are subject to the same hazards as all other workers in the metalworking trades. Specific additional hazards of welding include electrical shock, arc radiation, fumes and gases, fires and explosions, compressed gases, cutting and chipping operations, and high noise levels. Many of these also involve workers who share the welding environment. The increasing concern for health and safety in the industrial environment has resulted, in the case of welding, in ANSI Standard Z49.1, which covers safe practice for welding and cutting operations (19). An excellent review of the welding hazards and recommended practices are given in ref. 2. The American Welding Society provides information on safe welding practice and distributes information on welding hazards (20–22).

Fumes and gases associated with welding have been an area of growing concern (21–23). Fumes emanate from a number of sources, including electrodes, wires, base metals, coatings and contaminants, ozone produced by the ultraviolet radiation from the welding arc, and gases produced from the heat of the arc. These may lead to a number of health problems which could include acute poisoning under severe cases of ozone or nitrogen oxide concentration, chronic respiratory disease, a condition known as metal-fume fever involving zinc-containing fumes, skin disorders, and disorders of the nervous system, which can be caused by lead or manganese present in welding fumes.

Despite many studies of the health effects of welding, quantitative information is lacking, and although welding is known to cause some health problems, more research is needed.

BIBLIOGRAPHY

"Welding" in *ECT* 1st ed., Vol. 15, pp. 34–44, by S. A. Greenberg, American Welding Society; "Welding" in *ECT* 2nd ed., Vol. 22, pp. 241–252, by Edward A. Fenton, American Welding Society.

1. *Welding Handbook*, 7th ed., Vol. 1, American Welding Society, Miami, Fla., 1976; see also other volumes in this series.
2. H. B. Cary, *Modern Welding Technology*, Prentice-Hall, Inc., Englewood Cliffs, N.J., 1979.
3. *Welding and Brazing*, Vol. 6 of *Metals Handbook*, 8th ed., American Society for Metals, Metals Park, Ohio, 1971.
4. R. D. Simonson, *The History of Welding*, Monticello Books, Inc., 1969.
5. *Resistance Welding Manual*, 3rd ed., Vol. 2, Resistance Welder Manufacturer's Association, Philadelphia, Pa., 1961.
6. M. M. Schwartz, *Metals Joining Manual*, McGraw-Hill Book Company, New York, 1979.
7. H. H. Manko, *Solders and Soldering*, 2nd ed., McGraw-Hill Book Company, New York, 1979.
8. *Introductory Welding Metallurgy*, American Welding Society, Miami, Fla., 1968.
9. G. E. Linnert, *Welding Metallurgy*, 3rd ed., Vols. 1 and 2, American Welding Society, Miami, Fla., 1965.
10. J. F. Lancaster, *Metallurgy of Welding*, 3rd ed., George Allen & Unwin, London, 1980.
11. R. D. Stout and W. D. Doty, *Weldability of Steels*, Welding Research Council, New York, 1978; ref. 1, Vol. 4.
12. O. W. Blodgett, *Design of Weldments*, The James F. Lincoln Arc Welding Foundation, Cleveland, Ohio, 1963.
13. *Nondestructive Inspection and Quality Control*, Vol. 11 of ref. 3.
14. *Welding Inspection*, 2nd ed., American Welding Society, Miami, Fla., 1980.
15. "1982 Welding & Fabricating Buyers Guide," *Weld. Des. Fabr.*, (Jan. 1982).
16. T. B. Jefferson, *Weld. Des. Fabr.*, 71 (July 1982).
17. *The Welding Distributor*, 26 (Sept.–Oct. 1978).
18. *Moody's Industrial Manual* (American and Foreign), Moody's Investors Service, Inc., New York, 1981.
19. *Safety in Welding and Cutting*, ANSI Z49.1-1973, American Welding Society, Miami, Fla., 1973.
20. *Welding Safety and Health*, American Welding Society, Miami, Fla., 1982.
21. F. Y. Speight and H. C. Campbell, eds., *Fumes and Gases in the Welding Environment*, American Welding Society, Miami, Fla., 1979.
22. *Effects of Welding on Health*, American Welding Society, Miami, Fla., 1979.
23. *The Facts About Fume*, The Welding Institute, Abington, UK, 1976.

KARL F. GRAFF
The Ohio State University

WETTING AGENTS. See Surfactants and detersive systems.

Origins and History

The word cereal derives from the name of the Roman goddess, Ceres, in whose honor a spring festival, The Cerealis, was celebrated. This indicates the antiquity of these foods and the reverence with which they were esteemed. The latter probably reflected the important role cereals played in the diets of ancient peoples. Cereals are still an important dietary ingredient. As the world population continues to grow, cereals will become an increasing fraction of the diets for more and more people. That is recognized by the use of an ear of wheat as the symbol of the FAO. Below the wheat is the Latin inscription, *Fiat panis* (Let there be bread) (1).

Exactly where and when cereal grains were first cultivated has not been established. Most reports suggest that, for wheat, cultivation first occurred somewhere in the "fertile crescent" which extends from Egypt through Syria to and including the Tigris-Euphrates valley (2). It was in that region that the likely progenitors of wheat, emmer and einkorn, were found. These wild grains still grow in the area extending eastward from Asia Minor to Iran and Afghanistan.

The origins of the other cereals are less certain. Rice may have originated in Africa or Asia but probably was first cultivated somewhere between the southern People's Republic of China and southern Vietnam (3). Corn or maize originated in the Western Hemisphere. The Spanish conquistadors came in contact with corn during their early explorations of Central and South America (4). Corn originated perhaps in the lowlands of South America; others place its origin in Mexico or Guatemala (5). Wherever it was developed, by the time Columbus made his first voyage of exploration, corn as a food crop had spread over much of the Americas as well as to the West Indies (5). By 1492, the Indians had developed corn culture to such a high state that it is thought to have then ranked highest among cereals in efficiency of food production (6). The origin of rye is more difficult to ascertain. There appears to be no reference to it prior to the Christian era. It is still a principal ingredient in the diet of some peoples in northern Europe as it was in the UK until the late eighteenth century (7). Since barley grows wild in the Syria–Palestine area, it is likely to have originated there (7). Millet may have originated in the Sudan where pearl millet is still widely cultivated (3).

Botanically, cereals are simple, one-seed fruits. They include wheat, rice, corn (also called maize in some parts of the world), rye, barley, oats, sorghum, and millet. All cereals contain large amounts of starch and little fat; the fat is associated primarily with the germ and scutellum (single cotyledon or the first leaf). In most cases, the lipids in cereals contain a high concentration of unsaturated fatty acids, which are protected from oxidation by the presence of tocopherols. Actually, the oil from wheat germ was the starting material for the isolation of α-tocopherol (see Vitamins, vitamin E). As long as the antioxidants are in proximity to the lipids, oxidation of the lipids and the accompanying rancidity is minimized. This is one reason why wheat is stored in the kernel stage until it is milled. Generally, wheat is milled only when there is an order for flour. However, even in the intact kernel, the lipids are subject to hydrolytic rancidity (8), which together with oxidative rancidity can be estimated by determining the free fatty acid content. That is done by determining the amount in milligrams of potassium hydroxide required to neutralize the free fatty acids from 100 g of moisture-free grain (9). This is called the fat acidity value. One of the main purposes of milling, in addition to producing an acceptable flour, is to remove the wheat germ in as intact a state as possible. Thereby, the concentration of the lipids is reduced, the development of oxidative rancidity decreased, and the shelf life of the flour extended.

Cereals were among the earliest plants cultivated. They were related to some of the wild grasses indigenous to those parts of the world where civilizations had their origins. All of them are ideally suited for use as food under both primitive, and advanced conditions. For one thing, they can be stored for long periods under many conditions, thus providing a reserve against food shortages. This was done in ancient Egypt when Joseph advised the Pharaoh to store grain during the seven "plenteous" years to be used in the following seven "lean" years. In addition, cereals are nutritious foods that can be used in many ways, thus facilitating their incorporation in the diet at high levels over long periods of time. Even today, there are some areas, such as rural Iran, where wheat products, especially in the form of the flat breads indigenous to that region, provide as much as 70–90% of the daily caloric intake. Most of the cereals also respond well to primitive methods of agriculture with good yields and, with advanced technology and improved varieties, large yields per worker can be secured. Much more food is secured from fields planted in grains than can be obtained from cattle or other animals on the same land. The importance of this became evident during World War I. In the early stages of that conflagration, the German High Command made a conscious decision to continue the prewar level of meat, milk, and egg production to provide adequate nutrition for the men in the armed services and for the civilians who would be called on for arduous work in connection with the war. Had they decided instead on conversion of meadows and pastures to wheat fields and the direct consumption of the grain by human beings, the yield of food for the German people would have been far greater than it was when the land was given over to raising cattle, pigs, sheep, and poultry. This led a group of Scottish physicians to suggest that this mistake probably did more than any army general to lose the war (10). Finally, cereals grow in a wide variety of climatic and soil conditions, and they successfully compete with weeds for the limited amounts of nutrients and water where these plant factors are in short supply. This is an important reason why cereals have played, and continue to play, such an important role in the development of the human race.

Nutritional Value of Cereals

Deficiency Diseases. Not only did cereals make an important contribution to improving the general status of humankind, but they also were important dietary components of some groups of people who showed certain nutritional deficiencies. This observation led to the discovery of some of the vitamins. These deficiency diseases have been most prominently associated with use of rice, corn, and wheat.

Beriberi, Thiamine Deficiency. The recognition of vitamins and their importance to the health of human beings came about when Eijkman, a Dutch pathologist, was sent to Java in an attempt to cure an epidemic of beriberi that had appeared in one of the hospitals. Eijkman kept a flock of chickens on the hospital grounds to assist in discovering the disease agent he assumed was involved in the etiology of beriberi. These chickens were fed the scraps from the plates of the hospital patients—primarily polished rice, the common food in that part of the world (11).

Although Eijkman recognized the condition in his polyneuritic chickens as the analogue of beriberi, he misinterpreted its cause. He suggested on the basis of bacteriological concepts then dominating the medical field that beriberi resulted from the ingestion of a toxic substance associated with the starchy part of the rice. According to that theory, rice polishings contained an antitoxin which neutralized the toxin present in the polished rice. It was Eijkman's successor who established that polished rice lacked some substance essential for normal physiological functioning and that this substance occurred in rice polishings, beans, meat, and other foods (12). The critical ingredient is now known to be thiamine (vitamin B_1). The absence from the diet of similar substances is responsible for the development of scurvy, pellagra, rickets, etc (13).

Pellagra, Niacin Deficiency. It was 220 years after the first description of pellagra that nicotinic acid was discovered to be the cure for black tongue in dogs (14), a condition suggested by a veterinarian in North Carolina to be similar to human pellagra (15).

The contrast between a high incidence of pellagra among the inhabitants of the southeastern United States and its absence among the corn-eating people south of the Rio Grande was a puzzle for a number of years. It finally became evident that the lime water in which the corn kernels were steeped prior to being made into tortillas liberated nicotinic acid from the bound form (16). Untreated kernels of corn contain nicotinic acid as a complex from which it cannot be made available by human or animal digestive processes. However, a weak alkali, such as lime water, releases nicotinic acid from its complex and makes it available for absorption. Furthermore, the usual diet of the corn eaters south of the Rio Grande includes a large complement of beans. Beans are a rich source of tryptophan which can be converted to nicotinic acid by enzymes in the human body (see Vitamins, nicotinic acid and nicotinamide (B_3)).

Zinc Deficiency. A nutritional problem associated with consumption of large amounts of whole wheat products is the unavailability of dietary zinc, first observed in a patient with immature development and dwarfism in southern Iran (17). The patient showed marked improvement when placed on a well-balanced, nutritious diet for a year. In 1962, similar patients were observed in Egyptian villages. A deficiency of zinc was identified as the primary reason for the development of this condition. This deficiency of zinc results from the binding of that metal by the phytates present in whole wheat (18). Even when an excess of zinc is present in the diet, if conditions are right, that zinc may be complexed with the phytate and thus rendered unavailable to the body (see Mineral nutrients). Thus, a zinc deficiency may develop when its dietary level appears adequate but the diet contains large amounts of a food, such as whole wheat, that is high in phytates (see Vitamins, inositol).

Calcium Absorption. Phytates in cereal grains have also been reported to interfere with the absorption of calcium. However, a long-term study indicated a retention of calcium in subjects that consume large amounts of bread made with high extraction of flour (19).

Lysine Concentration. Many nutritionists are concerned about the low concentration of lysine in wheat proteins (see Amino acids). Initially, the insufficiency of lysine in wheat was noted because of the failure of weanling rats to grow when fed a ration in which white wheat flour provided most or all of the protein in the diet. The rats grew only when the white flour ration was supplemented with lysine or a protein source containing a relatively high concentration of that amino acid. Animal proteins were found to be good sources of lysine. That led nutritionists to advocate the use of wheat products in conjunction with animal proteins in human diets. The idea that human beings require not only cereals but also animal foods in their diet provided the theoretical basis for the development of American agriculture and much of the nutritional advice given the public.

However, the validity of rat studies for human nutrition has been called into serious question because of the vast difference in rates of growth of rats and humans (20). The daily weight gain of the infant rat is about 3% of its weight; the comparable rate for the human infant is 0.03%. The 100-fold difference in relative weight gains makes it impossible to evaluate food intended for human beings on the basis of its ability to provide good growth in rats (21).

Health Advantages of High Cereal Consumption

Blood Urea Levels and Kidney Function. One of the most intriguing observations made during a seven-week "bread" study was the reduction in blood urea level of the subjects (22). In normal individuals, the blood urea level was believed to be directly proportional to the amount of protein consumed (23). The normal blood urea concentration of individuals consuming 70 g of protein per day was projected as 32 mg/100 mL, and that was the level observed in the subjects at the end of the three-week control period. However, sometime after the "bread" diet was started, the blood urea level dropped to one half of the control value even though there had been no change in the level of dietary protein (22). Although there had been no change in the intake level of protein, the study did involve a complete elimination of animal protein during the "bread" diet period.

A marked reduction in the blood urea level of normal individuals implies an improvement in kidney function (see Urea; Diuretics). Such was observed in a subsequent study (24). Urea and creatinine clearances (clinical measures of kidney function) of six normal college men showed a 25% and 10% improvement, respectively, when they consumed a diet high in bread for seven weeks. This improvement occurred even though the tests at the end of the three-week control period were normal. Throughout the study, the protein level in the diet was maintained constant at 70 g/d. During the control period, the protein came from a mixture of plant and animal sources but during the "bread" period, ca 90% of the dietary protein came from wheat and animal protein was completely excluded. The improvement in kidney function associated with the consumption of a diet high in wheat products suggests a possible therapeutic value for wheat. That might be especially true for those patients who are starting to manifest aberrations in kidney function which ultimately would require renal dialysis or a kidney transplant.

Prevention of Osteoporosis. Another health benefit associated with the consumption of a diet high in cereals is the maintenance of bone-mineral content, especially among older women. After the menopause, women are subject to demineralization of bones, especially in the pelvis and spinal column. This condition, called osteoporosis, leaves them vulnerable to hip fractures. Despite operative procedures to correct the fracture, one sixth of all patients with hip fracture die within three months of injury. Some degree of deterioration in total life adjustment is a nearly inevitable occurrence in the five sixths who survive (25). Osteoporosis in women has been associated with a reduced estrogen production after the menopause. That plus the fact that animal studies indicated an involvement of estrogen in depositing calcium in the bones led to use of this hormone in treatment of osteoporotic women (26) until recent reports suggested a connection between estrogen therapy and endometrial cancer (27).

Since dietary cereals are low in sulfur-containing amino acids, they produce an alkaline urine which favors the retention of bone minerals. In postmenopausal women, there appears to be some interaction between the diet and the effect produced by estrogens on bone mineral content (28).

Prevention of Obesity. Cereal products may have an important role in obesity control because of the low concentration of fat in cereals and the fact that the starch in wheat flour acts, in the human body, somewhat like dietary fiber (29). Both the restriction in dietary fat and the presence of dietary fiber aid the individual in reducing caloric intake with a minimum of effort, and decrease the hunger reaction that plagues most people attempting to lose weight. Many people believe bread and cereal products are fattening, although bread frequently serves as the innocuous vehicle for the conveyance to the mouth of various fattening foods.

When the intake of fatty foods is curtailed, a large amount of bread in the diet is likely to produce a weight loss. For example, eight college men, eating 12 slices of commercial white bread per day, lost an average of 6.3 kg over an eight-week period (30). All of these men were from 10–15% overweight. Despite the continued loss of weight over the eight-week period, the few men in the group who experienced hunger reported it to be mild (30).

Dental Caries and Isomerase. Susceptibility to dental caries is a genetically determined trait. In susceptible individuals, the incidence of dental caries is directly related to the amount of sucrose in the diet (31–33); however, there are a number of factors that may modify the incidence. Although sucrose in the diet produces a high incidence of dental caries, its component monosaccharides used at an equivalent level in the diet result in a much lower incidence (see Sugar). It appears that the oxygen bridge linking glucose and fructose to form sucrose is needed as a source of energy by the bacteria (probably *Streptococcus mutans*) responsible for the development of the carious lesion.

The development of high fructose corn syrup (HFCS) may provide another health benefit attributable to cereal grains (see Syrups). These syrups are being used to an ever-increasing extent by the food industry. Shortly after HFCS came on the market, a dramatic increase in the price of sucrose acted as a spur to the production of the high fructose syrups. By 1976, 953 metric tons of corn syrup was being used in foods produced in the United States.

The high fructose syrups are produced from corn starch after the latter has been hydrolyzed to glucose by either acid or α-amylase. Treatment of the resulting fluid with an isomerase converts ca 42% of the glucose to fructose. The remaining solids consist almost exclusively of glucose and other saccharides (34). After clarification, this product is concentrated to a base of 71% solids, in which state it is bulk-distributed. The enzymes involved in the isomerization are of bacterial origin, most of them belonging to the *Streptomyces* group. The enzyme is immobilized by fixation on cellulose (see Enzymes, immobilized).

A few years ago, a solution became available with a 90% fructose content. Because of the sweetness of fructose, this syrup imparts the same sweetness to a product as a larger amount of sucrose (see Sweeteners). Since on a weight basis the monosaccharides have the same number of food calories as sucrose, this gives a means of reducing the caloric content of the food. Reduced caloric foods, such as jellies and preserves, salad dressing, and table syrup, are now being manufactured with 90% fructose corn syrup and are recognized as having quality improvements compared to the use of other sweeteners (34).

Health Problems Associated with the Consumption of Cereals

Celiac Disease. A disturbance of the lower gastrointestinal tract, celiac disease is a chronic disease characterized by loss of appetite and weight, depression and irritability, and diarrhea frequently followed by constipation (35). One of the more disturbing features of celiac disease is the large, frothy, foul-smelling stools. The disease may develop in childhood or later in life. Frequently, the patients who develop the disease in adulthood report having had some of the symptoms during childhood.

This disturbance was recognized shortly after World War II as being related to the ingestion of wheat. A group of physicians in the Netherlands was impressed by the fact that during the war they saw many cases of celiac disease. During that time, wheat was the primary staple of the diet. However, at the end of the war, other foods again became available and the number of children who developed celiac disease decreased. One Dutch group of investigators had a seven-year-old female patient who displayed extreme fluctuations in her symptoms. These changes were shown to be associated with the presence or absence of bread in her diet. Using that patient as the test subject, the group soon learned that her symptoms worsened shortly after she consumed foods containing wheat gluten (36). Although considerable work has been done attempting to identify the mechanism whereby gluten causes the disturbance characterizing celiac disease, little more can be said beyond that this appears to be an allergic reaction. For victims of celiac, the primary therapy involves the elimination of any dietary product that contains gluten. Since that protein occurs not only in wheat but also in rye and barley, the complete elimination of gluten from the diet becomes an onerous task.

Disturbances Associated with Flour-Aging Agents. Bread made with flour freshly milled from wheat has a different texture and appearance that we have come to associate with bread. Only after the flour has been aged by storage for a few months does the gluten develop the toughness that permits the bread to retain the shape it had when it was put in the oven. The aging of the flour likely involves a mild reaction of oxygen and some part of the gluten molecule. Additional evidence for this is the destruction of the yellowish pigment present in the flour prior to its storage. To hasten the reactions that occur during storage, chlorine gas and oxides of nitrogen were tried. Although they improved the freshly milled flour, they were not completely satisfactory. In 1921, nitrogen trichloride was introduced as a flour agenizing substance (37).

Although no untoward effects were observed that could be attributed to the long-term ingestion of agenized flour by people in the United States and the UK, and despite the absence of any symptoms in human subjects given large amounts of flour heavily treated with nitrogen trichloride prior to incorporation into foods, the use of nitrogen trichloride by the milling industry in the United States was prohibited after

Aug. 1, 1949 (38). That order recommended the use of chlorine dioxide for agenizing flour. Like nitrogen trichloride, chlorine dioxide is unstable and must be passed into the flour immediately after being generated. The chlorine dioxide is formed by passing chlorine mixed with air into a solution of sodium chlorite (39) (see Chlorine oxygen acids and salts).

Ergot. Ergotism is only indirectly related to health hazards associated with the consumption of cereals because the alkaloid which is responsible for ergotism is produced by a fungus (*Claviceps purpurea*) that grows primarily on rye. That the fungus may grow on other cereals than rye is implied in a report of this disturbance among the members of a British family who consumed no rye bread (40) (see Alkaloids).

The disease takes two different forms depending apparently upon whether ergotoxin, the alkaloid in ergot, attacks predominantly the nervous or the circulatory system. The former condition is characterized by severe convulsive seizures; the latter produces an intense burning and itching of the skin called St. Anthony's Fire.

Abortion is one of the prominent characteristics of ergot poisoning among both women and cattle. This is due to the action of ergotoxin in contracting the placental muscles. For that reason, small, regulated doses of ergotoxin are used to control bleeding following childbirth (see also Hormones, posterior pituitary hormones).

Colonic Cancer. Another area in which cereals may play an important role in maintaining the health of human beings is as a source of dietary fiber (qv). During the past decade, evidence has been collected that certain diseases, to which people in the more-developed countries are prone, owe their origin to the almost complete removal of the fibrous parts of cereals consumed in the diet. There is an almost complete absence of a variety of tumors in the lower gastrointestinal tract among the natives of African villages, whereas the incidence of these colonic tumors among those of European extraction living in the area is as high as in northern Europe (41). The primary, visible difference between these two groups of people is the nature of their diets. The native Africans consume a diet composed largely of unrefined foods and consequently high in fiber. On the other hand, the neighboring people of European origin rely on foods similar to those available in their countries of origin. The foods eaten by the latter group are either devoid of, or very low in, dietary fiber.

Prior to this work, dietary fiber, of which cellulose is one of the more important constituents, was considered important primarily as a means of preventing or overcoming constipation. Otherwise, dietary fiber was considered to be a metabolically inert substance. A large variety of diseases such as appendicitis, hiatus hernia, gallstones, ischemic heart disease, diabetes, obesity, dental caries, and duodenal ulcers are now suspected to be associated with the consumption of a highly refined diet (42).

Diabetes. Fiber has also been shown to be an aid in treating diabetes. Diabetic patients who required 20 or less units of insulin per day could do without their insulin injections provided they increased the amount of cereal bran in their diets (43). These patients had adult onset diabetes, a milder form of the disease than that which appears in children. Nevertheless, the fact that a high intake of bran relieves these patients of the need for daily insulin injections is a welcome new form of therapy. Practically all evidence indicates that dietary fiber decreases the maximum blood glucose level following a dose of glucose as in the typical glucose tolerance test. The lower level of blood glucose presumably occurs because its absorption from the intestinal tract is slowed. The mechanism whereby dietary fiber decreases the rate of absorption of glucose from the intestinal tract is not completely understood. One possibility is that the large molecules of dietary fiber physically impede the movement of the glucose molecules to the absorptive surface of the intestine. Another theory suggests that dietary fibers increase the viscosity of the intestinal contents and thereby slow the movement of glucose molecules (44).

Cardiovascular Diseases. For sometime it was believed that dietary fiber protected the individual from various heart disturbances, especially ischemic heart disease. That presumably occurred as a result of the action of dietary fiber in lowering serum cholesterol levels. This effect resulted, according to reports, from the increased rate of passage of food through the intestinal tract and consequent incomplete absorption of dietary cholesterol and some bile acids. Since some of the bile acids are secreted into the lower intestinal tract to be reabsorbed at a lower site and since cholesterol is the precursor of bile acids, this would lower blood cholesterol levels. However, some investigators now suggest that it may be other differences in dietary constituents or lifestyle that are responsible for the low blood cholesterol levels among primitive people whose diet is high in fiber (45). Inclusion of large amounts of high fiber foods in the diet may decrease the intake of cholesterol-containing foods and those high in saturated fats. The intake of these two dietary components is probably closely related to the level and nature of blood cholesterol (45).

The implications of the role that dietary fiber may play in the maintenance of human health should increase the demand for cereal products. It may well have an influence on the milling industry, especially if the demand for cereal products that are less highly milled becomes a reality.

Bread Consumption and Income

Many people believe that large amounts of bread and cereals in the diet are associated with inadequate nutrition and occur primarily in the economically impoverished segment of the population (46). Contrary to this general belief, the report of the McGovern Committee on Dietary Goals recommended for the United States a marked increase in the cereal products incorporated into the human diet (47). Such a dietary change would aid in decreasing the incidence of cardiovascular diseases as well as a number of others such as diabetes. If this recommendation has an effect on eating patterns in the United States, it will reverse the decreasing per capita consumption of flour that has been characteristic of this country for the past 50 yr. This reduction in flour consumption has been attributed to the improvement in the standard of living and the decreased amount of strenuous physical work performed by the average individual (48).

That high bread consumption is associated with poverty is a dictum that has been accepted by many. One treatise emphasizes that bread is the staple food of poverty (49). The statistics for bread consumption in the United States do not conform to that pattern. The USDA's *Decennial Survey of Food Consumption 1965–1966* indicates an increase in bread intake as the per capita amount of money spent on food goes up (assuming that the money spent on food is related to total income). On that basis, there is an almost linear increase in bread consumption from 0.27 kg per person per week for members of families spending less than $3.00 (in 1965) per person in a week for food, to 0.603 kg for those families spending $7.00–$8.99 (in 1965). Thereafter, the amount of bread eaten continues to increase but at a slower rate, to 0.81 kg per person in families spending ≥$20.00 (in 1965) per person per week for food (50). These trends in bread consumption hold true for both urban and rural families. Among the rural families, the change in bread consumption with the amount of money spent on food showed a similar trend, but the amount of bread eaten was less among the families spending the least money for food and considerably higher (1.195 vs 0.81 kg) for the families spending >$20.00 per person per week on food (in 1965). This is not a recent phenomenon since all of the reports starting with the first one in 1936 show similar trends in bread consumption.

This change in bread consumption is likely to be accentuated as the world's available resources become more limited. Although the primary emphasis to date has been on the impending energy shortage, freshwater supplies may become an even graver problem (see Water). The amount of water required for food production reaches stupendous values that vary tremendously for different kinds of foods. The water requirements for the production of different foods have been vividly depicted (51). The daily food consumed by the average person in the United States has been calculated to require ca 13.25 m^3 (3500 gal) water. Of that, 80.5% is needed for the production of animal foods such as meat, milk, eggs, etc and only 8.5% for the plant foods. As these foods are processed for human consumption, the water and energy requirements become even greater for animal foods; much less is needed for cereal grains. The grains are harvested in a semidehydrated state produced by the sun's energy. They can be transported and stored under conditions requiring much less energy than the water and refrigeration needs of animal foods. These may be some of the reasons why cereals constitute such a large proportion of the food consumed by people in the less-developed countries.

Nutritional Significance of Parboiling Rice

Rice has obviously played an important role in the diet of large numbers of people. The concentration of protein in white or polished rice is 5.3–13.4% (52). The value used most frequently in nutritional reports is 6.5% (53), a level well below the ≥12% of the food calories in the diet that are recommended to come from protein. A protein-rich animal product such as fish is important to the people who consume rice as the primary component of their diet. Although the concentration of protein in polished rice is low, rice protein does have a higher biological value than that assigned the protein in wheat flour because the essential amino acids in rice are present in proportions similar to their relative needs by human beings (52).

Parboiling involves soaking the whole rice kernels in warm water overnight and then drying the rice in the sun. During this process, the hulls are split open by the swelling of the grain. After drying, the endosperm can be separated easily from the hulls (54–55). This procedure overcomes the problems associated with previous methods of removing the hulls. Parboiling is also accomplished by steaming or cooking the intact rice grains. The higher nutritional value of the whole rice kernel versus polished rice was recognized long before the dawn of the vitamin era, but it is questionable whether that was the primary motive for the development of the parboiling process.

Parboiling rice has been practiced primarily by the Indians both in India and in those tropical countries to which they migrated. None of the other Asiatic people appear to have accepted parboiling as a step in rice milling. Apparently, the reason is the flavor of parboiled rice is offensive to many people (55). One nutritional advantage of parboiling is that thiamine and some other essential nutrients normally present in the bran migrate to the endosperm during the water soaking stage of the process (56). As a result of parboiling, the thiamine content of the rice is increased to 0.34 mg/100 g from 0.07 mg/100 g in regularly milled rice (57). The concentration of the other water-soluble vitamins is increased by a factor of 2–3 when rice is parboiled. Parboiling under well-controlled conditions is practiced by one or two rice millers in the United States (58). The product from the mill in Texas is sold throughout the country.

Beriberi was the primary deficiency disease formerly very common among those people who consumed large amounts of rice. The work on beriberi led ultimately to the isolation of thiamine. Not only is polished rice deficient in thiamine, but it contains practically no vitamin A, ascorbic acid, or vitamin D (57). One reason why a deficiency of thiamine becomes manifest before the disturbances with deficiencies of the other vitamins is that the body's stores of thiamine are more rapidly depleted than those of any of the other vitamins.

Enrichment of Cereal Products

During the 1930s, considerable prominence was given in the general press to the existence of various vitamin-deficiency diseases, especially beriberi, among various groups of people in the United States, such as alcoholics and prison inmates. There probably were various reasons for this, including the severe limitation placed on food purchase during the Depression. Furthermore, the emphasis on vitamins at that time very likely led to a diagnosis of vitamin deficiency among many people entirely on the basis of visual signs and easily described symptoms. This was understandable because during the period when many of the vitamins were being discovered there were practically no chemical tests to confirm or deny the presence of a deficiency disease.

With the entrance of the United States into World War II, one of the first presidential orders was for the enrichment of wheat flour and cereal products with certain vitamins and minerals. The standard of identity for enriched white flour became effective Jan. 1, 1942 (59). This order set minimum and maximum levels for thiamine, riboflavin, niacin, and iron added to flour and other cereal products. The addition of calcium and vitamin D were listed as optional ingredients. Since, at that time, synthetic riboflavin was not available in amounts to supply the requirement, the addition of that vitamin was delayed until adequate supplies became available. The standards, listed in the order promulgated by President Roosevelt early in World War II, were in effect throughout the war. In 1943, the FDA published a proposal according to which standards of identity were to be established for breads and cereal products. At the end of the war, lengthy hearings were held on the suggested standards. It was not until May 1952 that these standards were published to become effective in August of the same year (59). The FDA standards were incorporated into legislation passed by 34 states. The large number of states requiring compliance with the FDA regulations made it feasible to enrich all flour produced by a miller engaged in interstate commerce.

The standards of identity were established primarily to ensure that the name on the label of a product had a specific and uniform meaning. One of these terms was enriched. When that is used on the wrapper or container to describe a food, it signifies that the nutrients added to the food are in compliance with the standards for that food established by the FDA (60). The theoretical basis for the levels of nutrients in enriched foods is that the final levels in the processed food will be equal to but not exceed that in the food at the time of harvest.

On the basis of a number of dietary surveys and the results of some biochemical tests, the Food and Nutrition Board of the NAS in 1974 suggested increasing both the levels and number of nutrients added to cereals (59). The rationale for this suggestion was that both the dietary and biochemical data indicated a significant risk of deficiency of vitamin A, thiamine, riboflavin, niacin, vitamin B_6, folacin, iron, calcium, magnesium, and zinc among significant segments of the population (59). On that basis, the levels of nutrients in Table 1 were suggested for wheat, corn, and rice products where they were technically achievable.

The report justified fortification of cereal products since these foods constitute ca 26% of the daily caloric intake (ca 17% on a flour-equivalent basis). Percentages vary, but not significantly, relative to income and geographic region (59).

The addition of these nutrients to flour raised the question of their stability. Of primary concern would be the fate of some of the more labile vitamins when subjected to the various agents used to prepare flour for the commercial market. The results of such an investigation indicated that even when the vitamins are added to the flour prior to treatment with the various compounds, only vitamin B_6 underwent any significant destruction. Treatment of soft wheat flour in which it was incorporated destroyed 16% of the added vitamin B_6 (see Vitamins, vitamin B_6). Treatment of the flour with azodicarbonamide, potassium bromate, benzoyl peroxide, chlorine dioxide, or acetone peroxide produced no such destruction (62). In normal milling practice, the vitamins and minerals are added to the flour just before packaging. Thus, the vitamins do not come in direct contact with these compounds, so that vitamin destruction is minimized.

Although vitamin E is not among the nutrients suggested for addition to flour, when added as α-tocopherol acetate, it is more stable to the agents used in treating flour than the vitamin normally present therein. As much as 92% of the vitamin E originally present in flour is destroyed by chlorine and as much as 86% by chlorine dioxide.

Table 1. Proposed Enrichment of Cereals [a]

Nutrient	mg/100 g
vitamin A (retinol equivalent [b])	0.48
thiamine	0.64
riboflavin	0.40
niacin	5.29
vitamin B_6	0.44
folic acid	0.07
iron	8.81
calcium	198.20
magnesium	44.10
zinc	2.20

[a] Ref. 61.
[b] See Vitamins, vitamin A.

The Green Revolution

Beginning about 1940, efforts were initiated in various parts of the world to improve the yield of some of the basic food crops. The objective was to provide an adequate food supply for the large numbers of people in many of the less-developed countries (63). As a result, the food supplies in many countries have been markedly increased. For instance, in Mexico, importation of corn averaged 48,000 t/yr for the period 1941–1945. In the period of 1962–1966, domestic requirements were met and the surplus, averaging 496,000 t/yr, was exported (64). This was associated with doubling the yield of corn per hectare. A similar situation existed for wheat. For that cereal, Mexico imported an average of 253,000 t/yr in 1941–1945. That changed for the period 1962–1966, when an average of 276,000 t/yr was exported.

The high yielding strains of plants developed in this program are well adapted to the conditions where the crops are to be grown. These crops must be resistant to the indigenous diseases and insects; they must show favorable responses to fertilization and irrigation where that is necessary; they should be photoperiod-insensitive so they can be planted at various times throughout the year; they should be of improved nutritional value; and obviously they should be acceptable to the ultimate consumer. The latter was one of the problems of some of the early high yielding varieties of rice developed for consumption in the Orient.

The development of dwarf varieties of grains was necessary for a number of reasons. The long-stemmed cereals frequently suffered from lodging, especially if a high wind or heavy rain occurred just before harvest. The resulting loss of crop or the extra effort required to harvest the lodged grain made dwarf varieties which were not subject to lodging highly desirable. Furthermore, under conditions of limited plant nutrients, including water, a reduction in stem length would provide the developing grain with an extra supply of growth factors, thus facilitating a greater yield. One of the problems of fertilizing many of the long-stemmed varieties of cereals was that this frequently led to increased vegetative growth—and a greater possibility of lodging—rather than to increased yield (65).

Short varieties of wheat were known for many years but as long as straw was used as a bedding agent, especially for animals, and as a feed for ruminants, there was little interest in their use. However, in Japan, semidwarf wheat was recognized as early as 1800. Some of these varieties were imported into Italy in the next century and were used to produce improved varieties. One of these, Norin 10, was brought to the United States in 1946, where it was crossed with U.S. varieties, and taken to Mexico in the early 1950s. There the Norin-Brevor cross, in addition to some Italian varieties, was used by N. Borlaug (Nobel laureate, 1969) and associates to develop new Mexican varieties (66).

In a similar manner, the progenitors of dwarf varieties of rice were known to the Chinese as long ago as 1000 AD. Some of these were used at the International Rice Research Institute in developing the IR-8 and subsequent strains of dwarf rice (65).

Triticale

Until the development of triticale, no new cereal had been added to food crops for thousands of years. The ability of rye to grow under relatively adverse climatic and environmental conditions and the recognition that its genetic profile was not too different from that of wheat suggested the possibility of crossing these two cereals. The name triticale is derived from the class names for the two parent plants: *triti* from *triticum* for wheat and *cale* from *secale* for rye. An early attempt to cross-breed wheat and rye was reported in 1876 but the resulting hybrid was sterile (67). Shortly thereafter, a spontaneous hybrid of wheat and rye that was partly fertile was described in Germany. One reason for the infertility of these early hybrids was that the seed had only one set of the wheat and one set of the rye chromosomes. That problem was overcome when, in 1937, colchicine treatment of seedlings produced a doubling of their chromosomes, thus facilitating reproduction (67). Unless the chromosomes can be paired during meiosis, the seed is sterile.

Another technique used in the development of triticale has been embryo culture, which involves removing the embryo from the seed 10–12 d after pollination and transferring it to an agar medium containing various salts essential for growth and a number of plant hormones (67). This procedure permits culturing the plant cells in the presence of various compounds that may improve such functions as resistance to disease or insects, or may increase the concentration of certain plant constituents, such as the nutritionally essential amino acids that may be present in inadequate amounts in the parents (68). This procedure, which has been applied to rice, is also used with triticale.

Much of the developmental work on triticale has been done at the University of Manitoba, Canada. From there, several varieties were sent to CIMMYT (International Maize and Wheat Improvement Center) in Mexico for tests to see how these plants were adapted to semitropical regions. During that work, the pollen from an adjacent wheat plot was deposited on the stigma of a triticale plant. The genes in the wheat pollen converted the tall, semisterile triticale plant into a dwarf variety which was partially photoperiod-insensitive and left the plant fertile. This new strain, called Armadillo, has been developed into a strain of greater fertility, better test weight, higher yield, better insensitivity to the photoperiod, faster maturation, shorter straw length, higher resistance to lodging, and better nutritional qualities than older varieties of triticale (67).

Although triticale has some advantages over wheat and rye, certain problems will have to be overcome before this plant begins to compete with either of its parents. For one, the seed is shriveled to such an extent that it can be milled with modern machinery only at the expense of losing large amounts of the endosperm with the bran (69). The shriveling of triticale seeds has been related to the prolonged action of α-amylase, which breaks down the starch and produces shriveled seeds (67).

There is still some question as to whether triticale flour, by itself, will make an acceptable loaf of bread. However, there are reports that triticale is satisfactory for use in such products as bread, pastas, tortillas, chapattis, and infant foods (70). An increasing amount of farmland is being devoted to triticale especially in Ethiopia and some South American countries; in the latter region, the crop is used primarily for animal feed. In Manitoba, Canada, over 1200 ha (ca 3000 acres) of triticale has been raised for experimental use by the liquor industries (71).

Cereal Production

More than two thirds of the world's cultivated area is planted to cereal grains; in the less-developed countries, the fraction is even larger. Wheat, rice, and corn account for almost three fourths of the world's production (Table 2). Over the past 50 yr, except for oats and barley, there has been a steady increase in world production of all cereal grains. Part of this increase reflects the improvements in yields brought about by the Green Revolution.

The kind of cereal that predominates in any part of the world depends upon the adaptation of the plant to the climatic and environmental characteristics of that region and, to a certain extent, on consumer demand and political decisions. In some of the western African countries, for example, urbanization has been accompanied by an increased demand for rice. During the 1960s, the People's Republic of China was selling rice on the world market, not because it had a surplus, but for political and economic reasons. At that time, rice commanded a higher price than wheat, and thus by selling rice and buying wheat, the People's Republic of China increased the food available to its people. That was especially true of protein which is present in wheat at almost twice the level in rice (73).

Wheat. Wheat is adapted to the temperate zones and as such is an important crop in the United States. Even in the People's Republic of China and India it is grown in the northern parts of each country, whereas rice is limited largely to the southern regions. The difference in the protein content of these two cereals and their distribution were the basis for the explanation for the taller and sturdier Punjabi versus the smaller and leaner Madrasi peoples of India (74).

Although the yield of wheat per hectare in the United States is 1.5 times that in the USSR, the greater area devoted to wheat cultivation in the USSR leads to a total production of wheat 1.5 times that in the United States (75). The production of wheat in the USSR within the past decade has been plagued by both political and environmental problems. One of these was the decision to develop lands in the Asiatic segment for wheat cultivation and the lower-than-anticipated production in those areas. One of the primary environmental factors that limited wheat production in the USSR was a prolonged drought which culminated in the purchase of large amounts of wheat from the United States, Canada, and other western countries (76). This resulted in an

Table 2. World Production of Cereals, 10^6 Metric Tons per Year[a]

Crop	1948–1952 10^6 t	% Total	1965 10^6 t	% Total	1977 10^6 t	% Total	1980 10^6 t	% Total
wheat	171.2	25.1	265.8	26.3	390.7	26.7	444.5	28.6
rice	166.4	20.4	254.2	25.1	369.7	25.3	399.7	25.7
corn (maize)	139.4	24.4	226.2	22.4	346.2	23.7	392.2	25.2
oats	62.0	9.1	46.8	4.6	51.2	3.5	42.6	2.7
barley	59.0	8.6	104.8	10.4	176.3	12.0	162.4	10.4
millet–sorghum	47.4	6.9	78.0	7.7	103.6	7.1	87.3	5.6
rye	37.7	5.5	35.4	3.5	25.4	1.7	27.4	1.8
Total	683.1		1011.2		1463.1		1556.1	

[a] Ref. 72.

agreement with the USSR whereby that country agreed in 1975 that, during the succeeding five years, it would purchase a minimum of 6×10^6 metric tons of grain per year from the United States.

During the past 30 yr, world production of wheat, rice, corn, and barley has increased by factors of 2.4–2.8. In the same period, the production of millet and sorghum increased 1.8-fold, whereas that of rye and oats decreased (Table 2). In the last decade, cereal production in some countries has increased so much, especially as a result of high yielding strains, that nations have been converted from importers to exporters of cereal grains. Despite the contribution to the international wheat market by Canada, Australia, Argentina, and France, the United States still dominates that trade, as shown by the fact that in 1976, 40.5% and, in 1978, 46% of the wheat in international trade came from the United States (77). In 1979, the countries that imported large amounts of wheat from the United States were, in decreasing order: Japan, the Republic of Korea, Brazil, Chile, the Philippines, Venezuela, the Republic of China, Colombia, Peru, Portugal, and Italy. Each of these countries imported more than 500,000 t of U.S. wheat in 1979, with the first three countries importing 2,336,000–1,943,000 t. Japan purchased about 10% of the U.S. wheat sold abroad in 1978. In 1978, 14.6% of the wheat exported from the United States went for relief and charity, an increase over the 12.1% for the same purpose in 1977 (78).

In the next few years, the export of wheat by the United States may become even larger because some areas of the world such as the USSR and the countries in East Asia, the Middle East, and Africa, are increasingly dependent on food imports (79). These countries are projected to account, by 1988, for as much as two thirds of all grain imports (79). Of the principal wheat-exporting countries (United States, Canada, Australia, and France), Canada and the United States may have to provide 85% of the grain needed (79).

Different kinds of wheat are grown in various parts of the United States (Table 3) and those types are suited to different purposes. Most white flour is made from a mixture of different wheats. The proportion of each kind of wheat depends on a

Table 3. Wheat Varieties in the United States

Variety	Where grown	Primary use
Triticum aestivum (common wheat)		
hard red spring	Minnesota North and South Dakota Montana	flour for yeast breads
hard red winter	Kansas Nebraska Oklahoma Texas	flour for yeast breads
soft red winter	Ohio Indiana	flour for chemically leavened baked goods; all-purpose flour
white wheat	Pacific northwest	flour for pastries; all-purpose flour
Triticum durum		
red durum	North Dakota Minnesota	poultry and livestock feed
amber durum	North Dakota Minnesota	macaroni; semolina

number of factors, the principal one being the amount and properties of the protein it contains. The best flour for yeast-raised breads has a high protein content and a firm but elastic texture of the gluten proteins when wet. This flour is made primarily from the hard red wheats. These varieties are termed "hard" for the nature of the kernel which is vitreous; the hardness of the wheat is associated with the protein content of the wheat—the harder the wheat, the higher the level of protein. The redness of the wheat is apparent in the intact kernel, and the color is most prominent in the outer layers, but some carotenoid pigments are also present in the endosperm. These pigments are destroyed during the mild oxidation of the flour which is done primarily to "age" the gluten, thereby improving the appearance of the loaf of bread baked with that flour. Most millers make at least two kinds of flour intended primarily for breadmaking. The flour in the smaller packets (≤12 kg) usually has about 10.5% protein; it is milled to meet a variety of domestic needs. The flour sold in larger containers has 11.5% protein. The latter is used primarily by commercial bakers who need a flour that, in the dough stage, will withstand the mechanical agitation involved in large-scale breadmaking.

Years ago, wheat from certain parts of South Dakota was downgraded by the millers because of the garlic odor of the finished flour. That odor was due to the presence in the wheat of compounds containing selenium. The affected wheats were produced on soils that had a high concentration of this element. Although the selenium content of the soil was high, that element could not be absorbed by the wheat plant until the selenium had been cycled through certain members of the aster family (80).

Corn. Although corn or maize is considered to be principally a U.S. crop, fairly large amounts are produced in other countries. Brazil, the second largest producer, grew 18.7×10^6 t in 1979 (81) compared to the 192.7×10^6 t produced in the United States. The Eastern European countries in 1979 produced 32.9×10^6 t, with Romania, Hungary, and Bulgaria accounting for 96% of the total. Western European production was 22.3×10^6 t, with France, Italy, and Spain accounting for 86% of the total.

The warm, moist conditions required by corn for good growth probably define its production pattern in the United States. In decreasing order of production, the leading states are Iowa, Illinois, Nebraska, Indiana, and Minnesota. These five states accounted for 65% of the corn produced in the United States in 1979. Corn production in the United States has shown a continuous increase in yield for the past 40 yr, the largest increase occurring during the period of World War II. That was the time when hybrid corn seed rapidly replaced earlier varieties on increasing numbers of farms. Until that time, corn was judged principally on the size of the ears and the uniformity of the kernels rather than on yield per hectare. Earlier types of corn, when planted, produced lower yields than the hybrid corn which had been downgraded at the agricultural fairs.

The impact of hybrid corn on U.S. agriculture, the effect of the Green Revolution on world food production, and an interest in biotechnology attracted a number of oil, chemical, and pharmaceutical companies. They have entered this field by purchasing seed companies or expanding their research activities in closely related areas. They embarked on this venture to develop plants of improved yields, growth rates, and resistance to biological and environmental stresses (82). Besides increasing yields, it is anticipated that the concerted action of these companies will succeed in introducing nitrogen-fixing capability into plants such as cereals (see Nitrogen fixation). Should that become a reality, it will decrease the dependence of a large sector of the world on chemical fertilizers. Many of the less-developed countries can barely afford to purchase adequate amounts of fertilizer to increase yields of some of their crops as envisaged by the Green Revolution. If nitrogen fixation (qv) by cereal grains becomes a reality, the effect of fertilizers on their yields may be greatly reduced. The increases in yields among the cereals in the past 50 yr, which have had no parallel in the legumes, seems to substantiate this independence from fertilizer use. The legumes, largely because of their nitrogen-fixing abilities, are influenced to only a slight extent, if at all, by the application of fertilizers.

Rice. Most rice is grown on land that is flooded during the early part of the season. However, a small amount of rice is grown on nonirrigated plots located at higher elevations than the river bottoms or deltas where most rice is grown. Although total rice production has increased in the past 30 yr by a factor of 2.4, its relative contribution to total cereal production has not changed (Table 2).

Most of the world's rice is grown in the Far East where large amounts are consumed on the farms where it is raised. Most of the rice in the United States is grown in southwestern Louisiana, southeastern Texas, eastern Arkansas, the Sacramento and San Joaquin Valleys of California, and parts of the coastal area of South Carolina where rice culture was introduced to this country in the seventeenth century. The development of short-stemmed varieties which respond to fertilization with increased yield of kernels is largely responsible for the 1.5-fold increase in world production of rice from 1950 to 1965 (Table 2).

Oats. The world production of oats has shown an almost steady decrease during the past 30 yr (Table 2). Part of this can be attributed to the decrease in the number of horses on farms in the United States as they were replaced by tractors. This situation has had a profound influence on worldwide oat production since U.S. production of oats accounted for 20% of the world's supply in 1977–1979 (83). Canada produces less than half as much oats as the United States, with the FRG, France, Poland, and Sweden each contributing about 5% of the world production. The People's Republic of China reportedly produced about 3% total production in the same period.

Barley. Barley is adapted to grow fairly well in regions of cool weather with short growing seasons. Planted in the spring, barley grows farther north than most other cereals, maturing in only 2–3 mo (84). That helps to explain why Europe produced 43.5% of the world's supply of barley in 1980 (85). There has been little change in the production of barley from 1977 to 1980. One explanation for the flat production may be that barley does not make good bread—it is used by human beings principally in other foods and as a base for making alcoholic beverages, especially beer (qv). The annual consumption of barley as food in the United States has been stable for many years at 0.55 kg per capita (86). In the United States, about one fourth of the barley grown is fed to animals directly on the farm. Of that which is sold, over 30% is malted for brewing; the malt is also used as an ingredient in breakfast cereals, and in the production of malt syrups and confections (see Malts and malting).

The main producing countries are the USSR, which accounted for 29% of the world's production in 1980 (85); France, Canada, the UK, Spain, and the United States, in that order, followed the USSR. Within the United States, Idaho, North Dakota, California, Montana, Minnesota, and Washington, in that order, accounted for 73% of the country's 1980 production of barley (87). When barley is grown on land that contains an ample supply of nitrogenous compounds, it has a high protein content that is accompanied by a "flinty" starch which is not as desirable for malting as the "softer" barley which has a lower protein content (88).

Sorghum. Of the cereals grown for human use, sorghum is the best adapted to warm areas that may become dry during the growing season. Its well-developed root system makes it possible for sorghum to grow when rainfall is too limited for other cereals. Most of the world's production of sorghum is in Africa, where the seed is used extensively for human food. The latter is also true of India and the People's Republic of China (89). World annual production, including the United States is ca 54×10^6 t sorghum (90). In 1980, Kansas, Nebraska, and Texas accounted for 78% of the U.S. production of 6.45×10^6 t (91). The yield in the United States in 1980 ranged from 2.1 m^3/ha (24 bushels per acre) for Georgia to 6.5 m^3/ha (75 bushels per acre) for Arizona; the principal producing states produced 3.7–5.2 m^3/ha (43–60 bushels per acre). Practically all the sorghum raised in the United States is used for animal feed, one fourth of it used as silage or as forage, and the remainder is incorporated into poultry and livestock feeds (91). The composition of sorghum is similar to that of wheat except it must be low in antioxidants since the lipids become rancid if the intact seeds are kept for even a short period after harvesting (90). It is reported that a wax similar to carnauba can be extracted from the hulls (90).

Millet. Like sorghum, millet has small seeds in which the endosperm and hull layers are firmly attached which precludes both of these cereals from modern milling techniques. Millet grows well in poor soil where rainfall is scanty, probably one reason why millet has come to be regarded as the poor-man's food in India, Pakistan, and Africa just south of the Sahara where most millet is grown (92). About 85% of the millet produced in the world is used for human consumption (92).

Frequently, millet is listed with sorghum, and therefore it is difficult to secure estimates of the amount that is harvested. One source suggests that at least 27×10^6 t of millet annually is produced in the world (90,92). Some millet is grown in the United States, primarily as a forage crop for both stock animals and birds.

Storage of Cereals

One feature of cereal grains that makes them important as a year-round source of food for both people and animals is that they can be stored for long periods of time. The storage capability of the grain depends on low moisture content. It is reduced at harvest, almost always by solar energy. For most cereals, a maximum moisture content of ca 10% protects the grain during storage. In the developed countries, when fuel was cheap, less attention was paid to the moisture content of the grain when it was harvested. If the moisture content was too high, the grain was frequently dried prior to storage either on the farm or at the elevator.

Another factor that influences the respiratory rate is the oxygen in the air around the kernels. Normally, the air in a storage bin containing whole grain represents about one third of the volume of the bin. During respiration, the oxygen in the air is converted to carbon dioxide. Consequently, the concentration of that gas can be used to indicate the extent of respiration during storage. It may be only a crude indicator since, when the oxygen supply becomes limited, the seeds undergo fermentation (qv). Therefore, it is not worthwhile to fill the spaces between the seeds in an elevator with an inert gas or to attempt to store the grain under vacuum. Furthermore, under anaerobic conditions, the seeds produce alcohols, acids, and other compounds that reduce the value of the grain for seed, milling, and industrial purposes.

Storage may be associated with the formation of fatty acids, acid phosphates, and amino acids. The fatty acids result primarily from the hydrolysis of fats and are responsible for a rancid flavor; the phosphates arise principally from hydrolysis of the phytates; and the increase in the level of amino acids is a measure of protein hydrolysis.

Equally important to the respiration of the seeds during storage is the damage done by insects, molds, and animals, especially rodents. The loss during storage from each of these groups depends on various factors. In the underdeveloped countries where storage facilities are primitive, the loss from all three groups may represent as much as a third of the stored grain. Insects not only produce physical damage to the grain but also, since their respiratory rate is greater than that of the grain, increase the temperature in the storage bin and increase the respiratory rate of the grain itself. Thus, insects can produce primary and secondary damage to stored grain. Except where special precautions have been taken, grain-storage facilities are infested with the eggs of insects, mold spores, and rodents when the grain is stored. Bacteria are of minor importance in damaging stored grain because bacterial action usually requires more moisture than is present in grain-storage bins.

Milling of Cereals

Wheat. In 1979, the United States exported 62% of its wheat crop. Of the remainder, about two thirds was processed into human food of which flour was the principal component. Wheat that is unsuitable for high grade flour is often used for a number of other purposes, mainly animal feed. Some is malted and, as such, some is added to high grade flour to increase its diastasic activity; the rest is used primarily in making alcoholic beverages. Another product is starch (qv) used in making adhesives for, among other things, wallpaper and plywood. In the preparation of starch, the proteins of the endosperm are separated as gluten which has some use in the preparation of adhesives (qv), emulsifiers (see Emulsions), polishes (qv), and the meat-flavoring agent, monosodium glutamate (see Amino acids) (93).

Gluten is a complex of the two principal endosperm proteins, gliadin and glutenin. After the wheat has been milled and the resulting flour moistened with water, the endosperm proteins form a complex throughout the mass. The elasticity of the protein complex permits the entrapment of gas bubbles produced by the leavening agent (yeast or chemicals). The flour best suited for breadmaking contains proteins which form a gluten that retains the shape of the bread during baking as well as after the product cools. The quality of the flour protein in producing bread is assessed by a variety of techniques (94). None of the physical techniques by itself accurately predicts the baking qualities of a flour. For that reason, various procedures are used to evaluate a wheat for its breadmaking qualities. These include physical tests as well as the Kjeldahl determination of total protein. The protein determination is of value since the gluten content of a wheat increases with the total protein content of the wheat (see Bakery processes and leavening agents).

Within the past decade, instruments have been developed whereby the protein and moisture content of ground wheat can be determined in a very short time. This is based on the reflectance spectra of a near infrared (nir) beam that is focused on about 20 g of the grain sample. The sample is exposed to nir radiation from a number of wavelengths. The light reflected from the sample at each of these wavelengths is measured by instruments able to detect extremely low energy levels. These energy levels are converted into percent of moisture and protein by a microprocessor or minicomputer. In 1978, the USDA Federal Grain Inspection Service accepted the nir instruments for officially determining the protein content of grains (95) (see Analytical methods).

Wheat is difficult to mill because the large, deep crease that extends the length of the kernel precludes simply pearling the grain to obtain a white kernel for grinding (96). For centuries, millers have striven to get a clean separation of the endosperm, which ends up as flour, from the bran layers, which are used primarily as animal feed. To achieve the separation of endosperm from bran, flour millers subject the wheat, after it has been cleaned, to a breaking process that breaks open the kernel, releasing its contents so the endosperm particles can be separated from the bran. The wheat kernels are passed between a pair of spirally fluted, chilled iron rolls (97). The two rolls are run at different speeds to produce a shearing action which separates the endosperm from the bran particles. The endosperm particles are separated by screening, usually through silk. At the first break, the largest particles consist primarily of bran and are sent to the second set of rolls or the second break. The next smaller particles form semolina which is a mixture of clean endosperm and pieces of bran, some with adhering endosperm (98). The semolina is transferred to a purifier which is a slightly inclined sieve through which air currents are drawn. The meshes of the sieve become coarser down the slope. By that means, the dense, uncontaminated endosperm particles are separated from the lighter particles of endosperm with adhering bran; the lightest particles are pure bran which move on with the air blast. From the purifier, the semolina that contains no bran goes to the reduction rolls, which reduce the size of the particles. The reduction rolls have smooth surfaces and the two rolls at each stage move at about the same speed, crushing and grinding the endosperm particles. The sieving at that stage separates flour from the coarser endosperm particles. The latter pass through the second set of reduction rolls, after which the screening process is repeated. There may be as many as five reduction steps. At each one, the amount of flour secured becomes progressively less. Furthermore, the flour from the fourth or fifth reduction stage contains an increased amount of fine bran particles, which color the flour. During the reduction process, the germ is flattened but not disintegrated; at the end of the reduction process, the germ is separated from the bran by screening (98).

Toward the end of the operation, the flour is treated with a bleaching and aging agent (chlorine dioxide in the United States and the UK) and then the vitamin–mineral enrichment mixture is metered into the flour. After that, the flour is sifted and then bagged or put into a tank truck for bulk distribution to large bakeries. The flour distributed in bulk is delivered through large flexible pipes.

Durum Wheat. Some durum wheat is ground in mills that produce flour used specifically for the manufacture of macaroni, spaghetti, and noodles. The flour for such foods should be made from wheat with a high protein and carotenoid (yellow pigment) content. Before milling, the wheat is carefully cleaned. It is milled with machinery similar to that used to produce flour except that the breaking and sifting operations are more elaborate to permit a gradual reduction in particle size of the endosperm. This permits a decrease in the number of reduction stages.

Soft Wheat. The flour used in making cakes, pastries, cookies, doughnuts, crackers, pretzels, ice cream cones, and similar products should not have as much nor as strong gluten as the flour intended for the production of bread (99). Wheat intended for milling into flour for the cake and pastry industries should have kernels that are softer than the hard, flinty kernels of the wheat used for bread flour. The biochemical explanation for the adaptation of the soft wheats to these specialty foods is unknown. However, it has been recognized that the flour from soft wheats gives products that "are more tender, larger in size, greater in volume, and of superior internal structure than those from other classes of wheat" (98). Flour intended for cake production has a smaller particle size than that for bread and biscuits. Cake flour has passed through a silk screen with a very fine mesh. The flour milled from soft wheat has about 8% protein versus the 10–11% in bread flour and that protein has only a small amount of gluten. To weaken the gluten, the flour is treated with chlorine. The treatment lowers the pH of the flour and permits the incorporation of considerable amounts of lipid as shortening agents into the pastry mixture. After baking, the product is light and fine grained, largely because of the effect of the shortening in overcoming the toughening effect of the flour and egg proteins (100).

Bromated Flours. The FDA regulations permit the addition of potassium bromate at a level not exceeding 50 ppm of the finished flour, if the baking qualities of the flour are improved by the addition (101). Such flour is used in making biscuits. The disappearance of thiol (SH) groups resulting from the bromate oxidation may be a partial explanation for the improvement in gluten elasticity. This reaction occurs primarily at slightly elevated temperatures such as those during the early stages of baking (102).

Phosphated and Self-Rising Flours. In the southeastern part of the United States, the principal flour used in the home is the self-rising kind. This flour usually contains about 0.5% monocalcium phosphate monohydrate. Self-rising flour also contains sodium bicarbonate and salt. It is used in making biscuits, quick breads, and muffins, and requires as additional ingredients baking soda and sour milk. Since the biscuits prepared from self-rising flour have such a short shelf life, a process has been developed whereby dough in the form of individual biscuits is packaged in containers which have to be refrigerated until the biscuits are baked. Self-rising flour is made from soft wheat and, in most cases, has been treated with chlorine to slightly toughen the protein so as to produce a biscuit with increased volume and a correspondingly lighter texture (103).

Corn. Of the corn produced in the United States in 1980, about 62% was not used on the farms where it was grown (104). A large share of the corn sold by the farmers is incorporated into commercial feed for animals. Only about 4% of the harvested corn, but nevertheless, a very large amount on the basis of previous production figures ($>6 \times 10^6$ t), is subjected to dry milling (105). This involves increasing the moisture content to 21–24%, degerminating the corn, and breaking the endosperm with rollers similar to those used in milling wheat. The endosperm particles are dried to a moisture content of 14–16% and then are passed through a hominy separator consisting of cylindrical sieves that separate particles according to size. These endosperm fractions are passed through a series of reduction rolls. The material screened at the various reduction stages is used for different purposes. The coarsest particles are used in the preparation of corn flakes, the slightly smaller particles make up grits, and the next two groups constitute coarse and fine corn meal, respectively. Corn flour is composed of the smallest particles.

Corn starch is becoming an increasingly important product both as a food and as a starting material used in many industrial processes. It is prepared from corn kernels that have been steeped in dilute sulfurous acid for a few hours. The acid treatment softens the outer layers of the grain and dissolves certain soluble substances in the corn (106). The resulting solution, called corn steep liquor, is the primary nutrient for some of the molds and microorganisms involved in the production of antibiotics, primarily penicillin. The moist corn kernels are cracked in a special mill, which separates the intact germ from the rest of the kernel. Since the germ contains about 20% fat, it can be separated by flotation on water; the rest of the corn kernel falls to the bottom. The kernels are removed from the bottom of the pan and powdered in a burr mill. The hulls are removed by a series of siftings. The remaining mixture of starch and protein is separated from a slurry slowly run into long shallow channels. There, the starch settles out while the protein is carried to the end of the channel where it is recovered and used with the hulls to form animal feed. The starch is recovered from the channels, washed, and dried (106).

The food uses of corn starch are of relatively minor importance: primarily as a thickening agent in puddings and pie fillings. Reference has already been made to the production of high fructose syrups from corn starch (see High Fructose Corn Syrup above).

There are many industrial uses of corn starch such as adhesives, which usually involves modification of the starch by dextrinization, involving a partial splitting of the starch molecule either by enzymatic action or by heat in the presence or absence of chemicals (105). The dextrins so formed have wide use in the manufacture of multiwalled paper sacks, corrugated boards, paper tubes, laminated fiber boards, metal foil linings, bottle labels, and wallpaper and billposting (107).

Pregelatinization involves rupture of the hydrated starch granules. An aqueous suspension of this material has less thickening power than the original starch (107). It has extensive use in oil drilling where it controls water loss from the mud (see Petroleum, drilling fluids); it is also used as a bonding agent for foundry sand cores and for charcoal and coal briquettes, as an adhesive for the fibers in paper, as sizing for textiles, and in the manufacture of cold-water wallpaper pastes. Large amounts are used in gypsum wallboard and in cement joint formulations, and as a protective colloid in water-base paint (108).

A large array of modified starches have appeared during the past 50 yr. One of the more important of these is hydroxyethyl starch. It is made by adding ethylene oxide to a dilute alkaline slurry of starch. Hydroxyethyl starch is used as a coating and sizing agent for papers. It forms a continuous film on the surface of the paper and thus decreases the penetration into the paper of hydrophobic materials such as grease, wax, varnish, lacquer, and ink. It is used in textiles to form a smooth, flexible film which provides abrasion resistance and maximum weaving efficiency (109).

Oxidizing starch is another product extensively used in the paper, textile, and building-supplies industries. This product is made by treating starch in an alkaline, aqueous medium with hypochlorite (110).

Oxidized starch is used in making coated stock for printing and writing papers (111). Another oxidized starch is made with periodate which appears to split the glucose unit in the starch between C-2 and C-3. These two carbons are converted to aldehyde groups, which react with hydroxyl groups or with each other (112). This dialdehyde starch is used in making paper that retains its strength when wet, as for toweling, sanitary tissue, and maps. The improvement in wet strength imparted by the starch is temporary, so that the paper eventually disintegrates in water (112). The aldehyde groups in this starch are quite reactive and can combine with other molecules such as casein. When that happens, the protein in the adhesive becomes water insoluble and adheres more tenaciously to wood (113).

There are a number of commercially important substituted starches. The starch phosphates have considerable use in both foods and industrial products. When these starches are suspended in water, they form a very stable semisolid emulsion without cooking. Because of this property, they are replacing natural gums to a certain extent (114). They are being used to decrease rancidity in cottonseed and soybean oils, especially where rancidity may be catalyzed by such trace elements as iron, copper, nickel, or cobalt. These metals complex with the starch phosphates. The starch phosphates also enhance the emulsification of the oils (115). The presence of the orthophosphate group in the starch molecule permits cross bonding between two starch chains. The highly cross-bonded diesters are used in dry batteries where they provide a moisture-retentive filler and general adhesion and consistency (116).

Starch sulfates, nitrates, xanthates, and formates, as well as other derivatives, have been produced but they are only of experimental interest (115,117).

Rice. The rice kernel is covered with a tightly adhering hull which is relatively high in silicon, thus rendering it unsuitable for consumption. To remove the hulls, the cleaned rice is put into a shelling machine which consists of two horizontal circular steel plates or rubber rollers. The inner surface of the steel plates is coated with coarse carborundum embedded in cement. One plate remains stationary while the other rotates at such a height that the rice grains can assume a vertical position between the plates. The pressure thus exerted on the rice grain splits the hull and frees the endosperm with its adhering bran layers and germ; at that stage, it is known as brown rice. A blast of air is used to remove the light hulls from the brown rice and the remaining intact rice. The latter two are separated in a paddy machine where the lighter, unhulled rice rises to the top during shaking while the heavier endosperm moves downward. The unhulled rice emanating from the top of the paddy machine, passes through a second shelling machine, whose plates are set a little closer together than in the first machine. Separation of hulls from the brown rice and intact grains is again accomplished by a blast of air, after which the rice is again put through the paddy machine. The brown rice is separated from the bran layers in a cylindrical, rotating machine where the rice kernels scour each other and thus remove the outer bran layers and the germ. Final polishing of the rice occurs in a brush machine which is a vertical cylinder containing soft leather strips on its outer surface. This cylinder revolves within another stationary cylinder made of wire screen. As the brown rice flows down between these two cylinders, it is polished by the moving leather strips and the wire screen. The rice which emerges from the brush machine is screened to separate the intact endosperm from the broken pieces. By this procedure, 66–70% of the initial rice is recovered as white rice.

In the United States and some other countries, there is an additional step before the rice can be packaged which involves the addition of vitamins and minerals since practically all rice sold in the United States is enriched. Techniques had to be devised for enriching the grains without changing their shape.

There are two means whereby rice is enriched. One of these involves adding a large excess of the enrichment mixture as a coating to some of the rice grains and then diluting those grains with the polished rice in such proportions that the final mixture contains the legally prescribed vitamin and mineral dosage (118). This rice premix contains the enrichment ingredients as layers on the kernels which are then coated with a material that is relatively water insoluble. The penetrating yellow color of the riboflavin caused a problem with this enrichment technique since the person who was preparing the rice in the home frequently removed the yellow kernels on the assumption they were foreign matter. To overcome the problem, riboflavin is frequently omitted from the enrichment ingredients added to rice.

The other process involves adding the powdered premix with the milled rice. When the rice and premix are tumbled, the rice kernels become coated with the premix which adheres to the rice grains under ordinary handling. However, such rice cannot be washed and should be cooked in only enough water so that it is entirely absorbed by the rice. This rice must be packaged in containers that carry the admonition against washing the rice either before or after cooking.

Oats. Only 3–4% of the oat crop is destined for human consumption (119) and for that reason, oat milling is performed by only a few companies. The first step is the removal of foreign matter and the very lightest oat grains. To facilitate milling, the oats are dried to a moisture content of 6%; which increases the brittleness of the hulls and permits easier separation of the groats (the dehulled grain). The dried oats are separated into five or six groups on the basis of grain length. Each group of oats is conveyed to separate hullers which resemble those used in hulling rice. They consist of a lower, flat stationary "stone" and an upper one which is slightly conical and rotates rapidly. As in rice milling, the space between the two "stones" is adjusted to be slightly less than the length of the intact oat grain and slightly more than that of the groat. This exerts pressure on the oat grains, which assume a vertical position as they migrate from the center of the stones to the periphery. The pressure shatters the hull, thus releasing the groat. The mixture of unhulled oats, hulls, groats, and oat flour is sifted to remove the flour. The hulls are then removed by an air stream. Groats are separated from the unhulled grain in apron machines that have openings into which the smaller groats can pass while the unhulled oats pass through the machine for reprocessing.

For human use, the groats are treated with steam which partially cooks them and increases their moisture content. The latter decreases the loss of oat flour during the rolling process. Steam treatment of the intact groats produces the slower-cooking rolled oats. For the quick-cooking oats, the groats are cut so that each groat yields about three particles. These particles of oats are steamed and then rolled. The distance between the rollers is smaller for preparing the quick-cooking oats. The small particles in the quick-cooking oats explains why it requires less time to prepare oatmeal and related products. The yield of rolled oats is about 42% of the original cleaned oats.

Rye. Only about 20% of the rye crop in the United States is milled; about two thirds of the crop is used as stock feed (120). Rye milling is done in somewhat the same way as that for wheat. The principal difference arises from the tenacity with which rye bran adheres to the endosperm. Partially for that reason, the while flour is removed early in the milling process. It represents about 50–60% of the grain. As milling proceeds, the flour removed at successive stages becomes darker.

Since the gluten formed from rye flour is not well adapted to breadmaking, rye flour usually contains added wheat flour. The latter is milled from strong spring or hard winter wheats which are high in proteins that produce good loaf volumes when baked into bread. If the dark rye flour is used alone in breadmaking, as was the case in northern Europe, a black, soggy bread with a rather bitter flavor results (121).

BIBLIOGRAPHY

"Cereals" in *ECT* 1st ed., Vol. 3, pp. 591–634, by W. F. Geddes, University of Minnesota, and F. L. Dunlap, Wallace & Tiernan Co., Inc.; "Wheat" in *ECT* 2nd ed., Vol. 22, pp. 253–307, by Y. Pomeranz, U.S. Department of Agriculture, and M. M. MacMasters, Kansas State University.

1. W. R. Aykroyd and J. Doughty, *Wheat in Human Nutrition*, FAO Nutritional Studies, No. 23, Food and Agricultural Organization, Rome, 1970, p. 1.
2. J. Storck and W. D. Teague, *A History of Milling Flour for Man's Bread*, University of Minnesota Press, Minneapolis, Minn., 1952, p. 27.
3. W. J. Darby, P. Ghalioungiu, and L. Grivetti, *Food: The Gift of Osiris*, Vol. 2, Academic Press, New York, 1977, pp. 492, 493.
4. Ref. 3, p. 460.
5. H. A. Wallace and W. L. Brown, *Corn and Its Early Fathers*, Michigan State University Press, East Lansing, Mich., 1956, p. 35.
6. Ref. 5, p. 33.
7. Ref. 2, p. 37.
8. D. K. Mecham in Y. Pomeranz, ed., *Wheat, Chemistry and Technology*, American Association of Cereal Chemists, Inc., St. Paul, Minn., 1971, p. 395.
9. L. Zeleny in ref. 8, p. 34.
10. S. Davidson, R. Passmore, J. F. Brock, and A. S. Truswell, *Human Nutrition and Dietetics*, 7th ed., Churchill Livingstone, Edinburgh, 1979, p. 43.
11. R. R. Williams, *Toward the Conquest of Beriberi*, Harvard University Press, Cambridge, Mass., 1961, p. 38.
12. Ref. 11, p. 41.
13. K. Y. Guggenheim, *Nutrition and Nutritional Diseases*, D. C. Heath & Co., Lexington, Mass., 1981, p. 172.
14. C. A. Elvehjem, R. J. Madden, F. M. Strong, and D. W. Wooley, *J. Am. Chem. Soc.* **59**, 1767 (1937).
15. E. V. McCollum, *A History of Nutrition*, Houghton Mifflin Co., Boston, Mass., 1957, p. 306.
16. J. B. Mason, N. Gibson, and E. Kodicek, *Br. J. Nutr.* **30**, 297 (1973).
17. A. S. Prasad in A. S. Prasad, ed., *Trace Elements in Human Health and Disease*, Vol. 1, Academic Press, New York, 1976, pp. 1–20.
18. D. Oberleas and A. S. Prasad in ref. 17, pp. 155–162.
19. A. R. P. Walker, F. W. Fox, and J. T. Irving, *Biochem. J.* **42**, 452 (1948).
20. H. H. Mitchell, *Nutr. Rev.* **3**, 130 (1945).
21. A. O'Boyle, T. Watkins, and O. Mickelsen, *Fed. Proc. Fed. Am. Soc. Exp. Biol.* **41**, 394A (1982).
22. S. Bolourchi, J. S. Feurig, and O. Mickelsen, *Am. J. Clin. Nutr.* **21**, 836 (1968).
23. T. Addis, E. Barrett, L. J. Poo, and D. W. Yuen, *J. Clin. Invest.* **26**, 869 (1947).
24. D. D. Makdani, M. Ahmad, B. H. Selleck, D. R. Rovner, J. S. Feurig, and O. Mickelsen, to be published in *Am. J. Clin. Nutr.*
25. R. P. Heaney, *Clin. Obstet. Gynecol.* **19**, 791 (1976).
26. E. T. Jensen and E. Østergaard, *Am. J. Obstet. Gynecol.* **67**, 1094 (1954); D. C. Smith, *N. Eng. J. Med.* **293**, 1164 (1975); T. M. Mack and co-workers, *N. Eng. J. Med.* **294**, 1262 (1976).
27. E. C. Reifenstein and F. Albright, *J. Clin. Invest.* **26**, 24 (1947).
28. T. V. Sanchez, O. Mickelsen, A. G. Marsh, S. M. Garn, and G. H. Mayor, *Proceedings of the Fourth International Conference on Bone Measurement*, National Institute of Health Publication #80-1938, Public Health Services, U.S. Department of Health and Human Services, Washington, D.C., May 1980.
29. O. Mickelsen and D. D. Makdani, *Proceedings of the National Conference on Wheat Utilization Research*, ARS-NC-40, Agricultural Research Services, U.S. Department of Agriculture, Washington, D.C., 1975.
30. O. Mickelsen, D. D. Makdani, R. H. Cotton, S. T. Titcomb, J. C. Colmey, and R. Gatty, *Am. J. Clin. Nutr.* **32**, 1703 (1979).
31. *Pediatrics* **23**, 400 (1981).
32. J. H. Shaw, *J. Am. Med. Assoc.* **166**, 633 (1958).
33. B. E. Gustafsson and co-workers, *Acta Odontol. Scand.* **11**, 232 (1954).

34. B. H. Landis in *Products of the Corn Refining Industry in Food*, Seminar Proceedings, Corn Refiners Association, Washington, D.C., May 9, 1978, pp. 47–54.
35. J. S. Trier in M. H. Sleisenger and J. S. Fortran, eds., *Gastrointestinal Disease, Pathophysiology, Diagnosis, Management*, 2nd ed., Vol. II, W. B. Saunders Co., Philadelphia, Pa., 1978, pp. 1029–1053.
36. W. K. Dicke, H. A. Weijers, and J. H. van de Kamer, *Acta Paediat.* **42,** 34 (1953).
37. U.S. Pat. 1,367,530 (1921), J. C. Baker (to Wallace and Tiernan Co.).
38. *Fed. Regist.* **13,** 6969 (Nov. 27, 1948).
39. H. E. Magee, *Mon. Bull. Minist. Health Public Health Lab. Ser.*, 205 (Sept. 1950).
40. J. C. Drummond and A. Wilbraham, *The Englishman's Food*, Jonathan Cape, London, 1939, pp. 298–299.
41. D. P. Burkitt in D. P. Burkitt and H. C. Trowell, eds., *Refined Carbohydrate Foods and Disease, Some Implications of Dietary Fibre*, Academic Press, London, 1975, pp. 3–20.
42. D. P. Burkitt and H. C. Trowell in ref. 41.
43. T. G. Kiehm, J. W. Anderson, and K. Ward, *Am. J. Clin. Nutr.* **29,** 895 (1976).
44. D. J. A. Jenkins, A. R. Leeds, M. A. Gassull, B. Cochet, and K. G. M. M. Alberti, *Ann. Intern. Med.* **86,** 20 (1977).
45. M. Stasse-Wolthuis, *World Rev. Nutr. Diet.* **36,** 130 (1981).
46. W. H. Leonard and J. H. Marin, *Cereal Crops*, The MacMillan Co., London, 1963, p. 11.
47. *Dietary Goals for the United States*, U.S. Senate Select Committee on Nutrition and Human Needs, U.S. Government Printing Office, Washington, D.C., Feb. 1977.
48. Ref. 2, pp. 280–281.
49. Ref. 40, p. 353.
50. *Household Food Consumption Survey 1965–1966*, Report No. 17, *Food Consumption by Households by Money Value of Food and Quality of Diet*, Agricultural Research Service, U.S. Department of Agriculture, Washington, D.C., 1966, p. 13.
51. G. Borgstrom, *Too Many, A Study of Earth's Biological Limitations*, The Macmillan Co., New York, 1969, p. 154.
52. D. F. Houston and G. O. Kohler, *Nutritional Properties of Rice*, National Academy of Sciences, Washington, D.C., 1970, p. 11.
53. E. D. Wilson, K. H. Fisher, and M. E. Fuqua, *Principles of Nutrition*, 3rd ed., John Wiley & Sons, Inc., New York, 1975, p. 556.
54. Ref. 11, p. 113.
55. Ref. 11, p. 114.
56. Ref. 11, p. 112.
57. Ref. 52, p. 17.
58. Ref. 11, p. 115.
59. *Proposed Fortification Policy for Cereal-Grain Products*, Food and Nutrition Board, National Research Council, National Academy of Sciences, Washington, D.C., 1974, p. 4.
60. W. A. Krehl and J. J. Barboriak in R. S. Harris and H. von Loesecke, Jr., eds., *Nutritional Evaluation of Food Processing*, John Wiley & Sons, Inc., New York, 1960.
61. Ref. 59, p. 2.
62. P. M. Ranum, R. J. Loewe, and H. T. Gordon, *Cereal Chem.* **58,** 32 (1981).
63. M. S. Randhawa, *Green Revolution, A Case Study of Punjab*, Halsted Press, a division of John Wiley & Sons, Inc., New York, 1974; B. Sen, *Green Revolution in India: A Perspective*, Halsted Press, a division of John Wiley & Sons, Inc., New York, 1974.
64. R. Osoyo in *Strategy for the Conquest of Hunger*, Proceedings of a Symposium Convened by the Rockefeller Foundation, April 1–2, 1968, The Rockefeller Foundation, New York, 1968, p. 8.
65. D. G. Dalrymple, *Development and Spread of High-Yielding Varieties of Wheat and Rice in the Less Developed Nations*, USDA Foreign Agricultural Report No. 95, U.S. Department of Agriculture, Washington, D.C., Sept. 1978.
66. Ref. 65, p. ix.
67. J. L. Hulse and D. Spurgeon, *Sci. Am.* **231,** 72 (1974).
68. G. W. Schaefer and F. T. Sharpe, Jr., *In Vitro* **17,** 345 (1981).
69. J. H. Hulse and E. M. Laing, *Nutritive Value of Triticale Protein*, International Development and Research Centre, Ottawa, Canada, 1974.
70. J. L. Vetter, *L.I.F.E. Newsletter*, League for Food Education, Washington, D.C., Sept. 1975.
71. E. N. Larter, *Agric. Inst. Rev.* **23,** 12 (1968).

72. *Production Yearbook*, *Statistical Series*, Food and Agricultural Organization, Rome, Vol. 20, 1966; Vol. 33, 1979; Vol. 34, 1980.
73. O. Mickelsen in *Topics of Study Interest in Medicine and Public Health in the People's Republic of China, Report of a Planning Meeting*, (NIH) 72-395, National Institute of Health, U.S. Department of Health, Education, and Welfare, Washington, D.C., 1972, pp. 41–52.
74. R. McCarrison, *Studies in Deficiency Disease*, Oxford Medical Publishing, Henry Frowde and Hodder and Stoughton, London, 1921.
75. *Agricultural Statistics, 1981*, U.S. Department of Agriculture, U.S. Government Printing Office, Washington, D.C., 1981, p. 9.
76. R. D. Laird, J. Hajda, and B. A. Laird, *The Future of Agriculture in the Soviet Union and Eastern Europe, The 1976–1980 Five Year Plan*, Westview Press, Boulder, Colo., 1977, p. 2.
77. Ref. 75, p. 12.
78. Ref. 75, p. 13.
79. B. Bergland, *National Forum, The Phi Kappa Phi Journal* **69**(2), 3 (Spring 1979).
80. O. E. Olson in O. H. Muth, J. E. Oldfield, and P. H. Weswig, eds., *Selenium in Biomedicine*, AVI Publishing Co., Inc., Westport, Conn., 1967, p. 302.
81. Ref. 75, p. 35.
82. J. Walsh, *Science* **213**, 1339 (1981).
83. Ref. 75, p. 42.
84. R. W. Schery, *Plants for Man*, Prentice-Hall, Inc., Englewood Cliffs, N.J., 1972, p. 435.
85. Ref. 75, pp. 47–48.
86. Ref. 75, p. 49.
87. Ref. 75, p. 46.
88. Ref. 84, p. 437.
89. Ref. 84, p. 439.
90. Ref. 84, p. 440.
91. Ref. 75, p. 52.
92. Ref. 84, pp. 440–441.
93. Ref. 84, p. 426.
94. D. B. Pratt, Jr., in ref. 8, pp. 201–226.
95. R. D. Rosenthal, *An Introduction to Near Infrared Quantitative Analysis*, Neotec Instruments, Inc., Silver Springs, Md., 1978.
96. P. W. R. Eggitt and A. W. Hartley in A. Spicer, ed., *Bread, Social Nutritional and Agricultural Aspects of Wheaten Bread*, Applied Science Publishers, Ltd., London, 1975, p. 219.
97. E. Ziegler and E. N. Greer in ref. 8, p. 136.
98. Ref. 96, p. 220.
99. W. T. Yamazaki and D. D. Lord in ref. 8, p. 743.
100. Ref. 8, p. 755.
101. *Wheat and Corn Flour and Related Products—Definition and Standards*, Title 21, Pt 15.20, Food and Drug Administration, Washington, D.C., reprinted March 1953.
102. Y. Pomeranz in ref. 8, p. 615.
103. Ref. 99, p. 764.
104. Ref. 75, p. 34.
105. R. L. Whistler in G. E. Inglett, ed., *Corn, Culture, Processing Products*, AVI Publishing Co., Inc., Westport, Conn., 1970, p. 171.
106. Ref. 84, pp. 379–380.
107. Ref. 105, p. 175.
108. Ref. 105, p. 176.
109. Ref. 105, p. 179.
110. Ref. 105, p. 182.
111. Ref. 105, p. 185.
112. Ref. 105, p. 186.
113. Ref. 105, p. 187.
114. Ref. 105, p. 188.
115. Ref. 105, p. 189.
116. R. L. Whistler, Purdue University, West Lafayette, Ind., personal communication, 1982.
117. Ref. 105, pp. 190–193.
118. Ref. 52, pp. 30–31.

119. Ref. 84, p. 433.
120. Ref. 84, p. 439.
121. Ref. 84, p. 438.

Olaf Mickelsen
Consultant

WHEAT GERM OIL. See Fats and fatty oils.

WHEY. See Pet and livestock feeds.

WHISKEY. See Beverage spirits, distilled.

WHITE LEAD. See Pigments, inorganic.

WHITENING AGENTS. See Brighteners, optical.

WHITING, CaCO₃. See Pigments, inorganic.

WINE

The word wine was possibly first applied to the fermentation product of the sugars in the juice of grapes; this is its primary meaning. However, the fermented juices of many fruits are now called wine, and the term is also sometimes incorrectly applied to the alcoholic fermented juice of various plant materials containing sugars, eg, rice wine. It is thus used in contrast to fermented liquids made from starch-containing materials, such as beer (qv) and related beverages. The unmodified term wine applies only to the product obtained from fermented grapes. Orange, peach, cherry, blackberry, loganberry, currant, apple, strawberry, and other fruit wines are produced commercially in limited quantities in the United States, but in larger quantities in France, the UK, Poland, and Sweden. There is also a small but widespread home industry that produces wines from herbs and vegetables (dandelion, beans, rhubarb, or roses) by treating them with a sugar solution or from fermented honey (see also Beverage spirits, distilled).

An important aspect of wine is its intimate and long association with artistic, cultural, and religious activities. Thus, at the beginning of recorded history, wines were described or their production portrayed and their properties critically evaluated and praised. By 2500 BC, the Egyptians had evolved hieroglyphics that described various types of wines. Noah's reputed experience in planting a vineyard and making wine also indicates the early development of wine making. The Old Testament contains many references to wines and their properties, as does Greek literature. Greek wines frequently contained herbs, perfumes, and flavors, a practice that survives today (see also Flavors and spices). The reason for adding herbs and spices is not known, but it may have been an attempt to mask spoilage or prevent its development. Honey was probably added for the same reason and may have resulted in a secondary fermentation which produced enough alcohol ($\geq 15\%$) to prevent spoilage. Some herbs were probably added because of supposed medicinal or aphrodisiac values.

The religious and allegorical significance of wine was developed by the Greeks and has been utilized by many other religions including Christianity. It is unlikely that unfermented grape juice was more than a vintage-season beverage before the 19th century because of the lack of sterilization procedures and techniques.

In ancient Rome, wine production became increasingly organized and specialized, and the cultivation of grapes was highly developed. Various varieties of grapes and different methods of wine production were described by Columella and Pliny. Clay amphora of Greek origin and later wooden casks were used for aging. Although many Roman wines must have been very poor by modern standards, it is clear that the Romans had cultivated a taste for the beverage, lavished much care on its production, and gave it literary and artistic praise. The writings of Horace and Virgil on wines are well known.

During the Middle Ages, wines were produced in the Mediterranean countries and in northern France and Germany. The spread of Islam nearly destroyed the wine industry around the Mediterranean between 800 and 1400 AD. In the Moslem countries, only Christians and Jews produced small amounts for personal and sacramental purposes. The need for wine for religious ceremonies and the large number of monastic communities led to the production of wine by monasteries. Many important European vineyards from France to Yugoslavia and Cyprus owe their origins to Cistercian and other monastic orders which made wine for their own use as well as for sale. In the 17th century, the cooper's art improved, bottles were less expensive, and after corks became available, wines could be stored safely and for longer periods. Already in the pre-Christian era, wines were stored in caves under relatively constant and cool conditions; both are essential for minimizing secondary fermentation and for proper aging.

In the 19th century, the work of Pasteur and others not only demonstrated the role of yeasts in grape-juice fermentation, but also identified the various microorganisms responsible for spoilage; methods for their control were developed, and wine production changed from an uncertain art to a scientific industry. Wine making is still something of an art, as far as quality is concerned, but today most of the world's wine is produced by modern technology.

Definitions

Wine is the fermented juice of the fruit of one of several species of *Vitis*, most often of cultivars of *Vitis vinifera*, with or without the addition of sugar, grape concentrate, or reduced must (boiled-down grape juice), herbs, flavors, or alcohols. In the eastern United States, varieties of *V. labrusca* and hybrids of *V. vinifera* and *V. labrusca* are used. In the southeastern United States, varieties of *V. rotundifolia* are employed. Over 95% of the world's wine is made from varieties of *V. vinifera*, possibly because of the more subtle flavors of most *V. vinifera* wines. Grape breeders generally try to remove the strong labrusca flavor by complicated interspecific hybridization.

The present legal definition of wine in the United States (1) is:

United States Internal Revenue Code. Sec. 5381. Natural Wine. Natural wine is the product of the juice or must of sound, ripe grapes or other sound, ripe fruit, made with such cellar treatment as may be authorized under section 5382 and containing not more than 21 percent by weight of total solids. Any wine conforming to such definition except for having become substandard by reason of its condition shall be deemed not to be natural wine, unless the condition is corrected...

Sec. 5382. Cellar Treatment of Natural Wine. (a) General. Proper cellar treatment of natural wine constitutes those practices and procedures in the United States and elsewhere, whether historical or newly developed, of using various methods and materials to correct or stabilize the wine, or the fruit juice from which it is made, so as to produce a finished product acceptable in good commercial practice. Where a particular treatment has been used in a customary commercial practice, it shall continue to be recognized as a proper cellar treatment in the absence of regulations prescribed by the Secretary finding such treatment not to be a proper cellar treatment within the meaning of this subsection.

(b) Specifically Authorized Treatments. The practices and procedures specifically enumerated in this subsection shall be deemed proper cellar treatment for natural wine:

(1) The preparation and use of pure concentrated or unconcentrated juice or must. Concentrated juice or must reduced with water to its original density or to not less than 22 degrees Brix or unconcentrated juice or must reduced with water to not less than 22 degrees Brix shall be deemed to be juice or must, and shall include such amounts of water to clear crushing equipment as regulations prescribed by the Secretary may provide.

(2) The addition to natural wine, or to concentrated or unconcentrated juice or must, from one kind of fruit, of wine spirits (whether or not taxpaid) distilled in the United States from the same kind of fruit; except that (A) the wine, juice or concentrate shall not have an alcoholic content in excess of 24 percent by volume after the addition of the wine spirits, and (B) in the case of still wines, wine spirits may be added in any State only to natural wines produced by fermentation in bonded wine cellars located within the same State.

(3) Amelioration and sweetening of natural grape wines in accordance with section 5383.

(4) Amelioration and sweetening of natural wines from fruits other than grapes in accordance with section 5384.

(5) In the case of effervescent wines, such preparations for refermentation and for dosage as may be acceptable in good commercial practice, but only if the alcoholic content of the finished product does not exceed 14 percent by volume.

(6) The natural darkening of the sugars or other elements in juice, must, or wine due to storage, concentration, heating processes, or natural oxidation.

(7) The blending of natural wines with each other or with heavy-bodied blending wine or with concentrated or unconcentrated juice, whether or not such juice contains wine spirits, if the wines, juice, or wine spirits are from the same kind of fruit.

(8) Such use of acids to correct natural deficiencies and stabilize the wines as may be acceptable in good commercial practice...

Sec. 5383. Amelioration and Sweetening Limitations for Natural Grape Wines. (a) Sweetening of Grape Wines. Any natural grape wine may be sweetened after fermentation and before taxpayment with pure dry sugar or liquid sugar if the total solids content of the finished wine does not exceed 12 percent of the weight of the wine and the alcoholic content of the finished wine after sweetening is not more than 14 percent by volume; except that the use under this subsection of liquid sugar shall be limited so that the resultant volume will not exceed the volume which could result from the maximum authorized use of pure dry sugar only...

Sec. 5384. Amelioration and Sweetening Limitations for Natural Fruit and Berry Wines. (a) In General. To natural wine made from berries or fruit other than grapes, pure dry sugar or liquid sugar may be added to the juice in the fermenter, or to the wine after fermentation; but only if such wine has not more than 14 percent alcohol by volume after complete fermentation, or after complete fermentation and sweetening, and a total solids content not in excess of 21 percent by weight; and except that the use under this subsection of liquid sugar shall be limited so that the resultant volume will not exceed the volume which could result from the maximum authorized use of pure dry sugar only...

Sec. 5385. Specially Sweetened Natural Wines. (a) Definition. Specially sweetened natural wine is the product made by adding to natural wine of the winemaker's own production a sufficient quantity of pure dry sugar, or juice, or concentrated juice from the same kind of fruit, separately or in combination, to produce a finished product having a total solids content in excess of 17 percent by weight and an alcoholic content of not more than 14 percent by volume, and shall include extra sweet kosher wine and similarly heavily sweetened wines.

(b) Cellar Treatment. Specifically sweetened natural wines may be blended with each other, or with natural wine or heavy bodied blending wine in the further production of specially sweetened natural wine only, if the wines so blended are made from the same kind of fruit. Wines produced under this section may be cellar treated under the provisions of section 5382(a) and (c). Wine spirits may not be added to specially sweetened natural wine.

Sec. 5386. *Special Natural Wines.* (a) In General. Special natural wines are the products made, pursuant to a formula approved under this section, from a base of natural wine (including heavy-bodied blending wine) exclusively, with the addition, before, during or after fermentation, of natural herbs, spices, fruit juices, aromatics, essences, and other natural flavorings in such quantities or proportions as to enable such products to be distinguished from any natural wine not so treated, and with or without carbon dioxide naturally or artificially added, and with or without the addition, separately or in combination, of pure dry sugar or a solution of pure dry sugar and water, or caramel. No added wine spirits or alcohol or other spirits shall be used in any wine under this section except as may be contained in the natural wine (including heavy-bodied blending wine) used as a base or except as may be necessary in the production of approved essences or similar approved flavorings. The Brix degree of any solution of pure dry sugar and water used may be limited by regulations prescribed by the Secretary in accordance with good commercial practice.

(b) Cellar Treatment. Special natural wines may be cellar treated under the provisions of section 5382(a) and (c)...

Sec. 5387. *Agricultural Wines.* (a) In General. Wines made from agricultural products other than the juice of fruit shall be made in accordance with good commercial practice as may be prescribed by the Secretary by regulations. Wines made in accordance with such regulations shall be classed as "standard agricultural wines." Wines made under this section may be cellar treated under the provisions of section 5382(a) and (c)...

The standards of identity and quality in the California regulations generally conform to those of the Federal government.

Various states and foreign countries have slightly different definitions based on local needs (see Table 1). The regulations of the European Economic Community (EEC) provide for uniform regulations for trade in wines between its members and for imports from nonmember countries. Generally, the regulations have been written to allow legitimate local practices, eg, addition of sugar to low-sugar musts in the FRG and parts of France.

Table 1. EEC, Federal, and California Standards for Wines

Type	Alcohol, % max	Max volatile acidity[a], as % acetic acid	Min fixed acidity, as % tartaric acid	Max sulfur dioxide, mg/L	Min extract[b], g/100 mL
EEC					
red table		0.120		175[c]	
white and rosé		0.108		225[c]	
Federal					
red table	14	0.140	[d]	350[e]	none
white table	14	0.120	[d]	350[e]	none
dessert	17–21	0.120	[d]	350[e]	none
California					
red table		0.120	0.40	350	1.8
white table		0.110	0.30	350	1.7
dessert		0.110	0.25	350	none[f]

[a] Exclusive of sulfur dioxide.
[b] Minimum soluble solids content.
[c] For table wines with residual sugar the maximums are higher—up to 400 mg/L.
[d] No specification given.
[e] Not more than 70 of which may be free.
[f] Either minimum degree Balling or minimum percent reducing sugar is required, varying with the type.

Wines have been sold under regional names for many centuries. The Madrid and Lisbon agreements gave protection to regional names for many products including wine. However, the United States and some other countries did not participate in these conventions.

Codification for national protection to regional names began in France in 1935 under what is called *appelation d'origine contrôlée* (AOC or sometimes just AC) (2–3). In general, these regulations delimit the region protected, limit the varieties of grapes that can be used and the production per hectare, and restrict the enological practices. Larger geographical definitions may also contain smaller areas or even specific vineyards. Thus, Bordeaux includes all of the wines produced in the delimited Bordeaux district. Bordeaux supérior has higher standards. Médoc is a smaller regional appellation, and Pauillac is the smallest (a commune or roughly a township). Individual châteaux do not have an appellation of origin. In Burgundy, the subdivisions are Bourgogne or the area around a town, eg, Beaune or Meursault, and finally subdivisions such as Beaune Bressandes or Meursault Genevries. These smaller areas may belong to one or many growers. There are also some lesser appellations in France, *vins de consummation courant* (VCC) and subdivisions, and *vins delimitee qualité supérior* (VDQS).

Similar regulations have been developed for the FRG: Besitz, local area, and vineyards. Examples are Rheingau, Geisenheim (town), and Geisenheimer Mäuerchen (vineyard area). Again, a vineyard may have one or many owners. Furthermore, the sugar content of the fruit at harvest may permit quality designations, eg, Kabinet, Spätlese, Auslese, Beerenauslese, and Trockenbeerenauslese (for increasing sugar content and prices).

The Italian regulations provide for a *denominazione di origin controllata* (DOC) and *denominazione di origin controllata guarantie* (DOCG) (4). Several hundred DOCs have been defined but, so far, only a few DOCGs. The requirements for a DOCG are more restrictive than for a DOC, and presumably the quality and prices are higher.

In 1978, U.S. regulations were changed to give protection to regional names. These may be for large or small areas. So far, Augusta, Mo., Fennville, Mich., Finger Lakes, New York (and 8 other non-California regions), and Arrago Seco, Caramel Valley, Cienega Valley, Guenoc Valley, McDowell Valley, Napa Valley, Santa Cruz Mountains, San Pasqual, Santa Maria Valley, and Sonoma Valley (and 11 others), all in California, have been approved. Many others await approval. These are strictly geographical appellations with no indication of the types of wines that may be produced within the delimited area.

Proprietary wines are those with a name that is owned by a producer. All names of wines are in lower case except varietal types, proprietary types, and those with special regional significance. Thus, California burgundy is not capitalized since there is no regional significance to the name as used here. On the other hand, Burgundy designates a wine from the region of that name in France.

Standards

There are two types of standards for wines in the United States and most other countries. First, there are those based on taxes. In the United States, the tax is levied according to the alcohol content:

alcohol, %	≤14	>14–≤20	>21–24
tax, ¢/L	4.5	17.7	63.4

Very few wines have an alcohol content of 21% or more. The Federal government and some states have set up standards for spoiled wines to protect the consumer. These standards primarily limit the content of volatile acids, mainly acetic acid.

The international, Federal, and state public health authorities also set standards for wines on the basis of chlorides, sulfates, lead, arsenic, and other constituents. In the United States, it is seldom necessary to determine these constituents, since they are not known to be present in U.S. wines in excessive amounts.

Salicylic, benzoic, and monochloracetic acids may not be added to American wines, whereas clarification agents and sulfur dioxide, sorbic acid (qv), or sorbates are permitted. It can be presumed that additives not listed in the standard are prohibited. The maximum amounts permitted are specified. Other countries likewise generally prohibit certain practices and permit others, and those not specifically listed generally may not be employed without permission. Analytical procedures are discussed in refs. 5–8.

Classification

Wines may be classified in many ways: by alcohol (ethanol) content, place of production, color, method of production, or variety of grape or fruit. Wines with ≤14% alcohol are defined as table wines since they are normally consumed with meals. Wines with >14% are called dessert or aperitif wines since they are usually consumed after or before meals.

   I. Wines without added herbs or plant materials.
      A. Wines with excess carbon dioxide.
         1. From fermentation of added sugar. Usual pressure, 200–600 kPa (ca 2–6 atm).
            a. Containing anthocyanin and related (red) pigments.
               (1) Pink—pink or rosé sparkling types.
               (2) Red—sparkling burgundy or champagne rouge.
            b. Not containing red pigments.
               (1) With muscat flavor—sparkling muscat and muscato spumante
               (2) Without muscat flavor—Champagne, California (etc) champagne, or bulk-process champagne, Sekt, spumante, espumante, shampanski, etc.
                  (a) Below 1% sugar—brut type.
                  (b) Above 1.5% sugar, sec (dry), demi-sec, and doux types with increasing sugar.
         2. Wines with excess carbon dioxide, not from added sugar. Usual pressure, 20–200 kPa (ca 0.2–2 atm).
            a. Gassiness from fermentation of residual grape sugar. Includes occasional wines in Switzerland, France (some Vouvray), Italy, and the muscato amabile type of California.
            b. Gassiness from malo–lactic fermentation—vinhos verdes wines (white and red from northern Portugal) and some Italian wines.

3. Wines with added carbon dioxide.
    a. Containing anthocyanin and related (red) pigments—carbonated burgundy, and several proprietary red wines.
    b. Not containing red pigments—some Swiss and French types but increasingly rare.
B. Wines without obvious excess carbon dioxide.
  1. Wines 8–14% alcohol.
    a. Wines with anthocyanin and related (red) pigments.
      (1) Pink wines.
        (a) Dry—pink or rosé types and varietal types, such as Gamay, Grenache, Grignolino, and Tavel.
        (b) Sweet—Aleatico (see, however, 3a(1) below).
      (2) Full red color.
        (a) Dry (below 0.5% sugar).
          -1- With distinguishable (usually) varietal aromas.
            -a- With high acidity—Barbera.
            -b- With moderate acidity—Barolo, Beaujolais, Bordeaux (Médoc, St.-Émilion, etc), Burgundy, Cabernet Sauvignon or C. franc, Châteauneuf-du-Pape, Chianti, Fresia, Gamay, Hermitage, Petite Syrah, Pinot noir, P. St. George, Pinotage, Rioja, Ruby Cabernet, Syrah (Schiraz), Zinfandel, etc.
          -2- Without (normally) distinguishable varietal aromas—California (etc) burgundy, Carignane (Carignan), California (etc) claret, California (etc) dry red table, California (etc) chianti, Charbono, Cinsaut, Malvoisie, Mourestel, Nebbiolo (usually), Valdepenas.
        (b) Sweet. Proprietary types with or without Concord aroma (see also 2a below), California (etc) red table, and California sweet red table.
    b. Wines without anthocyanin and related (red) pigments.
      (1) Wines usually with distinguishable varietal aromas.
        (a) Containing sugar—German Auslese and California "Late Harvest" types, and many non-Auslese German wines from the Moselle, Rhine, etc, regions, Hungarian Tokay, light muscat, light sweet muscat, Late Harvest California Sauvignon blanc, White Riesling, etc, some Loire wines, Sauternes, sweet Catawba, sweet Sauvignon blanc, sweet Sémillon, and various proprietary wines.
        (b) Not containing sugar—Catawba, Chablis, Chardonnay, Chenin blanc (Steen), Delaware, Folle Blanche, Flora, some French Colombard (Colombard), Gewürztraminer, Graves, some Gray Riesling, some Loire wines, Moselle, Müller-Thurgau, Pinot blanc, Rhine (Rheingau, etc), Sauvignon blanc, Sémillon, Sylvaner, and White Riesling, etc.
      (2) Wines usually without distinguishable varietal aromas.
        (a) Containing sugar—California (etc) "château" types, California (etc) sweet white table, sauterne, sweet sauterne, and various proprietary labeled types. (See 2b below).

(b) Not containing sugar—California (etc) dry sauterne, California (etc) rhine, California (etc) white table, Burger, some French Colombard, some Gray Riesling, Green Hungarian, Ugni blanc (Trebbiano).
2. Wines with 14–17% alcohol.
   a. Containing anthocyanin or related (red) pigments—miscellaneous sweet red types—mainly with proprietary names.
   b. Not containing red pigments—miscellaneous sweet white types—mainly with proprietary names.
   c. Special types.
      (1) Blending—to increase the alcohol or to impart a special flavor to other wines or even other products, eg, whiskey.
      (2) Ecclesiastical, usually sweet (eg, *vino santo*) and various wines with dessert-type names but specially produced for special markets.
      (3) Fino and manzanilla types in Spain.
3. Wines with 17–21% alcohol.
   a. Containing anthocyanin and related (red) pigments.
      (1) With a muscat flavor
         (a) Pink—Aleatico (see also B1a(1) above).
         (b) Red—red or black muscatel.
      (2) Without a muscat flavor.
         (a) With a baked odor—California tokay.
         (b) Without a baked odor.
            -1- Brownish red—tawny port.
            -2- Red—port (including ruby, tawny, and vintage port).
   b. Not containing red pigments.
      (1) With a muscat aroma—Muscat blanc (ie, Muscat Frontignan and Muscato Canelli), Samos, Setúbal (Portugal), Sitges (Spain), and Australian and California muscatel.
      (2) Without a muscat aroma.
         (a) With a special odor due to treatment or aging.
            -1- With a raisin, cooked, reduced-must, or rancio odor.
               -a- Raisin odor—Malaga.
               -b- Baked odor—California (etc) sherry (dry, medium, or sweet), Madeira (usually sweet).
               -c- Reduced-must or burnt odor—Marsala.
               -d- Rancio (aged, slightly oxidized) odor—Banyuls (may have tawny color), Tarragona (such as Priorato), etc.
            -2- With a film yeast odor.
               -a- Dry (<1% sugar)—Spanish sherry (flor or fino, amontillado, etc), Australia, California, Cyprus, South Africa, and Soviet Union (various dry or cocktail sherries), Château Chalon (France).
               -b- Sweet—California medium or sweet flor sherry, some sweet Spanish sherry types.
         (b) Without a special odor due to treatment or aging.
            -1- With amber color—Angelica (California) white port

(Portugal), and some oloroso types (Spanish and others).
-2- Without amber color—white port (California).
II. Wines with added herbs or plant materials containing alkaloids or other flavoring materials.
  A. With a red color.
    1. Proprietary types—includes certain wines containing quinine and similar additives (Byrrh, Dubonnet, Campari, etc).
    2. Medicinal or home-produced types, such as iron- or herb-containing.
  B. Without a red color.
    1. Nearly dry—dry or French-type vermouth.
    2. Sweet (usually with a muscat flavor)—sweet or Italian-type vermouth.
    3. Proprietary types of fruit-flavored dessert and table wines (Thunderbird, Silver Satin, etc).
    4. Medicinal or home-produced types, such as gentian, rhubarb, dandelion, etc (see also A, 2 above).

Note. Only the more important types have been included in this classification with particular emphasis on the wines produced in, or imported into, the United States. Proprietary wines are those having a name that is owned by a producer. All wine types are in lower case except varietal types, proprietary types, and those having a specific legal regional significance. Thus, California burgundy is not capitalized as there is no geographical significance to the name as employed here. On the other hand, Burgundy means a wine from the region of that name in France. The "etc" following California usually represents other U.S. wines, such as New York or Ohio wines.

Legal Restrictions

Not only is the composition of the wines offered for sale subject to legal restrictions, but every producer of U.S. wines, except those produced in limited quantities for home consumption, must secure a Federal permit and take out a bond before beginning operations. Since this basic permit may be canceled for willful violation of Federal laws or regulations, the government possesses a powerful tool to deter violations. The regulations are frequently changed, and the regional office of the Bureau of Alcohol, Tobacco and Firearms of the Department of the Treasury should be consulted by anyone considering the wine business as to the necessary forms, bonds, and other papers which must be submitted. The bonded premises must be properly posted and protected by locks, and possess certain equipment (such as scales, ebulliometer) to facilitate inspection by government agents (commonly called gaugers). The capacity of all tanks must be recorded, and accurate records of production and movement of the wines and permissible additives must be maintained. The addition of sugar is strictly regulated and is forbidden in the production of most California wines. State laws are not as strict as Federal laws. In many countries, the regulations of production and labeling of wines are very detailed and require expert interpretation.

In most countries, the sale of wines is subject to permits, regulations, and taxes. The tax varies from state to state (or country to country) and is usually a function of the alcoholic content or type of wine. Most countries levy an import tax. In some cases, special tax stamps are purchased and affixed on the bottle or container. Special state taxes for state-sponsored advertising and research may also be assessed.

The addition of alcohol or herbs (fortification) to dessert wines and the production of vermouth and sparkling wines are subject to regulations. In some countries, the regulations are less stringent. Sweetening agents used for sparkling wines must be declared to the appropriate government agency.

A recent court decision in the United States requires the manufacturers of alcoholic beverages to disclose the ingredients of their products (9) but the case is on appeal.

Production

The special character of the fermentation of the various types of wines depends to a considerable extent on the composition of the fruit juice fermented (see also Fermentation).

Composition of Grapes. Grapes contain ca 15–25% sugar; partially dried grapes contain 30–40%. The percentage of sugar in the grapes, the extent of the fermentation, and the losses or additions of alcohol during treatment and storage determine the percentage of alcohol in the finished product. Since at least 9 vol % alcohol is usually necessary to prevent rapid acetification or spoilage, sugar must sometimes be added (chaptalization) to permit fermentation to reach the required alcohol content. The approximate amount of sugar necessary is calculated from the fermentation equation; ca 1% sugar fermented yields 0.55 vol % alcohol.

$$C_6H_{12}O_6 \rightarrow 2\ C_2H_5OH + 2\ CO_2$$

A minimum sugar content of ca 16.4% is therefore necessary to produce a wine of 9% alcohol. The soluble solids content of mature grape juice is >90% sugar. Usually, the sugar content of grape juice is determined with a hydrometer. The Brix or Balling hydrometer, used in the United States, reads in grams of sugar per 100 grams of liquid. In Europe, the specific gravity hydrometer (Oechsle) or other special hydrometers are employed; in Australia, the Baumé hydrometer is used. Sugar content can also be determined with a refractometer. For fruits, the hydrometer or refractometer is usually employed (see also Sugar, sugar analysis).

In the United States, it is permitted to add sugar to the must in all states except California. In Europe, sugar is added most frequently in Germany and occasionally in Switzerland and parts of France. Elsewhere, the sugar content of grapes is usually high enough to produce a wine with at least 9% alcohol. Although addition of sugar is prohibited in California, in the cool year of 1948 some grape concentrate (ca 70 wt % sugar) was added to musts (grape juice) from the cooler regions to ensure wines of >10% alcohol. Other fruits and vegetables contain smaller amounts of sugar, and supplementation is usually essential. Sucrose or invert sugar is commonly used. More sugar is added to fruit than to grapes, since the finished wine is usually not fermented dry, ie, without sugar, but is allowed to remain sweet. The chemical composition of wine and must is given in Table 2.

The organic acids are the second most important constituents. Grapes contain mostly tartaric and malic acids. The content varies with the state of maturity (decreasing as the grapes ripen), the variety, and the climatic conditions of the season or region (lower in warmer regions). It is always higher under cool conditions and may

Table 2. Composition of Grape Must[a] and Wines

Component	Must, mg/L	Wine, mg/L	Component	Must, mg/L	Wine, mg/L
Water	70–85[b]	60–85[b]	benzaldehyde [100-52-7]	trace	?
Carbohydrates	15–25[b]	0–20[b]	trans-2-hexenal [6728-26-3]	trace	?
glucose [50-99-7]	8–13[b]	0.1–10[b]	Acids	0.3–1.5[b]	0.3–1.2[b]
fructose [30237-26-4]	7–12[b]	0.1–10[b]	formic [64-18-6]	?	1.5–6.3
pentoses	10–200[b]	100–500[b]	acetic [64-19-7]	trace–0.02[b]	0.03–0.15[b,f]
arabinose [5328-37-0]	40–130[b]	trace	2-methylpropanoic [554-12-1]	?	trace
rhamnose [10485-94-6]	2–40[b]	trace	butanoic [107-92-6]	?	<0.5
xylose [58-86-6]	trace	trace	2-methylbutanoic [116-53-0]	?	>0.5
pectin [9000-69-5]	40–900[b]	30–500[b]	3-methylbutanoic [2835-39-47]	?	<0.5
inositol [87-89-8]	2–8[b]	trace			
Alcohols and related compounds			pentanoic [109-52-4]	?	trace
methanol [67-56-1]	trace	70–140	hexanoic [142-62-1]	?	1–>3
ethanol [64-17-5]	trace	5.6–17[b]	2-ethylhexanoic [149-57-5]	?	trace
1-propanol [71-23-8]	trace	15–50	cis-3-hexenoic [4219-24-3]	?	trace
2-propanol [67-63-0]	?	trace	heptanoic [111-14-8]	?	trace
1-methylthio-1-propanol [3877-15-4]	?	0.5–2	octanoic [124-07-2]	?	2–>4
2-methyl-1-propanol [78-83-1]	trace	33–150	nonanoic [112-05-0]	?	trace
			decanoic [334-48-5]	?	0.5–1
1-butanol [71-36-3]	trace	trace–5	9-decenoic [14436-32-9]	?	0.1–0.5
2-butanol [78-92-2]	trace	trace–40	undecenoic [1333-28-4]	?	trace
2-methyl-1-butanol [137-32-6]	trace	13–50	dodecanoic [143-07-7]	?	trace
3-methyl-1-butanol [123-51-3]	trace	50–160	lactic [50-21-5]	?	0.01–0.3[b]
			oxalic [144-62-7]	?	11–90
2,3-butanediol (threo) [513-85-9]	?	340–680	mesoxalic [473-90-5]	?	trace
			malonic [141-82-2]	?	trace
1-pentanol [71-41-0]	trace	trace–0.4	succinic [110-15-6]	?	0.01–0.20[b]
2,3-pentanediol [42027-23-6]	?	7–18	2,3-dimethyl succinic [13545-04-5]	?	trace
1-hexanol [111-27-3]	trace	0.5–3	3,3-dimethyl succinic [597-43-3]	?	trace
1-heptanol [111-70-6]	trace	trace			
1-octanol [111-87-5]	trace	0.2–1.5	ethyl acid succinate [636-48-6]	?	0.1–>0.5
benzyl alcohol [100-51-6]	?	0.1–0.3			
2-phenethanol [60-12-8]	trace	15–105	glutaric [110-94-1]	?	1–10
linalool [78-70-6]	trace	trace–0.4	adipic [124-04-9]	?	trace–30
nerol [106-25-2]	trace	trace	pimelic [111-16-0]	?	trace–20
geraniol [106-24-1]	trace	trace	azelaic [123-99-9]	?	trace–30
3,7-dimethyl-1,5,7-octatrien-3-ol [29957-43-5]	trace	trace–0.25	sebacic [111-20-6]	?	trace–10
			fumaric [119-17-8]	?	21–52
α-terpineol [98-55-5]	trace	trace–0.4	furan 3,4-carboxylic [3387-26-6]	?	trace–10
sorbitol [50-70-4]	trace	trace[b]			
glycerol[c] [56-81-5]	trace	0.4–2.5[b]	benzoic [65-85-0]	?	trace
Ketones and aldehydes[d]			2-phenylacetic [103-82-2]	?	trace
acetone [67-64-1]	trace	trace–0.04	2-phenylpropanoic [501-52-0]	?	trace
4,5-dimethyl-3-hydroxy-2(5H) furanone [28664-35-9]	?	trace	glycolic [79-14-1]	?	trace–40
3-hydroxy-2-butanone [513-86-0]	?	0.8–3.6	3-phenylpropenoic [621-82-9]	?	trace
			cis-aconitic [499-12-7]	?	trace–50
3-methyl-2-butanone [563-80-4]	trace	?	tricarballylic [99-14-9]	?	trace–10
2,3-butanedione [431-03-8]	?	0.1–3	2-methyl-2,3-dihydroxybutanoic		
2,3-pentanedione [123-54-6]	?	0.02–0.3	(threo) [14868-24-7]	?	40–200
3-hydroxy-2-pentanone [3142-66-3]	?	0.4–2.8	(erythro) [19774-31-3]	?	trace–95
			trans-geranoic [4698-08-2, 4613-38-1]	?	<0.1
β-ionone [14901-07-6]	trace	trace			
acetaldehyde [75-07-0]	trace	1–>300[e]	ascorbic [50-81-7]	10–180	?
propanal [123-38-6]	trace	?	malic [6915-15-7]	0.03–0.6[b,e]	0.001–0.5[b,e]
1-pentanal [110-62-3]	trace	?	2-methylmalic [2306-22-3]	?	1.9–15
butanal [123-72-8]	trace	?	tartaric [87-69-4]	0.5–1.0	0.4–0.8[b,e]
hexanal [66-25-1]	trace	trace	citric [77-92-9]	10–50	130–400[e]

Table 2 (continued)

Component	Must, mg/L	Wine, mg/L	Component	Must, mg/L	Wine, mg/L
isocitric [320-77-4]	?	trace–60	isopentyl acetate [123-92-2]	trace	trace–8
salicyclic [69-72-7]	?	<0.1	methyl acetate [79-20-9]	trace	trace
glyoxylic [298-12-4]	?	trace–6	propyl acetate [109-60-4]	trace	?
pyruvic [127-17-3]	?	8–50	pentyl acetate [628-63-7]	trace	?
oxalacetic [328-42-7]	?	trace–30	2-phenethyl acetate [103-45-7]	?	trace–2.6
2-ketoglutaric [328-50-7]	trace	30–60	phenyl acetate [122-79-9]	trace	?
levulinic [123-76-2]	?	trace–40	nonyl acetate [143-13-5]	trace	?
gluconic [133-42-6]	?	trace–3090[c]	octyl acetate [112-14-1]	trace	?
glucuronic [576-37-4]	trace	1–140[c]	hexyl acetate [142-92-7]	trace	0.2–2
galacturonic [14982-50-4]	?	10–>2000[c]	ethyl propanoate [105-37-3]	trace	trace–1.2
tetrahydroxyadipic [526-99-8]	?	trace–>650	pentyl propanoate [624-54-4]	trace	?
2-furoic [88-14-2]	?	trace–30	propyl propanoate [106-37-3]	trace	?
4-hydroxybenzoic [99-96-7]	?	<0.1	ethyl butanoate [105-54-4]	trace	trace
3,4-dihydroxybenzoic [99-50-3]	?	1–5	hexyl butanoate [1117-59-5]	trace	?
4-hydroxy-3-methoxybenzoic [121-34-6]	?	1–5	isobutyl butanoate [539-90-2]	trace	?
2,6-dihydroxybenzoic [303-07-1]	?	1–5	isobutyl isobutanoate [97-85-8]	trace	?
2,5-dihydroxybenzoic [490-79-9]	?	1–5	hexyl isobutanoate [2349-07-7]	trace	?
3,5-dimethoxy-4-hydroxy-benzoic [530-57-4]	?	1–5	isopentyl butanoate [106-27-4]	trace	?
3,4,5-trihydroxybenzoic [145-91-7]	?	1–5	ethyl pentanoate [539-82-2]	trace	trace
3,4,5-trihydroxycyclohexen-(1)-carboxylic [138-59-0]	?	1–5	hexyl pentanoate [1117-59-5]	trace	?
			isobutyl pentanoate	trace	?
1,3,4,5-tetrahydroxycyclo-hexen(1)-carboxylic [77-95-2]	?	1–5	ethyl hexanoate [123-66-0]	trace	0.1–2
			methyl hexanoate [106-70-7]	trace	?
			ethyl octanoate [106-32-1]	trace	0.2–1.5
eleagic [476-66-4]	?	1–5	ethyl nonanoate [123-29-5]	trace	?
2-hydroxycinnamic [583-17-5]	?	>0.1	ethyl decanoate [110-38-3]	trace	trace–0.3
			methyl anthranilate [134-20-3]	trace	trace–3
4-hydroxycinnamic [7400-08-0]	?	0.5–2.0	ethyl stearate [111-61-5]		
3,4-dihydroxycinnamic [331-39-5]	?	0.5–2.0	ethyl 2-hydroxypropanoate [97-64-3]	?	1–40
3-methoxy-4-cinnamic [1135-24-6]	?	0.1–0.5	isopentyl 2-hydroxypropanoate [19329-89-6]	?	trace–0.06
3,4-dimethoxy-4-cinnamic [530-59-6]	?	0.1–0.5	ethyl 2-methylbutanoate [7452-79-1]	?	trace
chlorogenic [327-97-9]	?	0.5–2.0	ethyl 3-methylbutanoate [108-64-5]	?	trace
isochlorogenic [14534-67-3]	?	<1	ethyl laurate [106-33-2]	trace	trace–0.4
neochlorogenic [906-33-2]			diethyl malonate [105-53-3]	trace	trace
nicotinic [59-67-6]	0.16–0.42	trace–10	diethyl succinate [121-75-5]	trace	0.8–2.0
2-carboxy-5-methoxy-indol [4382-54-1]	?	1–10	diethyl glutarate [818-38-2]	trace	trace
			ethyl phenylacetate [101-97-3]	trace	trace
3-indolylacetic [87-51-4]	?	1–10	isopentyl octanoate [2035-99-6]	?	trace
3-indolylacrylic [1204-06-4]	?	1–10			
pantothenic [79-83-4]	0.5–1.4	0.5–1.9	ethyl linoleate [544-35-4]	?	0.8
glyceric	?	trace	hexyl hexanoate [6378-65-0]	trace	?
sulfuric	0	0–350	isobutyl hexanoate [105-79-3]	trace	?
carbonic [463-79-6]	trace	<392			
Esters			isobutyl laurate [37811-72-6]	trace	?
ethyl formate [109-94-4]	trace	trace–8	hexyl laurate [34316-64-8]	trace	?
benzyl acetate [140-11-4]	?	trace–0.26	ethyl myristate [124-06-1]	trace	?
butyl acetate [105-46-4]	trace	?	dimethyl phthalate [131-11-3]	?	?
ethyl acetate [141-75-6]	trace	35–285			
geranyl acetate [16409-44-2]	trace	trace	pentyl octanoate [638-25-5]	?	?
isobutyl acetate [123-86-4]	?	trace–0.2			

Table 2 (continued)

Component	Must, mg/L	Wine, mg/L	Component	Must, mg/L	Wine, mg/L
3-(methylthio)-propyl acetate [13327-56-5]	?	trace	magnesium	100–250	100–200
ethyl lactate [97-64-3]	?	6.8–10	calcium	40–250	10–210
Polyphenol and related compounds			sodium	trace–200	trace–440
			iron	trace–30	trace–50
anthocyans[e]	trace	trace	aluminum	trace–30	trace–70
chlorophylls [479-61-8, 519-62-0]	trace	trace	manganese	trace–51	trace–50
			copper	trace–3	trace–5
xanthophyl [127-40-2]	trace	trace	boron	trace–70	trace–40
carotene [various]	trace	trace	rubidium	trace–1	trace–4
quercitin [117-39-5]	trace	trace	phosphate	200–500	30–900
quercitrin [522-12-3]	trace	trace	sulfate	30–3500	30–2200
kaempferol [520-18-3]	trace	trace	silicic acid	2–50	2–50
rutin [153-18-4]	?	4–10	chloride	10–100	10–600
catechin [154-23-4]	0.1–10	10–300	fluoride	trace	1–10
gallocatechin [970-73-0]	<1		iodide	trace	trace–10
gallocatechin gallate [27289-24-3]	trace		*Miscellaneous*		
			styrene [100-42-5]	?	0–0.1
epicatechin gallate	trace–<0.5		nerol oxide [1756-08-9]	trace	trace–0.1
gallic acid [149-91-7]	trace		*cis*-rose oxide [16409-43-1]	trace	trace
Nitrogenous compounds			*trans*-rose oxide [35598-65-3]	trace	trace
total	300–1700	100–900	*cis*-linalool oxide [5989-33-3]	trace	trace–0.26
protein	10–1000	10–30	*trans*-linalool oxide [34995-77-2]	?	trace–0.15
humin	10–20	10–20			
amide	10–40	10–80	ethanethiol [75-08-1]	?	trace
ammonia	10–12	0–200	dimethyl sulfide [75-18-3]	?	trace–0.4
residual	10–20	50–200	ethyl methyl sulfide [624-89-5]	?	trace
amino	170–1100	100–2000			
4,5-dimethyl-1,3-di-oxolane-2-propanamine [85236-72-2]	?	trace	diethyl sulfide [352-93-2]	?	trace
			diisopropyl sulfide [625-80-9]	?	trace
			ethyl-*n*-propyl sulfide [4110-50-3]	?	trace
1-pyrroline [5724-81-2]	?	trace	*tert*-butyl sulfide [592-65-4]	?	trace
Mineral compounds	0.3–0.5	0.15–0.40	isobutyl sulfide [592-65-4]	?	trace
potassium	0.15–0.25	0.045–0.175			

[a] Refs. 5–8.
[b] g/100 mL.
[c] Except more for botrytised grapes.
[d] Numerous acetals, lactones, secondary acetamides, and phenols have been identified in wines but quantitative data are lacking.
[e] Depends on variety and climatic conditions and cellar treatment.
[f] More in spoiled wines.

sometimes impart an unpleasantly tart taste. This occurs in the eastern U.S., Canada, Switzerland, and the FRG in some years. Regulations, different in each country, permit the use of water, calcium, or potassium carbonate or ion exchange to reduce the acidity. The pH of normal grape juice in moderate climatic zones is 3.0–3.6, and the titratable acidity is 0.5–1% (calculated as tartaric acid). In this range, most deleterious organisms grow slowly or not at all, thus allowing rapid growth of the desirable yeast. The relatively high titratable acidity and low pH of musts aid in the extraction of color from the skins and in wine clarification. In other fruits, the acidity is due to malic, citric, oxalic, and isocitric acids in varying proportions. This acidity is usually high enough to permit a disease-free fermentation and a stable product, unless diluted with too much water. Moreover, some fruit and berry wines must be sweetened to mask excessive acidity.

Only a small amount of nitrogenous material (0.3–1.0%) is found in grapes. However, this material is of considerable significance for yeast nutrition, bacterial stability, and flavor development, presumably because of the many amino acids present (see also Amino acids). During fermentation, the total amino acid content decreases, although the content of some acids may be higher in the finished wine than in the must because of their release by autolysis of yeast cells. The most important amino acids reported in grape juices or wines are alanine [56-41-7], arginine [74-79-3], aspartic acid [56-84-8], cystine [56-89-3], glutamic acid [56-86-0], glycine [56-40-6], histidine [71-00-1], isoleucine [73-32-5], leucine [56-87-13], lysine [56-87-1], methionine [63-68-3], phenylalanine [63-91-2], proline [147-85-3], serine [56-45-1], threonine [72-19-5], tryptophan [73-22-3], tyrosine [60-18-4], and 1-valine [72-18-4]. In other fruits, yeast propagation is limited by low nitrogen, and a nitrogen-containing compound is added to stimulate yeast growth (see also Yeasts). Apple and pear juices, for example, ferment slowly for this reason.

The pigments of grapes and fruits are usually located in the epidermal cells. A few varieties have red juice as well. During alcoholic fermentation, the cells are killed and these pigments are released. By separating the skin of red grapes from the juice before fermentation, it is possible to produce a white or nearly white wine. Anthocyanin and related red pigments are responsible for the color and probably also aid in clarification. Red wine (from red grapes) also contains considerable tannins, which affect taste, color, oxidation–reduction potential, and rate of aging.

The pectins of some fruits and grapes are a source of difficulty in juice clarification. They are rather insoluble in alcohol and precipitate during alcoholic fermentation.

The inorganic constituents are not of critical importance as they are usually present in sufficient amounts to catalyze yeast or enzyme functioning; excess iron or copper may cause turbidity. The high potassium–sodium ratio is believed to be of interest to persons with hypertension. The small amounts of copper and zinc may have some nutritional value.

Microorganisms. Wines are normally produced by fermentation with the yeast *Saccharomyces cerevisiae*, sometimes with *S. bayanus* or *S. oviformis* (10). Taxonomists are still reclassifying the genus *Saccharomyces*. These and other yeasts, found in grapes or fruit, multiply rapidly in the sweet juice, eventually causing fermentation. Although this process is adequate for grapes under most conditions, it may be inadequate for fruit. Under unfavorable climatic conditions, it may also be inadequate for poor quality grapes. For this reason, it is customary to add a pure culture of fermenting yeast. Numerous strains of *S. cerevisiae* are available, which, however, all produce very similar results. Pure yeast cultures are usually added at a rate of ca 1–3%. The actively fermenting culture is grown in sterilized must. Pressed wine yeasts grown in nongrape media have also been used.

To prevent growth and competition of undesirable organisms, 50–200 mg sulfur dioxide are usually added per liter ca 2 h before the pure yeast culture is added. The sulfur dioxide acts as a selective antiseptic and permits more or less unrestricted growth of the added yeasts. Originally, a piece of sulfur in the form of a wick was introduced into the cask and burned. When the must was introduced, as much as ≥25 mg of sulfur dioxide per liter might be absorbed from the air on the moist wooden walls of the cask. However, the amount of sulfur dioxide introduced is not easily controlled, because elemental sulfur sublimes onto the walls of the cask or pieces of the sulfur wick drop onto the bottom of the cask. During fermentation, this sulfur is reduced to hydrogen sulfide, imparting a very unpleasant odor. For this reason, salts that yield sulfur dioxide, such as potassium metabisulfite (pyrosulfite), aqueous solutions of sulfur dioxide, or the liquefied gas, are commonly used. The warmer the must and the poorer its quality, the more sulfur dioxide is needed (usually >150 mg/L). The sulfur dioxide kills or inhibits the growth and activity of undesirable bacteria and yeasts, increases the extraction of color and soluble material from the skins, and acts as an antioxidant. The resulting wines are thus of higher alcohol, extract, and total acid content, lower in volatile acidity, lighter in color in the case of white wines, and somewhat darker in the case of red wines, than those produced without sulfur dioxide. Other components, such as sugar, water, acid, or nitrogenous materials, eg, urea or ammonium phosphate, are added at the same time.

Equipment. Wine production can be very simple, but for large-scale operation, specialized equipment has been developed.

The grapes are crushed in combined stemmers and crushers. These remove the stems first by centrifugal force and then crush the berries. The actual crushing may be done by centrifugation or by passing the fruit through rollers or by both. Must pumps transfer the crushed grapes (must) to the fermentors or presses.

Fermentation tanks may be of wood, concrete, stainless steel, or iron lined with epoxy resins or a thin layer of stainless steel; they may be open or closed. Open wooden or concrete tanks were formerly used for red musts and closed containers for white musts. Today, stainless-steel or lined iron tanks are used for both red and white fermentation, for the storage of wines during clarification, for early maturing white table and dessert wines, and for blending and storage before bottling. Large open tanks may have coils, preferably of stainless steel, for cooling or heating, but most are now partially or wholly jacketed. In some wineries, the temperature is computer controlled.

After fermentation, the residue, ie, stems and skins, often called pomace or marc, must be transferred from the fermentor to the press. In some wineries, electric elevators are lowered into the tank and the pomace is raised to the top. From there, it is either dumped directly into the press or into a trough with a continuous belt or chain to carry it to the press. In other wineries, the fermentation tanks are raised above the floor and the tank's floor steeply slanted. The pomace is flushed with wine from the bottom into the conveyor and to the press. This wine can be used for distillation; water can also be used for the flushing, but the diluted wine can only be used for distillation.

The oldest type of press still in use is the screw-type basket press, usually operated vertically, either from above or below. It has been largely replaced by the horizontally operated hydraulic press. This press is more expensive to operate but produces a relatively clear juice from both red pomaces or white musts.

Another type has an inflatable rubber bag inside a cylinder with perforated holes. The cylinder is filled with crushed grapes. When the bag is inflated, the must in the cylinder is pressed. It is reported that computer-programmed operation extracts more juice with less pressure. Belt presses that crush and press are also available.

The continuous press, the most popular type, does not operate as well on fresh must, but is the cheapest to operate with fermented pomace and gives a high yield of liquid, albeit cloudy. When the press wine is distilled, the cloudiness is of no importance.

For some fruits, and occasionally for high pectin grape musts, the rack-and-cloth press was formerly used. This operation is now seldom seen, except possibly for apples and other fruits. The must is contained in cloth sacks which are placed between wooden racks. A pile of these is then pressed in a hydraulic press. The yield of clear juice is better than that obtained with the other types, but the operation is more expensive and the press cloths are difficult to clean.

White Table Wines. White wines of <14% alcohol are designated as white table wines. They are usually made from white grapes. Occasionally, a white wine is made from red grapes (as in the Champagne district of France) by separating the skins from the juice immediately after crushing.

Dry. A good dry white wine retains a light color and a fresh, unoxidized flavor. The key to its production is rapid transportation of the grapes from the vineyard to the winery. Crushing, stemming, and pressing should follow immediately after picking. The press juice is dark in color and high in tannins and may be settled before fermentation; that is, the juice is placed in a tank, and 50–150 mg of sulfur dioxide per liter are added and allowed to settle for 24–36 h. During this time, pieces of skin and other solid material settle at the bottom. The clear supernatant liquid is drawn off and used for the fermentation. Settling is carried out more satisfactorily in smaller containers (<7.5 m^3). It is most useful with varieties that yields a pulpy must and where a new wine is desired that clarifies rapidly. Musts, especially those from moldy fruit, are also clarified by centrifugation. Clear musts ferment slowly but more cleanly.

Some wine makers favor a skin maceration period of 6–24 h to increase flavor extraction. This procedure may result in darker-colored wines.

The juice is inoculated with a pure yeast culture. Fermentation is conducted in closed containers at ≤15.6°C. Many white wines are fermented at 6–10°C, but more time is required for fermentation at these temperatures. When fermentation is conducted in tanks of >100-hL capacity, the carbon dioxide given off by fermentation prevents the oxygen from reaching the wine. For smaller containers, fermentation traps permitting outflow of gas but preventing inflow are often employed.

The temperature and percentage of sugar (degree Brix with the hydrometer) should be determined daily during the fermentation. If the temperature rises much above 16°C, it may be necessary to cool the must by pumping a cooling liquid into the jacketed tanks or by pumping the must through a tubular heat exchanger or into a sump with cooled pipes. If the fermentation stops, refermentation may be necessary.

As the fermentation nears its end (the degree Brix is close to 0°), the tank should be filled with another wine of the same type, closed with a trap, or covered with a thick cloth to prevent air reaching the wine. When the Brix value reaches ca −1°, the wine is racked (drawn off or pumped) into another container. This container should be filled completely and the bung tapped in gently. Each day thereafter until fermentation ceases the bung can be loosened to release gas pressure. If the wine is stored at ca 10°C, it normally settles in ca 6 wk, ie, yeast cells, mucilaginous material, and cream of tartar (potassium acid tartrate) precipitate (see Tartaric acid). The settling is slower at higher temperature and with large containers. Whether the settling is complete or not, the wine must be racked from the sediment after ca 6 wk. This is particularly important if the wine is stored at somewhat higher temperatures since autolysis of the yeast cells may occur and hydrogen sulfide and other undesirable compounds are produced. New white table wines are often centrifuged for rapid clarification.

In transferring white wines from one tank to the other (racking), contact with oxygen must be avoided unless the wine has a yeasty or hydrogen sulfide odor, in which case some aeration is helpful. The second tank may be filled with carbon dioxide gas, or the wine may be pumped from the first tank to the bottom of the second, or, in small-scale operations, a sulfur wick may be burned in the second tank.

Dry white table wines are commonly bottled during the first year. The wine thus receives only one or two rackings before it is bottled. Stabilization procedures are necessary to ensure that the wines remain clear.

In general, the wine is passed through a filter press at the beginning of the year. Later in the spring, it is chilled to −4°C and held at this temperature for several days to precipitate the excess tartrates. Alternatively, the wine may be passed through a cation exchanger to exchange the potassium for sodium or hydrogen. This increases the sodium content, however, and is not favored. In some countries, eg, the FRG, the sodium content cannot exceed 60 mg/L. The wine is then racked and clarified (fined) with a slurry of bentonite or kieselguhr and is filtered, under exclusion of air, directly into the bottle. Modern and careful wineries usually make a test bottling which they subject to heat, cold, and sunlight to determine whether the wine remains brilliant. Wines with residual sugar are usually filtered through membrane filters. Just before bottling, the sulfur dioxide content should be adjusted. In the United States, enough sulfur dioxide is added to bring it to ca 150 mg/L.

Sweet. This class of wines is more difficult to produce and even more difficult to stabilize because the wines may contain less than 12% alcohol but 0.5–20 wt % sugar.

The production problem is twofold: First, the must has to be sufficiently high in sugar to produce ca 9–11% ethanol and yet have residual sugar. Second, the fermentation must be stopped before the ethanol exceeds 14%, and this is sometimes difficult. In Europe, some wines of this type contain slightly more than 14% alcohol, which should be avoided because of American tax regulations and the high alcohol.

In the United States, a high sugar content in the grapes is secured by delaying the harvest of suitable varieties as much as possible. A long delay, however, may cause sunburn or raisining on the vines, which is undesirable in white table wines. It is sometimes necessary to add grape juice, dessert wine, or grape concentrate to the finished wine in order to secure the necessary sugar content.

In Europe and in some parts of California, high humidity permits the growth of the fungus *Botrytis cinerea* on the ripe grapes. This fungus loosens the skins and allows loss of water; the grapes shrivel and increase in sugar content sometimes to as much as 40 wt %. The Sauternes of France, the sweeter Tokays of Hungary, the sweetest German (Auslese) wines, and many late harvest wines in California, are all produced from this type of grape. In another method, the grapes are dried between straw, suspended from strings, or put in shallow boxes. After two or three months, the grapes are crushed and pressed. The first procedure is used to a limited extent in France, whereas the other two are often employed in central and northern Italy, especially for sacramental wines.

The grapes are crushed, stemmed, and pressed as for dry white wines. The pressing may be more forceful in order to obtain the maximum yield. Settling or centrifuging is definitely advantageous with this type of must. The fermentation is conducted at 4.4–7.2°C. Various methods of stopping the fermentation are employed. Frequent rackings during fermentation keep the yeast population low, the successive propagation of the yeast cells reduces the nitrogen content, and the increasing alcohol content in the presence of sugar restricts further yeast growth. Some operators rack off the yeast and add a large amount of sulfur dioxide. Cooling at the same time is helpful, and this is the usual procedure in California. The wine may even be filtered or centrifuged to remove as much of the yeast as possible.

The sweet new wine must be watched constantly for signs of refermentation, which may be prevented by cooling, racking, filtration, and addition of sulfur dioxide or sorbates. As the wine ages, the tendency to referment is reduced. Sweet table wines are frequently held in the cask for one or two years for stabilization.

The wine should be bottled only after a trial bottling and a stability test. Various methods for ensuring bottle stability are employed. Least desirable is a very high sulfur dioxide content, which affects the sensory quality of the wine. In pasteurization of standard-quality wines, the wine is first heat-stabilized to a temperature slightly higher than the final pasteurization temperature and then cooled and filtered to remove precipitated solids. Finally, it is pasteurized hot in the bottles, and sealed with screwcaps. More recently, centrifugation and sterile filtration through pad or membrane filters has been employed. A low filter pressure reduces the volume; careful control of the prefiltration sterilization is required. Enologists generally frown on pasteurization of high quality wines.

Although sweet table wines produced by fermentation are generally considered the best, a similar effect can be obtained by adding a sweetening agent to a dry table wine. In the eastern United States, sucrose is used for both white and red wines, usually a blend of California table wine and eastern Concord flavored wine. The sugar content is often 13–14 wt %. Pasteurization is usually required because germproof filtration is difficult with such sweet wines. Producers that require a low sugar content may blend in a sweet dessert wine such as Angelica or port. A number of white, red, and rosé table wines with ca 1.5–2.5 wt % sugar are produced in California by adding a dessert wine or a concentrate. In Europe, grape juice is occasionally preserved with sulfur dioxide (called muté), and this is added to obtain a white sweet table wine.

Recently, so-called light or soft wines, usually white, with a low alcohol content have appeared on the U.S. market. They are made from low sugar grapes by removing the alcohol from part or all of the wine and blending. They are sold as moderate wines, ie, moderate with respect to alcohol and caloric content.

Red Table Wines. Much of the world's wine is red table wine with <14% alcohol. In France and Italy, particularly, it is an important part of the daily caloric intake. Red wines are relatively easy to produce compared with white table wines, and they are less subject to spoilage or clouding during aging.

The grapes should be harvested when they are sufficiently ripe to produce 11–13% alcohol. A sugar content of ca 21.5–23.5% is best. The grapes should be transported from the vineyard to the crusher without delay to prevent development of spoilage bacteria. Prompt crushing and destemming are essential.

Fermentation. The primary problem in red-wine production is the management of the so-called cap, ie, the floating mass of skins that rises above the liquid during fermentation. If the cap is allowed to dry, it may acidify; however, this seldom occurs in closed stainless-steel tanks. Furthermore, if the cap is not periodically submerged into the liquid, too little color is extracted from the skins. With small fermentors, the cap can be submerged manually. With very large containers (>7.6 m^3 or >2000 gal), the cap is too heavy to be forced down with wooden paddles. The juice is then pumped from the bottom of the container and sprayed over the cap. Instead of allowing the cap to float freely, various submerging systems have been devised. The oldest was a wooden latticework that fitted into the tank. The must was introduced under the framework, and when fermentation started, the cap was retained under the lattice while the juice rose and covered it.

In older systems, a permanent tank cover was constructed with a narrow opening into a basin. Sufficient must was introduced into the tank in such a way that when the fermentation started, the cap pressed against the cover and the juice rose through the narrow opening into the basin. With both these systems, good color and flavor were extracted by intimate contact of the cap and the liquid. The disadvantage of the submerged-cap systems is the management of the wooden latticework, the possibility of too much oxidation because of the free-floating surface of the fermenting must, and an excessively rapid fermentation.

Methods of making red wines include fermentations under pressure and continuous and automatic systems. For pressure fermentation, a metal tank lined with an inert material is used; a pressure of ca 300 kPa (3 atm) is maintained. This does not prolong the fermentation unduly and increases the color extraction. The cap, under a carbon dioxide atmosphere, cannot acetify. However, equipment for pressure fermentation is expensive.

Continuous systems are used in Argentina, Italy, the USSR, France, and Algeria, among others. A tall tank is used, and the sulfited must is introduced about midway. The pomace is continuously taken off the top, and the partially fermented wine is removed from the bottom. The fermentation may be contaminated by poor quality musts. Furthermore, it is difficult to provide grapes of the same variety for a sufficiently long time to make the system truly continuous. Automatic procedures provide automatic circulation of the must. The carbon dioxide pressure raises a portion of the must which then flows back over the cap. Construction costs are high, and the quality of the wine is not the best. In Burgundy and elsewhere, heat is used to extract the color of red grapes, in so-called thermovinification. The grapes are heated to 70–80°C and then pressed and cooled to 20–25°C before fermentation. Opinions differ as to the quality.

There is continuing interest in Europe in producing red wines from uncrushed grapes via the *macération carbonique* process. The uncrushed grapes are placed in closed tanks. Respiration occurs in the fruit and a high carbon dioxide atmosphere results in the death of the skin cells, releasing color. A slow fermentation occurs. The wine is light in color and tannins and has a special flavor which some find unpleasant and others pleasant. The process has not found wide favor in California because extra fermentors are required.

Red wines are normally fermented in contact with the skins until 70–90% of the sugar is used. Since the color pigments are extracted more rapidly than the tannin, skins and juice are separated as soon as the color extraction has reached its maximum, usually when ca 50–70% of the sugar has fermented. Moreover, where the grapes contain many shriveled or raisined berries, extended contact of the skins and juice increases the extraction of sugar from the high sugar fruit and may result in wines that contain too much alcohol.

Hydraulic basket presses are used for red wines, and if the pressure is not too high, operate very satisfactorily. At high pressures too much tannin, bitter-tasting material, and solids are extracted. The same is true for continuous presses. However, if the press juice is used for distillation, the residual solid material is not important; continuous presses are less expensive to operate.

After pressing, fermentation of the residual sugar requires one to six weeks, if the location is not too cold. The fermentation should be completed and no residual sugar left in the wine. Unless special precautions are taken, a low alcohol sweet wine is subject to spoilage by a variety of microorganisms, particularly if the pH is >3.6.

Aging. As soon as the sugar has fermented, the wine should be separated from the yeast deposit. Red table wines clarify more easily than white wines, and often a simple racking is sufficient; a rough filtration may also be necessary. In warm climates, where the total acidity is low, an early racking is very desirable. In cold climates, where the total acidity is high, the sediment (called lees) is kept in longer contact with the wine, which often contains bacteria capable of decarboxylating malic to lactic acid. This reduces the titratable acidity, and since lactic is a weaker acid than malic acid, a higher pH is obtained and the wine is less sour. Longer contact of sediment and wine also results in yeast autolysis which releases amino acids essential for the growth of the malolactic bacteria.

Because of their high tannin content, many red wines must be stabilized before bottling. The best red wines are aged for two or three years in wooden casks. The excess tannins are gradually oxidized or combine with aldehydes and precipitate; tartrate stability results. When red wines are bottled young, a special chilling removes excess tartrates. In addition, clarification (fining) with gelatin to remove excess tannin may be required. Just before bottling, a close (pad) filtration brings the wine to a perfect state of clarity. Even so, red wines frequently have a slight deposit when aged in the bottle for several years. If red table wines contain residual fermentable sugar, germ-proof filtration or hot bottling may be required.

Aging of red wines not only reduces excess sourness and tannins but also produces wines of special bouquets (by complicated oxidation–reduction–esterification reactions).

White table wines lose their yeasty odor during early aging. Many are then bottled and hardly change upon further aging. However, wines with a higher alcohol content, eg, Chardonnay, white Burgundy, Sauternes, Auslese, may be develop complicated and desirable bouquets in the bottle.

Sparkling Wines. Wines containing a permanent visible excess of carbon dioxide are called sparkling wines. The nomenclature of the sparkling wines is somewhat complicated. The most famous name is champagne, originally produced only in the region of that name in France where this appellation may be used only for wines produced and fermented in that region. However, the name is used in other countries as well, although local names are used in some countries, eg, Sekt in Germany and spumante in Italy. In the United States, the term champagne is used for most sparkling wines produced by a secondary fermentation of sugar in closed containers. If fermented in tanks, it must be so stated on the label.

In Europe, many wines are bottled with a slight residual sugar content. During aging this sugar may ferment, and the wines become slightly gassy; wines of Alsace, the Loire region, and Switzerland are frequently of this type. A certain amount of yeast growth is necessary to produce this gassiness, but the yeast deposit is often surprisingly small. At slightly higher temperature, this type of wine loses the gassiness that is one of its chief attractions; furthermore, the residual sugar may ferment, thereby producing a large yeast deposit which affects the odor.

Gassiness also results when the malic acid of the wine ferments, giving lactic acid and carbon dioxide. However, a high malic acid content is necessary for such fermentation. The vinho verde wines of Portugal are of this type, as are some gassy northern Italian wines.

Red wines must be a medium red without brown or violet overtones and have minimal tannin content. White sparkling wines must have a fresh fruity taste, have a light color, and be free of undesirable odors. Other requirements are given in the following list.

alcohol, %	10–11
pH	3.0–3.4
volatile acids	<0.04

Fermentation. The production of sparkling wines requires careful control of the harvesting, crushing, pressing, and fermentation. The harvest must be timed to obtain a sugar content adequate for at least 10% ethanol. However, it must not be delayed too long or the acidity is too low and the alcohol content will be too high. The crushing should be rapid and complete, followed immediately by pressing and settling. Where red grapes are used to produce white musts, as in the Champagne region of France, crushing may be omitted and the grapes sent straight to the presses. This gives a clearer and lighter colored must.

The must is fermented below 16°C and no sugar should remain. Pure yeast cultures of some agglutinating strains are employed. The new wines are clarified as rapidly as possible. They are cooled at 4.4°C for several weeks or chilled at ca −7°C for several days; tartrate stability is tested. The wines are fined and filtered.

Since only a rare single wine has all of the desired qualities, blending is common with wines of higher and lower alcohol content and acidity. However, low volatile acidity is essential as is absence of off-odors or off-tastes. Small amounts of citric acid are added to correct the total acidity. The free sulfur dioxide content should not exceed ca 5–10 mg/L in order not to interfere with the secondary fermentation or contribute an undesirable odor. The blended wine should be filtered and analyzed for sugar, iron, copper, and protein. Recommended maximums are 200, 5, 0.3, and 25 mg/L, respectively.

The total sugar content is then increased to 24 g/L. This is sufficient to give a pressure of ca 600 kPa (6 atm) in equilibrium with the wine if the carbon dioxide is not allowed to escape. In general, commercial invert sugar solutions are used, and occasionally sucrose. Rock candy is sometimes dissolved in wine containing ca 0.5% citric acid and heated to give a 50 vol % solution of invert sugar. An actively fermenting culture of *Saccharomyces cerevisiae* is then added at ca 1%. The so-called champagne strain is usually employed since it is an agglutinating type. To aid in later clarification, some producers also add a small amount of clarifying agent, such as bentonite, to the wine.

At this stage, the wine may either be transferred to special bottles (700–750 mL) or to stainless-steel pressure tanks of 20–1500 hL (500–40,000 gal) capacity for fermentation. In the tank process, the blending tank is used for the second fermentation. Although both procedures are used in the United States, the tank method is cheaper and less hazardous than the bottle procedure. The temperature is easily controlled, excess pressure can escape, and the wine can be stabilized in the tank. However, air comes in contact with the wine during the transfer from tank to tank since no counterpressure of carbon dioxide may be used in the United States. Fermentation in bottles is considerably more expensive; manual operations require great skill, and much equipment and time is needed. However, these wines have a lower aldehyde content and greater bouquet, presumably because of greater yeast autolysis during the longer aging period in bottles.

In the bottle operation, the blended, sweetened wine plus yeast are continuously agitated during bottling. Special bottles capable of withstanding a pressure of ca 800 kPa (8 atm) are used. They are closed with crown caps, which are cheap and conveniently attached and removed. The bottles are stacked in a constant-temperature room or, more commonly, held in bins. In the United States, the fermentation temperature is 16°C to ensure rapid and complete fermentation. However, the quality is enhanced at a lower temperature of 10°C. The fermentation takes from three weeks to six months, depending on the temperature and the particular wine. Wines with an alcohol content above 11.5% and lower nitrogen content may ferment more slowly below 10°C. Aging in the bottles in contact with the yeast for one to three years improves the quality (bouquet).

During fermentation, a yeast deposit forms on the lower side of the bottle, which must be removed before the wine is marketed. The bottles are placed upside down in special A racks to permit shaking the yeast deposit gradually down onto the cork. This procedure is called riddling and takes ca 3–6 wk. It is a highly labor-intensive hand manipulation which is being replaced by mechanized riddling racks or by mechanical shaking of large bins containing ca 500 bottles.

Then the deposit on the cork must be removed. The wine is under a pressure of at least 600 kPa (6 atm). The bottles are chilled to ca 2°C, and the necks are frozen solid by submersion in an ice–salt mixture or in special freezers. At this temperature the pressure is reduced considerably, and when the cork is removed the solid ice plug from the neck of the bottle containing the yeast deposit is ejected, carrying with it the yeast deposit. This procedure is called disgorging. Immediately after the plug is ejected, the bottle is temporarily closed to prevent loss of pressure. A small measured amount of sweetening agent and enough wine of the same kind to fill the bottle are then added. The sweetening agent (*liqueur d'expédition*) is usually a mixture of brandy, wine, and sugar. The formula is different for each company; however, American producers use little or no brandy. The best aged sparkling wines receive only enough of this mixture to bring the total sugar content to ca 1%. This is sold as brut or nature. The so-called dry (or sec) wines are sweeter (2–3% sugar), and the demi-sec may have ≥5% sugar. Very little sweet or doux sparkling wine, containing ca 10% sugar, is prepared. The bottle is closed with a large cork which is kept in place with a wire netting. The bottle is then ready for labeling and shipment. Losses during aging and disgorging reduce the pressure within the bottle of the finished wine to 300–400 kPa (3–4 atm). In many wineries, disgorging, sweetening, and corking operations are mechanized.

Instead of individually disgorging bottle-fermented wines, they are sometimes disgorged into a tank, sweetened, and then filtered into a bottle. No riddling is necessary. This procedure is called the transfer system. To counteract oxygen pick-up, a small amount of sulfur dioxide is usually added and a counterpressure of nitrogen applied. This procedure is less expensive.

In the tank system, two or more tanks are required. The first tank is jacketed for temperature control during fermentation and clarification. The fermentation is carried out at ca 16°C, since the main advantage of the tank system is rapid turnover of wine. Lower temperatures, ca 10°C, are reported to give a better quality. Fermentation is complete in 1–4 wk. The wine is then chilled and filtered to a second tank, where the required amount of sugar is added. Then the wine is bottled under pressure; sometimes the sugar solution is placed in the bottle first. The entire process may be completed in 4–6 wk. Sulfur dioxide is employed as an antioxidant to prevent fermentation of the added sugar and inhibit yeast growth. It may, however, impart an unpleasant odor. Since there is little time in the tank process for the yeast cells to die, a few viable yeasts may pass through the filter and into the wine. In the bottle process, when properly conducted, most of the yeast cells are dead after two or three years' storage.

Sparkling burgundy and other red or pink sparkling wines are produced by the same procedures. Because of their higher tannin content, they are frequently sweetened to 5% sugar to mask the astringency.

Carbonation. Instead of using costly fermentation processes to secure an excess of carbon dioxide, the gas may be added to the wine; good mixing is essential. Direct carbonation is conducted at a low temperature and with small bubbles. Carbonated wines are traditionally cheaper than fermentation-produced wines. Wines of lower quality and price may be used in their production. They may be older, but they should still retain a fruity flavor, have a light unoxidized color, and be free of off-odors. For the best flavor, the wines should have a relatively low redox potential; small additions of sulfur dioxide and ascorbic acid may be desirable. Before carbonation, the wine is clarified and the requisite amount of sugar added, ca 1–4%. Carbonation pressure is usually ca 400–500 kPa (4–5 atm). The same closure, cork or plastic, is employed as for other sparkling wines. Very few fully carbonated wines are now produced.

Dessert Wines. Less than 20% of all the Californian wines are fortified as dessert wines with an alcohol content of 17–21%. Both red and white dessert wines are produced; the white may be either dry or sweet (11). Worldwide, probably less than 10% of the wines are dessert types.

White Dessert Wines. Muscatel, Angelica, white port, and sweet sherry are the sweet dessert wines produced in the United States; dry sherry is the only dry type and it is often at least slightly sweet. Famous European dessert wines include Malaga (Spain), Marsala (Sicily), and Madeira (Madeira Islands).

For muscatel and related types, a must of high sugar content ($\geq 24\%$) is required. However, the pH should not be too high, and the grapes not raisined. They are crushed as for dry white table wines, but pressing is frequently dispensed with in California. Only the free-run juice is drawn off and used for wine. The sweet pomace is mixed with water and fermented. The wine thus produced is used for distillation. This procedure is economical as a large amount of alcohol is required in the production of dessert wines. Since only alcohol produced from wines may be used, considerable distilling wine is required. In the production of muscatel, the must and skins usually remain in contact for a day or two in order to increase the extraction of muscat flavor from the skins.

Normally ca 100 mg of sulfur dioxide per liter are added after crushing, and except at the beginning of the season, pure yeast cultures are seldom employed because the fermentation period is only 2–3 d. For example, a must of 25° Brix is fermented to ca 15° in order to produce a finished fortified wine of 12% sugar. At 15°, there is <5% alcohol. Since dessert wines are sold with 17–20% alcohol, ca 12–15% must be added. In the United States, this procedure is called addition of wine spirits or fortification; it is carried on under the strict control of the Bureau of Alcohol, Tobacco and Firearms of the U.S. Treasury Department. The fortifying brandy is ca 95% alcohol or 190 proof. Fortifications of 1000–4000 hL (25,000–100,000 gal) are commonly made in California.

The fortified wine is stored in tanks where the yeast is allowed to settle. The first racking and rough filtration follow a few weeks later. Dessert wines can be brought to brilliancy earlier than table wines because they are not very sensitive to the undesirable oxidative changes that occur with the repeated manipulations of table wines. Chilling, pasteurization, heavy fining, and repeated filtrations secure a brilliant wine. Some dessert wines reach the market within six months of the vintage and are stored in lined metal tanks. Others may be aged in casks or barrels for months to years. Although dessert wines profit by aging in wooden casks, this is done only where the original quality justifies it and when the demand for quality dessert wines is sufficient. Because of their high alcohol content, dessert wines are not as subject to acetification and thus can remain in the cask at higher temperatures for longer periods than table wines.

California white port is very young Angelica which has been partially decolorized with large amounts of charcoal or made from very light-colored free-run musts. It is seldom aged and production is small. Portuguese white port is not decolorized. California white port, Angelica, and muscatel are marketed with ca 12–14% sugar.

California sherry is a fortified wine of low sugar content that acquires its characteristic odor by being heated (baked) at ca 57°C for three or four months. It resembles the wines of Madeira much more than those of the Spanish sherry district, which are produced by different processes. This must is fermented as for sweet dessert wines, except that the fortification is postponed until only 1–7% sugar remains. The fortified wine is clarified and then placed in large redwood, lined metal, or stainless steel tanks. It is brought to the desired temperature by circulating hot water or steam in pipes placed in the tank or through the tank jacket. Not much additional heat is needed to maintain the temperature. Very little wine is baked in barrels exposed to the sun, since there is a large loss by evaporation. After the heating, the wine is clarified and marketed; it improves by aging in oak cooperage of 2–100 hL (50–2600 gal).

Sweet California sherry is usually a blend containing some slightly sweet sherry produced as described above plus some very sweet California Angelica. Baking a very sweet wine at 60°C results in a caramel flavor. If too pronounced, blending is necessary. A wine called California Tokay is produced by blending ca equal proportions of sherry, port, and Angelica. It has a brown-pink color and some of the slightly burnt flavor of sherry.

In the production of Spanish sherry, a surface film of yeast forms after the primary fermentation to ca 15% alcohol (12–13). The wine is kept in barrels that are only ca three-fourths full. Various oxidation–reduction changes occur during the formation of the film, including an increase in acetal and acetaldehyde and a decrease in acetic acid. This procedure gives the wine a characteristic and much appreciated bouquet. Variations of this process are now employed in Australia, California, Canada, Cyprus, South Africa, and the USSR.

Film-type yeasts have been successfully grown in submerged culture. A suitable dry white wine of 15.5% alcohol is pumped into a lined tank equipped with a stirrer. A rapidly fermenting culture of the film-type yeast is added and a small amount of air bubbled through the wine at a pressure of ca 101 kPa (1 atm). Periodically, the stirrer is operated to prevent the yeast from settling to the bottom of the tank. Under these conditions, the aldehyde content increases rapidly, often reaching 500–700 mg/L in 2–3 wk. For sale, the aldehyde content is reduced to 150–200 mg/L by blending. The process is being used commercially in Canada, the USSR, and California.

Red Dessert Wine. Port and port-type wines are the main red U.S. dessert wines, although a small amount of red muscatel is also produced. The prototype port is produced in a delimited district along the Douro river east of Oporto in Portugal. A number of red varieties are used, ie, the darker, stronger flavored wines are used for vintage or ruby port, whereas the lighter-colored wines are aged for tawny port.

The primary problem in the production of red dessert wines is to secure an adequate extraction of color during fermentation. This is made more difficult by the limited period of fermentation and the normally low color of grapes grown under warm climatic conditions. Most of these wines are produced in warm regions in order to secure grapes of high sugar content. The grapes are harvested at a 24° Brix or more and crushed and stemmed as for red table wine. For the best color, they should be harvested before the pH is too high. A compromise has to be made between 24° Brix and a pH of <3.5 or 3.6. In order to secure a good extraction of color from the skins, the juice and skins are heated very rapidly to ca 82°C, held for one or two minutes at this temperature, and then cooled. Previously, the mass of grapes was gradually heated to ca 70°C and then cooled, usually by circulating water in pipes in the tank. This procedure takes longer and may impart a cooked taste and a color more purple than desired. The cooled must is then fermented to the proper degree Brix.

In the absence of heating, the cap must be pressed down frequently into the fermenting liquid. In Portugal, the treading of grapes in shallow tanks had the same objective, but this custom is now disappearing. Whatever the system, pressing or drawing-off takes place in 2–4 d.

Ports are relatively simple to clarify and age. Some wines made from heated musts contain too much pectin and remain cloudy, and a pectin-splitting enzyme may be used to aid clarification. Ports may have the normal red color or be of the tawny type; it may require 3–5 yr to develop the proper tint. Alternatively, early maturing tawny ports may be produced by using tawny-colored grapes. Tawny ports are sometimes produced by heating the wine for three or four weeks at 60°C, but this procedure is illegal in Portugal and may produce an undesirable caramel flavor. A very limited amount of the finest Portuguese and port-types of other countries, including California, are bottled after two years' aging. After 10 to 20 years, this wine develops a special bouquet which is much appreciated. This wine is called vintage port and is very expensive.

Fruit and Berry Wines. The production of fruit wines is similar to that of grape wines, except that the sugar must always be added because of the low sugar content of most berries and fruit (13). Most fruits are soft and easily crushed. The juice is separated from the pulp of apples and pears, but most other fruits, particularly berries, are fermented with the pulp. Sufficient sugar is added to make a wine of 11–13% alcohol. For European fruit wines, only enough sugar to produce 9–11% alcohol is used. If fermented with the pulp, the juice is separated when color extraction is complete. The new wine is clarified in the usual way; most U.S. fruit wines are sweetened to 10% sugar or more before sale, although apple and pear wines are occasionally sold without sweetening. A few wines, eg, from peaches or apricots, are fortified to 18% alcohol as well as sweetened.

Grape and berry wines may not be mixed unless sold and labeled as a mixture. All berry and fruit wines are best when consumed young. After aging, the color fades and the characteristic odor of the fruit often disappears or is greatly diminished. These wines are easily stabilized and can safely be marketed soon after production. They are almost always pasteurized in the bottle to prevent fermentation. In the United States, demand appears to be limited.

Naturally Flavored Wines. In the United States, these wines were legally recognized in 1958. Natural herbs, spices, fruit juices, aromatics, essences, and other natural flavorings may provide the base, and sugar and caramel may be added; apple wine may be blended in. These wines are made both as table (<14% alcohol) and dessert wines. All are sold with proprietary names, eg, Bali Hai, Silver Satin, and Thunderbird. These wines are usually consumed with ice and soda. Some have a distinct citrus aroma. Sales have decreased since the 1960s.

Vermouth. Vermouths are nearly dry or sweet fortified wines to which herbs or herb extracts are added (10). The nearly dry or French type is used straight or for martini cocktails. The sweet or Italian type is used for Manhattan cocktails, but in Europe is more often consumed as a dessert wine or aperitif. The herbs in vermouth should be easily detectable, but the odor of no single herb should be allowed to predominate (see also Flavors and spices).

Dry. For dry vermouth, a light-colored wine of moderate total acidity is fortified to ca 17–18% alcohol. In the United States, dry table and dessert wines, such as low sugar sherry, are sometimes blended to yield a wine of the proper alcohol content and 1–4% sugar, to which a mixture of herbs is added, such as wormwood, gentian, orris, marjoram, centuary, bitter orange peel, pomegranate root, anise, nutmeg, vanilla, cinnamon, and others. The herbs and the proportions of each used are proprietary. Some typical formulas are given in refs. 11 and 14.

The herb mixture may be blended directly or placed in sacks that are submerged in the wine. The extraction may take from a few days to two weeks, depending on the flavor strength desired. If the extraction takes too long, tannins and other bitter substances are also extracted. Today, most wineries use herb extracts.

The wine is then filtered and stabilized. Some aging is desirable, but the color should be kept light. Much of the vermouth sold in the United States is partially decolorized with charcoal because bartenders prefer a very light-colored vermouth for martinis.

Sweet. The Italian sweet vermouth was originally produced in the 18th century with a muscatel wine as a base. At present, any sweet white dessert wine is used and grape concentrate or sucrose may be employed for sweetening. Furthermore, some caramel or dark-colored sherry is frequently added. Herb extracts rather than the herbs are generally employed. The appropriate mixture of herbs is extracted with alcohol, the extract filtered, and the proper amount added to the wine base. Sweet-vermouth herb mixtures often contain some vanilla. Clarification may be difficult if the herb extract was not clear or if the herbs were placed directly in the wine. Chilling, heating, and clarification with bentonite provide clarity.

Finishing. The aging, blending, clarification, and bottling practices in American wineries are carefully controlled by laboratory tests (14). Trained chemists taste and analyze the wines and prescribe the proper treatments to secure stability and the best quality.

Filtration. The wine is passed through a porous pad or layer of inert material with a fine pore size. For removing large particles, coarse pads are used; the filter plates can be ≥ 1 m in diameter. In order to extend the filtration, a filter aid of diatomaceous earth is mixed with the wine before filtration. Other filters consist of fine screens on which a precoat of porous inert material is placed. Smaller filters with pads of small pore size are employed for a polishing filtration or a prebottling filtration. For sweet table wines, sterile filtration may be required through pads with a pore size small enough to remove yeast cells. The entire bottling line must be sterile, and the filtration is conducted at a low, steady pressure. Membrane filters of small pore size are used for this purpose (see Membrane technology; Ultrafiltration).

Fining. Mechanical or chemical clarification is an old practice. A gelatin solution combines with the tannin in the wine, and the precipitate fines or clarifies the wine; a solution of egg white or isinglass may also be used. More recently, the organic fining agents have largely been supplanted by bentonite, a montmorillonite clay with excellent swelling properties and some adsorption capability. It may be used on hot or cold wines, and an excess does not result in cloudiness, as it may with organic clarifiers. It is especially effective in removing proteins. However, a large flocculant deposit is produced, and therefore the wine is tested to determine the minimum amount needed. A similar material, kieselguhr, is used in Europe.

Refrigeration. During the cold winters in Europe, wines lose their excess cream of tartar, especially when in small containers and after a long period of aging. In California, because of the mild winters, large containers, and relatively shorter aging period, artificial chilling is needed to remove tartrates. Of course, this is also practiced in Europe and elsewhere. Table wines are chilled to ca $-4°C$ and dessert wines to $<-5.5°C$. Chilled rooms, insulated tanks with internal circulating systems, and jacketed tanks are used. The process requires ca two weeks, and the tartrates should be filtered cold to prevent resolution.

Ion-exchange resins are now used by a few wineries to reduce the danger of cream of tartar precipitation and partially replace potassium by sodium or hydrogen. These must be used with caution. At least 500 mg/L of potassium should remain in the finished wine, and the sodium content should be <200 mg/L.

Pasteurization. Pasteurization stops microbial growth and aids in clarification; in the United States, the latter is more important. Contrary to common belief, pasteurization does not prevent reinfection; in fact, pasteurized wines may be more sensitive to infection. Pasteurization, however, when properly used, kills undesirable bacteria and is a useful practice for wines of moderate quality. Sweet table wines are sometimes hot-bottled above $66°C$ as a safety measure. When heating to aid clarification, even higher temperatures (but shorter periods) are used. The wine is then rapidly cooled and filtered.

Packaging. Automatic bottling, corking or capping, labeling, and casing lines are employed in large wineries. These are less labor intensive and give better results than manual procedures. The bottles are packed in cardboard boxes for shipment. Most wines are sold in glass bottles or plastic containers. There are standards for filling, and only certain bottle sizes are permitted in the United States.

Corks are employed for wines aged in the bottle, that is, for the better quality red and white table wines. Plastic or metal screw caps with inert liners are used for other table wines and most dessert wines (see Cork).

Spoilage

Wines are relatively immune to spoilage because of their alcohol content and low pH (10,14–15). Contact with air may result in acetification, which can be avoided by keeping the containers full or by pasteurization. Low acid wines may be spoiled by lactic acid bacteria, which can be prevented by a low concentration of sulfur dioxide. Yeast infection may be a problem.

Iron and copper originating from equipment occasionally cause cloudiness in white table wines; this problem does not occur with stainless-steel equipment. Iron cloudiness is partially inhibited by small amounts of citric acid which forms an iron complex.

Copper–protein cloudiness is partially inhibited by fining the wine while hot with bentonite, which removes the proteins responsible.

Economic Aspects

U.S. production and consumption of wines constitute only a small fraction of the world total (see Table 3). In 1980, the estimated world production was 351×10^5 m^3 (927×10^7 gal). The U.S. wine industry has been engaged for a number of years in a public relations campaign to increase per capita consumption. World production consists predominantly of red and white table wines.

The California wine industry dominates U.S. production. Some California grape concentrate and wine spirits are shipped to the east and then appear as eastern production. California produces more sparkling wine and vermouth than the other states combined. Californian wine is also shipped east and is converted there into sparkling wine or vermouth. The U.S. wine trade and production are given in Tables 4 and 5, respectively.

The prices of California wines have fluctuated very widely over the years, as bulk prices, in cents per liter, demonstrate:

	1967	1978
dessert wines, ¢/L	15	42
table wines, ¢/L	9–12	34

Table 3. Wine Production and Consumption, 1980

Country	Production, 10^5 m^3	Annual per capita consumption, L
France	69.20	95.4
Italy	84.75	93.1
Spain	43.52	60.0
United States	18.40	8.0

Table 4. U.S. Wine Trade, 1980–1981 Calendar Year Average [a], m^3 [b]

Type	Domestic production	Imports	Total
table	10,595	3,552	14,146
table, sparkling	991	226	1,217
dessert	1,521	106	1,628
vermouth	206	109	314
other special natural wines	1,225	119	1,344
Grand total	*14,538*	*4,112*	*18,649*

[a] Prepared from Reports of U.S. Treasury Department, Bureau of Alcohol, Tobacco and Firearms, and U.S. Department of Commerce, Bureau of the Census.
[b] To convert m^3 to gal (U.S. liquid), multiply by 264.

Table 5. U.S. Wine Production, 1980–1981 Crop Year Average[a], m³

Type	California	All other states	Total
gross[b]	15,596 (estd)	1,795 (estd)	17,391 (estd)
sparkling	922	155	1,078
vermouth	182	27	209
other special natural wines	1,155	107	1,262

[a] Prepared from reports of U.S. Treasury Department, Bureau of Alcohol, Tobacco and Firearms.
[b] Gross wine production is the quantity removed from fermentors plus the increase after fermentation by amelioration, sweeting, and addition of spirits, minus withdrawals for distillation. It includes wines subsequently used in producing sparkling wines, vermouth, or other special natural wines.

Sensory Evaluation

Consumers have definite preferences and tolerances for different types of wine. These originate largely from family and cultural traditions. In examining wines for their sensory differences or in serving wines with a meal, the drier (less sweet) wine with lower alcohol content precedes the sweeter, more alcoholic wines; white wines are generally served or tasted before red.

Anyone who so desires can become a wine connoisseur. Although some information is available (16–17), preference is highly subjective, and wines are judged on personal taste.

Professionals use score cards, ranking, paired, duo–trio, and triangular tests for identifying differences in wines. Statistical analysis of the data is recommended.

The results of local and regional judgings are often difficult to evaluate because of the varied ability of the participants and the usual lack of statistical evaluation of the results.

BIBLIOGRAPHY

"Wine" in *ECT* 1st ed., Vol. 15, pp. 48–72, by Maynard A. Amerine, University of California; "Wine" in *ECT* 2nd ed., Vol. 22, pp. 307–334, by M. A. Amerine, University of California, Davis, California.

1. U.S. Internal Revenue Service, *Federal Wine Regulations*, *26CFE (1954) Part 240*, Commerce Clearing House, Inc., Chicago, Ill., 1954, pp. 16001–16180, 16201–16240.
2. A. Lichine, *New Encyclopedia of Wines and Spirits*, 2nd ed., Alfred A. Knopf, Inc., New York, 1974.
3. D. Peppercorn, *Bordeaux*, Faber and Faber, London, 1982.
4. B. Anderson, *Vino. The Wines and Winemakers of Italy*, Little, Brown & Co., Boston, Mass., 1980.
5. M. A. Amerine, *Adv. Food Res.* **5**, 353 (1954); **8**, 133 (1958).
6. M. A. Amerine and M. A. Joslyn, *Table Wines, the Technology of Their Production*, 2nd ed., University of California Press, Berkeley, Calif., 1970.
7. M. A. Amerine and C. S. Ough, *Methods for Analysis of Musts and Wines*, John Wiley & Sons, Inc., New York, 1980.
8. P. Schreier, *CRC Crit. Rev. Food Sci. Nutr.* **12**, 59 (1979).
9. *Science* **219** (Feb. 25, 1983).
10. M. A. Amerine and R. E. Kunkee, *Ann. Rev. Microbiol.* **22**, 323 (1968).
11. M. A. Joslyn and M. A. Amerine, *Dessert, Appetizer and Other Flavored Wines*, Division of Agricultural Sciences, University of California, Berkeley, Calif., 1964.
12. J. C. M. Fornachon, *Studies on the Sherry Flor*, Australian Wine Board, Adelaide, Australia, 1953, reprinted 1972.

13. M. Gonzalez Gordon, *Jerez-Xeres-Scheris*, Jerez de la Frontera, Spain, 1948.
14. M. A. Amerine, H. W. Berg, R. E. Kunkee, C. S. Ough, V. L. Singleton, and A. D. Webb, *The Technology of Wine Making*, 4th ed., Avi Publishing Company, Westport, Conn., 1980.
15. J. C. M. Fornachon, *Bacterial Spoilage of Fortified Wines*, Australian Wine Board, Adelaide, Australia, 1943.
16. M. A. Amerine and E. B. Roessler, *Wines: Their Sensory Evaluation*, W. H. Freeman & Company, San Francisco, Calif., 230 pp., 1976, 2nd ed., xv, 432 pp., 1983.
17. M. Broadbent, *Wine Tasting: Enjoying; Understanding*, 5th ed., Christie Wine Publications, London, 1977.

General References

M. A. Amerine and V. L. Singleton, *Wine: An Introduction for Americans*, 2nd ed., University of California Press, Berkeley and Los Angeles, Calif., 1977.
Association of Official Analytical Chemists, *Methods of Analysis*, 13th ed., Washington, D.C., 1980.
G. Chappaz, *Le Vignoble et le Vin de Champagne*, Louis Larmat, Paris, France, 1948.
C. Chatfield and G. Adams, *Proximate Composition of American Food Materials*, U.S. Department of Agriculture circular, Washington, D.C.
P. G. Garoglio, *La Nuova Enologia*, Libreria LI. CO/SA, Florence, Italy, 1965.
E. Peynaud, *Connaissance et Travail du Vin*, Dunod, Paris, France, 1981.
J. Ribéreau-Gayon and E. Peynaud, *Analyse et Contrôle des Vins*, Librairie Polytechnique Ch. Béranger, Paris and Liège, France, 1972.
J. Ribéreau-Gayon, E. Peynaud, P. Ribéreau-Gayon, and P. Sudraud, *Traité d'Oenologie*, 4 vols., Dunod, Paris, 1977–1978.
G. Troost, *Die Technologie des Weines*, 5th ed., E. Ulmer Verlag, Stuttgart, FRG, 1980.
W. Younger, *Gods, Men and Wine*, World Publishing Co., Cleveland, Ohio, 1966.

M. A. AMERINE
University of California at Davis
Consultant, San Francisco Wine Institute

WINTERGREEN OIL. See Oils, essential.

WITHERITE, BaCO$_3$. See Barium compounds.

WOLFRAM AND WOLFRAM ALLOYS. See Tungsten and tungsten alloys.

WOOD

Wood is one of the most important natural resources, and one of the few that is renewable. It supplies material for many objects necessary to everyday living, ranging from homes and furniture to bridges and railroad ties. Wood yields fiber for pulp, paper, and fiberboard products and material for plywood, particleboard, pallets, and rayon textile fibers. It is also a source of energy and industrially important chemicals (see also Laminated wood-based composites; Paper; Pulp).

Production and consumption of wood, wood-derived products, and residues are given in various units of measure according to common usage (1–2). The cord, commonly used to measure log volume, refers to a stacked pile of wood containing 3.62 m^3 (128 ft^3) within its outside surfaces. Its standard dimensions are 1.2 × 1.2 × 2.4 m. Its weight, depending upon density and bark, ranges from 725 to 1500 kg. Sawn lumber is commonly measured in board feet (1 ft × 1 ft × 1 in.), a unit nominally containing 30 × 30 × 2.5 cm = 2360 cm^3 or $1/12$ ft^3. Roundwood equivalent may also be used and refers to the volume of logs or other roundwood products required to produce the lumber, plywood, wood pulp, or paper. Product weight is used to record the number of metric tons of paper and paperboard products produced.

Annual production of lumber in the United States has remained relatively constant for several decades at (36–38) × 10^9 board feet [(85–90) × 10^6 m^3]. The U.S. production of paper and paperboard, however, has been increasing steadily, as shown below (3):

Year	1950	1960	1970	1978
10^6 t	22	31	49	58

Timber demand for all forest products is projected to rise faster than supplies from U.S. forests. It is increasingly important to utilize the wood harvest efficiently, use currently underutilized hardwood species, and employ the considerable quantities of unused residues.

Structure

The anatomical structure of wood affects strength properties, appearance, resistance to penetration by water and chemicals, resistance to decay, pulp quality, and the chemical reactivity of wood (4). To use wood most effectively requires not only a knowledge of the amounts of various substances that make up wood, but also how those substances are distributed in the cell walls.

Woods are either hardwoods or softwoods. Hardwood trees (angiosperms, ie, plants with covered seeds) generally have broad leaves, are deciduous in the temperate regions of the world, and are porous, that is, they contain a vessel element. Softwood trees (conifers or gymnosperms, ie, plants with naked seeds) are cone bearing, generally have scalelike or needlelike leaves, and are nonporous, that is, they do not contain vessel elements. The terms hardwood and softwood have no direct relation to the hardness or softness of the wood. In fact, hardwood trees such as cottonwood, aspen, and balsa have softer wood than the western white pines and true firs; certain softwoods, such as longleaf pine and Douglas fir, produce wood that is much harder than that of basswood or yellow poplar.

Many mechanical properties of wood, such as bending and crushing strength, and hardness, depend upon the density of wood; denser woods are generally stronger (5). Wood density is determined largely by the relative thickness of the cell wall and by the proportions of thick-walled and thin-walled cells present.

The cells that make up the structural elements of wood are of various sizes and shapes and are firmly bonded together. Dry wood cells may be empty or partly filled with deposits such as gums, resins, or other extraneous substances. Long and pointed cells, known as fibers or tracheids, vary greatly in length within a tree and among species. Hardwood fibers are ca 1 mm long, softwood fibers ca 3 to 8 mm.

Just under the bark of a tree is a thin layer of cells, not visible to the naked eye, called the cambium. Here, cells divide and eventually differentiate to form bark tissue outside of the cambium and wood or xylem tissue inside. This newly formed wood contains many living cells and conducts sap upward in the tree, and hence is called sapwood. Eventually, the inner sapwood cells become inactive and are transformed into heartwood. This transformation is often accompanied by the formation of extractives that darken the wood, make it less porous, and sometimes provide more resistance to decay.

Because of the great structural variations in wood (6), there are many possibilities for selecting a species for a specific purpose. Some species, like spruce, combine light weight with relatively high stiffness and bending strength. Very heavy woods, like lignumvitae, are extremely hard and resistant to abrasion. A very light wood, like balsa, has high thermal insulation value; hickory has extremely high shock resistance; mahogany has excellent dimensional stability.

Composition

Wood is a complex polymeric structure consisting of lignin (qv) and carbohydrates (qv) (cellulose (qv) and hemicellulose), which form the visible lignocellulosic structure of wood (7–9). Also present, but not contributing to wood structure, are minor amounts of extraneous organic chemicals and minerals. The organic chemicals are extractable from the wood with neutral solvents, and are therefore called extractives. The minerals constitute the ash residue remaining after ignition at a high temperature.

The absolute composition of a species cannot be determined. Wood composition is affected by many variables including geographical location and weather conditions of the growth site. Enough is known to describe ranges of composition values within a species. Some generalizations are also possible to distinguish hardwood and softwood composition, eg, the average lignin content of softwood is slightly higher than that of hardwood. If the minerals and small amounts of nitrogen and sulfur (0.1–0.2%) are ignored, the elementary composition of dry wood averages 50% carbon, 6% hydrogen, and 44% oxygen.

Lignin. Lignin is an amorphous, insoluble, irregular organic polymer with a range of molecular weights. Its basic chemical structural unit is a methoxy-substituted propylphenol moiety, bonded in an irregular pattern of ether and carbon–carbon linkages. Lignin usually comprises 18–35% of the dry wood, most of it concentrated in the compound middle lamella and the layered cell wall. It imparts a woody, rigid structure to the cell walls, thus distinguishing wood from other fibrous plant materials of lesser lignin content. Analysis usually involves removing the carbohydrate material through acid hydrolysis and weighing the residue (9).

Carbohydrates. Carbohydrates are the principal components of the cell wall, usually 65–75 wt % of the dry wood; they are classified as cellulose or hemicellulose. Total hydrolysis yields simple sugars, primarily glucose and xylose in hardwoods and glucose and mannose in softwoods.

Cellulose is the main component of the wood cell wall, typically 40–50 wt % of the wood. Pure cellulose consists of glucose residues joined by 1,4-β-glucosidic bonds. The number of glucose units in the molecular chain is unknown, but probably exceeds 20,000. Wood cellulose is more resistant to dilute acid hydrolysis than hemicellulose. X-ray diffraction indicates a partial crystalline structure for wood cellulose. The crystalline regions are more difficult to hydrolyze than the amorphous regions as removal of the easily hydrolyzed material has little effect on the diffraction pattern.

Hemicellulose is a mixture of amorphous branched-chain readily hydrolyzable polysaccharides consisting of a few hundred sugar residues. The sugars include glucose, mannose, galactose, xylose, arabinose, and uronic acids. Many different hemicelluloses have been isolated from wood.

Extractives and Ash. The amount of extractives in wood varies from 5–20 wt % and includes a wide variety of complex organic chemicals. Many of these function as intermediates in tree metabolism, as energy reserves, or the tree's defense mechanism against microbiological attack. The extractives contribute to wood properties such as color, odor, and decay resistance.

Elemental analysis indicates the ash consists of calcium and potassium with lesser amounts of magnesium, sodium, manganese, and iron. Carbonates, phosphates, silicates, and sulfates are the most likely anions of the ash. Some woods contain significant amounts of silica.

The chemical compositions of North American hardwoods and softwoods are given in Table 1.

Wood–Liquid Relationship

Adsorption. Dry wood is highly hygroscopic. The amount of moisture adsorbed depends mainly on the relative humidity and temperature, as shown in Figure 1. Exceptions occur with species with high extractive contents (eg, redwood, cedar, and teak). The equilibrium-moisture contents of such woods are generally somewhat lower than those given in Figure 1.

In green wood, the cell walls are saturated, whereas some cell cavities are completely filled while others may be completely empty. Moisture in the cell walls is called bound, hygroscopic, or adsorbed water. Moisture in the cell cavities is called free or capillary water. The distinction is made because, under ordinary conditions, the removal of the free water has little or no effect on many wood properties. On the other hand, the removal of the cell-wall water has a pronounced effect.

In equilibrium with relative humidity below 100%, the moisture in wood is present primarily in the cell walls. The moisture content at which the cell walls would be saturated and the cell cavities empty is called the fiber-saturation point. Actually, such distribution is impossible. Beginning at ca 90% relative humidity, some condensation may occur in small capillaries. The determination of the fiber-saturation point is based on the fact that certain properties of wood (eg, strength and volume) change uniformly at first with increasing moisture content and then become independent of the moisture content, as shown in Figure 2. The equilibrium-moisture content (usually determined

582 WOOD

Table 1. Composition of North American Woods, Percent of Extractive-Free Wood[a]

	Glucan	Mannan	Galactan	Xylan	Arabinan	Uronic anhydride	Acetyl	Lignin	Ash
hardwoods									
trembling aspen, *Populus tremuloides*	57	2.3	0.8	16	0.4	3.3	3.4	16	0.2
beech, *Fagus grandifolia*	48	2.1	1.2	18	0.5	4.8	3.9	22	0.4
white birch, *Betula papyrifera*	45	1.5	0.6	25	0.5	4.6	4.4	19	0.2
yellow birch, *Betula alleghaniensis*	47	3.6	0.9	20	0.6	4.2	3.3	21	0.3
red maple, *Acer rubrum*	47	3.5	0.6	17	0.5	3.5	3.8	24	0.2
sugar maple, *Acer saccharum*	52	2.3	<0.1	15	0.8	4.4	2.9	23	0.3
sweet gum, *Liquidambar styraciflua*	39	3.1	0.8	18	0.3			24	0.2
American elm, *Ulmus americana*	53	2.4	0.9	12	0.6	3.6	3.9	24	0.3
southern red oak, *Quercus falcata*	41	2.0	1.2	19	0.4	4.5	3.3	24	0.8
softwoods									
balsam fir, *Abies balsamea*	47	12	1.0	4.8	0.5	3.4	1.5	29	0.2
eastern white cedar, *Thuja occidentalis*	45	8	1.5	7.5	1.3	4.2	1.1	31	0.2
eastern hemlock, *Tsuga canadensis*	45	11	1.2	4.0	0.6	3.3	1.7	32	0.2
jack pine, *Pinus banksiana*	46	11	1.4	7.1	1.4	3.9	1.2	29	0.2
eastern white pine, *Pinus strobus*	44	11	2.5	6.3	1.2	4.0	1.3	29	0.2
loblolly pine, *Pinus taeda*	45	11	2.3	6.8	1.7	3.8	1.1	28	0.3
Douglas-fir, *Pseudotsuga taxifolia*	44	11	4.7	2.8	2.7	2.8	0.8	32	0.4
black spruce, *Picea mariana*	48[b]	10[c]		8.0		4.1	1.1	28	0.4
white spruce, *Picea glauca*	46	12	1.2	6.8	1.6	3.6	1.3	27	0.3
tamarack, *Larix laricina*	46	13	2.3	4.3	1.0	2.9	1.5	29	0.2

[a] Ref. 8.
[b] Including galactan.
[c] Including arabinan.

by extrapolation), at which the property becomes constant at 25–30% moisture, is represented by the fiber-saturation point.

The density of wood substance is ca 1.5 g/cm^3. A species of a density of 0.5 g/cm^3 (based on ovendry weight and volume) has a void volume of 66.6%. Each 100 kg of the totally dry wood occupies 0.2 m^3 (7 ft^3) and contains 0.067 m^3 (2.4 ft^3) wood substance and 0.133 m^3 (4.7 ft^3) void. In waterlogged condition, the cell walls adsorb ca 0.028 m^3 (1 ft^3) water, whereas the cell cavities contain ca 0.133 m^3 (4.7 ft^3) water.

The average specific gravity of different species is given in Table 2 (5). The conventional way for expressing the specific gravity of wood is in terms of the ovendry weight and volume at 12% moisture content. The specific gravity based on the volume of the ovendry wood is ca 6% higher.

At low relative humidities, adsorption is due to interaction of water with accessible hydroxyl groups. These are present on the lignin and on the carbohydrates in the

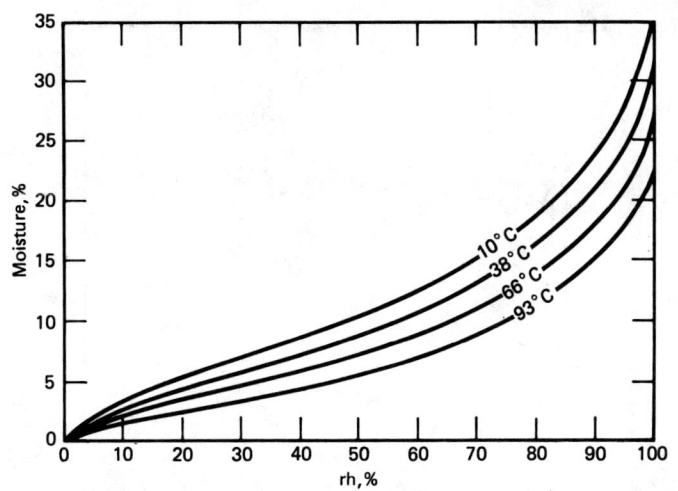

Figure 1. Relationship between the moisture content of wood (% of dry wood) and relative humidity at different temperatures.

Table 2. Specific Gravity of Some Common Woods Growing in the United States

Wood	Specific gravity[a]
aspen	0.40
birch, yellow	0.67
cottonwood, eastern	0.42
Douglas-fir, coast-type	0.51
fir, balsam	0.38
hemlock, western	0.44
maple, sugar	0.68
oak, white	0.73
pine	
lodgepole	0.43
ponderosa	0.42
longleaf	0.62
spruce, Engelmann	0.36
walnut, black	0.59

[a] Based on ovendry weight and volume at 12% moisture content.

noncrystalline or poorly crystalline regions. The high differential heat of adsorption by dry wood, ca 1.09 kJ/g (469 Btu/lb) water, reflects a very high affinity for moisture (10–12).

At high relative humidities, adsorption is believed to occur in response to a tendency for cellulose chains and lignin to disperse (solution tendency). Complete dispersion (dissolution) is prevented because of the strong interchain or interpolymer bonding at certain sites or regions. The differential heats of adsorption are much smaller than at low relative humidities.

Because the relative humidity of the atmosphere changes, the moisture content of wood undergoes corresponding changes. Effective protection against fluctuating atmospheric conditions is furnished by a surface coating of certain finishes, provided

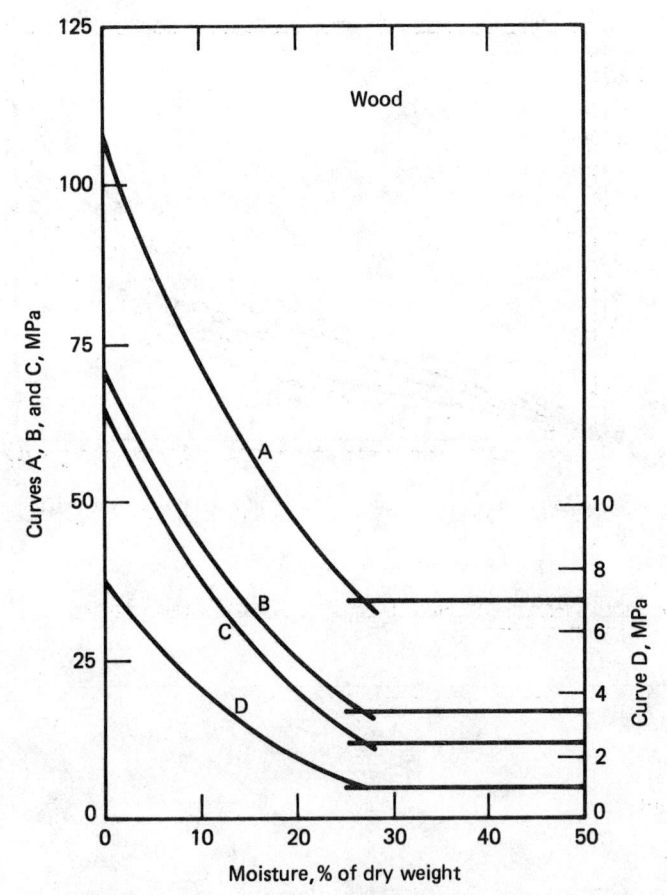

Figure 2. Relationship between various strength properties of wood and moisture content: A, modulus of rupture; B, fiber stress at elastic limit in static bending; C, maximum crushing strength parallel to the grain; D, fiber stress at elastic limit in compression perpendicular to the grain. To convert MPa to psi, multiply by 145.

the coating is applied to all surfaces of wood through which moisture gains access. However, no coating is absolutely moistureproof; coatings simply retard the rate at which moisture is taken up from or given off to the atmosphere. This means that coatings cannot be relied upon to keep moisture out of wood that is exposed to dampness constantly or for prolonged periods. Coatings vary markedly in their moisture-retarding efficiencies.

For some uses, it is important to protect wood against water, eg, doors, windows, door and window frames, and the lap and butt joints in wood siding. Water repellents and water-repellent preservatives, long used in the millwork industry, provide protection from wetting. They are designed to penetrate into wood, but they leave a very thin coating of wax, resin, and oil on the surface which repels water. However, they are not as effective in resisting water vapor (see Waterproofing).

Neither coatings nor water repellents alter the equilibrium-moisture content or equilibrium swelling of wood. This can only be accomplished by depositing bulking agents that block normal shrinkage within the cell walls, chemically replacing the hygroscopic hydroxyl groups of cellulose and lignin with less hygroscopic groups, or forming chemical cross-links between the structural units of wood.

Shrinking and Swelling. The adsorption and desorption of water in wood is accompanied by external volume changes. At moisture contents below the fiber-saturation point, the relationship may be a simple one, merely because the adsorbed water adds its volume to that of the wood, or the desorbed water subtracts its volume from the wood. The relationship may be complicated by the development of stresses. Theoretically, above the fiber-saturation point, no volume change should occur with a change in the moisture content. Actually, owing to the development of stresses, changes in volume or shape may occur. The magnitude of such stresses is minimized by drying wood under carefully controlled and empirically established conditions (11).

In the absence of drying stress (ie, with small specimens and extremely slow drying), the degree of shrinkage from the green to ovendry condition is, as a first approximation, proportional to the specific gravity of the wood. The value of the slope of the linear relationship is equal to the average fiber-saturation point of the wood. Serious deviations from the linear relationship may occur with species high in extractives.

Swelling or shrinking of wood is highly anisotropic. Tangential swelling (occurring tangent to the rings) is 1.5–3.5 times greater than radial swelling (occurring along a radius of the rings). Longitudinal swelling (occurring in the direction of tree growth) is usually very small. In certain abnormal woods, however, such as compression or tension wood, longitudinal swelling or shrinking may be relatively high (up to 1–2% for tension wood and 5–6% for compression wood) (12).

Permeability. Although wood is a porous material (60–70% void volume), its permeability (ie, flow of liquids under pressure) is extremely variable. This is due to the highly anisotropic shape and arrangement of the component cells and to the variable condition of the microscopic channels between cells. In the longitudinal direction, the permeability is 50–100 times greater than in the transverse direction (11). Sapwood is considerably more permeable than heartwood. In many instances, the permeability of the heartwood is practically zero. A rough comparison, however, may be made on the basis of heartwood permeability, as shown in Table 3.

Transport. Wood is composed of a complex capillary network through which transport occurs by capillarity, pressure permeability, and diffusion. A detailed study of the effect of capillary structure on the three transport mechanisms is given in ref. 11.

Table 3. Relative Permeability of the Heartwood of Some Common Species, Decreasing from Group 1 to Group 4

Group 1	Group 2	Group 3	Group 4
ponderosa pine	coastal Douglas-fir	eastern hemlock	alpine fir
basswood	jack pine	Engelmann spruce	Douglas-fir[a]
red oaks	loblolly pine	lodgepole pine	tamarack
slippery elm	longleaf pine	noble fir	western red cedar
tupelo gum	western hemlock	sitka spruce	black locust
white ash	cottonwood	western larch	red beech
	aspen	white fir	red gum
	silver maple	white spruce	white oaks
	sugar maple	rock elm	
	yellow birch	sycamore	

[a] Growing in interior regions; permeability practically zero.

Drying. The living tree holds much water in its cells. A southern pine log, 5 m long and 0.5 m in diameter, for example, may weigh as much as 1000 kg and contain ca 47% or 0.46 m^3 (16 ft^3) water.

There are a number of important reasons for drying: it reduces the likelihood of stain, mildew, or decay developing in transit, storage, or use; the shrinkage that accompanies drying can take place before the wood is put to use; wood increases in most of its strength properties as it dries below the fiber-saturation point (30% moisture content); the strength of joints made with fasteners, such as nails and screws, is greater in dry wood than in wet wood dried after assembly; the electrical resistance of wood increases greatly as it dries; dry wood is a better thermal insulating material than wet wood; and the appreciable reduction in weight that accompanies drying reduces shipping costs (see Drying).

Ideally, the temperature and relative humidity during drying should be controlled; if wood dries too rapidly, it is likely to split, check, warp, or honeycomb because of stresses.

Air-drying is a process of stacking sawmill products outdoors to dry (13). Control of drying rates is limited and great care must be taken to avoid degrading the wood. Drying time is a function of climatic condition; in damp coastal areas wood dries slowly, whereas in the arid regions of the Southwest it dries rapidly. Typical air-drying times for 2.5-cm thick lumber of various species are shown in Table 4.

Kiln-drying is a controlled drying process widely used for drying both hardwoods and softwoods. Dry-bulb temperatures seldom exceed 93°C and then only at the end of the drying schedule. In the initial stages, the relative humidity is maintained at a high level to control the moisture gradient in the wood and thus prevent splitting and checking. Modern kiln installations use forced-air circulation and are equipped with automatic controls for both dry- and wet-bulb temperatures. The kilns are vented to exhaust the moisture evaporated from the wood. Most installations are steam-heated, although furnace-type kilns fired with gas or oil are now under construction. Many species, especially hardwoods, are first air-dried to about 20% moisture content, then kiln-dried to the moisture content at which they will be used (Table 4). A typical time schedule for kiln-drying a softwood is shown in Table 5. Schedules for drying hardwoods are generally more complex (14).

Table 4. Approximate Air-Drying and Kiln-Drying Periods for 2.5-cm Lumber

Species	Air-dry to 20%	Kiln-dry to 6%[a]
baldcypress	100–300	10–20
hickory	70–200	7–15
magnolia	60–150	10–15
oak		
red	100–300	16–28
white	150–300	20–30
pine, southern	40–150	3–5
sweet gum	70–300	10–25
sycamore	70–200	6–12
tupelo	70–200	6–12
yellow poplar	60–150	6–10

(Days required to)

[a] From 20% moisture content.

Table 5. Typical Softwood Kiln-Drying Time Schedules for 2.5-cm Ponderosa Common Pine

Hours in kiln	Dry-bulb temperature, °C	Wet-bulb temperature, °C	Relative humidity, %
heartwood[a]			
1–8	54	43	52
8–16	58	44	48
16 until dry	60	44	41
sapwood[b]			
1–12	54	43	52
12–24	58	44	48
24 until dry	60	44	41

[a] Kiln-dried to average 12% moisture content in 24–36 h.
[b] Kiln-dried to average 12% moisture content in 48–72 h.

Special drying methods, such as with superheated-steam, solvent, vacuum, infrared radiation, and high frequency dielectric and microwave heating, are occasionally employed when accelerated drying is desired and the species being dried can withstand severe conditions without damage. None of these methods is of significant commercial importance.

Structural Material

Strength and Related Properties. In the framing of a building or the construction of an industrial unit, where wood is used because of its unique physical properties, strength and stiffness are primary requirements. Different species of wood have different mechanical properties that relate to the amount of wood substance per unit volume, ie, its specific gravity. Heavy wood such as oak tends to be stronger and stiffer than a light wood such as spruce. The strength of a piece of lumber depends also upon its grade or quality. The strength values of a lumber grade depend upon the size and number of such characteristics as knots, cross grain, shakes, splits, and wane (15). Wood free from these defects is known as clear wood.

Most strength properties of clear wood improve markedly as moisture is reduced below ca 30%, based on the ovendry weight (16) (Fig. 2). In structural lumber containing defects, however, the improvement in mechanical properties normally associated with wood as it dries may be largely offset by degradation during drying, particularly with lower grade material. There is no significant difference in the strength properties of wood that is conventionally kiln-dried compared with those of wood that is carefully air-dried. However, there is increasing interest in high temperature drying (110–120°C) of structural lumber to reduce processing time and energy consumption. These processes may reduce strength up to 20% depending upon species and property (17).

The mechanical properties of wood tend to deteriorate when it is heated and improve when it is cooled (5,17). Below ca 93°C, heat does not permanently affect the properties. The response of mechanical properties to temperature change is termed immediate or reversible effect.

The immediate effect of temperature on strength and modulus of elasticity, based upon several different loading modes, is illustrated in Figures 3–4 (5). The influence of moisture content and temperature is also shown.

High temperatures result in permanent degradation. This irreversible loss in

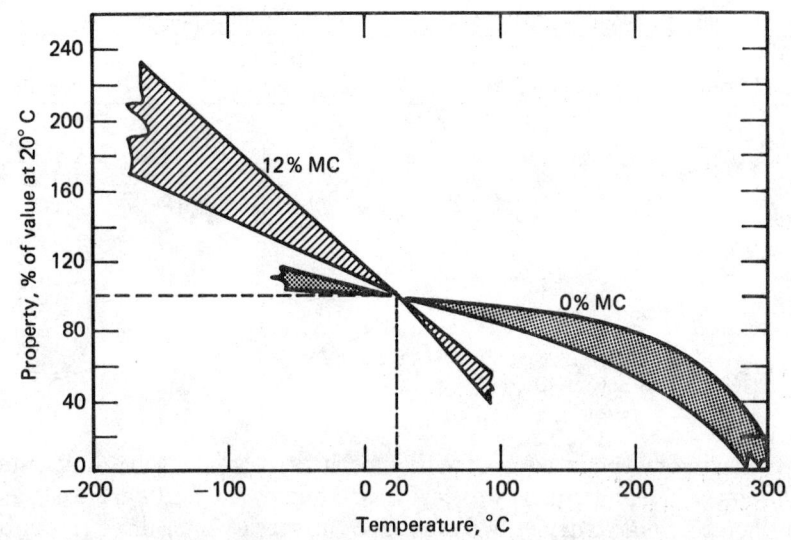

Figure 3. The immediate effect of temperature on the modulus of elasticity, relative to the value at 20°C. The plot is a composite of studies on the modulus as measured in bending, in tension parallel to the grain, and in compression parallel to grain. Variability in reported results is illustrated by the width of the bands. MC = moisture content.

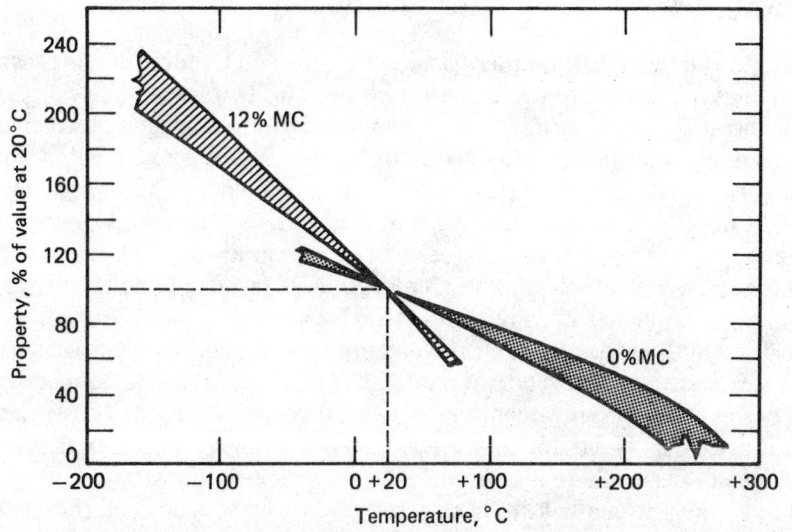

Figure 4. The immediate effect of temperature on strength properties, expressed as percent of value at 20°C. Trends illustrated are composites from studies on three strength properties: modulus of rupture in bending, tensile strength perpendicular to the grain, and compressive strength parallel to grain. Variability in reported results is illustrated by the width of the bands. MC = moisture content.

mechanical properties depends upon moisture content, heating medium, temperature, exposure period, and, to some extent, species. The effects of these factors on modulus of rupture, modulus of elasticity, and work-to-maximum load are illustrated in Figures 5–8 (5). The permanent property losses shown are based on tests conducted after specimens were cooled to ca 24°C and conditioned to a moisture content of 7–12%. If tested hot, presumably immediate and permanent effects would be additive. Thus, exposure to elevated temperature for extended periods has immediate and permanent

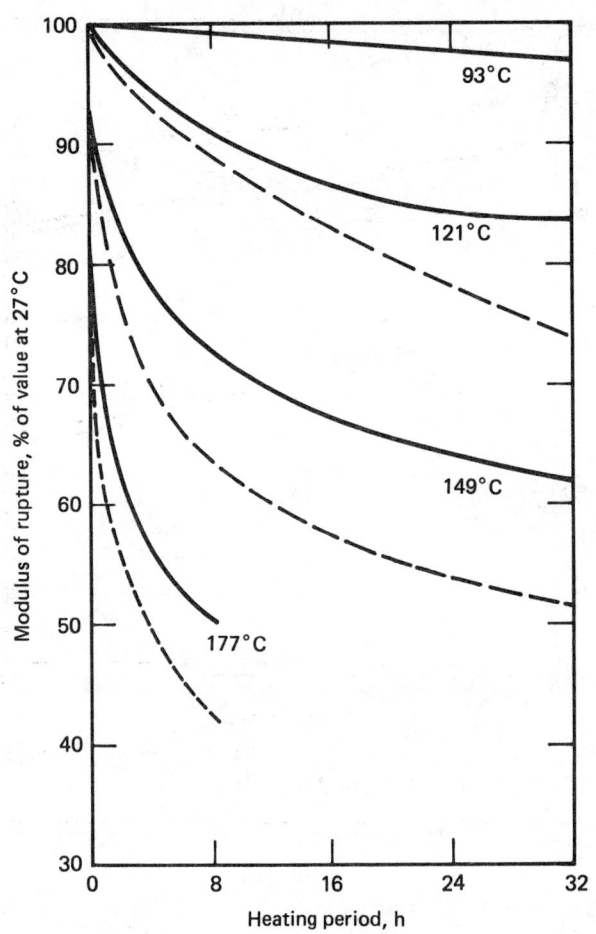

Figure 5. Permanent effect of heating in water (solid line) and in steam (dashed line) on the modulus of rupture. Data based on tests of Douglas-fir and Sitka spruce.

effects, and must be considered in the design of structures such as chemical storage tanks.

Repeated exposure to elevated temperature has a cumulative effect. For example, at a given set of exposure conditions the property losses are about the same after six exposure periods of one month each as it would be after a single 6-mo period.

The shape and size of wood pieces are important in analyzing the influence of temperature. If exposure is for only a short time and the inner parts of a large piece do not reach the temperature of the surrounding medium, the immediate effect on the strength of inner parts is less than for outer parts. The type of loading must also be considered. If the piece is to be stressed in bending, its outer fibers are subjected to the greatest load and ordinarily govern the ultimate strength; hence, under such a loading condition, the fact that the inner part is at a lower temperature may be of little significance.

For extended, noncyclic exposures, it can be assumed that the entire piece reaches the temperature of the heating medium and is, therefore, subject to permanent strength losses throughout the piece, regardless of size and mode of stress application. However, wood often does not reach the daily extremes in temperature of the air

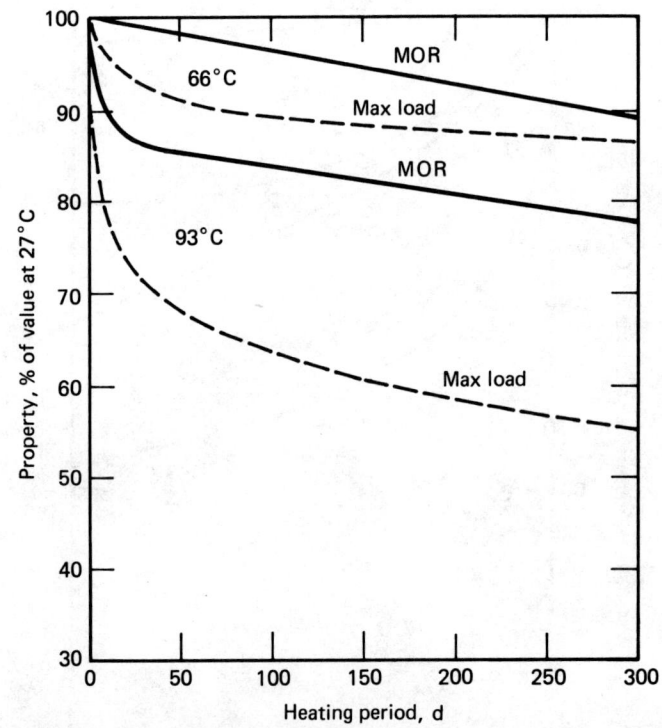

Figure 6. Permanent effect of heating in water on work-to-maximum load and on modulus of rupture (MOR). Data based on tests of Douglas-fir and Sitka spruce.

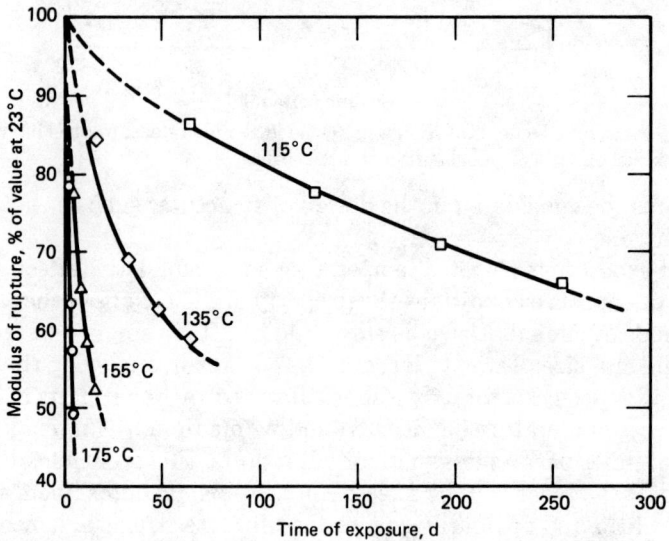

Figure 7. Permanent effect of oven heating at four temperatures on the modulus of rupture, based on four softwood and two hardwood species.

around it in ordinary construction; thus, long-term effects should be based on the accumulated temperature experience of critical structural parts.

The effect of absorption of various liquids upon the strength properties of wood

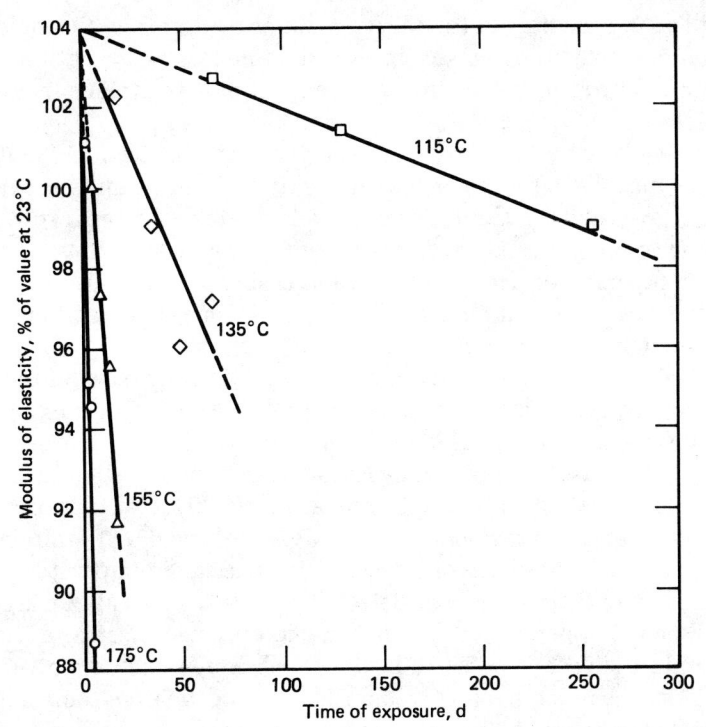

Figure 8. Permanent effect of oven heating at four temperatures on modulus of elasticity, based on four softwood and two hardwood species.

largely depends on the chemical nature and reactivity of the absorbed liquid. In general, neutral, nonswelling liquids have little if any effect upon the strength properties (18–19). Any liquid that causes wood to swell causes a reduction in strength. This effect may be temporary, existing only while the liquid remains in the wood. It may also be permanent, as in the case of chemically reactive liquids (strong acids and bases) and of a magnitude dependent upon time and temperature of exposure and concentration of the solution.

Wood preservatives are applied either from an oil system, such as creosote and petroleum solutions of pentachlorophenol, or a water system. Coal-tar creosote, creosote–coal-tar mixtures, creosote–petroleum oil mixtures, and pentachlorophenol dissolved in petroleum oils are practically inert to wood and have no chemical influence that would affect its strength.

The mechanical properties of wood can be damaged by preservatives applied from an aqueous system which can contain chromium, copper, arsenic, ammonia, zinc, and boron; in addition, these systems can corrode the mechanical fasteners. However, in the concentrations commonly used in most preservative treatments, their effect is minimal (15). Impact strength is an exception, as this property may be significantly reduced by inorganic preservatives.

Inorganic salts used in fire-retardant treatments are also reactive with wood. Because of the high concentrations required for effective protection, they have a more severe effect on mechanical properties than do wood-preserving salts (see Flame retardants). Bending strength may be reduced by 10–20% and impact strength by 30% or more (5,15).

Regardless of the effect of the preservative or fire-retardant chemicals, the impregnation process itself may result in considerable loss in strength because of the elevated temperatures and pressures applied. Careful treatment, however, causes minimal losses.

Heat and Fire Resistance. The physical and chemical properties of wood, like those of any organic material, are subject to deterioration. The rate and the extent of deterioration are governed by the interdependent factors of temperature, time, and moisture. In locations not conducive to decay or insect attack, wood is extremely stable at ordinary temperatures. However, with increasing temperature, the degradation of surface layers progresses into the interior layers. Prolonged heating at temperatures as low as 95°C may cause charring.

In general, the thermal degradation of wood and other cellulosic substances proceeds along two competing reaction pathways (20). At temperatures up to ca 200°C, carbon dioxide and traces of organic compounds are formed, in addition to the release of water vapor. The gases are not readily ignitable, but under certain conditions a pilot flame can ignite the volatiles after 14–30 min at 180°C (21). Exothermic reactions may occur near 200°C and, in situations where heat is conserved, self-ignition at temperatures as low as 100°C has been observed (22). Times and temperatures that might result in smoldering initiation can be determined (23). To provide a margin of safety, 77°C should be the upper limit in prolonged exposure near heating devices.

Temperatures in excess of 200°C lead to much more rapid decomposition. Under these conditions, the pyrolysis gases contain 200 or more different components (24–26), and the degradation is accompanied by reduction in weight, depending on temperature and duration of heating (27), as shown in Figure 9. Thermogravimetric analysis (tga) of wood, α-cellulose, and lignin (Fig. 10) indicates that a slow initial weight loss for lignin and wood begins at ca 200°C.

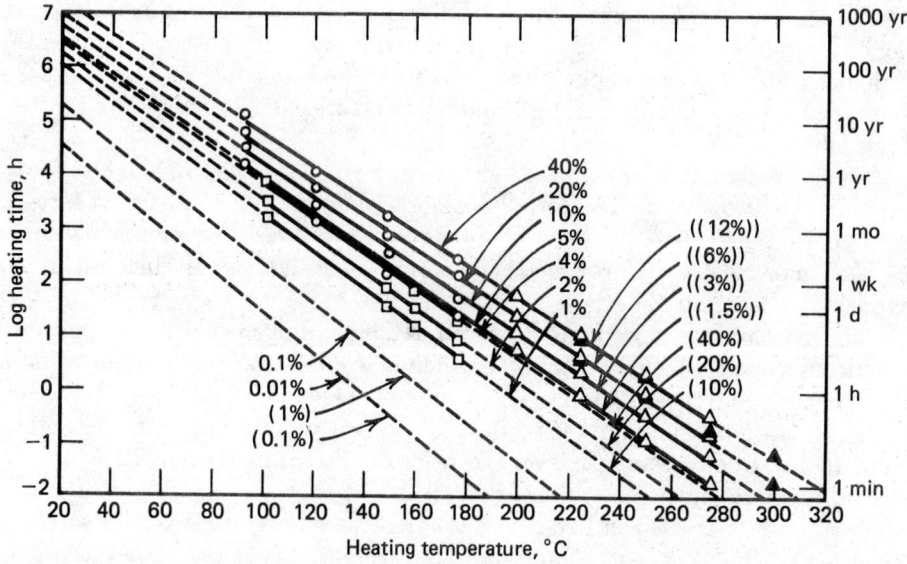

Figure 9. Logarithm of heating time versus temperature to attain various degrees of degradation of wood (27). No parenthesis indicates weight loss on oven heating; single parenthesis, modulus of rupture on oven heating; and double parenthesis, weight loss on heating beneath surface of molten metal. Courtesy of *Industrial and Engineering Chemistry*.

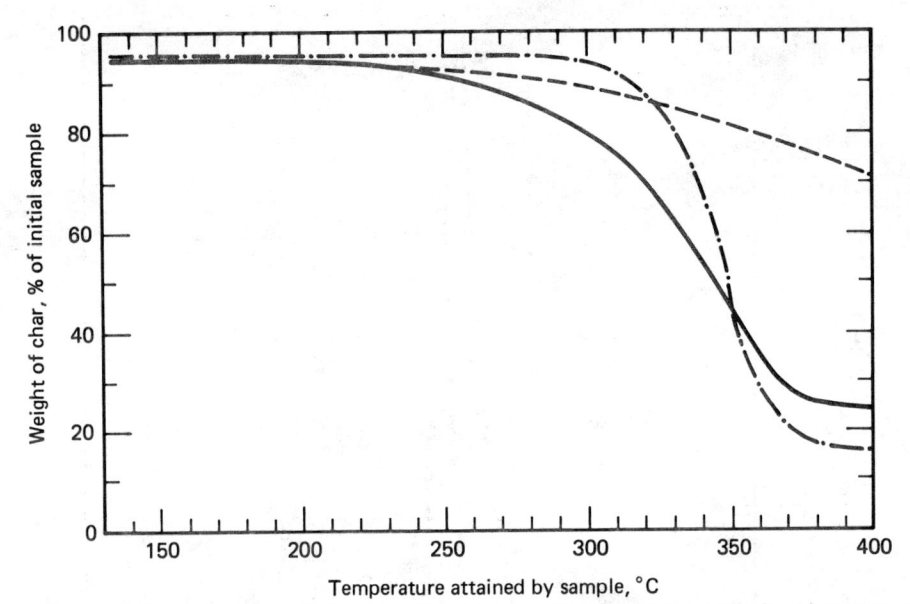

Figure 10. Portion of dynamic thermogravimetric curves between 130 and 400°C, with temperature rising 6°C/min, for wood, lignin, and α-cellulose. —— = wood; - - - = lignin; and - ·· - = α-cellulose (29).

Differential thermal analyses (dta) of wood and its components indicate that the thermal degradation reactions in an inert atmosphere release less than 5% of the heat released during combustion in air. The typical dta data given in Figure 11 (29) show two main exothermic peaks for wood. The peak near 320°C represents the flaming reaction resulting from combustion of the volatiles associated with cellulose pyrolysis; the other peak, near 440°C, represents glowing in place of the solid charcoal residue. The dta curves for α-cellulose and lignin in Figure 11 can be superimposed to approximate the curve for wood. Several extensive reviews on thermal degradation of wood are available (30–33).

Fire-performance codes for materials and constructions are mainly concerned with noncombustibility (34), fire endurance (35–36), fire-hazard classification (37), and smoke yield (38). Wood, even in its treated form, does not meet the requirements for noncombustibility (34). However, the safety of fire-resistant construction is controversial when the combustibility of the contents of the buildings is not regulated. Some codes permit materials with low flamespread characteristics to be classified as noncombustible, and, under this definition, certain treated wood products can be classified as noncombustible.

Wood in its untreated form has good resistance or endurance to fire penetration when used in thick sections for walls, doors, floors, ceilings, beams, and roofs. This endurance is due to low thermal conductivity which reduces the rate at which heat is transmitted to the interior. Typically, when the fire temperature at the surface of softwood is 870–980°C, the inner char-zone temperature is ca 290°C, and 6 mm further inward the temperature is 180°C or less. The penetration rate of this char line is about 0.6 mm/min, depending on the species, moisture content, and density (39–40). Owing to this slow penetration rate and low thermal conductivity, large wood members retain a substantial portion of their load-carrying capacity for considerable time during fire exposure (41). Chemical treatments reduce the charring rate (42).

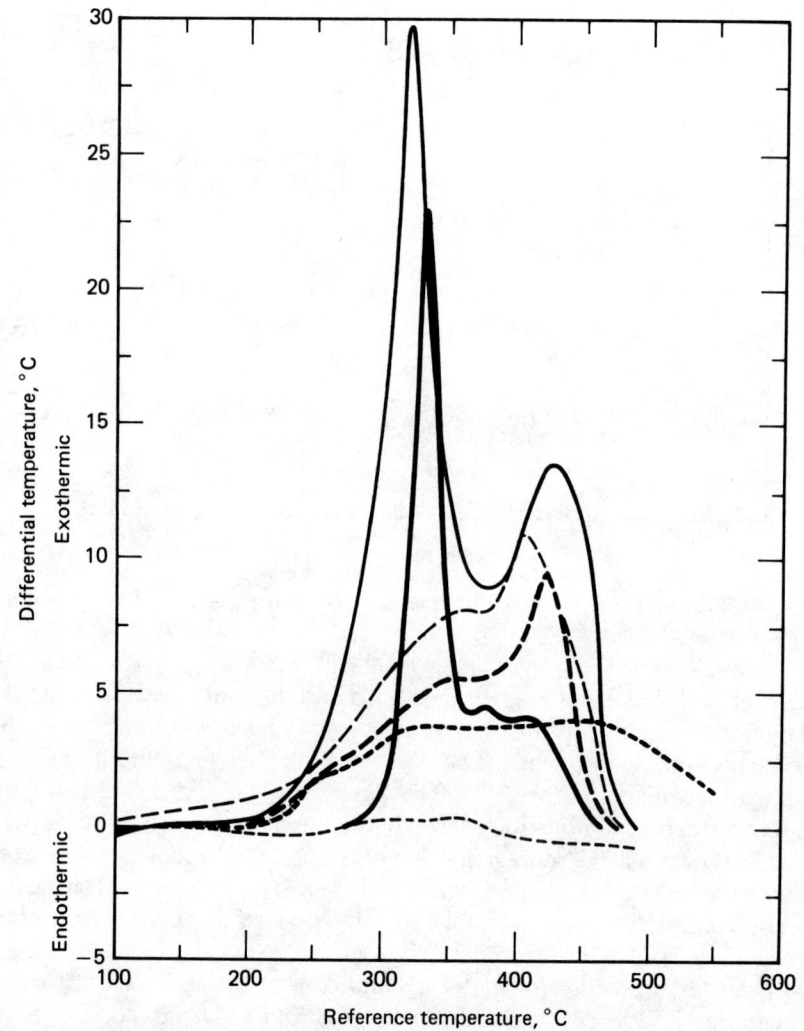

Figure 11. Differential thermal analysis of wood and its components at a heating rate of 12°C/min and a gas-flow rate of 30 cm³/min. Sample weight: wood, 40 mg; α-cellulose, 20 mg; lignin, 10 mg; charcoal, 12 mg. Treated wood sample contains 9% by weight of commercial fire retardant. —— = untreated wood in O₂; — — — = untreated wood charcoal in O₂; - - - = untreated wood in N₂; —— = untreated α-cellulose in O₂; — — — = untreated sulfate lignin in O₂; and - - - = wood treated with fire retardant in O₂.

The fire-hazard classification (flamespread, fuel contributed, and smoke development) for wood and wood products as measured by American Society for Testing and Materials (ASTM) E 84 (37) can be reduced with fire-retardant treatments, either chemical impregnation or coatings (43). In addition, the smoke yield measured by ASTM E 662 (38) can be lowered (44).

Fire-retardant chemicals, such as ammonium phosphate, ammonium sulfate, zinc chloride, dicyandiamide phosphate, borax, and boric acid, are often used in combinations. Borax and boric acid mixtures are moderately effective in reducing flamespread and afterglow without premature charring during severe drying operations. Although very hygroscopic, zinc chloride is an effective flame retardant; boric acid is often added to retard afterglow. The allowable design stresses of fire-retardant-treated members are reduced by ca 10%, but the manufacturer of the fire retardant should be contacted for a specific reduction factor for the particular treatment (45).

Solutions of these fire-retardant formulations are impregnated into wood under full cell pressure treatment to obtain dry chemical retentions of 65–95 kg/m^3; this type of treatment greatly reduces flamespread and afterglow. These effects are the result of changed thermal decomposition reactions that favor production of carbon dioxide and water (vapor) as opposed to more flammable components (26). Inhibition of char oxidation (glowing or smoldering) is also achieved.

Some of the chemicals mentioned above and others, such as chlorinated rubber or paraffin, antimony trioxide, calcium carbonate, calcium borate, pentaerythrithol, alumina trihydrate, titanium dioxide, and urea–melamine–formaldehyde resin, may be used to formulate fire-retardant coatings. Many of these coatings are formulated in such a way that the films intumesce (expand) when exposed to fire, thus insulating the wood surface from further thermal exposure. Fire-retardant coatings are mostly used for existing construction.

As a better understanding of fire phenomena has evolved (46), cost effective fire-retardants to resist the spreading of flames (47), and new formulations which give better resistance to leaching and weathering are being used (48). Combustible gases can, however, nullify the beneficial effects. These gases can flash ignite and spread fire to remote areas.

Resistance to Chemicals. Different species of wood vary in their resistance to chemical attack. The significant properties are believed to be inherent to the wood structure, which governs the rate of ingress of the chemical, and the composition of the cell wall, which affects the rate of action at the point of contact (49).

Wood is widely used as a structural material in the chemical industry because it is resistant to a large variety of chemicals. Its resistance to mild acids is far superior to that of steel but not as good as some of the more expensive acid-resistant alloys. Wood tanks used to store cold, dilute acid have a relatively long service life. However, increasing concentration or temperature causes the wood tank to deteriorate rapidly (5).

Softwoods are generally more resistant to acids than are hardwoods because they have high lignin and low hemicellulose contents. In general, heartwood is more resistant to acids than sapwood, probably because of higher extractive content and slower movement of liquid into the heartwood. For these reasons, the heartwood of certain conifers has been widely used in the chemical industry.

Oxidizing acids, such as nitric acid, attack wood faster than common mineral acids, although wood is frequently used in contact with dilute nitric acid. Oxidizing acids not only attack wood by hydrolysis of the polysaccharides but also degrade these polymers through oxidative reaction. Wood shows excellent resistance to organic acids, which gives it a distinct advantage over steel, concrete, rubber, and some plastics. Mild organic acids such as acetic acid have little effect on wood strength.

Alkaline solutions attack wood more rapidly than acids of equivalent concentrations, whereas strong oxidizing chemicals are harmful. Wood is seldom used where resistance to chlorine and hypochlorite solutions is required. These chemicals cause extensive degradation of cell-wall polymers. Wood tanks are, however, satisfactory for holding hydrogen peroxide solutions and give good service on contact with strong brine. Solutions of iron salts cause degradation, particularly of the polysaccharides.

In contact with iron under damp conditions, wood may show severe deterioration within a few years (50). Species high in acidic extractives seem especially prone to such attack.

Because traces of iron reduce the brilliance of many dyes, wood tanks have long been preferred to steel in the manufacture of dyes. Similarly, vinegar and sour foodstuffs are processed in wood tanks because common metals impart a metallic taste. Ease of fabrication may be the deciding factor in the use of wood tanks in less accessible areas to which ready-made tanks of other materials cannot be easily moved.

Resistance to chemical attack is generally improved by resin impregnation, which protects the underlying wood and reduces movement of liquid into the wood. Resistance to acids can be obtained by impregnating with phenolic resin, and to alkalies by impregnating with furfural resin (see Furan compounds; Phenolic resins).

Biodeterioration. Wood may be attacked by fungi, bacteria, insects, or marine borers. Decay fungi break down the various components enzymatically to forms that they readily assimilate. Such wood is said to be decayed, rotted, or doty. Wood susceptible to infection by decay fungi must have a moisture content slightly above its fiber-saturation point, ie, the moisture content at which the cell walls are saturated but there is no free water in the lumen. Conversely, if wood is too wet, ie, saturated, its oxygen supply is too limited for fungal growth. Decay can be prevented by keeping wood either too dry (below 20% moisture content) or too wet (lumens filled with water) for fungal development, by using naturally decay-resistant species, or by treating with preservatives.

Stainers attack wood and stain it deeply, lowering its grade as lumber and, hence, its value; stainers may also decrease resistance to impact loading (toughness). Molds discolor wood surfaces and may cause respiratory discomfort among workers and users. Wood is most likely to be infected by stain and mold fungi while air-drying in the green condition. These fungi are controlled by dipping the lumber, immediately after cutting, in an antistain (fungicidal) solution. Bacteria break down unlignified ray parenchyma cells and pit structures, thereby increasing the wood permeability. This can be detrimental, as increased absorbance may enhance susceptibility to fungal attack, or beneficial, where increased absorption of wood preservatives is desired in difficult-to-treat material (51).

Bacteria may impart a foul odor, rendering the wood unfit for indoor use. In addition, over many years, bacteria are capable of significantly reducing the strength of wood submerged in freshwater (52).

Termites are the most destructive insects attacking wood. Their attack can be prevented, or lessened, by using naturally resistant wood, or by treating wood with preservatives. For subterranean termites, which generally require contact with the ground to survive, poisoning the soil around the wood structure is the principal means of preventing infestation. A promising new approach to their control, entailing the use of insecticidal termite baits, is currently under investigation (53). The dry-wood termite flies directly to the wood into which it bores and does not require contact with the ground. Physical barriers, such as paint or screens, prevent infestation. Despite the great differences between fungi and termites, chemicals that inhibit fungi usually inhibit termites.

Marine borers inhabit saline or brackish waters where they are extremely destructive to untreated wood. The mollusks include the *Teredo* and *Bankia* borers; among the crustaceans, the *Limnoria* borers are the most widespread and destructive.

Protection against attack is afforded by preservatives or by using borer-resistant wood (see Coatings, marine).

For practical purposes, the sapwood of all species may be considered to be susceptible to biodeterioration. The heartwood of some species, however, contains toxic extractives that protect it against biological attack. Among the native species that have decay-resistant or highly decay-resistant heartwood are bald cypress, redwood, cedars, white oak, black locust, and black walnut (54). Douglas-fir, several of the pines, the larches, and honey locust are of intermediate decay resistance. Species low in decay resistance include the remainder of the pines, the spruces, true firs, ashes, aspens, birches, maples, hickories, red and black oaks, tupelo, and yellow poplar. Native woods considered somewhat resistant to termite attack include close-grained redwood heartwood and resinous heartwood of southern pine (5). Although several tropical woods show resistance to marine borers, no commercial native woods are sufficiently borer-resistant to be used untreated (5).

Wood slated for use under conditions conducive to decay is generally treated with preservative under pressure before installation (55–56). Both oil-type preservatives, such as creosote or petroleum solutions of pentachlorophenol, and waterborne preservatives, such as copper–chrome arsenate and ammoniacal-copper arsenate, are used when wood is to be in direct contact with the ground.

Creosote is applied to wood to be used in marine environments infested with teredine borers; however, where *Limnoria* abounds, treatment with water-soluble preservatives either alone or followed by impregnation with creosote (dual treatment) is recommended (57).

Where wood is to be used under low-to-moderate decay hazard conditions, eg, above ground, it may be protected by dip treatment with preservatives; pentachlorophenol in light oil has been found to afford some protection after 22 yr above ground (56). Additional protection to wood in use may be afforded through periodic brush treatment with a suitable wood preservative (58).

Redwood used for cooling towers sometimes undergoes premature deterioration due to either chemical or biological attack or a combination of both (50). Fungi play the more important role in such deterioration; chemicals in the water render redwood more susceptible to decay by accelerating the loss of fungicidal extractives (59).

Confidence in wood as a structural material in the chemical industry is well established. Today, wood treated with preservatives is used in construction under severe decay conditions formerly reserved for other building materials. Many buildings (ca 27,000) have been erected in the United States, for example, with all-wood foundations (60). In addition, treated posts have been recommended for low cost homes (61).

Modified Wood

In addition to preservation or fire protection, wood is modified to reduce shrinking and swelling under conditions of fluctuating relative humidity. Certain species with high extractives content, especially in the cell walls, have greater dimensional stability than species with low extractives content. This gives a clue to a means of obtaining still greater reductions in swelling and shrinking, that is, by deliberately adding large amounts of bulking agents to the cell walls.

Resin impregnation is a successful method of adding bulking agents, provided the resin can permeate the lumens and penetrate the cell walls. Resins of a very low degree of polymerization are used which polymerize after penetration has occurred. Thermosetting phenolic resins have been used successfully. The rate-determining step for successful treatment is the penetration of the resin into the cell wall. With green wood, this rate depends on the diffusion rate of the resin in the lumen-trapped water. With dry wood, pressure can be applied. In this case, the rate depends on the permeability of the wood. For both processes, the size of the object is very important; long treating times are used for large pieces and heartwood. The sapwood of some species is sufficiently permeable to admit resin fairly uniformly and in a reasonable time. Thin veneers are generally pressure impregnated (up to 1.4 MPa = ca 14 atm) with a 30% aqueous solution of a water-soluble resin. The wood is then slowly dried and heated at ca 150°C for 20 min to set the resin. Laminates are built up by gluing the individual sheets together. The product is called impreg. Its density is about 20% higher than that of the original wood, and its color is that of the original wood or slightly darker.

Antishrink efficiency (ASE), defined as

$$1 - \frac{\% \text{ swelling of treated specimen}}{\% \text{ swelling of control}} \times 100,$$

is a measure of the extent to which the swelling and shrinking tendency has been reduced. The antishrink efficiency of impreg increases with increasing content of phenolic resin and then tends to level off at ca 65% when the resin content reaches 30–35%, based on original wood. Impreg generally contains 25–35% resin (62).

The mechanical properties of resin-impregnated wood are improved or not affected except for toughness which is reduced by as much as 60% (63). Treatment with phenol–formaldehyde resins increases the decay resistance. Impreg stakes containing ca 30% resin had an average service life of 12 yr in ground tests (64). Biological resistance may be due to the fact that the cell walls of the treated wood resist moisture-supporting decay. It could also be due in part to toxic effects of partially polymerized phenolic resin on the destructive organism.

Heat resistance is improved markedly by resin impregnation. A block of impreg, subjected to 45 one-hour exposures at 204°C, showed no apparent loss in properties, although an untreated sample showed signs of deterioration after three one-hour exposures (65).

The largest industrial application of impreg is in die models for automobile body parts and other model dies. The dimensional stability and ease of shaping are the reasons that impreg, although expensive, has wide application.

If pressure is applied to dry, resin-treated veneers while they are being heat cured, a densified product (1.35 kg/m^3) is obtained. This material, called compreg, retains most of the advantages of impreg. In addition, owing to the two- to threefold increase in density, the mechanical properties are appreciably better than those of the original wood. The strength of compreg is increased in proportion to the compression. However, compreg is less tough than untreated wood but more tough than impreg.

Because of the plasticizing action of the resin-forming materials, the wood can be compressed under considerably lower pressures than dry, untreated wood. For example, treated spruce, cottonwood, and aspen veneer, dried to a moisture content of ca 6% but not cured, are compressed, when subjected to a pressure of only 1.72 MPa (ca 17 atm) at 149°C, to about half the original thickness and a specific gravity of ca 1.0.

In a 24-h water-soaking test, compreg has an ASE value of 95%. The rate of water pickup is so slow that complete swelling equilibrium of a 1.27-cm specimen is not achieved in a year at room temperature. Compreg is brown and acquires a high polish on buffing. It is made commercially in small quantities and is used for knife handles, gears, and certain musical instruments and decorative objects.

The wood cell wall can be bulked with leachable polyethylene glycol (PEG) to achieve dimensional stability (66). In this case, the wood is usually treated in a green condition and the PEG is exchanged for the cell-wall water. The green wood is soaked in a 30 wt % PEG-1000 solution for a length of time depending upon the thickness of the wood; two coats of polyurethane varnish are usually applied later to seal in the PEG and exclude water. Maximum ASE of 80% is achieved at PEG loadings of 45% weight gain. The strength properties of PEG-treated wood approximate those of untreated green wood (see Glycols; Polyethers).

Wood–plastic combination is another type of resin treatment. Wood is treated with a nonswelling vinyl-type monomer and cured by radiation or heat and a catalyst (67). The hygroscopic characteristics of the wood substance are not altered because little, if any, resin penetrates the cell walls. However, because of the high resin content (70–100%), the normally high void volume of wood is greatly reduced. With the elimination of this very important pathway for vapor diffusion, the response of the wood substance to changes in relative humidity is very slow, and moisture resistance is greatly improved. Hardness is increased appreciably. Wood–plastic materials are currently used in certain sporting equipment, musical instruments, and decorative objects.

It is possible to add an organic moiety to the hydroxyl groups on cell-wall components. This type of treatment also bulks the cell wall with a permanently bonded chemical (68). Many compounds modify wood chemically. Best results are obtained by the hydroxyl groups of wood reacting under neutral or mildly alkaline conditions below 120°C. The chemical system should be simple and must be capable of swelling the wood structure to facilitate penetration. The complete molecule must react quickly with wood components to yield stable chemical bonds while the treated wood retains the desirable properties of untreated wood. Anhydrides, epoxides, and isocyanates have ASE values of 60–75% at chemical weight gains of 20–30%.

Bending is another treatment process. Above 80°C, green wood becomes readily deformable. On cooling to room temperature and drying under restraint, the new shape persists. This is the basis for commercial bending of wood to various shapes. The deformation, however, is not a fully plastic one. An elastic component persists and produces some strain recovery at high or cyclic humidification. Ammonia is also used to bend wood (69). The wooden object is immersed in liquid ammonia for a period of time, depending on the dimensions, imparting appreciable plasticity to the wood. The ammonia is allowed to evaporate from the deformed wood and very little strain recovery occurs on humidification.

Chemical Raw Material

Wood is one of our most important renewable biomass resources. Unlike most biomass sources, wood is available year round and is more stable on storage. In the United States, residues from commercial forests, industrial wastes, urban wood wastes, and land clearing total 380×10^6 metric tons of dry wood (1). Increasingly, residues are incorporated into manufactured wood products and used as a fuel replacing petroleum, especially at wood-industry plants (70); some is converted to charcoal. Residues are also available for manufacturing chemicals, generally at a cost equivalent to the fuel value (71–73) (see Fuels from biomass; Fuels from waste).

The gasification of wood to produce methanol (qv) has received considerable recent attention. When wood is burned with limited oxygen, hydrogen and carbon monoxide are formed. The hydrogen and carbon monoxide can be used to produce low molecular weight chemical intermediates, such as methanol and ammonia. However, wood residues are inferior to coal as a feedstock because wood has a lower carbon content and therefore coal plants are more economical (see Coal chemicals and feedstocks, Supplement Volume). Wood is highly oxygenated, consisting predominately of polymeric carbohydrates, and is only 50% carbon. However, it contains only small amounts of sulfur and its ash content is lower than that of coal.

Because wood consists of two-thirds carbohydrates, considerable attention has been given to the potential of wood residues as a raw material for conversion to ethanol (qv) (74). Starch-rich sources such as corn are currently being used for the production of ethanol. Starch can be readily hydrolyzed in high yield to fermentable sugars. However, wood carbohydrates are difficult to hydrolyze. Hydrolysis with acids is impeded by the crystalline nature of cellulose, and hydrolysis with enzymes by the encrusting lignin which lowers the accessibility of the cellulose. Furthermore, only part of the sugars produced can be fermented by the currently used yeasts, the lignin interferes with processing, and hydrolytic by-products inhibit fermentation. Although the technology of wood hydrolysis is advancing, it remains to be demonstrated that the cost of ethanol from wood will be less than ethanol from starch and sugar-rich crops. Wood feedstocks are more economical when compared with crops because trees can be planted on land not suitable for food production and can be harvested throughout the year. However, the efficiency of photosynthesis is lower than for most crops.

The principal chemical industry based on wood is, of course, the pulp and paper industry which each year converts 80×10^6 t into almost 45×10^6 t of fiber products ranging from newsprint to pure cellulose. The latter is the raw material for a number of products, eg, rayon, cellulose acetate film base, cellulose nitrate explosives, cellophane, celluloid, and chemically modified cellulosic material with a wide range of properties.

Most of the over 36×10^6 t of organic chemicals removed from wood during pulping are burned for their energy content and to recover inorganic pulping chemicals. Some organic chemicals are recovered from this waste stream (75). Sulfite paper mills recover ca $\$100 \times 10^6$ worth of organic chemicals per year, half of which is for lignosulfonates, one third for vanillin, and the rest for small amounts of yeast, ethanol (from fermentation of sugars in the pulping liquor), and acetic acid. Sulfate (kraft) paper mills recover over $\$300 \times 10^6$ worth of organic chemicals per year, and this is expected to increase. It might be economically attractive for pulp mills to prehydrolyze hardwoods for the pentose sugars; isolate acetic, formic, lactic, and other hydroxy acids from the pulping liquor; and chemically utilize the lignin. The most valuable chemical by-products isolated at pulp mills are sulfate turpentine and tall oil (a mixture of fatty acids and rosin); dimethyl sulfide and dimethyl sulfoxide are also obtained from sulfate pulping liquors, and an arabinogalactan gum is extracted from larch before pulping. Thus, the pulp mill is evolving into a multiproduct chemical facility (see Paper; Pulp).

There are a few minor wood-based chemical industries. After chestnut blight wiped out the American chestnut, U.S. tannin production essentially ceased. The main natural tannins, wattle and quebracho, are now imported. High U.S. labor costs and the advent of synthetic tannins make re-establishment of a U.S. tannin industry unlikely. Tannins and by-product lignins are used in oil-well drilling muds (see Petroleum drilling fluids). Wax is isolated from Douglas fir bark. Tree exudates include rubber, true carbohydrate gums (eg, acacia gum), kinos (eg, the phenolic exudates from eucalypts), balsams (eg, Storax from *Liquidambar* spp.), and many different types of oleoresins (mixtures of a solid resin and a liquid essential oil). The most important oleoresin in the United States is pine gum (rosin plus turpentine), although Canada balsam is also well known.

Wood is the raw material of the naval stores industry. Naval stores, so named because of their importance to the wooden ships of past centuries, consist of rosin (diterpene resin acids), turpentine (monoterpene hydrocarbons), and associated chemicals derived from pine (see Terpenoids). These were obtained by wounding the tree to yield pine gum, but the high labor costs have substantially reduced this production in the United States. More important has been the production of rosin and turpentine obtained by the extraction of old pine stumps. This is a nonrenewable resource and this industry is in decline. The most important source of naval stores has been the sulfate liquors from pulping pine. In 1979, U.S. production of rosin from all sources was estimated at 320,000 metric tons and of turpentine at 74,000 metric tons. Distillation of tall oil (qv) provides, in addition to rosin, nearly 200,000 metric tons of tall oil fatty acids annually. In 1980, the value of naval stores primary products (rosin, turpentine, tall oil fatty acids, and processing by-products) was over 300×10^6 (76).

Hydrolysis

In the hydrolysis process (77–79), wood is treated with concentrated or dilute acid solution to produce a lignin rich residue and a liquor containing sugars, organic acids, furfural, and other chemicals. The process is adaptable to all species and all forms of wood waste. The liquor can be concentrated to a molasses for animal feed (80), used as a substrate for fermentation to ethanol or yeast (80), or dehydrated to furfural and levulinic acid (81–84). Attempts have been made to obtain marketable products from the lignin residue (85) rather than using it as a fuel, but currently only carbohydrate-derived products appear practical.

When concentrated acids are used, the carbohydrates are recovered in high yields, but the problem of economically recovering the large quantities of acid used has not been solved. At the present state of development, the dilute acid processes, especially percolating and two-stage, appear more promising.

A number of commercial plants using the Scholler percolation process (86) have been built in Germany, Switzerland, and the USSR; only the USSR plants are still in operation. In general, these plants were built to produce sugars for fermentation to ethanol and yeast (qv). Except under special circumstances, however, such a process has proved to be uneconomical. In the Scholler process, a hot dilute solution of sulfuric acid is percolated through wood chips; the solubilized sugars are carried with the solution withdrawn from the bottom of the digester. The acid concentration and temperature of the charged solution are continuously increased; the more labile hemicelluloses are hydrolyzed and removed in the early part of the cycle, whereas the resistant cellulose, yielding glucose, is hydrolyzed at the end.

A percolation process requiring less time was developed at the U.S. Forest Products Laboratory in Madison, Wisconsin, during World War II (87–89). For Douglas fir, this process gives a sugar yield of 40–45%; the sulfuric acid requirement is 5% based on the dry wood weight. The concentration of the resulting sugar solution is 4–5%. Yields per 100 kg of dry wood are 20–25 L of 100% ethanol.

In two-stage processes, the hemicellulose sugars are hydrolyzed in the first stage; the solubilized material is washed from the residue which is reimpregnated with acid and passed to the second stage where the cellulose is hydrolyzed. This process is much less energy- and capital-intensive than the percolation process, and gives a better fractionation of the hemicellulose sugars and glucose. Enzymatic saccharification of the cellulose requires two-stage processing. Sugar yields are about the same as those obtained by percolation, but concentrations of 10–12% are reached which significantly lower acid, energy, and equipment-capacity requirements. The two-stage process is not operated commercially but during World War II some pilot studies were done in Sweden (90). Much of the recent work on lignocellulose utilization has been directed toward the development of the two-stage process with either chemical or enzymatic hydrolysis of the cellulose (91–97).

High yields of pure glucose can be obtained by enzymatic cellulose hydrolysis. However, the enzyme is expensive and does not attack lignin-encrusted cellulose. Several processes have been studied using various organisms singly and in combination after subjecting the lignocellulose to different pretreatments (91,97–101). In the Gulf process, cellulose hydrolysis and glucose fermentation take place simultaneously in one vessel.

The Iotech steam explosion process has been used to prepare cellulosic substrates for enzymatic digestion (102–103). Wood, or other lignocellulose material, is subjected to a short prehydrolysis at high temperature and pressure and then rapidly decompressed. The combined chemical and mechanical treatment solubilizes the hemicellulose component in hot water and the lignin in alcohols. Freed of lignin encrustation, the cellulose is highly accessible to hydrolytic enzymes. The recovered lignin is thermoplastic and may be marketable as a resin component.

Although the hydrolysis of wood to produce simple sugars has not proved to be economically feasible, by-product sugars from sulfite pulping are used to produce ethanol and feed yeast (104). Furthermore, a hemicellulose molasses, obtained as a by-product in hardboard manufacture, can be used in cattle feeds instead of blackstrap molasses (105). Furfural can be produced from a variety of wood-processing by-products, such as spent sulfite liquor, liquors from the prehydrolysis of wood for kraft pulping, hardboard plants, and hardwood wastes (106).

Fuel Properties

The fuel properties of wood can be summarized by ultimate and proximate analyses and determination of heating value. The analytical procedures are the same as those for coal but with some modifications. Analytical results generally vary about as much within a species as they do between species, except that softwood species generally have a higher carbon content and heating values than hardwood species because of the presence of more lignin and resinous materials in softwood species (see Fuels from waste).

The higher heating value of wood and bark of softwood species is usually ca 21 kJ/g (ca 9000 Btu/lb) and slightly less for hardwood. The higher heating value includes the heating value of the condensed steam given off. These values are within ±5% of nearly all the values reported in the literature for specific samples. A systematic study of the heating value of a species has not been done because it requires measuring the heating value at various positions in a tree and from trees selected from the entire geographic range of the species.

Wood ash generally contains calcium, potassium, phosphorus, magnesium, and silica. Ashes recovered from burned wood are ca 25% water-soluble and the extract is strongly alkaline. The ash fusion temperature is in the range of 1300–1500°C.

The moisture content of freshly cut wood varies between species and portions of the tree. Between species, it can be 30–70% on a total weight basis (5); commonly, it is 45–50%. Within a tree, the heartwood generally has lower moisture content than the sapwood. For hardwood species, this difference is usually small; for softwood species such as Douglas-fir, the difference can be as great as 30% for heartwood compared to 50% for sapwood.

Charcoal Production

Charcoal is produced by heating wood under limited access of oxygen. When wood is heated slowly to ca 280°C, an exothermic reaction occurs. In the usual carbonization procedure, heating is prolonged to 400–500°C in the absence of air. The term charcoal also includes charcoal made from bark.

Charcoal is produced commercially from primary wood-processing residues and low quality roundwood in either kilns or continuous furnaces. A kiln is used if the raw material is in the form of roundwood, sawmill slabs, or edgings. In the United States, most kilns are constructed of poured concrete with a capacity of 40–100 cords of wood and operating on a 7–12-d cycle. Sawdust, shavings, or milled wood and bark are converted to charcoal in a continuous multiple-hearth furnace commonly referred to as a Herreshoff furnace. The capacity is usually at least one ton of charcoal per hour. The yield is ca 25 wt % on a dry basis.

The proximate analysis of charcoal is ca 20–25% volatile matter, 70–75% fixed carbon, and 5% ash. The higher heating value is ca 28 kJ/kg (12,000 Btu/lb). Charcoal briquettes have a higher heating value of ca 23–25 kJ/g (ca 9,900–10,800 Btu/lb) due to added ingredients.

To alleviate the air-pollution problem associated with charcoal kilns and furnaces, the gases from the kiln and furnaces are burned (see Air pollution control methods). They can be burned with additional fossil fuel to recover heat and steam (107–108), or in afterburners to nearly eliminate visible air pollution and odors (109).

Charcoal was an important industrial raw material in the United States for iron-ore reduction until it was replaced by coal in the early 1880s. Charcoal production increased, however, because of the demand for the by-product acetic acid, methanol, and acetone. In 1920, nearly 100 by-product recovery plants were in operation in the

United States, but the last plant ceased operation in 1969; U.S. production figures are given below (110).

Year	1909	1947	1956	1968	1980
Charcoal production, 10^3 t	450	190	240	450	725

The 1968 and 1980 figures are for the production of charcoal briquetts that contain a starch binder and possibly char from nut shells, lignite, clay, coal dust, and sawdust.

The increase in production since the 1940s reflects the use of charcoal briquetts for home and recreational cooking. The charcoal currently produced is nearly all consumed as briquetts for cooking. Some charcoal is used in certain metallurgical and filtration processes and horticultural uses. In Brazil, charcoal is produced in beehive-type kilns from natural and plantation-grown trees for use as a reducing agent for iron ore because Brazil does not have abundant supplies of coking coal. In many developing countries, charcoal is preferred for domestic cooking. It is made in pit-type kilns or portable sheet-metal kilns (111) (see also Carbon and artificial graphite).

Economic Aspects

In 1972, the wood industry contributed ca 48.5×10^9 to the U.S. gross national product (GNP), which is defined as the total value, at current prices, of the goods produced and services rendered in a country during a given period (generally one year) plus the balance of foreign trade (112). In terms of total U.S. employment, the wood industry employed one out of every 25 members of the workforce (4%).

The importance of wood in the U.S. economy is best evaluated in an aggregate sense, that is, through the assessment of wood products as an industrial raw material and the amount of GNP and employment originating in timber-based economic activities. In 1980, the USDA Forest Service published a detailed report on the timber economy (113). The compilation of the data took several years. A survey is given in Table 6 for 1972, the last year covered comprehensively in the report.

Timber Management. Timber management is defined as the process of improving, protecting, and otherwise managing forest lands for the production of timber and related products (113). Included are tree planting, timber-stand improvement, protection of forests from fire and insects, sales activities, and research.

The timber harvested from U.S. forests in 1976, ca 363×10^6 m^3 (11.9×10^9 ft^3), represents a 8% increase over 1972, and a 7% increase over 1967. Demand for hardwood timber, over half of which was growing in northern forests in the late 1960s, declined somewhat over the past three decades. Conversely, markets for softwood timber, largely a product of western and southern forests, have increased.

The stumpage value of the timber cut from U.S. forests in 1972 has more than doubled since 1963 and was 95% above 1967. It is assumed that the value of stumpage cut is the same as the total value added which is attributed to timber-management activities. Even though there are intermediate product costs, they are relatively small in comparison with the value of stumpage cut.

Table 6. Wood-Based Economic Activities in the United States, 1972[a]

Economic activity	Value of product or service, 10⁶	Value added[b], 10⁶ Total	Value added[b], 10⁶ Attributed to wood	Employment, in thousands Total	Employment, in thousands Attributed to wood
Timber management	2,864[c]	2,864	2,864	117	117
Harvesting	6,360	3,065	3,065	190	190
Primary manufacturing					
sawmills and planing mills	7,575	3,029	2,876	184	171
veneer and plywood plants	2,923	1,238	1,073	66	58
pulp, paper, and paperboard mills	11,705	5,417	4,583	218	184
all other	815	384	264	20	14
Total	23,018[d]	10,068	8,796	488	427
Secondary manufacturing					
millwork and prefabricated wood products	8,085	3,127	1,951	219	137
wooden containers	774	359	324	37	33
furniture	10,111	5,395	1,820	423	151
paper and paperboard products	16,553	7,605	5,063	413	278
fibers, plastics, and textiles	na	17,513	2,629	1,652	249
all others	na	na	718	na	52
Total	na	na	12,505	na	900
Construction	159,000	79,601	11,947	5,278	795
Transportation and marketing					
transportation	na	32,070	2,792	1,899	165
wholesale trade	694,300	70,456	2,997	4,310	181
retail trade	470,800	91,635	3,561	12,498	489
Total	na	194,161	9,350	18,707	835
Grand total			48,527[e]		3,264

[a] Ref. 1.
[b] Defined as the value of output less the cost of input which represents the contribution to the GNP (112).
[c] Value of stumpage cut.
[d] Increase of 83% over 1963 and 54% over 1967.
[e] Ca 4.1% of GNP.

Harvesting. Harvesting includes felling trees, cutting them into logs, cutting or collecting timber products such as Christmas trees and pine gum, and transporting these products to local delivery points such as rail yards, barge landings, and processing points (114).

The 1976 volume of round timber products is given in Table 7.

The total value of timber and related products harvested from U.S. forests in 1972 was almost double the value in 1963 and was 80% higher than in 1967. The total value of timber and related products harvested is derived from the value of stumpage plus the intermediate products used in harvesting and transportation. After these are deducted from the total value of the timber products harvested, the value added in harvesting (and the value added attributed to timber) amounted to an estimated 3.1×10^9 in 1972, double the 1963 amount of 1.5×10^9.

Primary Manufacturing. Primary manufacturing includes the activities involved in the processing of logs and related products into lumber, veneer, plywood, pulp and paper, turpentine, rosin, and other products (113).

Table 7. U.S. Timber Products, 1976, 1000 m³ [a]

Region	Total	Saw logs	Veneer logs	Pulpwood	Other products
north[b]	54,358	27,375	1,049	22,343	3,591
south[c]	158,644	64,175	15,895	74,055	4,519
west[d]	132,850	96,536	23,565	10,170	2,579
United States	345,852	188,086	40,509	106,568	10,689

[a] Ref. 113. To convert 1000 m³ to 10⁶ ft³, multiply by 28.32.
[b] Includes north central and northeast.
[c] Includes south central and southeast.
[d] Includes Mountain and Pacific states.

Some products harvested from the forests are ready for use and need only be marketed or transported to the consumer, eg, Christmas trees and fuelwood. In 1981, ca 40×10^6 cords of wood was consumed as fuelwood. Saw logs, veneer logs, and pulpwood are manufactured into lumber, plywood, wood pulp, and other similar items.

Primary manufacturing is carried out by sawmills and planing mills; the veneer and plywood industry; the pulp, paper, and paperboard industry; and manufacturers of diverse timber products such as excelsior, wood shingles, and gum and wood chemicals.

Softwood and hardwood lumbers were produced in 1982 at a rate of 22×10^9 and 5.0×10^9 board feet (52×10^6 and 12×10^6 m³), respectively. Lumber production was followed closely by softwood and hardwood plywood production on a 9.5-mm-thickness basis as they were produced at a 1.5×10^9 and 0.1×10^9-m² (16×10^9 and 1.1×10^9-ft²) rate in 1982. Also ca 49.5×10^6 t woodpulp produced in 1982.

The value of shipments data for the primary manufacturing industries is different from value added because of double counting since products of one plant or mill may be shipped as a raw material to other plants in the same industry.

Secondary Manufacturing. Secondary manufacturing includes activities involved in the processing of lumber, plywood, paper, and other products into finished goods such as furniture, toys, wearing apparel, and containers (113).

In the secondary manufacturing industries, the value added attributable to timber was $\$12.5 \times 10^9$, with ca two fifths originating in the paper and paperboard products industry. Another one fifth originated in the fibers, plastics, and textiles industries. Of the remainder, 16% was in the millwork and prefabricated wood-products industry, 15% in the furniture industry, 3% in the wooden-containers industry, and 6% in all other secondary manufacturing (1).

Construction. Construction activities involve the fabrication of lumber, plywood, and wood-based building supplies (113).

In the 1960s and early 1970s, three fourths of the softwood lumber and plywood, one tenth of the pulp products, and all poles, piling, and shingles were used in construction. In terms of value, residential construction represented the largest percentage, ca 44% in 1972.

Transportation and Marketing. Transportation and marketing involve the transportation of logs and related products from local delivery points to manufacturing plants or other consumers, transportation of primary and secondary products from points of manufacture to final consumers, and the marketing of these products through wholesale and retail channels (113).

Only a relatively small part of the activity of enterprises in transportation and marketing was based on timber products. In 1972, these products accounted for about 13.5% in railroad transportation, 10.3% in water transportation, and 5.7% in truck transportation.

BIBLIOGRAPHY

"Wood" in *ECT* 1st ed., Vol. 15, pp. 72–102, by E. G. Locke, R. H. Baechler, E. Beglinger, H. D. Bruce, J. T. Drow, K. G. Johnson, D. G. Laughnan, B. H. Paul, R. C. Rietz, J. F. Saeman, and H. Tarkow, U.S. Department of Agriculture; "Wood" in *ECT* 2nd ed., Vol. 22, pp. 358–387, by H. Tarkow, A. J. Baker, H. W. Eickner, W. E. Eslyn, G. J. Hajny, R. A. Hahn, R. C. Koeppen, M. A. Millet, and W. E. Moore, U.S. Department of Agriculture.

1. *An Analysis of the Timber Situation in the United States, 1952–2030*, Forest Resource Report 23, Forest Service, U.S. Department of Agriculture, Washington, D.C., 1982.
2. *Tree Biomass—A State-of-the-Art Compilation*, General Technical Report WO-33, Forest Service, U.S. Department of Agriculture, Washington, D.C., 1981.
3. A. Ulrich, *U.S. Timber Production, Trade, Consumption, and Price Statistics 1950–1980*, U.S. Department of Agriculture Miscellaneous Publication 1408, U.S. Department of Agriculture, Washington, D.C., 1981.
4. A. J. Panshin and C. deZeeuw, *Textbook of Wood Technology: Structure, Identification, Uses, and Properties of the Commercial Woods of the United States*, 4th ed., McGraw-Hill Book Co., Inc., New York, 1980.
5. *Wood Handbook*, Agriculture Handbook 72, U.S. Forest Products Laboratory, U.S. Department of Agriculture, Madison, Wisc., revised 1974.
6. R. L. Gray and R. A. Parham, *Chemtech* **12**(4), 232 (1982).
7. S. A. Rydholm, *Pulping Processes*, Interscience Publishers, a division of John Wiley & Sons, Inc., New York, 1965.
8. B. L. Browning, *The Chemistry of Wood*, Robert E. Krieger Publishing Company, Huntington, N.Y., 1975.
9. R. H. Farmer, *Chemistry in the Utilization of Wood*, Pergamon Press, Oxford, 1967.
10. C. Skaar and W. T. Simpson, *For. Prod. J.* **18**, 49 (1968).
11. A. J. Stamm, *Wood and Cellulose Science*, Ronald Press Company, New York, 1964.
12. F. F. P. Kollmann and W. A. Côté, Jr., *Principles of Wood Science and Technology*, Vol. 1, Springer-Verlag New York, Inc., New York, 1968.
13. R. C. Rietz and R. H. Page, *Air Drying of Lumber: A Guide to Industry Practices*, Agriculture Handbook 402, U.S. Department of Agriculture, U.S. Government Printing Office, Washington, D.C., 1971.
14. E. F. Rasmussen, *Dry Kiln Operators Manual*, Agriculture Handbook 188, U.S. Department of Agriculture, U.S. Government Printing Office, Washington, D.C., 1961.
15. *National Design Specifications for Wood Construction*, National Forest Products Association, Washington, D.C., 1977.
16. C. C. Gerhards, *Wood Fiber* **14**(1), 4 (1982).
17. C. C. Gerhards and J. M. McMillen, eds., *Proceedings of the Research Conference on High-Temperature Drying Effects on Mechanical Properties of Softwood Lumber*, U.S. Forest Products Laboratory, Forest Service, U.S. Department of Agriculture, Madison, Wisc., 1976.
18. H. D. Erickson and L. W. Rees, *J. Agric. Res. (Washington, D.C.)* **60**, 593 (1940).
19. Warren S. Thompson, *Effect of Chemicals, Chemical Atmospheres, and Contact with Metals on Southern Pine: A Review*, Research Report 6, Forest Products Utilization Laboratories, Mississippi State University, State College, Miss., 1969.
20. F. Shafizadeh and A. G. W. Bradbury, *J. Thermal Insulation* **2**, 141 (Jan. 1979).

21. *Ignition and Charring Temperature of Wood*, U.S. Forest Products Laboratory Report 1464, U.S. Forest Products Laboratory, U.S. Department of Agriculture, Madison, Wisc., 1958.
22. A. F. Matson, R. E. Dufour, and J. F. Breen, *Survey of Available Information on Ignition of Wood Exposed to Moderately Elevated Temperatures*, Bulletin of Research 51, Part II, Underwriters Laboratories, Inc., 1959.
23. E. L. Schaffer, *Fire Techn.* **16**(1), 22 (Feb. 1980).
24. V. W. Jahnsen, Ph.D. thesis, Purdue University, West Lafayette, Ind., 1961.
25. R. W. Porter, Ph.D. thesis, Michigan State University, East Lansing, Mich., 1963.
26. J. J. Brenden, *Effect of Fire-Retardant and Other Inorganic Salts on Pyrolysis Products of Ponderosa Pine at 250° and 350°C*, U.S. Forest Service Research Paper FPL 80, U.S. Forest Products Laboratory, U.S. Department of Agriculture, Madison, Wisc., 1967.
27. A. J. Stamm, *Ind. Eng. Chem.* **48**, 413 (1956).
28. F. L. Browne and W. K. Tang, *Fire Research Abstracts and Reviews* **4**(1-2), 76 (1962).
29. H. W. Eickner and W. K. Tang, *Proceedings of International Union of Forest Research Organizations*, Section 41, U.S. Forest Products Laboratory, U.S. Department of Agriculture, Madison, Wisc., 1963.
30. F. L. Browne, *Theories of the Combustion of Wood and its Control*, U.S. Forest Products Laboratory Report 2136, U.S. Forest Products Laboratory, U.S. Department of Agriculture, Madison, Wisc., 1958; reviewed and reaffirmed 1963.
31. F. C. Beall and H. W. Eickner, *Thermal Degradation of Wood Components: A Review of the Literature*, Forest Service Research Paper FPL 130, U.S. Forest Products Laboratory, U.S. Department of Agriculture, Madison, Wisc., 1970.
32. I. S. Goldstein in D. D. Nicholas, ed., *Degradation and Protection of Wood*, Vol. 1, Syracuse University Press, Syracuse, N.Y., 1973.
33. F. Shafizadeh, *Adv. Carbohydr. Chem.* **23**, 419 (1968).
34. *Standard Test Method for Behavior of Materials in a Vertical Tube Furnace at 750°C*, Standard E 136, American Society for Testing and Materials, Philadelphia, Pa., 1979.
35. *Standard Methods of Fire Tests of Building Construction and Materials*, Standard E 119, American Society for Testing and Materials, Philadelphia, Pa., 1981.
36. *Standard Methods of Fire Tests of Door Assemblies*, Standard E 152, American Society for Testing and Materials, Philadelphia, Pa., 1980.
37. *Standard Test Method for Surface Burning Characteristics of Building Materials*, Standard E 84, American Society for Testing and Materials, Philadelphia, Pa., 1981.
38. *Standard Test Method for Specific Optical Density of Smoke Generated by Solid Materials*, Standard E 662, American Society for Testing and Materials, Philadelphia, Pa., 1979.
39. E. L. Schaffer, *Review of Information Relating to the Charring Rate of Wood*, U.S. Forest Service Research Note FPL-0145, U.S. Forest Products Laboratory, U.S. Department of Agriculture, Madison, Wisc., 1966.
40. E. L. Schaffer, *Charring Rate of Selected Woods-Transverse to Grain*, U.S. Forest Service Research Paper FPL 69, U.S. Forest Products Laboratory, U.S. Department of Agriculture, Madison, Wisc., 1967.
41. E. L. Schaffer, *Wood Fiber* **9**(2), 145 (1977).
42. E. L. Schaffer, *J. Fire Flammability/Fire Retardant Chem.* **1**, 96 (1974).
43. H. W. Eickner, *ASTM Journal of Materials* **1**, 625 (1966).
44. J. J. Brenden, *How Nine Inorganic Salts Affected Smoke Yields from Douglas-Fir Plywood*, Forest Service Research Paper FPL 249, U.S. Forest Products Laboratory, U.S. Department of Agriculture, Madison, Wisc., 1975.
45. *National Design Specifications for Wood Construction*, National Forest Products Association, Washington, D.C., 1982.
46. J. J. Brenden, *Rate of Heat Release from Wood-base Building Materials Exposed to Fire*, Forest Service Research Paper FPL230, U.S. Forest Products Laboratory, U.S. Department of Agriculture, Madison, Wisc., 1974.
47. C. A. Holmes in I. S. Goldstein, ed., *American Chemical Society Symposium Series 43*, American Chemical Society, Washington, D.C., 1977.
48. C. A. Holmes and R. O. Knispel, *Exterior Weathering Durability of Some Leach-Resistant Fire-Retardant Treatments for Wood Shingles: A Five-Year Report*, Forest Service Research Paper FPL 403, U.S. Forest Products Research Laboratory, U.S. Department of Agriculture, Madison, Wisc., 1974.

49. R. H. Baechler, *J. For. Prod. Res. Soc.* **4,** 332 (1954).
50. R. H. Baechler and C. A. Richards, *Transactions American Society of Mechanical Engineers* **73,** 1055 (1951).
51. R. S. Smith in W. Liese, ed., *Economic Aspects of Bacteria in Wood, Biological Transformation of Wood by Microorganisms*, Springer-Verlag, Berlin, Heidelberg, New York, 1975, pp. 89–102.
52. W. E. Eslyn and J. W. Clark, *Mater. Org. Suppl.* **3,** 43 (1976).
53. G. R. Esenther and R. H. Beal, *Sociobiology* **4,** 215 (1979).
54. *Comparative Decay Resistance of Heartwood of Native Species*, U.S. Forest Service Research Note FPL-0153, U.S. Forest Products Laboratory, U.S. Department of Agriculture, Madison, Wisc., 1967.
55. *Wood Preservation: Treating Practices*, Federal Specification TT-W-5711, U.S. General Services Administration, Washington, D.C.
56. T. C. Scheffer and W. E. Eslyn, *For. Prod. J.* **28,** 25 (1978).
57. L. R. Gjovik and R. H. Baechler, Forest Service General Technical Report FPL-15, U.S. Forest Products Laboratory, U.S. Department of Agriculture, Madison, Wisc., 1977.
58. T. L. Highley, *For. Prod. J.* **30,** 49 (1980).
59. R. H. Baechler, J. O. Blew, and C. G. Duncan, *Causes and Prevention of Decay of Wood in Cooling Towers*, American Society of Mechanical Engineers Paper 16, Part 5, ASME, New York, 1961.
60. *All-Weather Wood Foundation System*, American Plywood Association, Tacoma, Wash., 1979.
61. L. O. Anderson, *Low-Cost Wood Homes for Rural America—Construction Manual*, Agriculture Handbook 364, U.S. Department of Agriculture, Washington, D.C., 1969.
62. R. M. Rowell and R. L. Youngs, *Dimensional Stability of Wood in Use*, Forest Service Research Note FPL-0243, U.S. Forest Products Laboratory, U.S. Department of Agriculture, Madison, Wisc., 1981.
63. E. C. O. Erikson, *Mechanical Properties of Laminated Modified Wood*, U.S. Forest Products Laboratory Report 1639, U.S. Forest Products Laboratory, U.S. Department of Agriculture, Madison, Wisc., 1965.
64. L. R. Gjovik and H. L. Davidson, *Comparison of Wood Preservatives in Stake Tests*, Forest Service Research Note FPL-02, U.S. Forest Products Laboratory, U.S. Department of Agriculture, Madison, Wisc., 1979.
65. R. M. Seborg and A. E. Vallier, *J. For. Prod. Res. Soc.* **4,** 305 (1954).
66. A. J. Stamm, *For. Prod. J.* **9,** 375 (1959).
67. J. A. Meyer, *Wood Sci.* **14**(2), 49 (1981).
68. R. M. Rowell, *Proc. Am. Wood-Preserv. Assoc.* **71,** 41 (1975).
69. C. Schuerch, *For. Prod. J.* **14,** 377 (1964).
70. D. A. Tillman, *Wood as an Energy Resource*, Academic Press, New York, 1978.
71. G. T. Maloney, *Chemicals from Pulp and Wood Waste, Production and Applications*, Noyes Data Corporation, Park Ridge, N.J., 1978.
72. F. Shafizadeh, K. V. Sarkanen, and D. A. Tillman, *Thermal Uses and Properties of Carbohydrates and Lignins*, Academic Press, New York, 1976.
73. K. V. Sarkanen and D. A. Tillman, *Progress in Biomass Conversion*, Vol. I, Academic Press, New York, 1979.
74. *Chem. Eng.*, 53 (Oct. 20, 1980); 51 (Jan. 26, 1981); 62 (June 15, 1981).
75. D. F. Zinkel in I. S. Goldstein, ed., *Organic Chemicals from Biomass*, CRC Press, Inc., Boca Raton, Fla., 1981, Chapt. 9.
76. D. F. Zinkel, *Pulp Chemicals Association 1980 Production Figures; Naval Stores Annual Summary, Preliminary 1979*, Crop Reporting Board SPCr 3(80), U.S. Department of Agriculture, Washington, D.C., 1980.
77. J. F. Harris, J. F. Saeman, and E. G. Locke in B. L. Browning, ed., *Wood as a Chemical Raw Material, The Chemistry of Wood*, Interscience Publishers, a division of John Wiley & Sons, Inc., New York, 1973, Chapt. 11, pp. 535–585.
78. J. A. Hall, J. F. Saeman, and J. F. Harris, *Unasylva* **10**(1), 7 (1956).
79. J. F. Harris, J. F. Saeman, and E. G. Locke, *For. Prod. J.* **8,** 248 (1958).
80. G. J. Hajny, *Biological Utilization of Wood for Production of Chemicals and Foodstuffs*, Forest Service Research Paper FPL 385, U.S. Forest Products Laboratory, U.S. Department of Agriculture, Madison, Wisc., 1981.
81. D. F. Root and co-workers, *For. Prod. J.* **9,** 158 (1959).
82. S. W. McKibbins and co-workers, *For. Prod. J.* **12,** 17 (1962).

83. J. F. Harris and J. M. Smuk, *For. Prod. J.* **11,** 303 (1961).
84. M. S. Feather and J. F. Harris, *Adv. Carbohydr. Chem. Biochem.* **28,** 161 (1973).
85. *Chem. Eng. News*, 35 (Nov. 3, 1980).
86. J. F. Saeman, E. G. Locke, and G. K. Dickerman, *Production of Wood Sugar in Germany and its Conversion to Yeast and Alcohol*, U.S. Department of Commerce Office Technical Service PB Report 7736, U.S. Department of Commerce, Washington, D.C., 1945.
87. E. E. Harris, *Adv. Carbohydr. Chem.* **4,** 154 (1949).
88. R. A. Lloyd and J. F. Harris, *Wood Hydrolysis for Sugar Production*, U.S. Forest Products Laboratory Report 2029, U.S. Forest Products Laboratory, U.S. Department of Agriculture, Madison, Wisc., 1955.
89. A. J. Panshin, *Wood Saccharification in Forest Products*, McGraw-Hill Book Co., Inc., New York, 1950, pp. 403–412.
90. K. N. Cederquist, *Some Remarks on Wood Hydrolyzation*, Report of a Seminar Held at Lucknow, India, 1952, on Production and Use of Power Alcohol in Asia and the Far East, Technical Assistance Administration, United Nations, New York, 1954, pp. 193–198.
91. A. J. Baker and T. W. Jeffries, *Status of Wood Hydrolysis for Ethanol Production*, Report of U.S. Department of Agriculture, Forest Service, for U.S. Agency for International Development Support Bureau, Office of Energy (TMR Authorization No. 81-89), Washington, D.C., June 1981.
92. U.S. Pat. 4,201,596 (May 6, 1980), J. A. Church and co-workers (to American Can Co.).
93. R. D. Ziminski, *paper presented at American Institute of Chemical Engineers Meeting, New Orleans, La.*, Nov. 1981.
94. D. R. Thompson and H. E. Grethlein, *Ind. Eng. Chem. Prod. Res. Dev.* **18**(3), 166 (1979).
95. G. T. Tsao and co-workers, *Lorre Biomass Conversion Conference*, Purdue University, West Lafayette, Ind., 1981.
96. R. S. Roberts, *Biotechnol. Bioeng. Symp.* **10,** 125 (1980).
97. *Enzymatic Hydrolysis of Cellulose to Glucose—A Report on the Natick Program*, U.S. Army, Natick Research and Development Command, Natick, Mass., Sept. 1981.
98. L. Spano, *J. Coatings Technol.* **50,** 71 (1978).
99. G. H. Emert, R. Katzen, R. E. Fredrickson, and K. F. Kaupisch, *Chem. Eng. Prog.* **76**(9), 47 (1980).
100. D. M. Jenkins and T. S. Reddy, *Economic Evaluation of the MIT Process for Manufacture of Ethanol*, DSE-3992-T1, National Technical Information Service, Washington, D.C., 1979.
101. R. Brooks, T.-M. Su, M. Brennan, and J. Frick, *Proceedings 3rd Annual Biomass Energy Systems Conference, Golden, Colo.*, SERI/TP-33-285, National Technical Information Service, Washington, D.C., June 1979.
102. R. H. Marchessault, S. Coulombe, T. Hanai, and H. Morikawa, *Monomers and Oligomers from Wood, Transactions of Technical Section* of the TR52, 1980. Published in *Pulp and Paper Canada* **81**(6), 1980.
103. L. Jurasek, *Devel. Ind. Microbiol.* **1978,** 177 (1979).
104. L. A. Underkofler and R. J. Hickey, *Industrial Fermentation*, Chemical Publishing Company, Inc., New York, 1954.
105. H. D. Turner, *For. Prod. J.* **14,** 282 (1964).
106. J. F. Harris, *Tappi* **61,** 41 (1978).
107. U.S. Pat. 4,280,878 (Oct. 30, 1979), G. E. Sprenger.
108. J. Rienks, *Forest Products Research Society Proceedings* No. P-75-13, Forest Products Research Society, Madison, Wisc., pp. 104–106, 1975.
109. J. Hartwig, *For. Prod. J.* **21** (1971).
110. *Charcoal and Charcoal Briquette Production in the United States*, Forest Service, Division of Forest Economics and Marketing Research, U.S. Department of Agriculture, Washington, D.C., 1963.
111. D. E. Earl, *A Report on Charcoal*, Food and Agriculture Organization of the United Nations, Rome, Italy, 1974.
112. *Terminology of Forest Science, Technology Practice and Products*, Society of American Foresters, Washington, D.C., 1971.
113. R. Phelps, *Timber in the United States Economy 1963, 1967, and 1972*, Forest Service General Technical Report WO-21, U.S. Department of Agriculture, Washington, D.C., 1980.

114. D. Hair, *The Economic Importance of Timber in the United States*, U.S. Department of Agriculture Miscellaneous Publication 941, U.S. Department of Agriculture, Washington, D.C., 1971.

THEODORE H. WEGNER
A. J. BAKER
B. A. BENDTSEN
J. J. BRENDEN
W. E. ESLYN
J. F. HARRIS
J. L. HOWARD
R. B. MILLER
R. C. PETTERSEN
J. W. ROWE
R. M. ROWELL
W. T. SIMPSON
D. F. ZINKEL
U.S. Department of Agriculture

WOOD-PLASTIC COMPOSITES. See Laminated wood-based composites.

WOOD PULP. See Pulp.

WOOL

Wool, the fibrous covering from the sheep (1), is by far the most important animal fiber used in textile manufacture. World production in 1979–1980 was 1.6×10^6 metric tons (clean) (2); other animal fibers, eg, mohair, alpaca, cashmere, and camel, totaled only 26,000–30,000 t (3) (see also Silk). However, the relative importance of wool as a textile fiber has declined over the decades (Table 1), as synthetic fibers have increasingly been used in textile consumption. Nevertheless, wool is still an important fiber in the middle and upper price ranges of the textile market. It is also an extremely important export for several nations, notably Australia, Argentina, New Zealand, and South Africa, and commands a price premium over most other fibers because of its outstanding natural properties of soft hand (the feel of the fabric), good moisture absorption (and hence comfort), and good drape (the way the fabric hangs) (see Fibers, chemical; Textiles).

Table 2 shows wool production and sheep numbers in the world's principal wool-producing countries.

Table 1. World Production of Cotton, Wool, and Synthetic Fibers, 10^6 t [a]

Year	Raw cotton	Wool (clean)	Synthetic filament	Staple	Cellulosic filament	Staple	Total	Total of all fibers
1900	3.162	0.73			0.001		0.001	3.893
1940	6.907	1.134	0.001	0.004	0.542	0.585	1.132	9.173
1950	6.647	1.057	0.054	0.015	0.871	0.737	1.677	9.381
1960	10.113	1.463	0.417	0.285	1.131	1.525	3.358	14.934
1970	11.784	1.602	2.397	2.417	1.391	2.187	8.393	21.779
1980	14.137	1.581	4.731	5.756	1.159	2.085	13.731	29.449

[a] Ref. 4.

Table 2. Wool Production and Sheep Numbers in Principal Wool-Producing Countries, 1979–1980 [a]

	Wool production (greasy basis) 1000 t	%	Sheep, 10^6
Australia	713	25.5	136
New Zealand	357	12.8	68
South Africa	103	3.7	24
Argentina	166	5.9	32
Uruguay	70	2.5	17
United States	48	1.7	13
United Kingdom	48	1.7	31
Turkey	57	2.1	46
Eastern Europe and The People's Republic of China	761	27.2	299
other	475	16.9	324
Total	2798 [b]	100	990

[a] Ref. 2.
[b] Clean equivalent is 1.602×10^6 t.

The three main distinctions of wool are fine, medium, and long (Table 3). The outstanding fine wools are produced by merino sheep, an animal first bred in Spain but now prevalent throughout the world. Over 75% of the sheep of the world's largest wool producer, Australia, are merino sheep, which are also bred in large numbers in South Africa, Argentina, and the USSR. Medium wools include types of English origin, eg, southdown, hampshire, dorset, and cheviot, as well as crossbreds, eg, columbia, targhee, corriedale, and polwarth, from interbreeding merino and long wool types. Coarse long wools come from sheep chiefly bred for mutton, eg, lincoln, cotswold, and leicester.

Raw wool from the sheep contains other constituents considered contaminants by wool processors. These can vary in content according to breed, nutrition, environment, and position of the wool on the sheep. The main contaminants are a solvent-soluble fraction called wool grease; protein material; a water-soluble fraction (largely perspiration salts collectively termed suint); dirt; and vegetable matter, eg, burrs and seeds from pastures. Table 4 gives some figures for percentages of fleece constituents in Australian wool, excluding the protein contaminant (5–6). The wools in Table 4 contain little or no vegetable matter, which in certain areas, can be as high as 20% of the raw-wool weight, but it is usually <2%.

Raw-Wool Specification. In buying raw wool, the wool processor is concerned about its quality and the quantity of pure fiber present. The particular properties of concern differ depending on the processing system used and the product. For example, for the main wool market, ie, fine and medium wools for apparel, the single-fiber and staple characteristics that are significant to processing are given in Table 5. The most important are the yield (the percentage content of pure fiber at a standard moisture content); the vegetable-matter content and type; the average fiber diameter; the average length of the fibers; the strength of the staples of fibers and the position of any weak spot along the fibers; the color of the clean wool; and the number (if any) of naturally colored fibers present. For carpet-type wools (long wools) the important properties (8) are yield; fiber diameter; fiber length; color; bulk (the volume occupied by the fibers in a yarn); medullation (see Fiber Morphology); and vegetable-matter content.

Table 3. Main Types of Wool

Type	Breed	Average length, cm	Average diameter, μm[a]	Grade or count[a]
fine	merino	3.7–10	10–30	90s–58s
medium	southdown hampshire dorset cheviot	5–10	20–40	60s–46s
crossbred	corriedale polwarth targhee columbia	7.5–15	20–40	60s–50s
long	lincoln	12.5–35	25–50	50s–36s

[a] Wool fineness is now largely described as the "micron," ie, micrometer. Traditionally, it has been expressed as grade or count which, in the worsted trade, represents the number of hanks, each 560 yd (1512 m) long, of the finest possible yarn that might be spun satisfactorily from 1 lb (0.454 kg) of the wool concerned. The higher the grade or count, the lower is the "micron."

Table 4. Fleece Composition of Australian Wool, %

Component	Sheep Merino	Sheep Crossbred	Skin[a]	Lambs, merino and crossbred
oven-dry wool				
max	66.8	72.2	70.0	68.0
min	29.4	49.3	50.6	54.0
av	48.9	61.0	63.0	60.5
wax				
max	25.4	19.3	20.0	23.5
min	10.0	5.3	9.9	6.4
av	16.1	10.6	15.8	16.0
suint				
max	13.0	13.6	1.4	7.2
min	2.0	4.4	0.1	3.4
av	6.1	8.2	0.6	5.4
dirt				
max	43.8	23.7	21.5	10.0
min	6.3	4.3	6.4	3.8
av	19.6	8.4	11.2	6.4
moisture				
max	12.6	14.2	9.6	14.2
min	8.1	9.5	7.2	9.0
av	9.6	12.0	8.0	11.2

[a] Skin wool is the term applied to fellmongered wool, usually removed from skins by a sweating process involving bacterial action or by action of lime–sulfide depilatory.

Table 5. The Significance of Single-Fiber and Staple Characteristics in Wool Textile Processing[a]

Characteristic	Importance
yield	primary
vegetable matter	primary
fiber diameter	primary
length	primary
strength and position of break	primary
color and colored fibers	primary
fiber diameter variability	secondary
length variability	secondary
cots[b]	secondary
crimp and resistance to compression	secondary
staple tip	minor
age, breed, and category	minor
style, character, and handle	minor

[a] Ref. 7.
[b] Cots are natural tangles of fibers in wool fleece.

Until the late 1960s and early 1970s, the characteristics of fine and medium wools (Table 5) were largely evaluated visually by wool valuers. However, with the development of sampling techniques and equipment capable of rapid and economical measurement of yield, fineness, and vegetable matter (9–10), the sale of wool by sample and objective measurement has become dominant in Australia and South Africa and has grown in use in New Zealand. In sale by sample, cores are drawn from each lot and tested for yield, fineness, and vegetable-matter content in accordance with international standards (11). In addition, a full-length display sample, representative of each lot and obtained by standard procedure (12), is available for buyers to appraise the other characteristics. Sale by sample has reduced costs by reducing the handling of bulk wool in wool-brokers' stores and selling operations, and has also led to more efficient utilization of wool in processing and more accurate alerting of the wool processors' requirements to the woolgrowers.

Techniques for efficiently and economically measuring the other important characteristics, ie, staple (fiber) length (10,13–14) and strength and position of any weakness in the fiber (10,15), are under development. Existing color-measuring equipment can be used to measure the color (whiteness or yellowness) of washed wool, but measuring colored-fiber content remains a problem (10). The overall objective of research underway is to develop a system of "sale by description" for fine and medium wools, whereby the buyer is presented only with measured data on the main characteristics of the raw wool plus an assessment of the less important characteristics by an independent skilled appraiser.

Fiber Growth

The skin of the sheep has two layers, an inner dermis and an epidermis. The fiber grows out of a tubelike structure in the epidermis known as the follicle. The two types of wool follicles are the primary, which develop first in the outerskin of the unborn lamb, and the secondary, which develops later. Figure 1 depicts the structure of the primary follicle. The follicle is formed by growth from the epidermis into the dermis (16). After a short while, the outgrowths of the sebaceous gland, which exudes the wool grease, and the sweat gland develop. There are ca 200×10^6 follicles in the skin of a merino sheep.

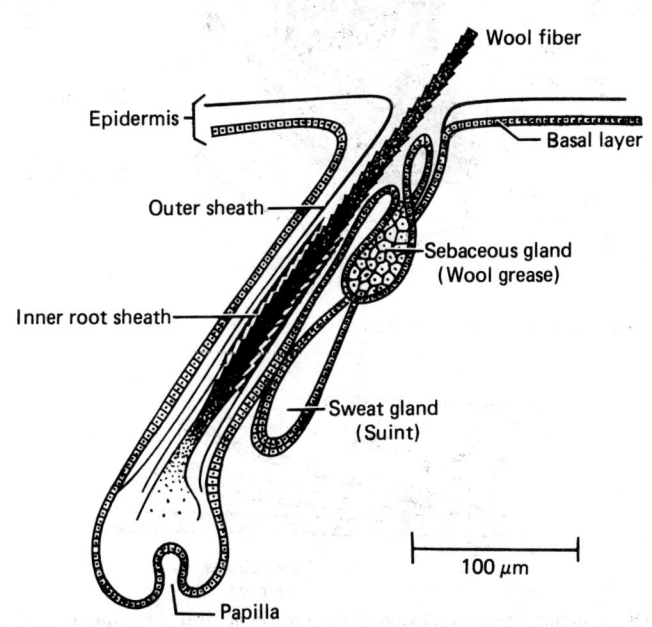

Figure 1. The primary wool follicle showing production of fiber, wax, and suint.

Fiber Morphology

The two types of cells in fine–medium wool fibers (Fig. 2) are corticle cells, which make up the center of the fiber, and cuticle cells, which form the outer protective cover around the cortex (16–17). The cuticle cells overlap like the tiles on a roof, with the protruding tips of the scales pointing toward the fiber tip (Fig. 3). The orientation of the scales in this way leads to a directional friction effect, ie, a difference in the coefficient of the friction of wool fiber when measured in the "with-scale" and "against-scale" direction. This effect is thought to be the main cause of wool felting (18) (see Shrink-Resistance Treatment). The dimensions of cuticle cells of merino wool are ca $20 \times 30 \times 0.5$ μm (19–20). They are made up of an enzyme-resistant exocuticle and an enzyme-digestible endocuticle (21–22), surrounded by a thin hydrophobic membrane, the epicuticle (19) (see Fig. 3).

Between the cuticle and the cortical cells, and separating the cortical cells themselves, is a cell membrane complex. This complex, ca 25 nm thick, basically "cements" the cells together, and has been called the intercellular cement. Some components of the complex are resistant to enzymes and chemicals, whereas others can easily be extracted with organic solvents or enzymes (23). Mechanically, the cell membrane complex is believed to be the weak link of the fiber (23–24).

The cortex consists of spindle-shaped cortical cells ca 100 μm long and is bilateral in structure with an orthocortex and a paracortex of different chemical reactivity

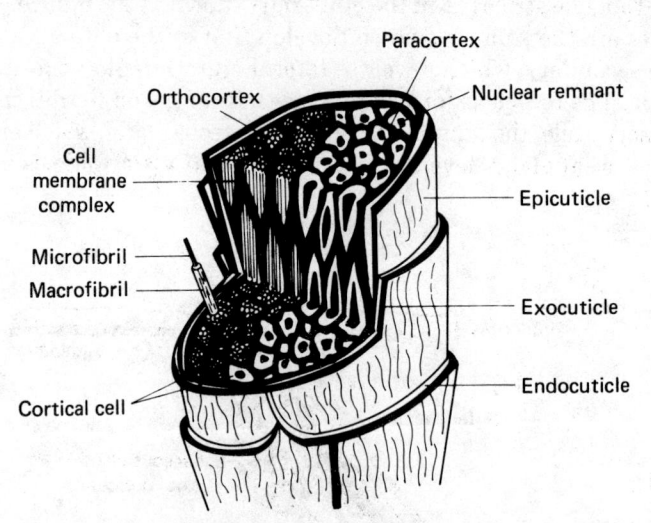

Figure 2. Diagrammatic representation of the morphological components of a wool fiber.

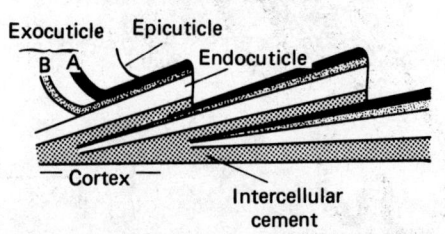

Figure 3. Diagrammatic representation of the morphological components of the cuticle of a wool fiber.

(25–29). In fine wools such as merino, the paracortex is always located on the inner concave side, and the orthocortex on the outer convex side of the fiber crimp curvature. In contrast, with coarse long fibers such as lincoln, the orthocortex forms a core surrounded by a tubular paracortex.

The cortical cells (Fig. 4) are made up of many highly organized fibrils (30–31). These fibrils in turn are composed of microfibrils of ca 8 nm diameter surrounded by an amorphous protein (32), the matrix. It is believed that these microfibrils are made up of smaller subunits, ie, protofibrils, each of which possibly consists of three α-helical polypeptide chains twisted together. Coarser fibers, such as some carpet wools, sometimes also contain a third component, the medulla, a hollow canal up the center of the fiber.

Physical Properties

That wool is still used as a textile fiber is to a large extent the result of its unique physical properties. In particular, it absorbs large volumes of moisture and hence is comfortable to wear. The absorption of moisture yields heat. The bulk (volume occupied by fibers in a yarn) given to it by its bilateral structure (and the resulting fiber crimp) adds to warmth as a greater volume of air is trapped than in a yarn of the same weight but of other fibers. Wool's unusual elastic properties give it outstanding resistance to and recovery from wrinkles, and outstanding drape. The principal physical properties of the fiber are given in Table 6.

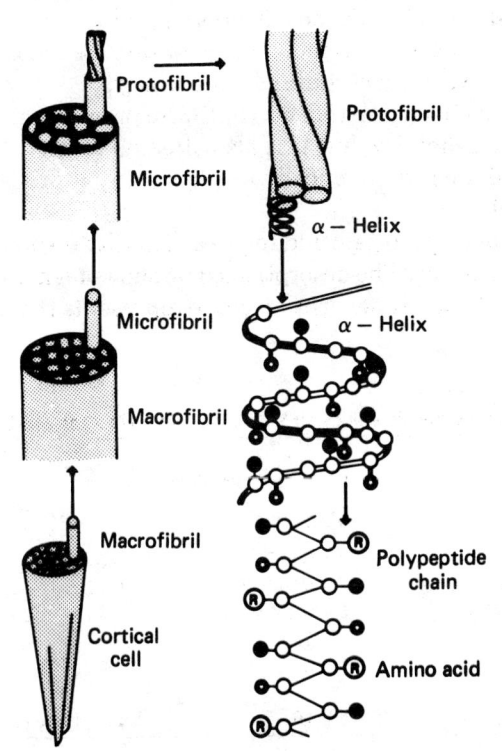

Figure 4. Cortical cell of a wool fiber, showing the various macro- and micro components.

Table 6. The Principal Physical Properties of Wool Fibers at 25°C[a]

Property	\multicolumn{8}{c}{Regain, %}
	0	5	10	15	20	25	30	33
relative humidity, %								
absorption	0	14.5	42	68	85	94	98	100
desorption	0	8	31.5	57.5	79	91.5	98	100
specific gravity	1.304	1.3135	1.3150	1.325	1.304	1.2915	1.2765	1.268
volume swelling, %	0	4.24	9.07	14.25	20.0	26.2	32.8	36.8
length swelling, %	0	0.55	0.93	1.08	1.15	1.17	1.18	1.19
radial swelling, %	0	1.82	4.00	6.32	8.88	11.69	14.57	16.26
heat of complete wetting, kJ/kg wool[b]	100.9	64.5	38.1	20.5	8.8	4.1	1.13	0
heat of complete absorption of liquid water, kJ/kg water[b]	854	624	431	276	142	100	42	33
relative Young's modulus	1.00	0.96	0.87	0.76	0.66	0.56	0.44	0.38
rigidity modulus (torsion), GPa[c]	1.76	1.60	1.26	0.90	0.50	0.28	0.16	0.1
electrical resistivity, 10^6 Ω·cm			4×10^4	800	40	6		
dielectric constant at 10^4 Hz	4.6	5.1	6.2	8.3	12.8			

[a] Ref. 33.
[b] To convert J to cal, divide by 4.184.
[c] To convert GPa to psi, multiply by 145,000.

Moisture Relations. Wool is the most hygroscopic of common textile fibers (Table 7). Its regain (moisture content) affects its ease of processing, especially in carding, combing, spinning, and weaving; accordingly, the relative humidity of the atmosphere in these mill areas needs to be controlled to optimize processing. The regain depends mainly on the relative humidity, but it is also affected by the state of the fiber, eg, whether it has been damaged chemically or physically in processing or whether it contains acid or alkali.

The relationship between regain and relative humidity is shown in Figure 5. There is a hysteresis effect on drying. The desorption curve shows higher regains at each value of relative humidity. The lower adsorption curve represents the behavior of dry wool

Table 7. Moisture Regain Values for Wool and Other Fibers at 21°C, %[a]

Fiber	At 65% rh	At 95% rh
Acrilan acrylic	1.5	5
Orlon acrylic	1.5	4
nylon	2.8–5	3.5–8.5
polypropylene	0.01–0.1	0.01–0.1
Dacron polyester	0.4–0.8	
viscose rayon	13	27
acetate	6.5	14
cotton	7	
wool	16	22.5 (90% rh, 25°C)

[a] Ref. 34.

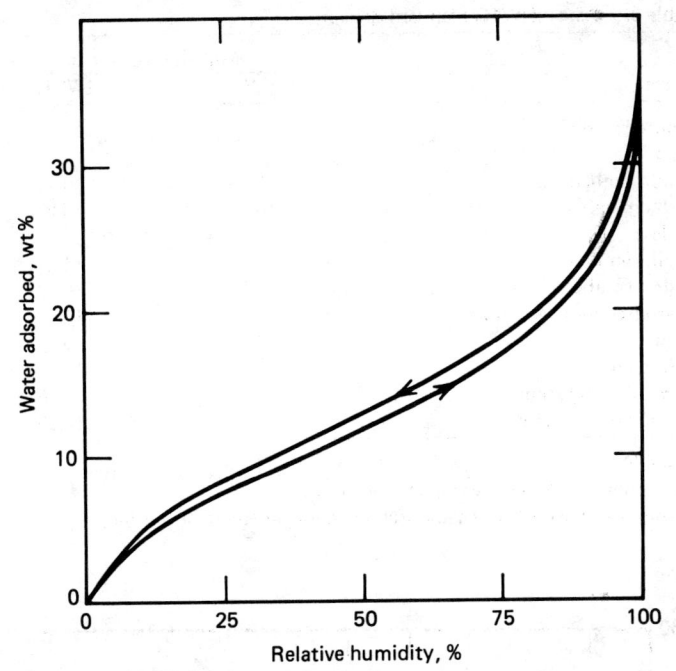

Figure 5. The moisture adsorption isotherm of 20-μm merino wool at 22°C.

of a given moisture content. In the normal range of relative humidities, there is a difference of ca 2% between the adsorption and desorption isotherm.

As can be seen from Table 6, most physical properties of the fiber are affected by the regain (see also Table 7). As the moisture content of the fiber increases, there is a marked radial swelling of the fiber attributable to its oriented structure. At high regains, fibers are easier to bend and twist and are much better conductors of electricity and less liable to generate static electricity. Accordingly, prior to spinning, wool tops should be stored to ensure the highest regain possible.

The regain of wool also has important bearing on commercial transactions where wool is sold by weight. Standard regains are, therefore, specified for this purpose, and these are given in Table 8.

Mechanical Properties. Both in the manufacturing processes of spinning, weaving, and knitting and in the behavior of the completed textile, the mechanical properties of wool fibers are the dominant physical factors. The fiber is characterized by high extensibility and relatively low breaking strength, and it has very unusual elastic properties particularly when wet. Figure 6 shows the stress–strain curves of wool at different relative humidities. The stress–strain properties can be considered in terms of three distinct regions of strain (36). Once the fiber has been straightened (this varies with the crimp in the fiber under test and for merino fibers can represent a considerable strain) the stress–strain curve has a nearly linear mechanically stiff region up to a few percent strain; this region is generally referred to as a Hookean region, although it does not closely approximate a Hookean spring in its properties (37). The strain of the fiber then increases rapidly with increase of stress up to strains of ca 25–30%, and this region is known as the yield region. For extensions beyond the yield region, ie, the postyield region, the fiber becomes stiffer with further strain.

Table 8. U.S. and IWTO Standard Regains

Form	Moisture regain, % U.S.[a]	IWTO[b]
scoured wool	13.63	17
tops		
oil, combed	15	19
dry, combed	15	18.25
noils[c]		
oil, combed		14
dry, combed		16
scoured and carbonized		17
yarn		
woolen	13	17
worsted, oil spun	13	18.25
worsted, dry spun	15	18.25

[a] Ref. 35.
[b] International Wool Testing Organization.
[c] Noils are shorter fibers separated from longer fibers in combing.

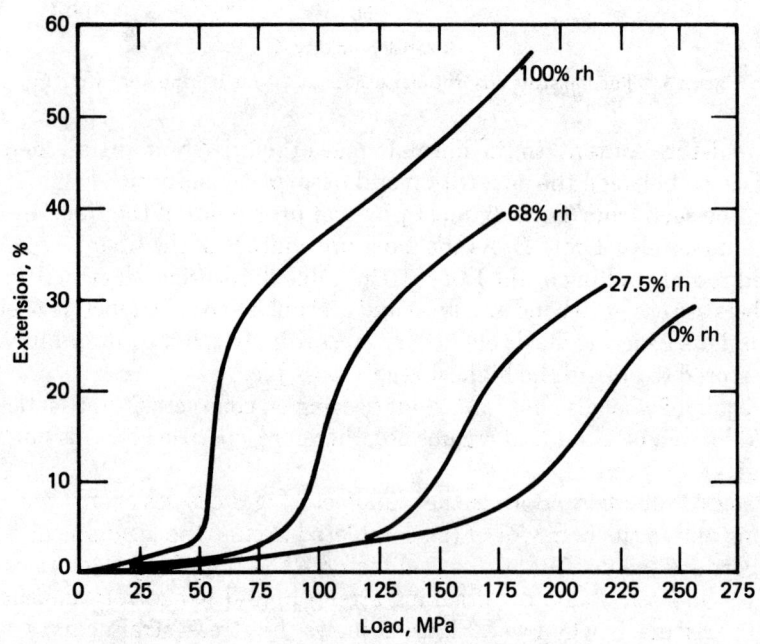

Figure 6. Load–extension curves for a wool fiber at different relative humidities. To convert MPa to psi, multiply by 145.

There is a hysteresis in the unloading curve as shown by a typical stress–strain curve for a wool fiber in water (Fig. 7). If the fiber is rested overnight in water at room temperature, the cycle can be repeated.

This mechanical behavior of a wool fiber has been explained as being the result of two systems acting in parallel—the microfibrils, which contain the organized α-helical material, and the matrix, the water-sensitive material outside the microfibrils (38).

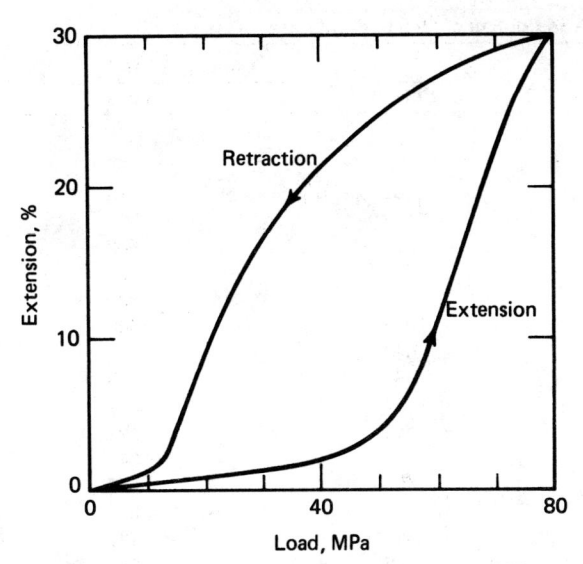

Figure 7. Load-extension curve for a wool fiber in water. To convert MPa to psi, multiply by 145.

Chemical Structure

Wool belongs to a family of proteins (qv) called the keratins. However, as explained above, morphologically, the fiber is a composite, and each of the components differs in chemical composition. Principally, the components are proteinaceous, but cleaned wool (after successive extraction with light petroleum, ethanol, and water) contains small amounts of lipid material and inorganic ions equivalent to an ash content of ca 0.5–1% after combustion of the fiber.

The simplest picture of the proteinaceous components of the wool fiber is one of polypeptides (qv) composed of α-amino acid residues. Eighteen amino acids (qv) are present in the hydrolysate of the fiber, but the relative amounts of the amino acids (qv) vary from one wool to another. Typical figures for three different samples of merino wool are given in Table 9. The ammonia produced in the hydrolysis of wool is presumed to arise from amide groups and gives a measure of the number of these groups originally present.

The side groups of the amino acids vary markedly in size and chemical nature. The nature of these groups is of course important to the chemical reactions of the fiber. For example, the basic groups (arginine, lysine, and histidine residues) can attract acid (anionic) dyes, and in addition, the side chains of lysine and histidine residues are important sites for the covalent attachment of reactive dyes (see Dyeing).

The main sulfur-containing amino acid present in wool, cystine, plays a very important role, as two of the residues form a disulfide cross-link between neighboring polypeptide chains or between amino acid residues in the same chain:

Table 9. The Amino Acid Content of Merino Wool[a]

Amino acid	Content, mol/g[b] Ref. 40	Ref. 41	Ref. 42
lysine	193	277	269
histidine	58	76	82
arginine	602	613	600
tryptophan	c	c	c
aspartic acid	503	602	560
threonine	547	564	572
serine	860	892	902
glutamic acid	1020	1046	1049
proline	633	561	522
glycine	688	815	757
alanine	417	512	469
½ cystine	943	1120	922
valine	423	546	486
methionine	37	47	44
isoleucine	234	318	275
leucine	583	721	676
tyrosine	353	380	349
phenylalanine	208	268	257
NH$_3$	887	855	

[a] Ref. 39.
[b] The results of three separate tests are reported from the indicated references.
[c] Tryptophan is destroyed under the conditions of hydrolysis used for these analyses; values in the range 35–44 mol/g have been obtained by alternative techniques (43–45).

The disulfide cross-links readily rearrange under influence of heat and moisture, facilitating conformational rearrangement of the wool proteins and so leading to relaxation of molecular stress in the fiber. Other cross-links, eg, isopeptide cross-links, are present in the fiber in small amount.

The extractable proteins of the wool fiber are of three different types that differ greatly in structure and properties. The predominant type (58%), SCMK-A (S-carboxymethylkerateine-A), contains a group of low sulfur proteins (sulfur content ca 1.7% (46) compared with the fiber overall at 3.5%). The second fraction (26%) SCMK-B, contains two groups of proteins (47): high sulfur (sulfur content 4–6%) and ultrahigh sulfur (sulfur content 8%) proteins. A third fraction (6%) contains another group of proteins, termed high Gly/Tyr proteins (48) because they are rich in glycine and tyrosine levels (sulfur content 0.5–2%).

The sequencing of amino acids in these protein fractions has been the subject of much research (49). There are at least four different families of proteins in the high sulfur fraction (50), and individual members from three of these families have been sequenced. The low sulfur groups contain at least three families. None of the constituent proteins has been sequenced completely, but partial sequences of helical regions of some of the proteins are now known (50). There are at least six families in the high Gly/Tyr group; one member, containing sixty-one residues, has been sequenced (51).

Wool Processing

The conversion of raw wool into a textile fabric or garment involves a long series of separate processes. There are two main processing systems, ie, worsted and woolen, although an appreciable volume of wool is also processed on the short-staple (cotton) system or on the semiworsted system for carpet use. The main stages in the woolen and worsted systems shown in Figure 8 are described in references 52 and 53, respectively. Nearly all the processing stages involve some aspect of chemistry. For example, although spinning is largely a mechanical operation, the lubrication of the fibers by the spinning oils involves surface chemistry.

Scouring. The first stage in wool processing is removal of the fleece impurities, principally grease, dirt, and suint, by scouring (54). This entails washing the raw wool in aqueous solutions (neutral or slightly alkaline) of nonionic detergent (or, less frequently, soap and soda), followed by rinsing in water. The process is carried out in a series of four or five bowls in which the wool is immersed in the wash liquor and transported through the liquor by mechanical rakes. The liquor flows countercurrent

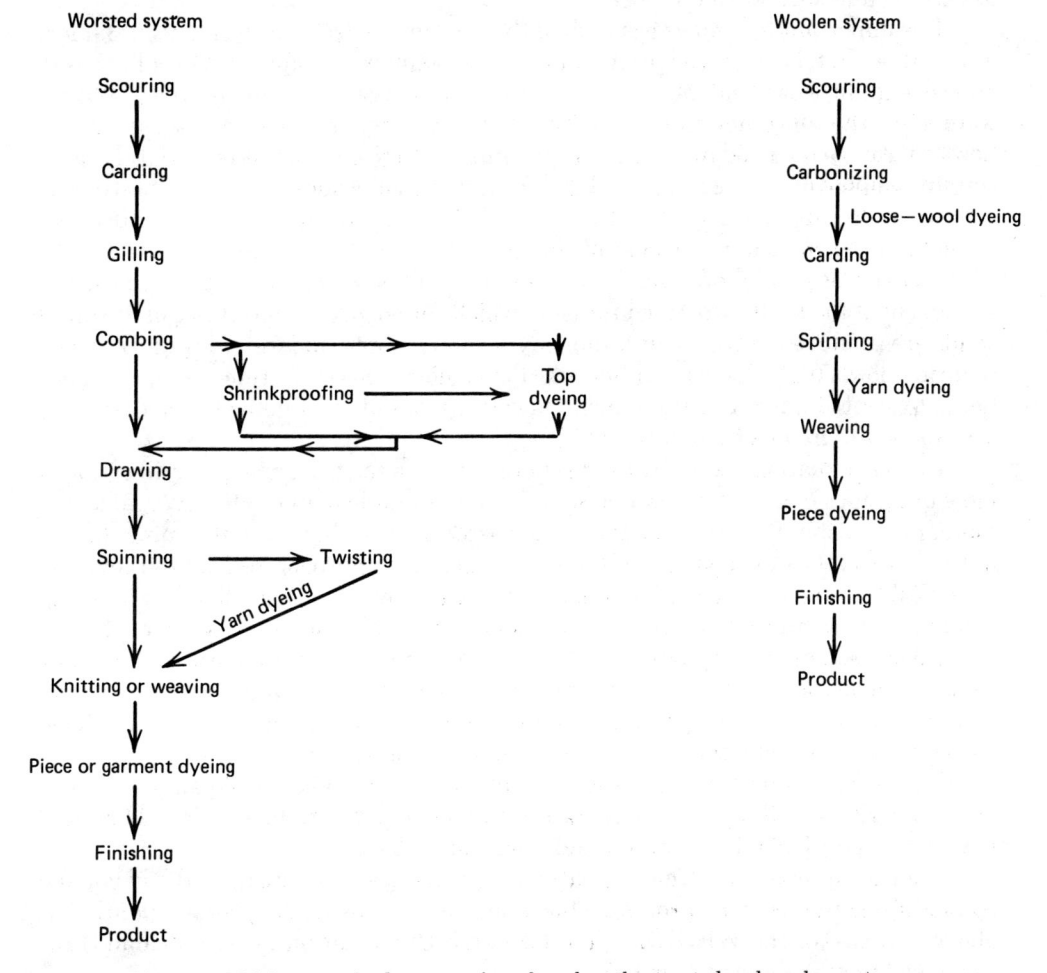

Figure 8. Main stages in the processing of wool on the worsted and woolen systems.

to the wool, and at the end of each bowl the wool passes through a roller squeeze. Other methods of fiber transport are used to a limited extent, eg, the wool is conveyed under jets of liquor (55) or passed around perforated drums (56), principally to limit the wool movement that leads to fiber felting. Agitation of wool in anhydrous solvents does not cause felting, and several plants (eg, 57–58) have been designed to take advantage of this property, but only a few are in operation commercially (54).

The traditional picture of the contaminants on the wool has been fiber covered with a layer of grease in which suint and dirt (both mineral and organic) are embedded. The principal action of the detergent or soap has been seen as one involving swelling and rolling up of the grease–suint and grease–dirt mixtures and suspension of these in the water (59–60). This simple picture has been questioned by a possible fourth main component, thought to be proteinaceous in nature (61–62), which is present as a layer on the wool fiber and also as small particles dispersed with the dirt and suint in the grease. Under certain scouring conditions (eg, neutral conditions or low ratios of liquor to wool), this material does not swell readily and hinders removal of dirt. As well as affecting the efficiency of scouring if not completely removed, this proteinaceous layer may also influence the color of scoured wool and subsequent processing stages such as carding and shrinkproofing (62).

It is important to remove as much of the wool grease from the waste scour liquors as possible, first, because the product has some commercial value (see Wool Grease), and second, because the pollution load of the discharge is thereby reduced. Wastewaters are therefore normally centrifuged before discharge. Not all of the grease, however, can be removed by centrifuging (normally up to 40% can be removed) because certain components of the grease and dirt form stable emulsions under normal scouring conditions. Because a high concentration of dissolved solids in the liquors reduces the stability of the emulsion, it is possible under certain conditions, to recover ca 85–90% of the grease originally present on the wool (63). This understanding led to the development of the Lo-flo scouring process in which very high concentrations of dissolved solids are produced in the scour liquors by severely restricting water input to special scouring bowls (64). The partial flocculation achieved eases the removal of contaminants by centrifuging. Grease removal is very good, but the removal efficiencies for dirt and suint are less than normal (54).

In most situations, wastewater discharged from a scouring machine still contains large quantities of grease and water-soluble and insoluble material (both organic and inorganic). Disposal or treatment of the wastes to comply with environmental requirements is thus expensive. Probably the cheapest approach to disposal involves biological treatment by irrigation of large land areas (not less than 20 ha per scouring machine is normally required) or maturation in very large shallow lagoons (54).

Physicochemical methods of wastewater treatment, eg, flocculation using inorganic or polymeric flocculants or sulfuric acid, have been investigated, as have physical techniques, eg, membrane processes and solvent extraction, but few processes have passed pilot-scale evaluation. One physical method that has attracted some commercial interest is evaporation; several evaporative plants were installed in Japan in the early 1970s, nearly all followed by incinerators for the sludge produced (65). They are, however, expensive in both capital and operating costs.

Another approach to effluent treatment is to reduce the effluent load and volume by modifying the scouring process. This is the basis of the Lo-flo process mentioned above and also of the WRONZ (Wool Research Organization of New Zealand) rationalized scouring system (66–68).

Carbonizing. Carbonizing is a process used to remove excessive contents of cellulosic impurities, eg, burrs and vegetable matter, from wool. It is carried out on loose wool (69), rags (70), and fabric (71). With loose wool and fabric, the wool is treated with aqueous sulfuric acid and then baked. The cellulosic matter is rapidly destroyed by the hot concentrated acid by which it is converted into friable hydrocellulose, whereas wool is scarcely affected by the treatment. For rags, hydrogen chloride gas is preferred as the mineral acid because it has less effect on the colors. After acid treatment, the carbonized vegetable matter is crushed to facilitate removal. The wool is then normally neutralized in alkali, although some mills omit this stage to facilitate subsequent dyeing with equalizing acid dyes.

Most industrial carbonizing is done on loose wool, and the technology has changed little over the last 40 years. Australian carbonizers now use surface-active agents to reduce fiber damage in the carbonizing process (72–73). The improved results depend on the use of high acid concentrations and rapid throughput (73). Factors important to the role of the surfactants are thought to be the increased efficiency of acid removal in squeezing (74) and the improved spreading of the acid over the fibers in the presence of the surfactant (75).

Attempts have been made to introduce processes and equipment for rapid acidizing, drying–baking, and crushing (76) and for carbonizing wool in sliver (rope) form (77), but neither has been adopted commercially.

Developments in the carbonizing of fabric (78) have been aimed at reducing the volume of acid used (79–80), in order to reduce energy in drying and save neutralization costs, and at the use of generally available textile processing equipment, eg, the pad-mangle, for application of the acid in a pad–dry–bake process (81).

Dyeing. Traditionally, wool is dyed in batch operations in which the textile is boiled in aqueous solutions of dye for periods of at least one hour. Techniques have also been developed for continuously dyeing wool in loose or sliver form (82–83). However, because of the need for large volumes of one color for economic operation, the number of installations of machines using such techniques has been limited.

The chemistry of wool dyeing is complex (84–85). The dyes used industrially are mainly anionic in nature (84). They adsorb on wool as the result of ionic attraction between sulfonic groups ($-SO_3^-$) on the dyes and charged amino ($-NH_3^+$) groups on the fiber and of nonpolar van der Waals forces between the hydrophobic dye anion and hydrophobic parts of the wool adjacent to $-NH_3^+$ groups. Other interactions, eg, hydrogen bonds, are probably also involved. Factors influencing these various interactions include pH and temperature of the dye solution; nature and concentration of electrolytes in the solution; hydration of dye and auxiliaries; effects of contaminants and pretreatments on the fiber surface; affinity of the dye for wool; and mode and depth of penetration of the dye into the fibers (85–86).

Three principal types of acid dyes are used for wool: equalizing acid dyes, milling acid dyes, and supermilling acid dyes.

Equalizing acid dyes have comparatively low affinity for wool, and it is easy to obtain uniform dyeings using them. They combine with the fiber largely because of ionic attraction. They are applied at ca pH 3 to increase the number of positively charged groups acting as sites of attraction. In general, these dyes are not particularly resistant to wet treatments.

Milling acid dyes are of higher molecular weight and possess fewer sulfonic groups than equalizing acid dyes. Because they depend less on ionic attraction for their affinity to wool, a higher pH is used to ensure uniformity. The pH required is usually obtained by adding a weak acid, eg, acetic acid.

Supermilling acid dyes are virtually independent of ionic attraction for good exhaustion and so are applied from an almost neutral dyebath.

Reactive dyes, usually anionic in character, are also important for wool (87–88). Because these dyes contain a group that reacts covalently with the wool fiber, they have outstanding resistance to removal by washing, etc. They also produce very bright hues. Because of these properties, they have significant application in dyeing wools that have been treated to withstand shrinkage in machine washing.

The resistance to wet treatments of some dyes can also be improved by treatment with metal salts, usually those of chromium, to reduce their solubility. These chrome dyes (89) are now supplemented by metal-complex dyes (90) of very high affinity for wool. In these, the metal complex is preformed; it contains one metal atom for each one or two dye molecules, the metal being either chromium or cobalt. The 1:1 complex dyes are applied to wool from strongly acid dyebaths; the 1:2 complexes, which are much more important than the 1:1 complexes, are applied from a very slightly acid bath (pH 5.5–7).

Wool is dyed in loose, sliver, yarn, fabric, or garment form (84), depending on the nature of the product, the equipment available, requirements of the market, and other factors. Traditionally, the process has involved movement of the dye solution through a compressed mass of the textile (eg, yarn package) or transport of the textile (eg, fabric) through the dye solution (see Dyes, application and evaluation). Both approaches entail the use of high volumes of dye solution relative to the volume of the textile being dyed and hence have tended to be expensive both in energy and in water (and consequently effluent disposal). General improvements in design of dyeing equipment to enable a reduction in solution volume have occurred. In package machines, this has involved the use of denser packages and higher rates of solution flow, but sometimes this damages the wool fiber (91). Early versions of jet-dyeing machines for fabric dyeing, originally developed for synthetic fabrics, presented problems in dyeing wool fabrics, but later versions that use gentler actions to achieve fabric movement have proved suitable (92).

Attempts to reduce energy usage, effluent levels, and damage to wool by prolonged boiling (dependent on pH (90)) have generally focused on dyeing below the boil or the use of continuous pad–steam dyeing (87,93). For example, the use of high concentrations of urea in the dye solution allows dyeing of tops or fabric in the cold (94–97). Several other methods have been developed involving the use of special auxiliaries in the dye solution below the boil, eg, ethoxylated nonylphenol surfactants (98) and benzyl and n-butyl alcohols (99–100) (see Dye carriers). CSIRO introduced two machines for pad–steam dyeing, one for loose wool (82), and one for sliver (83). A machine for continuous dyeing of yarns is under development (101). The use of dielectric heating to achieve dye fixation has also been introduced (102) (see Microwave technology).

Printing. The printing of wool is difficult technically, and this is thought to be the reason why less than 2% of total wool production is printed (103–104). The traditional procedure entails pretreatment with chlorine to improve both the color yield and the evenness of prints, followed by neutralization and drying. The design is then printed, usually by screen, and the fabric dried, steamed to fix the dyes, washed to remove print chemicals and unfixed dye, and dried again (see also Printing processes).

For chlorination, dichloroisocyanuric acid or its salts (105) are normally used (see Bleaching agents). Chlorination degrades the fabric, and when preserving the hand (feel) of the fabric is essential, the chlorination step is omitted, with consequent loss of color yield. To avoid the degradation due to chlorination, alternative means to improve color yield have been sought, among them the addition of certain surfactants (106) or fatty acid derivatives (107–108) to the print pastes and pretreatment of the fabric by a shrink-resistance process, eg, the Sirolan BAP process (109).

Incorporation of high concentrations (>300 g/L) of urea in a print paste greatly accelerates the fixation of low molecular weight dyes on wool. This discovery has been used in the development of a cold-print method (110). After being printed, the textile products are left wet and batched with interleaving polyethylene sheets for up to 24 h; they are then washed and dried. No steaming is required, and energy usage in the process is thereby minimized.

Pigment printing has not been widely adopted for wool because of the adverse influence of the encapsulating resins used on the hand of the printed area. Changes in hand can, however, be minimized by careful selection of binders and cross-linking agents (111).

Transfer-print papers marketed for polyester are unsuitable for wool because the sublimable dyes that are used have little affinity for the wool fiber (112). However, processes have been developed that enable this technique to be applied to wool (103). In the Fastran process (113–114), for example, conventional wool dyes are printed on the paper and are heat-transferred to wool (usually knitted garments) that has been padded with a thickened solution. After being printed, the garment is washed to remove thickening agent and unfixed dye. Another process (115–116), developed by the CSIRO in collaboration with the International Wool Secretariat, entails the transfer printing of certain sublimable metal-complexing dyes onto fabric pretreated with a mixture of an anionic surfactant, urea, lactic acid, and chromic chloride, followed by steaming. The process yields prints in a wide range of shades with good resistance to washing and light. The process requires neither prior chlorination nor washing of the fabric after printing and is the simplest method so far devised for printing wool.

Bleaching. Bleaching is carried out to reduce the natural yellowish tinge of normal wool (117–119). Either oxidation (with, principally, hydrogen peroxide, sodium peroxide, or sodium perborate) or reduction (with thiourea dioxide, sulfur dioxide, acidified sodium bisulfite, or sodium dithionite) is used, and sometimes both types of treatment are applied, in which case oxidation must precede reduction (see Bleaching agents).

Peroxide bleaching is the most common approach. The wool is treated at pH 9 (sodium silicate or a phosphate) with 0.6% hydrogen peroxide (0.6% H_2O_2 = 2 cm^3 O_2 liberated per mL of liquor). Treatment begins at 50°C, and the wool is soaked for 3–5 h or overnight. It is then rinsed, treated with dilute acetic acid, and rinsed again. It is important that no metal except stainless steel be present, since metals, particularly copper, act as catalysts for the decomposition of hydrogen peroxide and may cause localized degradation of the wool. Continuous processes utilizing hydrogen peroxide have also been introduced (78).

Whitening. One of the problems with bleached wool is its tendency to yellow on exposure to sunlight, particularly when wet. The mechanisms by which wool is bleached or yellowed by light are complex, and although considerable work (120–121) has been done on the chemical changes that occur on photoyellowing, much remains to be elucidated in this area.

Significant efforts have been made to develop chemical treatments that inhibit light degradation. For example, a sulfonated 2H-Z-(2-hydroxyphenyl)benztriazole retards phototendering but not photoyellowing (122). Ultraviolet absorbers (123) and antioxidants (qv) (124) do not significantly retard photoyellowing of wool (see Uv stabilizers). One of the most effective antiyellowing procedures is treatment with thiourea–formaldehyde mixtures or precondensates (125).

Insectproofing. The principal insects that attack wool are the case-bearing clothes moth (*Tinea metonella*, *Tinea dubiella*, and *Tinea pellionella*); the common clothes moth (*Tineola bisselliella*); the brown house moth (*Hofmannophila pseudospretalla*); and the variegated carpet beetle (*Anthrenus verbasci*). Wool is protected from attack of the larvae of these insects by application of an insecticide, usually from the dyebath (126) (see Insect-control technology).

The requirements for the insecticide are quite restrictive. For example, as well as being toxic to the insects at low levels of application, it must be stable to the application conditions (high temperature, aqueous solution), resistant to washing and light, and effective for prolonged periods on the textile (>15 yr for wool carpets). As a consequence, few general insecticides are suitable for insectproofing wool. The range commercially available (in 1982) is shown in Table 10 (126), and the structures of their active constituents in Figure 9 (126).

The most commonly used products are Eulan WA new and Eulan U33. Mitin LP and the pyrethroid-based formulations, Perigen and Mothproofing Agent 79, are more recently introduced products. Because of its high cost, Mitin FF has use only where high resistance to wet conditions is required since it is superior to the other products in this respect.

Conditions of application of insectproofing agents are dictated by the needs for satisfactory application of the dyes in the same bath (127–128). Conditions such as the presence of fibers other than wool, particularly nylon, and prolonged boiling at or above pH 5 can adversely affect the uptake of formulations based on polychloro-2-(chloromethylsulphamido)diphenyl ethers, eg, the Eulans, Mitin LP, and Molantin. Similarly, auxiliaries (particularly leveling agents) present to assist dye uptake and distribution may reduce application efficiency and increase residues in the dyebath effluent.

A wide variety of insecticides developed for general use have been investigated for their suitability for insectproofing, eg, organophosphorus, carbamate, and py-

Table 10. Currently Available Industrial Mothproofing Formulations[a]

Product	Manufacturer	Active constituent[b]
Eulan WA new	Bayer	1
Eulan U33	Bayer	1
Mitin LP	CIBA-GEIGY	1 and 2
Mitin FF high conc	CIBA-GEIGY	3
Mitin N	CIBA-GEIGY	2
Molantin P	Chemapol	1
Perigen	Wellcome	4
Mothproofing agent 79	Shell	4

[a] Ref. 126.
[b] See Figure 9.

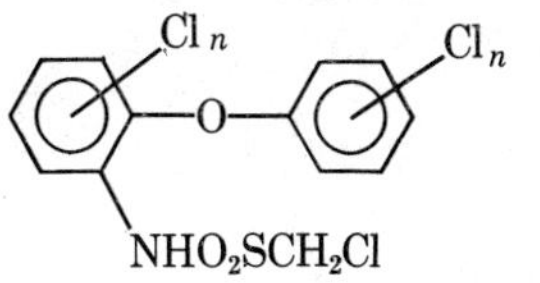

Figure 9. Chemical formulae of active constituents of some currently available industrial insect-proofing agents (see Table 10) (126).

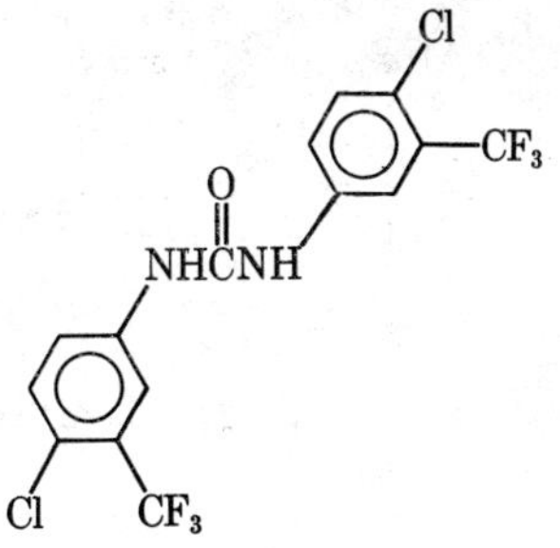

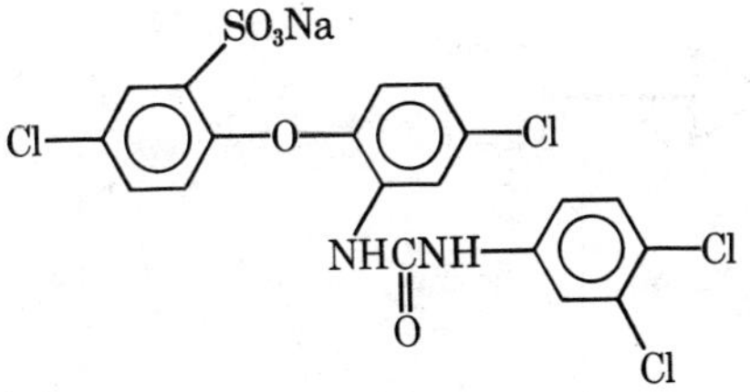

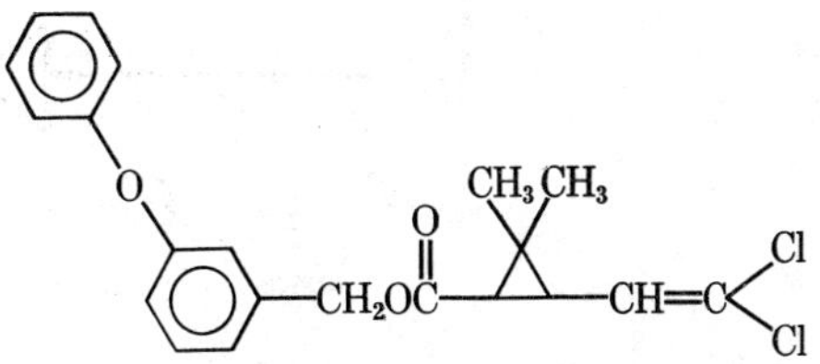

rethroid insecticides (126). Phosvel (129) has shown the best durability among the organophosphorus products. Carbamates, which are generally more active toward the carpet beetle than toward the common clothes moth, proved unsatisfactory because of their poor affinity for wool (130–131). The pyrethroids have proven to be particularly suitable, especially permethrin, fenpropanate, and fenvalerate (132).

Other chemicals investigated have included quaternary ammonium compounds, anionic surfactants, organotin compounds (eg, triphenyltin chloride and acetate), thioureas, and 2-thienylethylene (126). Some of these have shown activity against wool-eating insects, but all have disadvantages that have precluded their commercial use. Similarly, other forms of chemical control by compounds that affect juvenile-hormone activity have been investigated but have not met commercial requirements (126).

One novel approach has involved modification of certain organophosphorus insecticides that show activity against wool-eating pests but suffer from problems of volatility and poor affinity for wool. The modification entails attachment to the insecticide of a group that reacts covalently with the wool fiber (133). Once bonded to the fiber, the group becomes inactive, and losses owing to volatilization, photochemical degradation, and wet treatments are minimal. The insecticide is reactivated when an ester group strategically placed between the fiber-reactive and insecticidal portion of the compound is cleaved in the digestive tract of the wool-eating pests during digestion of the treated wool, as shown diagrammatically in Figure 10. The active organophosphorus insecticide is thus released and rapidly kills the insect.

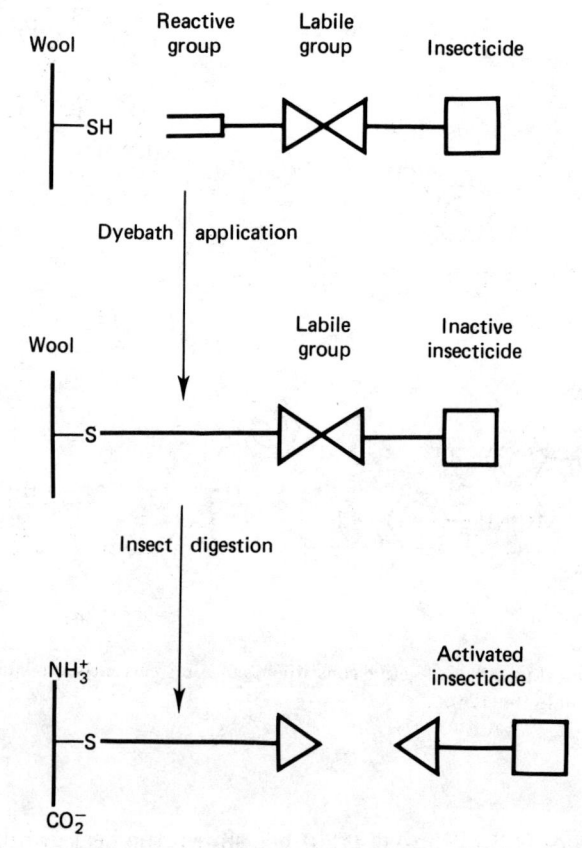

Figure 10. A diagrammatic representation of the principle of a fiber-reactive insectproofing agent (126).

Setting. Setting is an important part of the finishing of wool fabrics or garments to impart dimensional stability, shape, and hand appropriate to the product. Twist is set into yarns by steaming, and certain fabrics are set after weaving to prevent undue distortion during subsequent wet-finishing operations, a process known as crabbing. This is normally done by winding the fabric tightly on a roller and treating it in water at 60–100°C. Fabrics are often given a setting treatment to impart a certain amount of dimensional stability. This has traditionally been done by steaming again while the fabric is tightly wound on a roller and is known as blowing or decatizing. More recently, continuous machines, which are often integrated with other mechanical processes, eg, cutting, brushing, shrinking, and conditioning, have been adopted (78). Some of these machines operate at atmospheric pressure, and others at high pressure; some utilize chemicals (often sodium bisulfite or thioglycolate) to assist setting. The effectiveness of the setting treatments involving water or steam depends on fabric tension, the pH of the wool (optimum ca pH 8.5), the moisture content of the wool, the temperature, and the time (134).

The mechanism of setting wool involves bond fission and rebuilding; hydrogen bonds and disulfide and probably other covalent cross-links are involved (135–137). Disulfide rearrangement is promoted by thiols, which accounts for the efficiency in setting of agents that are capable of reducing cystine to cysteine residues, eg, bisulfite and thioglycolate. The rearrangement is catalyzed by the thiol anion (RS^-):

$$\text{WSSW} + \text{RS}^- \rightleftharpoons \text{WSSR} + \text{WS}^-$$

where W is wool. In acid solutions, thiols are largely nonionized, so that wool acquires little set under these conditions. Under alkaline or neutral conditions, thiols exist, at least partly, in their ionized state, and therefore wool sets well (138) (see also Hair preparations).

Commercial methods for setting durable creases in wool garments include (139): wet setting, eg, the Si-Ro-Set (140) and Immacula processes (141); hot-head pressing; dry creasing, eg, the Lintrak process (142); dry setting (143); and autoclave steaming. One of the techniques currently used for all-wool trousers and slacks of high wool content (>70%) blends is spraying the crease area with monoethanolamine sulfite and then steam pressing for approximately one minute (139). Alternatively, the Lintrak process, in which a thin line of silicone adhesive is laid into the inside of preformed trouser creases, may be used (142). For wool-blend slacks containing more than 30% polyester, acceptable crease durability can be obtained by hot-head or electric pressing at 170–180°C for 30 s. For wool and wool-blend skirts, the preferred method (139) is pleat formation followed by autoclave steaming at 110°C for 10 min.

Chemical setting alone will not stabilize wool garments sufficiently to maintain their shape during washing. Unless the garment is rendered shrink-resistant and the set shape stabilized, the garment will undergo felting shrinkage (see Durable-Press Wool).

Shrink-Resistance Treatment. Two main types of shrinkage can occur when wool fabrics or garments are washed: relaxation and felting.

Relaxation Shrinkage. Fabrics are often dried under tension because many drying machines are difficult to operate otherwise. The dimensions of most wet wool fabrics are greater than their dry dimensions. Hence, even when a cloth is dried at its exact wet dimensions, the fabric develops strains, which are released when the fabric is wet again and allowed to dry without tension. Such shrinkage is called relaxation shrinkage. Its control requires careful attention to finishing procedures (139).

Felting Shrinkage. This occurs when individual fibers move during laundering or other wet treatments. Because of the difference in the friction coefficient of a wool fiber with and against the direction of the scale, any movement of fibers always takes place towards the fiber root. How this leads to felting is explained simply in Figure 11. The mechanisms have been discussed in detail (144–145).

Two main types of shrink-resistance treatment are used industrially: degradative treatments and additive treatments. In degradative treatments, the proteins in the scales of the fiber (Fig. 12) are modified by treatment with, for example, chlorine (oxidizing), so that the scales are softened or sometimes removed (Fig. 13). Chlorination, the most common degradative treatment in use industrially, breaks both disulfide cross-links and peptide chains, and it has been postulated that water-soluble, high molecular weight peptides are produced from proteins in the cuticle without disrupting the epicuticle (146). Consequently, upon immersion of the wool in water, the cuticle absorbs water and becomes very soft. The differential friction effect is greatly reduced by the softening of the cuticle so that preferential movement of the fibers and hence felting shrinkage is reduced. Additive treatments entail application of a polymer to the fiber. The polymer masks the scale structure (Fig. 14), bonds adjacent fibers together (Fig. 15), or both.

Other chemical treatments that impart resistance to felting but have not had industrial acceptance include extraction with hot alcohol followed by treatment with

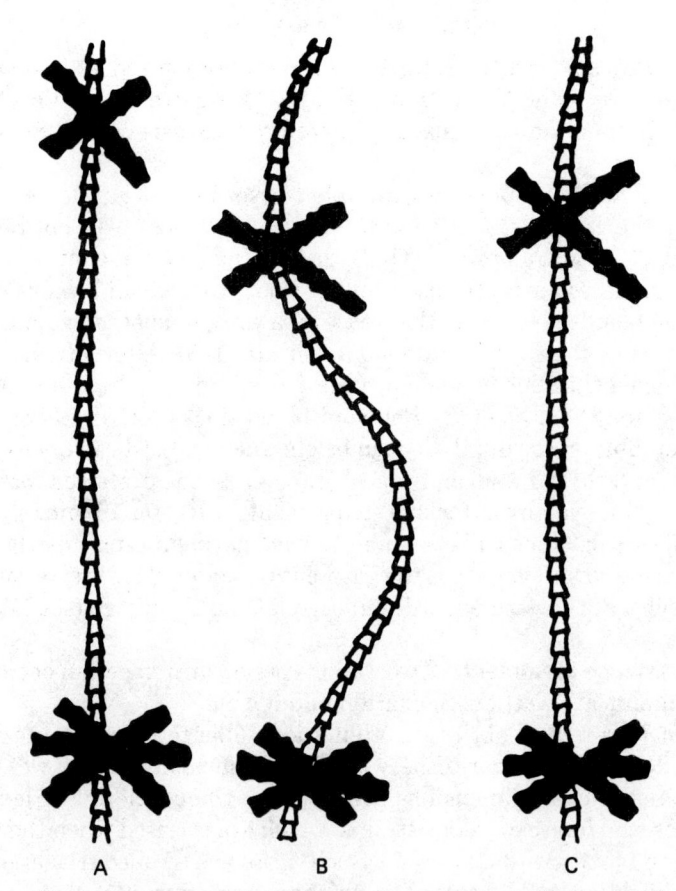

Figure 11. Schematic representation of fiber migration in a wool fabric which leads to felting shrinkage. A fiber is firmly entangled (A) with other fibers at its lower end but only partly entangled at its upper end. When the fabric is compressed during washing (B), the two ends of the fiber are brought together and it is temporarily bent. The scales are oriented so that the fiber can straighten elastically by sliding through the upper entanglement (C). The fibers in the two entanglements are now drawn closer together permanently because the scales permit fiber movement in one direction only. The fabric has shrunk.

aqueous ammonium thioglycolate (147); treatment with concentrated solutions of sodium hydroxide (148); and treatment with 1,4-benzoquinone or mercury(II) acetate (149) or aryl monoisocyanates (150). Significant research effort has been put into developing polymers and processes suitable for shrink-resistance treatment of wool (145,151–154). Industrially, shrink-resistance polymers are applied to wool mainly while it is in the form of top or of fabric, or garment form. Table 11 lists some of the polymers that have been used.

The effective shrink-resistance polymers on wool tops (slivers of parallel fibers), eg, those used in the chlorine–Hercosett (156–158) and Dylan GRC (159) processes, function by masking the scale structure (Fig. 14). For effective scale masking, these commercial polymers must spread over the surface of the wool fiber to encapsulate it completely. At present, a prerequisite for this is modification of the surface by low level chlorination to increase its surface energy (160). Because chlorination pretreatment requires careful control to avoid fiber damage and to give even treatment, development of polymers and processing conditions that enable encapsulation and

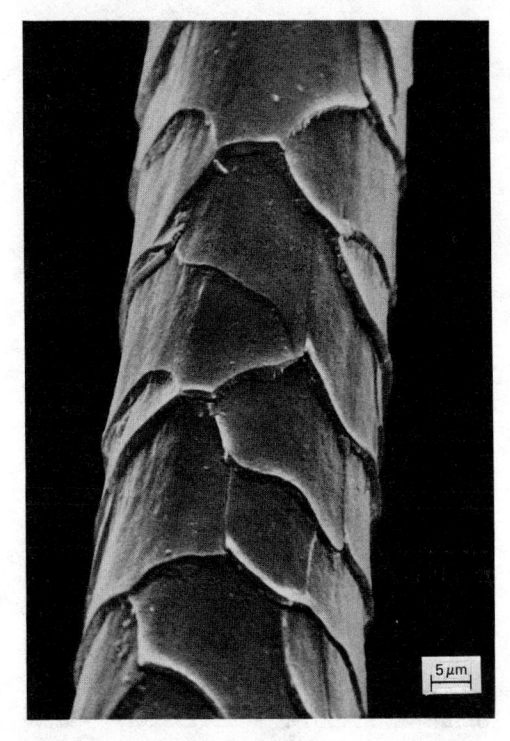

Figure 12. Electron micrograph of a merino wool fiber showing the scales on the surface.

Table 11. Some of the Polymers that Have Been Applied to Wool to Impart Shrink Resistance[a]

Trade name	Manufacturer	Backbone	Reactive group
Zeset TP	DuPont	poly(ethylene-co-vinyl acetate)	carboxylic acid chloride
Hercosett 57	Hercules	polyaminoamide	azetidinium
Primal K3	Rohm and Haas	polyacrylate	N-methylolamide
DC-109	Dow-Corning	dimethylpolysiloxane	silanol (in conjunction with an aminosilane cross-linker)
Oligan 500	CIBA-GEIGY	poly(propylene oxide)	thiol
Synthappret LKF	Bayer	poly(propylene oxide)	isocyanate
Synthappret BAP	Bayer	poly(propylene oxide)	carbamoyl sulfonate

[a] Ref. 155.

adhesion of the polymer to the fiber without pretreatment is an important research objective. Industrial chlorination has been improved by the introduction of the Kroy process, ie, the continuous chlorination of wool tops and sliver by passage of the material through a U-tube bath containing hypochlorous acid (161). This has been used both as a straight degradative process for shrink-resistance treatment and as a pretreatment before application of the Hercosett 57 (Hercules) resin.

Chlorination–resin processes are also used for shrink-resistance treatment of wool in fabric or garment form. However, for these products, there are also available processes that require no degradative pretreatment. The principal process used for fabric is the Sirolan BAP process (162–163), in which the fabric is padded with a mixture

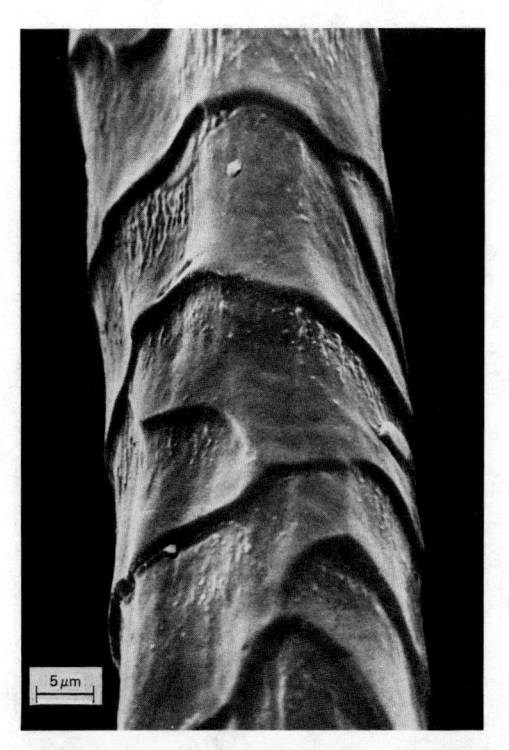

Figure 13. Electron micrograph of a merino wool fiber after treatment with chlorine, showing slight degradation of the scales.

of a reactive water-soluble polyurethane prepolymer and a polyurethane dispersion, with the polymers then cured by heating (see Urethane polymers).

Application of shrink-resistance polymers to garments or full-fashioned garment pieces is often done from organic solvents in modified drycleaning machines. Although several polymers have been developed for this purpose, only those based on silicones have received wide acceptance because of the soft hand they impart to the garment (164).

Three possible mechanisms by which polymers produce shrink-resistance in wool treated at the fabric or garment stage (145,165) are immobilization of fibers by fiber–fiber bonding which (Fig. 15) arises from small resin deposits cementing fibers together, thereby preventing fiber movement; scale masking or encapsulation of the fibers by a thin layer of polymer (Fig. 14); and deposition of large aggregates of polymer on the surface of fibers to prevent the scaly surfaces from interacting (166). The principal mechanism is believed to be fiber–fiber bonding, but probably all of these mechanisms operate to some extent depending on the type of polymer, the level of pretreatment (if any), the method of application, and subsequent processing (154).

Because of the high cost of polymer treatments for garments and the need for drycleaning equipment, research effort has been directed at developing processes in which polymers can be applied from aqueous solution by exhaustion (167–169).

Durable-Press Wool. Durable press is a description of garment performance implying that the garment is machine washable, retains any imparted creases or pleats, and requires minimum ironing after washing and drying. As already mentioned, chemical setting alone does not stabilize wool garments sufficiently to maintain their

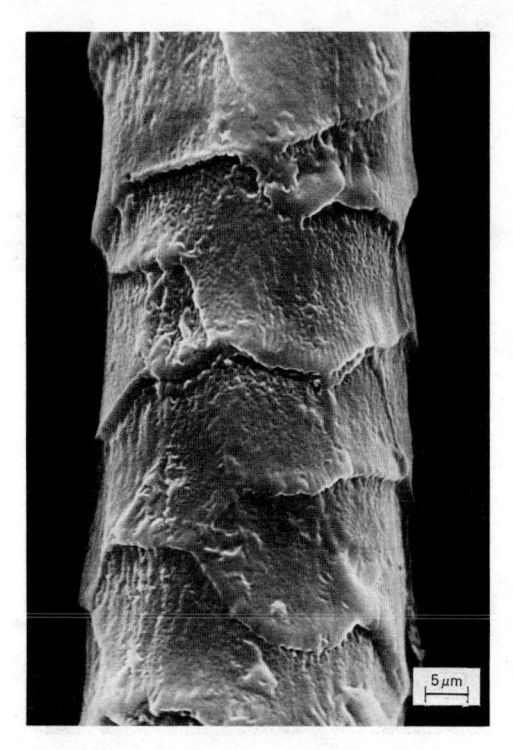

Figure 14. Electron micrograph of a merino wool fiber after treatment by the chlorine-Hercosett process, showing the polymer layer spread over the surface of the fiber.

shape during washing. The garment must also be rendered shrink-resistant. Thus, good durable-press effects on wool can be obtained in garments made from shrink-resistant wool fabric by a suitable setting treatment (139,170). However, careful selection of setting process is required to ensure the stability of the set to machine-wash conditions. Only two processes have, in fact, been applied commercially (139), ie, the CSIRO solvent–resin–steam process (171–172) and the International Wool Secretariat's durable-press process (173).

In the CSIRO process, a reactive polyurethane prepolymer is applied to the made-up garment from perchloroethylene. The garment is re-pressed to remove wrinkles and then hung in a steam oven. After a short warm-up period, saturated steam is introduced for 2 h. The polymer is cured on the garment in the required shape and then is capable of maintaining that shape while the free-hanging garment is set in steam.

The IWS process originally used a thiol-terminated prepolymer as the shrink-resistance agent, but this is no longer available commercially and may be replaced by the polymers used in the Sirolan BAP process (162–163). The polymer is applied to the fabric by padding and is cured. The fabric is then finished normally. After making-up, the garments are set using monoethanolamine sulfite.

Flame-Resistance Treatment. Wool is naturally a very flame-resistant fiber, and in most applications, no special treatment is needed. However, consumer demand for higher standards of flammability in, for example, aviation furnishings, drapes, and carpets has led to the development of treatments for improving this property. The treatments evolved from the observation that chrome mordanting of wool markedly

Figure 15. Electron micrograph of merino wool fibers in a fabric that has been treated with a typical shrink-resistance polymer, showing fiber–fiber bond formation.

improves its flame resistance (174). However, this approach did not prove practical for all situations because of the discoloration of undyed wool by the chrome mordant. Subsequently, it was shown that titanium or zirconium compounds give similar results (175), but the zirconium compounds did not share the discoloration disadvantage. The process, ie, the IWS Zirpro process, has been used extensively for wool in aircraft interiors to meet FAA requirements and for curtains and carpets in public buildings and similar areas where high flame resistance is required by legislation (176–177) (see Flame retardants in textiles).

Shrink-resistance processes compatible with Zirpro-treated wool have also been developed (178).

Wool Grease

In wool scouring, the contaminants on the wool, mainly grease, dirt, suint, and protein material, are washed off the fiber and remain in the wastewaters either in emulsion or suspension (grease, dirt, protein) or in solution (suint). Centrifugal extraction of the wastewaters produces a grease contaminated with detergent and suint. This product is called wool grease.

Lanolin is wool grease that has been refined to lighten its color and reduce its odor and free fatty acid content. Wool wax is the pure lipid material of the fleece, extractable with the usual fat solvents such as diethyl ether and chloroform. Wool grease is a mixture of compounds that are classed as waxes. However, it does not have the physical characteristics usually displayed by waxes. It is soft and slightly sticky with a greasy appearance. Some centrifugally recovered greases are light buff or ivory in color and practically odorless. Those recovered by other methods contain dark-colored and odorous impurities as well as free fatty acids. Grease recovered by acid-cracking of scour wastewaters can be used in lanolin production, but the centrifugally recovered products are more suited to the alkali refining, bleaching, and deodorizing required in preparing pharmaceutical-grade lanolin. Table 12 gives some physical and chemical data for wool wax.

Chemical Composition. Wool wax is a complex mixture of esters of water-soluble alcohols (180) and higher fatty acids (181) with a small proportion (ca 0.5%) of hydrocarbons (182). Considerable efforts have been made to identify the various components, but results are complicated by the fact that different workers use wool waxes from different sources and different analytical techniques. Nevertheless, significant progress has been made, and it is possible to give approximate percentages of the various components. The wool-wax acids (Table 13) are predominantly alkanoic, α-hydroxy, and ω-hydroxy acids. Each group contains normal, iso, and anteiso series of various chain length, and nearly all the acids are saturated.

The alcohol fraction is likewise a complex mixture of both aliphatic and cyclic compounds (Table 14). The principal components are cholesterol (34%) and lanosterol and dihydrolanosterol (38%). The aliphatic alcohols account for ca 22% of the unsaponifiable products. Sixty-nine components of aliphatic alcohols had been reported up to 1974 (latest reported work). The hydrocarbons (ca 0.5%) show structural similarity to the wool-wax acids or aliphatic alcohols and contain highly branched alkanes as well as cycloalkanes.

Wool-Grease Recovery. Current systems and research on wool-grease recovery have been reviewed comprehensively (187). The principal recovery process in use involves centrifuging in a cream-separator type of centrifuge modified by the addition of peripheral jets or other mechanical devices to remove dirt. Liquor is usually with-

Table 12. Some Physical and Chemical Data for Wool Wax[a]

color	yellow to pale brown
density (at 15°C)	0.94–0.97
refractive index (40°C)	1.48
melting point, °C	35–40
free acid content, %	4–10
free alcohol content, %	1–3
iodine value (Wijs)	13–30
saponification value	95–120
molecular weight (Rast method, in phenyl salicylate)	790–880
fatty acids, %	50–55
alcohols, %	50–45
acids	
melting point, °C	40–45
iodine value (Wijs)	10–20
mean molecular weight	330
alcohols	
melting point, °C	55–65
iodine value (Wijs)	40–50
mean molecular weight	370

[a] Ref. 179.

Table 13. Summary of the Average Composition of Wool-Wax Acids[a]

Acids[b]	Chain length[c]	Wool-wax acids, %
normal acids	C_8–C_{38}	10
iso acids	C_8–C_{40}	22
anteiso acids	C_7–C_{41}	28
normal α-hydroxy acids	C_{10}–C_{32}	17
iso α-hydroxy acids	C_{12}–C_{34}	9
anteiso α-hydroxy acids	C_{11}–C_{33}	3
normal ω-hydroxy acids	C_{22}–C_{36}	3
iso ω-hydroxy acids	C_{22}–C_{36}	0.5
anteiso ω-hydroxy acids	C_{23}–C_{35}	1
polyhydroxy acids		4.5[d]
unsaturated acids		2[d]

[a] Ref. 181.
[b] In the iso series, a branching methyl group is attached at the penultimate position, and in the anteiso series, at the antepenultimate position.
[c] Ref. 183.
[d] Tentative data.

Table 14. Summary of the Average Composition of Wool-Wax Alcohols[a]

Alcohol[b]	Chain length[c]	Approximate % of wool-wax alcohols
normal monoalcohols	C_{14}–C_{34}	2
iso monoalcohols	C_{14}–C_{36} }	
anteiso monoalcohols	C_{17}–C_{35} }	13
normal alkan-1,2-diols	C_{12}–C_{25}	1
iso alkan-1,2-diols	C_{14}–C_{30} }	
anteiso alkan-1,2-diols	C_{15}–C_{29} }	6
Total		22
cholesterol		34
lanosterol }		
dihydrolanosterol }		38
Total		72
hydrocarbons		1
autooxidation products undetermined		5

[a] Refs. 180, 183–186.
[b] See footnote [b], Table 13.
[c] Ref. 183.

drawn from the second bowl of a five-bowl scouring plant, heated to ca 90°C, and passed into a large settling tank to remove the heaviest dirt particles. The hot emulsion is then passed through a centrifuge designed to remove suspended solids, reheated, and run through the cream-separator type of centrifuge. This separates most of the grease remaining in the liquors. The liquor is cooled to ca 50°C and returned to the scouring plant or held in a reserve tank for later use.

With the introduction of scouring with nonionic detergents instead of soap and soda, grease-recovery rates increased (188); however, the quality of the grease decreased slightly because of increased recovery of the grease from the tip of the fiber (oxidized grease) (189). The recovery of grease by centrifuging has been optimized for normal scouring conditions in the WRONZ Comprehensive Scouring System (190), and the Lo-flo system of scouring (64) yields grease recoveries of ca 85% of the grease originally on the wool, although the quality of the grease is again reduced.

Other techniques aimed at improving grease recovery (and often attempting also to improve the scouring process itself) have included solvent degreasing of the wool (57–58), solvent extraction of the liquor or sludge (191), aeration (192–193), and physical and chemical destabilization (187).

Grease Refining and Fractionation. Lanolin to be used in pharmaceuticals and cosmetics must conform to strict requirements of purity, such as those in the U.S. and British Pharmacopoeias (194–195). These include specifications for the maximum allowable content of free fatty acids, moisture, ash, and free chloride. Lanolin intended for certain dermatological applications may have to meet further specifications in relation to free-alcohol and detergent contents (196–197).

The refining process most commonly used involves treatment with hot aqueous alkali to convert free fatty acids to soaps followed by bleaching, usually with hydrogen peroxide, although sodium chlorite, sodium hypochlorite, and ozone have also been used. Other techniques include distillation, steam stripping, neutralization by alkali, liquid thermal diffusion, and the use of active adsorbents (eg, charcoal and bentonite) and solvent fractionation (187).

Uses of Wool Grease. The uses of wool grease, lanolin, and lanolin derivatives are wide, ranging from pharmaceuticals and cosmetics (qv) to printing inks (qv), rust preventatives, and lubricants (see Lubrication and lubricants).

In pharmaceutical uses, the general inertness of lanolin and its derivatives, together with their ease of emulsification (198), have been important criteria. These properties are also important in cosmetics, but with emphasis also on the ability to absorb large quantities of water or, after suitable modification, to stabilize o/w emulsions.

The anticorrosion properties of lanolin have been utilized over a considerable period, and the product has the status of a temporary corrosion inhibitor (199) (see Corrosion and corrosion inhibitors). There is probably potential for greater use of the poorer qualities of lanolin for long-term storage and protection of machine parts.

Other uses have been reviewed in detail (187).

BIBLIOGRAPHY

"Wool" in *ECT* 1st ed., Vol. 15, pp. 103–134, by John Menkart, Textile Research Institute; "Wool" in *ECT* 2nd ed., Vol. 22, pp. 387–418, by M. Lipson, Division of Textile Industry, CSIRO, Geelong, Victoria, Australia.

1. *Textile Terms and Definitions* 7th ed., Textile Institute, Manchester, England, 1975.
2. *Wool Quarterly* (2), 19 (1981).
3. International Wool Secretariat, London, 1982.
4. *Organon World Cotton Statistics*, Textile Economics Bureau, Roseland, N.J., 1981.
5. M. Lipson and U. A. F. Black, *J. R. Soc. N.S.W.* **78,** 84 (1945).
6. R. B. Sweeten, *J. Text. Inst. Trans.* **40,** T727 (1949).
7. *Report to the Australian Wool Corporation by Specialist Working Group on Sale by Description*, Melbourne, Australia, 1978.
8. G. A. Carnaby and A. J. McKinnon, *Australas. Text.* **2**(2), 11 (1982).
9. *Objective Measurement of Wool in Australia*, Australian Wool Corporation, Melbourne, Australia, 1973.

10. B. H. Mackay, *Proc. Int. Wool Text. Res. Conf.*, (*Pretoria*) **I,** 59 (1980).
11. *Core Test Regulations, Test Methods, IWTO—19-76, 28-75, 3-73, 20-69*, International Wool Textile Organization, London, 1968.
12. *Australian Standard 1363—1976*, Standards Association of Australia, Melbourne, Australia, 1976.
13. A. Baumann, *A Wool Staple Length Meter*, CSIRO Division of Textile Physics, Ryde, New South Wales, Australia, in preparation.
14. J. Grignet and G. Bertoni, *Proc. Int. Wool Text. Res. Conf.* (*Aachen*) **V,** 393 (1975).
15. R. N. Caffin, *J. Text. Inst.* **71,** 65 (1980).
16. R. H. Peters, *The Chemistry of Fibres*, Vol. I of *Textile Science*, Elsevier, Amsterdam, The Netherlands, 1963, p. 263.
17. J. A. Maclaren and B. Milligan, *Wool Science: The Chemical Reactivity of the Wool Fibre*, Science Press, Sydney, Australia, 1981, p. 2.
18. K. R. Makinson, *Wool Shrinkproofing*, Marcel Dekker, Inc., New York, 1979, p. 147.
19. J. H. Bradbury and J. D. Leeder, *Aust. J. Biol. Sci.* **23,** 843 (1970).
20. H. M. Appleyard and C. M. Greville, *Nature* **166,** 1031 (1950).
21. M. S. C. Birbeck and E. H. Mercer, *J. Biophys. Biochem. Cytol.* **3,** 203, 215 (1957).
22. J. H. Bradbury and K. F. Ley, *Aust. J. Biol. Sci.* **25,** 1235 (1972).
23. H. Zahn, *Proc. Int. Wool Text. Conf.* (*Pretoria*) **I**(supplement), (1980).
24. C. A. Anderson, J. D. Leeder, and D. S. Taylor, *Wear* **21,** 115 (1972).
25. H. Horio and T. Kondo, *Text. Res. J.* **23,** 373 (1953).
26. J. H. Dusenbury and J. Menkart, *Proc. Int. Wool Text. Res. Conf.* (*Australia*), F-142 (1956).
27. J. H. Dusenbury, E. H. Mercer, and J. H. Wakelin, *Text. Res. J.* **24,** 890 (1954).
28. G. V. Chapman and J. H. Bradbury, *Arch. Biochem. Biophys.* **127,** 157 (1968).
29. P. Kassenbeck, *Proc. Int. Wool Text. Res. Conf.* (*Paris*) **I,** 367 (1965).
30. C. W. Hock, R. G. Ramsay, and M. Harris, *J. Res. Natl. Bur. Stand.* **27,** 181 (1941).
31. *Ibid.*, p. 234.
32. J. L. Farrant, A. L. G. Rees, and E. H. Mercer, *Nature* **159,** 535 (1947).
33. *Wool Research*, Vol. 2, Wool Industries Research Association, Leeds, UK, 1955.
34. "Man-Made Fiber Chart," *Text. World*, (1961).
35. W. von Bergen, ed., *Wool Handbook*, 3rd ed., Interscience Publishers, a division of John Wiley & Sons, Inc., New York, Vol. 1, 1963, p. 180.
36. J. B. Speakman, *J. Text. Inst. Trans.* **18,** T431 (1927).
37. E. G. Bendit, *J. Macromol. Sci-Phys.* **B17,** 129 (1980).
38. M. Feughelman, *Proc. Int. Wool Text. Res. Conf.* (*Pretoria*) **I,** 35 (1980).
39. Ref. 17, p. 6.
40. D. H. Simmonds, *Aust. J. Biol. Sci.* **8,** 537 (1955).
41. J. J. O'Donnell and E. D. P. Thompson, *Aust. J. Biol. Sci.* **14,** 740 (1962).
42. J. H. Bradbury, G. V. Chapman, and N. L. R. King, *Aust. J. Biol. Sci.* **18,** 353 (1965).
43. B. Milligan, L. A. Holt, and J. B. Caldwell, *Appl. Polym. Symp.* **18,** 113 (1971).
44. M. Cole, J. C. Fletcher, K. L. Gardner, and M. C. Corfield, *Appl. Polym. Symp.* **18,** 147 (1971).
45. L. A. Holt, B. Milligan, and L. J. Wolfram, *Text. Res. J.* **44,** 846 (1974).
46. I. J. O'Donnell and E. F. Woods, *J. Polym. Sci.* **21,** 397 (1956).
47. J. M. Gillespie in A. G. Lyne and B. F. Short, ed., *Biology of the Skin and Hair Growth*, Angus and Robertson, Sydney, Australia, 1965, p. 377.
48. H. Zahn and M. Biela, *Eur. J. Biochem.* **5,** 567 (1968).
49. W. G. Crewther and F. G. Lennox, *J. Proc. R. Soc. N.S.W.* **108,** 95 (1975).
50. W. G. Crewther, *Proc. Int. Wool Text. Res. Conf.* (*Aachen*) **I,** 1 (1975).
51. T. A. A. Dopheide, *Eur. J. Biochem.* **34,** 120 (1973).
52. A. Brearley and J. A. Iredale, *The Woollen Industry: An Outline of the Woollen Industry and Its Processes from Fibre to Fabric*, Wira, Leeds, UK, 1977.
53. A. Brearley and J. A. Iredale, *The Worsted Industry: An Account of the Worsted Industry and Its Processes from Fibre to Fabric*, Wira, Leeds, UK, 1980.
54. G. F. Wood, *Text. Prog.* **12**(1), (1982).
55. *Text. J. Aust.* **44**(6), 34 (1969).
56. *Text. Rec.*, 69 (Jan. 1964).
57. *CSIRO Division of Textile Industry Report No. G.10*, CSIRO Division of Textile Industry, Geelong, Australia, 1960.
58. J. Brach, *Wool Sci. Rev.* (36), 38 (May 1969).

59. A. S. C. Lawrence, *Nature* **183,** 1491 (1959).
60. R. P. Harker, *J. Text. Inst. Trans.* **50,** T189 (1959).
61. C. A. Anderson, *J. Text. Inst.* **73,** 289 (1982).
62. C. A. Anderson, *Australas. Text.* **2**(5), 28 (1982).
63. C. A. Anderson, J. R. Christoe, A. J. C. Pearson, J. J. Warner, and G. F. Wood, *Developments in Scouring and Carbonising 1980, Disc 80*, Australian Wool Corporation, Melbourne, 1980, p. 1.
64. G. F. Wood, A. J. C. Pearson, and J. R. Christoe, *CSIRO Division of Textile Industry Report No. G39*, CSIRO Division of Textile Industry, Geelong, Australia, 1980.
65. *Japanese Wool Textile Industry's Treatment of Effluent from Wool Scouring*, Japan Wool Spinners' Association, Tokyo, Jpn., 1973, p. 53.
66. R. G. Stewart, G. V. Barker, P. E. Chisnall, and J. L. Hoare, *WRONZ Report No. 25*, Wool Research Organization of New Zealand, Inc., Christchurch, New Zealand, 1974.
67. R. G. Stewart, *WRONZ Report No. 27*, Wool Research Organization of New Zealand, Inc., Christchurch, New Zealand, 1974.
68. R. G. Stewart, P. E. Chisnall, J. M. Flynn, and R. G. Jamieson, *WRONZ Communication No. 36*, Wool Research Organization of New Zealand, Inc., Christchurch, New Zealand, 1975.
69. T. A. Pressley and W. G. Crewther, *Wool Sci. Rev.* (30), 16 (1966); (31), 1 (1967).
70. N. C. Gee, *Shoddy and Mungo Manufacture*, Emmott & Co., Bradford, UK, 1950.
71. H. K. Rouette and G. K. Kittan, *Text. Praxis Int.* **36,** 784 (1981).
72. W. G. Crewther, *Proc. Int. Wool Textile Res. Conf.* (*Australia*) **E,** 408 (1955).
73. W. G. Crewther and T. A. Pressley, *Text. Res. J.* **28,** 67 (1958).
74. A. E. Davis, A. J. Johnson, and L. R. Mizell, *Text. Res. J.* **31,** 825 (1961).
75. M. S. Nosser, M. Chaikin, and A. J. Datyner, *J. Text. Inst.* **62,** 677 (1971).
76. M. S. Nosser and M. Chaikin, *Proc. Int. Wool Text. Res. Conf.* (*Aachen*) **IV,** 136 (1975).
77. H. J. Katz, D. E. A. Plate, and G. F. Wood, *Wool Sci. Rev.* (45), 28 (1973).
78. M. A. White, *Text. Prog.* **13**(2), (1983).
79. H. K. Rouette and A. Ohm, *Melliand Textilber.* **62,** 583 (1981).
80. H. K. Rouette, A. Gotz, and A. Ohm, *Int. Text. Bull. Dye. Print. Finish.* (3), 211 (1981).
81. M. A. White, *Proc. Conf. Recent Dev. in Wool and Wool Blend Processing*, CSIRO Division of Textile Industry, Geelong, Australia, 1983, p. 20.
82. I. B. Angliss, *CSIRO Division of Textile Industry Report No. G23*, CSIRO Division of Textile Industry, Geelong, Australia, 1973.
83. I. B. Angliss, P. R. Brady, J. Delmenico, and R. J. Hine, *Text. J. Aust.* **43**(4), 17 (1968).
84. C. L. Bird, *The Theory and Practice of Wool Dyeing*, Society of Dyers and Colourists, Bradford, England, 1972.
85. C. H. Giles in C. L. Bird and W. S. Boston, ed., *The Theory of Coloration of Textiles*, Society of Dyers and Colourists, Bradford, England, 1974, p. 78.
86. R. H. Peters, *The Physical Chemistry of Dyeing*, Vol. III of *Textile Chemistry*, Elsevier Publishing, Amsterdam, Oxford, England, and New York, 1975, p. 203.
87. D. M. Lewis, *Rev. Prog. Color.* **8,** 10 (1981).
88. Ref. 84, p. 127.
89. *Ibid.*, p. 106.
90. R. V. Peryman, *Proc. Int. Wool Text. Res. Conf.* (*Australia*) **E,** 17 (1955).
91. R. L. Holmes-Brown and G. A. Carnaby, *WRONZ Communication No. 67*, Wool Research Organization of New Zealand, Inc., Christchurch, New Zealand, 1980.
92. R. R. D. Holt and F. J. Harrigan, *Melliand Textilber.* **9,** 745 (1979).
93. I. B. Angliss, *Text. Prog.* **12**(4), (1982).
94. J. F. Graham and R. R. D. Holt, *Colourage* **25**(10), 45 (1978).
95. P. Spinaci, A. Panelli, and N. Cicchello, *Int. Dyer* **159,** 349 (1978).
96. J. F. Graham, *Int. Dyer* **150,** 558 (1973).
97. J. F. Graham and I. Seltzer, *Textilveredlung* **9,** 551 (1974).
98. J. R. Hine and J. R. McPhee, *Proc. Int. Wool Tex. Res. Conf.*, Paris III, 183 (1965).
99. L. Peters and C. B. Stevens, *J. Soc. Dyers Colour.* **72,** 100 (1956).
100. W. Beal and G. S. A. Corbishley, *J. Soc. Dyers Colour.* **87,** 329 (1971).
101. I. B. Angliss and R. J. Hine, *Proc. Int. Wool Text. Res. Conf.* (*Pretoria*) **V,** 629 (1980).
102. *Engineer* **244,** 28 (Apr. 7, 1977).
103. P. R. Brady, *Text. Prog.* **12**(4), (1982).
104. *Wool Statistics 1974–75*, Commonwealth Secretariat, London, 1975, p. 35.

105. K. Reincke, *Text. Praxis Int.* **25,** 419 (1970).
106. Brit. Pat. 1,377,085 (Aug. 19, 1966), (to Hoechst A.G.).
107. Brit. Pat. 1,165,253 (May 1, 1967), (to Sandoz Ltd.).
108. A. I. Matekskii, V. A. Kuz'menkov, and N. I. Volkova, *Tektil. Prom.* **30**(7), 69 (1970).
109. P. R. Brady and P. G. Cookson, *Proc. Int. Wool Text. Res. Conf. (Pretoria)* **V,** 517 (1980).
110. D. M. Lewis and I. Seltzer, *J. Soc. Dyers Colour.* **88,** 327 (1972).
111. P. R. Brady and R. J. Hine, *CSIRO Division of Textile Industry Report No. G35*, CSIRO Division of Textile Industry, Geelong, Australia, 1979.
112. P. R. Brady and P. G. Cookson, *J. Soc. Dyers Colour.* **97,** 159 (1981).
113. G. A. Smith, *Br. Knit. Ind.* **44,** 89 (Dec. 1971).
114. *Ibid.*, **46,** 77 (July 1973).
115. P. R. Brady, P. G. Cookson, K. W. Fincher, and D. M. Lewis, *J. Soc. Dyers Colour.* **96,** 188 (1980).
116. P. R. Brady, P. G. Cookson, K. W. Fincher, and D. M. Lewis, *Proc. Int. Wool Text. Res. Conf. (Pretoria)* **V,** 609 (1980).
117. Ref. 84, p. 54.
118. C. Frommelt, *Bayer Farben Rev.* (27), 46 (1977).
119. *Process Data Sheet CTE-054e*, BASF AG, Leverkusen, FRG, 1978.
120. B. Milligan, *Proc. Int. Wool Text. Res. Conf. (Pretoria)* **V,** 167 (1980).
121. C. H. Nicholls in N. S. Allen, ed., *Developments in Polymer Photochemistry*, Applied Science Publishers, London, 1980, p. 125.
122. P. J. Waters, N. A. Evans, L. A. Holt, and B. Milligan, *Proc. Int. Wool Text. Res. Conf. (Pretoria)* **V,** 195 (1980).
123. I. H. Leaver, P. J. Waters, and N. A. Evans, *J. Polym. Sci. Polym. Chem.* **17,** 1531 (1979).
124. L. A. Holt, B. Milligan, and L. J. Wolfram, *Text. Res. J.* **44,** 846 (1974).
125. B. Milligan and D. J. Tucker, *Text. Res. J.* **34,** 681 (1964).
126. R. J. Mayfield, *Text. Prog.* 11(4), (1982).
127. R. J. Mayfield, *CSIRO Division of Textile Industry, Report No. G40*, CSIRO Division of Textile Industry, Geelong, Australia, 1979.
128. R. J. Mayfield and G. J. O'Loughlin, *CSIRO Division of Textile Industry Report No. G41*, CSIRO Division of Textile Industry, Geelong, Australia, 1980.
129. R. M. Hoskinson, R. J. Mayfield, and I. M. Russell, *J. Text. Inst.* **67,** 19 (1976).
130. R. M. Hoskinson and I. M. Russell, *J. Text. Inst.* **65,** 387 (1974).
131. A. L. Black, R. M. Hoskinson, and I. M. Russell, *J. Text. Inst.* **67,** 68 (1976).
132. R. J. Mayfield and I. M. Russell, *J. Text. Inst.* **70,** 53 (1979).
133. F. W. Jones, R. J. Mayfield, and G. J. O'Loughlin, *Proc. Int. Wool Text. Res. Conf. (Pretoria)* **V,** 431 (1980).
134. C. S. Whewell, *J. Text. Inst. Proc.* **47,** P851 (1956).
135. J. B. Caldwell, S. J. Leach, S. J. Meschers, and B. Milligan, *Text. Res. J.* **34,** 627 (1964).
136. H. D. Weigmann, L. Rebenfeld, and C. Dansizer, *Proc. Int. Wool Text. Res. Conf. (Paris)* **II,** 319 (1965).
137. W. G. Crewther, *J. Soc. Dyers Colour.* **82,** 54 (1966).
138. Ref. 17, p. 294.
139. M. A. White, *CSIRO Division of Textile Industry Report No. G44*, CSIRO Division of Textile Industry, Geelong, Australia, 1982.
140. A. J. Farnworth, *Am. Dyest. Rep.* **49,** 996 (1960).
141. *Bull. Cent. Text. Contr. Rech. Sci. (Roubaix)* **43,** 39 (1959).
142. Technical brochures, IWS Technical Center, Ilkley, UK.
143. J. R. Cook and J. Delmenico, *Proc. Int. Wool Text. Res. Conf. (Paris)* **III,** 419 (1965).
144. K. R. Makinson, *Wool Sci. Rev.* **24,** 34 (1964).
145. Ref. 18, p. 233.
146. K. R. Makinson, *Text. Res. J.* **44,** 856 (1974).
147. A. J. Farnworth, *J. Soc. Dyers Colour.* **77,** 483 (1961).
148. M. Lipson, *J. Text. Inst. Proc.* **38,** P279 (1947).
149. T. Barr and J. B. Speakman, *J. Soc. Dyers Colour.* **60,** 783 (1944).
150. J. E. Moore, *Text. Res. J.* **26,** 936 (1956).
151. *Wool Sci. Rev.* **36,** 2 (1969); **37,** 37 (1969).
152. T. Shaw and J. Lewis, *Text. Prog.* 4(3), 1 (1972).
153. T. Shaw, *Wool Sci. Rev.* **46,** 44 (1973); **47,** 14 (1973).

154. J. Lewis, *Wool Sci. Rev.* **54**, 2 (1977); **55**, 23 (1978).
155. Ref. 17, p. 236.
156. J. R. McPhee, *Text. J. Aust.* **72**, 60 (1972).
157. P. Smith and J. H. Mills, *Chemtech*, 748 (1973).
158. J. Lewis, *Textilveredlung* **11**, 214 (1976).
159. *Hosiery Trade J.* **79**, 82 (1972).
160. R. H. Earle, R. H. Saunders, and L. R. Kangas, *Appl. Polym. Symp.* **18**, 707 (1971).
161. Brit. Pat. 1,524,392 (Sept. 13, 1978), (to Kroy Unshrinkable Wools Ltd.).
162. K. W. Fincher and M. A. White, *CSIRO Division of Textile Industry Report No. G30*, CSIRO Division of Textile Industry, Geelong, Australia, 1977.
163. M. A. Rushforth, *IWS Technical Information Bulletin No. 12*, International Wool Secretariat, Ilkley, England, 1977.
164. R. Bowrey, C. Brooke, and B. Robinson, *Textilia* **51**, 41 (1975).
165. A. Kershaw and J. Lewis, *Text. Month*, 40 (1976).
166. Ref. 18, p. 268.
167. D. Allanack, M. Rushforth, and T. Shaw, *Proc. Text. Inst. Conf., Edinburgh*, Textile Institute, Manchester, England, 1978.
168. A. Bereck, *Text. Res. J.* **49**, 233 (1979).
169. T. Jellinek, A. De Boos, and M. A. White, *Proc. Int. Wool Text. Res. Conf. (Pretoria)* **V**, 125 (1980).
170. A. G. De Boos and M. A. White, *Proc. Hung. Text. Conf., Budapest*, (1979).
171. M. A. White, *Text. J. Aust.* **48**(6), 20 (1973).
172. M. A. White, *CSIRO Division of Textile Industry Report No. G21*, CSIRO Division of Textile Industry, Geelong, Australia, 1977.
173. T. Shaw, *Wool Sci. Rev.* **47**, 14 (1973).
174. M. J. Koroskys, *Am. Dyest. Rep.* **60**(5), 48 (1971).
175. L. Benisek, *Text. Manuf.* **99**, 36 (Jan.–Feb. 1972); *Melliand Textilber.* **53**, 931 (1972).
176. L. Benisek, *J. Text. Inst.* **65**, 102, 140 (1974).
177. P. Gordon and L. Stephens, *J. Soc. Dyers Colour.* **90**, 239 (1974).
178. L. Benisek, G. Edmondson, and B. Greenwood, *Text. Inst. Ind.* **14**, 344 (1976).
179. E. V. Truter, *Wool Wax*, Cleaver-Hume Press Ltd., London, 1956, p. 32.
180. K. Motiuk, *J. Am. Oil Chem. Soc.* **56**, 651 (1979).
181. *Ibid.*, **56**, 91 (1979).
182. *Ibid.*, **57**(4), 145 (1980).
183. F. Fawaz, C. Miet, and F. Puisieux, *Ann. Pharm. Fr.* **31**(1), 63 (1973).
184. K. E. Murray and R. Schoenfeld, *Aust. J. Chem.* **8**, 424 (1955).
185. D. H. S. Horn, *J. Sci. Food Agric.* **9**, 632 (1958).
186. D. T. Downing, Z. H. Krantz, and K. E. Murray, *Aust. J. Chem.* **13**, 80 (1960).
187. R. G. Stewart and L. F. Story, *WRONZ Tech. Paper No. 4*, Wool Research Organization of New Zealand, Inc., Christchurch, New Zealand, 1980.
188. R. E. Wolfrom, *Am. Dyest. Rep.* **43**, P372 (1954).
189. C. A. Anderson and G. F. Wood, *Nature* **193**, 742 (1962).
190. R. G. Stewart, G. V. Barker, P. E. Chisnall, and J. L. Hoare, *WRONZ Report No. 25*, Wool Research Organization of New Zealand, Inc., Christchurch, New Zealand, 1974.
191. C. Moxhet, *Ind. Text. Belge* **7/8**, 51 (1971); **9**, 83 (1971); *World Text. Abstr.* 8105 (1971).
192. L. F. Evans and W. E. Ewers, *Aust. J. Appl. Sci.* **4**, 552 (1953).
193. USSR Pat. 104,212 (Dec. 1956), A. S. Salin and co-workers.
194. *The United States Pharmacopoeia XX (USP XX–NF XV)*, The United States Pharmacopeial Convention, Inc., Rockville, Md., 1980.
195. *British Pharmacopoeia*, H.M.S.O., London, 1980.
196. E. W. Clark, E. Cronin, and D. S. Wilkinson, *Contact Dermatitis* **3**, 69 (1977).
197. R. Elder, *J. Environ. Path. Toxicol.* **4**(4), 63 (1980).
198. A. Castillo and J. M. Sure, *Ars Pharm.* **9**, 367 (1968).
199. U.S. Pat. 2,473,614 (June 21, 1949), E. Snyder (to American Chemical Paint Co.).

General References

Refs. 17, 18, 35, 52, 53, 54, 84, 126, and 179 are also general references.

P. Alexander and R. F. Hudson in C. Earland, ed., *Wool: Its Chemistry and Physics*, 2nd ed., Chapman & Hall Ltd., London, 1963.

W. J. Onions, *Wool: An Introduction to its Properties, Varieties, Uses and Production*, Ernest Benn Ltd., London, 1962.

R. S. Asquith, ed., *Chemistry of Natural Protein Fibers*, Plenum Press, New York and London, 1977.

R. G. Stewart and L. F. Story, *Wool Grease: A Review of Its Recovery and Utilisation*, Wool Research Organization of N.Z. (Inc.), Christchurch, 1980; comprehensive review of the literature from 1947 to mid-1979.

Papers, 1st International Wool Textile Research Conference, Australia, 1955, CSIRO, Melbourne, Australia.

J. Text. Inst. Trans. **51**, T489 (1960).

Papers, 3rd International Wool Textile Research Conference, Paris, l'Institut Textile de France, Paris, 1965.

L. Rebenfeld, ed., *Proceedings of the 4th International Wool Textile Research Conference*, International Publishers, a division of John Wiley & Sons, Inc., New York, 1971.

H. Ziegler, ed., *Proceedings of the 5th International Wool Textile Research Conference*, Deutsches Wollforschungsinstitut an der Technischen Hochschule, Aachen, 1975.

Proceedings of the 6th International Wool Textile Research Conference, South African Wool and Textile Research Institute (SAWTRI) of the Council for Scientific and Industrial Research (CSIR), Pretoria, S. Africa, 1980.

Wool Science Review, International Wool Secretariat, London; published on an irregular basis.

W. S. BOSTON
CSIRO

WORKING FLUIDS. See Heat-exchange technology.

X

XANTHAN GUM. See Gums; Microbial polysaccharides.

XANTHATES

The salts of the O-esters of carbonodithioic acids and the corresponding O,S-diesters are xanthates. The free acids decompose on standing. Potassium ethyl xanthate was first prepared in 1822 by W. C. Zeise from potassium hydroxide, carbon disulfide, and ethanol. Most alcohols, including cellulose, undergo this reaction to form xanthates, but normally phenols do not (see Rayon). Potassium phenyl xanthate was prepared in 1960 from potassium phenoxide and carbon disulfide in dimethylformamide (1). Xanthates remained a laboratory curiosity until the turn of the twentieth century when the rubber industry developed a use for them in the curing and vulcanization of rubber (see Rubber chemicals; Rubber compounding). In 1925, xanthates were found to be excellent flotation reagents for the recovery of heavy-metal sulfides from ores (see Flotation). This is the principal use for the noncellulose xanthates; several of the alkali-metal xanthates are commercially available.

Nomenclature

The names xanthogenic acid and xanthogen refer to the acid C_2H_5OCSSH and the radical C_2H_5OCSS-, respectively. Xanthogenic is still used instead of xanthic and xanthogenate for xanthate, eg, sodium methyl xanthogenate [6370-03-2] for sodium methyl xanthate. The term xanthogen is sometimes used for the radical $HOCSS-$, eg, methyl xanthogen acetic acid for $CH_3OCSSCH_2COOH$. Examples of the current nomenclature are listed in Table 1 with the corresponding common names. The latter names are used in this article.

Table 1. Nomenclature of Some Xanthic Acids and Related Compounds

Formula	CAS Registry No.	CAS and IUPAC nomenclature	Common name
CH₃OC(S)SH	[2042-42-4]	O-methyl carbonodithioic acid	methyl xanthic acid
C₂H₅OC(S)SNa	[140-90-9]	sodium O-ethyl carbonodithioate, sodium O-ethyl dithiocarbonate	sodium ethyl xanthate
C₂H₅OC(S)S·CH₃	[623-54-1]	O-ethyl S-methyl carbonodithioate	methyl ethyl xanthate
(C₂H₅OC)₂S (S=)	[2905-52-4]	diethyl thiodicarbonate [(HO)C(S)]₂S)	diethyl xanthogen monosulfide
(C₂H₅OC)₂S₂ (S=)	[502-55-6]	diethyl thioperoxydicarbonate	diethyl dixanthogen
C₂H₅OC(S)SC(O)OC₂H₅	[3278-35-1]	diethyl thiodicarbonate, ((HO)C(O)SC(S)(OH)) 6-thioxo-3,7-dioxa-5-thio-4-nonane	diethyl xanthogen formate
C₂H₅OC(S)NHCH₃	[817-73-2]	O-ethyl methylcarbamothioate, O-ethyl methylthiocarbamate	ethyl methylthionocarbamate
CH₃OC(S)Cl	[2812-72-8]	O-methyl carbonochloridothioate	methyl chlorothionoformate
CH₃OC(S)NHC₆H₅	[13509-41-6]	O-methyl phenylcarbamothioate, O-methyl phenylthiocarbamate	methyl phenylthionocarbanilate

Properties

The free xanthic acids are unstable, colorless or yellow oils, and might decompose with explosive violence. They are soluble in the common organic solvents and are slightly soluble in water: methyl xanthic acid at 0°C, 0.05 mol/L; ethyl xanthic acid at 0°C, 0.02 mol/L; and n-butyl xanthic acid at 0°C, 0.0008 mol/L. Values for the dissociation constant for ethyl xanthic acid are $(2.0–3.0) \times 10^{-2}$ (2). Potentiometric determinations for C_1–C_8 xanthic acids show a decreasing acid strength with increasing molecular weight (3–4). The values for the ethyl derivative are $(1.82–3.4) \times 10^{-3}$. Similar values determined by the same method are reported in ref. 5 for a series in dimethylformamide. Values of 7.66×10^{-10} for ethyl xanthic acid and 9.68×10^{-10} for amyl xanthic acid are reported (6).

The alkali-metal salts, in contrast to the free acids, are relatively stable solids, are pale yellow when pure, and have a disagreeable odor. Pyrolysis studies of the potassium salts of ethyl, isopropyl, n-butyl, sec-butyl, and n-amyl xanthic acids with separation of the gaseous products by gas chromatography are reported in ref. 7. All salts produce carbonyl sulfide, carbon disulfide, thiols, alcohols, sulfides, disulfides, and some aldehydes. A comparison of melting points of xanthates and the temperature interval T_{dta} from the start to the end of the differential thermal analysis interval are given in Table 2 (7–8). The potassium salts of several xanthates derived from tertiary alcohols were pyrolyzed and the yield of olefins was superior to the Chugaev decomposition of the corresponding S-methyl esters (9). In most cases, the olefin product distributions were almost identical to those obtained from the esters.

Table 2. Decomposition Temperatures of Xanthates[a]

Compound	CAS Registry No.	T_{dta}, °C	Mp, °C
CH₃OC(=S)SK	[2667-20-1]	165–185	182–186
C₂H₅OC(=S)SK	[140-89-6]	210–225	225–226
n-C₃H₇OC(=S)SK	[2720-67-4]	220–240	233–239
i-C₃H₇OC(=S)SK	[140-92-1]	230–275	278–282
n-C₄H₉OC(=S)SK	[871-58-9]	235–255	255–256
i-C₄H₉OC(=S)SK	[13001-46-2]	250–265	260–270
s-C₄H₉OC(=S)SK	[141-96-8]	220–260	

[a] Refs. 7–8.

When exposed to air, the sodium salts tend to take up moisture and form dihydrates. The alkali-metal xanthates are soluble in water, alcohols, the lower ketones, pyridine, and acetonitrile. They are not particularly soluble in the nonpolar solvents, eg, ether or ligroin. The solubilities of a number of these salts are listed in Table 3. Potassium isopropyl xanthate is soluble in acetone to ca 6 wt %, whereas the corresponding methyl, ethyl, n-propyl, n-butyl, isobutyl, isoamyl, and benzyl [2720-79-8] xanthates are soluble to more than 10 wt % (8). The solubilities of the commercially available xanthates in water are plotted versus temperature in Figure 1 (11).

The heavy-metal salts, in contrast to the alkali metal salts, have lower melting

Table 3. Solubilities of Some Alkali-Metal Xanthates[a]

Xanthate	CAS Registry No.	Solvent	Solubility, g/100 g soln 0°C	35°C
potassium n-propyl		water	43.0	58.0
		n-propyl alcohol	1.9	8.9
sodium n-propyl	[14394-29-7]	water	17.6	43.3
		n-propyl alcohol	10.2	22.5
potassium isopropyl		water	16.6	37.2
		isopropyl alcohol		2.0
sodium isopropyl	[140-93-2]	water	12.1	37.9
		isopropyl alcohol		19.0
potassium n-butyl		water	32.4	47.9
		n-butyl alcohol		36.5
sodium n-butyl	[141-33-3]	water	20.0	76.2
		n-butyl alcohol		39.2
potassium isobutyl		water	10.7	47.7
		isobutyl alcohol	1.6	6.2
sodium isobutyl	[25306-75-6]	water	11.2	33.4
		isobutyl alcohol	1.2	20.5
potassium isoamyl	[928-70-1]	water	28.4	53.3
		isoamyl alcohol	2.0	6.5
sodium isoamyl	[2540-36-5]	water	24.7	43.5
		isoamyl alcohol	10.9	15.5

[a] Ref. 10.

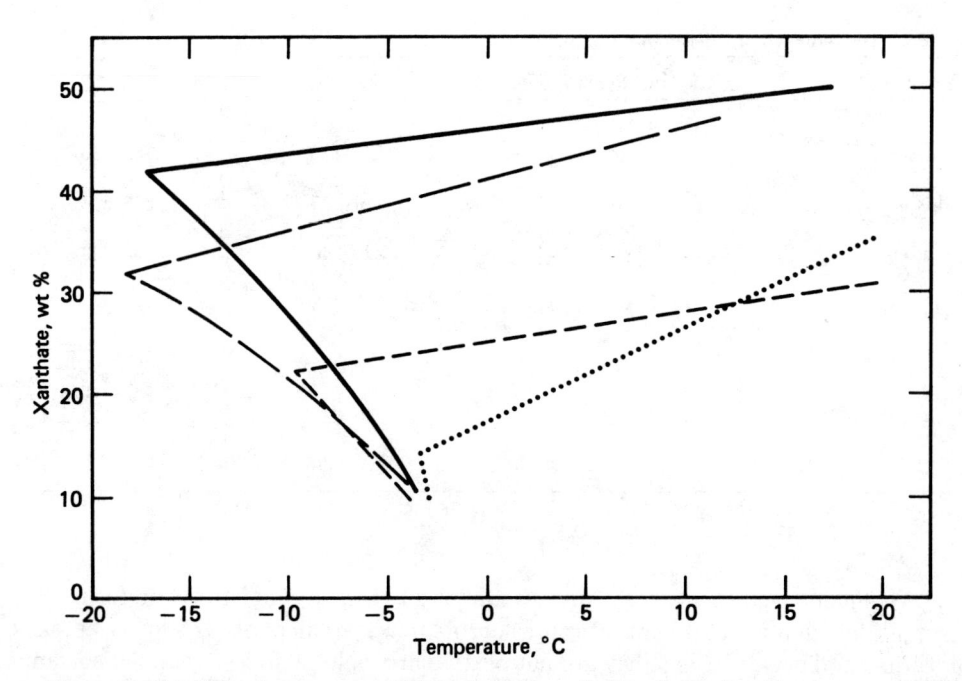

Figure 1. Solubility of some commercial xanthates. —— = Sodium isobutyl xanthate; — — = sodium ethyl xanthate; --- = sodium isopropyl xanthate; ··· = potassium amyl xanthate. Courtesy American Cyanamid Co.

points and are more soluble in organic solvents, eg, methylene chloride, chloroform, tetrahydrofuran, and benzene. They are slightly soluble in water, alcohol, aliphatic hydrocarbons, and ethyl ether (12). Their thermal decompositions have been extensively studied by dta and tga methods. They decompose to the metal sulfides and gaseous products, which are primarily carbonyl sulfide and carbon disulfide in varying ratios. In some cases, the dialkyl xanthate forms. Solvent extraction studies of a large number of elements as their xanthate salts have been reported (13).

The solubilities of the heavy-metal xanthates in water have been placed in the following orders of increasing solubility by two different authors (14–15): Hg^{2+}, Ag^+, Cu^+, Co^{3+}, As^{3+}, Pb^{2+}, Tl^+, Cd^{2+}, Ni^{2+}, Zn^{2+}; and Hg^{2+}, Hg_2^{2+}, Au^{3+}, Ag^+, Cu^+, Bi^{3+}, Pb^{2+}, Cd^{2+}, Ni^{2+}, Fe^{2+}, Zn^{2+}. The solubility products of many of these salts are reported in ref. 16. Surface tension depression increases with increasing molecular weight (17).

Alkalies stabilize xanthate solutions somewhat and the solutions readily decompose at acidic pHs (18–19). With a lower alkyl xanthate, the decomposition rate passes through a maximum at a low pH and then decreases smoothly as the acidity increases. However, this is not the case with salts of *tert*-butyl xanthic acid [110-50-9] (20). The decomposition accelerates at pH values above 10 or below 9; between these limits, no significant decomposition occurs during 24 h (21). The kinetic studies of the rate of decomposition of alkaline xanthate solutions show a minimum rate constant at pH 10 (22). The results of these studies agree with the report that over an eight-day period, 75% decomposition occurred at pH 6.5 and only 25% at pH 10.8 (23). In Figure 2, the effect of time, temperature, and concentration on aqueous solutions of sodium isopropyl xanthate is shown (11). The stability data for the other xanthates are similar;

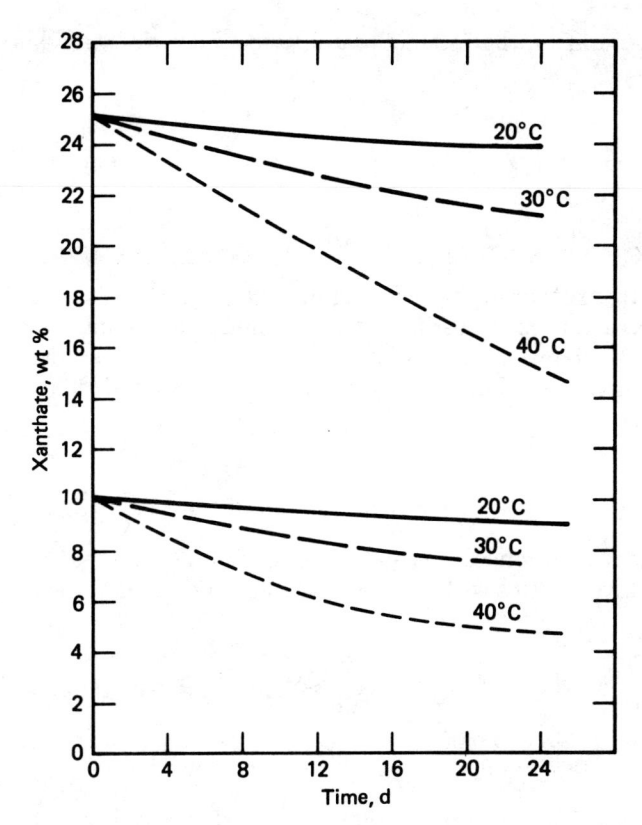

Figure 2. Decomposition of commercial sodium isopropyl xanthate solutions. Courtesy American Cyanamid Co.

xanthates prepared from secondary alcohols are more stable than those prepared from the primary alcohols. Branching of the chain and increasing the molecular weight also tends to increase the stability. The benzyl xanthates are less stable, both as solids and in solution.

Reactions. The chemistry of the xanthates is essentially that of the dithio acids. The free xanthic acids readily decompose in polar solvents, the rate being 10^6 times greater in methanol than in hexane. The acids decompose at room temperature to carbon disulfide and the corresponding alcohol; the resulting alcohol autocatalytically facilitates the decomposition.

$$\text{ROCSS}^- + \text{H}^+ \rightleftharpoons \text{ROC(=S)SH} \longrightarrow \text{ROH} + \text{CS}_2$$

The initial hydrolysis of the xanthate in aqueous solutions at room temperature is characterized by the following reaction involving potassium ethyl xanthate:

$$6\ \text{C}_2\text{H}_5\text{OC(=S)SK} + 3\ \text{H}_2\text{O} \longrightarrow 6\ \text{C}_2\text{H}_5\text{OH} + 2\ \text{KSC(=S)SK} + \text{K}_2\text{CO}_3 + 3\ \text{CS}_2$$

Further hydrolysis of the carbon disulfide and the trithiocarbonate produces hydrogen sulfide, etc (24).

The alkali metal xanthates react readily with the various alkylating reagents to form the S-esters:

$$C_2H_5O\overset{\underset{\|}{S}}{C}SNa + CH_3I \xrightarrow{alcohol} C_2H_5O\overset{\underset{\|}{S}}{C}SCH_3 + NaI$$

$$C_2H_5O\overset{\underset{\|}{S}}{C}SNa + ClCH_2\overset{\underset{\|}{O}}{C}ONa \xrightarrow{water} C_2H_5O\overset{\underset{\|}{S}}{C}SCH_2\overset{\underset{\|}{O}}{C}ONa + NaCl$$

The reactions are exothermic and cooling may be required.

The dialkyl esters react readily with ammonia or alkylamines to give the corresponding thionocarbamates and thiols. In general, heating is not required. In the following reaction, the thionocarbamate is the only water-insoluble material and separates as an oil:

$$C_2H_5O\overset{\underset{\|}{S}}{C}SCH_2\overset{\underset{\|}{O}}{C}ONa + C_2H_5NH_2 \rightarrow C_2H_5O\overset{\underset{\|}{S}}{C}NHC_2H_5 + HSCH_2\overset{\underset{\|}{O}}{C}ONa$$

The Chugaev reaction, or thermal decomposition of the S-substituted esters of the xanthates, gives olefins without rearrangement (25–26). For example:

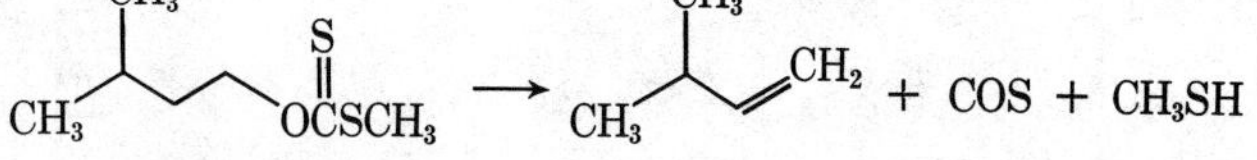

methyl isoamyl xanthate
[70061-61-9]

Esters derived from the primary alcohols are the most stable and those derived from the tertiary alcohols are the least stable. The decomposition temperature is lower in polar solvents, eg, dimethyl sulfoxide (DMSO), with decomposition occurring at 20°C for esters derived from the tertiary alcohols (27). Esters of benzyl xanthic acid yield stilbenes on heating, and those from neopentyl alcohols thermally rearrange to the corresponding dithiol esters (28–29). The dialkyl xanthate esters catalytically rearrange to the dithiol esters with conventional Lewis acids or trifluoroacetic acid (30–31).

Double bonds in or dialkylamino groups on the alkyl group of the S-methyl ester may facilitate isomerization to the dithiol ester (32). For example:

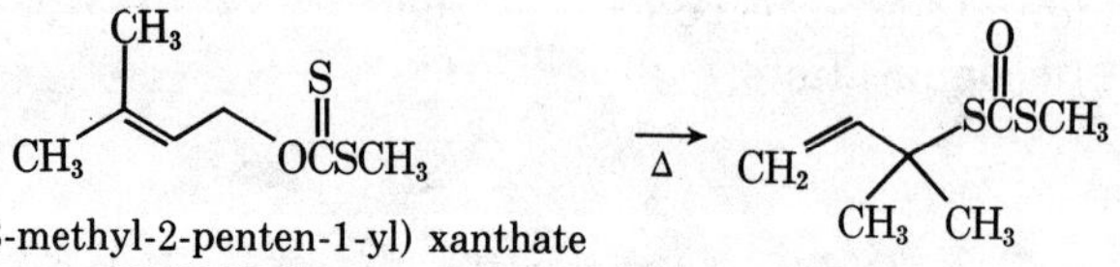

methyl (3-methyl-2-penten-1-yl) xanthate
[3817-83-2]

In a relatively low temperature procedure, olefins readily form from certain classes of xanthate esters (33):

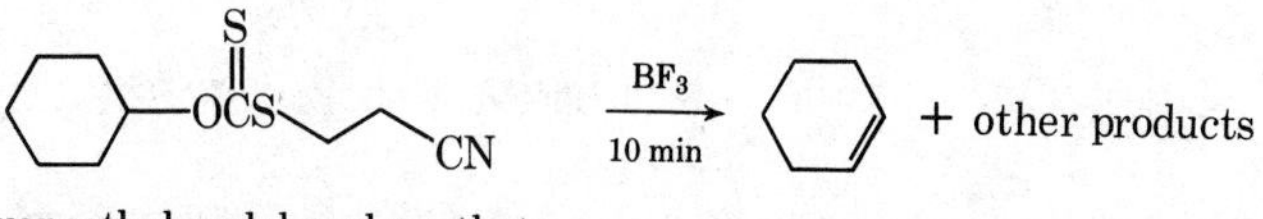

3-cyanoethyl cyclohexyl xanthate
[85909-63-3]

The reactions of the xanthate esters with some soft electrophiles proceed with good yields (34):

$$C_{18}H_{37}O\overset{S}{\overset{\|}{C}}SCH_3 + C_6H_5SCl \longrightarrow C_{18}H_{37}Cl + \left[CH_3\overset{O}{\overset{\|}{S}}CSSC_6H_5\right]$$

The reaction of phosgene with an aqueous solution of a xanthate and ether at 15–20°C gives the anhydrosulfide in good yield:

$$2\,RO\overset{S}{\overset{\|}{C}}SNa + COCl_2 \longrightarrow \left(RO\overset{S}{\overset{\|}{C}}\right)_2 S + COS + 2\,NaCl$$

The anhydrosulfide forms in the ether layer and is readily separated. The anhydrosulfides are useful in the preparation of thionocarbanilates, for example, by heating with an alcoholic solution of the arylamine:

$$\left(C_2H_5O\overset{S}{\overset{\|}{C}}\right)_2 S + C_6H_5NH_2 \longrightarrow C_2H_5O\overset{S}{\overset{\|}{C}}NHC_6H_5 + C_2H_5OH + CS_2$$

This method obviates the need to prepare the intermediate aryl isothiocyanate.

Many oxidizing agents, including sodium nitrite, convert the alkali-metal xanthates to the corresponding dixanthogen:

$$2\,RO\overset{S}{\overset{\|}{C}}SK + KI_3 \longrightarrow \left(RO\overset{S}{\overset{\|}{C}}\right)_2 S_2 + 3\,KI$$

$$2\,RO\overset{S}{\overset{\|}{C}}SK + K_2S_2O_8 \longrightarrow \left(RO\overset{S}{\overset{\|}{C}}\right)_2 S_2 + 2\,K_2SO_4$$

The reaction is generally carried out in water, and the resulting dixanthogen separates as a solid or oil. Copper salts also effect the oxidation:

$$4\,RO\overset{S}{\overset{\|}{C}}SNa + 2\,CuSO_4 \longrightarrow \left(RO\overset{S}{\overset{\|}{C}}\right)_2 S_2 + \left(RO\overset{S}{\overset{\|}{C}}S\right)_2 Cu_2 + 2\,Na_2SO_4$$

In the initial formation of the cupric xanthates, soluble xanthate complexes form prior to the precipitation of the cuprous xanthate with the concurrent formation of the dixanthogen (35). The dixanthogen can be separated by virtue of its solubility in ether. Older samples of alkali metal xanthates contain some dixanthogen, which is thought to form by the following reaction (24):

$$2\,C_2H_5O\overset{S}{\overset{\|}{C}}SK + 1/2\,O_2 + CO_2 \longrightarrow \left(C_2H_5O\overset{S}{\overset{\|}{C}}\right)_2 S_2 + K_2CO_3$$

The dixanthogens derived from the aryl xanthates are reported in ref. 36.

The sulfur chlorides give higher xanthate sulfides. For example, the following product is obtained from sulfur monochloride and an ether suspension at room temperature:

$$2\,RO\overset{S}{\overset{\|}{C}}SNa + S_2Cl_2 \longrightarrow \left(RO\overset{S}{\overset{\|}{C}}\right)_2 S_4 + 2\,NaCl$$

652 XANTHATES

Mixed anhydrosulfides can be readily obtained by adding an acetone solution of the xanthate to an acetone solution of the acyl halide at −35°C (37):

$$C_2H_5O\overset{S}{\underset{\|}{C}}SK + R\overset{O}{\underset{\|}{C}}Cl \longrightarrow C_2H_5O\overset{S}{\underset{\|}{C}}S\overset{O}{\underset{\|}{C}}R + KCl$$

The stability of these products varies (28). When R is methyl, the compound decomposes below room temperature, but when R is phenyl it is stable up to 40–45°C. Increasing the molecular weight of the alkyl group and introducing an electronegative group on the phenyl increases the stability. The aliphatic mixed anhydrosulfide can be decomposed by two different routes (37):

$$C_2H_5O\overset{S}{\underset{\|}{C}}S\overset{O}{\underset{\|}{C}}R \xrightarrow{\Delta} R\overset{O}{\underset{\|}{C}}OC_2H_5 + CS_2$$

$$\downarrow$$

$$C_2H_5O\overset{S}{\underset{\|}{C}}SR + CO$$

Thiol esters are obtained in 91% yield by the addition of an aroyl chloride to a slurry of an alkaline salt xanthate in acetone without cooling (38):

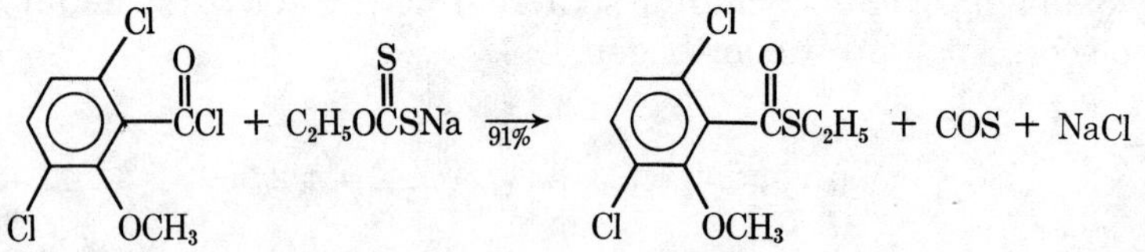

[2,6-dichloro-3-methoxybenzoyl chloride] + $C_2H_5O\overset{S}{\underset{\|}{C}}SNa \xrightarrow{91\%}$ [2,6-dichloro-3-methoxyphenyl thioester of C_2H_5] + COS + NaCl

The alkyl chloroformates react with cold ethereal dispersions of the xanthates to give the fairly stable xanthogen formates.

$$C_2H_5O\overset{S}{\underset{\|}{C}}SNa + C_2H_5O\overset{O}{\underset{\|}{C}}Cl \longrightarrow C_2H_5O\overset{S}{\underset{\|}{C}}S\overset{O}{\underset{\|}{C}}OC_2H_5 + NaCl$$

In the reaction between xanthates and sulfonyl chlorides, the xanthates convert to dixanthogens, and the sulfonyl chlorides reduce to sulfinic acids and other compounds (28):

$$2\,C_2H_5O\overset{O}{\underset{\|}{C}}SK + C_6H_5SO_2Cl \longrightarrow (C_2H_5O\overset{S}{\underset{\|}{C}})_2S_2 + C_6H_5SO_2K + KCl$$

The acid chlorides of the xanthic acids can be prepared by the reaction of chlorine with a dixanthogen (39):

$$(C_2H_5O\overset{S}{\underset{\|}{C}})_2S_2 + Cl_2 \longrightarrow 2\,C_2H_5O\overset{S}{\underset{\|}{C}}Cl + 2\,S$$

The S-alkoxycarbonylmethylxanthate esters treated with lithium diisopropylamide (LDA) and an aldehyde or ketone give excellent yields of α,β-unsaturated esters (40):

$$RO\overset{S}{\underset{\|}{C}}SCH_2\overset{O}{\underset{\|}{C}}OC_2H_5 + R'\overset{O}{\underset{\|}{C}}R'' \xrightarrow[\text{2. acid}]{\text{1. LDA}} \underset{\underset{R''}{|}}{R'C}=CH\overset{O}{\underset{\|}{C}}OC_2H_5$$

The Knoevenagel reaction is unsatisfactory for the direct preparation of these reaction products.

In the Leuckart thiophenol synthesis, the reaction of xanthates with diazonium compounds may be violent. The reaction can be controlled, however, by thermal decomposition of the intermediate azo compound as it forms; this provides a convenient and easy method of preparing the aryl esters, which can be saponified to the thiophenols. The crude reaction mixture is a more complex mixture than just an aryl alkyl xanthate (41):

$$\text{C}_2\text{H}_5\text{OCSK} + \text{C}_6\text{H}_5\text{N}_2\text{Cl} \longrightarrow \text{C}_2\text{H}_5\text{OCSC}_6\text{H}_5 + (\text{C}_6\text{H}_5\text{S})_2\text{CO} + \text{N}_2 + \text{KCl} + \text{other products}$$

Xanthates have been added to activated double bonds, eg, acrylic derivatives and α,β-unsaturated aldehydes and ketones (42–44):

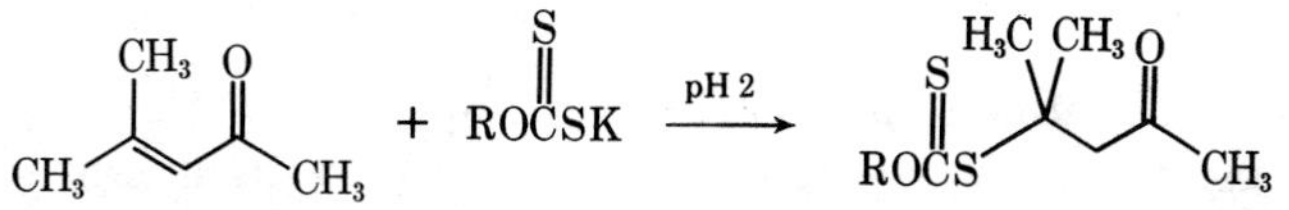

$$\underset{\text{CH}_3}{\overset{\text{CH}_3}{>}}\!\!\!\text{C}=\text{CH}-\overset{\text{O}}{\underset{\|}{\text{C}}}-\text{CH}_3 + \text{ROCSK} \xrightarrow{\text{pH 2}} \text{ROC(S)S}-\text{C}(\text{CH}_3)_2-\text{CH}(\text{OCH}_3)-\text{C}(\text{O})\text{CH}_3$$

The alkali-metal xanthates react with the lower alkylamines in the presence of catalytic amounts of nickel or palladium salts to give the dialkylthionocarbamates (45):

$$(\text{CH}_3)_2\text{CHOCSNa} + \text{C}_2\text{H}_5\text{NH}_2 \xrightarrow{\text{Ni}^{2+}} (\text{CH}_3)_2\text{CHOCNHC}_2\text{H}_5 + \text{NaSH}$$

Preparation and Manufacture

The alkali-metal xanthates are generally prepared from the reaction of sodium or potassium hydroxide with an alcohol and carbon disulfide. The initial reaction is the formation of the alkoxide, which reacts with carbon disulfide to give the xanthate:

$$\text{ROH} + \text{NaOH} \rightleftharpoons \text{RONa} + \text{H}_2\text{O}$$

$$\text{RONa} + \text{CS}_2 \longrightarrow \text{ROCSNa}$$

The overall heat of reaction for potassium isopropyl xanthate is 48.5 kJ/mol (11.6 kcal/mol) (46). The presence of water favors the principal side reaction:

$$6\,\text{NaOH} + 3\,\text{CS}_2 \rightarrow 2\,\text{Na}_2\text{CS}_3 + \text{Na}_2\text{CO}_3 + 3\,\text{H}_2\text{O}$$

When no water is present, very pure xanthates form (47). Ethanol reacts slowly with sodium trithiocarbonate to produce sodium ethyl xanthate (48):

$$\text{C}_2\text{H}_5\text{OH} + \text{NaSCSNa} \longrightarrow \text{C}_2\text{H}_5\text{OCSNa} + \text{NaHS}$$

The nature of the inorganic by-products in technical xanthates varies depending upon their exposure to air. Thus, Na_2S, Na_2SO_4, and $Na_2S_2O_3$ may be present in older samples exposed to air (49).

With tertiary and some of the more complex alcohols, the use of alkali hydroxides is not feasible, and it is necessary to use reagents such as sodium hydride, sodium amide, or the alkali metal to form the alkoxide:

$$2 \text{ROH} + 2 \text{K} \rightarrow 2 \text{ROK} + H_2$$

Powdered KOH in DMF (dimethylformamide) or DMSO reacts with carbon disulfide to give these xanthates.

Various xanthates can be readily prepared in the laboratory from alcohols (up to C_{16}), potassium hydroxide, and carbon disulfide (50).

Another and quite different procedure is the preparation of xanthates from ethers (51):

$$C_2H_5OC_2H_5 + 2 CS_2 + 2 NaOH \rightarrow 2 C_2H_5O\overset{\overset{S}{\|}}{C}SNa + H_2O$$

In a study of the influence of temperature (30–45°C) on the preparation of isopropyl xanthates, it was determined that increasing the temperature resulted in a decrease in the xanthate yield and an increase in by-products. Also, a decrease in the water content of the alcohol increases the xanthate yield (52).

Many of the heavy-metal xanthates have been prepared from aqueous solutions of the alkali-metal xanthates and the water-soluble compound of the heavy metal desired.

$$2 C_2H_5O\overset{\overset{S}{\|}}{C}SNa + ZnCl_2 \rightarrow (C_2H_5O\overset{\overset{S}{\|}}{C}S)_2Zn + 2 NaCl$$
zinc ethyl xanthate
[13435-48-8]

A modified and detailed procedure is given for the preparation of the corresponding Cr^{3+}, In^{3+}, and Co^{3+} xanthates in good yields and high purity (12).

Alkali-Metal Xanthates. The commercially available xanthates are prepared from various primary or secondary alcohols. The alkyl group varies from C_2 to C_5 and the alkali metal may be sodium or potassium. Not all of the commercially available alcohols in the C_2–C_6 range are available as their xanthates, but most could be made if there were sufficient demand for them. Except for a few foreign articles and the old reports of the U.S. Office of Technical Services, most information on the manufacture of xanthates is in the patent literature. Although the important patent on flotation uses of xanthates appeared in 1925, there are still articles and patents appearing on the manufacture of xanthates.

Steel (qv) generally is suitable as a material of construction for the apparatus used in the manufacture of xanthates. Cooling is required to minimize side reactions. The reaction temperature is generally maintained below 40°C. In one procedure, the reaction is carried out in an inert diluent, eg, petroleum ether (53–57). A plant in the FRG employed this method for the preparation of sodium amyl and sodium hexyl xanthate. The copper kettles were equipped with stirring and cooling apparatus. A petroleum ether slurry was made of equimolar quantities of powdered sodium hydroxide and the alcohol. Carbon disulfide was added and the temperature was maintained at ca 35–40°C. After three hours, the product was separated by filtration, washed with petroleum ether, and dried. The Dow Chemical xanthate plant, which is no longer in operation, in a similar manner used pentane as a diluent and the heat of reaction was dissipated by the refluxing of the pentane (bp 36°C), and this prevented overheating of the reaction mixture. The reactants and diluent were added to four steel reactors in series, which were equipped with stirrers and condensers. The product was fed from the last reactor to a spray dryer, then to a pelletizer, and finally to the drumming equipment. The spray dryer minimized the thermal decomposition of the product in the drying step. The volatile organic materials were recycled. The annual capacity of the plant was 13,600 metric tons of xanthates.

For the manufacturing of potassium ethyl xanthate, 400% excess of alcohol and equimolar quantities of 50 wt % aqueous potassium hydroxide were used (58). After 30 min at 40°C, the mixture was vacuum-drum-dried. The product was obtained in near quantitative yield and assayed at 95%. It is claimed that potassium amyl xanthate can be made with almost the same ratio of reactants and 60 wt % caustic potash (59).

Xanthates have been synthesized from C_6–C_8 oxo alcohols in a reactor with intensive stirring at 25°C. The water and unreacted alcohol are removed from the product at 40–50°C under vacuum to give a friable powder of 80% purity in 77% yield (60) (see Oxo process; Alcohols, higher aliphatic).

In another process, 50–150% excess carbon disulfide and a small excess of powdered alkali hydroxide are added with stirring and cooling to the lower alkyl alcohol. After completion of the reaction, the excess CS_2 and resulting water of reaction are removed by applying a vacuum to the reactor (61).

The most important hazard in the manufacturing of xanthates is the use of carbon disulfide (qv) because of its low flash point and ignition temperature and its toxicity. A report on the manufacture of sodium ethyl xanthate at Kennecott Nevada Mines Division has an excellent discussion of the various safety problems and the design of a facility. A plant layout and a description of the reagent preparations are also given (62).

One patent describes a continuous process involving an aqueous alkali-metal hydroxide, carbon disulfide, and an alcohol. The reported reaction time is 0.5–10 min before the mixture is fed to the dryer (63). The usual residence time is of the order of hours. A study in the USSR reports the use of the water–alcohol azeotrope for water removal from isobutyl or isoamyl alcohol and the appropriate alkali hydroxide to form the alkoxide prior to the addition of carbon disulfide (64).

Because of hydrate formation, the sodium salts tend to be difficult to dry. Excess water over that of hydration is believed to accelerate the decomposition of the xanthate salts. The effect of heat on the drying of sodium ethyl xanthate at 50°C has been studied (65):

Time of heating, h	Xanthate content, wt %
0	79.32
40	74.38
64	65.66
88	65.06
112	63.85
136	60.41

Economic Aspects

The two main producers of xanthates in the United States, ie, American Cyanamid and Dow, shut down their plants in the 1970s. American Cyanamid transferred its production to Canada. Xanthates in solution are available in some areas of the United States and Canada and are transported locally by truck. A few mines produce their own solution xanthates (62).

The U.S. consumption of xanthates for froth flotation was ca 3450 t/yr in 1965 and 2720 t/yr in 1980 (66–67). In another report, U.S. consumption of xanthates was ca 4900 t/yr in 1974 and 5900 t/yr in 1980 (68). Figures from the same source for noncommunist countries were 41,730 t/yr in 1974 and 47,180 t/yr in 1980. The consumption of xanthates just for the copper industry in the United States was ca 4850 t/yr for 1978 (69).

Standards

The alkali-metal xanthates are obtainable in the technical grade only, and they vary in particle size from powder to pellets. The color is any of various shades of yellow and the odor is unpleasant. There are no published specifications for the xanthates, but each manufacturer has its own standards. The assay on the commercial materials is usually ca 90–95%. The assay does not necessarily indicate the collector ability of a xanthate (see Flotation). Although all xanthates decompose to some extent on prolonged storage, such decomposition does not necessarily result in a corresponding reduction in mineral-collecting power. This lack of correlation between xanthate content and flotation-collecting ability has been noted in products produced by different methods. Apparently, some of the xanthate oxidation or decomposition products are effective as flotation agents.

The material is shipped in steel drums. The DOT regulations for the xanthates is the hazardous material description ORM-A (other regulated material—regulated by air), NOS.

Analytical Methods

Although there are several published analytical procedures, many are either tedious or inaccurate. For example, when assayed by the standard KI_3 method, the copper sulfate method, or the aqueous acid method, certain of the inorganic impurities assay as xanthate, thereby giving high results (70–71). When specifications are drawn up for xanthate content, the method of assay should be given.

In a satisfactory procedure developed by The Dow Chemical Company, the xanthate content is determined by dissolving the sample in acetone, removing the inorganic solids by filtration, decomposing the xanthate with an excess of standard acid, and back-titrating with a standard base. The components of the inorganic solid left on the filter can be determined by any standard procedure.

The water content can be determined by the Karl Fischer method (72). The nonxanthate organic material can be determined by extraction with ether. The uv absorption spectra of xanthates have been extensively used for the assay of dilute solutions of the latter (18,73). The interference of trithiocarbonates can be avoided through the formation of a nickel xanthate. The latter is extracted into heptane and the absorbance spectrum (425 nm) is measured (74).

The successful separation of xanthate-related compounds by high pressure liquid chromatography (hplc) methods have been reported (75–76). The thin-layer chromatography procedure has been used to determine the nature of the alcohols in a xanthate mixture. A short run of 3 cm at a development time of 25 min gives a complete separation of C_1–C_5 alkanol xanthates (77).

The analysis of solutions of technical xanthates by Ag^+ potentiometric titration, with the addition of ammonium hydroxide, has been successfully used at Dow (78).

Health and Safety Factors, Toxicology

Most of the published data on the toxicity of the xanthates is given in the foreign literature. The xanthates are low in acute and oral toxicity, as indicated in Table 4. Potassium amyl xanthate causes extensive pain and slight corneal injury to the eye and may burn the skin on prolonged contact. In a chronic toxicity study of this xanthate, an aqueous aerosol was applied to dogs, rabbits, rats, and mice. There were no adverse effects in the latter three at 23 mg/m^3, but liver damage occurred in dogs exposed to this level. A no-effect level was not determined. The recommended airborne industrial hygiene guide (IHG) is 1 mg/m^3 (83).

The alkali-metal xanthates are fairly safe to handle. The standard precautions of rubber gloves, dust mask, and goggles are sufficient when handling the solid or the solution. If xanthates contact the body, they may produce dermatitis, although susceptibility varies from person to person. Contaminated clothing should be laundered before being worn again. Internally, whether in the form of dust or fumes, the xanthates act similarly to carbon disulfide. The possible formation of carbon disulfide in xanthate solutions requires that proper care be taken in the preparation and handling of these solutions to avoid possible fires and explosions (11).

Under regulations for the enforcement of the Federal Insecticide, Fungicide, and

Table 4. Oral Toxicity of Xanthates

Xanthate	Species	LD$_0$, mg/kg	LD$_{50}$, mg/kg	Refs.
sodium ethyl	rat	500		79
potassium ethyl	rat	500	1700	79–80
	mouse		583	80
sodium isopropyl	rat	250		79
potassium isopropyl	rat		1700	80
	mouse		583	80
potassium n-butyl	mouse		411, 465	81–82
sodium isobutyl	rat	500		79
potassium isobutyl	rat		1290	80
	mouse		480	80
sodium sec-butyl	rat		>2000	79
potassium amyl (mixed)	rat	1000	1000–2000	79, 83
potassium isoamyl	rat		765	80
	mouse		470	80
C_5–C_6 mixture	rat		1500	84

Rodenticide Act, products containing over 50 wt % sodium isopropyl xanthate must bear the label "Caution. Irritating dust. Avoid breathing dust, avoid contact with skin and eyes" (85). Rubber goods in repeated contact with food may contain diethyl xanthogen disulfide not to exceed 5 wt % of the rubber products (86).

Xanthate drums should be kept as cool and dry as possible. Protection from moisture is the most important factor. A combination of moisture and hot weather causes sodium ethyl xanthate to ignite spontaneously (11).

Environmental Concerns. Concern for the well-being of the environment has resulted in studies on the effects of mining chemicals, including the xanthates, on various aquatic organisms worldwide (87–90). In a thorough and detailed study of three typical mill operations, it was concluded that residual organic flotation reagents do not seem to present widespread problems in effluent disposal. A large portion of the reagents used never reached the final discharge. Biodegradation appears to be the commonest method of reagent removal. At levels below 20–25 mg/L, the xanthates are biodegradable (91).

Uses

Outside of the importance of cellulose xanthates in the manufacture of rayon and cellophane, the primary use for the alkali metal xanthates is as collectors in the flotation of metallic sulfide ores. The xanthates are the principal metal-sulfide collectors. A great deal of experimental work on the production methods and potential uses of the starch xanthates has been reported (92). Allyl amyl xanthates are still useful reagents in the flotation of copper–molybdenum sulfide ores. Other uses of the xanthates are very minor. This is not so much the result of the lack of activity or performance of these compounds as it is the fact that superior products are available. Xanthates and derivatives have been recommended for the vulcanization of rubber, as herbicides (qv), insecticides, fungicides, high pressure lubricant additives, and for analytical procedures (see Insect control technology; Fungicides; Lubrication and lubricants). In addition to a continued increase in the number of use patents in the preceding fields, a new use of xanthates as inhibitors of fertilizer nitrogen transformation in soil has been reported as well as the use of certain metal xanthates as color developers for image-recording materials (93–94) (see Fertilizers; Color photography). For several years, sodium isopropyl xanthate was used as an intermediate in the manufacture of saccharin (see Sweeteners).

Derivatives

The principal derivatives are of three types. The dixanthogens,

$$(ROC)_2S_2,$$
$$\overset{S}{\underset{\|}{}}$$

were produced at one time as additives for polymers but are apparently no longer available. The mixed anhydrosulfides, ie, the xanthogen formates, are used in acid pulp flotation of sulfide ores. The most important one, diethyl xanthogenformate, has been used in Chile for over 50 yr. Dialkyl thionocarbamates are used in the flotation of copper and zinc sulfides, in particular where the rejection of iron pyrites is important.

The two principal products are isopropyl ethylthionocarbamate and butyl ethylthionocarbamate.

BIBLIOGRAPHY

"Xanthic Acids and Xanthates" in *ECT* 1st ed., Vol. 15, pp. 150–157, by G. H. Harris, The Dow Chemical Company; "Xanthates" in *ECT* 2nd ed., Vol. 22, pp. 419–429, by G. H. Harris, The Dow Chemical Company.

1. A. F. McKay, D. L. Garmaise, G. Y. Paris, S. Gelblum, and R. J. Ranz, *Can. J. Chem.* **38,** 2042 (1960).
2. I. Iwasaki and S. R. B. Cooke, *J. Phys. Chem.* **63,** 1321 (1959).
3. V. Hejl and F. Pechar, *Chem. Zvesti* **21,** 261 (1967).
4. A. P. Gavrish, *Ukr. Khim. Zh.* **29,** 900 (1963).
5. A. P. Sanzharova, *Ukr. Khim. Zh.* **35,** 91 (1969).
6. M. C. Fuerstenau, South Dakota School of Mines and Technology, Rapid City, S.D., private communication, April 1982.
7. I. Tyden, *Talanta* **13,** 1353 (1966).
8. I. S. Shupe, *J. Assoc. Offic. Agric. Chem.* **25,** 495 (1942).
9. K. G. Rutherford, R. M. Ottenbrite, and B. K. Tang, *J. Chem. Soc. (C),* 582 (1971).
10. I. Yu. Keskyula, S. B. Faerman, Ch. I. Kondrat'ev, E. L. Goncharova, R. M. Sorokina, and K. K. Chevychalova, *Trans. State Inst. Appl. Chem. (USSR)* **30,** 68 (1936).
11. *Xanthate Handbook*, American Cyanamid Company, Mining Chemicals Department, Wayne, N.J., 1972.
12. F. Galsbol and C. E. Schaffer, *Inorg. Synth.* **10,** 42 (1967).
13. E. M. Donaldson, *Talanta* **23,** 411, 417 (1976).
14. L. Malatesta, *Chim. Ind. (Milan)* **23,** 319 (1941).
15. A. T. Pilipenko, *Zh. Anal. Khim.* **4,** 227 (1949).
16. I. A. Kakovskii, *Proceedings of the International Congress on Surface Activity*, 2nd, London **4,** 225 (1957).
17. C. C. DeWitt, R. F. Makens, and A. W. Helz, *J. Am. Chem. Soc.* **57,** 796 (1935).
18. J. Dyer and L. H. Phifer, *Macromolecules*, 111 (1969).
19. R. J. Millican and C. K. Sauers, *J. Org. Chem.* **44,** 1964 (1979).
20. C. A. Bunton, P. Ng, and L. Sepulvada, *J. Org. Chem.* **39,** 1130 (1974).
21. P. J. C. Fierens, J. Adam, and E. Royers, *Ind. Chim. Belge Suppl. 1959* **2,** 777 (1959).
22. N. P. Finkelstein, *Colloid Polym. Sci.* **255,** 168 (1977).
23. I. Iwasaki and S. R. B. Cooke, *J. Am. Chem. Soc.* **80,** 285 (1958).
24. A. Pomianowski and J. Leja, *Can. J. Chem.* **41,** 2219 (1963).
25. G. L. O'Connor and H. R. Nace, *J. Am. Chem. Soc.* **74,** 5454 (1952); **75,** 2118 (1953).
26. H. R. Nace in A. C. Cope, ed., *Organic Reactions*, Vol. 12, John Wiley & Sons, Inc., New York, 1962.
27. P. Meurling, K. Sjoeberg, and B. Sjoeberg, *Acta Chim. Scand.* **26,** 279 (1972).
28. G. Bulmer and F. G. Mann, *J. Chem. Soc.*, 666, 677, 680 (1945).
29. K. G. Rutherford, B. K. Tang, L. K. M. Lam, and D. P. C. Fung, *Can. J. Chem.* **50,** 3288 (1972).
30. K. Komaki, T. Kawata, K. Harano, and T. Taguchi, *Chem. Pharm. Bull.* **26,** 3807 (1978).
31. M. W. Fichtner and N. F. Haley, *J. Org. Chem.* **46,** 3141 (1981).
32. T. Taguchi, Y. Kawazoe, K. Yoshihira, H. Kanayama, M. Mori, K. Tabata, and K. Harano, *Tetrahedron Lett.*, 2717 (1965).
33. *Chem. Eng. News* **47,** 41 (Sept. 22, 1969).
34. D. H. R. Barton, R. V. Stick, and R. Subramanian, *J. Chem. Soc. Perkin Trans. 1*, 2112 (1976).
35. G. Sparrow, A. Pomianowski, and J. Leja, *Sep. Sci.* **12,** 87 (1977).
36. H. W. Chen, J. P. Fackler, Jr., D. P. Schussler, and L. D. Thompson, *J. Am. Chem. Soc.* **100,** 2370 (1978).
37. D. H. R. Barton, M. V. George, and M. Tomoeda, *J. Chem. Soc.*, 1967 (1962).
38. J. J. D'amico, T. Shafer, and D. Jacobson, *Phosphorus Sulfur* **7,** 301 (1979).
39. D. Martin and W. Mucke, *Chem. Ber.* **98,** 2059 (1965).
40. K. Tanaka, R. Tanikaga, and A. Kaji, *Bull. Chem. Soc. Jpn.* **52,** 3619 (1979).
41. J. R. Cox, Jr., C. L. Gladys, L. Field, and D. E. Pearson, *J. Org. Chem.* **25,** 1083 (1960).

660 XANTHATES

42. O. Bayer, *Angew. Chem.* **61**, 229 (1949).
43. J. L. Garraway, *J. Chem. Soc.*, 4072 (1962).
44. S. V. Tsarenko, P. P. Gnatyuk, V. A. Ignatov, and V. A. Malii, *Izv. Vyssh. Uchebn. Zaved. Khim. Khim. Tekhnol.* **23**, 836 (1980).
45. U.S. Pat. 3,975,264 (Aug. 17, 1976), F. A. Booth, R. D. Crozier, and L. E. Strow (to Minerec Corporation).
46. V. P. Savel'yanov, I. P. Utrobin, R. T. Savel'yanova, and N. N. Evseev, *Russ. J. Phys. Chem.* **54**, 720 (1980).
47. U.S. Pat. 2,024,923 (Dec. 17, 1935), W. Hirschkind (to Great Western Electro-Chemical Co.).
48. G. Ingram and B. A. Toms, *J. Chem. Soc.*, 4328 (1957).
49. S. N. Danilov, N. M. Grad, and E. I. Geine, *Zh. Obshch. Khim.* **19**, 826 (1949).
50. L. S. Foster, *Technical Paper No. 2*, Department of Mining and Metallurgy Research, University of Utah, Salt Lake City, Utah, 1928.
51. U.S. Pat. 2,534,085 (Dec. 12, 1950), B. M. Vanderbilt and J. P. Thorn (to Standard Oil Development Co.).
52. T. T. Savel'yanova, I. P. Utrobin, N. N. Zaitseva, and U. P. Savel'yanova, *J. Appl. Chem. USSR* **53**, 1855 (1980).
53. W. Hensinger, *U.S. Office of Technical Services Report PB 74736*, U.S. Department of Commerce, Washington, D.C., 1934–1936.
54. U.S. Pat. 1,559,504 (Oct. 27, 1925), R. B. Crowell and G. F. Breckenridge (to Western Industries Co.).
55. U.S. Pat. 2,107,065 (Feb. 1, 1938), A. J. Van Peski (to Western Industries Co.).
56. Ger. Pat. 1,934,175 (Jan. 14, 1971), K. Baessler and G. Folz (to Farberwerke Hoechst A.G.).
57. USSR Pat. 727,641 (April 15, 1980), V. P. Savel'yanova, R. T. Savel'yanova, N. P. Utrobin, and T. V. Rudakova (to D. I. Mendeleev Chem-Tech. Inst.).
58. W. A. M. Edwards and J. H. Clayton, *U.S. Office of Technical Services Report PB 34028*, U.S. Department of Commerce, Washington, D.C., 1945, p. 28.
59. Ger. Pat. 765,196 (July 28, 1955), H. Gunther (to Deutsche Gold- und Silber-Scheideanstalt vorm Roessler).
60. G. A. Yefimova, Ve. I. Silina, and V. I. Ryaboy, *Sov. Chem. Ind.* **47**, 550 (1971).
61. Ger. Offen. 2,149,726 (April 6, 1972), G. J. Novak and A. J. Robertson (to American Cyanamid Company).
62. J. J. Harrington and P. B. Valenti, *1976 Fall Meeting, American Institute of Mining Engineers Paper No. 76B-358*, Engineering Societies Library, New York.
63. Czech. Pat. 97,929 (Jan. 15, 1961), O. Leminger, S. Novak, J. Chalupa, Z. Kubek, L. Kraus, F. Blechta, M. Morak, and J. Malek.
64. N. I. Gel'perin, E. M. Idel'son, A. K. Livshits, A. T. Borisenko, L. I. Gabrielova, and V. I. Zil'berg, *Sb. Nauchn. Tr. Gos. Nauchn. Issled. Inst. Tsvetn. Met.* 170 (1959); *Chem. Abstr.* **57**, 12313 (1962).
65. U.S. Pat. 1,724,549 (Aug. 13, 1929), T. W. Bartram and W. C. Weltman (to Rubber Service Laboratories Co.).
66. *Minerals Yearbook*, Vol. 1, U.S. Department of the Interior, Bureau of Mines, U.S. Government Printing Office, Washington, D.C., 1965, p. 75.
67. *Ibid.*, p. 34.
68. *Report on Ore Treatment and Extraction Metallurgy*, Roger Williams Technical and Economic Services (UK) Inc., London, 1974.
69. *Mining Industry Chemical Opportunities, U.S. and Canada, a Marketing-Technical Study*, Hochberg and Co., Inc., Chester, N.J.
70. W. S. Calcott, F. L. English, and F. B. Downing, *Eng. Min. J. Press* **118**, 980 (1924).
71. W. Hirschkind, *Eng. Min. J. Press* **119**, 968 (1925).
72. A. L. Linch, *Anal. Chem.* **23**, 293 (1951).
73. R. Joedodibroto, Ph.D. thesis, Syracuse University, Syracuse, N.Y., 1963.
74. D. Kyriacou, The Dow Chemical Co., private communication, 1969.
75. J. G. Eckhardt, K. Stetzenbach, M. F. Burke, and J. L. Moyers, *J. Chromatogr. Sci.* **16**, 510 (1978).
76. R. T. Honeyman and R. R. Schrieke, *Anal. Chim. Acta* **116**, 345 (1980).
77. A. Messina and D. Corradini, *J. Chromatogr.* **207**, 152 (1981).
78. H. O. Kerlinger, The Dow Chemical Co., private communication (1982).
79. The Dow Chemical Company, 1964, unpublished toxicological data to C. B. Shaffer, American Cyanamid; data from P. Avotin, American Cyanamid, private communication, 1982.

80. E. M. Trofinovich, S. M. Rykova, M. A. Molchanova, and L. B. Aleksandrovskaya, *Gig. Sanit.* (6), 95 (1976).
81. E. A. Babayan, *Materialy 2-oi[Vtoroi] Itog. Nauchn. Konf. Inst. Gigieny Truda i Prof. Zabolevan. Posvyashch. Vopr. Gigieny Truda i Prof. Patol. Erevan, Sb. 1963*, 75 (1964); *Chem. Abstr.* **64,** 8836e (1966).
82. E. A. Babayan, *Mater. Itogovoi Nauch. Konf. Vop. Gig. Tr. Profpatol. Khim. Gornorud. Prom.*, 3d 1966, 97 (1968); *Chem. Abstr.* **73,** 129111y (1970).
83. N. G. Fronk, The Dow Chemical Company, private communication, 1982.
84. A. Z. Buzina, A. I. Burkhanov, and Kh. B. Abeuev, *Zdravookhr. Kaz.*, 88 (1977).
85. *Fed. Regist.* **27,** 2267 (March 9, 1962).
86. *Fed. Regist.* **33,** 8338 (June 5, 1968).
87. M. C. Fuerstenau, *Trans. Soc. Min. Eng. AIME* **256,** 337 (1974).
88. G. Leduc, D. G. Dixon, and co-workers, *Arctic Land Use Research Report*, ALUR, 74-75-34, Ottawa, Canada, 1976.
89. A. Solak, A. Wernikowska-Ukleja, and B. Multan, *Cuprum* **6**(2), 32 (1979); *Chem. Abstr.* **91,** 205213x (1979).
90. M. V. Mosevich, N. M. Arshanitsa, and co-workers, *Izv. Gos. Nauchn. Issled. Inst. Ozern. Rechn. Rybn. Khoz.* **109,** 80 (1976).
91. A. D. Read and R. M. Manser, *Proc. Int. Miner. Process. Cong.*, 11th, 1323 (1975).
92. Numerous patents and publications of the Northern Regional Research Center, Peoria, Ill.
93. J. Ashworth, G. A. Rodgers, and G. G. Briggs, *Chem. Ind. (London)*, 90 (1979).
94. Jpn. Kokai Tokyo Koho 79 10,009 (Jan. 25, 1979), M. Satomura and K. Sato (to Fuji Photo Film Co., Ltd.).

General References

Reference 11 is also a general reference.
G. Gattow and W. R. Bahrendt, *Topics in Sulfur Chemistry, Vol. 2: Carbon Sulfides and Their Inorganic and Complex Chemistry*, Georg Thieme, Stuttgart, 1977.
J. Leja, *Surface Chemistry of Froth Flotation*, Plenum Press, New York, 1982.
S. R. Rao, *Xanthates and Related Compounds*, Marcel Dekker, Inc., New York, 1971.

GUY H. HARRIS
Consultant

XANTHENE DYES

Xanthene dyes are those containing the xanthylium (1) or dibenzo-γ-pyran nucleus as the chromophore with amino or hydroxy groups meta to the oxygen as the usual auxochromes.

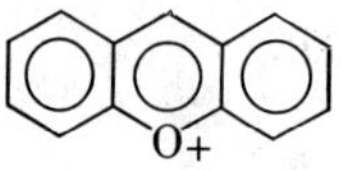

(1a) xanthylium [261-23-4]

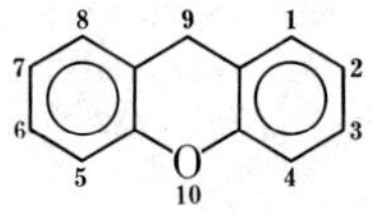

(1b) xanthene [92-83-1]

They have brilliant hues in the shade range of greenish-yellows to dark violets and blues, and they exhibit fluorescence but usually with inferior lightfastness compared with other chromophoric systems. They are used for the direct dyeing of wool and silk and mordant dyeing of cotton. Paper, leather, woods, food, drugs, and cosmetics are dyed with xanthenes (see Dyes, application and evaluation—application). Brilliant insoluble lakes are used in paints and varnishes.

Xanthenes date from 1871 when von Baeyer synthesized fluorescein by the condensation of two moles of resorcinol with one mole of phthalic anhydride in the presence of concentrated sulfuric acid (1).

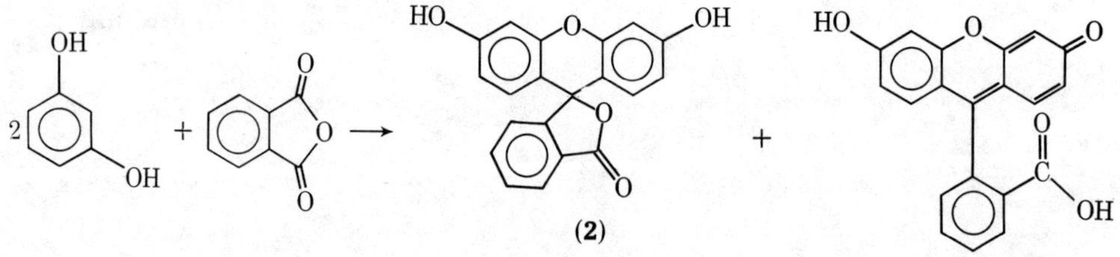

(2) Fluorescein A [2321-07-5]

(3) Fluorescein B [56503-30-1]

Depending upon the reaction conditions, the product can be isolated in either the lactoid form A (2) or the quinonoid form B (3). These 9-phenylxanthenes are closely related structurally to the triphenylmethane dyes (qv) and, like them, are cationic resonance hybrids:

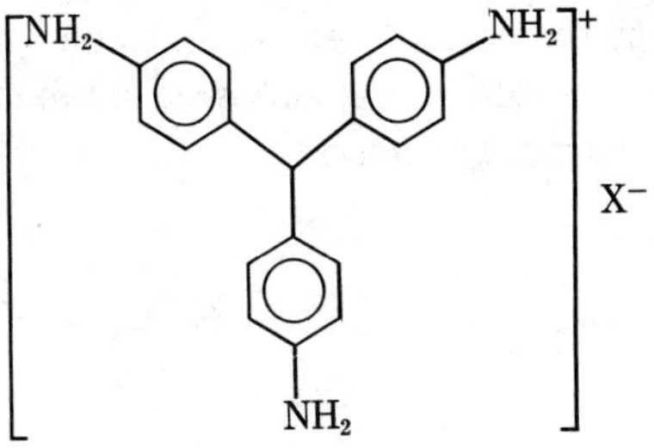

(4) pararosaniline [569-61-9] (triphenylmethane)

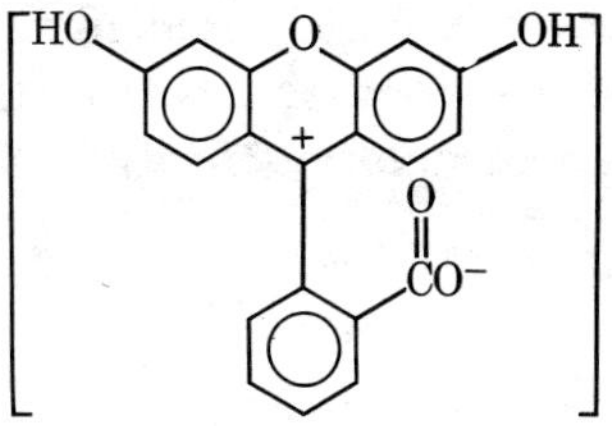

(5) fluorescein [2321-07-5] (hydroxyxanthene)

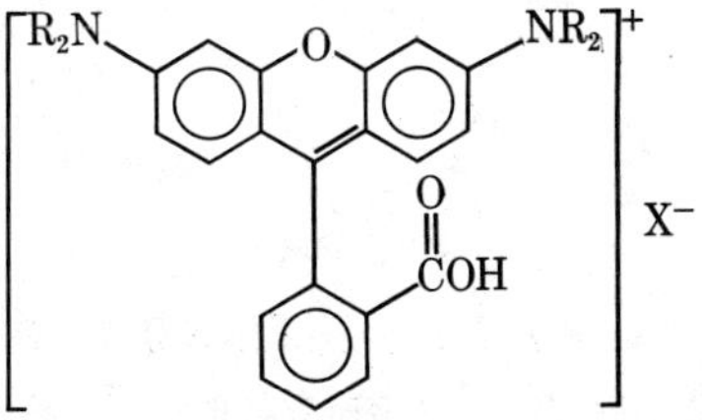

(6) a rhodamine (an aminoxanthene)

The main contributing resonance forms include the oxonium, carbonium, and ammonium structures:

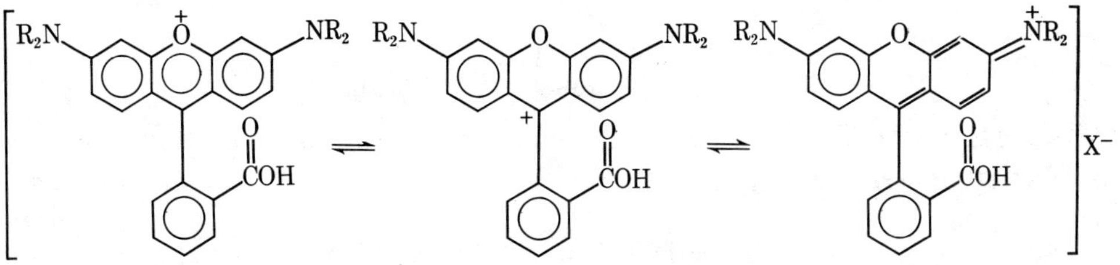

(7) oxonium ion   (8) carbonium ion   (9) ammonium ion

For uniformity with the structures given in the *Colour Index*, the ammonium radical is used for amino-substituted xanthenes and the keto form for hydroxy derivatives. This hybridization stabilization, including the positive charge on oxygen, probably accounts for the fluorescent properties of most xanthene dyes. As in other chromophoric systems, those dyes exhibiting fluorescence also exhibit notoriously poor lightfastness.

Xanthene dyes are classified into three groups according to the nature of the aromatic substitution: amino derivatives, hydroxy derivatives, and aminohydroxy derivatives.

Amino Derivatives

Pyronines. Pyronines are diphenylmethane derivatives synthesized by the condensation of m-dialkylaminophenols with formaldehyde, followed by oxidation of the xanthene derivative (**10**) to the corresponding xanthydrol (**11**) which, in the presence of acid, forms the dye (**12**).

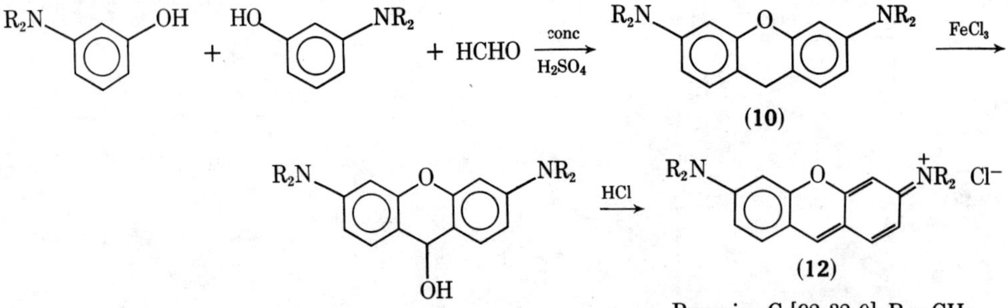

Pyronine G [92-32-0], R = CH₃
Pyronine B [2150-48-3], R = C₂H₅

If R is methyl, the dye produced is Pyronine G (CI 45005); if R is ethyl, Pyronine B (CI 45010) is obtained. If Pyronine G is oxidized with potassium permanganate, two methyl groups are eliminated with the formation of Acridine Red 3B (CI 45000) (**13**).

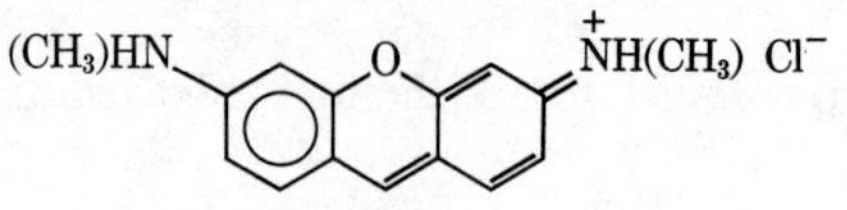

Acridine Red 3B
[2465-29-4]

Pyronine B and Acridine Red 3B are used as biological stains and are notable for their application in wet staining for the direct microscopic observations of living cells (2).

Succineins. If succinic anhydride instead of formaldehyde is used in the above reaction as the condensing agent with two moles of m-dialkylaminophenol, a succinein dyestuff forms. When m-dimethylaminophenol fuses with succinic anhydride and is isolated as the zinc double chloride, Basic Red 11 (CI 45050) (**14**) is obtained.

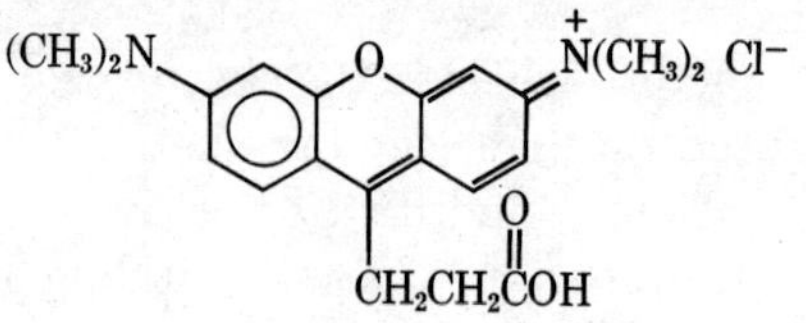

Basic Red 11 [72968-14-0]

Rhodamines. The rhodamines are economically the most important amino-substituted xanthene dyes. The total sales of Rhodamine B dyes in the United States in 1980 were over 10^7. If phthalic anhydride is used in place of formaldehyde in the above condensation reaction with *m*-dialkylaminophenol, a triphenylmethane analogue, 9-phenylxanthene, is produced. Historically, these have been called rhodamines. Rhodamine B (Basic Violet 10, CI 45170) (**15**) is usually manufactured by the condensation of two moles of *m*-diethylaminophenol with phthalic anhydride (3). An alternative route is the reaction of diethylamine with fluorescein dichloride [*630-88-6*] (3,6-dichlorofluoran) (**16**) under pressure:

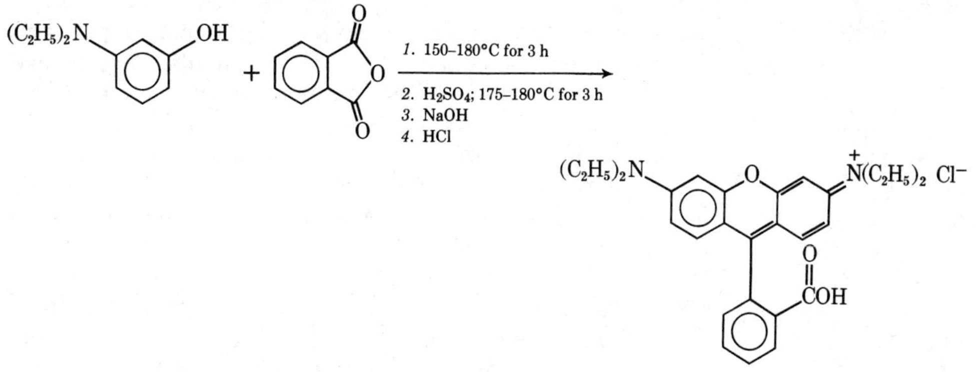
(C2H5)2N
OH
+
1. 150–180°C for 3 h
2. H2SO4; 175–180°C for 3 h
3. NaOH
4. HCl
(C2H5)2N
N(C2H5)2 Cl−
COH

(**15**)

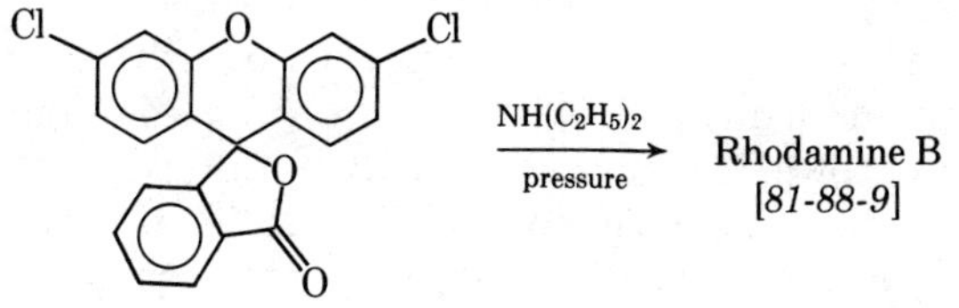
Cl
Cl
NH(C2H5)2
pressure
Rhodamine B
[81-88-9]

(**16**)

The free base of compound (**15**) is Rhodamine B Base [*509-34-2*] (Solvent Red 49; CI 45170:1). The phosphotungstomolybdic acid salt of (**15**) is Pigment Violet 1 [*1326-03-0*] (CI 45170:2). Pigment Red 173 [*12227-77-9*] (CI 45170:3) is the corresponding aluminum salt.

Esterification of the carboxyl groups also yields commercially useful dyes. If Rhodamine B is esterified with ethyl chloride or ethanol at 160–170°C under pressure, Basic Violet 11 (CI 45175) (**17**) forms.

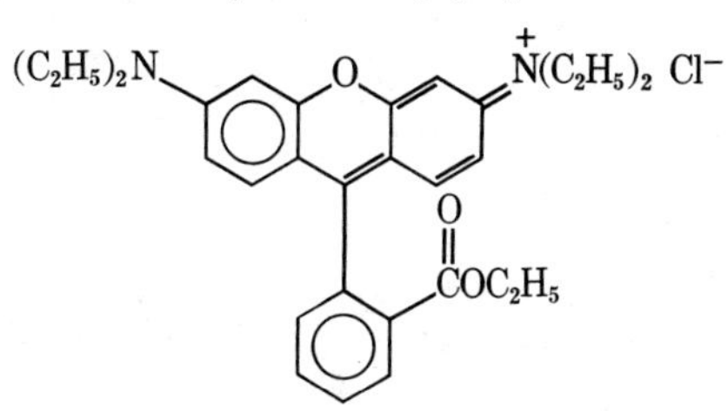
(C2H5)2N
N(C2H5)2 Cl−
COC2H5

(**17**)

Basic Violet 11
[*2390-63-8*]

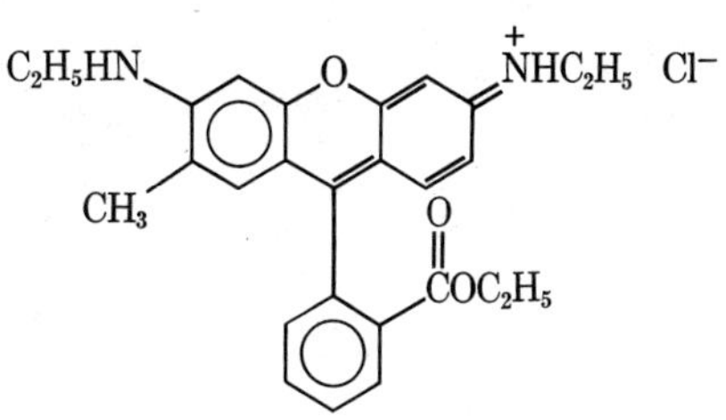
C2H5HN
NHC2H5 Cl−
CH3
COC2H5

(**18**)

Rhodamine 6G
[*989-38-8*]

Another commercially important esterified aminoxanthene is Rhodamine 6G (Basic Red 1; CI 45160) (**18**). This is manufactured by condensing 3-ethylamino-*p*-cresol with phthalic anhydride, then esterifying the product with ethanol and a mineral acid. The phosphotungstomolybdic acid salt of Rhodamine 6G is Pigment Red 81 [*12224-98-5*]

(CI 45160:1). The copper ferrocyanide complex of structure (**18**) is Pigment Red 169 [*12237-63-7*] (CI 45160:2). The 1980 sales for the Rhodamine 6G dyes in the United States were over 6×10^6. Recent manufacturing technology for the rhodamines has involved the preparation of stable, highly concentrated liquid forms (4). These are prepared by the reaction of the Rhodamine base with a dialkyl sulfate and a saturated aliphatic glycol at 100–160°C (5). These solutions are particularly suited for dyeing paper.

The rhodamines described thus far are basic rhodamines. They are used primarily for the dyeing of paper and the preparation of lakes for use as pigments. They are also used in the dyeing of silk and wool where brilliant shades with fluorescent effects are required, but where lightfastness is unimportant. Recently, many new uses for rhodamine dyes have been reported. For example, when vacuum-sublimed onto a video disk, Rhodamine B loses its color to form a clear stable film which becomes permanently colored upon exposure to uv light (6). This can be used in optical recording for computer storage or video recording (7). Fluorescent coloration to rigid or nonplasticized poly(vinyl chloride) (PVC) by the addition of selected rhodamines to the PVC resin before sheet or film formation has been reported (8). The addition of rhodamines to liquefied bis(hydroxyalkyl)aromatic dicarboxylic acid esters prior to their condensation to form polyesters to produce tinted polymer sheets is also used (9). Rhodamine 6G can be used in jet ink printing where fluorescence under ultraviolet radiation is a desired property (10). Selected xanthenes, including fluorescein, Rhodamine B, and Rhodamine 6G have been used as laser dyers (11).

Acid rhodamines are made by the introduction of the sulfonic acid group to the aminoxanthene base. However, these are usually not prepared by sulfonating basic rhodamines. The preferred route is the condensation of 3,6-dichlorofluoran (fluorescein dichloride) (**16**) with a primary aromatic amine in the presence of zinc chloride and quicklime. 3,6-Dichlorofluoran is made from the reaction of fluorescein (**2**) with phosphorus pentachloride. This product is then sulfonated. For example, if compound (**16**) is condensed with aniline and the product is sulfonated, Acid Violet 30 (CI 45186) (**19**) is produced.

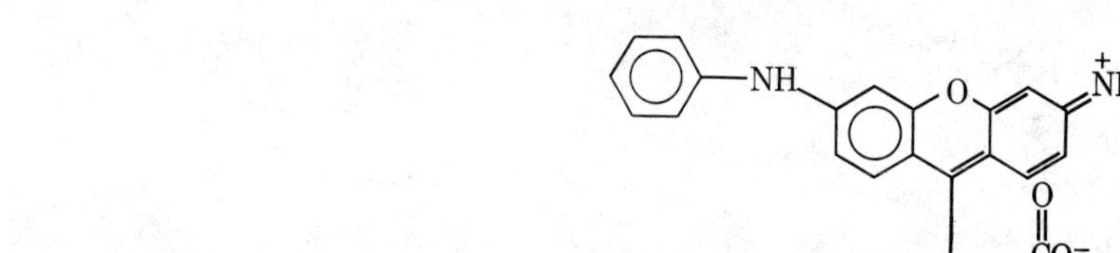

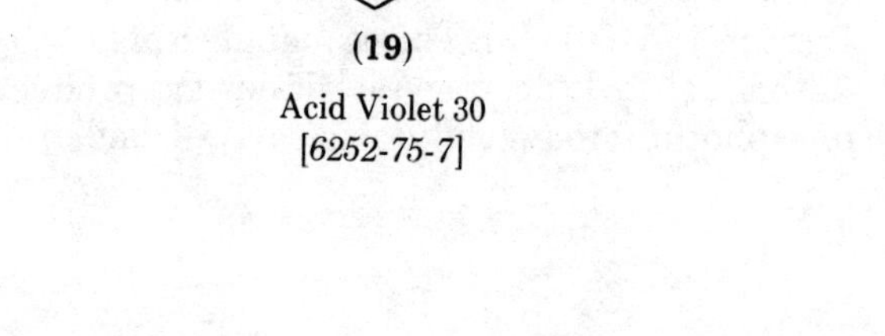

These acid rhodamines are usually used for dyeing silk and wool because they have level dyeing properties and show good fastness to alkali; however, they have poor lightfastness. Recent technology claims an improved process for manufacturing 3,6-diamino-substituted xanthenes by which the inner salts of 3,6-dihalo-9(2-sulfophenyl)xanthene-9-ols react with a primary or secondary amine in stoichiometric amounts in the presence of an inorganic acid-binding agent or acid-binding tertiary aliphatic or tertiary nitrogen-containing heterocyclic amine (12).

New, highly substituted acid rhodamines have been reported for fiber-reactive dye applications. For example, if 3,6-dichloro-9-(2-sulfophenyl)xanthene is first condensed with 1,4-phenylenediamine-3-sulfonic acid and then with N-methyltaurine and then acylated with cyanuric chloride, compound (20) is produced. This sulfurein derivative (20) dyes cellulosic fibers blue; the shades have good washfastness and improved lightfastness (13). Another route to fiber-reactive xanthenes is exemplified by the condensation of 4-nitro-2-sulfobenzaldehyde and 3-N-ethylamino-4-methylphenol. The product is reduced and then reacts with an acylating agent formed by the reaction of 1-aminobenzene-3,5-disulfonic acid and cyanuric chloride to produce compound (21), which gives brilliant red shades with good washfastness and moderate fastness to light (14).

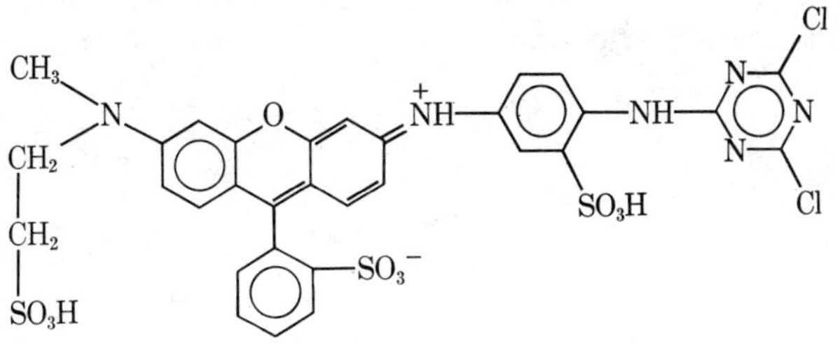

(20)
a sulfurein derivative

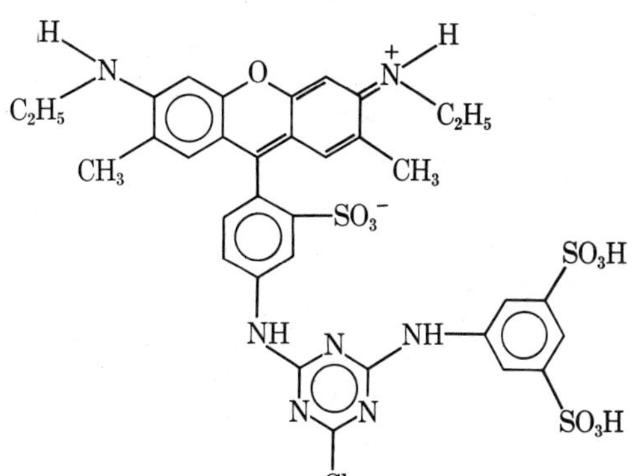

(21)
a sulfurein derivative

Reactive xanthene dyes with β-hydroxyethylsulfonyl groups, as exemplified by structure (22), provide brilliant shades and excellent washfastness on cotton (15).

The sulfurein derivative (22) is synthesized by condensing 3-aminophenyl-β-hydroxyethylsulfone with 3,6-dichloroxanthene-9-phenyl-2'-sulfonate at 90°C in N-

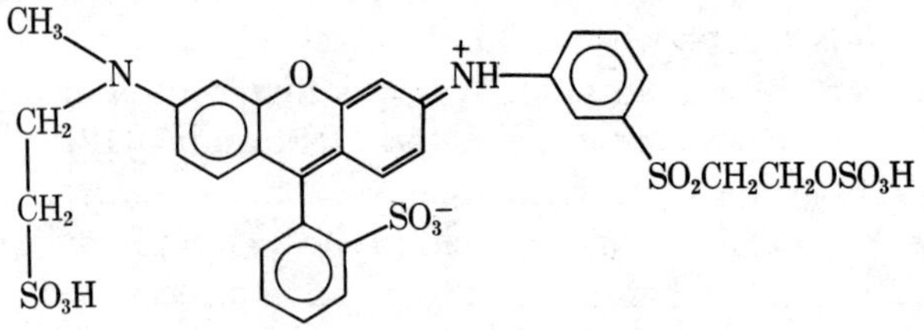

(22)
a sulfurein derivative

methylpyrrolidine or dimethylformamide, then condensing with N-methyltaurine, and finally esterifying with chlorosulfonic acid (16). A general route to asymmetrical acid xanthenes has been patented (17). Condensation of 3,6-dihaloxanthene-9-phenyl-2'-sulfonic acid, with an aromatic amine yields compound (23); a second condensation with an appropriately substituted aliphatic amine, usually with subsequent sulfonation, yields the acid rhodamine (24).

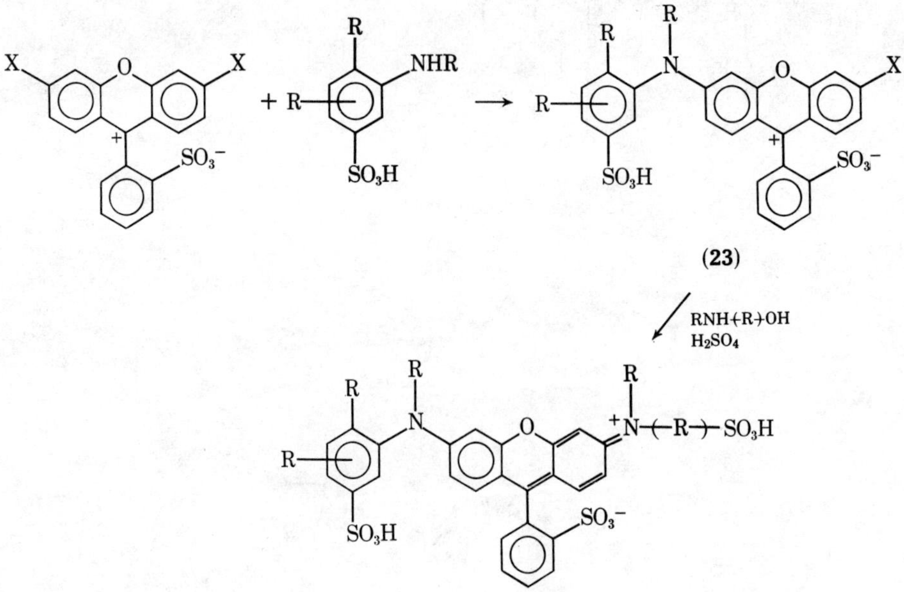

(23)

RNH(R)OH
H₂SO₄

(24)

Rosamines. Rosamines are 9-phenylxanthene derivatives prepared from substituted benzaldehydes instead of phthalic anhydride. Condensing benzaldehyde-2,4-disulfonic acid with m-diethylaminophenol, dehydrating the product with sulfuric acid, and oxidizing with ferric chloride yields Sulforhodamine B (Acid Red 52; CI 45100) (25), which is the most important rosamine.

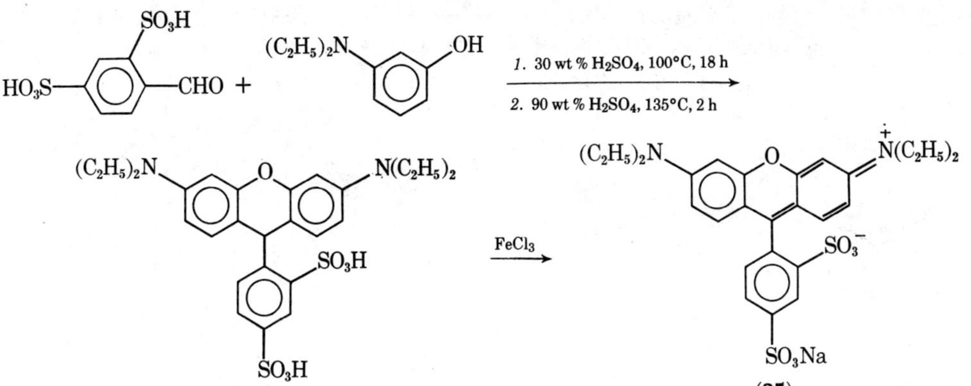

XANTHENE DYES

[Reaction scheme showing synthesis of Sulforhodamine B (25)]

(25)
Sulforhodamine B
[3520-42-1]

A series of fiber-reactive dyes have been made by the reaction of Sulforhodamine B with chlorosulfonic acid, an appropriately substituted diamine, and cyanuric chloride to yield dyes, eg, a Sulforhodamine B derivative (**26**), with good lightfastness (18).

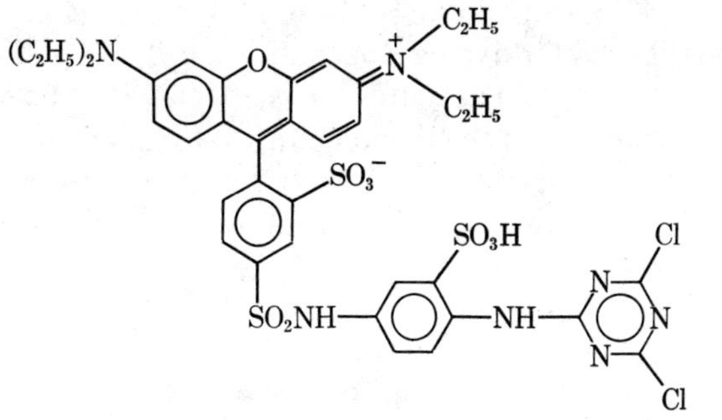

(26)
a Sulforhodamine B derivative

A group of substituted aminoxanthenes, ie, pyrazoloxanthenes, is used as color formers in pressure- or heat-sensitive imaging papers (19). These compounds are colorless, but, upon contact with an acidic electron-accepting material, are converted to resonance forms that are highly colored. An example is structure (**27**), which forms

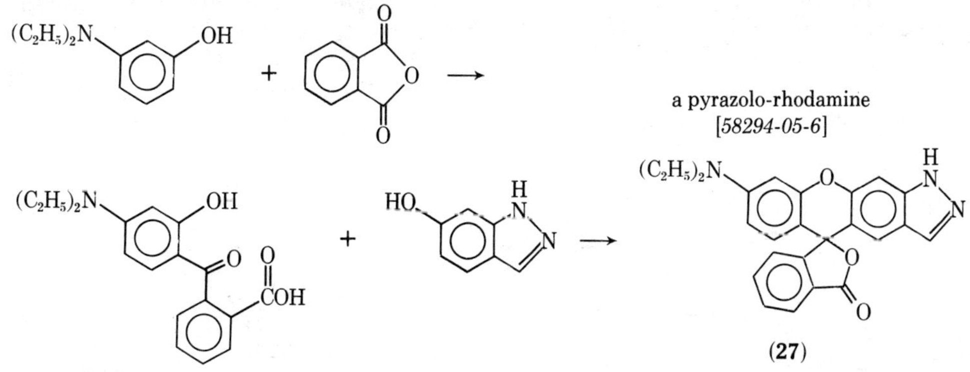

a pyrazolo-rhodamine
[58294-05-6]

(27)

upon the condensation of *N,N*-diethyl-*m*-aminophenol with phthalic anhydride, followed by addition of 6-hydroxyindazole in 80 wt % sulfuric acid (20).

Saccharein. Saccharein (**28**) is a basic dye prepared by the condensation of *m*-diethylaminophenol with saccharin at 165°C.

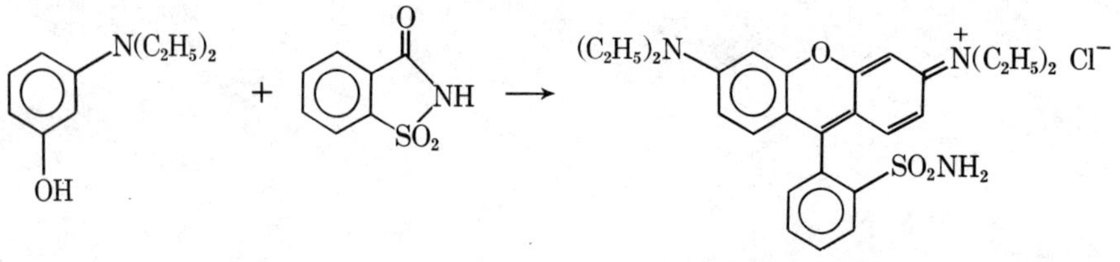

(**28**)

saccharein
[6837-69-0]

Hydroxyl Derivatives. The building block of most hydroxyl-substituted xanthenes, or fluorones, is fluorescein (**2**), (**3**). Although fluorescein itself is no longer used as a dye, its intense green fluorescence, even at extremely high dilutions, makes it ideal for tracing water flows to detect leakage and as sea markers for downed aircraft and missing ships. The sodium or potassium salt of fluorescein, commonly called uranine (CI 45350) (**29**), is still used for dyeing wool and silk brilliant yellow shades. The total domestic market for fluorescein and uranine is estimated to be over 0.5×10^6/yr.

A recent use of uranine is in the manufacture of luminescent laminates, eg, sheets, glass, and plastic films, that are transparent to electromagnetic waves and visible-light rays (21). Such material might be used in windows, viewing partitions, and optical lenses.

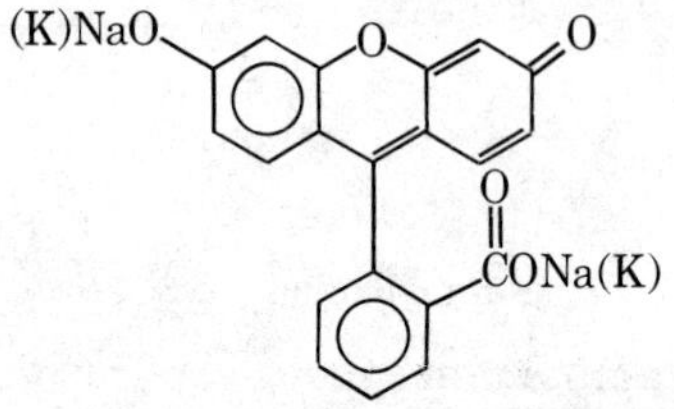

(**29**)

uranine [518-47-8] ([6417-85-2])

The principal use of fluorescein is as an intermediate for more highly substituted hydroxyxanthenes. When fluorescein is brominated in ethanolic solution and converted to the sodium salt with sodium chlorate, eosine (Acid Red 87; CI 45380) (**30**) forms. This has been shown to be the 2',4',5',7'-tetrabromo analogue. It is used for dyeing

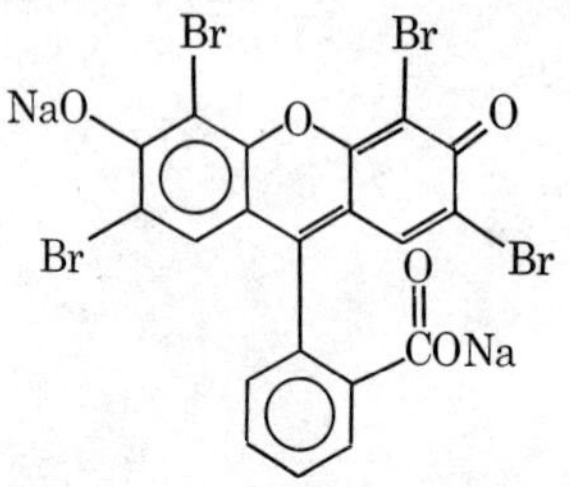

(**30**)

eosine [17372-87-1]

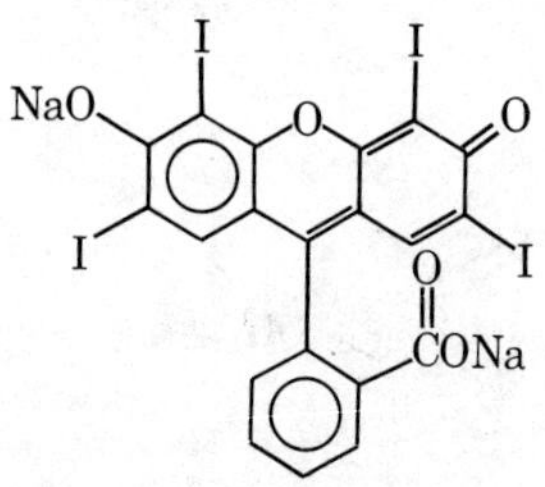

(**31**)

erythrosine [16423-68-0]

silk red with a brilliant yellow fluorescence and for coloring inks, dyeing paper, and coloring cosmetics. Another use is as an indicator in the analytical determination of polymeric biguanide concentrations in aqueous solutions (see Analytical methods). This is important in controlling the growth of bacteria and algae in swimming pools (22) (see Water, treatment of swimming pools, spas, and hot tubs). The lead salt of eosine is Pigment Red 90 [1326-05-2] (CI 45380:1), the free acid is Solvent Red 43 [15086-94-9] (CI 45380:2), and the aluminum salt is Pigment Red 90:1 [17372-87-1] (CI 45380:3). The value of the U.S. eosine market is estimated to be in excess of $600 × 10^6/yr.

When fluorescein reacts with iodine and potassium iodate in an ethanolic solution and converts to the sodium salt, the tetraiodo analogue erythrosine (Acid Red 51, CI 45430) (31) forms. This is used as a food coloring, a sensitizer in photographic plates, and in microscopical stains (see Colorants for food, drugs, and Cosmetics; Photography). Nitrated fluoresceins are used in dye applications. For example, dibromination of fluorescein in aqueous sodium hydroxide followed by treatment with mixed sulfuric–nitric acid yields the dibromo–dinitro analogue saffrosine (Acid Red 91; CI 45400) (32). Saffrosine is used to make fade-resistant electrophotographic sheet material by incorporating such nitro-substituted xanthenes in the photoconductive layer (23).

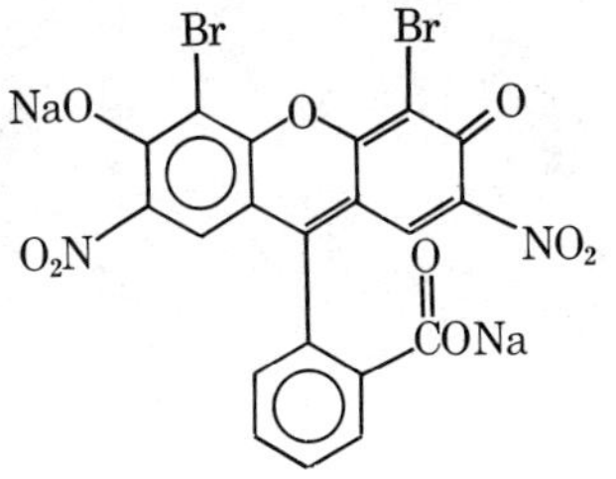

(32) saffrosine [548-24-3]

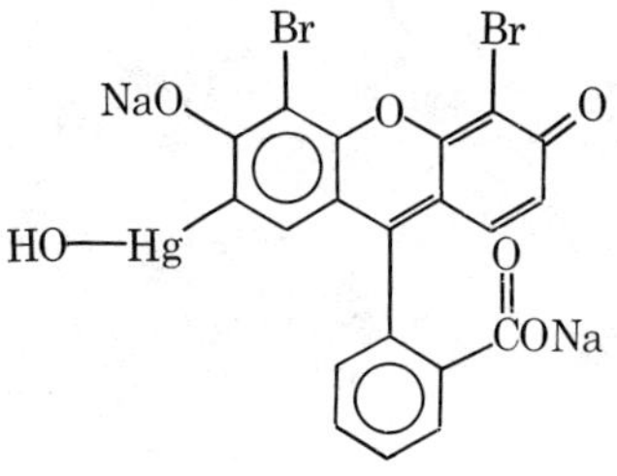

(33) merbromin [129-16-8]

Treatment of this same 4',5'-dibromofluorescein intermediate with mercuric acetate and conversion to the disodium salt yield the hydroxymercuric analogue merbromin or mercurochrome (33). It was once a widely used antiseptic, especially for skin disinfection, and was even administered internally. However, it has been replaced by more effective antibacterial agents.

Another group of halogenated fluorescein dyes is prepared by condensing chloro derivatives of phthalic anhydride with resorcinol, followed by bromination or iodination. Thus, Phloxine B (Acid Red 92, CI 45410) (34) is prepared by condensing tetrachlorophthalic anhydride with resorcinol followed by tetrabromination. Phloxine B undergoes ethylation to yield the yellowish-red acid dye Cyanosine B (35).

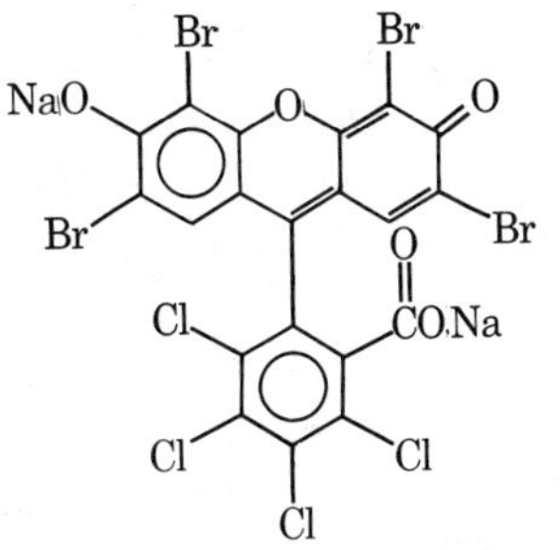

(34) Phloxine B [18472-87-2]

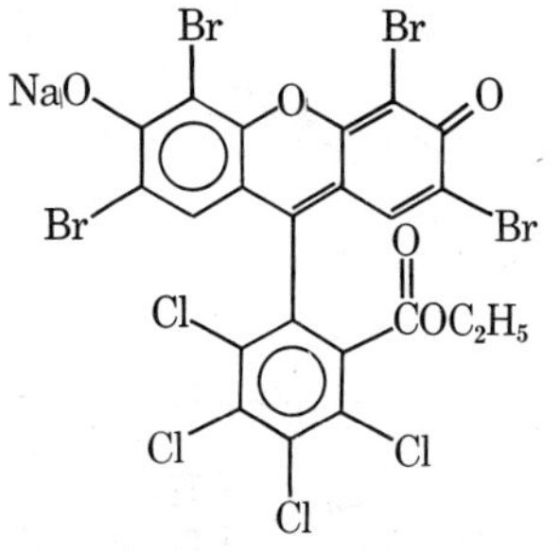

(35) Cyanosine B [6441-80-1]

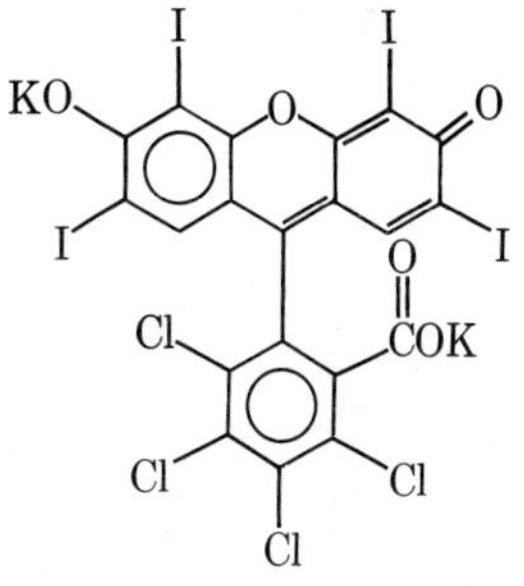

(36) Rose Bengal [632-68-8]

672 XANTHENE DYES

Another important polyhalogenated fluorescein dye is Rose Bengal (Acid Red 94, CI 45440) (36). This is synthesized by condensation of resorcinol with tetrachlorophthalic anhydride, tetraiodination, and conversion to the potassium salt. A new use for Rose Bengal and other tetrabromo- or tetraiodofluoresceins involves their chemiluminescent reaction with ozone. Measuring the intensity of the emitted light makes possible a quantitative determination of ozone concentration in the atmosphere (24) (see Chemiluminescence). Another recent use is as a nonsilver halide photographic system, particularly for use in making direct prints for microfilm enlargements (25).

Aminohydroxy Derivatives

Aminohydroxy-substituted xanthenes are of little commercial importance. They are synthesized by condensing one mole of m-dialkylaminophenol with phthalic anhydride, and then condensing that product with an appropriately substituted phenol. For example, Mordant Red 77 (CI 45300) (37) is prepared by condensing m-dimethylaminophenol with phthalic anhydride, and then condensing the product with 2,4-dihydroxybenzenesulfonic acid.

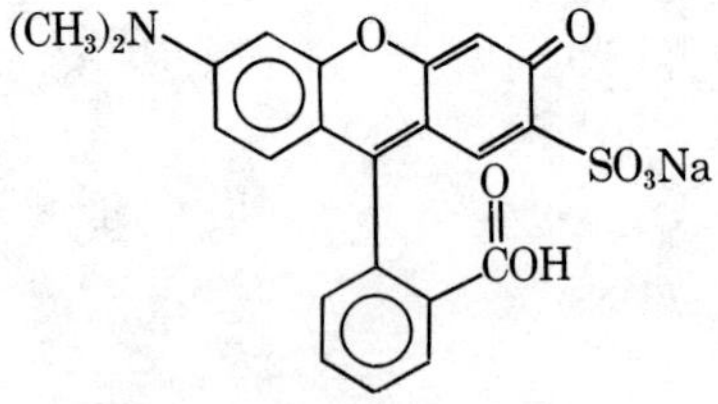

(37)
Mordant Red 77 [6528-43-4]

Miscellaneous Derivatives

Two additional xanthene analogues are termed fluorescent brighteners (see Brighteners, fluorescent). Condensation of two moles of p-cresol with one mole of phthalic anhydride yields 2′,7′-dimethylfluoran, which is cyclized with 24 wt % oleum and then reduced with zinc dust and ammonia under pressure to form Fluorescent Brightener 74 (CI 45550) (38). If the reduction is carried out with zinc dust and caustic soda in the presence of pyridine, followed by acetylation with acetic anhydride, Fluorescent Brightener 155 (CI 45555) (39) is obtained. These are used in the formulation

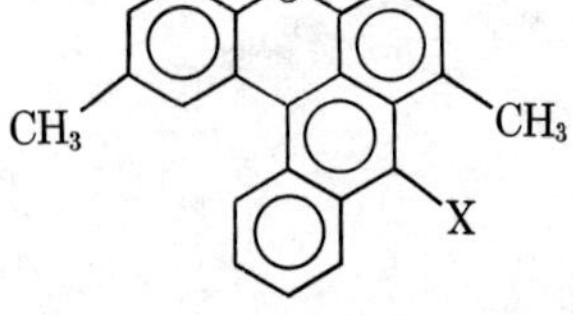

X = H   (38) Fluorescent Brightener 74 [81-37-8]

$$X = OCCH_3 \quad (39) \text{ Fluorescent Brightener 155 } [6250\text{-}49\text{-}3]$$

of solid polyolefin dielectric compositions for application in high voltage cables to prevent conductive treeing (26).

Another series of ring-closed xanthenes starting with benzoxanthene- and benzothioxanthenedicarboxylic acid hydrazides provides shades of bright yellow to red on acetyl cellulose, polyamides, and polyesters with excellent sublimation fastness and unusually good lightfastness. Because of their high fluorescence, they can be used for preparing daylight fluorescent pigments. For example, condensation of 8-methoxybenzo[d,e]xanthene-3,4-dicarboxylic acid hydrazide with acetylacetone in the presence of toluenesulfonic acid followed by cyclization of the hydrazone in N-methylimidazolidinone yields a mixture of isomers (40) and (41) (27):

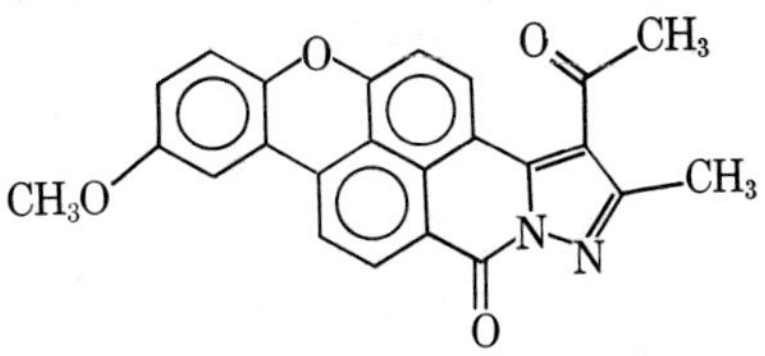

(40)

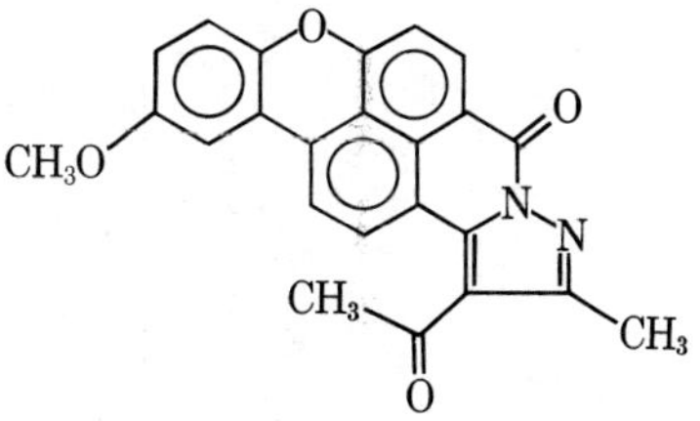

(41) [52204-14-5]

A series of water-soluble fiber-reactive xanthene dyes has been prepared from the reaction of benzoxanthenedicarboxylic acid anhydride disulfonic acid with, for example, 3-aminophenyl-β-hydroxyethyl sulfone to yield dyes, eg, compound (42), with high brilliancy and good fastness properties for dyeing of or printing on leather, wool, silk, or cellulosic fibers (28).

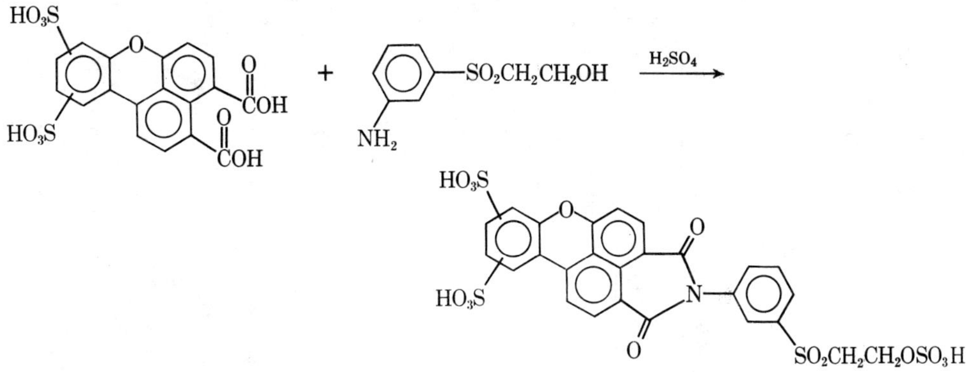

(42) [85682-00-4]

Economic Aspects

Xanthene dye production and import data and U.S. suppliers are listed in Table 1.

Table 1. Economic Aspects of Xanthene Dyes

CI name	CAS Reg. No.	CI number	Common name	U.S. Imports (U.S. Production), metric tons 1976	1977	1978	1979	1980	U.S. Suppliers, 1981
Acid Yellow 73	[2321-07-5]	45350	fluorescein	0.45	0.7	1.6	1.7	2.0	American Cyanamid Co.; Leeben Color, Division of Tricon Colors, Inc.; International Dyestuffs Corp.; Hilton-Davis Chemical Group of Sterling Drug, Inc.
Solvent Yellow 94	[518-47-8]	45350:1	uranine						
Acid Red 52	[3520-42-1]	45100	Sulforhodamine B	2.7	5.4	3.6	9.1	7.3	Atlantic Chemical Corp.; Carolina Color and Chemical Corp.; American Hoechst Corp.; Organic Chemical Corp.; Sandoz Colors and Chemicals, Inc.; Francolor Dyestuff, Inc.
Acid Red 87	[17372-87-1]	45380	eosine	0.45		0.7	3.2	2.7	Hilton-Davis Chemical Group of Sterling Drug, Inc.
Pigment Red 90	[85682-01-5]	45380:1							H. Kohnstann & Co.
Solvent Red 43	[15086-94-9]	45380:2							Hilton-Davis Chemical Group of Sterling Drug, Inc.; Sun Chemical Corp.
Acid Red 92	[18472-87-2]	45410	Phloxine B	3.4	1.1	0.2	0.2	0.2	Hilton-Davis Chemical Group of Sterling Drug, Inc.
Solvent Red 48	[13473-26-2]	45410:1							
Pigment Red 174	[15876-58-1]	45410:2							
Acid Red 94	[632-68-8]	45440	Rose Bengal	6.8	0.9	1.4	1.4	0.1	
Acid Red 289	[12220-28-9]					2.7	2.7	4.5	American Hoechst Corp.
Acid Violet 9	[6252-76-2]	45190		1.4	0.2	2.7	2.3	0.9	International Dyestuffs Corp.
Solvent Violet 10	[66225-66-9]								

674

Name	CAS	CI	Common name						Suppliers
Basic Red 1	[989-38-8]	45160	Rhodamine 6G	34	41	113	141	140	Atlantic Chemical Corp.; C. Lever Co.; International Dyestuffs Corp.; BASF Wyandotte Corp.; Dye Specialties, Inc.; Francolor Dyestuff, Inc.
Pigment Red 81	[12224-98-5]	45160:1		(263)	(233)	(254)	(227)	(186)	C. Lever Co.; Sun Chemical Corp.; Mobay Chemical Corp.; CIBA-GEIGY Corp.; BASF Wyandotte Corp.
Basic Violet 10	[81-88-9]	45170	Rhodamine B	91	82	102	154	75	Atlantic Chemical Corp.; American Cyanamid Co.; Buffalo Color Corp.; Sun Chemical Corp.; Leeben Color, Division of Tricon Colors, Inc.; International Dyestuffs Corp.; BASF Wyandotte Corp.; Dye Specialties, Inc; Mobay Chemical Corp.
Solvent Red 49	[509-34-2]	45170:1	Rhodamine B Base	6.8	11	21	2.7	14	American Cyanamid Co.; C. Lever Co.; Buffalo Color Corp.; International Dyestuffs Corp.; Dye Specialties, Inc.
Pigment Violet 1	[1326-03-0]	45170:2		29	(64)	(62)	(120)	(104)	CIBA-GEIGY Corp.; Sun Chemical Corp.
Basic Violet 11	[2390-63-8]	45175	Fanal Red 6BM (IG)		14	22	27	16	
Solvent Green 4	[81-37-8]	45550	Fluorescent Brightener 74	0.05	0.2	0.1			
Mordant Red 27	[6359-22-4]	45180	Chromoxane Brilliant Red				4.5	4.5	American Hoechst Corp.
Food Red 14	[16423-68-0]	45430	erythrosine						Warner Jenkinson Co.; Hilton-Davis Chemical Group of Sterling Drug, Inc.; Crompton and Knowles Corp.; Sun Chemical Corp.; Leeben Color, Division of Tricon Colors, Inc.

Table 2. Toxicological Properties of Selected Xanthene Dyes[a]

Compound	Structure	Property	Value, mg/kg
xanthene	(1b)	LD_{50} (mouse), subcutaneous	690
fluorescein	(5)	LD_{Lo} (rat), intraperitoneal	600
		LD_{Lo} (mouse)[b]	600
		LD_{Lo} (rabbit), oral	2,500
		LD_{Lo} (rabbit), intravenous	300
		LD_{Lo} (guinea pig)[b]	400
eosine	(30)	LD_{Lo} (rat), intraperitoneal	500
		LD_{Lo} (rat), subcutaneous	1,500
		TD_{Lo} (rat), subcutaneous	7,200
		LD_{50} (mouse), intravenous	550
		LD_{Lo} (rabbit), intravenous	300
erythrosine	(31)	LD_{50} (rat), intraperitoneal	300
		LD_{Lo} (rat), intravenous	200
		LD_{Lo} (mouse), oral	2,500
		LD_{50} (mouse), intravenous	700
Phloxine B	(34)	LD_{50} (mouse), intravenous	310
merbromin	(33)	LD_{Lo} (mouse), subcutaneous	20
		LD_{Lo} (rabbit), intravenous	15
uranine	(29)	LD_{50} (rat), intraperitoneal	1,700
		TD_{Lo} (rat), subcutaneous	17
		LD_{50} (mouse), intraperitoneal	1,800
		LD_{Lo} (guinea pig), intraperitoneal	1,000
Rhodamine B	(15)	LD_{Lo} (rat), oral	500
		TD_{Lo} (rat), subcutaneous	3,825
		LD_{50} (rat), intravenous	89,500
		LD_{Lo} (mouse), intraperitoneal	128
Rhodamine 6G	(18)	TD_{Lo} (rat), subcutaneous	132
		LD_{Lo} (mouse), intraperitoneal	2

[a] Ref. 29.
[b] Administration method unknown.

Health and Safety Factors, Toxicology

Xanthene dyes have not exhibited health or safety properties warranting special precautions; however, standard chemical labeling instructions are required. Toxicological properties of economically important dyes are listed in Table 2 (29).

BIBLIOGRAPHY

"Xanthene Dyes" in *ECT* 1st ed., Vol. 15, pp. 136–149 by W. G. Huey and S. K. Morse, General Aniline and Film Corporation; "Xanthene Dyes" in *ECT* 2nd ed., Vol. 22, pp. 430–437 by F. F. Cesark, American Cyanamid Co.

1. A. Baeyer, *Chem. Ber.* **4**, 558, 662 (1871).
2. U.S. Pat. 3,961,039 (June 1, 1976), R. Sternheimer.
3. *BIOS Report 959*, British Intelligence Objectives Subcommittees, England, 1946, pp. 8, 15.
4. U.S. Pat. 3,849,065 (Nov. 19, 1974), K. Schmeidl (to BASF).
5. U.S. Pat. 3,767,358 (Oct. 23, 1973), H. I. Stryker (to DuPont).
6. U.S. Pat. 3,690,889 (Sept. 12, 1972), S. E. Harrison and R. Drake (to RCA Corporation).
7. U.S. Pat. 3,767,408 (Oct. 23, 1973), S. E. Harrison and J. E. Goldmacher (to RCA Corporation).
8. U.S. Pat. 3,796,668 (Mar. 12, 1974), R. T. Hickcox (to Hercules, Inc.).

9. U.S. Pat. 3,644,270 (Feb. 22, 1972), G. D. Valiaveedam (to DuPont).
10. Brit. Pat. 1,494,768 (Dec. 14, 1977), D. M. Zabiak and K. S. Hwang (to A. B. Dick Co.).
11. U.S. Pat. 3,541,470 (Nov. 17, 1970), J. R. Lankard and P. P. Sorokin (to IBM Corp.).
12. Brit. Pat. 1,503,380 (Mar. 8, 1978), (to Hoechst AG).
13. U.S. Pat. 3,956,300 (May 11, 1976), P. W. Austin, A. T. Costello, and A. Crabtree (to ICI Ltd.).
14. Brit. Pat. 1,377,695 (Dec. 18, 1974), P. W. Austin and A. T. Costello (to ICI Ltd.).
15. Brit. Pat. 1,471,452 (Apr. 27, 1977), (to Hoechst AG); Brit. Pat. 1,471,453 (Apr. 27, 1977), (to Hoechst AG).
16. U.S. Pat. 3,772,335 (Nov. 13, 1973), F. Meininger and F. Kohlhass (to Hoechst AG).
17. Brit. Pat. 1,586,820 (Mar. 25, 1981), (to Hoechst AG).
18. U.S. Pat. 3,883,529 (May 13, 1975), P. W. Austin (to ICI Ltd.).
19. U.S. Pat. 3,988,492 (Oct. 26, 1976), S. M. Spatz (to Mead Corporation).
20. U.S. Pat. 3,929,825 (Dec. 30, 1975), S. M. Spatz (to Mead Corporation).
21. Brit. Pat. 1,429,597 (Mar. 24, 1976), G. O. Okikiolu (to Okikiolu Scientific and Industrial Organization).
22. Brit. Pat. 1,533,255 (Nov. 22, 1978), G. G. Barraclough, D. Myles, D. A. Reilly, and P. Tomlinson (to ICI Ltd.).
23. U.S. Pat. 3,951,655 (Apr. 20, 1976), B. Schoustra and H. Roncken (to Oce-van der Grinten N.V., Netherlands).
24. U.S. Pat. 3,975,159 (Aug. 17, 1976), S. van Heusden (to U.S. Philips Corporation).
25. U.S. Pat. 3,615,566 (Oct. 26, 1971), I. D. Robinson (to Arthur D. Little, Inc.).
26. U.S. Pat. 4,216,101 (Aug. 5, 1980), H. J. Davis (to Canada Wire and Cable Ltd.).
27. U.S. Pat. 3,853,884 (Dec. 10, 1974), H. Troster (to Hoechst AG).
28. U.S. Pat. 3,888,862 (June 10, 1975), F. Meininger and F. Kohlhaas (to Hoechst AG).
29. R. J. Lewis and R. L. Tatken, *Registry of Toxic Effects of Chemical Substances*, 1979 ed., U.S. Dept. of Health and Human Services, National Institute for Occupational Safety and Health, Cincinnati, Ohio, Sept. 1980.

General References

K. Venkataraman, *The Chemistry of Synthetic Dyes*, Vol. 2, Academic Press, Inc., New York, 1955, pp. 740–754.
H. A. Lubs, *The Chemistry of Synthetic Dyes and Pigments*, R. E. Krieger Publishing Company, Huntington, N.Y., 1955, pp. 291–301.
Colour Index, 3rd ed., The Society of Dyers and Colorists, Bradford, Yorkshire, UK, Vol. 4, 1971, pp. 4417–4430; Vol. 6, 1975, p. 6401.

RUSSELL E. FARRIS
Sandoz Colors & Chemicals

XENON. See Helium-group gases.

XENON COMPOUNDS. See Helium-group gases.

XEROGRAPHY. See Electrophotography.

X-RAY ANALYSIS. See X-ray technology.

X-RAY TECHNOLOGY

Materials Characterization

X rays are a form of electromagnetic radiation, which is manifested as continuous or white radiation and as characteristic radiation. X-ray photons are characterized by wavelength λ and energy E, which are related in the expression λ (nm) $\simeq 1.24/E(\text{keV})$. Continuous radiation is produced when a high energy electron beam decelerates as it approaches the electron clouds that surround the atomic nucleus. Characteristic radiation is produced following the ejection of an inner orbital electron by high energy particles and subsequent transition of atomic orbital electrons from states of high to low energy. When a monochromatic beam of x-ray photons falls onto a given specimen, absorption, scatter, or fluorescence may result. The coherently scattered photons may undergo subsequent interference, therein generating diffraction maxima. These three phenomena form the bases of three important x-ray methods: the absorption technique is the basis of radiographic analysis; the scattering effect is the basis of x-ray diffraction; and the fluorescence effect is the basis of x-ray fluorescence spectrometry (see Analytical methods; Nondestructive testing).

X rays were discovered by Wilhelm Roentgen in 1895 and the property of the atomic-number dependence of the absorption of x-ray photons was soon thereafter established and applied for medical diagnostic purposes (1) (see Medical diagnostic reagents). Radiographic techniques were also developed for industrial purposes and provide the means for the study of deformities and discontinuities in bulk materials. More recently, x-ray absorption methods have been used in security screening, eg, for high speed examination of luggage. Following the discovery of the diffraction of x rays in 1913 (2), two fields of materials analysis have developed. X-ray fluorescence spectrometry involves either the diffracting power of a single crystal to isolate narrow wavelength bands or a proportional detector to isolate narrow energy bands from the polychromatic-beam characteristic radiation excited in the sample. These methods are called wavelength-dispersive spectrometry and energy-dispersive spectrometry. Because of the known relationship between emission wavelength and atomic number, isolation of individual characteristic lines allows the unique identification of an element to be made, and elemental concentrations can be estimated from characteristic line intensities. Thus, this technique is a means of materials characterization in terms of chemical composition.

The second field of materials analysis involves characterization of atomic arrangement in the crystal lattice. Single-crystal x-ray diffraction allows the direct study of this atomic arrangement. An ordered array of atoms, ie, the crystal lattice, contains planes of high atomic density which imply planes of high electron density. A monochromatic beam of x-ray photons is scattered by these atomic electrons, and if the scattered photons interfere with each other, diffraction maxima may occur. In general, one diffracted line occurs for each unique set of planes in the lattice. X-ray powder diffractometry involves single or multiphase, ie, multicomponent, specimens comprising a random orientation of small crystallites; each is ca 1–50 μm in diameter. The powder pattern can be used as a unique fingerprint for a phase, and a large file of ca 38,000 standard single-phase patterns is available. Analytical methods based on manual- or computer-search techniques are available for unscrambling patterns of multiphase materials. These methods make possible qualitative and quantitative multiphase identification. Special techniques are also used for the study of stress, texture, topography, particle size, high and low temperature phase transformations, and other research applications. Use of computer-aided techniques is making general application of these specialized methods increasingly important in materials characterization.

Properties

When a beam of x-ray photons falls onto an absorber, a number of different processes may occur. The more important of these are illustrated in Figure 1. In this example, a monochromatic beam of radiation of wavelength λ and intensity I_o, is incident on an absorber of thickness x and density ρ. The fate of each x-ray photon is governed by absorption and scattering. In absorption, a certain fraction (I/I_o) of the radiation may pass through the absorber. Where this happens, the wavelength of the transmitted beam is unchanged and the intensity of the transmitted beam $I(\lambda_o)$ is given by

$$I(\lambda_o) = I_o \cdot \exp - (\mu \cdot \rho \cdot x) \qquad (1)$$

where μ is the mass absorption coefficient of the absorber for the wavelength λ_o. It is apparent from the above that a number of photons equal to $I_o - I$ are lost in the absorption process, and most of this loss results from the photoelectric effect. Photoelectric absorption, usually designated τ, occurs at each of the energy levels of the atom. Thus, the total photoelectric absorption is determined by the sum of each individual absorption within a specific shell.

Scattering σ occurs when an x-ray photon collides with one of the electrons of the absorbing element. Where this collision is elastic, ie, no energy is lost in the collision process, the scatter is said to be coherent or Rayleigh scatter. Since no energy change is involved, the coherently scattered radiation retains exactly the same wavelength as that of the incident beam. X-ray diffraction is a special case of coherent scatter,

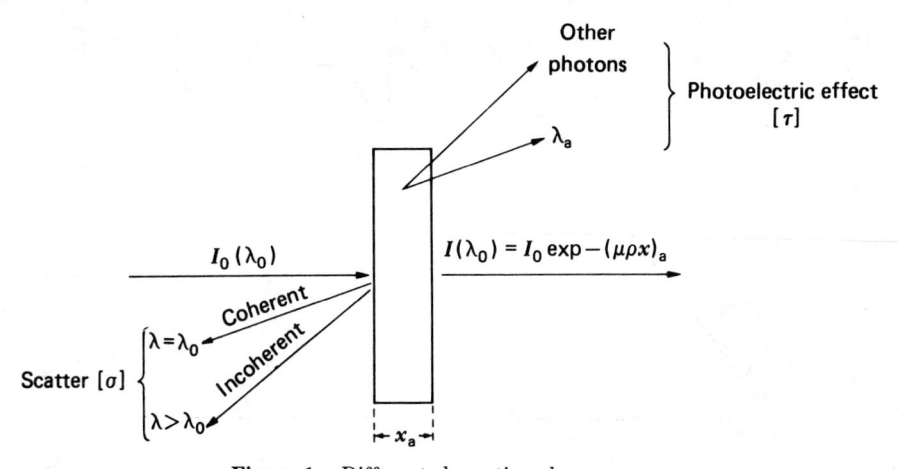

Figure 1. Different absorption phenomena.

where the scattered photons interfere with each other. The scattered photon may also give up a small part of its energy during the collision, especially where the electron with which the photon collides is only loosely bound. In this instance, the scatter is said to be incoherent or Compton scatter, and the wavelength of the incoherently scattered photon is greater than λ_o.

The value of the mass absorption μ referred to in equation 1 is a function of both the photoelectric absorption and the scatter:

$$\mu = f(\tau) + f(\sigma) \tag{2}$$

However, $f(\tau)$ is usually large compared with $f(\sigma)$. Because the photoelectric absorption is made up of absorption in the various atomic levels, it is an atomic-number-dependant function. As shown in Figure 2, a plot of μ against λ contains a number of discontinuities, called absorption edges, at wavelengths corresponding to the binding energies of the electrons in the various subshells. The absorption discontinuities are a main source of nonlinearity between x-ray intensity and composition in both x-ray fluorescence and x-ray powder diffractometry.

Under certain geometric conditions, wavelengths that are exactly in phase or exactly out of phase may cancel each other out or add to one another, as illustrated in Figure 3(**a**). The coherently scattered photons may thus constructively interfere with each other, thereby producing diffraction maxima. A crystal lattice consists of a regular arrangement of atoms, and when a monochromatic beam of radiation falls

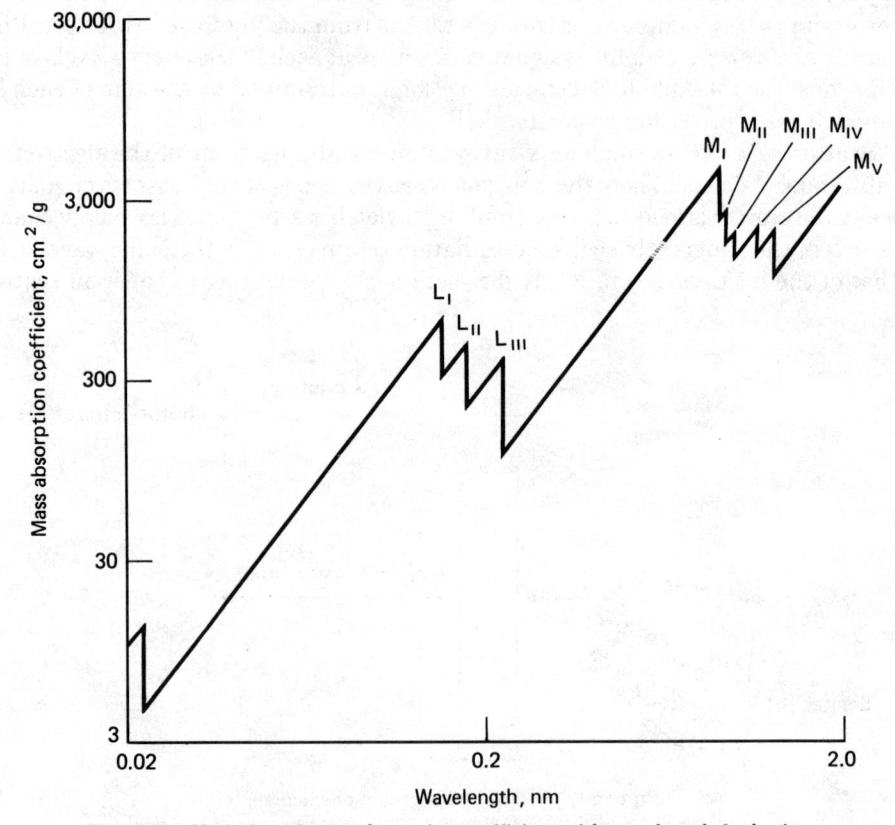

Figure 2. Variation of mass absorption coefficient with wavelength for barium.

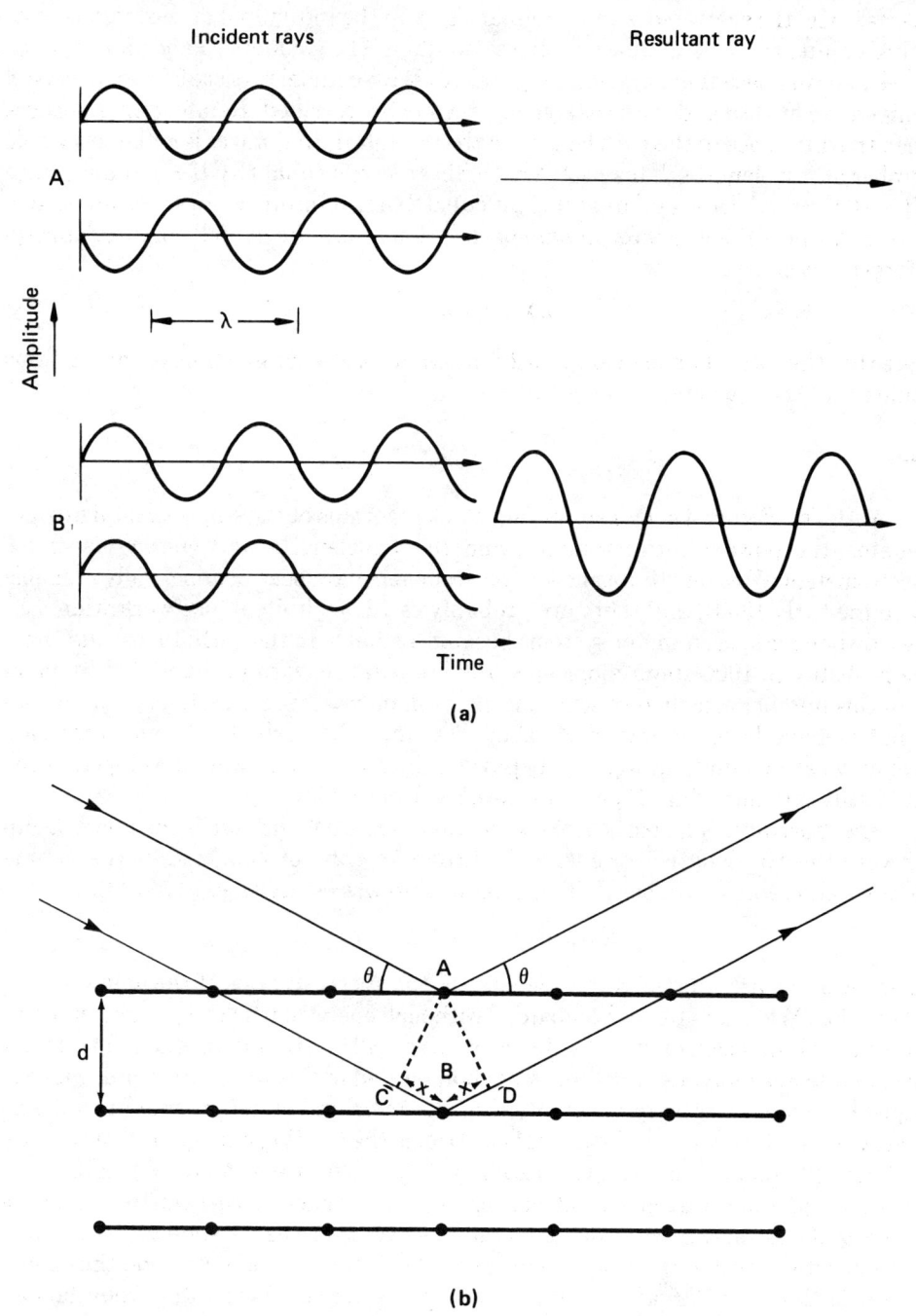

Figure 3. (a) Destructive (A) and constructive (B) interference. (b) Condition for diffraction of x rays.

onto these atomic layers, scattering occurs. In order to satisfy the requirement for interference, the scattered waves originating from the individual atoms, ie, the scattering points, must be in phase with one another. The geometric conditions for the waves to be in phase is illustrated in Figure 3(**b**). Two parallel rays strike a set of crystal planes at an angle θ and are scattered as previously described. Reinforcement occurs when the difference in the path lengths of the two interfering waves is equal to a whole number of wavelengths. This path length difference is equal to $CB + BD$ and, since $CB = BD = x$, $n\lambda$ (n = an integer) must equal $2x$ for reinforcement to occur. Also, $x = d \cdot \sin \theta$, where d is the interplanar spacing. Therefore the overall condition for reinforcement is Bragg's law:

$$n\lambda = 2\, d \cdot \sin \theta \tag{3}$$

Note that the total change in angle is 2θ and that a detector must move through this angle to see the scattered x rays.

Uses

Analysis. Figure 4 shows simplified block diagrams of the single-crystal diffractometer, the powder diffractometer, and the wavelength- and energy-dispersive spectrometers. Wavelength-dispersive spectrometers have been commercially available since the early 1950s, and there are probably ca 12,000 units in use worldwide. Energy-dispersive spectrometer systems became available in the early 1970s, and there are probably ca 1000 stand-alone spectrometers in use, with perhaps slightly more than this number attached to scanning electron microscopes. Single-crystal diffractometers have been commercially available since the early 1960s, and there are probably ca 1000 units in use. X-ray powder diffractometers were developed in the late 1940s and more than 15,000 units are used worldwide.

Spectroscopy. The basis of the x-ray fluorescence technique is the relationship between the wavelength λ or energy E of the x-ray photons emitted by the sample element and atomic number Z. This relationship was established in 1913 (3):

$$E/1.24 = 1/\lambda = K(Z - s)^2 \tag{4}$$

where K and s are constants that depend on the spectral series of the emission line in question. When an atom is bombarded with high energy particles, eg, x-ray photons, an inner-orbital electron may be displaced leaving the atom in an excited state. As an example, if a molybdenum atom were bombarded in this way, a K orbital vacancy might be created by the removal of one of the $1s$ orbital electrons (see Fig. 5). The energy required to remove the orbital electron is the binding energy and, in the case of the molybdenum K level, corresponds to 20 keV. The atom can regain stability by rearrangement of the atomic electrons, eg, by transference of a $2p$ electron from the L shell to the $1s$ vacancy. In so doing, the atom moves from the K^+ state to the L^+ state. The ionization level in the L shell is only ca 2.3 keV for molybdenum, and the energy excess, ie, $20-2.3 = 17.6$ keV, is emitted as a characteristic $K\alpha$ photon from the molybdenum. Not all vacancies result in the production of characteristic x-ray photons, since there is a competing internal rearrangement process known as the Auger effect. The ratio of the number of vacancies resulting in the production of characteristic x-ray photons to the total number of vacancies created in the excitation process is the fluorescent yield. The fluorescent yield is approximately one for the high atomic number

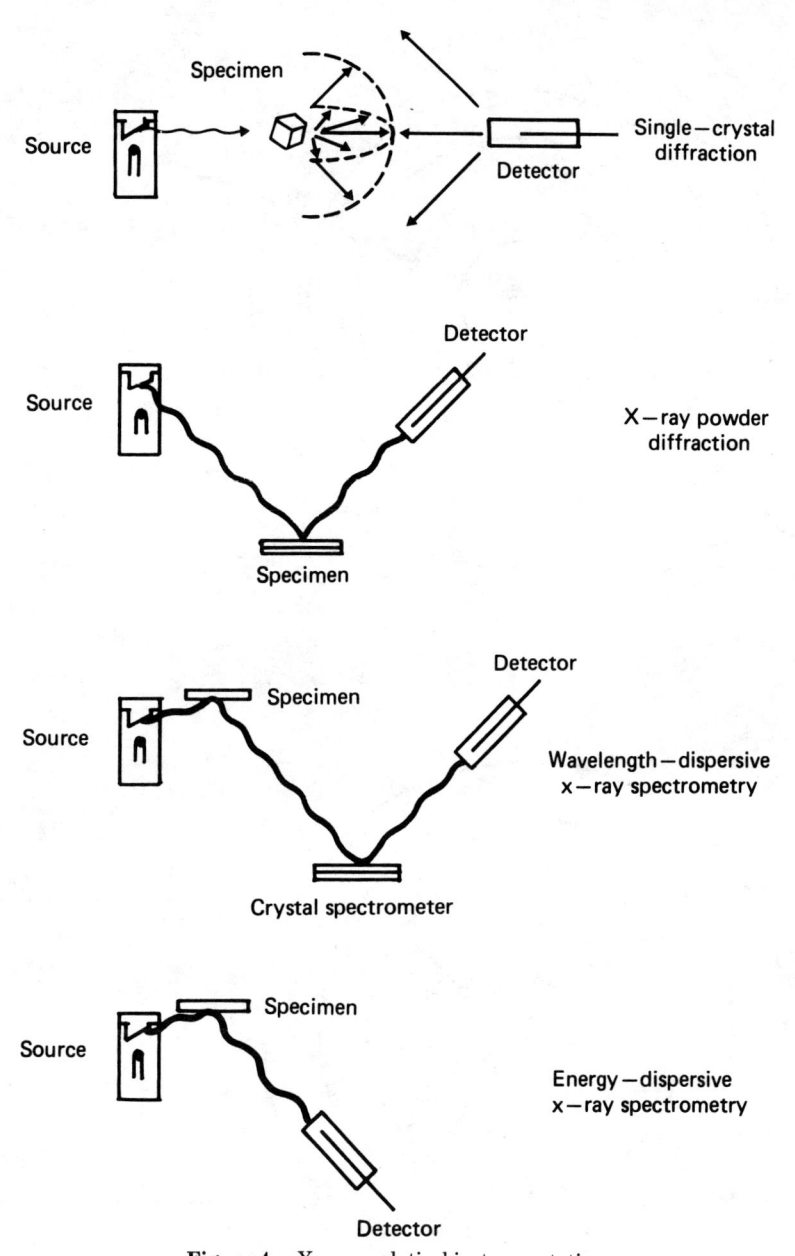

Figure 4. X-ray analytical instrumentation.

elements and may be as little as 0.01 for the low atomic elements, eg, Na, Mg, and Al. For those vacancies giving rise to characteristic x-ray photons, a series of very simple selection rules can be used to define which electrons can be transferred, namely, the principal quantum number n must change by 1, the angular quantum number ℓ must change by 1, and the vector sum of $\ell + s$ must be a positive number changing by 1 or 0. Therefore, for the K series, only p–s transitions are allowed, yielding two lines for each principle level change. Vacancies in the L level follow similar rules and give rise to L series lines. There are more of the L lines since p–s, s–p, and d–p transitions are allowed within the selection rules.

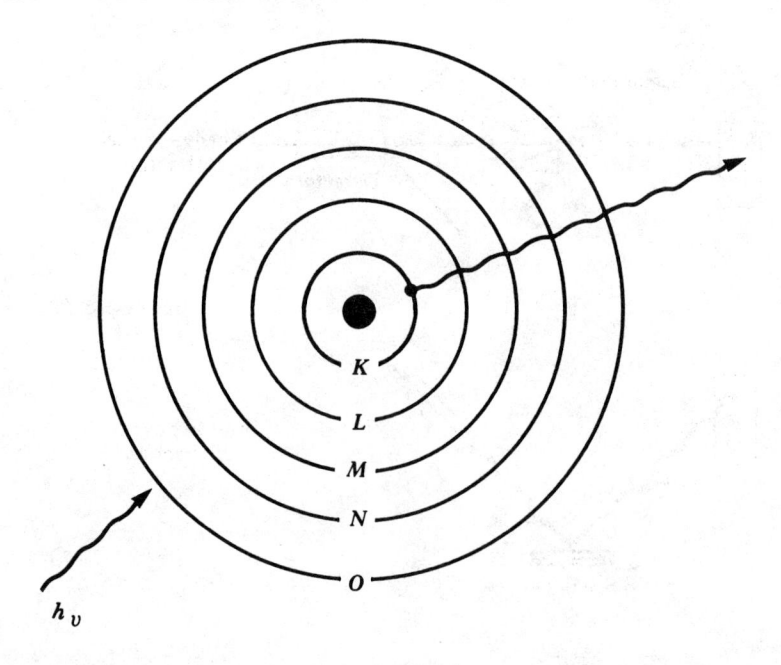

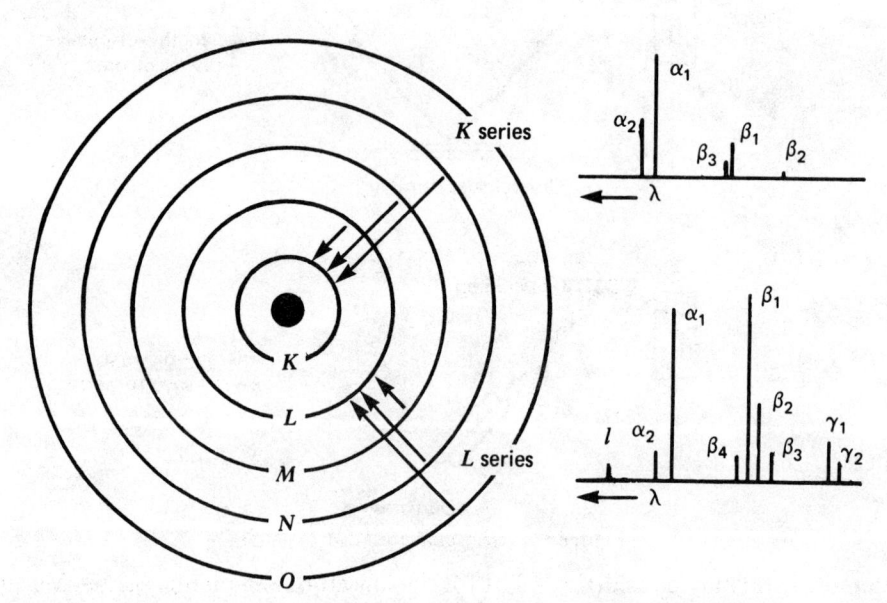

Figure 5. Production of characteristic K and L series radiation from molybdenum.

Most commercially available x-ray spectrometers have a range of ca 0.02–2.0 nm (60–0.6 keV), which allows measurement of the K series from F ($Z = 9$) to Lu ($Z = 71$) and for the L series from Mn ($Z = 25$) to U ($Z = 92$). Other line series can occur from the M and N levels, but these have little use in analytical x-ray spectrometry. Although in principle, almost any high energy particle can be used to excite characteristic radiation from a specimen, an x-ray source offers a reasonable compromise between efficiency, stability, and cost. Almost all commercially available x-ray spectrometers are based on such an excitation source. Since primary or source x-ray photons are used to excite secondary or specimen radiation, the technique is referred to as x-ray fluorescence spectrometry.

Instrumentation and techniques. All conventional x-ray spectrometers have three basic parts: the primary source unit, the spectrometer, and the measuring electronics. The primary source unit consists of a sealed x-ray tube and a very stable, high voltage generator capable of providing up to ca 3 kW of power at a potential of typically 60–80 kV. The sealed x-ray tube contains an anode of Cr, Rh, W, Ag, Au, or Mo and delivers an intense source of continuous radiation, which impinges on the analyzed specimen, where characteristic radiation is generated. A portion of the characteristic fluorescence radiation is then collected by the spectrometer where the beam is passed by a collimator or slit onto the surface of an analyzing crystal, where individual wavelengths are diffracted in accordance with the Bragg law. A photon detector, typically a gas-flow or a scintillation counter, converts the diffracted characteristic photons into voltage pulses, which are integrated and displayed as a measure of the characteristic line intensity. A goniometer maintains the required geometric conditions to ensure that the angle between source and crystal and that between the crystal and detector are the same.

The output from a wavelength-dispersive spectrometer may be either analogue or digital. For qualitative work, an analogue output is traditionally used, and in this instance, a rate meter is used to integrate the pulses over short time intervals, typically about a second. The output from the rate meter is fed to an intensity/time (x/t) recorder, which scans at a speed which is conveniently coupled with the goniometer scan speed. The recorder thus displays an intensity/time diagram, which becomes an intensity/2θ diagram. Tables are then used to interpret the wavelengths. For quantitative work, digital counting is more convenient, and a timer–scaler combination is provided, which allows pulses to be integrated over a period of several tens of seconds and then displayed as count or count rate. Most modern wavelength-dispersive spectrometers are controlled in some way by a minicomputer or microprocessor and, by means of specimen changers, are capable of very high specimen throughput. Once they are set up, the spectrometers run virtually unattended for several hours.

Like the wavelength-dispersive spectrometer, the energy-dispersive spectrometer also consists of the excitation source, spectrometer, and detection system. However, the detector acts as the dispersion agent. The detector is typically a lithium-drifted silicon, Si(Li), detector, which is a proportional detector of high intrinsic resolution. The resolution of the detector is typically 160–180 eV, compared to 20–200 eV for the wavelength- or crystal-dispersive system. The Si(Li) detector diode serves as a solid-state version of the gas-flow detector in the wavelength-dispersive system where the gas gain is unity. When an x-ray photon is stopped by the detector, an ionization cloud is generated in the form of electron–hole pairs. The numbers of electron–hole pairs created, ie, the total electric charge released, is proportional to the energy of the incident x-ray photons. The charge is swept from the diode by a high voltage applied across it. A preamplifier is responsible for collecting this charge on a feedback capacitor to produce a voltage pulse proportional to the original x-ray photon energy. Thus, when a range of photon energies is incident upon the detector, an equivalent range of voltage pulses is produced as a detector output. A multichannel analyzer is used to sort the arriving pulses at its input in the same fashion as in the production of a histogram representation of the x-ray energy spectrum. The output from an energy-dispersive spectrometer is generally displayed on a visual display unit. The operator is able to display dynamically the contents of the various channels as an energy spectrum, and provision is generally made to allow zooming in on portions of the spectrum of special interest, to overlay spectra, to subtract background, etc, in an interactive manner. As in the case of the modern wavelength-dispersive systems, nearly all energy-dispersive spectrometers incorporate some form of minicomputer which is available for spectral stripping, peak identification, quantitative analysis, and many other useful functions.

Both the sequential wavelength-dispersive spectrometer and the energy-dispersive spectrometer are well-suited for qualitative analysis of materials. As was shown in equation 4, there is a simple relationship between the wavelength or energy of a characteristic x-ray photon and the atomic number of the element from which the characteristic emission line occurs. Thus, each element emits a number of characteristic lines within a given series, eg, K, L, M, etc. Measuring the wavelengths or energies of a given series of lines from an unknown material makes possible the establishment of the atomic numbers of the excited elements. Because the characteristic x-ray spectra are so simple, the actual process of allocating atomic numbers to the emission lines is relatively simple, and the chance of making a gross error is small. As a comparison, the procedures for the qualitative analysis of multiphase materials with the x-ray powder diffractometer is much more complex. There are only ca 106 elements and, within the range of the conventional spectrometer, each element gives on an average only six lines. In diffraction, however, there are as many as several million (10^6) possible compounds, each of which can give on an average ca 50 lines. A further benefit of the x-ray emission spectrum for qualitative analysis is that, because transitions do arise from inner orbitals, the effect of chemical combination, ie, valence, is almost negligible.

The great flexibility and range of the various types of x-ray fluorescence spectrometer coupled with their high sensitivity and good inherent precision make them ideal for quantitative analysis. Like all instrumental methods of analysis, the high precision can be translated into high accuracy only if the various systematic errors in the analysis process are removed. The precision of a well-designed x-ray spectrometer is typically ca 0.1%, and the main source of this random error is the x-ray source, ie, the high voltage generator and the x-ray tube. In addition, there is a small count-time-dependent error arising from the statistics of the actual counting process. Systematic errors in quantitative x-ray spectrometry arise mainly from absorption- and specimen-related phenomena, including particle size and heterogeneity. Although these matrix effects are complicated, many excellent methods have been developed for handling them. The advent of the minicomputer-controlled spectrometer has done much to enhance the application of these correction procedures, and usually, most elements in the periodic table of atomic number 9 (F) and larger can be quantified to an accuracy of a few tenths of a percent (4). X-ray fluorescence is used in almost all areas of inorganic analysis.

X-ray spectrometers generally are one of three main types: scanning spectrometers for the sequential analysis of each element in a static sample; multichannel spectrometers, in which samples are analyzed for all elements simultaneously; and onstream spectrometers, in which a flowing slurry is analyzed for all elements simultaneously. In addition, various spectrometers are available with alternative primary sources, eg, nondispersive instruments with x-ray or isotope sources. The principal suppliers of the various types of commercially available spectrometers are listed in Table 1. The choice of instrumentation is very wide and, as shown in Table 2, over 20 basic instruments are available, each of which is usually obtainable at various stages of complexity. On-stream analyzer systems have special applications in ore beneficiation and similar industrial areas. Classical x-ray spectrometers are either wave-

Table 1. Principal Suppliers of X-Ray Instrumentation

Manufacturer	Instrument type
Bausch and Lomb (Sunnyvale, Calif., and Lausanne, Switzerland)	scanning, multichannel, on-stream, and energy-dispersive
Outokumpu Oy Research Laboratory (Tapiola, Finland)	on-stream
Philips Electronic Instruments (Mahwah, N.J.)	scanning, multichannel, portable, and energy-dispersive
Rigaku Denki Co., Ltd. (Tokyo, Japan)	scanning and multichannel
Siemens A.G. (Karlsruhe, FRG)	scanning and multichannel
Edax (Chicago, Ill.)	energy-dispersive
Kevex (Foster City, Calif.)	energy-dispersive

Table 2. Principal Features of Different Types of X-Ray Spectrometer

Type	Approximate number in use	Approximate cost, $	Typical application	Advantages	Disadvantages
scanning	10,000	80,000–150,000	general quantitative and qualitative analysis	versatile, relatively fast	slow compared to multichannel
multichannel	2,000	150,000–250,000	high speed replicate	very fast	inflexible
on-stream	100	200,000–300,000	ore-dressing process control	continuous analysis; excellent for rapid-trend analysis	sample presentation can be difficult; not very good for elements of low ($Z < 16$) atomic number
portable (x-ray)	100	15,000–30,000	rapid ore surveying, scrap-metal sorting, simple quantitative analysis	portable; crystal analyzer makes possible its use for almost all wavelengths	not as portable as isotope systems
portable (isotope)	2,000	5,000–15,000	rapid ore surveying, scrap-metal sorting, simple quantitative analysis	very portable	limited to fairly simple element systems; wavelength selected by means of filters

length-dispersive or energy-dispersive instruments. Probably ca 12,000 wavelength-dispersive instruments have been supplied commercially, and roughly half of these are in the United States. Energy-dispersive spectrometers became commercially available in the early 1970s, and there are ca 1000 units in use.

Single-channel wavelength-dispersive spectrometers are typically used for both routine and nonroutine analysis of a wide range of products, including ferrous and nonferrous alloys, oils, slags, sinters, ores, minerals, thin films, etc. These systems are very flexible but, relative to the multichannel spectrometers, are slow. The multichannel wavelength-dispersive instruments are used solely for routine, high throughput analyses where the principal need is for fast accurate analysis but where flexibility is unimportant. Energy-dispersive spectrometers have the advantage of displaying information about all of the elements simultaneously. They lack somewhat in resolution compared to the wavelength-dispersive systems, but they are very useful in quality control, trouble-shooting problems, etc. They have been particularly effective in the fields of scrap-alloy sorting, forensic science, and the provision of elemental data to supplement x-ray powder-diffraction data.

Crystallography. A crystal is a solid enclosed by symmetrically arranged plane surfaces that intersect at definite angles. Whether a substance is homogeneous or not can only be defined by the means that are available for measuring the crystallinity. In general, the shorter the wavelength, the smaller the crystalline region that is recognizable. Even noncrystalline materials have a degree of order, and each gives a diffraction pattern. For example, glassy materials and liquids generally give diffraction patterns in the form of one or more broad diffuse peaks or halos.

A crystalline substance has a definite form which is retained no matter what the physical size of the crystal. The macroscopic shape of a given crystal does not, however, necessarily reflect its microscopic structure, which is determined by the shape and packing of the individual atoms making up the compound. For example, a sodium chloride crystal has cubic symmetry, and by definition, the angles between all of the principal faces should be 90°. There are other crystal types that also have angles between faces of 90°, but the cube is unique in that the lengths of the sides are also equal. If a sodium chloride crystal were reduced in size until the smallest repeat unit or unit cell was determined, all sides of that unit would be equal. Therefore, a crystal type can be defined in terms of the lengths of the sides of its unit cell and the angles between the faces. Since every ordered material is made up of a unique arrangement and number of atoms, every ordered material gives a unique diffraction pattern.

In the single-crystal method, a monochromatic beam of radiation is allowed to fall onto a single crystal of the examined material, which is mounted in a holder that has at least two motions: rotation and tilting. A detector is mounted on the 2θ axis of a goniometer, and this detector also has two movements, one in the plane of the 2θ direction and one normal to this direction. By use of the movements of crystal and detector, all of the three-dimensional volume around the crystal can be searched for diffraction maxima. From the position and intensities of these maxima, it is possible to calculate the atomic arrangement of the crystal. The method is typically used for crystal-structure determination of organic materials.

Although the main use of x-ray diffraction for materials characterization is by use of the powder method, many special techniques have been developed for bulk crystalline and paracrystalline materials for the study of such effects as stress–strain, texture, topography, order–disorder transformations, etc (5).

Powder Diffraction. In x-ray powder diffraction, the specimen is typically a microcrystalline powder. A diffraction pattern of a single-phase material is typically in the form of a graph of diffraction angle or interplanar spacing vs diffracted line intensity. A pattern of a mixture of phases is thus made up of a series of superimposed diffractograms, one for each unique phase in the specimen. Each single-phase pattern can act as an empirical fingerprint for the identification of the various phases by pattern-recognition techniques based on a file of standard single-phase patterns. Quantitative phase analysis is also possible; however, it is difficult because of various experimental and other problems, including the many diffraction lines occurring from multiphase materials. A systematic means for unscrambling the superimposed diffraction patterns is based on the use of a file of single-phase patterns characterized in the first stage by their three strongest reflections and on a search technique used to match strong lines in the unknown pattern with the standard pattern lines (6). A potential match is then confirmed by comparison with the full pattern in question. The identified pattern is subtracted from the experimental pattern, and the procedure is repeated on the residue pattern until all lines are identified. The JCPDS (Joint Committee on Powder Diffraction Standards), International Centre for Diffraction Data in Swarthmore, Pa., maintains the powder data file, which currently stands at ca 38,000 patterns.

A Bragg-Brentano powder diffractometer allows a range of θ values to be scanned; the photon detector rotates at twice the angular speed of the specimen, thereby maintaining the required geometrical condition. The specimen consists of a random distribution of crystallites; thus, the appropriate planes are in the correct orientation to diffract the wavelength λ each time the Bragg condition is satisfied. Consequently, in x-ray diffraction, each peak angle value corresponds to a certain d spacing. With the wavelength-dispersive spectrometer, a single crystal of known d spacing is used to disperse the polychromatic beam of characteristic wavelengths of the sample, such that each wavelength diffracts at a discrete angle.

Instrumentation and techniques. Figure 6(**a**) is a photograph of a typical vertical powder diffractometer system, and Figure 6(**b**) illustrates the geometry of the system. This geometric arrangement is the Bragg-Brentano parafocusing system and is typified by a diverging beam from a line source F; the beam falls onto the specimen S, is diffracted, and passes through a receiving slit R to the detector. Distances FA and AR are equal. The amount of divergence is determined by the effective focal width of the source and the aperture of the divergence slit D. Axial divergence is controlled by two sets of parallel plate collimators (Soller slits) P and RP placed between the tube focus and the specimen, and the specimen and the scatter slit SS, respectively. Use of the narrower divergence slit gives less specimen coverage at a given diffraction angle, thus allowing the attainment of lower diffraction angles where the specimen has a larger apparent surface; thus, larger values of d are attainable. This is achieved, however, only at the expense of intensity loss. Choice of the divergence slit and its matched scatter slit is therefore governed by the angular range to be covered. The decision as to whether or not to increase the slit size at a given angle is determined by the available intensity. A scintillation detector is typically placed behind the scatter slit and this converts the diffracted x-ray photons into voltage pulses. These pulses may be integrated in a rate meter to give an analogue signal on an x/t recorder. Synchronizing the scanning speed of the goniometer with the recorder makes it possible to obtain a plot of degrees 2θ vs intensity, ie, the diffractogram. A timer–scaler is also provided for quantitative work and is used to obtain a measure of the integrated peak intensity of

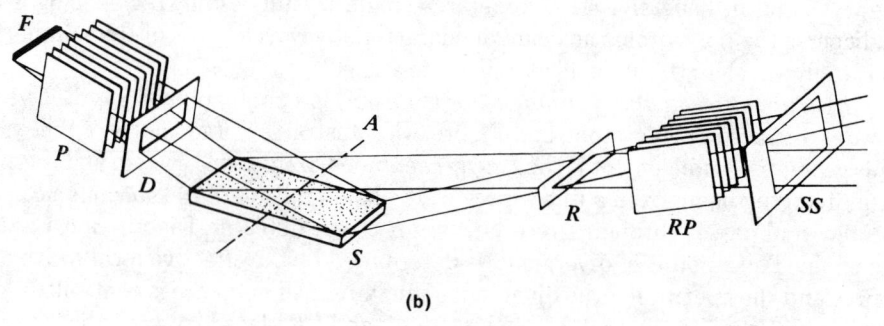

Figure 6. (a) The vertical diffractometer. (b) Geometry of the Bragg-Brentano diffractometer.

selected lines from each analyzed phase in the specimen. A diffracted beam monochromator may also be used to improve signal-to-noise characteristics.

Of all of the methods available to the analytical chemist, only x-ray diffraction provides general-purpose qualitative and quantitative information regarding the presence of phases, eg, compounds, in an unknown mixture. Although techniques such as dta provide some information about specific phase systems under certain circumstances, such methods are not general-purpose. A complication in the use of x-ray powder diffraction for the analysis of multiphase materials is that the patterns are superimposed one on top of the other, and consequently, there may be uncertainties as to which lines characterize the phases. A systematic search procedure is generally used and is based on the three strongest lines in the pattern. Once a potential match is indicated, the appropriate standard pattern is taken from the file and all lines are subtracted. This procedure is repeated until all significant lines in the unknown pattern are accounted for. The standard patterns are divided into those for mineral, inorganic, organic, and metallic substances, alloys, and common phases. X-ray powder diffractometry is neither a sensitive nor rapid means of analysis. The minimum detectable limit determined by routine qualitative procedures is several percent compared, for example, with a few ppm in x-ray fluorescence. A complete analysis would take several hours to complete, depending on the experience of the analyst and the complexity of the problem. The method is applicable to almost any crystalline material.

Once the presence of a phase has been established in a given specimen, the analyst should be able to determine how much of that phase is present by use of the intensities of one or more diffraction lines from the phase. However, it may be difficult to obtain an accurate value for these intensities. The intensities of the diffraction peaks are subject to various random and systematic errors which fall roughly into three categories. Those that are structure-dependent are a function of atomic size and atomic arrangement and depend on the scattering angle and temperature. Instrument-dependent errors are a function of diffractometer conditions, source power, slit widths, detector efficiency, etc. Specimen-dependent errors are a function of phase composition, specimen absorption, particle size, distribution, and orientation. For a given phase or selection of phases, all structure-dependent terms are fixed and have no influence on the quantitative procedure. Provided that the diffractometer terms are constant, the instrumental effects can also be ignored. If one calibrates the diffractometer with a sample of the pure phase of interest and then uses the same conditions for the analysis of the same phase in an unknown mixture, only the random errors associated with a given observation of intensity must be considered. The biggest problems in the quantitative analysis of multiphase mixtures are the specimen-dependent terms, specifically those dependent upon particle size and distribution, plus effects of absorption.

The applications of quantitative x-ray powder diffraction are many and varied. Some of the more common ones are ore and mineral analysis, quality control of rutile–anatase mixtures and retained austenite in steels, determination of phases in airborne particulates, various thin-film applications, study of catalysts, and analysis of cements. In the quantitative analysis of multiphase materials, accuracies of ca 1% can be obtained in those cases where the particle-orientation effect is either nonexistant or has been adequately compensated for.

Industrial. *On-Stream Analysis.* Approximately 200 x-ray spectrometry units for on-line process control are in regular use in mineral-beneficiation plants worldwide (7). Conventional x-ray spectrometry has been used for on-stream application where the normal fixed sample is replaced by a slurry cell. Special-purpose, isotope-excited systems have also been used successfully. The types of procedure considered for use in mineral extraction fall roughly into three categories: analysis of flowing slurries, eg, in most ore-dressing and mineral-beneficiation plants; analysis of dry, flowing solids, eg, crushed ores on conveyor belts or dry, raw cement mix en route to a kiln; and analysis of drill cores and insides of bore holes. Probably the main disadvantage of the on-stream analyzer-based system on a conventional x-ray-source–crystal-dispersion arrangement is that its inherent high cost limits the size of the installation to one, or at most two, units. This requires some extremely complex plumbing arrangements for the sample streams, which can cause delays in getting the analysis to the control center. A cycling system is generally used to allow sampling of multiple lines. However, even if certain priority channels are allowed, the cycle time can be 3–15 min. Although this may be satisfactory in some installations, the growing trend towards use of on-line computer control requires almost immediate use of analytical data and delays of minutes are unacceptable. In addition, there are problems in the determination of the lower atomic number elements ($Z < 20$), partially because of the attenuation of the fluorescent signal by the plastic windows of the flow cell but also because of particle effects which arise because individual solid particles in the slurry may be heterogeneous over the same order of magnitude as the penetration depth of the x-ray beam (see also Instrumentation and control).

The potential use of radioisotope sources for on-stream analysis is also well-established, and a survey includes an impressive list of applications and feasibility studies involving the use of radioisotopes for the on-stream analysis of slurries (8) (see Radioisotopes). Successful applications include the determination of elements including Pb, Sn, Ca, Ba, Zn, Cu, Mo, and Nb in flotation feeds and slurries. A great advantage of the radioisotope system is its low cost and the fact that sources can be obtained or specially fabricated for almost any special application. Unfortunately, the photon yield is generally much lower than with x-ray-tube-excited sources, which precludes the use of the relatively inefficient crystal-dispersive spectrometers. Use of energy-dispersive spectrometers has been important in this area, and additional techniques, including use of filters and secondary sources, have enhanced their applicability. Typical of these special sources is the γ–x source, the geometry of which is illustrated in Figure 7 (9). The source comprises three essential parts: the γ-source (in this case ^{153}Gd), the target material in the shape of a cone, and a lead shield. The γ source excites K x radiation from the target cone, and this secondary x radiation is directed onto the sample. Some commonly used γ–x sources and their energy ranges are as follows: ^{241}Am, 20–51 keV; ^{109}Cd, 8.6–19 keV; ^{153}Gd, 19–77 keV; ^{155}Eu, 71–77 keV; and ^{145}Sm, 14–32 keV.

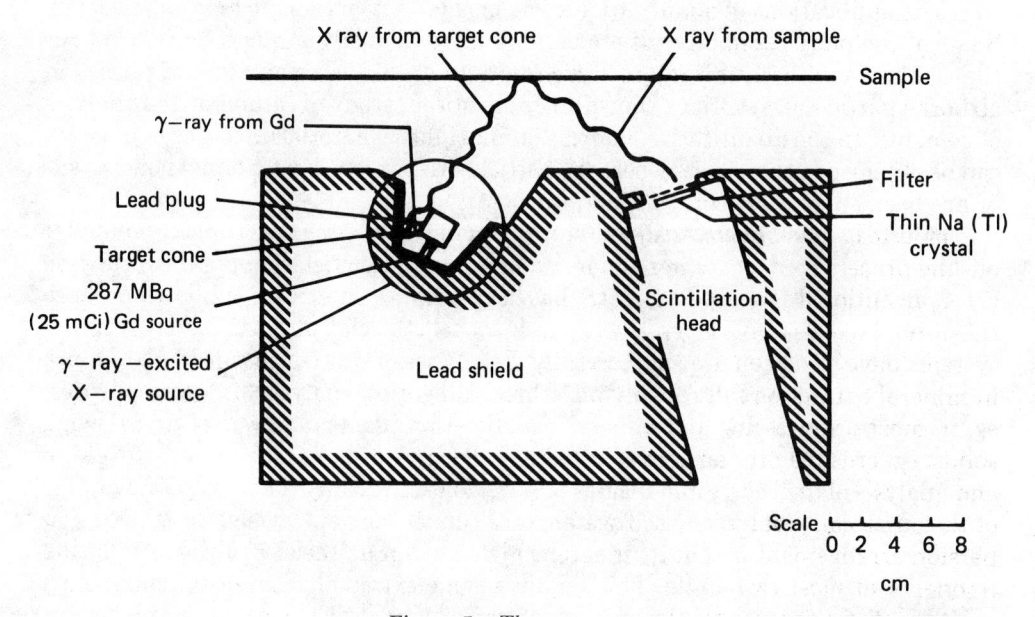

Figure 7. The γ–x source.

The conventional x-ray spectrometer modified for the analysis of a flowing sample has not been used successfully in the analysis of dry, flowing solids, mainly because of the particle-size effect. However, there has been much success in the use of successive sampling rather than continuous irradiation (see Fig. 8). In continuous irradiation, a sample depth equivalent to tens of micrometer is continuously irradiated for an integration time related to the number of count required; this time is typically ca 60 s. After a short delay for the data readout, the process is repeated. In successive sampling, a sample is taken at time intervals required to prepare and analyze the sample, usually ca 200 s. In some instances, the entire sample may be analyzed continuously, but in all cases, a homogeneous specimen of ca 2–5 g is analyzed. In the first case, homogeneity over the cross section of the stream is critical, but this is not so in the second method where a larger analyzed volume is taken and homogenized before analysis. The successive-sampling technique combined with high pressure pelletizing or borax fusion of the sample, followed by conventional x-ray fluorescence spectrometry, has been applied with great success, for example, to the analysis of cements (see Cement). Special instruments based on the use of semiconductor counters have been used for certain applications in the analysis of the insides of drill-core holes and specifically for the determination of high atomic-number elements in low grade ores. For example, there is a system for the determination of gold and silver in a silicon matrix at concentrations as low as 20 ppm (10). In this example, an annular source of ^{125}I is used with a silicon semiconductor counter for the determination of silver and a germanium detector for the analysis of the higher energy gold K spectrum.

Level and Thickness Gauging. The thickness of a solid, homogeneous material is readily measurable by x-ray absorptiometry, and this method can be applied for the estimation of bulk thickness as well as for the measurement of layer thickness. Figure

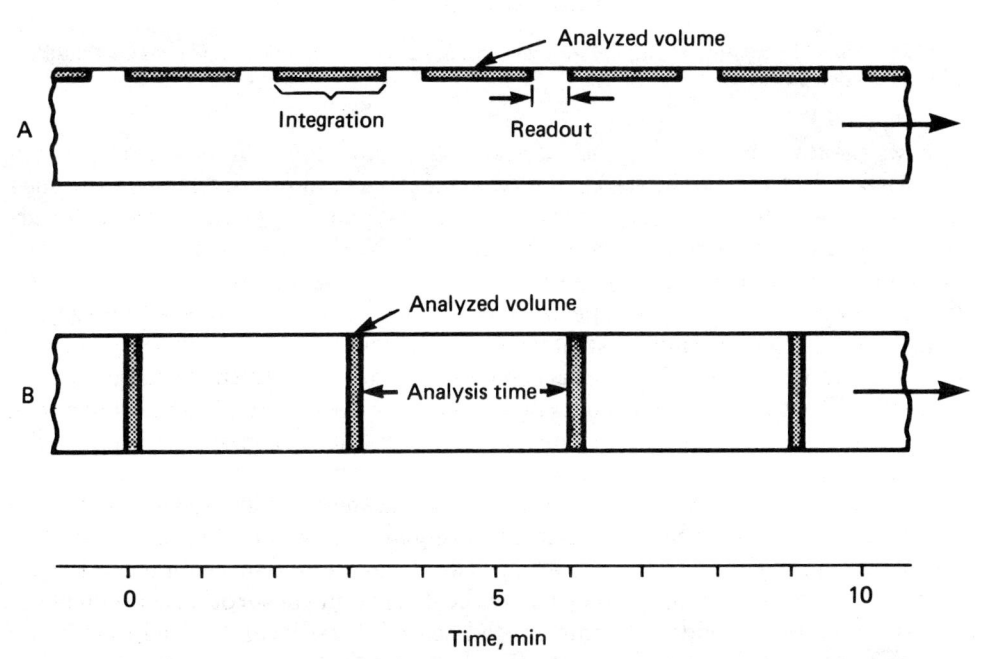

Figure 8. Methods for the analysis of a flowing, dry solid. A, Continuous irradiation. B, Continuous sampling.

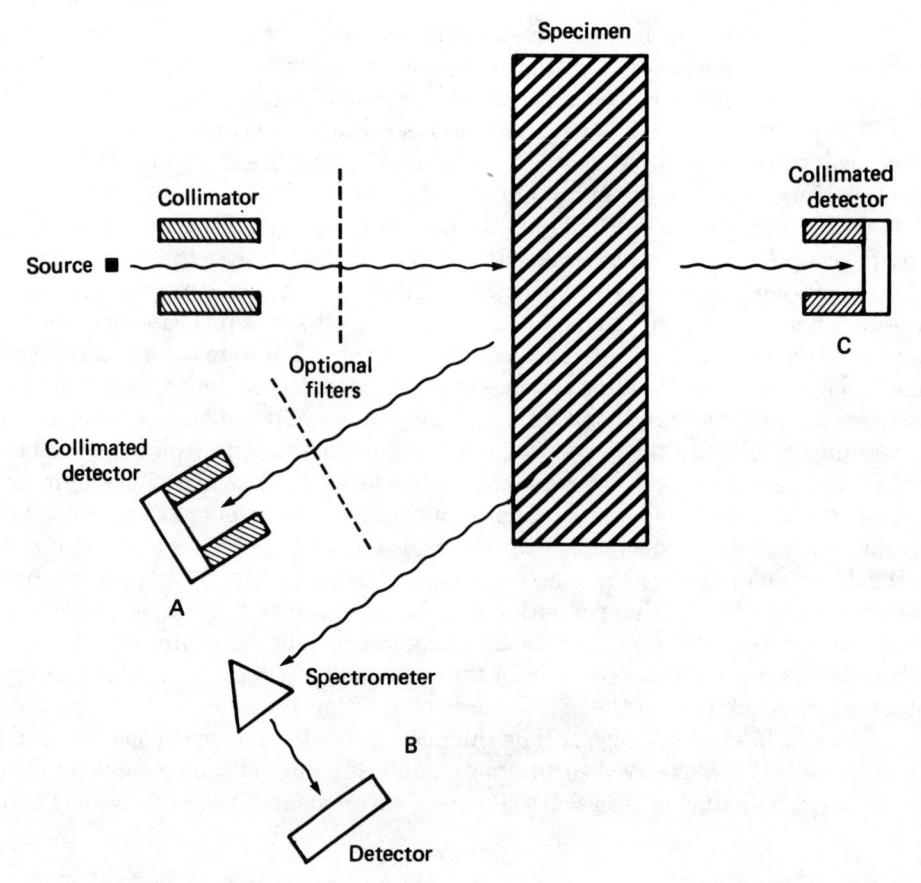

Figure 9. Three basic methods for x-ray gauging. A, Backscatter position. B, Fluorescence position. C, Transmission position.

9 shows the three basic techniques used for x-ray gauging and indicates a specimen being irradiated by a collimated primary beam of x-ray photons, which may or may not have been filtered to change the spectral distribution. In the backscatter arrangement A, a collimated detector with optional filter is placed at an angle from the incident beam but on the same side of the specimen as that beam. Both scattered and fluorescent radiation from the specimen can enter the detector. In the fluorescence arrangement B, a spectrometer is interposed between the specimen and the detector, such that only selected wavelengths can enter the detector. In the transmission arrangement C, a collimated detector is placed on the opposite side of the specimen than the source, such that only radiation transmitted by the specimen can enter the detector.

The most common method of measuring the thickness or consistency of a material is using the transmission mode C. As shown in equation 1, the attenuation of the x-ray beam depends upon specimen density, thickness, and mass-absorption coefficient. Thus, for a given material, eg, steel, where the density and absorption coefficient are known, the ratio of incident to transmitted beam intensity can be used to measure the thickness of the steel, for example, the completely automatic measurement of the thickness of sheet steel (11). Alternatively, where the apparent thickness is known, the same intensity ratio can be used to monitor the variations in absorption coefficient and, therefore, average atomic number of the material within this layer. As an example, this technique has been used to determine the plutonium content and distribution in flat reactor fuel plates (12). Where thickness and mass-absorption coefficient are known, the intensity ratio can be used to measure changes in bulk density caused by, for instance, the presence of voids or holes. This latter technique has been used successfully to measure porosity in the manufacture of materials, eg, leather, textiles, and storage batteries. Where solution contained within a pipe, for example, has a measurable effect on the total attenuation of an x-ray beam, the absorption method can be used to establish the level of liquid within the closed pipe.

X-ray methods have also been used with great success for the measurement of layer and coating thicknesses. Methods based on backscatter and fluorescence arrangement have been used. Figure 10 illustrates the different schemes that are typically used. A polychromatic beam of radiation falling onto a specimen excites characteristic radiation from the layer and substrate elements and is scattered by the sample. The total backscatter signal is atomic-number dependant and thus provides a simple and inexpensive means of estimating layer thickness or layer average atomic number.

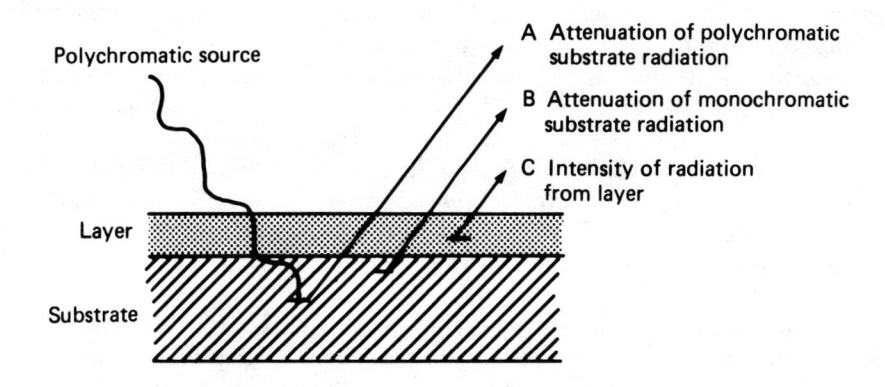

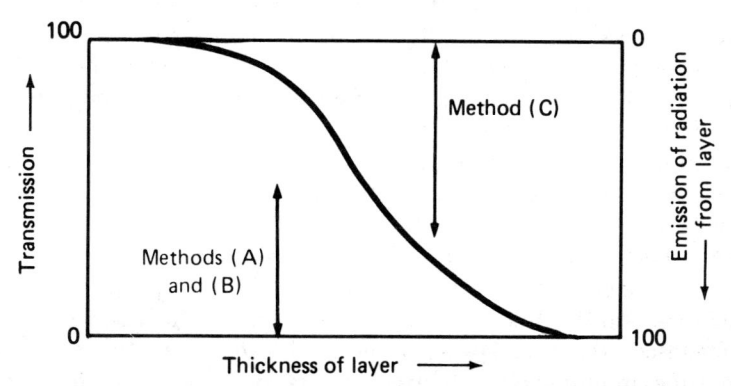

Figure 10. Methods for the measurement of coating thickness. A, Attenuation of polychromatic substrate radiation. B, Attenuation of monochromatic substrate radiation. C, Intensity of radiation from layer.

However, this technique is somewhat empirical, and methods based on the selection of the fluorescence signal from layer or substrate wavelength are usually more successful. The attenuation of the radiation intensity from the substrate layer by the surface layer can be used to measure layer thickness, ie, method A in Figure 10, provided that the layer thickness is within critical depth. An increase in layer thickness still gives a measurable change in the transmitted intensity. Where a single wavelength can be selected from the substrate layer radiation by means of filters or a spectrometer channel, the calculation of layer thickness becomes much easier and much more precise, since only one value of the mass-absorption coefficient has to be considered (see method B, Figure 10). Where the layer thickness radiation is within critical depth, the intensity of characteristic radiation from a layer element can be used to estimate the layer thickness (method C, Figure 10). These techniques have been applied to a wide variety of quality control problems, including the measurement of thin nickel coatings, the cladding thickness of fuel elements, the thickness of oxide layers on aluminum, and the thickness of aluminum on silicon (13–16).

Nondestructive Testing. X-ray imaging tests are widely used to examine interior regions of metal castings, fusion weldments, composite structures, and brazed components (see Nondestructive testing). Radiographic tests are made on pipeline welds, pressure vessels, nuclear fuel rods, and other critical materials and components that may contain three-dimensional voids, inclusions, gaps, or cracks that are aligned so that the critical areas are parallel to the x-ray beam. Since penetrating radiation tests depend upon the absorption properties of materials on x-ray photons, the tests can reveal changes in thickness and density and the presence of inclusions in the material. Essentially, two basic techniques are used. In x-ray radiography, the sample is placed between an x-ray source and a film. Exposure produces an absorption shadowgraph of the object being irradiated. In x-ray fluoroscopy, the film is replaced by some form of x-ray transducer, which converts the transmitted x-ray signal into a voltage level.

Radiography is an effective means of nondestructive testing inspection, and it is limited only by the time required for exposure and development of the film. Evaluation must, in this instance, take place off-line. This can cause complications, particularly if many objects are being examined and the analyst must correlate many exposures with these objects. X-ray fluoroscopy allows direct on-line examinations to be made. In this case, defects, eg, porosity or shrink holes, show as contrasting areas on the fluorescent screen. A simple fluoroscopic setup is shown in Figure 11. The fluoroscopic technique also allows the object to be manipulated so as to bring it into an optimum viewing position. It can thus be observed in motion on a screen in front of the operator, providing three-dimensional observation of the size and shape of a defect or inclusion. In most commercially available systems, the object on the screen can be enlarged. Fluoroscopy can be integrated into production or quality control in such a way that each inspection takes only a few minutes and no intermediate file of paperwork is required.

For the inspection of heavy or awkward castings, manipulation of the object may be impossible. In such cases, the x-ray tube must be transported to the object and maneuvered to obtain the desired angle for inspection. Castings that can be inspected within an enclosed area, eg, a specially designed lead-lined room, can be examined by means of tubes mounted on mobile or fixed stands or on overhead suspension systems. Such an arrangement provides an inspection system of maximum flexibility and accessibility.

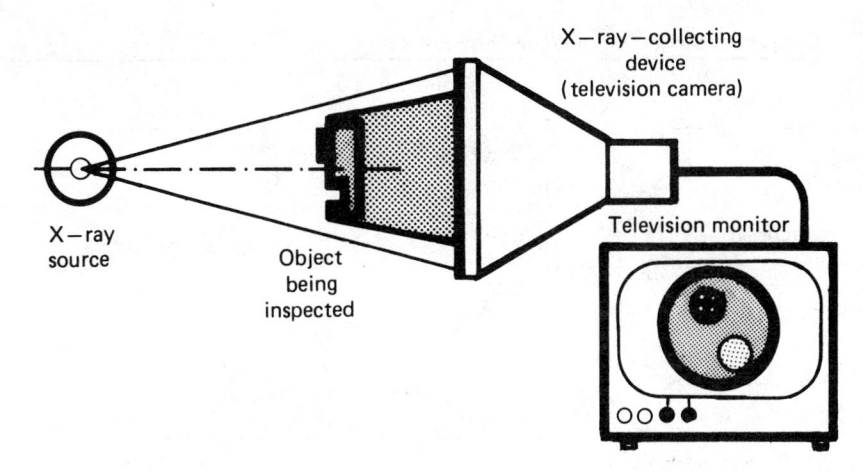

Figure 11. Basic hardware for x-ray fluoroscopic inspection.

Technological advances in fluoroscopy have greatly reduced inspection and interpretation times and provided a level of picture quality and resolution that meets extremely rigid specifications for light alloy and iron–steel castings at thicknesses used in most cast components. However, the incentive to tighten inspection criteria for castings has increased. It is often necessary to establish what happens inside a given sample, and fluoroscopic techniques allow such examinations to be made, eg, examinations of a nuclear-fuel-rod container or the behavior of a projectile inside a gun barrel. In the food processing and packaging industry, x-ray inspection is becoming increasingly important. With growing output and greater use of mass production, the chances of accidentally introducing foreign bodies, eg, glass or metallic particles, increase. X-ray techniques are also used in the search for new food products.

Security Screening Systems. X-ray imaging is used by customs and security personnel to examine the contents of packages and parcels. X-ray machines became practical for security applications in the late 1970s because of the ability to generate an x-ray image with a very low dose rate of x rays that would not even damage photographic film (17). The dose rate is ca 50,000 times less than that from high dose systems with direct-viewing fluorescent screens. Table 3 lists the typical characteristics of high and low dose rate systems.

The requirements of security x-ray systems generally demand that they are enclosed, self-contained units (called cabinet x-ray systems) that can be operated by unskilled personnel. The units must also be absolutely radiation safe under all operating conditions. Low dose systems are usually designed in terms of a radiation-shielded cabinet, in which the radiation source is contained within a radiation-safe enclosure. A conventional system has a control panel, x-ray generator, viewing system, and door or port through which the luggage to be inspected is loaded. Because low dose systems require much less shielding than high dose systems, items to be inspected can be transported through them at high speed via flexible lead curtains, which act as baffles.

There are essentially three basic techniques used in low dose systems: continuous x-ray systems, pulsed x-ray systems, and scanning x-ray beam flying-spot scanning. A typical unit is illustrated in Figure 12. The continuous system operates with a low level x-ray beam which illuminates a fluorescent screen. A multistage light amplifier

698   X-RAY TECHNOLOGY

Table 3. Comparison of High Dose and Low Dose Systems

Property	Searching, high dose and medium dose	Screening, low dose and film safe
weight	heavy	light
image quality	good	fair to good
voltage, kV	100–200	50–100
power consumption	high	low
throughput	slow	fast
operating controls	many	few
primary beam dose rate, Gy/h[a]	0.005–5.0	low dose: 0.005–0.05 film safe: 0.01 mGy per inspection
mobility	poor	good
size	medium to large	small to medium
maintenance	medium	medium

[a] To convert Gy/h to rad/h, divide by 0.01.

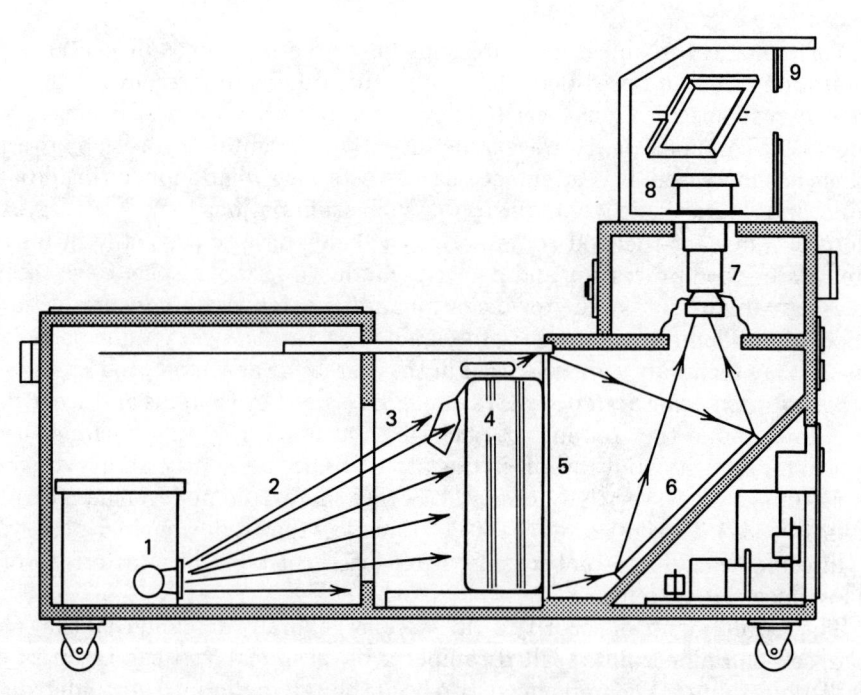

Figure 12. Typical x-ray security system. 1, x-ray tube; 2, x-ray beam; 3, x-ray chamber; 4, item being inspected; 5, fluorescent screen; 6, mirror; 7, light amplifier or television camera; 8, adjustable viewing mirror; and 9, adjustable viewing hood.

is attached to a closed-circuit television (TV) camera for viewing on a TV monitor or to an output lens for direct viewing of the phosphor of the light-amplifier module. The pulsed x-ray system produces a single x-ray pulse which momentarily illuminates the screen. During this time, a closed-circuit TV camera views the screen and the video signal is stored. The image is then repeatedly displayed on a TV monitor. The storage device is cleared before each inspection cycle. The scanning x-ray beam system operates on a two-dimensional raster scan, in which the beam moves vertically while a conveyor belt moves the parcel being examined horizontally. The vertical beam is divided and stored in an X–Y raster, which is displayed on the TV monitor; thus, the x-ray image is presented as a series of closely spaced dots of differing brightness. The storage device repeatedly retrieves information from the TV monitor picture in the same manner as the pulsed x-ray system.

In the United States, there are ca 1000 low dose systems at transportation and public sites and ca 100–200 high dose systems at governmental and private security locations. All of the low dose systems are film safe, since many travelers carry film and cameras on their persons. In the United States, several thousand prohibited weapons and prohibited articles are detected by these screening systems each year. In addition to their ability to detect such items, they also are a psychological deterrent to criminals.

Lithography. X-ray and uv-based lithographic techniques have provided the fine-line instrumentation required for the manufacture of microelectronic circuits (18). Lithography is the transfer of an image from one surface to another and is used industrially to make micropatterns for high speed reproduction of components. X-ray lithography is essentially proximity printing with soft x-rays (19). It is used for the manufacture of the replicate pattern; chemical etching then reproduces this pattern on the required substrate. In x-ray microscopy, an absorption shadowgraph is produced on a photographic plate and the developed film is viewed through an optical microscope. Although the use of x-ray lithography is a relatively new technique dating back only to the early 1970s, there have been rapid advances in the field, and today the technique is widely accepted (20–21) (see also Photoreactive polymers; Integrated circuits).

Figure 13 is a simplified diagram of an x-ray lithographic exposure station and indicates the three essential parts: a very intense x-ray source; a mask consisting of the device pattern to be transferred and made on some highly absorbing medium mounted on a thin-membrane substrate; and an x-ray resist of high resolution suitable for the subsequent device fabrication steps. The maximum absorption for most useful resist materials is at ca 50 nm, but the best replica resolution is at ca 7.5 nm. The sources that deliver this highly intense x-ray beam might be a rotating anode x-ray tube at 15–25 kW, an electron storage ring emitting synchrotron radiation, or multi-million (10^6) degree plasmas heated by lasers or electric discharges (22) (see Plasma technology in Supplement Volume). Some success has also been achieved with specially designed, stationary x-ray sources (23). The material for the production of the absorbing mask usually is gold supported on a plastic substrate, eg, mylar, poly(methyl methacrylate) (PMMA), beryllium, or a polyimide.

Figure 14 illustrates the stages in the production of a device by the x-ray lithographic technique. A mixture of ethylenediamine, catechol, and hydrofluoric acid etches all types of silicon, except that which is heavily doped with boron. The absorber pattern is first transferred to a thin silicon wafer (B), which is doped with boron. The silicon is then etched through the rear side through windows in the protected silicon layer, leaving the boron-doped membrane stretched over the frame of silicon which was protected during etching. Membranes produced in this way are transparent to red light and are useful for the wavelengths 0.7–1.5 nm (21). Typical thicknesses are 1–5 μm.

The principal advantage of x-ray lithography relative to other techniques results primarily from the fact that x rays are not reflected and are only slightly refracted; that is, practically, x rays travel in straight lines and accurately reproduce pattern

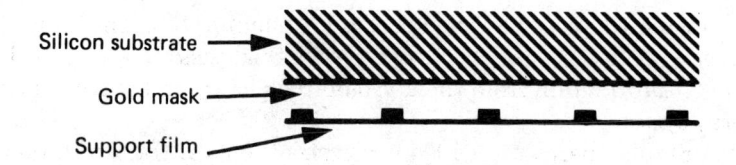

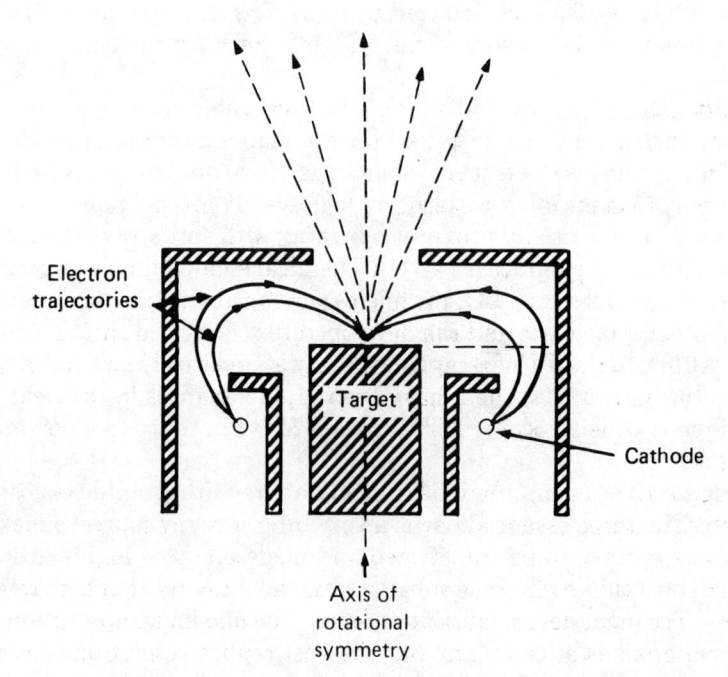

Figure 13. X-ray exposure station with a rotating anode x-ray tube.

characteristics. The high aspect ratios obtained with x-ray lithography cannot be obtained with any other lithographic technique. It also appears that mask life with x rays is better than with electron-beam methods. However, a disadvantage of the x-ray method is the low overall efficiency of x-ray sources, requiring expensive, high power generators. Also, precise alignment with x rays is difficult, and the first mask is invariably produced with an electron beam. It is hoped that future development of focusing x-ray systems will avoid this costly and inconvenient intermediate step.

Medical and Scientific Research. *Diagnostic.* One of the first practical applications of the use of x rays was for the radiographic analysis of the human body. The absorption of x rays varies as a fourth power of atomic number. Since human bones are high in calcium ($Z = 20$), they absorb x rays to a greater extent than do flesh and soft tissue. Thus, if a limb is placed between a polychromatic beam of x rays and a piece of photographic film, transmitted photons falling onto the film produce a shadowgraph of the bone structure of the limb.

Factors important to the design of systems for radiographic analysis are the energy distribution of the source, ie, the voltage on the x-ray tube and any filtration of the beam, and the response characteristics of the film, which depends on the graininess

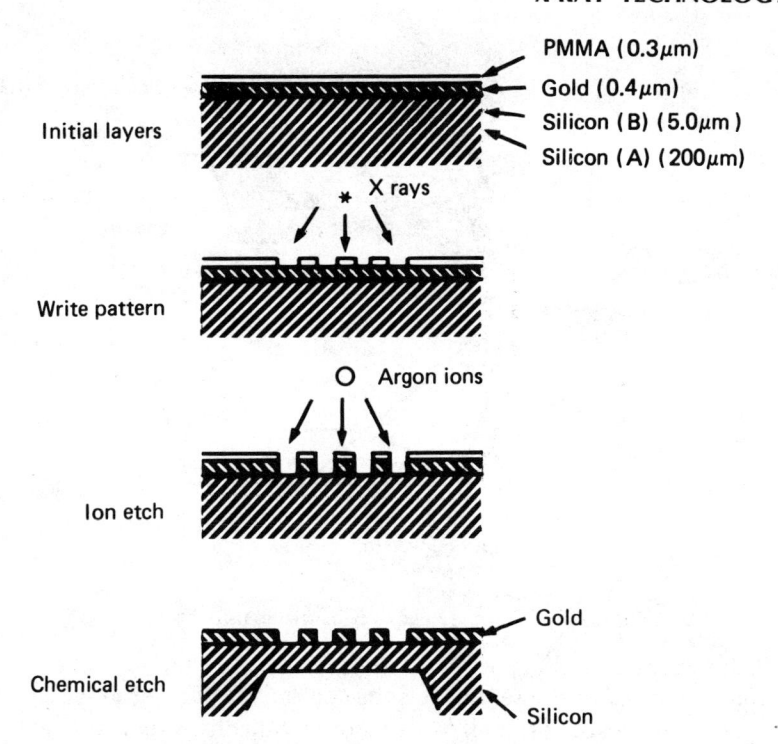

Figure 14. Steps in the fabrication of a device by x-ray lithography.

of the emulsion. Contrast and sharpness of the film image are important. Exposure time and image quality can be improved by use of a sandwich arrangement of a double-sided emulsion between phosphor screens. A main disadvantage of film is that it represents a passive rather than an active detection device. If a given exposure fails to reveal the detail sought by the radiographer, the exposure may have to be repeated several times. By direct use of a fluorescence screen in place of the film, it is, in principle, possible to obtain a real-time image. However, in practice, an intense primary source would be required to give an observable signal. Use of various solid-state amplifiers as an image intensifier obviates this problem (24).

In its simplest form, the solid-state amplifier comprises an x-ray sensitive photocathode and an electroluminescent screen. The photoconductor changes the local field strength of the electroluminescent screen as a function of x-ray flux, thereby producing light and dark areas. Figure 15 is a schematic of an image intensifier, in which primary electrons produced when x-ray photons fall onto the photosensitive layer are accelerated and focused onto the output screen. A closed-circuit TV system usually is coupled to the output port of the image intensifier for fluoroscopic examination. An important development in real-time imaging was the vidicon tube (25). The term vidicon was first applied to the tubes that replaced the photoelectric-charge-storage–TV tube combination. The original vidicons were based on an Sb_2S_3 photoconductor, which was scanned over its rear surface by an electron beam. In these and in modern instruments, the beam scans the surface of the photoconductor and deposits a small negative charge, thereby bringing the scanned area to the same potential as the cathode. X-ray photons falling onto the front surface induce conductivity, producing a positive potential proportional to the quantity of radiation. When the

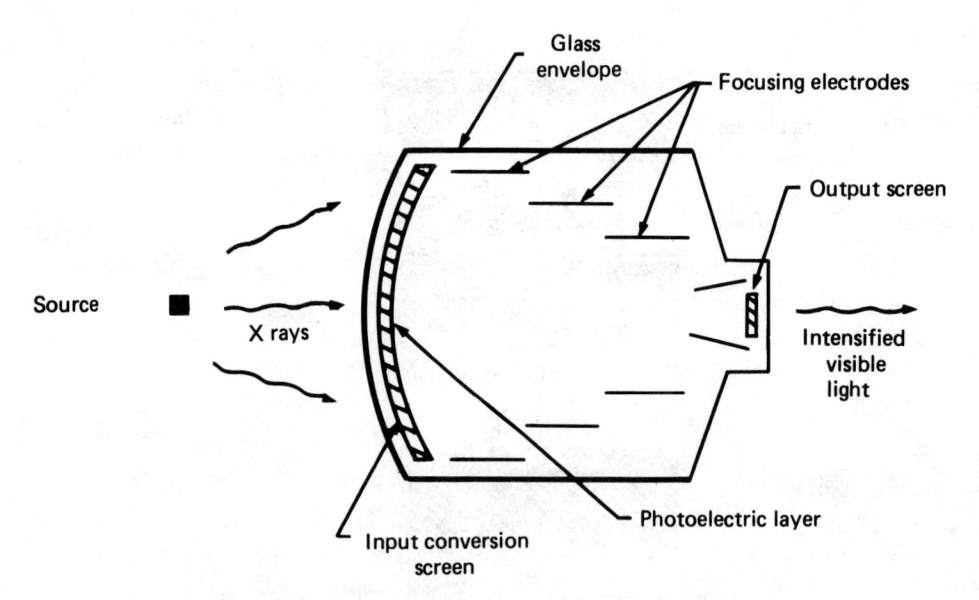

Figure 15. Conventional x-ray image intensifier tube.

scanning beam returns to this particular area, a certain negative charge is required to return the area to cathode potential; it is the deposition of this negative charge which is used to generate a display signal. Most modern vidicon systems use lead oxide or amorphous selenium as the photoconductive surface.

Conventional real-time fluoroscopy involves a vidicon-type detector and allows viewing of the image on a TV monitor. Use of such a system enables real-time visualization of dynamically changing situations in the living body. Figure 16 is a comparison of several different experimental configurations as used in image-intensifier systems. An experimental problem is obtaining sufficient contrast between the volume of interest and the immediate surroundings. Unfortunately, blood and soft tissue have very similar absorption characteristics, since their average atomic numbers are almost identical. Contrast can sometimes be enhanced by ingestion or injection of contrast agents, eg, barium, immediately before or even during fluoroscopic examination. In arteriography, image enhancers are injected into the bloodstream.

Another method of image improvement is the use of data-processing techniques, eg, image subtraction, or image enhancement. In the former method, the significant intrinsic noise associated with the vidicon–TV tube system can be almost completely removed by storing the experimental image in a suitable digitized form and then subtracting a similar digitized stored image representing the average dark field pixel pattern. The reticon systems have about a sixfold better signal–noise ratio than the vidicon systems. For example, one system includes a reticon camera with a 32×32 photodiode array, which collects an image that is then fed by an analogue device to a digital converter to a magnetic disk (26). This system has been used on an experimental basis to detect pulmonary pulsations in monkeys. The signal from the x-ray detector is stored as a digital image; computer-based data processing is applied, and the treated image is redisplayed. Image enhancement is a high speed, noise-reduction, edge-enhancement process that acts as an interface between the detector system and the TV image. In many ways, it is similar to the sequence used in analytical x-ray instrumentation, where raw data are collected, stored, and smoothed; peak maxima are

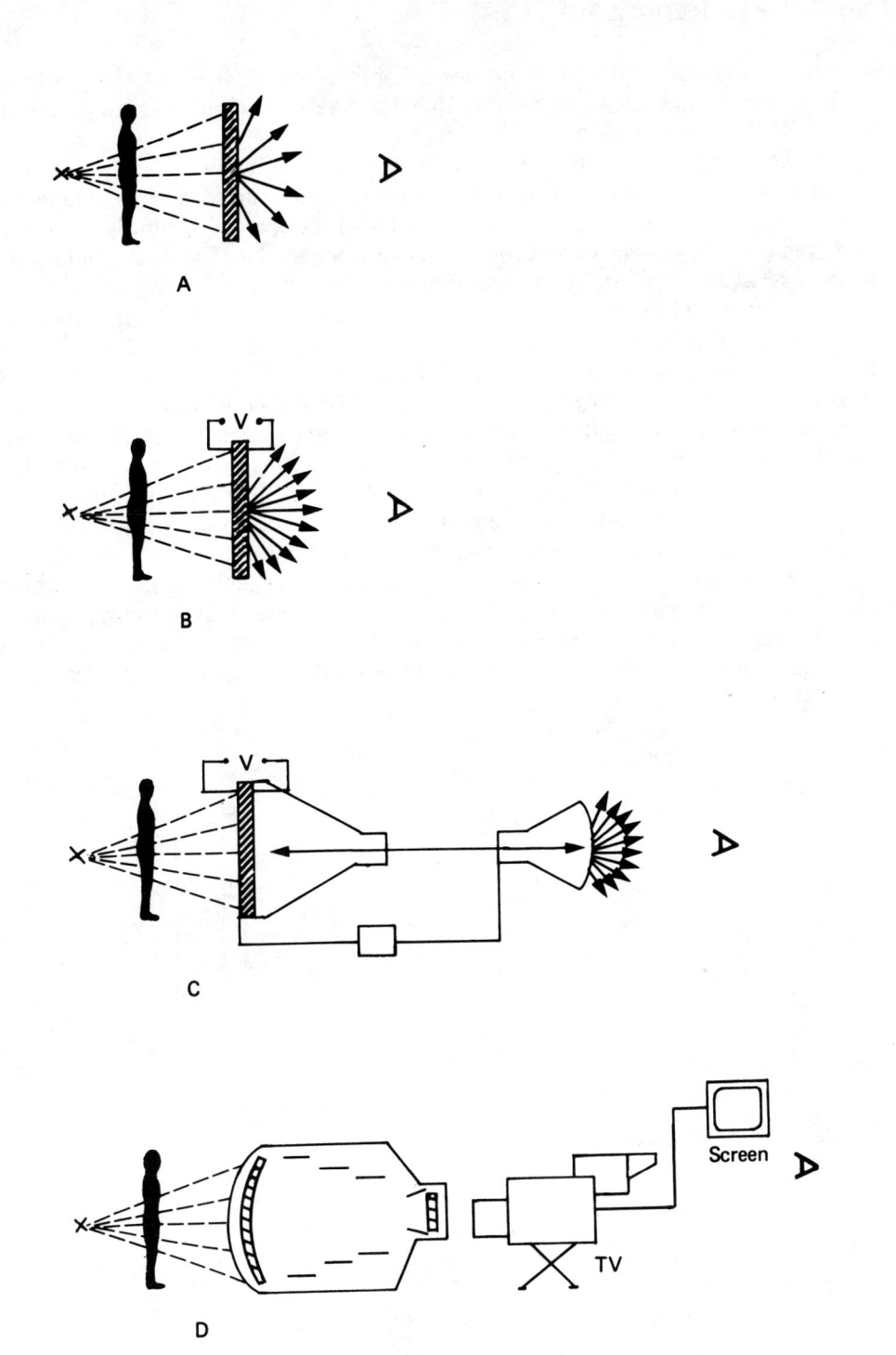

Figure 16. Image intensifier systems. A, Fluoroscopic screen. B, Solid-state amplifier. C, Vidicon system. D, Intensifier tube with television monitor.

sought; and the treated data are redisplayed. For real-time applications, this processing of fluoroscopic data allows rates of ca 10 MHz pixels, which is equal to processing 262,144 (512^2) pixel words on a TV frame in ca 0.033 s.

Tomography. In normal radiography, the image is maintained as sharp as possible during an exposure by minimizing relative movement of source, object, and detection medium during exposure. In tomography, suitable relative movement is introduced in all planes of the object except for the one of interest. This blurs all nondesirable planes except for the plane of interest. Figure 17 compares normal projective imaging with tomographic imaging. In the former case, the two orthogonally placed objects produce an image of similar sharpness. In tomography, movement of the detector during the course of a contiguous series of exposures focuses just one of the orthogonal members. Tomography may be described as sectional roentgenography. The x-ray tube moves curvilinearly during exposure but is synchronized with the recording medium in the opposite direction; consequently, the shadow of the selected plane remains stationary on the film while all other planes are displaced relative to it and appear blurred or obliterated. The tomographic method gives three-dimensional detail, which would be completely masked by conventional fluoroscopy.

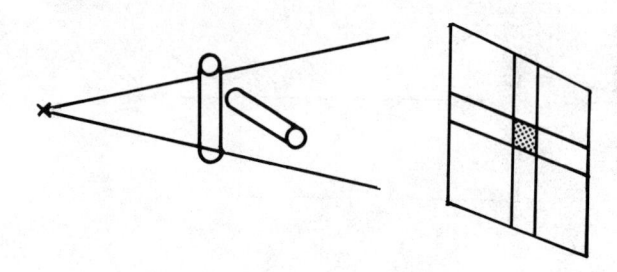

The hardware for tomographic work typically allows multiple exposures. Generally, this is done with a single x-ray source and a multiple-array detector. Figure 18 is a photograph of a commercial instrument. This has a rotating anode x-ray tube, which operates at 280 kW and generates a pulse width of 2 ms and a pulse frequency

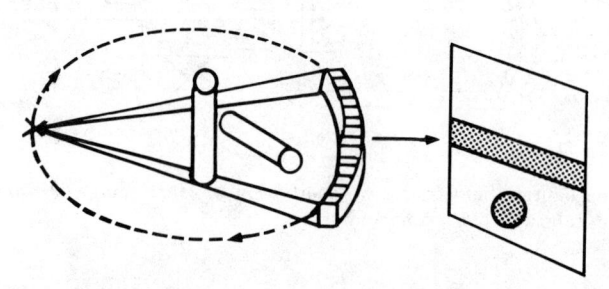

Figure 17. X-ray fluoroscopy.

Figure 18. Philips Tomoscan 310.

of 8 ms. The detector is a xenon-gas ionization chamber with 576 individual detector elements. It produces excellent quality pictures in extremely short exposure times, typically in a few seconds. Advanced computer software is also available with these systems; they provide great flexibility in terms of pattern storage, enhancement, and redisplay called computerized tomography (CT) (see Superconducting materials). Figure 19 is a tomographic section of a thorax. All of the three-dimensional information is stored, allowing the display of different sections or slices through the area of interest. Newer developments in tomographic instrumentation utilize multiple x-ray sources, which permit an even higher degree of flexibility (27). One such system consists of twenty-eight x-ray sources arranged in a semicircle with twenty-eight corresponding image intensifiers (28). This entire system permits mathematical reconstruction imaging of a cylindrical, three-dimensional volume (23 cm axially, 30 cm transaxially). This system should be applicable to the study of dynamic relationships of anatomic structure and to the function of moving organs, especially the heart, lungs, and circulation. This technique is invaluable for the study of patients with cardiopulmonary disabilities or abnormalities of vascular anatomy.

In vascular radiography, the principle of geometric magnification is applied to demonstrate the smallest possible anatomical structures. This technique essentially involves a fan-type x-ray beam which diverges through the object under examination. Details inadequately imaged in a normal radiograph become strikingly visible by means of this approach. The same idea can be applied to tomography but with even greater advantage, in that the inefficient use of detectors in conventional fan-beam geometry is avoided by coupling the source and detector. A direct fan-beam scanner with a rotating x-ray source–detector assembly attached to it is mounted on a rigid frame, which can be moved radially with respect to the isocenter. Mechanical movements and ad-

706 X-RAY TECHNOLOGY

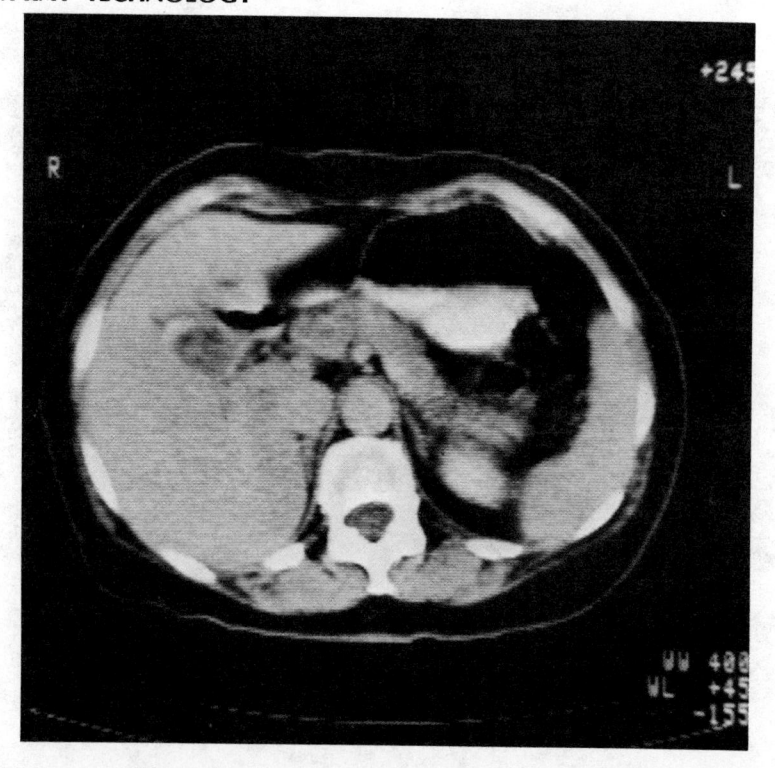

Figure 19. Tomographic section of a human thorax.

justments can be made automatically without repositioning the patient. Reconstructive zoom and variable-image regeneration capability provides increased visibility of the area without rescanning.

Astronomy. X-ray astronomy allows exploration of the universe far beyond that made possible with optical and radioastronomy. In 1962, an Aerobee rocket equipped with three Geiger counters was launched in New Mexico, and the first measurement was made of x-rays from a source beyond the solar system (29). The most modern detectors can detect sources many orders of magnitude weaker than were observed with the detectors used in 1962 (30). Cosmic x rays are generated in regions of extreme violence, either where electrons have been accelerated to very high speeds or, more commonly, where there is gas at a temperature of millions (10^6) of degrees Celsius. X-ray astronomy has been applied to view clouds of ultrahot gases from the atmospheres of stars to enormous pools that envelop whole clusters of galaxies, as well as streams of gas apparently about to disappear from the universe into black holes.

The earth's atmosphere absorbs most of the cosmic x-ray photons before they reach the earth's surface, the absorption being almost complete at an altitude of ca 30 km. Consequently, observations can only be made from rockets or orbiting satellites. A very intense x-ray source (Sco X-1) has been determined in the constellation Scorpius (31). The strongest x-ray sources in our galaxy are double-star systems; in each such system, a compact neutron star draws gas from a normal companion star. As this gas spirals down onto the neutron star's surface, it heats to ca 10^7 °C and produces copius quantities of x-ray photons. For example, the gas streams in Sco X-1 emit 10^{10}

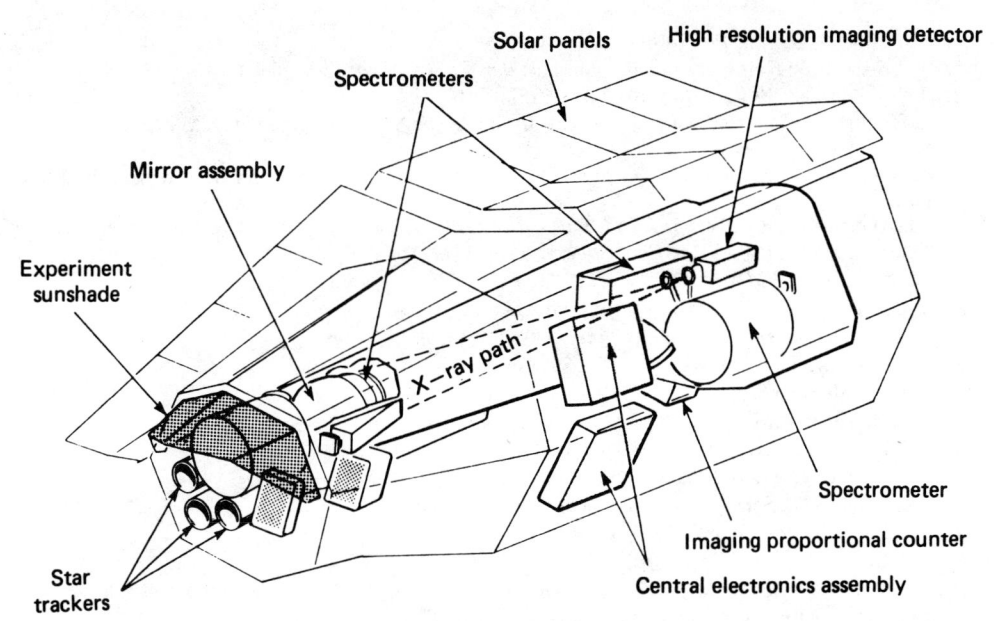

Figure 20. The Einstein observatory.

times as much power in x rays as a star like the sun. Therefore, the x-ray output from Sco X-1 is one thousand times the sun's power output at all wavelengths.

The development of detector arrays allows directional information to be obtained, and collimators make possible observations from a single source without influence from others. Since x-ray photons cannot be bent with a lens, as can visible light, grazing-incidence techniques are used. X rays can be reflected only if they strike a polished metal surface at a very shallow angle. Thus, x-ray telescopes have very different configurations from optical and radio reflectors. An x-ray telescope used simply to concentrate x rays has a slightly tapered cylinder, which is highly polished on the inside. An imaging x-ray telescope adds a confocal hyperboloid of revolution. X rays entering parallel to the axis of the telescope and near the edge of the tube strike the inner surfaces at grazing incidence, reflecting to reach the focus further along the axis. Figure 20 shows a sketch of the Einstein observatory, which was launched in 1978. This satellite is equipped with imaging x-ray mirrors and is able to pinpoint sources with great accuracy. All orbiting observatories have a limited lifetime, which is predicted by the lifetime of batteries or electronics package, or until the gas used for driving mechanical movements to sight on one source or another is exhausted. The longest lived satellite was Copernicus, whose x-ray telescope was working well when the satellite was turned off after eight years of use. The Einstein observatory was closed down after two and one half years of use owing to exhaustion of the gas supply (see also Space chemistry).

BIBLIOGRAPHY

"X-Ray Fluorescence Spectrography" in *ECT* 1st ed., Vol. 15, pp. 176–185, W. Parrish, Philips Laboratories; "X-Ray Analysis" in *ECT* 2nd ed., Vol. 22, pp. 438–467, W. Parrish, International Business Machines Corporation.

1. W. C. Roentgen, *Ann. Phys. Chem.* **64,** 1 (1898).
2. W. Friedrich, P. Knipping, and M. von Laue, *Ber. Bayer. Akad. Wiss.*, 303 (1912).
3. H. G. J. Moseley, *Philos. Mag.* **26,** 1024 (1912).
4. R. Jenkins, R. W. Gould, and D. A. Gedcke, *Quantitative X-Ray Spectrometry*, Marcel Dekker, Inc., New York, 1981.
5. H. P. Klug and L. E. Alexander, *X-Ray Diffraction Procedures*, 2nd ed., John Wiley & Sons, Inc., New York, 1974.
6. J. D. Hanawalt and H. W. Rinn, *Ind. Eng. Chem.* **8,** 244 (1936).
7. R. Jenkins and J. L. de Vries, *Can. Spectrosc.* **16,** 3 (1971).
8. J. S. Watt, *Aust. I.M.M. Proc.* **233,** 69 (1970).
9. J. A. Hope and J. S. Watt, *Int. J. Appl. Rad. Isot.* **16,** 9 (1965).
10. P. G. Burkhalter, *Instrum. Ind. Geophys.* **20,** 353 (1969).
11. S. Berstein, *J. Soc. Nondestr. Test.* **16,** 305 (1958).
12. M. C. Lambert, *U.S. A.E.C. Rep.* **HW-57941** (1958).
13. E. P. Bertin and R. J. Longobucco, *Met. Finish.* **60,** 42 (1962).
14. B. J. Lowe, P. D. Sierer, and R. B. Ogilvie, *Nucleonics* **17,** 70 (1959).
15. F. Lihl, H. Ebel, and R. Struszkiewicz, *Z. Metallkd.* **59,** 63 (1968).
16. J. E. Cline and S. Schwartz, *J. Electrochem. Soc.* **114,** 605 (1967).
17. D. J. Haas, *Security World*, 20 (Aug. 1976).
18. M. P. Lepselter, *IEEE Spectrum*, 26 (May 1981).
19. G. A. Garretson, *X-Ray Lithography Primer, IEEE X-Ray Standards Workshop*, Oct. 1981.
20. D. L. Spears and H. I. Smith, *Electron. Lett.* **8,** 102 (1972).
21. E. Spiller and R. Feder, *Topics Appl. Phys.* **22,** 35 (1977).
22. D. J. Nagel, *SPIE Newsletter* **297** (1981).
23. J. R. Maldonado and co-workers, *J. Vac. Sci. Technol.* **16,** 1942 (1979).
24. W. Kuhl, *ASTM STP-716*, American Society of Testing and Materials, Philadelphia, Pa., 1980, p. 33.
25. U.S. Pat. 2,817,781 (Dec. 1957), E. E. Sheldon.
26. M. C. Ziskin and C. M. Phillips, *ASTM STP-716*, American Society of Testing and Materials, Philadelphia, Pa., 1980, p. 294.
27. E. L. Ritman and co-workers in ref. 26, p. 277.
28. E. L. Ritman and co-workers, *Proc. Mayo Clinic* **53,** 3 (1978).
29. H. Freidman, *Space Research*, Vol. 4, N. Holland Publishing Co., Amsterdam, The Netherlands, 1964, p. 966.
30. N. Henbest, *New Sci.*, 720 (Mar. 1982).
31. R. Giacconi, H. Gursky, F. R. Paolini, and B. Rossi, *Phys. Rev. Lett.* **9,** 439 (1962).

RONALD JENKINS
Philips Electronic Instruments, Inc.

XYLENES AND ETHYLBENZENE

Xylenes and ethylbenzene are C_8 benzene homologues with the molecular formula C_8H_{10}. The three xylene isomers are o-xylene, m-xylene, and p-xylene, which differ in the positions of two methyl groups on the benzene ring. The term mixed xylenes describes a mixture of ethylbenzene and the three xylene isomers. Mixed xylenes are largely derived from petroleum (see Table 1).

p-xylene   m-xylene   o-xylene   ethylbenzene

Ethylbenzene is always present, except in the small amount of xylenes produced by toluene disproportionation. Ethylbenzene is a diluent which can accumulate in recycle processing schemes and, hence, has a strong impact on the separation of the individual xylene isomers. Ethylbenzene is commercially important as a precursor to styrene; it is usually made from benzene and ethylene (see also Styrene).

The demand for gasoline far surpasses that for petrochemicals (see Gasoline and other motor fuels). In 1979, gasoline production was 356×10^6 metric tons, whereas that of benzene, xylene, and toluene was ca 6×10^6 t, 2.7×10^6 t, and 0.9×10^6 t, respectively (2). Gasoline contains ca 24% aromatic material. Thus, the demand for mixed xylenes for petrochemicals use is strongly influenced by the demand for gasoline. By-product credits have a decisive impact on the production economics of the individual isomers.

U.S. mixed-xylenes production is distributed roughly as follows: p-xylene, 50–60%; to gasoline blending, 10–25%; o-xylene, 10–15%; solvents, 10%; ethylbenzene, 3%; and m-xylene, 1% (3).

p-Xylene is oxidized to terephthalic acid or dimethyl terephthalate as the first step in a sequence that produces polyesters (qv). o-Xylene is oxidized to phthalic anhydride which is converted to plasticizers (qv). m-Xylene is oxidized to isophthalic acid for use in polyesters. Ethylbenzene is dehydrogenated to styrene.

Table 1. U.S. 1980 Supply Sources for Mixed Xylenes[a]

Source	Thousand metric tons	%
catalytic reformate	3507	94.5
pyrolysis gasoline	163	4.4
toluene disproportionation	33	0.9
coke-oven light oil	10	0.2
Total	3713	

[a] Ref. 1.

Physical Properties

Because of their similar structure, the three xylenes and the isomeric ethylbenzene exhibit similar properties (see Table 2).

The distillation characteristics of the C_8-aromatic compounds are of considerable practical importance. Ethylbenzene can be separated from the remaining isomers by means of a column with more than 300 trays and a high reflux ratio. In commercial practice, two or three columns are used in series. This procedure is highly energy intensive and has become less competitive as the cost of energy has increased. Furthermore, advances in the technology of the alkylation of benzene with ethylene makes distillation even less useful (9–10). However, the removal of ethylbenzene simplifies the subsequent separation of the C_8-aromatic compounds.

o-Xylene is more readily separated from m-xylene because of a 5°C difference

Table 2. Physical Properties for C_8-Aromatic Compounds[a]

Property	p-Xylene [106-42-3]	m-Xylene [108-38-3]	o-Xylene [95-47-6]	Ethylbenzene [100-41-4]
molecular weight	106.167	106.167	106.167	106.167
density at 25°C, g/cm³	0.8610	0.8642	0.8802	0.8671
boiling point, °C	138.37	139.12	144.41	136.19
freezing point, °C	13.263	−47.872	−25.182	−94.975
refractive index at 25°C	1.4958	1.4971	1.5054	1.4959
surface tension[b], mN/m (= dyn/cm)	28.27	31.23	32.5	31.50
dielectric constant at 25°C	2.27	2.367	2.568	2.412
dipole moment of liquid, C·m[c]	0	0.30	0.51	0.36
critical properties[b]				
critical density, mmol/cm³	2.64	2.66	2.71	2.67
critical volume, cm³/mol	379.0	376.0	369.0	374.0
critical pressure, MPa[d]	3.511	3.535	3.730	3.701
critical temperature, °C	343.05	343.90	357.15	343.05
thermodynamic properties[e]				
C_s at 25°C, J/(mol·K)[f]	181.66	183.44	188.07	185.96
S_s at 25°C, J/(mol·K)[f]	247.36	253.25	246.61	255.19
$H_o - H_o$ at 25°C, J/mol	44.641	40.616	42.382	40.219
$-(G_s - H_o/T)$ at 25°C, J/(mol·K)[f]	97.633	117.03	104.46	120.29
heats of transition, J/(mol·K)[f]				
vaporization at 25°C	42.036	42.036	43.413	42.226
fusion	17.112	11.569	13.598	9.164
formation at 25°C	−24.43	−25.418	−24.439	−12.456
entropy of formation	247.4	252.2	246.5	255.2
vapor pressure, Antoine equation[g]				
A	6.1155	6.1349	6.1239	6.0821
B	1453.430	1462.266	1474.679	1424.255
C	215.307	215.105	213.686	213.206

[a] Convenient graphic representations of the change in property with temperature are given in refs. 4 and 5.
[b] Ref. 6.
[c] To convert C·m to D, divide by 3.336×10^{-30}.
[d] To convert MPa to psi, multiply by 145.
[e] Refs. 7–8.
[f] To convert J to cal, divide by 4.184.
[g] $\log P_{kPa} = A - B/(C + t) (\log P_{mm\ Hg} = \log P_{kPa} + \log 7.50)$.

in boiling point. This procedure is used commercially with one or two columns with a total number of ca 150 trays and a high reflux ratio. First, m-xylene and o-xylene are separated by fractional distillation and then o-xylene is recovered by redistillation from the bottoms $C_{>9}$-aromatic compounds. p-Xylene and m-xylene cannot be separated by distillation because their boiling points are too close. Azeotropic separations of C_8-aromatic compounds have been studied extensively but are of little practical importance.

The difference in freezing point between p-xylene and the other C_8-aromatic compounds is utilized for p-xylene separation (see below).

Since xylenes are important components of gasoline, their combustion characteristics are of interest. The critical compression ratios are 14.2, 13.6, and 9.6 for p-, m-, and o-xylene, respectively. o-Xylene oxidizes much faster than either the meta or para isomers.

The research octane values are 113, 116.4, 117.5, and 107.4 for ethylbenzene, p-, m-, and o-xylene, respectively (11). The motor octane numbers are 6–8 units lower.

Chemical Properties

Reactions Involving the Position of the Alkyl Substituents. These reactions include isomerization, disproportionation, and dealkylation. The interconversion of the three xylene isomers is catalyzed by acids. Xylenes isomerize to near equilibrium levels in a hydrogen fluoride–boron trifluoride system with low boron trifluoride concentrations (12). At high boron trifluoride concentrations, the m-xylene concentration is nearly 100%, because of the formation of the xylene–hydrogen fluoride–boron trifluoride complex (1:1:1). Both o- and p-xylene isomerize in toluene solution under the influence of hydrogen bromide and aluminum bromide (13); the reactions are first order with respect to both xylene and hydrogen bromide–aluminum bromide.

The mechanism involves the rapid and reversible addition of a proton to the aromatic ring followed by a 1,2-intramolecular methyl shift (14).

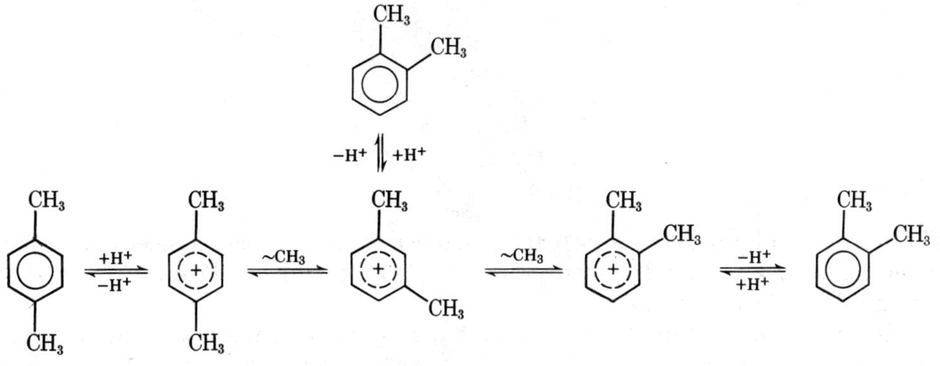

The equilibrium concentration is affected slightly by temperature. Isomerization at lower temperatures produces more p-xylene, the preferred isomer.

The concentration of the three xylene isomers is affected by the reaction rate and the initial concentration of each isomer. Mild deviations from equilibrium can be achieved owing to shape-selectivity effects in zeolite pores (see Molecular sieves).

The thermal isomerization of the three xylenes was studied at 1000°C (15). Side reactions predominated, and only a small percentage of xylenes was interconverted.

Transalkylation of xylenes occurs under more severe conditions and is catalyzed by acid. In this case, the methyl migration is intermolecular and ultimately produces benzene and hexamethylbenzene (see Alkylation).

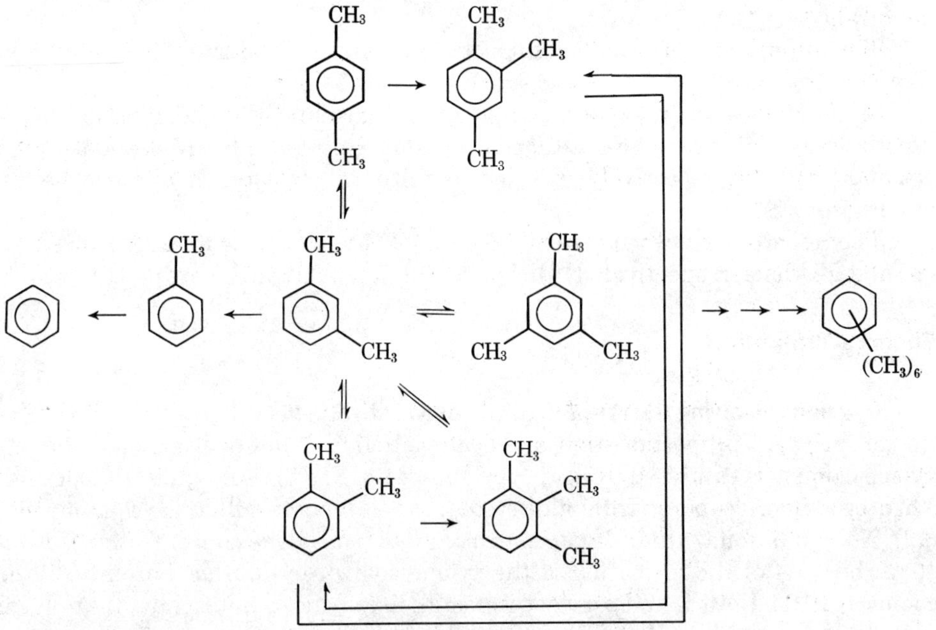

Isomerization over a zeolite catalyst system was accompanied by transalkylation (16). The transalkylation reaction of trimethylbenzenes with benzene to xylenes involved diphenylalkane-type intermediates:

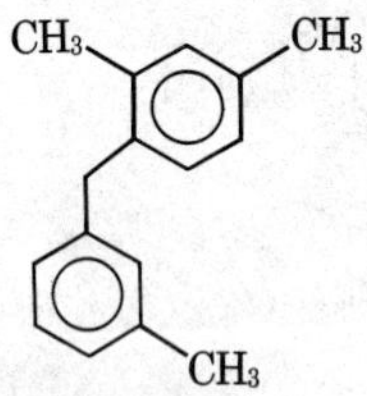

The overall equilibrium constants for all possible methylbenzenes has been determined experimentally and calculated theoretically (17) (see Fig. 1). Overall equilibrium constants are given in Table 3.

In the isomerization of xylenes, some disproportionation always occurs. Furthermore, the intermolecular movement of methyl groups generates xylenes from toluene or from toluene and C_9-aromatic compounds. When all four C_8-aromatic isomers are subjected to disproportionation, the ethyl group of ethylbenzene migrates faster than the methyl of xylenes. Thus, the final product includes dimethylethylbenzenes, trimethylbenzenes, and diethylbenzenes.

Dealkylation usually occurs in the presence of hydrogen and is generally of little importance. Attempted hydrogenolysis of ethylbenzene to toluene (22) was not a commercial success.

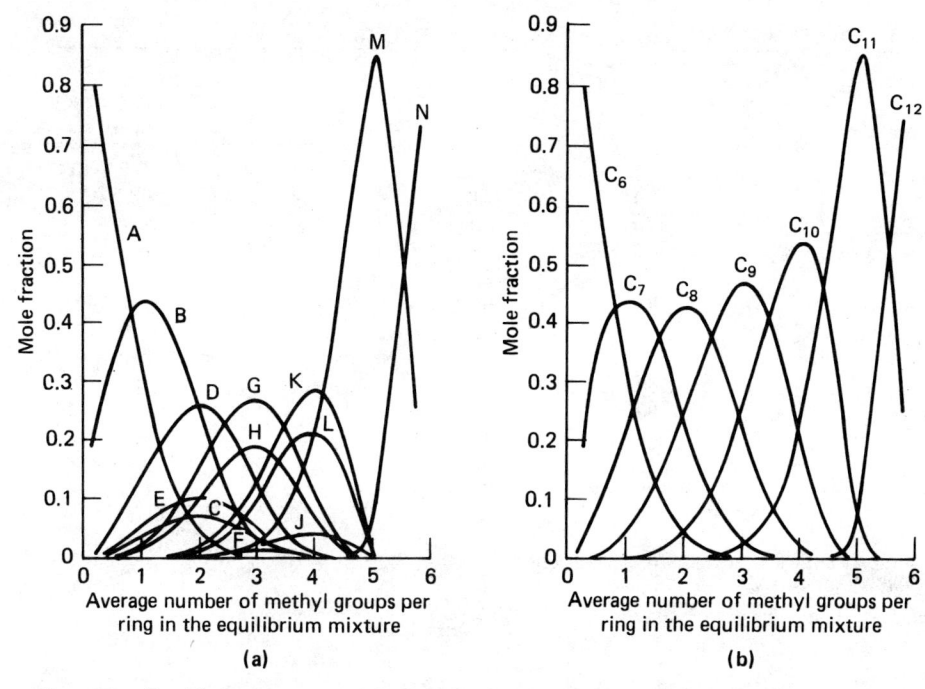

Figure 1. Equilibrium concentration of methyl benzenes in ideal gas state at 25°C. A, Benzene; B, toluene; C, o-xylene; D, m-xylene; E, p-xylene; F, hemimellitene; G, pseudocumene; H, mesitylene; J, prehnitene; K, isodurene; L, durene; M, pentamethylbenzene; and N, hexamethylbenzene (17).

Reactions of the Alkyl Groups. p-Xylene is oxidized to either terephthalic acid or dimethyl terephthalate which is then condensed with ethylene glycol to form polyesters. Oxidation of o-xylene yields phthalic anhydride, another important commodity chemical. Phthalic anhydride is used in the production of its esters, mainly 2-ethylhexyl and higher esters, which are used as plasticizers for synthetic polymers. m-Xylene is oxidized to isophthalic acid, which is also converted to esters and eventually used in plasticizers and resins (see Phthalic acids and other benzenepolycarboxylic acids).

In a study of the slow combustion of the three xylenes, it was observed that o-xylene is much more reactive toward oxygen than the meta and para isomers (23). Under identical conditions, o-xylene is approximately ten times as reactive as its isomers. It was proposed that the initial steps in the mechanism of formation of each isomer are the same.

Table 3. Overall Equilibrium Constants of Methylbenzenes[a,b]

Hydrocarbon	K_{eq}[c] Ref. 19	Ref. 20	Ref. 21	HCJ[d] and inductive model[a]
benzene		0.09	2×10^{-4}	$1-2 \times 10^{-4}$
toluene	ca 0.01	0.63	0.25	0.11–0.13
o-xylene	2	1.1		1.8
m-xylene	20	26	300	90–110
p-xylene	1	1	1	1
pseudocumene	40	63		110–140
hemimellitene	40	69		200–310
durene	120	140		510–810
prehnitene	170	400		960–1700
mesitylene	2800	1.3×10^4	2×10^5	$2-3.6 \times 10^4$
isodurene	5600	1.6×10^4		$0.6-1 \times 10^5$
pentamethylbenzene	8700	2.9×10^4		$1.1-2.5 \times 10^5$
hexamethylbenzene	8.0×10^4	9.7×10^4	1×10^7	$0.7-1.5 \times 10^6$

[a] Ref. 18.
[b] Column headings indicate sources of various test results.
[c] Relative to p-xylene.
[d] HCJ = hyperconjugative.

In the oxidation of p-xylene and m-xylene, formaldehyde is a degenerate branching intermediate, whereas phthalan is formed from o-xylene.

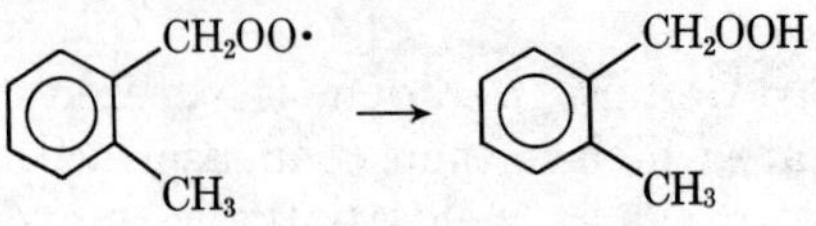

p-Xylene forms p-xylylene when heated above 1200°C. Its structure is represented by a p-quinoid structure or as a p-benzenoid biradical.

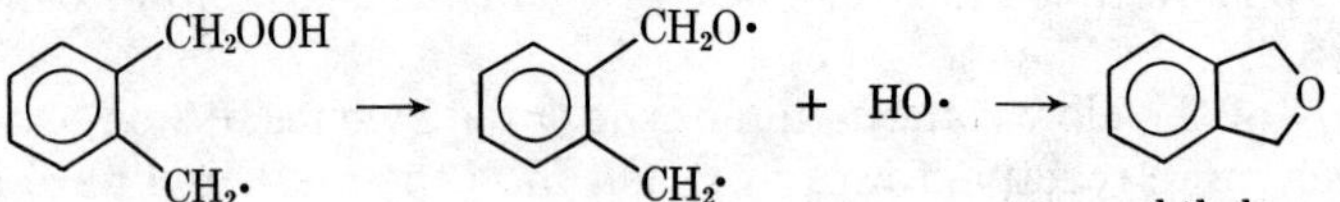

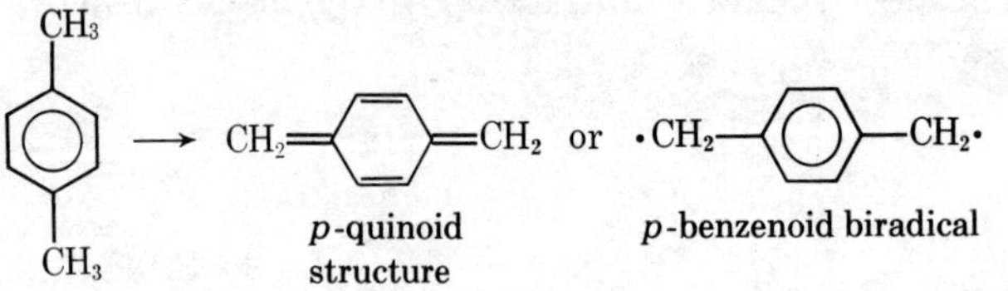

p-quinoid structure

p-benzenoid biradical

Condensation gives poly(p-xylylene) (24–27) (see Xylylene polymers).

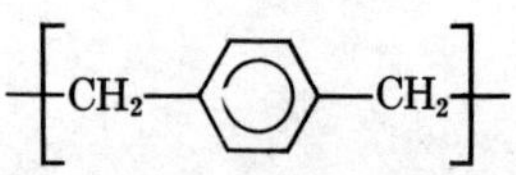

poly(-p-xylylene)

The difference in acidity has been proposed as a basis for separation of p-xylene–m-xylene and ethylbenzene–p-xylene (28–31). The process first requires the formation of the carbanion of cumene in excess cumene containing a chelating agent such as N,N,N',N'-tetramethyl-1,2-cyclohexanediamine. The equilibrium constants for the reaction of p-xylene and m-xylene with the carbanion are 125 and 1160, respectively. The equilibrium constant for metalation of p-xylene with metalated ethylbenzene is 5.4. The p-xylene–m-xylene mixture is usually enriched in p-xylene, which is distilled overhead because the artificially induced separation factor is raised from about one to three. It is unlikely that this process will be commercialized.

Ammoxidation, the reaction of an organic compound with ammonia and oxygen, can be written (32):

$$2\ RCH_3 + 2\ NH_3 + 3\ O_2 \rightarrow 2\ RCN + 6\ H_2O$$

The reaction can be carried out at low hydrocarbon concentration. The Showa Denka Co. practices this reaction with a p-xylene–m-xylene mixture (33), whereas Mitsubishi Gas-Chemical Company uses high purity m-xylene to first form the dicyanide (34); in each process, hydrogenation to the diamine follows.

m-Xylenediamine and phosgene give m-xylene diisocyanate, which is used in urethane resins (35–37).

Reactions of the Aromatic Ring. The reactions of the aromatic ring of the C_8-aromatic isomers are generally electrophilic substitution reactions.

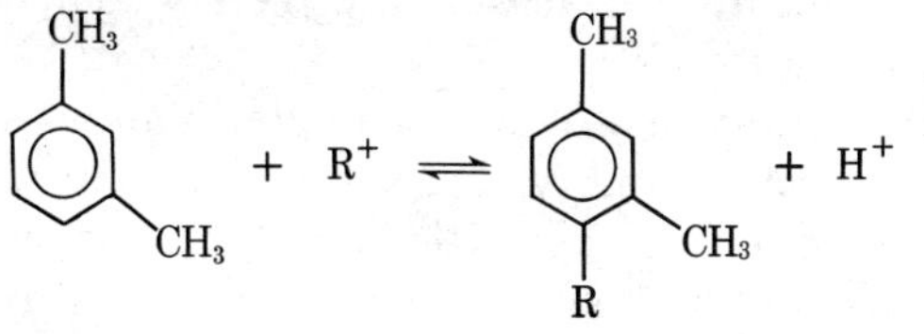

All the classical electrophilic substitution reactions may be considered, but in most instances they are of little practical significance (see Friedel-Crafts reactions). Under acid conditions, formaldehyde yields a mixture of low molecular weight polymers.

The relative nuclear chlorination rates of polymethylbenzenes have been studied (38–39). The higher the substitution, the higher is the rate of chlorination.

As in most electrophilic reactions, the ability to stabilize the positive charge generated by the initial addition strongly affects the relative rates. m-Xylene reacts faster than its two isomers because both methyl groups work in conjunction to stabilize the charge on the next-but-one carbon.

Sulfonation was, at one time, used to separate m-xylene from the other C_8-aromatic isomers. The meta-isomer reacts most rapidly to form the sulfonic acid which remains in the aqueous phase. The sulfonation reaction is reversible, and m-xylene can be regenerated.

In a study of the sulfonation reactions of several methylbenzenes, two mechanisms were suggested depending on acid strength, one involving $H_3SO_4^+$ and the other $H_2S_2O_7$ (40). The degree of sulfonation at the sterically most hindered position increases with increasing sulfuric acid concentration. The activation entropy data show that the $H_3SO_4^+$ mechanism is more affected by steric hindrance than the other mechanism.

Hydrogenation of the aromatic ring has been proposed as a route to facilitate separation of the C_8-aromatic isomers which boil over a range of ca 8°C, whereas the hydrogenated products boil over a range of ca 12°C (41). However, cycloparaffinic (naphthenic) products obtained from o-xylene and ethylbenzene boil only 3°C apart, which impedes separation.

Generally, ethylbenzene is hydrogenated more rapidly than the xylenes. It was proposed that partial hydrogenation of a C_8-aromatic mixture over a selective catalyst could yield a mixture of cycloparaffins high in ethylcyclohexane and unreacted C_8-aromatic compounds. This stream could be passed over an Octafining-type catalyst which converts ethylcyclohexane to xylenes (see below). Of the many catalysts examined, 0.5 wt % rhodium on alumina provided the greatest selectivity for ethylbenzene hydrogenation compared to xylenes (42). Selectivities greater than 10:1 were achieved, but these were considered insufficient to be of practical value.

Complex Formation. All four C_8-aromatic isomers have a strong tendency to form several different types of complexes. Complexes with electrophilic agents are utilized in xylene separation. Equimolar complexes of m-xylene and HBr (mp −77°C) and ethylbenzene and HBr (mp −103°C) have been reported (43–44). Similarly, HCl complexes undergo rapid formation and decomposition at −80°C (45).

Complexes of the type $HX–AlX_3–ArH$ are common, and the formation of the $HF–AlF_3$-m-xylene complex is the key to the Mitsubishi Gas-Chemical Company (MGCC) process for m-xylene recovery (see below).

An entirely different type is the Werner complex, which can be used to separate aromatic isomers (46–48). Both p-xylene and ethylbenzene form Werner complexes with Ni (tetrakis-γ-picoline dithiocyanate) (49). Werner complexes with xylenes and β- and γ-picoline were studied (50).

The molecular configuration of Werner complexes and the size and shape of the guest isomers determine compatibility. The thermal decomposition of Werner complexes has been studied (51). The mole ratio of guest to host was 0.4, 0.8, and 1.0 for m-xylene, ethylbenzene, and p-xylene, respectively, in their complexes with Ni(tetrakis-γ-picoline dithiocyanate). When heated, the complex releases 4-methylpyridine and the guest molecules, but the order in which this occurs depends on the guest. Since complexes of the Werner type have different selectivities for the four C_8-aromatic isomers, they have been suggested as a means of separating the isomers (52–53).

Inclusion compounds of the C_8-aromatic compounds with tris(o-phenylenedioxy)cyclotriphosphazene were used to separate the individual isomers (54–58). The Schardinger dextrins, such as α-cyclodextrin, β-cyclodextrin, and γ-cyclodextrin, are used for clathration (qv); α-cyclodextrin is particularly useful for recovering p-xylene from a C_8-aromatic mixture (59–60). Pyromellitic dianhydride (61) and beryllium oxybenzoate (62) also form complexes, and procedures for separations were developed.

A 1:1 complex melting at 24.8°C is formed between p-xylene and carbon tetrachloride (63). The other C_8-aromatic compounds do not form these complexes. Carbon tetrabromide and chloral, CCl_3CHO, form addition compounds with p-xylene.

Manufacture

A schematic of a process for benzene, o-xylene, and p-xylene is shown in Figure 2. The 65–175°C cut from a straight-run petroleum fraction or from an isocracker is used as feed to a reformer (unit A). This is followed by heart-cutting and extraction in units B, C, and D. If benzene is a required product, its concentration can be increased via toluene recovery (unit E), dealkylation (unit F), and benzene recovered by distil-

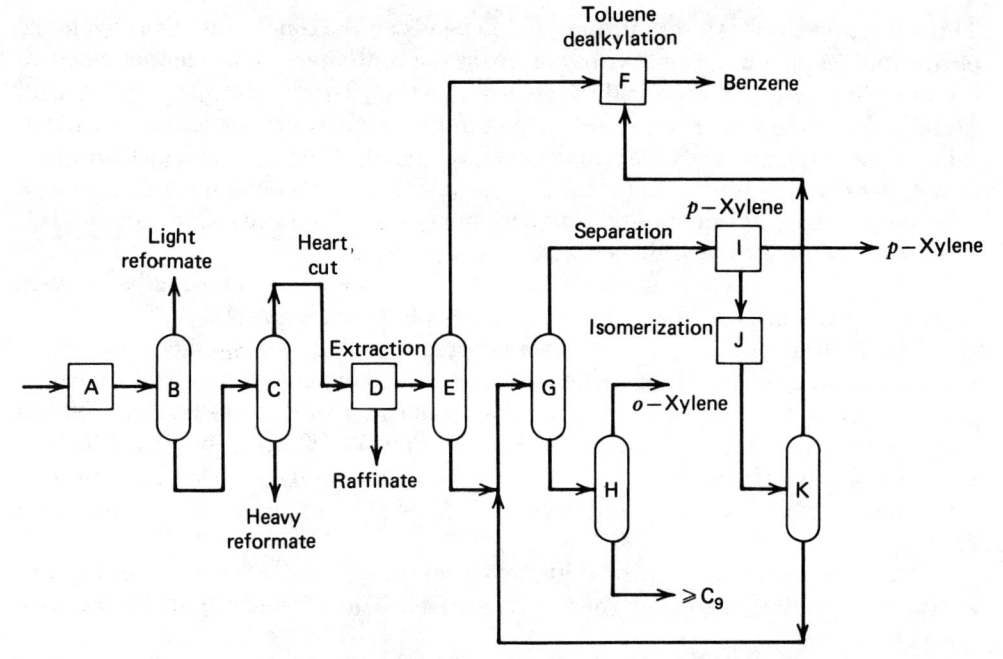

Figure 2. A general scheme for producing benzene, o-xylene, and p-xylene.

lation. o-Xylene and m-xylene are separated in unit G, and o-xylene is recovered by rerunning in unit H (see BTX processing).

p-Xylene is separated in unit I, and the mother liquor or raffinate is sent to unit J for reequilibration of xylenes. Any toluene and lighter materials made in this process are rejected (unit K) and sent to toluene dealkylation. The isomerate is recycled to unit G. Since p-xylene and o-xylene recovery is followed by reequilibration of xylenes in the isomerization unit, a recycle system or loop is established. Under a constant set of conditions, each stream eventually equilibrates both in volume and composition. However, impurities or diluents accumulate. In xylene processing, ethylbenzene accumulates unless a reaction with greater selectivity for ethylbenzene than xylenes occurred, or ethylbenzene was rejected by distillation.

In most processes, ethylbenzene is not recovered because of high energy costs. Where ethylbenzene is recovered, the sequence of steps after extraction (unit D, Fig. 2) is toluene rejection (unit E) followed by ethylbenzene removal and m-xylene–o-xylene separation (unit G).

Xylenes are also obtained by transalkylation, eg, disproportionation of toluene to give benzene and xylenes or from a mixture of toluene and C$_9$-aromatic compounds.

These reactions reach an equilibrium that depends on the conditions. The advantage of this route is that a stream free of ethylbenzene is produced. The commercial processes include Atlantic Richfield's Xylenes Plus (64), Toray's Tatoray (65–67), and Mobil's TDP. The first two can be adapted for toluene disproportionation or transalkylation of toluene and C_9-aromatic compounds, the latter for toluene disproportionation only. The ability to perform these reactions adds flexibility to a refiner's capability to match products and production to the market needs. Toluene could alternatively be used to provide benzene via dealkylation.

The Xylenes Plus process uses a nonnoble-metal catalyst in a moving-bed system, and hydrogen is not required. Liquid yield is 95–97% (see Table 4).

The Tatoray process uses a catalyst referred to as T-81 with high conversion per pass and good stability over several months. It is used in pellet form in a fixed bed in the presence of hydrogen and is regenerable by burning off the coke formed. Typical operating conditions are 1–5 MPa (10–50 atm), 350–530°C, and a hydrogen–hydrocarbon mol ratio of 5:12. Per pass conversion is increased by increasing pressure or temperature, which, however, reduces yield. A lower hydrogen ratio (<5) increases coke formation.

The Tatoray process shown in Figure 3 uses toluene feed. The effect of the proportion of trimethylbenzene in the feed is shown in Table 5; make-up hydrogen is ca 0.4 wt %.

The TDP process. More recently, Mobil developed and commercialized, at their Naples, Italy plant, a toluene disproportionation process that is referred to as the TDP process. A ZSM-5 zeolite catalyst is employed that is regenerable by burning off coke (see Molecular sieves).

Crystallization Processes for *p*-Xylene. Until the development of molecular-sieve adsorption, crystallization (qv) was the only practical method for *p*-xylene production. The feed to the crystallizer contains ca 20% *p*-xylene, and the bulk of the remainder is ethylbenzene and *o*- and *m*-xylenes. Upon cooling, *p*-xylene crystallizes first. Eventually a temperature is reached where crystals of another isomer form. This usually occurs at the *p*-xylene–*m*-xylene eutectic, although theoretically it could occur at the *p*-xylene–*o*-xylene eutectic. The latter limits the efficiency of the crystallization process. *p*-Xylene always has a certain solubility in the remaining C_8-aromatic isomers at whatever temperature the eutectic is reached. In practice, the *p*-xylene–*m*-xylene eutectic is reached at ca −68°C; crystallization usually begins at ca −4°C.

The key to good solid–liquid separation is crystal size; the larger the crystal, the better the separation. Crystal size is affected by the degree of supersaturation and nucleation, which is affected by temperature, agitation, or presence of crystal growth sites, etc. Crystal sizes varied between 200 × 1000 µm and 20 × 50 µm, depending on the method used (68).

Table 4. Xylenes Plus Transalkylation of Toluene and $C_{>9}$-Aromatic Compounds, m^3/d

Feed		Product	
Toluene	$C_{>9}$-aromatic compounds	Benzene	Xylene
12,580		4,650	6,920
12,580	6,290	2,360	15,750
12,580	12,580	2,170	21,950

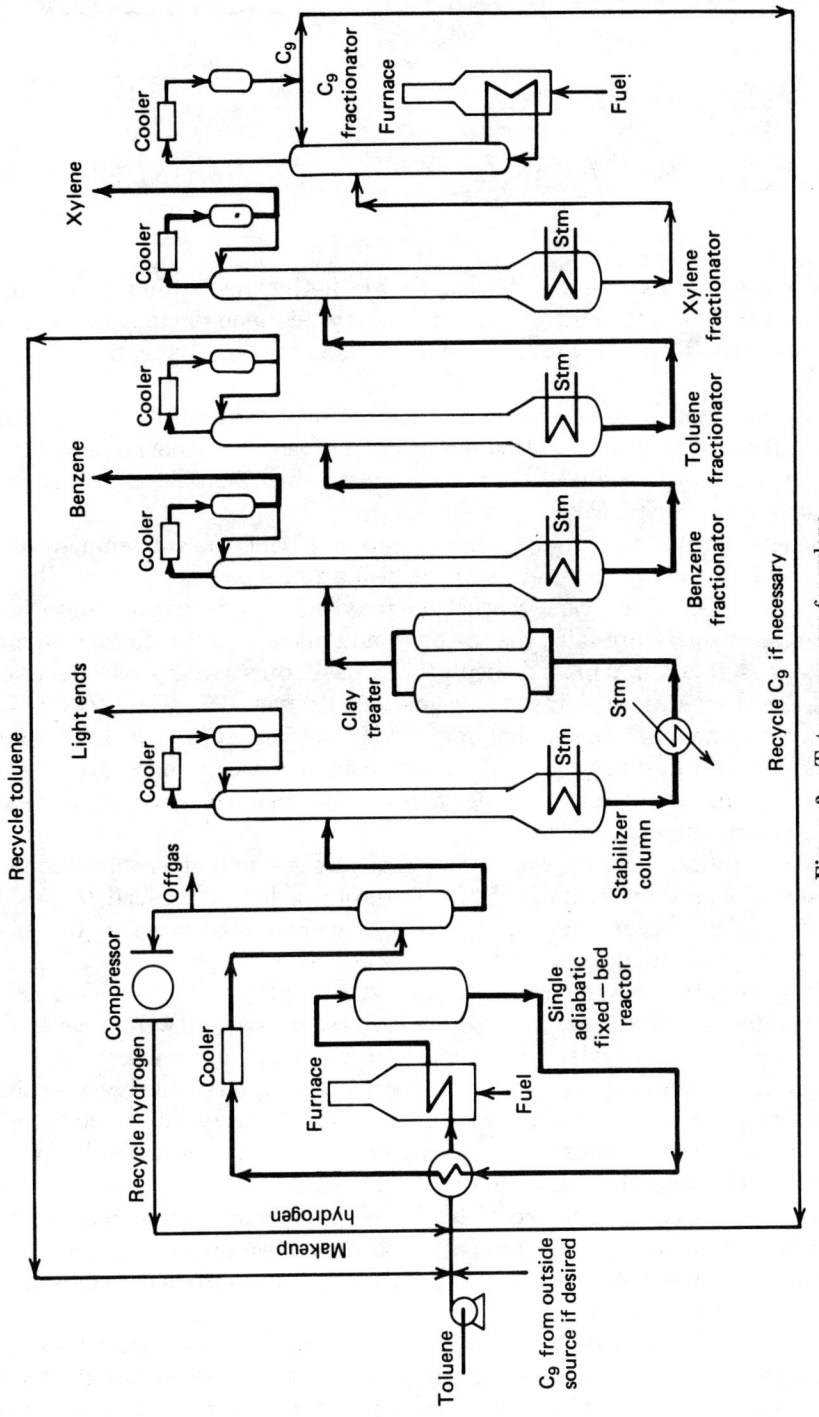

Figure 3. Tatoray process for xylene.

Table 5. Tatoray Transalkylation of Toluene and C$_{>9}$-Aromatic Compounds, Relative Weight Units

Feed		Product	
Toluene	C$_{>9}$-aromatic compounds	Benzene	Xylene
1000		414	561
750	250	252	716
489	511	209	871

p-Xylene crystals have a 6:1:0.5 length–width–thickness ratio. The thickness is hard to discern. Under the microscope, p-xylene crystals upon rotation disappear when viewed sideways. Thus, the width-to-thickness ratio may be less than 1:0.5, perhaps 1:0.3 (68).

p-Xylene crystals are generally produced in two or more stages of crystallization and are separated by centrifugation (69–70). The first-stage cake has a purity of ca 80–90%. The impurity is due to the mother liquor which wets the crystal surface or is occluded in the crystal cake.

The efficiency of the solid–liquid separation depends on the temperature and the loading of the centrifuges. A feed comprising 20% p-xylene, 20% ethylbenzene, 40% m-xylene, and 20% o-xylene reaches the p-xylene–m-xylene eutectic at −73.3°C. However, as temperature falls, the viscosity and density of the mother liquor rise sharply. Thus, it becomes more difficult for the centrifuges to achieve effective separation. The theoretical solubility of p-xylene in xylenes at −73.3°C is 6.24%; however, >8% p-xylene remains in the mother liquor. This inefficiency is owing to fine crystals being carried over in the mother liquor, and to an increase in temperature between the final crystallizer and the centrifuge and the conversion of mechanical work into heat in the centrifuges.

The theoretical recovery, when there is no increase in temperature and perfect liquid–solids separation occurs, is 15 of 20 units of p-xylene in the feed, ie, 75%. If 8% p-xylene remains in the mother liquor, 13.4 units are recovered of the original 20 units or 67%. Most processes operate in the 60–70% efficiency range.

In the event that the feed composition is 20% p-xylene, 50% m-xylene, 20% ethylbenzene, and 10% o-xylene, as might be the case if o-xylene were recovered as a separate product, the p-xylene–m-xylene eutectic temperature would be −66.9°C and theoretical solubility 8.04%. In this case, crystal separation should be easier because the viscosity and density difference compared to that of the crystals would be greater. With an increase in temperature, p-xylene in the mother liquor would probably exceed 9%, and thus the separation efficiency would be ≤64%.

The cake is usually reslurried with a higher purity stream from a later stage of purification. At this stage, p-xylene crystals have a less well-defined shape, are more chunky, and drain well. A second stage of centrifugation is sufficient in most cases to give p-xylene of ≥99% purity.

The commercial processes for producing p-xylene by crystallization use either direct-contact or indirect refrigeration. The latter has the disadvantage that the walls of the vessel tend to foul which reduces heat transfer. The scrapers used to remove the crystal incrustation can damage the crystals causing overall reduction in crystal size.

The Chevron p-Xylene Crystallization process (71) is the most widely used in the world; the FRG, Mexico, Japan, the UK, and the United States have plants using this method (see Fig. 4). It is characterized by the use of circulating magma; CO_2 is used as the direct-contact coolant. The crystallization plant follows a distillation column which removes toluene and lighter materials; the bottoms from this column are dry and become feed to the crystallization unit.

The feed is exchanged against mother liquor from the first-stage centrifuge to a temperature close to the crystal point. Liquid CO_2 may be injected either into the circulating magma or directly into the feed before entering the crystallizer. The pressure in the lower part of the circulation loop permits dispersion of CO_2 and prevents flashing, which would otherwise cause local supersaturation resulting in nucleation and formation of fine crystals.

As the slurry rises in the circulation leg, some CO_2 vaporizes and cools the system. Cooling is gradual; the degree of supersaturation is low, and thus crystal growth occurs on the existing crystals. The vaporized CO_2 is removed at the top of the crystallizer and sent to the refrigeration unit. The slurry passes to the vacuum crystallizer which operates at a reduced pressure, and further cooling occurs as CO_2 is flashed from solution. The crystals are separated from mother liquor in two stages, the first using screen-bowl centrifuges and the second pusher centrifuges. The Chevron process offers the advantages of using inert CO_2 and the growth of relatively large crystals which permit good solid–liquids separation in the centrifuges.

The Krupp process for *p*-xylene crystallization is reported to be used in eight plants with a total capacity of 140,000 t/yr (72). The feed, after cooling against mother liquor, is cooled in a scraped chiller and sent to a holding tank to encourage growth, preferably with a system in which a classifier permits recirculation of fines. The cake is separated by a rotary drum filter and washed with a second-stage recycle stream. The resulting slurry is sent to a final scraped chiller and centrifugation.

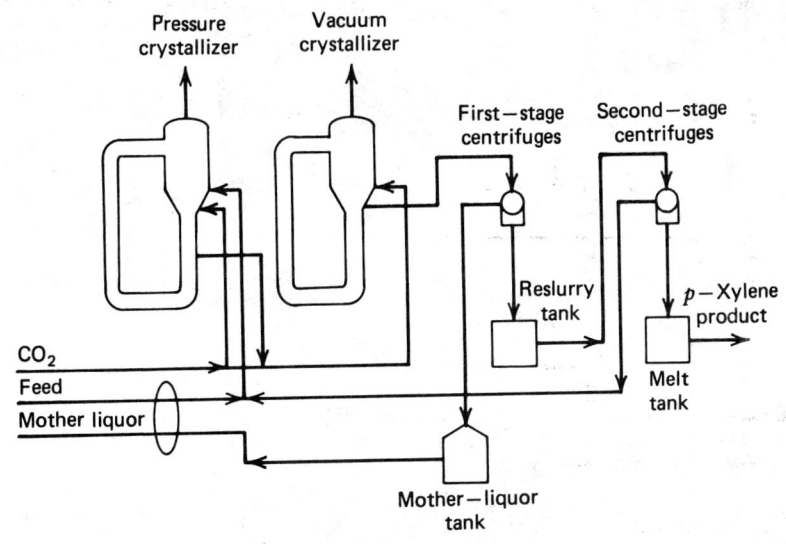

Figure 4. Chevron *p*-xylene recovery process.

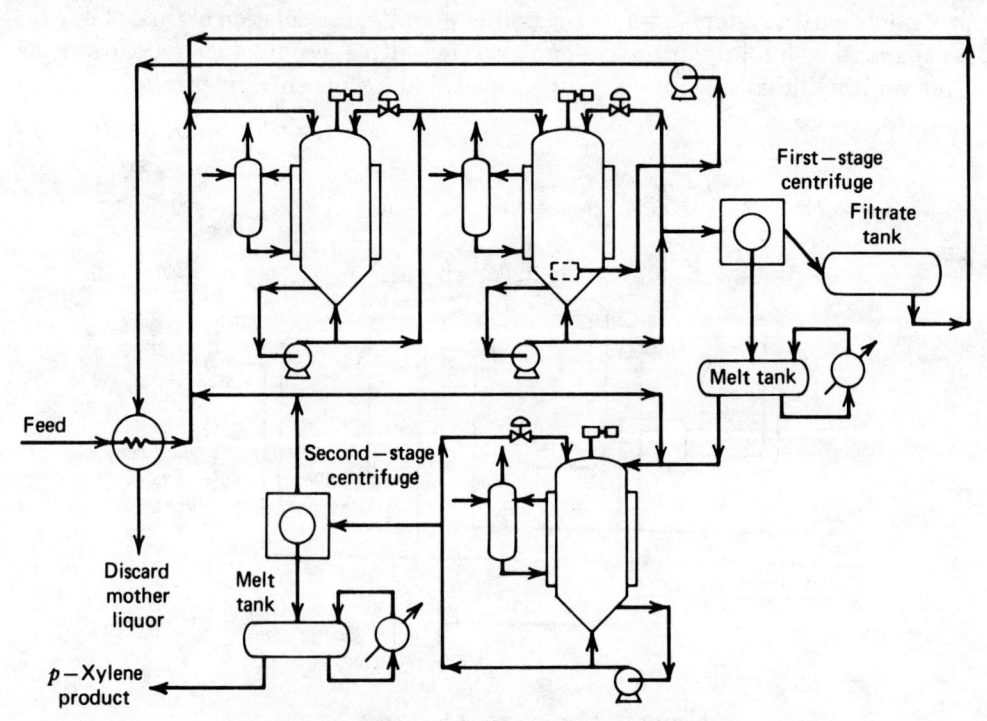

The Amoco p-Xylene Crystallization process (73–74) is used by over half of the current U.S. capacity (see Fig. 5). Indirect refrigeration, with ethylene as refrigerant, is employed in the first stage of crystallization.

In two or more first-stage crystallizers, the temperature is brought down in stages to approximately the *p*-xylene–*m*-xylene eutectic. The crystallizers are fitted with scrapers mounted on a central shaft, which provide agitation and maintain a good heat-exchange surface. The slurry in the crystallizer is recirculated at such a rate that the larger crystals are deposited near the bottom of the crystallizer where the slurry is removed. The residence time in each of the two crystallizers is ca 3 h in order to encourage crystal growth.

Sufficient mother liquor is withdrawn through a micrometallic filter or similar element, situated within the final first-stage crystallizer. It is critical to withdraw the mother liquor at the lowest possible temperature and to maintain an overall mass balance around the plant. The mother liquor may be sent for isomerization.

The slurry from the final first-stage crystallizer is transferred to a centrifuge wherein the mother liquor is separated from the first-stage cake and recycled to the first crystallization stage.

The first-stage cake is melted and passes to a second-stage crystallizer designed like those in the first stage but cooled by propane. The product from the second-stage crystallizer is pumped to a final centrifugation stage where mother liquor is separated and recycled to the second-stage crystallizer feed to control solids concentration. Melting the cake gives *p*-xylene, the purity of which depends on the temperature of the final crystallization stage. In an overall balance, 26.5 metric tons of feed per hour yields 3.3 t of 99.1% *p*-xylene per hour if the filtration step is omitted. The same feed

Figure 5. Amoco *p*-xylene crystallization process (73).

yields 3.7 t of 99.1% p-xylene per hour in the process described above. This represents an 11% yield improvement and an overall recovery of 71%.

The Maruzen process for p-xylene crystallization uses ethylene gas as a direct refrigerant in a two-stage process (75–76); two commercial plants with a total capacity of ca 100,000 t/yr have been built. The crystals are separated by centrifugation.

The Arco process resembles the Amoco process; in both, cooling is provided by ethylene in two first-stage crystallizers fitted with scraper blades (77–78). The slurry is centrifuged and the mother liquor rejected. The cake is remelted, cooled indirectly by propane in a second-stage crystallizer, and sent to a second-stage centrifuge. The mother liquor is recycled to the first-stage crystallizers and the cake washed on the centrifuge with toluene. The p-xylene is recovered as a bottoms product after toluene is removed by distillation. By late 1981, six plants with a total design capacity of 375,000 t/yr had been built in the United States and four other countries.

Arco has published descriptions of a single-stage crystallization process which still incorporates the toluene-wash step. It is not known whether any plants have been constructed or converted to the single-stage technology.

A process was developed by Institut Français du Pétrole but has not been commercialized (79–80). p-Xylene is passed countercurrent to an immiscible refrigerant. The p-xylene crystals are separated from liquid in a conical-shaped column equipped with a filter.

p-Xylene crystallization by increasing pressure has been described but is not used commercially (81).

The Phillips process applies pulsed-column technique (68,82). Although commercialized, no plants are currently in operation. A continuous multistage melt-purification process has been developed for p-xylene crystallization (83).

Chlorotrifluoromethane has been patented as direct refrigerant for a crystallization process (84–85). Although this material offers a number of the benefits associated with direct refrigeration, it is too expensive for commercialization of this process.

Adsorption Processes for *p*-Xylene. In the late 1960s, UOP introduced the Parex process. By 1981, sixteen Parex units were on-stream, and production exceeded 10^6 t/yr. To date, fifteen plants are either being designed or are under construction (86). p-Xylene recovery per pass is 90–95% compared to 60–70% for crystallization. Thus, to produce an equal amount of p-xylene, the feed to the separation unit is ca 30% less for Parex than for crystallization, and the feed to the isomerization unit is 25% less. A similar reduction in volume to the distillation columns occurs, reducing the operating costs of the processing loop (see Adsorptive separation, liquids).

The Parex process is based on the principle of continuous selective adsorption in the liquid phase, employing a fixed bed of solid adsorbent (87–97). This solid adsorbent, made from zeolite material that has undergone exchange with barium and potassium, permits entry into the pore structure of the main feed components. Therefore, the shape–size selectivity is not critical to the separation technique. Advantage is taken of small differences in affinity for the different feed components. In addition, the combination of the unique adsorbent properties with UOP's simulated moving-bed procedure, previously used in their Molex process, became the key to success. The desorbent material is neither weakly nor strongly adsorbed with respect to the C_8-aromatic feed components. Preferred desorbents are toluene or p-diethylbenzene which can be readily distilled from p-xylene.

The open structure of the solid adsorbent, exposing a relatively large surface area to the feed, gives access to more adsorptive sites than if the adsorption was limited to the exterior surface alone. p-Xylene is the most readily adsorbed material.

The desorbing liquid must be readily separated from the feed components by distillation. Both lower and higher boiling desorbents have been used commercially. The desorbent is adsorbed by the adsorbent to about the same extent as the feed hydrocarbons. It must be capable of being desorbed by feed hydrocarbons which it displaces by mass action.

Although the adsorption process could be conducted in both batchwise or continuous fashion, only the latter is used commercially. The continuous process operates with a fixed bed, which appears to move in the opposite direction from the liquid streams.

In zone I in Figure 6(**a**), the feed component A is completely adsorbed from the upwardly flowing liquid stream. The solid at the top of zone I contains only feed component B and desorbent D. Zone II is a rectification section. Since the solid entering this zone has been immediately in contact with feed, it contains feed components A and B, and desorbent D, whereas the liquid entering zone II contains only A and D. As the adsorbent passes downward, it adsorbs more of A and D, and consequently, B is desorbed since A is preferentially adsorbed. Zone III is a desorption section; A is removed from the adsorbent by the desorbent D which enters the bottom of this zone. Zone IV is a secondary rectification zone; D is partially desorbed by the rising stream of liquid containing B and D. The liquid composition along the length of the bed is shown in Figure 6(**b**).

The commercial fixed-bed Parex process is shown in Figure 7 where 12 access lines connect a large rotary valve to distributors in the adsorbent bed; four lines transport the three process streams between the valve and the distillation columns and admit the feed. A liquid circulating pump passes fluid from the top of the column to the bottom via another line. This system is capable of producing all the process aspects of moving the solid past fixed positions of fluid feed, inlet product, and withdrawal. Moving these positions in the direction of liquid flow is equivalent to moving

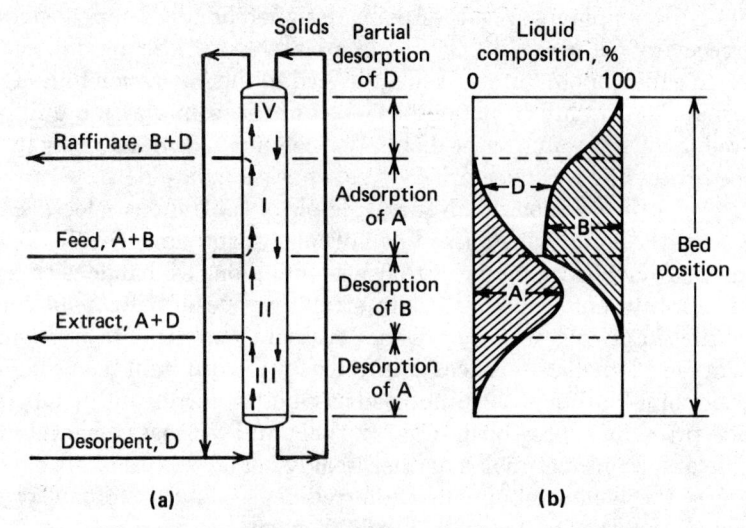

Figure 6. Parex concentration profile. Courtesy of UOP.

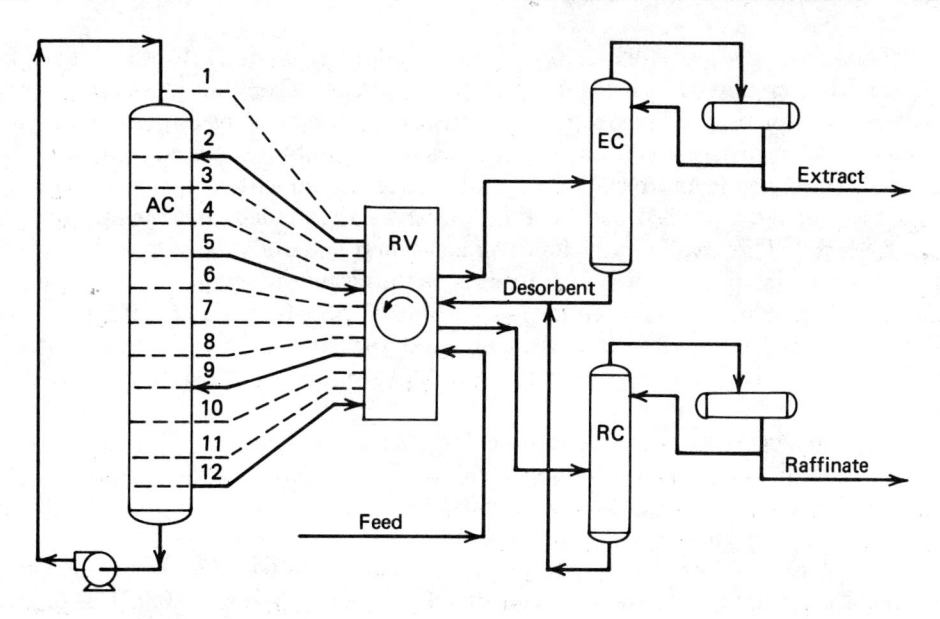

Figure 7. Parex-simulated moving bed for adsorptive separation. AC = adsorbent chamber; RV = rotary valve; EC = extract column; and RC = raffinate column. Lines: 2, desorbent; 5, extract; 9, feed; 12, raffinate; 1, 3, 4, 6, 7, 8, 10, and 11 are inactive access lines. Courtesy of UOP.

the solid in the opposite direction. All lines between valve and column are inactive, except for lines 2, 5, 9, and 12, which are performing as indicated in the diagram. When the valve is rotated counterclockwise by one notch, lines 10 (feed), 6 (extract), 3 (desorbent), and 1 (raffinate) become active, and 2, 5, 9, and 12 become inactive. The greater the availability of bed access lines, the more closely the system resembles a continuous countercurrent system.

The flow rates of fluids in the bed must be adjusted to compensate for the rate at which positions of feed and withdrawal points move past void volumes in the bed. The pump is programmed to appropriately change the flow rate each time the valve is moved.

The operating conditions are 250–300°C and moderate pressures. p-Diethylbenzene, which can be separated from its isomers by the Parex process, is a preferred high-boiling solvent. Small amounts of toluene present in the process feed may contaminate the product p-xylene and must be rejected in a finishing column.

The typical p-xylene product of 99.5% purity contains 0.3% ethylbenzene, 0.1% m-xylene, and 0.1% o-xylene. The relatively high content of ethylbenzene is owing to ethylbenzene being more readily adsorbed than m-xylene or o-xylene. Since benzoic acid, the oxidation product of ethylbenzene, is a chain terminator in the polymerization step to form polyesters, the purity of p-xylene must be slightly higher for material produced by the Parex process than the purity of p-xylene produced by crystallization.

The Parex process is generally combined with an isomerization process, and thus Parex feed normally contains ca 20% p-xylene. In one instance, this process has been used on a mother-liquor stream from a crystallization process which contained ca 10% p-xylene (98). Normally, this is not a cost-effective approach because Parex unit size and attendent equipment are geared more to liquid throughput than to p-xylene concentration.

The Aromax process (99–103) was developed by Toray Industries in Japan and has much in common with the UOP Parex process. A desorbent liquid is used to assist in the separation of *p*-xylene from C_8-aromatic compounds using a zeolite-based adsorbent. The adsorbent is contained in a series of horizontally situated chambers that are isolated from one another. Each chamber is equipped with a valve in such a way that it can be fed and desorbed. A series of valves are opened, and the appropriate chambers fed. These valves are closed and a second bank of valves opened simultaneously ca 5 min later. Progression of this type fulfills all the requirements of feeding and desorbing. The operating conditions are 140°C, 785–980 kPa (7.7–9.6 atm), and flow rates of 5–13 L/h with diethylbenzene as desorbent. *p*-Xylene is preferentially adsorbed and recovered together with desorbent from which it can be recovered by distillation.

p-Xylene yield per pass is better than 90%, and typical purity is 99.5%. Impurities, in one example, contained ethylbenzene, 0.37%; *m*-xylene, 0.06%; *o*-xylene, 0.3%; toluene, 0.1%; and nonaromatic compounds, 0.1%. As in the Parex process, ethylbenzene is, after *p*-xylene, preferentially adsorbed.

The Asahi process has yet to be commercialized (104–107). Development is complete, and pilot results are being confirmed in larger equipment (108). This process appears to have been thoroughly studied and covered by patents and may achieve commercial significance (see Fig. 8). It is based on displacement chromatography with the acid of an improved zeolite adsorbent and a specially chosen desorbent material. The amount of xylene adsorbed depends on the ionic radius of the exchanged cation; it is larger for monovalent cations. Other important factors are the silica–alumina ratio and the moisture content.

The Asahi process produces ethylbenzene as well as *p*-xylene. Furthermore, the amount of zeolite and the heat consumption are about half of that required for the

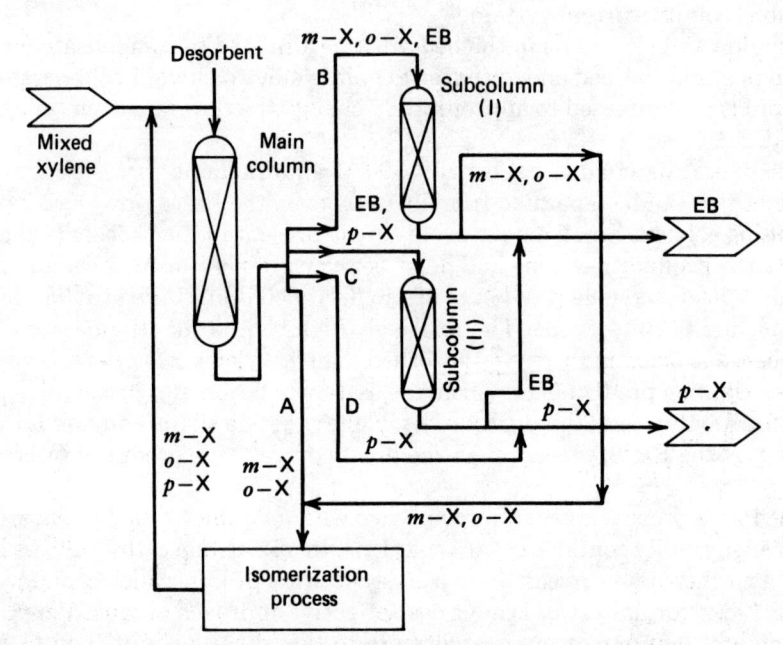

Figure 8. Asahi separation process. EB = ethylbenzene; *o*-X = *o*-xylene; *p*-X = *p*-xylene; and *m*-X = *m*-xylene. Courtesy of Asahi Chemical Industry Co. Ltd. (104).

Parex process. Investment cost is relatively low because of the simplicity of the process. Projections for commercial utilization show a requirement of 170 t of adsorbent to recover 70,000 t of p-xylene and >15,000 t of ethylbenzene. The fuel consumption is 9.2 MJ/kg (3960 Btu/lb) of combined product or 7.1 MJ/kg (3060 Btu/lb) for p-xylene only.

Mitsubishi Gas-Chemical Company (MGCC) process. The MGCC process is the only efficient commercial method to separate m-xylene (109–119). In addition to m-xylene production, this process greatly simplifies and reduces the cost of separating the other C_8-aromatic isomers.

When a C_8-aromatic mixture is treated with HF–BF_3, two layers are formed. In the absence of BF_3, the hydrocarbon and acid layers have <1% mutual solubility. In the presence of BF_3, m-xylene selectively dissolves in the HF phase. After stirring, the two layers separate rapidly. The selectivity of separation of m-xylene is increased by the addition of a diluent, normally a C_8 paraffin.

m-Xylene dissolves in HF–BF_3 because of the formation of a 1:1 molecular complex, xylene–HBF_4. m-Xylene is the most basic xylene, and its complex with HF–BF_3 is the most stable. The relative basicities of ethylbenzene and p-, m-, and o-xylene are 0.14, 1, 100, and 2, respectively (120).

m-Xylene of >99% purity can be obtained with the MGCC process with <1% m-xylene left in the raffinate by phase separation of hydrocarbon layer from the complex–HF layer. The latter undergoes thermal decomposition which liberates the components of the complex.

The m-xylene–HF–BF_3 complex isomerizes by increasing the temperature to below 100°C. The HF and BF_3 are removed to yield xylenes which approach equilibrium at the operating temperature. This favors the para isomer over the ortho isomer because of the lower temperature of operation compared to other xylene isomerization processes.

The MGCC process is shown schematically in Figure 9. The mixed C_8-aromatic feed is sent to an extractor (unit A) where it is in contact with HF–BF_3 and hexane. The m-xylene–HF–BF_3 complex is sent to the isomerization section (unit D) or the decomposer (unit B). In the decomposer, BF_3 is stripped and taken overhead from condenser–separator (unit C), while HF in hexane is recycled from the bottom of C. Recovered m-xylene is isomerized in unit D and returned to the extractor A, or sent for further purification to column E. Meanwhile, the remaining C_8-aromatic compounds and hexane are sent to raffinate column F where residual BF_3 and HF are separated as well as hexane for recycle. Higher boiling materials are rejected in column H, and ethylbenzene and o-xylene are recovered in columns I and J. The overhead from J becomes feed to p-xylene separation, unit K. The raffinate or mother liquor is then recycled for isomerization.

The MGCC process is used in Japan, the United States, and Spain in plants with a total capacity of 164,000 t/yr.

Isomerization. The flow sheet of Figure 2 reveals that an enriched xylenes stream receives considerable processing before entering the xylenes loop. Therefore, a once-through procedure for xylenes recovery is generally not economic. The introduction of the isomerization step to reequilibrate the xylenes enables efficient utilization of the enriched xylenes stream.

The three xylene isomers can be interconverted with the aid of an acid catalyst. Examination of the patent literature reveals that claims for xylene isomerization have

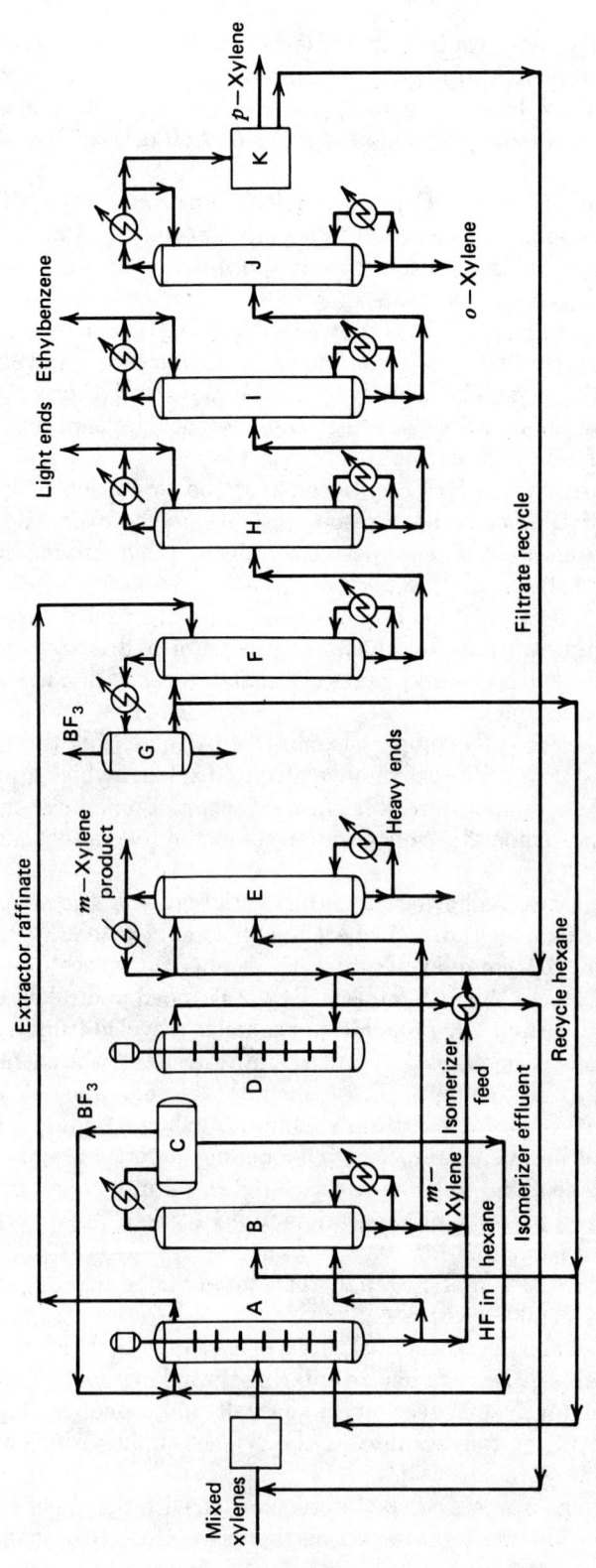

Figure 9. Xylenes separation via Mitsubishi Gas-Chemical Company's HF–BF$_3$ extraction–isomerization process. A = extractor; B = decomposer; C = separator; D = isomerization reactor; E = heavy ends tower; F = raffinate tower; G = separator; H = light-ends fractionator; I = ethylbenzene fractionator; J = o-xylene fractionator; and K = p-xylene crystallizer (114).

been made for innumerable catalysts, such as the Friedel-Crafts type and silica–alumina, both amorphous and zeolitic.

Unless ethylbenzene is removed by distillation or the feed originates entirely from toluene transalkylation, ethylbenzene is always present in the feed to a xylenes processing loop. Ethylbenzene is, therefore, a diluent in the feed to the xylenes-separation process. In all xylenes isomerization processes, ethylbenzene should undergo some partial transformation to assure that its concentration in the separation unit feed is kept to a minimum. The greater tendency of ethylbenzene to undergo disproportionation and hydrogenolysis has been used to reduce ethylbenzene content as has its isomerization to xylenes.

Disproportionation is difficult to suppress and occurs to some extent in all isomerization processes. Molecular-shape selectivity can be applied to reduce disproportionation to very low levels. Isomerization requires a cycloparaffinic intermediate to effect this conversion. However, any conversion to xylenes at least partially offsets any xylenes lost to side reactions.

Most isomerization processes are controlled by a thermodynamic equilibrium which is a function of temperature. This usually involves just the three xylene isomers, but for processes in which ethylbenzene can be converted to xylenes, it is included in the equilibrium (Fig. 10). The m-xylene concentration is higher than that of the ortho isomer because of the lower entropy of the latter owing to a restriction to rotation of the two methyl groups (122). The para isomer also has a lower entropy value because of its greater symmetry.

The xylene isomerization proceeds as if the methyl group moves in a sequence of 1,2-steps, ie, there is no direct conversion between o- and p-xylene (123–124).

Dual-Function Processes. *The Octafining process* was developed by Engelhard Industries and Atlantic Richfield Company and has been employed successfully on a commercial scale for many years (125–127). The catalyst is prepared by mixing equal amounts of a silica–alumina cracking catalyst with platinized alumina; the platinum content of the mixture is ca 0.5 wt %. The reaction conditions are as follows: pressure of 1.14–2.51 MPa (150–350 psig), with 1.31–1.65 MPa (175–225 psig) preferred; hydrogen–hydrocarbon ratio <10:1, with 4–6:1 preferred; temperature at start of run near 425°C, and liquid hourly space velocity (LHSV) 0.6–1.6/h. The reaction tem-

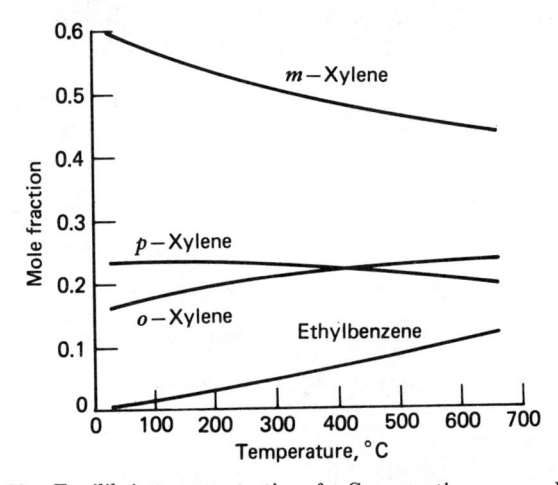

Figure 10. Equilibrium concentrations for C_8-aromatic compounds (121).

perature is gradually increased to ca 480°C. The catalyst can be regenerated at least once by controlled combustion of the carbon which gradually accumulates over the life of the run.

The key to the Octafining process is its ability to convert ethylbenzene to xylenes and to interconvert the isomeric xylenes. The catalyst promotes carbonium ion-type reactions as well as hydrogenation–dehydrogenation. In this respect, the catalyst is very similar to reforming catalysts. A proper balance of the following reactions has to be achieved: hydrogenation of aromatic compounds and dehydrogenation and isomerization of cycloparaffins. Side reactions include disproportionation and hydrodealkylation of aromatic compounds and hydrodecyclization and hydrodealkylation of cycloparaffins. The mechanism for the conversion of ethylbenzene and xylenes is shown in Figure 11, together with some side reactions.

The cycloparaffinic intermediates provide a link between xylenes and ethylbenzene, and a certain concentration of these compounds has to be maintained. This concentration is determined by the operating conditions. Higher pressure favors hydrogenation, higher temperatures favor dehydrogenation, and the conditions usually represent a compromise. Higher temperatures generally favor side reactions. Effluent from the reactors includes the C_8-cycloparaffinic intermediates. These are separated together with toluene and recycled and hence do not represent a yield loss (128). Therefore, it is more appropriate to judge the efficiency of the Octafining process by C_8-ring loss rather than C_8-aromatic loss.

The Octafining process has been used commercially since 1960. Over 30 units have been built in the United States and 16 other countries.

The Octafining catalyst has recently been improved by depositing platinum on alumina and combining the mixture with a form of hydrogen mordenite (zeolite) in such proportions that the final platinum content is <0.5 wt % (129–132).

In many instances, the original Octafining process has been supplanted by other processes or Octafining II.

Octafining II is a substantial improvement over the original process. The space velocity is at least twice that of the original process with the consequence that only half the volume of catalyst is required. The isomerization activity, C_8-ring retention, and catalyst stability is higher. The catalyst can be regenerated, but there has yet to be sufficient commercial use to determine its ultimate life; it has operated without loss in performance for over four years.

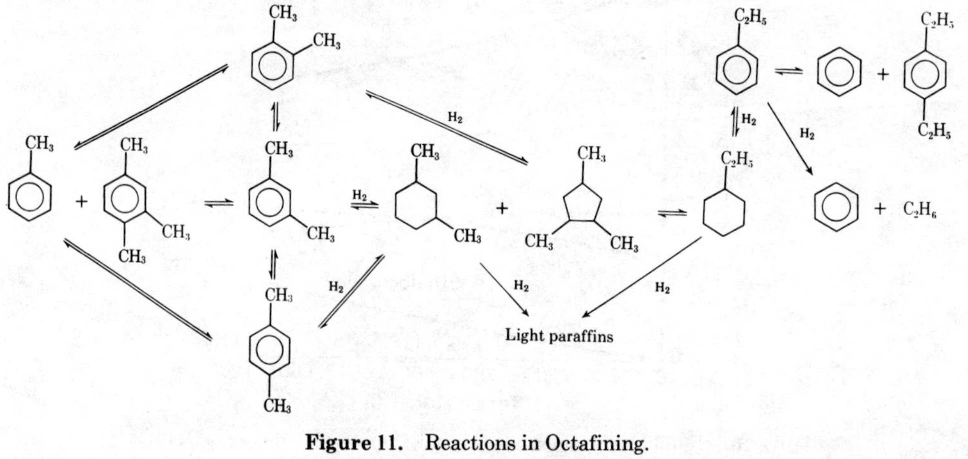

Figure 11. Reactions in Octafining.

The UOP Isomar process is similar in many respects to Octafining (133–134). A platinum-containing catalyst is used to both isomerize xylenes and convert ethylbenzene to xylenes by way of hydrogenated intermediates. Conditions of temperature, pressure, space rate, and hydrogen–hydrocarbon ratio are similar to those of the original Octafining process. Cycle length before regeneration is generally about six months. At one time six Isomar units were in operation, and 13 others were under design or construction. None are known to be in operation in the United States at present.

An example of an overall material balance using the Parex process for *p*-xylene separation and Isomar as the isomerization process is given in Table 6 for the production of *o*-xylene and *p*-xylene. The net overall effect is the complete conversion of ethylbenzene and *m*-xylene to *p*- and *o*-xylene.

The Isolene II process, developed by Toray Industries, converts ethylbenzene to xylenes (99–100,102). The catalyst is again platinum on an acidic support; operating conditions are similar to those of Octafining. The catalyst is regenerable with cycle lengths of several months and a total life of over two years. A 110,000-t/yr plant was built at Toray's Kawasaki complex; two other plants were licensed and built. The Isolene II process is used in conjunction with the Aromax process at Kawasaki, Japan.

Amorphous Silica–Alumina Catalysts. Amorphous silica–alumina catalysts have long been used for xylene isomerization, eg, in the Chevron (135), Maruzen (136), and ICI (137–140) processes. The advantage of these processes is their simplicity. No hydrogen is required, and the only side reaction of any importance is disproportionation. However, in the absence of hydrogen, coke buildup on the catalyst is rapid, and regeneration is required every 3–30 d. An additional swing reactor is needed that can be taken off-stream for regeneration. A duplicate reactor or, preferably, three reactors one-half cycle out of phase serve this purpose.

Space rates are generally lower, usually 0.5–1.0/h LHSV. Thus, the volume of catalyst, which is inexpensive, is large. The disproportionation losses may be high, up to 5–10% xylene and 10–15% ethylbenzene per pass under conditions at which xylenes are isomerized to near equilibrium values.

In each process, *p*-xylene equilibrium is reached to the extent of 95–100%. The freshly regenerated catalyst is brought on-stream at 370–420°C, and temperature is gradually increased to maintain a >95% approach to equilibrium. The length of cycle depends on feed; it is longer when the *o*-xylene concentration in the feed is low. The catalysts are very stable to regeneration by burnoff of coke and may last several years.

Table 6. Material Balance for Parex-Isomar Loop Producing *p*-Xylene and *o*-Xylene [a]

C_8-aromatic compound	Feed t/d	Feed wt %	Products t/d	Products wt %
ethylbenzene	50.0	17.3		
p-xylene	51.0	17.6	157.6	54.5
m-xylene	125.1	43.3		
o-xylene	63.0	21.8	101.4	35.1
Total [b]	289.1	100.0	259.0	89.6

[a] Ref. 133.
[b] Net C_8-aromatic compounds process loss: 8.2 wt %.

Generally, the catalyst-activity loss is due to a decrease in surface area which occurs slowly at regeneration conditions.

The ICI process is used in four UK plants in conjunction with separation facilities to produce 200,000 t of p-xylene per year; there are also plants in Poland, Japan, and the USSR. The ICI process operates without any feed diluent, and the catalyst can undergo at least one rejuvenation *ex situ*. The operating range is 430–450°C, with carbon deposition increasing rapidly at higher temperatures. Cycle lengths tend to be three to four days before regeneration. In most instances, ethylbenzene is rejected from the fresh feed although concentrations of 20–25% in the isomerization unit can be tolerated. The feed rate is ca 0.5/h LHSV over a compressed pellet catalyst with 800 g/cm^3 bulk density.

The Chevron process was used in two U.S. plants. Cycle lengths range from 6 to 30 days, depending on catalyst age and o-xylene content of the feed; temperature is 370–470°C and space velocity ca 0.5/h. Steam dilution of ca 5 wt % reduces disproportionation losses.

Zeolite-Based Xylene Isomerization. In the 1970s, research performed at Mobil Oil Corporation on a versatile series of zeolite materials, referred to as ZSM, initiated a new era in xylene isomerization. From this work emerged three xylene isomerization processes called LTI (low temperature isomerization), MVPI (Mobil's vapor phase isomerization), and MLPI (Mobil's low pressure isomerization).

Conditions, feed, and product of an LTI reactor (129–131,141) are shown in Table 7. Under these mild conditions, the product p-xylene concentration is at least 95% of theoretical, yet the loss to disproportionation is negligible. In general, the process is facilitated if the feed is low in p-xylene and if 10–20% toluene is used as diluent. The operation takes place in the liquid phase and normally operates at ca 2.17 MPa (300 psig) and 200–260°C. The catalyst is regenerated when the reactor temperature reaches 275°C; it is expected to be good for more than two years. As far as is known, the LTI process has never been used commercially. However, it may be considered important in its role as a stepping stone to the widely used MVPI and MLPI processes.

The development of the series of ZSM-5 catalysts revolutionized commercial xylene isomerization (132,142–144). The silica–alumina ratio is at least 10 and preferably higher; ratios as high as 60 have been reported. It is suggested that the catalyst used in the MVPI Process is one in which ZSM-5 is exchanged with ammonium and nickel salts to the NiHZSM-5 form, which is incorporated with an alumina binder and

Table 7. Reactor Feed and Product Compositions in the LTI Process[a,b], wt %

Component	Feed	Product
benzene	<0.1	0.2
toluene	15.0	15.0
ethylbenzene	3.0	3.0
p-xylene	11.5	18.5
m-xylene	66.3	43.9
o-xylene	4.2	19.1
C$_{>9}$-aromatic compounds	<0.1	0.3

[a] Ref. 130.
[b] LTI = low temperature isomerization. Conditions: Weight hourly space velocity (WHSV) = 3.0; temperature = 238°C; pressure = 1.48 MPa (200 psig).

extruded (145). This procedure is described in Mobil patents, and it is not clear whether other patented refinements are included in the commercial catalysts (129–132,141).

Patents describe a broad range of conditions for what is believed to be the MVPI process (145–146). Temperatures are in the range 255–290°C, pressure is generally ca 1480 kPa (200 psig), the hydrogen–hydrocarbon ratio is ca 6:1, and WHSV is dependant on temperature but in the range of 2–50, normally 5–10. The process is characterized by low xylene losses to side reactions, by a p-xylene equilibrium slightly higher than theoretical, by a strong preference for ethylbenzene disproportionation over xylene disproportionation, and by extreme stability of performance (145–151).

The primary disproportionation reactions between ethylbenzene (EB) and xylenes (XYL) are shown in Table 8 with relative rate constants (131).

Reactions 1 and 2 involve transfer of an ethyl group, and reactions 3 and 4 of a methyl group; the relative rate constants are typical of a feed containing 20% ethylbenzene and 80% xylenes. It is typical in acid-catalyzed reactions that ethyl groups transfer faster than methyl groups, but the factor is usually two to four times instead of the approximate factor of 10 for the Mobil processes. Shape-selective catalysis in ZSM systems may be the key to the unusual selectivity (152).

The patent which appears to most closely reflect the MLPI process describes the use of an HZSM-5 catalyst and the hydrogen form of the ZSM-5 material in a blend containing 35 wt % alumina (153). The conditions described included temperatures of 290–380°C, pressure ca 274 kPa (25 psig), and WHSV of 5–8.5/h based on total composite catalyst. No hydrogen is used. Ethylbenzene conversion is maintained at ca 25% but declines with time. Temperature is gradually raised to compensate. Cycle lengths of over six months were described in patents (153) and could be well over a year in commercial practice. Total catalyst life of at least four years can be anticipated. Typical feed and product compositions are shown in Table 9. Both the MLPI and MVPI catalysts can be regenerated by controlled combustion to remove carbon. Regeneration is required more frequently for the MLPI process.

The MVPI and MLPI processes have been rapidly accepted and implemented because they can be installed in existing isomerization plants with little more than a change in catalysts. The difference in space velocity and temperature of operation has to be accommodated. This is particularly true for conversion from an amorphous silica–alumina catalyst to the MLPI catalyst. The space-rate difference between these two processes could be a factor of 10–16. With such a reduction in catalyst volume, the reactor may have to be modified in such a way that the height–diameter ratio is at least 0.2 to ensure good feed distribution.

The success of the MVPI and MLPI processes is indicated by the fact that 14 units

Table 8. Reactions between Ethylbenzene and Xylenes

Reaction[a]	Relative rate constant
1. EB + EB ⇌ B + DEB	125
2. EB + XYL ⇌ B + DMEB	17
3. EB + XYL ⇌ T + MEB	4
4. XYL + XYL ⇌ T + TMB	1

[a] B = benzene; T = toluene; DEB = diethylbenzene; DMEB = dimethylethylbenzene; MEB = methylethylbenzene; and TMB = trimethylbenzene.

Table 9. Typical Feed and Product Composition for MLPI[a] Process, Wt %

Component	Feed	Product
≤C$_5$	0	0.1
benzene	0	1.9
toluene	0	0.5
ethylbenzene	20.5	15.5
p-xylene	8.7	19.0
m-xylene	50.7	42.2
o-xylene	20.0	17.2
≥C$_9$-aromatic compounds	0.0	3.6
Total	100.0	100.0
ethylbenzene loss, %		24.4
xylene loss, %		1.3

[a] Mobil's low pressure isomerization (153).

have been converted to MVPI and three to MLPI, representing ca half of the world's p-xylene capacity (154). Almost all U.S. manufacturers have converted to one of these two processes.

The commercial impact of the MVPI and MLPI processes on p-xylene production capacity has been substantial. The reduction of ethylbenzene concentration reduces the loop volumes considerably. In the event that an existing plant was hydraulically limited or limited by separation capacity, eg, by refrigeration capacity in a crystallization plant, additional feed could be processed, thereby increasing the throughput of an existing operation. Consequently, some existing plant capacities were increased by as much as 35% and overall operating costs reduced by up to 25%. Conversion in the United States to the Mobil processes increases U.S. capacity in the same way that two new facilities would do.

Process Choice. Currently, the MGCC process is preferred for m-xylene production. Crystallization processes are well established for p-xylene production and will remain so for many years to come as large plants are already paid off and proprietary technology eliminates royalty fees. The reliability of crystallization caused hesitation to switch to other processes, and at least one large crystallization plant is under construction (155).

At this point, the Parex process is well established and would appear to have advantages over crystallization because the higher p-xylene recovery results in less loop traffic. If operating and construction costs for a Parex plant and a crystallization plant to produce the same amount of p-xylene were identical, Parex would offer the advantage that the other parts of the processing loop would be at least 15% smaller with correspondingly lower operating costs.

There is not enough known about Aromax versus Parex to determine whether there is any technical or economic advantage to either process. However, in the United States, UOP appears to have an overwhelming patent-position advantage, and Aromax would probably be excluded from use in this country.

A p-xylene separation process would certainly be accompanied by an isomerization process. If fresh feed to the loop is expensive or limited, the need exists to efficiently convert it to p-xylene or p- and o-xylene, the choice might be Octafining II. In other cases, the choice might be to MVPI or MLPI with the latter appearing to have some economic advantages.

The Asahi process could have the lead in new technology, but more recently selective toluene methylation for p-xylene has been reported (156–157). However, a final crystallization or adsorption process has to be added to attain the required commercial purity.

Economic Aspects

Since World War II, the main source of xylenes has been from reforming petroleum fractions. Earlier, xylenes were produced from coal. It is possible that coal will again become an important source of xylenes as bountiful supplies of coal in the U.S. are developed for fuel and petrochemical uses.

The demand for the individual C_8-aromatic isomers has grown rapidly, particularly since the mid-1960s; worldwide capacities are shown in Table 10, U.S. economic data in Table 11. The U.S., Western Europe, and Japan account for the bulk.

The U.S. capacity and production figures for o-xylene, p-xylene, and ethylbenzene are shown in Figure 12 and Table 11. Growth started in the mid-1960s and continued into the 1980s. However, capacity and production are well below the industry estimates for 1982 (159). The 1980 supply-and-demand picture and the principal uses are shown

Table 10. Worldwide Capacity for C_8-Aromatic Isomers in 1981 (Year End), 1000 t[a]

Location	o-Xylene	p-Xylene	m-Xylene	Ethylbenzene
United States	576	2,543	79	4,411
Canada and Mexico	49	40	0	599
South America	108	123	0	264
Western Europe	732	1,157	30	4,598
Eastern Europe	>171	>387	0	>669
Middle East	10	0	0	29
Far East	379	1,180	55	2,307
Oceania	0	0	0	138
Africa	0	0	0	23
Total	>2,025	>5,430	164	>13,038

[a] Ref. 1.

Table 11. U.S. Mixed Xylenes Economic Data, 1000 t[a]

	1979	%	1980	%
production	3877		3713	
imports	183		153	
exports	131		166	
consumption	3929		3700	
in gasoline	654	16.6	490	13.2
o-xylene	490	12.5	454	12.3
m-xylene	45	1.1	33	0.9
p-xylene	2273	57.9	2283	61.7
ethylbenzene	108	2.7	114	3.1
solvents	359	9.1	326	8.8
year-end capacity	5764		5614	
production–capacity	0.67		0.66	

[a] Ref. 1.

736 XYLENES AND ETHYLBENZENE

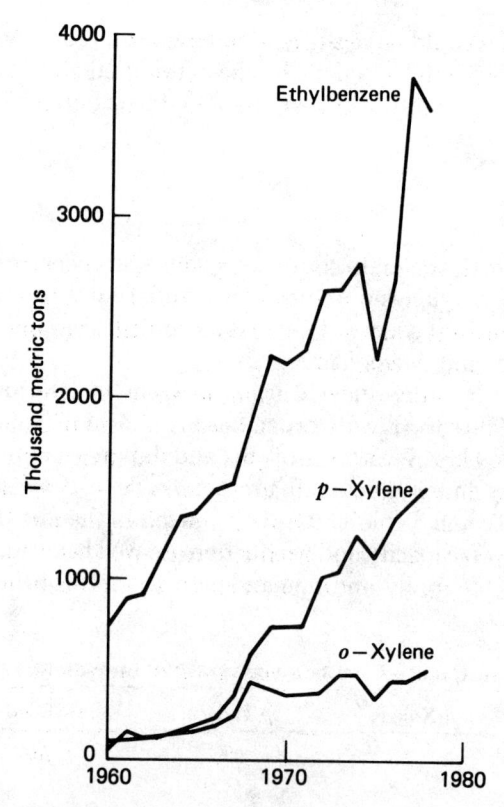

Figure 12. U.S. capacity (year end) for o-xylene, p-xylene, and ethylbenzene (158).

Table 12. U.S. 1980 Supply and Demand for Xylene Isomers, 1000 t[a]

	o-Xylene	m-Xylene	p-Xylene
production	452	41	1922
imports	0	7	22
exports	215	9	376
total U.S. consumption	237	39	1568
phthalic anhydride	261		
isophthalic acid		36	
dimethyl terephthalate			900
terephthalic acid			638
other	6	3	9
year-end capacity	590	79	2542
production–capacity	0.77	0.52	0.74

[a] Ref. 1.

in Table 12. The projected 1982–1987 growth rates for o-xylene, m-xylene, p-xylene, and ethylbenzene are 6.5%, 5.5%, 5%, and 3.5%, respectively (1).

The U.S. producers and capacities for p-xylene, o-xylene, and ethylbenzene are shown in Table 13; companies using the Octafining II and their production are given in Table 14 (see also Petroleum refining). Most of the ethylbenzene produced is used captively for styrene (qv). Almost all o-xylene is sold on the open market, whereas p-xylene consumption is divided between the open market and captive use in a 40:60 ratio.

Table 13. U.S. Capacity for *p*-Xylene (Jan. 1982), *o*-Xylene (1981 year end), and Ethylbenzene (Jan. 1983)

Company	Location	Process Separation	Process Isomerization	*p*-Xylene, 1000 t/yr	*o*-Xylene, 1000 t/yr	Ethylbenzene[a], 1000 t/yr
American Hoechst	Baton Rouge, La.					315
	Bayport, Texas					469
Amoco	Decatur, Ala.	Amoco	Amoco	590		
	Texas City, Texas	Amoco	Amoco	408		313[c]
Arco	Houston, Texas	Arco	Mobil	177	109	
	Channelview, Texas					526[c]
Charver	Houston, Texas					16[b]
Chevron	Pascagoula, Miss.	Chevron	Mobil	185		
COS-MAR	Carville, La.					785
Dow	Freeport, Texas					839
DuPont-Conoco	Alvin, Texas					27[b]
El Paso Natural Gas	Odessa, Texas					125[c]
Exxon	Baytown, Texas	Esso/Parex	Mobil	190	131[d]	
Gulf Oil	Donaldsville, La.					308[c]
Koch[e]	Corpus Christi, Texas	Amoco/Parex	Mobil	177	72	50[b,c]
Monsanto	Chocolate Bayou, Texas				11[f]	
	Texas City, Texas					708[c]
Phillips	Guayama, Puerto Rico	Parex	Octafining	213	59	
St. Croix	St. Croix, Virgin Islands	Parex	Mobil	272		
Shell	Deer Park, Texas				57	
Tenneco	Chalmette, La.	Amoco	Mobil	57	54	
U.S. Steel	Houston, Texas					61[b]
Total				2269[g]	493	4542

[a] Ref. 161.
[b] Distillation from mixed xylenes.
[c] From benzene and ethylene.
[d] Some used captively.
[e] Formerly Sun.
[f] Captive use only.
[g] Total is ca 400 t less than peak capacity a few years ago.

Amoco is the only U.S. producer of *m*-xylene with an annual capacity of 79,000 t. The 1980 production was 39,000 t.

In recent years, many small xylene plants have shut down because operations on a small scale are not economical or the technology has become obsolete. Ethylbenzene from mixed xylenes is highly energy intensive, and therefore, fewer units using this approach are being operated.

The price histories for the four isomers are given in Table 15. The dramatic price increases since 1974 reflect the increase in crude oil prices. The profitability of any particular plant depends on many factors, such as age, size, and ability to adapt to more efficient current technology.

Table 14. Units Using O-750 Catalyst[a]

Licensee	Location	Start up	Production, 1000 t/yr p-Xylene	o-Xylene
BP California[b]	Grangemouth, Scotland	Nov. 1977	100	130
China Petrochemical Development Corporation	Kaohsiung, Republic of China	July 1978	25	0
Chinese Petroleum Corporation	Lin Yuan, Republic of China Train 1	Feb. 1979	110	40
Chinese Petroleum Corporation	Lin Yuan, Republic of China Train 2	Jan. 1980	110	40
Slovnaft	Bratislava, Czechoslovakia	Apr. 1979	40	0
Centrala Industriala de Rafinarii Si Petrochemie	Brazi, Romania	Apr. 1980	20	0
Mazowieckie Zaklady Raffineryinei Petrochemiczne	Plock, Poland	June 1980	40	0
Liao Yang Petrochemical and Fibers Corporation	Liao Yang, People's Republic of China	Apr. 1980	60	0
Petroquimica General Mosconi	Ensenada, Argentina	Apr. 1981	40	26

[a] Ref. 160.
[b] Closed.

Table 15. C_8-Aromatic Price History, $/metric ton[a]

Year	p-Xylene	o-Xylene	m-Xylene	Ethylbenzene
1960	308.7			275.5
1961	286.7		507.2	
1965	198.5		419.0	220.5
1970	132.3		319.7	132.3
1974	324.1		352.8	308.6
1975	352.8	209.5		
1976	374.9	253.6		
1978	275.6	264.6	485.1	
1979	573.3	485.1		
1980	595.4	507.2		
1981	683.6	551.3		

[a] Ref. 1.

Specifications and Analytical Methods

Typical analyses for C_8-aromatic isomers are given in Table 16. The traditional freezing-point method for p-xylene purity is preferred over chromatographic methods. The latter do not detect water or carbon dioxide, which though present at low volume concentrations can be present in appreciable molar quantities. However, chromatographic separation of the C_8 aromatic isomers is utilized in research and manufacture. The main problem is to obtain good separation of the p- and m-xylene peaks. In chromatographic analysis of >99% p-xylene, p-xylene peak tends to tail over the m-xylene peak, another reason to prefer the freezing-point method.

Table 16. Typical Analyses of Commercial C$_8$-Aromatic Isomers

Assay	ASTM method	Ethylbenzene	p-Xylene	m-Xylene	o-Xylene
purity, mol %	D 1016	99.5	99.1	98.5	96.0
distillation range, °C	D 850		137.9–138.4	138.5–139.7	144.0–145.0
specific gravity, 15.5°C/15.5°C	D 891	0.871	0.866	0.869	0.884
color Saybolt	D 156	>25	>30	>30	20
flash point, °C	D 56	15	27	29	31

Conditions for chromatographic separation of C$_8$-aromatic isomers are given in Table 17. This procedure readily separates benzene, toluene, and the C$_8$-aromatic isomers. By raising the temperature to ca 140°C, reasonably good separation of C$_9$- and C$_{10}$-aromatic compounds can be achieved. Di-n-propyl tetrachlorophthalate is also a useful column coating because it reverses the positions of p-xylenes and m-xylenes. Thus, the m-xylene peak elutes ahead of p-xylene and provides better analysis of >99% p-xylene.

Bentone-34 is commonly used in packed columns (162–163). The retention indexes of many benzene homologues on squalane were determined (164). Gas chromatography of C$_6$–C$_{11}$-aromatic compounds using a Ucon B550X-coated capillary column is discussed in ref. 165. Other procedures use phthalic acids (166), liquid crystals (167), and Werner complexes (168) as the separation media. Gel permeation chromatography of alkylbenzenes and the separation of the C$_8$ aromatics with treated zeolites are described in refs. 169–172.

Health and Safety Factors

The xylene isomers are designated as flammable liquids and as such should be stored in approved closed containers with a red label out of doors and away from heat or open flames. Limits for transportation by air are one liter (passenger planes) and 40 L (cargo planes).

Table 17. Chromatographic Separation of C$_8$-Aromatic Isomers

column[a]	30.5 m × 0.5 mm id
sample size, µL	0.2
detector	flame ionization
temperature, °C	70
pressure, kPa (psig)	156 (8)
helium flow, mL/min	8

[a] Stainless-steel column coated with m-bis(m-phenoxyphenoxy)benzene.

Table 18. Flash Points and Autoignition Temperatures of the C$_8$-Aromatic Compounds, °C

Compound	Flash point	Autoignition temperature
p-xylene	27	530
m-xylene	29	530
o-xylene	31	460
ethylbenzene	15	432

The flash points and autoignition temperatures are given in Table 18. The vapor can travel along the ground to an ignition source. In the event of fire, water reduces the temperature, but foam, carbon dioxide, or dry chemical are preferred extinguishers. The lower and upper flammability limits are 1% and 7%.

The field detection limit is ca 10 ppm, and the laboratory detection limit is 0.003 ppm. The odor threshold limits (lower, medium, and upper) are 0.26, 2.21, 4.13 ppm, respectively.

The xylenes are not very toxic. They are mild skin irritants, and skin protection and the cannister-type face masks are recommended. The oral LD_{50} value for rats is 4000 ppm. Prolonged (8 h) exposure by humans should be limited to 200 ppm.

Xylenes show only mild toxicity to fish, and the threshold limit for crop damage is 800–2400 ppm. Biodegradation with activated seed is slow, and sewage digestion is impaired by 0.1% concentrations. In the event of a spill, oil-skimming equipment, adsorbent foam, and charcoal may be used for cleanup.

Uses

The bulk of xylenes, which are mostly produced by reforming petroleum fractions, is used in gasoline (see Gasoline and other motor fuels). Mixed xylenes, which are obtained under severe reforming conditions or, more likely, from reformate extracts, are used as solvents in the paint and coatings industry (see also Solvents, industrial). This use is likely to decline, particularly in the United States, as efforts are increased to reduce hydrocarbon emissions into the air. The paint industry is moving more and more toward water-based latex paints for environmental reasons and ease in use and cleanup.

Ethylbenzene is mainly used as a precursor to styrene (qv).

o-Xylene is used almost entirely as the feedstock for phthalic anhydride manufacture. Although *o*-xylene is the preferred feed, naphthalene is competing as a starting material because of its currently low price. Phthalic anhydride is a basic building block for plasticizers (qv). *o*-Xylene is also used for the preparation of phthalonitrile which is converted to the copper phthalocyanine, a pigment.

m-Xylene is used for the manufacture of isophthalic acid and to a lesser extent, isophthalonitrile. This latter is the starting material of the fungicide tetrachloroisophthalonitrile. Isophthalic acid is the base of unsaturated polyester resins which have good corrosion resistance and greater strength and higher modulus than resins derived from phthalic anhydride.

Commercially, *p*-xylene is the most important isomer. Almost all is converted via terephthalic acid or dimethylterephthalate and reaction with ethylene glycol to poly(ethylene terephthalate) for use in fibers, films, or resins. Poly(ethylene terephthalate) fibers are well established and used in blends with cotton for household fabrics, carpets, and apparel (see Phthalic acid and other benzene polycarboxylic acids; Polyesters).

BIBLIOGRAPHY

"Xylenes and Ethylbenzene" in *ECT* 1st ed., Vol. 15, pp. 186–194, by Maury Lapeyrouse, Esso Research and Engineering Company; "Xylenes and Ethylbenzene" in *ECT* 2nd ed., Vol. 22, pp. 467–507, by Harry E. Cier, Esso Research and Engineering Co.

1. *Chemical Economics Handbook*, SRI International, Menlo Park, Calif., 1981.
2. *Chem. Week* **128**(1), 42 (1981).
3. J. D. Anderson, *Chem. Eng. Prog.*, 39 (Dec. 1980).
4. C. L. Yaws, *Chem. Eng.*, 113 (July 21, 1975).
5. C. L. Yaws, *Chem. Eng.*, 73 (Sept. 29, 1975).
6. J. Chao and co-workers, *Hydrocarbon Process. Int. Ed.* **59**(8), 117 (1980).
7. J. Chao, *Hydrocarbon Process.*, 295 (Nov. 1979).
8. D. R. Stull and co-workers, *The Chemical Thermodynamics of Organic Compounds*, John Wiley & Sons, Inc., New York, 1968.
9. F. G. Dwyer, P. J. Lewis, and F. M. Schneider, *Chem. Eng.*, 90 (Jan. 6, 1976).
10. P. J. Lewis and F. G. Dwyer, *Oil Gas J.*, 55 (Sept. 26, 1977).
11. W. L. Nelson, *Oil Gas J.*, 68 (Mar. 19, 1970).
12. D. A. McCaulay and A. P. Lien, *J. Am. Chem. Soc.* **74**, 6246 (1950).
13. H. C. Brown and H. Jungk, *J. Am. Chem. Soc.* **77**, 5579 (1955).
14. R. H. Allen and L. D. Yats, *J. Am. Chem. Soc.* **81**, 5289 (1959).
15. W. D. Crow and C. Wentrup, *Tetrahedron Lett.* **27**, 3111 (1968).
16. M. A. Lanewalla and A. P. Bolton, *J. Org. Chem.* **34**(10), 3107 (1969).
17. C. J. Egan, *J. Chem. Eng. Data* **5**(3), 298 (1960).
18. S. Ehrenson, *J. Am. Chem. Soc.* **84**, 2681 (1962).
19. A. McCaulay and A. P. Lien, *J. Am. Chem. Soc.* **73**, 2013 (1951).
20. M. Kilpatrick and F. E. Luborsky, *J. Am. Chem. Soc.* **75**, 577 (1953).
21. E. L. Mackor, A. Hofstra, and J. H. van der Waals, *Trans. Faraday Soc.*, 186 (1958).
22. U.S. Pat. 3,919,339 (Nov. 11, 1975), D. L. Ransley (to Chevron Research Company).
23. J. Barnard and B. M. Sankey, *Combust. Flame* **12**(4), 345 (1968).
24. L. Errede and M. Swarc, *Q. Rev.* **12**, 301 (1958).
25. L. Errede, R. Gregorian, and J. Hoyt, *J. Am. Chem. Soc.* **82**, 5218 (1960).
26. L. Errede and N. Kroll, *J. Polym. Sci.* **60**, 33 (1962).
27. U.S. Pat. 3,412,167 (Nov. 19, 1968), J. W. Lewis (to Union Carbide Company).
28. *Chem. Eng. News*, 30 (June 14, 1971).
29. U.S. Pat. 3,849,261 (Nov. 19, 1974), G. Gau (to Agence Nationale de Valorisation de la Recherche).
30. S. Veracini and G. Gau, *Nouv. J. Chem.* **2**(5), 523 (1978).
31. G. Gau and S. Marques, *J. Am. Chem. Soc.* **98**(6), 1538 (1976).
32. D. J. Hadley, *Chem. Ind.*, 238 (Feb. 25, 1961).
33. *Hydrocarbon Process.*, 252 (Nov. 1969).
34. *Hydrocarbon Process.*, 238 (Nov. 1981).
35. E. Kamegaya, *Chem. Econ. Eng.* **2**(2), 30 (1970).
36. S. Ariki and A. Ohira, *Chem. Econ. Eng.* **5**(7), (1973).
37. N. E. Ockerbloom, *Hydrocarbon Process.*, 101 (Feb. 1972).
38. P. B. de la Mare and J. Robertson, *J. Chem. Soc.*, 279 (1943).
39. F. Condon, *J. Am. Chem. Soc.* **70**, 1963 (1948).
40. K. Lammertsma, C. J. Verlaan, and H. Cerfontain, *Chem. Soc. J. Perkins Trans.* **2**(8), 719 (1978).
41. U.S. Pat. 2,282,231 (May 5, 1942), W. J. Mattox (to UOP).
42. D. L. Ransley, Chevron Research, Richmond, Calif., unpublished work, 1971.
43. O. Maas and J. Russell, *J. Am. Chem. Soc.* **40**, 1561 (1918).
44. O. Maas, E. H. Boomer, and D. M. Morrison, *J. Am. Chem. Soc.* **45**, 1433 (1923).
45. H. C. Brown and H. W. Pearsall, *J. Am. Chem. Soc.* **74**, 191 (1952).
46. W. D. Schaeffer and W. S. Dorsey, *Advances in Petroleum Chemistry and Refining*, Vol. 6, Wiley-Interscience, New York, 1962.
47. W. D. Schaeffer and co-workers, *J. Am. Chem. Soc.* **79**, 5870 (1957).
48. U.S. Pat. 2,798,103 (July 2, 1957), and U.S. Pat. 2,951,104 (Aug. 30, 1960), W. D. Schaeffer and J. D. Wordie (to Union Oil Company).
49. M. J. Minton and co-workers, *J. Phys. Chem.* **71**(11), 3618 (1967).
50. J. Menyhart, *Magy. Kem. Lapja* **22**(1), 18 (1967).
51. F. Casellato, *Erdoel Kohle* **22**, 2 (1969).
52. E. Ger. Pat. 56,236 (June 6, 1967), W. Jugel.
53. Brit. Pat. 1,151,606 (May 14, 1969), J. Lumsden (to Imperial Smelting Company).
54. H. R. Allcock and L. A. Siegel, *J. Am. Chem. Soc.* **86**, 5140 (1964).
55. U.S. Pat. 3,472,762 (Oct. 14, 1969), A. Goldrup and M. T. Westaway (to British Petroleum).

56. U.S. 3,484,500 (Dec. 16, 1969), A. Goldrup, A. B. Morrison, and M. T. Westaway (to British Petroleum).
57. Brit. Pat. 1,183,524 (Mar. 11, 1970), (to British Petroleum Company).
58. U.S. Pat. 3,504,047 (Mar. 31, 1970), J. N. Haresnape (to British Petroleum Company).
59. U.S. Pat. 3,456,028 (July 15, 1969), C. G. Gerhold and D. B. Broughton (to UOP).
60. U.S. Pat. 3,465,055 (Sept. 2, 1969), W. K. T. Gleim, R. C. Wackler, and R. C. Ranquist (to UOP).
61. U.S. Pat. 2,900,428 (Aug. 18, 1959), C. D. Heaton and W. E. Toland (to Chevron Research Company).
62. USSR Pat. 176,277 (Nov. 2, 1965), V. S. Bogdanov and co-workers.
63. C. J. Egan and R. V. Luthy, *Ind. Eng. Chem.* **47,** 250 (1955).
64. J. A. Verdol, *Oil Gas J.*, 63 (1969).
65. T. Iwamura, S. Otani, and M. Sato, *Bull. Jpn. Pet. Inst.* **13**(1), 116 (1971).
66. S. Otani, *Chem. Eng.*, 118 (July 20, 1970).
67. *Hydrocarbon Process.*, 133 (Nov. 1977).
68. D. L. McKay, G. H. Dale, and D. C. Tabler, *Chem. Eng. Prog.* **69**(11), 104 (1966).
69. C. M. Ambler, *Chem. Eng.*, 96 (Oct. 20, 1969).
70. D. K. Baumann and D. B. Todd, *Chem. Eng. Prog.*, 61 (Sept. 1973).
71. U.S. Pat. 3,467,724 (Sept. 16, 1969), S. A. Laurich (Chevron Research Company).
72. *Hydrocarbon Process.*, 253 (Nov. 1979).
73. U.S. Pat. 3,177,265 (Apr. 6, 1965), G. C. Lammers (to Standard Oil Company, Indiana).
74. Belg. Pat. 617,795 (Sept. 14, 1962), (to Standard Oil Company, Indiana).
75. Y. Hatanaka and T. Nakamura, *Oil Gas J.* **70**(47), 60 (1972).
76. *Hydrocarbon Process.*, 254 (1979).
77. R. J. Desiderio and co-workers, *Hydrocarbon Process.* **53**(8), 81 (1974).
78. *Hydrocarbon Process.*, 239 (1981).
79. H. F. Wiegandt and R. Lafay, *7th World Pet. Conf.* **4,** 47 (1967).
80. R. Lafay, *Chem. Ind.* **99**(10), 1555 (1968).
81. M. Moritoki, *Chem. Econ. Eng. Rev.* **8**(9), 20 (1976).
82. D. L. McKay and H. W. Goard, *Chem. Eng. Prog.* **61**(11), 99 (1966).
83. J. A. Brodie, *Mech. Chem. Eng. Trans.* **7,** 37 (1971).
84. A. G. Duncan and R. H. Phillips, *Trans. Inst. Chem. Eng.* **54,** 153 (1976).
85. Brit. Pat. 1,326,903 (Aug. 15, 1971), A. G. Duncan (to UK Atomic Energy Authority).
86. Private communication, UOP, Des Plaines, Ill., 1981.
87. D. B. Broughton and co-workers, *The Separation of p-Xylene from C_8–Hydrocarbon Mixtures by the Parex Process*, American Institute of Chemical Engineers, Puerto Rico, May 17–20, 1970.
88. D. B. Broughton and co-workers, *Chem. Eng. Prog.* **66**(9), 70 (1970).
89. D. P. Thornton, Jr., *Hydrocarbon Process.*, 151 (Nov. 1970).
90. F. H. Adams, *Environ. Chem. News*, 62 (Oct. 13, 1972).
91. D. B. Broughton, *Chem. Eng. Prog.* **73**(10), 49 (1977).
92. *Hydrocarbon Process.*, 256 (Nov. 1979).
93. U.S. Pat. 3,663,638 (May 16, 1972), R. W. Neuzil (to UOP).
94. U.S. Pat. 3,665,046 (May 23, 1972), A. J. De Rosset (to UOP).
95. U.S. Pat. 3,686,342 (Aug. 22, 1972), R. W. Neuzil (to UOP).
96. U.S. Pat. 3,636,180 (Jan. 18, 1972), D. B. Broughton (to UOP).
97. U.S. Pat. 3,696,107 (Oct. 3, 1972), R. W. Neuzil (to UOP).
98. H. J. Biezer, G. R. Winter, and G. Barbu, *Oil Gas J.*, (Aug. 11, 1975).
99. S. Otani and co-workers, *Chem. Econ. Eng. Rev.*, 56 (June 1971).
100. S. Otani and co-workers, *Pac. Chem. Eng. Congr. Proc.* **20**(1), 550 (1973).
101. S. Otani, *Chem. Eng.*, 106 (Sept. 17, 1973).
102. S. Otani and co-workers, *Jpn. Pet. Inst. Bull.* **16**(1), 60 (1974).
103. U.S. Pat. 3,761,533 (Sept. 25, 1973), S. Otani and co-workers (to Toray Industries, Inc.).
104. M. Seko, T. Miyake, and K. Inada, *New Process for p-Xylene and Ethylbenzene Separation from Mixed Xylene*, presented to the ACS/CSJ Chemical Congress, Honolulu, Hawaii, Apr. 1979.
105. M. Seko, *Oil Gas J.*, 81 (July 2, 1979).
106. M. Seko, T. Miyake, and K. Inada, *Ind. Eng. Chem. Prod. Res. Dev.* **18**(4), 263 (1979).
107. M. Seko, T. Miyake, and K. Inada, *Hydrocarbon Process.*, 133 (Jan. 1980).
108. Private communication, Asahi Chemical Industry Company, Ltd., Tokyo, Jpn., 1981.
109. Y. Igarashi and T. Ueno, *American Chemical Society Meeting A.7*, Atlantic City, N.J., 1968.

110. *Hydrocarbon Process.*, 254 (Nov. 1969).
111. Y. Igarashi, *Jpn. Chem. Q.* **IV**(4), 27 (1969).
112. T. Ueno, *Bull. Jpn. Pet. Inst.* **12,** (May 1970).
113. G. R. Herrin and E. H. Martel, *Chem. Eng. (London)*, 319 (Sept. 1971).
114. J. C. Davis, *Chem. Eng.*, 77 (Aug. 9, 1971).
115. T. Ueno and T. Nakano, *Eighth World Pet. Congr. Proc.* **4,** 187 (1971).
116. G. R. Herrin and E. H. Martel, *Pet. Petrochem. Int.* **12**(7), 74 (1972).
117. S. Ariki and A. Ohira, *Chem. Econ. Eng. Rev.* **5**(7), 30 (1973).
118. J. J. H. Masseling, *Chem. Technol.*, 714 (Nov. 1976).
119. *Hydrocarbon Process.*, 255 (Nov. 1979).
120. *Selected Values of Properties of Hydrocarbons and Related Compounds*, Carnegie Institute of Technology, Pittsburgh, Pa., 1956.
121. *Selected Values of Physical and Thermodynamic Properties of Hydrocarbons and Related Compounds*, API Research Project 44, Carnegie Press, 1953.
122. M. H. Everdell, *J. Chem. Ed.* **44**(9), 538 (1967).
123. K. G. Harbison, *J. Chem. Ed.* **47**(12), 837 (1970).
124. D. A. McCaulay in G. A. Olah, ed., *Friedel-Crafts and Related Reactions*, Vol. 2, Wiley-Interscience, a division of John Wiley & Sons, Inc., New York, 1964, pp. 1049–1073.
125. H. F. Uhlig and W. C. Pfefferle, *Am. Chem. Soc., Div. Pet. Chem. D*, 154 (Sept. 1969).
126. P. M. Pitts, Jr., J. E. Connor, Jr., and L. N. Leum, *Ind. Eng. Chem.* **47**(4), 770 (1955).
127. U.S. Pat. 2,976,332 (Mar. 21, 1961), L. N. Leum and J. E. Connor, Jr. (to Atlantic Refining Co.).
128. U.S. Pat. 4,139,571 (Feb. 13, 1979), R. A. Riehm (to Atlantic Richfield Company).
129. P. Grandio and co-workers, *Oil Gas J.*, 63 (Nov. 1971).
130. P. Grandio and co-workers, *Hydrocarbon Process.* **51,** 85 (1972).
131. U.S. Pat. 3,856,871 (Dec. 24, 1974), W. O. Haag and D. H. Olson (to Mobil Oil Corp.).
132. U.S. Pat. 3,702,886 (Nov. 14, 1972), R. J. Argauer and G. R. Landolt (to Mobil Oil Corp.).
133. C. V. Berger, *Hydrocarbon Process.*, 173 (Sept. 1973).
134. *Hydrocarbon Process.*, 239 (Nov. 1977).
135. H. C. Ries, *Process Economics Program 25A*, Stanford Research Institute, Menlo Park, Calif., July 1970, pp. 41–53.
136. *Hydrocarbon Process.*, 240 (Nov. 1981).
137. *Hydrocarbon Process.*, 109 (Aug. 1969).
138. *Hydrocarbon Process.*, 253 (Nov. 1969).
139. U.S. Pat. 3,793,384 (Feb. 19, 1974), J. G. Chenoweth and co-workers (to Imperial Chemical Industries).
140. U.S. Pat. 3,860,668 (Jan. 14, 1975), J. K. January and A. Marchant (to Imperial Chemical Industries).
141. P. Grandio and co-workers, *Am. Chem. Soc. Div. of Pet. Chem.* **16**(3), B70 (1971).
142. N. Y. Chen and W. E. Garwood, *J. Catal.* **52,** 453 (1978).
143. U.S. Pat. 3,790,471 (Feb. 5, 1974), R. J. Argauer and G. R. Landolt (to Mobil Oil Corp.).
144. D. H. Olson and co-workers, *J. Phys. Chem.* **85,** 2238 (1981).
145. U.S. Pat. 3,856,873 (Dec. 24, 1974), G. T. Burress (to Mobil Oil Corp.).
146. U.S. Pat. 3,856,872 (Dec. 24, 1974), R. A. Morrison (to Mobil Oil Corp.).
147. U.S. Pat. 4,159,283 (June 20, 1979), M. P. Nicoletti and J. F. van Kirk (to Mobil Oil Corp.).
148. U.S. Pat. 4,158,676 (June 19, 1979), F. A. Smith, L. L. Beckenridge, and A. B. Schwartz (to Mobil Oil Corp.).
149. U.S. Pat. 4,218,573 (Aug. 19, 1980), S. A. Tabak and R. A. Morrison (to Mobil Oil Corp.).
150. U.S. Pat. 4,224,141 (Sept. 23, 1980), R. A. Morrison and S. A. Tabak (to Mobil Oil Corp.).
151. U.S. Pat. 3,699,182 (Dec. 5, 1969), J. Cattanach (to Mobil Oil Corp.).
152. P. B. Weisz, *7th International Congress of Catalysis*, Tokyo, Jpn., 1980.
153. U.S. Pat. 4,101,596 (July 18, 1978), K. M. Mitchell and J. J. Wise (to Mobil Oil Corp.).
154. Private communication, Mobil Research and Development Corp., New York, 1981.
155. *Hydrocarbon Process.*, 25 (Oct. 1981).
156. W. W. Kaeding and co-workers, *J. Catal.* **67,** 159 (1981).
157. U.S. Pat. 4,283,306 (Aug. 11, 1981), F. E. Herkes (to DuPont).
158. *Chemical Economics Handbook*, SRI International, Menlo Park, Calif., 1979.
159. *Chem. Eng. News*, 23 (Nov. 22, 1982).
160. Private communication, Engelhard Industries, Newark, N.J., 1981.

161. *Chemical Economics Handbook*, SRI International, Menlo Park, Calif., 1983.
162. J. M. Vergnaud, *J. Chromatogr.* **27**, 54 (1967).
163. C. F. Raley and J. W. Kaufman, *Anal. Chem.* **40**(8), 1371 (1968).
164. M. Dimov and D. Papazova, *J. Chromatogr.* **137**, 265 (1977).
165. F. Baumann and S. M. Csicsery, *J. Chromatogr.* **26**, 262 (1967).
166. M. F. Burke and L. B. Rogers, *J. Chromatogr. Sci.* **6**(1), 75 (1968).
167. A. B. Richmond, *J. Chromatogr. Sci.* **9**(9), 571 (1971).
168. A. C. Bhattacharyya and A. Bhattacharjee, *J. Chromatogr.* **41**(3-4), 446 (1969).
169. M. Popl, J. Coupek, and S. Pokorny, *J. Chromatogr.* **104**, 135 (1975).
170. U.S. Pat. 3,653,184 (Apr. 4, 1972), B. M. Drinkard, P. T. Allen, and E. H. Unger (to Mobil Oil Corp.).
171. U.S. Pat. 3,656,278 (Apr. 18, 1972), B. M. Drinkard, P. T. Allen, and E. H. Unger (to Mobil Oil Corp.).
172. U.S. Pat. 3,724,170 (Apr. 3, 1973), P. T. Allen, B. M. Drinkard, and E. H. Unger (to Mobil Oil Corp.).

DEREK L. RANSLEY
Chevron Research Company

XYLYLENE POLYMERS

Poly(p-xylylene) [25722-33-2] coatings achieved economic importance in the 1960s, although the polymers were synthesized over seventy-five years ago by Thiele (1). Later, p-xylylene, which resembles p-benzoquinone, was used as a model for stability and property prediction and was thought to be too reactive to exist in the condensed phase (2–3). However, it was actually synthesized by fast-flow pyrolysis of p-xylene in the gas phase at very low pressures (4). Subsequent studies showed that this extremely reactive monomer can be stabilized by collecting the pyrolyzed gas stream in a solvent at −78°C (5). This observation enabled researchers to elucidate the chemistry of this diradical-like monomer under controlled reaction conditions.

The p-xylylene monomers are considered prototypes of the Chichibabin hydrocarbons and can be represented by a quinoid (1) or benzenoid (2) structure. If the structure exists in the quinoid form, it would have diamagnetic properties, whereas the benzenoid form would have a diradical structure with paramagnetic properties (6). It was established that the monomer is diamagnetic and therefore the p-xylylenes probably exist in the quinoid form (1).

Because of their unique properties, the xylylene polymers are used as special coatings or films. These polymers are highly crystalline, straight–chain organic compounds with excellent dielectric characteristics and molecular weights of ca 500,000.

The poly(p-xylylene) polymers are usually prepared by the pyrolysis of di-p-xylylene. A patent has been issued that describes the electrolytic preparation of poly(p-xylylene) (7).

Synthesis

The starting material for the synthesis of the commercial xylylene polymers is a dimer, di-p-xylylene (**3**) or 2,5′-dichlorodi-p-xylylene (**4**).

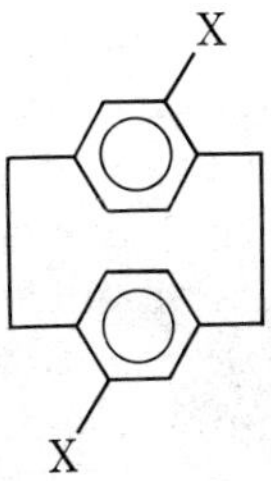

(3) X = H
(4) X = Cl

Dimer. Preparation. Di-p-xylylene is prepared by a number of synthetic routes. The most direct route is by pyrolysis of p-xylene in the presence of steam at atmospheric pressure, followed by quenching of the pyrolysis gases in an aromatic hydrocarbon such as xylene or toluene (eq. 1). This route produces di-p-xylylene in reasonable yields (8) and is used commercially by Union Carbide Corporation.

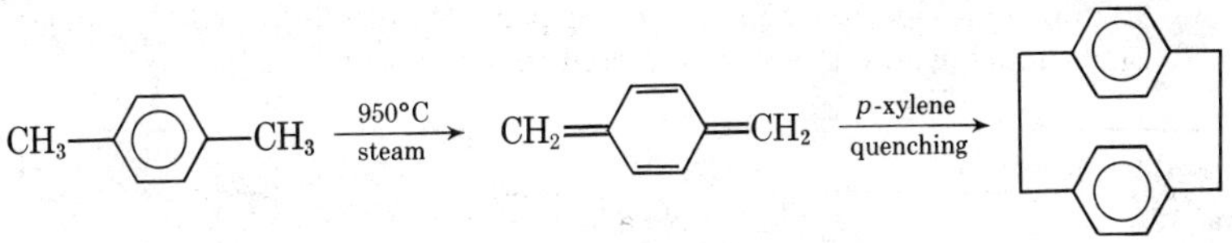

$$CH_3-\langle\bigcirc\rangle-CH_3 \xrightarrow[\text{steam}]{950°C} CH_2=\langle\bigcirc\rangle=CH_2 \xrightarrow[\text{quenching}]{p\text{-xylene}} [\text{dimer}] \quad (1)$$

Another convenient laboratory approach for preparation of small quantities is given below (9):

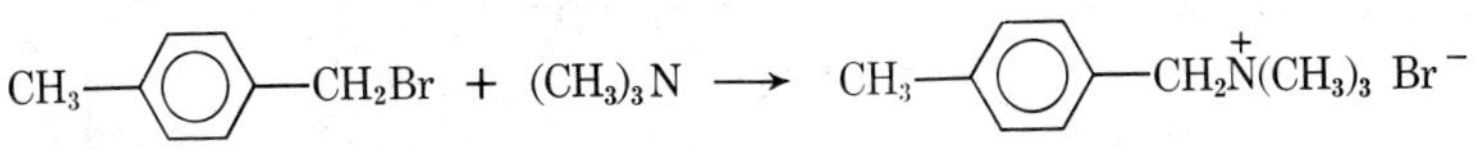

$$CH_3-\langle\bigcirc\rangle-CH_2Br + (CH_3)_3N \rightarrow CH_3-\langle\bigcirc\rangle-CH_2\overset{+}{N}(CH_3)_3\ Br^- \quad (2)$$

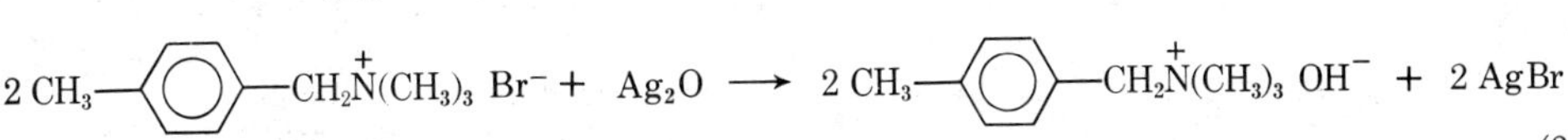

$$2\ CH_3-\langle\bigcirc\rangle-CH_2\overset{+}{N}(CH_3)_3\ Br^- + Ag_2O \rightarrow 2\ CH_3-\langle\bigcirc\rangle-CH_2\overset{+}{N}(CH_3)_3\ OH^- + 2\ AgBr \quad (3)$$

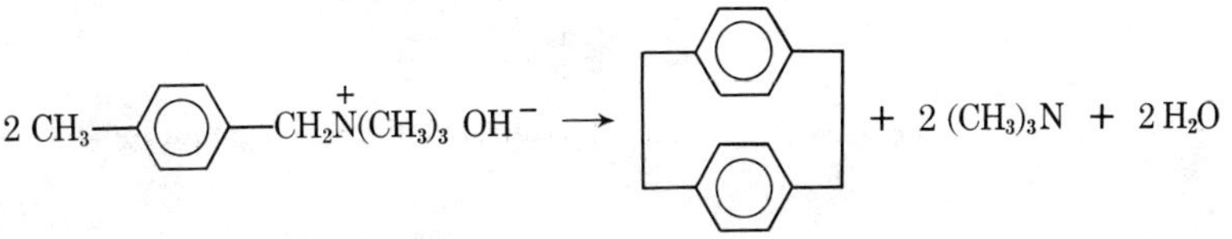

$$2\ CH_3-\langle\bigcirc\rangle-CH_2\overset{+}{N}(CH_3)_3\ OH^- \rightarrow [\text{dimer}] + 2\ (CH_3)_3N + 2\ H_2O \quad (4)$$

Properties. Di-*p*-xylylene and dichloro-*p*-xylylene dimers are white, high-melting crystalline solids. Melting points and densities are given in Table 1.

Purification. Unsubstituted di-*p*-xylylene is readily purified by recrystallization from xylene. Dichlorodi-*p*-xylylene is a mixture of isomers as prepared by chlorination of di-*p*-xylylene. It is not necessary to separate these isomers since chloro-*p*-xylylene is the only pyrolysis product, regardless of which isomeric dimer is used as starting material. Hence, the primary purpose of purification is to remove residue and chlorinated dimers. Separation of residue and some fractionation can be accomplished by distillation. Dichlorodi-*p*-xylylene has a boiling point of approximately 185°C at ca 5.3 Pa (0.04 mm Hg). The dimers can be recrystallized from carbon tetrachloride or ethanol.

Gas phase chromatography is used routinely to check the purity of di-*p*-xylylene and its derivatives; thermal conductivity detection is generally used with helium as the carrier gas.

Polymerization. Di-*p*-xylylene is vaporized at ca 133 Pa (1 mm Hg) and 200°C and the vapors are conducted through a pyrolysis or cleavage zone at 600°C to yield monomeric *p*-xylylene diradicals. At ambient temperature, these stable *p*-xylylene diradicals instantaneously polymerize quantitatively upon contact with any solid surface without the formation of any intermediate phase (see Fig. 1) (11). This is often referred to as the Gorham process.

Parts to be coated are rotated in the vacuum chamber by a controlled variable-

Table 1. Some Properties of Xylylene Dimers[a]

Dimer	Mp, °C	Density, g/cm³
di-*p*-xylylene[b]	284 (decomp)	1.22
dichlorodi-*p*-xylylene	140–160[c]	1.3

[a] From Ref. 10.
[b] DPX.
[c] Mixture of isomers, the highest melting point of which is 212–213°C, which can be isolated after repeated recrystallizations from ethanol.

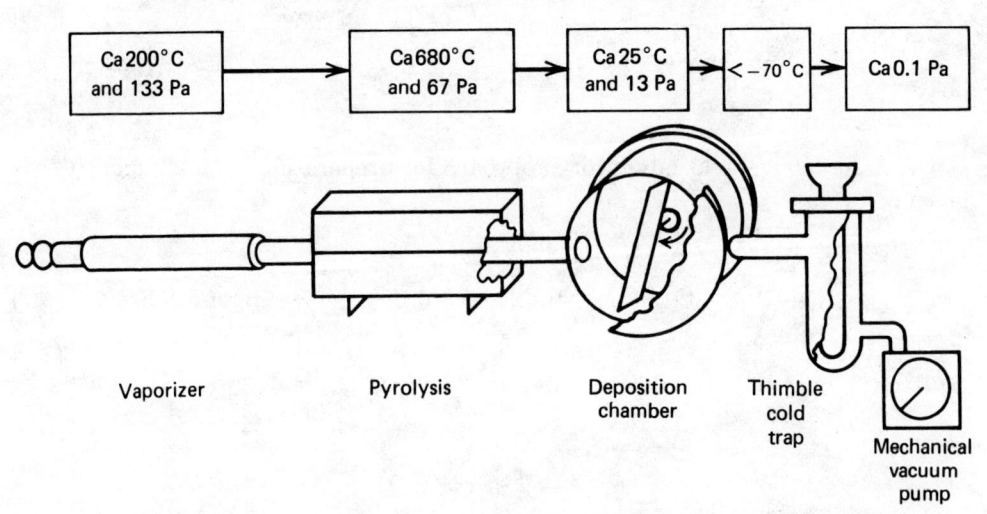

Figure 1. Manufacture of poly(*p*-xylylene). To convert Pa to mm Hg, multiply by 0.0075.

speed motor to allow even distribution of the polymeric coating. This is not a line-of-sight process as in vacuum metallizing; the vapor is pervasive, but it coats without briding and holes can be jacketed evenly. Moreover, the object to be coated remains at ambient temperature, eliminating the risk of thermal damage. Coating thickness is controlled easily and very accurately by regulating the amount of dimer vaporized. Deposition chambers of virtually any size can be constructed. Large parts (up to 152 cm long and 46 cm high) have been processed in equipment similar to that shown in Figure 2 (see Film deposition techniques).

Mechanism. The polymerization process takes place in two completely distinct and separate steps:

$$\text{di-}p\text{-xylene} \longrightarrow \cdot CH_2-\bigcirc-CH_2\cdot \longrightarrow \left[CH_2-\bigcirc-CH_2\right]_n \quad (5)$$

$$\updownarrow$$

$$CH_2=\bigcirc=CH_2$$

di-*p*-xylylene   *p*-xylylene   poly(*p*-xylylene)

The first stage involves the cleavage of the two methylene–methylene bonds in di-*p*-xylylene. This molecule is stable in the vapor phase, but spontaneously polymerizes in the second step upon condensation to form high molecular weight poly(*p*-xylylene).

Figure 2. Model 1050 Parylene Coater. Courtesy of Union Carbide Corporation.

748   XYLYLENE POLYMERS

Pyrolysis. The pyrolysis of di-p-xylylene produces two molecules of p-xylylene rather than one long–chain reactive intermediate such as (5).

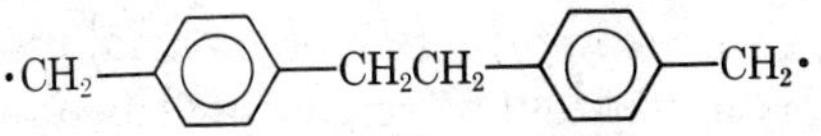

$$\cdot CH_2-\phi-CH_2CH_2-\phi-CH_2\cdot$$
(5)

This was shown by quenching the pyrolysis gases in the deposition zone with an excess of iodine vapor. p-Xylylene diodide was isolated in 50% yield, unequivocally indicating the presence of p-xylylene in the pyrolysis vapor.

This experiment did not shed light on the question of whether the methylene–methylene bonds in di-p-xylylene are broken sequentially or simultaneously in the pyrolysis step. However, during the attempted pyrolytic polymerization of α-hydroxydi-p-xylylene (6), the product was unexpectedly found to be the long–chain aldehyde (7). Compound (6) undoubtedly pyrolyzes in the fashion shown in equation 6 involving the initial pyrolysis of only one methylene–methylene bond to form (8). The intermediate (8) then rearranges to form the stable product (7) which was isolated in >90% yield.

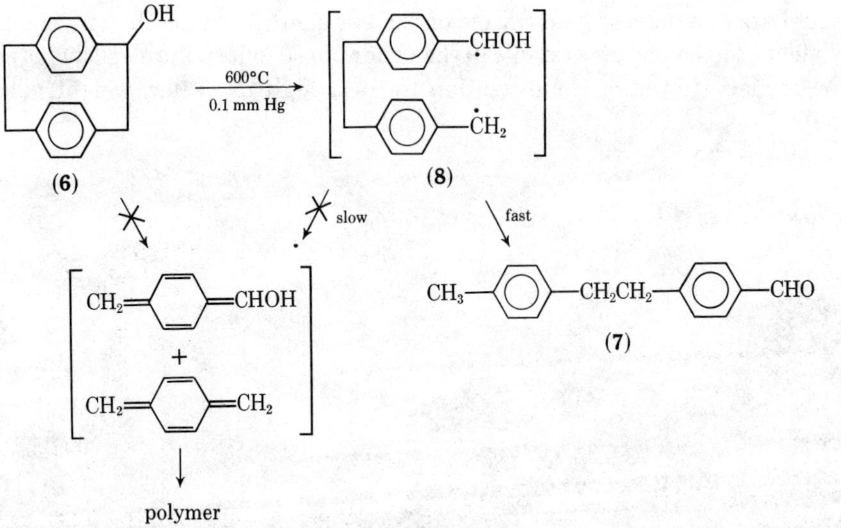

The normal cleavage of the second double bond to form p-xylylene species is thus interrupted. The transformation indicates a stepwise rather than simultaneous cleavage of the two benzylic bonds.

This example, however, is the only known case in which abnormal pyrolysis occurs. Over thirty other substituted di-p-xylylenes, including ring-substituted and other derivatives, such as ar-chlorodi-p-xylylene, have been pyrolyzed to yield polymerizable p-xylylenes as the main product. However, the case cited above, provides an interesting clue in favor of sequential cleavage of the methylene–methylene bonds.

Pyrolysis of monosubstituted di-p-xylylenes such as ar-acetyldi-p-xylylene (9) results in the formation of two reactive p-xylylenes polymerizing at different temperatures. The two species were separated in the form of their polymers using the principle of threshold–condensation temperature. The pyrolysis vapors containing the two monomers (10) and (1) were passed initially through a zone maintained at a temperature low enough to permit rapid condensation and polymerization of *ar*-acetyl-*p*-xylylene. The temperature was, however, substantially above the T_c of *p*-xylylene, which passed through the first zone and polymerized in a final zone maintained at ambient temperature. In a sense, the monomers were fractionated on the basis of volatility and isolated in the form of their polymers. These transformations are illustrated by equation 7. The results are explained only by the formation of both *p*-xylylene and *ar*-acetyl-*p*-xylylene in the pyrolysis step. This provides conclusive evidence that both bridge bonds in the di-*p*-xylylene are broken during pyrolysis.

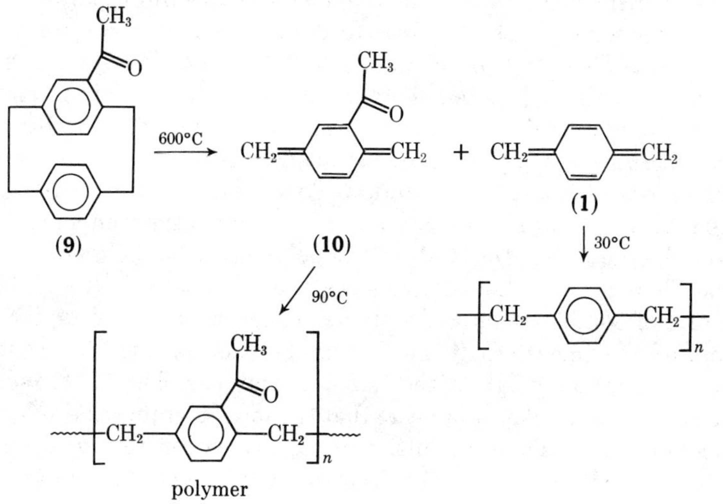

If the pyrolysis step is interrupted and then continued again, laminated poly(*p*-xylylene) films are produced.

Polymerization Reaction. The polymerization step is proceeded by a free-radical mechanism in which, as a first step, two molecules of *p*-xylylene condense on a surface and react to form the diradical intermediate (5) (eq. 8). The first step, forming (5), is probably reversible.

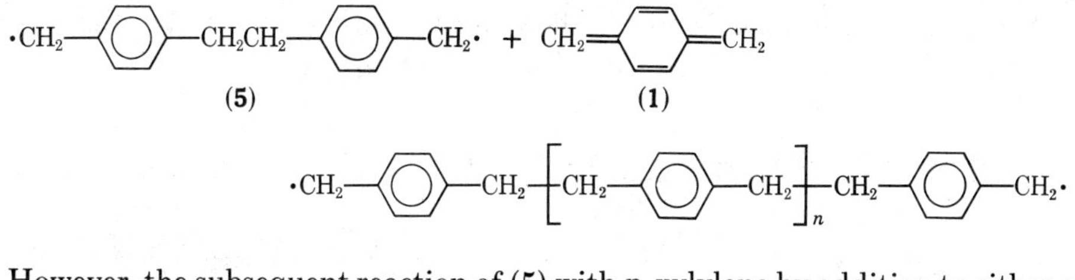

However, the subsequent reaction of (5) with *p*-xylylene by addition to either end of the reactive diradical of the intermediate gives the stable species (11) (in which *n* is 1, 2, or 3) as shown in equation (8). Growth then progresses by addition of *p*-xylylene to each end of the diradical and is terminated by reaction of the radical end groups with reactive sites in other growing polymer chains, by reaction of the free-radical sites with chain-transfer agents (eg, oxygen or mercaptans), or by the reactive sites becoming buried in the polymer matrix.

This proposed mechanism of polymerization of *p*-xylylenes suggests that the rate of polymerization should be markedly increased by lowering the temperature of the deposition surface to raise the rate of condensation and, therefore, the concentration of molecules of p-xylenes in the condensed phase. This has been shown to be the case, and relative polymerization rates of 1, 10, and 100 were observed for p-xylylene on surfaces maintained at 30, 0, and $-40°C$, respectively, and at equivalent monomer concentrations in the vapor phase. These data provide strong evidence that the rate-determining step in the polymerization is condensation of a p-xylylene molecule in the vicinity of growing free radical, and that addition of the condensed molecules to the reactive site is very rapid in comparison.

This mechanism suggests that the polymer should exhibit paramagnetic properties, which is indeed the case. Polymers formed by this technique have been shown to contain radical concentrations of $(5-10) \times 10^{-4}$ mol of free electrons per mol of xylylene. These so-called living polymers retain their free-radical content for several months when stored at ambient temperatures in an inert oxygen-free atmosphere.

The polymerization mechanism described above requires the initial condensation of p-xylylene species before polymerization. This implies that monomers of different volatility exhibit different condensation and polymerization tendencies, which has been found to be the case in the study of the polymerization behavior of substituted p-xylylenes. There is a threshold condensation temperature, T_c (see Table 2), above which the rate of condensation and polymerization is very slow under the system pressure conditions of 6.5–13 Pa (0.05–0.10 mm Hg) normally used. This temperature is related to the molecular weight of the monomer and the polarity of the substituent(s) attached to the aromatic ring. It appears that the monomer must be able to condense on a surface and be available in sufficiently high concentration to react with its neighbors to form initially diradical intermediates, such as (12), containing at least three and probably four monomer units.

$\cdot CH_2-\bigcirc_X-CH_2CH_2-\bigcirc_X-CH_2CH_2-\bigcirc_X-CH_2\cdot$

(12)

Once this intermediate is formed, growth appears to be very rapid through addition of monomer units that condense in the vicinity of either radical site. High mo-

Table 2. Threshold–Condensation Temperature of Substituted p-Xylylenes[a]

Monomer	CAS Reg. No.	T_c, °C	Polymer, CAS Reg. No.
p-xylylene	[502-86-3]	30	[25722-33-2]
2-methyl-p-xylylene	[10366-12-8]	60	[67076-70-4]
2-ethyl-p-xylylene	[10366-13-9]	90	[9069-95-8]
2-chloro-p-xylylene	[623-25-6]	90	[9052-19-1]
2-acetyl-p-xylylene	[10366-11-7]	130	[67076-72-6]
2-cyano-p-xylylene	[10366-10-6]	130	[85567-63-1]
2-bromo-p-xylylene	[91-13-4]	130	[65863-37-8]
dichloro-p-xylylene	[28347-13-9]	130	[52261-45-7]

[a] Ref. 10.

lecular weight polymers are then formed by a free-radical addition mechanism (12–14).

Kinetic studies, using 0.05–0.1 M solutions of p-xylylene at $-78°C$, elucidated the mechanisms of initiation, propagation, and termination (15). In addition, conditions for the selective formation of the dimer, trimer or high molecular weight polymer from the monomer were established.

The poly(p-xylylene) film is transparent when thin, but white or opaque when deposited in thick layers. It is formed by condensation of gaseous p-xylylene, started by p-xylylene units combining into large diradicals, which continue to spontaneously grow at both ends. The remarkable lack of polymerization in the gas phase (6) was explained by cyclization which, particularly in the early growth stage, is competitive with diradical polymerization (16–17). However, the concentration of monomer is low in the gas phase and so cyclization is more probable than propagation, whereas in the condensed phase, cyclization is hindered by polymer crystallization. It has been postulated that the central dimer CH_2CH_2 bond is weak, because a great amount of binding energy is lost when opening two C—C double bonds forming a C—C single bond with two uncoupled electrons (6). Such a diradical is expected to revert to the more stable xylylene units. If, however, four units combine into a linear tetrameric diradical, a comparatively stable species results. By breaking the central CH_2CH_2 bond on the tetramer, two diradicals can be produced from one tetrameric diradical with no additional driving force available to facilitate the process.

Numerous xylylene polymers have been prepared by the pyrolysis reaction and characterized starting with appropriately substituted di-p-xylylenes (12–14). Some of the unchanged starting material is still present in poly(chloro-p-xylylene) [9052-19-1]. However, by optimizing deposition conditions, films have been reported containing only 0.15% of the dichloro-p-xylylene dimer (18).

Copolymerization. Attempts to copolymerize p-xylylene with styrene or butadiene either in the vapor or liquid phase resulted only in the formation of poly(p-xylylene) (6,19–20). Copolymerization with maleic anhydride and chloroprene was successful in solution at $-78°C$ using oxygen (21), PCl_3, and $RPCl_2$ (22), or SO_2 (23) as catalysts.

p-Xylylene copolymerizes with other monomers, such as chloro-p-xylylene, pseudocumene, and 2,5-dimethylpyrazine. A mixture of p-xylene and the other monomer was pyrolyzed or each component was pyrolyzed separately, followed by mixing the gaseous products and condensation. Alternatively, the monomers are mixed in the desired proportion at a low temperature; heating results in polymerization (5). Copolymerization was confirmed by x-ray diffraction studies (21,24).

Properties

Molecular Weight. Measurement of the soluble polymer fraction with a high temperature viscometer gave a molecular weight of 20,000 (24). This was confirmed by an osmotic pressure measurement of the chloroform solution giving a value of ca 24,000.

The high molecular weight of poly(p-xylylene) polymers is demonstrated by the fact that the stretched polymer yields highly oriented fibers (19). Other evidence included the swelling of sulfonated poly(p-xylylene) in water and the failure of the polymer solution to pass through a sintered glass membrane which is designed to prevent diffusion of polymers of molecular weight above 20,000.

Contrary to early reports (4–5,19–20,24–26), poly(p-xylylene) is a linear, high molecular weight polymer that is completely soluble at high temperatures. Paramagnetic studies reveal a molecular weight above 2×10^5 (27). Similar values were obtained when the polymer was prepared by the Wurtz reaction (28).

Structure. The high birefringence of β-poly(p-xylylene) reveals that the molecules are all parallel to the c axis and are distributed nearly isotropically about it (29). For example, molecules in the benzene crystal lattice are arranged in two sets with their planes approximately at right angles to one another and parallel to the b axis (30).

It had been predicted that the parallel coplanar structure would correspond to the α cell, considering the morphology of the elongated pseudorectangular α single crystals (29,31–32). It was shown that the benzene rings in α-poly(p-xylylene) are indeed oriented in the same direction and are parallel (33). They are parallel to the b axis, which is preferentially oriented along the direction perpendicular to the substrate surface when p-xylylene monomer polymerizes by deposition from the vapor phase. It is more probable that the parallel coplanar structure corresponds to the α cell, considering the morphology of the elongated pseudorectangular α single crystals (29,31–32).

The observed fiber-repeat distance for α and β poly-p-xylylene is 6.55 and 6.58 $\times 10^{-4}$ µm, respectively, compared to a calculated value of 6.52×10^{-4} µm for a regular staggered configuration (5). Similarly, the bond angles and interatomic distances for 4,4'-dimethyldibenzyl show the length of the unit

$$-CH_2CH_2-\!\!\left\langle\!\!\bigcirc\!\!\right\rangle\!\!-$$

(13)

to be 0.66 nm, in agreement with the repeat distance determined for both the α and β modifications (34). Therefore, the molecules in both polymorphs are most likely organized in an extended configuration.

A polymer formed during deposition between -20 and $80°C$ crystallizes almost entirely in the α modification. Both the α and β polymorphs are deposited between -20 and $-60°C$ (35–36). Below $-100°C$ the polymer is deposited entirely in the β modification. Films deposited below $-20°C$ were originally believed to crystallize in a third crystal form, γ, on the basis of x-ray diffraction studies (36). However, these diffraction effects are due to a mixture of β diffraction lines and tri-p-xylylene (35). Similarly, white blocks which formed on films deposited at $50°C$ were proposed to be a fourth polymorph of poly(p-xylylene) (37), but x-ray reflections of this polymorph were ascribed to the physical carryover of the di-p-xylylene (dimer) crystals (38). Above the T_g, both the alpha and β forms codeposit to $220°C$. Between 220 and $270°C$ the beta form predominates, and at $270°C$ a reversible smectic mesomorphic transition of β_1 to β_2 occurs on heating (see Liquid crystals). The latter transition is the temperature where the crystal order is lost along the molecular chain direction, ie, the β_2 form is a two-dimensional crystal. This transition is very similar to the room-temperature transition ($25°C$) in Teflon, polytetrafluoroethylene, which accounts, in part, for its low coefficient of friction (39). The complex transformation of poly(p-xylylene) is illustrated in Figure 3.

The thermal behavior of poly(p-xylylene) is unusual. Apparently, the α modification is the most stable thermodynamic form near room temperature, but at low and

```
|   β   |  β+α  |   α   |  α+β  |   β   |  β₂  | Liquid
|-------|-------|-------|-------|-------|------|--------
-196    -60    -20     80      220     270    420
                      T_g, °C                       T_m
```

|← Deposition temperature, °C →|

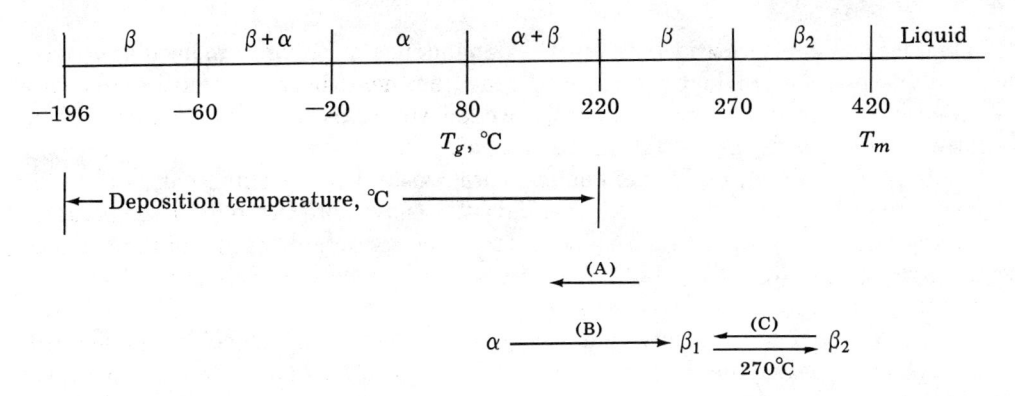

$$\alpha \xrightarrow{(B)} \beta_1 \underset{270°C}{\overset{(C)}{\rightleftarrows}} \beta_2$$

$$\xleftarrow{(A)}$$

Figure 3. The transformation of poly(p-xylylene). (A) Observed for crystals suspended in α-chloronaphthalene (74). (B) The α → β transition not reversible in films. (C) Spontaneously reversible transition.

high deposition temperatures, and above 220°C, the β polymorph is the more stable form.

Molecular Orientation. Early studies indicated that the molecular orientation in vapor-deposited poly(p-xylylene) films was anisotropic (19). Later, a careful diffraction study of films deposited on various crystalline and amorphous substrates was undertaken (40). Wide-angle x-ray diffraction patterns of films >1 μm in thickness, and electron-diffraction patterns of thin films (0.01–0.02 μm), presented a typical random-oriented powder pattern. This occurred when the beam of radiation was directed perpendicular to the plane of the film and revealed that the polymer molecules were randomly oriented. However, when the beam was inclined at some angle other than 90° to the film, the continuous Debye rings broke up into arcs, establishing the presence of preferred orientation.

Analysis of complex oriented electron–diffraction patterns of films annealed to the hexagonal β modification revealed that most of the crystallites are oriented with the (10.0) prism faces lying in the plane of the film. Since the polymer chains are aligned in the c-axis direction [00.1], it follows that the molecules within oriented crystallites are also lying in the plane of the film. These conclusions were qualitatively extended on the basis of x-ray data to include films as thick as 100 μm.

The degree of orientation is essentially independent of the type of substrate upon which the films are deposited from the vapor phase. However, orientation is inversely proportional to the partial pressure of the p-xylylene monomer in the vapor phase above the substrate. Growth rates were 6×10^{-4} μm at 25°C, with a sticking coefficient (p-xylylene radicals striking the surface) of 0.0025 ± 0.0010.

The condensation of p-xylylene on a cold surface could be represented as a simultaneous polymerization and crystallization process (41). By analogy with other polymer systems, this should afford extended-chain crystallization. However, on the basis of small-angle x-ray diffraction studies of lamellar structures observed in vapor-deposited films, the crystals in these films must contain folded chains (42).

A mathematical model developed for the vapor deposition and polymerization of p-xylylene links process variables (temperature, pressure, rate of growth) with molecular rate constants, diffusional mass transport, and molecular weight (43). This model explains the lack of gas–phase reactivity of p-xylylene, and its conformality. It also provides an as yet untested prediction of the pressure dependence of growth rate, and the pressure and temperature dependence of polymer molecular weight. Qualitatively, the polymer is formed in a medium near the surface of the substrate which is best described as a solid slightly swollen with an as yet reacted monomer. The latter is on the order of several tenths of a percent.

Solubility. Poly(p-xylylene) polymers are resistant to solvents, acids, and bases. Early workers reported that the polymer did not dissolve until the solution temperature approached the crystalline melting point (24). The intractable nature of the material was ascribed to extreme crystallinity or presence of cross-links. Poly(p-xylylenes) are also unaffected by stress-cracking agents.

Poly(p-xylylene) and poly(chloro-p-xylylene) dissolve in high boiling liquids such as α-chloronaphthalene or benzyl benzoate above 200°C. In an extensive solubility study, the effects of organic solvents on poly-p-xylylene and its chloro and dichloro derivatives were determined (44). The solvents included alcohols, ketones, aliphatic and aromatic hydrocarbons, chlorinated aliphatic and aromatic nitrogen heterocyclic solvents and fluorinated aliphatic solvents. A minor swelling effect with a 3% maximum increase in film thickness was observed. Vacuum drying reversed this effect.

The solubility in deionized water, dilute bases, and dilute and concentrated nonoxidizing and oxidizing acids was also investigated; no effect was observed at room temperature. However, at 75°C for 90 minutes, acids caused swelling ranging from 0.7% with hydrochloric to 8.2% with chromic acid. Nitric acid under these conditions caused severe degradation, but sulfuric acid did not have an appreciable effect (45). However, in the presence of traces of silver ion, sulfonation results, yielding a product containing one sulfonic acid group per benzene ring (19,20,26).

Reactivity. Chlorination with a pyridine solution of sulfuryl chloride under uv light results mainly in chlorinated ethylene groups (20).

Concentrated nitric acid at 50°C for a prolonged period gives poly(dinitro-p-xylylene) [85586-90-9], an explosive material with the sensitivity of pentaerythritol tetranitrate and the power of trinitrotoluene (24).

Boiling the polymer for 24 h in a chromic acid–acetic acid solution gives terephthalic acid as the main product (25).

Oxidation takes place slowly at high temperatures in the presence of atmospheric oxygen. Once ignited, the polymer burns fiercely (24). Heating in a vacuum at 415°C gave random chain scission: benzene, toluene, xylene, p-methylstyrene, and p-ethyltoluene were identified as the volatile products of decomposition (46). The reported activation energy is 318 kJ/mol (76 kcal/mol) and relative thermal stability is similar to that of Teflon.

Both vacuum and oxidative pyrolysis of the vapor-deposited polymer gives low molecular weight products (47); presumably dimers and pentamers produced by an unzippering mechanism. Hydrogen evolution is also noted, because of the breakdown of conjugated unsaturated polymer structures. A small percentage (ca 1–2%) of unreacted cyclic dimer, present in the original polymer, is recovered. The initial slow evolution of decomposition products during vacuum pyrolysis may be due to the small number of branch points in the polymer.

Physical and Mechanical Properties. The physical and mechanical properties of several xylylene polymers are given in Table 3 and Table 4. In general, the properties were measured on 25–50 μm films that had been deposited on glass and removed by stripping the film from the surface. These films are relatively high modulus, tough polymers which compare favorably with other coating resins (48).

Table 3. Mechanical and Thermal Properties of Several p-Xylylene Polymers[a]

Properties	Poly(p-xylylene)	Poly(chloro-p-xylylene)	Poly(bromo-p-xylylene)	Poly(dichloro-p-xylylene)	Poly(cyano-p-xylylene)	Poly(methyl-p-xylylene)	Poly(ethyl-p-xylylene)
tensile properties at room temperature[b]							
tensile strength, MPa[c]	46.9	69.0	55.2	41.3	60.0	65.5	145.0
tensile modulus, MPa[c]	2414	3172	2759	2759	3000	2759	1207
elongation at break, %	10–15	220	30	5–10	7–12	230	275
thermal properties[d]							
crystalline melting point, °C	420	290	270	380	270	200–210	160–170
glass-transition temperature, °C	80	80	80	110	90	50–60	25
tensile modulus at 200°C, MPa[c]	172.4	172.4	137.9	172.4	137.9	<7.0	<0.7

[a] Ref. 10.
[b] Measured on 0.025–0.050 mm films in Instron tensile tester at 10% strain/min.
[c] To convert MPa to psi, multiply by 145.
[d] From secant modulus–temperature curve. Melting points also obtained from x-ray data and from melting behavior in sealed melting point tubes.

Table 4. Properties of Commercial Xylylene Polymers[a]

Properties[b]	ASTM method or conditions	Poly(chloro-p-xylylene)	Poly(p-xylylene)[c]
secant modulus, MPa[d]	D 882-56T at 1% strain	2759	2414
tensile strength, MPa[d]	D 882-56T at 10% strain/min	69.0	41.4–75.9
yield strength, MPa[d]	D 882-56T at 10% strain/min	55.2	42.1
elongation to break, %	D 882-56T at 10% strain/min	200	20–250
yield elongation, %	D 882-56T at 10% strain/min	2.9	2.5
density, g/cm³	D 1505-57T	1.289	1.10–1.12
index of refraction, n_D^{23}	Abbe refractometer	1.639	1.661
water absorption, 24 h, %	D 570-57T	0.06	0.01
Rockwell hardness	D 785-65	R80	R88
coefficient of friction	D 1894-63		
static		0.29	0.25
dynamic		0.29	0.25

[a] Ref. 10.
[b] Measured on films 0.025–0.075 mm thick, except where specified.
[c] Properties depend on deposition conditions.
[d] To convert MPa to psi, multiply by 145.

Poly(chloro-p-xylylene) is an excellent barrier to permanent gases and moisture vapor, whereas the barrier properties of poly(p-xylylene) are only average (see Table 5) (see Barrier polymers). The water–vapor permeability of unsupported poly-(chloro-p-xylylene) is constant throughout a range of thicknesses (49). The uniaxial and biaxial mechanical properties of very thin poly(p-xylylene) coatings are uniform to thicknesses as low as 0.12 μm (50). Abrasion, hardness, and impact resistance were measured on thin coatings (below 25 μm). The bulk wear properties are comparable to those of epoxies and urethanes and in several instances very thin coatings are superior to thicker ones.

Poly(p-xylylene) films have low coefficients of friction, which are slightly higher than those for Teflon, but about the same as those for nylon.

After 1000 h at 210°C, samples of poly(p-xylylene) and poly(chloro-p-xylylene) exhibited significant changes only in elongation, density, and tensile strength. Elongation decreased to ca 5%, density to ca 1.13, g/cm³ and tensile strength to ca 75.9 MPa (11,000 psi). Modulus remains unchanged at 2410 MPa (350,000 psi) for poly(p-xy-

Table 5. Barrier Properties of Xylylene Polymers[a]

Property	Poly(chloro-p-xylylene)	Poly(p-xylylene)
gas permeability, (cm³·cm)/(100 cm²·d·GPa)[b]		
to nitrogen	2.4×10^{-4}	3.0×10^{-3}
to oxygen	2.0×10^{-3}	1.5×10^{-2}
to carbon dioxide	3.0×10^{-3}	8.4×10^{-2}
to hydrogen	4.3×10^{-2}	21.3×10^{-2}
to chlorine	1.6×10^{-4}	2.9×10^{-2}
moisture–vapor transmission, g·cm/100 cm²·d[c]	2.4×10^{-4}	6.3×10^{-4}

[a] Refs. 10, 53.
[b] ASTM D 1434-63T. To convert GPa to atm, multiply by 9872.
[c] ASTM E 96-63T.

lylene), but increased from 2760–3450 MPa (400,000–500,000 psi) for the chloro analogue.

Mechanical Relaxations. For mechanical-relaxation measurements a free-oscillating inverted torsion pendulum is used (54). The various loss peaks are grouped by the letters α, β, γ, and δ in descending order of the temperature regions where they generally occur (55).

The α-loss process occurs at the highest temperature and is generally the most pronounced mechanical relaxation. It is commonly called the glass-transition temperature. Poly(dichloro-p-xylylene) exhibits such a transition in the 300–420 K region which is centered at 386 K (0.325 Hz) and agrees well with the value obtained from secant modulus–temperature data.

The β-relaxation transition generally occurs in the 220–320 K region and is associated with the rotation of side groups appended in the main chain. None of the poly(p-xylylenes) show a transition in this region.

The γ transition is a local oscillation mode centered at ca 150 K, which may occur either in side chains or in the polymer backbone. The γ–peak position is essentially independent of crystallinity, contrary to the α peak, which is shifted to higher values with increasing crystallinity, but is inversely proportional to the amount of crystallinity. Poly(p-xylylene), poly(chloro-p-xylylene), and poly(dichloro-p-xylylene) have large γ–loss peaks at 159 K (0.54 Hz), 254 K (0.4 Hz), and 156 K (0.34 Hz), respectively. These transitions have been ascribed to local motions of the phenylene groups (54).

Electrical Properties. The poly(p-xylylenes) exhibit good electrical properties; as shown in Table 6, dielectric strength and volume and surface resistivity are high. The dielectric constants and dissipation factors, however, are very low and better than for most epoxies, silicones, or polyurethanes, but not as good as for Teflon and some other fluorocarbons.

The best electrical properties are exhibited by the unsubstituted polymer. However, the monochloro and disubstituted chloro analogues display higher dielectric constants, and dissipation factors, as is usually the case with chlorinated polymers which exhibit a high dipole moment.

The volume resistivity, dielectric constant, and dissipation factor as a function

Table 6. Electrical Properties[a]

Property	Poly(p-xylylene)	Poly(chloro-p-xylylene)	Poly(dichloro-p-xylylene)
dielectric strength[b], V/10 μm	2756	2205	2165
volume resistivity at 23°C, 50% rh, $\Omega \cdot$cm	1×10^{17}	6×10^{16}	2×10^{16}
surface resistivity at 23°C, 50% rh, Ω	10^{13}	10^{14}	5×10^{16}
dielectric constant			
at 60 Hz	2.65	3.15	2.84
10^3 Hz	2.65	3.10	2.82
10^6 Hz	2.65	2.95	2.80
dissipation factor			
at 60 Hz	0.0002	0.020	0.004
10^3 Hz	0.0002	0.019	0.003
10^6 Hz	0.0006	0.013	0.002

[a] Refs. 10–11.
[b] For a 25.4-μm film.

of temperature are plotted in Figure 4 (11). These properties combined with application at ambient temperature contribute to the suitability of poly(p-xylylene) coatings as a primary insulation and for high frequency application.

After heating poly(p-xylylene) and poly(chloro-p-xylylene) films at 210°C for 1000 h, no significant changes were noted in electrical properties; the dielectric strength remains at 275 V/10 μm, the dissipation factor at 0.0002, and the dielectric constant at 2.65.

Even under increased thermal stress, the electrical properties of poly(p-xylylene) remain unaffected. For example, after 1000 h at 265°C, when test specimens begin to embrittle, and after 1500 h, when most samples are too fragile for physical testing, these electrical properties are not significantly altered with the exception of the dissipation factor which decreased from the control range of 0.02–0.01 for poly(chloro-p-xylylene).

High Temperature and Related Properties. The high crystalline melting points of the poly(p-xylylenes) suggest the possibility of a useful service life at elevated temperatures. However, the thermal endurance in air is not exceptional. The short-term (1000 h) use temperatures of poly(p-xylylene), poly(chloro-p-xylylene), and poly(dichloro-p-xylylene) are ca 95, 115, and 130°C, respectively. Extrapolation of failure time (50% loss in tensile strength) versus reciprocal of absolute temperature (Fig. 5) indicates that the corresponding long-term (10-yr) use temperatures are 60, 80, and 100°C, respectively. The problem is due to oxidative degradation and the presence of the aliphatic ethylene groups attached to the phenyl ring which enables oxidation and thermal cleavage to occur readily at these sites at elevated temperatures. Although the unsubstituted poly(p-xylylene) has the highest melting point, it has the lowest continuous use temperature because of lack of stability imparted by the presence of chlorine atoms in the chloro-substituted polymer.

The performance of both poly(p-xylylene) and poly(chloro-p-xylylene) greatly improves in the absence of air or in inert atmospheres, as shown in Figure 6. Oxidative degeneration does not take place and degradation is chiefly due to thermal cleavage of carbon–carbon bonds.

Thermogravimetric and differential scanning calorimetric data for halogenated poly(p-xylylenes) have been published (56). In-depth studies have also been reported on pyrolysis (47), oxidative stability (57), as well as kinetics and mechanisms of thermal degradation (58) of poly(p-xylylene).

Total weight loss at 121°C and 133 μPa (0.001 mm Hg) is 0.12% for poly(chloro-p-xylylene) and 0.03% for poly(p-xylylene). Volatile condensible losses were less than 0.1% for both.

In measurement of weathering properties by accelerated weathering tests, 0.05 mm films of poly(p-xylylene) and poly(chloro-p-xylylene) become brittle in less than 100 h. This is apparently the result of oxidative degradation of the polymer chain catalyzed by uv radiation. These data indicate that poly(p-xylylene) cannot be recommended for extended exposure to sunlight.

Optical. Both poly(p-xylylene) and poly(chloro-p-xylylene) absorb very little in the visible region and are, therefore, transparent and colorless. Below about 280 nm, both polymers absorb strongly. The uv and ir spectra have been published (10) as well as the Raman spectra (59–60).

Films of poly(p-xylylene) develop a pronounced birefringence during stretching. The stretching birefringence constant is +0.2, positive and very high as compared to

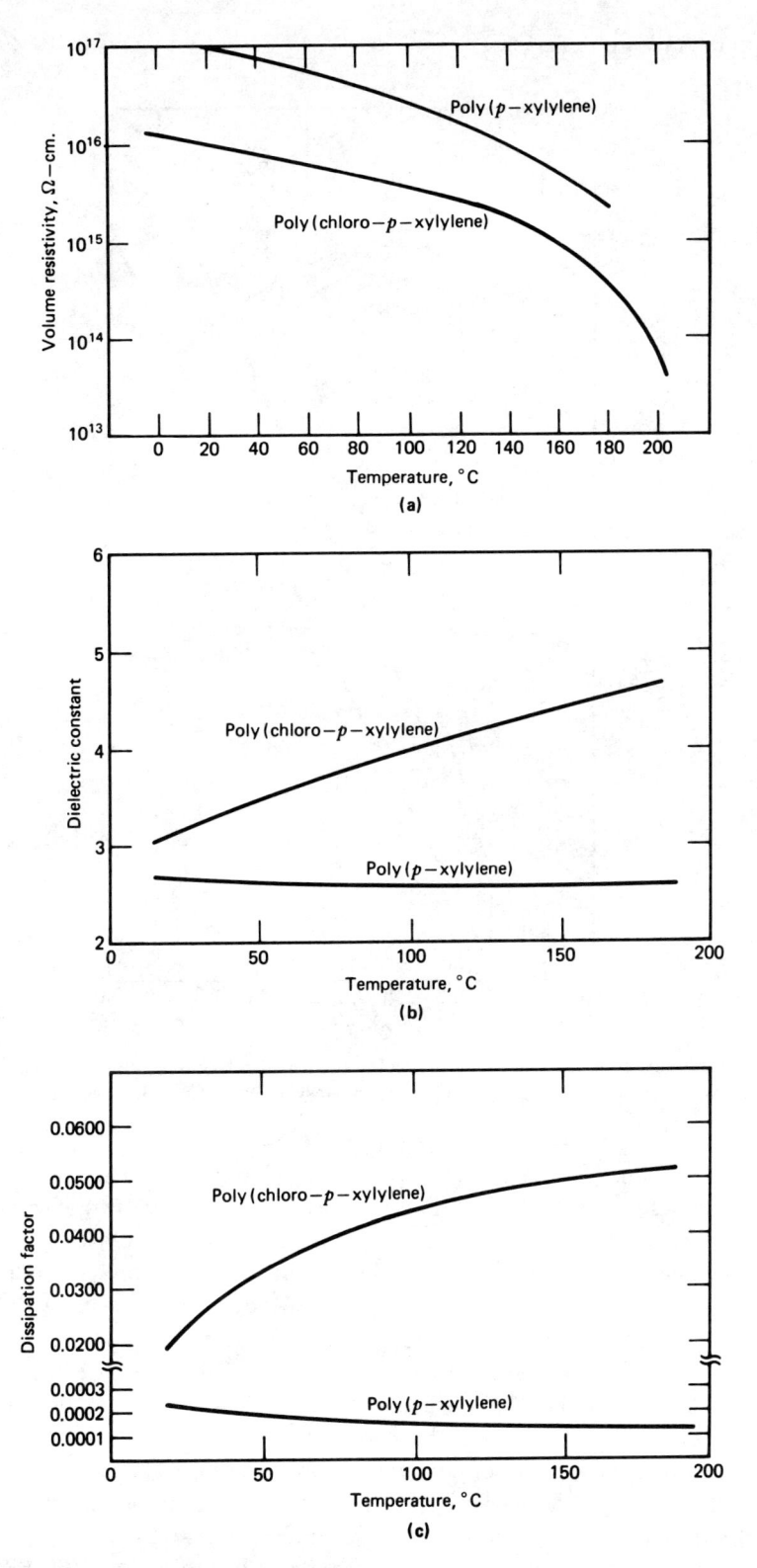

Figure 4. (a) Volume resistivity vs temperature (11). (b) Dielectric constant 1 kHz. (c) Dissipation factor vs temperature at 1 kHz.

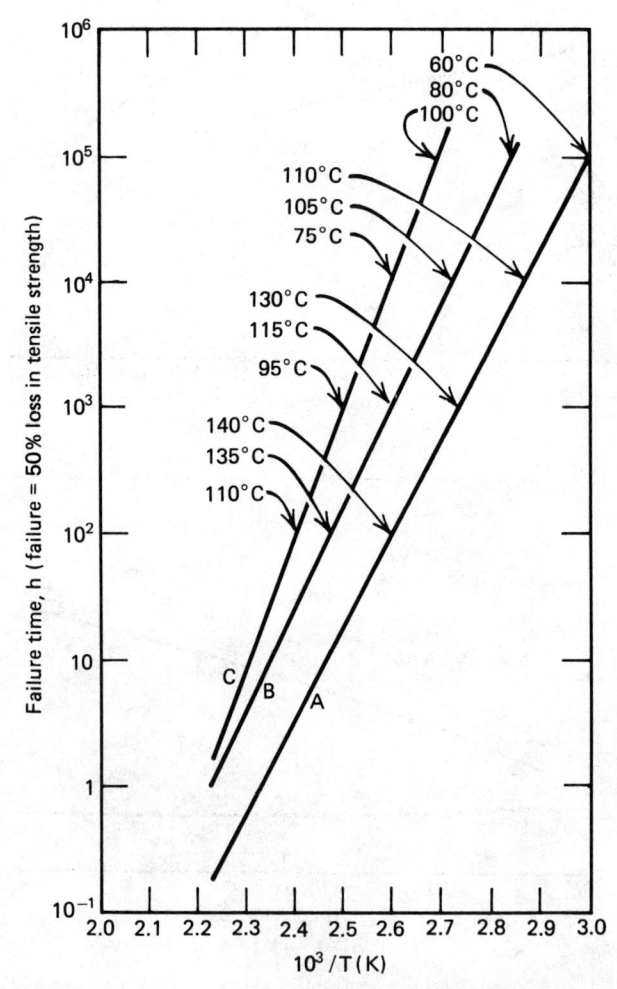

Figure 5. Failure time vs. reciprocal of aging temperature (in air); A = poly(p-xylylene), B = poly-(chloro-p-xylylene), C = poly(dichloro-p-xylylene). Courtesy of Union Carbide Corporation.

other polymeric films (17). In addition, poly(p-xylylene) films exhibit a positive and very large Brewster constant (20). The high and positive birefringence stretching constant and Brewster constant would indicate excellent photoelastic properties.

The average refractive index in the plane of unannealed poly(p-xylylene) films as measured in an Abbe refractometer is $n_D^{23} = 1.661$. Similarly, the value determined by the Becke line method is $n_D^{23} = 1.660 \pm 0.005$. The index remains the same for films annealed at 190°C, but increases slightly to $n_D^{23} = 1.665 \pm 0.005$ for films annealed 4 h at 255°C (29). The deviations represent the uncertainty in matching the index of the polymer with immersion oils in the Becke method.

The positive birefringence of poly(p-xylylene) supports the contention that the 270°C transition is a smectic mesomorphic transition, since the optical characteristics of a smectic mesophase are those of a positive uniaxial crystal (61).

Luminescence from poly(p-xylylene) films has been reported (62–63). Photoconductivity has been investigated (64).

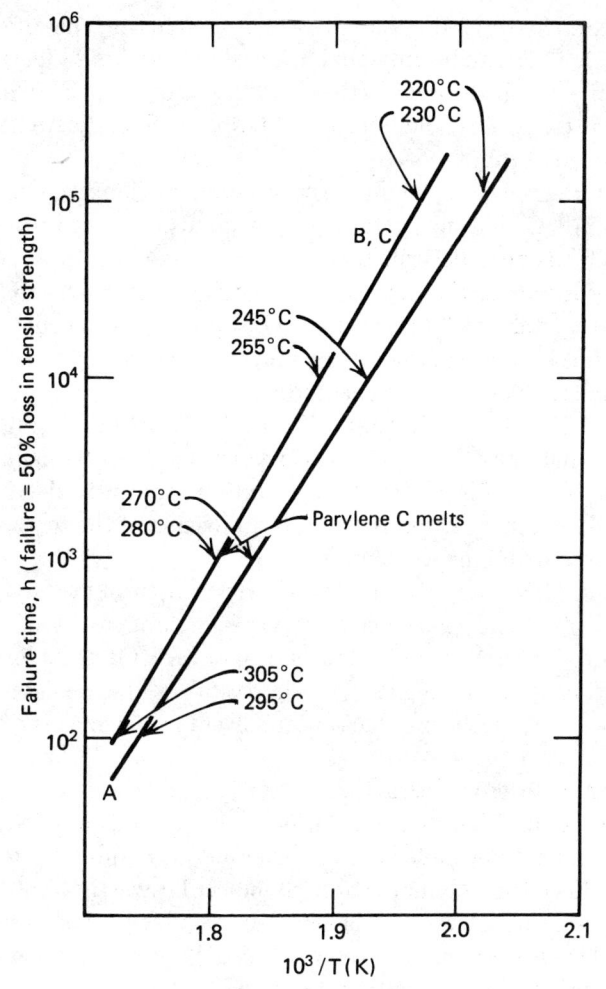

Figure 6. Failure time vs reciprocal of aging temperature (inert atmosphere); A = poly(p-xylylene), B = poly(chloro-p-xylylene), C = poly(dichloro-p-xylylene). Courtesy of Union Carbide Corporation.

Other Properties. Both poly(p-xylylene) and poly(chloro-p-xylylene) exhibit a high degree of radiation resistance as measured by gamma-ray degradation in vacuum. Tensile and electrical properties remained unchanged after a dosage of 1 MGy (100 Mrad) at a rate of 1.6 kGy/h (160 krad/h). Exposure in air, however, leads to rapid embrittlement. The effect of electrons, glow discharge, protons, uv, and x-rays on poly(p-xylylene) has been studied (65). In addition, the Lawrence Livermore Laboratory has published many papers on laser effects on poly(p-xylylene) films (66–70).

Poly(chloro-p-xylylene) has outstanding dimensional and low temperature properties. Steel panels coated with this polymer and chilled in liquid nitrogen at −165°C withstood impacts of more than 11.3 J (8.33 ft·lbf) in a modified Gardiner falling-ball impact test, compared with values of over 22.6 J (16.66 ft·lbf) at ambient temperature. Unsupported 0.05-mm films can be flexed 180° six times at −165°C before failure occurs. Neither electrical nor physical properties are affected by changing the temperature from ca −271°C to ambient.

Films prepared by vapor deposition are deposited in the α modification at ambient temperature (71). They can be annealed below 200°C without phase transformation, but above 220°C are transformed to the β modification (26). Both annealed and unannealed films are extensible, and afford oriented x-ray fiber diagrams (65). Extensions of 1000% have been obtained.

The highest degree of crystalline perfection and orientation is realized in films annealed above 270°C. This temperature corresponds to a third reversible crystalline transition, which has been described as a smectic mesomorphic transition (29). The polymer loses order along the direction of the chains above 270°C, and becomes a two-dimensional crystal (72). This conclusion is supported by mechanical creep studies, in which an abrupt increase in the rate of elongation occurs above 270°C, and by differential thermal analysis and optical studies.

If a solution of poly(p-xylylene) (0.05-wt % in α–chloronaphthalene) is slowly cooled from ca 240 to 209°C, a Tyndall effect is noted after 0.5 h, indicating the formation of a precipitate. If isothermal precipitation is permitted to continue for 16 h, and the suspension is then cooled to 25°C, examination of the precipitate reveals two entirely different morphological structures.

Above 250°C, dried single α crystals slowly transform to the β modification, as shown in x-ray diffraction patterns. However, the appearance of the crystal is not preceptibly altered (73). Although films of α–modification transform irreversibly to the β–modification upon heating (24), single crystals of the hexagonal β modification transform to the α form in the presence of the solvent from which they were crystallized (74).

Single-crystal electron-diffraction patterns correspond to the azimuthal (hko) projection of the x-ray fiber diagrams of both the α and β modifications (31). This indicates that the polymer molecules are oriented perpendicular to the plane of the single crystals. Since the polymer is of high molecular weight and the single crystals are only ca 0.01 μm thick, the crystals must fold regularly on the growth faces (edges of the crystals) (75–76). Virtually all polymer single crystals are now known to exhibit chain folding (see also Crystallization; Polymers).

Since the single-crystal electron-diffraction patterns present an undistorted projection of the crystal lattice in a direction perpendicular (or nearly so) to the chain axis, it can be seen by inspection that the β modification is hexagonal. By suitable calibration it can be determined that the recurring spacing is 2.0 nm, which is the approximate lattice dimension in a direction perpendicular to the chain axis. The various levels of the fiber diagram reveal that the molecular repeating unit is 0.66 nm. This corresponds to a fully extended planar zigzag-ordered sequence based on known bond angles and distances (29,41,45). By applying these observations to accurate x-ray powder camera data, it was found that the lattice constants of the β unit cell are a = 2.052 ± 0.005 nm and c = 0.6581 ± 0.0020 nm.

A number of different unit cell parameters were proposed by early investigators for the α modification (25,29,35,74,77). The cell finally proposed in ref. 33 proved to be correct, as shown by the crystalline structure determination of poly(p-xylylene) (78). The cell is monoclinic with a = 5.92, b = 10.64, c (fiber axis) = 0.655 nm, β = 134.7°, ρ (theory) = 1.18 g/cm^3, ρ (observed) = 1.13 g/cm^3). Hence, poly(p-xylylene) deposited at or near room temperature is about 60% crystalline (79).

Health and Safety Factors

The Union Carbide Corporation has issued safety data sheets for di-p-xylylene (DPX-N, Parylene dimer), dichloro-p-xylylene (DPX-C, Parylene C dimer), and tetrachlorodi-p-xylylene (DPX-D, Parylene D dimer). The autoignition temperatures are 257 and 310°C, respectively, for the first two; the tetrachloro derivative does not ignite. Carbon dioxide or dry chemical extinguishers are recommended for small fires, foam or water or large fires. Self-contained breathing equipment should be available for firemen. Dispersion of dust should be avoided as an explosion hazard; data are not available on the effects of exposure. The hazardous products of thermal decomposition and burning are carbon dioxide, carbon monoxide and, for the chlorinated derivatives, hydrogen chloride. Hazardous polymerization does not occur. Waste disposal by burial rather than incineration is recommended. The usual respirators may be employed as needed when working with these compounds.

Applications

Poly(p-xylylene) and poly(chloro-p-xylylene) have numerous commercial applications. Most utilize the combination of excellent properties and some feature of the vapor-deposition polymerization process where the reactive monomers condense and polymerize on any solid surface placed in the condensation zone. The chemical nature of the surface is unimportant, and many materials, including strong acids, bases, alkali metals or explosives like RDX (cyclotrimethylenetrinitramine) (80) or ammonium nitrate (81) have been coated in a deposition chamber. Levitating microballoons have also been coated in a similar fashion (30).

The polymerization of p-xylylenes on condensation is extremely rapid and appears to proceed from gaseous monomer to solid polymer without passing through a viscous stage. The monomer behaves as a reactive medium which surrounds solid objects placed in the deposition chamber. For these reasons, films of uniform thickness are deposited on all surfaces, including sharp edges, crevices, and the inside of holes (see Film deposition techniques; also Film and sheeting materials; Coating processes; Coatings, resistant).

Commercial applications involve mainly poly(p-xylylene), poly(chloro-p-xylylene), and lesser amounts of poly(dichloro-p-xylylene). Other derivatives have been synthesized, eg, the tetrachloro by vapor deposition (82) and the tetrafluoro (83) and hexafluoro by an electrolytic process (7).

Some of the halogenated derivatives exhibit superior temperature resistance, eg, the useful properties of poly-$\alpha,\alpha,\alpha',\alpha'$-tetrafluoro-$p$-xylylene [74952-03-7] were maintained for over 3000 h at 250°C in air (84–85).

Surface Preparation and Adhesion Enhancement. For maximum adhesion, the surfaces are treated before coating. Silane adhesion promoters establish a bridge-type chemical bond between the inorganic and organic materials because of the unique affinity of both portions of the molecule. For example, the surface is treated with a dilute solution of (γ-methacryloxypropyl)trimethoxysilane. The silicon part of the molecule can bond to the inorganic, metallic, or glass surface, whereas the organic portion is tailored to combine with the polymeric surface.

Another technique employs plasma pretreatment. Plasmas of argon and oxygen promote adhesion of poly(p-xylylene) coatings on many difficult to bond substrates such as gold, nickel, Kovar, Teflon (FEP), Kapton, silicon, tantalum, titanium, and tungsten (86). Without plasma treatment, 180° peel tests show strengths of a few g/cm, which can be increased by one or two orders of magnitude by d-c plasma treatment in a deposition chamber.

Film Stabilization. Oxidation resistance is imparted by the application of an antioxidant to the deposited film (87) or the incorporation of an antioxidant with the dimer feed stock (88–89). Heating freshly deposited films of poly(dichloro-p-xylylene) and poly(p-xylylene) above the T_g improves polymer properties (90).

Electrical and Electronic Applications. The unique poly(p-xylylene) vapor-deposition process has widespread application in the field of electronics. As a true conformal coating, it can be deposited on and under complex delicate circuitry with a minimum of handling. In addition, the comparatively good electrical properties, fungus resistance, and ability to protect its barrier properties from hostile environments have made poly(p-xylylene)s the preferred choice.

Circuit Boards. Poly(chloro-p-xylylene) is used for coatings on circuit boards, as shown in Figure 7, where protection of the components and circuitry from moisture or reactive gases in the atmosphere is required. Circuit boards are an excellent example of a substrate that can be readily coated by vapor deposition. It is protected only with difficulty by conventional coatings applied in liquid form by spraying, dipping, or brushing. Conventional coatings result in surface tension pulling the coating away from the corners, filling low areas and, increasing the chance of pinholes. Poly(p-xylylene) films, on the other hand, grow from the surface of the substrate upward covering everything with a uniform thick coating, even closely spaced components on all sides without bridging.

Tests demonstrate that 0.025–0.050 mm of poly(chloro-p-xylylene) on circuit boards provides protection superior to that of 0.13 mm coatings of other compositions, such as epoxy resins, silicones, and polyurethanes (91).

When coating circuit boards, a small amount of anthracene is frequently added to the dimer starting material. The anthracene is not affected by the deposition and codeposits uniformly with the poly(p-xylylene) polymer. A visual inspection of the final coating for continuity using uv illumination also detects pinholes.

Above 150°C, coated multilayer circuit boards have equal bond strength and higher bond resistances than single-layer boards. The poly(p-xylylene) coating actually doubles the wire-bond strength (92).

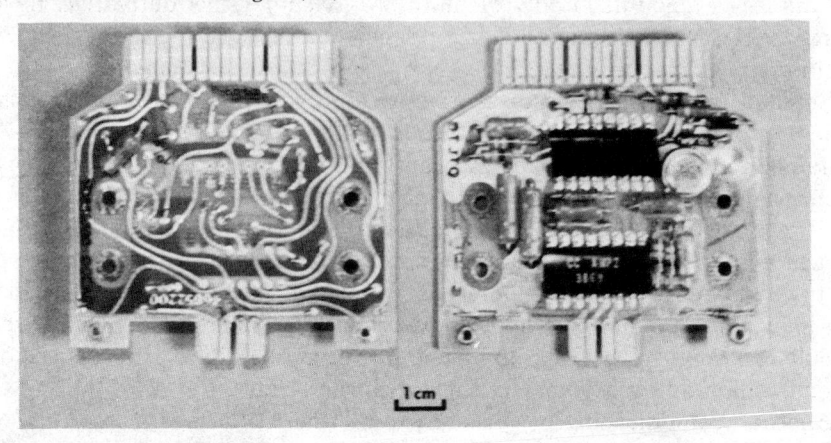

Figure 7. Circuit boards coated with poly(p-xylylene). Courtesy of Union Carbide Corporation.

Capacitors. An important application of vapor deposited poly(p-xylylene) is in thin-film capacitors (Parylene Flat–Kap capacitors, Union Carbide Corp.). In these units, 2–3 μm poly(p-xylylene) films serve as the dielectric (93). Poly(p-xylylene) is deposited in uniform, pinhole-free films on aluminum foil at 2-μm thickness with a dielectric strength >500 V, and a very low dissipation factor over a wide range of temperatures and frequencies. The aluminum foil is slit into tapes and fabricated into capacitors. Because of the extremely thin dielectric, these capacitors provide volumetric efficiencies up to twenty times the efficiencies of glass, mica, and porcelain capacitors. They are used in filters, integrators, and timing circuits. Microelectronic printed capacitors coated with poly(chloro-p-xylylene) were exposed to humidity for 10 d without adverse effects (94).

Thermistors. Thermistors, which are protected by conformal coatings, are typically used in expendable bathythermograph systems which measure and record the temperature of ocean waters to depths of 460 m. The temperature–sensing element is a miniature thermistor encased in a thin protective coating of poly(chloro-p-xylylene).

Surface Passivation. Poly(p-xylylene) coatings function as a barrier on semiconductor device surfaces by preventing lowering of insulation resistance (95–96). Extended electrical stressing of coated devices at 175°C indicates the absence of failures. In addition, changes in the electrical parameters are negligible; the average threshold voltage shifts to <0.1 V, the breakdown voltage changes to <0.1 V, and leakage currents drift within a few picoamperes (see Semiconductors).

Solid-State Relay. A hybrid microelectronic solid-state relay coated with poly(p-xylylene) coating provided d-c 2.5 kV input and switched a load of d-c 400 V at 2 mA. The poly(p-xylylene) coating was the only material tested which was still operational after 200 cycles of 45 min each from −120 to 80°C (97).

Hybrid Circuits. Uncoated hybrid microcircuits are vulnerable to chemical and particulate contamination as well as loose conductive particles (98–104). These small, usually microscopic, pieces of metal such as solder balls and weld splatter can remain inside hermetically sealed devices during manufacturing. The application of a poly-(chloro-p-xylylene) coating to the inside of hybrid circuits as shown in Figure 7 protects the surface from contaminants and handling, and immobilizes trapped particles within the device by coating them. This technique has become known as particle-tie down or closed-package control (105–106) and is used by hybrid manufacturers.

Poly(p-xylylene) vapor penetrates under a silicon chip that is mounted ca 12 μm above the surface where other coatings cannot flow. In addition, by adhering to the flying leads on hybrids, it increases the bond strength 5–10 times.

Discrete Core Memories. Ferrites (qv) are readily tumble-coated which results in total encapsulation of the toroid (107). These ferrites, molded to close tolerances, retain their dimensions because the ultrathin poly(p-xylylene) sheath provides protection without altering the dimension, shape, or magnetic properties. It also improves the mechanical crush strength of the core and prevents cutting of wire insulation or the introduction of stresses which might cause magnetostriction.

Up to 10,000 6-mm ferrite toroids can be coated in a single run using a mold-tumbling process giving a 2.5 μm coating for a negligible cost.

Field-Effect Transistors. A pH–insensitive poly(p-xylylene) polymer has been prepared for use as a chemically modified poly(p-xylylene) gate field effect transistor (108–109).

Nichrome Resistors. Poly(p-xylylene) coated nichrome resistors are protected from moist environments for periods of at least a week up to 82°C (94,110).

Electrets. A recent paper has reported a poly(p-xylylene) electret, or a dielectric body containing a permanent state of electric polarization, that was usable at high temperatures (111). Poly(dichloro-p-xylylene) and poly($\alpha,\alpha,\alpha',\alpha'$-tetrafluoro-p-xylylene) electrets can be used at 90 and 170°C, respectively, well over the 60°C use temperature of Teflon. The 10-year use temperature for most stable electrets is projected to be above 100°C (84).

Plastic-Packaged Microelectronics. The realiability of plastic packaged devices increases when coated with poly(chloro-p-xylene). The barrier coating is applied internally or externally similar to circuit boards (112).

Transformers and Coils. Poly(p-xylylene) film is used in a unique transformer with a smaller and less expensive hybrid-isolation amplifier (113). The insulation of wire-wound coils with a poly(p-xylylene) conformal coating has been reported (114–115).

Transducer. Standard metallic bellows and strain gauges are susceptible to attack from hostile environments. Poly(p-xylylene) coatings have provided total conformal protection without significantly affecting the sensitivity or mechanical properties.

Biomedical. The use of poly(chloro-p-xylylene) polymers in the biomedical fields is increasing rapidly (116–124). Long-term implantation devices such as cardiac pacemakers require protection from corrosive tissue fluids (see Fig. 8). Electronic circuitry within the pacemakers upon contact with tissue fluids results in a corrosive breach in the seal leading to almost immediate failure of the coatings. A secondary moisture barrier provided by a 0.25-mm coating of poly(chloro-p-xylylene) performed remarkably well over a thirty day period as contrasted to only 58 h prior to failure of the second-best candidate coating (125) (see Prosthetic and biomedical devices).

Particle Encapsulation. Poly(p-xylylene) vapor deposition is used for encapsulation of particulate solids (126). The particles, granules, or objects to be encapsulated are placed in a bottle in the deposition chamber and the nozzle from the pyrolysis tube

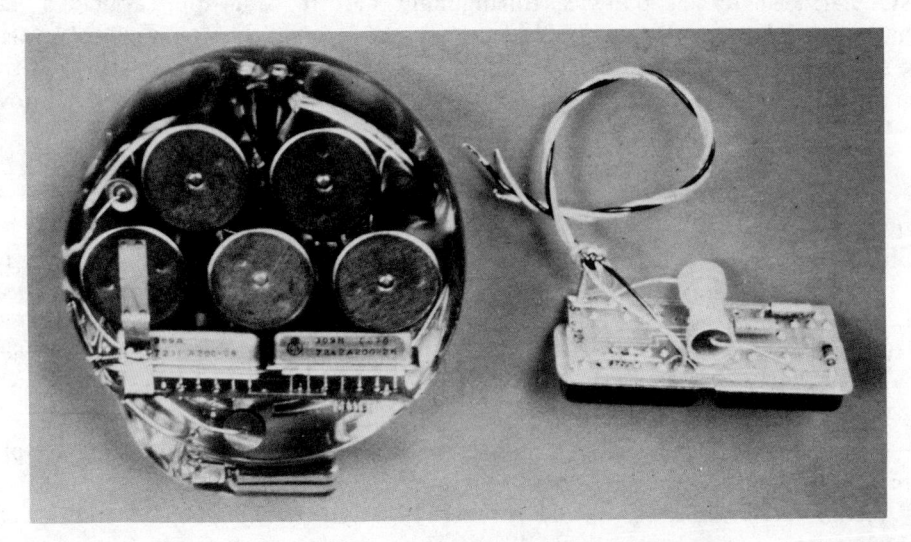

Figure 8. Cardiac pacemaker coated with poly(p-xylylene). Courtesy of Union Carbide Corporation.

is inserted in the mouth of the bottle. The monomers pass from the pyrolysis zone through the nozzle into the bottle and polymerize on the surface of the tumbling particles or granules. The inner surface of the rotating bottle is also coated. Particles as small as 150–300 µm are encapsulated by this method. It is essential to keep the particles in continuous motion to prevent agglomeration and sticking. Since no solvents are involved and no liquid viscous stage is encountered during the polymerization, coherent, uniform coatings are deposited around each particle (see also Embedding).

Particulate materials that have been studied in the encapsulation process include sodium dichromate, lithium, lithium hydride, zinc, lithium aluminum hydride, ferrite cores, and electrical and electronic connectors (see Electrical connectors).

Pellicles, Thin Membranes, and Electrooptical Applications. Ultrathin pellicles (unsupported membranes), 0.1–5 µm in thickness have been prepared from both poly(p-xylylene) and poly(chloro-p-xylylene) (see Membrane technology). Pellicles are normally produced in diameters up to 10 cm; they can be vacuum metallized with gold, aluminum, or other materials. Pellicles are used for micrometeorite-detector instruments, dielectric supports for planar capacitors, and extremely fast-responding, low-mass thermistors and thermocouples. Such pellicles have also been used as beam splitters in optical instruments and as windows for devices for measuring nuclear radiation. Both poly(p-xylylene) and poly(chloro-p-xylylene) are transparent (127); the latter has better clarity. Total transmission through the visible region is >95% for poly(chloro-p-xylylene) and >90% for poly(p-xylylene), and both polymers are virtually nonabsorbing in most regions up to the far infrared. These films are used as optical waveguides (128).

Lubricants. Coatings of poly(p-xylylene) are ideal dry film lubricants because of their low static and dynamic coefficients of friction. A 2.54 µm coating on a 225 µm motor shaft has performed for 10,000 continuous hours without fault (see Lubrication and lubricants).

Optical Devices. Poly(p-xylylene) coatings have been applied to mirrors, reflectors, beam splitters, display panels, and other optical devices in thicknesses of less than 1 µm. These transparent and colorless coatings exhibit little absorption in the visible-light region.

Other Applications. Numerous other applications for vapor-deposited poly(p-xylylenes) are under investigation. It has been shown that very thin ($\cong$ 1 µm) coatings of poly(p-xylylene) on steam condenser tube bundles increase the overall heat-transfer coefficient from 20 to 50%. This is achieved by promoting condensation of the steam in droplets rather than as a film owing to the hydrophobic nature of poly(p-xylylenes) (129). Steam can be condensed more quickly in the presence of bare surfaces between drops than when an insulating film of water covers the surface. A very thin polymer film is essential to prevent reduction of the heat-transfer rate which may be caused by the insulating nature of the polymer.

In addition, applications include antifouling or corrosion-resistant coatings and the possible use as a coating to reduce the permeability of plastic materials. Polyethylene bottles coated internally with 6 µm thick poly(chloro-p-xylylene) film have been filled with lemon oil for three years and found to have lost only two grams through permeation, whereas lemon oil diffuses extremely rapidly through uncoated polyethylene.

Deposited poly(p-xylylene films) achieve surface replication to a detail as fine as 0.01 μm (130). Poly(p-xylylene) coatings have been applied to the surfaces of syntactic composites (131–132) and polyurethane baffle materials (133) to increase moisture resistance as well as a critical-thickness coating spacer for high speed computer printers. A patent has been issued for a poly(p-xylylene) coated fishing lure (134).

Poly(p-xylylene) spent polymeric film and the dimer have been fabricated into target material and sputtered (see also Film deposition techniques). However, only a rudimentary similarity between the infrared spectrum of poly(p-xylylene) produced by the standard vapor-phase synthetic technique and the sputtered materials was in evidence. Even with differences in electrical and physical properties the use of poly(p-xylylene) coatings prepared by sputtering is not precluded (135–137).

BIBLIOGRAPHY

1. J. Thiele and H. Balhorn, *Chem. Ber.* **37,** 1463 (1904).
2. A. J. Namiot, M. E. Dyatkina, and Ya. K. Syrkin, *Compt. Rend. Acad. Sci. U.R.S.S.* **48,** 267 (1945).
3. M. E. Djatkina and Ya. K. Syrkin, *Acta Physiochim. U.R.S.S.* **21,** 23 (1946).
4. M. Szwarc, *Disc. Faraday Soc.* **2,** 46 (1947).
5. L. A. Errede and B. F. Landrum, *J. Am. Chem. Soc.* **79,** 4952 (1957).
6. L. A. Errede and M. Szwarc, *Quart. Rev. Chem. Soc.* **12,** 301 (1958).
7. U.S. Pat. 3,399,124 (Oct. 24, 1967), H. G. Gilch (to Union Carbide Corp.).
8. U.S. Pat. 3,149,175 (Sept. 15, 1964), D. F. Pollart (to Union Carbide Corp.).
9. H. E. Winberg and E. S. Fawcett, *Org. Syn.* **42,** 83 (1962).
10. W. F. Gorham and W. D. Niegisch in N. M. Bikales, ed., *Encyclopedia of Polymer Science and Technology*, Vol. 15, Wiley-Interscience, New York, 1971, p. 99.
11. Union Carbide Corporation, *Parylene Environmentally Compatible Coatings*, Sales Brochure.
12. W. F. Gorham, *J. Polym. Sci. Part A–1* **4,** 3027 (1966).
13. U.S. Pat. 3,342,754 (Sept. 19, 1967), W. F. Gorham (to Union Carbide Corp.).
14. U.S. Pat. 3,288,728 (Nov. 29, 1966), W. F. Gorham (to Union Carbide Corp.).
15. L. A. Errede, R. S. Gregonan and J. M. Hoyt, *J. Am. Chem. Soc.* **82,** 5218 (1960).
16. R. N. Howard, *Trans. Faraday Soc.* **46,** 204 (1950).
17. B. H. Zimm and J. K. Bragg, *J. Polym. Sci.* **9,** 476 (1952).
18. S. L. Bagdasarian, G. L. Fix, and J. S. Judge, *Electron. Prog.* **20**(4), 17 (1978).
19. M. H. Kaufman, H. F. Mark, and R. B. Mesrobian, *J. Polym. Sci.* **13,** 3 (1954).
20. R. S. Corley, H. C. Haas, M. W. Kane, and D. I. Livingston, *J. Polym. Sci.* **13,** 137 (1954).
21. L. A. Errede and S. L. Hopwood, Jr., *J. Am. Chem. Soc.* **79,** 6507 (1957).
22. L. A. Errede and W. A. Pearson, *J. Am. Chem. Soc.* **83,** 954 (1961).
23. L. A. Errede and J. M. Hoyt, *J. Am. Chem. Soc.* **82,** 436 (1960).
24. L. A. Auspos, C. W. Burham, L. A. R. Hall, J. K. Hubbard, W. Kirk, Jr., J. R. Schaefgen, and S. B. Speck, *J. Polym. Sci.* **15,** 19 (1955).
25. D. J. Brown and A. C. Farthing, *J. Chem. Soc.* 3270 (1953).
26. L. A. Errede and R. S. Gregorian, *J. Polym. Sci.* **60,** 21 (1962).
27. W. F. Gorham, *Am. Chem. Soc. Div. Polym. Chem. Preprints* **6**(1), 73 (1965).
28. M. Kobayashi, *J. Polym. Sci. Polym. Lett. Ed.* **8,** 823 (1970).
29. W. D. Niegisch, *J. Appl. Phys.* **37,** 4041 (1966).
30. N. H. Hartshorne and A. Stuart, *Crystals and the Polarizing Microscope*, 3rd ed., Edward Arnold and Company, London, 1960, p. 168.
31. W. D. Niegisch, *Polym. Lett.* **4,** 531 (1966).
32. A. Kajiura, M. Fujii, K. Kikuchi, S. Irie, and H. Watase, *Kolloid Z. Z. Polym.* **224,** 128 (1968).
33. S. Kubo and B. Wunderlich, *Makromol. Chem.* **162,** 1 (1972).
34. R. W. Wyckoff, *Crystal Structures*, Vol. 5, Interscience Publishers, New York, 1960, Fig. XIV–a–31.
35. S. Kubo and B. Wunderlich, *Makromol. Chem.* **157,** 299 (1971).
36. U.S. Pat. 3,509,075 (April 28, 1970), W. D. Niegisch and W. E. Loeb (to Union Carbide Corp.).
37. R. Iwamoto, R. C. Bopp, and B. Wunderlich, *J. Polym. Sci., Polym. Phys. Ed.* **13,** 1925 (1975).

38. W. D. Niegisch, Union Carbide, Bound Brook, N.J., private communication, 1982.
39. C. W. Bunn and E. R. Howells, *Nature* **174,** 549 (1954).
40. W. D. Niegisch, *J. Appl. Phys.* **38,** 4110 (1967).
41. B. Wunderlich, *Advan. Polym. Sci.* **5,** 568 (1968).
42. S. Kubo and B. Wunderlich, *J. Appl. Phys.* **42,** 4558 (1971).
43. W. F. Beach, *Macromolecules* **11,** 72 (1978).
44. Union Carbide Corporation Technology Letter, No. 12, May 1975.
45. M. Szwarc, *J. Polym. Sci.* **6,** 319 (1951).
46. S. L. Madorsky and S. Strauss, *J. Res. Nat. Bur. Stand.* **55,** 223 (1955).
47. H. H. G. Jellinek and S. N. Lipovac, *J. Polym. Sci., Part A-1* **8,** 2517 (1970).
48. N. Ya. Marusii, B. V. Tkachuk, and V. I. Pavlov, *Vysokomol. Soedin. Ser. A* **23,** 1091 (1981).
49. M. A. Spivack and G. Gerrante, *J. Electrochem. Soc.* **116,** 1592 (1969).
50. M. A. Spivack, *Rev. Sci. Instrum.* **43,** 985 (1972).
51. Union Carbide Corporation Technology Letter, No. 3, July 1972.
52. M. A. Spivack and C. E. White, *Electron. Packag. Prod.* **12,** 84 (1972).
53. Union Carbide Corporation Technology Letter, No. 6, Oct. 1977.
54. C. Chung, *The Effects of Orientation and Chemical Structure on the Mechanical Behavior of Polymer Solids*, Ph.D. Thesis, Rutgers State University, New Brunswick, N.J., 1969. *Diss–Ab Str. Int.* **B30,** 338–9 (1970); C. Chung and J. A. Sauer, *Polymer* **11,** 454 (1970), C. I. Chung and J. A. Sauer, *J. Polym. Sci., Part A-2,* **9,** 1097 (1971).
55. K. Deutsch, E. A. W. Hoff, and W. Reddish, *J. Polym. Sci.* **13,** 565 (1954).
56. B. L. Joesten, *J. Appl. Polym. Sci.* **18,** 439 (1974).
57. T. E. Nowlin, D. F. Smith, Jr., and G. S. Cieloszyk, *J. Polym. Sci. Polym. Chem. Ed. USSR* **18,** 2103 (1980).
58. S. Ya. Lazareva, A. V. Osysov, and Y. E. Malkov, *Polym. Sci. USSR* **21,** 1654 (1979).
59. M. S. Mathur and N. A. Weir, *J. Mol. Struct.* **15,** 459 (1973).
60. M. S. Mathur and G. C. Tabisz, *J. Cryst. Mol. Struc.* **4**(1), 23 (1973).
61. G. W. Gray, *Molecular Structure and the Properties of Liquid Crystals*, Academic Press, New York, 1962, p. 22.
62. Y. Takai, J. H. Caldewood, and N. S. Allen, *Makromol. Rapid Commun. Chem.* **1**(1), 17 (1980).
63. Y. Takai, T. Mizutanic, and M. Ieda, *Jpn. J. Appl. Phys.* **17,** 651 (1978).
64. H. T. Coffey, J. E. Nanvicz, and R. C. Adamo, *Photoconductivity of High–Voltage Space Insulating Materials*, NASA-CR-134995, Oct. 1975.
65. Union Carbide Corporation Technology Letter, No. 10, Jan. 1974.
66. K. R. Manes, H. G. Ahlstrom, R. A. Haas, and J. F. Holzrichter, *J. Opt. Soc. Am.* **67,** 717 (1977).
67. R. A. Haas, D. W. Phillion, M. J. Boyle, H. N. Kornblum, and V. C. Rupert, Rept. No. CONF-751130-21, University of California, Lawrence Livermore Lab, Dec. 10, 1975.
68. R. A. Haas, W. C. Mead, W. L. Kruer, D. W. Phillion, H. N. Kornblum, J. D. Lindl, D. McQuigg, V. C. Rupert, and K. G. Tirsell, *Phys. Fluids* **20**(2), 322 (1977).
69. R. A. Haas, M. J. Boyle, K. R. Manes, and J. E. Swain, *J. Appl. Phys.* **47,** 1318 (1976).
70. W. C. Mead, R. A. Haas, W. L. Kruer, D. W. Phillion, H. N. Kornblum, J. D. Lindl, D. R. McQuigg, and V. C. Rupert, *Phys. Rev. Lett.* **37,** 489 (1976).
71. Ref. 59, footnote 10.
72. C. A. Sperati and W. H. Starkweather, *Fortschr. Hochpolym. Forsch.* **2,** 456 (1961).
73. W. D. Niegisch, Union Carbide, Bound Brook, N.J., unpublished results, 1982.
74. G. Lieser, Thesis, Naturwissenschaftliche Fakultät der Johannes Gutenberg Universität zu Mainz, Mainz, 1971.
75. K. H. Storks, *J. Am. Chem. Soc.* **60,** 1753 (1938).
76. A. Keller, *Phil. Mag.* **2,** 1171 (1957).
77. S. Kubo and B. Wunderlich, *J. Appl. Phys.* **42,** 4565 (1971).
78. R. Iwamoto and B. Wunderlich, *J. Polym. Sci. Polym. Phys. Ed.* **11,** 2403 (1973).
79. S. Kubo and B. Wunderlich, *J. Polym. Sci. Part A-2* **10,** 1949 (1972).
80. A. F. Smetana and T. C. Castorina, paper presented at Compat. Propellants, Explosives and Pyrotechnics Plastic Additive Conference, 11-E, 1975, pp. 1–26.
81. T. C. Castorina and A. F. Smetana, *J. Appl. Polym. Sci.* **18,** 1373 (1974).
82. H. V. Pebalk, I. Ye. Kardash, N. V. Kozlova, and A. N. Pravednikov, *Polym. Sci. USSR* **21,** 824 (1979).
83. S. W. Chow, W. E. Loeb, and C. E. White, *J. Appl. Polym. Sci.* **13,** 2325 (1969).
84. W. R. Norris, *J. Org. Chem.* **37**(1), 147 (1972).

XYLYLENE POLYMERS

85. B. L. Joesten, *J. Appl. Polym. Sci.* **18,** 439 (1974).
86. T. Riley, T. Cobo Mahuson, and K. Seibest, *Investigation Into the Effect of Plasma Pretreatment On the Adhesion of Parylene To Various Substrates*, American Society For Metals Aerospace Activity of the Cleaning, Finishing and Coating Division and the Society for the Advancement of Material & Process Engineering, Los Angeles, Calif., Feb. 5–6, 1980, pp. 95–113.
87. U.S. Pat. 4,173,664 (Nov. 6, 1979), G. S. Cieloszyk (to Union Carbide Corp.).
88. U.S. Pat. 4,176,209 (Nov. 27, 1979), T. E. Baker, G. L. Fix, and J. S. Judge (to Raytheon Company).
89. T. E. Baker, S. L. Bagadasarian, G. L. Fix, and J. S. Judge, *J. Electrochem. Soc.* **124,** 897 (1977).
90. U.S. Pat. 3,503,903 (Mar. 31, 1970), R. G. Shaw, Y. L. Yeh, and J. W. Lewis (to Union Carbide Corp.).
91. W. E. Loeb and C. E. White, *Proceedings of the Technical Program, National Electronic Packaging Convention*, NEP/CON '68 West, Los Angeles, Calif., Jan. 30–Feb. 1, 1968; NEP/CON '68 East, New York, June 4–6, 1968, pp. 228–240.
92. R. G. Oswald, W. R. R. Demiranda, and C. W. White, *Proc. Int. Microelectron. Symp.*, 1975, pp. 324–339.
93. R. C. Greer, *Development of Thin Rolled Film Capacitor*, Final Report, Bureau of Ships, Contract No. BSR-89519, Union Carbide Corporation, May 14, 1968.
94. H. F. Windsor, *NASA Workshop/Conference*, NASA Lewis Research Center, Cleveland, Ohio, Oct. 19–20, 1976.
95. S. M. Lee, *Polymeric Films for Semiconductor Passivation*, NASA Contract NAS 12–2011, March 1969.
96. S. M. Lee, J. J. Licari, and I. Litant, *Trans. Metall. Soc.* **1,** 701 (1970).
97. B. L. Sater and T. J. Riley, paper presented at *IEEE Power Electronics Specialists Conf.*, California Institute of Technology, Pasadena, Calif., June 11–13, 1973, pp. 163–169.
98. W. M. Jayne, Jr., *International Microelectronics Symposium*, San Francisco, Calif., Oct. 22–24, 1973, Session pp. 4B–6.
99. D. L. Kinser, *Insulation/Circuits* **26**(13), 88 (1980).
100. F. Z. Keister, *Study of Methods for Protecting Hybrid Microcircuits from Contaminating Particles*, NASA TM X-68587, 1972.
101. F. W. Oberin, *Development for Application of Parylene Coatings*, NASA CR-120536, June 1974.
102. J. R. Szedon, T. A. Temofonte, and T. R. Kiggins, *Protective Coating for Hybrid Microcircuits*, U.S. Army Report, ECOM-72-0217-F, July 1974.
103. J. W. Adolphsen, W. A. Kagdis, and A. R. Timmins, *A Survey of Particle Contamination in Electronic Devices*, NASA TM X-71245, Dec. 1976.
104. V. S. Kale, and T. J. Riley, *IEEE Trans. Parts Hybrids Packag.* **PHP-13,** 273 (1977).
105. W. E. Loeb, *Spacecraft Contam. Conf.*, NASA Conf. Publ., NASA–CP–2039, 1978, pp. 863–879.
106. D. M. Wylie, *Final Report NASA CR 135071*, Teledyne Microelectronics Division, Aug. 1976.
107. J. H. Magee and R. D. Fisher, *IEEE Transactions on Magnetics* **Mag–6**(i), 34 (1970).
108. M. Fujihara, M. Fukui, and T. Osa, *J. Electroanal. Chem. International Electrochem.* **106,** 413 (1980).
109. M. Fujihara and T. Osa, *Prog. Batteries Sol. Cells* **3,** 298 (1980).
111. C. R. Rashkee and T. E. Nowlin, *J. Appl. Polym. Sci.* **25,** 1639 (1981).
112. *NASA Tech. Briefs*, winter 1976, p. 638.
113. B. Olschewski, *Electronics* **51**(15), 105 (1978).
114. R. Olson, *Insulation/Circuits* **24**(11), 33 (1978).
115. Union Carbide Corp., *Insulation Circuits* **24**(1), 38 (1978).
116. W. E. Loeb, M. Bak, M. Saliman, E. M. Schmidt, *IEEE Trans. Bio–Med. Eng.* **BMF–24**(2), 121 (1977).
117. A. Thornton, *Conformal Coatings for Pacemaker Applications*, NBS Spec. Publ. **109,** 400 (1979).
118. F. R. Tittman and W. F. Beach, *Synth. Biomed. Polym: Concepts Appl.*, 117 (1980).
119. H. L. Lee, Jr., A. L. Cupples, and R. J. Schubert, *Improved Encapsulation Materials for Implantation*, National Heart Institute Report, June 1969, pp. 69–154.
120. J. S. Byck, S. Chow, L. J. Gonsior, W. A. Miller, W. P. Mulvaney, L. M. Robeson, and M. A. Spivack, *Proc. of Artificial Heart Program Conf.*, June 1969, Chapt. 12, p. 123.
121. S. G. Eskin, C. D. Armeniades, J. T. Lie, L. Trevino, and J. H. Kennedy, *Biomed. Mat. Res.* **10**(1), 113 (1976).
122. W. A. Miller, M. A. Spivack, F. R. Tittman, and J. S. Byck, *Textile Res. J.* **43,** 728 (1973).

123. A. W. Hahn, H. K. Yasuda, W. J. James, M. F. Nichols, R. K. Sadhir, A. K. Sharma, O. A. Pringle, D. H. Hork, and E. J. Carlson, *Biomed. Sci. Instrum.* **17,** 109 (1981).
124. W. A. Spivack, M. A. Tittman, and J. S. Byck, *J. Text. Res. J.* **43,** 728 (1973).
125. D. Devanathan and R. Carr, *IEEE Trans. Biomed. Eng.* **BME–29,** 671 (1980).
126. W. E. Loeb, *SPE. J.* **27**(9), 46 (1971).
127. M. A. Spivack, *Rev. of Sci. Instrum.* **41,** 1614 (1970).
128. O. I. Szentesi and E. A. Noga, *Appl. Opt.* **13,** 2458 (1974).
129. P. J. Capitano, W. E. Loeb, M. A. Spivack, and F. R. Tittmann, *Final Report to Office of Saline Water*, (Contract No. 14–01–0001–1102), July 5, 1967.
130. W. Stokes, D. Rey, and J. L. Thompson, *Govt. Rept. Announce. Index (U.S.)* **81,** 1695 (1981).
131. A. H. Schloman, *Bendix Report*, NTIS, HC A03/MF A01, Nov. 1980.
132. R. J. McWhirter, *Energy Res. Abstr.* **6**(2), Abstr. No. 2627 (1981).
133. N. Bilo and P. M. Sawko, *J. Cell. Plast.* **11,** 207 (1975).
134. U.S. Pat. 4,225,647 (Sept. 30, 1980), R. A. Parent.
135. U.S. Pat. 3,663,265 (May 16, 1972), S. M. Lee and W. A. Bailey (to North American Rockwell Corp.).
136. U.S. Pat. 3,666,533 (May 30, 1972), S. M. Lee (to North American Rockwell Corp.).
137. D. T. Morrison and A. Taylor, *Proceedings of The Second German Sputtering School and Congress*, Tegernsee, FRG, May 3–5, 1971, p. 134.

STUART M. LEE
Ford Aerospace and Communications Corp.

Y

YEASTS

Yeasts are probably the oldest cultivated plants; their use dates to 2000 BC. The recognition of yeasts as a group of microorganisms and of their role in fermentation preceded their taxonomic classification. Yeasts are eukaryotic organisms, ie, they have a well-developed nucleus enclosed in a nuclear membrane. Eukaryotes have more than one chromosome. Yeasts are fungi that exist as single cells during at least some part of their life cycle. They usually reproduce vegetatively by budding or fission, but parasexual and sexual propagation also occurs in most yeasts. They have no photosynthetic ability and are not motile. The most extensive and authoritative texts on yeast taxonomy are references 1–2. A listing of yeast genera is given in ref. 3. Among those listed, the genus *Saccharomyces* is of greatest practical and economic importance for the baking, beer (qv), and wine (qv) industries, as well as for the production of biomass (see Bakery processes and leavening agents, yeast-raised products). Other yeasts participate in alcoholic fermentations and occur as food-spoilage organisms or pathogens (see Beverage spirits, distilled; Fermentation).

Morphology, Reproduction, and Life Cycles

The shape of a single yeast cell is usually spherical to ellipsoidal but may be cylindrical, ogival, pyramidal, or apiculate. Often the shape results from specific sites of bud formation. The apiculate or lemon-shaped form arises from bipolar bud formation, ie, buds form at the ends of the long polar axis of an originally elliptical cell. The size of yeast cells varies. For ellipsoidal cells, eg, of *Saccharomyces cerevisiae*, a length of the long axis of 5–10 μm and of the short axis of 3–5 μm is common. The volume of such a cell may be 40 μm^3 and its weight 10 pg. The density of the cells is 1.03–1.1 g/cm^3 (4). Yeasts may form a pseudomycelium consisting of single or branched chains of cells if the daughter cells do not separate from the mother cells.

The yeast cell is surrounded by a strong and mechanically refractory cell wall which may account for 20–30 wt % of the cell solids (5). The cell wall consists of alkali-soluble β-glucan, alkali-insoluble glucan, and glycoproteins. The linkages between the glucan molecules are β-(1→3) or β-(1→6). The cell wall surrounds the plasma membrane, which regulates the transport of chemical compounds into and out of the cell either by simple diffusion or by active transport. Some yeasts form viscous and sticky capsular material on the outside of the cell wall. Fimbriae (small, hairlike structures) are on the outside of the cell wall and may play a role in cell flocculation.

The large nucleus of yeast cells is surrounded by a membrane, or tonoplast, which has many pores with an average diameter of ca 0.085 μm and which contains the genetic material of the cell. In haploid cells of *S. cerevisiae*, there are probably 17 chromosomes, and the nucleus contains 90% of the deoxyribonucleic acid (DNA) of the cell. Other species may have smaller numbers of chromosomes. Yeast vacuoles can be seen through a phase-contrast microscope. There usually are one to four vacuoles in the protoplasm, and they appear to contain hydrolytic enzymes and have a storage function for other chemical compounds. Mitochondria are small bodies within the protoplasm and contain ≥5 wt % of the cell DNA. They serve the respiratory function of the cell and contain cytochromes. Under anaerobic conditions, the mitochondria degenerate but can reform if a cell suspension is aerated. Lipid globules are also in the protoplasm. The cytoplasm or cytoplasmic ground substance of the cell contains most of the ribonucleic acid (RNA) which may account for 7–12 wt % of the total cell solids. It also contains the carbohydrate storage materials of the cell, namely, yeast glycogen, which resembles liver glycogen, and a specific nonreducing disaccharide, trehalose. The enzymes of the glycolytic pathway are cytoplasmic.

Vegetative reproduction in yeasts occurs mostly by budding and, in the instance of *Schizosaccharomyces* and *Endomycopsis*, by fission. In these asexual processes, the nucleus of the mother cell divides and one portion enters the newly formed bud. The bud grows until it has reached a size almost as large as the mother cell and then separates from it, forming a bud scar on the surface of the mother cell. The number of bud scars can be counted, and the oldest cells of a population in the stationary phase typically have 12–15 bud scars, which is apparently the maximum number of daughter cells that can be produced by a mother cell.

Fission in *Schizosaccharomyces* and *Endomycopsis* is by formation of a cross wall without constriction of the cell. This cross wall or septum divides, forming two individual, separate walls. At this time, the two vegetative cells separate.

In *Ascomycetes*, sexual reproduction is through the formation of spores in a cell which serves as an ascus or spore sac. Generally, four (but sometimes eight or more) spores form per ascus. This is sometimes reflected in the name of the species, eg, *Schizosaccharomyces octosporus* or *Kluyveromyces polysporus*. The yeasts belonging to *Basidiomycetes* form external spores. Karyogamy, the fusion of the nuclei of two cells, and meiosis, the reductive division of the chromosomes of the nucleus, are the bases of sexual reproduction. Sexual reproduction of yeasts consists of an alternation between the haploid stage (one set of chromosomes) and the diploid stage (two sets of chromosomes). Figure 1 shows schematically the life cycle of yeasts existing predominantly in the haploid state, eg, for *Saccharomyces rouxii*; in the diploid state, eg, for *S. cerevisiae*; or in both states, eg, for *Saccharomycopsis lipolytica*. An alternation between the haploid and the diploid state may also occur without the production of sexual spores (parasexuality).

Genetic Aspects

Yeasts have translation and transcription mechanisms similar to those of higher eukaryotic organisms. Accordingly, higher eukaryotic genes are expressed in yeasts when they are inserted by transformation. However, most of the work on genetic improvement of industrial yeast strains is performed by the more traditional methods of hybridization and mutation (7) (see Genetic Engineering, Vol. 11 and Supplement Volume).

Hybridization. It is possible to isolate single ascospores of a yeast, eg, *S. cerevisiae*, by means of a micromanipulator. If individual spores from one strain are placed next to the spores of another strain, then conjugation of spores may occur and a hybrid can form. Such hybrids can also be produced by conjugation of spores of different species of the same genus. These techniques have led to improved strains of bakers' yeasts (8–9).

In many instances, protoplast fusion of the cells occurs. This is the fusion of two cells whose cell walls have been digested with enzymes. Such naked protoplasts can often be induced to fuse. Fusion has been achieved between species of *Saccharomyces* and of *Kluyveromyces*. Even intergeneric fusion has been achieved between *Candida* and *Pichia* strains as well as several other types. Protoplast fusion between wine strains

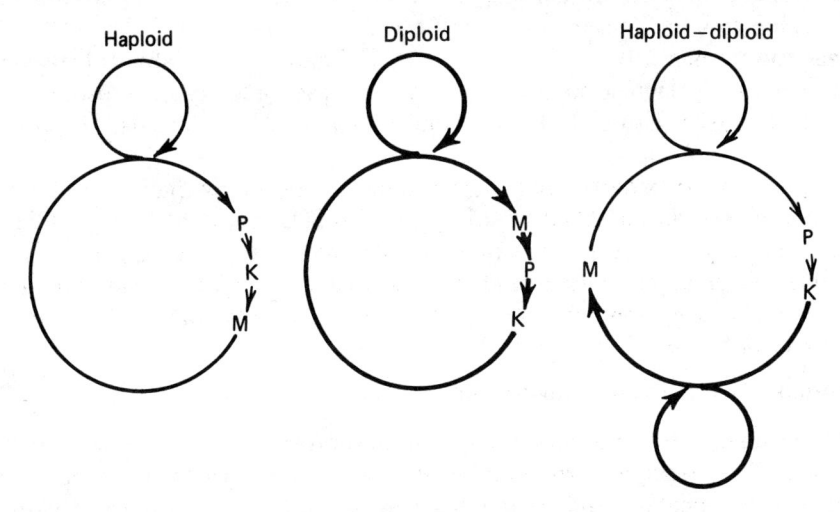

Figure 1. Scheme of essential life cycles (6). Thin line = haploid phase; heavy line = diploid phase; K = karyogamy; M = meiosis; and P = plasmogamy. The smaller circles represent the possibility of asexual propagation.

of *S. cerevisiae* and *Schiz. pombe* has been achieved, but results have not yet been published.

Mutation. A normal population of bakers' yeast cells usually contains 1% of respiratory-deficient mutant cells, which are so named because they lack one or several of the respiratory enzymes. They can grow fermentatively but do not grow on carbon sources they cannot ferment. They are easily recognized by their formation of very small, or petite, colonies when grown on nutrient agar (10).

For industrial work, mutations are induced by x rays, uv light irradiation, or chemicals, eg, nitrosoguanidine. Examples of useful mutants are strains of *Candida membranefaciens*, which produce 1-threonine; *Hansenula anomala*, which produce tryptophan; or strains of *Candida lipolytica*, which produce citric acid (11–12). An auxotrophic mutant of *S. cerevisiae* that requires leucine for growth has been produced for use in wine fermentations. This yeast produces only minimal quantities of isoamyl alcohol, a fusel oil fraction derived from leucine by the Ehrlich reaction.

Fermentative and Respiratory Metabolism

Most yeasts are strict aerobes or ferment weakly. Strongly fermenting yeasts, eg, *S. cerevisiae*, grow fairly well under anaerobic conditions but with a lower yield. The yield of yeast based on the weight of sugar consumed is greater in the presence of oxygen, and oxygen inhibits the rate of glucose utilization. However, under aerobic conditions, alcohol is produced if the concentration of glucose exceeds a threshold value of 0.1–0.2 wt %. This is the reverse Pasteur or Crabtree effect and is also called glucose inhibition or catabolite repression. In the presence of higher concentrations of glucose, synthesis of respiratory enzymes, eg, cytochromes, is inhibited.

The anaerobic pathways for glucose utilization by yeasts and other microbes lead to the formation of various alcohols and organic acids (see Fig. 2). Ethanol is the principal product of yeasts used in the beverage industry. It forms from pyruvate via acetaldehyde. Pyruvate forms in the Embden-Meyerhof-Parnass pathway (EMP) and the related hexose monophosphate pathway (HMP) (13). In either case, anaerobic fermentation results in 2 mol ethanol and 2 mol CO_2 from 1 mol glucose. The reaction produces a net gain of 2 mol adenosine triphosphate (ATP), which supplies the energy for anaerobic growth. This anaerobic pathway from sugars to alcohol and CO_2 characterizes the fermentative activity of yeasts. Common fermentable substrates are glucose, fructose, galactose, mannose, maltose, lactose, cellobiose, sucrose, and some trisaccharides (see Ethanol; Fuels from biomass; Alcohol fuels, Supplement Volume).

Respiratory, or oxidative, pathways produce more energy than anaerobic pathways. Complete oxidation of one mole of glucose to CO_2 may yield 36 mol ATP in the tricarboxylic acid cycle or related oxidative pathways. Many more substrates can be respired than fermented; these include pentoses, eg, by *Candida* species; ethanol, eg, by *Saccharomyces* sp; methanol, eg, by *Hansenula* sp; and alkanes, eg, by *Saccharomycopsis lipolytica*.

Composition, Nutrients, and Growth Rate

The elemental and vitamin compositions of some representative yeasts are listed in Table 1. The principal carbon and energy sources for yeasts are carbohydrates (usually sugars), alcohols, and organic acids, as well as a few other specific hydrocarbons. Nitrogen is usually supplied in the form of ammonia, urea, or amino acids. The

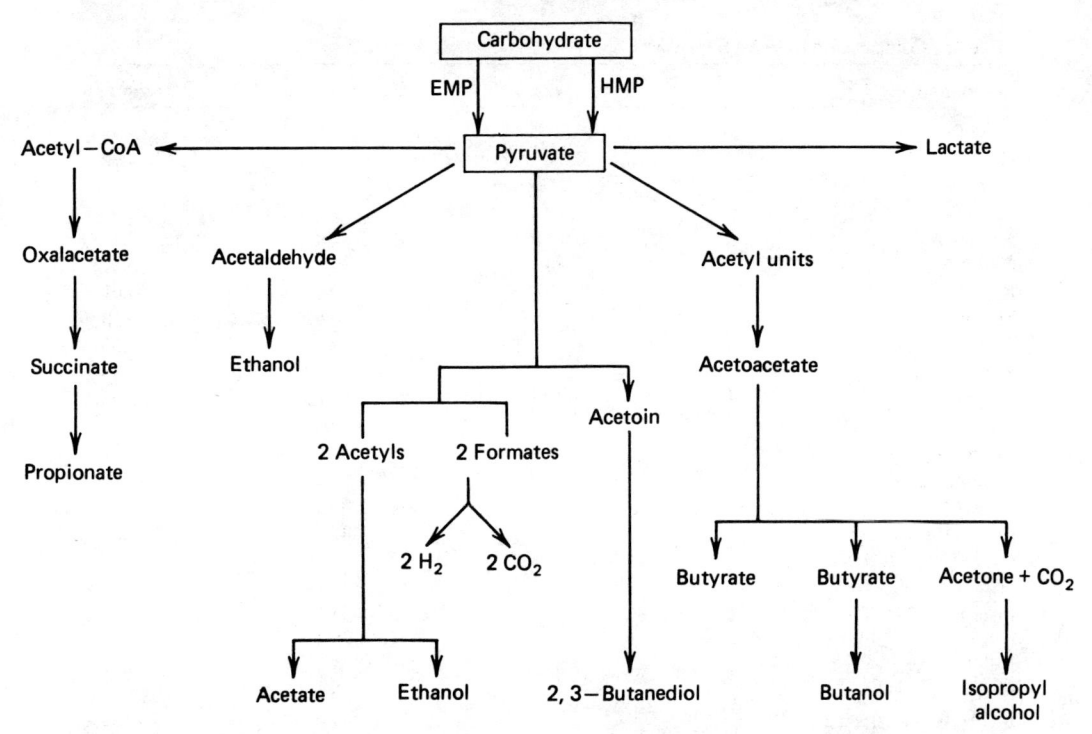

Figure 2. Fermentation products of carbohydrate metabolism (13). EMP = Embden-Meyerhof-Parnass pathway; HMP = hexose monophosphate pathway.

main essential mineral elements are phosphorus supplied as phosphoric acid and potassium, with smaller amounts of magnesium and trace amounts of copper, zinc, and iron. These requirements are characteristic of all yeasts. The vitamin requirements, however, differ among species. For laboratory and many industrial processes, a commercial yeast extract is assumed to contain all of the required nutrients (see also Mineral nutrients).

The specific growth rate constant for exponential growth μ is defined by the equation $\mu \times dt = dM/M$, where M is the mass of yeast and t is time. Growth is also often expressed as the generation time, ie, the time required for a doubling of the yeast population, or as the hourly growth rate. The relationship of these expressions of exponential growth is shown in Table 2. Exponential growth ceases when at least one nutrient becomes limiting. This deficiency may also include oxygen, which is difficult to supply in saturating amounts because of its poor solubility.

Yeasts that have been grown in clear media, eg, clarified molasses or brewers' worts, can be recovered by centrifuging or filtration followed by pressing. Yeasts that have grown in distillers' mashes or grape musts, which contain insoluble particles, cannot be recovered economically.

Yeast-Fermented Foods and Beverages

Table 3 shows U.S. production of alcoholic beverages, baked goods, and yeast biomass and the production of industrial fuel alcohol by fermentation. All of the yeast produced in wine fermentations is either discarded as waste or spread with the lees

Table 1. Composition of Yeast (Dry Mass)[a]

Component	Bakers' yeast	Brewers' yeast	Candida sp
C, wt %	47.0		45.94
H, wt %	6.0		6.72
N, wt %	8.5		7.31
O, wt %	32.5		32.08
ash, wt %	6.0	6.4	7.75
Ca, wt %		0.13	0.57
Fe, wt %		0.01	0.01
Mg, wt %		0.23	0.13
P, wt %		1.43	1.68
K, wt %		1.72	1.88
Na, wt %		0.07	0.01
Co, mg/kg		0.2	
Cu, mg/kg		33.0	13.4
Mn, mg/kg		5.7	38.7
Zn, mg/kg		38.7	99.2
dry matter, wt %		93.0	93.0
crude fiber, wt %		3.0	2.0
ether extract, wt %		1.1	2.5
protein (N × 6.25), estd wt %		44.6	48.3
protein, digestible, wt %		38.4	41.5
thiamine, mg/kg		91.7	6.2
riboflavin, mg/kg		35.0	44.4
nicotinic acid, mg/kg		447.5	500.3
vitamin B_6, mg/kg		43.3	29.5
biotin, mg/kg			1.1
pantothenic acid, mg/kg		109.8	82.9
folic acid, mg/kg		9.7	23.3
choline, mg/kg		3885.2	2910.6

[a] Refs. 14–15.

Table 2. Relation of the Generation Time to the Hourly Growth Rate and the Specific Growth Rate Constant

Generation time, h	Hourly growth rate, C_t/C_o[a]	Specific growth rate constant, μ, h^{-1}
1	2.00	0.693
2	1.41	0.345
3	1.26	0.230
4	1.19	0.173
5	1.15	0.139

[a] C_o = concentration of yeast solids at time 0. C_t = concentration of yeast solids at time t; in this case, t = 1 h.

or still residues on agricultural fields. All of the excess yeast produced during the fermentation of distilled beverages or in fuel-ethanol production is recovered with the spent grains and is sold as feed (see Pet and livestock feeds). Yeast produced during beer fermentations is largely combined with the spent grains and sold as feed. However, much is recovered separately by centrifugation and is used for the production of food-grade, or inactive, brewers' yeast or is used in the production of yeast extracts.

Table 3. 1981 U.S. Production of Yeast-Fermented Foods, Beverages, and Fuel Alcohol and Production of Yeast

Product	Food, beverage, and fuel production, 10^6 m^3 (except where noted)	Production of yeast[a], 1000 metric tons
beer	18.52	40
wine	1.61	4
baked goods (yeast-raised)	7.5×10^6 t	60
yeast biomass (inactive)		7–9
distilled alcohol beverages (50 vol % ethanol)	2.5	25
fuel and industrial alcohol (100 vol % ethanol)	1.5	30

[a] Estimates based on production of 2–2.5 kg of yeast solids per cubic meter of the alcohol beverage (at 8–12 vol % ethanol).

Bakers' yeast is supplied as a moist press cake or an active dried yeast. There is little or no growth of yeasts during the dough fermentation, and therefore, the estimate shown in Table 3 is based on production in yeast factories. Some bakers' yeast is also used for the production of food-grade, inactive dried yeast. Yeast biomass refers to the production of torula (*Candida utilis*) yeast grown on waste sulfite liquor or ethanol and smaller amounts of *Kluyveromyces fragilis* grown on whey. These dead yeast products are generally used as inactive, dried, food-grade yeasts (see Foods, nonconventional; Microbial transformations).

Table 4 is a comparison of the conditions under which yeasts are used in various industries. The fermentative activity of the yeasts is almost the same in all industries. For alcoholic beverages, there is a five-to-tenfold multiplication of yeast cells during fermentation. Differences arise mainly from differences in temperature, which are reflected in the fermentation times. For the baking industry, the temperature is high;

Table 4. Conditions for Commercial Use of Yeast[a]

	Bread dough	Beer Lager	Beer Ale	Wine	Whiskey
raw material	flour, sugar	wort	wort	grape or fruit juice	cereal mash
time of fermentation	1–3 h	8–10 d	2–6 d	5–10 d	3 d
fermentation temperature, °C	30–35	10	20	15–27	35
pH	5.2–4.8	5.2–4.2	5.2–4.2	3.5–3.5	4.9–4.0
ethanol, wt %	2–3	4	4	11–13	7–9
yeast cells[b]					
start	275	6–10	6–10	5–10	5–10
end	300	30–70	30–70	50–150	50–150
CO$_2$ production rate, mmol CO$_2$/(h·g yeast) solids	18–35	16–25	16–25	25–32	

[a] Ref. 16.
[b] Million (10^6) cells per gram (bread dough) or 10^6 cells/mL (alcohol beverages).

therefore, the fermentation time is short, and with the very high cell counts at the start, there is little or no growth during the fermentation.

Bakers' Yeast Production. Bakers' yeast is grown aerobically in fed-batch fermentations and under conditions of carbohydrate limitation. This assures the growth of yeast and prevents or minimizes the formation of ethanol. Such yeasts have excellent dough-leavening ability. They perform better in the bakery than yeasts grown anaerobically.

Strains. All bakers' yeast strains are *S. cerevisiae*. Suitable strains may be obtained from all culture collections. Since yeast sold to bakers is alive, it is possible to isolate the strains used for commercial production. Strains used for the production of compressed yeast in the United States or for the so-called fast compressed yeasts in Europe are similar. The cells are ovoid with diameters of 4–6 μm. In contrast, strains used for the production of active dry yeast are larger, ie, 5–8 μm. They are generally of slightly lower fermentation activity on an equivalent solids basis but are more tolerant to drying conditions.

There are few proprietary strains, but some hybrids of bakers' yeast strains have been patented (1,8–9). Pure cultures of bakers' yeast strains may be kept on agar slants with frequent transfers or in lyophilized form.

Raw materials. Until the 1930s, grain worts were used as the principal carbon and energy sources. Since then, cane and beet molasses have been used in all countries because they are the least expensive sources of fermentable sugars (see Sugar, beet sugar; Sugar, cane sugar). Molasses contains ca 50–55 wt % fermentable sugar in the form of sucrose, glucose, and fructose (see Syrups). It also is a main source of potassium and other minerals, and cane molasses is the principal source of biotin, which is a required nutrient for bakers' yeast production (see Vitamins, biotin). The basic source of nitrogen is NH_3 added as ammonium sulfate or as liquid or anhydrous ammonia. Urea may also be used, and many of the amino acids are sources of assimilable nitrogen (17).

Molasses contains sufficient potassium for yeast growth, but phosphate must be added. Bakers' yeast contains ca 3 wt % phosphorus as P_2O_5, based on yeast solids. Phosphate is added as phosphoric acid or as phosphate salts. Calcium and magnesium salts in smaller concentrations are also added. Iron, zinc, copper, manganese, and molybdenum are required as trace minerals. The vitamin requirements for biotin, inositol, and pantothenic acid are usually in molasses. Thiamine beyond that required for yeast growth is usually added, because it contributes to fermentation activity of the final compressed yeast.

Fermentation. The molasses of 80° Brix (% wt/vol sugars) is diluted to ca 36° Brix and adjusted to pH 5. It is clarified by centrifuging and is sterilized by brief, high temperature procedures (18). The other nutrients are either included in the molasses wort or added separately during the fermentation. Ammonia is generally added separately, since it is used as a nutrient and as a means of pH adjustment. The fermentation is carried out at 30°C and pH 4.5–6.5. Higher pH levels are conducive to faster growth, but lower pH values protect better against contaminants.

The actual fermentation is carried out as a fed-batch procedure (formerly known as the Zulauf process); ie, media are added to the seed yeast at a slow and programmed rate so that sugar concentrations in the fermentor never exceed 0.1 wt %. At higher sugar concentrations in the fermentor, alcohol forms which lowers the yield of yeast and increases the biological oxygen demand (BOD) of the effluent. With sufficient aeration, it is possible to greatly extend the exponential phase, but in practice this is rarely the case. Figure 3 shows some practical feed curves for a 12-h fermentation as well as an experimental feed curve for exponential growth.

Oxygen is a critical nutrient for yeast growth. One kilogram of oxygen is required for the growth of one kilogram of yeast solids. This oxygen is supplied by bubbling air through a series of pipes with small holes arranged at the bottom of the fermentor (19). The distribution of air in the fermentor and the formation of very small bubbles is often aided by agitators. Oxygen is only sparingly soluble in water, and at saturation, the concentration of oxygen at 30°C is ca 7 ppm. During uptake by concentrated yeast suspensions, the oxygen level may decrease to ca 5% of saturation. In large fermentors, the rate of air supply is about one volume of air per fermentor volume per minute (VVM). Alcohol forms if the supply of air is limited.

Finally, the rate of yeast growth must be limited to a specific growth rate constant of 0.23. This is defined as $dM/M \times dt$, where M is the mass of yeast and t is time. For an hourly specific growth rate of 0.23, the hourly increase in cell mass is 1.26, and the generation time, ie, the doubling time of cell mass, is three hours. At higher growth rates, alcohol again forms. During yeast growth, heat is liberated and must be removed from the fermentor by internal cooling tubes or by pumping of the fermentor contents through external heat exchangers. In practice, the yield of yeast cell mass is ca 50 wt % of the sugar used (20–22).

The fermentation period is usually 12–18 h. For a 15-h fermentation, the generation time may increase from three hours, to five hours, and then to seven hours, allowing for an eightfold increase in cell mass. Generally, aeration continues briefly after feed additions have been stopped to allow the yeast to mature. This results in a reduction of the number of budding cells, and it increases storage stability of the com-

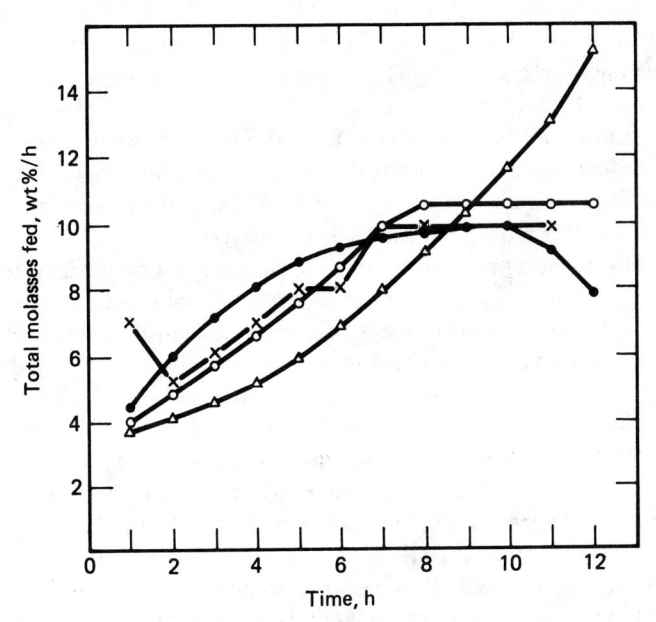

Figure 3. Molasses feed curves for bakers' yeast fed-batch fermentations. ·—·—·, Experimental, exponential feed curve. O—O—O—O, Experimental, exponential followed by constant feed. X—X—X—X, commercial, pragmatic curve with constant feed from seventh hour. ●—●—●—●, Commercial, pragmatic curve with constant feed from eighth hour and reduced feed from the eleventh hour.

pressed yeast. At the end of the growth period, the concentration of yeast solids in the fermentor is ca 5 wt %. Experimentally, the concentration of cell solids may be increased to 10 wt % or higher, but in commercial practice, the limitation of oxygen transfer limits the cell mass to 5–5.5 wt %.

The foregoing comments on yeast fermentations apply only to the final trade fermentation which is carried out in large (ca 200 m^3 or 52,800 gal) fermentors. Inoculum seed for this fermentation is prepared in a series of fermentations, beginning with those in small shake flasks followed by transfers to larger and larger vessels to the final trade fermentation. There may be as many as eight stages in this sequence, producing perhaps 0.8 kg, 3.5 kg, 25 kg, 120 kg, 2.5 t, 15 t, and 100 t of bakers' yeast from a single pure-culture slant (23).

The earlier stages of the series may be carried out under conditions approaching those of pure-culture fermentations, ie, with sterile media, sterile air, and presterilized vessels. Fermentations in larger vessels and the final trade fermentation are conducted under quasisterile conditions, and yeast growth is accompanied by some growth of contaminant bacteria. These are generally lactic acid-producing organisms and are sometimes coliform bacteria. The occurrence of *Salmonella* in fermentor liquids has not been reported. Massive and detrimental infections with *Oidium lactis* or wild yeasts have been reported (24–25) (see also Sterilization techniques).

Harvesting and packaging of compressed yeast. At a 5% yeast solids concentration in the fermentor, the yeast occupies ca 12% of the fermentor volume. It is harvested by centrifuging in nozzle centrifuges and is washed several times with water (18). The final centrifugate is cooled and stored in refrigerated tanks. It is called yeast cream because of its off-white appearance. A moist press cake is obtained by pressing the cream in plate-and-frame filter presses or filtration through a rotary vacuum filter. The press cake is mixed with emulsifiers (0.1–0.2 wt %), and its moisture content is adjusted to 70 wt %. It is then extruded through Teflon nozzles in the shape of thick strands with rectangular cross sections and is cut into 0.45-kg cakes, which are wrapped with waxed paper.

During pressing, mixing, and extruding, the temperature of the yeast cakes is 10–15°C. They must be cooled in cartons for about two days before shipment in refrigerated containers. Compressed yeast is also shipped in the form of irregularly shaped pieces packed in bulk containers as crumbled yeast.

Compressed yeast is a perishable commodity and must be refrigerated at all times. At a storage temperature of 5–8°C, there is a 3–5% loss of fermentation activity per week. Compressed yeast can also be kept frozen for several months. However, after thawing, some browning and some softening of the cake occur.

Active dry yeast. The production of active dry yeast (ADY) is very similar to that of compressed yeast. However, a selected strain of *S. cerevisiae* is used, and the nitrogen concentration (based on yeast solids) is kept at ca 7 wt % compared with 8–9 wt % for compressed yeast. The press cake made with the active dry yeast strain is extruded through a perforated plate in the form of fine strands with diameters of 2–3 mm. The strands are chopped into lengths of 3–10 mm and then are dried on endless belts of steel mesh in drying chambers, in roto-louvre dryers, or in fluid-bed (air-lift) dryers. During drying, it is necessary to keep the temperature of the yeast cell mass ≤40°C. Drying times in air-lift dryers are 10 min to 2 h; in drying chambers, 3–4 h; and in roto-louver dryers, ≥6 h. The drying process involving an endless belt and drying chambers is continuous, whereas the roto-louver process is a batch process. Air-lift drying is usually carried out as a batch process, although continuous air-lift dryers are available.

Regular ADY is dried to a moisture content of 7.5–8.5%; the instant ADY contains ≤6 wt % moisture. Table 5 is a comparison of the gross chemical compositions of compressed yeast and active dry yeast. Table 6 is a summary of the fermentative activity of these yeast preparations in various doughs. Lean doughs are doughs with <1% added sugar, and sweet doughs contain 15–25% sugar; these percentages are based on the weight of the flour.

Yeasts in Baked Foods. Baked goods can be leavened with yeast, chemicals, the foam of egg whites, or steam. There is a preference for yeast-leavened baked goods because of their desirable flavor. Therefore, only those products in which yeasts do not perform well are leavened by chemical leavening, which is the result of osmotic pressure. The osmotic pressure of doughs depends inversely on moisture content and directly on concentration of salt and sugars. Cookies and highly sweetened cake doughs have an osmotic pressure too high for adequate fermenting activity of yeasts. Some attempts have been made to grow yeasts with exceptional tolerance to high osmotic pressures, but such strains have not been used commercially (26). There are some products in which either yeasts or chemical leavening, or both, are used, eg, doughnuts and pizza doughs.

Another requirement for proper leavening is the presence of a protein matrix that is sufficiently elastic to entrap small bubbles of CO_2, eg, wheat protein gluten. The protein of rye is less suitable, and the proteins of other cereals, eg, rice, oats, or corn, are practically useless for this purpose.

Sourdoughs. The microflora of flour contains large numbers of lactic acid-producing bacteria. In past centuries, fermentations of bread doughs were mixed yeast and bacterial fermentations. This is still true for most doughs made exclusively from rye flour. In the United States, rye bread contains generally small concentrations of rye flour and large concentrations of wheat flour. Sour rye doughs and sour wheat

Table 5. Composition of Compressed Yeast and Active Dry Yeast (ADY), wt % of Yeast Solids

Component	Compressed yeast	ADY	Instant ADY
crude protein (N × 6.25)	50–56	41–47	44–53
phosphorus (as P_2O_5)	2.8–3.0	2.2–2.5	2.3–2.8
ash	6–8	6–8	6–8
lipids[a]	4–5	4–5	4–5
water	69–70	7.5–8.5	4.0–6.0

[a] Extracted after hydrolysis.

Table 6. Fermenting Power of Bakers' Yeasts, Grams of Sugar Fermented per Grams of Yeast Solids per Hour

	Regular dough	Lean dough	Sweet dough
compressed yeast	2.22–2.34	2.36–2.58	0.95–1.12
active dry yeast (ADY)	1.42–1.54	1.22–1.28	0.83–0.89
instant ADY	1.60–1.83	1.84–1.97	0.77–0.86

doughs are inoculated with bacterial starter cultures and with yeast. The lactic acid bacteria usually in sourdough fermentations are *Lactobacillus plantarum*, *Lactobacillus brevis*, and *Lactobacillus fermenti*. Figure 4 shows the drop in pH of a rye dough as a function of the development of yeasts and lactic acid bacteria (27).

The production of soda crackers is also based on a mixed fermentation. Doughs for cracker production are inoculated with very small amounts of bakers' yeast. During the first 3–5 h of the 18-h sponge fermentations, yeast fermentation predominates. Thereafter, bacterial fermentation causes a rapid decrease in the pH of the sponge through formation of lactic acid.

The production of San Francisco sourdough bread is a typical sponge-dough fermentation, and a portion of each sponge is retained for inoculation of the succeeding sponge. The yeast responsible for this fermentation is *Torulopsis holmii* and the lactic acid-producing bacterium is *Lactobacillus sanfrancisco* (28). The same yeast is also responsible for the fermentation of the Italian pannetone, a traditional Christmas fruitcake (29).

White Pan Bread. More than 50% of all of the flour used in the production of baked foods is made into bread, rolls, and buns, and white bread is the predominant dough type. The basic method of production is the sponge-dough method. The sponge generally contains 60–70 wt % of the flour, all of the yeast, and yeast food. It undergoes a 3–4 h sponge fermentation. The sponge, the remainder of the flour, and all of the other ingredients, ie, salt, sugars, fat, etc, are then mixed into a dough, which is permitted to rest for 20–30 min. Then it is divided, rounded, molded, and panned. The pans are placed into a proofing cabinet where the final fermentation takes place before baking. Mixing of the sponge, the sponge fermentation, and mixing of the dough are

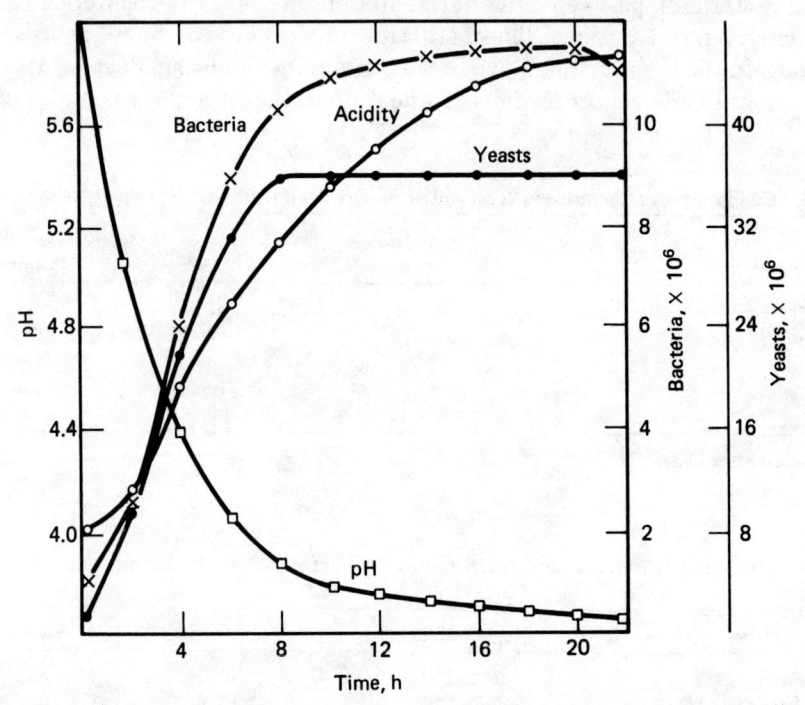

Figure 4. Relationship of yeast and bacteria count and pH of rye sourdough. Yeasts and bacteria are quantified as the number of cells per gram of dough (27).

batch processes. However, from the dividing step to the final baking, the process is generally continuous.

Since the 1950s and 1960s, other production methods have been used. For example, the sponge can be replaced by a liquid sponge or preferment, which contains very little or no flour but most or all of the water, yeast, yeast nutrients, sugar, and a suitable buffer. It must remain sufficiently fluid so that it can be pumped. Addition of the buffer prevents the pH from dropping below 4.5. The preferment is generally set at 25–30°C, and the fermentation time is 1–2 h. Another method is the use of continuous dough mixing, which must be combined with the use of liquid preferments. Continuous, high speed mixing produces a very extensible, slack dough with a uniform distribution of fine bubbles of carbon dioxide. The resulting bread has a very fine grain and is very soft.

All of the preceding methods have the advantage of flexibility. Depending on the course of the sponge fermentation, alterations can be made in dough-mixing time or in the addition of ingredients, eg, oxidants, to the dough. In case of equipment breakdown, sponges can be maintained for several hours without spoiling, and preferments can be cooled and stored overnight. Wholesale bakers also use straight dough methods, in which all of the ingredients of a bread formula and the water are mixed in a single operation. The dough is then fermented for 2–4 h. The operations following fermentation are the same as those for sponge and dough methods. For more detailed descriptions of actual bakery operations, see references 30–31.

The function of yeast in bread baking is threefold: it serves as a leavening agent, it contributes flavor, and it matures the dough. In contrast to chemical leavening, which is a fast reaction, yeast leavening is slow and proceeds as long as fermentable sugar is available. Flour contains ca 1 wt % fermentable sugar, which is generally fermented within 1–1.5 h. However, flour also contains ca 6–12 wt % of damaged starch granules, which are hydrolyzed by amylases to maltose. These amylases are principally β-amylase, which is inherent to the flour, and added α-amylase. The latter may be derived from barley malt or from fungal sources. The fermentation pattern of a dough to which no sugar has been added is shown in Figure 5. It is apparent that glucose is fermented faster than fructose. No conclusion can be drawn on the rate of maltose fermentation, since maltose is continuously produced while it is fermented.

Most bread doughs contain 4–8 wt % of added sugars, ie, sucrose, dextrose, corn

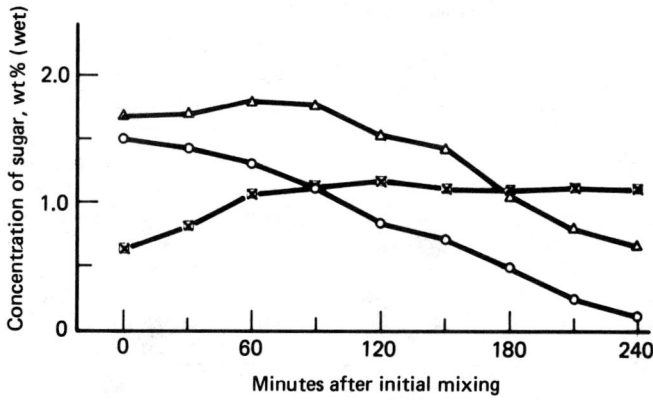

Figure 5. Changes in sugar concentration in fermenting dough (32). △, Fructose. ○, Glucose. x, Maltose.

syrup, or high fructose corn syrup. Most of these sugars are used for fermentation, but there is some residual sugar in the bread. For instance, with the addition of 6.7 wt % sucrose, the residual sugars in the baked bread are 2.6 wt % fructose, 1.7 wt % glucose, and 0.7 wt % maltose (33). Lactose (milk sugar) is not fermented by bakers' yeast, and any residual lactose in bread is derived from added milk solids or whey.

The pH of doughs or preferments has little effect on yeast fermentation rate unless the pH drops below 4.5. Prior to fermentation, the pH of a dough is 5–6. During fermentation, it decreases to ca 4.8 because of the production of bicarbonate and some organic acids. During baking, there is a slight rise in pH as CO_2 is removed. Temperature has a considerable effect on yeast activity: up to 38°C, yeast activity increases rapidly; at higher temperatures, the activity diminishes; and at 52°C, bakers' yeast cells die within 10 min. Bakery sponges, straight doughs, and preferments are generally set at 24–26°C. During fermentation, there is a rise of 2–5°C resulting from the metabolic activity of the yeast. In the case of continuously mixed doughs, the temperature may reach 35°C outside the mixer. Proof-box temperatures are usually 30–40°C. During baking, the temperature of the dough rises rapidly and yeast fermentation ceases after ca 10 min in the oven. However, during this time, production of additional CO_2 results in some expansion of the dough in the pan, the so-called oven spring. The reduced solubility of CO_2 in the dough water and the expansion of CO_2 bubbles at higher temperatures also cause oven spring.

The effect of osmotic pressure on yeast activity is of great importance, and is often overlooked. At salt concentrations up to 1.5 wt %, the effect is slight; salt concentrations of 2–2.5 wt %, which are common in bread, inhibit yeast activity considerably. Similarly, at sugar concentrations above 4 wt %, the inhibiting effect is apparent. Therefore, yeast-raised sweet doughs, eg, doughnut, coffee-cake, or Danish-roll doughs, contain very high concentrations of yeast (see Table 7).

Ethanol inhibits yeast fermentation. This is important in wine fermentations where alcohol levels may reach 11–12 vol %. In dough fermentations, alcohol concentrations rarely exceed 3 vol % and are not inhibitory. Mold inhibitors are commonly added to bread doughs. These may be propionates (0.25 wt % of the flour) or small

Table 7. Usual Levels of Compressed Yeast in Doughs[a]

Type	Yeast, wt %[b]	Type	Yeast, wt %[b]
breads		sweet doughs	
sponge and dough	2.5	yeast-raised doughnuts	4–7
continuously mixed dough	2.5–3.5	Danish and coffee cakes	5–10
breads from mixed grains	2.5–4.0	specialties	
whole wheat	2.75	partially baked rolls	2
high fiber breads	3–5	frozen, unbaked doughs	5–5.5
no-time doughs[c]	3–4	pretzels	0.5
sour rye bread	1.2	soda crackers	0.2
buns and rolls		traditional breads	
hard rolls	2	habab (Syria)	1.5
hamburger buns	4	creola (Cuba)	1.0
English muffins	2–8	barbery (Syria)	0.25

[a] Ref. 34.
[b] Based on the weight of the flour.
[c] No fermentation occurs.

concentrations of vinegar. These mold inhibitors retard leavening to a slight but noticeable degree.

There is little or no growth of yeast during dough fermentation. However, budding of cells can be observed during sponge fermentation or during the proof stage of a straight dough fermentation (35). The addition of yeast nutrients in yeast foods is common; ca 0.05 wt % ammonium sulfate or ammonium chloride are usually added (36).

The effectiveness of yeast leavening depends on the development of a protein matrix, which can be obtained with wheat flour which contains 10–14 wt % protein. It cannot be developed with corn flour, and consequently, corn breads have a course, grainy crumb. A portion of the CO_2 produced during the fermentation escapes and is not available for leavening. This is important during the proof when the volume of the loaf is determined. The amount of high speed mixing is critical to development of the elastic protein matrix. Undermixed or overmixed doughs do not retain the gas well.

Yeast fermentation is also central to flavor development. This flavor is derived from the fermentation by-products, eg, higher alcohols, esters, aldehydes, and other carbonyl compounds. However, in contrast to alcoholic beverages, the flavor of baked bread is produced by the interaction of these fermentation by-products with other dough constituents. A considerable part of bread flavor is formed in the crust during the browning. After the bread has been baked and cooled this flavor diffuses into the bread crumb. Actually, very little is known about these reactions despite the identification of more than 100 flavor compounds (37–38).

The role of yeast in the maturation of fermenting doughs is even less clear. The alcohol and carbon dioxide, which develop during fermentation, probably influence the elastic properties of the protein matrix. No experimental procedure that would permit this to be checked in the absence of yeast has been developed.

Bakers' Compressed Yeast. Compressed yeast is received by the baker in 0.45-kg cakes or as crumbled material in 22.7-kg bags and refrigerated while stored. Deliveries are received on alternate days or once a week. The stock must be rotated so that the yeast is not kept longer in the bakery than 1–2 wk. The yeast cakes may be placed directly into the dough mixer on top of the flour. Crumbled yeast is preferred for the production of preferments. In many bakeries, compressed yeast is slurried with a small portion of the dough water and other ingredients, eg, oxidants, enrichment tablets, and yeast nutrients, in a premix tank. The content of this tank is then pumped into the mixer.

Compressed yeast is also sold through grocery stores in 18-g and 56-g packages; the product has a lower protein content and a lower fermentative activity. It contains approximately 10 wt % added starch for improved shelf life. Most of this type of compressed yeast has been replaced by active dry yeast.

Active Dry Yeast (ADY). In several applications, ADY has replaced compressed yeast, eg, where storage stability and convenience are key considerations. The yeast is packaged in hermetically sealed, aluminum-foil pouches for domestic use and in 0.45–0.91-kg cans for institutional use. The yeast is protected in a nitrogen gas atmosphere or under vacuum. For these uses, the integrity of the package is at least as important as the fermentative activity of the yeast in that a minimum storage life of one year can be guaranteed. Ten-kilogram quantities can also be vacuum-packed in foil pouches. Active dry yeast for use in wholesale bakeries is usually delivered in polyethylene-lined fiber drums and is not protected by an inert atmosphere. Accordingly, it should be used within 2–3 mo.

The active dry yeast that has been available in the United States for the past 35 yr and accounts for most of active dry yeast usage is referred to as regular active dry yeast. This yeast requires rehydration in water, preferably at 35–40°C, before addition to the other dough ingredients. A period of 5–10 min is adequate for full rehydration and optimum activity. Instant ADY does not require separate rehydration. It is always packaged in a protective atmosphere or under vacuum. On an equivalent solids basis, the fermentative activity of the instant ADY is greater than that of the regular ADY, but it is not as great as that of compressed yeast.

All active dry yeasts make doughs softer, more extensible, and less elastic. The slackening effect is more pronounced with the regular ADY than with the instant ADY. Slackening can be counteracted with higher concentrations of oxidants.

Brewers' Yeast. The basic raw materials for the production of beer are sweet worts formed by enzymatic hydrolysis of various cereal starches. The principal cereal is barley which, after malting, is also the source of enzymes that hydrolyze starches, glucans, and proteins. In some European countries, eg, the FRG, the mash bill consists entirely of malted barley. In other countries, adjuncts, eg, corn, rice, corn syrup, or glucose, are common. In the United States, the mash bill typically contains 60 wt % barley malt and 40 wt % corn or rice. In the UK, sugar syrups are often used.

Malt is produced by steeping barley in water for several days. This initiates germination, and during sprouting, several enzymes are activated. These are important in the digestive process that occurs in malt. The germinated barley is dried and then kilned at ca 80°C or at higher temperatures for very dark malts.

Brewers' wort is made by mashing barley malt or malt and adjuncts with water. In the infusion method, the temperature of the mash is gradually raised to 60–65°C. Heating is gradual, and the mash is maintained at ca 52°C to permit proper action by proteolytic enzymes. At the final temperature of 60–65°C, amylases catalyze the hydrolysis of starches to maltose, oligosaccharides, and dextrins. Adjuncts are generally heated to boiling which gelatinizes their starch. This can be done because the adjuncts contain no essential enzymes that must be preserved for mashing. Gelatinized starch is readily hydrolyzed by the amylases of the malt mash. The boiled adjuncts are added slowly to the malt mash to raise its temperature in the desired sequence. Finally, the mash is filtered to separate the insoluble constituents, eg, barley hulls, from the soluble constituents, ie, the extract. The filtrate or sweet wort is then pumped into the brew kettles, where the wort is boiled with hops. This inactivates the enzymes, extracts the bitter hops flavor, and sterilizes the wort.

Brewery fermentations are carried out with pure-culture yeasts. The yeasts are grown in the brewery from a pure culture on a slant of agar-fortified sterile wort (see Fig. 6). Often, yeast from a preceding commercial beer fermentation is inoculated into one or several succeeding fermentations. The fermentations are not carried out under strictly aerobic conditions. Notwithstanding some aeration with sterile air, the yeasts' metabolism is not respiratory because the sugar concentration is too high. During the aseptic propagation of the seed yeast, there is a tenfold increase in cells from one step to the next. Stock cultures of brewers' yeasts may be maintained on solid media, ie, agar-fortified malt extract, yeast extracts, glucose, and peptone, or in a lyophilized form.

Two species of pure-culture yeasts are used in brewing: *S. cerevisiae*, the top-

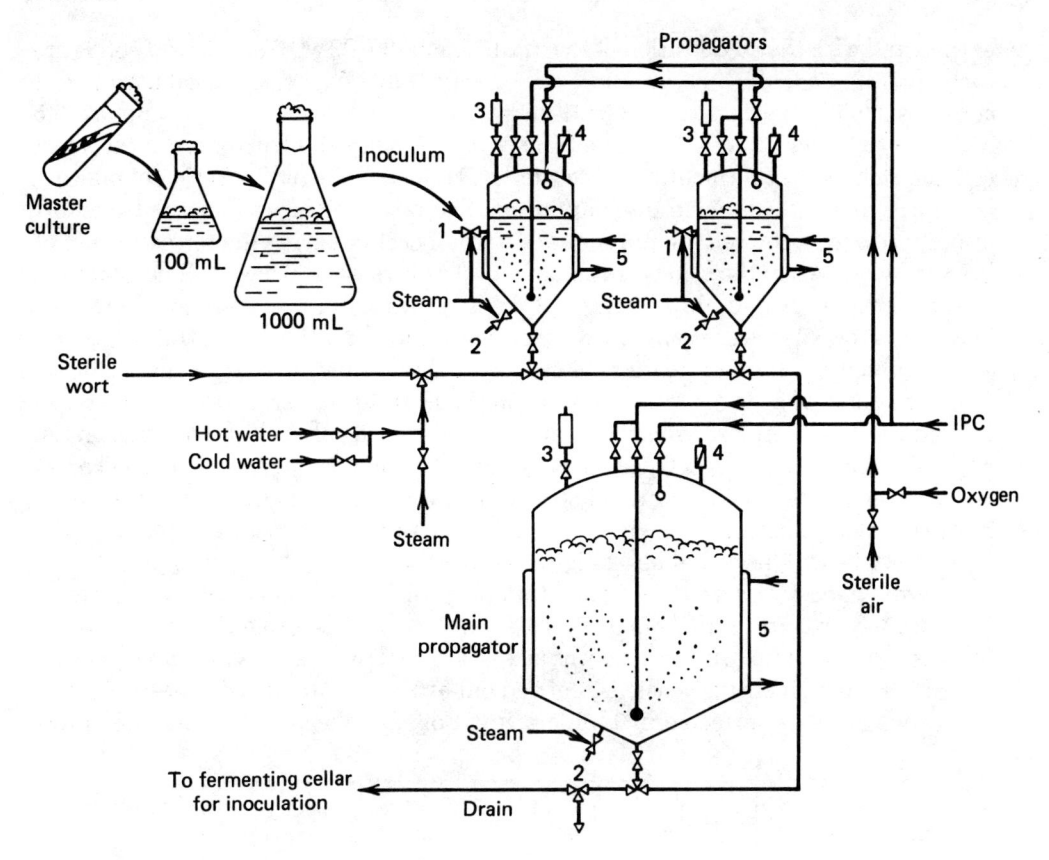

Figure 6. Pure yeast culture propagation system (39). 1, Inoculating cock; 2, sampling cock; 3, vent; 4, safety valve; and 5, attemperation.

fermenting yeast, is used for the production of ale, and *S. uvarum* (synonym *S. carlsbergensis*), the bottom-fermenting yeast, is used for the production of lager beer. The yeasts are indistinguishable microscopically. The presence of melibiase activity characterizes *S. uvarum*. The *Saccharomyces* strains, which ferment glucose, maltose, fructose, and raffinose completely and which form in malt agar cultures round-to-oval cells with a ratio between length and width of 1–2, are represented in *S. carlsbergensis* (40–41). The presence of melibiase (α-galactosidase) in *S. uvarum* permits complete fermentation of the trisaccharide raffinose, whereas *S. cerevisiae* ferments only the fructose portion. This diagnostic feature is of no importance for brewing, since worts contain little or no raffinose. In general, the top-fermenting yeasts have stronger respiratory systems than the bottom-fermenting yeasts. The former are also often less flocculant.

The suitability of brewers' strains depends largely on the rate and extent of growth, rate and extent of fermentation, degree of flocculence, and influence on the flavor and aroma of beer. Fermentation rates of brewers' yeasts are of the same order of magnitude as wine, bakers', distillers', or sake yeasts. For instance, in glucose and maltose media at 25°C, fermentation rates are 600 and 700 mL of CO_2 per hour per gram of yeast solids, respectively (42). In a wort at 25°C, the rate is 265 (mL/hr)/g (43). The concentration of fermentable sugars in worts varies with the concentration of the extract and with the composition of the mash. About 90–92 wt % of the wort solids are carbohydrates, and of these, 70–80 wt % are fermentable sugars. A typical barley wort contains 3.5–6% maltose, 1.1–1.8 wt % maltotriose, 0.5–1.0 wt % glucose, and 0.1–0.5 wt % fructose and sucrose (44). The transport of glucose and fructose into the yeast cell and their rate of fermentation are similar. However, the fermentation of maltose is subject to adaptation and de-adaptation. The rate-limiting step is the transport of maltose across the cell membrane. Adaptation occurs during the fermentation of brewers' worts when the concentration of glucose decreases ≤0.6 wt %. The synthesis of the transport system for maltose uptake is repressed in the medium by glucose; however, a few strains are constitutive maltose fermentors. However, strains that are grown on maltose ferment maltose as rapidly as, and sometimes faster than, glucose (45). The changes in sugar concentrations in the wort during fermentation are shown in Figure 7. Maltotetraose, isomaltose, and higher glucose polymers are not fermented. Maltose fermentation takes place after the concentration of glucose has reached a low level, and maltotriose fermentation takes place only after the concentration of maltose has reached a low level. Some strains do not ferment maltotriose, and consequently, beer brewed with this strain has a high concentration of this trisaccharide.

Fermentation. Lager beer fermentations with *S. uvarum*, which is a bottom-fermenting yeast, are carried out at 10–16°C for 6–10 d. This fermentation is followed by a lager period of up to several months at 2–5°C. Ale fermentations with *S. cerevisiae*, which is a top-fermenting yeast, are carried out at 15–22°C for 3–5 d. The rate of fermentation is greatly affected by temperature. For example, at 25°C, the rate of fer-

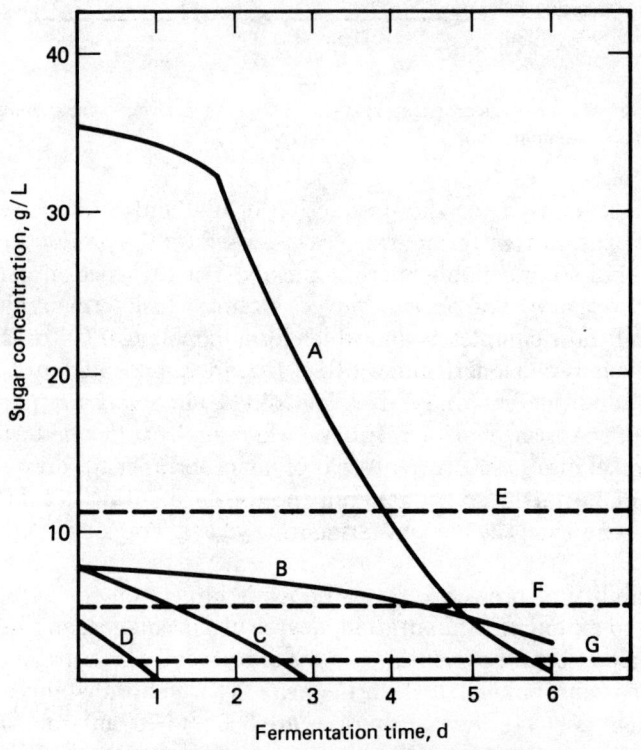

Figure 7. Concentration of various sugars in the primary fermentation of lager beer (46). A, Maltose; B, maltotriose; C, glucose; D, sucrose; E, glucosans; F, maltotetraose; and G, isomaltose.

mentation is about three to four times faster than at 10°C. The pH drops during beer fermentation from ca 5.5 to ca 4.3–4.5. However, this does not greatly affect the rate of fermentation. At pH 3.5–5, the rate of maltose fermentation is almost constant, but it drops sharply at pH values outside that range. The optimum pH for maltotriose fermentation is 4.3–5.4. Yeast fermentations and growth are inhibited by high concentrations of ethanol and dissolved carbon dioxide. The alcohol content of beer may vary from 3.4–3.8 wt % in U.S. lager beer to 8.7 wt % in British stout. At ethanol levels up to 5–6 wt %, the inhibiting effect is slight. At CO_2 concentrations of 0.4–0.5 wt %, the inhibiting effect is negligible.

Brewers' fermentations usually contain $(5-10) \times 10^6$ yeast cells per milliliter at first. Depending on the size of the cells, this may correspond to ca 0.75 g yeast solids/L. During fermentation, there is considerable growth resulting in a four-to-eightfold increase in cell mass. The basic carbon source for this growth is carbohydrate. Some oxygen is supplied at the start of the fermentation by aeration of the wort. The nitrogen source is amino acids of the wort. Some amino acids, eg, asparagine, serine, threonine, and lysine, are assimilated rapidly; others are absorbed more slowly. Proline is not absorbed at all, and the amino acid composition of the final beer is relatively high in proline. Brewers' worts usually contain sufficient minerals and vitamins to support yeast growth.

Flocculation. The degree of flocculence of a brewers' yeast is very important. This is true for both top- and bottom-fermenting yeasts. If the yeast is too flocculent, a large portion rapidly settles, thereby delaying fermentation. If it is too powdery, the fermentation is rapid but the yeast does not settle at the end of the fermentation.

Flocculation results from the clumping or aggregation of yeast cells. Such clumps, eg, of lager beer yeasts, settle in the fermentor and some, eg, ale yeasts, rise to the top. Flocculence of yeast strains is genetically determined, but it is also greatly influenced by physical and chemical factors, eg, the flocculating effect of divalent cations (Ca and Mg); the pH value, temperature, and electric charge on the cell surface; and fimbriae (hairlike appendages) on cell walls of some yeast strains.

Flavor and Aroma. The flavor compounds, which are by-products of the alcoholic fermentation, are higher alcohols (fusel oils), esters, diketones, aldehydes, organic acids, and sulfur compounds. Although many of these compounds can be determined analytically, it is not always possible to determine their effect on flavor. For example, all of the higher alcohols are present in concentrations below the threshold of perception of individual compounds. However, the combination of the higher alcohols affects flavor. Their total concentration may be 50–100 ppm. About 50–66 wt % of this fraction consists of isoamyl alcohol and, in decreasing concentration, active amyl alcohol, isobutyl alcohol, and propanol. The amounts of amyl and butyl alcohols are negligible. The yeast strain has a large effect on the formation of higher alcohols; the same is true for esters. The concentration of esters in U.S. lager beer is 25–50 ppm. Ethyl acetate, isoamyl acetate, and phenethyl acetate probably contribute a fruity flavor to beer. Ethyl acetate forms from the reaction of acetyl coenzyme A with ethanol, and its formation depends greatly on temperature. Maximum formation occurs at 20–25°C. Therefore, top-fermented beers contain higher amounts of this ester.

Diacetyl, acetoin, and diketones form during yeast fermentation. Diacetyl has a pronounced effect on flavor, with threshold of perception values at 0.1–0.2 ppm. At 0.45 ppm, it gives a cheesey flavor. The formation of diacetyl during lager beer fermentation is shown in Figure 8 (47). It probably forms by the decarboxylation of

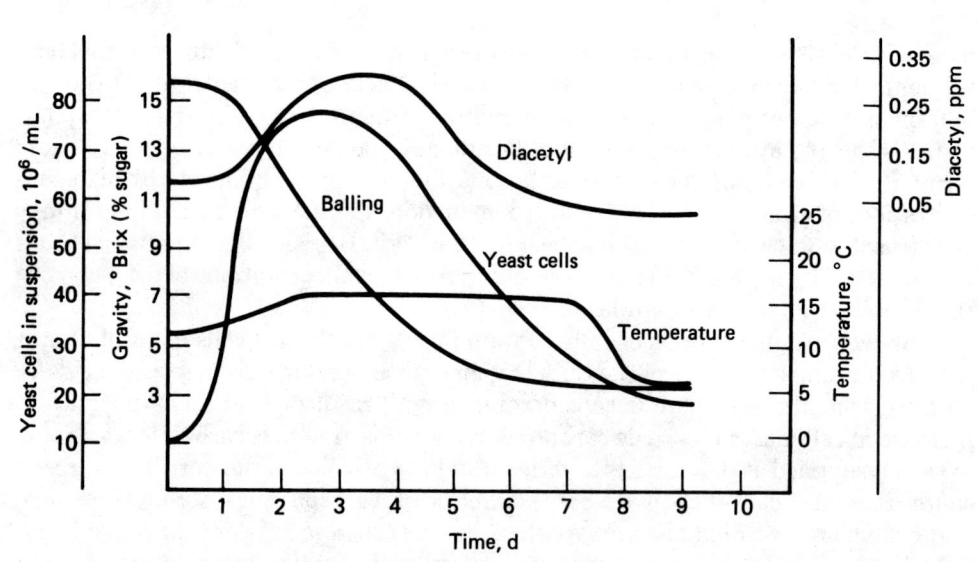

Figure 8. Some characteristic parameters of a typical high gravity fermentation (47).

α-ethyl acetolactate to acetoin and oxidation of acetoin to diacetyl. Yeast cells in suspension irreversibly reduce diacetyl to acetoin by diacetyl reductase. U.S. lager beer has a very mild flavor and generally has lower concentrations of diacetyl than ale. Concentrations of aldehyde are usually 10–20 ppm. Aldehyde effect on flavor must be minor, since the threshold perception value is ca 25 ppm.

Organic acids and carbon dioxide lower the pH of beer during fermentation. The principal acids formed are lactic, pyruvic, citric, malic, and acetic acids. The concentration of the individual acids is 100–200 ppm. Some of the sulfur compounds which form during fermentation may have a desirable flavor effect at extremely low levels. At higher levels, their presence is undesirable. The main volatile sulfur compounds and their taste detection threshold values are as follows: H_2S, 5–10 ppb (parts per billion (10^9)); ethanethiol, 5–10 ppb; dimethyl sulfide, 35–60 ppb; and diethyl sulfide, 3–30 ppb. Sulfur dioxide also forms during fermentation at 5–50 ppm. Its presence at levels above 50 ppm can be tasted.

The taste of beer can be spoiled by contaminating bacteria or yeasts. The most common spoilage bacteria are lactic acid producers, acetic acid producers, and *Zymomonas* species. Wild yeasts are species other than the pure-culture strain used. For example, *S. uvarum* is a wild yeast in ale fermentations, whereas *S. cerevisiae* is classified as wild in lager beer fermentations. Other common spoilage yeasts are *S. diastaticus* and species of *Pichia*, *Candida*, and *Brettanomyces*. Yeasts also affect beer flavor by partial adsorption of bitter substances, ie, humolones and isohumolones, extracted from hops during kettle boil.

Kaffir Beer. The best known beer native to Africa is Kaffir beer, the traditional beverage of the Bantus of South Africa. Kaffir beer is brewed from sorghum and corn. Conversion to fermentable sugars is commonly achieved by the addition of malt. Enzymatic conversion and fermentation proceed simultaneously as they do in distillers' mashes. The final beer is neither filtered nor sterilized, and it contains the yeast and the insoluble grain residues. Its keeping quality probably results from its low pH of 3.5 (33,48).

Wine Yeasts. Wine yeast describes many genera that occur naturally in grape musts and that participate in spontaneous fermentation in wine. The number of yeast cells on intact young grape berries is not large; however, it increases with increasing maturity of the grapes, and there are many yeasts on injured grapes where the grape juice is exposed. Nonetheless, most yeast cells that participate in spontaneous fermentations are probably derived from yeasts adhering to crushers, presses, and other cellar equipment.

Spontaneous Fermentations. The percentages of various yeast species occurring in must before, during, and after completion of fermentation are listed in Table 8 (49). Compilations by other authors often show a different pattern, but in general, the start of the fermentation is dominated by *Kloeckera apiculata*, *Metschnikowia pulcherrima*, and *Torulopsis stellata*. *Saccharomyces rosei* appears between the start of the fermentation and the main fermentation period. The main fermentation is dominated by *S. cerevisiea* and *S. uvarum*, and the end of the fermentation by *S. cerevisiae* and *S. bayanus* (50). The succession of species during the fermentation depends largely on their alcohol tolerance. As the alcohol concentration increases, the less tolerant yeasts die.

Spontaneous fermentations are used for the production of wines in France, in some other European countries, and in South America. They generally start more slowly than inoculated fermentations (51). They also make control of the fermentation difficult and may encourage the growth of undesirable contaminants. However, it has been claimed that spontaneous fermentations produce wines with better organoleptic properties, in particular, more complex flavor and aroma.

Pure-culture yeasts are strains of *S. cerevisiae* and *S. bayanus*, which are available as pure cultures on slants or in lyophilized form. These yeasts are propagated in wineries in equipment similar to that shown in Figure 9. Such equipment has been described by several authors (52–53). The requirement for complete asepsis is not as stringent as for the propagation of pure yeast cultures in breweries because of the lower

Table 8. Occurrence of Yeast Species during the Fermentation of White Wines, % of the Wines Tested[a]

Yeast	Before fermentation (213 samples)	Half-fermented (208 samples)	End of fermentation (165 samples)
Kloeckera apiculata	36.1	0	0
Torulopsis bacillaris	23.4	1.0	0
Saccharomyces cerevisiae (*S. vini*)	31.4	87.5	50.3
Saccharomyces oviformis (*S. bayanus*)	0.5	2.4	38.1
Saccharomyces chevalieri	0.5	3.8	3.6
Saccharomyces fructuum	0.9	0	0.6
Saccharomyces carlsbergensis	2.3	1.0	0
Saccharomyces uvarum	0.5	0	0
Saccharomyces steineri	2.3	1.9	1.2
Saccharomyces heterogenicus	0	0.5	1.8
Saccharomyces acidifaciens	0.5	0.5	1.2
Saccharomyces elegans	0	0.5	0
Saccharomyces veronae	0	0.5	0
Saccharomyces rosei	0.9	0.5	2.4
Torulopsis delbrueckii	0.5	0	0

[a] Ref. 49.

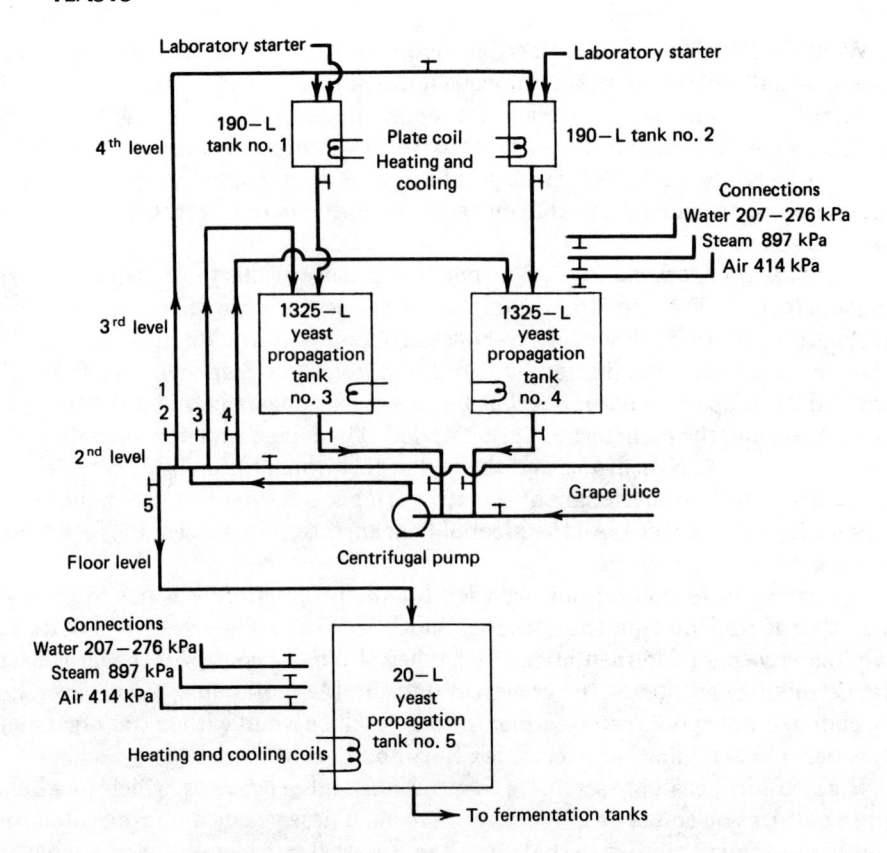

Figure 9. Schematic for wine yeast propagation with grape must (52). To convert L to gal, divide by 3.785; to convert kPa to psi, multiply by 0.145.

pH of must (pH 3–4) and because the final wine fermentations are not carried out under pure-culture conditions.

It is customary to add up to 200 ppm SO_2 to musts to inhibit the natural flora of the must. Strains of *S. cerevisiae* and *S. bayanus* may be acclimatized to fairly high concentrations of SO_2 by growing them in the presence of sulfite. The cultures that are propagated in the winery are added in liquid suspension to musts for the production of wines, usually at ca 1–2% of the must volume. The use of pure cultures is advantageous in that many strains are available. Such properties as degree of flocculence, absence of foaming, fast fermentation, no H_2S formation, no SO_2 formation, resistance to SO_2 and other inhibitors, and production of particular flavor compounds may then be selected. However, there are no ideal strains possessing all of the desired properties.

The production and use of active dry wine yeasts began in the 1960s. Such yeasts are produced by companies producing bakers' yeast and by methods resembling those used for bakers' yeast production, ie, the yeasts are grown aerobically under conditions of carbohydrate limitation (54). The dry yeasts have excellent storage stability, and if they are packaged under an inert atmosphere, ie, N_2, CO_2, or vacuum, their shelf life is one year or longer. They also can be acclimatized to high levels of SO_2. Their use in the United States and Australia is common. These yeasts have also been introduced successfully into the European winemaking countries. About ten strains of

S. cerevisiae, five strains of *S. bayanus*, and one strain of *S. fermentati* are commercially available. Although the total number of strains is smaller than is available as liquid, pure-culture yeasts, it satisfies the needs of the industry.

A comparison between wines produced by spontaneous fermentation and those produced by inoculation with selected yeasts, either as liquid cultures or as active dry yeasts, is not very meaningful since wines made by spontaneous fermentation are variable in composition as well as in their organoleptic properties. However, a comparison between wines made with liquid, pure-culture yeasts and active dry yeasts of the same strain is meaningful and is presented in Table 9, which indicates that the composition of the wines is identical. The table also shows the assays that are routinely done in wineries to ascertain the composition of wines. Compounds other than those shown contribute greatly to the flavor of wines.

Saccharomyces. *Saccharomyces* sp are strong fermentors. Their cells are generally ovoid to spherical, elongated, or ellipsoid. Sporulation can be induced and the asci have 1–4 spores. They propagate vegetatively by multilateral budding. *S. cerevisiae* (*S. vini*) with ellipsoidal cells, which measure 3–7 µm by 4–12.5 µm, is the most frequently encountered wine yeast. It may produce 18–20 vol % ethanol under suitable conditions, but only one half of the 44 strains tested produced >14 vol % ethanol (56). Another important species, *S. bayanus* (*S. oviformis*) is an equally good producer of alcohol. Several strains produced >19 vol % ethanol (57). *S. uvarum* (*S. carlsbergensis*) includes wine strains and strains used in the production of lager beer. The wine strains occur frequently during the main fermentation and are then superseded by *S. cerevisiae* and *S. bayanus*. *S. rosei* forms lesser concentrations of ethanol (6–8 vol %) and, like *S. uvarum*, often occurs during the main fermentation. *S. chevalieri* has properties similar to *S. cerevisiae*. *S. italicus*, a less frequently occurring yeast, also produces high concentrations of ethanol, but more slowly than *S. cerevisiae*. Both *S. chevalieri* and *S. italicus* are considered synonymous to *S. cerevisiae*.

Table 9. Analysis of Wines Made from the Same Must with the Same Strain of Yeast as a Liquid Culture and as Active Dry Yeast[a]

	70-13 Strain[b] Liquid	70-13 Strain[b] Dry	70-13 Strain[c] Liquid	70-13 Strain[c] Dry
density, g/L	999.7	999.7	995.8	995.8
alcohol concentration, vol %	11.29	11.38	11.58	11.58
residual sugars, g/L	1.50	1.50	1.37	1.35
volatile acidity (as H_2SO_4), g/L	0.14	0.19	0.23	0.22
total acidity (as H_2SO_4), g/L	4.60	4.70	3.40	3.30
tartaric acid, meq/L	51.3	51.0	34.3	34.7
malic acid, meq/L	44.5	43.4		
pH	3.5	3.5	3.63	3.67
glycerol, g/L			7.4	7.9
SO_2, total, mg/L			71	41

[a] Initial must: 192.1 g sugar per liter; addition of 50 mg SO_2/L; fermented in 500-L vessels; inoculated with 4.1×10^6 cells/mL with liquid-yeast culture or 7.5×10^6 cells/mL with active dry yeast (55).
[b] Grapes crushed on Sept. 30.
[c] Grapes crushed on Mar. 11.

Kloeckera. *Kloeckera* is the imperfect form of *Hanseniaspora*. The small cells are pointed or lemon-shaped and show bilateral budding. They do not form ascospores. *K. apiculata*, with cell dimensions of (2–4.5) μm $\times$ (5–8) μm, often dominates the early phases of spontaneous wine fermentations. It stops growing and fermenting at ethanol concentrations of 4–5 vol %. Other species in musts are *K. corticis* and *K. africana*.

Metschnikowia. The only important *Metschnikowia* species occurring in wines is *M. pulcherrima* (*Candida pulcherrima*). It propagates by multilateral budding or by formation of large asci with needle-shaped ascospores. It forms less than 4 vol % ethanol. With *K. apiculata* and *T. stellata*, it dominates the early stages of spontaneous fermentation.

Torulopsis. The common species in musts is *T. stellata* (*T. bacillaris*), which does not form sexual spores. The cells are oval to elongated oval and propagate vegetatively by multilateral budding. Its alcohol-producing ability varies.

Pichia. The most prominent species, *P. membranefaciens*, is a film-forming yeast. It is a strict aerobe and does not produce ethanol, but it tolerates high concentrations of this alcohol. The cells are ellipsoid to cylindrical with a cell size of 2.5–4.5 μm $\times$ 5–14 μm. It reproduces vegetatively by multilateral budding or sexually by formation of 2–4 spores. It grows on the surface of young wines and forms a continuous film.

Hansenula. *Hansenula* yeasts have a medium cell size, ie, 2.5–5 μm $\times$ 5.5–20 μm. They propagate by multilateral budding or by forming 2–4 spores, which are hat-shaped or Saturn-shaped. Some species, eg, *H. anomala*, may produce up to 10 vol % ethanol and large amounts of ethyl acetate and other esters.

Candida. The asporogenous cells propagate by budding. A pseudomycelium and sometimes a true mycelium may form. *C. krusei*, *C. vini*, and *C. valida* are wine-spoilage, film-forming yeasts. The last two species used to be called *C. mycoderma*.

Saccharomycodes. *Saccharomycodes ludwigii*, the only species of the genus, has lemon-shaped, 5–8 μm $\times$ 10–30 μm cells. The ascosporogenous cell produces four spherical spores. It is a wine-spoilage yeast and is highly resistant to SO_2.

Schizosaccharomyces. *Schizosaccharomyces* species reproduce vegetatively by fission, not by budding, as a wall forms across the middle section of the elongated or cylindrical cells. The yeasts also produce spores. *Schiz. pombe* is noted for its ability to degrade L-malic acid. Under anaerobic conditions, malic acid is degraded to ethanol and CO_2. The yeast is not often used in spontaneously fermenting musts but has been used as a pure-culture yeast.

Brettanomyces. *Brettanomyces* cells are ovoid or pointed at one end. The asporogenous yeast propagates by budding. *B. intermedius* is a wine-spoilage yeast. A strong characteristic odor forms during fermentation, and under aerobic conditions, the yeast produces large concentrations of acetic acid. It can tolerate up to 12 vol % ethanol but is very sensitive to SO_2.

Distilled Beverages. Distilled alcoholic beverages are made by fermentation of sugars from grains or fruits followed by distillation. The cereal grains used for whiskey are wheat, corn, and rye. Potatoes are used for the production of vodka or neutral spirits, molasses for rum, and grape or other fruit juices for brandy. The most significant difference is between the starchy cereal grains or potatoes, whose starch must be converted to fermentable sugars, and the sugar-containing substrates.

With the exception of some Scotch and Irish whiskeys, the distilled beverage industry ferments whole mashes of grains; that is, fermentation proceeds simultaneously with the conversion of starch to fermentable sugars. For preparation of the mash, the cereal grains, except barley malt, are cooked with water to gelatinize the starch. The cooking temperature often exceeds 100°C in pressurized equipment. The mash is then cooled to ca 63°C, and a slurry of barley malt is added. The amount of malt is ca 5–10% of the total mash. Malt is a source of amylases for conversion of the starch to fermentable sugars, and it gives the whiskey some flavor. Fungal enzymes are sometimes used for the conversion (3). The combined mash is further cooled to 20–25°C in preparation for inoculation with yeast.

Distillers' yeast is made up of selected strains of *S. cerevisiae*. Many distillers use their own, proprietary strains. Commercial bakers' yeast may be used for the production of grain neutral spirits. Yeast propagation starts with pure-culture slants. It is carried through several stages of propagation with increasing volume and with regular, cooled distillers' mash as the medium. Inoculation into commercial fermentors is at 3–5 vol %. Backstocking is the inoculation of a fermentor with portions of an actively fermenting mash. In the UK, residual brewers' yeast is sometimes used for the fermentation.

Distillers' fermentations are generally set at pH 4.8–5.0; the pH should not drop below 4.0 during the fermentation. The set temperature should be sufficiently low so that the final temperature remains below 30°C, and the fermentation time may be 2–5 days. The latter stages of the fermentation are very slow because starch conversion is the limiting factor after ca 30 h (58). Alcohol concentrations at the end of the fermentation may be 7–9 vol % (59–60).

Certain lactic acid-producing bacteria are common contaminants in distillers' fermentations. *S. lactis* may produce excessive amounts of volatile fatty acids. Some *Lactobacillus* species convert glycerol to β-hydroxypropionaldehyde. During distillation, this breaks down to acrolein, which gives the distilled beverage an acrid odor.

Fermentation of fruit juices other than from grapes, principally apple juice, is the same as that in the production of grape wine. For the production of rum, cane molasses is diluted to a sugar concentration of 15–20%. The sucrose, glucose, and fructose of such diluted molasses is completely fermented within 36 h at 30–32°C.

Microbial Biomass and Single-Cell Protein

In all fermented foods, microbes contribute to the food supply by their preservative action, eg, by lowering pH and producing ethanol, or by making such foods more palatable. However, the use of microbes as a direct source of food started with the drying of spent brewers' yeast in ca 1910 and with the production of *Candida utilis* yeast during the Second World War. There are, however, some notable exceptions: mushrooms (macrofungi) have always been used as food; algae have been used as part of the human diet by the Aztecs of Central America; and small beer, ie, the sediment of beer, has been used as a vitamin supplement for infants (see Foods, nonconventional).

During the 1960s and 1970s, the world protein shortage stimulated the production of microbial biomass from traditional substrates, eg, carbohydrates, and from alternative raw materials, eg, *n*-paraffins, ethanol, and methanol. Present industrial processes for the production of biomass use yeast cultures (see Fuels from biomass). Exceptions are the methanol-based, aerobic process for growing *Pseudomonas methy-*

lotropha in the UK and the production of the fungal biomass *Paecilomyces varioti* on sulfite liquor in Finland. In general, yeasts are preferred because their cell size is larger than that of bacteria and permits recovery by centrifugation, and also because of their historically safe use in the food industry.

Substrates and Nutrient Requirements. Traditional carbon sources for the production of biomass are fermentable sugars. The preferred source is beet or cane molasses because of their low cost and because they contribute some nitrogenous nutrients, minerals, and vitamins. Starchy materials, ie, corn, rice, potatoes, and cassava, may be used only after the carbohydrate has been hydrolyzed to fermentable sugar. Cellulose may be used as a carbon source after acid or enzymatic hydrolysis to glucose. Alkanes of sulfite waste liquor, ethanol, and methanol are used. These materials serve as the principal carbon and energy sources for the growth of yeasts. In addition, yeasts require large amounts of nitrogen, phosphorus, potassium, calcium, and magnesium as well as trace minerals. The inorganic nutrient requirements for *C. utilis* are listed in Table 10. The carbon source, ie, ethanol, for this particular fermentation contributes neither vitamins nor minerals. The vitamin requirements of different yeasts vary. For example, bakers' yeast (*S. cerevisiae*) requires the addition of biotin, which is usually contributed by cane molasses, whereas *C. utilis* does not require biotin.

Fermentation. Yeast biomass need not be produced under conditions of complete sterility. Most yeast fermentations are carried out at pH ≤4.5, which limits bacterial growth. The most common contaminants are lactic acid-producing bacteria and vinegar organisms. All of the processes designed specifically for yeast growth are highly aerobic and thus require costly heat removal. The supply of oxygen in the form of air is also costly, since oxygen-transfer rates from the gas to the liquid are low.

The simplest fermentor is a cylindrical vessel equipped with air-sparger tubes at the bottom, internal cooling coils, and perhaps an agitator. A schematic of a loop fermentor, in which directional flow of the fermentor liquid is induced by draft tubes, is shown in Figure 10. The energy requirement for cooling and aeration is high. In many industrial fermentations, the rate of aeration is one volume of air per volume of fer-

Table 10. Inorganic Nutrient Requirements for *Candida utilis* Grown on Ethanol[a]

Nutrient element	Nutrient input per 100 grams of cells produced	Typical source
Macronutrients, g		
phosphorus	2–4	H_3PO_4
potassium	2–3	KCl
magnesium	0.3–0.6	$MgCl_2 \cdot 6H_2O$; $MgSO_4$
calcium	0.001–0.2	$CaCl_2$
sodium	0.01–0.2	Na_2CO_3; NaCl
Micronutrients, mg		
iron	6–13	$Fe(CH_3COCH_2CO)_3$
manganese	4–8	$MnSO_4 \cdot H_2O$
zinc	2–6	$ZnSO_4 \cdot 7H_2O$
molybdenum	1–2	$NaMoO_4 \cdot 2H_2O$
iodine	1–3	KI
copper	0.5–1	$CuSO_4 \cdot 5H_2O$

[a] Ref. 61.

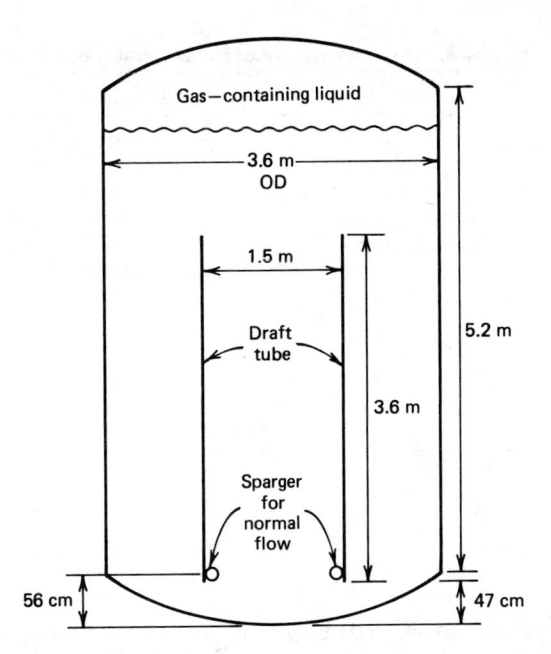

Figure 10. Air-lift fermentor with 1.5-m draft tube; the vessel is air-sparged, and there is no agitator (62).

mentor liquid per minute. A high aeration rate lowers the density of the gas-containing liquid and favors the formation of foam, and this restricts the useful liquid volume of the fermentor.

Continuous processes are used for the production of yeast biomass. Raw material as liquid feed is added continuously to the fermentor and an equal amount of fermentor liquid is withdrawn for harvesting of the yeast cells. These are homogeneous fermentations in a stirred fermentor or in two fermentors in series. The growth rate in such fermentors is high. A typical dilution rate for the production of *C. utilis* on sulfite liquor is 0.25, ie, one-fourth of the fermentor volume is withdrawn hourly. The temperature is usually held at 30°C.

Nutritional Value. Microbial biomass is either recovered as a by-product of a food fermentation, eg, of beer, or grown specifically as a food or feed supplement. In either case, its value is mainly in its contribution to protein nutrition. There is an additional contribution of minerals and vitamins, but this is less important. Yeasts contain 7.5–9 wt % nitrogen, ie, N × 6.25 = 47–56% crude protein; 6–12 wt % nucleic acids; 5–9 wt % ash; and 2–6 wt % lipids (63).

The value corresponding to N × 6.25 may be called crude protein since it includes the nucleic acid fraction, which accounts for ca 12–15 wt % of the crude protein. Table 11 is a list of the amino acid compositions of three yeast species used for food or feed. The essential amino acids compositions are similar, ie, a high lysine content and a low content of sulfur-containing amino acids. The protein efficiency ratio of bakers' yeast is 2.02. The addition of 0.16 wt % D,L-methionine increases this value to 2.27, and that of 0.5 wt % D,L-methionine to 2.77. These figures are based on the weight of true protein solids and on a protein efficiency ratio (PER) for casein of 2.5. A protein fraction isolated from bakers' yeast and essentially free from nucleic acids has a PER of 2.2 (65). The protein efficiency ratio is the ratio of the weight gain of young rats to the weight of protein fed.

Table 11. Amino Acid Composition of Commercial Yeast Biomass Products, g/g Nitrogen

Component	C. lipolytica[a]	C. utilis[b]	S. cerevisiae[c]
lysine	0.46	0.448	0.527
valine	0.37	0.333	0.417
leucine	0.46	0.436	0.508
isoleucine	0.32	0.27	0.324
threonine	0.31	0.294	0.33
methionine	0.11	0.0638	0.102
cystine	0.69	0.038	0.10
phenylalanine	0.27	0.231	0.28
tryptophan	0.88	0.003	0.069
histidine	0.13	0.128	0.196
tyrosine	0.23	0.205	0.304
arginine	0.32	0.448	0.334

[a] Ref. 31.
[b] Ref. 61.
[c] Ref. 64.

Brewers' and bakers' inactive dried yeasts are used as dietary supplements. They contribute some protein and some B vitamins but no vitamin C, vitamin B_{12}, or fat-soluble vitamins. These yeasts also contain trace minerals. The discovery of selenium as an essential nutrient was first made with yeast (66). The glucose tolerance factor (GTF) in yeast is a chromium-containing organic compound and mediates the effect of insulin. It seems to be important for older persons who cannot synthesize GTF from the inorganic chromium in the diet (67). Also, the cell wall fraction of bakers' yeast reduces cholesterol levels in rats that are fed a hypercholesterolemic diet (68).

Production. Microbial biomass is frequently referred to as single-cell protein (SCP). However, it is best to refer to the entire microbial cell mass as microbial biomass and to reserve the designation SCP to the proteins isolated from it. The most widely available biomass is a by-product of the brewing industry. The multiplication of yeast cells during brewing results in a surplus of yeast cells. For every 117 L (barrel) of beer brewed, 0.2–0.3 kg of beer-yeast solids is recovered. In the United States, this results in the availability of ca 40×10^6 kg/yr of brewers' yeast. Most of this yeast is dried with the spent brewers' grains and is sold as feed. About 10,000 metric tons is dried separately as dried brewers' yeast. The yeast is recovered from the beer by centrifuging and is dried on roller drum dryers or spray dryers. The dried yeast, which is bitter from residual hops, is sold as animal feed or a pet-food supplement. It can be debittered by alkaline extraction of the bitter substances, and the product is marketed mainly by the health-food industry. It is available as tablets, powder, or flakes, and is often fortified with extra vitamins. Distillers' yeasts cannot be readily separated from the fermented mash and are sold with the spent distillers' grains. Wine yeast is not recovered.

Bakers' inactive dried yeast is also used widely in the health-food industry. It is possible to produce high levels of nicotinic acid and thiamine in bakers' yeast during growth. Since this yeast may be grown as a nutritional yeast, there is greater choice in its composition than in that of brewers' yeast. For instance, its crude protein content can be raised to 50–55%, and it can be used as a vehicle for the incorporation of microelements, eg, selenium or chromium, into the human diet.

Candida utilis yeast is grown on waste sulfite liquor in western countries, on molasses in Cuba and Taiwan, and on cellulose acid hydrolysates in the USSR and other Eastern European countries. The carbon sources for *C. utilis* are hexoses, pentoses, and many organic acids. Sulfite liquor from hardwood trees contains 2–3 wt % fermentable sugars. Of these, 20% are hexoses and 80% are pentoses; in sulfite liquor from softwoods, this ratio is reversed. For production of this yeast, the sulfite liquor must be stripped of SO_2. The fermentation is carried out in very large, highly aerated fermentors. The dilution rate of the continuous fermentation is 0.27–0.30, and the fermentation is carried out at pH 4 and 30°C (69). The first U.S. installation of this process has been described (34,69–71).

Cheese whey contains 70–75 wt % milk sugar, which can serve as the carbon source for the growth of lactose-fermenting yeasts; *Kluyveromyces fragilis* is generally used for this purpose. The total volume of *K. fragilis* yeast produced is considerably smaller than that of the other yeasts, but it is a continuous process characterized by recycling of yeast cells (72).

Of the petrochemical substrates, both *n*-alkanes and gas oil can be used as carbon and energy sources. The yeasts that have been used commercially are *Candida tropicalis* and *Candida lipolytica* (62,73–74). The water-insoluble alkane is one of the phases in the fermentor. There are two immiscible liquid phases, ie, *n*-alkanes and water; a semisolid phase, ie, yeast; and a gaseous phase, ie, air. For yeasts grown on carbohydrates, yields are ca 50%, whereas yeasts grown on *n*-alkanes generally give yields of 95–115%, based on the weight of the alkane. The cells can easily be recovered by centrifugation. If the yeasts are grown on gas oil, they must be extracted with solvents to remove residual hydrocarbons. During the 1970s, some plants operated with *n*-alkanes as the substrate. The steep increase in oil prices and problems with regulatory agencies in several countries have stopped this development.

The following yeast genera contain species that are facultative methylotrophs: *Candida*, *Hansenula*, *Pichia*, *Torulopsis*, and *Trichoderma*. Growth substrates for such yeasts are methanol, carbohydrates, or organic acids. However, commercial production of biomass is also carried out with the bacterium *Methylophilus methylotropha* (75). Ethanol may be used as a carbon source for the production of certain species of the following genera: *Candida*, *Debaromyces*, *Endomycopsis*, *Hansenula*, *Mycoderma*, *Pichia*, *Rhodotorula*, and *Saccharomyces*. In the United States, *C. utilis* is grown on ethanol, and in Finland, ethanol is used with molasses for the growth of bakers' yeast (61).

One of the most promising substrates for future production of microbial biomass and single-cell protein is cellulose (qv) from agricultural residues, eg, wood pulp, sawdust, feed-lot waste, corn stover, rice hulls, nut shells, and bagasse. All of these materials contain cellulose as the principal carbon source (76–78). Cotton fibers with 90 wt % cellulose are one of the purest materials, whereas dicotyledenous leaves contain only ca 15–20 wt %. Woody residues and grasses generally contain mixtures of cellulose, hemicellulose, and lignin in varying proportions. The main problem preventing large-scale use of cellulose is the difficulty of converting it to glucose by enzymatic or acid hydrolysis. Acid hydrolysis, which is practiced in Eastern Europe, is costly because the acid must be neutralized and the resulting salts must be disposed of. Nevertheless, some investigators have found acid hydrolysis to be more economical than enzymatic hydrolysis (79–80).

Enzymatic hydrolysis with cellulase and cellobiase is the subject of intensive research (81). The most commonly used enzymes are cellulases and cellobiases derived from *Trichoderma reesei*. In many instances, 50% of the cellulose can be converted to glucose, and higher conversion rates can be achieved but require pretreatment of the cellulose with alkali, amines, or ammonia or by mechanical comminution (82). Newsprint is one of the more attractive raw materials for the production of glucose syrups, ethanol, and microbial biomass from the glucose (83–84). At present, it is not economical compared to traditional substrates, but progress in this area of research is expected.

Processing. Recovery of microbial biomass from the fermentor requires centrifugation, washing, concentration, and drying. *C. utilis* yeast grown on sulfite liquor contains ca 10 wt % lignosulfonic acids if it is not washed. This crude material is suitable for feed, but extensive washing results in a yeast suitable for food use.

Extraction of protein for the production of single-cell protein requires cracking of the cell wall to release the cytoplasmic contents. This can be achieved by high speed ball mills or colloid mills. For high pressure extrusion of cell suspensions, the most practical method is homogenization at high pressure (50–60 MPa (7250–8700 psi)) (85). This is usually followed by alkaline extraction and precipitation with acids. The alkaline extract may be adjusted to a mildly acidic pH and incubated to permit hydrolysis of the nucleic acids with nucleases. Precipitation of the protein at a more acid pH then results in a single-cell protein with a negligible concentration of nucleic acid (65).

Health and Safety Factors. The presence of nucleic acids in yeast is one of the main obstacles to their use in human foods (86). The ingestion of purines by humans and some other primates increases plasma levels of uric acid, which may cause gout. No more than two grams of nucleic acids should be in a daily human diet (87). This limits the daily intake of inactive dry yeasts by humans to ca 20 g/d. Other animals readily convert uric acid to allantoin, which is excreted in the urine.

The use of higher concentrations of yeast proteins in the human diet requires the removal of the nucleic acid fraction (88). A reduction of the nucleic acid content can be achieved by alkaline extraction followed by precipitation of the protein with acids. However, the yields of yeast protein by such procedures are low, and therefore, the cost of yeast protein may double. This and other procedures for the removal of yeast nucleic acids have been reviewed and reported (41,89–92).

Lysinoalanine, a nephrotoxic factor, has been determined in alkali-treated soybean protein (93). Since some processes for the extraction of yeast protein from whole cells involve an alkaline treatment, the possible presence of lysinoalanine must be considered (94). Treatment at a pH >8 at 100°C or at a pH >10.5 at 25°C leads to the formation of lysinoalanine. However, small concentrations of lysinoalanine occur in almost all heat-treated protein.

Small concentrations of tyramine (0.1–1.6 mg/g) and histamine (0.2–2.8 mg/g) have been determined in yeast autolysates. These form from the decarboxylation of the corresponding amino acids (95). Such compounds also form in other fermented foods. The use of yeast extracts as condiments limits the amounts in the diet to a very small percentage of the food intake.

Uses. Inactive dried yeasts are used as ingredients in many formulated foods, eg, baby foods, soups, gravies, and meat extenders; as carriers of spices and smoke flavor; and in baked goods. Inclusion in dough formulas generally results in greater extensibility of the dough because of the reducing action of sulfhydryl groups leached from the dried yeasts. Usually, yeasts used in the health food industry are fortified with higher concentrations of B vitamins, principally thiamine, riboflavin, and niacin.

There has been much research on extraction of the protein of yeasts and on their characterization (94). These proteins can be spun into fibers or texturized (96–98). However, such isolated yeast proteins are more expensive than isolated soy proteins, and there is no market for them.

Dried yeasts are used extensively in animal feeds. Most of the distillers' and brewers' yeasts are incorporated in feeds by codrying with spent grains. Approximately 15,000 t of brewers' yeast is dried as such and sold for feed use. The main outlet is for rations of monogastric animals. U.S. specifications for feed yeast are listed in Table 12 (99). Appreciable amounts of brewers' yeast and torula (*C. utilis*) are also used in pet foods, principally in dog and cat foods. They are also included in feeds for birds, fish, mink, and bees.

Yeast-Derived Products

Enzymes. Invertase (β-fructofuranosidase) is commercially produced from either *S. cerevisiae* or *S. uvarum*. Since the enzyme is not excreted by the yeasts, it must be isolated by autolysis of the cells. The autolysate is filtered, and the enzyme precipitates from the filtrate upon addition of ethanol or isopropyl alcohol. Invertase is available in dry form or in solution containing 50 vol % glycerol as stabilizer. The enzyme is used for sucrose hydrolysis in the production of high test molasses and in the production of cream-center candies (100).

Lactase (β-galactosidase) is produced commercially from a lactose-fermenting yeast, *Kluyveromyces fragilis*. The yeast enzyme has an optimal pH of 6–7 and is suitable for the hydrolysis of lactose in milk or skim milk (see Enzymes, industrial).

Table 12. Specifications for Feed Yeasts of the American Association of Feed Control Officials[a]

Product	Fermentative	Genus	Protein[b]	Comment
primary dried yeast	no	*Saccharomyces*	40%	separated from the mash or medium
active dry yeast	yes	*Saccharomyces*		15×10^6 line cells per gram; no added cereal or filler permitted
irradiated dried yeast	no	*Saccharomyces*		source of vitamin D_2, ie, calciferol
brewers' dried yeast	no	*Saccharomyces*	40%	nonextracted
grain distillers' dried yeast	no	*Saccharomyces*	40%	separated from the mash or medium
molasses distillers' dried yeast	no	*Saccharomyces*	40%	separated from the mash or medium
torula dried yeast	no	*Torulopsis* (now *Candida*)	40%	separated from the mash or medium
yeast culture	yes	*Saccharomyces*		yeast plus medium

[a] Ref. 99.
[b] Crude protein (N $\times$ 6.25).

Yeast Extracts. Autolysis in yeast cells can be induced by raising the temperature of a cell suspension to 45–55°C, at which the yeast cells die but their hydrolytic enzymes remain active. During the process of protein autolysis, carbohydrates and nucleic acid compounds are hydrolyzed and solubilized. Figure 11 shows the rate of protein solubilization and amino acid formation as a function of time. Commercial processes for yeast autolysis generally last ≥10 h. At the end of autolysis, the insoluble cell-wall material can be separated from the solubilized solids by centrifugation or filtration. The resulting extract is evaporated to a paste of ca 75% solids, or it can be spray dried to a powder. The powder contains 50–80% proteinaceous material consisting of peptides and amino acids; the latter is expressed as total nitrogen × 6.25. Amino nitrogen may account for 25–45 wt % of the total nitrogen.

Yeast extracts made from brewers' yeast (*S. uvarum* or *S. cerevisiae*), from bakers' yeast (*S. cerevisiae*), from alcohol-grown yeast (*C. utilis*), and from whey-grown yeast (*K. fragilis*) are commercially available. These extracts are used in almost all fermentation media for the production of antibiotics, in cheese-starter culture, in the production of vinegar, etc. They are also used widely in the food industry as condiments providing savory flavors for soups, gravies, bouillon cubes, and as flavor potentiators for cheese products (see Flavor and spices).

Nucleotides. The two nucleotides 5'-inosinic acid and 5'-guanylic acid, which are produced from yeast RNA, are potent flavor potentiators for meat products. They act synergistically with monosodium glutamate and are usually used in conjunction with this amino acid (see Amino acids, monosodium glutamate).

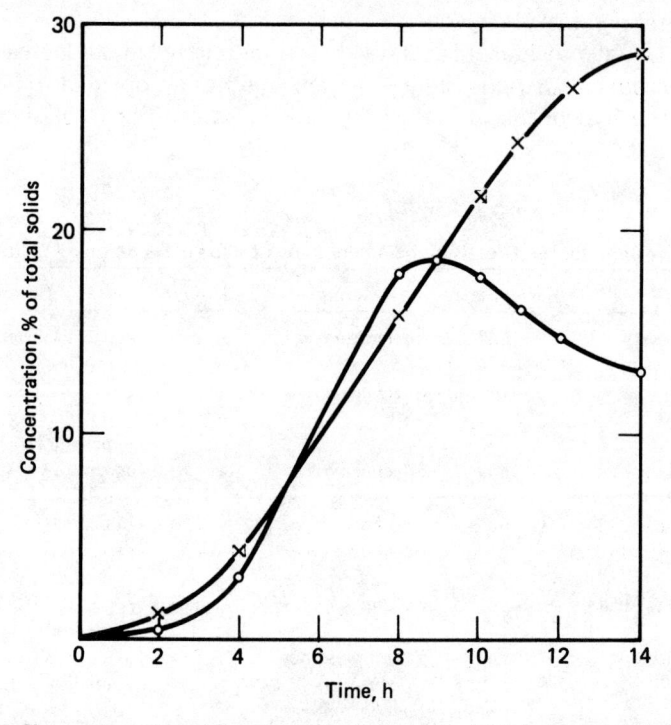

Figure 11. Changes in protein and amino acid levels in yeast extract during autolysis (101). O—O—O, Released protein as % of total yeast solids. x—x—x, released amino acids as % of total yeast solids. To convert kPa to psi, multiply by 0.145; to convert L to gal, divide by 3.785.

Food Preservation and Food Spoilage

Many foods are preserved by fermentation, including alcoholic beverages, pickles, cheese, and fish sauce. Usually, spontaneous fermentations by mixed populations of yeasts and bacteria are involved. The preservative effect results from a lowering of the pH or the formation of ethanol. Yeasts do not produce any potent antibiotics.

Yeasts also act as spoilage organisms. Jams, jellies, and honey can be fermented by osmophilic yeasts, eg, *S. mellis* or *S. rouxii*. Wild yeasts may also spoil wines or beers. Film-forming yeasts, eg, *P. membranefaciens*, may grow on the surface of fermented vegetables, such as sauerkraut or pickles. *K. fragilis* and other lactose-fermenting species frequently occur in milk products. Butter and oleomargarine are occasionally spoiled by *C. lipolytica*. Such spoilage always results in the development of undesirable flavors with serious economic losses.

Pathogenic Yeasts

Few yeasts are pathogenic. However, *Candida albicans* is the best known of the potentially pathogenic species. It may cause infections of the skin or of the mucous membranes, eg, oral cavity, intestinal tract, and vagina, called candidiasis or moniliasis. Occasionally, lesions of the skin may be caused by other species of *Candida*, eg, *C. tropicalis*. *Cryptococcus neoformis* may cause serious infections of a number of organs, particularly the meninges, and sometimes with fatal results. Some fungi can grow in the manner of yeasts at body temperatures and in laboratory cultures at 37°C but grow as mycelial fungi at lower temperatures. Of these, *Histoplasma capsulatum* invades the lungs and may spread to other parts of the body. The systemic disease, histoplasmosis, can be fatal (see Chemotherapeutics, antibacterial and antimycotic).

Outlook

During the second half of the 19th century and the early 20th century, yeasts have been used as experimental subjects for the elucidation of metabolic pathways. The fruit fly *Drosophila*, the bacterium *Escherichia coli*, and the yeasts *Saccharomyces cerevisiae* and *Neurospora crassa* have been the main species genetically studied. Yeasts will probably play a principal role in future genetic work. Commercial fermentations will be increasingly important in the production of foods, energy, and pharmaceuticals (qv). Continuous fermentations have been practiced for some time, but the continuous, heterogeneous fermentation with immobilized yeast cells is new and promises to bring greatly increased productivity to alcoholic fermentations (7,65,80) (see Enzymes, immobilized).

BIBLIOGRAPHY

"Yeasts" in *ECT* 1st ed., Vol. 15, pp. 195–219, by W. J. Nickerson, Institute of Microbiology, Rutgers University, and A. H. Rose, University of Birmingham, England; "Yeasts" in *ECT* 2nd ed., Vol. 22, pp. 507–554, by George I. de Becze, St. Thomas Institute for Advanced Studies.

1. J. Lodder, ed., *The Yeasts: A Taxonomic Study*, 2nd ed., North Holland Publishing Co., Amsterdam, The Netherlands, 1970.
2. A. Kockova-Kratochvilova, *Yeasts and Yeast-Like Organisms* (in Slovak), ALFA Publishing House, Bratislava, Czech., 1981.

3. A. Kockova-Kratochvilova in H. J. Rehm and G. Reed, eds., *Biotechnology*, Vol. 1, Verlag Chemie, Weinheim, FRG, 1981.
4. C. L. Cooney in ref. 3, p. 73.
5. W. N. Arnold, ed., *Yeast Cell Envelopes—Biochemistry, Biophysics, and Ultrastructure*, Vols. I and II, CRC Press, Boca Raton, Fla., 1981.
6. K. Esser and U. Stahl in ref. 3, p. 305.
7. A. Hinnen and co-workers in O. K. Sebek and A. I. Laskin, eds., *Genetics of Industrial Microorganisms*, American Society of Microbiology, Washington, D.C., 1979.
8. U.S. Pat. 3,993,783 (Nov. 23, 1976), A. Langejan and B. Khoudokormoff.
9. Belg. Pat. 862,191 (June 22, 1978), (to Lesaffre Cie).
10. H. J. Phaff, M. W. Miller, and E. M. Mrak, *The Life of Yeasts*, 2nd ed., Harvard University Press, Cambridge, Mass., 1978.
11. K. K. Kapoor, K. Chaudhary, and P. Tauro in G. Reed, ed., *Prescott and Dunn's Industrial Microbiology*, 4th ed., AVI Publishing Co., Westport, Conn., 1982.
12. K. Nakayama in ref. 11, p. 748.
13. H. W. Doelle in ref. 3, p. 113.
14. P. L. Altmann and D. S. Dittmar in *Metabolism, Handbook of FASEB*, FASEB, Bethesda, Md., 1968.
15. J. White, *Yeast Technology*, Academic Press, London, 1954.
16. G. Reed, *Process Biochem.* **9**(9), 11 (1974).
17. R. Kautzmann, *Branntweinwirtsch.* **9,** 215 (1969).
18. K. Rosen, *Process Biochem.* **12**(3), 10 (1977).
19. G. de Becze and A. J. Liebmann, *Ind. Eng. Chem.* **36,** 882 (1944).
20. H. Dellweg, W. K. Bronn, and W. Hartmeier, *Kem. Kemi* **4**(12), 611 (1977).
21. S. L. Chen and F. Gutmanis, *Biotechnol. Bioeng.* **18,** 1455 (1976).
22. E. Oura, *Biotechnol. Bioeng.* **16,** 1197 (1974).
23. H. Suomalainen, *Pure Appl. Chem.* **7,** 639 (1963).
24. M. S. Fowell, *J. Appl. Bacteriol.* **28,** 373 (1965).
25. M. S. Fowell, *Process Biochem.* **2**(12), 11 (1967).
26. S. Windisch and B. A. Schubert, *Gordian* **73**(7/8), 288 (1973).
27. M. Rohrlich and J. Stegeman, *Brot Gebaeck* **12,** 41 (1958).
28. T. F. Sugihara, L. Kline, and N. W. Miller, *Appl. Microbiol.* **21,** 456 (1971).
29. T. F. Sugihara, *Baker's Dig.* **51**(5), 76 (1977).
30. J. G. Ponte, Jr., V. A. de Stefanis, and S. T. Titcomb, unpublished data, ITT Continental Baking Co., Rye, N.Y., 1970.
31. E. J. Pyler, *Baking Science and Technology*, Seibel Publishing, Chicago, Ill., 1973.
32. R. T. Tang and co-workers, *Baker's Dig.* **46**(4), 48 (1972).
33. B. S. Platt, *Food Technol.* **18,** 662 (1964).
34. Ref. 11, p. 279.
35. A. J. Thorn and J. W. Ross, *Cereal Chem.* **37,** 415 (1960).
36. G. Reed, *Baker's Dig.* **46**(6), 16 (1972).
37. J. A. Maga, *CRC Crit. Rev. Food Technol.* **5,** 55 (1974).
38. C. D. Magoffin and A. C. Hoseney, *Baker's Dig.* **48**(6), 22 (1974).
39. J. G. Piesley and T. Lom in H. M. Broderick, ed., *The Practical Brewer*, 2nd ed., Master Brewers Association of the Americas, Madison, Wisc., 1977.
40. U.S. Pat. 3,394,008 (July 23, 1968), J. Lodder (to Koninklijke Gist-Stiritusfabriek).
41. G. Reed and H. J. Peppler, *Yeast Technology*, AVI Publishing Co., Westport, Conn., 1973.
42. S. R. Griffin, *J. Inst. Brewing* **76,** 45 (1970).
43. M. Amaha, *Bull. Brewing Sci. (Japan)* **12**(1), 43 (1966).
44. I. C. MacWilliam and J. F. Chapperton, *Eur. Brew. Conv. Proc.*, 271 (1969).
45. C. A. Masschelein and co-workers, *Eur. Brew. Conv. Proc.*, 381 (1963).
46. J. Montreuil, S. Mullet, and R. Scriban, *Wallerstein Lab. Commun.* **24,** 301 (1961).
47. J. R. Helbert in ref. 11, p. 403.
48. L. Novellie, *Wallerstein Lab. Commun.* **31,** 17 (1968).
49. S. Domerq, *Ann. Technol. Agric.* **6,** 5 (1957).
50. I. Benda, *Bayer. Landwirtsch. Jahrb.* **39,** 595 (1962).
51. A. Schmitt, K. Curschmann, and H. Koehler, *Rebe Wein* **32**(9), 364 (1979).
52. R. T. de Soto, *Am. J. Enol. Vitic.* **6,** 26 (1955).

53. B. C. Rankine, *Ann. Technol. Agric.* **27**(1), 189 (1978).
54. G. Thoukis, G. Reed, and J. R. Bouthilet, *Am. J. Enol. Vitic.* **14**, 148 (1963).
55. P. Bidan and J. Maugenet, *Bull. Off. Int. Vin* **54**(601), 241 (1981).
56. B. C. Rankine, *J. Appl. Sci.* **4**, 590 (1953).
57. E. Minarik, *Mitt. Rebe Wein Ser. A (Klosterneuburg)* **13**, 185 (1963).
58. S. C. Pan, A. A. Andreasen, and P. Kolachov, *Ind. Eng. Chem.* **42**, 1783 (1950).
59. D. Brandt in ref. 11, p. 468.
60. W. F. Maisch, M. Sobolov, and A. J. Petricola in H. J. Pepper and D. Perlamn, eds., *Microbial Technology*, 2nd ed., Vol. II, Academic Press, Inc., New York, 1979.
61. U.S. Pat. 3,865,691 (Feb. 11, 1975), J. A. Ridgeway and co-workers (to Standard Oil at Indiana).
62. R. G. Cooper, R. S. Silver, and J. P. Boyle in S. R. Tannenbaum and D. I. C. Wang, ed., *Single Cell Protein*, Vol. II, MIT Press, Cambridge, Mass., 1975.
63. R. Kihlberg, *Ann. Rev. Microbiol.* **26**, 427 (1972).
64. Universal Foods, Milwaukee, Wisc., private communication.
65. R. D. Seeley, *Tech. Master Brew. Assoc. Am.* **14**(1), 35 (1977).
66. K. Schwarz and C. M. Foltz, *J. Am. Chem. Soc.* **79**, 3292 (1957).
67. W. Mertz, *Nutr. Rev.* **33**(5), 129 (1975).
68. E. A. Robbins and R. D. Seeley, *J. Food Sci.* **42**(3), 694 (1977).
69. H. J. Peppler in A. H. Rose and J. S. Harrison, ed., *The Yeasts*, Vol. 3, Academic Press, New York, 1970.
70. G. Butschek and G. Krause in R. Reiff and co-eds., *The Yeasts* (in German), Verlag Hans Carl, Nürnberg, FRG, 1962.
71. A. Anderson, R. B. Weisbaum, and K. Robe, *Food Process.* **35**(7), 58 (1974).
72. S. Bernstein, C. H. Tzeng, and D. Sisson, *Biotechnol. Bioeng. Symp.* **7**, (1977).
73. K. Allison in F. Wagner, ed., *Symposium Microbial Production of Protein 1975*, Verlag Chemie, Weinheim, FRG, 1975, p. 73.
74. R. Knecht and co-workers, *Process Biochem.* **12**(4), 11 (1977).
75. H. Sahm and F. Wagner in ref. 73, p. 17.
76. J. L. Cooper in *Biotechnol. Bioeng. Symp.* **6**, 251 (1976).
77. J. H. Slonecker in ref. 76, p 235.
78. R. N. Stone in ref. 76, p. 223.
79. H. E. Grethelein, *Biotechnol. Bioeng.* **20**, 503 (1978).
80. G. T. Tsao, *Process Biochem.* **13**(10), 12 (1978).
81. *Biotechnol. Bioeng. Symp.* **6**, (1976).
82. M. A. Millet, A. J. Baker, and L. D. Satter in ref. 81.
83. G. R. Cysewski and C. R. Wilke, *Biotechnol. Bioeng.* **18**, 1297 (1976).
84. C. R. Wilke, R. D. Yang, and U. von Stokar in ref. 81.
85. P. Dunnil and M. D. Lilly in ref. 62.
86. C. I. Waslien and co-workers, *J. Food Sci.* **35**, 294 (1970).
87. J. C. Edozien and co-workers, *Nature* **228**, 180 (1970).
88. L. Viikari and M. Linko, *Process Biochem.* **12**(4), 17 (1977).
89. *Chem. Eng. News* **56**(17), 18 (1978).
90. G. Hedenskog and H. Mogren, *Biotechnol. Bioeng.* **15**, 129 (1973).
91. A. J. Sinskey and S. R. Tannenbaum in ref. 62.
92. W. E. Trevelyan, *J. Sci. Food Agric.* **27**(3), 225 (1976).
93. J. C. Woodart and D. D. Short, *J. Nutr.* **103**(4), 569 (1973).
94. M. Lindblom, *Biotechnol. Bioeng.* **16**, 1495 (1974).
95. B. Blackwell, L. A. Mabbit, and W. Marley, *J. Food Sci.* **34**, 47 (1969).
96. C. K. Rha in ref. 62.
97. U.S. Pat. 3,781,204 (Dec. 11, 1973), C. Akin (to Standard Oil of Indiana).
98. S. R. Tannenbaum in J. R. Whitaker and S. R. Tannebaum, ed., *Single Cell Protein*, AVI Publ. Co., Westport, Conn., 1977.
99. H. J. Peppler and C. W. Stone, *Feed Manage.* **27**(8), 17 (1976).
100. G. Reed, *Enzymes in Food Processing*, Academic Press, Inc., New York, 1966.
101. J. S. Hough and I. S. Maddox, *Process Biochem.* **5**(5), 50 (1970).

General References

Refs. 1, 2, 3, 11, 15, and 41 are general references.

C. C. Lindegren, *The Yeast Cell. Its Genetics and Cytology*, Educational Publishing, St. Louis, Mo., 1949.

L. J. Wickerham, *Taxonomy of Yeasts*, technical bulletin 1029, U.S. Dept. of Agriculture, Washington, D.C., 1951.

A. H. Cook, ed., *The Technology of Yeast*, Academic Press, Inc., New York, 1958.

G. I. de Becze, "Classification of Yeasts," *Wallerstein Lab. Commun.* **22**(77), 103 (1959); **22**(78), 199 (1959); **23**(81), 99 (1960); **25**(86), 43 (1962).

F. Reiff and co-eds., *Die Hefen* (*The Yeasts*), Vols. I and II, Verlag Hans Carl, Nurnberg, FRG, 1962.

C. Rainbow and A. H. Rose, ed., *Biochemistry of Industrial Microorganisms*, Academic Press, Inc., New York, 1963.

H. J. Peppler, ed., *Microbial Technology*, Reinhold Publishing Corporation, New York, 1967.

A. H. Rose and J. S. Harrison, ed., *The Yeasts*, Academic Press, London, Vol. I, 1969; Vol. II, 1971; Vol. III, 1970.

J. A. Washington, ed., *Laboratory Procedures in Clinical Microbiology*, Little, Brown & Co., Boston, Mass., 1974.

A. H. Rose, ed., *Alcoholic Beverages*, Vol. 1 of *Economic Microbiology*, Academic Press, London, 1977.

H. J. Peppler and D. Perlman, ed., *Microbial Technology*, 2nd ed., Vols. I and II, Academic Press, Inc., New York, 1979.

W. K. Joklik, H. P. Willett, and N. B. Amos, eds., *Zinsser Microbiology*, 17th ed., Appleton-Century-Crofts, New York, 1980.

GERALD REED
Amber Laboratories, Inc.

YOUNG-HELMHOLTZ COLOR-VISION THEORY. See Color.

YTTERBIUM. See Rare-earth elements.

YTTRIUM. See Rare-earth elements.

Z

ZEOLITES. See Molecular sieves.

ZIEGLER-NATTA CATALYSTS. See Catalysis; Olefin polymers; Organometallics.

ZINC AND ZINC ALLOYS

Zinc [7440-66-6] is a relatively active metal and its compounds are stable. Since it is not found free in nature, it was discovered much later than less-reactive metals such as copper, gold, silver, iron, and lead. In early times, the smelting of copper ores containing zinc resulted in brasses, which were known to the Romans before 200 BC. Later, brasses were made by heating copper with zinc oxide or carbonate and charcoal. The oldest piece of zinc extant is an idol found in a prehistoric Dacian site in Transylvania which analyzed at 87.52% zinc, 11.41% lead, and 1.07% iron (1). Zinc smelting is thought to have originated in China, where it was known in the seventh century AD how to make malleable zinc. In India, zinc was produced from ore mined at Zawar before 1380. By the seventeenth century, zinc was imported into Europe from Asia and in 1743 a zinc smelter for zinc oxide ore was erected in Bristol, UK. By the early nineteenth century, zinc smelting was well established in Germany and Belgium. Zinc was first produced in the United States at the arsenal in Washington, D.C., in 1835, and by 1860 The New Jersey Zinc Company had well-established smelting operations at Bethlehem, Pennsylvania.

In 1758 in England, John Champion was granted a patent for making zinc from roasted zinc sulfide, the main ore of zinc. This route is still the principal method of production today. Commercial zinc metallurgy started with batchwise horizontal retorts, which have had many variations, and progressed to continuous vertical-retort processes. Although some of the latter are important today, electrowinning dominates the field, and the Imperial Smelting blast furnace for zinc–lead remains competitive.

The main application of zinc is to protect iron and other metals from corrosion.

Zinc in contact with iron and other metals, as a coating or attached anode, corrodes sacrificially and protects the iron. Most commonly, zinc is applied in the molten state, ie, galvanizing, but is also applied by various mechanical procedures using zinc dust or powder (see Metallic coatings; Metal surface treatments). Zinc dust paints are growing in importance (2). Another important use is in alloys for die casting. These alloys are used extensively because of their high quality and low cost. Brass and bronze products account for the third largest usage.

The monograph on zinc is a valuable general reference on zinc technology (3). Furthermore, detailed descriptions of extractive processes, resource data, and environmental- and energy-related papers from symposia of the Metallurgical Society of the AIME are a rich source of information (4–7).

Occurrence

Zinc, like most metals, is found in all natural waters and soils as well as the atmosphere and is an important trace element in plant and animal life (see Mineral nutrients). Rocks of various kinds contain 20–200 ppm zinc and normal soils 10–30 ppm (average ca 50 ppm) in uncontaminated areas. The average zinc content of coal is 33 ppm. Seawater contains 1–27 μg/L (median ca 8 μg/L), and uncontaminated freshwater usually <10 μg/L.

Zinc ores are widely distributed throughout the world; 55 zinc minerals are known (8–10). However, only those listed in Table 1 are of commercial importance. Of these, sphalerite provides ca 90% of the zinc produced today. Sulfide ores usually occur in the range of 2–12% zinc (average ca 4%) as mined.

The oxygen-containing ores in Table 1 are often found in richer deposits ranging up to 35% zinc, most commonly smithsonite and hemimorphite. They are thought to have originated from sulfide mineralization and are minor sources of zinc. However, in the beginning, the U.S. zinc industry was based on the deposit of franklinite, zincite, and willemite in Sussex County, New Jersey. Zinc deposits are classified as contact metamorphic, irregular and associated fissure fillings, vein, stratabound in metamorphic rocks, stratabound in carbonate rocks, strataform, and deposits formed by supergene enrichment of laterization (10). Carbonate rocks, eg, limestone and dolomite, are common host-rocks of zinc ore.

In the United States, the richest zinc district is the Mississippi Valley; however, Canada leads the world in estimated reserves and the United States is second (see Tables 2–3).

Table 1. Common Zinc Minerals

Name	Composition	% Zn
sphalerite[a]	ZnS	67.0
hemimorphite[b]	$Zn_4Si_2O_7(OH)_2 \cdot H_2O$	54.2
smithsonite	$ZnCO_3$	52.0
hydrozincite	$Zn_5(OH)_6(CO_3)_2$	56.0
zincite	ZnO	80.3
willemite	Zn_2SiO_4	58.5
franklinite	$(Zn,Fe,Mn)(Fe,Mn)_2O_4$	15–20

[a] Zinc blend, wurtzite.
[b] Calamine.

Table 2. World Zinc Reserves, 10^6 Metric Tons[a]

Country	Reserves	Total[b]
Canada	28	56
United States	22	45
Australia	19	40
USSR	11	22
Republic of South Africa	4	17
Peru	7	11
Brazil	5	11
Spain	5	10
Ireland	8	9
Japan	5	7
Iran	4	6
People's Republic of China	1	5
India	3	4
Mexico	3	4
Yugoslavia	2	4
Zaire	1	2
Total	128	253

[a] Ref. 11.
[b] Includes inferred reserves and some identified subeconomic resources.

Table 3. U.S. Zinc Reserves[a]

District	States	Reserves
Mississippi Valley	Illinois, Missouri, Wisconsin, mid-Tennessee	8.3
South Appalachian	Virginia, east Tennessee	6.5
Northeastern	Maine, New York, New Jersey, Pennsylvania	3.4
Southwestern	Arizona, New Mexico, Utah, Colorado, California	2.1
Northwestern	Idaho, Montana, Washington	1.7
Total		22.0

[a] Ref. 11.

Zinc minerals tend to be associated with those of other metals; the most common are zinc–lead or lead–zinc, depending upon the dominant metal, zinc–copper or copper–zinc, and base metal such as silver. Zinc does occur alone, most often in the Northeastern district, and here, as elsewhere, recoverable amounts of cadmium (up to 0.5%) are present. Other minor metals recovered from zinc ores are indium, germanium, and thallium (see under By-Product Metals).

The exploration, evaluation, and development of zinc and lead ore bodies in North and Central America are discussed in ref. 12. A survey of world zinc production in ref. 13 gives all operating mines and mills, and their methods, production, and chemical analysis of the products; zinc smelters are included.

Open-pit zinc mining is not common, since most mines are below the surface. The Kidd Creek Mine in Ontario, Canada, is a combination open-pit–underground mine. It is one of the richest deposits in the world with an estimated 62.5×10^6 t grading 7.08% zinc, 1.33% copper, and 151 g silver (14). Underground mining methods include room-and-pillar, shrinkage, cut-and-fill, and square set (11). In the United States, ca 30 mines are worked for zinc.

Physical Properties

Zinc is a lustrous, blue-white metal, which can be formed into virtually any shape by the common metal-forming techniques such as rolling, drawing, extruding, etc. The hexagonal close-packed crystal structure governs the behavior of zinc during fabrication. Physical properties are given in Table 4.

Mechanical history, heat, and impurities greatly affect the mechanical properties. Pure zinc is ductile at room temperature and does not have a definite yield point as do most structural metals. Rather, it creeps under sufficient constant load. The impurities of commercial zinc and alloying metals are carefully controlled to achieve the desired mechanical properties.

Table 4. Physical Properties of Zinc[a]

Property	Value
ionic radius, Zn^{2+}, nm	0.074
covalent radius, nm	0.131
metallic radius, nm	0.138
ionization potential, eV	
first	9.39
second	17.87
third	40.0
density	
solid, g/cm³	
at 25°C	7.133
at 419.5°C	6.830
liquid, g/mL	
at 419.5°C	6.620
at 800°C	6.250
melting point, °C	419.5
boiling point, °C	907
heat of fusion at 419.5°C, kJ/mol[b]	7.387
heat of vaporization at 907°C, kJ/mol[b]	114.8
coefficient of thermal expansion, mm/(m·K)	
volume	8.9
linear, polycrystalline	39.7
thermal conductivity, W/(m·K)	
solid	
at 18°C	113.0
at 419.5°C	96.0
liquid	
at 419.5°C	60.7
at 750°C	56.5
electrical resistivity, nΩ/m	
polycrystalline	$R = 54.6\,(1 + 0.0042\,t)$[c]
liquid at 423°C	369.55
heat capacity, J/(mol·K)[b]	
solid	$22.39 + 10^{-2}\,T$[d]
liquid	31.39
gas	20.80

[a] Ref. 15.
[b] To convert J to cal, divide by 4.184.
[c] $t = 0\text{--}100°C$.
[d] $T = 298\text{--}692.7$ K.

Chemical Properties

The most significant chemical property of zinc is its high reduction potential. Zinc, which is above iron in the electromotive series, displaces iron ions from solution and prevents dissolution of the iron. For this reason, zinc is used extensively in coating steel, eg, by galvanizing and in zinc dust paints, and as a sacrificial anode in protecting pipelines, ship hulls, etc.

In batteries, a zinc anode undergoes the oxidation reaction,

$$Zn \rightarrow Zn^{2+} + 2\,e \;(+0.763\text{ V})$$

to provide a flow of electrons to the external circuit. For instance, in the familiar dry cell, the zinc is oxidized to zinc chloride while manganese dioxide is reduced at the cathode. The alkaline zinc battery with manganese dioxide cathode uses potassium hydroxide electrolyte and forms zincate ions or precipitates zinc oxide at the anode:

$$Zn + 4\,OH^- \rightarrow Zn(OH)_4^{2-} + 2\,e \;(+1.216\text{ V})$$

$$Zn + 2\,OH^- \rightarrow ZnO + H_2O + 2\,e \;(+1.245\text{ V})$$

Other alkaline primary cells couple zinc with oxides of mercury or silver and some even use atmospheric oxygen (zinc–air cell). Frequently, zinc powder is used in the fabrication of batteries because of its high surface area. Secondary (rechargeable) cells with zinc anodes under development are the alkaline zinc–nickel oxide and zinc–chlorine (see Batteries and electric cells).

The capability of zinc to reduce the ions of many metals to their metallic state is the basis of important applications. However, metals are removed from zinc solutions by displacement with finely divided zinc before winning by electrolysis. Gold and silver are displaced from cyanide leach solutions with zinc and the following metals are similarly recovered from various solutions: platinum group, cadmium, indium, thallium, and sometimes copper.

Zinc hydrosulfite (zinc dithionite) is a powerful reducing agent used in bleaching paper and textiles; it is prepared from zinc dust and sulfur dioxide:

$$Zn + 2\,SO_2 \rightarrow ZnS_2O_4$$

Another hydrosulfite reducing agent is zinc formaldehyde sulfoxylate, Zn-(HSO$_2$.CH$_2$O)$_2$ (see Bleaching agents).

Pure zinc displaces hydrogen from acidic solutions slowly because of its high hydrogen overvoltage. This fact allows zinc to be electrodeposited by reduction from highly purified acidic solutions with only slight evolution of hydrogen. Surface amalgamation with mercury is useful in alkaline-battery applications because it increases the hydrogen overvoltage and reduces hydrogen gassing. Conversely, some metals reduce overvoltage and increase the rate of hydrogen evolution. Zinc dust, because of its high surface area, must be kept bone dry to avoid slow displacement of hydrogen from water. Massive zinc, on the other hand, reacts with steam rapidly above 350°C (see also Electrochemical technology; Electroplating).

Oxidizing elements such as oxygen, sulfur, and halides react with zinc at room temperature in the presence of moisture, but do not in its absence. At higher temperature, the reactions can be vigorous even when dry. For instance, a powdered mixture of zinc and sulfur explodes if warmed and zinc reacts rapidly with oxygen at

225°C. Atmospheric corrosion of zinc results in hydrated basic carbonates, $x\text{ZnO} \cdot y\text{CO}_2 \cdot z\text{H}_2\text{O}$, of variable stoichiometry which form a surface film.

Zinc does not react with nitrogen, even at elevated temperatures but zinc nitride, Zn_3N_2, forms with ammonia at red heat. Zinc sulfide, the most common form of zinc in nature, is not reduced directly in commercial practice because of reactions of the zinc vapor during condensation. Rather, the sulfide is burned (roasted) to the oxide plus sulfur dioxide before reduction. However, zinc can be reduced to the metal at ca 1300°C with carbon or iron.

In the Parkes desilvering process, 1–2% zinc is added to molten lead where it reacts with any gold, silver, and copper to form intermetallic compounds which float as crusts or dross that is skimmed (see Lead and lead alloys).

Processing

Concentration. Zinc ores are too low in zinc content for direct reduction and must be concentrated. They are first crushed, usually underground, in jaw, gyratory, or cone crushers, and then ground to 75–150 µm (100–200 mesh) by ball or rod milling to a degree necessary to separate the zinc minerals from other minerals and gangue in later flotation (qv). In some ores, the zinc and lead minerals are too intimately mixed to be separated by flotation and are separated by leaching for subsequent electrowinning or in the Imperial Smelting furnace (ISF). The beneficiation of zinc ores is usually more complex than for other nonferrous ores because of the diversity of minerals present. Not only are copper or lead often associated, but the sulfide ore is sometimes partially oxidized.

Beneficiation before grinding can be done by gravity or magnetic methods (16) (see Gravity concentration). Wetherill high intensity magnetic separators were developed by The New Jersey Zinc Company in 1896 to separate franklinite from their Franklin and Sterling ores (see Magnetic separation). Jigging and tabling recovered the zincite and willemite. The concentrates were further upgraded to crude zinc oxide (17).

Zinc ores are generally floated at the mine (18). In the case of simple zinc sulfide ores, flotation is carried out by treatment with copper sulfate to activate the sphalerite causing it to be wet by the organic collector (eg, xanthate). The now-hydrophobic zinc ore particles attach themselves to the rising bubbles. Oxidized ore particles present must be sulfidized with sodium sulfide to be floated (19). Flotation produces concentrates which are ca 50–60% zinc. In mixed ore, the lead and copper are usually floated after depressing the sphalerite with cyanide or zinc sulfate. The sphalerite is then activated and floated.

Roasting. Copper and lead sulfides are directly smelted but not zinc sulfide. However, theoretical calculations are encouraging (20) and, if an efficient means of condensing zinc rapidly from 1600 K in the presence of carbon dioxide, sulfur dioxide, and steam can be devised, the process may be feasible. The reaction of zinc vapor to yield zinc oxide or zinc sulfide presents the main difficulty.

The zinc sulfide in the concentrate is always converted to oxide by roasting. An exception is the direct leach process described below. The principal overall roasting reaction is strongly exothermic and provides excess heat which is recovered.

$$\text{ZnS(s)} + \tfrac{3}{2}\,\text{O}_2(\text{g}) \rightarrow \text{ZnO(s)} + \text{SO}_2(\text{g}) + \text{ca } 5.02 \text{ MJ/kg (1200 kcal/kg)} \tag{1}$$

Under certain conditions, the oxide is sulfatized:

$$\text{ZnO} + \text{SO}_2 + \tfrac{1}{2}\,\text{O}_2 \rightarrow \text{ZnSO}_4 \tag{2}$$

In a usual roaster gas at equilibrium, the sulfate decomposes at ca 860°C, but it is difficult to avoid some sulfatization during cooling and indeed some plants require a degree of sulfation to maintain sulfate balance.

Another important reaction accounts for sulfur trioxide in the roaster off-gases and is a component of reaction 2.

$$SO_2 + \tfrac{1}{2} O_2 \rightarrow SO_3 \tag{3}$$

Reaction 3 also occurs on cooling since the concentration of SO_3 is very low at roaster temperatures of 950°C and approaches zero at 1000°C. Another important reaction that occurs during roasting is the formation of zinc ferrite, $ZnO.Fe_2O_3$ above 650°C (see Ferrites). Zinc ores contain 5–12% iron. Zinc ferrite forms solid solutions with other spinels, such as $FeO.Fe_2O_3$, and therefore the zinc–iron compositions formed are of indefinite stoichiometry. Ferritic zinc is difficult to solubilize in hydrometallurgical leaching but several recovery processes are discussed below.

The sulfur dioxide of reaction 1 is cooled in a waste-heat boiler, freed from calcine, and converted to trioxide. The oxidation and conversion to sulfuric acid is conducted in a conventional acid plant (see also Sulfuric acid and sulfur trioxide).

For environmental and economic reasons, the early practice of roasting zinc sulfide and discharging the sulfur dioxide to the atmosphere gave way to plants where the sulfur dioxide is converted to sulfuric acid. Desulfurization takes place while the ore particles are suspended in hot gases. Called flash- and fluid-bed roasters, these processes are described below. Some plants use combinations of roasters and sintering for desulfurization.

Multiple-Hearth Roasters. The circular types consist of a series of hearths arranged vertically in such a way that the ore entering the top is rabbled and dropped down from hearth to hearth, until it is completely oxidized. The hearths are usually stationary and the plows revolve, such as in the Wedge, Herreshoff, Ord, Skinner, and other roasters (21). In other furnaces, the hearths revolve and the rabbles are fixed, eg, the deSpirlet and its modification, the Barrier.

A conventional circular-wedge roaster consists of a brick-lined steel shell with hearths arched gently upward from the periphery to a central shaft. The brick hearths may number from eight to 16 and are ca 1 m apart. The central steel shaft (ca 1.2 m in diameter) revolves at 1 rpm or less carrying two rabble arms per hearth. These rabbles, cooled with air or water, plow the ore from the outside to the center of the hearth where it is dropped to the next hearth for plowing in the opposite direction. The calcine thus proceeds to the bottom where it is dropped into a conveyor. The sulfide sulfur at this point is ca 3.5% (22).

The uppermost hearth serves to dry the damp ore in the hot (ca 500°C) gases exiting the top of the roaster. These gases may contain up to 15% of the total crude oxide and up to 6% sulfur dioxide, high enough to be fed to a sulfuric acid plant. In some pyrometallurgical operations, desulfurization is continued in a sintering step.

Multiple-hearth roasting offers ease of operation, ability to handle a wide variety of ores or blends, and little downtime. On the other hand, these furnaces are no longer being built because of their high capital and labor costs, relatively low sulfur dioxide off-gas, need for added fuel, and marginal opportunity for waste-heat recovery.

Flash Roasting. As more and more zinc concentrates were produced by flotation, which requires finely ground ore, more rapid roasting was possible. Instantaneous oxidation was noted in multiple-hearth furnaces when particles were dropped from one hearth to another. The Consolidated Mining and Smelting Company of Canada took advantage of this observation to develop the flash or suspension roaster at Trail, B.C., Canada, in the early 1930s (23). An 8-m Wedge roaster with eight hearths was modified by removing all but the top two and bottom two hearths. Ore, fed to the top, is dried and then removed for grinding. After classification, it is conveyed back to the roaster where it is blown into the central chamber in a turbulent fashion using high pressure air jets. Oxidation occurs at ca 1000°C and ca 40% of the product falls to the bottom hearths. This coarser material is further desulfurized as it is rabbled out of the roaster. Some or all of the dust leaving with the gas stream is usually returned to the bottom hearths for further oxidation or decomposition of sulfate. The dust is removed from the gas by means of a cyclone following the waste-heat boiler. An electrostatic precipitator, wet scrubber, and dryer (sulfuric acid tower) clean up the gas before it passes to the acid plant.

New flash roasters dry on the bottom hearth; the ore is introduced in two opposed burners for increased turbulence (24). Such roasters with combustion chambers of 8–9 m high are capable of dead roasting (sulfide removal to <0.5%) over 300 t of zinc concentrates per day with 10% sulfur dioxide in the off-gas.

The flash roaster is flexible in handling various flotation concentrates and reaching the degree of desulfurization desired, ie, 0.5–3.0% sulfate sulfur. Waste heat is easily recovered. However, grinding and rabbling must be done mechanically.

Fluidization Roasting. Fluid-bed roasting offers the advantages of reduced machinery, high throughput, good heat recovery, and rich sulfur dioxide off-gas (25–26) (see Fluidization). Most electrolytic zinc plants use fluid-bed roasting because of the ease of controlling the temperature to achieve proper sulfide and sulfate levels. Compared to flash roasting, zinc leachability is better and sulfation is increased while the formation of zinc ferrite is minimized (see also Extractive metallurgy).

Slurry fed. Dorr-Oliver, Inc., pioneered fluidization roasting and developed their FluoSolids process originally for treating arsenopyrite gold ores. The reactor (27) for roasting up to 155 t of concentrates per day is an insulated cylinder of brick-lined steel construction and with a conical bottom 9.5 m high with a diameter of 6.3–7.5 m. The windbox is separated from the reaction chamber by a distribution plate containing 780 nozzles designed to keep calcine from falling into the windbox. Fluidization air is supplied at 20,000 m³/h at 34.3 kPa (257 mm Hg), ie, 30% excess over stoichiometric. In operation, the bed is 1.5 m deep and at 1.5 kPa (11 mm Hg), under a slight positive pressure.

After start-up, the temperature is raised to the bed ignition temperature of 900°C. The feed slurry at 80% solids is pumped from an agitated tank and injected by atomization with compressed air. Evaporation of water from the slurry, or added supplementally, controls roasting at 900°C. The coarse calcine (40% of total) exists the top of the bed and flows to an evaporative cooler followed by cyclone and bag filter. The gases carry 60% of the solids to a waste-heat boiler where 15% of the total calcine is collected. Cyclones recover 40%, and 5% is picked up by an electrostatic precipitator. The gas at this point is ca 10% sulfur dioxide. The collected calcine is screened and the coarse fraction (>2 mm) is hammer-milled before being sent to leaching in the electrolytic plant.

The coarse calcine cooler operates at 300°C, while the waste-heat boiler cools the gas to 350°C. The tubes in the boilers have a chain-shaking arrangement operated by pneumatic hammers. Steam production is 0.78 kg/kg of dry concentrate. The only trouble with dust reported is in the connection between the reactor and waste-heat boiler. It is necessary to cool the gas stream quickly to avoid sulfation, but even so the carry-over calcine contains on the order of four times more sulfate than the coarse overflow. In this plant, the composite calcine is 0.1% sulfide and 2.2% sulfate sulfur.

A similar FluoSolids roaster handles 155 t/d of zinc concentrates through a roaster which is 1 m higher than standard (28).

The Dorr-Oliver slurry-fed roaster allows easy introduction of the concentrate, relatively low dust carry-over, good sealing, and freedom from cooling coils in the bed. It is usually operated at negative pressures to avoid sulfur dioxide and dust leakage but the in-leakage of air causes sulfation which may be undesirable. A temperature >900°C aids desulfurization by reducing carry-over; that is, the higher temperature induces particle sintering and agglomeration which diverts more of the calcine to bed overflow where it avoids sulfation. The principal disadvantage of the slurry-feed system is that the added water reduces the amount of heat that can be recovered.

Dry fed. The dry feeding of fine concentrate to fluid-bed roasting was developed at S.A. Vieille-Montagne (VM) in Balen, Belgium, using a modified turbulent-layer process (BASF pyrites roaster). This process was sold to the zinc industry by Lurgi whose largest installation treats 800 metric tons of concentrates per day (29). In this roaster, unlike the Dorr-Oliver FluoSolids, so-called dry concentrates with 5–10% moisture are fed by mechanical belt slingers. In addition, the freeboard diameter is larger. Heat is removed from the bed by cooling coils and adding water. The VM roaster is the best for heat recovery and has a relatively low gas throughput.

The newest zinc plant in the United States at Clarksville, Tennessee, owned by the Jersey Miniere Zinc Co., uses a VM roaster (30). A single roaster treats 382 t of concentrate per day and recycles dross producing 333 t of calcine, while the acid plant produces 338 t of 100% sulfuric acid. The grate area of the roaster is 64.7 m^2 and normally would have five cooling coils in the bed. The low heat value (low iron sulfide) of the concentrates allows removal of three coils. Calcine leaves the roaster as bed overflow (ca 60%) or carry-over into the waste-heat boiler (ca 40%). The latter generates 18,160 kg of steam per hour at 3.9 MPa (ca 39 atm) pressure. Gases leaving the electrostatic precipitator contain 9.5% SO$_2$. Calcine from the roaster overflow and waste-heat boiler are cooled to ca 100°C in a rotary-drum cooler and combined with cyclone and precipitator collects for dry ball-milling to <120 μm. Sulfur levels in the calcine are 0.25% sulfide and <1.5% sulfate. The clean, cool gas passes through a Norzinc mercury-removal tower (31) before entering the acid plant. Mercury reacts with soluble mercury(II) chloride to precipitate mercury(I) chloride, ensuring <1 ppm mercury content.

Canadian Electrolytic Zinc, Ltd., at Valleyfield, Quebec, uses two VM roasters to process 360 t of concentrates per day (32). Each roaster has 3,300 tuyeres in its grate of 34 m^2 area, to fluidize a bed of 1 m depth; gas exits at 990°C. This plant was the first to use a water-cooled settling chamber in the waste-heat boiler for reducing the degree of sulfation. The gas exits the boiler at 320°C. An attempt was made to use spent electrolyte from the cells for roaster temperature control. This was designed to control the total sulfate in the plant but it caused excessive bed agglomeration and increased sulfation.

Agglomerate fed. Fluid-bed roasters of this type were developed in order to avoid the sintering stage in preparing feed for pyrometallurgical zinc operations where agglomerate strength and freedom from volatile components (sulfates, cadmium, etc) is essential.

Metallurgie Hoboken-Overpelt S.A. (MHO) in Belgium uses hard pellets as a suitable feed for horizontal retorts (33). The MHO roaster, started in 1954, was the first to recover waste heat and was developed in collaboration with Dorr-Oliver. Concentrates and recycled fines are homogenized with sulfuric acid from scrubbing and pelletized to 0.5–4-mm size. The pellets are dried and fed to the rectangular roaster where the bed flow is longitudinal to the discharge port. The furnace is 5.5 m high inside. For 85% of the length, fluidization air is introduced and the balance of the bottom is an aperture where the pellets fall into a finishing bin. Crude pellets are less dense than calcine and tend to float in the upper portion of the 1-m-deep bed. Calcine enters the finishing bin at 980°C where incoming air completes the oxidation and sweeps away the sulfur dioxide to prevent sulfation. Sulfur content of the calcine is ca 0.05% sulfide and 0.35% sulfate. The dust problem is reduced by carrying over only 20% of the calcine to the boiler. A relatively long retention and high temperature (up to 1060°C) tend to eliminate chlorides, fluorides, mercury, and selenium. Ores containing low melting materials such as silica and magnesia do not create sticking problems. A disadvantage is the need for considerable grinding before leaching.

The New Jersey Zinc Company patented a fluidized-pellet roaster which was installed in several zinc plants. Called a fluid-column roaster, it resembles a shaft furnace and can handle 370 t of concentrate per day. This roaster can be operated at 1080–1100°C to eliminate 90% of the cadmium and 92% of the lead. The fluid-column roaster has the same advantages as the MHO roaster; the pelletizing cost is a disadvantage for both systems.

St. Joe Minerals Corporation uses a fluid-bed roaster to finish the roasting at 950°C of material that has been deleaded in a modified multiple-hearth furnace operated with insufficient oxidation (34). First, sulfur is reduced from 31 to 22% and lead from 0.5 to 0.013%. Somewhat aggregated, the product is hammer-milled before final roasting. Half of the calcined product is bed overflow and special hot cyclones before the boiler remove the other half; total sulfur is ca 1.5%. Boiler and precipitator dusts are higher in sulfur, lead, etc, and are separated.

Sintering. Sintering before further pyrometallurgical smelting completes the roast, eliminates volatile material, and aggregates fine calcine. Aggregates need considerable strength and porosity in vertical retorts, electrothermic furnaces, and the Imperial Smelting process (ISF). Sintering practice can be classified as follows (22): on low sulfur calcine with carbonaceous fuel in one or two stages; on unroasted concentrates with high sinter recycle; and on 7–9% sulfur calcine without fuel. Zinciferous material is mixed with carbonaceous fuel, if necessary, and water. The mix is pelletized, or partly so, and fed onto a traveling endless grate. After ignition, air is drawn through the bed to burn the carbon and zinc sulfide.

Sintering machines, usually of the Dwight-Lloyd downdraft type, are used. In downdraft operations, metallic impurities are usually not eliminated in one pass because they tend to condense in the bottom portion of the bed. For this reason, the product is sliced from the top of the 10–20-cm bed and the bottom fraction (20–40%) recycled or used where the impurity can be tolerated. Chloride may be added to improve the transfer of cadmium and lead from the bed into the windbox and downstream dust and fume-collection system. Since the electrothermic zinc process requires very strong pellets, double sintering is practiced with silica flux added to the second stage. Heat is not recovered from downdraft sintering operations because of high gas volumes, moisture in the charge, and large sinter recycle loads, and a considerable energy debit is incurred. In addition, it is expensive to scrub the dilute sulfur dioxide in the off-gases.

Updraft sintering is used on unroasted concentrates and the off-gases are built to a high enough sulfur dioxide content (5–6.5%) to allow sulfuric acid production. This is accomplished by preventing gas leakage around the bed with grease-sealed slide rails and by gas recirculation. In order to avoid excessive bed temperatures and maintain porosity, sinter recycle (ca 80%) is high. Some heat is recovered to heat incoming air.

Reduction. *Electrolytic Process.* The electrolytic process, where zinc is deposited from an aqueous solution onto the cathode, treats complex ores that did not lend themselves to pyrometallurgical recovery (see also Electrochemical processing). For instance, some abundant ores containing combinations of lead, copper, silver, iron, and zinc could be concentrated to only 30–40% zinc, an unsuitable level for pyrometallurgy. The first successful electrolytic zinc plant was built by the Anaconda Company at Anaconda, Montana, and began production in 1915. The sulfide concentrate is oxidized to crude oxide and leached with return acid from the cells. The zinc sulfate solution is purified and electrolyzed. Some plants also feed crude zinc oxide from pyrometallurgical recovery operations. At present, there are four electrolytic zinc plants in the United States with annual capacities from 50,000 to 100,000 metric tons. A simplified flow diagram is shown in Figure 1 and detailed descriptions of plants and processes are given in the literature (35–39).

Leaching. Crude zinc oxide in the form of roasted concentrate and, at some plants, fume from slag-fuming is fed to the leach circuit where zinc is solubilized with 100–200 g/L sulfuric acid which has been returned from the cells. Grinding is integrated into the leaching step. Zinc ferrite formed in roasting is not soluble in the dilute acid and this zinc is recovered. Residues also contain lead and often silver. The pH of the

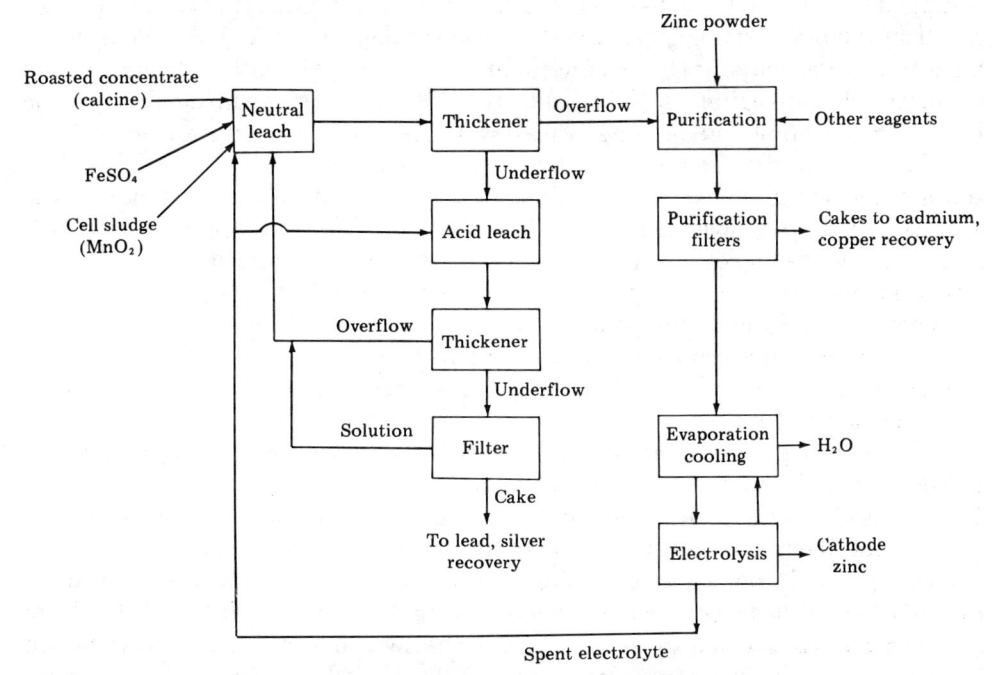

Figure 1. Simplified flow diagram, electrolytic zinc.

solution from which the residue is separated is ca 5.0 where hydrous iron(III), aluminum, indium, and silicon oxides are precipitated but basic zinc sulfate is not. The hydrous oxides adsorb arsenic, antimony, germanium, and other impurities. Thus, leaching is important in purification.

Leaching is a one- or two-step procedure, batch or continuous. In one-step leaches, calcine is added to neutralize return acid until the soluble zinc is in solution and the pH is high enough to precipitate iron(III) and other impurity hydroxides. The final pH adjustment may be made with lime or limestone instead of calcine to decrease zinc loss to the residue. Although the one-step procedure is simple and requires less capital equipment than the two-step, it must be carefully controlled to ensure zinc dissolution and avoid precipitation of basic zinc sulfate above pH 5.5. Residues containing 15–20% zinc (primarily ferrite) are sent for pyrometallurgical recovery. In the two-step procedure, more impurities are removed. New zinc plants use various modifications of this procedure because it includes recovery of the ferritic zinc. The first leach of the two-step procedure is called the neutral leach. Excess calcine is added to ca half the return acid plus acidic solution from the second leach step to precipitate iron and other impurities and ensure a pH (5.0–5.2) high enough for proper flocculation of the residue. The leachate is sent to the purification section and the residue, containing ca 50–75% of the available zinc, is leached with acid to dissolve the remaining available zinc. The final acid contains 3–5 g zinc per liter and, although some of the impurities remain in the solution returning to neutral leach, most remains with the residue. Depending on the ore, the residue may also contain lead, copper, silver, and gold. In many plants, the hot acid leach (95°C) dissolves zinc ferrites and other values which otherwise would result in the loss of up to 15% of the zinc and higher proportions of cadmium and copper.

The leaching is 50–60°C without external heating. The cone-bottom tanks are equipped with a pipe from just above the solution level to near the bottom through which air is blown forming bubbles which lower the density of the slurry. These leach tanks, called Pachucas, are fairly efficient and are still popular although many plants employ mechanical agitators. Most plants use 3–5 tanks in series with acid and calcine being fed to the first and, in some cases, downstream from the first tank as well.

Recovery of zinc from leach residue. The residues from dilute acid leaching contain zinc ferrite, entrained zinc sulfate, undissolved zinc oxide, and other values, such as lead, copper, silver and gold. Although hot acid leaching improves copper and zinc recovery, the process was not widely used in the past because of the difficulty of separating iron which forms a gelatinous precipitate. Usually the residue was treated pyrometallurgically and often still is.

In addition, three commercially successful methods have been developed in which iron(III) precipitates are formed that filter and wash well, and which increase the recovery of zinc by 10–15%.

The jarosite process separates iron(III) from zinc in acid solution by precipitation of $MFe_3(OH)_6(SO_4)_2$, where M is an alkali metal (usually sodium) or ammonium (see Fig. 2) (40–41). Other monovalent and hydronium ions also form jarosites which are found in the precipitate to some degree. Properly seeded, the relatively coarse jarosite can be separated from the zinc-bearing solution efficiently. The reaction is usually carried out at 95°C by adding ammonia or sodium hydroxide after the pH has been adjusted with calcine and the iron oxidized. The neutral leach residue is leached in hot acid (spent + makeup) with final acidity >20 g/L and essentially all the zinc, in-

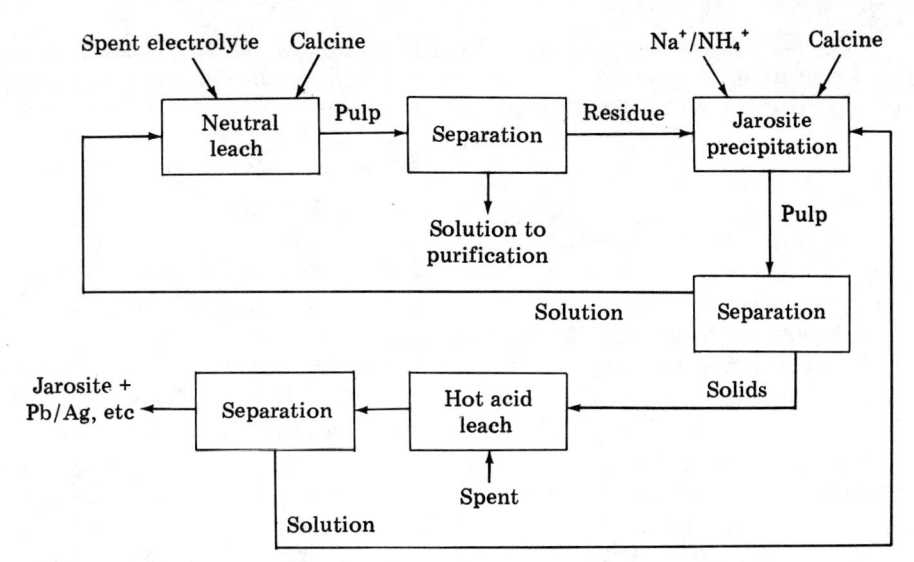

Figure 2. Flow sheet for modified jarosite precipitation process.

cluding ferrite, is solubilized. Ammonium jarosite is then precipitated in the presence of the residue or after separating it. If the residue contains appreciable lead or silver, they are first separated to avoid loss to the jarosite waste solids. Minimum use of calcine in jarosite neutralization is required for maximum recovery of lead and silver as well as zinc and other metals.

In a modification, the conversion process, the jarosite residue is hydrothermally decomposed to hematite by autoclaving at 220–250°C. This solubilizes zinc and other metal values and the hematite has a potential for iron recovery. Hematite stockpiles are less of a problem than jarosite because hematite is denser and holds up less of the soluble metals.

The jarosite process controls both alkali metals and sulfate in the zinc-plant circuit and consumes little neutralizing agent. It is used in 16 plants worldwide, which account for ca 25% of the noncommunist world's primary zinc output.

The goethite process precipitates crystalline αFeO.OH (goethite) as well as βFeO.OH, αFe$_2$O$_3$, and amorphous phases. The reaction is carried out at 90°C and pH 3.0, for 4–6 h in either batch or continuous fashion, and the iron(III) ion must be kept <1 g/L. Both jarosite and goethite solids are usually lagooned.

The hematite process is used at the Iijima plant of the Akita Zinc Co. in Japan, where it started in 1972, and at Ruhr-Zink, Datteln, FRG. In the Iijima process, neutral leach residue is leached with hot acid and the iron(III) in the leachate is reduced with sulfur dioxide at ca 100°C. Copper is precipitated with hydrogen sulfide although it is not essential to the process. Ruhr-Zink reduces with concentrate, zinc sulfide, giving elemental sulfur as product. Sulfur and unreacted zinc concentrate are coagulated at elevated pressure and temperature and separated by classification from finer lead–silver residue. The solution is neutralized to pH 4.5 with limestone to yield gypsum which is filtered and sold. This leaves iron(II) sulfate solution from which αFe$_2$O$_3$ is precipitated by oxidizing at ca 190°C under an oxygen pressure of 8.4 kPa (63 mm Hg). The hematite product is denser than jarosite or goethite; it can be sold to steel and cement plants.

Direct leaching of concentrates. Sherritt Gordon Mines, Ltd., has adapted the process first used on nickel sulfide ores to zinc sulfide oxidation with air in aqueous slurry under pressure (42–43). The concentrates are leached directly with return acid from the cells and the sulfide is converted to free sulfur:

$$ZnS + Fe_2(SO_4)_3 \rightarrow 2\ FeSO_4 + ZnSO_4 + S$$
$$\underline{2\ FeSO_4 + H_2SO_4 + \tfrac{1}{2}\ O_2 \rightarrow Fe_2(SO_4)_3 + H_2O}$$
$$ZnS + H_2SO_4 + \tfrac{1}{2}\ O_2 \rightarrow ZnSO_4 + S + H_2O$$

Thus, roasting is avoided. The process, especially amenable to high iron and copper concentrates, has been installed by Cominco, Ltd. (44) at Trail, B.C., Canada, and will be installed at the Kidd Creek Mines, Ltd., plant at Timmins, Ontario.

Purification. Purification of the electrolyte is extremely important since the presence of impurities, even in trace amounts, can seriously impede or halt production (45–46). Impurities may lower the hydrogen overvoltage causing the electrical energy to be consumed in depositing hydrogen instead of zinc. The reversible deposition potential of hydrogen from acid onto platinum is 1.70 V but on zinc, because of overvoltage, it is 2.4 V. The standard potential needed for zinc reduction is 2.35 V (45). Therefore, if the hydrogen potential is lowered by a mere 0.05 V, it is deposited strongly in competition with zinc. Very low impurity levels are usually very harmful and some exhibit synergism. For instance, although cobalt is tolerated in some cells at up to 10 mg/L, it increases the damage caused by germanium. Therefore, it is difficult to assess the true effect of various impurities especially since temperature, current density, and electrolyte concentration are also factors. The effects of various impurities are given in Table 5.

Purification actually starts with the precipitation of the hydrous oxides of iron, alumina, silica, and tin which carry along arsenic, antimony, and, to some extent, germanium. Lead and silver sulfates coprecipitate but lead is reintroduced into the electrolyte by anode corrosion, as is aluminum from the cathodes; and copper by bus-bar corrosion.

Purification is primarily based upon displacement from solution, ie, so-called cementation with metallic zinc of the metals below zinc on the electromotive series. For this purpose, zinc is melted and atomized with an air jet to so-called zinc dust, which is screened into fractions in the range 70–300 μm (ca 50 to 200 mesh) for use. Zinc is added in large excess over stoichiometric and about 5% of the cathode zinc is recycled for this purpose. Cobalt is removed in some plants by precipitation with α-nitroso-β-naphthol. In one plant, dimethylglyoxime is used to insolubilize nickel. Most importantly, magnesium, alkali metals, chloride, and fluoride are not removed by iron precipitation or cementation but by stripping the zinc from a portion of the electrolyte and discarding the solution. The plant solutions are saturated with calcium sulfate and incoming calcium is precipitated as gypsum at relatively cool points, eg, the cooling towers and coils. Gypsum is an important outlet for sulfate which enters with the calcine in small amounts.

Purification is highly specific for any given plant and ore composition. In general, the zinc-dust particle size, pH, temperature, agitation and presence of additives (such as copper, arsenic, and antimony) are important in determining the reaction rate. Both batch and continuous procedures are employed. Final polishing stages using minimal zinc are sometimes added. The three most popular purification flow sheets are described in Table 6 (47), but many variations are practiced, eg, new three-stage procedures (48–49).

Table 5. Effect of Impurities in Zinc Electrowinning

Impurity	Reported range—11 plants, mg/L	Class[a]	Current efficiency	Resolution	Deposit	Other
germanium	0.005–0.2	4	lowers	high	spongy	worse with cobalt
tellurium	<0.001[b]	4	lowers	high	uneven	worse with cobalt
selenium	<0.002[b]	4	lowers	high	uneven	
arsenic	0.003–0.02	4	lowers	high	corrugated	worse with cobalt
antimony	0.01–0.03	4	lowers	high	beady poor adhesion	worse with cobalt or germanium
copper	0.05–0.2	4	lowers	yes		
nickel	<0.01–0.5	3	lowers	mild	holes	
cobalt	0.03–2.0	3	lowers	mild	holes	reduces lead deposition
tin	<0.02[b]	3	lowers	yes	filmy	
iron	0.2–25	3	lowers			may reduce anode corrosion
cadmium	0.01–5	2				deposits with zinc
lead	1[b]	2				deposits with zinc
thallium	0.5–5[b]	2				increases lead deposition, deposits
aluminum	10[b]	1				increases electrolyte resistivity
magnesium	6.5–12 g/L	1				increases electrolyte resistivity
manganese	3–3.5 g/L	1				oxide forms on anode, lowers corrosion, may insulate anode
chloride	20–100					corrodes lead anodes
fluoride	2[b]					corrodes electrodes, causes sticking of zinc to cathode

[a] Class 1. Higher reduction potential than zinc. Will not deposit on cathode but will affect conductivity. 2. Lower reduction potential than zinc. Hydrogen overvoltage above 0.65 V. Deposits with zinc but does not affect zinc deposition. 3. Reduction potential between zinc and hydrogen. May deposit and redissolve, affecting current efficiency. Causes local lowering of hydrogen overvoltage, making pits, holes, etc. 4. Reduction potential below hydrogen. Hydrogen overvoltage below 0.65 V. Deposits and causes hydrogen evolution. Tends to form hydrides.
[b] One plant reporting.

Copper sulfate, in small amounts, activates the zinc dust by forming zinc–copper couples. Arsenic(III) and antimony(III) oxides are used to remove cobalt and nickel; they activate the zinc and form intermetallic compounds such as CoAs (49). Antimony is less toxic than arsenic and its hydride, stibine, is less stable than arsine and does not form as readily. Hydrogen, formed in the purification tanks, may give these hydrides and venting and surveillance is mandatory. The reverse antimony procedure gives a good separation of cadmium and cobalt.

Aeration must be avoided since it can oxidize and resolubilize the cemented (precipitated) impurities. Filter presses are used after each step and the cakes are leached to recover various values. For example, cadmium is dissolved, recemented with zinc, and recovered on site either electrolytically or by distillation. A copper residue of 25–60% copper is sold for recovery elsewhere. The other impurities cannot be recovered economically with the exception of cobalt in some plants.

Electrolysis. The net electrochemical reaction is $ZnSO_4 + H_2O \rightarrow Zn + H_2SO_4 + 0.5\ O_2$. Electrolysis is carried out in cells with 17–44 aluminum cathodes and 18–45 lead anodes per cell (30,50). Cell tanks are usually concrete lined with lead, rubber, or plastic. Anodes and cathodes in a given cell are connected in parallel with an imposed

Table 6. Purification of Zinc Liquor

Stage	Conventional arsenic[a]	Conventional antimony[a]	Reverse antimony[b]
stage 1			
remove	Co, Cu, Ni, As, Sb	Co, Cu, Ni, Cd	Cu, Cd, Tl
reagent	coarse Zn, CuSO$_4$[c], As$_2$O$_3$	coarse Zn, Sb[d]	Zn
temperature, °C	90	65–75	ambient
pH	4.0		
stage 2			
remove	Cd, Tl	Co, Cu, Ni, Cd	Co, Ge, Ni
reagent	Zn	fine Zn	Zn, Sb
temperature, °C	70–80	ambient	90
pH	3.0		

[a] To remove Co and Ni in stage 1 (hot).
[b] Reverse antimony: stage 2, Sb (hot) for Co, Ni removal.
[c] CuSO$_4$ added if Cu is <400 mg/L.
[d] α-Nitroso-β-naphthol may be substituted for Sb to precipitate Co.

voltage of 3.3–3.8 V which is higher than the standard 2.35 V because of the resistance of the electrolyte, gaseous, and solid films on the electrodes, and contact resistance. Energy consumed is 3.3 kW·h/kg zinc at slightly over 50% efficiency. Cells are arranged in groups connected in series with direct current supplied by rectifiers which are controlled by load-optimizing equipment in many plants. Thus, the power consumption varies, lower during peak demand period and higher during off-hours.

Increasing the current density increases the zinc-output rate per cell. Electrolytic plants were formerly classified as either low current density (200–450 A/m^2)–low acid (<6%) or high current density (1000 A/m^2)–high acid (22–28%). The ratio of zinc to acid in solution must be kept high enough to avoid resolution of cathode zinc and therefore high acid plants operate at higher zinc levels. At higher current density, more heat must be removed and solution purity is more critical but the return acid can more readily dissolve zinc ferrite formed in roasting. The choice of current density to use is therefore determined by many factors. Modern plants have current densities of 300–600 A/m^2 and achieve efficiencies of slightly over 90% in good operation (see Table 7). Glue ($\leq$100 mg/L) improves the smoothness of the cathode deposit and moderates the effect of impurities (52).

The purified zinc solution is fed to the electrolyte recirculating stream at a rate that holds the composition of the electrolyte constant, commonly ranging from 100–200 g H$_2$SO$_4$ and 45–70 g Zn/L. Continuous monitoring of the density and conductivity of the spent acid aids in control. This is important since the range of acidity is narrow for maximum current efficiency at any given current density.

The cells are fed individually but cascading is practiced in some older plants. In this system, electrolyte overflows from one cell to the next in a series of 3–9 cells. More commonly, cells are placed in rows, with each cell overflowing into a common spent-acid launder.

Cooling by means of evaporative cooling towers is required to maintain a constant temperature of 30–40°C. At higher temperatures, the deposit is rougher, impurity effects are more pronounced, lead codeposition is favored, and the manganese dioxide formed at the anode increases and tends to adhere rather than fall to the bottom of the cell.

Table 7. Cellhouse Operating Parameters[a]

Parameters	AMAX Sauget, Illinois	Electrolytic Zinc of Australasia Risdon, Tasmania	Canadian Electrolytic Zinc Valleyfield, Quebec	Vieille Montagne Balen, Belgium[b]	Jersey Miniere Clarksville, Tennessee	Mitsubishi Akita, Japan	Mitsui Hikoshima, Japan
electrolyte							
Zn, g/L	55	46	70	50	90 (65)[c]	47	na
H_2SO_4, g/L	200	100	200	190	128 (175)[c]	115	na
temperature, °C		35	35	30	32	36	36
cell							
arrangement	cascade	row[d]	row	row	row	row	row
cathodes/cell	27	48	36	44	49	39	33
cathode area, m^2	0.65	0.65	1.1	2.6	2.6	1.6	1.6
current density, A/m^2	807	540	500–700	300–400	375	370–430	300–600
voltage	3.8	3.5	3.5	3.3–3.5	3.3	3.55	
liner	lead	lead	lead	PVC	PVC	rubber	rubber
deposition, h	24	72	24				
stripping							
type	manual	manual	manual[e]	auto	auto	auto	auto
removal				knives	knives	hinged rubber edge-knives	high frequency shake
cycle, h				48	48	48	32–48

[a] Older plants are described in ref. 51.
[b] The National Zinc Company plant at Bartlesville is, like the Jersey Miniere plant, based on the Vieille-Montagne design.
[c] Secondary cells.
[d] New section.
[e] Old section.

Oxygen evolved from the anodes as well as some hydrogen from the cathodes produces a mist which is trapped by a froth maintained by adding cresylic acid, sodium silicate, and gum arabic, or glue plus cresol. Alkaline-earth carbonates prevent lead contamination of the cathode zinc. Most of the lead is deposited in the cell sludge as insoluble carbonate–sulfate.

Cells and anodes are cleaned on a cycle of 40–90 d when the sludge, mostly manganese dioxide, is pumped out. The manganese dioxide coating on the anodes is removed with water jets and the anodes mechanically straightened. Aluminum cathodes are welded to aluminum header bars which have copper contact points welded or cast on. The aluminum is polished at a frequency depending upon the corrosiveness of the electrolyte. Cathodes are often anodized to minimize corrosion by brief contact with the "live" anodes upon re-entering the cell after stripping. Chlorides and fluorides are corrosive and make the zinc adhere to the cathode. This is especially critical in mechanized stripping operations which are rapidly replacing the manual method. Antimony chemicals are added in some plants (≤0.08 mg Sb/L) to combat sticking. Manual stripping of the deposit from the cathodes has always been an expensive and onerous task. Several plants adapted automatic mechanical strippers to existing cellhouses and, in 1969, Vieille-Montagne started a fully automated cellhouse at Balen, Belgium, featuring jumbo cathodes of large area (2.6 m^2).

Melting, casting, zinc dust. Cathode zinc is melted, usually by induction heating, and distributed by pump to a casting machine which produces 26-kg slabs of special high grade zinc. Larger castings of 1090 kg and 450 kg, for instance, are made in preheated molds. Some of the zinc may be pumped to an alloying furnace for the addition of metals such as lead, aluminum, and copper. These alloys are cast into slabs, jumbo blocks, and special shapes.

Some of the melted zinc is fed to the zinc-dust unit where the molten zinc may be dropped from a crucible through a small orifice (2.5 mm) to be atomized in a blast of air. Solidified droplets are collected in a chamber and screened to the proper size for purification and cadmium plant cementation. Frequently, coarse (+70–200 μm) and fine (−70 μm) fractions are required.

Ammonium chloride is used as a flux in the melting furnace because the large surface of the cathodes favors the formation of dross, ie, oxide-coated globules of zinc. The dross is separated by liquation or air-swept milling into metal and oxide fractions. In the latter, the oxide fraction is swept out of the mill and can be returned to roasting for the elimination of chloride. Metallic zinc is recycled. Overall melting efficiency is 96–98%.

Pyrometallurgical Processes. Zinc pyrometallurgy is based upon reduction of zinc oxide (53).

$$ZnO(s) + CO(g) \leftrightarrows Zn(g) + CO_2(g)$$

$$CO_2(g) + C(s) \leftrightarrows 2\,CO(g)$$

$$ZnO(s) + C(s) \leftrightarrows Zn(g) + CO(g)$$

The lowest temperature for the reaction is 857°C but 1100–1300°C is required for acceptable rates. In this range, the two component reactions proceed at about the same rate and the reduction is diffusion controlled (54). Both reactions are reversible and the overall reaction is endothermic, requiring about 5.5 GJ/t Zn (1.2 × 10^9 cal/short ton Zn) at 1200 K.

Carbon monoxide and dioxide oxidize zinc vapor below 1100–1300°C although only the carbon dioxide reaction is significant. Rapid condensation of the zinc vapor avoids the formation of zinc-oxide-coated droplets, so-called blue powder.

At first, batchwise horizontal retorts were used for smelting, and later continuous vertical retorts, both externally fired. Continuous, internally heated furnaces such as the electrothermic furnace followed, and the last important development was the Imperial Smelting blast furnace.

Imperial Smelting furnace. In the ISF process, zinc vapor is produced in a blast furnace along with lead. Although air is introduced to provide heat through the burning of coke, the atmosphere in the upper portion of the furnace is not oxidizing despite the carbon dioxide content (CO$_2$/CO = 10%/25%) because of the high temperature, 1000–1050°C. To prevent reaction of carbon dioxide with zinc vapor, the gases leaving the furnace are rapidly cooled to 550–600°C in splash condensers (see Figure 3).

The Imperial Smelting Corporation, Ltd., brought its blast furnace on-stream at Avonmouth, UK, in 1952 (55), and in 1980 there were 13 operating ISF plants around the world ranging in capacity from 30,000 to 105,000 metric tons per year. There are none in the United States, Mexico, or Canada since the Brunswick Mining & Smelting Corporation plant at Belledune, N.B., Canada, was converted to straight lead in 1972. The ISF can handle complex zinc–lead–copper ores, zinc-containing dust from

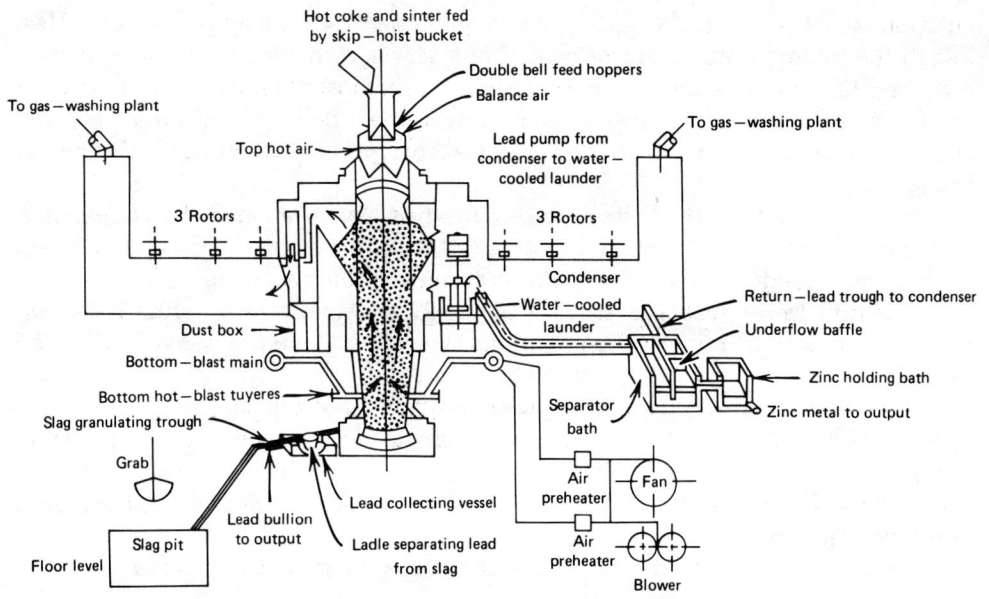

Figure 3. Imperial Smelting furnace.

steelmaking, drosses, leach plant residues, etc, in a large, economical unit. The ISF is more energy-efficient (37.9 GJ/t or 32.6 × 10⁶ Btu/short ton Prime Western Zinc) than either the vertical retort (60.2 GJ/t or 51.8 × 10⁶ Btu/short ton) or the electrothermic (72.2 GJ/t or 62.1 × 10⁶ Btu/short ton) processes (56). In fact, the figures favor the ISF over electrolytic (49.8 GJ/t or 42.9 Btu/short ton) for Prime Western and 44.8 vs 50.1 GJ/t (38.6 vs 43.1 × 10⁶ Btu/t) for special high grade (57). The electrolytic process, however, uses lower cost fuel in some locations and has other cost advantages including labor and environmental.

Feeds have lead–zinc ratios from 0.45 to 0.82 (58). The feed mix, primarily sulfide concentrate, is updraft-sintered and oxidized with the off-gases passing to an acid plant. Most of the cadmium is eliminated in the sintering step and the lead oxide, if above 7% lead, acts as a flux to produce the hard sinter required for the blast furnace (58). Only a small portion of the sintering heat is recovered compared to the waste heat recovered in fluid-bed roasting. Hot, briquetted, roasted concentrate or oxidized ore is an alternative to sintering, especially where lead is low. This approach is being used to supplement sintered feed up to 25%. Successful trials of 100% briquetted feed have been made, opening the possibility of using the ISF for zinc alone.

Sinter and metallurgical coke are preheated to 400 and 800°C, respectively, and fed into the top of the blast furnace through a double-bell hopper. Preheated air (800–1000°C) is blown in through tuyeres near the bottom providing heat and carbon monoxide by oxidation of the coke. Lead, silver, and copper are deposited in the molten lead bullion tapped from the furnace bottom. Slag floats on top of the bullion and is tapped periodically. Zinc vapor exiting the top is heated to 1000°C by combusting the off-gas with injected air and is rapidly condensed in a spray of molten lead after which the molten metals flow to a cooler where a zinc-rich layer overflows and the high lead bottom recirculates to the splash condenser. Zinc recoveries range from 92 to 95% and, with lead at 1.1–1.3%, the zinc is in the Prime Western class.

Since a large amount of heat must be absorbed at the condenser, the lead recirculation rate is very high, 300–500 t/h per condenser or 400 t lead per ton zinc. Heat lost to the cooling water is not recovered but heat is recovered from the condenser off-gases (LCV or low calorific value gas) at 450°C in most plants to preheat air, coke, etc. Consideration is being given to using a waste-heat boiler at this point. The Hachinohe smelter in Japan has reported on heat recovery from LCV gas and other energy savings (59).

Grades purer than the Prime Western product have been made by vacuum dezincing the liquid alloy to recover 99.9% zinc at 0.02% lead. Distillation (New Jersey Zinc process) produces a purer product at a considerable cost in energy.

The ISFs range in shaft area from 15.3 to 27.1 m^2 and produce from 112 to 334 t/d. The standard furnace is 17.2 m^2 and its typical performance is given in Table 8 (58).

Efficiency is improved and cost lowered by the following changes:

Hot (500–700°C) binderless briquetting of oxidized zinc and lead feed materials to replace sintering (see Size enlargement).

Changes in furnace design. More uniform charge distribution, new furnace shapes, and tuyere design.

Drying of the air blast lowers coke usage 3% for each 1% of water vapor removed.

Heat recovery from furnace off-gas (LCV gas) normally used to preheat coke and blast air; LCV gas has also been burned to heat melting baths and generate steam and power.

Vertical retort. The vertical retort process was developed by The New Jersey Zinc Company (60). Production started in 1929 at Palmerton, Pennsylvania. A charge of 25% bituminous coal, 10% anthracite, 10% recirculated fines, and 55% sinter is mixed with suitable binders and densified in a chaser mill. The mix is briquetted and coked autogenously, that is, the volatiles from the coking process burn in the bed of briquettes to supply the needed heat. Briquettes leave the coker at ca 700°C and are transferred to the top of the retort where they are charged on a periodic cycle. Radiant heat from the walls drives the reduction as the briquettes descend. The loaf-shaped briquettes, strong enough to withstand breakage, are extracted from the bottom through a water

Table 8. Imperial Smelting Furnace Performance[a]

Characteristic	Value
production[b], t/d	
Zn	256
Pb	128
input ratio	
Pb:Zn	0.5
C:Zn	0.75
Zn in slag, %	7.0
slag:slab Zn	0.65
recovery, %	
Zn	94.5
Pb	95.0
carbon burned per metric ton slab Zn	0.82

[a] Ref. 58.
[b] Full blast.

seal. Carbon monoxide forces the zinc vapor up the column; ca 95% of the metal is condensed and the balance scrubbed out.

The Palmerton, Pennsylvania, plant had 43 retorts with an output of ca 8 t/d per retort. Recovery was ca 94% when the plant was shut down in 1980. The zinc contained ca 0.3% lead, 0.10% cadmium, and 0.01% iron plus minor impurities. Lead and aluminum are added to produce galvanizer's zinc.

Zinc of 99.995% purity is produced in fractional-distillation columns which were developed along with the vertical retorts. Refining is a two-stage procedure in which higher boiling lead and iron are separated in the first column, whereas lower boiling cadmium is distilled in a second column (61).

Electrothermic process. Electric-arc furnaces with molten iron-slag baths have never been successful despite many attempts (62). However, a process developed by the St. Joseph Lead Company (now St. Joe Zinc Company), based upon resistance heating, was employed in four plants in 1981 (34). There is relatively little dusting and zinc condensation is efficient.

Sintering in two stages, the second with sand, produces a high purity, hard sinter which is required for the oxide and high grade zinc furnaces. Single-stage sinter is fed to the Prime Western and intermediate-grade circuits. Both sinters are sized to 9–25 mm. Coke is fed at ca 300%. Small furnaces have daily outputs of 16–25 t, large ones 100 t. Current consumption is 1250 kW per electrode pair at 200–230 V. Vapor exits through an annular vapor ring and is led to the condenser.

The Weaton-Najarian zinc condenser was commercialized in 1936. The condenser and cooling well of the electrothermic furnace hold 48 t of molten zinc. Hot zinc-laden gases bubble through the zinc in the condenser and cause rapid circulation through the cooling well which is kept at 480–500°C by water coils. The off-gases are scrubbed and burned for fuel value. Scrubber water is ponded to recover blue powder.

Horizontal retort. In 1800, the first commercial zinc process made use of the horizontal retort. In 1980, only three such plants remain because they are not competitive in terms of labor and fuel costs. Furthermore, the dust produced presents a serious pollution problem. Nevertheless, in 1956, the tonnage of zinc produced from horizontal retorts was above that of any previous year; today, it is mostly of historical interest (1,63).

Banks of Belgian horizontal cylindrical retorts are arranged in a furnace containing 500–700 retorts. Heat is applied externally by natural or producer gas. The retorts are charged with a damp mixture (35–40% zinc) of sinter, blue powder, and carbonaceous reducing agent (50%). The cycle is ca 48 h, and 40 kg zinc is produced per retort. The zinc vapor is liquefied in a clay condenser. Metal is drawn periodically in 10-kg lots into a ladle which is skimmed to remove blue powder and impurities. Slab zinc is cast in molds directly from the ladle and the casting is skimmed.

Recoveries are 88–95%, and the quality of the metal depends upon the composition of the charge. At the Asarco Mexicana plant at Rosita, Coahuila, slightly over 10% of the furnace production exceeds the Prime Western specifications.

Secondary Recovery. Zinc is recovered as metal, dust, and chemicals (including oxide) from secondary sources, mostly scrap (see also Recycling). So-called old scrap originates from die castings and engraver's plates, whereas new scrap, such as drosses, skimmings, flue dust, clippings, and residues, originates in various processes. The amount of old scrap has remained at ca 5–6% of total production for the last two decades but it may be possible to increase recovery from die castings and steelmaking dust.

In 1980, secondary slab zinc was produced in the United States at nine plants as well as by two primary producers. Zinc dust was produced at eight plants, mostly from secondary materials. Recovery practices vary greatly, but zinc products are usually made from metallic scrap by melting and distillation. Dust is made by rapid condensation of zinc vapor in oxygen-free atmospheres; the spherical particles have diameters of 1–20 μm (2).

A novel process, called Zincex, was brought on-stream in 1976 by Metalquimica del Nervion in Bilbao, Spain, where 8000 t/yr is produced from pyrite cinders (64). The cinders are given a chloridizing roast and leached to produce a solution containing 20–30 g/L zinc, 18–25 g/L iron, 60–100 g/L chloride, and 120–155 g/L sulfate. Solvent extraction in two stages recovers 98% of the zinc. The first stage uses a secondary amine to extract $ZnCl_4^{2-}$ which is stripped with water. Di-2-ethylhexyl phosphoric acid absorbs zinc ion in the second stage where spent electrolyte is the stripping agent. Thus, zinc is well-separated from chloride, iron, and other harmful impurities.

Economic Aspects

The United States, USSR, Canada, and Australia have the largest known reserves of zinc ore which should permit mining at current levels into the next century (see Tables 2 and 3). World mine production of recoverable zinc between 1970 and 1980 is given in Table 9 (65–66). Mine production in the United States in the 1970s was lower than in the 1960s. The U.S. share of world production has fallen from ca 9% in 1970 to ca 5.5% in 1980.

Tennessee is leading in mine production followed by Missouri; five of the eight principal producer states are located in the eastern part of the United States with a combined output in 1980 of over 50% of U.S. production.

Zinc consumption is broken down into slab and all-classes classifications. The latter includes zinc-containing materials made from ores and residues as well as from secondary (scrap) zinc. Consumption and production figures of slab zinc show that a substantial difference exists which is made up by imports (see Table 10). The imports consist of ore and concentrates based on zinc content, slab zinc in the form of blocks and pigs, zinc dust by gross weight, waste and scrap on a gross weight basis, and dross and skimmings based on zinc content. In 1970, ore and concentrate imports were about equal to U.S. mine production and slab zinc imports represented 30% of U.S. production. In the late 1970s, imports of ore and concentrates decreased below U.S. production and imports of slab zinc increased to exceed U.S. production from 1976

Table 9. United States, Canada, North America, South America, and World Mine Production, 1000 t

Year	United States	Canada	North America[a]	South America[b]	World	World/United States ratio
1970	484.6	1239.2	1903.6	397.3	5464.0	11.28
1975	425.8	1055.2	1738.4	544.2	5988.1	14.06
1977	448.3	1300.2	2052.8	635.2	6608.2	14.74
1979	293.8	1204.4	1766.4	643.2	6335.1	21.56
1980	343.9	1058.7	1661.1	649.1	6214.1	18.07

[a] United States, Canada, Mexico, Guatemala, Honduras, Nicaragua.
[b] Argentina, Bolivia, Brazil, Peru, Chile, Colombia, Ecuador.

Table 10. U.S. Production, Consumption, and Prices of Slab Zinc, 1000 t

Use	1970	1976	1978	1980
production				
domestic ores, recoverable Zn content	484.6	439.6	302.7	313.1
slab zinc				
from domestic ores	366.5	346.4	267.4	234.4
from foreign ores	429.9	106.1	139.4	113.7
from scrap	70.0	62.2	34.8	28.8
Total	*866.4*	*514.7*	*441.6*	*376.9*
imports				
ores and concentrates (Zn content)	477.0	88.1	187.4	129.9
slab zinc	245.3	648.2	617.8	410.6
other[a]	9.9	18.4	19.6	11.5
consumption				
slab zinc	1076.8	1135.8	1063.3	817.9
all classes	1420.6	1479.0	1449.2	1153.6
price[b], Prime Western, yearly average, ¢/kg				
United States	33.778	81.607	68.291	82.529
London (LME)[c]	29.035	71.230	59.248	76.033

[a] Zinc dust, gross weight; waste and scrap, gross weight; dross and skimmings = zinc content.
[b] Ref. 67.
[c] Ref. 68.

to 1980. Imports of ore and concentrates in 1980 came from Canada (48.5%), Peru (30.9%), and Mexico (12.2%) for a combined total of about 92%. Canada supplied ca 68% of slab zinc and Mexico and Australia each supplied ca 6%.

For 1980, the figures for U.S. exports are given in Table 11.

Production processes are given in Table 12. Electrolytic processes have increased because of lower cost and fewer environmental problems. In 1970, Montana was the third largest producer of slab zinc, but in August, 1972, the Anaconda Company terminated operations of its electrolytic plant at Great Falls, Montana, which had a rated annual capacity of 162,000 t. Production of slab zinc in Tennessee commenced in 1978 when Jersey Miniere Zinc Company began operation of its new 90,000 t/yr electrolytic smelter at Clarksville. Today, Tennessee is the leading producer of slab zinc. Although the U.S. demand for zinc metal in the past ten years has remained fairly stable, smelting capacity has declined by almost 50%. Plants closed because they were obsolete and could not meet environmental standards or obtain sufficient concentrate. Consequently, slab zinc has replaced concentrates as the principal import form. This situation is expected to prevail up to the year 2000 (69–71).

Table 11. U.S. Zinc Exports[a]

Material	1000 t
ores and concentrates	54.5
zinc pigs and slabs	0.3
zinc sheet and strip	3.1
dross, scrap, ashes, and skimmings	34.1
Total	*92.0[b]*

[a] Refs. 65–66.
[b] Ca one sixth of imports.

Table 12. Distribution of Zinc Production Processes

Process	Date of commercialization	Noncommunist production, % 1958[a]	1968[a]	1980[b]
electrolytic	1915	50	56	76
horizontal retort	1800	32	15	2
vertical retort	1930	7	14	7
electrothermic	1936	3	4	4
Imperial Smelting (ISF)	1952	8	11	11

[a] Ref. 13.
[b] Based on capacity.

Electrolytic and distillation processes are used to produce primary slab zinc at smelters from ores and concentrates, whereas redistillation is used to recover zinc from secondary zinc materials at both primary and secondary smelters (see Table 13) (65–66). In 1965, nearly 60% of the primary slab zinc was manufactured by the distillation process. However, in 1975, electrolytic slab zinc became dominant and by 1980 its share increased to 85%.

In 1980, the 310,000 t obtained from secondary smelters was 89% of primary zinc production. Slab zinc made from scrap was down to about 29,000 tons or 9.3% of secondary output caused by reduced production at the primary smelters. The balance of the zinc-containing scrap utilized in 1980 went into the production of zinc dust (9%); zinc, magnesium, and aluminum alloys (3%); brass and bronze alloys (59%); and pigments and chemicals (19%). As a result of the closing of U.S. primary production, greater reliance will have to be placed on secondary zinc and imports (69–71).

The world production of mined zinc ore and of zinc-slab production are given in Table 14. The 12 producing countries listed accounted for 75–80% of the zinc mined in the world. In many cases, ore is smelted in the country in which it is mined, but an active market in zinc concentrates exists. The most notable exceptions are the FRG, France, Belgium, Italy, and the Netherlands, which produce metal from imported ore in varying degrees. Mining countries that have little or no smelting capacity are Peru, Ireland, and Sweden.

Zinc consumption is categorized in five semifabricating markets (see Table 15). Galvanizing was the main market for zinc in the 1970s followed by zinc-base casting alloys and brass and bronze. Depressed construction and automotive industries caused a decline from 1979 to 1980 of ca 18%, and the die-casting business declined 34% and galvanizing 24%.

Table 13. Slab-Zinc Production in the United States by Various Reduction Methods, 1000 t[a]

	Method of reduction				
	Primary		Secondary redistilled		
Year	Electrolytic	Distilled	At primary smelters	At secondary smelters	Grand total
1970	356.8	439.6	59.7	10.3	866.4
1975	211.7	185.8	31.7	20.8	450.0
1977	210.5	197.9	26.4	19.5	454.3
1980	295.2	52.9	15.3	13.4	376.8

[a] Refs. 65–66.

Table 14. World Mine Production[a] and Slab-Zinc Production, 1000 t[b]

Country	1970 Mine production	1970 Slab zinc	1976 Mine production	1976 Slab zinc	1978 Mine production	1978 Slab zinc	1980 Mine production	1980 Slab zinc
Australia	484.2	260.6	461.9	249.2	473.3	294.3	470.6	306.0
Belgium		234.8		244.4		232.9		238.0
Canada	1135.7	417.9	1145.0	472.3	1245.3	495.4	1058.7	586.7
France		223.7		233.5		231.2		252.8
FRG		301.6		304.8		306.8		368.8
Ireland	96.5		62.8		176.0		228.8	
Italy		142.0		191.2		177.6		206.4
Japan	279.7	676.3	260.0	742.1	274.6	767.9	236.9	743.9
Mexico	266.4		259.2		244.9		239.3	
Netherlands		46.2		123.2		135.3		168.1
People's Republic of China	99.8	99.8	135.0	150.0	150.8	160.0	155.0	160.6
Peru	299.1		413.7		457.5		492.0	
Poland	242.0	209.0	215.0	237.0	231.0	222.0	237.2	215.0
Spain	98.1		82.3		143.2		170.4	
Sweden	93.4		128.3		167.4		170.0	
United States	484.6	804.5	483.0	514.7	337.0	441.5	343.9	358.6
USSR[c]	557.9	557.9	1020.1	1000.0	1030.0	1055.0	1020.1	1088.6
other	1050.3	999.9	1586.2	1327.1	1509.7	1521.3	1391.3	1502.4
world	5187.7	4974.2	6252.5	5789.5	6440.7	6041.2	6214.2	6195.9

[a] Based on zinc content.
[b] Refs. 65–66.
[c] Estimated.

Die-Casting Alloys. Consumption of die-casting alloys increased rapidly in the 1940s because of usage in the automotive industry. Zinc tonnage used in die castings reached a peak in 1965 with 578,766 metric tons. The decline in the zinc die-casting market can be attributed to reduced automotive sales and competition from aluminum, magnesium, and plastics, particularly where weight limitations and reductions are required. However, the development of thin-walled die castings and improved alloys and finishing techniques indicates that die casting will continue to consume large quantities of zinc in the transportation and appliance industries. The price of zinc relative to that of aluminum and plastics and public preference for metal parts because of their durability are important factors affecting the use of zinc. Markets and consumption for zinc die castings are given in Table 16.

Zinc Dust. U.S. production and imports are given in Table 17 (66), distribution of consumption in Table 18 (2). The slight decline since 1970 can be attributed to a reduced demand for chemical uses since the demand for zinc dust for coatings has increased.

Specifications and Grades

The chemical requirements for three slab-zinc grades made from ore or other material by distillation or electrolysis are specified by ASTM as given in Table 19 (74). Zinc produced by sweating or remelting of secondary zinc is not covered. Purity ranges from 98.0 to 99.99%; the main impurities are Pb, Fe, and Cd. The impurities strongly

Table 15. Distribution of U.S. Slab-Zinc Consumption, 1000 t [a]

Use	1970	1976	1978	1980
galvanizing				
sheet and strip	229.7	223.8	268.7	218.9
wire and wire rope	28.0	24.3	22.8	20.6
tube and pipe	58.5	44.4	47.4	34.8
fittings (for tube and pipe)	8.6	5.9	6.9	5.3
tanks and containers	3.6	3.0	2.9	3.8
structural shapes	17.0	31.4	33.3	19.4
fasteners	4.8	3.7	4.8	2.7
pole-line hardware	9.0	4.3	4.9	3.8
fencing, wire cloth, and netting	16.4	20.0	25.0	12.1
other and unspecified uses	54.6	32.2	37.4	23.6
Total	*430.2*	*393.0*	*454.1*	*345.0*
zinc-base alloys				
die-casting alloys	411.4	380.8	346.0	202.6
dies and rod alloys	0.8	0.9	0.5	0.2
slush and sandcasting alloys	9.1	5.7	7.6	5.6
Total	*421.3*	*387.4*	*354.1*	*208.4*
brass and bronze[b]				
sheet, strip, and plate	55.9	82.7	70.2	37.6
rod and wire	37.6	49.5	46.3	32.4
tube	8.2	6.7	6.8	4.6
castings and billets	4.2	3.8	4.4	1.8
copper-base ingots	9.0	7.0	6.6	17.0
other copper-base products	0.9	1.1	7.2	3.0
Total	*115.8*	*150.8*	*141.5*	*96.4*
rolled zinc	37.2	27.1	24.9	21.1
zinc oxide	39.8	35.4	37.2	29.9
other[c]	33.0	35.3	38.9	25.0
estimated undistributed consumption				91.2
Grand total	*1077.3*	*1029.0*	*1050.7*	*817.0*

[a] Refs. 65–66, 72.
[b] For zinc in brass objects, see Copper alloys.
[c] Zinc dust, light metal alloys, desilvering lead, bronze and brass powders, zinc chemicals.

affect the physical and chemical properties. Over the years, the intermediate grades, brass special, and selected grades of slab zinc, with purities between high grade and Prime Western, have been deleted from the specification because use requirements have changed. Nevertheless, the use determines the quality standards (75).

The ability of the zinc industry to produce special high grade by zinc distillation and electrolytic processes led to the development of modern zinc die-casting alloys which represent the second largest market for zinc. Because zinc die castings contain aluminum, this extra-pure grade must be extremely low in iron, lead, cadmium, and tin in order to avoid damaging intercrystalline corrosion. Such rigorous control of composition is essential for stable zinc die castings.

In galvanizing, the largest use of slab zinc, the iron or steel is dipped in molten zinc or electroplated. For hot-dip galvanizing, the lower grades are satisfactory. For some applications, such as wire, which must withstand bending without flaking of the coating, high grade or special high grade is employed. In electrogalvanizing, zinc is obtained either from zinc anodes or by direct solution from ore and subsequent puri-

Table 16. Markets for Zinc Die Castings, 1000 t[a]

Year	Total consumption	Automotive components	Builders hardware	Electrical components	Domestic appliances	Industrial agricultural, and commercial machinery	Miscellaneous industries	Sport goods and toys	Scientific, sound, and television equipment
1974	411.7	208.9	77.9	27.2	34.5	31.4	9.0	11.4	11.4
1976	377.8	187.4	79.5	33.5	27.4	21.5	9.8	8.9	9.6
1978	377.7	174.6	84.4	39.3	26.9	23.8	10.0	8.6	10.0
1980	284.7	94.0	76.0	37.8	22.8	25.4	11.6	9.3	7.8

[a] Refs. 65, 73.

Table 17. U.S. Production and Imports of Zinc Dust, 1000 t

Zinc dust	1970	1975	1978	1980
production	46.0	38.2	38.5	42.6
imports	8.5	5.2	9.0	

Table 18. Distribution of Zinc-Dust Consumption, 1974–1977 Average

Use	1000 t/yr	Percent
coatings		
industrial	17.1	35.2
automobile	10.0	20.5
marine	3.4	6.9
chemical uses	18.2	37.4
Total	*48.7*	*100.0*

Table 19. Compositions of the Commercial Grades of Zinc[a], %

Grade	Lead, maximum	Iron, maximum	Cadmium, maximum	Zinc, minimum by difference
special high grade[b]	0.003	0.003	0.003	99.990
high grade	0.03	0.02	0.02	99.90
Prime Western[c]	1.4	0.05	0.20	98.0

[a] When specified for use in the manufacture of rolled zinc or brass, aluminum max = 0.005%.
[b] Tin max = 0.001%.
[c] Aluminum max = 0.05%.

fication of the electrolyte. High grade and special high grade are used for anodes in electrogalvanizing.

To meet the specification for Prime Western, zinc is debased with lead. This may increase the cost, when the price of lead is higher than that of zinc. Generally, lead, the principal impurity in Prime Western, aids the galvanizing process by increasing the fluidity of the bath and defining spangle boundaries. The solubility of lead in molten zinc at galvanizing temperatures (425–460°C) is ca 1–1.5% by weight. If the lead content is higher, the lead settles at the bottom of the galvanizing kettle, protecting it from attack by molten zinc and facilitating the removal of iron–zinc alloy dross.

All grades of zinc slab are used to some degree in brasses and bronzes. In many leaded brass-mill products, the lead originates from the slab zinc; the accompanying cadmium is usually acceptable.

All grades may be used in rolled zinc alloys, although special high grade and high grade are most commonly employed. When aluminum is added to superplastic zinc alloys (Zn-22Al), special high grade is required plus protective alloying elements to prevent intergranular corrosion caused by moist service conditions, as is the case with zinc die castings.

Analytical Methods

Zinc in ores at the concentrating mill is often determined polarographically and mill slurries are commonly monitored continuously for zinc by x-ray fluorescence to control addition of flotation reagent (76). Low zinc concentrations in solution are analyzed polarographically or by atomic absorption spectroscopy (AAS) with a detection limit as low as 0.01 µg/mL (77). This method is used by most modern mills for determining other metals such as lead, cadmium, copper, cobalt, nickel, magnesium, and calcium. Silver and gold are fire-assayed. In concentrates where zinc is high and great precision is required, wet methods are often used, such as the titration with potassium ferrocyanide (78).

Zinc smelters use x-ray fluorescence spectrometry to analyze for zinc and many other metals in concentrates, calcines, residues, and trace elements precipitated from solution, such as arsenic, antimony, selenium, tellurium, and tin. X-ray analysis is also used for qualitative and semiquantitative analysis. Electrolytic smelters rely heavily on AAS and polarography for solutions, residues, and environmental samples.

Analysis of zinc solutions at the purification stage before electrolysis is critical and several metals present in low concentrations are monitored carefully. Methods vary from plant to plant but are highly specific and usually capable of detecting 0.1 ppm or less. Colorimetric process-control methods are used for cobalt, antimony, and germanium, turbidimetric methods for cadmium and copper. Alternatively, cadmium, cobalt, and copper are determined polarographically, arsenic and antimony by a modified Gutzeit test, and nickel with a dimethylglyoxime spot test.

Finished zinc and zinc alloys are usually analyzed for metals other than zinc by emission spectroscopy and the zinc determined by difference. ASTM method E 27 describes a technique using a dissolved sample and photographic detection. The internal standard is the zinc line at 267.0 nm. However, procedures using solid samples are generally preferred and photoelectric detection often replaces optical detection. Samples are cast and machined on the surface where the arc is struck. Up to 15 elements can be determined in a few minutes by modern automatic spectrometers. ASTM gives wet chemical methods for metals other than zinc (79).

The Committee on Medical and Biologic Effects of Environmental Pollutants of the National Research Council presents a well-referenced review of zinc sampling and analysis as they pertain to physiological media, water, air, foods, soils, and plants (80). Standard methods have not yet evolved in the clinical field but some are recommended. Most analyses today are by AAS which uses the 213.8-nm wavelength for zinc with reported sensitivities of 0.01–0.025 µg/mL. Electrothermal atomization (graphite furnace) lowers the detection limit by up to three orders of magnitude (77). Spectrophotometric procedures are popular, commonly employing dithizone (diphenylthiocarbazone) and zincon to develop red (535 nm) and blue (620 nm), respectively. Those methods detect zinc concentrations as low as 1 µg/mL (80). Polarography is still employed for water samples containing 0.01–0.1 mg Zn/L, but is being replaced by AAS.

Health and Safety Factors; Environmental Aspects

Recent review articles describe the role of zinc in ecological cycles, zinc as pollutant, and the biological importance and hazards of zinc (80–81).

Airborne zinc from automobiles, fuel combustion, incineration, and soil erosion is generally <1 µg/m^3 even in urban areas. At this concentration, zinc in the air is not a problem but fallout increases the concentration in waterways near cities; yet drinking water rarely exceeds the standard 5 mg/L. In the vicinity of zinc production facilities, lead smelters, brass works, etc, airborne zinc, cadmium, and lead are much higher but even here problems are created by fallout.

Because of modern agricultural practices, zinc deficiency is recognized in the soil of 39 U.S. states and supplementary additions are recommended. Deficiency occurs in some plants when tissue content drops below 20 ppm; the normal range is 25–150 ppm. The toxic effect of zinc has been seen at 400 ppm, although plants vary widely in their tolerance.

The use of municipal sludge in agriculture must be monitored carefully to avoid excess zinc. The source of zinc in such sludges is galvanized piping, electroplating and other industrial waste, zinc-containing products, and human excrement.

Zinc is not toxic to humans; ca one gram per day may be ingested without ill effects. Doses on the order of 10 g cause nausea and diarrhea. Recommended dietary allowance is 15 mg/d for adults. Inhalation of fresh zinc oxide may cause a temporary illness called metal fume fever (see also Zinc compounds). Animals have a high tolerance for zinc, although horses are the most sensitive and intakes of 3000 mg/kg body weight have shown toxic effects. Less is known about the toxic effects of zinc on marine organisms but they are generally regarded as more sensitive. The median lethal concentration in water for some fish has been set at 100–300 g Zn/L. Conversely, bivalve mollusks (eg, oysters) and algae tend to accumulate zinc, and >4000 mg Zn/kg dry weight has been found in the soft tissue of some oysters.

Zinc in the diet is necessary for growth and wound healing in animals and human beings. Zinc and cadmium intake are generally related, especially in contaminated areas, and the interaction of the two in the body is complex. The effects of zinc and cadmium vary greatly, but some of the adverse effects of the latter can be mitigated by sufficient zinc (see Mineral nutrients).

Hazards of Production. In most zinc mines, zinc is present as the sulfide and coexists with other minerals, especially lead, copper, and cadmium. Therefore, the escape of zinc from mines and mills is accompanied by these other, often more toxic, materials. Mining and concentrating, usually by flotations, does not present any unusual hazards to personnel. Atmospheric pollution is of little consequence at minesites but considerable effort is required to flocculate and settle fine ore particles which would find their way into receiving waters. Atmospheric zinc loss is estimated at 100 g/t of zinc mined, mostly from handling dry ore and concentrate and wind erosion of tailing piles. Effluent waters from well-managed concentration plants should be on the order of 5 µg Zn/L.

Occupational risks caused by chemicals in electrolytic zinc plants are described in ref. 82. Pyrometallurgical plants dealing with zinc have much the same hazards as electrolytic plants except that arsenic and antimony are problems only if present in the concentrate. On the other hand, danger from carbon oxides at the reduction step requires monitoring. Hygienic rules should be carefully observed.

In recent years, zinc plants have expended large sums to control in-plant dusts and environmental pollution. Liquid effluents are limed and settled to precipitate metals as hydroxides. Flocculants are used to reduce the total suspended solids and, in some instances, filtration of thickener overflow is practiced. Typically, zinc can be held to 1 mg/L and cadmium 0.1 mg/L on average by good control, and many procedures involve increased reusage of water. Particulate emissions are controlled mainly through venting, baghouses, and water scrubbers. Sulfur dioxide emissions have been reduced by installing double-absorption acid plants and improved containment of dilute gases.

Uses

Metallic Coatings. Zinc coatings on iron and steel for corrosion protection represented the largest market for zinc in the 1970s (see Table 15). Methods in general include hot-dip galvanizing, continuous-line galvanizing, electrogalvanizing, zinc plating, zinc spraying, and painting with zinc-bearing paints (83).

Die-Casting Alloys. The principal application of zinc as a structural material is in alloys for pressure die casting (see Table 16). The principal consumer is the automotive industry where uses include handles and locks, mechanical components, electrical components, body hardware and trim, lamp and lighting fittings, instruments, and other components. The zinc die castings per average car, reflecting the need for lighter weight and thin-walled castings, dropped from ca 22.6 kg per car in 1974 and 1975 to 13.6 kg in 1978, and 12.8 kg on the 1979 models. Builder's hardware, the second largest market for zinc die castings, includes covered door and window hardware, locks and keys, furniture and cabinet hardware, hand tools and cutlery, bathroom and plumbing fittings, general hardware, marine hardware, and luggage hardware.

The three zinc alloys, designated as Nos. 3, 5, and 7, are used for the hot-chamber die-casting process. Their compositions are given in Table 20 along with a cold-chamber zinc die-casting composition, No. 16 alloy (84–86). The No. 16 alloy was developed by the International Lead Zinc Research Organization (ILZRO) and has superior creep resistance at both ambient and elevated temperatures (87).

The zinc alloys 3, 5, and 7 are somewhat similar in properties and can generally be used interchangeably. The No. 3 alloy has excellent retention of impact strength and dimensions; No. 5 has greater hardness, tensile strength, and more resistance to creep, but suffers some loss of impact strength when used continuously at elevated temperatures; No. 7, the most recent addition to the series, has properties similar to No. 3, but is slightly softer and more ductile. Because of its lower magnesium content, it exhibits better fluidity and castability than either 3 or 5 (88).

Aluminum is the principal alloying addition varying from 3.5 to 4.3%. The Zn–Al phase diagram is shown in Figure 4 (89). Most die-cast and foundry alloys are Zn–Al

Table 20. Compositions of Zinc Die-Casting Alloys, wt %[a]

Element	No. 3 ASTM AG40A Ingot[b]	No. 3 ASTM AG40A Die casting[c,d]	No. 5 ASTM AC41A Ingot[b]	No. 5 ASTM AC41A Die casting[c,d]	No. 7 Ingot[b]	No. 7 Die casting[c]	No. 16 Ingot	No. 16 Die casting
copper	0.10 max	0.25 max[e]	0.75–1.25	0.75–1.25	0.10 max	0.25 max[e]		1.0–1.5
aluminum	3.9–4.3	3.5–4.3	3.9–4.3	3.5–4.3	3.9–4.3	3.5–4.3		0.01–0.04
magnesium	0.03–0.06	0.03–0.08[f]	0.03–0.06	0.03–0.08[f]	0.010–0.020	0.005–0.020		0.02
iron, max	0.075	0.100	0.075	0.100	0.075	0.100		0.04
lead, max	0.005	0.007	0.005	0.007	0.0020	0.0030		0.005
cadmium, max	0.004	0.005	0.004	0.005	0.0020	0.0020		0.004
tin, max	0.002	0.005	0.002	0.005	0.0010	0.0010		0.003
nickel					0.005–0.020	0.005–0.020		
titanium								0.15–0.25
chromium								0.10–0.20

[a] Zinc is the remainder.
[b] ASTM B 240.
[c] May contain nickel, chromium, silicon, and manganese in amounts of 0.02, 0.02, 0.035, and 0.5%, respectively. No harmful effects have ever been noted due to the presence of these elements in these concentrations; therefore, analyses are not required for these elements.
[d] ASTM B 86.
[e] For most commercial applications, a copper content in the range of 0.25 to 0.75% does not adversely affect the serviceability of die castings and should not serve as a basis for rejection; ASTM B 240-79 (82).
[f] Magnesium may be as low as 0.015% provided that the lead, cadmium, and tin do not exceed 0.003, 0.003, and 0.001%, respectively.

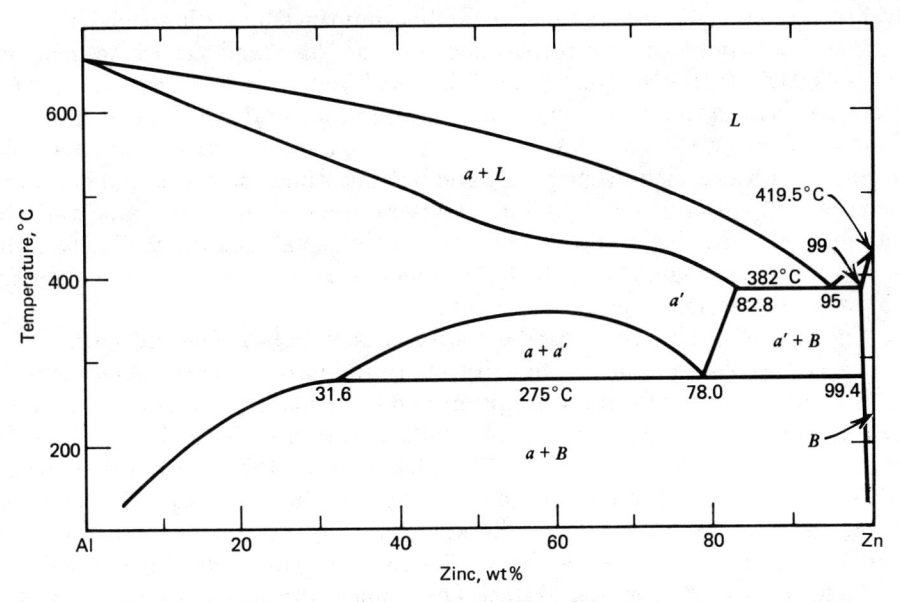

Figure 4. Aluminum–zinc phase diagram.

compositions. Aluminum improves die-cast strength, reduces grain size, and decreases the solution rate of iron and steel parts to the extent that a submerged plunger-type die-casting machine can be used. The spread between 3.5 and 4.3% allows sufficient latitude for production without decreasing optimum properties. An aluminum content below 3.5% promotes hot shortness, impairs the surface finish and lowers the mechanical properties. On the high aluminum side, a loss of impact strength begins at 4.5% aluminum, and at 5% the alloy is brittle. Aluminum in zinc, in the presence of excessive amounts of certain impurities (eg, Pb, Sn), can result in intergranular corrosion in die castings exposed to moist, warm atmospheres, causing them to lose strength, crack, and sometimes even disintegrate. Specifications place limits on these harmful impurities to avoid this problem, and special high grade zinc is employed for die-casting alloys (90–91).

Copper increases tensile strength and hardness and offers some protection against elements that promote intergranular corrosion. However, copper reduces impact strength and dimensional stability owing to aging and is therefore kept at 1.25% max.

Magnesium prevents intergranular corrosion by counteracting the harmful effect of small quantities of impurities such as tin and lead (92). The minimum values for magnesium are necessary for the level of impurities allowed in the casting specification. The maximum limit for magnesium in the alloy specifications is set arbitrarily to allow some spread for commercial production. Operating close to the minimum values lessens the hot shortness and improves castability. The No. 7 alloy allows the lowest minimum magnesium content in the presence of nickel which neutralizes elements that promote intergranular corrosion. Magnesium also increases the strength and the hardness.

Maximum limits are placed on iron, lead, cadmium, and tin. The maximum limits for lead and tin are set at amounts that do not promote subsurface network corrosion at the minimum levels of magnesium allowed in the die castings. At concentrations of ca 0.1% and higher, cadmium is detrimental to mechanical properties. Cadmium adversely affects hot shortness and castability, but the amount in special high grade zinc is too low to produce a noticeable effect. For the Nos. 3 and 5 alloys, the maximum limit is 0.004%, but for the No. 7 alloy, it is 0.002%, and special high grade zinc must be selected to meet this requirement in the preparation of this alloy. Iron has no detrimental effects on the permanence or properties, but an excess of iron may affect finishing operations such as machining and buffing. Zinc castings should not contain much over 0.02% iron. The ASTM limit of 0.100% iron is an arbitrary value which may be on the high side. The ASTM specification allows small quantities of nickel, chromium, silicon, and manganese which may come from aluminum used in alloying or from remelting of electroplated scrap (90).

The melting of zinc alloys for die casting entails (91) the melting down of virgin ingot; remelting of scrap from the foundry and trimming operations and reconstitution of this melt by means of suitable additions; and the holding of quantities of molten metal at a closely controlled temperature adjacent to the die-casting machine. The furnaces are generally of the pot type, although immersion-tube and induction furnaces are also used. Heat is supplied by gas, oil, or electrical resistance. The capacity of melting furnaces is between 450 and 9000 kg; total capacity is usually five to seven times the amount of metal required per hour. The pots are made from gray or ductile cast iron. Ladles are cast iron or pressed steel. The temperature of zinc is kept at 475–500°C in the melting and alloying pots, whereas holding furnace temperatures are between 390 and 425°C, depending on the composition of the alloy being cast and part size.

Die castings are produced under pressure in permanent metal molds (93–95). Die casting with zinc-base alloys is one of the most efficient and versatile production methods which can be used for the manufacture of accurate, complex metal components. The machine used is of the so-called hot-chamber or submerged-plunger type because the alloys melt at low temperature and do not attack the injection-pump material (gray or ductile cast iron). The die-casting die or mold consists of a cover die and an ejector die which meet at the parting line. Machines range in size (clamping force) from 25 to 2,500 tons.

Injection pressures for hot-chamber machines are between 6.9 and 20.7 MPa (1000–3000 psi). The velocity with which the metal enters the die is controlled by the pressure and die design. The optimum may differ from each casting and in the past it has been determined primarily by trial and error. In recent years, techniques have been developed for the use of instrumentation to calibrate the shot system and die design programs are available for optimizing gate and runner systems in dies (96–97). The P–Q^2 technique (relationship between pressure and flow), developed by the Commonwealth Scientific and Industrial Research Organization, is utilized to characterize the rate with which the die-casting machine-shot system delivers zinc to the die. Using this information, metal flow during die filling can be controlled for castings of various sizes (98). Computer programs are available to aid the die caster in applying these rules and guidelines (99).

For casting zinc alloys, die temperatures are generally between 150 and 250°C. The lower temperatures are employed for heavy castings, higher temperatures for thin-wall castings. Hot-chamber machines can be operated rapidly with a high degree of automatic control, generally at rates of 50–500 shots per hour. Special machines greatly exceed these rates with 2000–5000 shots per hour up to 18,000 per hour for a zipper-casting machine. Castings may weigh from a few grams to over 22.5 kg.

Zinc alloys can be cast in single-cavity, multiple-cavity, combination, or unit dies. Combination or family dies consist of a series of cavities in one die for casting two or more parts. Unit dies are separate, small dies, usually with a single cavity, which are inserted in a single master-holding die and operate several at a time in large machines. Because temperatures for die casting are relatively low, 150–250°C, hot-worked tool steels are not generally required. For short-run dies, low alloy steels, like 4140 steel, are suitable. However, for very long runs, particularly when high dimensional accuracy is required, hot-worked tool steels, such as H11, H12, and H13, give the longest die life. Ejector pins of nitrided H11 tool steel or of 7140 alloy steel are available as stock items for insertion in the dies. Hardenable grades of stainless steel, such as type 440B, are often used for cores. Hot-worked tool steels such as H11, H12, and H13 can be used for both cores and slides. Lubricating the slides and cores with molybdenum disulfide or colloidal graphite in oil helps to ensure smooth action and to minimize wear.

The life of a casting die depends on the temperature of the metal being cast, thermal gradients within the die, and frequency of exposure to high temperature. The die life for casting zinc alloys is generally much longer than that for casting aluminum, magnesium, or copper alloys; it is not unusual for dies to last for one million (10^6) shots or more.

Zinc die castings may be machined, bent, swaged, or coined for finishing or for slight changes of shape (90). They can be joined by riveting, spinning, welding, and soldering in assembly operations. Corrosion resistance against atmospheric and seawater conditions is good, particularly if first treated with chromate using a weak chromic acid solution to form a passive film on the surface. Anodized coatings protect zinc against severe corrosive environments and improve wear resistance. Paint, lacquer, or enamel finishes can be applied for decorative purposes. A large number of parts are chromium plated. For such treatment, the castings are usually buffed to remove traces of parting lines and flow marks, and then plated with copper, nickel, or chromium. Other metals can also be electroplated on zinc.

During the 1970s, casting techniques for the production of thin-wall zinc die castings were developed by the ILZRO to allow the faster production of components containing less zinc at lower cost (96). Because of the resulting weight saving per part, zinc became cost- and weight-competitive with aluminum and plastic (100–101).

Castings with exceptionally thin wall sections are produced by faster cooling rates, controlled metal injection, and a meticulous standard of die construction which is necessarily associated with automatic machine operation. Since die casting is a heat-transfer process, any reduction in the mass of metal to be chilled in each cycle allows faster operation. For the user, metal saving has not only a financial advantage, but gives a component with better strength-to-weight ratio since thin sections have a fine-grained structure with superior properties. Thus, castings up to ca 3 kg can be made with walls ca 0.75 mm thick (102–103).

Purging the die cavity with a reactive gas before metal injection eliminates gas porosity which is found in die castings because of trapped air. In the pore-free process, the air in the die cavity is displaced with oxygen or other reactive gas. When the die is charged with the molten zinc alloy, the zinc reacts with the gas to form tiny, solid particles that are dispersed throughout the pore-free casting. The absence of porosity in the castings results in improved mechanical properties. This process is still in the developmental stage for zinc die castings (104).

Foundry Alloys. The tonnage of slab zinc used in foundry applications is small compared with that used for die casting, 5334 t in 1979 and 5662 t in 1980.

High Strength Alloys. Before 1967, the only zinc gravity-casting alloys of note were the slush and forming-die alloys. In that year, the first high strength zinc gravity-casting alloy, No. 12, was introduced with 11% Al. This alloy was originally marketed as a prototype alloy to be used in sand or permanent-mold casting of parts that would later be made as die castings, since many of its mechanical properties approximate those of die-cast alloy No. 3. However, much wider application of this alloy in the foundry industry has developed (105). Subsequently, two additional alloys were developed containing 8% aluminum (No. 8) and 27% aluminum (No. 27) (106); compositions are given in Table 21 (86,107–108).

Physical and mechanical properties are given in Table 22 (105–106,108–110). The densities reflect the effect of aluminum: the Zn–27% Al alloy is ca 30% lighter than

Table 21. Compositions of High Strength Zinc Foundry Alloys, wt %[a]

Element	No. 8	No. 12[b]	No. 27
aluminum	8.0–8.8	10.5–11.5	25–28
copper	0.8–1.3	0.5–1.25	2.0–2.5
magnesium	0.015–0.03	0.015–0.03	0.01–0.02
iron	0.10	0.075	0.10
lead	0.004	0.004	0.004
cadmium	0.003	0.003	0.003
tin	0.002	0.002	0.002
zinc	balance	balance	balance

[a] Single units indicate maximum amounts permitted.
[b] Covered by ASTM B 669-80 (105). A revision of this specification is under consideration to include Nos. 8 and 27.

Table 22. Properties of High Strength Zinc Foundry Alloys[a]

Property	Alloy No. 8, Permanent-Mold Cast	Alloy No. 12 Sand Cast	Alloy No. 12 Permanent-Mold Cast	Alloy No. 27 Sand Cast[b]	Alloy No. 27 Sand Cast, H.T.[c]
physical					
density, g/cm^3	6.37	6.03	6.03	5.01	5.01
melting or solidification range, °C	375–404	377–432	377–432	375–487	375–487
mechanical					
tensile strength, MPa[d]	221–225	276–310	310–345	400–441	310–324
yield strength, 0.2% MPa[d]	207	207	214	365	255
elongation, %	1–2	1–3	4–7	3–6	8–11
Brinell hardness[e]	85–90	92–96	105–125	110–120	90–100

[a] Refs. 105–106, 108–110.
[b] Primary purpose.
[c] Heat treated.
[d] To convert MPa to psi, multiply by 145.
[e] 500 kg for 30 s.

zinc, 17% lighter than the Zn–11% Al alloy, and 21% lighter than the Zn–8% Al alloy. All three alloys possess a good combination of mechanical properties. The sand-cast Zn–27% Al alloy is superior to the other two alloys with a tensile strength close to 50% higher. Ductility of the Zn–27% Al alloy is slightly higher; however, if ductility is required, the heat-treated condition should be specified. The Zn–27% Al alloy is also the most creep-resistant (87,109). Mechanical properties of these alloys are equal or superior to those of many brass, bronze, aluminum, and iron alloys.

The Zn–Al system permits manipulation of the mechanical properties by suitable heat treatment. The aluminum-rich alpha phase is especially suitable for solution hardening since it can be supersaturated by as much as 30 wt % zinc. Furthermore, both alpha and beta phases can be strengthened by precipitation because of decreasing solute solubility with decreasing temperature.

Alloys Nos. 8, 12, and 27 should not be prepared and stored in cast-iron crucibles since excessive iron pickup decreases fluidity. Many types of refractory crucibles can be used for molten zinc–aluminum alloys. Silicon carbide crucibles are generally recommended. However, unlike other alloys, zinc alloys can be cast in many types of molds, including sand, plaster, silicone rubber, graphite, and cast iron (105,110), as well as bronze, aluminum, and beryllium–copper molds. Silicone rubber and graphite molds offer some special advantages. Parts with reverse taper can be cast in silicone rubber molds because of their flexibility. Graphite is easy to machine, does not distort or warp, and promotes rapid solidification because of high thermal conductivity; over 20,000 zinc alloy castings have been made with the same graphite mold (111). Castings made in graphite molds show superior surface finish and mechanical properties. Alloy No. 12 can be sand- and permanent-mold cast and graphite molds are particularly successful. Alloy No. 8 has excellent finishing characteristics and presents a low cost replacement for brass and bronze castings.

The gravity-casting alloys are gaining widespread acceptance for industrial structural parts; business machines, office equipment, builder's hardware, and marine applications (110–112). They have replaced cast iron, copper-based alloys, and aluminum alloys in many applications, eg, Alloy No. 12 is used in bearing applications, where bushings have been tested in mine elevators, drill motors, 50-ton dump trucks, and drill-track rollers. The alloy bushings showed less wear than bronze when both were used in tandem on the same equipment. It is expected that zinc-based bearings and bushings will find wider acceptance as a substitute for bronze because of their lower cost and superior mechanical properties, particularly for low speed, high-load-type service (108,113).

Because zinc-based alloys have low melting points, energy savings in the melting operation are substantial and the foundry operation is essentially free of fume. With the current trend of increasing energy costs and pollution control, cost benefits can be considerable (110).

Slush Alloys. Slush casting is limited generally to the production of hollow casting (114), eg, lamp bases, lighting fixtures, and casket hardware. In slush casting, a molten metal is poured into a split metal mold, generally made of bronze, until the mold is filled; the mold is immediately inverted and the liquid metal is allowed to run out, leaving a thin-shell casting behind, whose shape is a replication of the cavity walls. The thickness of the wall of the casting depends on the time interval between the filling and the inverting of the mold, as well as on the chemical and physical properties of the alloy and the temperature and composition of the mold. Skilled operators produce good castings with remarkable uniformity in weight.

Slush-casting alloys must be fairly low melting and freeze over a temperature range, that is, the liquidus and solidus temperatures must be significantly different to provide a slushy range.

Zinc slush-casting alloy compositions are based on the Zn–Al system. The two commonly used alloys have nominal aluminum contents of 4.75% Al and 5.5% Al, which fall on either side of the zinc–aluminum eutectic at 5% Al with an invariant melting point of 382°C (see Fig. 4). Special high grade zinc is employed to control impurity levels in order to avoid corrosion problems. The lower aluminum alloy with 0.25% Cu produces castings with improved mechanical properties and longer service life, because copper affords protection against corrosion. Slush casting with copper-containing alloy, however, is more difficult and yields castings with thicker walls than the straight binary Zn–5.5% Al alloy.

The cast zinc–aluminum–copper slush alloy, after aging for ten years indoors, shows a tensile strength of 238 MPa (34,500 psi) with Charpy impact strengths of 1.4–4.1 J (1–3 ft·lbf) (87).

Forming-Die Alloys. The tonnage of slab zinc used in this application is small. The use of zinc alloy dies started in the aircraft industry during World War II (115). Zinc-based alloys cast in sand and plaster molds continue to be used for short-run dies for steel and aluminum stampings in the automotive and aircraft industries (116). Considerable cost savings are realized with these low melting zinc-based alloys which are easy to polish, machine, weld, and remelt.

The common composition is 4% Al and 3% Cu, with or without a small amount of magnesium. Other elements, such as Ni or Ti, increase die life (117). The base alloy has tensile strengths of 210–280 MPa (30,450–40,600 psi) and Charpy impact strengths of ca 20–27 J (15–20 ft·lbf) (87).

Rolled Zinc. Rolled-zinc products constitute ca 2–3% of slab zinc consumed in the United States over the past ten years. The U.S. production has declined by ca 50% since 1965 (66) and in 1980 only three companies rolling zinc products were left.

The ASTM B 69-66 (1979) gives typical compositions of rolled zinc for informational purposes and general guidance and not for specification purposes (118). The maximum solid solubilities of other metals in zinc are always small, 0.5–1% for manganese and aluminum; ca 2–3% for cadmium, palladium, and copper; ca 8% for silver; and 10–15% for gold; solubilities at room temperature are lower. Only fractional percentages, or minute amounts, of other metals enter the structure of the zinc crystal, and only copper, cadmium, and aluminum are added to modern zinc rolling alloys. The commercial grades of rolled zinc contain the natural impurities lead, cadmium, and iron. Rolled zinc compositions developed to meet various applications are given in Table 23 (86,119). Except for the superplastic Zn–Al, the total alloying content in the other alloys is <2%. Limits are placed on impurities like cadmium, lead, iron, tin, arsenic, bismuth, and indium which cause hot shortness or edge cracking. Lead, tin, cadmium, and bismuth must be held at very low concentrations in aluminum containing alloys because they promote intercrystalline corrosion, which takes the form of exfoliation of the surface and general embrittlement; therefore, special high grade zinc is used in the preparation of these alloys.

The zinc is normally melted in a gas, oil, or coal-fired reverberatory furnace with a capacity up to 100 tons or in a low frequency induction furnace with a capacity of a few tons. The more highly alloyed compositions are more effectively melted and mixed in low frequency induction furnaces. The furnace must be refractory-lined to eliminate iron pickup by the molten metal. The metal temperature is maintained below 500°C to minimize loss by oxidation. A ladle is used to transfer the metal for casting into molds; the pouring temperature is usually ca 440°C. Zinc scrap is not generally suitable for remelting because it may contain undesirable impurities.

Although more and more zinc sheet and strip are produced in continuous mills, some is still produced by rolling slabs cast in open or closed book-type molds made of cast iron (120–123). The casting temperatures are between 440 and 510°C, mold temperatures between 80 and 120°C. The contact surfaces of the mold must be smooth and clean to allow unrestricted shrinkage of the cast slab. Mold lubricant is not necessary, but if used should be held to a minimum. Slabs cast in open molds must be skimmed immediately to remove surface oxide. Rolling slabs are cast 1.87–10 cm thick.

Table 23. Rolled Zinc Alloys

Alloy	Pb	Cd	Fe	Cu	Ti	Mg	Al	Zn
zinc, pure	0.003–0.10 max	0.003–0.007 max	0.0014–0.012 max	0.001 max				99.88–99.99%
Zn–Pb–Cd–Fe	0.04–0.40	0.035–0.35	0.006–0.030 max	0.002–0.005 max				99.2%
Zn–Cu	0.02–0.10 max	0.007–0.08	0.012 max	0.12–0.90				98.9%
Zn–Cu–Mg	0.10 max	0.04 max	0.012 max	0.65–0.85		0.006–0.016		97.98%
Zn–Cu–Ti	0.003–0.30	0.007–0.02 max	0.002–0.012 max	0.50–0.90	0.08–0.16			98.61%
Zn–Al–Mg	0.002–0.003 max	0.002–0.003 max	0.002–0.003 max	0.002 max		0.03–0.07	0.08–0.11	99.81%
Zn–Hi Al, superplastic	0.003 max	0.003 max	0.003 max	0–0.6		0–0.03	20–24	75.36%

Zinc rolling slabs have been cast successfully by semicontinuous direct-chill casting methods. This is the preferred method for superplastic zinc alloys which, because of their large freezing range, display unacceptable surface shrinkage when cast in open molds.

The open-mold cast slabs are preheated to 150–260°C for initial rolling in rough or breakdown mill. Composition of slab influences the temperature employed. Special high grade or high grade is used to avoid the harmful effect of impurities. Tin should be absent altogether, because its presence at concentrations of 0.004% can cause the zinc to crumble at hot working temperatures. Slabs for strip are reduced by longitudinal rolling to a thickness convenient for coiling, generally <2.5 mm. For sheet, slabs are rolled laterally until the desired width is obtained. Then the sheets are turned 90° and rolled to a thickness two to four times the final gauge. By cross rolling, the slab is worked in two directions, reducing to some degree the effects of anisotropy.

Strip is produced as wide as 2 m and in thicknesses as low as 0.1 mm in regular mills. Foil in thicknesses of 0.025 mm or less is produced in special mills. To obtain strip with a bright surface, high ductility, and low hardness, finish rolling is performed hot at 120–150°C.

Zinc sheet is produced by the pack-rolling process, in which 2–40 rough-rolled sheets are stacked together in packs and rolled simultaneously. Packs must be split frequently, interchanging inner and outer sheets to equalize temperatures and reductions. With care, good quality sheets can be produced by this technique but considerable variations in properties can occur. If a bright ductile product is desired, rolls are held at 120–150°C; the reduction on the last pass is 20–40%.

Lubrication of sheet and strip is necessary for all operations. Although for special operations vegetable and mineral oils may be employed, a mixture of paraffin and tallow oil is normally perferred in rough rolling. Requirements for finish-roll lubricant are more strict because of staining caused by breakdown of the oil or reaction with the zinc. Strip zinc is usually finish-rolled with cotton seed or mineral oil.

In recent years, much attention has been given by the zinc industry to the development of methods for casting and rolling the metal continuously (124–126). The Hazelett casting machine employed consists essentially of two continuous steel belts, supported and driven through a system of pulleys (127). These form the walls of the mold, which is closed at the sides by sets of small steel blocks linked together by cables forming two endless chains sandwiched between the belts. Molten metal is fed from a tundish or funnel into the cavity between the belts, which are cooled by water sprays. The belts are rotated at a controlled speed and, as the metal solidifies, the slab is drawn downward, thus producing a continuous cast ingot; its thickness can be varied from 10 to 75 mm by adjusting the belt separation. Owing to the continuous nature of the casting operation, the grain size of the solid slab is much finer than when static molds are used, and subsequent reduction in the rolling mill is therefore easier, giving a more uniform product.

After leaving the casting machine, the slab is cooled by water sprays to 180–240°C and fed into the mill, which is generally of the four-high roll type, where a 60% reduction is taken in one pass. The strip is then coiled and, when the casting run has been completed, is fed back through the mill and rolled at a temperature of 80–90°C, with a reduction up to 50%, to give the required final thickness, finish, and properties.

Zinc sheet and strip can be easily fabricated by drawing, spinning, bending, stamping, coining, embossing, blanking, and impact extrusion. Rolled-zinc products in the form of strip, sheet, wire, and rod have many and varied commercial applications. Strip is formed into dry-cell battery cans, mason jar covers, organ pipes, grommets, eyelets, and many other objects, some of which are subsequently brass or chromium plated (jewelry, medallions, bathroom accessories, etc) (128). The zinc–carbon dry-cell application accounts for about one half the rolled-zinc consumption in the United States (see Batteries and electric cells). Sheet zinc is used in photoengraving and also in the construction of roofing and other architectural uses. Special high grade zinc with a maximum iron content of 0.0014% in the form of plate and rods gives cathodic protection to steel in marine and pipeline applications.

The low creep resistance of unalloyed zinc sheet and strip is one of its most serious defects and restricts its application. Strip made with Zn–0.6 Cu–0.10 Ti composition has superior creep resistance and is used in structural applications where high, continuously applied stresses are encountered in service.

A new large tonnage application for strip zinc has recently developed in the coinage field. The U.S. Mint has decided to replace the solid copper (95% Cu–5% Zn) penny with a less expensive copper-plated penny made from zinc strip (Zn–0.8% Cu alloy) (129–130). Zinc blanks stamped from the Hazelett-cast rolled zinc are barrel-plated with copper prior to coining to produce the finished penny. This application has a potential annual market of 40,000–45,000 t of zinc.

Another commercial development of the 1970s is the application of superplasticity which is exhibited by a number of zinc alloys (131–134). Under the right conditions, the material becomes exceptionally soft and ductile and, under low stresses, extensions exceeding 1000% can be obtained without fracture. The grain size must be extremely small (about 1 micrometer) and stable. This grain size is less than one tenth that of common metals in the wrought condition.

Extremely fine-grained two-phase structures can be developed in zinc–aluminum eutectoid alloys (ca 22% aluminum with and without minor alloying additions) by rapid quenching from above the eutectoid temperature of 275°C or by rolling to appreciable degrees of deformation at somewhat lower temperatures, or both. If sheet with this fine-grained structure is heated to just below the transformation temperature of 245–270°C, optimum superplastic behavior is obtained. Because of the ductility and low strength in this condition, the material is similar to thermoplastic polymers and, similarly, can be molded into complex shapes by various processes. When cooled to room temperature, its strength and hardness are many times greater than those of any thermoplastic. Tensile strengths of the superplastic zinc alloys at room temperature range from 190 MPa (28,000 psi) for the binary alloy to 380 MPa (55,000 psi) with the higher alloy compositions. A simple postforming heat treatment can increase strength to 350–440 MPa (51,000–64,000 psi). These alloys can be molded easily like thermoplastics at moderately elevated temperatures, but at room temperature have normal metallic properties, including strength, stiffness, conductivity, and stability, which no plastic can equal. The tooling cost is generally considerably lower than for conventional pressing, which is particularly beneficial for limited production runs (10,000 parts or less).

Parts made from these materials are formed from sheet metal, or solid, three-dimensional metal preforms (135). Complex housings, bezels, and equipment covers are typical of the former. Applications involving solid shapes include gears, pulleys, flywheels, and similar parts.

Zinc Dust and Powder. Zinc dust and powder are particulate forms of zinc which have many interesting and useful applications (2). The terms dust and powder have been used more or less indiscriminately to designate particulate zinc materials. Here, the term zinc dust designates material produced by condensation of zinc vapor, whereas zinc powder indicates the product obtained by atomizing molten zinc. Zinc dust is manufactured by vaporizing zinc from a suitable source material. The vapor is condensed under conditions which permit control over purity and fineness (136–138). The zinc vapor eminating from a boiler is led into a condensing chamber where it is diluted with inert gases and quickly chilled. The frozen droplets or dust are collected at the bottom of the condenser, screened, and packaged. The purity of the zinc dust depends on the source material or charge to the zinc boiler.

In the atomizing process, a stream of molten zinc is broken into tiny droplets by the force of a pressurized fluid impinging on the stream. The fluid can be any convenient material, although air is normally used. The atomized drops cool and solidify rapidly in a collection chamber. The powder is screened to specified sizes. Particulate zinc is also produced by other methods such as electrolytic deposition and spinning-cup techniques, but these are not of commercial importance.

Zinc dust is smaller in particle size and spherical in shape, whereas zinc powder is coarser in size and irregular in shape. The particle size of zinc dust, important in some applications, is controlled by adjusting the rate of condensation. Rapid cooling produces fine dust, slower condensation coarse dust. In the case of zinc powder, changes in the atomization parameters can be employed to change particle size to some degree. The particle size distributions for commercial zinc powders range from 44 to 841 μm (325–20 mesh). The purity of zinc powders is 98–99.6%.

ASTM recognizes two types of zinc dust in specification ASTM D 520-51 (reapproved 1976) (139), which includes permissible impurity concentrations. The metallic content of most commercial grades is 95–97%. The zinc oxide content is between 3 and 5%; finer dusts contain higher concentrations because of high surface areas. Zinc dusts are manufactured in various size ranges, and a typical commercial dust has an average

Table 24. Other Metals in North American Zinc Concentrates

Metal	Percent[a]	Process step Electrolytic	Pyrometallurgical
lead	0.2–6.0	leach residue	sinter fume, refining column, ISF lead bullion and copper dross
copper	0.1–2.8	leach residue, purification residue	retort residue, ISF lead bullion and copper dross
cadmium	0.1–0.8	purification residue	sinter fume, refining column
silver	27–1900[b]	leach residue	retort residue, ISF lead bullion and copper dross
gold	0.3–12[b]	leach residue	retort residue, ISF lead bullion and copper dross
mercury	0.001–0.03	roaster off-gas	roaster off-gas
germanium	0.004–0.04	leach residue	sinter fume
gallium	0.005–0.02	leach residue	retort residue
indium	0.01–0.03	leach residue	sinter fume, refining column
thallium	trace	leach residue	sinter fume

[a] Unless otherwise indicated.
[b] g/t.

particle diameter between 4 and 8 μm. Usually, dusts are screened to be essentially free of particles coarser than 75 μm (200 mesh).

The principal use of zinc dust is in paint coatings (see Table 18). Zinc-dust paints can be classified into those containing zinc oxide as a substantial portion of the pigmentation and those in which zinc dust represents most of the pigment (see Paint). In the latter, the zinc-dust concentration in the dry paint film is high enough to provide galvanic protection to an iron or steel substrate similar to a galvanized coating (140–141). These paints are classified as zinc-rich paints, and represent one of the fastest growing new markets for zinc in recent years. Such paints are used to supplement galvanized steel for automotive underbody protection against corrosion (142), bridges and other structures, ship hulls, and buildings (2,143–144). Zinc-rich paint is recommended whenever conventional galvanizing is not suitable.

Zinc dust is used in the sherardizing process where work pieces are tumbled with zinc dust in rotating steel drums which are heated electrically or by gas to 370–420°C (145). The steel parts are uniformly coated with zinc. In the chemical and metallurgical industries, zinc dust is used as a reducing agent, in the manufacture of hydrosulfite compounds for the textile and paper industries, and to enhance the physical properties of plastics and lubricants (2).

By far, the largest application of zinc powder is for solution purification in electrolytic zinc plants. This application consumed an estimated 17,700 t of powder in 1980. Zinc powder is also used in primary batteries, frictional materials, spray metallizing, mechanical plating, and chemical formulations.

By-Product Metals

Ores that are exploited primarily for zinc invariably contain one or more other valuable metals. Various ores contain different combinations of such other metals (see Table 24). For instance, eastern ores commonly contain only cadmium as a significant recoverable value.

Cadmium and mercury are usually recovered in separate processes at the zinc plant. The others are shipped as enriched residues to plants that specialize in their recovery.

BIBLIOGRAPHY

"Zinc and Zinc Alloys" in *ECT* 1st ed., Vol. 15, pp. 224–275, by A. Paul Thompson and H. W. Schultz, The Eagle-Picher Co.; "Zinc and Zinc Alloys" in *ECT* 2nd ed., Vol. 22, pp. 555–603, by A. W. Schlechten, Colorado School of Mines, and A. P. Thompson, Eagle-Picher Industries, Inc.

1. H. O. Hoffman, *Metallurgy of Zinc and Cadmium*, McGraw-Hill Book Co., New York, 1922.
2. B. C. Hafford, W. E. Pepper, and T. B. Lloyd, *Zinc Dust and Zinc Powder*, International Lead Zinc Research Organization, New York, 1982.
3. C. H. Mathewson, ed., *Zinc*, Reinhold Publishing Co., New York, 1959.
4. D. O. Rausch and B. C. Mariacher, eds., *AIME-World Symposium on Mining and Metallurgy of Lead and Zinc*, Vol. I, American Institute of Mining Engineers, New York, 1970.
5. C. H. Cotterill and C. M. Cigan, eds., *AIME Symposium on Mining and Metallurgy of Lead and Zinc*, Vol. II, American Institute of Mining Engineers, New York, 1970.
6. J. M. Cigan, T. S. Mackey, and T. J. O'Keefe, eds., *Lead-Zinc-Tin '80*, American Institute of Mining Engineers, Warrendale, Pa., 1979.
7. J. N. Anderson and P. E. Queneau, eds., *Pyrometallurgical Processes in Nonferrous Metallurgy*, Gordon and Breach Science Publishing, New York, 1967.

848 ZINC AND ZINC ALLOYS

8. E. S. Dana in W. E. Ford, ed., *A Textbook of Mineralogy*, 4th ed., John Wiley & Sons, Inc., New York, 1945, p. 851.
9. A. M. Bateman in ref. 3, Chapt. 3.
10. H. Wedow and co-workers, *U.S. Geol. Sur. Prof. Pap.* **820,** 697 (1973).
11. V. A. Cammarota, Jr., in J. O. Nriagu, ed., *Zinc in the Environment*, John Wiley & Sons, Inc., New York, 1980, pp. 1–38.
12. J. W. Chandler in ref. 4, Chapt. 2.
13. A. L. Hatch, K. C. Hendrick, and W. Seis in ref. 5, Chapt. 2.
14. *Canadian Mines Handbook*, Northern Miner Press, Toronto, 1970–1971, p. 350.
15. E. W. Horvick in H. Baker, ed., *Metals Handbook*, 9th ed., Vol. 2, American Society for Metals, Metals Park, Ohio, 1979, pp. 824–826.
16. E. N. Doyle in ref. 4, Chapt. 40.
17. J. S. Pellett, *Mining Eng. (N.Y.)* **5,** 1211 (1953).
18. T. G. White and K. L. Clifford in D. O. Rausch and co-workers, eds., *Lead-Zinc Update*, American Institute of Mining Engineers, New York, 1977, pp. 119–147.
19. M. Rey and co-workers, *Trans. AIME* **199,** 416 (1954).
20. T. R. A. Davey in ref. 6, pp. 48–65.
21. W. Gowland, *The Metallurgy of Non-Ferrous Metals*, 2nd ed., C. Griffen, London, 1918.
22. K. D. McBean in ref. 3, Chapt. 6.
23. U.S. Pats. 1,884,348 (Oct. 25, 1932); 1,963,288 (June 19, 1934); 2,089,306 (Aug. 10, 1937), B. A. Stimmel, K. D. McBean, and G. Cruicshank (to Consolidated Mining and Smelting Co. of Canada).
24. J. H. Reid in ref. 7, Chapt. 5.
25. M. M. Avedesian, preprint, *Conference of the Canadian Society of Chemical Engineers*, Ottawa, Paper No. 5d, Oct. 21, 1974.
26. K. Natesan and W. O. Philbrook, *Metall. Trans.* **1,** 1353 (1970).
27. G. Stankovic in ref. 5, Chapt. 4.
28. M. Okozaki, Y. Nakane, and H. Noguci in ref. 7, Chapt. 2.
29. A. Berg and F. Pape, *Erzmetall* **31,** 224 (1978).
30. L. A. Painter in ref. 6, Chapt. 8.
31. G. Steintveit in ref. 6, pp. 85–96.
32. K. H. Heino and co-workers in ref. 5, Chapt. 5.
33. R. Denoiseux and co-workers in ref. 6, Chapt. 5.
34. R. E. Lund and co-workers in ref. 5, Chapt. 20.
35. *Eng. Min. J.* **182**(9), 99 (1981).
36. R. R. Knobler and co-workers, *Erzmetall* **32,** 109 (1979).
37. S. Tsunoda and co-workers, *AIME-TMS Paper A73-65*, 1973.
38. F. S. Gaunce and co-workers, *CIM Bull.* **67,** 116 (1974).
39. K. Shimokawa and T. Takesue, *MMIJ-AIME Joint Meeting Paper T IVCZ*, May 1972.
40. V. Arregui, A. R. Gordon, and G. Steintveit in ref. 6, pp. 97–123.
41. J. E. Dutrizac in ref. 6, pp. 32–64.
42. Can. Pat. 238,439 (Oct. 22, 1975); U.S. Pat. 4,004,991 (Jan. 25, 1977), H. Veltman and co-workers (to Sherritt Gordon Mines, Ltd.).
43. H. Veltman and G. L. Bolton, *Erzmetall* **33,** 76 (1980).
44. E. G. Parker, *CIM Bull.* **74,** 145 (1981).
45. F. S. Weimer, G. T. Wever, and R. J. Lapee in ref. 3, Chapt. 6.
46. H. H. Fukubayashi, T. J. O'Keefe, and W. C. Clinton, *U.S. Bur. Mines Rep. Invest.*, 7966 (1974).
47. G. M. Meisel, *J. Metals* **26**(8), 25 (1974).
48. D. E. Rodier in ref. 6, Chapt. 9.
49. S. Fugleberg and co-workers in ref. 6, Chapt. 10.
50. N. G. Freeman and A. Pyatt in ref. 6, pp. 222–246.
51. C. L. Mantell, *Electrochemical Engineering*, McGraw-Hill, Inc., New York, 1960.
52. R. C. Kerby and C. J. Krauss in ref. 6, pp. 187–203.
53. C. G. Maier, *U.S. Bur. Mines Bull.*, 324 (1930).
54. E. C. Truesdale and R. M. Waring, *Trans. AIME* **159,** 97 (1944).
55. R. M. Sellwood in ref. 5, Chapt. 21.
56. H. H. Kellogg in ref. 6, pp. 28–47.
57. W. Hopkins and A. W. Richards, *J. Met.* **30,** 12 (1978).
58. C. F. Harris, A. W. Richards, and A. W. Robson in ref. 6, pp. 247–260.

59. H. Nakagawa and co-workers in ref. 6, Chapt. 16.
60. L. D. Fetterolf and co-workers in ref. 5, Chapt. 19.
61. E. L. Miller in ref. 3, Chapt. 7.
62. G. F. Weaton in ref. 3, pp. 272–273.
63. K. A. Phillips in ref. 3, pp. 225–251.
64. E. D. Nogueira, J. M. Regife, and A. M. Arcocha, *Eng. Min. J.* **180**(10), 92 (1979).
65. *Minerals Year Book*, U.S. Bureau of Mines, Washington, D.C., 1970–1980.
66. *Non-Ferrous Metal Data*, American Bureau of Metal Statistics, Inc., New York, 1926–1980.
67. Ref. 66, 1970, p. 90 and 1980, p. 86.
68. Ref. 66, 1970, p. 91 and 1980, p. 87.
69. H. H. Landsberg and co-workers, *Resources in America's Future—Patterns of Requirements and Availabilities, 1960–2000*, The Johns Hopkins Press, Baltimore, Md., 1963.
70. *1981 U.S. Industrial Outlook for 200 Industries with Projections for 1985*, Bureau of Industrial Economics, U.S. Government Printing Office, Washington, D.C., Jan. 1981.
71. W. Malenbaum, *World Demand for Raw Materials in 1985 and 2000*, McGraw-Hill, Inc., New York, 1978.
72. *Annual Review, U.S. Zinc and Cadmium Industries*, Zinc Institute, Inc., New York, 1970–1980.
73. *U.S. Automotive Market for Zinc Die Casting*, Zinc Institute Inc., New York, 1974–1980.
74. *ASTM Standards, Specification B 6-77*, American Society for Testing and Materials, Philadelphia, Pa., 1981, Part 8, pp. 11–14.
75. C. H. Mathewson in ref. 3, Chapt. 9, pp. 386–390.
76. T. G. White, *Min. Congr. J.* **63**(5), 54 (1977).
77. J. C. Van Loon, *Analytical Atomic Absorption Spectroscopy*, Academic Press, New York, 1980.
78. N. H. Furman, ed., *Standard Methods of Chemical Analysis*, Vol. I, D. Van Nostrand Co., Princeton, N.J., 1962, Chapt. 52.
79. *Annual Book of Standards*, American Society for Testing and Materials, Philadelphia, Pa., 1980, Parts 12 and 42.
80. *Zinc*, National Research Council, Subcommittee on Zinc, University Park Press, Baltimore, Md., 1979, pp. 273–296, 327–341.
81. J. D. Nriagu, ed., *Zinc in the Environment*, John Wiley & Sons, Inc., New York, 1980, Parts 1 and 2.
82. P. Krick in ref. 6, pp. 693–699.
83. *Zinc Coatings for Corrosion Protection*, Zinc Institute, Inc., New York, 1977, 52 pp.
84. *ASTM Standard, Specification B 240-79*, American Society for Testing and Materials, Philadelphia, Pa., 1980, Part 7, pp. 332–334.
85. *ASTM Standard, Specification B 86-76*, American Society for Testing and Materials, Philadelphia, Pa., 1980, Part 7, pp. 45–49.
86. E. W. Horvick in ref. 15, pp. 629–637.
87. *Engineering Properties of Zinc Alloys*, International Lead Zinc Research Organization, New York, 1980, pp. 1–36.
88. F. R. Sauerwine, T. P. Groeneveld, and F. C. Bennett, *Tech Data '80*, The American Die Casting Institute–Die Casting Research Institute, Des Plaines, Ill., Technical Seminar, March 20, 1980.
89. M. Hansen, *Constitution of Binary Alloys*, McGraw-Hill Book Co., Inc., New York, 1958, p. 149.
90. E. A. Anderson and G. L. Werley, *Zamak Alloys for Zinc Die Casting*, 12th ed., The New Jersey Zinc Company, Palmerton, Pa., 1965.
91. G. L. Werley in T. Lyman, ed., *Metals Handbook*, 1948 ed., American Society for Metals, Cleveland, Ohio, 1948, pp. 1078–1082.
92. C. W. Roberts, *Metallurgia* **64**, 57 (Aug. 1961).
93. T. Lyman, ed., *Metals Handbook*, 8th ed., Vol. 5, American Society for Metals, Metals Park, Ohio, 1970, pp. 444–448.
94. Ref. 93, pp. 285–333.
95. A. Street, *The Die Casting Book*, Portcollis Press, Ltd., Surrey, UK, 1977.
96. T. P. Groeneveld and W. D. Kaiser, *Designing for Thin-Wall Zinc Die Casting*, International Lead Zinc Research Organization, New York, 1975.
97. W. D. Kaiser and T. P. Groeneveld, *Trans. Soc. Die Cast. Eng.* (March 1975).
98. A. J. Davis, H. Siauw, and G. N. Payne, *Trans. Soc. Die Cast. Eng.* (June 1977).
99. D. C. H. Nevison, *Using Computer Aids in Die Casting*, 10th International Pressure Die Casting Conference, May 10–14, 1981, Zinc Institute Inc., New York, 1981, 16 pp.

100. D. C. H. Nevison, *Precis. Met.* **32**(3), 27 (March 1974).
101. D. C. H. Nevison, *Ind. Design* **21**(7), 51 (Sept. 1974).
102. E. T. Foster, *Appliance Engineer* **9**, 67 (April 1975).
103. F. R. Sauerwine and A. D. Behler, *10th Society of Die Casting Engineers Congress*, Paper No. G-T79-101, St. Louis, Mo., March 1979, 16 pp.
104. J. T. Winship, *Am. Mach.* (Aug. 1979).
105. *Design Applications of Gravity Cast Zinc*, International Lead Zinc Research Organization/Zinc Institute, New York, 1972, 36 pp.
106. E. Gervais, H. Levert, and M. Bess, *84th Casting Congress and Exposition of the American Foundrymen's Society*, St. Louis, Mo., April 21–25, 1980, American Foundrymen's Society, Des Plaines, Ill.
107. *ASTM Standards, Specification B 669-80*, American Society for Testing and Materials, Philadelphia, Pa., 1981, Part 8, pp. 789–792.
108. D. Apelian, M. Paliwal, and D. C. Herrschaft, *J. Met.*, 12 (Nov. 1981).
109. E. Gervais, A. Y. Kandeil, and H. Levert, *Die Cast. Eng.* **25**, 44 (Sept.–Oct. 1981).
110. *Zinc Foundry Alloys*, Zinc Institute, Inc., New York, 1980, 8 pp.
111. W. Mihaichuk, *Mod. Cast.* **71**, 39 (July 1981).
112. T. Riedlinger, *Mod. Met.* **37**, 12, Chicago, Ill. (Jan. 1981).
113. J. W. Kissel and K. F. Dufrane, *The Bearing Properties of Zinc Gravity Casting Alloys*, ILZRO Project ZM-298, Progress Report No. 2, International Lead Zinc Research Organization, New York, 1980.
114. G. L. Werley in ref. 91, p. 334.
115. J. C. Fox in ref. 3, Chapt. 9, pp. 423–453.
116. Ref. 93, pp. 222–236, 265–284.
117. T. Lyman, ed., *Metals Handbook*, 8th ed., Vol. 4, American Society for Metals, Metals Park, Ohio, 1969, pp. 145–153.
118. *ASTM Standard, Specification B 69-66*, reapproved 1979, American Society for Testing and Materials, Philadelphia, Pa., 1981, Part 8, pp. 35–40.
119. Ref. 87, Chapt. 3, pp. 57–60.
120. R. K. Martin in ref. 3, Chapt. 9, pp. 523–527.
121. E. H. Kelton in ref. 91, pp. 1082–1084.
122. C. W. Roberts and B. Walters, *J. Inst. Met.* **76**, 557 (1949–1950).
123. E. W. Horvick in ref. 15, pp. 629–638.
124. S. W. K. Morgan, *Zinc and Its Alloys*, MacDonald and Evans, Ltd., London, UK, 1977, p. 155.
125. *Steel (USA)* **161**(3), 83 (July 17, 1967).
126. B. D. Wakefield, *Iron Age* **204**(21), 122 (Nov. 20, 1969).
127. U.S. Pat. 3,865,176 (Feb. 11, 1975), J. Dampas and R. W. Hazelett (to Hazelett Strip Casting Corp.).
128. W. J. Mosbach, *Met. Prog.*, 60 (May 1975).
129. *Plating and Surface Finishing* **68**(7), 36 (July 1981).
130. A. Collart and C. P. Tombras, Sr., *Ind. Research and Development* **24**(2), 170 (Feb. 1982).
131. *Superplastic Metal Manual*, The New Jersey Zinc Company, Palmerton, Pa., 1970.
132. R. H. Johnson, *Met. Mater.* **4**(9), 115 (1970).
133. C. B. Tennant, Paper No. A70-48, *TMS-AIME Annual Meeting*, Denver, Colo., Feb. 1970.
134. W. Showak and S. R. Dunbar in T. Lyman, ed., *Metals Handbook*, 8th ed., Vol. 7, American Society for Metals, Metals Park, Ohio, 1972, pp. 335–340.
135. D. R. Dreger, *Mach. Des.*, 110 (Aug. 8, 1974).
136. J. N. Pomeroy and co-worker in ref. 3, Chapt. 6, pp. 314–321.
137. A. Pass, *J. Oil and Colour Chem. Assoc.* **35**, 241 (June 1952).
138. D. S. Newton in ref. 5, pp. 99–1005.
139. *ASTM Standards, Specification D 520-51*, reapproved 1976, American Society for Testing and Materials, Philadelphia, Pa., 1981, Part 28.
140. U. R. Evans, *Metal Ind. (London)* **67**, 114, 118 (1945).
141. J. E. O. Mayne and U. R. Evans, *J. Soc. Chem. Ind. London Rev. Sec.* **22**, 109 (1944).
142. A. W. Kennedy, *Modern Paint and Coatings* **66**(9), 21 (1976).
143. A. C. Elm, *Zinc Dust Metal Protective Coatings*, The New Jersey Zinc Company, Palmerton, Pa., 1968.

144. C. G. Munger, *Mater. Prot.* **2**(3), 8 (March 1963).
145. C. T. Flachbarth and W. H. Taylor in ref. 3, Chapt. 9, pp. 511–513.

THOMAS B. LLOYD
Gulf and Western National Resources Group

WALTER SHOWAK
The New Jersey Zinc Company, Inc.

ZINC COMPOUNDS

Zinc usually occurs as the sulfide but significant quantities of the oxide, carbonate, silicate [14374-77-7], and basic compounds of the latter two are also mined (see Zinc and zinc alloys).

Properties

Zinc is in Group IIB of the periodic table and exhibits a valence of +2 in all its compounds. Being high on the electromotive series, zinc forms quite stable compounds and, as such, resembles magnesium. Bonding in zinc compounds tends to be covalent, as in the sulfide and oxide. With strongly electropositive elements, eg, chlorine, the bond is more ionic. Zinc also tends to form stable covalent complex ions, eg, with ammonia $[Zn(NH_3)_2]^{2+}$, cyanide $[Zn(CN)_4]^{2-}$, and hydroxyl $[Zn(OH)_4]^{2-}$. The coordination number is usually four, to a lesser degree six, and in some cases five. A good review of zinc compounds is given in ref. 1.

Zinc forms salts with acids but since it is amphoteric, it also forms zincates, eg, $[Zn(OH)_3.H_2O]^-$ and $[Zn(OH)_4]^{2-}$. The tendency of zinc to form stable hydroxy complexes is also important because some basic zinc salts are only slightly soluble in water. Examples are $3Zn(OH)_2.ZnSO_4$ [12027-98-4] and $4Zn(OH)_2.ZnCl_2$ [11073-22-6], which may precipitate upon neutralization of acidic solutions of the salts.

Properties of zinc salts of inorganic and organic salts are listed in Table 1 with other commercially important zinc chemicals. In the dithiocarbamates, 2-mercaptobenzothiazole, and formaldehyde sulfoxylate, zinc is covalently bound to sulfur. In compounds such as the oxide, borate, and silicate, the covalent bonds with oxygen are very stable. Zinc–carbon bonds occur in diorganozinc compounds, eg, diethylzinc [557-20-0]. Such compounds were much used in organic synthesis prior to the development of the more convenient Grignard route (see Grignard reaction).

The water solubility of zinc compounds varies greatly, as shown in Table 1. Water-soluble compounds not listed are zinc formate [557-41-5], chlorate [10361-95-2], fluorosilicate [16871-71-9], and thiocyanate [557-42-6]. Also, the water-soluble amino and cyanide complexes have many uses.

Table 1. Properties, Prices, and Uses of Zinc Compounds

Zinc compound	Formula, synonym	CAS Registry No.	Sp gr	Mp, °C	Solubility[a], g/100 g solvent Water	Other	Price[b], $/kg (Aug. 1981)	Uses
acetate	$Zn(C_2H_3O_2)_2 \cdot 2H_2O$	[5970-45-6]	1.735	237	$40^{25°C}$	$67^{100°C}$ $3^{25°C}$ alcohol	3.53	wood preservative, mordant antiseptics, catalyst, waterproofing
ammonium chloride	$ZnCl_2 \cdot 2NH_4Cl$	[52628-25-8]	1.88	150 dec	$66^{0°C}$	$69^{30°C}$	0.71	galvanizing, solder flux, adhesives
diborate	$ZnO \cdot B_2O_3 \cdot 2H_2O$	[27043-84-1]	3.64		$0.007^{25°C}$	sl sol HCl	1.21	fireproofing, ceramics, fungicide
dodecaborate	$2ZnO \cdot 3B_2O_3 \cdot 3.5H_2O$	[12513-27-8]	4.22	980	insol		1.54	fire retardant
bromide	$ZnBr_2$	[13550-22-6]	4.21	394	$471^{25°C}$	$675^{100°C}$ sol alcohol, ether	62.00[c]	photographic paper, catalyst, batteries
carbonate	$ZnCO_3$	[3486-35-9]	4.40	$-CO_2$ at 300	$0.001^{15°C}$	insol alcohol	45.40[c]	ceramics, rubber, astringent (lotions)
chloride	$ZnCl_2$	[7646-85-7]	2.91	275	$432^{25°C}$	$614^{100°C}$ sol ether	0.41	textiles, adhesives, flux, wood preservative, antiseptic, astringent
cyanide	$Zn(CN)_2$	[557-21-1]	1.85	800 dec	$0.005^{20°C}$	sol alkali, CN$^-$	3.63	electroplating, gold extraction
dithiocarbamates								
dibutyl	$Zn\left[\begin{array}{c}S\\SCN(C_4H_9)_2\end{array}\right]_2$	[136-23-2]	1.21	106	insol	sol C_6H_6, CS_2, $CHCl_3$	3.17	vulcanization accelerator, lube oil
diethyl	$Zn\left[\begin{array}{c}S\\SCN(C_2H_5)_2\end{array}\right]_2$	[14324-55-1]	1.48	176	insol	sol C_6H_6, CS_2, $CHCl_3$	3.32	vulcanization accelerator
ethylenebis	$Zn\left[\begin{array}{c}S\\SCNHCH_2\end{array}\right]_2$, zineb	[12122-67-7]			insol	sol C_6H_6, CS_2, $CHCl_3$	5.50	fungicide, insecticide
dimethyl	$Zn\left[\begin{array}{c}S\\SCN(CH_3)_2\end{array}\right]_2$, ziram	[137-30-4]	1.71	249	$0.0065^{25°C}$	sol CS_2, acetone, alkali	3.06	vulcanization accelerator, fungicide
2-ethylhexanoate	$Zn(C_8H_{16}O_2)_2$	[136-53-8]	0.90		insol	sol hydrocarbon	1.87	paint drier, silicone rubber cure
fluoroborate	$Zn(BF_4)_2 \cdot 6H_2O$	[13826-88-5]		$-H_2O$ at 60°C	$>100^{25°C}$	sol alcohol	0.60	plating, bonderizing, textile resin cure
fluoride	ZnF_2	[7783-49-5]	4.95	872	$1.6^{18°C}$	sol hot acid, NH_4OH	100.00[c]	ceramics, impregnating wood, galvanizing
formaldehyde sulfoxylate	$Zn(HSO_2 \cdot CH_2O)_2$	[24887-06-7]		90 dec	$60^{25°C}$	insol alcohol	2.31	reducing agent, drying, polymerization

name	formula	CAS Registry Number	density	mp, °C	bp, °C	solubility	price, $/kg	uses
hydrosulfite	$ZnS_2O_4 \cdot 2H_2O$	[7779-86-4]		200 dec	$40^{20°C}$		2.25	bleach, especially textile, paper, reducing agent
iodide	ZnI_2	[10139-47-6]	4.70	446	$432^{18°C}$	sol alcohol	182^c	medicine, photography
2-mercaptobenzo-thiazole	$Zn(SC_6H_4NCS)_2$	[155-04-4]	1.70	300 dec	insol	sol C_6H_6, CS_2, $CHCl_3$	3.17	vulcanization accelerator for latex
naphthenate	$Zn[(C_2H)_5CHCOO]_2$	[12001-85-3]			insol	sol hydrocarbon, acid	1.17	paint film improver, rot proofer
nitrate	$Zn(NO_3)_2 \cdot 6H_2O$	[10196-18-6]	2.065	−18	93	sol alcohol	0.75	textiles as resin catalyst, mordant latex coagulant
oxide	ZnO	[1314-13-2]	5.47, 5.61	1800 sublimes	$0.0004^{18°C}$	sol acid, alkali, NH_4OH	1.05–1.12	vulcanization accelerator, mildewstat, pigment, supplement in feed and fertilizer, catalyst, ceramics, intermediate
peroxide	ZnO_2	[1314-22-3]	1.57	212 explodes	insol	sol acid	5.50	cosmetic powders as antiseptic
phosphate	$Zn_3(PO_4)_2$	[7779-90-0]	4.00	900	$2.6^{25°C}$	sol NH_4OH, insol alcohol	1.85	metal coatings, dental cement
potassium chromate	$4ZnO \cdot K_2O \cdot 4CrO_3 \cdot 3H_2O$, zinc yellow	[37300-23-5]	3.36–3.46		$0.24^{25°C}$		2.47	rust-inhibiting pigment
resinate	$Zn(C_{20}H_{30}O_2)_2$	[9010-69-9]	1.24	205	insol	sol hydrocarbon	0.99	inks, paint drier
selenide	$ZnSe$	[1315-09-9]	5.42	>1100	insol	sol acid	330.0	phosphor
silicofluoride	$ZnSiF_6 \cdot 6H_2O$	[18433-42-6]	2.10	100 dec	$77^{10°C}$		0.37	laundry sour, wood preservative, plaster additive
stearate	$Zn(C_{17}H_{35}COO)_2$	[557-05-1]	1.09	120	insol	sol hydrocarbon	1.90	lubricant, mold release, vinyl stabilizer, anticake, water repellant
sulfate	$ZnSO_4$	[7733-02-0]	3.54	680 dec	$41.9^{0°C}$	sol glycerol	0.77	rayon bath, agriculture, zinc plating, intermediate, flotation, mordant
sulfate	$ZnSO_4 \cdot H_2O$	[7446-19-7]	3.28	238 dec	$101^{70°C}$		0.58	rayon bath, agriculture, zinc plating, intermediate, flotation, mordant
sulfide	ZnS	[1314-98-3]	3.98, 4.10	1185 sublimes	$0.0007^{18°C}$	sol acid	1.76^d	phosphor, white pigment, dental materials
tetroxychromate	$4Zn(OH)_2 \cdot ZnCrO_4$	[13530-65-9]	3.87–3.97		insol		2.57	wash primer
undecylenate	$Zn[CH_2=CH(CH_2)_8COO]_2$	[557-08-4]		115	insol	sol hydrocarbon	9.70	dermal fungicide

[a] Insol = insoluble, v sol = very soluble, dec = decompose.
[b] Ref. 2.
[c] Reagent price.
[d] Pigment price.

Zinc compounds are generally colorless unless the other component, eg, chromate, is colored. The lack of color of most zinc compounds in visible light is a great advantage in that they do not color paint films, plastics, rubber, cosmetics, etc. However, when excited by various types of radiation and at various temperatures, zinc oxide, sulfide, selenide [1315-09-9], and related compounds exhibit luminescence, ie, they emit colored light (see Luminescent materials). Zinc-based phosphors can be produced in many colors, depending upon the added dopants. They are used in television tubes, luminescent glasses, and various specialty products.

Zinc Oxide

Physical Properties. Some of the physical properties of zinc oxide are listed in Table 2. Of great importance is the fact that it completely absorbs ultraviolet light below 366 nm and, thus, is unique among white pigments (15) (see Uv stabilizers). Its high refractive indexes make it a good white pigment where its mean diameter for maximum light scattering is 0.25 µm. The crystal structure of zinc oxide is likely to stabilize defects, eg, zinc excess or deficiency and inclusion of foreign ions, and therefore has useful semiconductor properties. Variously doped oxides are used for photocopying, catalysts, and phosphors (16–17) (see Reprography; Catalysts; Semiconductors).

Chemical Properties. Zinc oxide, as an amphoteric material, reacts with acids to form zinc salts and with strong alkalies to form zincates. In the vulcanization of rubber, the chemical role of zinc oxide is complex and the free oxide is required, probably as an activator (see Rubber chemicals). Zinc soaps result from the oxide's interaction with organic acids and their interaction with the accelerator. The oxide is used alone as an accelerator in certain elastomers, eg, neoprene and thiokols, which contain

Table 2. Selected Physical Properties of Zinc Oxide

Property	Value	Ref.
mp, °C	ca 1975 (subl)	3
color	white in finely divided form	
refractive index, 0.5 µm	2.015, 2.068	4
specific gravity	5.68	
water solubility, minimum at pH	9.7	5
K_{sp} Zn(OH)$_2$	4.5×10^{-17}	
heat capacity (at 25°C), J/(mol·°C)^a	40.26	3
$\Delta H_{formation}$ (at 419.5–907°C), kJ/mol^a	−356.1	6
$\Delta F_{formation}$ (at 419.5°C), kJ/mol^a	−281.6	6
$\Delta F_{formation}$ (at 907°C), kJ/mol^a	−229.0	
S formation (at 25°C), J/mol^a	43.65	7
coefficient of expansion, $\times 10^{-6}$/°C	4.0	8
conductivity, W/(m·K)	25.2	9
crystal structure	hexagonal, wurtzite	
conductivity (n-type), S/cm	10^{-7}–10^3	10
piezoelectricity (lithium-doped)	ca 4× that of quartz	11
magnetic susceptibility (at 196°C), $\times 10^{-6}$ Hz units	0.20	12
pyroelectric current density, MA/(m^2·s·K)	6.8	13
$E°$ of Zn + ½ O$_2$ = ZnO (at 25°C), V	1.649	14

^{a} To convert J to cal, divide by 4.184.

chlorine and sulfur in the polymer molecules. Zinc oxide reacts with carbon dioxide in moist air to form oxycarbonate. Acidic gases, eg, hydrogen sulfide, sulfur dioxide, and chlorine, react with zinc oxide, and carbon monoxide or hydrogen reduce it to the metal. At high temperatures, zinc oxide replaces sodium oxide in silicate glasses. An important biochemical property of the oxide is its fungicidal/mildewstatic action (see Fungicides) (18). It is also soluble in body fluids and soils (19).

Production and Processing. Primary zinc oxide is manufactured by oxidizing zinc vapor in burners wherein the concentration of zinc vapor and the flow of air is controlled so as to develop the desired particle size and shape. The hot gases and particulate oxide or fume pass through tubular coolers and then the ZnO is separated in a baghouse. The purity of the zinc oxide depends upon the source of the zinc vapor. Zinc oxide of great purity is required for pharmaceutical, photoconductive, and certain other grades, and these are made by the indirect (French) process which accounted for 34% of U.S. zinc production in 1976–1980. Zinc vapor from previously purified zinc metal is burned (20–21). Less-pure zinc oxide is manufactured by the direct (American) process, by which impure zinc oxide is reduced to zinc vapor which is then burned (20). The American process accounted for 58% of U.S. zinc oxide production in 1976–1980. Certain impurities in the original crude ore and carbonaceous reductant are inseparable from the product oxide. Nevertheless, the uses dictate that certain impurities be at low levels. For instance, cadmium, lead, iron, sulfur, copper, and manganese can be deleterious in rubber if they exceed certain concentrations (22–23). Some scrap zinc materials undergo both French and American processes; other processes are used to produce secondary zinc oxide from zinc scrap, zinc-containing sludges, and metallurgical slags.

Direct (American) Process. *Grate furnaces.* In the eastern Wetherhill furnace, four or more firebrick furnaces (called a block) having common walls are charged in cyclic fashion. Coal that is hot from the previous charge is first spread on the grate and, after ignition, a damp, well-blended mixture of zinciferous material and coal is added. The bed is maintained in a reducing condition with carbon monoxide to produce zinc and lead, if present. Metal vapors are drawn into a chamber above the furnace, where combustion air oxidizes them to pigment. The hot pigment–gas stream enters a cooling duct common to the whole block and, in this way, the product becomes a uniform blend. The raw material contains 15–75 wt % zinc, but usually over 60 wt %, and the coal is adjusted accordingly. For best results, anthracite or coke is used (see Coal). Calcium, iron, and silica influence the nature of the clinker and the elimination of zinc and must, therefore, be minimized.

The western Wetherill furnace is like the eastern furnace except that a block of 12 or more furnaces has a common combustion chamber. This results in improved control and uniformity of product. Clinker usually contains 8–15 wt % zinc, which is often recovered in a Waelz kiln (24).

The traveling-grate furnace requires less labor, increases the output per unit of grate area, and produces more uniform product than the Wetherill furnaces. The traveling grate is an endless chain of cast-iron bars, driven by sprockets, which traverses a firebrick chamber. Anthracite briquettes are fed to a depth of ca 15 cm. After ignition by the previous charge, the coal briquettes are covered by 15–16.5 cm of ore/coal briquettes. The latter are dried with waste heat from the furnace. Zinc vapor evolves and burns in a combustion chamber and the spent clinker falls into containers for removal (24–25).

Rotary kilns. The pigment-grade oxide kiln, because of its high temperature, produces pigment-quality zinc oxide and makes possible higher recovery as compared to the grate furnaces. The kilns, which are 2.4-m dia and 12.1–15.2-m long, are fed in the firing end with a mixture of 65 wt % sinter and 35 wt % anthracite. Very slow rotation and small slope give sufficient retention time for good zinc elimination. The atmosphere in the kiln is reducing. Air is then control-fed into a firebrick combustion chamber, where the zinc is burned to the desired size and shape. The exiting solid residue is water-quenched.

The Waelz process is used to beneficiate many metals, including zinc, and produces a relatively impure oxide (26). This is true because the loose feed is charged at the end of the kiln where metal vapors are exiting and because the feed materials are usually quite impure. Where the process is used to make crude zinc oxide, feeds include oxidized and sulfide zinc ores, residues, zinc-bearing iron ores and flue dusts, lead furnace slags, mill slimes, electrolytic-zinc leach residues, and ores of various other metals, eg, tin and gold (27). The metallurgy of the process is similar to the preceding processes, but it is complicated by the presence of other materials in the feed, eg, sulfates, sulfides, carbonates, and silicates. Conditioners, eg, silica, lime, and iron oxide, are often added to control the fluidity of the charge (28). Also, lime can be used to fix sulfur as calcium sulfide, so that materials containing up to 20 wt % sulfur can be introduced into the kiln. Any sulfur dioxide formed tends, because of the high temperature and long retention time, to be oxidized to trioxide and reacts with metallic oxides to form sulfate.

The charge, which contains ca 25 wt % carbon, progresses through the 11–30-m long kiln, countercurrent to the hot gases from the fuel burner and the burning carbon in the bed. Lifters give the bed a considerable mixing action, as the kiln rotates at 1–1.5 rpm. The particles are reduced when submerged in the bed and are oxidized on the surface. Conditions are adjusted so that the products in the exiting gas are zinc and cadmium vapors, stannous oxide, germanium monoxide, carbon monoxide, and often further oxidized states of these compounds. They are fully oxidized in the combustion chamber which may also serve as a settling chamber for fly ash from the kiln. Settled material is recycled to the kiln. The fume is cooled and collected in a baghouse. Very often, the crude ZnO is sintered to remove cadmium, sulfur, and lead and used for the production of zinc oxide or zinc metal.

Electrothermic process. Electrothermic zinc smelting is described in refs. 29–30. The oxide furnaces in the one U.S. plant are 11-m high, three are 1.75-m dia, and the fourth is 2.44-m dia with eight graphite electrodes at each of two levels. The electrical resistance path through the descending charge of 70 wt % sinter and 30 wt % coke is 9-m long, and it supplies the energy for smelting. The zinc vapor and carbon monoxide pass through exit ports at four levels between the upper and lower electrodes into manifolds where pigment forms by oxidation with added air. After cooling and passing through a cleaning cyclone, the oxide is collected in a baghouse (31).

Indirect (French) Process. Zinc metal vapor for burning is produced in several ways, one of which involves horizontal retorts. Since all the vapor is burned in a combustion chamber, the purity of the oxide depends on that of the zinc feed. Oxide of the highest purity requires special high grade zinc and less-pure products are made by blending in Prime Western and even scrap zinc.

Another method involves an electric-arc vaporizer which is >2000°C before burning (25,32). One of the features of the process is a rapid quench of the hot gas flow to yield very fine oxide particles (≤0.15 nm). This product is quite reactive and imparts accelerated cure rates to rubber. Internally fired rotary kilns are used extensively in Canada and Europe and, to a limited extent, in the United States (24). The burning occurs in the kiln and the heat is sufficient to melt and vaporize the zinc. Because of the lower temperatures, the particles are coarser than those produced in the other processes. In a fourth process, zinc metal which is purified in a vertical refining column is burned. In essence, the purification is a distillation and impure zinc can be used to make extremely pure oxide. Also, a wide range of particle sizes is possible (33).

Secondary Zinc Oxide. Secondary oxide is that made from scrap, eg, galvanizer's dross, trimmings, automotive zinc scrap, etc, and is largely used by rubber and ceramics manufacturers. About one fifth of the oxide produced in the United States is from secondary material, although some of it is made by primary producers. Most secondary oxide is of the French type in that the starting zinc is metallic. The zinc is usually vaporized in upright retorts in such a way that the higher boiling impurities, eg, lead, do not vaporize and relatively pure oxide is produced (34). Another procedure for manufacturing secondary oxide is precipitation of zinc carbonate from waste zinc solution, followed by drying, calcining, and grinding to a coarse oxide suited for ceramics. A similar product results from the calcination of zinc hydrosulfite [7779-86-4] sludge. Such sludges are by-products of the use of zinc hydrosulfite as a bleaching agent.

Leaded Zinc Oxide. Oxides containing more than 5 wt % basic lead sulfate are classified as leaded and are made in the American process from high lead materials, usually lead sulfide mineral, or by blending zinc oxide and basic lead sulfate. There is only one manufacturer in the United States and the product contains 20–28 wt % basic lead sulfate. Leaded oxides are used only in rubber in the United States.

Slag-Fuming Process. Three plants in the United States, two in Canada, and one in Mexico process slag from blast furnaces (usually lead) and residues to make zinc oxide fume. The oxide is impure and most is sent to zinc smelters for conversion into metal; however, a small amount is sold as pigment. The process consists of blowing powdered coal into a bath of molten slag via multiple double-inlet tuyeres (35). Zinc is eliminated from the 10–18 wt % zinc charge in ca 2 h, and the coal is the fuel and the reductant, ie, carbon monoxide, source. The direct method of introduction of the coal combines drying and grinding with the feeding of coal into the furnace feed lines. Indirect feeding uses previously dried and ground coal from a bin. In either case, the coal–air mixture must be carefully controlled to minimize excess coal yet avoid too low a temperature. The slag temperature is maintained at 1150–1200°C. Recovery is ca 90% or better for both lead and zinc, and results in fume of ca 70 wt % zinc oxide and 7–10 wt % lead.

Zinc Oxide Treatments. There are a number of treatment procedures in addition to the customary screening and pulverizing. Various chemical materials are added to improve ease of incorporation and dispersibility of the oxide in rubber and paint (see Paint). Commonly, these are fatty acids, eg, propionic and stearic, and oils (36–37). Phosphoric acid is added to reduce chemical activity and trialkyl phosphates to combat dusting (38–39). American-process zinc oxide may contain acidity in the form of oxysulfur compounds; this can be removed by washing with solutions of ammonia or ammonium salts (40). However, improved American-process oxides, which do not need washing, are being produced. Unless recalcined, zinc oxide tends to be bulky, dusty, and have poor dry flow. These characteristics are especially objectionable to those who use the oxide in ceramics and rubber. Therefore, some grades are pelleted in rotating drums or densified by passing through rolls.

Economic Aspects. Table 3 shows that rubber production is the largest market for zinc oxide; the downturn in 1980 resulted from a drop in tire production because of the production trend to smaller tires, more importation of tires, and a recession. The drop in paint usage reflects the trend to water-base paints, which originally contained no zinc oxide. However, its growing use in such paints is based upon improved formulations based on zinc oxide. The increased use in agriculture is a result of the realization of the importance of zinc as a trace element. The rise in use of zinc-oxide-coated paper for photocopying is followed by a slackening in use because of a shift to plain-paper copiers.

Zinc oxide prices are listed in Table 4.

Health and Safety Factors. Zinc oxide is considered nontoxic, but inhalation of freshly formed fume can cause zinc chills also known as brass-founder's ague. The symptoms are fever and cough followed by chills after ca 4–8 h (43). No aftereffects have been noted and workers who are continually exposed quickly develop a resistance (44). So-called chronic zinc poisoning is caused by toxic impurities, eg, lead, cadmium, arsenic, and antimony, which commonly contaminate zinc ores (45). The ACGIH has classified zinc oxide as a nuisance particulate and set the threshold limit value for air-borne oxide at 5 mg/m^3 (46) (see also Uses, Nutrition).

Table 3. U.S. Zinc Oxide Shipments by Industry, Metric Tons[a]

Industry	1965	1970	1975	1978	1979	1980
rubber	93,518	101,079	87,279	97,989	93,075	61,796
paints	27,500	19,862	9,994	13,237	12,503	12,165
ceramics	9,083	8,175	5,715	9,245	9,236	5,702
chemicals	10,313	17,631	15,916	27,057	27,710	17,551
agriculture	887	2,038	1,676	4,847	4,397	6,930
photocopying	8,000[b]	28,894	22,359	19,096	16,148	9,604
other	20,052	15,809	10,815	9,981	16,700	22,028
Total	169,353	193,488	153,754	181,452	179,769	135,776

[a] Ref. 41.
[b] Estimated.

Table 4. Year-End U.S. Prices of Zinc Oxide, $/kg[a]

Grade	1970	1975	1978	1979	1980
American process, lead-free	0.37	0.89	0.90	0.98	0.97
French process, lead-free					
lead-free	0.39	0.91	0.93	1.01	1.01
high purity	0.40	0.95	0.97	1.04	1.05
electrophotographic	0.44	0.97	0.99	1.07	1.05
leaded[b]					
12 wt % lead	0.31	0.78	0.82	0.87	0.87
18 wt % lead	0.33				
35 wt % lead	0.37				

[a] Ref. 42.
[b] As basic lead sulfate.

Uses. The uses of zinc oxide can be divided into two groups based on the chemical and physical properties of the compound (47). The largest user, the rubber industry, uses it chemically as a vulcanization activator and accelerator and to slow rubber aging by neutralizing sulfur and organic acids formed by oxidation. Fine oxides are used for fast cures and coarse, sulfated grades for slow cures. Physically, it is a reinforcing agent, a heat conductor, a white pigment, and an absorber of uv light (see Uv stabilizers).

In paints, zinc oxide serves as a mildewstat and acid buffer as well as a pigment. The oxide also is a starting material for many zinc chemicals. The oxide supplies zinc in animal feeds and is a fertilizer supplement used in zinc-deficient soils. Its chemical action in cosmetics (qv) and drugs is varied and complex but, based upon its fungicidal activity, it promotes wound healing. It is also essential in nutrition. Zinc oxide is used to prepare dental cements in combination with eugenol and phosphoric and poly(acrylic acid)s (48) (see Dental materials).

Zinc oxide functions in ceramics in several ways. Added to glasses, it imparts low thermal expansion, low melting, and increased chemical resistivity. The semiconducting property of a great variety of glasses and ceramics is based on their zinc oxide content. Zinc ferrites are basically zinc–ferrite oxide spinels, which are highly magnetic. Usually they also contain other oxides, eg, nickel and manganese oxides. Ferrites (qv) are used in many electrical and electronic devices. The oxide is used as a catalyst in alkylation, oxidation, hydrogenation, and dehydrogenation. The oxide is also used in coated photocopy paper (see Electrophotography).

Nutrition. Zinc is essential to the proper functioning of plants and animals and, as zinc sulfate and oxide, it is used as a feed supplement (49–51) (see Mineral nutrients; Pet and livestock feed). Most crops use less than a kilogram of zinc per 1000 m^2 per year, so that zinc salts added at 1.3–4.5 kg/ha gradually build up the zinc reserve (52). Animals, including humans, store relatively little available zinc and, thus, require a constant supply in the diet. For instance, beef cattle require 10–30 mg/kg dry feed, dairy cattle 40 mg/kg, and breeding hens 65 mg/kg. Zinc from plants is considered less available to monogastric animals than zinc from animal protein.

Zinc is second only to iron as a trace metal in humans. A 70-kg human body has ca 4.0 g iron, 2.0 g zinc, 0.2 g manganese, 0.1 g copper, and <0.1 g of all other elements combined (53). The recommended dietary allowance of zinc is 15 mg/d for adults and 25 mg/d for lactating females (54). Supplemental zinc is administered in the form of sulfate, chloride, acetate, gluconate [4468-02-4], stearate, or oxide but the anion is not involved in the processes of utilization. Supplementary zinc is prescribed for zinc deficiency and certain disorders. Deficiency is thought to be more widespread in the United States than previously assumed (55). Although excessive zinc produces toxic symptoms, such symptoms rarely occur. Large doses or continued ingestion of zinc may cause gastrointestinal distress; for example, 2 g of zinc sulfate is recommended as an emetic. However, 18 patients given 660 mg zinc daily for 16–26 weeks showed no ill effects but a 16-year-old boy who ingested 12 g of elemental zinc in two days did develop noticeable ill effects (56).

Insufficient zinc results in slowed growth, delayed wound healing, poor appetite, mental lethargy, and sexual immaturity and it interferes with the immune response. The main function of zinc in metabolism is enzymatic and there is evidence of other physiologic roles, eg, in stabilization of membrane structure (57).

Zinc Chloride

Zinc chloride melts at 275°C, boils at 720°C, and is stable in the vapor phase up to 900°C. It is very hygroscopic, extremely water-soluble, and soluble in organic liquids such as alcohols, esters, ketones, ethers, amides, and nitrides. Hydrates with 1, 1.5, 2.5, 3, and 4 molecules of water have been identified and great care must be exercised to avoid hydration of the anhydrous form. Aqueous solutions of zinc chloride are acidic (pH = 1.0 for 6 M) and, when partially neutralized, can form slightly soluble basic chlorides, eg, $ZnCl_2 \cdot 4Zn(OH)_2$ [11073-22-6] and $Zn(OH)Cl$ [14031-59-5]. Many other basic chlorides have been reported (58).

Anhydrous zinc chloride can be made from the reaction of the metal with chlorine or hydrogen chloride. It is usually made commercially by the reaction of aqueous hydrochloric acid with scrap zinc materials or roasted ore, ie, crude zinc oxide. The solution is purified in various ways depending upon the impurities present. For example, iron and manganese precipitate after partial neutralization with zinc oxide or other alkali and oxidation with chlorine or sodium hypochlorite. Heavy metals are removed with zinc powder. The solution is concentrated by boiling, and hydrochloric acid is added to prevent the formation of basic chlorides. Zinc chloride is usually sold as a 47.4 wt % (sp gr 1.53) solution, but is also produced in solid form by further evaporation until, upon cooling, an almost anhydrous salt crystallizes. The solid is sometimes sold in fused form.

The fumes of zinc chloride are highly toxic and can damage mucous membranes and cause pale gray cyanation. It can also ulcerate the skin of workers using it as a soldering flux or those handling wood impregnated with it (59).

The largest use of zinc chloride in the United States is in wood preservation, fluxes, and batteries (see Batteries and electric cells). Zinc chloride solution dissolves vegetable fiber and is widely used in mercerizing cotton (qv), swelling fibers, as a mordant in dyeing, parchmentizing paper, etc (see Fibers, vegetable; Dyes, application and evaluation–application; Papermaking additives). It dissolves metal oxides and is used as a flux, especially in galvanizing (see Solders). Zinc electroplating is often done with a chloride bath (see Electroplating). In medicine, it is used in antiseptics, deodorants, dental cements, and disinfectants (qv). Zinc chloride solutions preserve wood and textiles and are used in adhesives (qv) and embalming fluids (see Textiles). Other uses are in organic synthesis, eg, in the preparation of methyl chloride and diethylzinc, as a dehydrant, in rubber vulcanization, and in oil refining. The consumption of zinc chloride is declining, as shown in Table 5.

Table 5. U.S. Production and Importation of Zinc Oxide, Sulfate, and Chloride, Metric Tons[a]

Zinc compound	1960 Production	1960 Import	1970 Production	1970 Import	1975 Production	1975 Import	1980 Production	1980 Import
zinc oxide	125,309	11,517	202,059	10,952	150,050	11,963	145,509	29,843
zinc sulfate	25,064	1,581	48,236	5,713	21,223	2,895	35,159	3,871
zinc chloride[b]	23,995	806	20,139	1,044	na[c]	696	11,676	1,008

[a] Ref. 41.
[b] Includes zinc chloride in zinc ammonium chloride and chromated zinc chloride.
[c] Na = not available.

Zinc Sulfate

Anhydrous zinc sulfate forms when its hydrates are heated above 238°C. At ca 680°C, sulfur trioxide separates from the compound, forming 3ZnO.2SO$_3$ [12037-14-8] and above 930°C the compound is decomposed to zinc oxide. The three stable hydrates are ZnSO$_4$.H$_2$O [7446-19-7], ZnSO$_4$.6H$_2$O [13986-24-8], and ZnSO$_4$.7H$_2$O (orthorhombic) [7446-20-0]. The latter heptahydrate occurs in a few small deposits as the mineral goslarite [7446-20-0]. Three unstable hydrates are ZnSO$_4$.4H$_2$O [33309-49-8], ZnSO$_4$.2H$_2$O [80867-26-1], and ZnSO$_4$.7H$_2$O (monoclinic).

The solubility of zinc sulfate increases almost linearly with temperature from 27.6 wt % (as ZnSO$_4$) at −7°C to 41.4 wt % at 39°C. In this range, the heptahydrate is the solid phase. As the temperature rises, the solid phase becomes the hexahydrate and its solubility increases to a maximum of 47.7 wt % at 70°C. Above this temperature, the solid phase is the monohydrate and solubility declines with temperature to 44.0 wt % at the boiling point (105°C). Many basic zinc sulfates have been reported but probably the only true compounds are hydrates of 3Zn(OH)$_2$.ZnSO$_4$ [12027-98-4] (60).

Zinc sulfate was made by 15 companies in 1980 from secondary materials (93%) and from roasted ore, ie, zinc oxide (7%). The zinciferous material reacts with sulfuric acid to form a solution, which is purified. After filtration, the solution is heated to evaporation and heptahydrate crystals are separated. It is sometimes sold in this form but usually as the monohydrate [7446-19-7], which is made by dehydration at ca 100°C. Very pure zinc sulfate solution is made in the manufacture of the pigment lithopone [1345-05-7], ZnS.BaSO$_4$, and of zinc by electrowinning (see Zinc and zinc alloys).

Zinc sulfate is used in fertilizers (qv), sprays, and animal feeds in which it serves as a valuable trace element and disease-control agent. In the manufacture of rayon, it is a crenulating agent in the precipitation bath. It is also the starting material for the manufacture of many zinc chemicals and is used in textile dyeing and printing, flotation (qv) reagents, electrogalvanizing, paper bleaching, and glue (qv). In 1950, rayon accounted for 46% of zinc sulfate consumption, in 1960 55%, and in 1966 39%. Statistics for rayon are not available beyond 1966 but the use in agriculture grew from 39% in 1966 to 49% in 1977, to 72% in 1979, and to 78% in 1980. The actual tonnage used in agriculture also has increased steadily.

BIBLIOGRAPHY

"Zinc Compounds" in *ECT* 1st ed., Vol. 15, pp. 275–281 by A. P. Thompson and H. W. Schutz, the Eagle-Picher Co.; "Zinc Compounds" in *ECT* 2nd ed., Vol. 22, pp. 604–613 by A. P. Thompson, Eagle-Picher Industries, Inc.

1. M. Farnsworth and C. H. Kline, *Zinc Chemicals*, Zinc Institute, Inc., New York, 1973.
2. *Chem. Mark. Rep.* **220**(9), 40 (1981).
3. F. D. Rossini and co-workers, *Circular 500*, U.S. Bureau of Standards, Washington, D.C., 1952.
4. W. Bond, *J. Appl. Phys.* **36**, 1974 (1964).
5. L. Blok and P. L. DeBruyn, *J. Colloid Interface Sci.* **32**, 518 (1970).
6. T. C. Wilder, *Trans. Metall. Soc. AIME* **245**, 1370 (1969).
7. K. K. Kelley, *U.S. Bur. Mines Bull.*, 584 (1960).
8. D. D. Wagman and co-workers, *U.S. Bureau of Standards Technical Note*, U.S. Bureau of Standards, Washington, D.C., 1968, pp. 270–273.
9. P. Pascal, *Nouveau Traite de Chemic Minerals*, Vol. V, Masson et Cie, Paris, 1956.

10. M. Seitz and D. Whitmore, *Phys. Chem. Solids* **29,** 1033 (1968).
11. *Ceram. Ind. (Chicago)* **84,** 157 (1965).
12. J. Turkevich and P. W. Selwood, *J. Am. Chem. Soc.* **63,** 1077 (1941).
13. H. Ibach, *Solid State Commun.* **4,** 353 (1966).
14. W. J. Hamer, *J. Electroanal. Chem.* **10,** 140 (1965).
15. G. F. A. Stutz, *J. Franklin Inst.* **202,** 89 (1926).
16. H. E. Brown, *Zinc Oxide, Properties and Applications*, International Lead Zinc Organization, Inc., New York, 1976, pp. 56–59.
17. *Ibid.*, pp. 88–89.
18. S. B. Salvin, *Ind. Eng. Chem.* **36,** 336 (1944).
19. A. M. van Rij and W. J. Pories in J. O. Nriagu, ed., *Zinc in the Environment*, John Wiley & Sons, Inc., New York, 1980, Part 2, pp. 215–236.
20. E. H. Bunce and H. M. Haslam, *Trans. AIME* **121,** 678 (1936).
21. C. D. Holley, *The Lead and Zinc Pigments*, John Wiley & Sons, Inc., New York, 1909, pp. 152–153.
22. M. Morton, ed., *Introduction to Rubber Technology*, Reinhold Publishing Co., New York, 1959, p. 222.
23. A. P. Thompson in C. H. Mathewson, ed., *Zinc*, Reinhold Publishing Co., New York, 1959, p. 646.
24. J. H. Calbeck in ref. 23, pp. 344–367.
25. U.S. Pat. 1,522,097 (Jan. 6, 1925), F. G. Breyer, E. C. Gaskill, and J. A. Singmaster (to The New Jersey Zinc Co.).
26. U.S. Pat. 959,924 (May 31, 1910), E. Dedolph.
27. W. E. Harris, *Trans. AIME* **121,** 702 (1936).
28. C. W. Morrison in ref. 23, pp. 298–306.
29. U.S. Pats. 1,743,886 (Jan. 14, 1930); 1,773,779 (Aug. 26, 1930); 1,775,591 (Sept. 9, 1930), E. C. Gaskill (to St. Joseph Lead Co.).
30. J. J. Rankin in ref. 23, pp. 374–384.
31. R. E. Lund and co-workers in C. H. Cotterill and J. M. Cigan, eds., *Extractive Metallurgy of Lead and Zinc*, Vol. II, AIME, New York, 1970, pp. 549–580.
32. U.S. Pats. 1,522,096; 1,522,098 (Jan. 6, 1925), F. G. Breyer, E. C. Gaskill, and J. A. Singmaster (to The New Jersey Zinc Co.).
33. W. A. Thomas and W. A. Handwerk, *Min. Eng. (N.Y.)* **5,** 1203 (1953).
34. T. R. Janes in ref. 23, p. 385.
35. H. E. Lee and W. T. Isbell in ref. 23, pp. 307–314.
36. U.S. Pat. 2,303,330 (Dec. 1, 1942), B. R. Silver and E. R. Bridgewater (to The New Jersey Zinc Co. and E. I. du Pont de Nemours & Co., Inc.).
37. U.S. Pat. 1,997,925 (April 16, 1935), A. C. Eide (to American Zinc Lead and Smelting Co.).
38. U.S. Pat. 2,251,869 (Aug. 5, 1941), D. L. Gamble and J. H. Haslam (to The New Jersey Zinc Co.).
39. U.S. Pat. 4,270,955 (June 2, 1981), D. M. Eshelman (to The New Jersey Zinc Co.).
40. U.S. Pat. 2,372,367 (March 27, 1945), H. A. Depew (to American Lead Zinc & Smelting Co.).
41. *Minerals Yearbook*, U.S. Bureau of Mines, Washington, D.C., 1960, 1965, 1970, 1975, 1980.
42. *Chem. Market. Rep.* **198,** 43 (Dec. 28, 1970); **208,** 35 (Dec. 29, 1975); **214,** 37 (Dec. 25, 1978); **216,** 37 (Dec. 31, 1979); **218,** 37 (Dec. 29, 1980).
43. *Zinc—Medical and Biologic Effects of Environmental Pollutants*, Subcommittee on Zinc, National Research Council, University Park Press, Baltimore, Md., 1979, pp. 251–253.
44. P. Drinker and co-workers, *J. Ind. Hyg.* **3,** 98 (1927).
45. R. P. Batchelor and co-workers, *J. Ind. Hyg.* **8,** 322 (1926).
46. W. D. Kelly, *Nat. Saf. News* **119,** 65 (1979).
47. Ref. 16, pp. 6–7.
48. R. G. Silvey and G. E. Myers in ref. 19, Part 2, pp. 237–253.
49. Ref. 19, Parts 1 and 2.
50. Ref. 43, Chapt. 7.
51. G. J. Brewer and A. S. Prasad, eds., *Progress in Clinical and Biological Research*, Vol. 14, A. R. Liss, New York, 1977.
52. Ref. 43, pp. 73–83.
53. Ref. 43, p. 123.
54. Ref. 19, Part 2, pp. 4–5.
55. Ref. 43, p. 187.
56. Ref. 43, p. 249.

57. Ref. 19, Part 2, pp. 45–53.
58. J. W. Hoffman and I. Lander, *Aust. J. Chem.* **21,** 1439 (1968).
59. N. L. Sax, *Dangerous Properties of Industrial Materials*, 5th ed., Van Nostrand Reinhold, New York, 1979, pp. 1100–1104.
60. L. C. Copeland and O. A. Short, *J. Am. Chem. Soc.* **62,** 3285 (1940).

THOMAS B. LLOYD
Gulf and Western Natural Resources Group

ZIRCONIUM AND ZIRCONIUM COMPOUNDS

Zirconium [7440-67-7] is classified in subgroup IVB of the periodic table with its sister metallic elements titanium and hafnium. Zirconium forms a very stable oxide. The principal valence state of zirconium is +4, its only stable valence in aqueous solutions. The naturally occurring isotopes are given in Table 1. Zirconium compounds commonly exhibit coordinations of 6, 7, and 8. The aqueous chemistry of zirconium is characterized by the high degree of hydrolysis, the formation of polymeric species, and the multitude of complex ions that can be formed.

Zirconium occurs naturally as a silicate in zircon [1490-68-2], the oxide baddeleyite [12036-23-6], and in other oxide compounds. Zircon is an almost ubiquitous mineral, occurring in granular limestone, gneiss, syenite, granite, sandstone, and many other minerals, albeit in small proportion, so that zircon is widely distributed in the earth's crust. The average concentration of zirconium in the earth's crust is estimated at 220 ppm, about the same abundance as barium (250 ppm) and chromium (200 ppm) (2).

Zircon has been known as a gem mineral since Biblical times and was known as jargon in Sri Lanka (Ceylon) and as hyacinth in France. The name zircon possibly comes from the Arabic *zargon* for the gold or dark amber color of the more common gemstone. Zircons may be colorless, amber, red, reddish brown, blue, green, or black. In 1789, Klaproth announced that in analyzing jargon from Sri Lanka he had found 68% of an unknown earth which he called zirkonerde (3). In 1797, Vauqelin studied

Table 1. Naturally Occurring Zirconium Isotopes

Isotope	CAS Registry No.	Occurrence, %[a]	Thermal neutron capture cross section, 10^{-28} m^2 [b]
^{90}Zr	[13982-15-5]	51.45	0.03
^{91}Zr	[14331-93-2]	11.32	1.14
^{92}Zr	[14392-15-5]	17.19	0.21
^{94}Zr	[14119-12-1]	17.28	0.055
^{96}Zr	[15691-06-2]	2.76	0.020

[a] Ref. 1.
[b] To convert m^2 to barns, multiply by 10^{28}.

this new earth, to which the name zirconia was given, and published the preparation and properties of some of its compounds (4). In 1824, Berzelius prepared the first crude zirconium metal, a black powder, by heating potassium and potassium hexafluorozirconate [16923-95-8] in a closed pot (5). In 1914, the first relatively pure zirconium was prepared by the reduction of zirconium tetrachloride [10026-11-6] with sodium in a bomb.

High purity zirconium was first produced by van Arkel and de Boer in 1925. They vaporized zirconium tetraiodide [13986-26-0] into a bulb containing a hot tungsten filament which caused the tetraiodide to dissociate, depositing zirconium on the filament.

Zirconium is used as a containment material for the uranium oxide fuel pellets in nuclear power reactors (see Nuclear reactors). Zirconium is particularly useful for this application because of its ready availability, good ductility, resistance to radiation damage, low thermal-neutron absorption cross section 18×10^{-30} m^2 (0.18 barns), and excellent corrosion resistance in pressurized hot water up to 350°C. Zirconium is used as an alloy strengthening agent in aluminum and magnesium, and as the burning component in flash bulbs. It is employed as a corrosion-resistant metal in the chemical process industry, and as pressure-vessel material of construction in the ASME Boiler and Pressure Vessel Codes.

Occurrence and Mining

Zirconium is found in at least 37 different mineral forms (6) but the predominant commercial source is the mineral zircon, zirconium orthosilicate. Other current mineral sources are baddeleyite and eudialyte [12173-26-1].

Zircon occurs worldwide as an accessory mineral in igneous, metamorphic, and sedimentary rocks. Weathering has resulted in segregation and concentration of the heavy mineral sands in layers or lenses of placer deposits in river beds and ocean beaches. All commercial sources of zircon are derived from the mining of these ancient, unconsolidated beach deposits, the largest of which are in Kerala State in India, Sri Lanka, the east and west Coasts of Australia, on the Trail Ridge in Florida, and at Richards Bay in the Republic of South Africa. These heavy mineral sands are processed for the recovery of the titanium-bearing minerals ilmenite, rutile, and leucoxene, and the zircon is obtained as a co-product. The output of zircon depends largely then on the market for these titanium minerals used in producing titanium oxide white pigment and titanium metal.

The deposits, usually ca 4% heavy minerals, are mined with front-end loaders or sand dredges. Typically, the overburden is bulldozed away, the excavation is flooded, and the raw sand is handled by a floating sand dredge capable of dredging to a depth of 18 m. The material is broken up by a cutter head to the bottom of the deposit and the sand slurry is pumped to a wet-mill concentrator mounted on a floating barge behind the dredge. Initial wet concentration using screens, Reichert cones, spirals, and cyclones removes the coarse sand, slimes, and light-density sands to produce a 40 wt % heavy-mineral concentrate. The tailings are returned to the back end of the excavation and used for rehabilitation of worked-out areas. The concentrate is dried and iron oxide and other surface coatings are removed; various combinations of gravity separation (qv), magnetic separation (qv), and electrostatic separation yield individual concentrates of rutile, ilmenite, leucoxene, zircon, monazite, and xenotime. Other heavy minerals such as staurolite, tourmaline, sillimanite, corundum, and magnetite may be recovered as local situations warrant. Typical analyses of zircon sand samples are given in Table 2.

Baddeleyite, a naturally occurring zirconium oxide, has been found in the Poco de Caldas region of the states of Sao Paulo and Minas Geraes in Brazil, the Kola Peninsula of the USSR, and the northeastern Transvaal of the Republic of South Africa. Brazilian baddeleyite occurs frequently with zircon, and ore shipments are reported to contain 65–85% zirconium oxide, 12–18% silica, and 0.5% uranium oxide. Very little of this ore is exported now because all radioactive minerals are under close control of the Brazilian government.

The Phalaborwa complex in the northeastern Transvaal is a complex volcanic orebody. Different sections are mined to recover magnetite, apatite, a copper concentrate, vermiculite, and baddeleyite, listed in order of annual quantities mined. The baddeleyite is contained in the foskorite ore zone at a zirconium oxide concentration of 0.2%, and at a lesser concentration in the carbonatite orebody. Although baddeleyite is recovered from the process tailings to meet market demand, the maximum output could be limited by the requirements for the magnetite and apatite. The baddeleyite concentrate contains ca 96% zirconium oxide with a hafnium content of 2% Hf/Zr+Hf. A comminuted, chemically beneficiated concentrate containing ca 99% zirconium oxide is produced also.

Eudialyte, $(Na,Ca)_6ZrOH(Si_3O_9)_2$, from a large deposit near Narssaq in southwest Greenland, is the source of pure zirconium oxide. The hafnium ratio in the ore is 2.2% Hf/Zr+Hf.

Recently, two zirconium-containing minerals were discovered in North America, namely, welognite [55659-01-3], $(Sr_{2.8}Ca_{0.2})ZrNa_2(CO_3)_6 \cdot 3H_2O$, and gittinsite [75331-27-0], $CaZrSi_2O_7$ (7–8). Unlike zircon, the zirconium content of these minerals and eudialyte can be dissolved by strong acid.

Table 2. Typical Analysis of Zircon Sands, %

Assay	Sri Lanka	India	Nigeria	Republic of South Africa	United States, Florida	Australia East	Australia West
constituents							
$(Zr+Hf)O_2$	64.9	64.4	58.2	65.0	65.3	65.7	64.6
Fe_2O_3	0.16	0.23	0.90	0.18	0.07	0.06	0.28
Al_2O_3	0.20	0.60	0.65	0.15	0.17	0.14	0.25
TiO_2	0.71	0.16	0.05	0.08	0.10	0.17	0.32
P_2O_5	0.18	0.13	0.39	0.13	0.05	0.07	0.08
U_3O_8	0.03	0.04	0.11	0.03	0.03	0.03	0.03
Nb_2O_5	<0.01	<0.01	2.0	<0.01	<0.01	<0.01	<0.01
Hf/Hf+Zr	2.2	2.3	6.8	2.3	2.2	2.2	2.2
size, μm (mesh)							
>149 (+100)	0.1	41.8	97.2	5.6	0.4	83.8	20.2
<74–149 (−100–200)	40.7	48.6	2.7	92.7	91.7	14.2	77.9
<74 (−200)	59.2	9.6	0.1	1.7	7.9	2.0	1.9

Table 3. Physical Properties of Zirconium

Property	Value	Refs.
atomic weight	91.22	
density at 298.15 K, g/cm^3	6.5107	3, 5
crystal structure		
αZr		9
close-packed hexagonal space group	P6$_3$/mmc	10
a, nm	0.3231	
c, nm	0.5146	
c/a	1.5927	
βZr		11
body-centered cubic space group	Im3m	9
α–β transition temperature, K	1136 ± 5	12
melting temperature, K	2125 ± 10	13
boiling temperature, K	4577 ± 100	13
vapor pressure, $T < 2125$ K, $\log_{10} P_{kPa}$[a]		14
βZr	$8.956 \pm 0.080 \cdot (30810 \pm 240) \, T^{-1}$	
liquid Zr	$8.547 \pm 0.080 \cdot (29940 \pm 240) \, T^{-1}$	
heat of transition, kJ/mol[b]	3.89 ± 0.08	13
heat of melting, kJ/mol[b]	18.8 ± 2.1	14
heat of boiling, kJ/mol[b]	573.2 ± 4.6	15
heat of sublimation at 298 K, kJ/mol[b]	600.8	15
heat capacity, $T = 298$–1136, J/(mol·K)[b]		12
αZr	$22.857 \pm 8.970 \times 10^{-3} \, T - 0.69 \times 10^5 \, T^{-2}$	
βZr	$21.493 \pm 6.586 \times 10^{-3} \, T + 36.718 \times 10^5 \, T^{-2}$	
entropy at 298.15 K, J/(mol·K)[b]	38.99 ± 0.46	12
thermal expansion		16
single crystal		
perpendicular to c-axis, $L_{TC} = L_{0°C}$	$1 + 5.145 \times 10^{-6} \, T$	
parallel to c-axis, $L_{TC} = L_{0°C}$	$1 + 9.213 \times 10^{-6} \, T - 6.385 \times 10^{-9} \, T^2 + 18.491 \times 10^{-12} \, T^3 - 9.856 \times 10^{-15} \, T^4$	
volumetric, $V_{TC} = V_{0°C}$	$1 + 19.756 \times 10^{-6} \, T - 7.023 \times 10^{-9} \, T^2 + 19.146 \times 10^{-12} \, T^3 - 9.980 \times 10^{-15} \, T^4$	
polycrystalline		
linear, for random orientation, $L_{TC} = L_{0°C}$	$1 + 6.499 \times 10^{-6} \, T - 2.096 \times 10^{-9} \, T^2 + 6.108 \times 10^{-12} \, T^3 - 3.259 \times 10^{-15} \, T^4$	
thermal conductivity, W/(m·K)		17[c]
0–10 K	$1.08189 \times 10^{-1} \, T + 1.65524 \times 10^{-3} \, T^2 - 2.63239 \times 10^{-4} \, T^3$	
10–25 K	$9.48650 \times 10^{-1} + 3.44316 \times 10^{-1} \, T - 1.79663 \times 10^{-2} \, T^2 + 2.82821 \times 10^{-4} \, T^3$	
25–80 K	$1.80754 - 5.50950 \times 10^{-2} \, T - 7.61839 \times 10^{-4} \, T^2 - 3.72006 \times 10^{-6} \, T^3$	
80–500 K	$5.18009 \times 10^{-1} - 2.36738 \times 10^{-3} \, T + 6.28905 \times 10^{-6} \, T^2 - 5.58159 \times 10^{-9} \, T^3$	
500–1900 K	$2.44486 \times 10^{-1} - 2.3982 \times 10^{-4} \, T + 2.27218 \times 10^{-7} \, T^2 - 6.24923 \times 10^{-11} \, T^3$	
electrical resistivity, Zr, at 5°C, Ω·cm	43.74 0.08 × 10^{-6}	18
temperature coefficient, 0–200°C, per °C	42.5×10^{-4}	19
elastic moduli, Zr, GPa[d]		20–21
single-crystal adiabatic, 19.7°C, 6.505 g/cm^3		
C_{11}	143.5 ± 0.2	
C_{12}	72.5 ± 0.2	
C_{13}	65.4 ± 0.2	

Table 3. (continued)

Property	Value	Refs.
C_{33}	164.9 ± 0.2	
C_{44}	32.07 ± 0.03	
polycrystalline adiabatic[e]		20
Young's modulus, GPa[e]	97.1	
shear modulus, GPa[e]	36.5	
bulk modulus, GPa[e]	954	
Poisson's ratio	0.33	22
Brinell hardness number, HB[f]	90–130	

[a] To convert kPa to mm Hg, multiply by 7.5.
[b] To convert J to cal, divide by 4.184.
[c] Tabulated selected experimental values were regressed to give the equation quoted. Estimated values tabulated were not used in the regression. All equations represent tabulated, nonestimated values within ±2% or better.
[d] To convert GPa to psi, multiply by 145,000.
[e] Calculated from single-crystal adiabatic moduli by the Voight method.
[f] At good Kroll-process purity, lower for iodide zirconium.

Physical Properties

Zirconium is a hard, shiny, ductile metal, similar to stainless steel in appearance. It can be hot-worked to form slabs, rods, and rounds from arc-melted ingot. Further cold-working of zirconium with intermediate annealings produces sheet, foil, bar wire, and tubing. Physical properties are given in Table 3.

Chemical Properties

Zirconium forms anhydrous compounds in which its valence may be 1, 2, 3, or 4, but the chemistry of zirconium is characterized by the difficulty of reduction to oxidation states less than four. In aqueous systems, zirconium is always quadrivalent. It has high coordination numbers, and exhibits hydrolysis which is slow to come to equilibrium, and as a consequence zirconium compounds in aqueous systems are polymerized.

Zirconium is a highly active metal which, like aluminum, seems quite passive because of its stable, cohesive, protective oxide film which is always present in air or water. Massive zirconium does not burn in air, but oxidizes rapidly above 600°C in air. Clean zirconium plate ignites spontaneously in oxygen of ca 2 MPa (300 psi); the autoignition pressure drops as the metal thickness decreases. Zirconium powder ignites quite easily. Powder (<44 μm or −325 mesh) prepared in an inert atmosphere by the hydride–dehydride process ignites spontaneously upon contact with air unless its surface has been conditioned, ie, preoxidized by slow addition of air to the inert atmosphere. Heated zirconium is readily oxidized by carbon dioxide, sulfur dioxide, or water vapor.

Zirconium reacts more slowly with nitrogen than with oxygen. Heating in nitrogen for 3 min gives a 0.3 μm layer of zirconium nitride [25658-42-8] at 700°C or a 1.2-μm layer at 900°C. The nitriding rate is enhanced by the presence of oxygen in the nitrogen or on the metal surface. Clean zirconium in ultrapure nitrogen reacts more slowly. Although the nitride reaction occurs at 900°C or higher, diffusion of nitrogen into zirconium is slow, and temperatures of 1300°C are needed to fully nitride the metal.

868　ZIRCONIUM AND ZIRCONIUM COMPOUNDS

Heated zirconium is readily chlorinated by ammonium chloride, molten stannous chloride, zinc chloride, and chlorinated hydrocarbons and the common chlorinating agents. It is slowly attacked by molten magnesium chloride in the absence of free magnesium, which is always present in the Kroll process.

Zirconium is readily attacked by acidic solutions containing fluorides. As little as 3 ppm fluoride ion in 50% boiling sulfuric acid corrodes zirconium at 1.25 mm/yr. Solutions of ammonium hydrogen fluoride or potassium hydrogen fluoride have been used for pickling and electropolishing zirconium. Commercial pickling is conducted with nitric–hydrofluoric acid mixtures (see also Metal surface treatments).

Corrosion Resistance. Zirconium is resistant to corrosion by water and steam, mineral acids, strong alkalies, organic acids, salt solutions, and molten salts (28) (see also Corrosion and corrosion inhibitors). This property is attributed to the presence of a dense adherent oxide film which forms at ambient temperatures. Any break in the film reforms instantly and spontaneously in most environments.

Zirconium is completely resistant to sulfuric acid up to boiling temperatures, at concentrations up to 70 wt %, except that the heat-affected zones at welds have lower resistance in >55 wt % concentration acid (see Fig. 1). Fluoride ions must be excluded from the sulfuric acid. Cupric, ferric, or nitrate ions significantly increase the corrosion rate of zirconium in 65–75 wt % sulfuric acid.

Zirconium resists attack by nitric acid at concentrations up to 70 wt % and up to 250°C. Above concentrations of 70 wt %, zirconium is susceptible to stress-corrosion cracking in welds and points of high sustained tensile stress (29). Otherwise, zirconium is resistant to nitric acid concentrations of 70–98 wt % up to the boiling point.

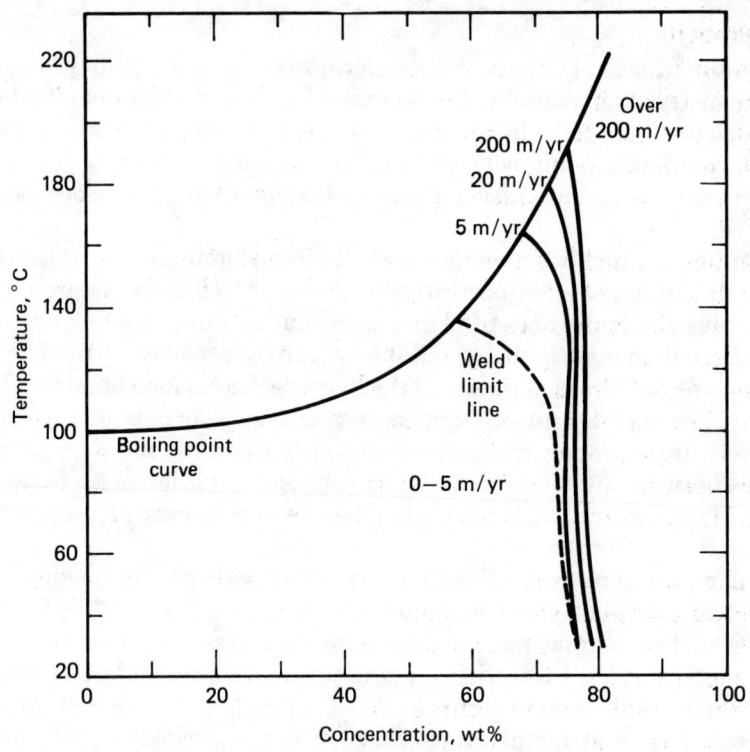

Figure 1. Corrosion of zirconium in sulfuric acid.

Zirconium is not attacked by caustics up to boiling temperatures. It is resistant to molten sodium hydroxide to 1000°C, but is less resistant to potassium hydroxide.

Zirconium is totally resistant to corrosion by organic acids. It has been used in urea-production plants for more than two decades.

Zirconium is totally resistant to attack of hydrochloric acid in all concentrations to temperatures well above boiling (see Fig. 2). Aeration has no effect, but oxidizing agents such as cupric or ferric ions may cause pitting. Zirconium also has excellent corrosion resistance to hydrobromic and hydriodic acid.

Processing

Decomposition of Zircon. Zircon is a highly refractory mineral as shown by its geological stability; the ore is cracked only with strong reagents and high temperature.

Electric Furnace. Zircon and coke have reacted in an electric arc furnace to produce a crude zirconium carbide nitride [*12713-24-5*] (ca 6 wt % C, 2 wt % N, 1 wt % O):

$$ZrSiO_4 + 3\,C \xrightarrow{air} Zr(C,N,O) + SiO\uparrow + 3\,CO\uparrow$$

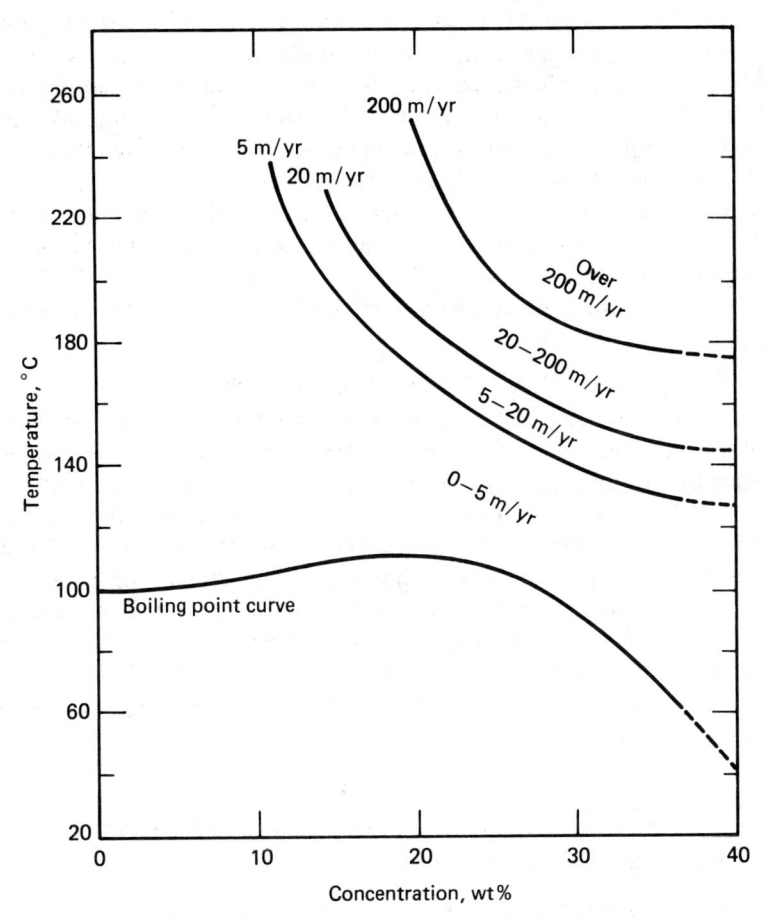

Figure 2. Corrosion of zirconium in hydrochloric acid.

With a deficiency of carbon, the silica is not reduced to carbide but converted into silicon monoxide which is vaporized at the reaction temperature, estimated to be 2500°C. In actual operation, the above proportions are optimal. The additional carbon is derived from the graphite electrodes. If more coke is added, more silicon is retained as the carbide in the fused ingot. When feed of the batch charge is completed, the fused ingot is separated from the unreacted charge which had insulated the ingot from the furnace shell, and the hot ingot is allowed to oxidize to crude zirconia. Alternatively, the entire charge is cooled and separated and the ingot is broken into lumps which are subsequently chlorinated.

Mixed zircon, coke, iron oxide, and lime reduced together produce zirconium ferrosilicon [71503-20-3], 15 wt % Zr, which is an alloy agent. Fused zirconia [1314-23-4] has been made from zircon but baddeleyite is now the preferred feed for the production of fused zirconia and fused alumina–zirconia by electric-arc-furnace processing.

Caustic Fusion. Fusion of finely ground zircon with caustic soda at 600°C or with soda ash produces a frit containing sodium silicate, sodium zirconate [12201-48-8], and some sodium silicozirconate (25–26). Water removes most of the sodium and silica, leaving a hydrous zirconium oxide which is soluble in most mineral acids. If the fusion is conducted with less alkali, the resulting frit is essentially Na_2ZrSiO_5 [12027-83-7] which can be ground and treated with a strong mineral acid to solubilize and extract the zirconium. These two methods are most commonly used to produce aqueous zirconium solutions, hydrated zirconium compounds, and zirconium oxide.

Similarly, fusion of milled zircon with dolomite or lime forms $CaSiO_3$ and $MgZrO_3$ [12032-31-4], $CaZrO_3$ [12013-47-7], and $CaO.Ca_2SiO_4$ or $CaSiO_3$ and ZrO_2, and is used to prepare zirconium oxide, usually as calcia-stabilized cubic zirconia because of the calcia left in solid solution in the zirconia (27–29).

Fluorosilicate Fusion. The fusion reaction of milled zircon with potassium hydrogen fluoride was used to prepare potassium hexafluorozirconate [16923-95-8] for studies leading to the first separation of hafnium and zirconium (30). Similar reactions using potassium hexafluorosilicate have been used (31–32) commercially in the United States and USSR:

$$K_2SiF_6 + ZrSiO_4 \rightarrow K_2ZrF_6 + 2\ SiO_2$$

The use of potassium hexafluorosilicate is preferred over sodium hexafluorosilicate because of the lower tendency of the potassium compound to dissociate and lose silicon tetrafluoride by sublimation. The addition of potassium carbonate or chloride to the fusion mix further reduces this tendency and promotes completion of the reaction. The reaction is conducted in a rotary furnace operating at 700°C. The product is crushed prior to leaching with acidified hot water. The hot slurry is filtered to remove the silica, and potassium hexafluorozirconate crystallizes as the solution cools.

Chlorination. Historically, the production of zirconium tetrachloride from zircon sand involved first a reduction to carbide nitride (see above) followed by the very exothermic reaction of the crushed carbide nitride with chlorine gas in a water-cooled vertical shaft furnace:

$$Zr(C,N,O) + 2\ Cl_2 \xrightarrow{700-1200°C} ZrCl_4 + (CO + N_2)\uparrow$$

In the current practice, milled zircon and coke are chlorinated in fluidized beds using chlorine as the fluidizing medium:

$$ZrSiO_4 + 4\ C + 4\ Cl_2 \xrightarrow{1100°C} ZrCl_4 + SiCl_4 + 4\ CO$$

Additional energy to sustain the endothermic reaction is provided chemically by the addition of silicon carbide grain or electrically by use of electrothermal fluidized beds (33–34), induction heating, or resistance heating. Chlorine efficiencies are typically 98% or better.

The product gases are first cooled below 200°C to selectively condense so-called zirconium tetrachloride snow in a large space condenser. The silicon tetrachloride subsequently is condensed in a quench condenser wherein the warm gases are countercurrently scrubbed with liquid silicon tetrachloride at −20°C. The silicon tetrachloride is purified by stripping and distillation.

Thermal Dissociation. The thermal dissociation of zircon into zirconia and cristobalite or liquid silica above 1650°C has been studied as a means of producing zirconia, but the process was hampered by the partial recombination of the two phases during cooling. Passing particulate zircon through the intense heat of a plasma and immediately thereafter quenching the molten droplet results in the formation of dissociated zircon particles of generally spheroidal configuration (35). The spheroids consist of a fused, amorphous silica mass in which radiating crystallites of monoclinic zirconia are dispersed.

This dissociated zircon is amenable to hot aqueous caustic leaching to remove the silica in the form of soluble sodium silicate. The remaining skeletal structure of zirconia is readily washed to remove residual caustic. Purity of this zirconia is directly related to the purity of the starting zircon since only silica, phosphate, and trace alkalies and alkaline earth are removed during the leach. This zirconia, and the untreated dissociated zircon, are both proposed for use in ceramic color glazes (36) (see Colorants for ceramics).

Separation of Hafnium. Zirconium and hafnium always occur together in natural minerals and therefore all zirconium compounds contain hafnium, usually about 2 wt % Hf/Hf+Zr. However, the only applications that require hafnium-free material are zirconium components of water-cooled nuclear reactors.

Zirconium and hafnium have very similar chemical properties, exhibit the same valences, and have similar ionic radii, ie, 0.074 mm for Zr^{4+}, 0.075 mm for Hf^{4+} (see Hafnium and hafnium compounds). Because of these similarities, their separation was difficult (37–40). Today, the separation of zirconium and hafnium by multistage counter-current liquid–liquid extraction is routine (41) (see Extraction, liquid–liquid extraction).

In the initial thiocyanate-complex liquid–liquid extraction process (42–43), the thiocyanate complexes of hafnium and zirconium were extracted with ether from a dilute sulfuric acid solution of zirconium and hafnium to obtain hafnium. This process was modified in 1949–1950 by an Oak Ridge team and is still used in the United States today. A solution of thiocyanic acid in methyl isobutyl ketone (MIBK) is used to preferentially extract hafnium from a concentrated zirconium–hafnium oxide chloride solution which also contains thiocyanic acid. The separated metals are recovered by precipitation as basic zirconium sulfate and hydrous hafnium oxide, respectively, and calcined to the oxide (44–45). This process is being used by Teledyne Wah Chang Albany Corp. and Western Zirconium Division of Westinghouse, and was used by Carborundum Metals Co., Reactive Metals Inc., AMAX Specialty Metals, Toyo Zirconium in Japan, and Pechiney Ugine Kuhlmann in France.

In the tributyl phosphate extraction process developed at the Ames Laboratory, Iowa State University (46–48), a solution of tributyl phosphate (TBP) in heptane is used to preferentially extract zirconium from an acid solution (mixed hydrochloric–nitric or nitric acid) of zirconium and hafnium (45). Most other impurity elements remain with the hafnium in the aqueous acid layer. Zirconium recovered from the organic phase can be precipitated by neutralization without need for further purification.

High molecular weight primary, secondary, and tertiary amines can be employed as extractants for zirconium and hafnium in hydrochloric acid (49–51). With similar aqueous-phase conditions, the selectivity is in the order tertiary > secondary > primary amines. The addition of small amounts of nitric acid increases the separation of zirconium and hafnium but decreases the zirconium yield. Good extraction of zirconium and hafnium from ca 1-M sulfuric acid has been effected with tertiaryamines (52–54), with separation factors of 10 or more. A system of this type, using trioctylamine in kerosene as the organic solvent, is used by Nippon Mining of Japan in the production of zirconium (55).

Zirconium and hafnium are separated by fractional distillation of the anhydrous tetrachlorides in a continuous molten solvent salt KCl–$AlCl_3$ system at atmospheric pressure (56–57). Zirconium and hafnium tetrachlorides are soluble in KCl–$AlCl_3$ without compound formation and are produced simultaneously.

Pure zirconium tetrachloride is obtained by the fractional distillation of the anhydrous tetrachlorides in a high pressure system (58). Commercial operation of the fractional distillation process in a batch mode was recently proposed by Ishizuka Research Institute (59). The mixed tetrachlorides are heated above 437°C, the triple point of zirconium tetrachloride. All of the hafnium tetrachloride and some of the zirconium tetrachloride are distilled, leaving pure zirconium tetrachloride. The innovative aspect of this operation is the use of a double-shell reactor. The autogenous pressure of 3–4.5 MPa (30–45 atm) inside the heated reactor is balanced by the nitrogen pressure contained in the cold outer reactor (60). However, previous evaluation in the USSR of the binary distillation process (61) has cast doubt on the feasibility of also producing zirconium-free hafnium tetrachloride by this method because of the limited range of operating temperature imposed by the small difference in temperature between the triple point, 433°C, and critical temperature, 453°C, of hafnium tetrachloride.

Reduction. Brezelius attempted the first reduction of zirconium in 1824 by the reaction of sodium with potassium fluorozirconate. However, the first pure ductile metal was made in 1925 by the iodide thermal-dissociation method. The successful commercial production of pure ductile zirconium via the magnesium reduction of zirconium tetrachloride vapor in an inert gas atmosphere was the result of the intense research efforts of Kroll and co-workers at the U.S. Bureau of Mines in 1945–1950 (62–67).

Obtaining pure ductile zirconium by reduction of the oxide is particularly difficult because of the tendency of hot zirconium to dissolve considerable amounts of oxygen, making the metal brittle at room temperature. Therefore, it is common practice to reduce oxygen-free zirconium tetrachloride.

Kroll Process. Hafnium-free zirconium dioxide is mixed with pulverized coke and fed into an induction-heated chlorinator where the mixture is fluidized by chlorine gas; reaction at 900°C gives zirconium tetrachloride and carbon dioxide. The product gas passes through a nickel-lined condenser where zirconium tetrachloride powder is formed by cooling below 200°C. It is purified by subliming and recondensing in a nitrogen–hydrogen atmosphere to reduce the aluminum and phosphorus contents.

The tetrachloride powder is charged into the upper chamber of a vertical cylindrical steel retort which contains cast ingots of magnesium in a stainless-steel liner within the lower chamber. The retort is sealed and evacuated and backfilled with argon several times at 200°C. Heat applied to the lower retort chamber melts the magnesium which in turn reduces the zirconium tetrachloride vapors as they sublime out of the upper chamber. The retort is cooled and unloaded. The stainless-steel liner is peeled from the reduction mass and the bulk of the magnesium chloride salt is physically separated from the Zr–Mg regulus. Several reduction reguli are stacked and loaded into a furnace for removal of residual magnesium chloride. The distillation is conducted in a vacuum <1.3 Pa (<10 µm Hg). As the temperature is gradually increased to 980°C, the magnesium chloride melts and drains and the magnesium metal is distilled and condensed on the cold lower retort wall, leaving porous zirconium sponge. After cooling and conditioning, the sponge is removed from the retort and broken into chunks using a hydraulic chisel. The chunks are graded, crushed to ca 1 cm dia, and sampled before blending and melting.

Other Reductions. Ductile, pure zirconium has been made by a two-stage sodium reduction of zirconium tetrachloride (68) in which the tetrachloride and sodium are continuously fed into a stirred reactor to form zirconium dichloride [13762-26-0]; heating with additional sodium yields zirconium metal. Leaching with water removes the sodium chloride from the zirconium. Bomb reduction of pure zirconium tetrafluoride with calcium also produces pure metal (69).

Finely divided zirconium powder is made by bomb reduction of zirconium oxide with magnesium or calcium. The powder is separated by leaching with cold hydrochloric acid. Because of its large surface area, it is extremely pyrophoric and hazardous to produce. Zirconium powder made by this technique is used in ordnance fuses and incendiaries, but it contains considerable oxygen and therefore cannot be converted into ductile mill products.

Electrolysis. Electrowinning of zirconium has long been considered as an alternative to the Kroll process, and at one time zirconium was produced electrolytically in a prototype production cell (70). Electrolysis of an all-chloride molten-salt system is inefficient because of the stability of lower chlorides in these melts. The presence of fluoride salts in the melt increases the stability of Zr^{4+} in solution, decreasing the concentration of lower-valence zirconium ions, and results in much higher current efficiencies. The chloride–electrolyte systems and electrolysis approaches are reviewed in refs. 71–72. The recovery of zirconium metal by electrolysis of aqueous solutions is not thermodynamically feasible, although efforts in this direction persist.

Refining. Zirconium sponge produced by the Kroll process has adequate purity and ductility for most uses. For applications requiring extremely soft metal and for research studies on the properties of the pure metal, it can be further purified by the van Arkel–de Boer (iodide-bar) process using a selective vapor transport (73–76). Zirconium sponge is loaded into a cylindrical Inconel vessel. The vessel lid contains insulated electrical lines from which a zirconium wire filament in a hairpin shape is suspended. Iodine is added to the evacuated vessel which then is heated to 250°C in a molten-salt bath. Volatile zirconium tetraiodide forms and diffuses to the central zirconium filament which is resistance-heated to 1200–1500°C. The tetraiodide thermally dissociates at the hot filament, depositing zirconium and releasing iodine to react again with the sponge feed. The deposition rate is controlled by the feed-bed temperature, filament temperature, iodine concentration, filament-to-bed distance, and the presence of other gases. Ordinarily, bars of 40 mm dia are grown from 3-mm filament wire. Under conditions optimized for zirconium transfer, the impurity metals and metallic oxides, carbides, and nitrides transfer poorly so that the zirconium filament is much softer and purer than the starting sponge.

Electron-beam melting of zirconium has been used to remove the more volatile impurities such as iron, but the relatively high volatility of zirconium precludes effective purification. Electrorefining in fused-salt baths (77–78) and purification by d-c electrotransport (79) have been demonstrated but are not in commercial use.

Economic Aspects

Normally, zircon sand is readily available as a by-product of rutile and ilmenite mining at ca $150 per metric ton. However, zircon and baddeleyite are obtained as by-products of their operations, and therefore, the supply is limited by the demand for other minerals. In 1974, when a new use for zircon in tundish nozzles developed in the Japanese steel industry, a resulting surge in demand and stockpiling raised zircon prices to $500/t. Worldwide production by country is given in Table 4, U.S. consumption between 1972 and 1981 in Table 5.

Zirconium oxide is made in a wide variety of grades which sell for $1.10–15.40/kg. The principal U.S. producers of zirconium oxides and zirconium chemicals are Norton, Carborundum, Magnesium Elektron, Harshaw, Ferro Corp., Corning, TAM Ceramics, Associated Minerals Consolidated, Ltd., and Teledyne Wah Chang Albany.

Table 4. Estimated Zircon Production, 1000 t [a]

Country	1980	1981
Australia	540	470
Republic of South Africa	88	110
USSR	80	80
United States	70	80
India	16	17
People's Republic of China	14	15
Brazil	4	4
Sri Lanka	3	4

[a] Ref. 80.

Table 5. U.S. Zircon Consumption, 1000 t [a]

Consumption	1972	1975	1978	1980	1981
foundry	92	46	72	55	75
zircon refractories	12	28	27	25	25
AZS refractories [b]	13	15	11	8	5
ZrO_2 + AZ abrasives [c]	18	12	17	18	13
alloys, <90% Zr	3	3	3	2	5
other [d]	30	18	34	32	27
Total	168	122	164	140	150

[a] Ref. 80.
[b] Alumina–zirconia–silica refractories.
[c] Alumina–zirconia abrasives.
[d] Other includes zirconium chemicals, primary metal, welding rod, miscellaneous.

Zirconium metal, hafnium-free, is being produced in volume by Teledyne Wah Chang Albany and Western Zirconium in the United States, CEZUS (Pechiney Ugine Kuhlmann) in France, and in the USSR, primarily for use in nuclear energy programs. Zirconium metal production programs are under development in Japan, India, and Brazil. The total annual U.S., French, USSR, and Japanese zirconium metal production capacity is estimated at 6800 t of sponge.

The current market prices for standard zirconium sheet and strip are between $44 and 66/kg. Since zirconium metal production methods are energy and labor intensive, escalation in the cost of zirconium continues to reflect these relationships.

Specifications and Standards

The U.S. specifications for zirconium are listed in Table 6. For nuclear power use, each reactor vendor issues particular, detailed specifications which usually include the pertinent ASTM nuclear specifications.

The four ASTM-specified grades of commercial zirconium and zirconium alloys are given in Table 7. The unalloyed Zr 702, the most commonly used commercial grade, has the best overall chemical corrosion resistance but it is the lowest in strength; Zr 705 has similar corrosion resistance in most environments but its strength is almost double that of Zr 702. It has better formability in applications requiring bending through a sharp radius. Both Zr 702 and 705 are approved for use in the construction of pressure vessels according to the American Society of Mechanical Engineers Boiler and Pressure Vessel Code, Section VIII.

Table 6. U.S. Specifications for Zirconium

Form	ASTM Nuclear	ASTM Commercial	ASME	DOE	AWS[a]
sponge	B 349	B 494			
ingot	B 350	B 495		M 10-1T	
bars, rod, and wire	B 351	B 550	SB550	M 7-9T	
flat rolled products	B 352	B 551	SB551	M 5-6T	
tubing	B 353	B 523	SB523	M 3-8T	
forging and extrusions		B 493	SB493	M 2-9T	
bare welding rods			SFA5.24	M 1-16T	A5.24
descaling and cleaning		B 614			
welding fittings		B 653			
seamless and welded pipe		B 658	SB658		
aqueous corrosion testing	G 2	G 2			

[a] American Welding Society.

Table 7. ASTM Grades and Commercial Designations of Commercial Zirconium and Alloys

Nominal composition, wt %	R 60702, Zr 702	R 60704, Zr 704	R 60705, Zr 705	R 60706, Zr 706
Zr + Hf, min	99.2	97.5	95.2	95.5
Fe + Cr	<0.2	0.3	<0.2	<0.2
Sn		1.5		
Nb			2.5	2.5
Hf	2.0	2.0	2.0	2.0

The chemical corrosion resistance of Zr 704 is slightly less than that of Zr 702 in some environments but Zr 704 is superior in high temperature, high pressure water, and steam. A softer version of Zr 705 is Zr 706, developed specifically for severe forming applications such as panel-type heat exchangers.

There are no industry-wide specifications for zirconium metal castings, or for zirconium chemicals.

For nuclear applications, hafnium-free zirconium is used, mostly as Zircaloy 2 and Zircaloy 4, in alloys developed in the U.S. Naval Nuclear Propulsion Program. The nickel-free Zircaloy 4 absorbs less of the hydrogen generated by steam corrosion, and is used increasingly. Zr–2.5Nb was adapted by Atomic Energy of Canada, Ltd., for use in Candu reactor pressure tubes (see Nuclear reactors). Excel is a newer, stronger, creep-resistant alloy developed in Canada as a pressure-tube material. Ozhennite 0.5 is another zirconium alloy from the USSR for nuclear applications. A UK zirconium alloy, A.T.R., is used in carbon dioxide cooled reactors. The alloying compositions are given in Table 8.

Analytical Methods

Zirconium is often determined gravimetrically. The most common procedure utilizes mandelic acid (81) which is fairly specific for zirconium plus hafnium. Other precipitants, including nine inorganic and 42 organic reagents, are listed in ref. 82. Volumetric procedures for zirconium, which also include hafnium as zirconium, are limited to either EDTA titrations (83) or indirect procedures (84). X-ray fluorescence spectroscopy gives quantitative results for zirconium, without including hafnium, for concentrations from 0.1% to 50% (85). Atomic absorption determines zirconium in aluminum in the presence of hafnium at concentrations of 0.1–3% (86).

Emission spectroscopy is used for lower concentrations and trace levels. Methods, as outlined in ASTM procedures (87), include zirconium in aluminum and aluminum alloys, ceramics, sand, magnesium alloys, and titanium.

Colorimetric methods always include hafnium. Most methods employ a separation step such as solvent extraction. The three reagents used successfully are 8-hydroxyquinoline (88), alizarin red S (89), and catechol violet (90).

Impurities in zirconium and zirconium alloys and compounds are often determined by emission spectroscopy. Both carrier distillation techniques and point-to-plane methods are available (91–92). Several metallic impurities can be determined instantaneously by this method. Atomic absorption analysis has been used for iron, chromium, tin, copper, nickel, and magnesium (93). The interstitial gases, hydrogen,

Table 8. Hafnium-Free Zirconium Alloys For Nuclear Service

Element	Zircaloy 2	Zircaloy 4	Zr–2.5 Nb	Excel	A.T.R.	Ozhennite 0.5
Sn	1.5	1.5	<0.02	3.5	<0.01	0.02
Fe	0.14	0.22	<0.08	<0.08	<0.05	0.1
Cr	0.1	0.1	<0.02	<0.02	<0.01	<0.02
Ni	0.05	<0.004	<0.007	<0.007	<0.004	0.1
Nb			2.5	0.8		0.1
Cu					0.55	
Mo				0.8	0.55	

nitrogen, and oxygen are most often determined by chromatography (81). Procedures for carbon, chloride, fluoride, phosphorus, silicon, sulfur, titanium, and uranium in zirconium are given in the literature (81,94–96).

Health and Safety Factors

Zirconium is generally nontoxic as an element or in compounds (97–98). At pH normally associated with biological activity, zirconium chiefly exists as the dioxide which is insoluble in water and in this form zirconium is physiologically inert.

Chelated complexes such as sodium zirconium lactate [15529-67-6] or ammonium zirconium carbonate [22829-17-0], and acidic forms such as zirconium hydroxy oxide chloride [18428-88-1] have been used in preparations in deodorants or for treatment for poison oak and poison ivy dermatitis. In such occasions, when the skin had been cut or abraded, a few users developed granulomas which have been identified as a delayed hypersensitivity to zirconium (99). These may take several weeks to develop, and commonly persist for six months to over a year.

The oral toxicity is low; OSHA standards for pulmonary exposure specify a TLV of 5 mg zirconium per m^3.

In massive form, zirconium is commonly brought to red heat before forging, swaging, or other hot working with only slight surface oxidation of the ingot or slab. Conversely, zirconium metal with a high surface area such as sponge, foil, fine powder, fine machining chips, and grinding dust is extremely flammable and often pyrophoric, especially when slightly damp. Many unexpected ignitions have occurred caused by improper handling of finely divided forms of zirconium metal, resulting in more than one incident of fatal flash burns to metal workers. Very fine dusts of zirconium ignite when dispersed in air. Fires can be extinguished with a blanket of argon or a layer of dry salt or sand. Water is hazardous unless the zirconium can be quenched below its ignition temperature. A settled 25-kg mass of 20-μm particles ignited in the bottom of a water-filled 55-gal (208-L) drum burns for hours. A stream of water results in the generation of hydrogen and steam which disperse the burning fragments and spread the fire (100). Safe handling of zirconium has been discussed in several publications (100–102).

Zirconium powder reacts exothermically with many other elements, including hydrogen, boron, carbon, nitrogen, and the halogens, although the ignition temperature is usually above 200°C. The reaction between zirconium powder and platinum is especially violent.

Finely divided zirconium is classified as a flammable solid and shipping regulations are prescribed accordingly (103). Metal powder finer than 74 μm (270 mesh) is limited to 2.26 kg per individual container.

Uses

The largest use is foundry sand, where zircon is used as the basic mold material, as facing material on mold cores, and in ram mixes. Because of its chemical inertness and high melting point, zircon is wetted less easily by molten metal, producing smoother surfaces on iron, high alloy steel, aluminum, and bronze castings. Zircon-sand molds have greater thermal shock resistance and better dimensional stability than quartz-sand molds. The zircon grains are usually bonded with sodium silicate in the formation of the molds and cores, but other binders, eg, clay, lignin-containing liquors, and furan resin, are used. The consumption of zircon in foundry applications was significantly reduced by the 1974 price increases. Alternatives include olivine, chromite, and quartz as well as increased cleaning and recycling of zircon foundry sand.

Fired zircon and AZS (alumina–zirconia–silica) refractories, including brick, crucibles, and saggers, have their main use in the production and handling of molten glasses (see Refractories). Blends of zircon with alumina or magnesia improve spalling resistance, reduce thermal conductivity, withstand thermal cycling, and have better corrosion resistance in specific glass-furnace environments. Other uses for zircon refractories include hearths and nozzles for handling molten metal and in extrusion dies. Finely milled zircon is used as an opacifier in glazes for tiles and sanitary wares. The use of zircon in these applications is ca 60% of the U.S. consumption of zircon and baddeleyite; the remaining 40% are used in the conversion into other forms of zirconium.

Zirconium oxide is fused with alumina in electric-arc furnaces to make alumina–zirconia abrasive grains for use in grinding wheels, coated-abrasive disks, and belts (104) (see Abrasives). The addition of zirconia improves the shock resistance of brittle alumina and toughens the abrasive. Most of the baddeleyite imported is used for this application, as is zirconia produced by burning zirconium carbide nitride.

Zirconium oxide is used in the production of ceramic colors or stains for ceramic tile and sanitary wares. Zirconia and silica are fired together to form zircon in the presence of small amounts of other elements which are trapped in the zircon lattice to form colors such as tin–vanadium yellow, praseodymium–zircon yellow [68187-15-5], vanadium–zircon blue [12067-91-3], iron–zircon pink [68412-79-3], indium–vanadium orange (105–108).

Zirconium oxide increases the refractive index of some optical glasses, and is used for dispersion hardening of platinum and ruthenium. Very fine zirconium oxide has been used for polishing glass but ceria seems to be preferred.

Yttria or calcia-stabilized cubic zirconias are used extensively as the solid electrolyte in oxygen sensors in automotive and boiler exhausts, and oxygen-content probes for molten copper or iron in smelters. Stabilized zirconia has been proposed for high temperature fuel cells operating with natural gas, or converting ammonia to nitric acid (109). Stabilized zirconias are used for diverse applications such as ceramic tubing, extrusion dies, ball-milling media, high temperature heating elements (110), insulating fibers, thermal barrier coatings, ceramic diesel engines, and a nonlubricated ball-bearing assembly for use in a space vehicle. Single-crystal cubic zirconias of high purity are being produced as a substitute for diamonds in jewelry (111) (see Gems, synthetic).

Zirconate compounds exhibit several interesting properties. Lead zirconate–titanate [12626-81-2] compositions display piezoelectric properties which are utilized in the production of FM-coupled mode filters, resonators in microprocessor clocks, photoflash actuators, phonograph cartridges, gas ignitors, audio tweeters and beepers, and ultrasonic transducers. Lanthanum-modified lead zirconate–titanate ceramics have been studied for photoferroelectric image storage (112) (see Ferroelectrics). Alkaline-earth zirconate dielectrics are used in ceramic capacitors.

The uses of other zirconium compounds result primarily from the ability of zirconium to complex with carboxyl groups and to form an insoluble organic compound. Ammonium zirconium carbonate enhances the fungicidal action of copper salts on cotton cloth (113) and zirconium acetate has been used as an algicide. Both ammonium zirconium carbonate and zirconium acetate [14311-93-4] have been used in the waterproofing of fabrics (114). Freshly prepared zirconium sulfate solutions are preferred over chromium solutions in the tanning of leather to prepare white leathers, although other mixed sulfate solutions containing zirconium, chromium, and aluminum have been proposed. In the retanning of chrome-treated leather, an intermediate treatment with ammonium zirconium carbonate improves the results obtained during the second chrome tanning (115). Zirconium carbonate produces durable acrylic-emulsion floor polishes which are easily stripped by treatment with household ammonia (see Polishes). Both zirconium sulfate [14644-61-2] and potassium hexafluorozirconate have been used for flame-resistant treatment of wool fabric. Very dilute fluoride solutions containing zirconium and gluconic acids are used for the preparation of clean aluminum surfaces to be subsequently painted or overcoated (116). Similarly, an aqueous solution of zirconium nitrate [13746-89-9] and poly(vinyl alcohol) is applied as a precoating on glass surfaces before the application of phosphor photobinder in the production of cathode-ray tubes.

Zirconium tetrafluoride [7783-64-4] is used in some fluoride-based glasses. These glasses are the first chemically and mechanically stable bulk glasses to have continuous high transparency from the near uv to the mid-ir (0.3–6 μm) (117–118). Zirconium oxide and tetrachloride have use as catalysts (119), and zirconium sulfate is used in preparing a nickel catalyst for the hydrogenation of vegetable oil. Zirconium 2-ethylhexanoate [22464-99-9] is used with cobalt driers to replace lead compounds as driers in oil-based and alkyd paints (see Driers and metallic soaps).

Sodium hydrogen zirconium phosphate [34370-53-1] is an ion-exchange material used in portable kidney dialysis systems which regenerate and recirculate the dialysate solution. The solution picks up urea during the dialysis. The urea reacts with urease to form ammonia, which is absorbed by the sodium hydrogen zirconium phosphate.

Zirconium phosphate [13772-29-7] also absorbs cesium and other radioactive-decay daughter products, and has been proposed as part of permanent disposal systems for nuclear fuel waste processing.

Zirconium metal is marketed in three forms: zirconium-containing silicon–manganese, iron, ferrosilicon, or magnesium master alloys; commercially-pure zirconium metal; and hafnium-free pure zirconium metal. The use of zircon for the production of zirconium metal of all three types is ca 5–8% of the total U.S. zircon consumption.

Silicon–manganese–zirconium, ferrozirconium, and ferrosilicon–zirconium (and some pure zirconium) are used in the steel industry for deoxidizing. Magnesium–zirconium is added to magnesium and aluminum alloys for grain refining and strengthening. Most magnesium alloys used at elevated temperatures contain zirconium.

Because of its flammability, zirconium is used in military ordnance including percussion-primer compositions, delay fuses, tracers, and pyrophoric shrapnel, in getters for vacuum tubes, inert-gas glove boxes, and sodium-filled hollow-shaft exhaust valves, and as shredded foil in flashbulbs for photography and excitation of lasers. Alloying applications of pure zirconium include zirconium–niobium superconductors, titanium aircraft alloys, and strengthening of copper alloys. Pure zirconium is being increasingly used as a corrosion-resistant metal in fabricating columns, pumps, pipe, valves, heat exchangers, and tanks for severe chemical environments, particularly sulfuric and hydrochloric acids, except those containing fluorides. In this capacity, it is also used in facilities producing urea, hydrogen peroxide, methyl methacrylate, or acetic acid.

Hafnium-free zirconium alloys containing tin or niobium are used for tubing to hold uranium oxide fuel pellets inside water-cooled nuclear reactors. Zirconium–niobium alloys are used for pressure tubes and structural components in Canadian, USSR, and FRG reactor designs.

Hafnium-free zirconium is particularly well-suited for these applications because of its ductility, excellent oxidation resistance in pure water at 300°C, low thermal neutron absorption, and low susceptibility to radiation. Nuclear fuel cladding and reactor core structural components are the principal uses for zirconium metal.

Compounds

The numerous intermetallic compounds of zirconium, from $ZrAl_3$ to $ZrZn_6$, are reviewed in refs. 120–123.

Hydrides. Zirconium hydride [7704-99-6] in powder form was produced by the reduction of zirconium oxide with calcium hydride in a bomb reactor. However, the workup was hazardous and many fires and explosions occurred when the calcium oxide was dissolved with hydrochloric acid to recover the hydride powder. With the ready availability of zirconium metal via the Kroll process, zirconium hydride can be obtained by exothermic absorption of hydrogen by pure zirconium, usually highly porous sponge. The heat of formation is 167.4 J/mol (40 kcal/mol) hydrogen absorbed.

Zirconium absorbs or desorbs hydrogen reversibly. Provided the metal surface is clean, equilibrium is attained quickly above 400°C, and very slowly below 250°C. The hydrogen solubility limit in alpha zirconium has been expressed (124) as solubility, ppm = $1.61 \times 10^5 \exp(-8959/RT)$. The addition of hydrogen lowers the $\alpha \rightarrow \beta$ transition from 876 to 550°C at 6 at. % hydrogen. Further hydrogen addition causes the formation of hydride phases, culminating in the body-centered tetragonal epsilon phase as the hydrogen content approaches the limiting concentration of ZrH_2 (125–126). Hydrogen is removed by heating and evacuating. To reduce the hydrogen content to 10 ppm, the specimen must be heated to 900°C at 1 Pa (ca 10^{-2} mm Hg) (see Hydrides).

Zirconium hydride is not a true compound of fixed stoichiometry but rather a series of crystalline phases through which zirconium metal transforms with changing hydrogen concentration and temperature. The γ phase hydride exists below 250°C in the narrow composition range $ZrH_{0.9}$ to $ZrH_{1.1}$, the δ phase has a compositional range $ZrH_{1.5}$ to $ZrH_{1.7}$, and the ϵ phase has a compositional range $ZrH_{1.8}$–ZrH_2. Most commercial hydride powder contains δ and ϵ phase.

Above 40 wt % hydrogen content at room temperature, zirconium hydride is brittle, ie, has no tensile ductility, and it becomes more friable with increasing hydrogen content. This behavior and the reversibility of the hydride reaction are utilized in preparing zirconium alloy powders for powder metallurgy purposes by the hydride–dehydride process. The mechanical and physical properties of zirconium hydride, and their variation with hydrogen content of the hydride, are reviewed in ref. 127.

Intermetallic compounds of zirconium with iron, cobalt, and manganese absorb and desorb considerable amounts of hydrogen, up to $ZrMn_{2.4}H_3$ [68417-38-9] (128) and $ZrV_2H_{5.3}$ [63440-37-9] (129). These and other zirconium intermetallic compounds are being extensively studied for possible hydrogen storage applications (130).

The metallic monohalides zirconium chloride [14989-34-5], ZrCl, and zirconium bromide [31483-18-8], ZrBr, reversibly absorb hydrogen up to a limiting composition of ZrXH (131). These hydrides are less stable than the binary hydride ZrH_2, and begin to disproportionate above 400°C to ZrH_2 and ZrX_4 in a hydrogen atmosphere (see also Hydrides).

Carbide. Zirconium carbide [12020-14-3], nominally ZrC, is a dark-gray brittle solid. It is made typically by a carbothermic reduction of zirconium oxide in a induction-heated vacuum furnace. Alternative production methods, especially for deposition on a substrate, consist of vapor-phase reaction of a volatile zirconium halide, usually $ZrCl_4$, with a hydrocarbon in a hydrogen atmosphere at 900–1400°C.

Once initiated, zirconium and carbon powders react exothermically in a vacuum or inert atmosphere to form zirconium carbide. With the greater availability of relatively pure metal powders, this technique is coming into common use for the production of several refractory carbides. Zirconium carbide is not a fixed stoichiometric compound, but a defect compound with a single-phase composition ranging from $ZrC_{0.6}$ to $ZrC_{0.98}$ at 2400°C.

Zirconium carbide is inert to most reagents but is dissolved by hydrofluoric acid solutions which also contain nitrate or peroxide ions, and by hot concentrated sulfuric acid. Zirconium carbide reacts exothermically with halogens above 250°C to form zirconium tetrahalides, and with oxidizers to zirconium dioxide in air above 700°C. Zirconium carbide forms solid solutions with other transition-metal carbides and most of the transition-metal nitrides without the formation of compounds. When heated in a vacuum, zirconium carbide, $ZrC_{0.6-0.98}$, moves to the congruent composition $ZrC_{ca\ 0.85}$, with a melting point of 3420°C.

As a hard, high melting carbide and possible constituent of UC-fueled reactors, zirconium carbide has been studied extensively. The preparation, behavior, and properties of zirconium and other carbides are reviewed in ref. 132, temperature-correlated engineering property data in ref. 133 (see also Carbides).

Nitride. Cubic zirconium nitride, ZrN, a brittle, yellow solid, is prepared by heating zirconium turnings or loose sponge to 1100–1500°C in a nitrogen or ammonia atmosphere. The reaction is slow because of the slow diffusion of nitrogen through the protective nitride layer, and nitrogen contents approaching stoichiometric require time or higher temperatures such as those of a plasma torch. Zirconium tetrachloride and nitrogen in a hydrogen atmosphere above 1000°C yield zirconium nitride films or powders; impurity effects cause the growth of nitride whiskers (134). Ammonia and zirconium tetrachloride forms adducts which on heating to 750°C give $ZrN_{1.3}$ [25658-42-8], which in turn yields ZrN at 1200°C (135). Unlike its analogues, TiN and HfN, which are being used as wear-resistant surface layers on cemented-carbide tool bits, no significant commercial applications for ZrN have been developed (see also Nitrides).

Zirconium nitride is dissolved by concentrated hydrofluoric acid, dissolved slowly by hot concentrated sulfuric acid, and oxidizes to zirconium oxide above 700°C in air.

Borides. Zirconium forms two borides: zirconium diboride [12045-64-6], ZrB_2, and zirconium dodecaboride [12046-91-2], ZrB_{12}. The diboride is synthesized from the elements, by vapor phase coreduction of zirconium and boron halides, or by the carbothermic reduction of zirconium oxide and boron carbide; boric oxide is avoided because of its relatively high vapor pressure at the reaction temperature.

$$2\ ZrO_2 + B_4C + 3\ C \xrightarrow{1400°C} 2\ ZrB_2 + 4\ CO$$

The diboride has a hexagonal structure and melts at 3245°C (136); it is considered to have the best oxidation resistance of all the refractory "hard metals." The dodecaboride has a cubic structure.

Phosphides. Zirconium forms several phosphides: ZrP_3 [39318-19-9], ZrP_2 [12037-80-8], and $ZrP_{0.6}$ [12066-61-4]; they are part of the Zr–P phase diagram (137). The solubility of phosphorus in zirconium metal is low, ca 50 ppm, and at higher concentrations it collects as separate globules at the metal grain boundaries. Analysis indicates that this material is Zr_3P.

Chalcogenides. The reactions of pure zirconium turnings with threefold quantities of elemental sulfur, selenium, or tellurium give ZrS_3 [12166-31-3], $ZrSe_3$ [12166-53-9], and $ZrTe_3$ [39294-10-5] (138). Zirconium disulfide [12039-15-5] is made from the elemental powders and by the action of carbon disulfide on zirconium oxide above 1200°C (139); some ZrOS [12164-95-3] is usually also obtained. The higher sulfides disproportionate at ca 700°C; synthesis reactions at 900–1000°C with S:Zr ratios between 0.2 and 2.3 produced crystals that were identified as Zr_9S_2 [12595-12-9], $Zr_{21}S_8$ [12595-19-6], Zr_2S [12334-07-5], ZrS_{1-x}, ZrS [12067-18-4], and $ZrS_{1.3}$ [37244-09-0] (140). Zirconium disulfide is a semiconductor with a cadmium-iodide-type layered structure consisting of stacked sandwiches each containing single sheets of metal cations between two sheets of anions.

Several compounds such as $BaZrS_3$ [12026-44-7], $SrZrS_3$ [12143-75-8], and $CaZrS_3$ [59087-48-8], have been made by reacting carbon disulfide with the corresponding zirconate at high temperature (141), whereas $PbZrS_3$ [12510-11-1] was produced from the elements zirconium and sulfur plus lead sulfide sealed in a platinum capsule which was then pressurized and heated (142). Lithium zirconium disulfide [55964-34-6], $LiZrS_2$, was also synthesized. Zirconium disulfide forms organometallic intercalations with a series of low-ionization (<6.2 eV)-sandwich compounds with parallel rings (143).

The dichalcogenides are hexagonal, the diamagnetic trichalcogenides are monoclinic. The compound sulfides $BaZrS_3$ [12026-44-7], $SrZrS_3$ [12143-75-8], and $CaZrS_3$ [59087-48-8] have an orthorhombic distorted perovskite structure, although $BaZrS_3$ prepared at 1000°C displayed a tetragonal perovskite structure which was attributed to a sulfur deficiency (141).

Oxides. Zirconium dioxide [1314-23-4], ZrO_2, is the most important oxide of zirconium. It melts at 2710 ± 35°C, and in the pure state exists in four solid phases: monoclinic, tetragonal, orthorhombic, and a cubic fluorite structure. The monoclinic phase is stable up to ca 1100°C, and transforms to tetragonal as the temperature increases to 1200°C. The proportion of tetragonal during the transformation is temperature dependent and not time-related (144). The transformation temperature exhibits a large hysteresis, and the tetragonal-to-monoclinic transition occurs between 1000 and 850°C; it is a martensitic transformation. The volume expansion resulting from the tetragonal-to-monoclinic transition can be restrained by high external pressure, increasing the stability of the tetragonal phase at lower temperatures (145). At higher pressures, an orthorhombic (cotunnite-type) structure has been identified (146) and it has been suggested that at 1000°C the zirconium oxide pressure-transformations are

$$\text{monoclinic} \xleftrightarrow{\text{GPa (ca } 10^4 \text{ atm})} \text{tetragonal} \xleftrightarrow{10 \text{ GPa (ca } 10^5 \text{ atm})} \text{orthorhombic}$$

The tetragonal-to-cubic reversible transformation occurs at 2370°C. The existence of cubic zirconia was proven by high temperature x-ray diffraction studies but little is known about the transformation. The cubic structure can be stabilized at lower temperatures by the introduction of vacancies in the anion lattice. The existence of cubic zirconia down to 1525°C has been shown in substoichiometric ZrO_{2-x} (147–149) and zirconium nitride-stabilized cubic zirconia (150). Vacancies are also generated by adding alkaline-earth oxides or rare-earth oxides to form stable solid solution cubic zirconia. Although these materials were initially considered stable at room temperature, more recent studies indicate that the kinetics of change are slow and that the lower temperature stability ranges from 280 to 1400°C, depending on the stabilizing oxide (151). Fully stabilized cubic magnesia–zirconia ceramics do not withstand thermal cycling, whereas CaO- and Y_2O_3-stabilized oxides exhibit higher thermal stability.

Partially stabilized zirconias are formed by adding insufficient yttria, calcia, or magnesia to completely stabilize the zirconia, firing the mix, and then slowly cooling the sintered material in order to precipitate tetragonal zirconia within the grains of stabilized zirconia. The improved toughness of partially stabilized zirconias is the result of stress-induced martensitic transformation of the tetragonal particles to the monoclinic form in the stress field of a propagating crack. Their commercial application is based on their high strength and thermal shock resistance.

Zirconia prepared by the thermal decomposition of zirconium salts is often metastable tetragonal, or metastable cubic, and reverts to the stable monoclinic form upon heating to 800°C. These metastable forms apparently occur because of the presence of other ions during the hydrolysis of the zirconium; their stability has been ascribed both to crystallite size and surface energy (152–153) as well as strain energy and the formation of domains (154).

Zirconium oxide is stable to most reagents but dissolves slowly in hot concentrated sulfuric acid and in concentrated hydrofluoric acid. Carbon tetrachloride or phosgene above 300°C, or chlorine and carbon above 600°C give $ZrCl_4$. Above 1400°C, zirconium oxide is reduced by carbon to zirconium carbide. At high temperatures, zirconium oxide reacts with many metal oxides, including BaO, CaO, MgO, SrO, Y_2O_3, Re_2O_3, and PbO, to form solid solution oxides and, with the exception of magnesium, zirconate compounds. Molten-salt synthesis procedures give uniform, stoichiometric, finely divided alkaline-earth zirconate powders for use as a dielectric (155).

Although lower oxides do not occur terrestrially, they can be produced at high temperature. Zirconium monoxide [*12036-01-0*], ZrO, has been observed in the sun's spectra, and in the spectra of zirconium oxide being evaporated from the surface of a tungsten filament (156). As oxygen is added to alpha zirconium, additional diffraction lines appear at Zr_3O [*12059-93-7*] and higher oxygen contents (157–158); Zr_3O occurs as an ordered super lattice in zirconium, representing the limit of oxygen solubility in zirconium. Other lower oxide compounds that are considered stable are Zr_8O [*53801-45-9*] and Zr_2O [*12412-49-6*] (159).

Silicates. Zirconium silicate, $ZrSiO_4$, occurs naturally as zircon. Zircon is tetragonal, with the zirconium and silicon linked through oxygen atoms to form edge-sharing alternating SiO_4 tetrahedra and ZrO_8 triangular dodecahedra (160). Zircon is isomorphous with xenotime which occurs in solid solution in zircon crystals (161).

Upon heating to 900°C, zircon is transformed to a scheelite-type structure at 12 GPa (ca 120,000 atm), and a further transformation to the $KAlF_4$ structure at 17 GPa (ca 170,000 atm) has been predicted (162).

Zircon is synthesized by heating a mixture of zirconium oxide and silicon oxide to 1500°C for several hours (163). The corresponding hafnium silicate, hafnon, has been synthesized also. Zircon can be dissociated into the respective oxides by heating above 1540°C and rapidly quenching to prevent recombination. Commercially, this is done by passing closely sized zircon through a streaming arc plasma (38).

Zircon silicate is highly stable. Decomposition of zircon is accomplished only by very aggressive chemical attack, usually at high temperature.

Many other silicate minerals contain zirconium as a constituent (164–165). These minerals may be altered or metamict zircons such as cyrtolite or malacon.

Halides. *Quadrivalent.* Zirconium tetrafluoride is prepared by fluorination of zirconium metal, but this is hampered by the low volatility of the tetrafluoride which coats the surface of the metal. An effective method is the halogen exchange between flowing hydrogen fluoride gas and zirconium tetrachloride at 300°C. Large volumes are produced by the addition of concentrated hydrofluoric acid to a concentrated nitric acid solution of zirconium; zirconium tetrafluoride monohydrate [14956-11-3] precipitates (69). The recovered crystals are dried and treated with hydrogen fluoride gas at 450°C in a fluid-bed reactor. The thermal dissociation of fluorozirconates also yields zirconium tetrafluoride.

The physical properties of zirconium tetrafluoride are listed in Table 9.

Zirconium tetrafluoride dissolves in dilute acid without hydrolysis, and can be

Table 9. Physical Properties of Zirconium Tetrahalides

Property	ZrF_4	$ZrCl_4$	$ZrBr_4$	ZrI_4
color	white	white	white	orange–yellow
melting point, °C	932	437	450	499
sublimation temperature at 101.3 kPa (= 1 atm), °C	903	331	357	431
density, g/cm³	4.43	2.80		4.85
critical temperature, °C		503.5	532	686
critical pressure, MPa[a]		5.7	4.3	4.1
critical density, g/cm³		0.76	0.97	1.13
vapor pressure $\log P_{kPa} = A - B/T$[b]				
A	12.682	10.891	11.393	8.695
B	12430	5400	5945	6700
range t, °C	408–640	207–416	216–332	137–307
crystal structure	monoclinic (B)	monoclinic	cubic	cubic
space group, nm	I 2/a	P 2/c	P a3	P a3
a	0.957	0.6361	1.095	1.179[e]
b	0.993	0.7407		
c	0.773	0.6256		
β, °	99.47	109.30		
Zr–X bond distance, nm	0.210 avg	2.307[c]		
		2.498[d]		
		2.665[d]		

[a] To convert MPa to atm, divide by 0.101.
[b] To convert kPa to mm Hg, multiply by 7.5. $T = K$.
[c] Between terminal chlorine atoms.
[d] Between bridging chlorine atoms.
[e] Ref. 167.

recovered as the monohydrate [14956-11-3] by crystallization from nitric acid solutions. If the solution is acidified with hydrofluoric acid, $ZrF_4.3H_2O$ [14517-16-9] crystallizes at 10–30 wt % HF; $HZrF_5.4H_2O$ [18129-16-9] crystallizes at 30–35 wt % HF, and at higher HF concentrations $H_2ZrF_6.2H_2O$ [12021-95-3] can be recovered.

Potassium hexafluorozirconate, K_2ZrF_6, can be crystallized from a zirconium oxide chloride [7699-43-6] solution by addition of excess hydrofluoric acid and a stoichiometric amount of potassium fluoride. Industrially, it is produced by the fusion of zircon flour with potassium fluorosilicate as described above (31). Sodium fluorozirconates can be produced by equivalent processes, or by adding sodium fluoride to a warm acidified solution of potassium fluorozirconate to precipitate the less soluble sodium salt. Several different sodium salts can be precipitated, depending on the sodium fluoride concentration: $NaZrF_5.H_2O$ [20982-58-5] is recovered from sodium fluoride concentration less than 0.21%, Na_2ZrF_6 [16925-26-1] is recovered from concentrations between 0.21 and 0.4%, and $Na_5Zr_2F_{13}$ [12022-20-7] from concentrations between 0.47 and 1.18%. At higher sodium fluoride concentrations, Na_3ZrF_7 [17442-98-7] precipitates. All of these solutions must remain acidic to prevent the formation of oxide fluorides. The solubilities and equilibrium phases for $ZrF-Na(K,Rb,Cs)F-H_2O$ systems are given in ref. 168. For comparison with the aqueous system, the molten $NaF.ZrF_4$ salt system had the following stable phases: $Na_3Zr_4F_{19}$ [12140-35-1], $Na_7Zr_6F_{31}$ [12140-37-3], $Na_3Zr_2F_{11}$ [12140-29-3], Na_2ZrF_6, $Na_5Zr_2F_{13}$, and Na_3ZrF_7 (169). (In addition, Na_4HfF_8 has been found in the hafnium system.)

Zirconium tetrachloride, $ZrCl_4$, is prepared by a variety of anhydrous chlorination procedures. The reaction of chlorine or hydrogen chloride with zirconium metal above 300°C, or phosgene or carbon tetrachloride on zirconium oxide above 450°C, or chlorine on an intimate mixture of zirconium oxide and carbon above 700°C are commonly used.

Zirconium tetrachloride is a tetrahedral monomer in the gas phase, but the solid is a polymer of $ZrCl_4$ octahedra arranged in zigzag chains in such a way that each zirconium has two pairs of bridging chlorine anions and two terminal or t-chlorine anions. The octahedra are distorted with unequal Zr–Cl bridge bonds of 0.2498 and 0.2655 nm. The physical properties of zirconium tetrachloride are given in Table 9.

Zirconium tetrachloride is instantly hydrolyzed in water to zirconium oxide dichloride octahydrate [13520-92-8]. Zirconium tetrachloride exchanges chlorine for oxo bonds in the reaction with hydroxylic ligands, forming alkoxides from alcohols (see Alkoxides, metal). Zirconium tetrachloride combines with many Lewis bases such as dimethyl sulfoxide, phosphorus oxychloride and amines including ammonia, ethers, and ketones. The zirconium organometallic compounds are all derived from zirconium tetrachloride.

Zirconium tetrachloride forms additional compounds with phosphorus pentachloride: $ZrCl_4.PCl_5$ and $ZrCl_4.2PCl_5$. However, the alleged $2ZrCl_4.PCl_5$ has been found to be a low boiling azeotrope of $ZrCl_4$ and $ZrCl_4.PCl_5$ (170).

Zirconium tetrachloride forms hexachlorozirconates with alkali-metal chlorides, eg, Li_2ZrCl_6 [18346-96-8], Na_2ZrCl_6 [18346-98-0], K_2ZrCl_6 [18346-99-1], Rb_2ZrCl_6 [19381-65-8], and Cs_2ZrCl_6, and with alkaline-earth metal chlorides: $SrZrCl_6$ [21210-13-9] and $BaZrCl_6$ [21210-12-8]. The vapor pressure of $ZrCl_4$ over these melts as a function of the respective alkali chlorides and of $ZrCl_4$ concentration were studied as potential electrolytes for the electrowinning of zirconium (72). The zirconium tetrachloride vapor pressure increased in the following sequence: Cs < Rb < K < Na <

Li. The stability of a hexachlorohafnate is greater than that of a comparable hexachlorozirconate (171), and this has been proposed as a separation method (172).

Zirconium tetrabromide [13777-25-8], $ZrBr_4$, is prepared directly from the elements or by the reaction of bromine on a mixture of zirconium oxide and carbon. It may also be made by halogen exchange between the tetrachloride and aluminum bromide. The physical properties are given in Table 9. The chemical behavior is similar to that of the tetrachloride.

Zirconium tetraiodide [13986-26-0], ZrI_4, is prepared directly from the elements, by the reaction of iodine on zirconium carbide, or by halogen exchange with aluminum triiodide. The reaction of iodine with zirconium oxide and carbon does not proceed. The physical properties are given in Table 9.

Zirconium tetraiodide is the least thermally stable zirconium tetrahalide. At 1400°C, it disproportionates to Zr metal and iodine vapor. This behavior is utilized in the van Arkel–de Boer process to refine zirconium. As with the tetrachloride and tetrabromide, the tetraiodide forms additional adducts with gaseous ammonia which, upon heating, decompose through several steps ending with zirconium nitride.

Lower Valent Halides. Zirconium trichloride [10241-03-9], $ZrCl_3$, tribromide [24621-18-9], $ZrBr_3$, and triiodide [13779-87-8], ZrI_3, are produced from zirconium metal and tetrahalide in sealed tantalum tubes at 500–700°C. The hafnium trihalides are made in the same manner. These reactions are slow, and may not reach equilibrium for days. Zirconium monohalides as the reductant increase the reaction rates. Another procedure uses quartz tubes with double bulbs; the halide is kept at the sublimation temperature and the metal at 700°C. In a procedure employing comparatively low temperatures, the tetrahalides are dissolved in the corresponding liquid aluminum halide at the eutectic and reduced by metallic zirconium or aluminum between 220 and 300°C. The zirconium trihalides and $ZrI_{3.40}$ [29950-62-7] were prepared in this manner (173). Zirconium trichloride, tribromide, and triiodide have been made from their respective tetrahalides by atomic hydrogen reduction (174).

The trichloride and tribromide disproportionate above 200°C at or below 101.3 kPa (1 atm) (175):

$$12\ ZrCl_3 \xrightarrow{200°C} 10\ ZrCl_{2.8} + 2\ ZrCl_4$$

$$10\ ZrCl_{2.8} \xrightarrow{300°C} 5\ ZrCl_{1.6} + 5\ ZrCl_4$$

$$5\ ZrCl_{1.6} \rightarrow 4\ ZrCl + ZrCl_4$$

$$4\ ZrCl \xrightarrow[vac]{600°C} 3\ Zr + ZrCl_4$$

Other researchers have reported the trihalides to be nonstoichiometric with composition ranges of $ZrCl_{2.94}$–$ZrCl_{3.03}$, $ZrBr_{2.87}$–$ZrBr_{3.23}$, and $ZrI_{2.83}$–$ZrI_{3.43}$ (176). The composition ranges of the lower iodides were reported to be $ZrI_{1.9}$–$ZrI_{2.1}$, and $ZrI_{1.1}$–$ZrI_{1.3}$ (177).

Zirconium dichloride, $ZrCl_2$, has been made by a month-long reaction of ZrCl and $ZrCl_3$ at 650–750°C. The product has a three-layer sheet structure of close-packed Cl–Zr–Cl (178).

The first known cluster ion for group IVB elements is $ZrCl_{2.5}$, produced similarly to $ZrCl_2$: $(Zr_6Cl_{12})^{3+}(Cl^-)_3$, with nine delocalized bonding electrons in the metal cage (178). ZrI_2 also contains the 12-electron cage Zr_6I_{12}, as does the isostructural Zr_6Cl_{12}.

Attempts to prepare zirconium trifluoride, ZrF$_3$, by the zirconium reduction of zirconium tetrafluoride were unsuccessful, but it has been made by heating zirconium hydride to 750°C in a stream of hydrogen and hydrogen fluoride (179).

Monovalent Halides. Zirconium monochloride [14989-34-5], ZrCl, was discovered during electrorefining studies of zirconium in a SrCl$_2$–NaCl–ZrCl$_4$ melt intended to produce pure ductile hafnium-depleted zirconium from crude zirconium anodes (180–181). The monochloride is also called Zirklor. It is obtained as black flakes with a graphite slip-plane behavior and was proposed as a lubricant (182–183).

Zirconium monochloride and zirconium monobromide [31483-18-8] are prepared in better purity by equilibration of mixed lower halides with zirconium foil at 625°C (184–185) or by slowly heating zirconium tetrahalide with zirconium turnings at 400–800°C over a period of 2 wk and holding at 800–850°C for a few additional days (186). Similar attempts to produce zirconium monoiodide [14728-76-8] were unsuccessful; it was, however, obtained from the reaction of hydrogen iodide with metallic zirconium above 2000 K (187).

Zirconium chloride and bromide have closely related but dissimilar structures. Both contain two metal layers enclosed between two nonmetal layers which both have hexagonal structure. In ZrCl, the four-layer sandwich repeats in layers stacked up according to /abca/bcab/cabc/, whereas the ZrBr stacking order is /abca/cabc/bcab/ (188). Both are metallic conductors, but the difference in packing results in different mechanical properties; the bromide is much more brittle.

Zirconium monochloride is reported to disproportionate at 610°C in a high vacuum (184) and to exert a ZrCl$_4$ pressure of 101.3 kPa (1 atm) at 900°C and 293.7 kPa (2.9 atm) at 975°C, and to melt at 1100°C in an autogenous ZrCl$_4$ pressure of 1.5–2 MPa (15–20 atm) (185). The disproportionation proceeds at a lower temperature in the presence of copper or silver (182).

Zirconium monochloride reacts with sodium ethoxide to form additional adducts which hydrolyze in water. The monochloride does not react with benzene in a Friedel-Crafts reaction, and does not enter into intercalation reactions similar to those of zirconium disulfide. Both monohalides add hydrogen reversibly up to a limiting composition of ZrXH (131).

Hydroxyl Compounds. The aqueous chemistry of zirconium is complex, and in the past its understanding was complicated by differing interpretations. In a study of zirconium oxide chloride and zirconium oxide bromide, the polymeric cation [Zr$_4$(OH)$_8$(H$_2$O)$_{16}$]$^{8+}$ was identified (189); the earlier postulated moiety [Zr=O]$^{2+}$ was discarded. In the tetramer, the zirconium atoms are connected by double hydroxyl bridges (shown without the coordinating water molecules):

$$\begin{array}{c} \diagdown \mid \text{OH} \mid \text{OH} \mid \text{OH} \mid \text{OH} \\ \text{Zr} \quad \text{Zr} \quad \text{Zr} \quad \text{Zr} \\ \diagup \mid \text{OH} \mid \text{OH} \mid \text{OH} \mid \diagdown \text{OH} \end{array}$$

The tetramer exists in two-molal zirconium chloride and nitrate solutions, but it polymerizes into cross-linked chains on hydrolysis (190–191); in strong acid solutions, the hydroxyl bridges can be replaced by other anions to form trimers (192) and monomers (192–193).

Hydrous Oxides and Hydroxides. Hydroxide addition to aqueous zirconium solutions precipitates a white gel formerly called a hydroxide, but now commonly considered hydrous zirconium oxide hydrate [12164-98-6], $ZrO_2 \cdot nH_2O$. However, the behavior of this material changes with time and temperature.

The freshly precipitated material with four hydroxyl groups titratable with potassium fluoride for each zirconium atom can be considered a true hydroxide (194–195). When the freshly precipitated hydroxide was kept in water for two to three days before titration, two hydroxyl groups were titrated quickly, and two slowly on mixing for 30 h. These were called α and γ hydroxides. An intermediate β hydroxide with three hydroxyl groups per zirconium could be produced only by precipitation from solutions of zirconium oxide chloride or nitrate in methanol, with extremely high zirconium concentrations. Drying these hydroxides in air gave a powder, which no longer contained any titratable hydroxyl groups; the material was very sparingly soluble in acid.

The hydroxides as precipitated are amorphous, but if they are refluxed in a neutral or slightly acidic solution they convert to a mixture of cubic and monoclinic hydrous zirconia crystallites; on continued refluxing, only the monoclinic form persists (196). If the refluxing is conducted in an alkaline solution, metastable cubic zirconia is formed (197).

Oxide Chlorides. Zirconium oxide dichloride, $ZrOCl_2 \cdot 8H_2O$ [13520-92-8], commonly called zirconium oxychloride, is really a hydroxyl chloride, $[Zr_4(OH)_8 \cdot 16H_2O]\text{-}Cl_8 \cdot 12H_2O$ (189). Zirconium oxychloride is produced commercially by caustic fusion of zircon, followed by water washing to remove sodium silicate and to hydrolyze the sodium zirconate; the wet filter pulp is dissolved in hot hydrochloric acid, and $ZrOCl_2 \cdot 8H_2O$ is recovered from the solution by crystallization. An aqueous solution is also produced by the dissolution and hydrolysis of zirconium tetrachloride in water, or by the addition of hydrochloric acid to zirconium carbonate.

Tetragonal prisms of $ZrOCl_2 \cdot 8H_2O$ crystallize from hot hydrochloric acid solutions by cooling or by increasing the acidity; in >32% hydrochloric acid transparent hexagonal plates are formed probably as $ZrOCl_2 \cdot 2HCl \cdot 10H_2O$ (198).

Zirconium oxychloride is an important intermediate from which other zirconium chemicals are produced. It readily effloresces, and hydrates with 5–7 H_2O are common. The salt cannot be dried to the anhydrous form, and decomposes to hydrogen chloride and zirconium oxide.

Zirconium hydroxy oxychloride [18428-88-1], nominally $ZrO(OH)Cl$, is produced by dissolving hydrous zirconia in hydrochloric acid in an equal molar proportion, and is available only in solution. Other oxychlorides with Cl:Zr ratios <2 are discussed in ref. 199.

Anhydrous zirconium oxide chloride, $ZrOCl_2$ [7699-43-6], has been prepared by the reaction of dichlorine oxide with a zirconium tetrachloride suspension in carbon tetrachloride starting at $-30°C$ and slowly rising to room temperature. The white solid is extremely hygroscopic and decomposes to $ZrCl_4$ and ZrO_2 at $250°C$ (200).

Nitrates. Anhydrous zirconium tetranitrate [12372-57-5], $Zr(NO_3)_4$, is prepared from zirconium tetrachloride and nitrogen pentoxide (201). The hydrated compounds are obtained from aqueous nitric acid (165); $ZrO(NO_3)_2 \cdot 2H_2O$ [20213-65-4] is most commonly used; $Zr(NO_3)_4 \cdot 5H_2O$ [12372-57-5] can be produced from strong nitric acid.

Aqueous solutions of zirconium oxide dinitrate [13826-66-9] (zirconium oxynitrate) and zirconium oxychloride behave very similarly; these two compounds have been cocrystallized in solid solution (202) where $ZrO(NO_3)_2.5H_2O$ was the stable hydrate.

Carbonates. Basic zirconium carbonate [37356-18-6] is produced in a two-step process in which zirconium is precipitated as a basic sulfate from an oxychloride solution. The carbonate is formed by an exchange reaction between a water slurry of basic zirconium sulfate and sodium carbonate or ammonium carbonate at 80°C (203). The particulate product is easily filtered. Freshly precipitated zirconium hydroxide, dispersed in water under carbon dioxide in a pressure vessel at ca 200–300 kPa (2–3 atm), absorbs carbon dioxide to form the basic zirconium carbonate (204). Washed free of other anions, it can be dissolved in organic acids such as lactic, acetic, citric, oxalic, and tartaric to form zirconium oxy salts of these acids.

Basic zirconium carbonate is nominally $2ZrO_2.CO_2.xH_2O$ but the $ZrO_2:CO_2$ ratio may range from 4:1 to 1:1 depending on the methods and techniques of preparation.

Basic zirconium carbonate reacts with sodium or ammonium carbonate solutions to give water-soluble double carbonates. The ammonium double carbonate is nominally $NH_4[Zr_2O(OH)_3(CO_3)_3]$. These solutions are stable at room temperature, but upon heating they lose carbon dioxide and hydrous zirconia precipitates.

In a study of zirconium double carbonates (205), a family of carbonates with $CO_3:Zr$ ratios of 1, 1.5, 2, 2.5, 3, 3.5, and 4:1 were identified. None of these compounds combined with ammonia, confirming the absence of HCO_3^-.

Sulfates. Sulfate ions strongly complex zirconium, removing hydroxyl groups and forming anionic complexes. With increasing acidity, all hydroxyl groups are replaced; zirconium sulfate [7446-31-3], $Zr(SO_4)_2.4H_2O$, with an orthorhombic structure (206), can be crystallized from a 45% sulfuric acid solution. Zirconium sulfate forms various hydrates, and 13 different crystalline $Zr(SO_4)_2.nH_2O$ [14644-61-2] systems are described in ref. 207. The tetrahydrate loses three hydration waters above 100°C and becomes anhydrous at 380°C. Many complex sulfates are formed with alkali-metal sulfates.

Basic zirconium sulfates are formed by hydrolysis of zirconium sulfate, which is broken up into fragments that undergo further hydrolysis to yield a series of basic sulfates with the generic formula $Zr_n(OH)_{2n+2}(SO_4)_{n-1}$ (190).

Basic zirconium sulfate is also considered to be strands of $[Zr(OH)_2^b]_n^{2n+}$ joined by bridging sulfates (208). The resulting formula is $Zr(OH)_2^b(OH)^t(SO_4)_{0.5}^b.nH_2O$, where b are bridging anions and t are terminal anions.

The most common basic sulfate is $5ZrO_2.3SO_3.nH_2O$ [84583-91-5] which is precipitated in good yield when a zirconium oxychloride solution is heated with the stoichiometric amount of sulfate ion. It is used to prepare high purity oxides and ammonium zirconium carbonate.

The stability of zirconium sulfate solutions to spontaneous precipitation when heated to 70°C for 2 h was studied as a function of $SO_3:ZrO_2$ ratio and metal concentrations (209). The zirconium solutions were considered unstable at metal concentrations below 0.64 M or at $SO_3:ZrO_2$ ratios <1.2.

Phosphates. Phosphate ions precipitate group IVB metals from strongly acid solutions. This ability is used in analytical procedures to separate zirconium from other elements, and to prepare zirconium phosphates. The precipitate is a gelatinous amorphous solid of variable composition. However, when refluxed in strong phosphoric acid, a crystalline, stoichiometrically constant compound forms of composition $Zr(HPO_4)_2 \cdot H_2O$, known as zirconium bis(monohydrogen phosphate) [*13772-31-1*] or α zirconium phosphate (210), or αZP. This compound is also obtained by gradual precipitation of the phosphate from a heated zirconium fluoride solution (211). On heating, the αZP undergoes three transformations, losing water and finally converting to cubic zirconium pyrophosphate [*33712-62-8*], ZrP_2O_7, at 1000°C (212).

The main interest in zirconium phosphates relates to their ion-exchange properties. If amorphous zirconium phosphate is equilibrated with sodium hydroxide to pH 7, one hydrogen is displaced and $ZrNaH(PO_4)_2 \cdot 5H_2O$ [*13933-56-7*] is obtained. The spacing between the zirconium layers is increased from 0.76 to 1.18 nm, which allows this phosphate to exchange larger ions.

The gels precipitated as described above are not useful in ion-exchange systems because their fine size impedes fluid flow and allows particulate entrainment. Controlled larger-sized particles of zirconium phosphate are obtained by first producing the desired particle size zirconium hydrous oxide by sol–gel techniques or by controlled precipitation of zirconium basic sulfate. These active, very slightly soluble compounds are then slurried in phosphoric acid to produce zirconium bis(monohydrogen phosphate) and subsequently sodium zirconium hydrogen phosphate pentahydrate with the desired hydraulic characteristics (213–214).

Although zirconium phosphate is insoluble in acids, it is easily hydrolyzed in excess caustic to give hydrous zirconium oxide. Zirconium phosphate forms soluble complexes with a large excess of zirconium oxide chloride, and therefore separation of phosphorus from zirconium oxide chloride solutions is difficult (215).

The properties and behavior of double phosphates such as $Na_{1+x}Y_xZr_{2-x}(PO_4)_3$ (216), sodium–zirconium phosphate–silicates (217) and alkali-modified zirconium pyrophosphates (218) are under study as potential solid electrolytes in high temperature batteries.

Alkoxides. Zirconium alkoxides are part of a family of alcohol-derived compounds (219). The binary zirconium compounds have the general formula $ZrX_4-n(OR)_n$. They are easily hydrolyzed and must be prepared under anhydrous conditions. They are prepared by the reaction of zirconium tetrahalides and alcohols:

$$ZrCl_4 + 3\ C_2H_5OH \rightarrow ZrCl_2(OC_2H_5)_2 \cdot C_2H_5OH + 2\ HCl\uparrow$$
[*87227-58-5*]

All four chlorines can be substituted if anhydrous ammonia is added to combine with the hydrogen chloride:

$$ZrCl_4 + 4\ CH_3OH + 4\ NH_3 \rightarrow Zr(OCH_3)_4 + 4\ NH_4Cl$$
[*28469-78-5*]

Alkoxides of other alcohols are formed by alcohol exchange. The general stability of the alcohols in exchange is primary > secondary > tertiary, although the reaction can be driven in the opposite direction by removal of the more volatile alcohol:

$$Zr(OC_2H_5)_4 + 3\ (CH_3)_3COH \rightarrow Zr(OC_2H_5)[OC(CH_3)_3]_3 + 3\ C_2H_5OH$$
[*18267-08-8*] [*87227-52-9*]

Alkoxides can be formed also by reaction of zirconium dialkylamines with alcohols, and alkoxides can be exchanged also by transesterification reactions.

Double alkoxides of zirconium with alkali metals of the type $MZr_2(OR)_9$ have been obtained by reaction of alkali metal alkoxides with zirconium alkoxides (220). Although these usually are monomeric derivatives, the reaction between zirconium tetra-*t*-butoxide [*1071-76-7*] and sodium *t*-butoxide was found (221) to form dimeric [NaZr(OC(CH$_3$)$_3$)$_5$]$_2$.

Zirconium alkoxides readily hydrolyze to hydrous zirconia. However, when limited amounts of water are added to zirconium alkoxides, they partially hydrolyze in a variety of reactions depending on the particular alkoxide (222). Zirconium tetraisopropoxide [*2171-98-4*] reacts with fatty acids to form carboxylates (223), and with glycols to form mono- and diglycolates (224).

Carboxylates. Zirconium hydroxy carboxylates of the generic type

$$\text{Zr(OH)}_3(\text{OCR})$$
$$\overset{\text{O}}{\underset{\|}{}}$$

and

$$\text{Zr(OH)}_2(\text{OCR})_2$$

are precipitated by adding carboxylic salts to hydrochloric acid solutions of zirconium. With larger organic ligands, these compounds are water insoluble. They have been used for gravimetric determination, such as trihydroxyzirconium mandelate [*87227-59-6*] or for industrial purification precipitations, such as hydrogen trihydroxyzirconium phthalate [*62313-97-7*]:

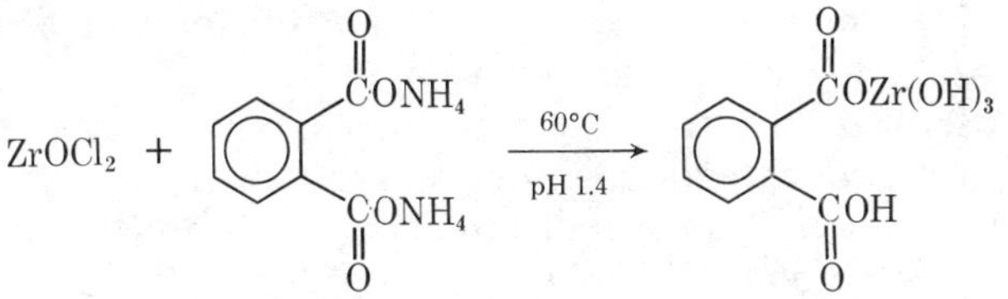

ZrOCl$_2$ + [benzene-1,2-dicarboxamide: C$_6$H$_4$(CONH$_4$)$_2$] $\xrightarrow[\text{pH 1.4}]{60°C}$ [benzene ring with –COZr(OH)$_3$ and –COH groups ortho]

Zirconium oxalates exist as compounds, double compounds, and mixed oxalato complexes (165,195,225–226). When the carboxylate ligand is a longer alkyl chain, the materials often are called zirconium soaps.

Zirconium tetracarboxylates,

$$\text{Zr(OCR)}_4$$

are prepared by reactions of zirconium tetraalkoxides or zirconium tetrachloride with anhydrous carboxylic acids (223).

Amides, Imides, Alkamides. When zirconium tetrachloride reacts with liquid ammonia, only one chloride is displaced to form a white precipitate, insoluble in liquid ammonia (227):

$$\text{ZrCl}_4 + (x+2)\,\text{NH}_3 \rightarrow \text{ZrCl}_3(\text{NH}_2)\cdot x\,\text{NH}_3 + \text{NH}_4\text{Cl}$$
[*87319-91-3*]

Upon heating in vacuum to 100°C, this material loses several ammonia moieties to give ZrCl$_3$(NH$_2$)·NH$_3$.

The presence of NH$_4$Cl causes the formation of a complex which is soluble in excess liquid ammonia (228):

$$(\text{NH}_4)_2\text{ZrCl}_6 + \text{NH}_3 \rightarrow (\text{NH}_4)_2[\text{ZrCl}_5(\text{NH}_2)] + \text{NH}_4\text{Cl}$$
[*19381-66-9*]   [*87247-93-2*]

Additional halides can be removed from the above insoluble ammonolysis product by reaction with potassium imide to form zirconium imide, $Zr(NH_2)_2$ [87227-54-1]:

$$ZrBr_3NH_2 + KNH_2 \rightarrow ZrBr_2(NH_2)_2 + KBr$$
$$ZrBr_2(NH_2) + 2 KNH_2 \rightarrow Zr(NH_2)_4 + 2 KBr$$
$$Zr(NH_2)_4 \rightarrow Zr(NH_2)_2 + 2 NH_3$$

Zirconium dichlorobis(dimethylamide) [87227-57-4], $ZrCl_2[N(CH_3)_2]_2$ is made directly from methylamine and zirconium tetrachloride but all of the halogens can be substituted by treating zirconium tetrachloride with the appropriate lithium alkylamide (229). Zirconium arylamines are made from zirconium alkoxides which first are converted to aryloxides (230).

The zirconium–nitrogen bond is weaker than the zirconium–oxygen bond even under anhydrous conditions. When zirconium tetrachloride reacts with carbonyl-containing amides such as

$$HC(O)NH_2, \quad CH_3C(O)NH_2, \quad HC(O)N(CH_3)_2,$$

or

$$CH_3C(O)N(CH_3)_2$$

all the bonding to zirconium is through carbonyl groups (231).

Organometallic Compounds. Certain zirconium organometallic compounds are highly reactive toward low molecular weight unsaturated molecules. Some of these compounds are useful in various organic syntheses; others function effectively as catalysts for polymerization, hydrogenation, or isomerization.

Some simple zirconium organometallic compounds, such as tetramethylzirconium [6727-89-5], are known. In general, these compounds are very unstable. It appears that zirconium must be π-bonded to at least one moderately large ligand, such as a cyclopentadienyl group, for the compound to be stable. The abbreviation Cp is used here for the cyclopentadienyl group (C_5H_5), and Cp' for [$C_5(CH_3)_5$] (see Organometallics—metal π complexes).

The metallocene complexes of M = Ti, Zr, and Hf are most stable when the two Cp groups are not parallel, in contrast to most other transition metal–Cp complexes. The most stable angle for the zirconium metallocenes is ca 40°, which partially accounts for the more interesting chemistry of these compounds compared to other transition metallocenes.

Cyclopentadienyl zirconium compounds are similar in structure and behavior to their titanium analogues. The increasing strength of the M–H and M–C bonds in the series M = Ti, Zr, and Hf makes the zirconium and hafnium compounds slightly more stable and thus somewhat less reactive than their titanium analogues. For example, ferrocenyl lithium reacts with Cp_2MCl_2 to give $Cp_2M(Fc)_2$. The colors of these compounds, green for Ti and red for Zr and Hf, are the results of electron transitions from a ferrocene orbital to a Cp_2M orbital (232), and is a measure of the relative bond energies. In general, the zirconium compounds are more complex and less reactive than the titanium analogues.

Hydrides. Zirconium hydrides react easily with unsaturated molecules. This process, termed hydrozirconation, replaces the hydrogen with the unsaturated group:

$$Cp_2ZrHCl + R \rightarrow Cp_2ZrRCl$$
[37342-97-5]

RCO₂H, R₂C=O, and RC≡N react to give Cp₂ZrXCl complexes where X = RCO₂—, R₂CHO—, and RCH=N—, respectively. Esters RCO₂R' give a mixture of

$$Cp_2ZrCl(OCH_2R)$$

and Cp₂ZrCl(OR'); CO₂ is reduced to formaldehyde, with formation of (Cp₂ZrCl)₂O [12097-04-0]; cyclopentadiene produces Cp₃ZrCl [69005-93-2]; and ethylene gives Cp₂ZrCl(CH₂CH₃) [12109-84-1] (233).

The hydrides can also be used to form primary alcohols from either terminal or internal olefins. The olefin and hydride form an alkenyl zirconium, Cp₂ZrRCl, which is oxidized to the alcohol. Protonic oxidizing agents such as peroxides and peracids form the alcohol directly, but dry oxygen may also be used to form the alkoxide which can be hydrolyzed (234).

Zirconium hydrides undergo 1,2-addition with 1,3 dienes to give γ,δ-unsaturated complexes in 80–90% yield. Treatment of these complexes with CO at 20°C and 345 kPa (50 psi) followed by hydrolysis gives γ,δ-unsaturated aldehydes (235).

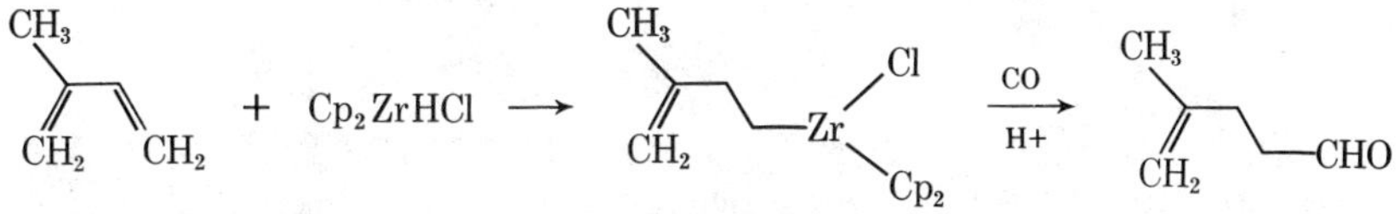

[58079-71-3]

Cyclopropanes can be generated in high yield by treatment of halogen-containing alkenes with Cp₂ZrHCl; eg,

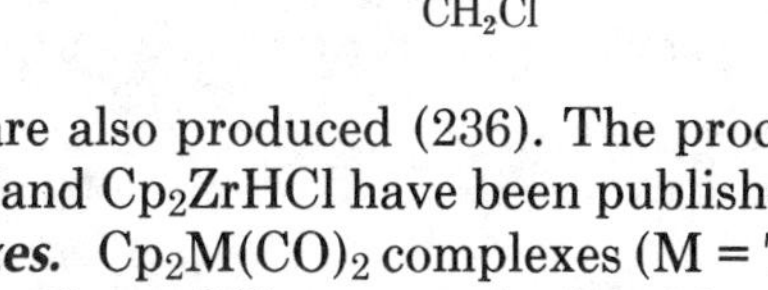

(41%)

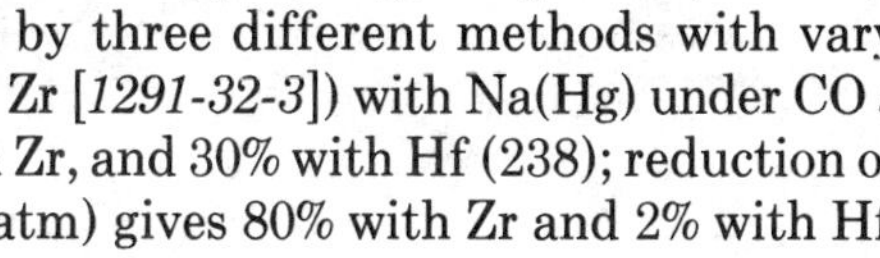

(31%)

Alkenes and alkanes are also produced (236). The procedures for the synthesis of Cp₂ZrH₂ [37342-98-6] and Cp₂ZrHCl have been published in detail (237).

Carbonyl Complexes. Cp₂M(CO)₂ complexes (M = Ti, Zr [59487-85-3], and Hf) have been prepared by three different methods with varying yields: reduction of Cp₂MCl₂ (when M = Zr [1291-32-3]) with Na(Hg) under CO at 101.3 kPa (1 atm) gives 80% with Ti, 9% with Zr, and 30% with Hf (238); reduction of Cp₂MCl₂ with Li under CO at 20 MPa (200 atm) gives 80% with Zr and 2% with Hf (239); and treatment of Cp₂MBH₄ with (C₂H₅)₃N under CO at 101.3 kPa (1 atm) gives 80% with Ti and 15% with Zr (240). The nmr and ir spectra are similar for all three analogues, with the CO stretching frequencies decreasing in the order Ti > Zr > Hf (238).

An x-ray study of the structure of Cp₂Hf(CO)₂ revealed the expected tetrahedral disposition of ligands with OC–Hf–CO and (centroid Cp)–Hf–(centroid Cp) angles of 89.3° and 141°, respectively, and mean bond lengths for both bond types of 0.216 nm (241). The Zr analogue is isomorphous with bond lengths of 0.2187 nm and a OC–Zr–CO bond angle of 89.2° (242).

894   ZIRCONIUM AND ZIRCONIUM COMPOUNDS

The oxidative addition reactions of $Cp_2M(CO)_2$ show some interesting differences between the Ti and Zr analogues. For $Cp_2Ti(CO)_2$, both

$$\underset{\text{RCCl}}{\overset{\overset{\displaystyle O}{\|}}{}}$$

(R = CH_3, C_6H_5) and RX (R = CH_3, C_2H_5, Pr^i, Bu^s) react to give acyls,

$$\underset{Cp_2Ti(CR)X.}{\overset{\overset{\displaystyle O}{\|}}{}}$$

In contrast, $Cp_2Zr(CO)_2$ [59487-85-3] reacts with CH_3I, without CO insertion, to give Cp_2ZrCH_3I [63643-49-2]. Reaction with $P(CH_3)_3$ gives $Cp_2Zr(CO)[P(CH_3)_3]$ [63637-45-6]; diphenylacetylene gives the metallocycle $Cp_2Zr[C_4(C_6H_5)_4]$ [63637-34-3] (243–244).

$$\underset{Cp_2Zr(CR)}{\overset{\overset{\displaystyle O}{\|}}{}}$$

R is formed by carbonylation of Cp_2ZrR_2 (R = CH_3 [12636-72-5], $CH_2C_6H_5$ [37206-41-0], and C_6H_5 [51177-89-0]) and is reversible for the alkyls, but not for aryls. Equilibrium and thermodynamic data were published for the alkyls. A crystal structure

$$\underset{Cp_2Zr(CCH_3)CH_3}{\overset{\overset{\displaystyle O}{\|}}{}}$$

for

$$\underset{RCR,}{\overset{\overset{\displaystyle O}{\|}}{}}$$

[60970-97-0] is used as support for interpreting the failure to eliminate the ketone as occurs in the Ti analogue, in terms of stabilization resulting from the side-bonded acyl group (245).

A Zr(IV) carbonyl, $Cp'_2ZrH_2(CO)$ [61396-35-8], was formed by exposing a solution of Cp'_2ZrH_2 [61396-34-7] to CO at −80°C. The structure was inferred by observation of $^{13}C-H$ coupling when ^{13}CO was used. No ir spectra could be obtained. Warming $Cp'_2ZrH_2(CO)$ gives the dimer [61396-37-0] (246):

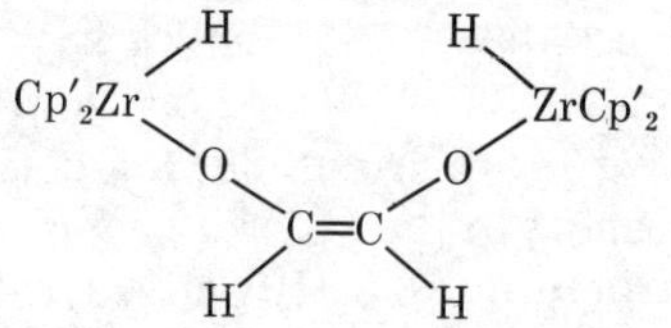

$$Cp'_2Zr \underset{O}{\overset{H}{\diagdown}} \underset{C=C}{} \underset{O}{\overset{H}{\diagup}} ZrCp'_2$$

The CO stretching frequency of $Cp'_2ZrH_2(CO)$ was observed by other workers at 2044 cm^{-1} (the Hf analogue was observed at 2036 cm^{-1}) (247).

Cp$_2$Zr(CO)$_2$ in hot toluene reacts with CO to give a cyclic trimer (Cp$_3$ZrO)$_3$. The Zr$_3$O$_3$ ring in this unusual compound is nearly planar, with Zr–O–Zr and O–Zr–O bond angles of 142.5° and 97.5°, respectively. The mean Zr–O bond length (0.196 nm) seems to indicate a large degree of Zr–O double bonding (243).

Dinitrogen Complexes. The relative inertness of molecular nitrogen is well known, however, some Cp'–Zr compounds coordinate dinitrogen and substantially increase its reactivity. The nitrogen molecule can be coordinated either in a terminal position or as a bridge in dimeric structures.

Bridging and terminal nitrogens have been compared in [Cp$'_2$Zr(N$_2$)]$_2$N$_2$. The bridging N$_2$ has a longer N–N distance (0.118 vs 0.1115 nm) and shorter Zr–N distances. In addition, the bridging N$_2$ stretching frequency is very low, 1578 cm^{-1} (248).

The increase in reactivity of coordinated N$_2$ has been assumed to be associated with increased bond length and decreased stretching frequency. A labeling study has shown that this is an oversimplification. In the protonolysis of [Cp$'_2$Zr(N$_2$)]$_2$N$_2$, the hydrazine produced comes equally from terminal and bridging N$_2$. An intermediate, such as Cp$'_2$Zr(N$_2$H)$_2$ [86165-22-2], was proposed where the bridging and terminal N$_2$ have become equivalent. Furthermore, careful carbonylation of the dimer produces [Cp$'_2$Zr(CO)]$_2$N which on protonolysis gives no reduced form of N$_2$, indicating that both bridging and terminal N$_2$ are required for reduction (249).

A similar process has been patented covering Cp$'_2$ZrR [86165-24-4], Cp$'_2$ZrR(N$_2$) [86165-25-5], and (Cp$'_2$ZrR)$_2$N$_2$ [86165-23-3], where R = CH[Si(CH$_3$)$_3$]$_2$. Protonolysis of the dinitrogen complexes gives hydrazine and ammonia (250).

Alkyl and Aryl Complexes. Cp$_2$MR$_2$ and Cp$_2$MRCl, [R = CH(C$_6$H$_5$)$_2$, CH(Si(CH$_3$)$_3$)$_2$] have been prepared and studied by nmr, and crystal structures for R = CHC$_6$H$_5$ have been reported. For Zr, the metal–carbon distance (0.2388 nm) is substantially longer than for R = CH$_3$, whereas no increase is observed with Hf (251).

The thermal decomposition of several Cp$_2$ZrR$_2$ compounds has been studied; RH is formed and the hydrogen derives from either a Cp or another R group and not from a solvent molecule (252).

Mixed-Metal Systems. Mixed-metal systems, where a zirconium alkyl is formed and the alkyl group transferred to another metal, are a new application of the hydrozirconation reaction. These systems offer the advantages of the easy formation of the Zr–alkyl as well as the versatility of alkyl–metal reagents. For example, Cp$_2$ZrRCl (R = alkyl or alkenyl) reacts with AlCl$_3$ to give an Al–alkyl species which may then be acylated with

$$\underset{\underset{\text{R'CCl}}{\|}}{\text{O}}$$

to give ketones,

$$\underset{\underset{\text{R'CR}}{\|}}{\text{O}}$$

in up to 98% yield. In contrast, direct acylation of the Zr compounds is difficult (R = alkyl) or impossible (R = alkenyl). The carbon configuration is retained during the transmetallation step which is faster for alkenyls, suggesting a bridged Zr–R–Al intermediate (253). Acyl–aluminum compounds,

$$(R\overset{O}{\overset{\|}{C}})AlCl_2,$$

which have potential use in organic synthesis as acyl anion equivalents, can be prepared by a similar reaction (254).

Alkenyl groups have also been transferred to Cu or Pd, leading to alkenyl dimers RCH=CH—CH=CHR in high yield (255). Alkenyls undergo conjugate addition to enones in the presence of catalytic amounts of nickel acetylacetonate in high yield and selectivity (256). Nickel tetralithium catalyzes the reaction of Zr–alkenyls with aryl halides to give the cross-coupling products RCH=CHAr (257). The reaction proceeds much more efficiently if the Ni catalyst is first reduced with diisobutyl aluminum hydride (258).

Alkenyl zirconium complexes derived from alkynes form C–C bonds when added to allylic palladium complexes. The stereochemistry differs from that found in reactions of corresponding carbanions with allyl–Pd in a way that suggests the Cp_2ZrRCl alkylates first at Pd, rather than by direct attack on the allyl group (259).

Catalysts. Several types of zirconium organometallic compounds are useful catalysts. In addition to the catalytic properties of the molecules, the fact that they can be bound to a relatively inert substrate increases their utility. A polymer-bound Cp-complex can be obtained through Cp–ring exchange. Treatment of Cp_2MCl_2 (M = Ti, Zr, Hf) with two equivalents of NaC_5D_5, followed by HCl, gives a mixture of $(C_5H_5)_2$—, $(C_5H_5)(C_5D_5)$—, and $(C_5D_5)_2$—MCl_2 (260). Although this indicates that ring exchange in Cp_4M is fast, other studies have shown that the ratio of exchange products depends upon whether H- or D-substituted Cp_2MCl_2 is used as a reactant (261). Though Cp-ring exchange is not simply a matter of a Cp_4M intermediate, the products may nonetheless be used to bind Cp–Zr complexes to various substrates.

Polymer–Cp–MCl_3 complexes have been formed with the Cp-group covalently bound to a polystyrene bead. The metal complex is uniformly distributed throughout the bead, as shown by electron microprobe x-ray fluorescence. Olefin hydrogenation catalysts were then prepared by reduction with butyl lithium (262).

Ziegler polymerization catalysts may be prepared from Cp–Zr complexes and trialkylaluminum. The molecular weight of the polymers can be controlled over a wide range by varying the temperature. The activity of these catalysts is considerably increased by the addition of small amounts of water (263–264) (see Olefin polymers).

Zirconium–allyl complexes also have catalytic properties. Tetraallylzirconium [12090-34-5] on a silica substrate catalyzes ethylene polymerization (265). Supported on silica, ZrR_4 (R = allyl or neopentyl) catalyzes olefin isomerization (266).

BIBLIOGRAPHY

"Zirconium and Zirconium Alloys" in *ECT* 1st ed., Vol. 15, pp. 282–290, by W. B. Blumenthal, National Lead Company; "Zirconium Compounds" in *ECT* 1st ed., Vol. 15, pp. 290–312, by W. B. Blumenthal, National Lead Company; "Zirconium and Zirconium Compounds" in *ECT* 2nd ed., Vol. 22, pp. 614–679, W. B. Blumenthal and J. D. Roach, National Lead Company.

1. N. E. Holden, *Pure Appl. Chem.* **52**, 2349 (1980).
2. B. Mason, *Principles of Geochemistry*, John Wiley & Sons, Inc., New York, 1952.

3. M. H. Klaproth, *Ann. Chim. Phys.* **6**(1), 1 (1789).
4. L. N. Vauquelin, *Ann. Chim. Phys.* **22**(1), 179 (1797).
5. J. J. Berzelius, *Ann. Chim. Phys.* **26**(2), 43 (1824).
6. H. D. Hess, *USBM R.I. 5856*, U.S. Bureau of Mines, Washington, D.C., 1962.
7. J. D. Grice and G. Perrault, *Can. Mineral.* **13**, 209 (1975).
8. H. G. Ansell, A. C. Roberts, and A. G. Plant, *Can. Mineral.* **18**, 201 (1980).
9. W. B. Pearson, *Handbook of Lattice Spacings and Structures of Metals and Alloys*, International Series on Metal Physics and Physical Metallurgy, Pergamon Press, New York, 1958, p. 130.
10. D. L. Douglass, *The Metallurgy of Zirconium*, International Atomic Energy Agency, Vienna, 1971, p. 4.
11. Ref. 10, p. 5.
12. O. Kubaschewski, ed., *Zirconium: Physico-Chemical Properties of its Compounds and Alloys*, International Atomic Energy Agency, Vienna, 1976, p. 8.
13. Ref. 12, p. 9.
14. Ref. 12, p. 10.
15. Ref. 12, p. 11.
16. Ref. 10, p. 11.
17. Y. S. Touloukian, R. W. Powell, C. Y. Ho, and P. G. Klemens, *Thermophysical Properties of Matter*, Vol. 1, *Thermal Conductivity, Metallic Elements and Alloys*, IFI/Plenum, New York, 1970.
18. B. Lustmann and F. Kerze, Jr., eds., *The Metallurgy of Zirconium*, McGraw-Hill, New York, 1955, p. 365.
19. Ref. 18, p. 370.
20. Ref. 10, p. 77.
21. Ref. 10, p. 78.
22. Ref. 18, p. 375.
23. J. H. McClain, *USBM IC 7686*, U.S. Bureau of Mines, Washington, D.C., 1954.
24. T. L. Yau, *Paper No. 102 presented at NACE Corrosion/81 April 1981*, Toronto, Canada.
25. H. L. Gilbert, C. Q. Morrison, A. Jones, and A. W. Henderson, *USBM RI 5091*, U.S. Bureau of Mines, Washington, D.C., 1954.
26. H. S. Choi, *Can. Min. Metall. Bull.* **67**, 65 (1965).
27. *Chem. Eng.* **61**, 124 (1954).
28. U.S. Pat. 2,721,115 (Oct. 18, 1955) (to Zirconium Corp. of America); 2,721,117 (Oct. 18, 1955); 2,578,748 (Dec. 18, 1951) (to Sylvester & Co.); 3,832,441 (Aug. 27, 1974), R. A. Schoenlaub.
29. U.S. Pat. 3,109,704 (1963), J. K. Olby.
30. G. von Hevesy, *Chem. Rev.* **2**, 1 (1925).
31. U.S. Pat. 2,653,855 (1953), H. C. Kawecki.
32. N. P. Sajin and E. A. Pepelyaeva, in *Proceedings of the International Conference on Peaceful Uses of Atomic Energy*, Vol. 8, United Nations, New York, 1958, p. 559.
33. A. A. Manieh and D. R. Spink, *Can. Metall. Q.* **12**, 331 (1973).
34. A. A. Manieh, D. S. Scott, and D. R. Spink, *Can. J. Chem. Eng.* **52**, 507 (1972).
35. P. H. Wilks, *Chem. Eng. Prog.* **68**, (April 1974).
36. A. M. Evans and J. P. H. Williamson, *J. Mater. Sci.* **12**, 779 (1977).
37. R. H. Nielsen in D. E. Thomas and E. T. Hayes, eds., *Metallurgy of Hafnium*, U.S. Atomic Energy Commission, Washington, D.C., 1960, pp. 49–82.
38. I. V. Vinarov, *Russ. Chem. Rev.* **36**, 522 (1967).
39. D. Royston and P. G. Alfredson, *AAEC.TM* 570, Australian Atomic Energy Commission, Oct. 1970.
40. P. Pascal, *Nouveau traité de chemie minérale*, Tome IX, Masson, Paris, 1963.
41. O. A. Sinegribova and G. A. Yagodin, *At. Energy Rev.* **4**(1), 93 (1966).
42. W. Fischer and W. Chalybaeus, *Z. Anorg. Allg. Chem.* **255**, 79 (1947).
43. W. Fischer, W. Chalybaeus, and M. Zumbusch, *Z. Anorg. Allg. Chem.* **255**, 277 (1948).
44. J. H. McClain and S. M. Shelton in C. R. Tipton, Jr., ed., *Reactor Handbook*, Vol. 1, *Materials*, 2nd ed., Interscience Publishers, Inc., New York, 1960.
45. J. M. Googin in F. R. Bruce, J. M. Fletcher, and H. H. Hyman, eds., *Progress in Nuclear Energy*, Series III: Process Chemistry, Vol. 2, Pergamon Press, New York, 1958.
46. U.S. Pat. 2,753,250 (July 3, 1956), H. A. Wilhelm, K. A. Walsh, and J. V. Kerrigan (to the United States of America as represented by the U.S. Atomic Energy Commission).

47. G. H. Beyer and H. C. Peterson, *AEC report ISC-182*, U.S. Atomic Energy Commission, Washington, D.C., Dec. 3, 1951.
48. R. P. Cox, H. C. Peterson, and G. H. Beyer, *Ind. Eng. Chem.* **50**(2), 141 (1958).
49. F. L. Moore, *National Academy of Sciences Report NAS-NS 3101*, Office of Technical Services, Department of Commerce, Washington, D.C., Dec. 1960.
50. E. Cerrai and C. Testa, *Energ. Nucl. (Milan)* **6**, 707 (1959).
51. Ref. 50, p. 768.
52. J. G. Moore, C. A. Blake, and J. M. Schmitt, *USAEC Report ORNL 2346*, U.S. Atomic Energy Commission, Washington, D.C., July 1957.
53. K. B. Brown, *USAEC Report ORNL-TM-107*, U.S. Atomic Energy Commission, Washington, D.C., Feb. 1962.
54. D. J. McDonald, *Sep. Sci. and Technol.* **16**, 1355 (1981).
55. M. Takahashi, H. Miyazaki, and Y. Katon, paper presented at the 6th International Conference on Zirconium in the Nuclear Industry, Vancouver, B.C., Canada, June 1982, to be published as ASTM Special Technical Publication 824 in 1984.
56. L. Moulin, P. Thouvenin, and P. Brun in ref. 55.
57. U.S. Pat. 4,021,531 (May 3, 1977), P. Besson, J. Guerin, P. Brun, and M. Bakes (to Ugine Aciers (Pechiney Ugine Kuhlman, France).
58. U.S. Pat. 2,852,446 (Sept. 16, 1958), M. L. Bromberg (to E. I. du Pont de Nemours & Co., Inc.).
59. *Chem. Eng.*, 106 (Jan. 26, 1981).
60. W. W. Minkler and E. F. Baroch in J. K. Tien and J. F. Elliott, eds., *Metallurgical Treatises*, Metallurgical Society of AIME, New York, 1981.
61. L. A. Nisel'son, T. D. Sokolova, and V. I. Stolyarov, *Dokl. Akad. Nauk SSSR* **168**, 385 (1966).
62. W. J. Kroll, A. W. Schlechten, and L. A. Yerkes, *Trans. Electrochem. Soc.* **89**, 263 (1976).
63. W. J. Kroll, A. W. Schlechten, W. R. Carmody, L. A. Yerkes, H. P. Holmes, and H. L. Gilbert, *Trans. Electrochem. Soc.* **92**, 99 (1947).
64. W. J. Kroll, C. T. Anderson, H. P. Holmes, L. A. Yerkes, and H. L. Gilbert, *J. Electrochem. Soc.*, 94 (1948).
65. W. J. Kroll, W. W. Stephens, and H. P. Holmes, *Trans. AIME* **188**, 1445 (1950).
66. W. W. Stephens and J. H. McClain, *Chem. Eng.*, 116 (March 1951).
67. S. M. Shelton, E. D. Dilling, and J. H. McClain, in ref. 24 in *Proc. Int. Conf. on The Peaceful Uses of Atomic Energy* **8**, 505–550, United Nations (1957).
68. *Chem. Eng.*, 126 (Oct. 9, 1967).
69. W. J. S. Craigen, E. C. Joe, and G. M. Ritcey, *Can. Metall. Q.* **9**, 485 (1970).
70. *ZrHf Newsletter*, Amax Specialty Metals Corp., Akron, N.Y., Sept. 1973.
71. P. Pint and S. N. Flengas, *Trans. Instit. Min. Metallurgy* **87**, 29 (1978).
72. S. F. Flengas and P. Pint, *Can. Metall. Q.* **8**, 151 (1969).
73. Z. M. Shapiro in ref. 18, Chapt. 5.
74. E. M. Sherwood and I. E. Campbell in ref. 37, Chapt. 4.
75. V. S. Emelyanov, P. D. Bystrov, and A. I. Eustyukhin, *J. Nucl. Energy* **4**, 253 (1957).
76. R. F. Rolston, *Iodide Metals and Metal Iodides*, The Electrochemistry Society, Inc., New York, 1961.
77. D. H. Baker, J. R. Nettle, and H. Knudsen, *USBM R.I. 5758*, U.S. Bureau of Mines, Washington, D.C., 1961.
78. J. C. Sehra, I. G. Sharma, P. L. Vijay, and A. Unnikrishnan, *Trans. Indian Inst. Met.* **30**, 14 (1977).
79. F. A. Schmidt, O. N. Carlson, and C. E. Swanson, Jr., *Metall. Trans.* **1**, 1371 (1970).
80. *Minerals Yearbook*, U.S. Bureau of Mines, Washington, D.C, 1972–1982.
81. F. D. Snell and L. S. Ettre, eds., *Encyclopedia of Industrial Chemical Analysis*, Vol. 14, Interscience Publishers, a division of John Wiley & Sons, Inc., New York, 1971, pp. 103–152.
82. R. B. Hahn in I. M. Kolthoff and P. J. Elving, eds., *Treatise on Analytical Chemistry*, Vol. 5, Interscience Publishers, a division of John Wiley & Sons, Inc., New York, 1961, Part II, pp. 61–138.
83. J. S. Fritz and M. Johnson, *Anal. Chem.* **27**, 1653 (1955).
84. P. R. Subbaraman and K. S. Rajan, *J. Sci. Ind. Res.* **13B**, 31 (1963).
85. C. L. Luke, *Anal. Chim. Acta* **37**, 284 (1967).
86. R. J. Jaworowski, R. P. Weverling, and D. J. Bracco, *Anal. Chim. Acta* **37**, 284 (1967).
87. *Methods for Emission Spectrochemical Analysis*, 7th ed., American Society for Testing and Materials, Philadelphia, Pa., 1982.

88. R. T. Van Santen, J. H. Schlewitz, and C. H. Toy, *Anal. Chim. Acta* **33,** 593 (1965).
89. E. B. Sandell, *Colorimetric Determination of Trace Metals*, 3rd ed., Interscience Publishers, New York, 1956, p. 976.
90. J. P. Yound, J. R. French, and J. C. White, *Anal. Chem.* **30,** 422 (1958).
91. Ref. 87, p. 766.
92. Ref. 87, p. 770.
93. J. H. Schlewitz and M. Shields, *At. Absorpt. Newsl.* **10,** 39 (1971).
94. S. V. Elinson and K. L. Petrov, *Zirconium, Chemical and Physical Methods of Analysis*, Main Administration on the Use of Atomic Energy at the Council of Ministers, Moscow, 1960; English translation, Office of Technical Services, Department of Commerce, Washington, D.C., 1962.
95. E. B. Read and H. M. Read, *The Chemical Analysis of Zirconium and Zircaloy Metals*, Nuclear Metals, Cambridge, Mass., 1957.
96. *Chemical Analysis of Zirconium and Zirconium-base Alloys*, ASTM E 146-68, American Society for Testing and Materials, Philadelphia, Pa., 1969.
97. W. B. Blumenthal, *J. Sci. Ind. Res.* **35,** 485, July 1976.
98. I. C. Smith and B. L. Carson, *Trace Metals in the Environment*, Vol. 3, *Zirconium*, Ann Arbor Science, Ann Arbor, Mich., 1978.
99. W. B. Shelley and H. J. Hurley in M. Sauter, ed., *Immunological Diseases*, 2nd ed., Little, Brown & Co., Boston, Mass., 1971.
100. W. W. Allison, *Zirconium, Zircaloy, and Hafnium Safe Practice Guide for Shipping, Storing, Handling, Processing, and Scrap Disposal*, WAPD-TM-17, Bettis Atomic Power Laboratory, Pittsburgh, Pa., 1960.
101. *National Fire Codes*, Vol. 13, NFPA No. 482M-1974, National Fire Protection Association, Boston, Mass., 1978.
102. J. Schemel, *ASTM Manual on Zirconium and Hafnium*, ASTM STP 639, American Society for Testing and Materials, Philadelphia, Pa., 1977.
103. *Bureau of Explosives Tariff No. BOE-6000-B*, Bureau of Explosives, Association of American Railroads, Washington, D.C.
104. R. A. Rowse and J. E. Patchett, *Sagamore Army Materials Research Conference Proceedings, 1979*, 1981, pp. 215–228.
105. C. A. Seabright and H. C. Draker, *Ceramic Bulletin* **40,** 1 (1961).
106. D. N. Crook, *J. Aust. Ceram. Soc.* **4**(2), 31 (1968).
107. C. L. Hackler, *Ceramic Bulletin* **54,** 283 (1975).
108. R. A. Eppler, *Ceramic Bulletin* **56,** 213 (1977).
109. R. D. Fare and C. G. Yayenas, *J. Electrochem. Soc.* **127,** 1478 (1980).
110. A. M. Anthony in A. H. Heuer and L. W. Hobbs, eds., *Science and Technology of Zirconia*, The American Ceramic Society, Columbus, Ohio, 1981, pp. 437–454.
111. K. Nassau, *Lapidary J.* **35,** 1210 (1981).
112. C. E. Land, *IEEE Trans. Electron Devices* **ED-26,** 1143 (Aug. 1979).
113. C. J. Conner, A. S. Cooper, L. W. Mazzeno, and W. A. Reaves, *Am. Dyest. Rep.*, 62 (April 1972).
114. W. B. Blumenthal, *Ind. Eng. Chem.* **42,** 640 (1950).
115. U.S. Pat. 3,551,089 (Dec. 1970), W. O. Dawson, S. C. O'Connor, and F. S. Fellowes.
116. U.S. Pat. 4,273,592 (June 1981), T. L. Kelly.
117. M. P. Brassington, T. Hailing, A. J. Miller, and G. A. Saunders, *Mater. Res. Bull.* **16,** 613 (1981).
118. R. N. Brown, B. Bendow, M. G. Drexhage, and C. T. Moynihan, *Appl. Opt.* **21,** 361 (1982).
119. N. W. Connon, *Eastman Org. Chem. Bull.*, Vol. 44 (1972).
120. M. Hansen, *Constitution of Binary Alloys*, McGraw-Hill, New York, 1958.
121. R. P. Elliott, *Constitution of Binary Alloys*, 1st Suppl., McGraw-Hill, New York, 1965.
122. F. A. Shunk, *Constitution of Binary Alloys*, 2nd Suppl., McGraw-Hill, New York, 1969.
123. O. Kubaschewski-von Goldbeck in ref. 12, Chapt. 2.
124. Ref. 10, p. 158.
125. W. M. Mueller, J. P. Blackledge, and G. G. Libowitz, *Metal Hydrides*, Academic Press, New York, 1968, pp. 241–335.
126. E. Fromm and E. Gebhardt, *Gase und Kohlenstoff in Metallen*, Springer, Berlin, 1976.
127. R. L. Beck and W. M. Mueller in ref. 125, pp. 286–321.
128. R. M. van Essen and K. H. J. Busehow, *Mater. Res. Bull.* **15,** 1149 (1980).
129. D. Shaltiel, I. Jacob, and D. Davidov, *J. Less Common Met.* **53,** 117 (1977).

130. D. G. Ivey, R. I. Chittim, K. J. Chittim, and D. O. Northwood, *J. Materials for Energy Systems* **3**, 3 (Dec. 81).
131. A. W. Struss and J. P. Corbett, *Inorg. Chem.* **16**, 360 (1977).
132. E. K. Storms, *The Refractory Carbides*, Academic Press, New York, 1967.
133. *Engineering Property Data on Selected Ceramics, Vol. II Carbides*, Metals and Ceramics Information Center, Battelle Columbus Laboratory, Columbus, Ohio, 1979.
134. S. Motojima, E. Kani, Y. Takahashi, and K. Sugiyama, *J. Mater. Sci.* **14**, 1495 (1979).
135. Y. Okabe, Hojo, and Kato, *Yogyo Kyokai Shi* **85**, 173 (1977).
136. E. Rudy, *Compendium of Phase Diagram Material*, Air Force Materials Laboratory, Wright-Patterson Air Force Base, Ohio, 1969.
137. Ref. 12, p. 105.
138. L. Brattas and A. Kuekshus, *Acta Chem. Scand.* **26**, 3441 (1972).
139. H. Schafer, *Chemical Transport Reactions*, Academic Press, New York, 1964.
140. B. R. Conard and H. F. Franzen, *High Temp. Sci.* **3**, 49 (1971).
141. A. Clearfield, *Acta Crystallogr.* **16**, 135 (1963).
142. S. Yamoaka, *J. Am. Ceram. Soc.* **55**, 111 (1972).
143. W. B. Davies, M. L. H. Green, and A. J. Jackson, *J. Chem. Soc. Chem. Comm.* **19**, 781 (1976).
144. C. F. Grain and R. C. Garvie, *USBM R.I. 6619*, U.S. Bureau of Mines, Washington, D.C., 1965.
145. E. D. Whitney, *J. Electrochem. Soc.* **112**, 91 (1965).
146. L. G. Liu, *J. Phys. Chem. Solids* **41**, 331 (1980).
147. E. Gebhardt, H. Sechezzi, and W. Durrshnabel, *J. Nucl. Mater.* **4**, 255 (1961).
148. R. Ruh and H. Garrett, *J. Am. Ceram. Soc.* **50**, 257 (1967).
149. R. J. Ackermann, S. P. Carg, and E. G. Rauh, *J. Am. Ceram. Soc.* **61**, 275 (1978).
150. N. Claussen, R. Wagner, L. J. Gauckler, and G. Petzow, *J. Am. Ceram. Soc.* **61**, 369 (1978).
151. U. S. Stubican and J. R. Hellman in ref. 110.
152. R. G. Garvie in A. M. Alper, ed., *High Temperature Oxides*, Academic Press, New York, 1970, p. 127.
153. R. G. Garvie, *J. Phys. Chem.* **82**, 218 (1978).
154. H. T. Rijnten, *Zirconia*, Drukkerij Gebr. Janssen N. V. Nijmegen, 1971, pp. 67–70.
155. U.S. Pats. 4,293,534; 4,293,535 (Oct. 6, 1981), R. H. Arendt (to General Electric).
156. E. G. Rauh and S. P. Garg, *J. Am. Ceram. Soc.* **63**, 239 (1980).
157. D. L. Douglass, *At. Energy Rev.* **1**, 75 (1963).
158. H. M. Chung and F. L. Yagee in ref. 55.
159. Ref. 12, p. 104.
160. J. A. Speer in P. A. Ribbe, ed., *Orthosilicates*, Minerological Society of America, Washington, D.C., 1980.
161. P. A. Romans, L. Brown, and J. C. White, *Am. Mineralogist* **60**, 475 (1975).
162. R. M. Hazen and L. W. Finger, *Am. Mineral.* **64**, 196 (1979).
163. C. E. Curtis and H. G. Sowman, *J. Am. Ceram. Soc.* **36**, 190 (1953).
164. O. I. Lee, *Chem. Rev.* **5**, 22 (1928).
165. W. B. Blumenthal, *The Chemical Behavior of Zirconium*, D. Van Nostrand Co., Princeton, N.J., 1958.
166. H. D. Hess, *USBM R.I. 5856*, U.S. Bureau of Mines, Washington, D.C., 1962.
167. M. S. Ferrante and R. S. McClune, *BM RI 8418*, U.S. Department of the Interior, Washington, D.C., 1980.
168. I. V. Tananaev and L. S. Guzeeva, *Russ. J. Inorg. Chem.* **11**, 587 (1966).
169. C. J. Barton and co-workers, *J. Phys. Chem.* **62**, 665 (1958).
170. M. K. Chikanova, *J. Gen. Chem. USSR* **46**, 2088 (1976).
171. H. S. Ray, B. G. Bhat, G. S. Reddy, and A. K. Biswas, *Trans. Indian Inst. Met.* **32**, 177 (1978).
172. Can. Pat. 863,258 (Feb. 9, 1971), J. E. Dutrizag and S. N. Flengas.
173. E. M. Larsen, J. S. Wrazel, and L. G. Hoard, *Inorg. Chem.* **21**, 2619 (1982).
174. I. E. Newnham and J. A. Watts, *J. Am. Chem. Soc.* **82**, 2113 (1960).
175. R. A. J. Shelton, *J. Less Common Met.* **61**, 51 (1978).
176. R. L. Daake and J. D. Corbett, *Inorg. Chem.* **17**, 1192 (1978).
177. D. Cubicciotti, R. L. Jones, and B. C. Syrett in D. G. Franklin, ed., *Zirconium in the Nuclear Industry*, ASTM Special Technical Publication 754, American Society for Testing and Materials, Philadelphia, Pa., 1982.

178. J. D. Corbett, R. L. Daake, A. Cisar, and D. H. Guthrie in D. L. Hildenbrand and D. D. Cubicciotti, eds., *High Temperature Metal Halide Chemistry*, The Electrochemical Society, Princeton, N.J., 1978.
179. P. Ehrlich, F. Ploger, and E. Koch, *Z. Anorg. Allg. Chem.* **333,** 209 (1964).
180. U.S. Pat. 2,941,931 (June 21, 1960), R. S. Dean.
181. Ref. 37, p. 69.
182. R. S. Dean, *Ind. Lab.* **45,** (April 1959).
183. F. L. Scott, H. Q. Smith, and I. Mockrin, *Technical Report No. 7, Office Naval Research Contract Nonr 2687(00)*, Pennsalt Chemicals Corp., Research and Development Dept., Wyndmoor, Pa., Sept. 10, 1959.
184. A. W. Struss and J. D. Corbett, *Inorg. Chem.* **9,** 1373 (1970).
185. R. L. Daake and J. D. Corbett, *Inorg. Chem.* **16,** 2029 (1977).
186. D. G. Adolfson and J. D. Corbett, *Inorg. Chem.* **15,** 1820 (1976).
187. P. D. Kleinschmidt, D. Cubicciotti, and D. L. Hildebrand, *J. Electrochem. Soc.* **125,** 1543 (1978).
188. J. F. Marchiando, B. N. Harmon, and S. H. Kiu, *Physica* **99B,** 259 (1980).
189. A. Clearfield and P. A. Vaughan, *Acta Crystallogr.* **9,** 555 (1956).
190. A. Clearfield, *Rev. Pure Appl. Chem.* **14,** 91 (1964).
191. J. R. Fryer, J. L. Hutchison, and R. Paterson, *J. Colloid Interface Sci.* **34,** 238 (1970).
192. J. C. Mailen, D. E. Horner, S. E. Dorris, N. Pih, S. M. Robinson, and R. G. Yates, *Sep. Sci. Technol.* **15,** 959 (1980).
193. D. H. Devia and A. G. Sykes, *Inorg. Chem.* **20,** 910 (1981).
194. L. M. Zaitsev and G. S. Bochkarev, *Russ. J. Inorg. Chem.* **7,** 711 (1962).
195. L. M. Zaitsev, *Russ. J. Inorg. Chem.* **11,** 900 (1966).
196. A. Clearfield, *Inorg. Chem.* **3,** 146 (1964).
197. U.S. Pat. 3,334,962 (Aug. 8, 1967), A. Clearfield (to National Lead Co.).
198. I. G. Kamaeva, L. A. Mel'Nik, and V. V. Serebrennikov, *Russ. J. Inorg. Chem.* **13,** 1026 (1968).
199. Ref. 165, pp. 132–135.
200. K. Dehnicke and J. Weidlein, *Angew. Chem. Int. Ed.* **5,** 1041 (1966).
201. B. O. Field and C. J. Hardy, *Proc. Chem. Soc.*, 76 (Feb. 1962).
202. F. M. Farhan and H. Nedjat, *J. Chem. Eng. Data* **21,** 441 (1976).
203. U.S. Pat. 3,510,254 (May 5, 1970), R. N. Bell (to Stauffer Chemical Co.).
204. U.S. Pat. 2,316,141 (April 6, 1943), E. Wainer (to the Titanium Alloy Manufacturing Co.).
205. L. A. Pospelova and L. M. Zaitsev, *Russ. J. Inorg. Chem.* **11,** 995 (1966).
206. J. Singer and D. T. Cromer, *Acta Crystallogr.* **12,** 719 (1959).
207. I. J. Bear and W. G. Mumme, *Rev. Pure Appl. Chem.* **21,** 189 (1971).
208. F. Farnworth, S. L. Jones, and I. McAlpine in R. Thompson, ed., *Specialty Inorganic Chemicals*, The Royal Society of Chemistry, London, 1981, pp. 248–284.
209. M. A. Kolenkova and co-workers, *J. Appl. Chem. USSR* **50,** 156 (1977).
210. A. Clearfield and J. A. Stynes, *J. Inorg. Nucl. Chem.* **26,** 117 (1964).
211. G. Alberti and E. Torracca, *J. Inorg. Nucl. Chem.* **30,** 317 (1968).
212. A. Clearfield and S. P. Pack, *J. Inorg. Nucl. Chem.* **37,** 1983 (1975).
213. A. Clearfield, *Inorganic Ion Exchange Materials*, CRC Press, Boca Raton, Fla., 1982.
214. U.S. Pat. 4,025,608 (May 24, 1977), D. S. Tawil, M. H. Clubley, and F. Farnworth (to Magnesium Elektron Ltd.).
215. U.S. Pat. 4,256,463 (Mar. 17, 1981), D. P. Carter (to Teledyne Industries, Inc.).
216. S. Fusitsu, M. Nagai, and T. Kanazawa, *Solid State Ionics* **3/4,** 233 (1981).
217. H. Y.-P. Hong, *Mater. Res. Bull.* **11,** 173 (1976).
218. R. Sacks, Y. Avigal, and E. Banks, *J. Electrochem. Soc.* **129,** 726 (1982).
219. D. C. Bradley, R. C Mehrotra, and D. P. Gaur, *Metal Alkoxides*, Academic Press, New York, 1978.
220. W. G. Bartley and W. Wardlaw, *J. Chem. Soc.*, 421 (1958).
221. R. C. Mehrotra and M. M. Agrawal, *J. Chem. Soc.*, 1026 (1967).
222. Ref. 219, pp. 163–165.
223. R. N. Kapoor and R. C. Mehrotra, *J. Chem. Soc.*, 422 (1959).
224. U. B. Saxena, A. K. Rai, and R. C. Mehrotra, *Inorg. Chim. Acta* **7,** 681 (1973).
225. L. M. Zaitsev, *Russ. J. Inorg. Chem.* **9,** 1279 (1964).
226. L. M. Zaitsev, G. S. Bockharev, and N. V. Kozhenkova, *Russ. J. Inorg. Chem.* **10,** 590 (1965).
227. J. E. Drake and G. W. A. Fowles, *J. Less Common Met.* **2,** 401 (1960).
228. *Ibid.*, **3,** 149 (1961).

229. M. Allbutt and G. W. A. Fowles, *J. Inorg. Nucl. Chem.* **25,** 67 (1963).
230. U.S. Pat. 3,626,010 (July 31, 1968), C. J. Sorrel (to Ethyl Corp.).
231. A. Clearfield and E. J. Malkiewich, *J. Inorg. Nucl. Chem.* **25,** 237 (1963).
232. G. A. Razuvauv, G. A. Domrachev, V. V. Sharutin, and O. N. Suvarova, *J. Organometal. Chem.* **141,** 313 (1977).
233. P. Etievant, G. Tainturier, and B. Gautheron, *C. R. Acad. Sci. Ser. C* **283,** 233 (1976).
234. T. F. Blackburn, J. A. Labinger, and J. Schwartz, *Tetrahedron Lett.*, 3041 (1975).
235. C. A. Bertelo and J. Schwartz, *J. Am. Chem. Soc.* **98,** 262 (1975).
236. W. Tam and M. F. Rettig, *J. Organometal. Chem.* **108,** (1976).
237. P. C. Wailes and H. Weigold, *Inorg. Synth.* **19,** 223 (1979).
238. J. L. Thomas and K. T. Brown, *J. Organometal. Chem.* **111,** 297 (1976).
239. B. Demerseman, G. Bouquet, and M. Bigorgne, *J. Organometal. Chem.* **107,** c19 (1976).
240. G. Fachinetti, G. Fochi, and C. Floriani, *J. Chem. Soc. Chem. Commun.*, 230 (1976).
241. D. J. Sikora, M. D. Rausch, R. D. Rogers, and J. L. Atwood, *J. Am. Chem. Soc.* **101,** 5079 (1979).
242. J. L. Atwood, R. D. Rogers, W. E. Hunter, C. Floriano, G. Fachinetti, and A. Chiesi-Villa, *A. Inorg. Chem.* **19,** 3812 (1980).
243. G. Fachinetti, C. Floriani, and H. Stoecki-Evans, *J. Chem. Soc. Dalton Trans.* 2297 (1977).
244. B. Demerseman, G. Bouquet, and M. Bigorgne, *J. Organometal. Chem.* **132,** 223 (1977).
245. G. Fachinetti, G. Fochi, and C. Floriani, *J. Chem. Soc. Dalton Trans.*, 1946 (1977).
246. J. M. Manriquez, D. R. McAlister, R. D. Sanner, and J. E. Bercaw, *J. Am. Chem. Soc. Chem. Commun.*, 230 (1976).
247. J. A. Marsella, C. J. Curtis, J. E. Bercaw, and K. G. Caulton, *J. Am. Chem. Soc.* **102,** 7244 (1980).
248. R. D. Sanner, J. M. Manriquez, R. E. Marsh, and J. E. Bercaw, *J. Am. Chem. Soc.* **98,** 8351 (1976).
249. J. M. Manriquez, R. D. Scanner, R. E. Marsh, and J. E. Bercaw, *J. Am. Chem. Soc.* **98,** 3042 (1976).
250. Brit. Pat. 2,018,267 (1979), M. F. Lappert, M. J. S. Gynane, and J. Jeffrey.
251. J. L. Atwood, G. K. Barker, J. Holton, W. E. Hunter, M. F. Lappert, and R. Pearce, *J. Am. Chem. Soc.* **99,** 6645 (1977).
252. G. A. Razuvaev, V. N. Latyaeva, L. I. Vyshinskaya, G. A. Vasileva, V. I. Khruleva, and L. I. Smirnova, *Dokl. Akad. NAUK SSSR* **231,** 114 (1976).
253. D. B. Carr and J. Schwartz, *J. Am. Chem. Soc.* **99,** 638 (1977).
254. D. B. Carr and J. Schwartz, *J. Organometal. Chem.* **139,** c21 (1977).
255. M. Yoshifuju, M. J. Loots, and J. Schwartz, *Tetrahedron Lett.*, c21 (1977).
256. M. J. Loots and J. Schwartz, *J. Am. Chem. Soc.* **99,** 8045 (1977).
257. E. Negishi and D. E. Van Horn, *J. Am. Chem. Soc.* **99,** 3168 (1977).
258. J. Schwartz, M. J. Loots, and J. Kosugi, *J. Am. Chem. Soc.* **102,** 1333 (1980).
259. J. S. Temple and J. Schwartz, *J. Am. Chem. Soc.* **102,** 7382 (1980).
260. J. G. Lee and C. H. Brubaker, *J. Organometal. Chem.* **135,** 115 (1977).
261. O. Khan, A. Dormaond, and J. P. Letourneux, *J. Organometal. Chem.* **132,** 149 (1977).
262. E. S. Chandrasekaran, R. H. Grubbs, and C. H. Brubaker, Jr., *J. Organometal. Chem.* **120,** 49 (1976).
263. A. Anderson, H.-G. Cordes, J. Herwig, W. Kaminshky, A. Merck, R. Mottweiler, J. Pein, H. Sinn, and H.-J. Vollmer, *Angew. Chem.* **88,** 649 (1976).
264. J. Sinn, W. Kanimshky, H.-J. Vollmer, and R. Wolot, *Angew. Chem. Int. Ed. Engl.* **19,** 390 (1980).
265. V. A. Zakharov, V. K. Dudchenko, E. A. Paukshtis, L. G. Karakchiev, and Y. I. Yermakov, *J. Mol. Catal.* **2,** 421 (1977).
266. J. Schwartz and M. D. Ward, *J. Mol. Catal.* **8,** 465 (1980).

RALPH H. NIELSEN
JAMES H. SCHLEWITZ
HENRY NIELSEN
Teledyne Wah Chang Albany

ZONE REFINING

Zone refining is one of a class of techniques known as fractional solidification, in which a separation is brought about by crystallization of a melt without solvent being added (see also Crystallization) (1–8). Solid–liquid phase equilibria are utilized, but as described later, phenomena are much more complex than in separation processes utilizing vapor–liquid equilibria. In most of the fractional-solidification techniques described in the article on crystallization, small separate crystals are formed rapidly in a relatively isothermal melt. In zone refining, on the other hand, a massive solid is formed slowly with a sizable temperature gradient imposed at the solid–liquid interface.

Zone refining was developed in the early 1950s at Bell Telephone Laboratories in response to the need for extremely pure germanium. It was an essential step for the development of the transistor and thus for the entire electronics revolution. Before that time, purification was carried out by normal freezing, also called progressive freezing (see Semiconductors). A melt was slowly frozen, causing more impurities to be concentrated in the last melt to freeze. In progressive freezing, however, a single solidification did not give sufficient purification. Simply remelting the entire ingot would redistribute the impurities, thereby eliminating the purification of the first solidification. The impure end could be removed, but handling and cutting introduce new impurities. The problem was solved by melting only a small part of the material and passing this molten zone down the ingot. Subsequent zone passes increased purification without requiring handling or cutting or permitting excessive backmixing.

Zone refining can be applied to the purification of almost every type of substance that can be melted and solidified, eg, elements, organic compounds, and inorganic compounds. Because the solid–liquid phase equilibria are not favorable for all impurities, zone refining often is combined with other techniques to achieve ultrahigh purity.

The high cost of zone refining has thus far limited its application to laboratory reagents and valuable chemicals such as electronic materials. The cost arises primarily from the low processing rates, handling, and high energy consumption owing to the large temperature gradients needed.

Actually, zone refining is only one of a class of techniques known as zone melting in which a molten zone is passed down a solid rod. Zone melting is used routinely to collect impurities in high purity materials, eg, silicon, in preparation for chemical analysis. Growth of bulk single crystals is an important application that includes commercial float-zoning of silicon crystals for the semiconductor industry (7). Floating-zone melting is being considered for use aboard the U.S. Space Shuttle because the absence of gravity might permit larger crystals to be grown with greater homogeneity (see Space chemistry).

Addition of solvent to the zone allows crystals to be grown that either decompose before melting (incongruent melting) or have a very high vapor pressure at the melting point. This technique has been called both temperature-gradient zone melting and the traveling-heater method of crystal growth. It has also been used for fabrication of semiconductor devices (9).

Continuous zone-refining techniques have been developed, both theoretically and experimentally, but they have not been commercialized (1,4,10–11).

Theory

Early zone-refining theory attempted to correlate the concentration of impurities with location.

Progressive Freezing. Although a rod geometry is not required, directional solidification in the Bridgman configuration is easiest to visualize. It is widely used for single-crystal growth and preparation of dendritic and eutectic composite materials.

To derive the concentration profile for progressive freezing, a material balance is employed for solidification of a small fraction dg of melt, as shown in Figure 1. Integration from the beginning of solidification gives (1,4,8):

$$\int_0^g (1-g)^{-1} dg = -\ln(1-g) = \int_{w_o}^{w_l} (w_l - w_s)^{-1} dw_l \qquad (1)$$

where g is mass fraction already solidified, w_o is the original mass fraction of impurity, w_s is the mass fraction of impurity in the solid at g, and w_l is the average impurity concentration in the liquid melt. It has been assumed that no mass transfer occurs in the solid, which is nearly always true because of the absence of convective mixing and the very low diffusion coefficients. Without solid-state diffusion, the solid, once it is formed, does not know the melt is there. This is analogous to batch evaporation with the removal of the vapor as it forms.

In order to integrate equation 1, it is necessary to have a relationship between w_l and w_s. If there is equilibrium between the bulk melt and the solid freezing out, phase-diagram data may be used. As shown in Figure 2, in the limit of very small impurity contents, w_s is often proportional to w_l, ie, $w_s = kw_l$, where k is the distribution or segregation coefficient. Equation 1 may then be integrated:

$$w_s/w_o = k(1-g)^{k-1} \qquad (2)$$

This relationship is plotted in Figure 3 for several values of k; for $k < 1$ (the usual situation for impurities), $w_s \to \infty$ as $g \to 1$. Clearly, this assumption cannot be correct; w_s is no longer proportional to w_l at high concentrations.

Zone Melting. A similar material balance may be made for a zone of mass m_z moving a short distance in such a manner that the mass dm of solid is frozen out and an identical mass melts into the zone (see Fig. 4). For the first zone pass, it is assumed that the rod is initially at uniform composition w_o to obtain the following (1,4,8):

$$\int_0^m m_z^{-1} dm = m/m_z = \int_{w_o}^{w_l} (w_o - w_s)^{-1} dw_l \qquad (3)$$

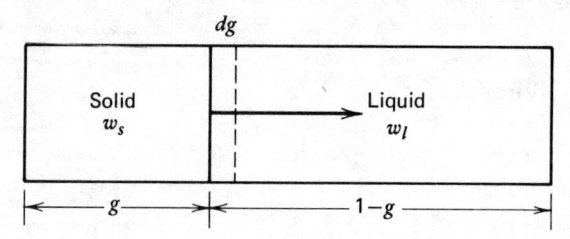

Figure 1. Solidification of differential mass fraction dg of a melt. Mass fraction of impurity in melt is w_l and in solid freezing out is $w_s(g)$.

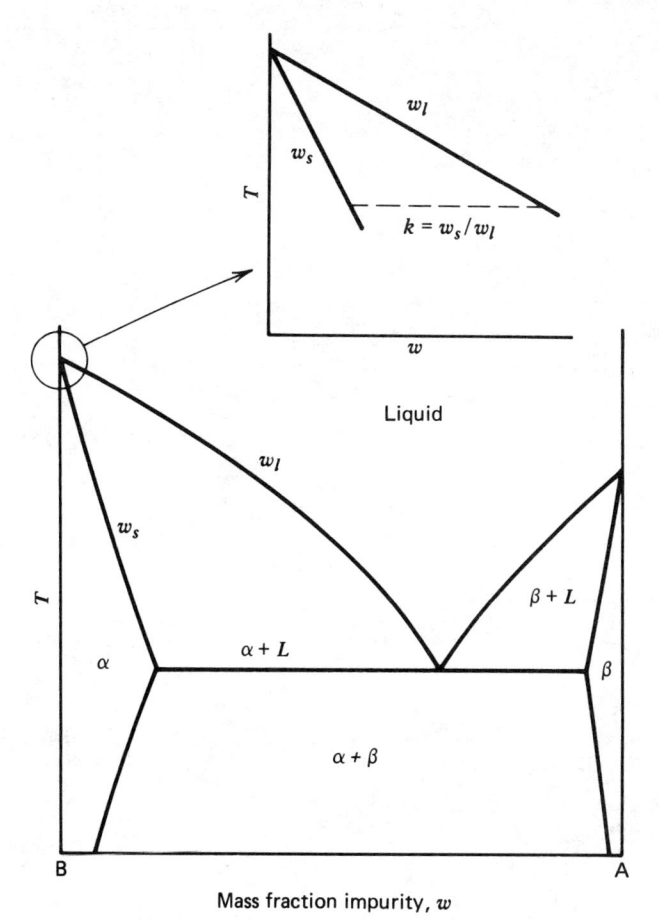

Figure 2. Typical binary phase diagram for host and impurity, showing a constant distribution coefficient if impurity content is low. L = liquid composition after some solidification, α = B and small amount of A, β = A and small amount of B, w_l = liquidus, and w_s = solidus.

For constant $k = w_s/w_l$, this may be integrated to yield

$$w_s = w_o[1 - (1 - k)\exp(-km/m_z)] \qquad (4)$$

Note that m/m_z is approximately the number of zone lengths down the rod.

Computations are more difficult for subsequent zone passes, since the starting composition of the rod is no longer uniform. Nevertheless, a variety of numerical and analytical results has been obtained for infinite and for finite rods (1,4,12–16). A typical result is shown in Figure 5. Substantial purification can be attained even when k is not significantly different from 1.

With subsequent zone passes, back-mixing becomes increasingly significant. When the number of zone passes becomes very large, the concentration profile reaches the ultimate distribution and subsequent passes have no influence. The number of zone passes required to approach the ultimate distribution is about twice the number of zone masses in the rod for small effective distribution coefficients. Thus, longer zones permit the attainment of the ultimate distribution sooner. On the other hand, the attainable separation increases as the zone size becomes smaller (permitting less back-mixing) (1,4).

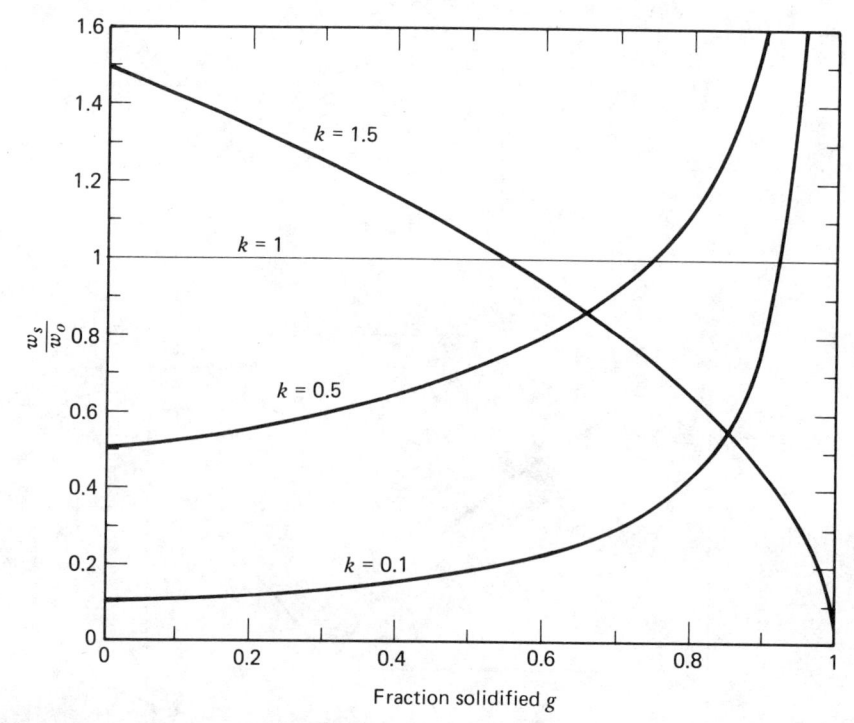

Figure 3. Impurity concentration profiles resulting from progressive freezing with different values of distribution coefficient k (from eq. 2).

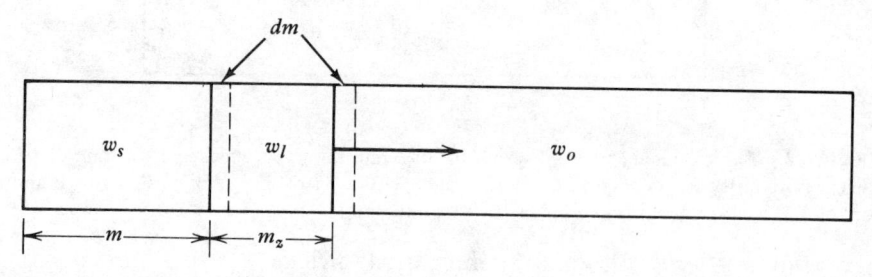

Figure 4. Movement of molten zone by differential amount dm for initial zone pass; original solid has uniform impurity content w_o.

Although most impurities have $k < 1$, there are exceptions, and it cannot be automatically assumed that the purest material is at the front end of the ingot; sometimes it is in the middle.

Theoretical treatments have also been given to other situations of interest. In zone leveling, the impurity level is homogenized either by moving the zone back and forth along a linear rod, or by movement around a ring (1). Evaporation and condensation of a volatile impurity (17–18), as well as decomposition of the material itself (13), have also been treated.

Nonideal Separations. In numerous instances, the ideal equations 2 and 4 have been verified experimentally. However, in other experiments different results were obtained, reflecting failure of one or more of the assumptions made in deriving equations 2 and 4. Likewise, much theoretical work is concerned with modified assumptions, including varying distribution coefficient k (19), eutectic-forming phase behavior (4,20–21), varying mass m_z of zone (22), and solid-state diffusion (23).

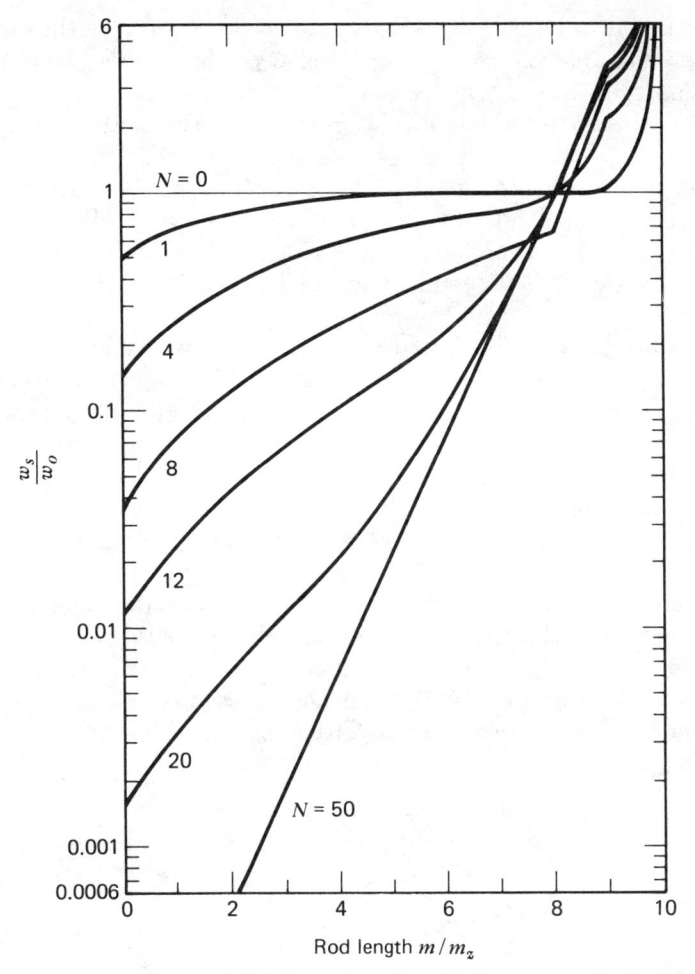

Figure 5. Concentration profiles for different numbers N of zone passes for $k = 0.5$ and a rod containing 10 zone masses. Obtained by numerical methods of ref. 13 using Texas Instruments 99/4 Home Computer.

The assumption of equilibrium between solid w_s and bulk melt w_l is frequently violated because of lack of complete mixing in the melt. A steady-state fictitious stagnant-film treatment may be employed to arrive at an effective distribution coefficient,

$$k_{\text{eff}} \equiv \frac{w_s}{w_l} = \frac{k}{k + (1-k)\exp(-\delta V \rho_s / D \rho_l)} \qquad (5)$$

where k is the equilibrium distribution coefficient from the phase diagram, δ is the film thickness (erroneously called the boundary-layer thickness), V the linear freezing rate, ρ_s the density of the solid, D the binary diffusion coefficient in the melt, and ρ_l the density of the liquid (1,4–5,24–26). The film thickness δ is related to the mass-transfer coefficient k_c and the Sherwood number N_{Sh} (see also Mass transfer) by

$$\delta = D/k_c = L/Sh \qquad (6)$$

It has been found to be only weakly dependent on V (25). As shown by Figure 6, equation 5 predicts that $k_{\text{eff}} \to k$ as $V \to 0$, and $k_{\text{eff}} \to 1$ as $V \to \infty$. Thus, the separation obtained diminishes as the zone travel rate increases.

In practice, the freezing rate V is rarely constant. Fluctuating convective currents in the melt (27–32) lead to a fluctuating freezing rate, which causes k_{eff} to be somewhat nearer unity than predicted by equation 5. A fluctuating freezing rate also leads to impurity striations, ie, bands of varying impurity content parallel to the freezing interface. Irregular free convection in liquid metals has been damped successfully by means of a constant magnetic field, about 0.4 T (4 kG) for a melt of 1 cm dia (27,33–36).

The equilibrium distribution coefficient k varies with distance down the ingot because of compositional stress in the crystal (37–38), departures from linearity on the phase diagram at large impurity concentrations, and interactions with other impurities. The actual interfacial-distribution coefficient k_i can depart from equilibrium because of adsorption on the growing interface (5,39–40), differences in incorporation kinetics of impurity and host atoms (41–43), and interface-field effects (44). Deviation of k_i from k in such cases increases as the growth velocity V increases. For $k < 1$, k_i can even exceed 1 at moderate V. The well-known facet effect generally is thought to arise from such phenomena, and k_i is much greater on the faceted portion of the interface than on a nonfaceted part of the same interface (4–5).

As the zone-travel rate is increased, a breakdown of the freezing interface from smooth to dendritic is eventually observed. Dendrites are treelike structures that trap melt and dramatically reduce separation. The onset of interface breakdown is predicted by the elementary theory of constitutional supercooling (1,4–5). As shown in Figure

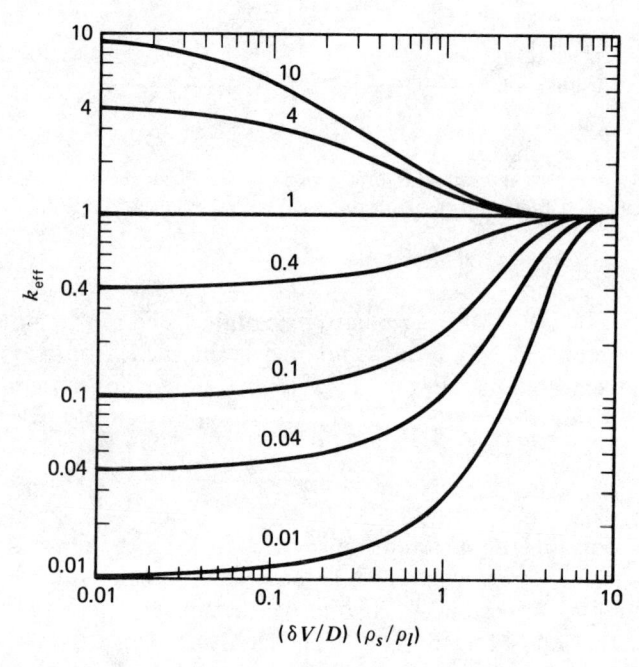

Figure 6. Effective distribution coefficient k_{eff} vs interfacial distribution coefficient k and dimensionless zone-travel velocity; k_{eff} is to be used in place of k in expressions for concentration profiles, such as equations 2 and 4.

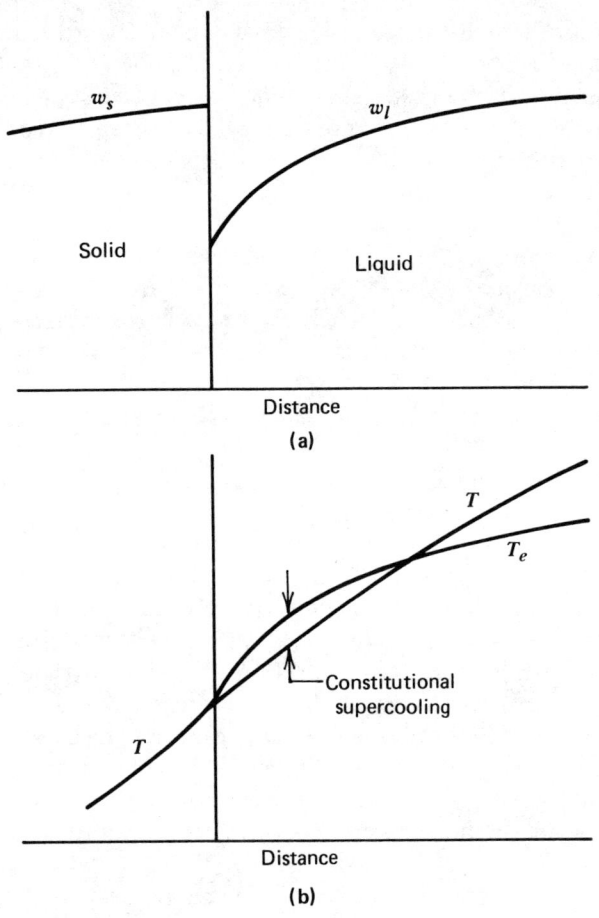

Figure 7. Constitutional supercooling. (a) Impurity concentration profile during solidification; (b) actual temperature T and equilibrium freezing temperature T_e during solidification.

7(a), $k < 1$ causes impurity to accumulate in the melt at the freezing interface, thereby lowering the freezing point. Thus, the actual temperature may be below the melting point for some distance into the melt, even though the two are identical at the freezing interface (see Fig. 7(b)). This is known as constitutional supercooling because it is brought about by a change in constitution or composition. The interface is unstable with respect to perturbation in shape, and cells or dendrites are expected.

At the onset of constitutional supercooling, the melting-point gradient exceeds the temperature gradient. Equating these gradients leads to the criterion for constitutional supercooling:

$$V/G > D\rho_l/(\partial T_e/\partial w_i)\rho_s(w_s - w_i) = Dk\rho_l/(\partial T_e/\partial w_i)\rho_s w_s(k-1)$$
$$= D\rho_l[k + (1-k)\exp(-\delta V \rho_s/\rho_l D)]/(\partial T_e/\partial w_i)\rho_s(k-1)w_l \quad (7)$$

where G is the imposed interfacial-temperature gradient and $\partial T_e/\partial w_i$ is the slope of the liquidus on the phase diagram (change of freezing point with impurity concentration). Much more advanced and complicated treatments of morphological stability have been made (45–54). However, most experimental results are adequately predicted by the constitutional-supercooling theory outlined here (55–57).

From equation 7, it may be seen that the tendency toward constitutional supercooling increases as the freezing rate V increases, the temperature gradient G decreases, the impurity content w increases, the separation ($w_s - w_i$) between liquidus and solidus in the phase diagram increases, and the stirring decreases (δ increases). This explains why zone melting is limited to purification of materials with low impurity contents, and why substantial temperature gradients and low zone-travel rates are necessary.

Under the microscope, the onset of constitutional supercooling first manifests itself by the formation of a grooved interface, rather than dendrites. Dendritic growth occurs at higher freezing rates. In strongly faceted materials (high entropy of fusion), needles and plates may occur rather than branched dendrites. The separation is reduced whenever the interface is no longer smooth. Nevertheless, k_{eff} does not become precisely one even with severely dendritic growth, ie, some separation still occurs (58–61).

Insoluble particles may also be removed during zone refining. Most particles are pushed by a freezing solid–liquid interface, provided that the freezing rate is sufficiently slow. In the absence of stirring, a particle is pushed up to a critical freezing rate V_c, beyond which it is trapped (62–65). The critical freezing rate is roughly inversely proportional to the particle size for smooth particles and less dependent on particle size for rough particles. For 15-μm diameter smooth spheres, V_c is about 10 mm/h.

Stirring dramatically enhances separation of foreign particles by zone melting. If the particles are not allowed to rest on the freezing interface, they are pushed at much larger velocities than V_c in the absence of stirring, although there is no sharp demarcation velocity between pushing and engulfment. The probability of trapping increases as particle size and stirring decrease and the amount of particles increases (66–67).

A variety of other phenomena influence fractional solidification of organic compounds (68).

Optimization. Zone travel rate, sample geometry and orientation, zone length and spacing, stirring in the zone, and the number of zone passes can all be controlled. These variables can be optimized either with respect to purification or to the purification rate.

For maximum purification, the zone should move as slowly as possible, the sample should be many zone lengths long, the melt should be stirred, and more zone passes should be made than several times the number of zone lengths in the sample. Stirring and slow zone travel lower $\delta V/D$, which in turn increases $|1 - k_{\text{eff}}|$, as shown by equation 5 and Figure 6. In practice, it is sufficient for $\delta V/D \lesssim 0.1$.

Stirring is essential to maximize the rate of purification, and the travel rate V should be adjusted in such a way that $\delta V/D \approx 1$ (1,4). In addition, the narrow zones should be as close together as possible, a rod should be only about 10 zone lengths long, and there should be about as many zone passes as there are zone lengths in the rod. In the absence of tube rotation, heat transfer limits the minimum zone length and zone spacing to approximately the sample diameter, which is typically about 1 cm. The value $\delta V/D \approx 1$ derives from a consideration of equation 5, which is only valid in the absence of constitutional supercooling. In no case should the zone be moved so fast that dendritic growth occurs. Thus, whenever possible, the freezing interface should be examined with a microscope during zoning (69).

Dendritic growth can be avoided by keeping the zone travel rate low for the first few passes. As impurity is removed (for $k < 1$), higher zoning rates are possible. If dendritic growth is not a problem, $\delta V/D \approx 1$ at the beginning, followed by a gradual decrease of V for subsequent passes, in order for $|1 - k_{\text{eff}}|$ to increase. Otherwise, the possible ultimate purification is limited, and the purification rate of later zone passes is low.

Since the separation occurs at the solid–liquid interface, the purification rate is maximized by having many zones close together. If this is not possible and only one zone can be passed at a time, the rate is maximized by having the first zone as long as possible and decreasing the length of subsequent zones.

Equipment and Techniques

Containers. The ideal container for zone melting should not contaminate the melt nor be damaged by the melt or subsequent contraction of the solid. For organic materials, borosilicate glasses are especially suitable, although metals and fluorocarbon and other polymers have also been successfully employed.

For many metals and semiconductors, fused silica (often erroneously called quartz) is ideal (5). Sometimes, the substance sticks to the glass and causes a break. Such breakage is prevented by coating the inside of the vessel with a film of carbon, conveniently deposited by passing acetone vapor through the glass tube at ca 700°C. However, at high temperatures such coating may result in contamination with carbon or silicon. Metals such as aluminum, with a high free energy of oxidation, react with silica glasses. In such cases, an oxide container of still higher free energy of formation is used, such as alumina or even sapphire. Zirconia and boron nitride have also been employed.

High melting inorganic salts and oxides are conveniently, albeit expensively, treated in noble-metal containers, platinum, rhodium, and iridium. Refractory containers, such as graphite, molybdenum, and tungsten in inert atmospheres, are also successfully employed for such salts. Low melting salts are refined in fused silica. Glassy carbon and pyrolytic graphite are good substitutes for ordinary graphite when carbon particles are a problem (see Refractories; Silica).

It is prudent to perform zone melting in a dry inert atmosphere. Oxygen causes most organic melts to oxidize slowly. Oxygen and moisture not only oxidize metals and semiconductors, but often enhance sticking to the container. Molten salts attack silica more rapidly in the presence of moisture. Oxygen and water are considered impurities in some inorganic compounds.

Moisture is removed from fused silica by heating, preferably in vacuum. Prior to this, silica ampuls are sometimes rinsed with a detergent solution, hydrofluoric acid, nitric acid, or methanol. After cleaning, drying, and loading, the ampul may be alternately evacuated and back-filled with inert gas. Sealing is easier if the inert-gas pressure is only slightly below atmospheric.

The container may either be in the form of a tube or a horizontal boat. A boat is advantageous for removal of the substance after zone melting, but may not be feasible if the substance is volatile. Most organic compounds, for example, are too volatile to be zone refined in an open boat. After zone melting, a silica glass ampul may be removed by etching with hydrofluoric acid, carefully cutting lengthwise with a diamond saw, or tapping with a ball-peen hammer to propagate cracks.

Because the solid and melt differ in density, repeated zone passes can transport material along the container. If the substance expands upon melting, it tends to travel toward the front, and vice versa. With a horizontal boat, this can be avoided by tilting. The tilt can be adjusted in such a way that whenever a zone reaches the end of the ingot some melt overflows, thereby removing some impurity and preventing back-mixing upon subsequent zone passes. Precautions must be taken with a sealed tube to prevent rupture of the container caused by material transport. With organic compounds that expand upon melting, provisions must be made for this expansion when the zone forms at the front end of the ampul by leaving an empty space at the front end, separated from the substance by a fluorocarbon polymer plug. Each time a new zone is formed, the plug is forced to move to take up the expansion.

Drive Mechanisms. Either the heaters or the sample may be moved. The optimal zone travel rates are typically rather slow, ie, ca 1 cm/h. Thus, electric-motor drives require gearing systems, resulting in an undesirably jerky stick-slip movement. This is avoided with double-rod supports with linear ball bearings and with low backlash gears, where tension on the moving piece pushes in the direction of motion. To ensure a smooth drive, it is useful to take time-lapse films during zoning. Even with the most stringent precautions, both low level mechanical-drive fluctuations and temperature fluctuations persist, and a zone-travel rate >1 mm/h is generally not considered practical. If constitutional-supercooling problems require a rate below this, other purification methods are preferable.

Pneumatic and hydraulic drive systems have also been used, although not widely.

If many zones are present in the sample simultaneously, it may be advantageous to move the ampul (or heaters) slowly by one zone spacing, and then rapidly backward to catch the next zone. By such a reciprocating action, many zones can be moved continuously through the sample without a bank of heaters longer than the sample.

Heating and Cooling. Heat must be applied to form the molten zones, and this heat must be removed from the adjacent solid material (4,70). In principle, any heat source can be used, including direct flames. However, the most common method is to place electrical resistance heaters around the container. In air, nichrome wire is useful to ca 1000°C, Kanthal to ca 1300°C, and platinum–rhodium alloys to ca 1700°C. In an inert atmosphere or vacuum, molybdenum, tungsten, and graphite can be used to well over 2000°C.

Conductors can be heated to very high temperatures by induction heating wherein the sample is surrounded by a water-cooled coil carrying high currents at high frequencies. A frequency of 450 kHz is typical for good conductors with a diameter >1–2 cm. Floating-zone melting requires frequencies of several MHz. Basically, the high frequency electromagnetic field induces a current within the sample itself, which is then heated by the resistance losses. The lower the frequency and the lower the electrical resistance, the deeper the field penetrates. Contact with the specimen is not required, and there are no limits to the temperatures attainable. If the substance is not a conductor, it is still possible to use induction heating. A conducting ring, eg, graphite, is placed between the sample and the high frequency coil. This susceptor is heated by the field, and in turn transmits heat to the sample by radiation, conduction, and convection (see Furnaces, electric).

Molten zones are also formed by radiant heating (71). The light source may be focused carbon arcs, xenon lamps, sunlight, or lasers. Very high temperatures have been achieved with all of these. For example, sapphire has been float-zoned in this manner, at over 2000°C.

An electron beam can be used for floating-zone melting of a conducting rod. For example, a heated tungsten ring is placed around the zone in a vacuum and made strongly negative. Electrons are emitted from the tungsten and deliver the entire voltage-drop energy to the zone. A positive connection is made to the end of the rod to drain the current.

Electric heaters have also been directly immersed in the molten zone. Zone refining has been accomplished with a single helical heater rotating in an annular sample space (71).

It is important to control temperature and power input of the heaters. A fluctuating heat input leads to a fluctuating freezing rate, thereby reducing separation. It can also cause breakage of the container because $\rho_l \neq \rho_s$. A variable autotransformer (eg, Powerstat or Variac) fed by a constant voltage transformer (eg, Sola) affords a convenient and cheap source of energy. The voltage is increased until the zone is the desired size. An oversized constant-voltage transformer does not give constant voltage. It is not necessary to actually know the temperature anywhere in the system.

Heat is often removed by simply allowing it to escape by convection, radiation, and conduction. However, such uncontrolled escape can lead to very large temperature fluctuations. It is better to surround the entire container, heaters and all, with a controlled-temperature cooled chamber. Even then, buoyancy-driven free convection from the ampul can lead to small temperature fluctuations. Jets of air or cooling water applied directly onto the ampul adjacent to the heater have been employed. Both temperature and flow rate of the coolant should be controlled.

Floating-Zone Melting. No completely satisfactory container material exists for many high melting materials. In such cases, floating-zone melting may be employed (1,5,7). Heat is applied to a vertical rod, usually by induction, electron-beam, or radiant methods. In early applications, the roughly cylindrical molten zone was held by surface tension, and rods of only a few mm in diameter could be zoned thereby. The demand for large-diameter silicon for semiconductor devices led to a modification. Typically, an induction-heating pancake coil is used with a diameter smaller than that of crystal. As shown in Figure 8, the upper feed rod melts to a conical shape, with the melt running down through the so-called eye-of-the-needle induction heating coil. The melt collects in a puddle and freezes onto the growing crystal below. The electromagnetic field of the induction heating coil supports the molten zone. In this way, dislocation-free single silicon crystals of >10 cm dia and almost 1 m long are grown commercially.

The primary application for floating-zone melting is crystal growth rather than purification. Semiconductor-grade silicon is not purified by zone refining; silicon chlorides are distilled and then reduced with hydrogen (see Silicon).

Stirring. As noted earlier, stirring increases the optimal zone-travel rate by lowering δ and aids significantly in removal of foreign particles. Even in the absence of stirring, convection motion occurs in the melt (27). In a vertical sealed ampul, the driving force is buoyancy, ie, the interaction of gravity with density gradients owing to temperature and compositional variations. If the zone is shorter than the heater and the impurity content is low, the most vigorous convection is along the top interface. If that were the sole consideration, the top interface should be the freezing interface, ie, the zone should be moved downward. However, a gas bubble frequently forms at the top of the zone caused by release of the gas bubbles trapped in the solid during the casting of the feed ingot. Material evaporates across the gas space at the top of the zone, leading to a dramatically altered separation. Furthermore, the processes of

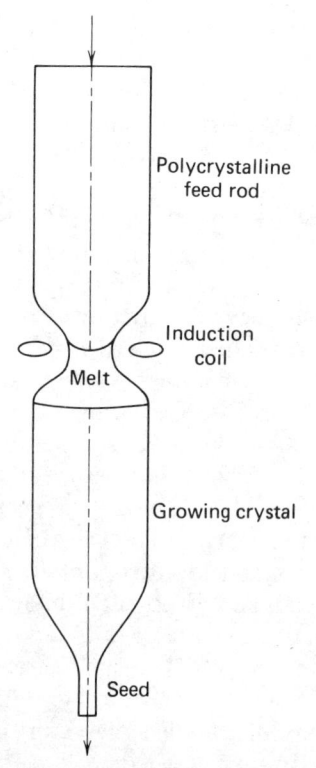

Figure 8. Geometry for float-zoning large-diameter conducting materials.

evaporation, condensation, freezing, and dripping of condensate back into the melt are so complicated that the separation is difficult to predict and may be irreproducible (68). Therefore, the zone is frequently moved up so that the freezing interface is on the bottom.

The buoyancy-driven natural convection along the freezing interface in horizontal operation tends to be fairly vigorous. However, it also leads to spreading of the zone at the top owing to convection transport of heat upward.

With natural convection, δ is between 0.1 and several mm (4).

It is frequently very helpful to rotate the tube about its axis. In horizontal zone melting with a gas bubble along the top of the zone, rotation causes good stirring and enhances removal of foreign particles. Some insoluble particles are still trapped as long as the bubble contacts the freezing interface. Tilting the tube or making subsequent zone passes without a bubble can lead to virtually complete elimination of foreign particles. Although particles are very effectively removed if no gas bubble is present, δ is then about the same as with free convection (66–67).

With vertical zone melting and horizontal zone melting without a gas bubble, simple tube rotation at a constant moderate velocity does not significantly influence δ. In those cases, accelerated crucible rotation or spin up–spin down could be used (72–75). The tube is spun more rapidly than described above, but not at constant velocity. It may, for example, be spun rapidly, suddenly stopped, spun rapidly, etc, resulting in very vigorous stirring.

In recent years, it has been shown theoretically and experimentally that surface-tension gradients can give rise to very vigorous stirring (76–80). A free melt–vapor surface is required, as with floating-zone melting or horizontal processing in an open boat. Surface tension depends on temperature and composition, both of which vary along the melt surface. The surface moves from areas of low surface tension to areas of high surface tension, carrying the underlying melt along via viscosity. This is called Marangoni convection or thermocapillary convection. It occurs only if a surfactant does not immobilize the surface, which occurs surprisingly often. For example, most organic melts contain surfactant impurities which strongly inhibit Marangoni convection (81). Surface segregation also occurs in metal-alloy melts.

Induction heating is thought to cause vigorous convection because of the spatially varying average force field imposed along the melt surface.

Deliberate stirring can be imposed on conductors with a transverse rotating magnetic field or by passage of electric current axially with a transverse magnetic field. Conversely, a constant magnetic field with no current imposed greatly reduces natural convection.

Oscillators (82) and ultrasonic vibrations (83) have also been used to lower δ.

Economic Aspects

Mere specification of zone refined is no guarantee of the purity of a chemical and has fallen into disuse. Specification is now more often by % purity, which has a strong influence on price. The price of 99.9999% pure zone-refined metals range is $150–1100/kg.

The primary commercial product is the dislocation-free float-zoned single-crystal silicon wafer for the semiconductor industry. These wafers are ca 0.4 mm thick and 5–10 cm in diameter. Prices are 1–3¢/cm^2, depending on the uniformity of impurity doping, surface preparation, and diameter. Trade names include Mon-X, Lopex, Perfex, Topsil, and Waso. Production is estimated as 100–200 metric tons per year (7).

Commercial zone-melting equipment ranges in cost from ca $2000 for a simple apparatus for organic compounds up to $600,000 for highly automated equipment to float-zone silicon 10 cm in diameter. It is relatively inexpensive to construct a zone refiner for organic compounds (1–4,84–85).

BIBLIOGRAPHY

"Zone Refining" in *ECT* 2nd ed., Vol. 22, pp. 680–702, by Alan Lawley and D. Robert Hay, Drexel University.

1. W. G. Pfann, *Zone Melting*, 2nd ed., John Wiley & Sons, Inc., New York, 1966.
2. E. F. G. Herington, *Zone Melting of Organic Compounds*, John Wiley & Sons, Inc., New York, 1966.
3. H. Schildknecht, *Zonenschmelzen*, Verlag Chemie, Weinheim, FRG, 1964.
4. M. Zief and W. R. Wilcox, *Fractional Solidification*, Marcel Dekker, Inc., New York, 1967.
5. J. C. Brice, *The Growth of Crystals from Liquids*, North-Holland Publishing Company, Amsterdam, The Netherlands, 1973.
6. J. S. Shah in B. R. Pamplin, ed., *Crystal Growth*, Pergamon, Oxford, UK, 1975, Chapt. 4.
7. W. Keller and A. Mühlbauer, *Floating-Zone Silicon*, Marcel Dekker, Inc., New York, 1981.
8. W. R. Wilcox, *CHEMI Module SMT-50*, American Institute of Chemical Engineers, in press.
9. M. Chang and R. Kennedy, *J. Electrochem. Soc.* **128,** 2193 (1981).
10. G. R. Atwood, *Chem. Eng. Prog. Symp. Ser.* **65,** 112 (1969).

11. A. M. Nazar and T. W. Clyne in T. N. Veziroglu, ed., *Proceedings Multi-Phase Flow and Heat Transfer Symposium*, Hemisphere, Washington, D.C., 1980.
12. N. W. Lord, *Trans. AIME* **197,** 1531 (1953).
13. W. R. Wilcox, *Sep. Sci. Technol.* **17,** 1117 (1982).
14. L. Burris, Jr., C. H. Stockman, and I. G. Dillon, *Trans. AIME* **203,** 1017 (1955).
15. E. Helfand and R. L. Kornegay, *J. Appl. Phys.* **37,** 2484 (1966).
16. I. Brown, *Brit. J. Appl. Phys.* **8,** 457 (1957).
17. J. R. Gould, *Trans. AIME* **221,** 1154 (1961).
18. Sh. I. Peizulaev, *Inorg. Mater.* **3,** 1329 (1967).
19. G. Matz, *Chem. Ing. Techn.* **4,** 381 (1964); Sh. I. Peizulaev, *Inorg. Mater.* **3,** 1329 (1967).
20. B. V. Ramarao, W. R. Wilcox, and W. E. Briggs, *J. Cryst. Growth* **59,** 557 (1982).
21. P. S. Ravishankar and W. R. Wilcox, *J. Cryst. Growth* **43,** 480 (1978).
22. H. M. Yeh and W. H. Yeh, *Sep. Sci. Technol.* **14,** 795 (1979).
23. D. Fischer, *J. Appl. Phys.* **44,** 1977 (1973).
24. J. A. Burton, R. C. Prim, and W. P. Slichter, *J. Chem. Phys.* **21,** 1987 (1953).
25. W. R. Wilcox, *Mater. Res. Bull.* **4,** 265 (1969).
26. W. R. Wilcox, *J. Appl. Phys.* **35,** 636 (1964).
27. J. R. Carruthers in W. R. Wilcox and R. A. Lefever, eds., *Preparation and Properties of Solid State Materials*, Vol. 3, Marcel Dekker, Inc., New York, 1977, Chapt. 1.
28. J. R. Carruthers, *J. Cryst. Growth* **32,** 13 (1976).
29. W. R. Wilcox and L. D. Fullmer, *J. Appl. Phys.* **36,** 2201 (1965).
30. M. A. Azouni, *J. Cryst. Growth* **47,** 109 (1979).
31. H. C. Gatos and A. F. Witt in A. Bishay, ed., *Recent Advances in Science and Technology of Materials*, Vol. 1, Plenum Publishing Corp., N.Y., 1973.
32. W. R. Wilcox, *Sep. Sci.* **2,** 411 (1967).
33. S. Sen, R. A. Lefever, and W. R. Wilcox, *J. Cryst. Growth* **43,** 526 (1978).
34. W. R. Wilcox and S. Sen, *Mater. Res. Bull.* **13,** 293 (1978).
35. H. P. Utech and M. C. Flemings, *J. Appl. Phys.* **37,** 2021 (1966).
36. A. F. Witt, C. J. Herman, and H. C. Gatos, *J. Mater. Sci.* **5,** 822 (1970).
37. W. R. Wilcox, *Mater. Res. Bull.* **2,** 121 (1967).
38. J. C. Brice, *J. Cryst. Growth* **28,** 249 (1975).
39. V. V. Voronkov and A. A. Chernov, *Sov. Phys.-Crystallogr.* **12,** 186 (1967).
40. D. E. Temkin, *Sov. Phys.-Crystallogr.* **17,** 405 (1972).
41. J. D. Weeks and G. H. Gilmer, *Adv. Chem. Phys.* **40,** 157 (1979).
42. H. Pfeiffer, *J. Cryst. Growth* **52,** 350 (1981).
43. J. C. Brice, *J. Cryst. Growth* **10,** 205 (1971).
44. W. A. Tiller and K.-S. Ahn, *J. Cryst. Growth* **49,** 483 (1980).
45. R. F. Sekerka in P. Hartman, ed., *Crystal Growth: An Introduction*, North-Holland Publishing Co., Amsterdam, The Netherlands, 1973, Chapt. 15.
46. D. J. Wollkind in W. R. Wilcox, ed., *Preparation and Properties of Solid State Materials*, Vol. 4, Marcel Dekker, Inc., New York, 1979, Chapt. 4.
47. S. R. Coriell and R. F. Sekerka, *J. Cryst. Growth* **34,** 157 (1976).
48. J. S. Langer, *Acta Met.* **25,** 1121 (1977).
49. T. Sato, K. Shibata, and G. Ohira, *J. Cryst. Growth* **40,** 69 (1977).
50. D. E. Temkin, *Sov. Phys. Crystallogr.* **22,** 529 (1977).
51. W. W. Mullins and R. F. Sekerka, *J. Appl. Phys.* **35,** 444 (1964).
52. A. L. Conlet, B. Billia, and I. Capella, *J. Cryst. Growth* **51,** 106 (1981).
53. S. R. Coriell, M. R. Cordes, W. J. Boettinger, and R. F. Sekerka, *J. Cryst. Growth* **49,** 13 (1981).
54. S. R. Coriell, D. T. J. Hurle, and R. F. Sekerka, *J. Cryst. Growth* **32,** 1 (1976).
55. D. E. Holmes and H. C. Gatos, *J. Appl. Phys.* **52,** 2971 (1981).
56. J. P. Dismukes and W. M. Yim, *J. Cryst. Growth* **22,** 287 (1974).
57. D. J. Morantz and K. K. Mathur, *J. Cryst. Growth* **16,** 147 (1972).
58. J. Verhoeven, *Met. Trans.* **2,** 2673 (1971).
59. D. D. Edie and D. J. Kirwan, *Ind. Eng. Chem. Fundam.* **12,** 100 (1973).
60. B. Ozum and D. J. Kirwan, *AIChE Symp. Ser.* **72**(153), 1 (1976).
61. N. Streat and F. Weinberg, *Met. Trans.* **7B,** 417 (1976).
62. A. A. Chernov, D. E. Temkin, and A. M. Mel'nikova, *Sov. Phys. Crystallogr.* **21,** 652 (1976).
63. S. N. Omenyi and A. W. Neumann, *J. Appl. Phys.* **47,** 3956 (1976).

64. R. R. Gilpin, *J. Colloid Int. Sci.* **74,** 44 (1980).
65. P. F. Aubourg, *Interaction of Second Phase Particles with Crystal Growing from Melt*, Ph.D. thesis, MIT, Cambridge, Mass., 1978.
66. J. E. Coon and W. R. Wilcox, *Sep. Sci. Technol.* **15,** 1401 (1980).
67. R. B. Fedich and W. R. Wilcox, *Sep. Sci. Technol.* **15,** 31 (1980).
68. W. R. Wilcox, *Sep. Sci.* **4,** 95 (1969).
69. C. E. Chang and W. R. Wilcox, *J. Cryst. Growth* **21,** 182 (1974).
70. N. Kobayashi, *J. Cryst. Growth* **43,** 417 (1978).
71. A. R. McGhie, P. J. Rennolds, and G. J. Sloan, *Anal. Chem.* **52,** 1738 (1980).
72. N. J. G. Bollen, M. J. Van Essen, and W. M. Smit, *Anal. Chim. Acta* **38,** 279 (1967).
73. G. J. Sloan, *Mol. Cryst.* **1,** 161 (1966).
74. K. D. Wolter, P. L. Carella, G. A. Moebus, and J. F. Johnson, *Sep. Sci. Technol.* **14,** 805 (1979).
75. I. N. Anikin and Kh. S. Bagdasarov, *Sov. Phys. Crystallogr.* **25,** 81 (1980).
76. P. A. Clark and W. R. Wilcox, *J. Cryst. Growth* **50,** 461 (1980).
77. C. E. Chang, W. R. Wilcox, and R. A. Lefever, *Mater. Res. Bull.* **14,** 527 (1979).
78. D. Schwabe and A. Scharmann, *J. Cryst. Growth* **46,** 125 (1979).
79. *Ibid.*, **52,** 435 (1981).
80. Ch.-H. Chun, *J. Cryst. Growth* **48,** 600 (1980).
81. W. R. Wilcox, R. S. Subramanian, J. M. Papazian, H. D. Smith, and D. M. Mattox, *AIAA J.* **17,** 1022 (1979).
82. B. K. Jindal, *J. Cryst. Growth* **16,** 280 (1972).
83. O. V. Abramov, I. I. Teumin, V. A. Filonenko, and G. I. Eskin, *Sov. Phys. Acoust.* **13,** 141 (1967).
84. G. F. Needham, G. Boehme, R. D. Willett, and D. D. Swank, *J. Chem. Ed.* **59,** 63 (1982).
85. D. Fischer, *Mater. Res. Bull.* **8,** 385 (1973).

WILLIAM R. WILCOX
Clarkson College of Technology

ZYMURGY. See Beer; Fermentation; Wine.